8 Transistor Modeling Hybrid equivalent: $V_i = h_i I_i + h_r V_o$, $I_o = h_f I_i + h_o V_o$;
Approximate CE hybrid equivalent: $h_{re} = h_{oe} = 0$, h_{ie}, $h_{fe} I_b$;
r_e model: $r_e = 26\text{ mV}/I_E\text{ (mA)}$, $h_{ib} = r_e$, $h_{fb} = -1$, $\beta = h_{fe}$, $\beta r_e = h_{ie}$;
Graphical determination of h parameters: $h_{ie} \cong \Delta v_{be}/\Delta i_b$ with V_{CE} = constant, $h_{re} \cong \Delta v_{be}/\Delta v_{ce}$ with I_B = constant, $h_{fe} \cong \Delta i_c/\Delta i_b$ with V_{CE} = constant, $h_{oe} \cong \Delta i_c/\Delta v_{ce}$ with I_B = constant

9 BJT Small-Signal Analysis CE fixed bias: $Z_i = R_B \| h_{ie} = R_B \| \beta r_e$, $Z_o = R_C$,
$A_v = -h_{fe} R_C/h_{ie} = -R_C/r_e$, $A_i \cong h_{fe} \cong \beta$;
Effect of h_{oe}: $Z_o = R_C \| 1/h_{oe}$, $A_v = -h_{fe}(R_C \| 1/h_{oe})/h_{ie}$, $A_i = h_{fe}/(1 + h_{oe} R_C)$;
Voltage-divider bias: $R_{BB} = R_{B1} \| R_{B2}$, $Z_i = R_{BB} \| h_{ie} = R_{BB} \| \beta r_e$, $Z_o = R_C$,
$A_v = -h_{fe} R_C/h_{ie} = -R_C/r_e$, $A_i = R_{BB} h_{fe}/(R_{BB} + h_{ie}) = R_{BB}\beta/(R_{BB} + \beta r_e)$;
Effect of h_{oe}: $Z_o = R_C \| 1/h_{oe}$, $A_v = -h_{fe}(R_C \| 1/h_{oe})/h_{ie}$,
$A_i = R_{BB} h_{fe}/[(1 + h_{oe} R_C)(R_{BB} + h_{ie})]$;
CE unbypassed emitter bias: $Z_b = h_{ie} + (1 + h_{fe})R_E$, $Z_b = \beta(r_e + R_E)$, $Z_i = R_B \| Z_b$,
$Z_o = R_C$, $A_v = -h_{fe} R_C/Z_b = -\beta R_C/(r_e + R_E)$, for $Z_b \cong h_{fe} R_E$, $A_v = -R_C/R_E$,
$A_i = R_B h_{fe}/(R_B + Z_b)$;
Emitter follower: $Z_i = R_B \| Z_b$, $Z_b = R_E \| [h_{ie}/(1 + h_{fe})]$, $A_v = R_E/[R_E + h_{ie}/(1 + h_{fe})]$,
$A_i = (1 + h_{fe})R_B/(R_B + Z_b)$;
Common-base: $Z_i = R_E \| h_{ib} = R_E \| r_e$, $Z_o = R_C$, $A_v = -h_{fb} R_C/h_{ib} = -R_C/r_e$,
$A_i = -h_{fb} = -1$;
Collector feedback: $A_v \cong -h_{fe} R_C/h_{ie}$, $A_i = h_{fe} R_F/(R_F + h_{fe} R_C)$ for $h_{fe} R_C \gg R_F$;
$A_i \cong R_F/R_C$, $Z_i = h_{ie} \| R_F/|A_v|$, $Z_o \cong R_C \| R_F$;
Loading effects of R_S and R_L: $Z_{\text{in}} = R_S + Z_i$, $A_{v_s} = [V_o/V_i][V_i/V_s]$, $|A_i| = (Z_i/R_L)(|A_v|)$;
Complete hybrid: $A_i = h_f/(1 + h_o R_L)$, $A_v = -h_f R_L/[h_i + (h_i h_o - h_f h_r)R_L]$,
$Z_i = h_i - h_f h_r R_L/(1 + h_o R_L)$, $Z_o = 1/\left(h_o - \dfrac{h_f h_r}{h_i + R_s}\right)$,
$A_p = h_f^2 R_L/[(1 + h_o R_L)(h_i + (h_i h_o - h_f h_r)R_L] = |A_v| \cdot |A_i|$;
Systems approach: $R_{\text{th}} = Z_o = R_o$, $E_{\text{th}} = V_o = A_v V_i$, $A_{v_L} = R_L A_v/(R_L + R_o)$,
$|A_i| = \dfrac{R_L}{Z_i}|A_{v_L}|$

10 FET Small-Signal Analysis $g_m = g_{mo}(1 - V_{GS}/V_P)$, $g_{mo} = 2I_{DSS}/|V_P|$;
Basic configuration: $A_v = -g_m R_D$, $R_i = R_G$, $R_o = R_D$, $r_m = 1/g_m$, $A_v = -R_D/r_m$;
Unbypassed source resistance: $A_v = -g_m R_D/(1 + g_m R_{S1}) = -R_D/(r_m + R_{s1})$;
Loading effects: $A_v = -(R_D \| R_L)/(r_m + R_S)$;
Source resistance: $V_i = R_i V_S/(R_S + R_i)$;
Source follower: $A_v = g_m R_S/(1 + g_m R_S) = R_S/(r_m + R_S)$, $R_i = R_G$, $R_o = R_S \| r_m$;
Common-gate: $A_v = g_m R_D = R_D/r_m$, $R_i = R_S$, $R_o = R_D$

11 Multistage Systems and Frequency Considerations
$A_{v_T} = \pm A_{v_1} A_{v_2} A_{v_3} \cdots A_{v_n}$, $A_{i_T} = \pm A_{i_1} A_{i_2} A_{i_3} \cdots A_{i_n}$, $|A_{v_T}| = |A_{i_T}| \cdot |Z_L/Z_{i1}|$,
$|A_{P_T}| = |A_{v_T}||A_{i_T}|$;
Transformer-coupled: $V_1/V_2 = N_1/N_2 = a$, $I_1/I_2 = N_2/N_1 = 1/a$, $Z_i = a^2 Z_L$;
Darlington: $A_i = h_{fe_1} h_{fe_2}/[1 + h_{oe_1}(h_{fe_2} R_E)]$; for $h_{fe_1} = h_{fe_2} = h_{fe}$ and $h_{oe_1} = h_{oe_2} = h_{oe}$,
$A_i \cong h_{fe}^2/(1 + h_{oe} h_{fe} R_E)$; for $h_{oe} h_{fe} R_E \le 0.1$, $A_i \cong h_{fe}^2 = \beta^2$,
$Z_{i_1} = h_{fe_1} h_{fe_2} R_E/(h_{oe_1} h_{fe_2} R_E + 1)$; for $h_{fe_1} = h_{fe_2} = h_{fe}$ and $h_{oe_1} = h_{oe_2} = h_{oe}$,
$Z_{i_1} \cong h_{fe}^2 R_E/(1 + h_{oe} h_{fe} R_E)$; for $h_{oe} h_{fe} R_E \le 0.1$, $Z_{i_1} \cong h_{fe}^2 R_E = \beta^2 R_E$,

ELECTRONIC DEVICES AND CIRCUIT THEORY

Robert Boylestad

Louis Nashelsky

4th Edition

ELECTRONIC AND CIRCUIT

PRENTICE-HALL, INC., Englewood Cliffs, NJ 07632

DEVICES THEORY

Library of Congress Cataloging-in-Publication Data

Boylestad, Robert L.
Electronic devices and circuit theory.

Includes index.
1. Electronic circuits. 2. Electronic apparatus and appliances. I. Nashelsky, Louis. II. Title.
TK7867.B66 1987 621.38 86–22661
ISBN 0–13–250556–8

Editorial/production supervision: Eileen M. O'Sullivan
Interior design: Maureen Eide and Janet Schmid
Cover design: Janet Schmid
Cover photograph: C. Paul Ambrose/FPG
Manufacturing buyers: John Hall and Carol Bystrom

ELECTRONIC DEVICES AND CIRCUIT THEORY
4TH. EDITION
ROBERT BOYLESTAD, LOUIS NASHELSKY

Printed in the United States of America

10 9 8 7 6 5 4 3 2 1

0-13-250556-8 025

Prentice-Hall International (UK) Limited, *London*
Prentice-Hall of Australia Pty. Limited, *Sydney*
Prentice-Hall Canada Inc., *Toronto*
Prentice-Hall Hispanoamericana, S.A., *Mexico*
Prentice-Hall of India Private Limited, *New Delhi*
Prentice-Hall of Japan, Inc., *Tokyo*
Prentice-Hall of Southeast Asia Pte. Ltd., *Singapore*
Editora Prentice-Hall do Brasil, Ltda., *Rio de Janeiro*

Dedicated to:

ELSE MARIE, ERIC, ALISON and STACEY

and to

KATRIN, KIRA, LARREN, and THOMAS

Contents

Preface

In addition to the variety of improvements normally associated with a new edition it became obvious in recent years that there were two specific areas that required special attention. One was to address the confusion developing due to the use of two models to analyze the ac response of transistors. The other was to improve the pedagogical treatment of a number of fundamental areas.

With regard to the question of the r_e vs. hybrid model, the development describes the similarities that exist between both models and which model should be used in a particular area of application or investigation. A student should be familiar with both models since both approaches are used and appear in industry and educational literature. In essence, sufficient exposure should be provided to each model.

In response to the second concern there was a complete revision of some chapters to provide an improved pedagogical order to the presentation. Fundamental concepts are highlighted with an expanded set of examples from the simple to the more complex. A number of sections in the advanced chapters were completely rewritten to insure a clearer understanding of the analysis and design procedures of some relatively complex material. Material was updated throughout the text to reflect industry advances since the last edition. The problem set has been carefully reviewed and expanded with an effort to provide a more practical set of exercises. In particular, standard resistor values were used throughout the problem section to develop an awareness of typical industry values.

Computer solutions are provided to demonstrate how the computer can be used to support the analysis and design effort. No attempt was made to develop computer techniques; instead, through the use of BASIC language, an attempt was made to demonstrate how equations appearing in a chapter would appear in computer format and how a program can provide an almost instantaneous response to lengthy analysis and design procedures.

A second color was added to this edition primarily to support the learning process and not simply for esthetic reasons. In particular, note its use in the ac analysis section to identify the ac model and its use in identifying curves provided on the same set of axis.

A major support item was added to this edition in the form of a set of laboratory experiments closely tied to the content and sequence of the text material. An effort was made to limit the component and instrumentation requirements and avoid a "cook-book" approach.

This text is designed primarily for use in a two-semester or three-trimester sequence in the basic electronics area. It is expected that the student has taken a course in dc circuit analysis and has either taken or is taking a course in ac circuit analysis.

In an effort to aid the student, the text contains extensive examples that stress the main points of each chapter. There are also numerous illustrations to guide the student through new concepts and techniques. Important conclusions are emphasized by boxed equations or boldface answers to make the student aware of the essential points covered.

The text is a result of a two-semester electronics course sequence that both authors were actively involved in teaching over a period of years. However, the 18 chapters actually contain more material than can be covered in two 15-week semesters (or three 10-week trimesters). This preface will suggest how the authors feel the material can be best organized.

Essentially, the first 7 chapters provide the basic background to electronic devices from a dc viewpoint—including construction, biasing, and operation as single stages. The material in these chapters can be included in the first semester with the option left to the teacher of stressing some areas more than others, or some not at all. The course could begin with the theory and operation of the two terminal semiconductor diode. Since the theory is usually taught in conjunction with laboratory exercises, the material has been organized with regard to provide practical circuit examples for examination in the laboratory.

Chapters 1 and 2 include a description of the construction, characteristics, and application of two-terminal semiconductor diodes. Chapter 2 is a new chapter designed to develop an improved sense of the practical behavior of semiconductor diodes in a variety of configurations.

Chapter 3 includes a detailed description of a number of two-terminal devices such as the Zener diode, LED, LCD, solar cell, Schottky diode, etc., that have become increasingly important. There was a serious attempt to expand on the coverage to ensure that some practical considerations associated with their use be more fully understood and appreciated.

The construction and theory of operation of the BJT transistor is covered in detail in Chapter 4. The operation of the transistor is presented both mathematically and graphically; the amplifying action of the transistor is defined and demonstrated. Actual current directions are used in this introductory area, since both authors feel that the material is better absorbed using this approach. The investigation of the data sheet was expanded to ensure its readability and impact on the application of the device.

It is the authors' experience that the student can better comprehend the operation of the BJT transistor device if, initially, the dc bias and ac operation are treated separately. Thus, Chapter 5 deals only with the dc bias of the BJT transistor. This is done for common-emitter, common-base, and common-collector for a variety of bias current types. Numerous examples help demonstrate the theory presented. Also, some design problems are included to provide a well-rounded treatment.

If possible, the material on the field-effect transistor (FET) should also be covered in the first semester of electronics. After the presentation and development of the concepts of dc bias of the BJT, Chapters 6 and 7 then cover a number of practical FET circuits.

Chapters 8 and 9 are two of the most important in the basic coverage area and should be given sufficient time in any course. The development of the BJT transistor ac equivalent model is covered in detail, followed by analysis of the ac operation of the full small-signal circuit. The mathematics are kept short and direct, with a generous number of examples provided so that the student should be able to follow the ideas presented. The hybrid and r_e model equivalent circuits of the transistor are introduced and the usual engineering simplifications are applied to provide a more practical treatment.

Chapter 10 presents the ac small-signal analysis of the FET with numerous examples. Stage loading, overall gain calculations, and use of decibels are all covered in Chapter 11. A number of examples help to emphasize the main points of the chapter. Increased emphasis has been placed on the use of the approximate analysis techniques for multi-stage amplifiers. The material on frequency has been revised for increased clarity. In addition, computer analysis of some configurations is included.

Chapter 12 covers the operation of power transistors in a few basic power amplifier circuits. Most important is the operation of the push-pull circuit. Transistor push-pull networks with and without a transformer (transformerless circuits) are covered. Additional material on quasi-complementary push-pull amplifiers and on class-B power and efficiency is also included.

Chapter 13 is a "catch-all" of a number of *pnpn* devices—covering their construction, operation, and circuit applications. It can be covered quickly or even passed over, if desired, without loss of continuity. Also covered are the UJT and PUT.

The fabrication and construction of integrated circuits (ICs) is provided in Chapter 14. The content has again been updated as a result of a recent visit to the Phoenix branch of the Motorola Corporation and includes many of the advances made in this area. If desired, the content of this chapter can be assigned mainly as student reading.

Chapters 15–17 deal with a range of popular linear integrated circuits—their basic fabrication, operation, and most important, their practical application. The topics covered in these chapters are representative of the concepts and developments taking place in the electronics field.

Chapter 18 has been extensively rewritten to clarify the terminology employed and better describe the analysis techniques applied to feedback systems.

We wish to thank Professors Aidala and Katz for their continued help and encouragement. Thanks are also extended to Eileen O'Sullivan of Prentice-Hall for her diligent efforts to produce the best possible text and Greg Burnell and Alice Barr of Prentice-Hall for their encouragement and support in getting this fourth edition out so quickly and painlessly. Special thanks go to Professor Sidney Sonsky for checking the numerical calculations and proofreading the manuscript in its critical galley stage. Thanks are also due to Janet Schmid for her creativity in designing the book and for her assistance on working with the technical art; Melissa Halverstadt, Marketing Manager; Carol Bystrom, Manufacturing Buyer; and, Debbie Kesar, Schedule Coordinator.

ROBERT BOYLESTAD
Westport, Conn.

LOUIS NASHELSKY
Great Neck, N.Y.

Acknowledgments

We want to express our gratitude to those individuals who took the time, interest, and effort to review our text—*Electronic Devices and Circuit Theory*. Their comments and suggestions contributed to the development of the fourth edition to ensure that the content and approach would reflect recent trends and developments.

We are anxious to hear from current adopters of the text to ensure that the text fulfills their needs and is as technically accurate as possible. Comments are always appreciated and will receive a personal reply at the earliest opportunity.

Ernest Lee Abbott, Napa Valley College
J. L. Brockbank, DeVry Institute of Technology
John W. Coons, RCA Bloomington
Kenneth Dunn, Pensacola Jr. College
Bryan L. Gagnon, Indiana Vocational Technical College
Trevor Glave, British Columbia Institute of Technology
Thurman Grass, Lima Technical College
Harold Hambrock, DeVry Institute of Technology
D. Hunchak, Northern Alberta Institute of Technology
Robert W. James, DeVry Institute of Technology
John Jellema, Eastern Michigan University
Albert Koon, Tidewater Community College
Ken Kramer, DeVry Institute of Technology
David Krispinsky, Rochester Institute of Technology
Paul Maini, Suffolk Community College
Jim Pannell, DeVry Institute of Technology
Lester C. Peach, Illinois Institute of Technology
Fred Pirkey, DeVry Institute of Technology
Raghuwanshi Pravin M., DeVry Institute of Technology
P. K. Rastogi, Case Western Reserve University

H. C. Schenkenberger, DeVry Institute of Technology
Sheau-Yong Sheu, Brookdale Community College
Ted Stulp, DeVry Institute of Technology
Paul T. Svatik, Owens Technical College
Basil M. Taraleshkov, DeVry Institute of Technology
Edward Troyan, Lehigh County Community College
James R. Williams, Brevard Community College
O. S. Zemlak, Algonquin College
Sidney Sonsky, Queensborough Community College
Kelvin Shih, Lawrence Institute of Technology
Marcela Rydlora-Erlich, Vermont Technical College

ELECTRONIC DEVICES AND CIRCUIT THEORY

1

Semiconductor Diodes

1.1 INTRODUCTION

The few decades following the introduction of the semiconductor transistor in the late 1940s have seen a very dramatic change in the electronics industry. The miniaturization that has resulted leaves us to wonder about its limits. Complete systems now appear on a wafer thousands of times smaller than the single element of earlier networks. The advantages associated with semiconductor systems as compared to the tube networks of prior years are, for the most part, immediately obvious: smaller and lightweight, no heater requirement or heater loss (as required for tubes), more rugged construction, more efficient, and not requiring a warm-up period.

The miniaturization of recent years has resulted in semiconductor systems so small that the primary purpose of the container is simply to provide some means of handling the device and ensuring that the leads remain properly fixed to the semiconductor wafer. The limits of miniaturization appear to be limited by three factors: the quality of the semiconductor material itself, the network design technique, and the limits of the manufacturing and processing equipment.

1.2 GENERAL CHARACTERISTICS

The label *semiconductor* itself provides a hint as to its characteristics. The prefix *semi* is normally applied to anything midway between two limits. The term *conductor* is applied to any material that will permit a generous flow of charge due to the application of a limited amount of external pressure. A semiconductor, therefore, is a material that has a conductivity level somewhere between the extremes of an insulator (very low conductivity) and a conductor, such as copper, which has a high level of conductivity. Inversely related to the conductivity of a material is its resistance to the flow of charge, or current. That is, the higher the conductivity level, the lower the resistance level. In tables, the term *resistivity* (ρ, Greek letter rho) is often used when comparing the resistance levels of materials. The resisitivity of a material can be examined by noting the resistance of a sample having a length of 1 cm and a cross-sectional area of 1 cm^2, as shown in Fig. 1.1. Recall that the equation for the resistance of a material (at a particular temperature) is determined by $R = \rho l/A$, where R is measured in ohms, l is the length of the sample, A is its incident surface area, and ρ is the resistivity. If $l = 1$ cm and $A = 1\ cm^2$, then $R = \rho$, as indicated above. The magnitude of the resistance of a 1-cm^3 sample of the material is therefore determined by the resistivity. Or, in other words, the higher the resistivity, the more the magnitude of the resistance for such a sample. The units for ρ as defined by the equation are as follows:

$$\rho = \frac{RA}{l} \Rightarrow \frac{\Omega \cdot cm^2}{cm} = \boxed{\Omega \cdot cm} \tag{1.1}$$

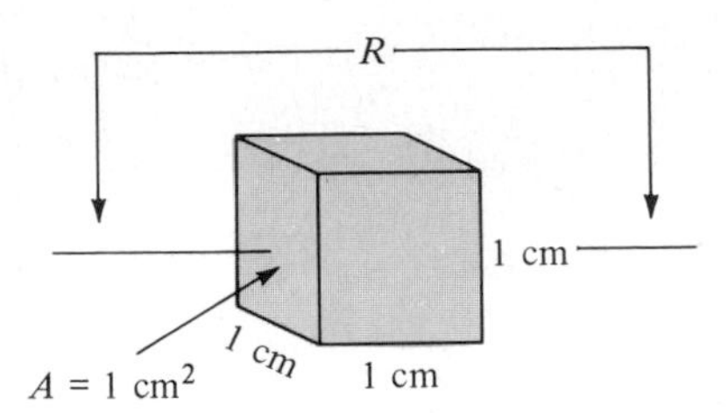

Figure 1.1

Please be assured that this book is not heavily bent toward mathematical developments and complex algebraic techniques. The authors feel that a clear understanding of dimensions is an absolute necessity for proper development of an engineering background. We would assume that the somewhat abstract unit of measure for resistivity now has a measure of understanding and clarity. Incidentally, the actual resistance of a semiconductor material as measured above is called its *bulk* resistance. The resistance introduced by connecting the leads to the bulk material is called the *ohmic contact* resistance. These terms will appear in the description of devices to be introduced throughout.

In Table 1.1, typical resistivity values are provided for three broad categories of materials. Although you may be familiar with the electrical properties of copper and mica from your past studies, the characteristics of the semiconductor materials of germanium (Ge) and silicon (Si) may be relatively new. As you will find in the chapters to follow, they are certainly not the only two semiconductor materials. They are, however, the two materials that have received the broadest range of interest in the development of semiconductor devices. In recent years the shift has been steadily toward silicon and away from germanium, but germanium is still in modest production.

Note in Table 1.1 the extreme range between the conductor and insulating materials for the 1-cm length of the material. Eighteen places separate the placement of the

TABLE 1.1 Typical Resistivity Values (At 300 K—Room Temperature)

Conductor	*Semiconductor*	*Insulator*
$\rho \cong 10^{-6}\ \Omega \cdot \text{cm}$ (copper)	$\rho \cong 50\ \Omega \cdot \text{cm}$ (germanium) $\rho \cong 50 \times 10^{3}\ \Omega \cdot \text{cm}$ (silicon)	$\rho \cong 10^{12}\ \Omega \cdot \text{cm}$ (mica)

decimal point for one number from the other. Ge and Si have received the attention they have for a number of reasons. One very important consideration is the fact that they can be manufactured to a very high purity level. In fact, recent advances have reduced impurity levels in the pure material to 1 part in 10 billion (1:10, 000,000,000). One might ask if these low impurity levels are really necessary. They certainly are if you consider that the addition of one part impurity (of the proper type) per million in a wafer of silicon material can change that material from a relatively poor conductor to a good conductor of electricity. We are obviously dealing with a whole new spectrum of comparison levels when we deal with the semiconductor medium. The ability to change the characteristics of the material significantly through this process, known as "doping," is yet another reason why Ge and Si have received such wide attention. Further reasons include the fact that their characteristics can be altered significantly through the application of heat or light—an important consideration in the development of heat- and light-sensitive devices.

Some of the unique qualities of Ge and Si noted above are due to their atomic structure. The atoms of both materials form a very definite pattern that is periodic in nature (i.e., continually repeats itself). One complete pattern is called a *crystal* and the periodic arrangement of the atoms a *lattice*. For Ge and Si the crystal has the three-dimensional diamond structure of Fig. 1.2. Any material composed solely of repeating crystal structures of the same kind is called a *single-crystal* structure. For semiconductor materials of practical application in the electronics field, this single-crystal feature exists, and, in addition, the periodicity of the structure does not change significantly with the addition of impurities in the doping process.

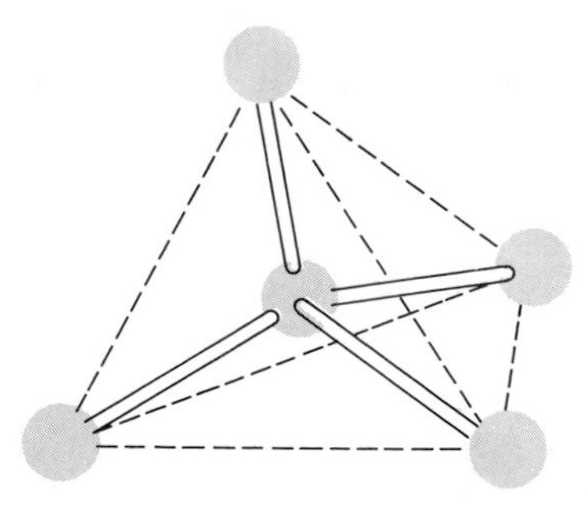

Figure 1.2 Ge and Si single-crystal structure.

Let us now examine the structure of the atom itself and note how it might affect the electrical characteristics of the material. As you are aware, the atom is composed of three basic particles; the *electron*, the *proton*, and the *neutron*. In the atomic lattice, the neutrons and protons form the *neucleus*, while the electrons revolve around the nucleus in a fixed *orbit*. The Bohr models of the two most commonly used semiconductors, *germanium* and *silicon,* are shown in Fig. 1.3.

As indicated by Fig. 1.3a, the germanium atom has 32 orbiting electrons, while silicon has 14 orbiting electrons. In each case, there are 4 electrons in the outermost (*valence*) shell. The potential (*ionization potential*) required to remove any one of these 4 valence electrons is lower than that required for any other electron in the structure. In a pure germanium or silicon crystal these 4 valence electrons are bonded to 4 adjoining atoms, as shown in Fig. 1.4 for silicon. Both Ge and Si are referred to as *tetravalent* atoms because they each have four valence electrons.

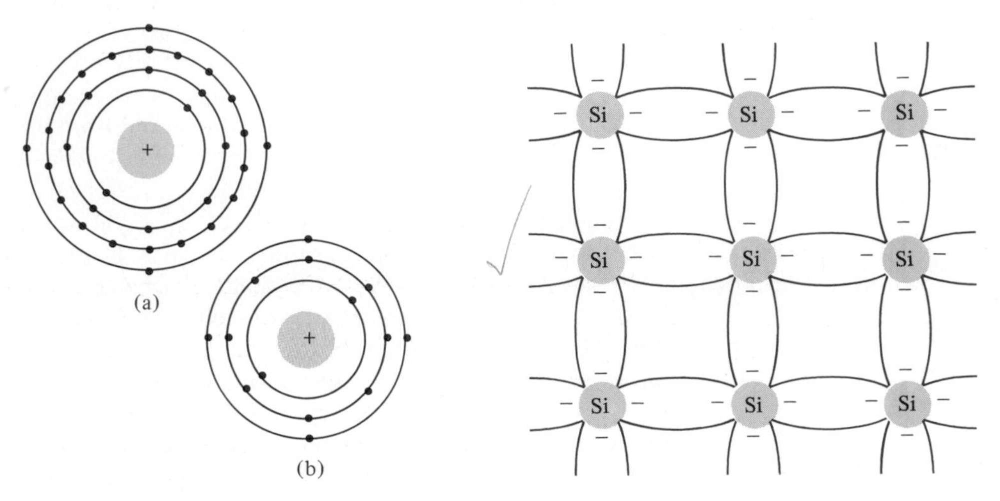

Figure 1.3 Atomic structure: (a) germanium; (b) silicon.

Figure 1.4 Covalent bonding of the silicon atom.

This type of bonding, formed by *sharing* electrons, is called *covalent bonding*. Although the covalent bond will result in a stronger bond between the valence electrons and their parent atom, it is still possible for the valence electrons to absorb sufficient kinetic energy from natural causes to break the covalent bond and assume the "free" state. These natural causes include effects such as light energy in the form of photons and thermal energy from the surrounding medium. At room temperature there are approximately 1.5×10^{10} free carriers in a cubic centimeter of intrinsic silicon material. *Intrinsic* materials are those semiconductors that have been carefully refined to reduce the impurities to a very low level—essentially as pure as can be made available through modern technology. The free electrons in the material due only to natural causes are referred to as *intrinsic carriers*. At the same temperature, intrinsic germanium material will have approximately 2.5×10^{13} free carriers per cubic centimeter. The ratio of the number of carriers in germanium to that of silicon is greater than 10^3 and would indicate that germanium is a much better conductor at room temperature. This may be true, but both are still considered poor conductors in the intrinsic state. Note in Table 1.1 that the resistivity also differs by a ratio of about 1000 : 1, with silicon having the larger value. This should be the case, of course, since resistivity and conductivity are inversely related.

A change in the temperature of a semiconductor material can increase the number of free electrons quite substantially. As the temperature rises from absolute zero (0 K), an increasing number of valence electrons absorb sufficient thermal energy to break the covalent bond and contribute to the number of free carriers as described above. This increased number of carriers will increase the conductivity index and result in a lower resistance level. Semiconductor materials such as Ge and Si that show a reduction in resistance with increase in temperature are said to have a *negative temperature coefficient*. You will probably recall that the resistance of most conductors will increase with temperature. This is due to the fact that the numbers of carriers in a conductor will not increase significantly with temperature, but their vibration pattern above a relatively fixed location will make it increasingly difficult for electrons to pass through. An increase in temperature therefore results in an increased resistance level and a *positive temperature coefficient*.

1.3 ENERGY LEVELS

In the isolated atomic structure there are discrete (individual) energy levels associated with each orbiting electron as shown in Fig. 1.5a. Each material will, in fact, have its own set of permissible energy levels for the electrons in its atomic structure. The more distant the electron from the nucleus, the higher the energy state, and any electron that has left its parent atom has a higher energy state than any electron in the atomic structure. Between the discrete energy levels are gaps in which no electrons in the isolated atomic structure can appear. As the atoms of a material are brought closer together to form the crystal lattice structure, there is an interaction between atoms that will result in the electrons in a particular orbit of one atom having slightly different energy levels from electrons in the same orbit of an adjoining atom. The net result is an expansion of the discrete levels of possible energy states for the valence electrons to that of bands as shown in Fig. 1.5b. Note that there are still boundary levels and maximum energy states in which any electron in the atomic lattice can find itself, and there remains a *forbidden* region between the valence band and the ionization level. Recall that ionization is the mechanism whereby an electron can absorb sufficient energy to break away from the atomic structure and join the "free" carriers in the conduction band. You will note that energy is measured in *electron volts* (eV). The unit of measure is appropriate, since

$$W(\text{energy}) = P(\text{power}) \cdot t(\text{time})$$

but

$$P = VI$$

resulting in

$$W = VIt$$

but

$$I = \frac{Q}{t} \quad \text{or} \quad Q = It$$

$$\boxed{W = QV} \quad \text{joules} \tag{1.2}$$

Substituting the charge of an electron and a potential difference of 1 volt into Eq. (1.2) will result in an energy level referred to as one *electron volt*. Since energy

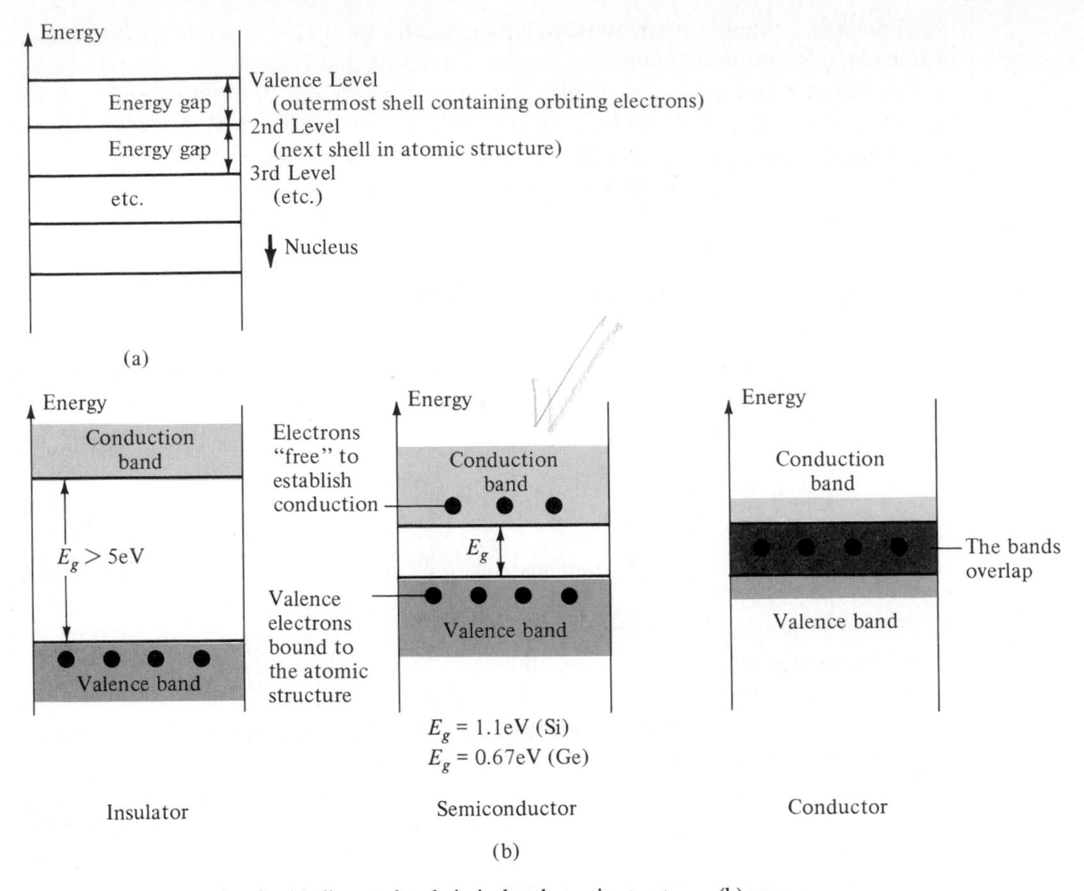

Figure 1.5 Energy levels: (a) discrete levels in isolated atomic structures; (b) conduction and valence bands of an insulator, semiconductor, and conductor.

is also measured in joules and the charge of one electron = 1.6×10^{-19} coulomb,

$$W = QV = (1.6 \times 10^{-19}\text{C})(1\text{ V})$$

$$\boxed{1\text{ eV} = 1.6 \times 10^{-19}\text{ J}} \qquad (1.3)$$

The small unit of measure is required to avoid reference to very small numbers in the discussion to follow.

At 0 K or absolute zero, all the valence electrons of the semiconductor materials are in the valence bands. However, at room temperature (300 K) a large number of electrons have acquired sufficient energy to enter the conduction band, that is, to bridge the 1.1-eV gap for silicon and 0.67 eV for germanium. The obviously lower E_g for germanium accounts for the increased number of carriers in that material as compared to silicon at room temperature. Note for the insulator that the energy gap is typically 5 eV or more. Very few electrons can acquire the required energy at room temperature, with the result that the material remains an insulator. The conductor has electrons in the conduction band even at 0 K. Quite obviously, therefore, at room temperature there are more than enough free carriers to sustain a heavy flow of charge, or current.

We see in Section 1.4 that if certain impurities are added to the intrinsic semiconductor materials, the result will be permissible energy states in the forbidden band and a net reduction in E_g for both semiconductor materials—consequently increased carrier density in the conduction band at room temperature!

1.4 EXTRINSIC MATERIALS—*n*- AND *p*-TYPE

The characteristics of semiconductor materials can be altered significantly by the addition of certain impurity atoms into the relatively pure semiconductor material. These impurities, although only added to perhaps 1 part in 10 million, can alter the band structure sufficiently to totally change the electrical properties of the material. A semiconductor material that has been subjected to this *doping* process is called an *extrinsic* material. There are two extrinsic materials of immeasurable importance to semiconductor device fabrication: *n*-type and *p*-type. Each will be described in some detail in the following paragraphs.

n-Type Material

Both the *n*- and *p*-type materials are formed by adding a predetermined number of impurity atoms into a germanium or silicon base. The *n*-type is created by adding those impurity elements that have *five* valence electrons (*pentavalent*), such as *antimony, arsenic,* and *phosphorus*. The effect of such impurity elements is indicated in Fig. 1.6 (using antimony as the impurity in a silicon base). Note that the four covalent bonds are still present. There is, however, an additional fifth electron due to the impurity atom, which is *unassociated* with any particular covalent bond. This remaining electron, loosely bound to its parent (antimony) atom, is relatively free to move within the newly formed *n*-type material. Since the inserted impurity atom has donated a rela-

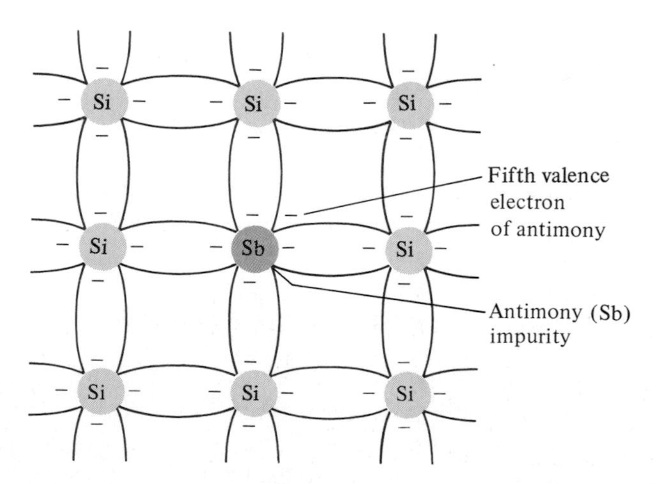

Figure 1.6 Anitmony impurity in *n*-type material.

tively "free" electron to the structure, impurities with five valence electrons are called *donor* atoms. It is important to realize that even though a large number of "free" carriers have been established in the *n*-type material, it is still electrically *neutral* since ideally the number of positively charged protons in the nuclei is still equal to the number of "free" and orbiting negatively charged electrons in the structure.

The effect of this doping process on the relative conductivity can best be described through the use of the energy-band diagram of Fig. 1.7. Note that a discrete energy level (called the *donor* level) appears in the forbidden band with an E_g significantly less than that of the intrinsic material. Those "free" electrons due to the added impurity sit at this energy level and have absolutely no difficulty absorbing a sufficient measure of thermal energy to move into the conduction band at room temperature. The result is that at room temperature, there are a large number of carriers (electrons) in the conduction level and the conductivity of the material increases significantly. At room temperature in an intrinsic Si material there is about one free electron for every 10^{12} atoms (1 to 10^9 for Ge). If our dosage level were 1 in 10 million (10^7), the ratio ($10^{12}/10^7 = 10^5$) would indicate that the carrier concentration has increased by a ratio of 100,000:1.

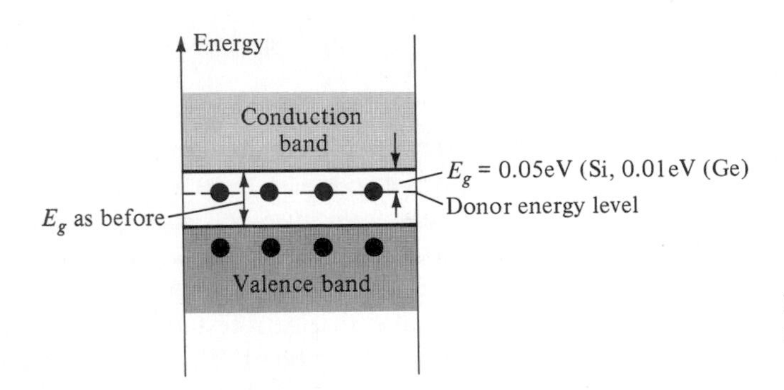

Figure 1.7 Effect of donor impurities on the energy band structure.

p-Type Material

The *p*-type material is formed by doping a pure germanium or silicon crystal with impurity atoms having *three* valence electrons. The elements most frequently used for this purpose are *boron, gallium,* and *indium*. The effect of one of these elements, boron, on a base of silicon is indicated in Fig. 1.8.

Note that there is now an insufficient number of electrons to complete the covalent bonds of the newly formed lattice. The resulting vacancy is called a *hole* and is represented by a small circle or positive sign due to the absence of a negative charge. Since the resulting vacancy will readily *accept* a "free" electron, the impurities added are called *acceptor* atoms. The resulting *p*-type material is electrically neutral, for the same reasons as for the *n*-type material.

The effect of the hole on conduction is shown in Fig. 1.9. If a valence electron acquires sufficient kinetic energy to break its covalent bond and fills the void created by a hole, then a vacancy, or hole, will be created in the covalent bond that released the electron. There is therefore a transfer of holes to the left and electrons to the right,

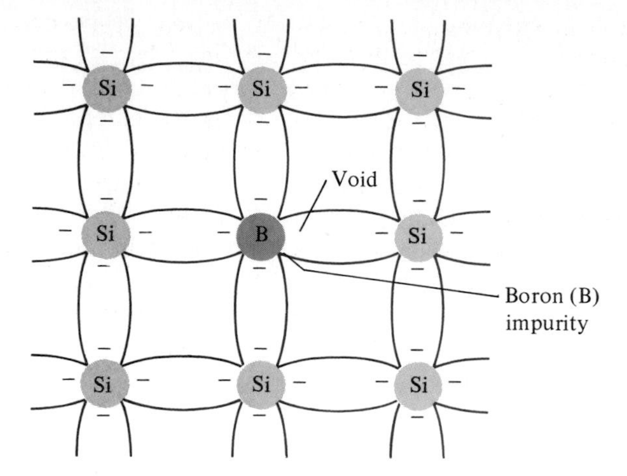

Figure 1.8 Boron impurity in *p*-type material.

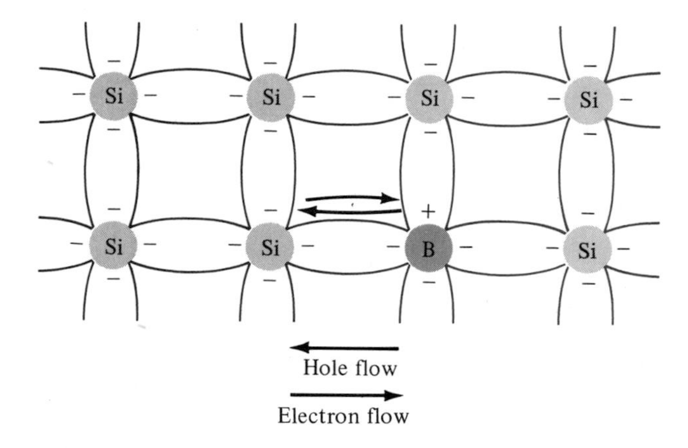

Figure 1.9 Electron versus hole flow.

as shown in Fig. 1.9. The direction to be used in this text is that of *conventional* flow, which is indicated by the direction of hole flow.

In the intrinsic state, the number of free electrons in Ge or Si is due only to those few electrons in the valence band that have acquired sufficient energy from thermal or light sources to break the covalent bond or to the few impurities that could not be removed. The vacancies left behind in the covalent bonding structure represent our very limited supply of holes. In an *n*-type material, the number of holes has not changed significantly from this intrinsic level. The net result, therefore, is that the number of electrons far outweighs the number of holes. For this reason the electron is called the *majority carrier* and the hole the *minority carrier*, as shown in Fig. 1.10a. Note in Fig. 1.10b that the reverse is true for the *p*-type material. When the fifth electron of a donor atom leaves the parent atom, the atom remaining acquires a net positive charge: hence the positive sign in the donor-ion representation. For similar reasons, the negative sign appears in the acceptor ion.

The *n*- and *p*-type materials represent the basic building blocks of semiconductor devices. We will find later in this chapter that the joining of a single *n*-type material with a *p*-type material will result in a semiconductor element of considerable importance in electronic systems.

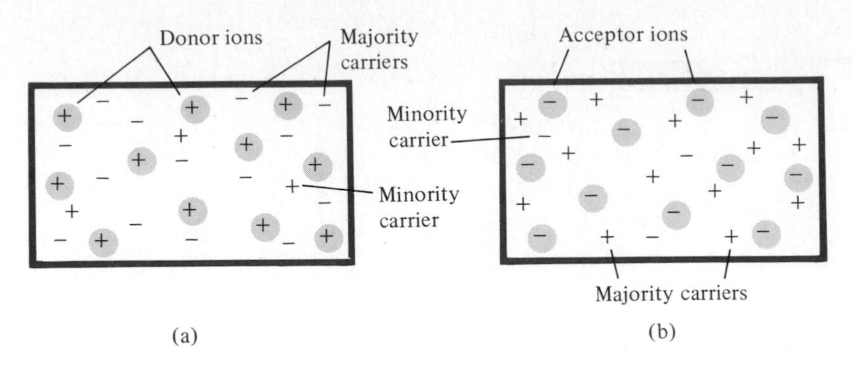

Figure 1.10 (a) n-type material; (b) p-type material.

1.5 IDEAL DIODE

The first electronic device to be introduced is called the *diode*. It is the simplest of semiconductor devices but plays a very vital role in electronic systems, with its characteristics that closely match those of a simple switch. It will appear in a range of applications, extending from the simple to the very complex. In addition to the details of its construction and characteristics, the very important data and graphs to be found on specification sheets will also be covered to ensure an understanding of the terminology employed and to demonstrate the wealth of information typically available from manufacturers.

Before examining the construction and characteristics of an actual device, we first consider the ideal device, to provide a basis for comparison. The *ideal diode* is a *two-terminal* device having the symbol and characteristics shown in Fig. 1.11a and b, respectively.

In the description of the elements to follow, it is critical that the various *letter symbols, voltage polarities,* and *current directions* be defined. If the polarity of the applied voltage is consistent with that shown in Fig. 1.11a, the portion of the characteristics to be considered in Fig. 1.11b, is to the right of the vertical axis. If a reverse voltage is applied, the characteristics to the left are pertinent. If the current through the doide has the direction indicated in Fig. 1.11a, the portion of the characteristics to be considered is above the horizontal axis, while a reversal in direction would require the use of the characteristics below the axis. For the majority of the device

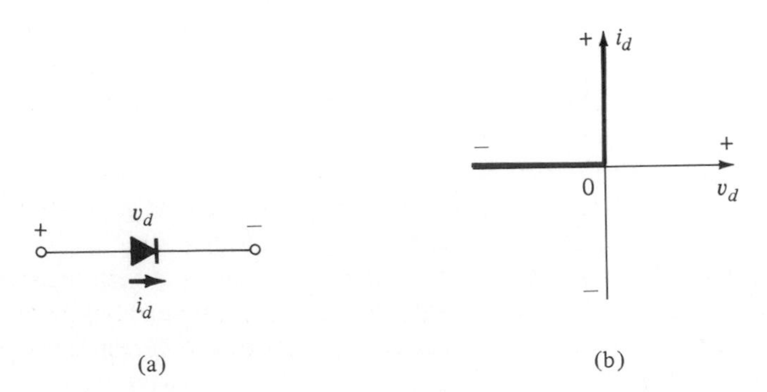

Figure 1.11 Ideal diode: (a) symbol; (b) characteristics.

characteristics to appear in this text the *ordinate* will be the *current* axis, while the *abscissa* will be the *voltage axis*.

One of the important parameters for the diode is the resistance at the point or region of operation. If we consider the region defined by the direction of i_d and polarity of v_d in Fig. 1.11a (upper-right quadrant of Fig. 1.11b), we shall find that the value of the forward resistance, R_f, as defined by Ohm's law is

$$R_f = \frac{V_f}{I_f} = \frac{0}{2,\ 3,\ \text{mA},\ \ldots,\ \text{or any positive value}} = 0\ \Omega$$

where V_f is the forward voltage across the diode and I_f is the forward current through the diode.

The ideal diode, therefore, is a short circuit for the forward region of conduction ($i_d \neq 0$).

If we now consider the region of negatively applied potential (third quadrant) of Fig. 1.11b,

$$R_r = \frac{V_r}{I_r} = \frac{-5,\ -20,\ \text{or any reverse-bias potential}}{0}$$

$$= \text{very large number, which for our purposes we shall consider to be infinite } (\infty)$$

where V_r is the reverse voltage across the diode and I_r is the reverse current in the diode.

The ideal diode, therefore, is an open circuit in the region of nonconduction ($i_d = 0$).

In review, the conditions depicted in Fig. 1.12 are applicable.

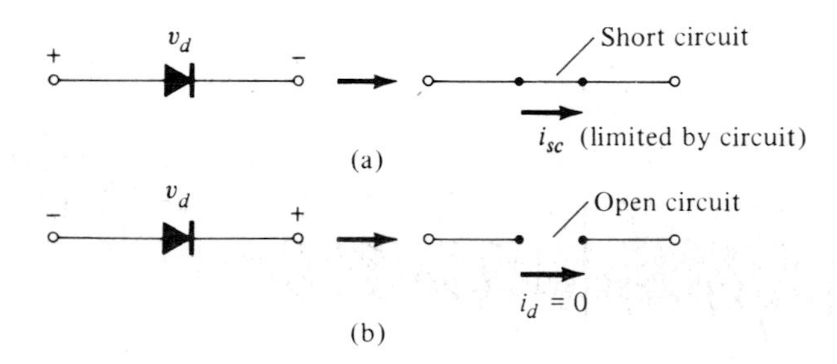

Figure 1.12 (a) Conduction and (b) nonconduction states of the ideal diode as determined by the applied bias.

In general, it is relatively simple to determine whether a diode is in the region of conduction or nonconduction by simply noting the direction of the current i_d to be established by an applied voltage. For convential flow (opposite to that of electron flow), if the resultant diode current has the same direction as the arrowhead of the diode symbol, the diode is operating in the conducting region. This is depicted in Fig. 1.13.

The actual construction and characteristics of a practical diode are introduced in Chapter 2. Compare the levels of forward and reverse resistance with those just obtained and judge whether the ideal diode is one that closely "fits" the real world.

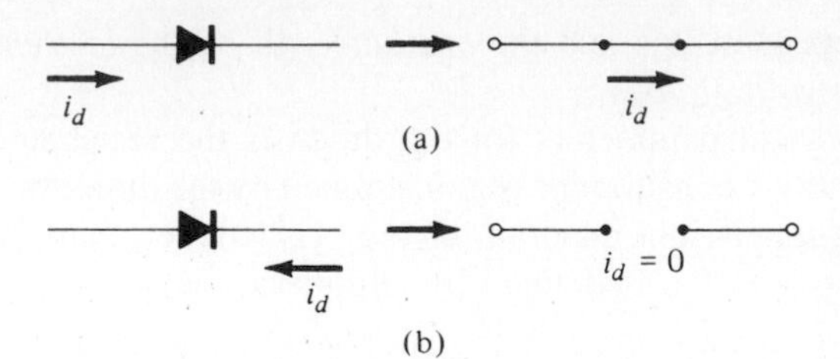

Figure 1.13 (a) Conduction and (b) nonconduction states of the ideal diode as determined by current direction of applied network.

1.6 BASIC CONSTRUCTION AND CHARACTERISTICS

Earlier in this chapter both *n*- and *p*-type materials were introduced. The semiconductor diode is formed by simply bringing these materials together (constructed from the same base—Ge or Si), as shown in Fig. 1.14, using techniques to be described later in the chapter. At the instant the two materials are "joined" the electrons and holes in the region of the junction will combine resulting in a lack of carriers in the region near the junction. This region of uncovered positive and negative ions is called the *depletion* region due to the depletion of carriers in this region.

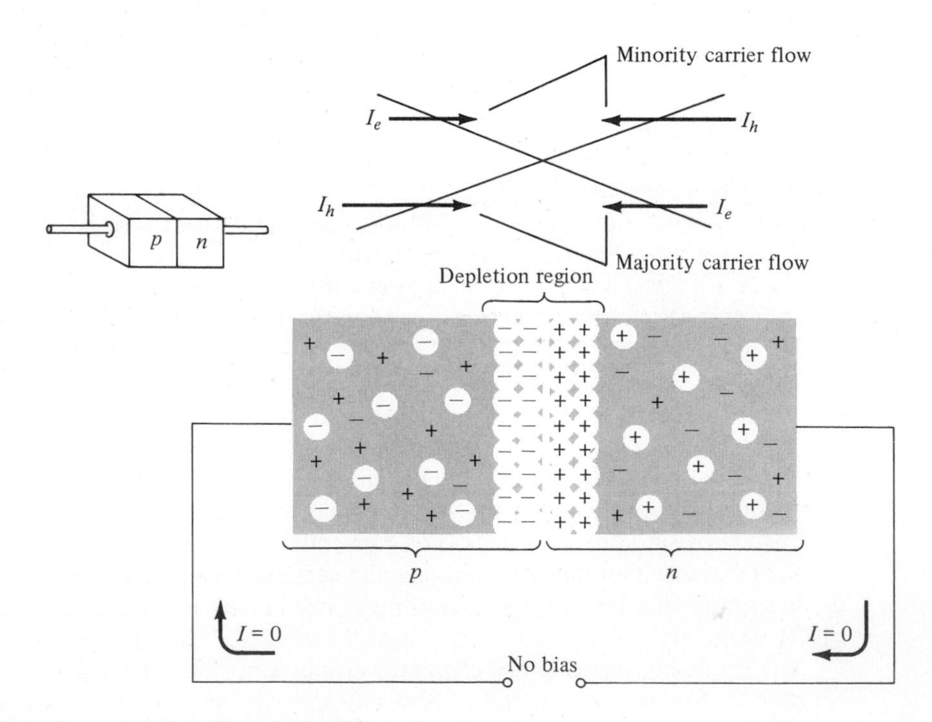

Figure 1.14 *p-n* junction with no external bias.

No Applied Bias

The minority carriers in the *n*-type material that find themselves within the depletion region will pass directly into the *p*-type material. The closer the minority carrier is to the junction, the greater the attraction for the layer of negative ions and the less the opposition of the positive ions in the depletion region of the *n*-type material. For the purposes of future discussions we shall assume that all the minority carriers of the *n*-type material that find themselves in the depletion region due to their random motion will pass directly into the *p*-type material. Similar discussion can be applied to the minority carriers (electrons) of the *p*-type material. This carrier flow has been indicated in Fig. 1.14 for the minority carriers of each material.

The majority carriers in the *n*-type material must overcome the attractive forces of the layer of positive ions in the *n*-type material and the shield of negative ions in the *p*-type material in order to migrate into the neutral region of the *p*-type material. The number of majority carriers is so large in the *n*-type material, however, that there will be invariably a small number of majority carriers with sufficient kinetic energy to pass through the depletion region into the *p*-type material. Again, the same type of discussion can be applied to the majority carriers of the *p*-type material. The resulting flow due to the majority carriers is also shown in Fig. 1.14.

A close examination of Fig. 1.14 will reveal that the relative magnitudes of the flow vectors are such that the net flow in either direction is zero. This cancellation of vectors has been indicated by crossed lines. The length of the vector representing hole flow has been drawn longer than that for electron flow to demonstrate that the magnitude of each need not be the same for cancellation and that the doping levels for each material may result in an unequal carrier flow of holes and electrons. In summary, *the net flow of charge in any one direction with no applied voltage is zero.*

Reverse-Bias Condition

If an external potential of *V* volts is applied across the *p-n* junction such that the positive terminal is connected to the *n*-type material and the negative terminal is connected to the *p*-type material as shown in Fig. 1.15, the number of uncovered positive ions in the depletion region of the *n*-type material will increase due to the large number of "free" electrons drawn to the positive potential of the applied voltage. For similar reasons, the number of uncovered negative ions will increase in the *p*-type material. The net effect, therefore, is a widening of the depletion region. This widening of the depletion region will establish too great a barrier for the majority carriers to overcome, effectively reducing the majority carrier flow to zero (Fig. 1.15).

The number of minority carriers, however, that find themselves entering the depletion region will not change, resulting in minority-carrier flow vectors of the same magnitude indicated in Fig. 1.14 with no applied voltage. The current that exists under these conditions is called the *reverse saturation current* and is represented by the subscript *s*. It is seldom more than a few microamperes in magnitude except for high-power devices. The term "saturation" comes from the fact that it reaches its maximum value quickly and does not change significantly with increase in the reverse-bias potential. The situation depicted in Fig. 1.15 is referred to as a *reverse-bias* condition.

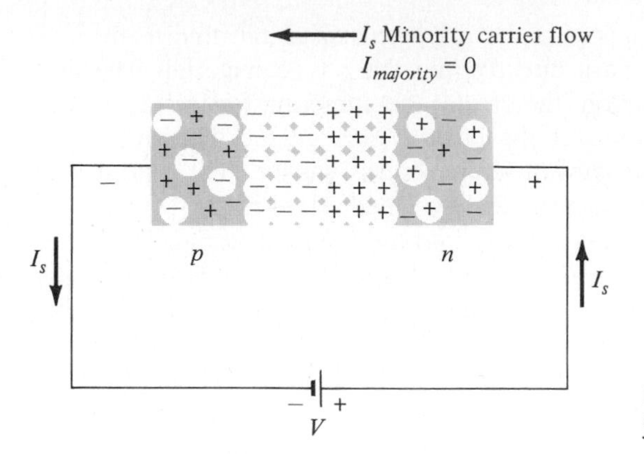

Figure 1.15 Reverse-biased *p-n* junction.

Forward-Bias Condition

A *forward-bias* condition is established by applying the positive potential to the *p*-type material and the negative potential to the *n*-type material (for future reference, note that the forward-bias condition is defined by the corresponding first letter in *p*-type and *p*ositive or in *n*-type and *n*egative), as shown in Fig. 1.16. Note that the minority-carrier flow has not changed in magnitude, but the reduction in the width of the depletion region has resulted in a heavy majority flow across the junction. The magnitude of the majority-carrier flow will increase exponentially with increasing forward bias, as indicated in Figs. 1.17 and 1.18. Note in Fig 1.17 the similarities with the ideal diode except at a very large negative voltage. The offset in the first quadrant and the sharp drop in the third quadrant will be examined in this chapter. To reiterate, the first quadrant represents the forward-bias region, and the third quadrant the reverse-bias region. Note the extreme change in scales for both the voltage and current in Fig. 1.18. For the current it is a 5000 : 1 change. The vertical scale for the majority

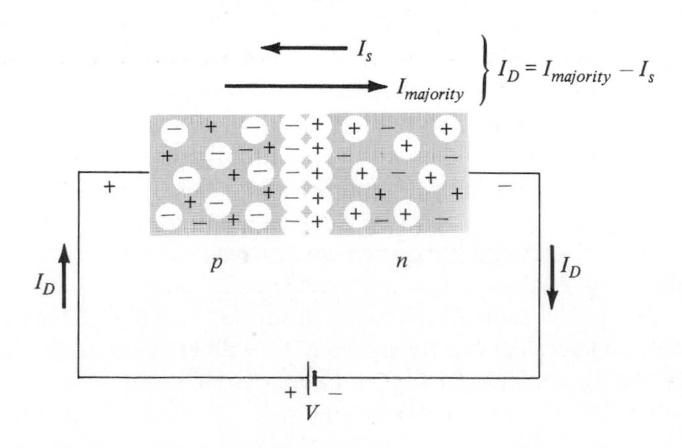

Figure 1.16 Forward-biased *p-n* junction.

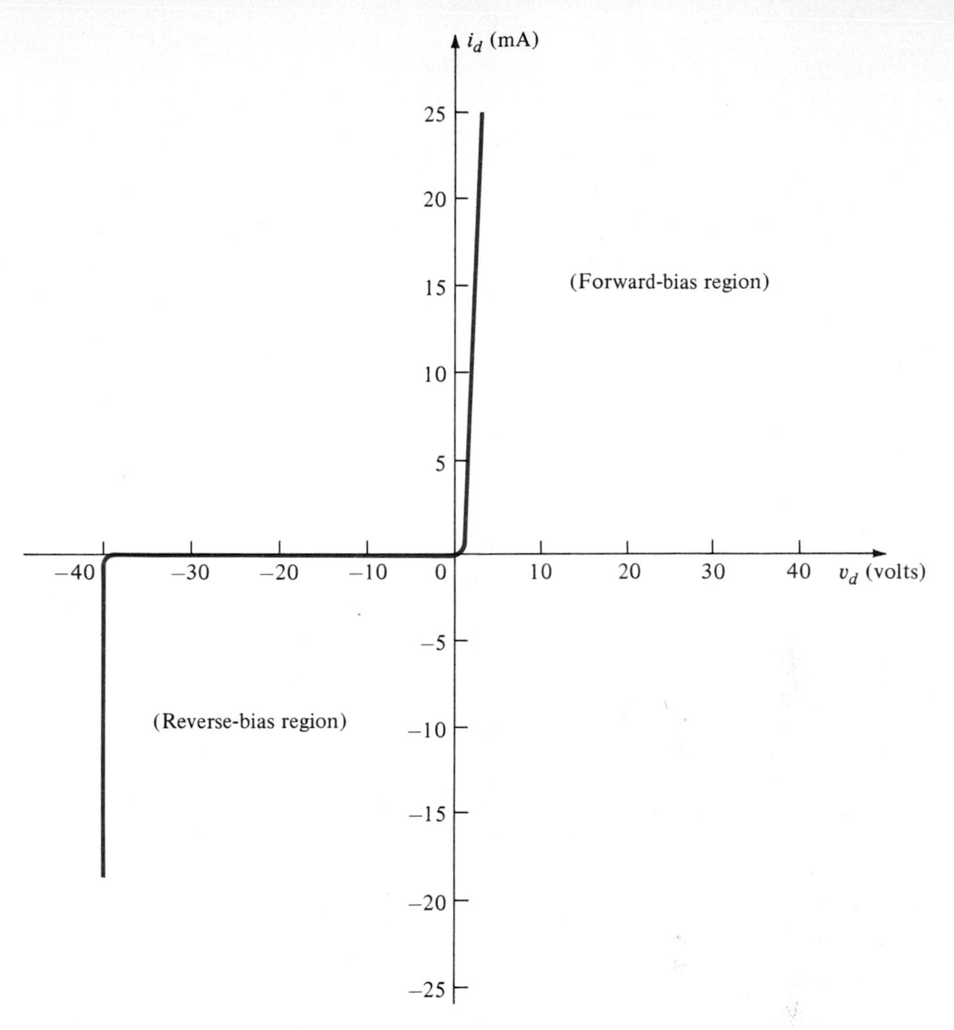

Figure 1.17 Semiconductor diode characteristics—continuous scale on the vertical and horizontal axes.

of smaller units is the milliampere, as shown in Fig. 1.18. However, semiconductor diodes are available today with ampere in the vertical scale, even though the diameter of the casing may not be greater than $\frac{1}{4}$ in.

It can be demonstrated through the use of solid-state physics that this diode current can be mathematically related to temperature (T_K) and applied bias (V) in the following manner:

$$\boxed{I = I_s(e^{kV/T_K} - 1)} \tag{1.4}$$

where I_s = reverse saturation current
$k = 11{,}600/\eta$
with $\eta = 1$ for Ge and 2 for Si for small values of i_d

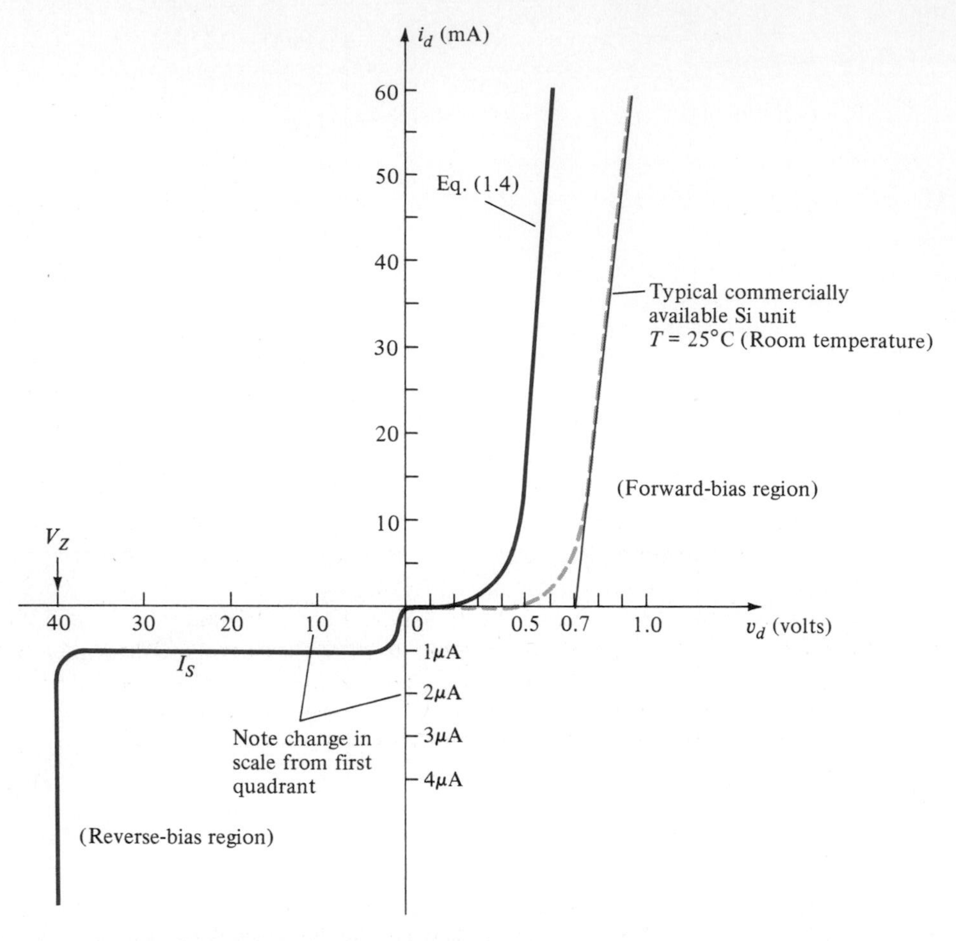

Figure 1.18 Semiconductor diode (Si) characteristics.

and $\eta = 1$ for Ge and Si in the vertical section of the characteristics

$T_K = T_C + 273°$ (T_K—degrees Kelvin, T_C—degrees Celsius)

Note the exponential factor that will result in a very sharp increase in I with increasing levels of V. The characteristics of a commercially available silicon (Si) diode will differ slightly from the characteristics of Fig. 1.18 because of the *body* or *bulk* resistance of the semiconductor material and the *contact* resistance between the semiconductor material and the external metallic conductor. They will cause the curve to shift slightly in the forward-bias region, as indicated by the dashed line in Fig. 1.18. As construction techniques improve and these undesired resistance levels are reduced, the commercially available unit will approach the characteristic defined by Eq. (1.4).

In an effort to demonstrate that Eq. (1.4) does in fact represent the curves of Fig. 1.18, let us determine the current I for the forward-bias voltage of 0.5 V at room temperature (25°C).

$$I_S = 1\ \mu A = 1 \times 10^{-6}\ A$$

$$T_K = T_C + 273° = 25° + 273° = 298°$$

$$k(\text{Si}) = \frac{11,600}{2} = 5800$$

$$\frac{kV}{T_K} = \frac{(5800)(0.5)}{298} = 9.732$$

and $$I = I_s(e^{9.732} - 1) = (1 \times 10^{-6})(16848 - 1) = 16.848 \times 10^{-3}$$

so that $$I \cong \mathbf{16.8\ mA}$$

as verified by Fig. 1.18.

Temperature can have a marked effect on the diode current. This is clearly demonstrated by the factor T_K in Eq. (1.4). The effect of varying T_K will be determined for the forward-bias condition in the problems appearing at the end of the chapter. In the reverse-bias region it has been found experimentally that the *reverse saturation current I_s will almost double in magnitude for every 10°C change in temperature*. It is not uncommon for a germanium diode with an I_s in the order of 1 or 2 μA at 25°C to have a leakage current of 100 μA = 0.1 mA at a temperature of 100°C. Current levels of this magnitude in the reverse-bias region would certainly question our desired open-circuit condition in the reverse-bias region. Typical values of I_s for silicon are much lower than that of germanium for similar power and current levels. The result is that even at high temperatures the levels of I_s for silicon diodes do not reach the same high levels obtained for germanium—a very important reason that silicon devices enjoy a significantly higher level of development and utilization in design. Fundamentally, the open-circuit equivalent in the reverse-bias region is better realized at any temperature with silicon than with germanium.

Zener Region

Note the sharp change in the characteristics of Fig. 1.19 at the reverse-bias potential V_Z (the subscript Z refers to the name Zener). This constant-voltage effect is induced by a high reverse-bias voltage across the diode. When the applied reverse potential becomes more and more negative, a point is eventually reached where the few free minority carriers have developed sufficient velocity to liberate additional carriers through ionization. That is, they collide with the valence electrons and impart sufficient energy to them to permit them to leave the parent atom. These additional carriers can then aid the ionization process to the point where a high *avalanche* current is established and the *avalanche breakdown* region determined.

The avalanche region (V_Z) can be brought closer to the vertical axis by increasing the doping levels in the p- and n-type materials. However, as V_Z decreases to very low levels, such as −5 V, another mechanism, called *Zener breakdown*, will contribute to the sharp change in the characteristic. It occurs because there is a strong electric field in the region of the junction that can disrupt the bonding forces within the atom and "generate" carriers. Although the Zener breakdown mechanism is only a significant contributor at lower levels of V_Z, this sharp change in the characteristic at any level is called the *Zener region* and diodes employing this unique portion of the characteristic of a p-n junction are called *Zener diodes*. They are described in detail in Chapter 3.

The Zener region of the semiconductor diode described must be avoided if the

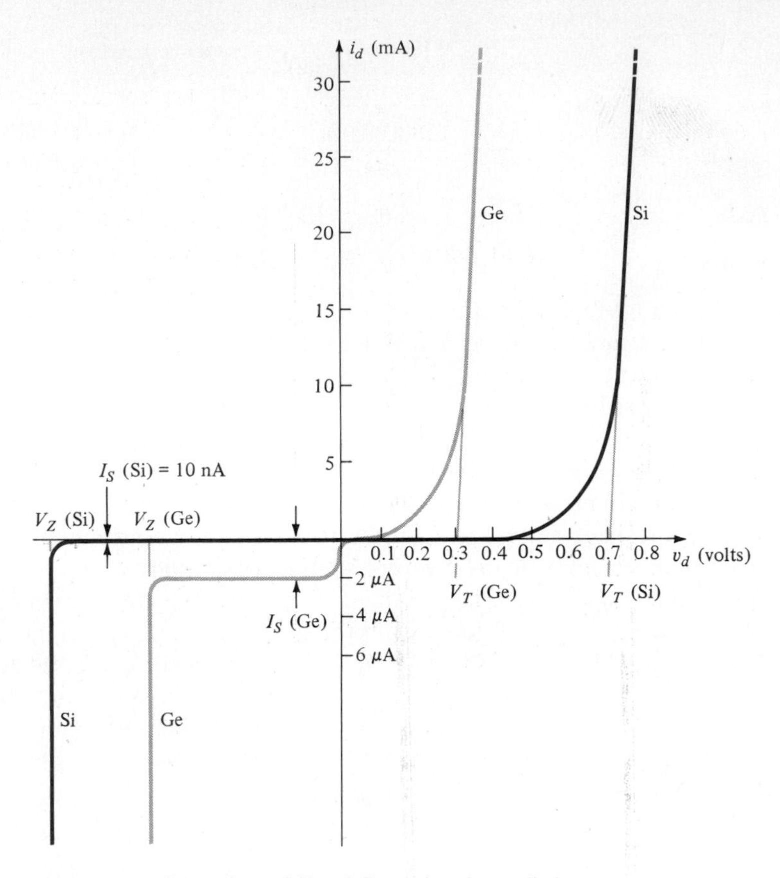

Figure 1.19 Comparison of Si and Ge semiconductor diodes.

response of a system is not to be completely altered by the sharp change in characteristics in this reverse-voltage region. The maximum reverse-bias potential that can be applied before entering this region is called the *peak inverse voltage* (referred to simply as the PIV rating), or the *peak reverse voltage* (denoted by PRV rating). If an application requires a PIV rating greater than that of a single unit, a number of diodes of the same characteristics can be connected in series. Diodes are also connected in parallel to increase the current-carrying capacity.

Silicon versus Germanium

Silicon diodes have, in general, higher PIV and current ratings and wider temperature ranges than germanium diodes. PIV ratings for silicon can be in the neighborhood of 1000 V, whereas the maximum value for germanium is closer to 400 V. Silicon can be used for applications in which the temperature may rise to about 200°C (400°F), whereas germanium has a much lower maximum rating (100°C). The disadvantage of silicon, however, as compared to germanium, as indicated in Fig. 1.18, is the higher forward-bias voltage required to reach the region of upward swing. It is typically of the order of magnitude of 0.7 V for *commercially* available silicon diodes and 0.3 V for germanium diodes. The increased offset for silicon is due primarily to the factor η in Eq. (1.4). This factor only plays a part in determining the shape of the curve at very low current levels. Once the curve starts its vertical rise, the factor η drops to 1 (the continuous value for germanium). This is evidenced by the similarities in the curves once the offset potential is reached. The potential at which this rise occurs is commonly referred to as the *offset, threshold,* or *firing potential*. Frequently, the first letter of a term that describes a particular quantity is used in the notation for that quantity. However, to ensure a minimum of confusion with other terms, such as output voltage (V_o) and forward voltage (V_F), the notation V_T has been adopted for this text, from the word "threshold."

In review:

$$V_T = 0.7 \text{ (Si)}$$
$$V_T = 0.3 \text{ (Ge)}$$

Obviously, the closer the upward swing is to the vertical axis, the more "ideal" the device. However, the other characteristics of silicon as compared to germanium still make it the choice in the majority of commercially available units.

1.7 DC OR STATIC RESISTANCE

The resistance of a diode at a particular operating point is called the *dc* or *static* resistance of the diode. It is determined by

$$R_{dc} = \frac{V_D}{I_D} \quad \text{ohms} \tag{1.5}$$

For the ideal diode of Fig. 1.20 the dc resistance at $i_d = 20$ mA is

$$R_{dc} = \frac{V_D}{I_D} = \frac{0}{20 \text{ mA}} = 0\ \Omega$$

as expected, while the dc resistance of the silicon diode is

$$R_{dc} = \frac{V_D}{I_D} = \frac{0.8}{20 \text{ mA}} = 40\ \Omega$$

At $i_d = 2$ mA the resistance of the ideal diode remains as 0 Ω, but the resistance of the silicon diode is now

$$R_{dc} = \frac{V_D}{I_D} = \frac{0.5}{2 \text{ mA}} = 250\ \Omega$$

The results clearly indicate that the dc resistance of a diode in the forward-bias region decreases as you approach the region of higher currents and voltages.

In the reverse-bias region at $v_d = -10$ V the resistance of the ideal diode is

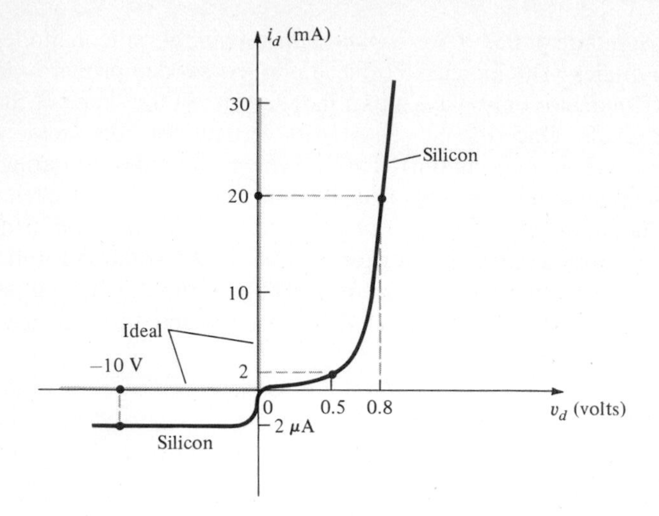

Figure 1.20

essentially infinite resistance (open-circuit equivalent), as determined by

$$R_{dc} = \frac{V_D}{I_D} = \frac{-10}{0} = \longrightarrow \Omega$$

while the resistance of the silicon diode is

$$R_{dc} = \frac{V_D}{I_D} = \frac{-10}{-2\ \mu A} = 5\ M\Omega$$

which is certainly equivalent to an open circuit for many applications.

Once the dc resistance is known at a particular operating point, the diode can be replaced by a resistive element as shown in Fig. 1.21 and the analysis can continue.

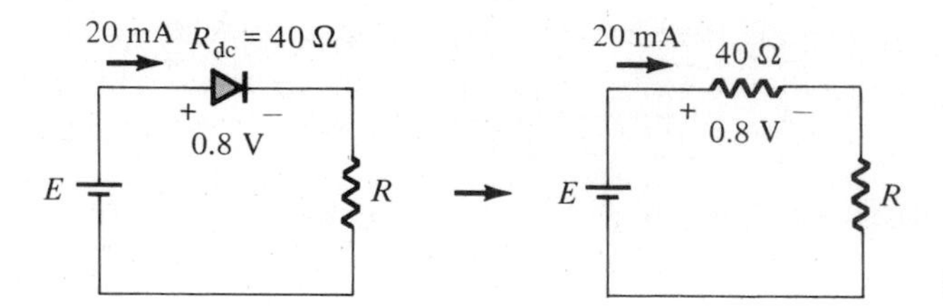

Figure 1.21

1.8 AC OR DYNAMIC RESISTANCE

It is obvious from Fig. 1.20 that the dc resistance of a diode is independent of the shape of the characteristic in the region surrounding the point of interest. If a sinusoidal rather than dc input is applied, the situation will change completely. The varying input will move the instantaneous operating point up and down a region of the characteristics and define a specific change in current and voltage as shown in Fig. 1.22. With no

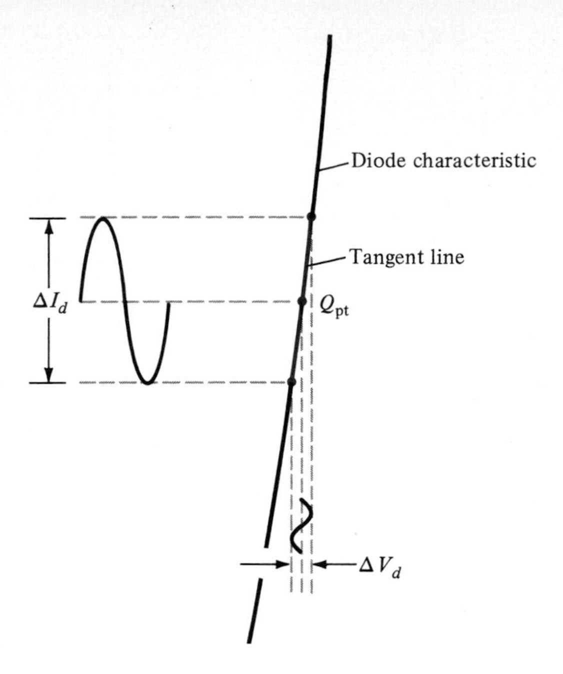

Figure 1.22

applied varying signal, the point of operation would be the Q-point appearing on Fig. 1.22 determined by the applied dc levels. The designation *Q-point* is derived from the word *quiescent,* which means "still of unvarying level."

A straight line drawn tangent to the curve through the Q-point will define a particular change in voltage and current that can be used to determine the *ac* or *dynamic* resistance for this region of the diode characteristics. In equation form,

$$r_{ac} = r_d = \frac{\Delta V_d}{\Delta I_d} \tag{1.6}$$

The steeper the slope, the less the value of ΔV_d for the same change in ΔI_d and the less the resistance. The ac resistance in the vertical-rise region of the characteristic is therefore quite small, while the ac resistance is much higher at low current levels.

EXAMPLE 1.1

For the characteristics of Fig. 1.23:
(a) Determine the ac resistance for region 1.
(b) Determine the ac resistance for region 2.
(c) Compare the results of parts (a) and (b).

Solution:

(a) For region 1:

$$\Delta V_d \cong 0.72 - 0.57 = 0.15 \text{ V} \qquad \Delta I_d \cong (6 - 2) \text{ mA} = 4 \text{ mA}$$

and
$$r_{d_1} = \frac{\Delta V_d}{\Delta I_d} = \frac{0.15 \text{ V}}{4 \text{ mA}} = \mathbf{37.5\ \Omega}$$

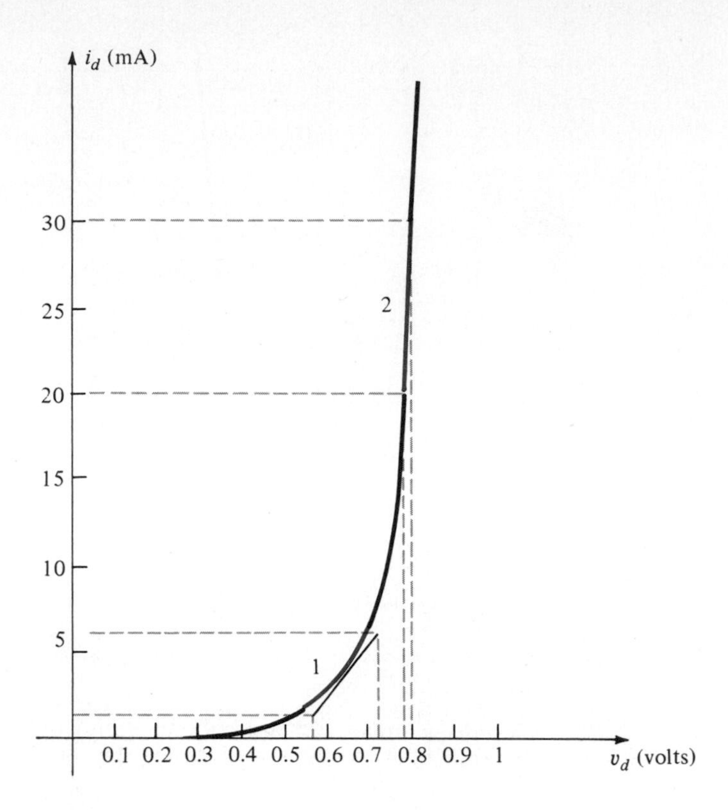

Figure 1.23

(b) For region 2:

$$\Delta V_d \cong 0.8 - 0.78 = 0.02 \text{ V} \qquad \Delta I_d \cong (30 - 20) \text{ mA} = 10 \text{ mA}$$

and
$$r_{d_2} = \frac{\Delta V_d}{\Delta I_d} = \frac{0.02 \text{ V}}{10 \text{ mA}} = \mathbf{2\ \Omega}$$

(c) The ratio $r_{d_1}:r_{d_2} = 37.5:2 =$ **18.75:1**.

There is a basic definition in differential calculus that states that *the derivative of a function at a point is equal to the slope of a tangent line drawn at that point.* Equation (1.6), as defined by Fig. 1.22, is, therefore, essentially finding the derivative of the function at the Q-point of operation. If we find the derivative of the general equation [Eq. (1.4)] for the semiconductor diode with respect to the applied forward bias and then invert the result, we will have an equation for the dynamic or ac resistance in that region. That is, taking the derivative of Eq. (1.4) with respect to the applied bias will result in

$$\frac{d}{dV}(I) = \frac{d}{dV}[I_s(e^{kV/T_K} - 1)]$$

and
$$\frac{dI}{dV} = \frac{k}{T_K}(I + I_s)$$

following a few basic maneuvers of differential calculus. In general, $I \gg I_s$ and

$$\frac{dI}{dV} = \frac{k}{T_K} I$$

Substituting $\eta = 1$ for Ge and Si in the vertical-rise section of the characteristics, we obtain

$$k = \frac{11{,}600}{\eta} = \frac{11{,}600}{1} = 11{,}600$$

and at room temperature,

$$T_K = T_C + 273° = 25° + 273° = 298°$$

so that

$$\frac{k}{T_K} = \frac{11{,}600}{298} \cong 38.93$$

and

$$\frac{dI}{dV} = 38.93I$$

Flipping the result to define a resistance ratio ($R = V/I$) gives us

$$\frac{dV}{dI} \cong \frac{0.026}{I}$$

or

$$\boxed{r_d = \frac{dV}{dI} = \frac{26 \text{ mV}}{I_D \text{ (mA)}}}_{\text{Ge, Si}} \qquad (1.7)$$

The significance of Eq. (1.7) must be clearly understood. It implies that the dynamic resistance can be found by simply substituting the quiescent value of the diode current into the equation. There is no need to have the characteristics available or to worry about sketching tangent lines as defined by Eq. (1.6).

We already realize from Eq. (1.6) that the shape of the curve will have an effect on the dynamic resistance. The fact that the silicon and germanium curves in Fig. 1.19 are almost identical after they begin their vertical rise would suggest that the equation for the dynamic resistance of each might be the same as indicated by Eq. (1.7).

It was already noted on Fig. 1.18 that the characteristics of the commercial unit are slightly different from those determined by Eq. (1.4) because of the bulk and contact resistance of the semiconductor device. This additional resistance level can be included in Eq. (1.7) by adding a factor denoted r_B as appearing in Eq. (1.8).

$$\boxed{r'_d = \frac{26 \text{ mV}}{I_D \text{ (mA)}} + r_B} \qquad \text{ohms} \qquad (1.8)$$

The factor r_B (measured in ohms) can range from typically 0.1 for high-power devices to 2 for some low-power, general-purpose diodes. As construction techniques improve, this additional factor will continue to decrease in importance until it can be dropped and Eq. (1.7) applied. For values of I_D in mA, the units of the first term are like those of r_B: ohms. For low levels of current, the first factor of Eq. (1.8) will certainly predominate. Consider

$$I_D = 1 \text{ mA}$$

with $$r_B = 2\ \Omega$$

Then $$r_d' = \frac{26}{1} + 2 = \mathbf{28\ \Omega}$$

At higher levels of current the second factor may predominate. Consider

$$I_D = 52\text{ mA}$$

with $$r_B = 2\ \Omega$$

Then $$r_d' = \frac{26}{52} + 2 = 0.5 + 2 = \mathbf{2.5\ \Omega}$$

For Example 1.1 the ac resistance at 25 mA was calculated to be 2 Ω. Using Eq. (1.7), we have

$$r_d = \frac{26\text{ mV}}{I_D\text{ (mA)}} = \frac{26\text{ mV}}{25\text{ mA}} = 1.04\ \Omega$$

The difference of about 1 Ω could be treated as the contribution of r_B.

At $I_D = 4$ mA, the ac resistance was calculated to be about 37.5 Ω. Using Eq. (1.7) yields

$$r_d = \frac{26\text{ mV}}{I_D\text{ (mA)}} = \frac{26\text{ mV}}{4\text{ mA}} = 6.5\ \Omega$$

which is quite different from 37.5 Ω. Recall, however, that Eqs. (1.7) and (1.8) are defined only for the vertical-rise section of the characteristics, where $\eta = 1$.

The question of how one is to determine which value to choose for r_B will probably arise. For some devices 2 Ω will be an excellent choice, while for others the approximate average of 1 Ω will perhaps be more appropriate. Certainly, 2 Ω could always be used as a worst-case design approach. However, it would appear that technology is reaching the point where an average value of 1 Ω would, in general, be more appropriate. Of course, the problem of choosing a correct value only arises in the intermediate range of current levels. At low levels of current either choice of r_B would be an insignificant factor. At higher levels the resistance level is so low in comparison to the other series elements that it can probably be ignored. For the purposes of this text the value of r_B chosen for an example will be directly related to the current level; it will extend from a minimum value at high currents of 0.1 Ω to a maximum value of 2 Ω at low levels. Experience will develop a sense for what value to choose, and, indeed, whether it is a factor of significance at all.

In summary, keep in mind that the static or dc resistance of a diode is determined solely by the point of operation, while the dynamic resistance is determined by the shape of the curve to the region of interest.

1.9 AVERAGE AC RESISTANCE

If the input signal is sufficiently large to produce the type of swing indicated in Fig. 1.24, the resistance associated with the device for this region is called the *average ac resistance*. The average ac resistance is, by definition, the resistance determined by

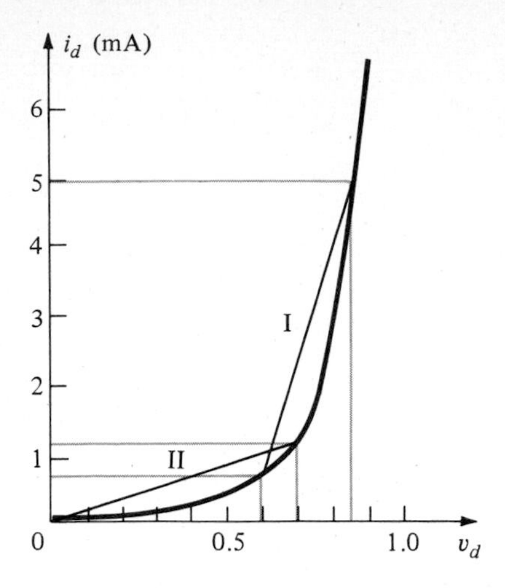

Figure 1.24 Average ac resistance.

a straight line drawn between the two intersections determined by the maximum and minimum values of input voltage. In equation form (note Fig. 1.24)

$$r_{av} = \left.\frac{\Delta V_d}{\Delta I_d}\right|_{\text{pt. to pt.}} \tag{1.9}$$

For the situation indicated by Fig. 1.24 in region I,

$$r_{av} = \left.\frac{\Delta V_d}{\Delta I_d}\right|_{\text{pt. to pt.}} = \frac{0.85 - 0.6}{(5 - 0.75) \times 10^{-3}} = \frac{0.25}{4.25 \times 10^{-3}} = 58.8\ \Omega$$

For region II,

$$r_{av} = \left.\frac{\Delta V_d}{\Delta I_d}\right|_{\text{pt. to pt.}} = \frac{0.7 - 0}{(1.2 - 0) \times 10^{-3}} = 583.3\ \Omega$$

Note the significant increase in resistance as you progress down the curve. For curves in which the current is the vertical axis and the voltage the horizontal, it is useful to remember that the more horizontal the region, the higher the resistance. Or, if preferred, the more vertical the region, the less the resistance.

It is important to note in this discussion of average ac resistance that the resistance to be associated with the element is determined *only* by the region of interest, *not* by the entire characteristic.

1.10 EQUIVALENT CIRCUITS—DIODE MODELS

An equivalent circuit is a combination of elements properly chosen to best represent the actual terminal characteristics of a device, system, and so on. That is, once the equivalent circuit is determined, the device symbol can be removed from a schematic and the equivalent circuit inserted in its place without severely affecting the behavior of the overall system.

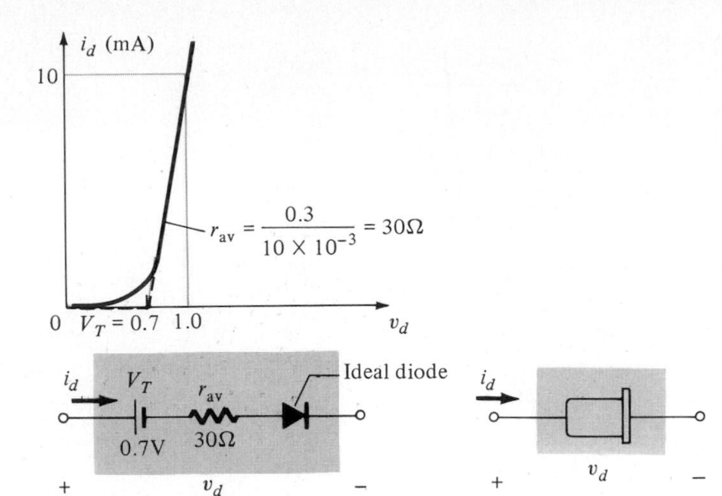

Figure 1.25 Piecewise linear equivalent circuits for a semiconductor diode.

One technique for obtaining an equivalent circuit for a diode is to approximate the characteristics of the device by straight-line segments, as shown in Fig. 1.25. This type of equivalent circuit is called a *piecewise-linear equivalent circuit*. It should be obvious from each curve that the straight-line segments do not result in an exact equivalence between the characteristics and the equivalent circuit. It will, however, at least provide a *first approximation* to its terminal behavior. In each case the resistance chosen is the average ac resistance as defined by Eq. (1.9). The equivalent circuit appears below the curve in Fig. 1.25. The ideal diode was included to indicate that there is only one direction of conduction through the device and that the reverse-bias state is an open-circuit state.

Since a silicon semiconductor diode does not reach the conduction state ("fire") until approximately 0.7 V, an opposing battery V_T of this value must appear in the equivalent circuit. This indicates that the total forward voltage V_D across the diode must be greater than V_T before the ideal diode in the equivalent circuit will be forward biased.

Keep in mind, however, that V_T is not an independent source of energy in a system. You will not measure a voltage $V_T = 0.7$ V across an isolated silicon diode using simply a voltmeter. It is simply a useful way of representing the horizontal offset of the semiconductor diode.

The value of r_{av} can usually be determined purely from a few numerical values given on a specification sheet. The complete characteristics, therefore, are usually unnecessary for this calculation. For instance, for a semiconductor diode, if $I_F = 10$ mA, at 1 V, we know that for silicon a shift of 0.7 V is required before the characteristics rise and

$$r_{av} = \frac{1 - 0.7}{10 \text{ mA}} = \frac{0.3}{10 \text{ mA}} = 30 \ \Omega$$

For most applications, the resistance r_{av} is sufficiently small to be ignored in comparison to the other elements of the network. The removal of r_{av} from the equivalent circuit is the same as implying that the characteristics of the diode appear as shown in Fig. 1.26. Indeed, this approximation is frequently employed in semicon-

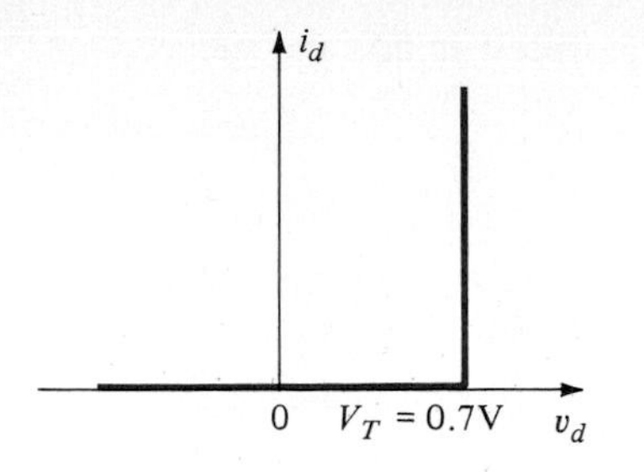

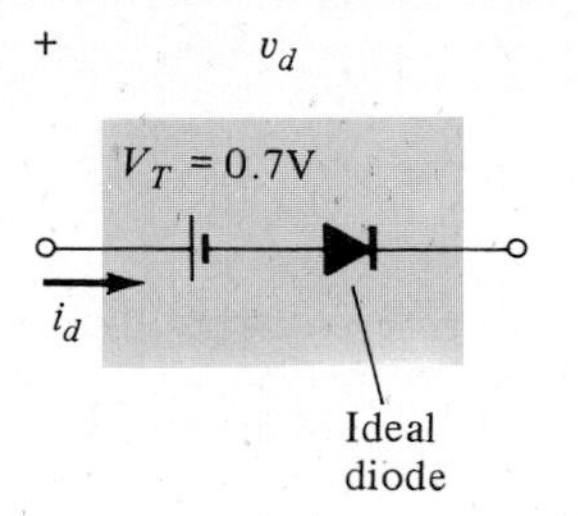

Figure 1.26 Approximate equivalent circuit for the silicon semiconductor diode.

ductor circuit analysis. The reduced equivalent circuit appears in the same figure. It states that a forward-biased silicon diode in an electronic system under dc conditions has a drop of 0.7 V across it in the conduction state no matter what the diode current (within rated values, of course).

In fact, we can now go a step further and say that the 0.7 V in comparison to the applied voltages can often be ignored, leaving only the ideal diode as an equivalent for the semiconductor device. It is for this very reason that many of the applications to follow in later sections use ideal diodes rather than the complete equivalent. Except for small applied voltages or series resistances, it is never too far from the actual response and it does not cloud the application with a great deal of mathematical exercises.

In industry a popular substitution for the phrase "diode equivalent circuit" is diode *model*—a model by definition being a representation of an existing device, object, system, and so on. In fact, this substitute terminology will be used almost exclusively in the chapters to follow.

For clarity, the diode models employed for the range of circuit parameters and applications are provided in Fig. 1.27 with their piecewise-linear characteristics. Each

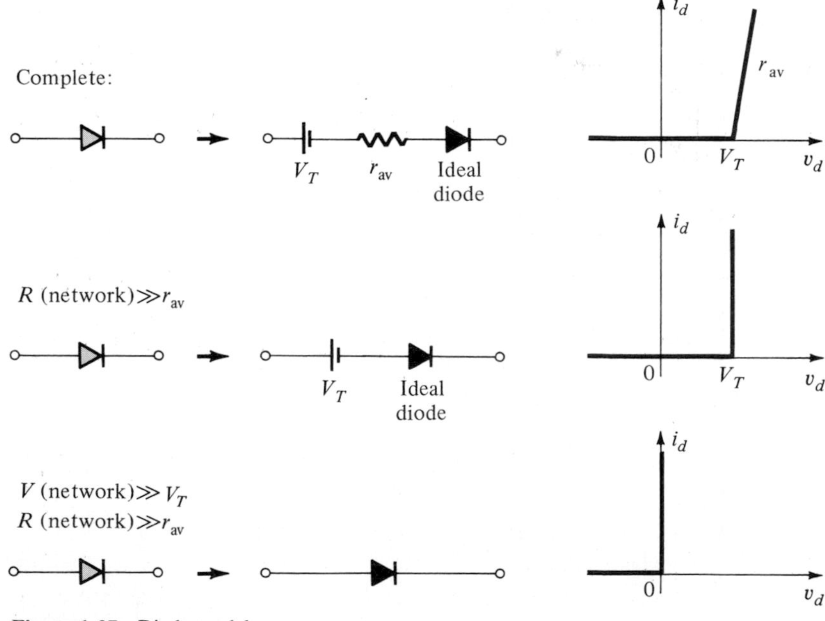

Figure 1.27 Diode models.

will be applied in an application or two in the analysis to follow in this and succeeding chapters.

EXAMPLE 1.2

Given the network of Fig. 1.28:
(a) Determine which model for the silicon diode appears to be the most appropriate for the level of circuit parameters.
(b) Calculate the resulting current and voltage for the resistor R.

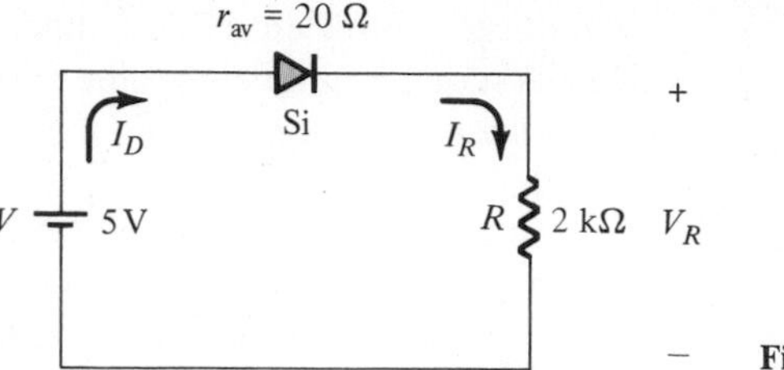

Figure 1.28

Solution:

(a) Since R is much greater than r_{av} of the diode, r_{av} can be dropped on an approximate basis. V_T, however, is 14% of V and therefore should be included. The chosen model appears in Fig. 1.29.

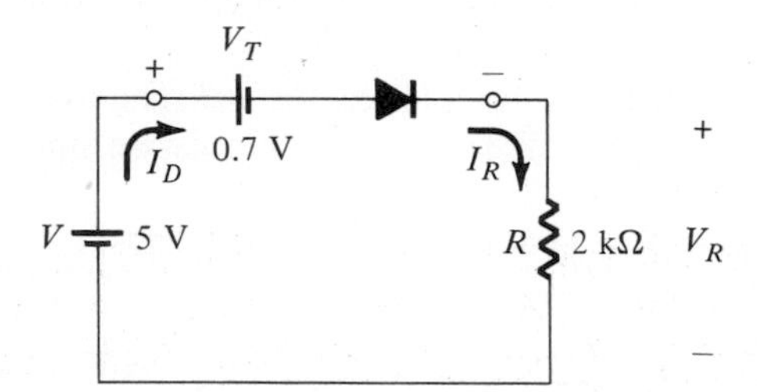

Figure 1.29

(b) The applied voltage has established a voltage across the diode that places the diode in the short-circuit state. Replacing the diode by the short-circuit equivalent will result in the network of Fig. 1.30, where it is clear that

$$V_R = V - V_T = 5 - 0.7 = \mathbf{4.3\ V}$$

and

$$I_D = I_R = \frac{V_R}{R} = \frac{4.3}{2\ k\Omega} = \mathbf{2.15\ mA}$$

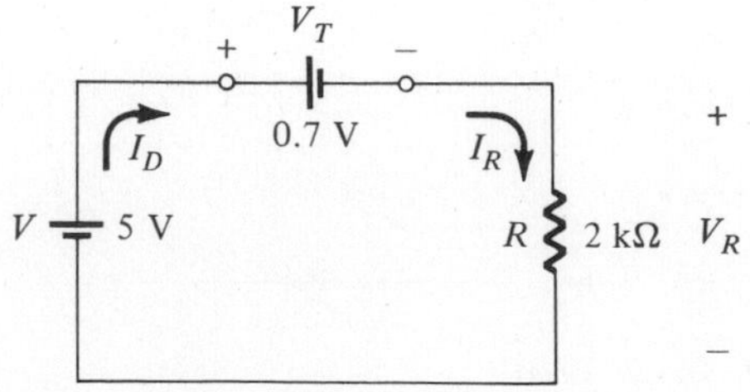

Figure 1.30

Chapter 2 includes a wide variety of applications of the diode employing the models of Fig. 1.27. This chapter is devoted primarily to introducing the basic operation, characteristics, models, and construction techniques of semiconductor diodes.

1.11 DRIFT AND DIFFUSION CURRENTS

The flow of charge, or current, through a semiconductor material is normally referred to as one of two types: drift and diffusion. *Drift current* relates directly to the mechanism encountered in the flow of charge in a conductor. When a voltage is applied across the material as shown in Fig. 1.31, the electrons are naturally drawn to the positive end of the sample. However, collisions with the other atoms, ions, and carriers encountered in their movement may result in an erratic path, as shown in the figure. The net result, however, is a drift of carriers to the positive end.

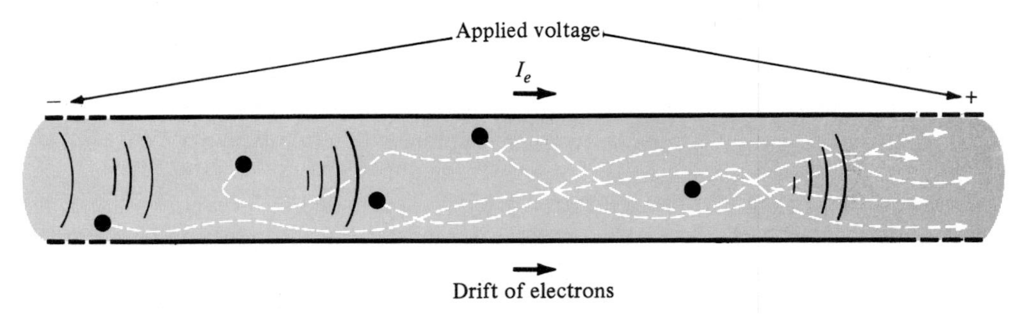

Figure 1.31 Drift current.

The concept of *diffusion current* is best described by considering the effect of placing a drop of dye into a clear pool of water. The heavy concentration of dye will eventually diffuse through the clear water. The darker color of the heavy concentration of dye will give way to a much lighter shade as it spreads out through the liquid. The same effect will take place in a semiconductor material if a heavy concentration of carriers is introduced to a region as shown in Fig. 1.32a. In time they will distribute themselves evenly through the material, as shown in Fig. 1.32b. This movement is due

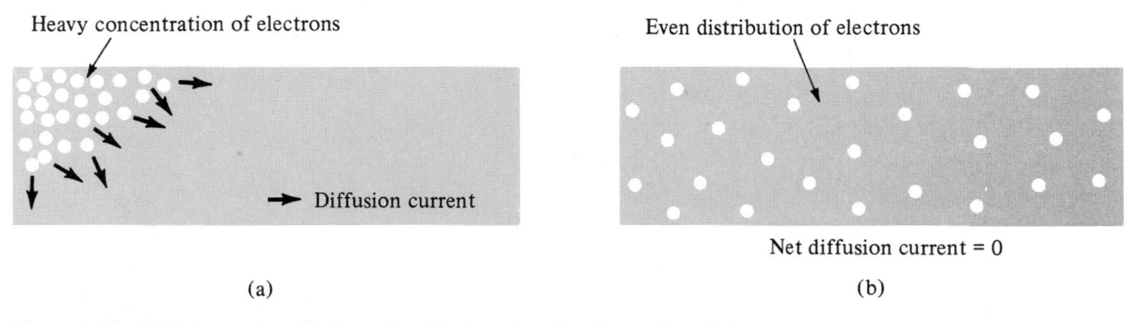

Figure 1.32 Diffusion current: (a) heavy introduction of carriers in a region of the semiconductor material; (b) steady-state conduction.

only to interaction between neighboring atoms; there is no applied source of energy as is required for drift current. Diffusion current is an important consideration in the examination of minority-carrier flow in *n*- and *p*-type materials. The diffusion phenomenon is also important as a technology for the doping process and must be carefully investigated when models are constructed for semiconductor devices (diffusion capacitance, etc.).

1.12 TRANSITION AND DIFFUSION CAPACITANCE

Electronic devices are inherently sensitive to very high frequencies. Most shunt capacitive effects that can be ignored at lower frequencies because the reactance $X_C = 1/2\pi fC$ is very large (open-circuit equivalent) cannot be ignored at very high frequencies. X_C will become sufficiently small due to the high value of f to introduce a low-reactance "shorting" path. In the *p-n* semiconductor diode, there are two capacitive effects to be considered. Both types of capacitance are present in the forward- and reverse-bias regions, but one so outweighs the other in each region that we consider the effects of only one in each region. In the reverse-bias region we have the *transition-* or *depletion-*region capacitance (C_T), while in the forward-bias region we have the *diffusion* (C_D) or *storage* capacitance.

Recall that the basic equation for the capacitance of a parallel-plate capacitor is defined by $C = \epsilon A/d$, where ϵ is the permittivity of the dielectric (insulator) between the plates of area A separated by a distance d. In the reverse-bias region there is a depletion region (free of carriers) that behaves essentially like an insulator between the layers of opposite charge. Since the depletion region will increase with increased reverse-bias potential, the resulting transition capacitance will decrease, as shown in Fig. 1.33. The fact that the capacitance is dependent on the applied reverse-bias potential has application in a number of electronic systems. In fact, in Chapter 3 a diode will be introduced whose existence is wholly dependent on this phenomenon.

Although the effect described above will also be present in the forward-bias region, it is overshadowed by a capacitance effect directly dependent on the rate at which charge is injected into the regions just outside the depletion region. In other words, directly dependent on the resulting current of the diode. Increased levels of

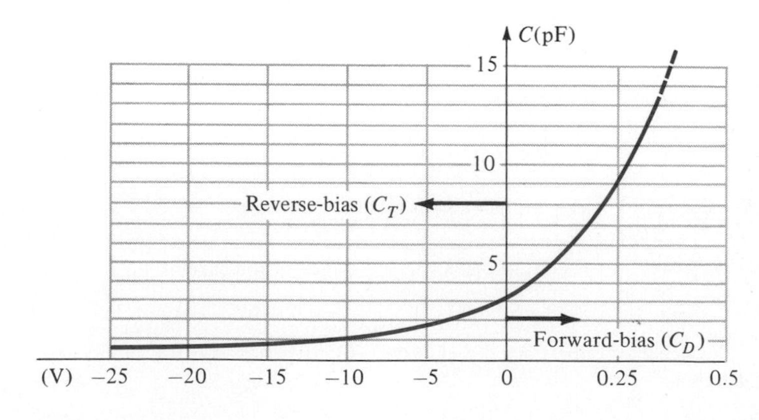

Figure 1.33 Transition and diffusion capacitance versus applied bias for a silicon diode.

current will result in increased levels of diffusion capacitance. However, increased levels of current result in reduced levels of associated resistance (to be demonstrated shortly), and the resulting time constant ($\tau = RC$), which is very important in high-speed applications, does not become excessive.

The capacitive effects described above are represented by a capacitor in parallel with the ideal diode, as shown in Fig. 1.34. For low- or midfrequency applications (except in the power area), however, the capacitor is normally not included in the diode symbol.

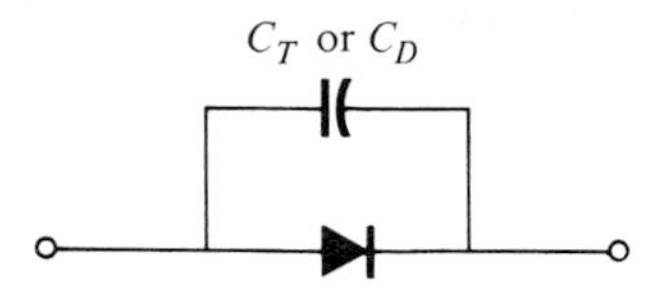

Figure 1.34 Including the effect of the transition or diffusion capacitance on the semiconductor diode.

1.13 REVERSE RECOVERY TIME

There are certain pieces of data that are normally provided on diode specification sheets provided by manufacturers. One such quantity that has not been considered yet is the reverse recovery time denoted by t_{rr}. In the forward-bias state it was shown earlier that there are a large number of electrons from the n-type material progressing through the p-type material and a large number of holes in the n-type—a requirement for conduction. The electrons in the p-type and of holes progressing through the n-type material establish a large number of minority carriers in each material. If the applied voltage should be reversed to establish a reverse-bias situation, we would ideally like to see the diode change instantaneously from the conduction state to the nonconduction state. However, because of the large number of minority carriers in each material, the diode will simply reverse as shown in Fig. 1.35 and stay at this measurable level for the period of time t_s (storage time) required for the minority carriers to return to their majority-carrier state in the opposite material. In essence, the diode will remain in the short-circuit state with a current I_{reverse} determined by the network parameters. Eventually, when this storage phase has passed, the current will reduce in level to that associated with the nonconduction state. This second period of time is denoted by t_t (transition interval). The reverse recovery time is the sum of these

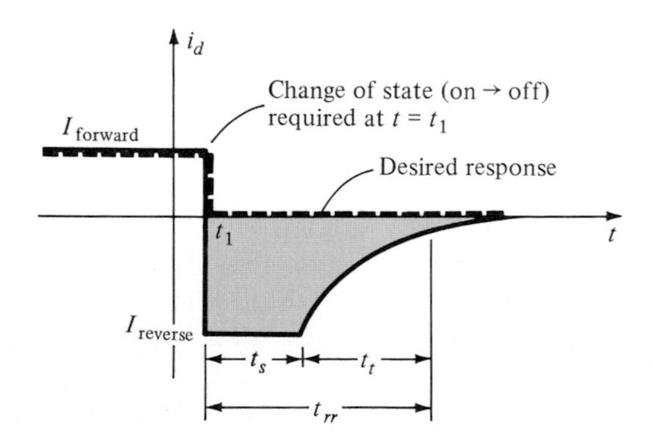

Figure 1.35 Defining the reverse recovery time.

two intervals: $t_{rr} = t_s + t_t$. Naturally, it is an important consideration in high-speed switching applications. Most commercially available switching diodes have a t_{rr} in the range of a few nanoseconds to 1 μs. Units are available, however, with a t_{rr} of only a few hundred picoseconds (10^{-12}).

1.14 TEMPERATURE EFFECTS

Temperature is a very important consideration in the design or analysis of electronic systems. It will affect virtually all of the characteristics of any semiconductor device. The change in characteristics of a semiconductor diode due to temperature variations above and below room temperature (25°C) is shown in Fig. 1.36. Note the reduced levels of forward voltage drop but increased levels of saturation current at 100°C. The Zener potential is also experiencing a pronounced change in level.

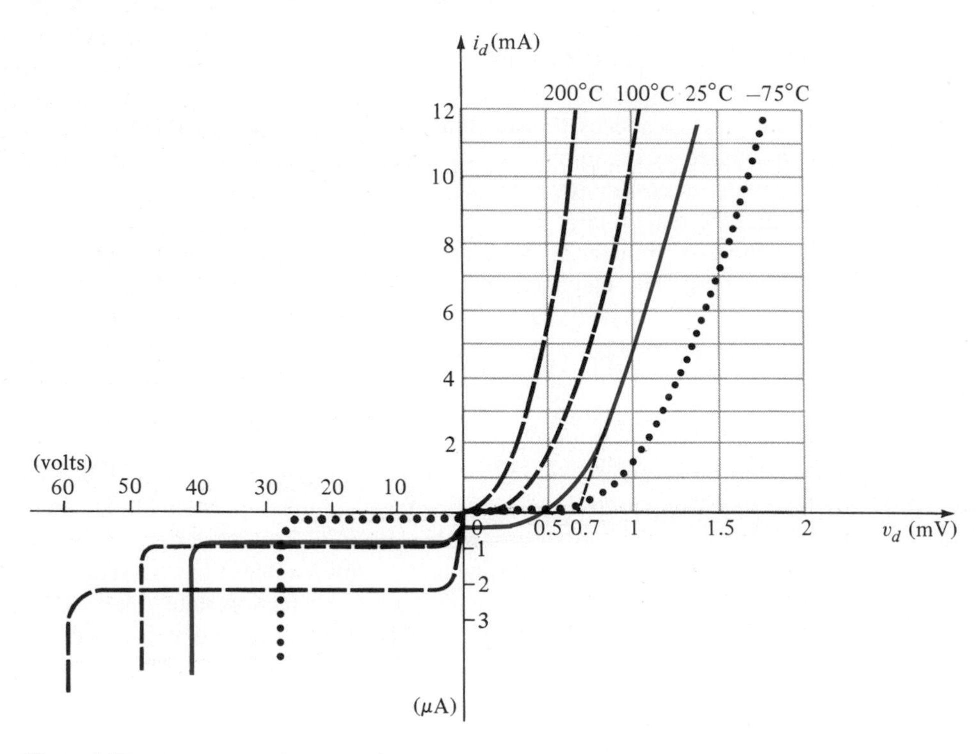

Figure 1.36 Variation in diode characteristics with temperature change.

1.15 DIODE SPECIFICATION SHEETS

Data on specific semiconductor devices are normally provided by the manufacturer in two forms. One is a very brief description of a device that will permit a quick review of all devices available within a few pages. The other is a thorough examination of a device, including graphs, applications, and so on, which usually appears as a separate entity. The latter is normally only provided when specifically requested.

There are certain pieces of data, however, that normally appear on either one. They are included below:

1. The maximum forward voltage $V_{F(\max)}$ (at a specified current and temperature)
2. The maximum forward current $I_{F(\max)}$ (at a specified temperature)
3. The maximum reverse current $I_{R(\max)}$ (at a specified temperature)
4. The reverse-voltage rating (PIV) or PRV or V(BR), where BR comes from the term "breakdown" (at a specified temperature)
5. Maximum capacitance
6. Maximum t_{rr}
7. The maximum operating (or case) temperature

Depending on the type of diode being considered, additional data may also be provided, such as frequency range, noise level, switching time, thermal resistance levels, and peak repetitive values. For the application in mind, the significance of the data will usually be self-apparent. If the maximum power or dissipation rating is also provided, it is understood to be equal to the following product:

$$\boxed{P_{D_{\max}} = V_D I_D} \tag{1.10}$$

where I_D and V_D are the diode current and voltage at a particular point of operation, each variable not to exceed its maximum value. The information in Table 1.2 was taken directly from a Texas Instruments, Inc., data book. Note that the forward voltage drop does not exceed 1 V, but the current has maximum values of 1 to 200 mA.

For the 1N463, if we establish maximum forward voltage and current conditions:

$$P_D = V_D I_D = (1)(1 \times 10^{-3}) = 1 \text{ mW} \quad \text{(a low-power device)}$$

Of course, a device may have a maximum dissipation less than that established by the maximum values. That is, if the voltage is a maximum, the current may have to be less than rated maximum value. Note the increase in I_R for each device with temperature. For the 1N463, it is 30/0.5 = 60 times larger.

An exact copy of the data provided by Fairchild Camera and Instrument Corporation for their BAY 73 and BA 129 high-voltage/low-leakage diodes appears in Figs. 1.37 and 1.38. This example would represent the expanded list of data and

TABLE 1.2 General-Purpose Diodes

	Forward Current			Maximum I_R			
				25°C		150°C	
Device Type	I_F (mA)	V_F (V)	V_{BR} (V)	V	μA	V	μA
1N463	1.0	1.0	200	175	0.5	175	30
1N462	5.0	1.0	70	60	0.5	60	30
1N459A	100.0	1.0	200	175	0.025	175	5
T151	200.0	1.0	20	10	1	—	—

DIFFUSED SILICON PLANAR

- **BV . . . 125 V (MIN) @ 100 μA (BAY73)**
- **BV . . . 200 V (MIN) @ 100 μA (BA129)**

ABSOLUTE MAXIMUM RATINGS (Note 1)

Temperatures

Storage Temperature Range	−65°C to +200°C
Maximum Junction Operating Temperature	+175°C
Lead Temperature	+260°C

Power Dissipation (Note 2)

Maximum Total Power Dissipation at 25°C Ambient	500 mW
Linear Power Derating Factor (from 25°C)	3.33 mW/°C

Maximum Voltage and Currents

WIV	Working Inverse Voltage	BAY73	100 V
		BA129	180 V
I_O	Average Rectified Current		200 mA
I_F	Continuous Forward Current		500 mA
i_f	Peak Repetitive Forward Current		600 mA
$i_{f(surge)}$	Peak Forward Surge Current		
	Pulse Width = 1 s		1.0 A
	Pulse Width = 1 μs		4.0 A

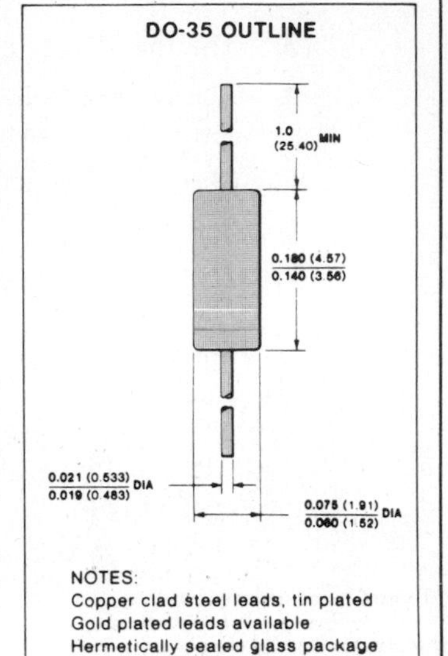

NOTES:
Copper clad steel leads, tin plated
Gold plated leads available
Hermetically sealed glass package
Package weight is 0.14 gram

ELECTRICAL CHARACTERISTICS (25°C Ambient Temperature unless otherwise noted)

SYMBOL	CHARACTERISTIC	BAY73		BA129		UNITS	TEST CONDITIONS
		MIN	MAX	MIN	MAX		
V_F	Forward Voltage	0.85	1.00			V	I_F = 200 mA
		0.81	0.94			V	I_F = 100 mA
		0.78	0.88	0.78	1.00	V	I_F =50 mA
		0.69	0.80	0.69	0.83	V	I_F = 10 mA
		0.67	0.75			V	I_F = 5.0 mA
		0.60	0.68	0.60	0.71	V	I_F = 1.0 mA
				0.51	0.60	V	I_F = 0.1 mA
I_R	Reverse Current		500			nA	V_R = 20 V, T_A = 125°C
			5.0			nA	V_R = 100 V
			1.0			μA	V_R = 100 V, T_A = 125°C
					10	nA	V_R = 180 V
					5.0	μA	V_R = 180 V, T_A = 100°C
BV	Breakdown Voltage	125		200		V	I_R = 100 μA
C	Capacitance		8.0		6.0	pF	V_R = 0, f = 1.0 MHz
t_{rr}	Reverse Recovery Time		3.0			μs	I_f = 10 mA, V_r = 35 V R_L = 1.0 to 100 KΩ C_L = 10 pf, JAN 256

NOTES:
1. These ratings are limiting values above which the serviceability of the diode may be impaired.
2. These are steady state limits. The factory should be consulted on applications involving pulses or low duty-cycle operation.

Figure 1.37 Electrical characteristics of the Fairchild Bay 73 · BA 129 high-voltage, low-leakage diodes. (Courtesy Fairchild Camera and Instrument Corporation.)

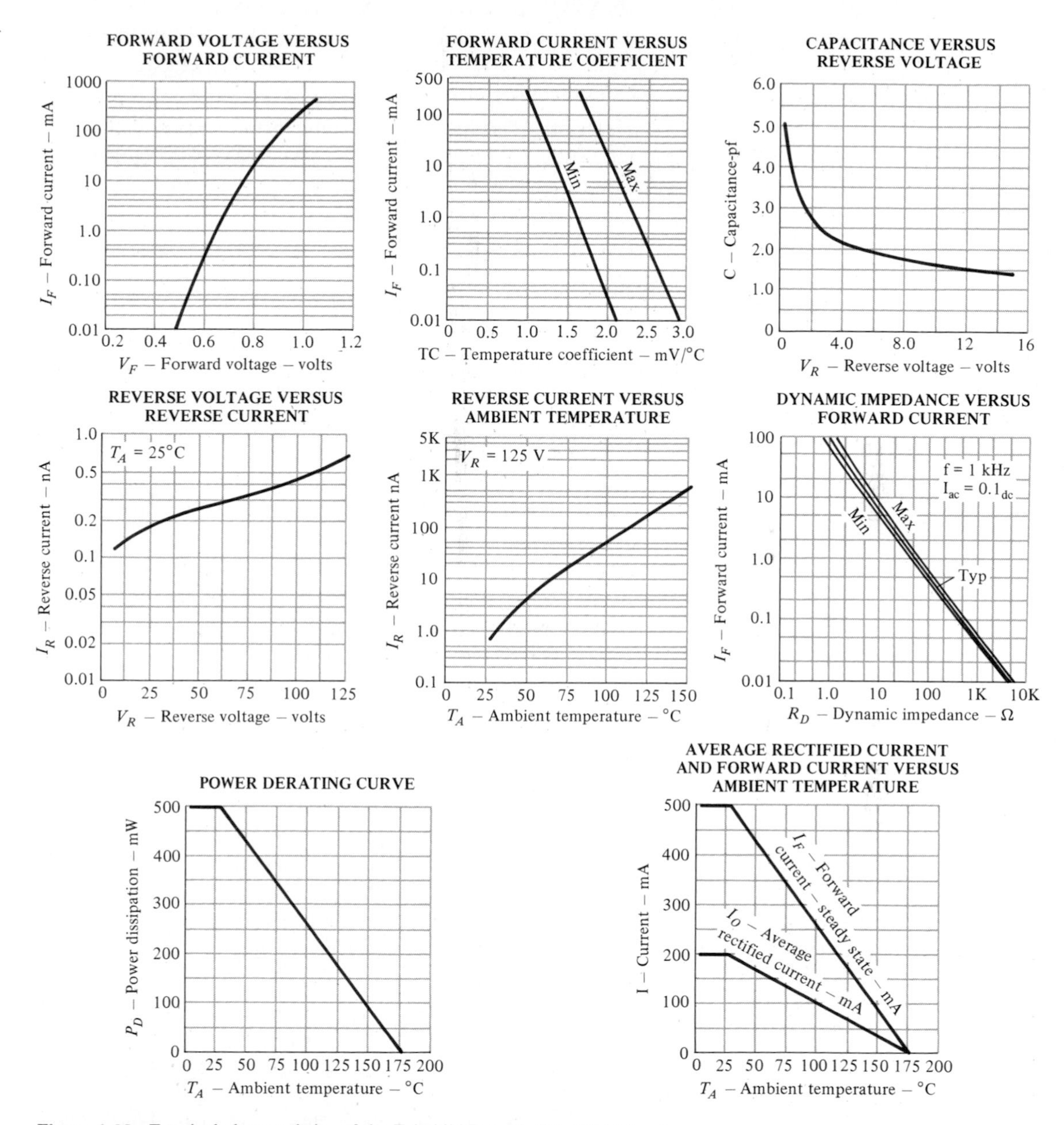

Figure 1.38 Terminal characteristics of the Fairchild Bay 73 · BA 129 high-voltage diodes. (Courtesy Fairchild Camera and Instrument Corporation.)

characteristics. Note that all but the average rectified current, peak repetitive forward current, and peak forward surge current have been defined in this chapter. The significance of these three quantities is as follows:

1. *Average rectified current.* A half-wave-rectified signal (described in Section 2.5) has an average value defined by $I_{av} = 0.318I_{peak}$. The average current rating is lower than the continuous forward current because a half-wave current waveform will have instantaneous values much higher than the average value.
2. *Peak repetitive forward current.* This is the maximum instantaneous value of repetitive forward current. Note that since it is at this level for a brief period of time, its level can be higher than the continuous level.
3. *Peak forward surge current.* On occasion during turn-on, malfunctions, and so on, there will be very high currents through the device for very brief intervals of time (that are not repetitive). This rating defines the maximum value and the time interval for such surges in current level.

Note the logarithmic scale appearing on some of the curves of Fig. 1.38. Each region is bisected such that the value of each horizontal line should be fairly obvious. For I_F versus V_F the horizontal lines between 1.0 and 10.0 mA are 2 mA, 4 mA, 6 mA, and 8 mA. Again, most of the axis variables on the graphs provided have been introduced, resulting in a set of curves that have some recognizable meaning. The temperature coefficient defines the change in voltage with temperature at different current levels. A range of values for the temperature coefficient is provided at each current level. The dynamic impedance (actually, simply the resistance of the device at that forward current) will be discussed in a later section. Note the effect of temperature in the power rating and current ratings of the device in the bottom-right figure.

1.16 SEMICONDUCTOR DIODE NOTATION

The notation most frequently used for semiconductor diodes is provided in Fig. 1.39. For most diodes any marking such as a dot or band, as shown in Fig. 1.39, appears at the cathode end. The terminology anode and cathode is a carryover from vacuum-tube notation. The anode refers to the higher or positive potential and the cathode refers to the lower or negative terminal. This combination of bias levels will result in a forward-bias or "on" condition for the diode. In general, the maximum current-carrying capacity of the diodes of Fig. 1.39 increases from the left to the right. For

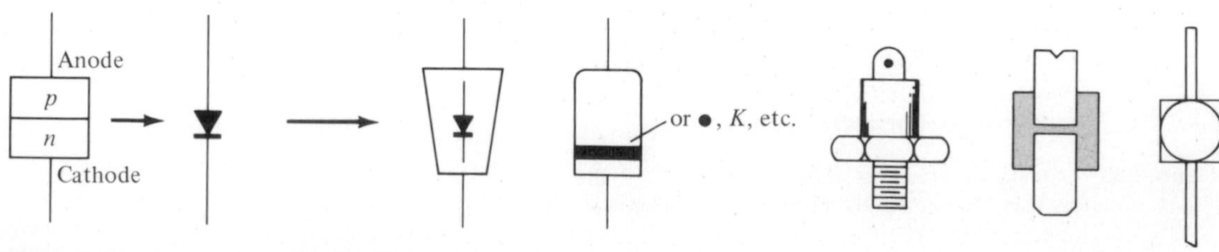

Figure 1.39 Semiconductor diode notation.

each the size will increase with the current rating to ensure that it can handle the additional power dissipation. All but the stud type are limited to a few amperes.

1.17 DIODE OHMMETER CHECK

The condition of a semiconductor diode can be quickly determined by using an ohmmeter such as is found on the standard VOM. The internal battery (often 1.5 V) of the ohmmeter section will either forward- or reverse-bias the diode when applied. If the positive (normally the red) lead is connected to the anode and the negative (normally the black) lead to the cathode, the diode is forward-biased and the meter should indicate a low resistance. The R × 1000 or R × 10,000 setting should be suitable for this measurement. With the reverse polarity the internal battery will back bias the diode and the resistance should be very large. A small reverse-bias resistance reading indicates a "short" condition while a large forward-bias resistance indicates an "open" situation. The basic connections for the tests appear in Fig. 1.40.

Most digital multimeters (DMMs) have a built-in diode check represented by a diode symbol —▶|— as a range selection. A forward-bias connection will provide the firing voltage while a reverse-bias connection will indicate O.L. to represent the open-circuit state.

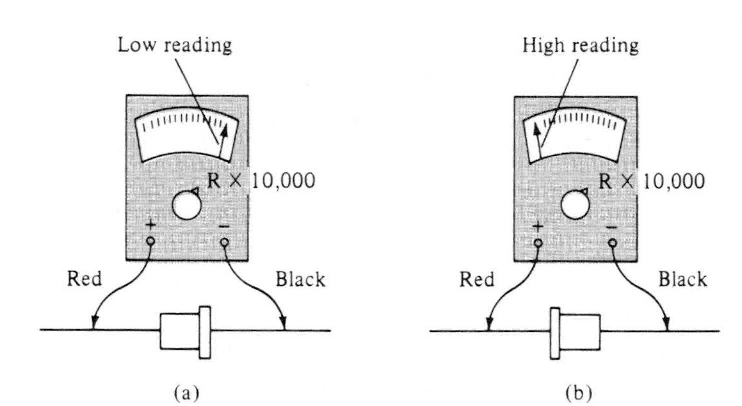

Figure 1.40 Ohmmeter testing of a semiconductor diode: (a) forward bias; (b) reverse bias.

1.18 MANUFACTURING TECHNIQUES

The first step in the manufacture of any semiconductor device is to obtain semiconductor materials, such as germanium or silicon, of the desired purity level. Impurity levels of *less* than *one* part in *one billion* (1 in 1,000,000,000) are required for most semiconductor fabrications today.

The raw materials are first subjected to a series of chemical reactions and a zone-refining process to form a polycrystalline crystal of the desired purity level. The atoms of a polycrystalline crystal are haphazardly arranged, while in the single crystal desired the atoms are arranged in a symmetrical, uniform, geometrical lattice structure.

Zone-refining apparatus is shown in Fig. 1.41. It consists of a graphite or quartz boat for minimum contamination, a quartz container, and a set of RF (radio-frequency) induction coils. Either the coils or boat must be movable along the length of the quartz container. The same result will be obtained in either case, although moving coils are discussed here since it appears to be the more popular method. The interior of the quartz container is filled with either an inert (little or no chemical reaction) gas, or vacuum, to reduce further the chance of contamination. In the zone refining process, a bar of germanium is placed in the boat with the coils at one end of the bar as shown in Fig. 1.41. The radio-frequency signal is then applied to the coil, which will induce a flow of charge (eddy currents) in the germanium ingot. The magnitude of these currents is increased until sufficient heat is developed to melt that region of the semiconductor material. The impurities in the ingot will enter a more liquid state than the surrounding semiconductor material. If the induction coils of Fig. 1.41 are now slowly moved to the right to induce melting in the neighboring region, the "more fluidic" impurities will "follow" the molten region. The net result is that a large percentage of the impurities will appear at the right end of the ingot when the induction coils have reached this end. This end piece of impurities can then be cut off and the entire process repeated until the desired purity level is reached.

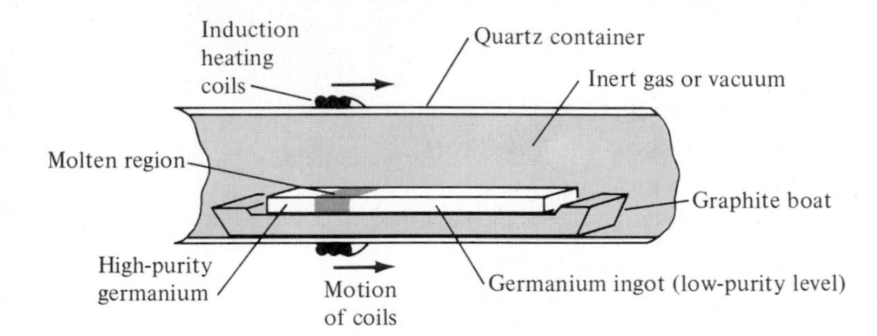

Figure 1.41 Zone refining process.

The final operation before semiconductor fabrication can take place is the formation of a single crystal of a germanium or silicon. This can be accomplished using either the *Czochralski* or the *floating-zone* technique, the latter being the more recently devised. The apparatus employed in the Czochralski technique is shown in Fig. 1.42a. The polycrystalline material is first transformed to the molten state by the RF induction coils. A single-crystal "seed" of the desired impurity level is then immersed in the molten germanium and gradually withdrawn while the shaft holding the seed is slowly turning. As the "seed" is withdrawn, a single-crystal germanium lattice structure will grow on the "seed" as shown in Fig. 1.42a. The resulting single-crystal ingots are typically 6 to 36 in. in length and 1 to 5 in. in diameter (Fig. 1.42b). Ingots having a length of 48 in. and a diameter of 3 in. have been grown. The weight of such a structure is about 28.5 lb.

The floating-zone technique eliminates the need for having both a zone refining and single-crystal forming process. Both can be accomplished at the same time using this technique. A second advantage of this method is the absence of the graphite or quartz boat, which often introduces impurities into the germanium or silicon ingot. Two clamps hold the bar of germanium or silicon in the vertical position within a set

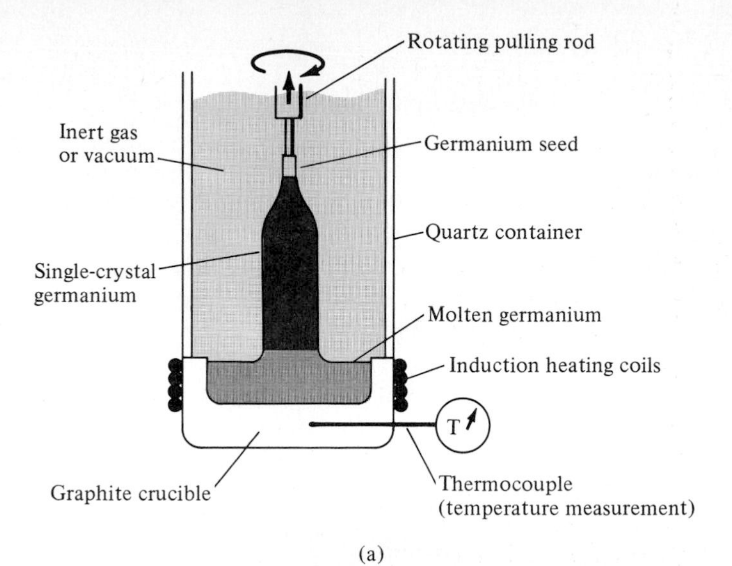

(a)

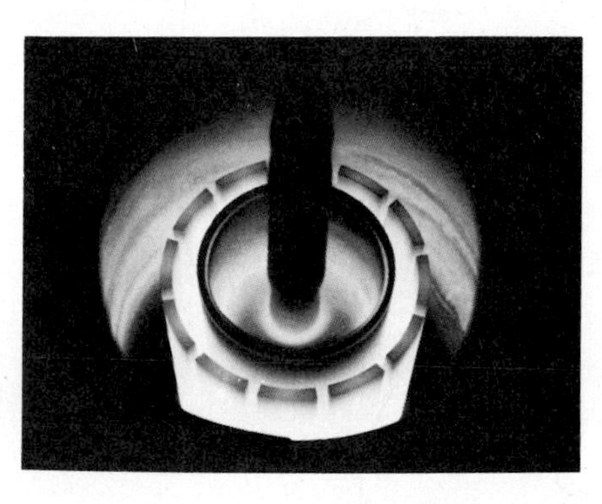

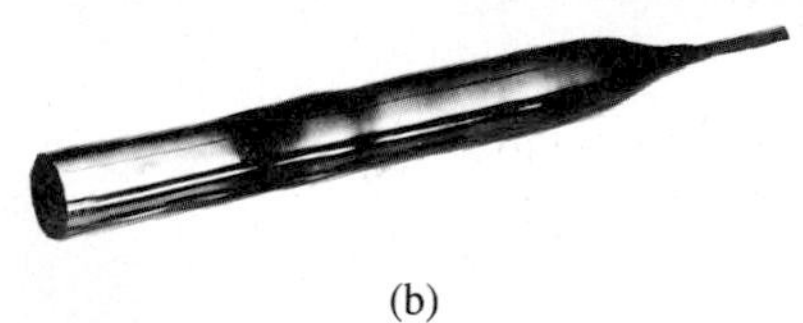

(b)

Figure 1.42 Czochralski technique and ingots. [(b) *Top,* courtesy Texas Instruments, Inc.; *bottom,* courtesy Motorola, Inc.]

of movable RF induction coils as shown in Fig. 1.43. A small single-crystal "seed" of the desired impurity level is deposited at the lower end of the bar and heated with the germanium bar until the molten state is reached. The induction coils are then slowly moved up the germanium or silicon ingot while the bar is slowly rotating. As before, the impurities follow the molten state, resulting in an improved impurity level single-crystal germanium lattice below the molten zone. Through proper control of the process, there will always be sufficient surface tension present in the semiconductor material to ensure that the ingot does not rupture in the molten zone.

The single-crystal structure produced can then be cut into wafers sometimes as thin as $\frac{1}{1000}$ (or 0.001) of an inch ($\cong \frac{1}{5}$ the thickness of this paper). This cutting process can be accomplished using the setup of Fig. 1.44a or b. In Fig. 1.44a, tungsten wires (0.001 in. in diameter) with abrasive deposited surfaces are connected to supporting

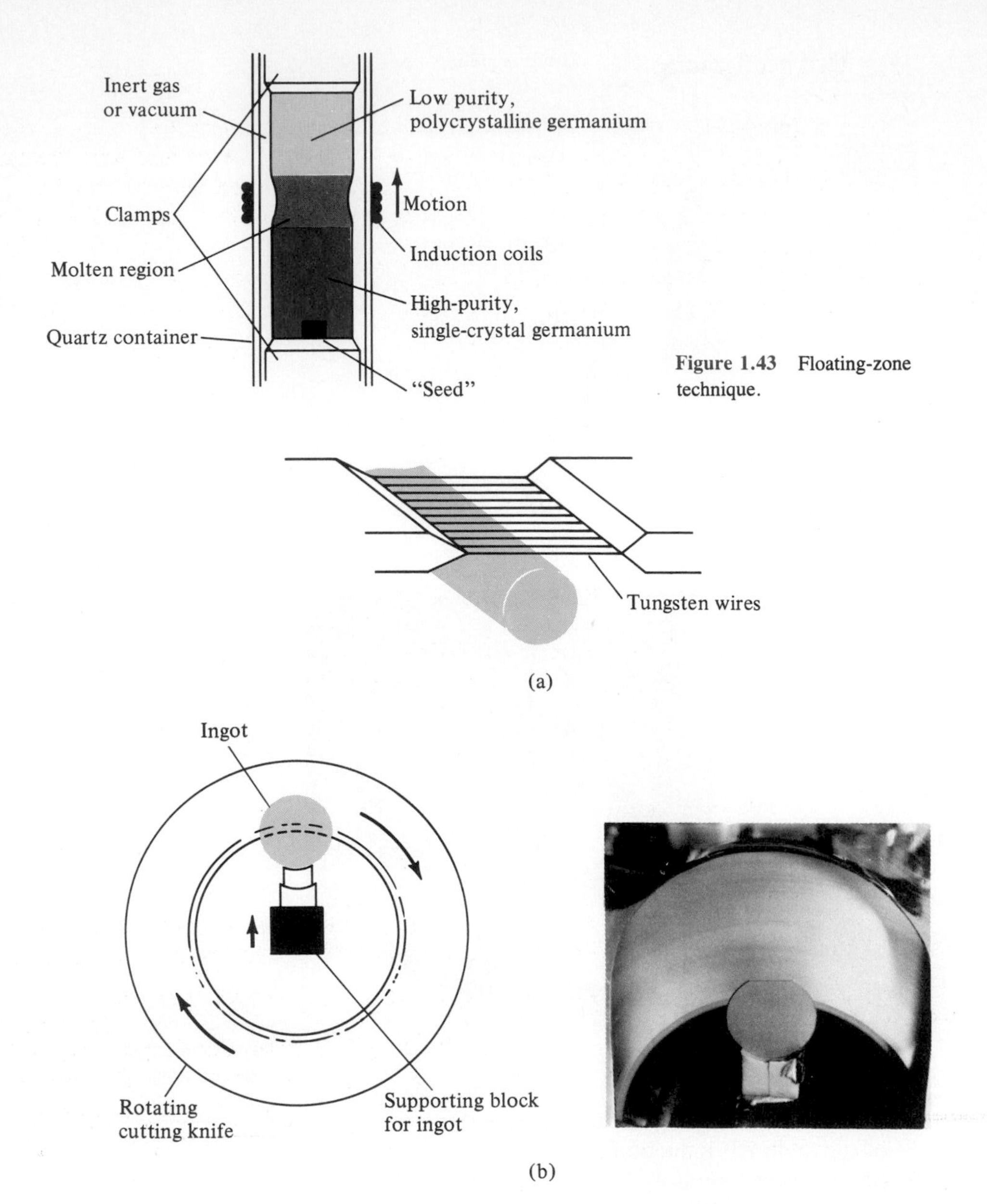

Figure 1.43 Floating-zone technique.

Figure 1.44 Slicing the single-crystal ingot into wafers. (Courtesy Texas Instruments Inc.)

blocks at the proper spacing and then the entire system is moved back and forth as a saw. The system of Fig. 1.44b is self-explanatory.

Other semiconductor materials will be introduced as their area of application is considered.

Semiconductor diodes are normally one of the following types: grown junction, alloy, diffusion, epitaxial growth, or point contact. Each is described in some detail in this section.

Grown Junction

Diodes of this type are formed during the Czochralski *crystal pulling* process. Impurities of p- and n-type can be alternately added to the molten semiconductor material in the crucible, resulting in a *p-n* junction, as indicated in Fig. 1.45 when the crystal is pulled. After slicing, the large-area device can then be cut into a large number (sometimes thousands) of smaller-area semiconductor diodes. The area of grown-junction diodes is sufficiently large to handle high currents (and therefore have high power ratings). The large area, however, will introduce undesired junction capacitive effects.

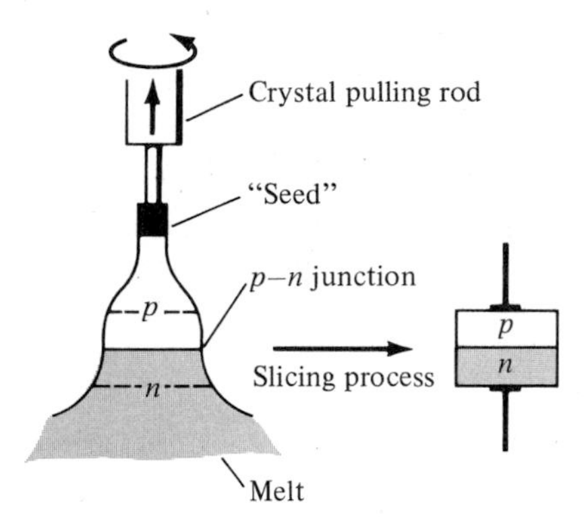

Figure 1.45 Grown junction diode.

Alloy

The alloy process will result in a junction-type semiconductor diode that will also have a high current rating and large PIV rating. The junction capacitance is also large, however, due to the large junction area.

The *p-n* junction is formed by first placing a p-type impurity on an n-type substrate and heating the two until liquefaction occurs where the two materials meet (Fig. 1.46). An alloy will result that, when cooled, will produce a *p-n* junction at the boundary of the alloy and substrate. The roles played by the n- and p-type materials can be interchanged.

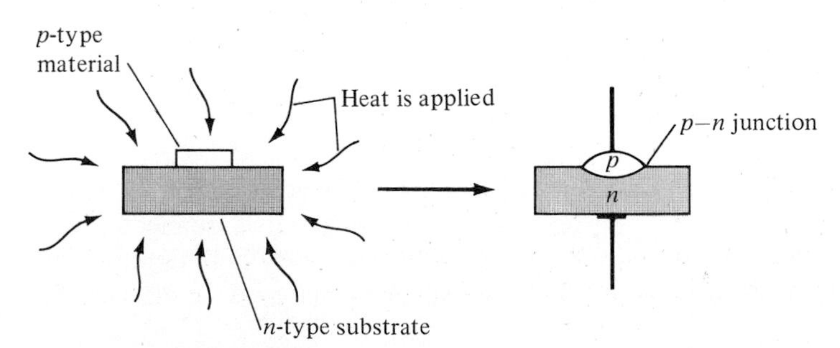

Figure 1.46 Alloy process diode.

Diffusion

The diffusion process of forming semiconductor junction diodes can employ either solid or gaseous diffusion. This process requires more time than the alloy process but it is relatively inexpensive and can be very accurately controlled. Diffusion is a process by which a heavy concentration of particles will "diffuse" into a surrounding region of lesser concentration. The primary difference between the diffusion and alloy process is the fact that liquefaction is not reached in the diffusion process. Heat is applied in the diffusion process only to increase the activity of the elements involved.

The process of solid diffusion commences with the "painting" of an acceptor impurity on an n-type substrate and heating the two until the impurity diffuses into the substrate to form the p-type layer (Fig. 1.47a).

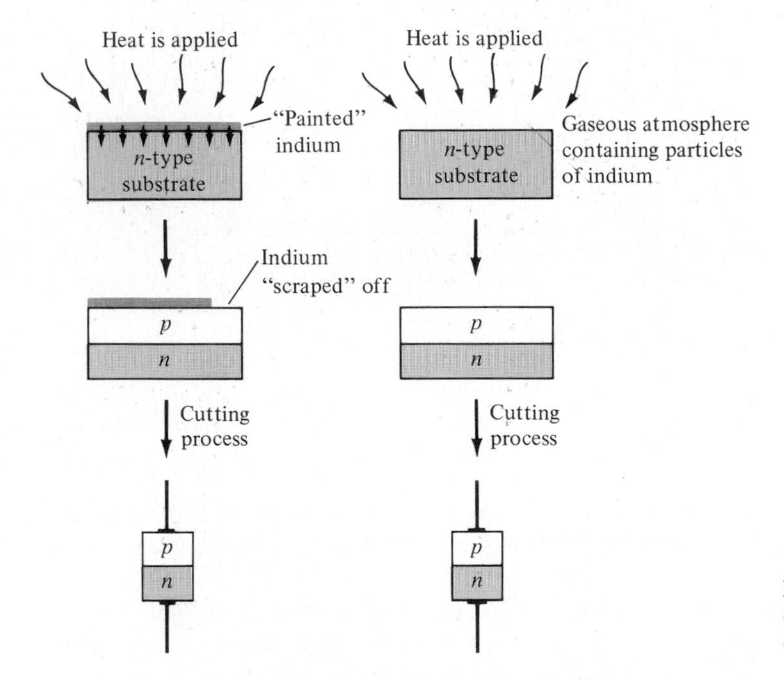

Figure 1.47 Diffusion process diodes: (a) solid diffusion; (b) gaseous diffusion.

In the process of gaseous diffusion, an n-type material is submerged in a gaseous atmosphere of acceptor impurities and then heated (Fig. 1.47b). The impurity diffuses into the substrate to form the p-type layer of the semiconductor diode. The roles of the p- and the n-type materials can also be interchanged in each case. The diffusion process is the most frequently used today in the manufacture of semiconductor diodes.

Epitaxial Growth

The term "epitaxial" has its derivation from the Greek terms *epi* meaning "upon" and *taxis* meaning "arrangement." A base wafer of n^+ material is connected to a metallic conductor as shown in Fig. 1.48. The n^+ indicates a very high doping level for a reduced resistance characteristic. Its purpose is to act as a semiconductor

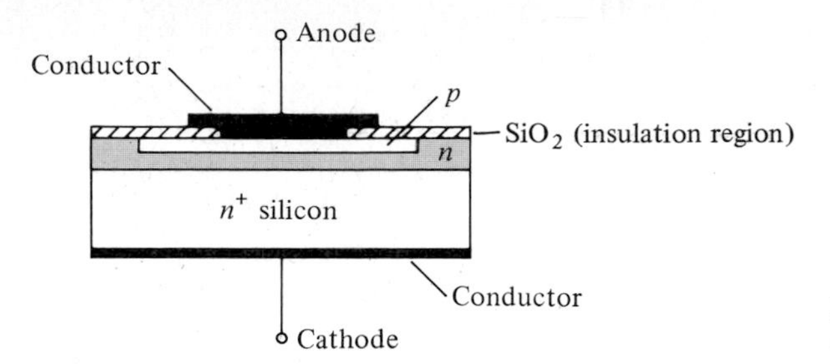

Figure 1.48 Epitaxially grown semiconductor diode.

extension of the conductor and not the n-type material of the p-n junction. The n-type layer is to be deposited on this layer as shown in Fig. 1.48 using a diffusion process. This technique of using an n^+ base gives the manufacturer definite design advantages. The p-type silicon is then applied by using a diffusion technique and the anode metallic connector added as indicated in Fig. 1.48.

Point Contact

The point-contact semiconductor diode is constructed by pressing a phosphor-bronze spring (called a cat whisker) against an n-type substrate (Fig. 1.49). A high current is then passed through the whisker and substrate for a short period of time, resulting in a number of atoms passing from the wire into the n-type material to create a p-region in the wafer. The small area of the p-n junction results in a very small junction capacitance (typically 1 pF or less). For this reason, the point-contact diode is frequently used in applications where very high frequencies are encountered, such as in microwave mixers and detectors. The disadvantage of the small contact area is the resulting low current ratings and characteristics less ideal than those obtained from junction-type semiconductor diodes. The basic construction and photographs of point-contact diodes appear in Fig. 1.50. Various types of junction diodes appear in Fig. 1.51.

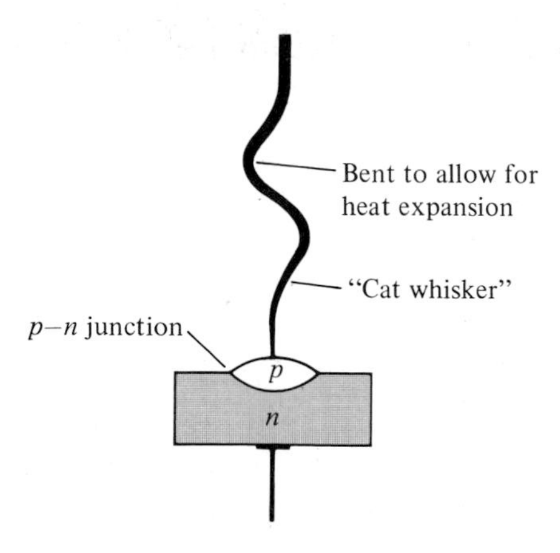

Figure 1.49 Point-contact diode.

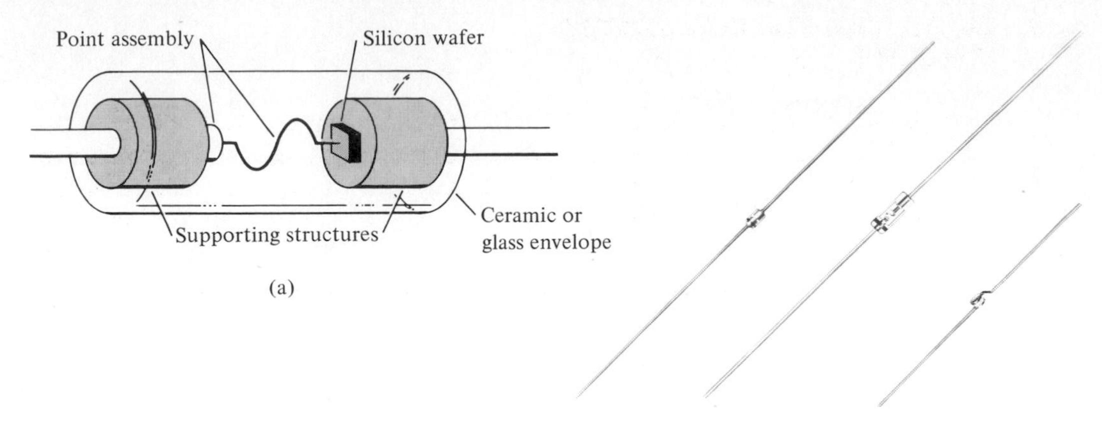

Figure 1.50 Point-contact diodes: (a) basic construction; (b) various types. (Courtesy General Electric Company.)

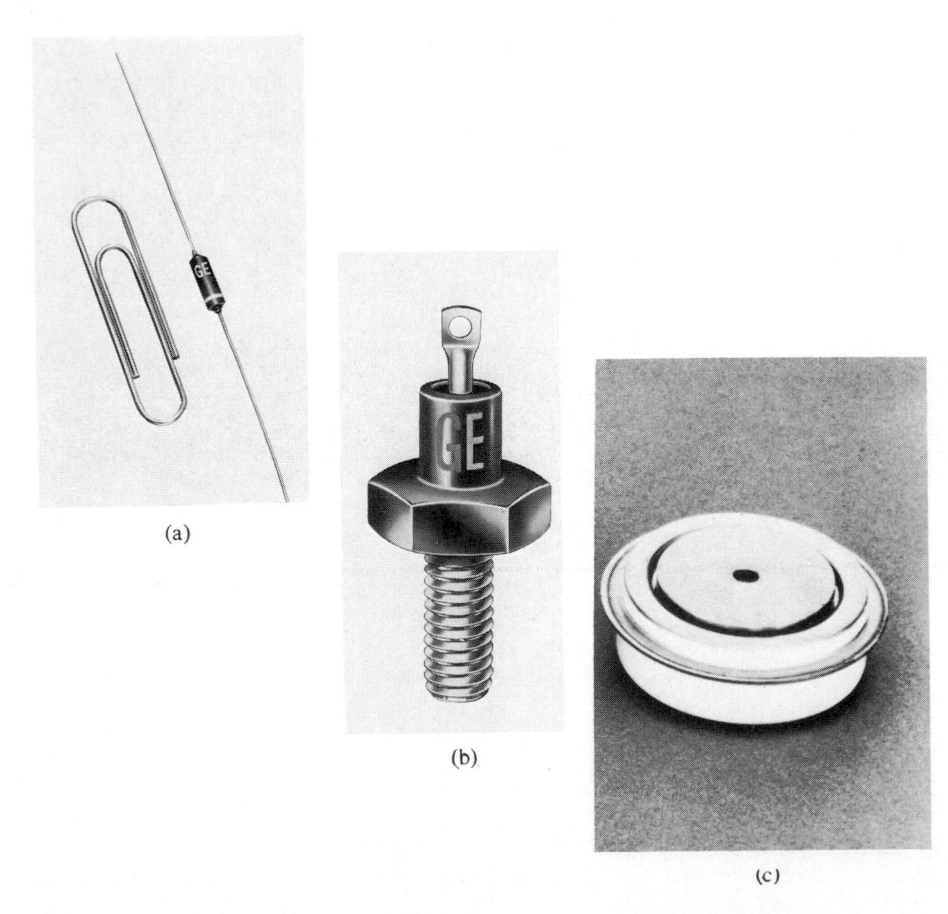

Figure 1.51 Various types of junction diodes. [(a) and (b) Courtesy General Electric Company; (c) courtesy International Rectifier Corporation.]

1.19 DIODE ARRAYS—INTEGRATED CIRCUITS

The unique characteristics of integrated circuits will be introduced in Chapter 14. However, we have reached a plateau in our introduction to electronic circuits that permits at least a surface examination of diode arrays in the integrated-circuit package. You will find that the integrated circuit is not a unique device with characteristics

FSA1410M

PLANAR AIR-ISOLATED MONOLITHIC DIODE ARRAY

- C . . . 5.0 pF (MAX)
- ΔV_F . . . 15 mV (MAX) @ 10 mA

ABSOLUTE MAXIMUM RATINGS (Note 1)

Temperatures

Storage Temperature Range	−55°C to +200°C
Maximum Junction Operating Temperature	+150°C
Lead Temperature	+260°C

Power Dissipation (Note 2)

Maximum Dissipation per Junction at 25°C Ambient	400 mW
per Package at 25°C Ambient	600 mW
Linear Derating Factor (from 25°C) Junction	3.2 mW/°C
Package	4.8 mW/°C

Maximum Voltage and Currents

WIV	Working Inverse Voltage	55 V
I_F	Continuous Forward Current	350 mA
$i_{f(surge)}$	Peak Forward Surge Current	
	Pulse Width = 1.0 s	1.0 A
	Pulse Width = 1.0 μs	2.0 A

CONNECTION DIAGRAM

FSA1410M

1 2 3 4 5 6 7 8 9

See Package Outline TO-96

ELECTRICAL CHARACTERISTICS (25°C Ambient Temperature unless otherwise noted)

SYMBOL	CHARACTERISTIC	MIN	MAX	UNITS	TEST CONDITIONS
B_V	Breakdown Voltage	60		V	I_R = 10 μA
V_F	Forward Voltage (Note 3)		1.5	V	I_F = 500 mA
			1.1	V	I_F = 200 mA
			1.0	V	I_F = 100 mA
I_R	Reverse Current		100	nA	V_R = 40 V
	Reverse Current (T_A = 150°C)		100	μA	V_R = 40 V
C	Capacitance		5.0	pF	V_R = 0, f = 1 MHz
V_{FM}	Peak Forward Voltage		4.0	V	I_f = 500 mA, t_r < 10 ns
t_{fr}	Forward Recovery Time		40	ns	I_f = 500 mA, t_r < 10 ns
t_{rr}	Reverse Recovery Time		10	ns	$I_f = I_r$ = 10−200 mA
					R_L = 100 Ω, Rec. to 0.1 I_r
			50	ns	I_f = 500 mA, I_r = 50 mA
					R_L = 100 Ω, Rec. to 5 mA
ΔV_F	Forward Voltage Match		15	mV	I_F = 10 mA

NOTES:
1. These ratings are limiting values above which life or satisfactory performance may be impaired.
2. These are steady state limits. The factory should be consulted on applications involving pulsed or low duty cycle operation.
3. V_F is measured using an 8 ms pulse.

Figure 1.52 Monolithic diode array. (Courtesy Farichild Camera and Instrument Corporation.)

totally different from those we will examine in these introductory chapters. It is simply a packaging technique that permits a significant reduction in the size of electronic systems. In other words, internal to the integrated circuit are systems and discrete devices that were available long before the integrated circuit as we know it today became a reality.

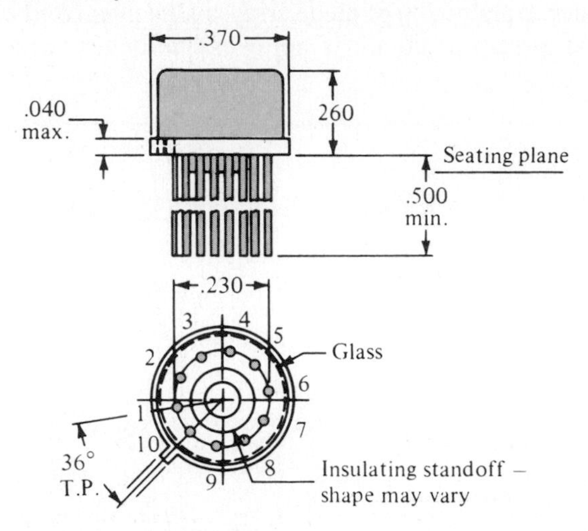

Figure 1.53 Package outline TO-96 for the FSA 1410M diode array. All dimensions are in inches. (Courtesy Fairchild Camera and Instrument Corporation.)

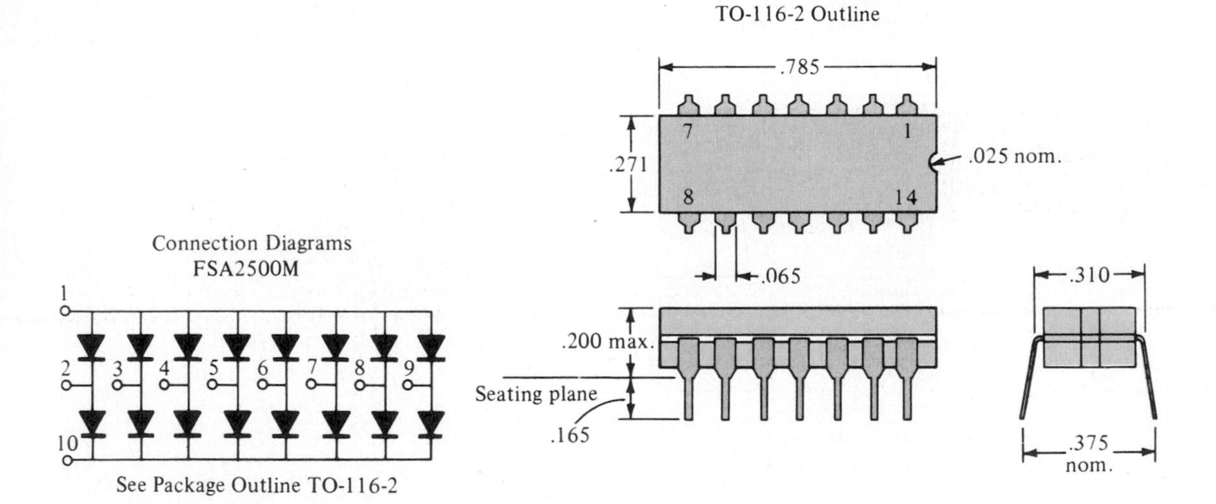

Figure 1.54 Monolithic diode array. All dimensions are in inches. (Courtesy Fairchild Camera and Instrument Corporation.)

One possible array appears in Fig. 1.52. Note that eight diodes are internal to the Fairchild FSA 1410M diode array. That is, in the container shown in Fig. 1.53 there are diodes set in a single silicon wafer that have all the anodes connected to pin 1 and the cathodes of each to pins 2 through 9. Note in the same figure that pin 1 can be determined as being to the left of the small projection in the case if we look from the bottom toward the case. The other numbers then follow in sequence. If only one diode is to be used, then only pins 1 and 2 (or any number from 3 to 9) would be used. The remaining diodes would be left hanging and not affect the network that pins 1 and 2 are connected to.

Another diode array appears in Fig. 1.54. In this case the package is different but the numbering sequence appears in the outline. Pin 1 is the pin directly above the small indentation as you look down on the device.

PROBLEMS

§ 1.2

1. In your own words, define semiconductor, resistivity, bulk resistance, and ohmic contact resistance.

2. (a) Using Table 1.1, determine the resistance of a silicon sample having an area of 1 cm^2 and a length of 3 cm.
(b) Repeat part (a) if the length is 1 cm and the area 4 cm^2.
(c) Repeat part (a) if the length is 8 cm and the area 0.5 cm^2.
(d) Repeat part (a) for copper and compare the results.

3. Sketch the atomic structure of copper and discuss why it is a good conductor and how its structure is different from germanium and silicon.

4. Define, in your own words, an intrinsic material, a negative temperature coefficient, and covalent bonding.

5. Consult your reference library and list three materials that have a negative temperature coefficient and three that have a positive temperature coefficient.

§ 1.3

6. How much energy in joules is required to move a charge of 6 C through a difference in potential of 3 V?

7. If 48 eV of energy is required to move a charge through a potential difference of 12 V, determine the charge involved.

8. Consult your reference library and determine the level of E_g for GaP and ZnS, two semiconductor materials of practical value. In addition, determine the written name for each material.

§ 1.4

9. Describe the difference between n-type and p-type semiconductor materials.

10. Describe the difference between donor and acceptor impurities.

11. Describe the difference between majority and minority carriers.

12. Sketch the atomic structure of silicon and insert an impurity of arsenic as demonstrated for germanium in Fig. 1.6.

13. Repeat Problem 12 but insert an impurity of indium.

14. Consult your reference library and find another explanation of hole versus electron flow. Using both descriptions, describe in your own words the process of hole induction.

§ 1.5

15. Describe, in your own words, the characteristics of the *ideal* diode and how they determine the on and off state of the device. That is, describe why the short-circuit and open-circuit equivalents are appropriate.

16. What is the one important difference between the characteristics of a simple switch and those of an ideal diode? (*Hint*: Consider the conduction state and the direction of charge flow.)

§ 1.6

17. Describe in your own words the conditions established by a forward- and reverse-bias condition on a *p-n* junction diode and how it effects the resulting current.

18. Describe how will you remember the forward- and reverse-bias states of the *p-n* junction diode. That is, how will you remember which potential (positive or negative) is applied to which terminal?

19. Referring to Fig. 1.18, determine the average difference in voltage between the typical commercially available Si unit and the characteristic determined by Eq. (1.4) for the range i_d = 10 mA to 50 mA.

20. Using Eq. (1.4), determine the diode current at 20°C for a silicon diode with I_s = 50 μA and an applied forward bias of 0.6 V.

21. Repeat Problem 20 for T = 100°C (boiling point of water). Assume that I_s has increased to 5.0 μA.

22. In the reverse-bias region the saturation current of a silicon diode is about 0.1 μA (T = 20°C). Determine its approximate value if the temperature is increased 40°C.

23. Compare the characteristics of a silicon and germanium diode and determine which you would prefer to use for most practical applications. Give some detail. Refer to a manufacturer's listing and compare the characteristic of a germanium and silicon diode of similar maximum ratings.

§ 1.7

24. Determine the static or dc resistance of the diode of Fig. 1.20 at a forward current of 5 mA.

25. Repeat Problem 24 at a forward current of 30 mA and compare results.

26. Determine the static or dc resistance of the diode of Fig. 1.20 at a reverse voltage of −5 V. How does it compare to the value determined at a reverse voltage of −10 V?

§ 1.8

27. Determine the dynamic (ac) resistance of the diode of Fig. 1.23 at a forward current of 10 mA.

28. Determine the dynamic (ac) resistance of the diode of Fig. 1.23 at a forward current of 10 mA using Eq. (1.7) and compare with the results of Problem 27. What is the r_B contribution?

29. Calculate the dc and ac resistance for the diode of Fig. 1.23 at a forward current of 20 mA and compare their magnitudes.

§ 1.9

30. Determine the average ac resistance for the diode of Fig. 1.24 for the region between 0.4 and 0.8 V.

31. Determine the dc and ac resistance for the diode of Fig. 1.24 at 0.6 V and compare to the average ac resistance obtained in Problem 30.

§ 1.10

32. Find the piecewise-linear equivalent circuit for the diode of Fig. 1.24. Assume that the straight-line segment for the semiconductor diode intersects the horizontal axis at 0.7 V (Si).

33. Repeat Example 1.2 if $r_{av} = 32\ \Omega$ and $R_L = 220\ \Omega$.

§ 1.11

34. Describe another example of the diffusion process.

35. How is drift current different from diffusion current?

§ 1.12

36. (a) Referring to Fig. 1.33, determine the transition capacitance at reverse-bias potentials of -25 V and -10 V. What is the ratio of the change in capacitance to the change in voltage?
(b) Repeat part (a) for reverse-bias potentials of -10 V and -1 V. Determine the ratio of the change in capacitance to the change in voltage.
(c) How do the ratios determined in parts (a) and (b) compare? What does it tell you about which range may have more areas of practical application?

37. Referring to Fig. 1.33, determine the diffusion capacitance at 0 V and 0.25 V.

38. Describe in your own words how diffusion and transition capacitances differ.

39. Determine the reactance offered by a diode described by the characteristics of Fig. 1.33 at a forward potential of 0.2 V and a reverse potential of -20 V if the applied frequency is 6 MHz.

§ 1.13

40. Sketch the waveform for i of the network of Fig. 1.55 if $t_t = 2t_s$ and the total reverse recovery time is 9 ns.

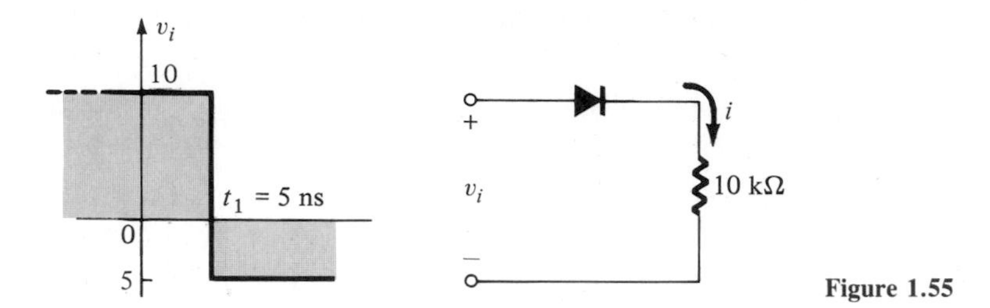

Figure 1.55

§ 1.14

41. Determine the forward voltage drop across the diode whose characteristics appear in Fig. 1.36 at temperatures of −75°C, 25°C, 100°C, and 200°C and a current of 10 mA. For each temperature, determine the level of saturation current. Compare the extremes of each and comment on the ratio of the two.

§ 1.15

42. Determine the maximum power dissipation for the T151 diode in the forward-bias region. What is the maximum reverse-bias dissipation at $V = -10$ V ($T = 25$°C)?

43. Repeat Problem 42 for the 1N459A diode. Assume that I_R does not change significantly with reverse bias.

44. Using the data of Fig. 1.37, sketch the power derating curve for the BAY73 diode and find its rated power level at 50°C.

45. Plot a curve of I_F (ordinate) versus V_F (max)(abscissa) for the BAY73 diode from the data of Fig. 1.37 and make any noteworthy comments.

46. Repeat Problem 45 for the reverse current ($T_A = 25$°C).

47. What is the capacitive reactance of the BA129 diode at a frequency of 1 MHz?

48. If $t_t = t_s$, sketch a curve of the reverse recovery period for the BAY73 diode.

49. Determine the peak current associated with a rated average rectified level of $I_O = 200$ mA (half-wave-rectified signal).

50. How does the curve of V_F versus I_F in Fig. 1.38 compare with the results of Problem 45?

51. (a) Referring to Fig. 1.38, determine the temperature coefficient at a forward current of 1.0 mA under maximum conditions.
(b) Using the results of part (a), determine the change in forward voltage if the temperature should increase 20°C.

52. Compare the power derating curve of Fig. 1.38 with the results of Problem 44.

53. What is the change in dynamic impedance if the current should be reduced from 10 mA to 0.1 mA under maximum conditions? Refer to Fig. 1.38.

§ 1.18

54. Describe the Czochralski method for fabricating a single crystal of germanium or silicon.

55. How is the floating-zone technique different from the Czochralski method?

56. What is the induction heating? Describe in your own words.

§ 1.19

57. What is the maximum power dissipation at 75°C for each diode in the FSA 1410M array? Sketch the power derating curve.

58. Referring to Fig. 1.52, list the detrimental effects of increasing the forward current (per diode) beyond 100 mA and the temperature above room temperature (25°C).

59. If each diode of the FSA 1410M has a current of 40 mA, what is the current through terminal 1? What is the forward voltage drop from pin 1 to 9 if diodes 1, 5, and 9 are active as forward currents (I_F) of 100 mA?

60. If each diode in Fig. 1.54 has a V_T of 0.7 V and an I_F equal to 30 mA, determine the current through terminals 1 and 10 and the voltage from across pins 1 and 10.

2

Diode Applications

2.1 INTRODUCTION

The construction, characteristics, and models of semiconductor diodes were introduced in Chapter 1. The primary goal of this chapter is to develop a working knowledge of the diode in a variety of configurations using models appropriate for the area of application. By chapter's end, the fundamental behavior pattern of diodes in dc and ac networks should be clearly understood. The concepts learned in this chapter will have significant carryover in the chapters to follow. For instance, diodes are frequently employed in the description of the basic construction of transistors and in the analysis of transistor networks in the dc and ac domains.

The content of this chapter will reveal a positive side of the study of a field such as electronic devices and systems—once the basic behavior of a device is understood, its function and response in an infinite variety of configurations can be determined. The range of applications is endless, yet its characteristics and models remain the same.

2.2 SERIES DIODE CONFIGURATIONS WITH DC INPUTS

For the analysis to follow, the approximate models for the diode will be employed using the notation defined in Section 1.10 and appearing in Fig. 2.1. Keep in mind for all models of Fig. 2.1 that an applied or resulting voltage across the diode with the polarity of Fig. 2.2 of *any magnitude* will result in an open-circuit equivalent—the

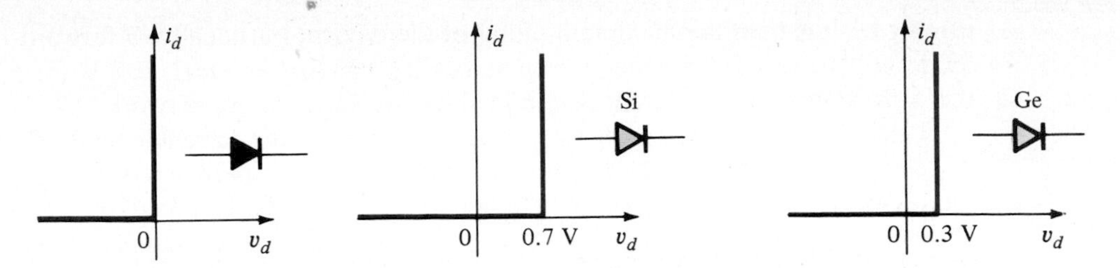

Figure 2.1 Ideal and approximate characteristics for the diode.

v_d
− +
$i_d = 0$ A
$i_d = 0$ A

Figure 2.2 "Off" state of any diode.

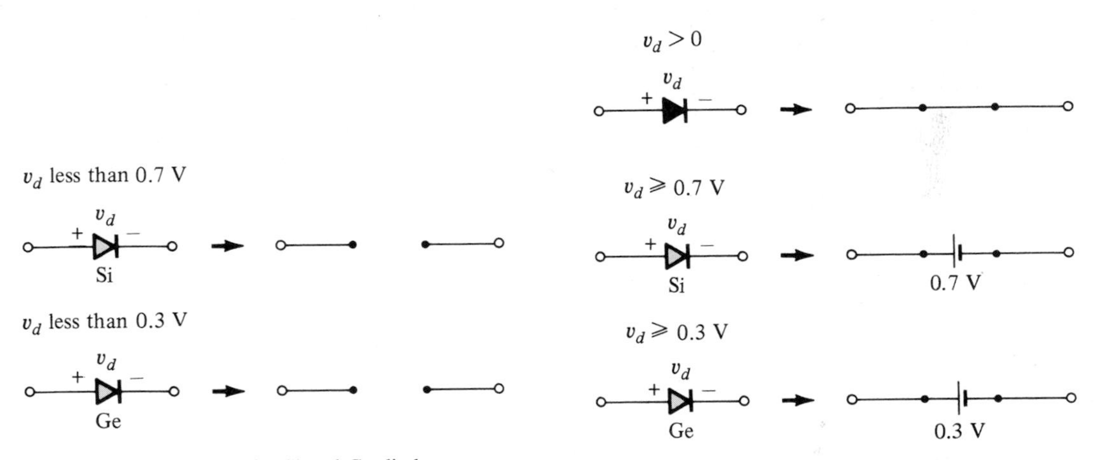

Figure 2.3 "Off" states for Si and Ge diodes.

Figure 2.4 "On" states for ideal and approximate equivalent diodes

"off" condition. For the approximate models (not including the ideal), voltages less than 0.7 for silicon and 0.3 for germanium with the polarity of Fig. 2.3 will also result in an open-circuit equivalent. For the ideal diode, any positive voltage with the polarity of Fig. 2.3 will result in the equivalent "on"-state short-circuit equivalent of Fig. 2.4.

For the analysis to follow, be sure to keep in mind that the 0.7- and 0.3-V sources appearing in the equivalent circuits are not *independent* sources of energy. For instance, an isolated diode on a table will not register 0.7 or 0.3 V if a voltmeter is connected across its terminals. It is simply a mechanism, method—whatever—to include the effects of the offset voltage of the diode that separates it from the ideal characteristics.

The magnitude of the applied forward bias obviously has a pronounced effect on the behavior of the diode. However, what about the resulting current level? For the open-circuit situation, it is obviously 0 A. When the short-circuit state is appropriate, the *current is determined by the network to which the diode is connected*. Of course, it must be less than the maximum rating of the device, but ideally, a forward-biased diode will have a fixed voltage drop across it (0 V for the ideal, 0.7 V for Si, and 0.3 V for Ge), with a current determined by the surrounding network.

Let us begin by determining the various voltages and the current level for the series dc configuration of Fig. 2.5. The first question of major importance is the state of the diode—should the short-circuit or the open-circuit state be assumed? For most situations, simply removing the diode from the picture entirely and determining the direction of the resulting conventional current will provide the hint we need regarding the diode state. If the current direction (as in Fig. 2.6) has the same direction as the arrow in the diode symbol, the diode is in the "on" state as long as the network has sufficient voltage to compensate for the V_T back-bias voltage of the diode.

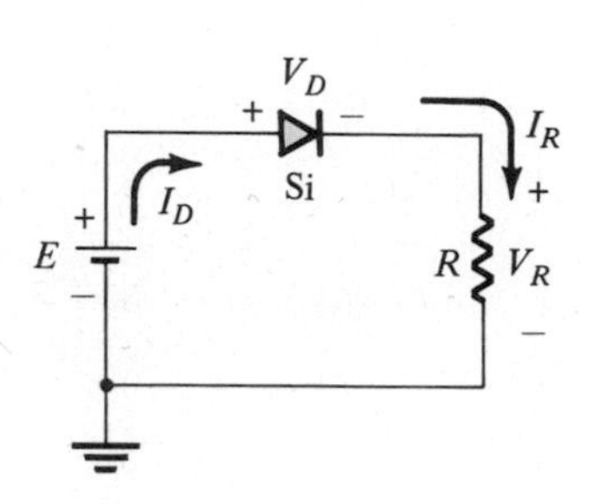

Figure 2.5 Series diode configuration.

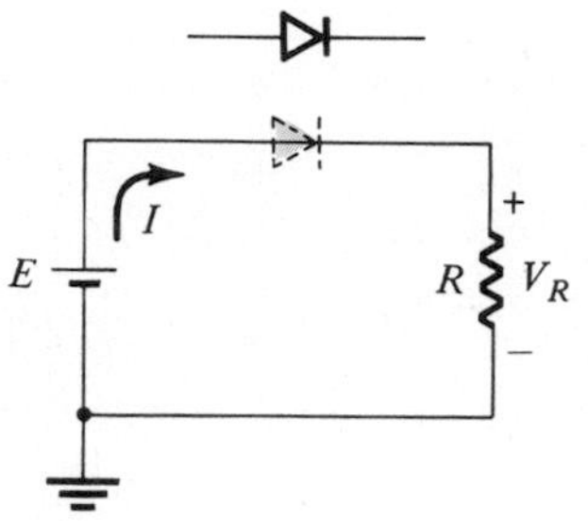

Figure 2.6 Determining the state of the diode.

Assuming that $E > V_T$, the diode is in the "on" state and the equivalent network of Fig. 2.7 results. Now

$$\boxed{V_D = V_T} \tag{2.1}$$

$$\boxed{V_R = E - V_T} \tag{2.2}$$

and

$$\boxed{I_D = I_R = \frac{V_R}{R}} \tag{2.3}$$

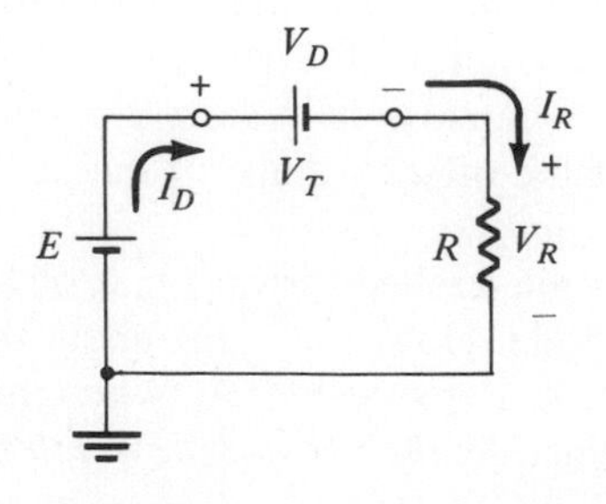

Figure 2.7 Substituting the "on"-state equivalent for the Si diode.

EXAMPLE 2.1

For the series diode configuration of Fig. 2.8, determine V_D, V_R, and I_D.

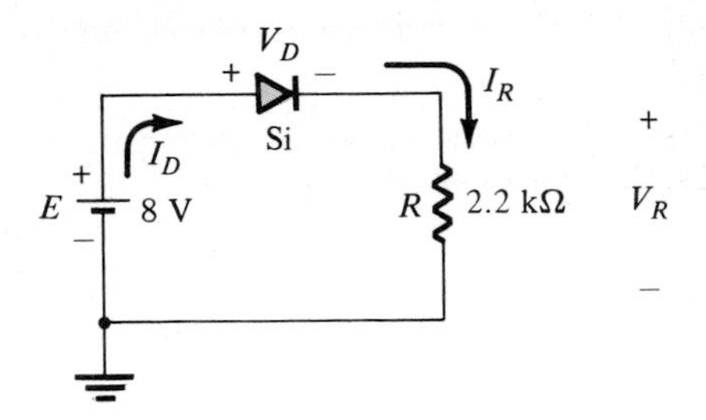

Figure 2.8

Solution:

Since the applied voltage establishes a current in the clockwise direction to match the arrow of the symbol and the diode is in the "on" state,

$$V_D = \mathbf{0.7\ V}$$

$$V_R = E - V_T = 8 - 0.7 = \mathbf{7.3\ V}$$

$$I_D = I_R = \frac{V_R}{R} = \frac{7.3}{2.2\ \text{k}\Omega} \cong \mathbf{3.32\ mA}$$

EXAMPLE 2.2

Repeat Example 2.1 with the diode reversed.

Solution:

Removing the diode, we find that the direction of I in Fig. 2.9 is opposite to the arrow in the diode symbol and the diode equivalent is the open circuit no matter which model is employed. The result is the network of Fig. 2.10, where I_D = **0 A** due to the open circuit. Since $V_R = I_R R$, $V_R = (0)R = 0$ V. Applying Kirchhoff's voltage law around the closed loop yields

$$E - V_D - V_R = 0$$

and

$$V_D = E - V_R = E - 0 = E = \mathbf{8\ V}$$

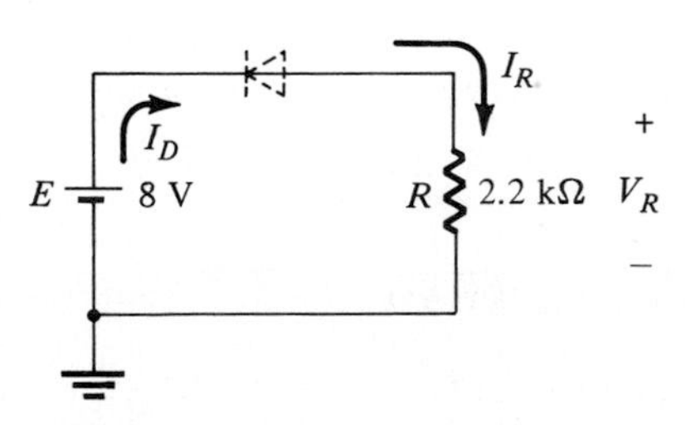

Figure 2.9

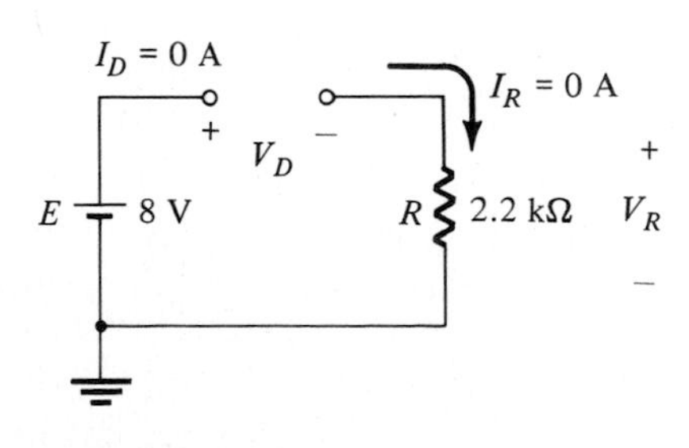

Figure 2.10

In particular, note in Example 2.2 the high voltage across the diode even though it is an "off" state. The current is zero, but the voltage is significant. For review purposes, keep the following in mind for the analysis to follow:

1. *An open circuit can have any voltage across its terminals, but the current is always 0 A.*
2. *A short circuit has a 0-V drop across its terminals, but the current is limited only by the surrounding network.*

In the next example the notation of Fig. 2.11 will be employed for the applied voltage. It is a common industry notation and one with which the reader should become very familiar. Such notation and other defined voltage levels are treated further in Chapter 5.

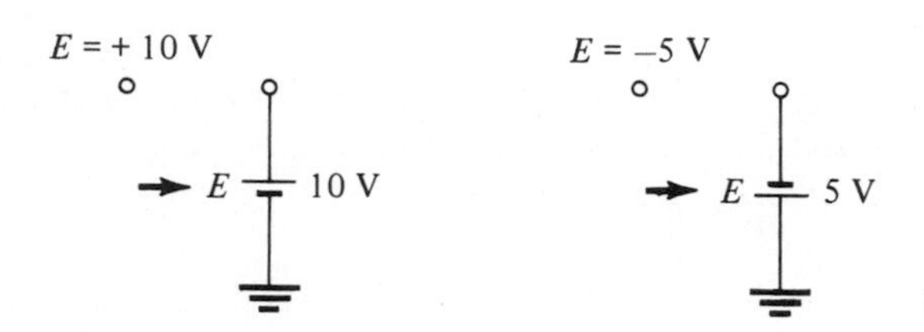

Figure 2.11 Source notation.

EXAMPLE 2.3

Determine V_o and I_D for the series circuit of Fig. 2.12.

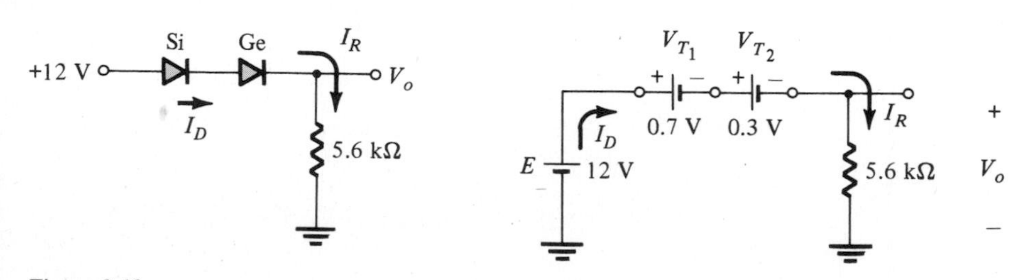

Figure 2.12

Figure 2.13

Solution:

An attack similar to that applied in Example 2.1 will reveal that the resulting current has the same direction as the arrowheads of the symbols of both diodes, and the network of Fig. 2.13 results because $E = 12\text{ V} > (0.7 + 0.3) = 1\text{ V}$. Note the redrawn supply of 12 V and the polarity of V_o across the 5.6-kΩ resistor. The resulting voltage

$$V_o = E - V_{T_1} - V_{T_2} = 12 - 0.7 - 0.3 = \mathbf{11\ V}$$

and

$$I_D = I_R = \frac{V_R}{R} = \frac{V_o}{R} = \frac{11}{5.6\text{ k}\Omega} \cong \mathbf{1.96\ mA}$$

EXAMPLE 2.4

Determine I_D, V_{D_2}, and V_o for the circuit of Fig. 2.14.

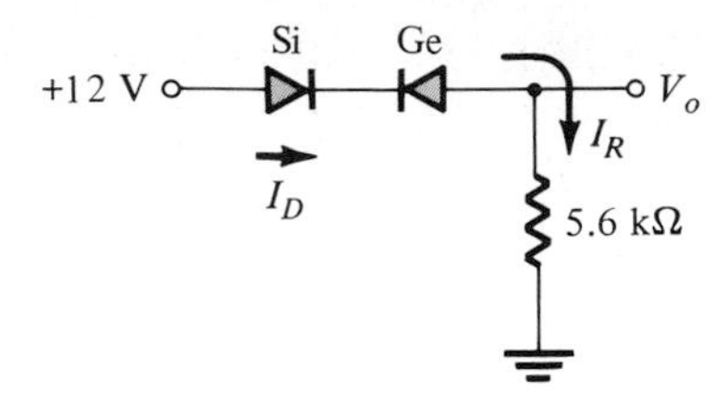

Figure 2.14

Solution:

Removing the diodes and determining the direction of the resulting current I will result in the circuit of Fig. 2.15. There is a match in current direction for the silicon diode but not for the germanium diode. The combination of a short circuit in series with an open circuit always results in an open circuit and I_D = **0 A**, as shown in Fig. 2.16.

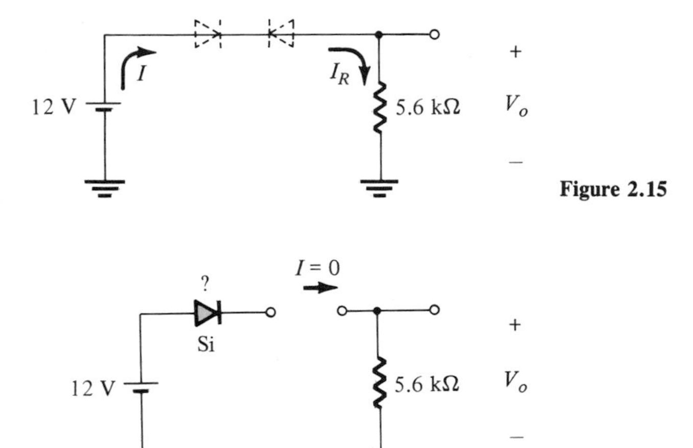

Figure 2.15

Figure 2.16

The question remains as to what to substitute for the silicon diode. For the analysis to follow in this and succeeding chapters, simply recall for the actual practical diode that when $V_D = 0$ V, $i_D = 0$ A (and vice versa), as described for the no-bias situation in Chapter 1. The conditions described by $V_{D_1} = 0$ V and $I_D = 0$ A are indicated on Fig. 2.17.

$$V_o = I_R R = I_D R = (0)R = 0 \text{ V}$$

and

$$V_{D_2} = V_{\text{open circuit}} = E = \mathbf{12\ V}$$

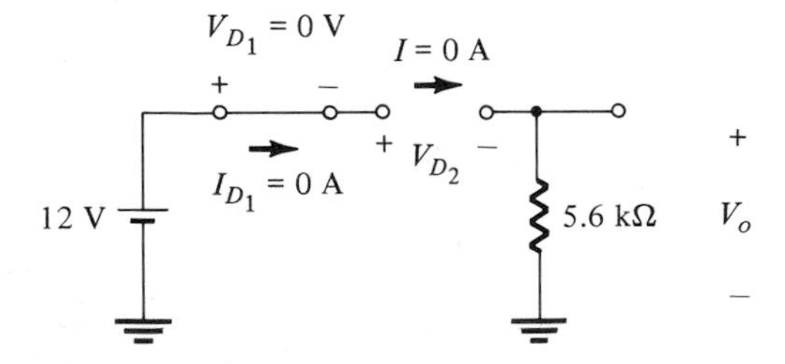

Figure 2.17

Applying Kirchhoff's voltage law in a clockwise direction gives us

$$E - V_{D_1} - V_{D_2} - V_o = 0$$

and

$$V_{D_2} = E - V_{D_1} - V_o = 12\text{ V} - 0 - 0$$

$$= \mathbf{12\ V}$$

$$V_o = \mathbf{0\ V}$$

EXAMPLE 2.5

Determine I, V_1, V_2, and V_o for the series dc configuration of Fig. 2.18.

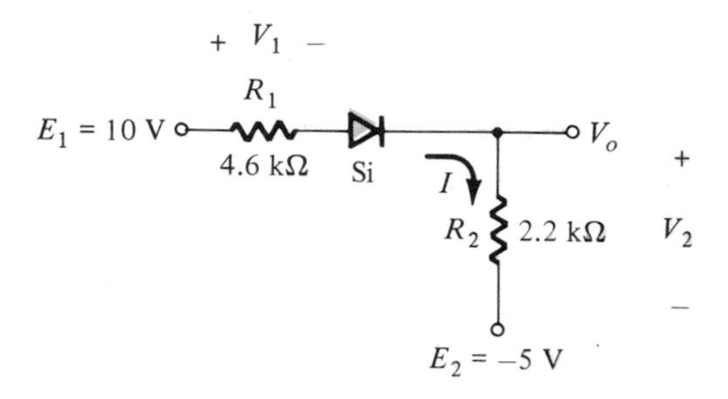

Figure 2.18

Solution:

The sources are drawn and the current direction determined as indicated in Fig. 2.19. The diode is in the "on" state and the approximate equivalent model is substituted in Fig. 2.20.

$$I = \frac{E_1 + E_2 - V_D}{R_1 + R_2} = \frac{10 + 5 - 0.7}{4.6\text{ k}\Omega + 2.2\text{ k}\Omega} = \frac{14.3}{6.8\text{ k}\Omega}$$

$$\cong \mathbf{2.1\ mA}$$

$$V_1 = IR_1 = (2.1\text{ mA})(4.6\ k\Omega) = \mathbf{9.66\ V}$$

$$V_2 = IR_2 = (2.1\text{ mA})(2.2\text{ k}\Omega) = \mathbf{4.62\ V}$$

Applying Kirchhoff's voltage law to the output section in the clockwise direction, we have

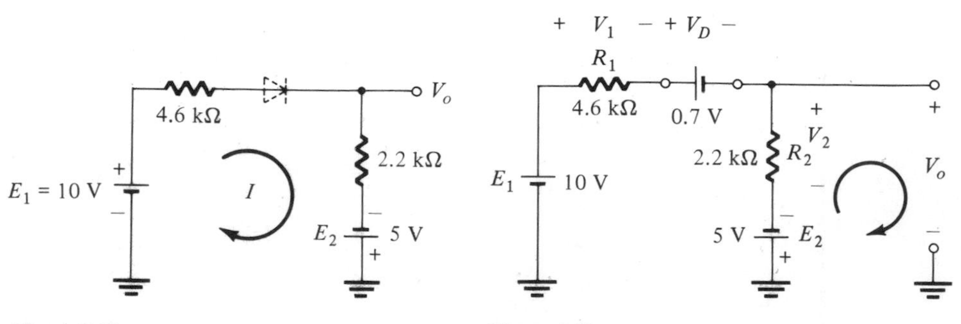

Figure 2.19

Figure 2.20

$$-E_2 + V_2 - V_o = 0$$

and
$$V_o = V_2 - E_2 = 4.62 - 5 = \mathbf{-0.38\ V}$$

The minus sign indicating that V_o has a polarity opposite to that appearing in Fig. 2.20.

2.3 PARALLEL AND SERIES–PARALLEL CONFIGURATIONS

The methods applied in Section 2.2 can be extended to the analysis of parallel and series–parallel configurations. For each area of application, simply match the sequential series of steps applied to series diode configurations.

EXAMPLE 2.6

Determine V_o, I_1, I_{D_1}, and I_{D_2} for the parallel diode configuration of Fig. 2.21.

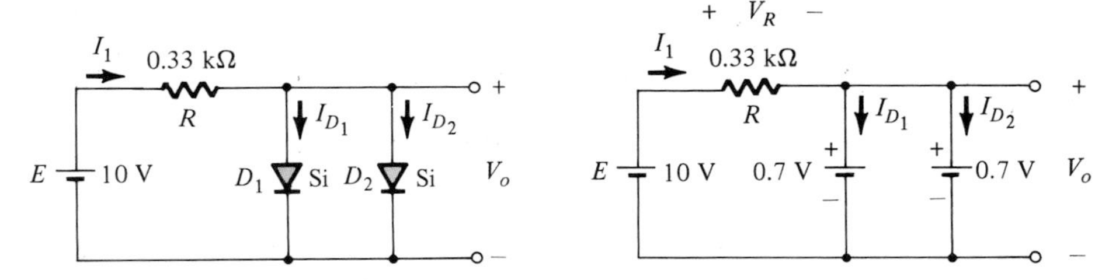

Figure 2.21

Figure 2.22

Solution:

For the applied voltage the "pressure" of the source is to establish a current through each diode in the same direction as shown in Fig. 2.22. Since the resulting current direction matches that of the arrow in each diode symbol and the applied voltage is greater than 0.7 V, both diodes are in the "on" state. The voltage across parallel elements is always the same and

$$V_o = \mathbf{0.7\ V}$$

The current

$$I_1 = \frac{V_R}{R} = \frac{E - V_D}{R} = \frac{10 - 0.7}{0.33\ \text{k}\Omega} = \mathbf{28.18\ mA}$$

Assuming diodes of similar characteristics, we have

$$I_{D_1} = I_{D_2} = \frac{I_1}{2} = \frac{28.18\ \text{mA}}{2} = \mathbf{14.09\ mA}$$

Example 2.6 demonstrated one reason for placing diodes in parallel. If the current rating of the diodes of Fig. 2.21 is only 20 mA, a current of 28.18 mA would damage the device if it appeared alone in Fig. 2.21. By placing two in parallel, the current is limited to a safe value of 14.09 mA with the same terminal voltage.

EXAMPLE 2.7

Determine the current I for the network of Fig. 2.23.

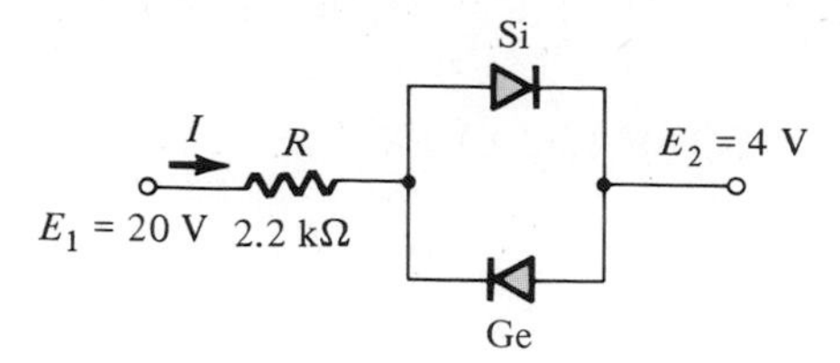

Figure 2.23

Solution:

Redrawing the network as shown in Fig. 2.24 reveals that the resulting current direction is such as to turn on the silicon diode and turn off the germanium diode. The resulting current I is then

$$I = \frac{E_1 - E_2 - V_D}{R} = \frac{20 - 4 - 0.7}{2.2\ k\Omega} \cong \mathbf{6.95\ mA}$$

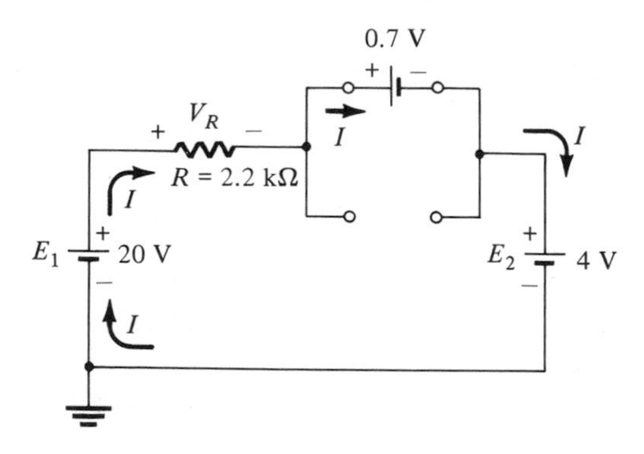

Figure 2.24

EXAMPLE 2.8

Determine the currents I_1, I_2, and I_{D_2} for the network of Fig. 2.25.

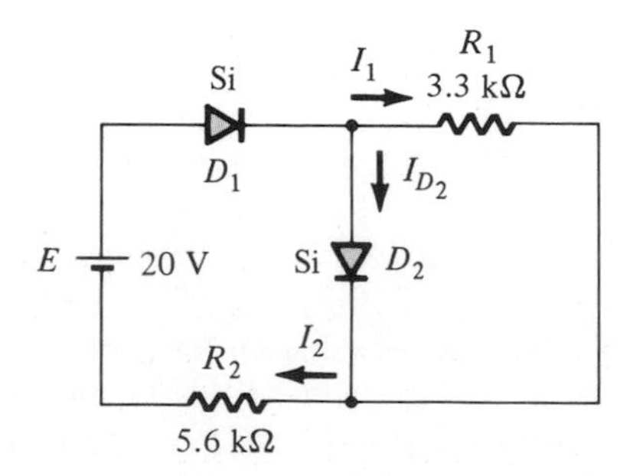

Figure 2.25

Solution:

The applied voltage (pressure) is such as to turn both diodes on, as noted by the resulting current directions in the network of Fig. 2.26. Note that once the equivalent network is drawn, the solution is obtained through an application of techniques applied to dc series–parallel networks.

$$I_1 = \frac{V_{T2}}{R_1} = \frac{0.7}{3.3\text{ k}\Omega} = \mathbf{0.212\ mA}$$

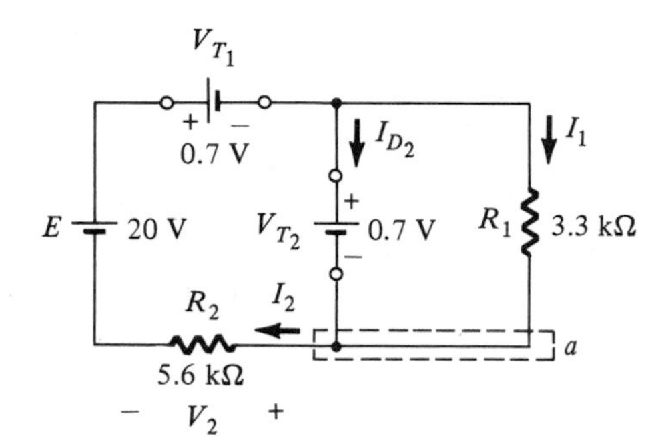

Figure 2.26

Applying the Kirchhoff's voltage law around the indicated loop in the clockwise direction yields

$$-V_2 + E - V_{T_1} - V_{T_2} = 0$$

and
$$V_2 = E - V_{T_1} - V_{T_2} = 20 - 0.7 - 0.7 = 18.6\text{ V}$$

with
$$I_2 = \frac{V_2}{R_2} = \frac{18.6}{5.6\text{ k}\Omega} = \mathbf{3.32\ mA}$$

At the bottom node (a),

$$I_{D_2} + I_1 = I_2$$

and
$$I_{D_2} = I_2 - I_1 = 3.32\text{ mA} - 0.212\text{ mA} = \mathbf{3.108\ mA}$$

2.4 AND/OR GATES

The tools of analysis are now at our disposal and the opportunity to investigate a computer configuration is one that will demonstrate the range of applications of this relatively simple device. Our analysis will be limited to determining the voltage levels and will not include a detailed discussion of Boolean algebra or positive and negative logic.

The network to be analyzed in Example 2.9 is an OR gate for positive logic. That is, the 10-V level of Fig. 2.27 is assigned a "1" for Boolean algebra while the 0-V input is assigned a "0." An OR gate is such that the output voltage level will be a 1 if either *or* both inputs is a 1. The output is a 0 if both inputs are at the 0 level.

The analysis of AND/OR gates is made measurably easier by using the approximate equivalent for a diode rather than the ideal because we can stipulate that the voltage across the diode must be 0.7 V (0.3 V for Ge) postive for the silicon diode to switch to the "on" state.

In general, the best approach is simply to establish a "gut" feeling for the state of the diodes by noting the direction and the "pressure" established by the applied potentials. The analysis will then verify or negate your initial assumptions.

EXAMPLE 2.9

Determine V_o for the network of Fig. 2.27.

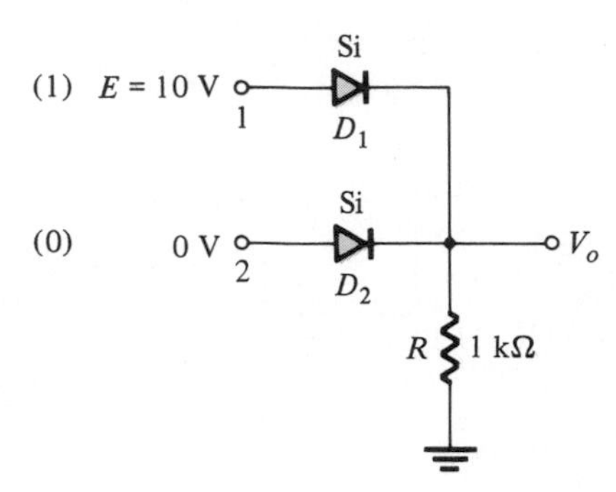

Figure 2.27 Positive logic OR gate.

Solution:

First note that there is only one applied potential—10 V at terminal 1. Terminal 2 with a 0 V input is essentially at ground potential, as shown in the redrawn network of Fig. 2.28. Figure 2.28 "suggests" that D_1, is probably in the "on" state due to the applied 10 V while D_2 with its "positive" side at 0 V is probably "off." Assuming these states will result in the configuration of Fig. 2.29.

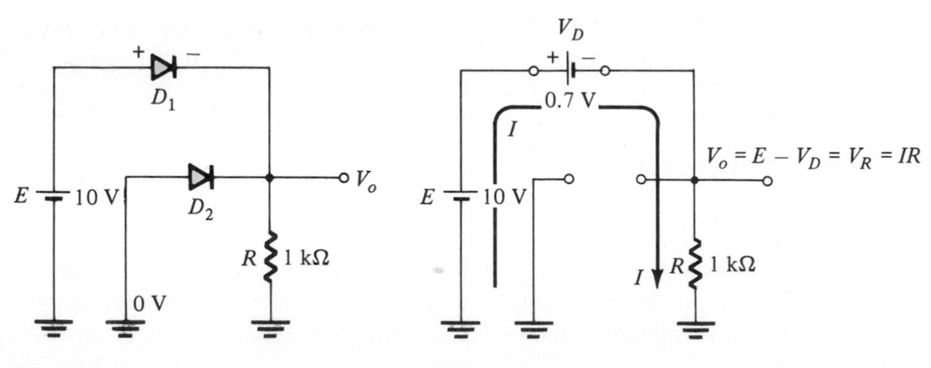

Figure 2.28

Figure 2.29

The next step is simply to check that there is no contradiction to our assumptions. That is, note that the polarity across D_1 is such as to turn it on and the polarity across D_2 is such as to turn it off. For D_1 the "on" state establishes V_o at $V_o = E - V_D = 10 - 0.7 = \mathbf{9.3\ V}$. With 9.3 V at the cathode (−) side of D_2 and 0 V at the anode (+) side, D_2 is definitely in the "off" state. The current direction and the resulting continuous path for conduction further confirm our assumption that D_1 is conducting. Our assumptions seem confirmed by the resulting voltages and current, and our initial analysis can be assumed to be correct. The output voltage level is not 10 V as defined for an input of 1, but the 9.3 V is sufficiently large to be considered a 1 level. The output is therefore at a 1 level with only one input, which suggests that that the gate is an OR gate. An analysis of the same network with two 10-V inputs will result in both diodes being in the "on" state and an output of 9.3 V. A 0-V input at both inputs will not provide the 0.7 V required to turn the diodes on and the output will be a 0 due to the 0-V output level. For the network of Fig. 2.29 the current level is determined by

$$I = \frac{E - V_D}{R} = \frac{10 - 0.7}{1\ k\Omega} = \mathbf{9.3\ mA}$$

EXAMPLE 2.10

Determine the output level for the positive logic AND gate of Fig. 2.30.

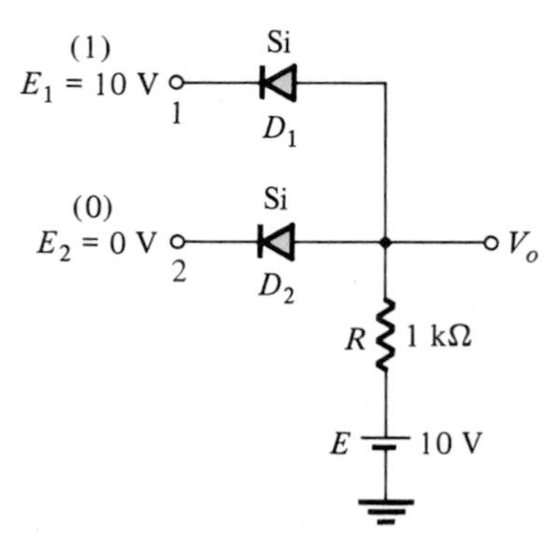

Figure 2.30 Positive logic AND gate.

Solution:

Note in this case that an independent source appears in the grounded leg of the network. For reasons soon to become obvious it is chosen at the same level as the input logic level. The network is redrawn in Fig. 2.31 with our initial assumptions regarding the state of the diodes. With 10 V at the cathode side of D_1 it is assumed that D_1 is in the "off" state even though there is a 10-V source connected to the anode of D_1 through the resistor. However, recall that we mentioned in the introduction to this section that the use of the approximate

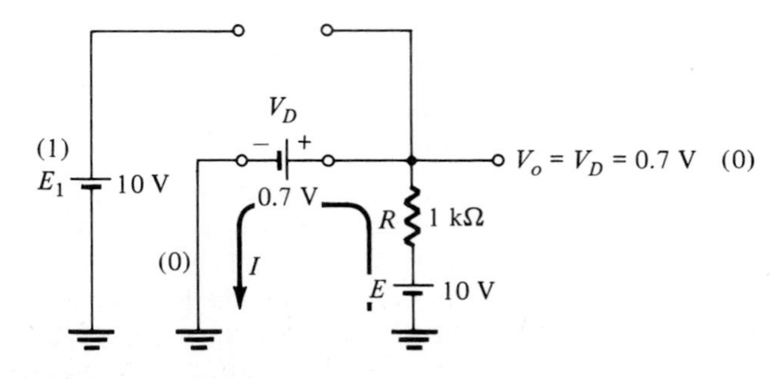

Figure 2.31

model will be an aid to the analysis. For D_1, where will the 0.7 V come from if the input and source voltages are at the same level and creating opposing "pressures"? D_2 is assumed to be in the "on" state due to the low voltage at the cathode side and the availability of the 10-V source through the 1 kΩ resistor.

For the network of Fig. 2.31 the voltage at V_o is 0.7 V due to the forward-biased diode D_2. With 0.7 V at the anode of D_1 and 10 V at the cathode, D_1 is definitely in the "off" state. The current I will have the direction indicated in Fig. 2.31 and a magnitude equal to

$$I = \frac{E - V_D}{R} = \frac{10 - 0.7}{1\ k\Omega} = \mathbf{9.3\ mA}$$

The state of the diodes is therefore confirmed and our earlier analysis was correct. Although not 0 V as earlier defined for the 0 level, the output voltage is sufficiently small to be considered a 0 level. For the AND gate, therefore, a single input will result in a 0-level output. The remaining states of the diodes for the possibilities of two inputs and no inputs will be examined in the problems at the end of the chapter.

2.5 HALF-WAVE RECTIFICATION

The diode analysis will now be expanded to include time-varying functions such as the sinusoidal waveform and the square wave. There is no question that the degree of difficulty will increase, but once a few fundamental maneuvers are understood, the analysis will be fairly direct and follow a common thread.

The simplest of networks to examine with a time-varying signal appears in Fig. 2.32. For the moment we will use the ideal model to ensure that the approach is not clouded by additional mathematical complexity.

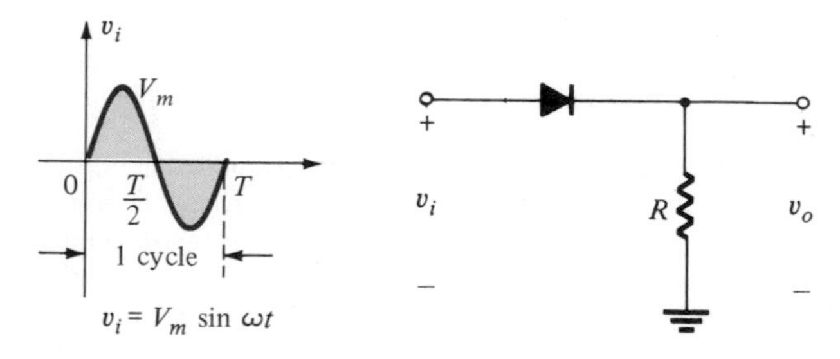

Figure 2.32 Half-wave rectifier.

Over one full cycle, defined by the period T of Fig. 2.32, the average value (the algebraic sum of the areas above and below the axis) is zero. The circuit of Fig. 2.32, called a *half-wave rectifier*, will generate a waveform v_o that will have an average value of particular use in the ac-to-dc conversion process.

During the interval $t = 0 \rightarrow T/2$ in Fig. 2.32, the polarity of the input voltage v_i is shown in Fig. 2.33. The result is the polarity shown across the diode, resulting in the short-circuit equivalent appearing in the adjoining figure. The output is now connected directly to the input, with the result that for the period $0 \rightarrow T/2$, $v_o = v_i$.

For the period $T/2 \rightarrow T$, the polarity of the input v_i is as shown in Fig. 2.34 and the resulting polarity across the ideal diode produces an "off" state with an open-circuit equivalent. The result is the absence of a path for charge to flow and $v_o = iR = (0)R = 0$ V for the period $T/2 \rightarrow T$. The input v_i and the output v_o were

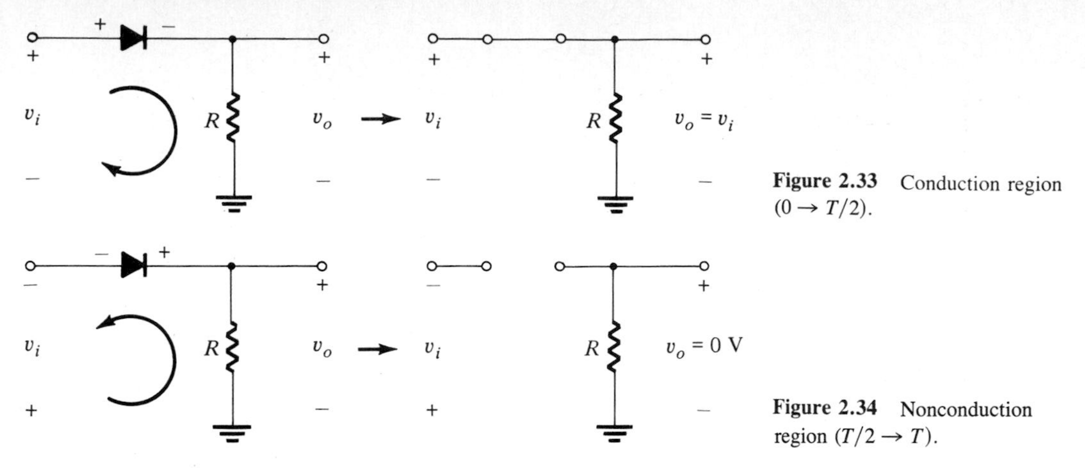

Figure 2.33 Conduction region $(0 \rightarrow T/2)$.

Figure 2.34 Nonconduction region $(T/2 \rightarrow T)$.

sketched together in Fig. 2.35 for comparison purposes. The output signal v_o now has a net positive area above the axis over a full period and an average value determined by

$$\boxed{\text{average (dc value)} = 0.318V_m} \tag{2.4}$$

The process of removing one-half the input signal to establish a dc level is aptly called *half-wave rectification*. The term "rectification" comes from the use of the term "rectifier" for diodes employed in power supplies for the ac-to-dc conversion process. Later we expand on the use of this pulsating voltage to establish a steady dc voltage.

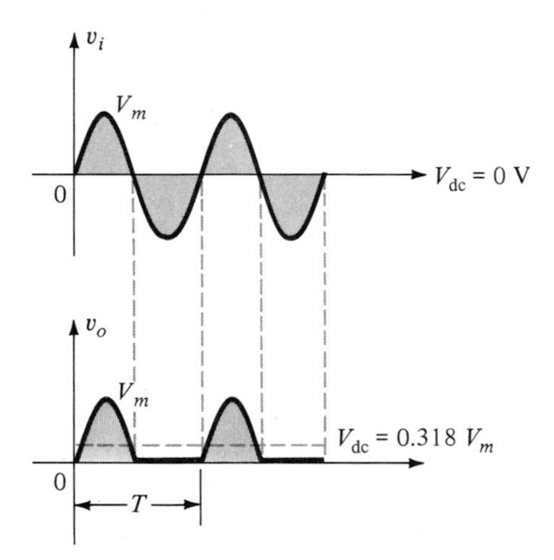

Figure 2.35 Half-wave rectified signal.

The effect of using a silicon diode with $V_T = 0.7$ V is demonstrated by Fig. 2.36 for the forward-bias region. The input must now be at least 0.7 V before the diode conducts resulting in a zero output level until the transition occurs. When conducting the difference between v_o and v_i is a fixed level of $V_T = 0.7$ V and $v_o = v_i - V_T$, as shown in the figure. The net effect is a reduction in area above the axis, which naturally reduces the resulting dc voltage level. If V_m is much greater than V_o, the difference can be ignored and Eq. (2.4) applied. That is,

$$\boxed{V_{dc} \cong 0.318V_m}_{\,V_m >> V_T} \tag{2.5}$$

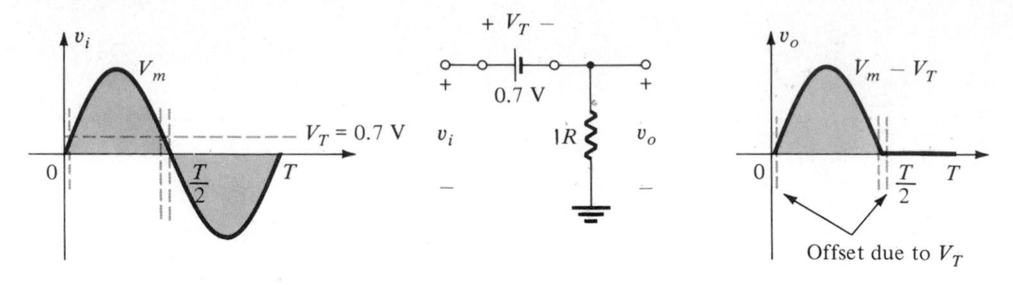

Figure 2.36 Effect of V_T on half-wave-rectified signal.

If V_m is relatively close to V_T, a good approximation is simply to use Eq. (2.4) but with a peak value reduced by the offset voltage. The difference in dc level due to neglecting the fringe areas on the left and right is normally small enough to be of little consequence. Therefore,

$$V_{dc} \cong 0.318(V_m - V_T) \quad \text{(} V_m \text{ close to } V_T\text{)} \tag{2.6}$$

EXAMPLE 2.11

(a) Sketch the output v_o and determine the dc level of the output for the network of Fig. 2.37.
(b) Repeat part (a) if the ideal diode is replaced by a silicon diode.

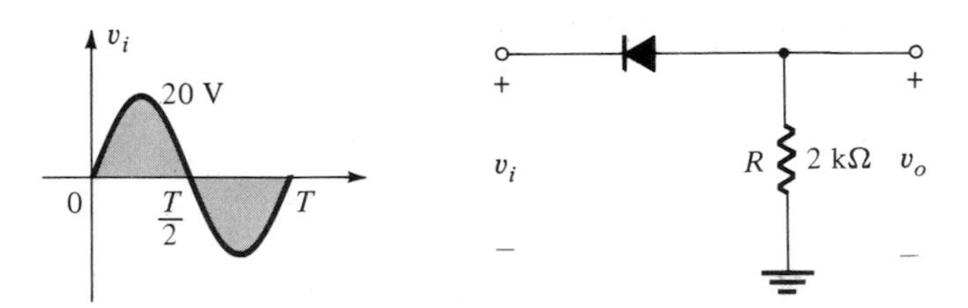

Figure 2.37 Network for Example 2.11.

Solution:

(a) In this situation the diode will conduct during the negative part of the input as shown in Fig. 2.38, and v_o will appear as shown in the same figure. For the full period, the dc level is

$$V_{dc} = -0.318V_m = -0.318(20) = \mathbf{-6.36\ V}$$

The negative sign indicates that the polarity of the output is opposite to the defined polarity of Fig. 2.37.
(b) Using a silicon diode, the output has the appearance of Fig. 2.39 and

$$V_{dc} \cong -0.318(V_m - 0.7) = -0.318(19.3) \cong \mathbf{-6.14\ V}$$

The resulting drop in dc level is 0.22 V or about 3.5%.

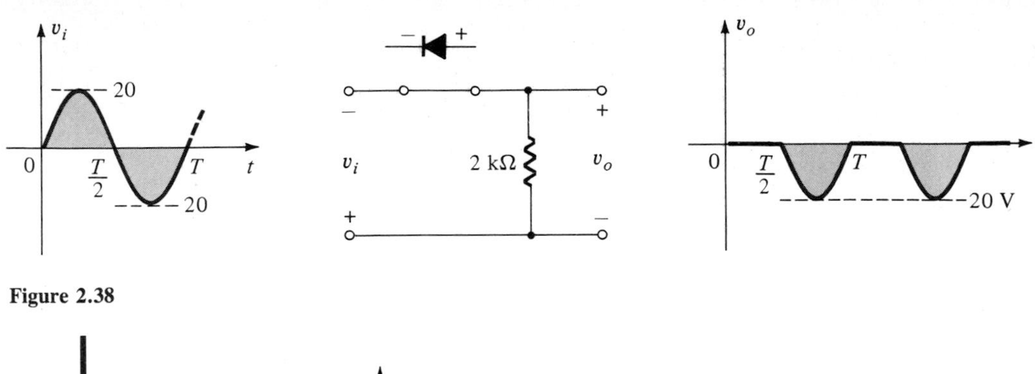

Figure 2.38

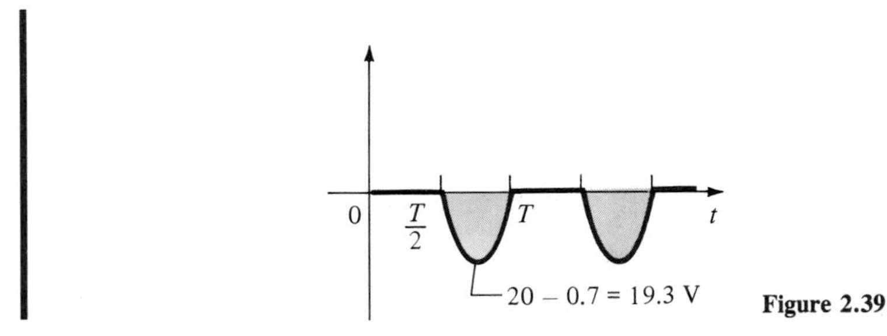

Figure 2.39

The peak-inverse-voltage (PIV) rating of the diode is of primary importance in the design of rectification systems. Recall that it is voltage rating that must not be exceeded in the reverse-bias region or the diode will enter the Zener avalanche region. The required PIV rating for the half-wave rectifier can be determined from Fig. 2.40, which displays the reverse-biased diode of Fig. 2.32 with maximum applied voltage. Applying Kirchhoff's voltage law, it is fairly obvious that the PIV rating of the diode must equal or exceed the peak value of the applied voltage. Therefore,

$$\boxed{\text{PIV rating} = V_m}_{\text{half-wave rectifier}} \tag{2.7}$$

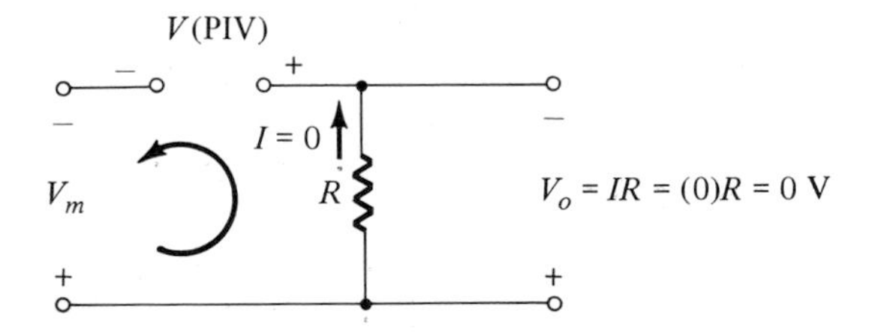

Figure 2.40 Determining the required PIV rating for the half-wave rectifier.

2.6 FULL-WAVE RECTIFICATION

The dc level obtained from a sinusoidal input can be improved 100% using a process called *full-wave rectification*. The most familiar network for performing such a function appears in Fig. 2.41 with its four diodes in a *bridge* configuration. During the period $t = 0$ to $T/2$ the polarity of the input is as shown in Fig. 2.42. The resulting

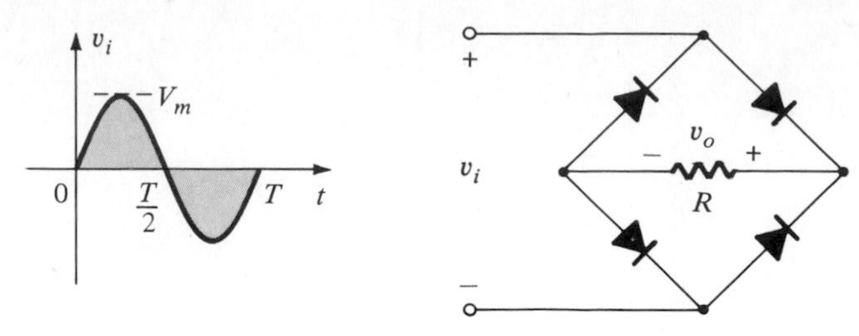

Figure 2.41 Full-wave bridge rectifier.

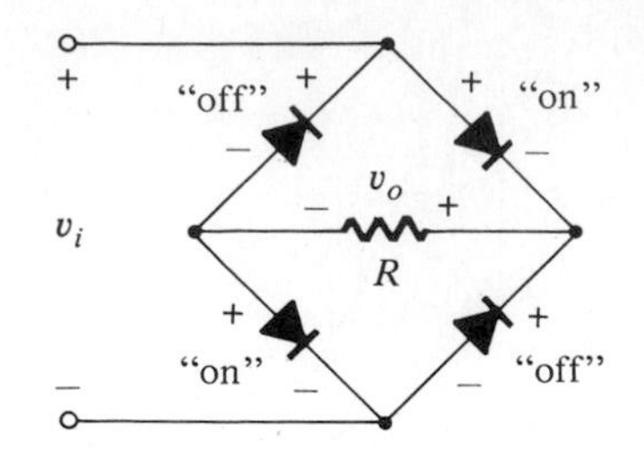

Figure 2.42 Network of Fig. 2.41 for the period $0 \rightarrow T/2$ of the input voltage V_i.

polarities across the ideal diodes are also shown in Fig. 2.42 to reveal that D_2 and D_3 are conducting while D_1 and D_4 are in the "off" state. The net result is the configuration of Fig. 2.43, with its indicated current and polarity across R. Since the diodes are ideal the load voltage $v_o = v_i$, as shown in the same figure.

For the negative region of the input the conducting diodes are D_1 and D_4, resulting in the configuration of Fig. 2.44. The important result is that the polarity across the load resistor R is the same as in Fig. 2.42, establishing a second positive pulse, as shown in Fig. 2.44. Over one full cycle the input and output voltages will appear as shown in Fig. 2.45.

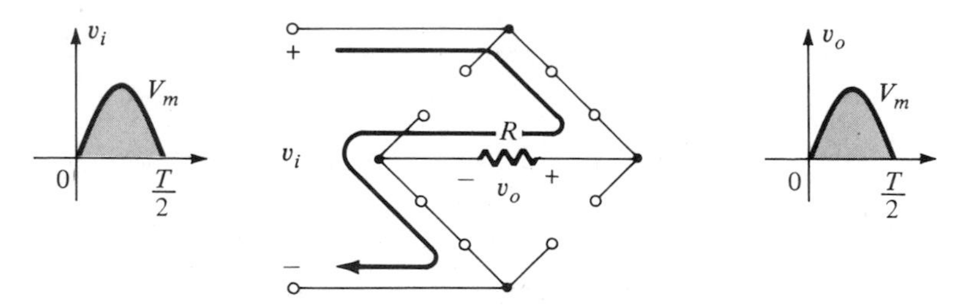

Figure 2.43

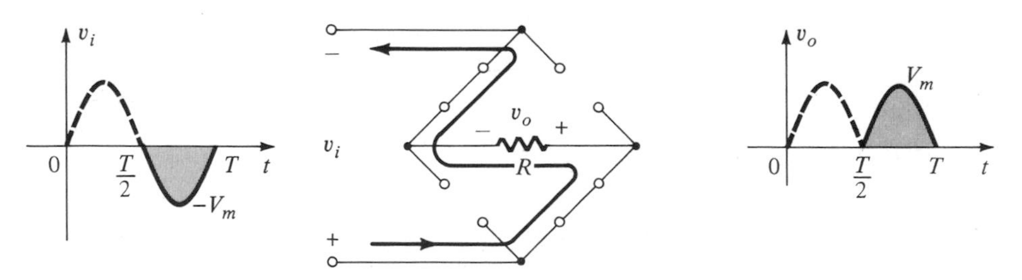

Figure 2.44

Since the area above the axis for one full cycle is now twice that obtained for a half-wave system, the dc level has also been doubled and

$$\boxed{\text{average (dc value)} = 0.636\ V_m} \tag{2.8}$$

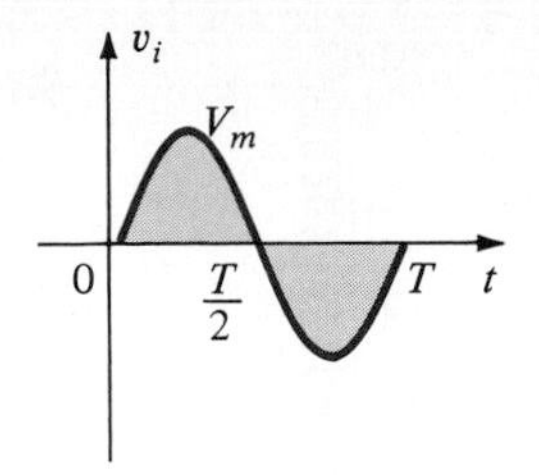

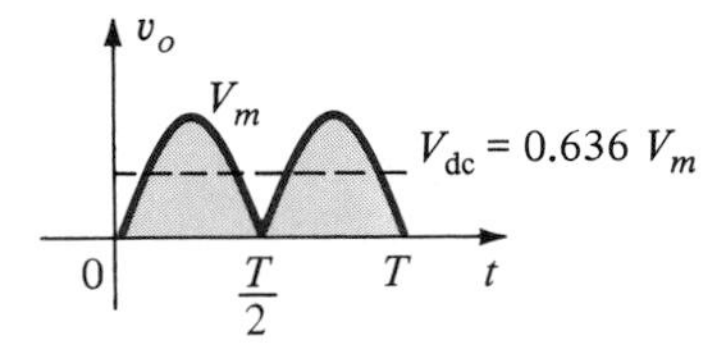

Figure 2.45

The effect of V_o has also doubled, as shown in Fig. 2.46 for silicon diodes during the positive conduction phase. However, if $V_m >> 2V_T$ then

$$\boxed{V_{dc} \cong 0.636V_m} \quad _{V_m \gg 2V_T} \tag{2.9}$$

and if V_m is close to $2V_T$,

$$\boxed{V_{dc} \cong 0.636(V_m - 2V_T)} \quad _{V_m \text{ close to } 2V_T} \tag{2.10}$$

The required PIV of each diode (ideal) can be determined from Fig. 2.47 obtained at the peak of the positive region of the input signal. For the indicated loop the maximum voltage across R is V_m and

$$\boxed{\text{PIV} = V_m} \quad _{\text{full-wave bridge rectifier}} \tag{2.11}$$

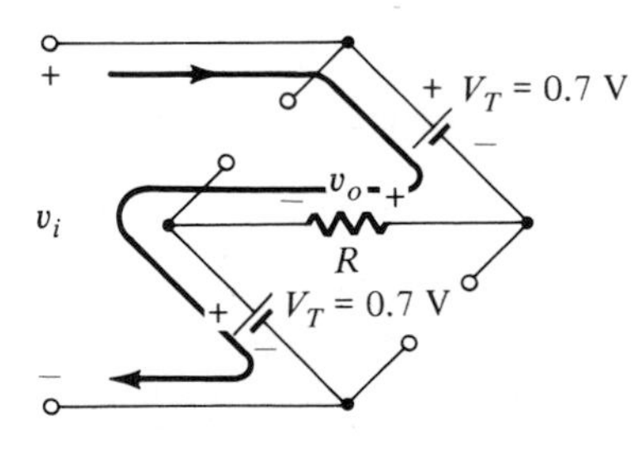

Figure 2.46

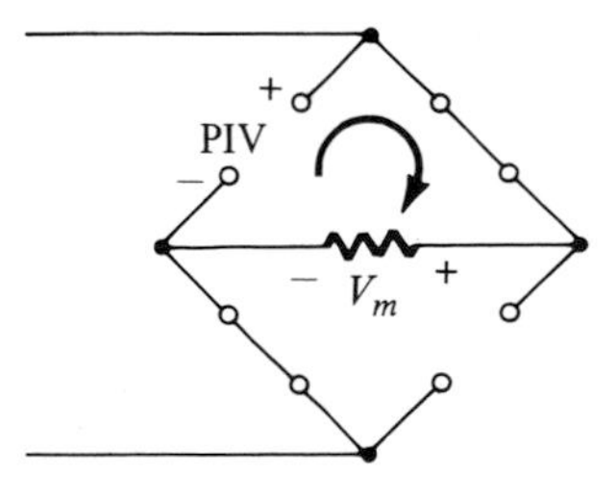

Figure 2.47

A second popular full-wave rectifier appears in Fig. 2.48 with only two diodes but requiring a center-tapped (CT) transformer to establish the input signal across each section of the secondary of the transformer.

During the positive portion of v_i applied to the primary of the transformer, the network will appear as shown in Fig. 2.49. D_1 assumes the short-circuit equivalent

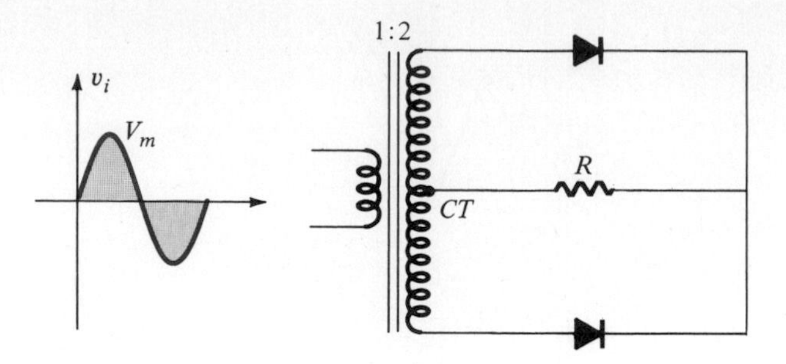

Figure 2.48

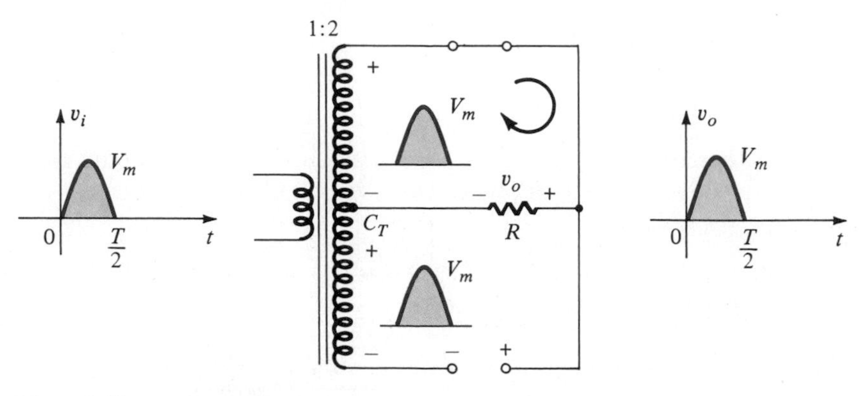

Figure 2.49

and D_2 the open-circuit equivalent, as determined by the secondary voltages and the resulting current directions. The output voltage appears as shown in Fig. 2.49.

During the negative portion of the input the network appears as shown in Fig. 2.50, reversing the roles of the diodes but maintaining the same polarity for the voltage across the load resistor R. The net effect is the same output as that appearing in Fig. 2.45 with the same dc levels.

The network of Fig. 2.51 will help us determine the net PIV for each diode for this fiull-wave rectifier. Inserting the maximum voltage for the secondary voltage and V_m

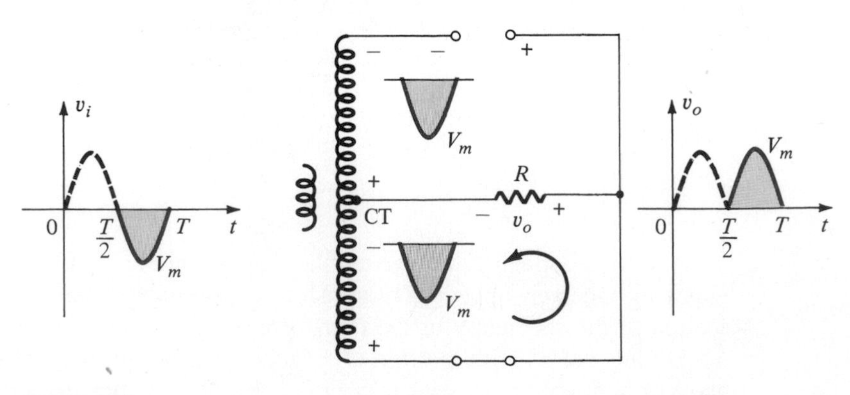

Figure 2.50

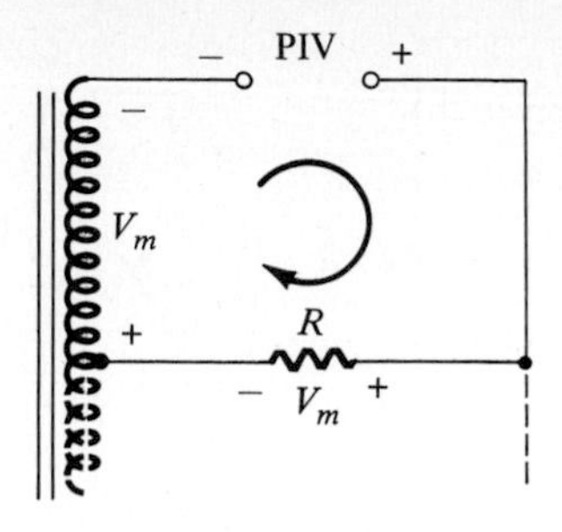

Figure 2.51

as established by the adjoining loop will result in

$$\text{PIV} = V_{\text{secondary}} + V_R$$
$$= V_m + V_m$$

$$\boxed{\text{PIV} = 2V_m} \quad \text{transformer full-wave rectifier} \qquad (2.12)$$

EXAMPLE 2.12

Determine the output waveform for the network of Fig. 2.52 and calculate the output dc level and the required PIV of each diode.

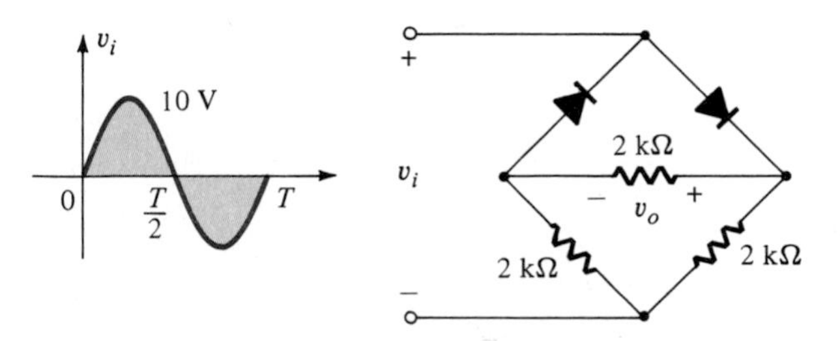

Figure 2.52

Solution:

The network will appear as shown in Fig. 2.53 for the positive region of the input voltage. Redrawing the network will result in the configuration of Fig. 2.54, where $v_o = \frac{1}{2}v_i$ or $V_{o\,(\text{max})} = \frac{1}{2}V_{i\,(\text{max})} = \frac{1}{2}(10) = 5$ V, as shown in Fig. 2.54. For the negative part of the input the roles of the diodes will be interchanged and v_o will appear as shown in Fig. 2.55.

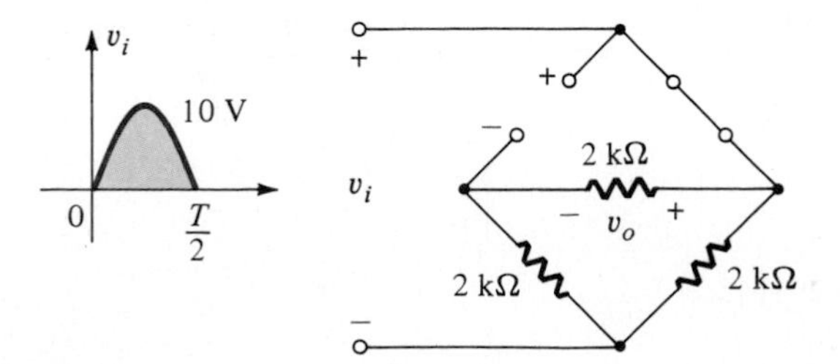

Figure 2.53

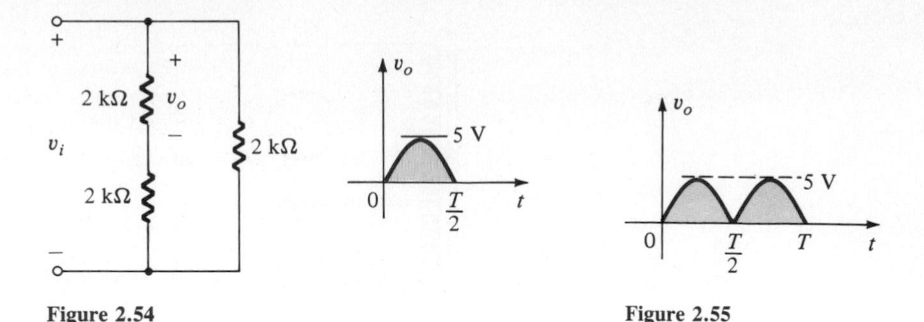

Figure 2.54 **Figure 2.55**

The effect of removing two diodes from the bridge configuration was therefore to reduce the available dc level to the following:

$$V_{dc} = 0.636(5) = \mathbf{3.18\ V}$$

or that available from a half-wave rectifier with the same input. However, the PIV as determined from Fig. 2.56 is equal to the maximum voltage across R, which is 5 V or half of that required for a half-wave rectifier with the same input.

2.7 CLIPPERS

There are a variety of diode networks called *clippers* that have the ability to "clip" off a portion of the input signal without distorting the remaining part of the alternating waveform. The half-wave rectifier of Section 2.5 is an example of the simplest form of diode clipper—one resistor and diode. Depending on the orientation of the diode the positive or negative region of the input signal is "clipped" off.

There are two general categories of clippers: series and parallel. The series configuration is defined as one where the diode is in series with the load, while the parallel variety has the diode in a branch parallel to the load.

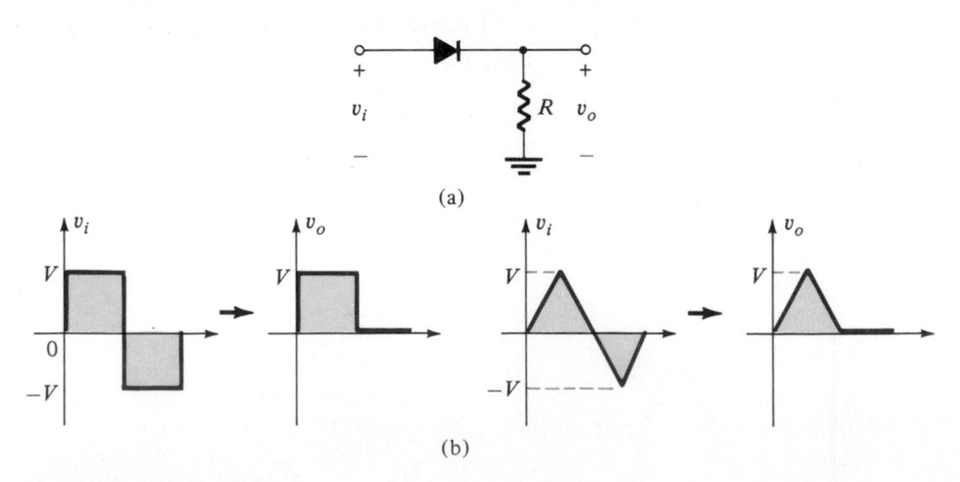

Figure 2.56 Series clipper.

Series

The response of the series configuration of Fig. 2.56a to a variety of alternating waveforms is provided in Fig. 2.56b. Although first introduced as a half-wave rectifier (for sinusoidal waveforms), there are no boundaries on the type of signals that can be applied to a clipper.

The addition of a dc supply such as shown in Fig. 2.57 can have a pronounced effect on the output of a clipper. Our initial discussion will be limited to ideal diodes, with the effect of V_o reserved for a concluding example.

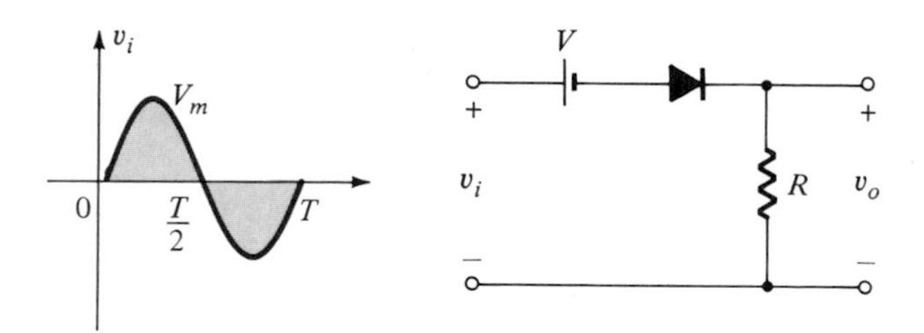

Figure 2.57

There is no general procedure for analyzing networks such as the type in Fig. 2.57, but there are a few thoughts to keep in mind as you work toward a solution.

1. *Make a mental sketch of the response of the network based on the direction of the diode and the applied voltage levels*. For the network of Fig. 2.57, the direction of the diode suggests that the signal v_i must be positive to turn it on. The dc supply further requires that the voltage v_i be greater than V volts to turn the diode on.

   The negative region of the input signal is "pressuring" the diode into the "off" state, supported further by the dc supply. In general, therefore, we can be quite sure the diode is an open circuit ("off" state) for the negative region of the input signal.

2. *Determine the applied voltage (transition voltage) that will cause a change in state for the diode*. For the ideal diode the condition $i_d = 0$ or $v_d = 0$ will be employed to determine the level of v_i to effect a transition.

   Applying the condition $i_d = 0$ at $v_d = 0$ to the network of Fig. 2.57 will result in the configuration of Fig. 2.58, where it is recognized that the level of v_i that will cause a transition in state is

$$\boxed{v_i = V_{dc}} \tag{2.13}$$

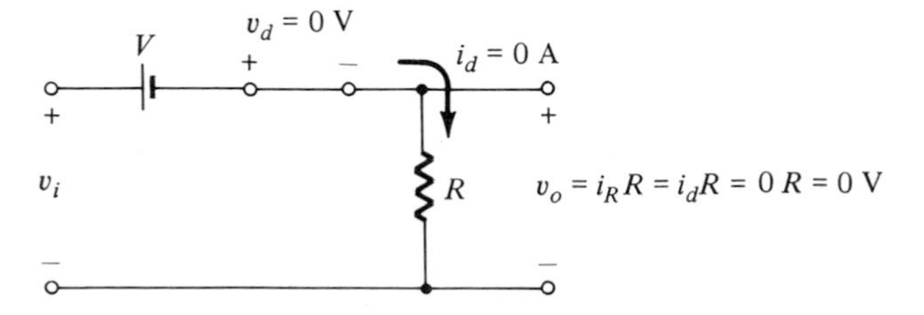

Figure 2.58

For an input voltage greater than V volts the diode is in the short-circuit state, while for input voltages less than V volts it is in the open-circuit or "off" state.

3. *Be continually aware of the defined terminals and polarity of v_o.* When the diode is in the short-circuit state, such as shown in Fig. 2.59, the output voltage v_o is the same as from part (a) to (b) as noted on the figure, and applying Kirchhoff's voltage law, we have

$$v_i - V - v_o = 0 \quad \text{(CW direction)}$$

and

$$\boxed{v_o = v_i - V} \tag{2.14}$$

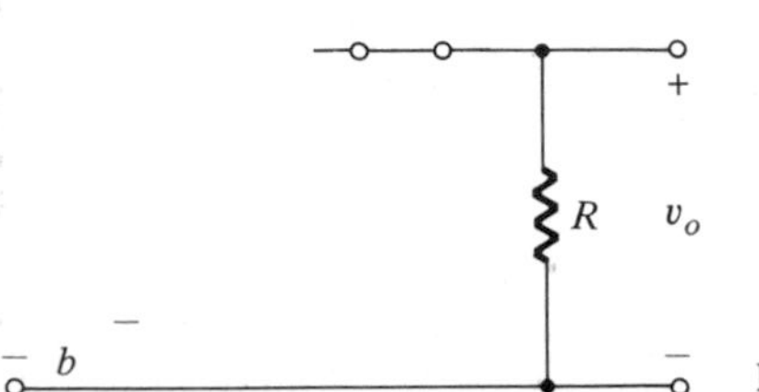

Figure 2.59

4. *It can be helpful to sketch the input signal above the output as shown in Fig. 2.60 and determine the output at instantaneous values of the input.* It is then possible that the output voltage can be sketched from the resulting data points of v_o. Keep in mind that at an instantaneous value of v_i the input can be treated as a dc supply of that value and the corresponding dc (the instantaneous value) value of the output determined. For instance, at $v_i = V_m$ for the network of Fig. 2.57, the network to be analyzed appears in Fig. 2.61. For $V_m > V$ the diode is in the short-circuit state and $V_o = V_m - V$, as shown in Fig. 2.60.

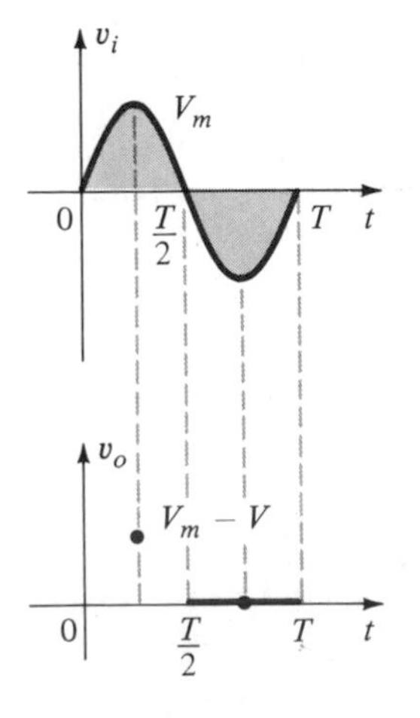

Figure 2.60

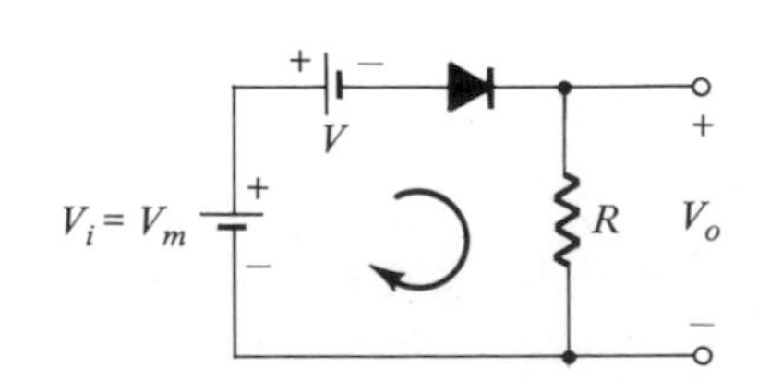

Figure 2.61

The same procedure can be applied to any number of points of the input voltage until a sufficient number of output points are available to sketch the continuous curve for v_o. For the network of Fig. 2.57 the output has the appearance of Fig. 2.62.

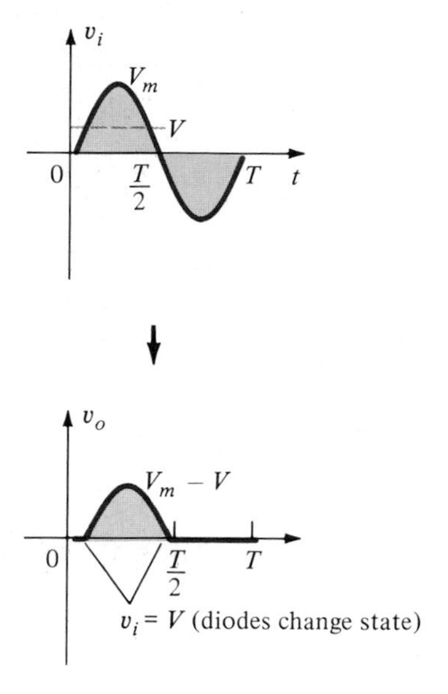

Figure 2.62

EXAMPLE 2.13

Determine the output waveform for the network of Fig. 2.63.

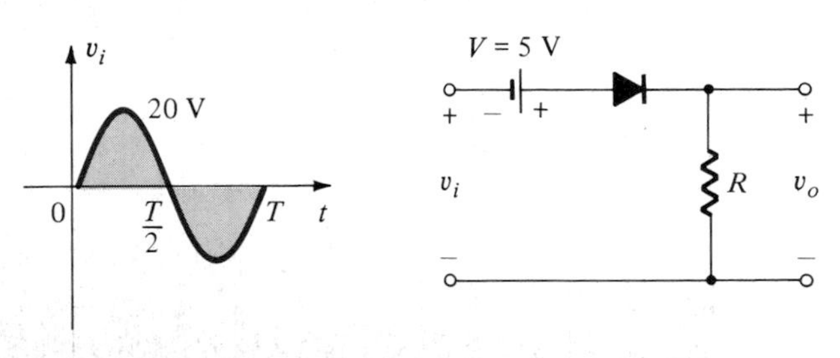

Figure 2.63

Solution:

Past experience suggests that the diode will be in the "on" state for the positive region of v_i—especially when we note the aiding effect of $V = 5$ V. The network will then appear as shown in Fig. 2.64 and $v_o = v_i + 5$. Substituting $i_d = 0$ at $v_d = 0$ for the transition voltage, we obtain the network of Fig. 2.65 and $v_i = -5$ V.

For voltages more negative than -5 V the diode will enter its open-circuit state, while for voltages more positive than -5 V the diode is in the short-circuit state. The input and output voltages appear in Fig. 2.66.

The analysis of clipper networks with square-wave inputs is actually easier to analyze than with sinusoidal inputs because only two levels have to be considered. In other words,

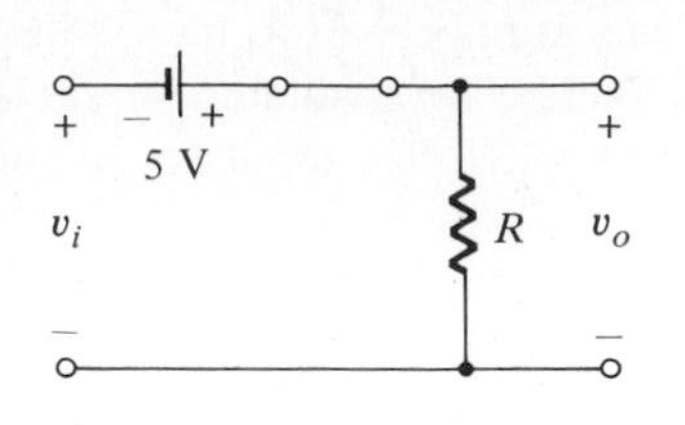

Figure 2.64

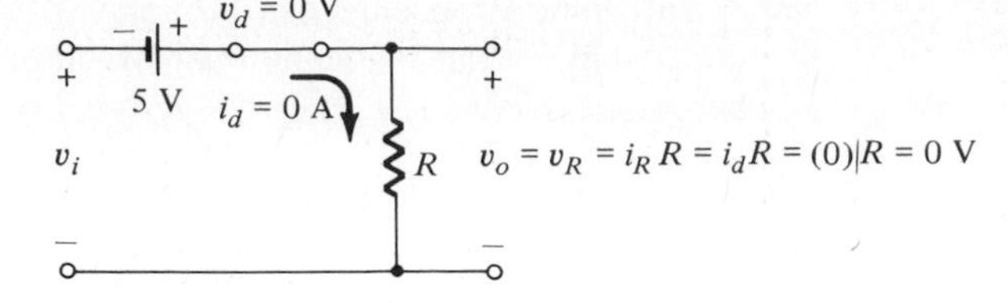

Figure 2.65

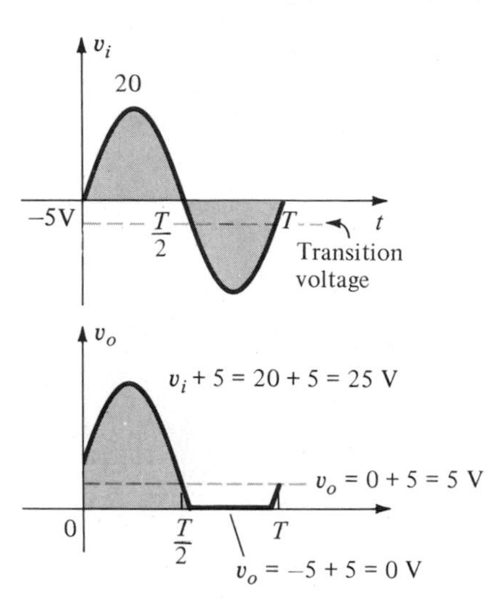

Figure 2.66

the network can be analyzed as if it had two dc level inputs with the resulting output v_o plotted in the proper time frame.

EXAMPLE 2.14

Repeat Example 2.13 for the square-wave input of Fig. 2.67.

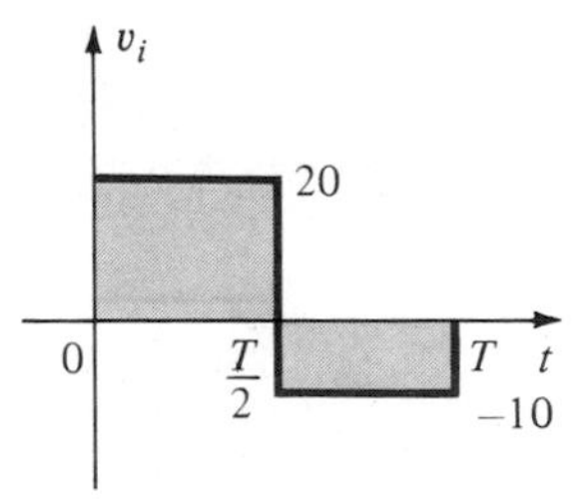

Figure 2.67

Solution:

For $V_i = 20$ V $(0 \rightarrow T/2)$ the network of Fig. 2.68 will result. The diode is in the short-circuit state and $V_o = 20 + 5 = 25$ V. For $v_i = -10$ V the network of Fig. 2.69 will result, placing the diode in the "off" state and $v_o = i_R R = (0)R = 0$ V. The resulting output voltage appears in Fig. 2.70.

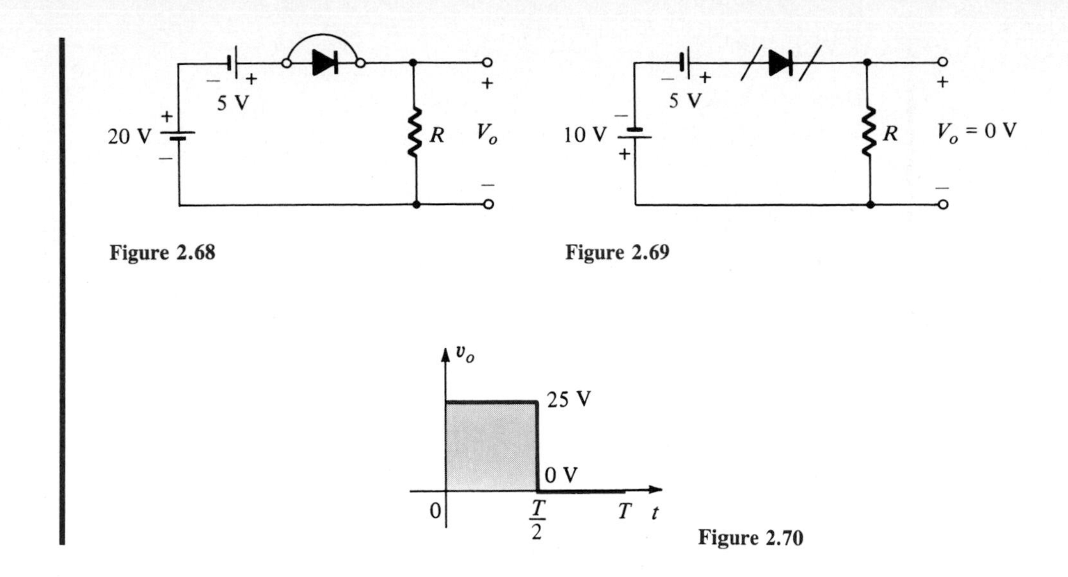

Figure 2.68

Figure 2.69

Figure 2.70

Note in Example 2.14 that the clipper not only clipped off 5 V from the total swing but raised the dc level of the signal by 5 V.

Parallel

The network of Fig. 2.71 is the simplest of parallel diode configurations with the output for the same inputs of Fig. 2.56. The analysis of parallel configurations is very similar to that applied to series configurations, as demonstrated in the next example.

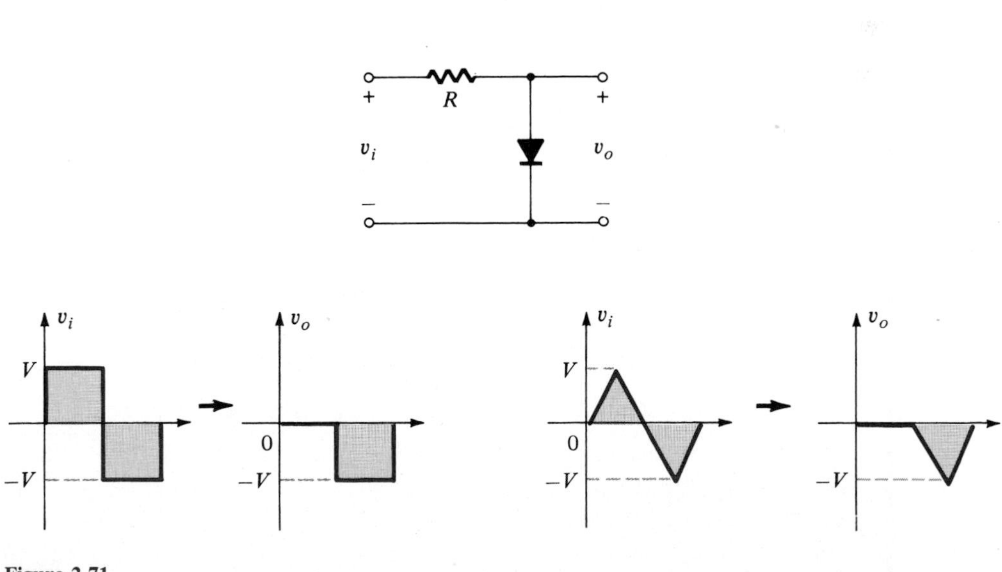

Figure 2.71

EXAMPLE 2.15

Determine v_o for the network of Fig. 2.72.

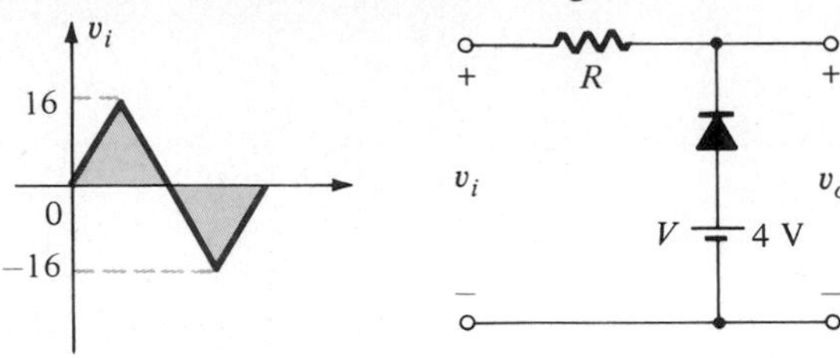

Figure 2.72

Solution:

The polarity of the dc supply and the direction of the diode strongly suggest that the diode will be in the "on" state for the negative region of the input signal. For this region the network will appear as shown in Fig. 2.73, where the defined terminals for v_o require that $v_o = V = 4$ V.

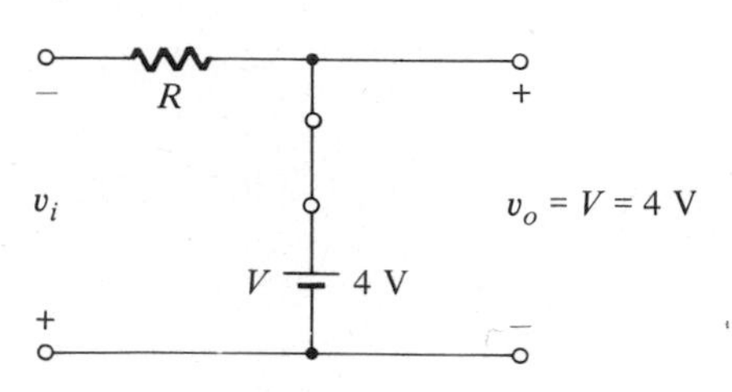

Figure 2.73

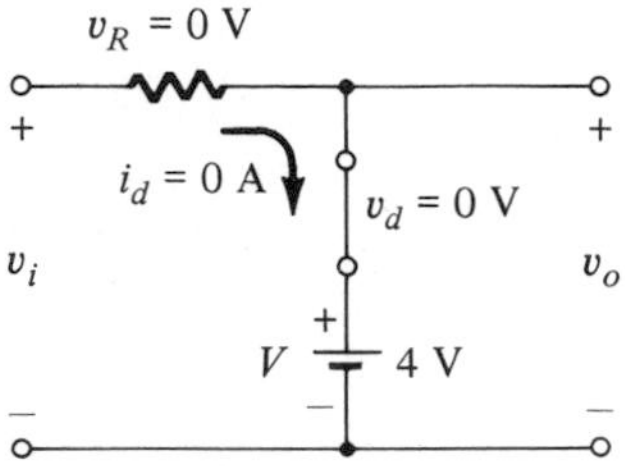

Figure 2.74

The transition state can be determined from Fig. 2.74, where the condition $i_d = 0$ A at $v_d = 0$ V has been imposed. The result is v_i (transition) $= V = 4$ V.

Since the dc supply is obviously "pressuring" the diode to stay in the short-circuit state, the input voltage must be greater than 4 V for the diode to be in the "off" state. Any input voltage less than 4 V will result in a short-circuited diode.

For the open-circuit state the network will appear as shown in Fig. 2.75, where $v_o = v_i$. Completing the sketch of v_o results in the waveform of Fig. 2.76.

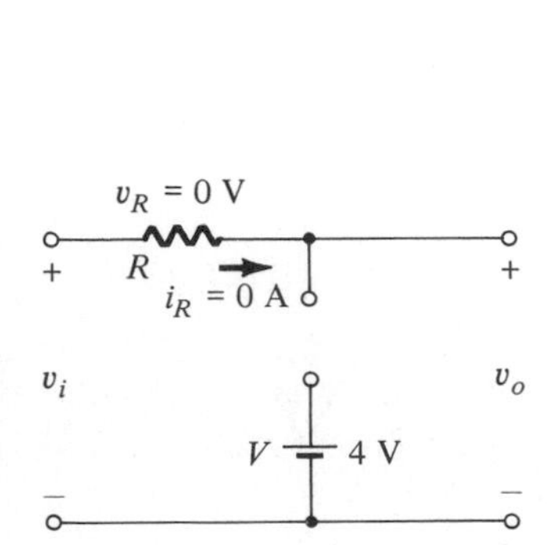

Figure 2.75

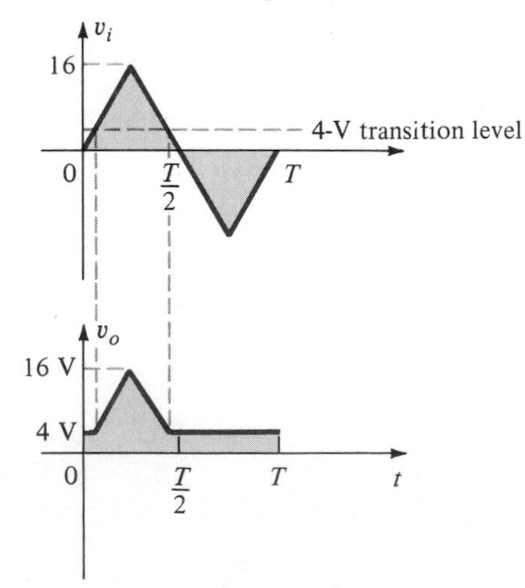

Figure 2.76

To examine the effects of V_T on the output voltage, the next example will specify a silicon diode rather than an ideal diode equivalent.

EXAMPLE 2.16

Repeat Example 2.15 using a silicon diode with $V_T = 0.7$ V.

Solution:

The transition voltage can first be determined by applying the condition $i_d = 0$ A at $v_d = V_D = 0.7$ V and obtaining the network of Fig. 2.77. Applying Kirchhoff's voltage law around the output loop in the clockwise direction, we find

$$v_i + V_T - V = 0$$

and

$$v_i = V - V_T = 4 - 0.7 = 3.3 \text{ V}$$

For input voltages greater than 3.3 V, the diode will be an open circuit and $v_o = v_i$. For input voltages of less than 3.3 V, the diode will be in the "on" state and the network of Fig. 2.78 results, where

$$v_o = 4 - 0.7 = 3.3 \text{ V}$$

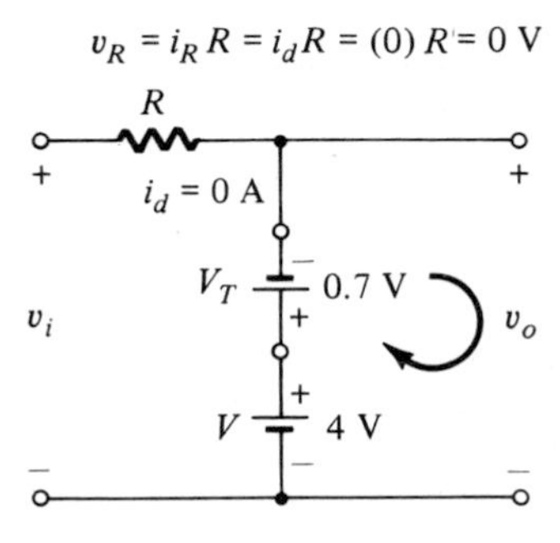

Figure 2.77

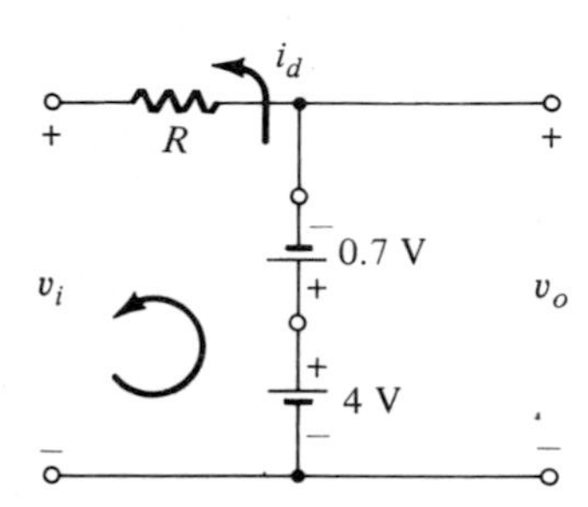

Figure 2.78

The resulting output waveform appears in Fig. 2.79. Note that the only effect of V_T was to drop the "on"-state level to 3.3 V from 4 V.

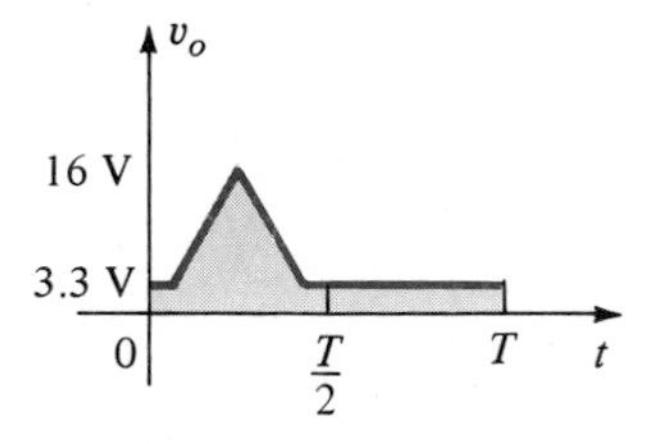

Figure 2.79

There is no question that including the effects of V_T will complicate the analysis somewhat, but once the analysis is understood with the ideal diode, the procedure, including the effects of V_T, will not be that difficult.

A variety of series and parallel clippers with the resulting output for the sinusoidal input are provided in Fig. 2.80. In particular, note the response of the last configur-

SIMPLE SERIES CLIPPERS

POSITIVE

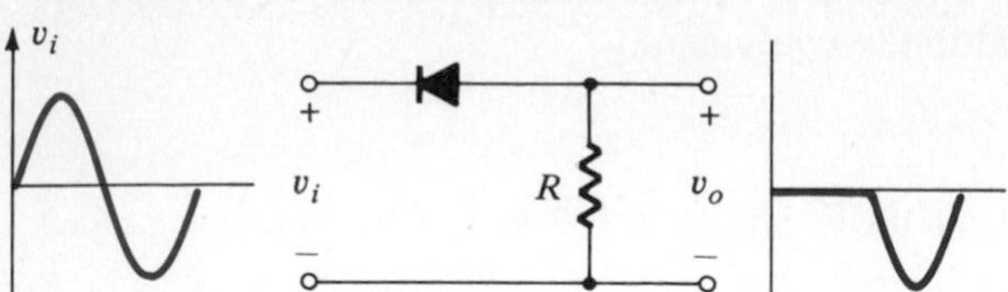

NEGATIVE

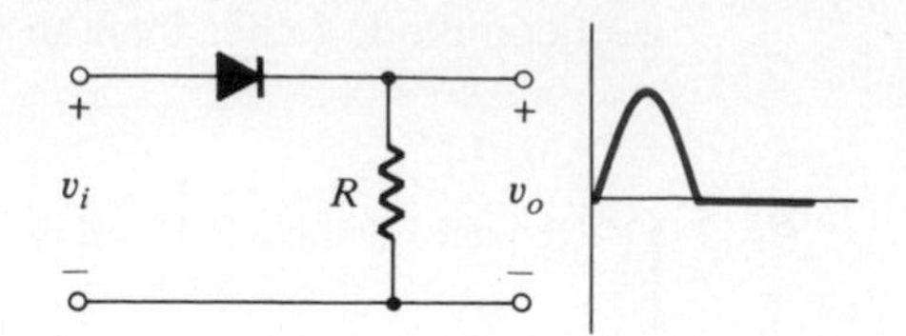

BIASED SERIES CLIPPERS

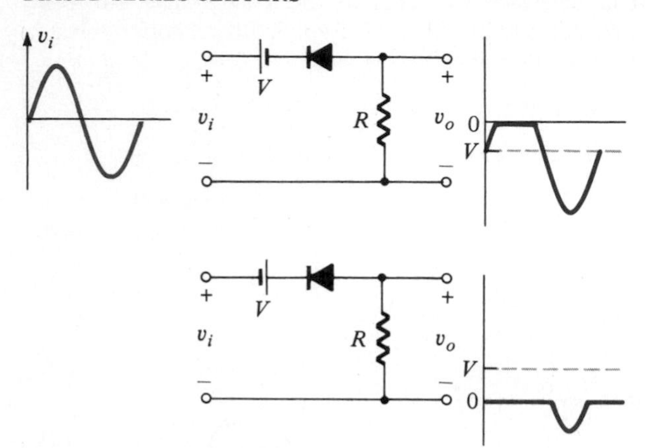

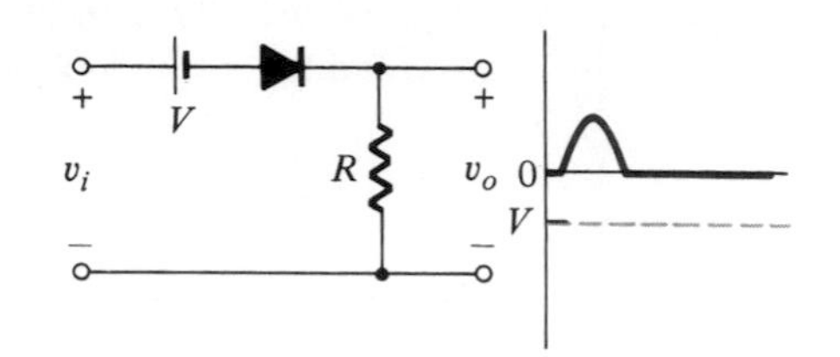

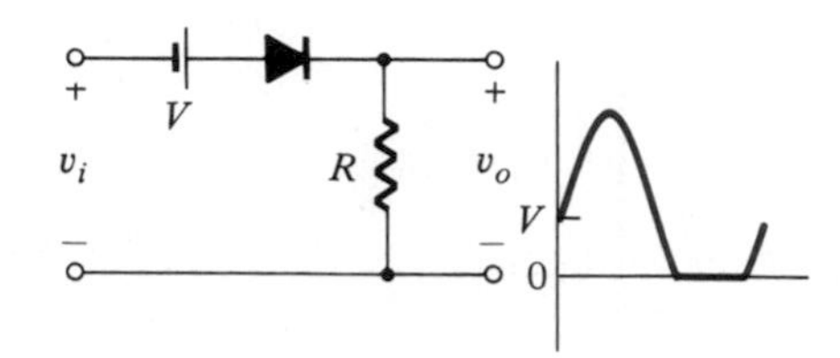

SIMPLE PARALLEL CLIPPERS

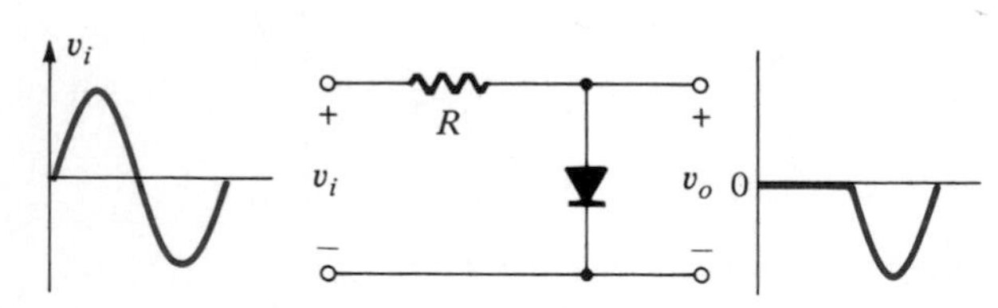

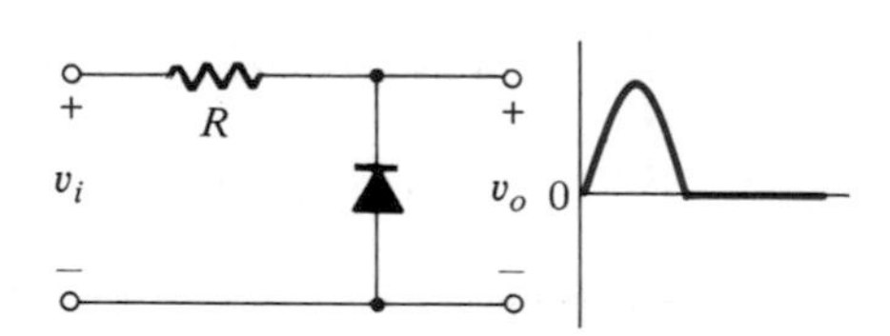

BIASED PARALLEL CLIPPERS

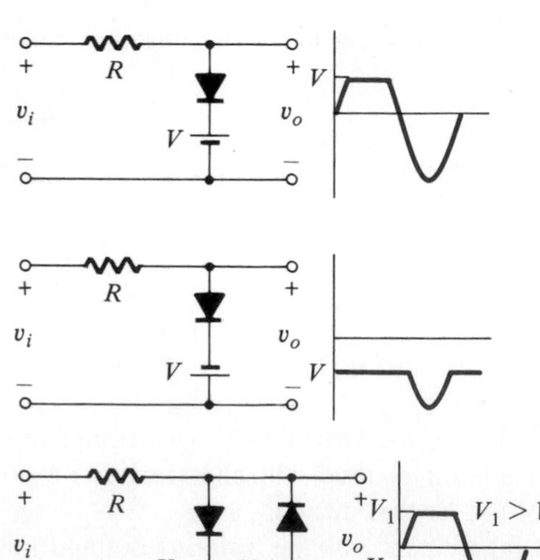

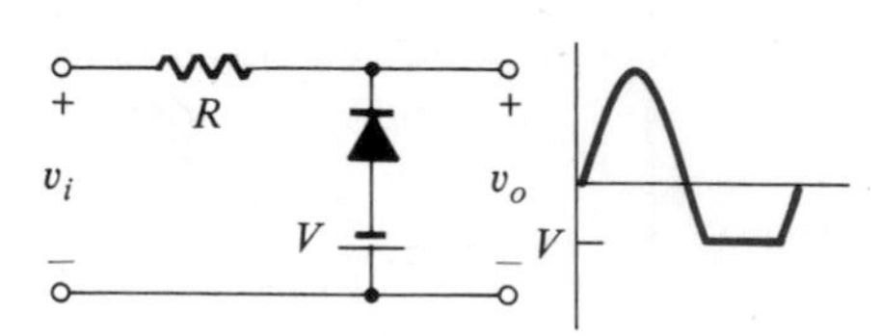

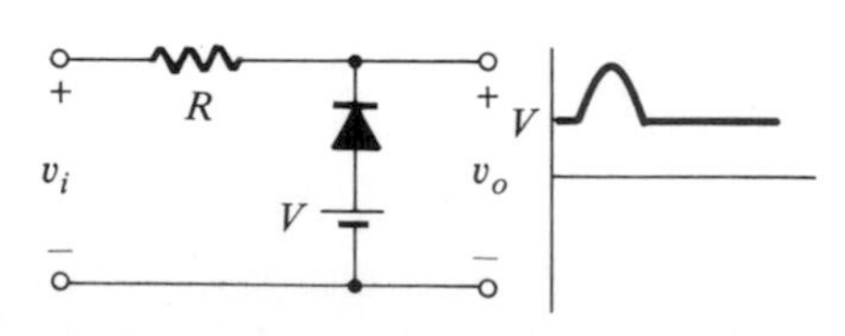

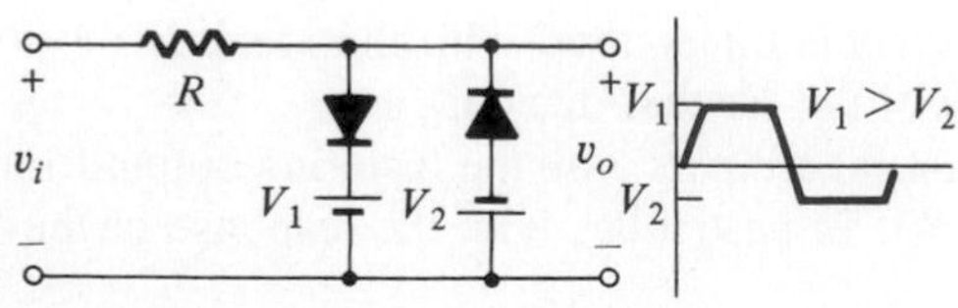

Figure 2.80 Clipping circuits.

ation, with its ability to clip off a positive and a negative section as determined by the magnitude of the dc supplies.

2.8 CLAMPERS

The *clamping* network is one that will "clamp" a signal to a different dc level. The network must have a capacitor, a diode, and a resistive element, but it can also employ an independent dc supply to introduce an additional shift.

The magnitude of R and C must be chosen such that the time constant $\tau = RC$ is large enough to ensure that the voltage across the capacitor does not discharge significantly during the interval the diode is nonconducting. Throughout the analysis we will assume that for all practical purposes the capacitor will fully charge or discharge in five time constants.

The network of Fig. 2.81 will clamp the input signal to the zero level (for ideal diodes). The resistor R can be the load resistor or a parallel combination of the load resistor and a resistor designed to provide the desired level of R.

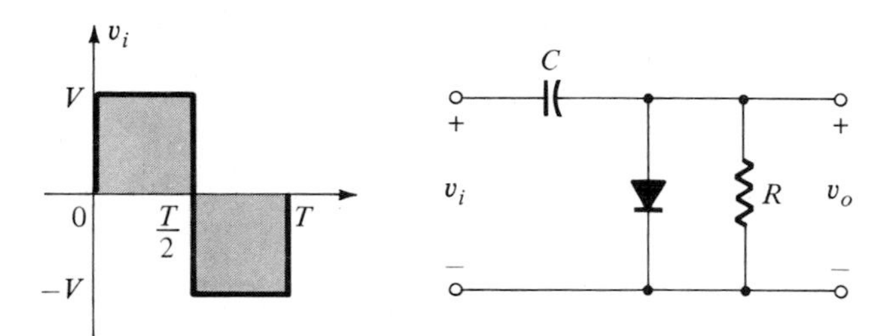

Figure 2.81 Clamper.

During the interval $0 \rightarrow T/2$ the network will appear as shown in Fig. 2.82, with the diode in the short-circuit state "shorting out" the effect of the resistor R. The resulting RC time constant is so small (R determined by the inherent resistance of the network) that the capacitor will charge to V volts very quickly. During this interval the output voltage is directly across the short circuit and $v_o = 0$ V.

When the input switches to the $-V$ state, the network will appear as shown in Fig. 2.83, with the open-circuit equivalent for the diode determined by the applied signal and stored voltage across the capacitor—both "pressuring" current through the diode from anode to cathode. Now that R is back in the network the time constant determined by the RC product is sufficiently large to establish a discharge period 5τ much greater

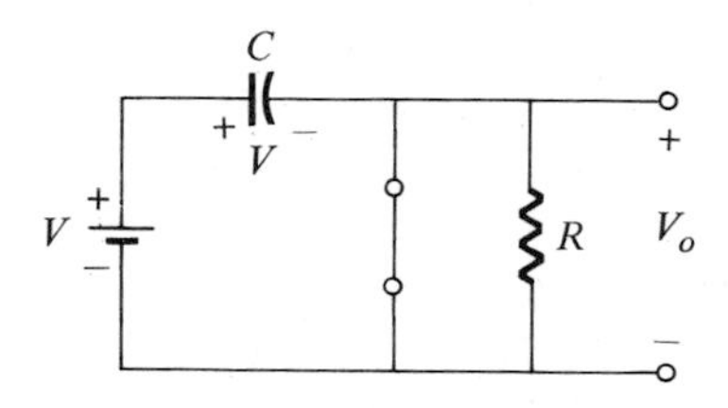

Figure 2.82

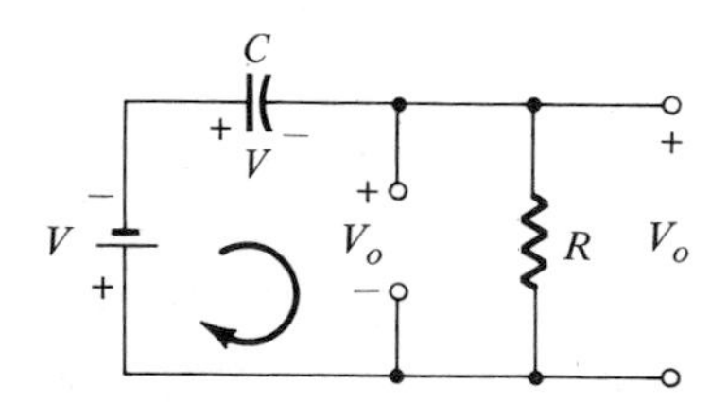

Figure 2.83

than the period $T/2 \rightarrow T$ and it can be assumed on an approximate basis that the capacitor holds onto all its charge and therefore voltage (since $V = Q/C$) during this period.

Since v_o is in parallel with the diode and resistor, it can also be drawn in the alternate position shown in Fig. 2.83. Applying Kirchhoff's voltage law around the input loop will result in

$$-V - V - V_o = 0$$

and

$$V_o = -2V$$

The negative sign resulting from the fact that the polarity of $2V$ is opposite to that defined for v_o. The resulting output waveform appears in Fig. 2.84 with the input signal. The output signal is clamped to 0 V for the interval 0 to $T/2$ but maintains the same total swing ($2V$) as the input.

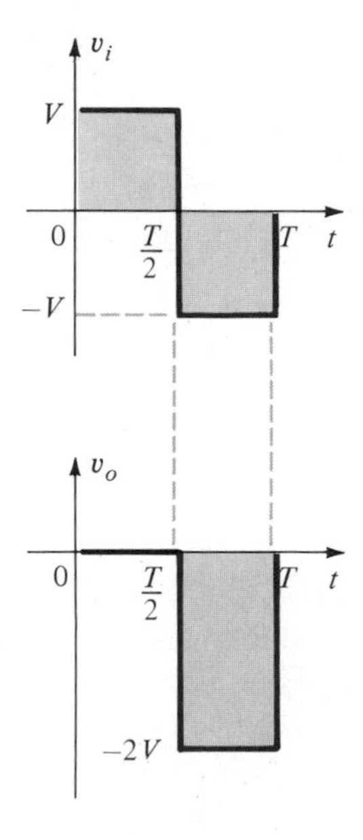

Figure 2.84

For clamping networks, *the total swing of the output is equal to the total swing of the input signal*. This fact is an excellent checking tool for the result obtained.

In general, the following may be helpful when analyzing clamping networks:

1. Always start the analysis of clamping networks by *considering that part of the input signal that will forward bias the diode*. This may require skipping an interval of the input signal (as demonstrated in an example to follow), but the analysis will not be extended by an unnecessary measure of investigation.
2. During the period that the diode is in the short-circuit state, assume that the capacitor will charge up instantaneously (an approximation) to a level determined by the voltage across the capacitor in its equivalent open-circuit state.

3. Assume that during the period the diode is an open-circuit ("off" state), the capacitor will hold on to all its charge and therefore voltage.
4. Throughout the analysis maintain a continual awareness of where v_o is defined to ensure that the proper levels for v_o are obtained in the analysis.
5. Keep in mind the general rule that the swing of the output must match the swing of the input signal.

EXAMPLE 2.17

Determine v_o for the network of Fig. 2.85 for the indicated input.

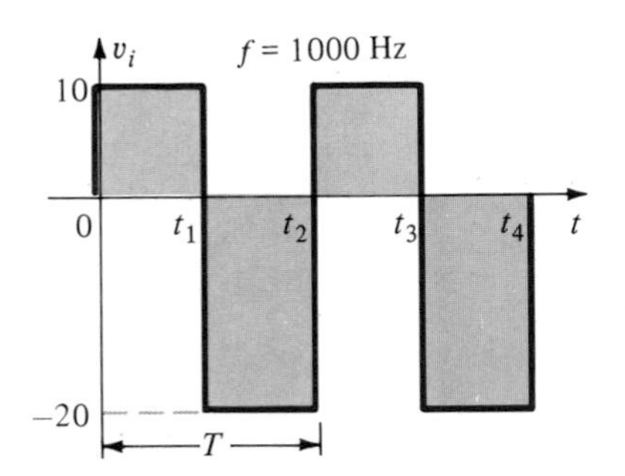

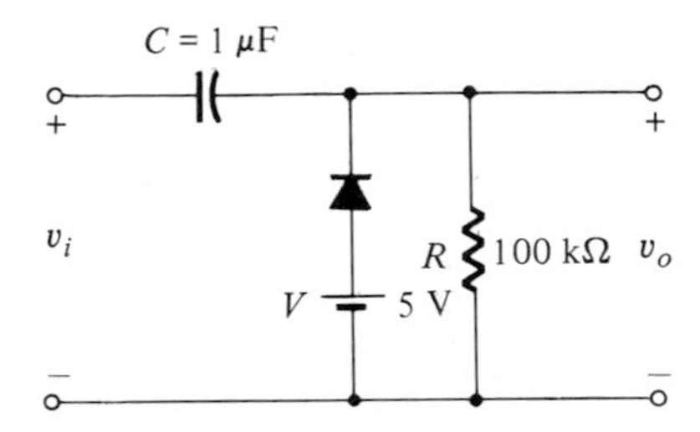

Figure 2.85

Solution:

Note that the frequency is 1000 Hz, resulting in a period of 1 ms and an interval of 0.5 ms between levels. The analysis will begin with the period $t_1 \rightarrow t_2$ of the input signal since the diode is in its short-circuit state as recommended by comment 1. For this interval the network will appear as shown in Fig. 2.86.

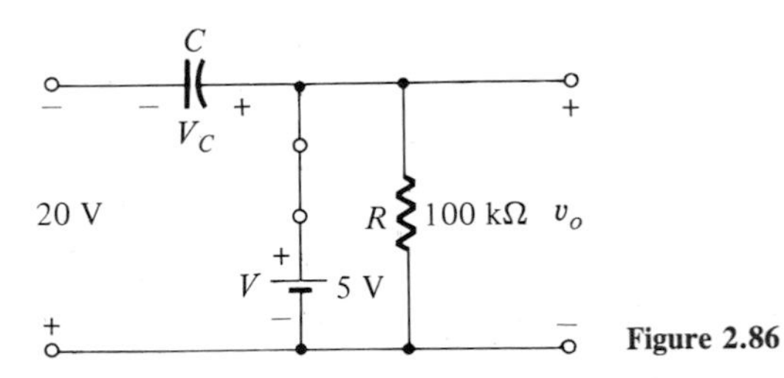

Figure 2.86

The output is across R, but it is also directly across the 5-V battery if you follow the direct connection between the defined terminals for v_o and the battery terminals. The result is $v_o = 5$ V for this interval.

Applying Kirchhoff's voltage law around the input loop will result in

$$-20 + V_C - 5 = 0$$

and

$$V_C = 25 \text{ V}$$

The capacitor will therefore charge up to 25 V, as stated in comment 2. In this case the resistor R is not shorted out by the diode but a Thévenin equivalent circuit of that portion of the network which includes the battery and the resistor will result in $R_{Th} = 0\ \Omega$ with $E_{Th} = V = 5$ V. For the period $t_2 \rightarrow t_3$ the network will appear as shown in Fig. 2.87.

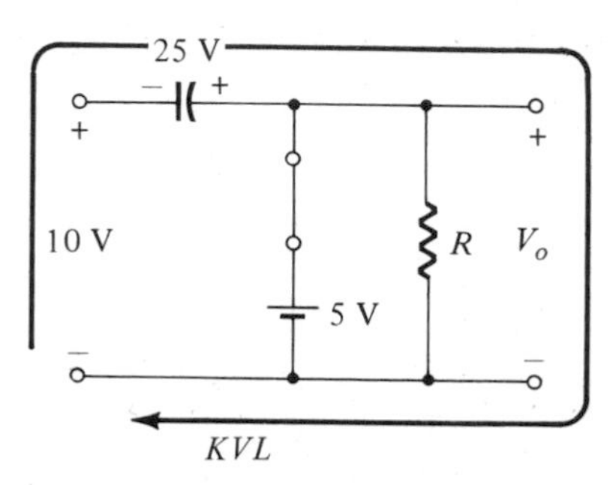

Figure 2.87

The open-circuit equivalent for the diode will remove the 5-V battery from having any effect on v_o, and applying Kirchhoff's voltage law around the outside loop of the network will result in

$$+10 + 25 - V_o = 0$$

and

$$V_o = 35 \text{ V}$$

The time constant of the discharging network of Fig. 2.87 is determined by the product RC and has the magnitude

$$\tau = RC = (100 \text{ k}\Omega)(0.1\ \mu\text{F}) = 0.01 \text{ s} = 10 \text{ ms}$$

The total discharge time is therefore $5\tau = 5(10 \text{ ms}) = 50$ ms.

Since the interval $t_2 \rightarrow t_3$ will only last for 0.5 ms, it is certainly a good approximation to assume that the capacitor will hold its voltage during the discharge period between pulses of the input signal. The resulting output appears in Fig. 2.88 with the

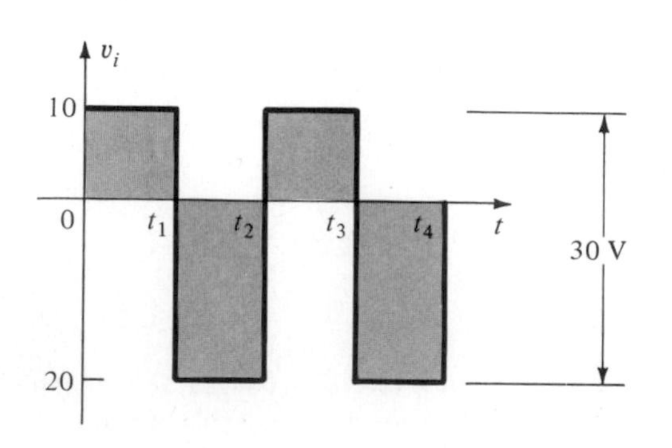

Figure 2.88

input signal. Note that the output swing of 30 V matches the input swing as noted in comment 5.

EXAMPLE 2.18

Repeat Example 2.17 using a silicon diode with $V_T = 0.7$ V.

Solution:

For the short-circuit state the network now takes on the appearance of Fig. 2.89 and v_o can be determined by Kirchhoff's voltage law in the output section.

$$+5 - 0.7 - V_o = 0$$

and

$$V_o = 5 - 0.7 = 4.3 \text{ V}$$

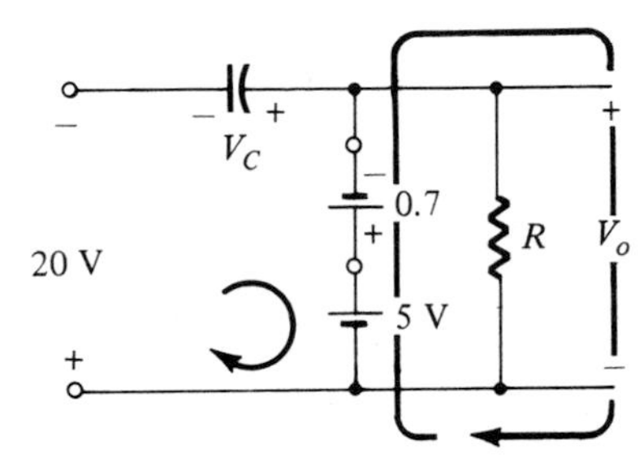

Figure 2.89

For the input section Kirchhoff's voltage law will result in

$$-20 + V_C + 0.7 - 5 = 0$$

and

$$V_C = 25 - 0.7 = 24.3 \text{ V}$$

For the period $t_2 \rightarrow t_3$ the network will now appear as in Fig. 2.90 with the only change being the voltage across the capacitor. Applying Kirchhoff's voltage law yields

$$+10 + 24.3 - V_o = 0$$

and

$$V_o = 34.3 \text{ V}$$

The resulting output appears in Fig. 2.91, verifying the statement that the input and output swings are the same.

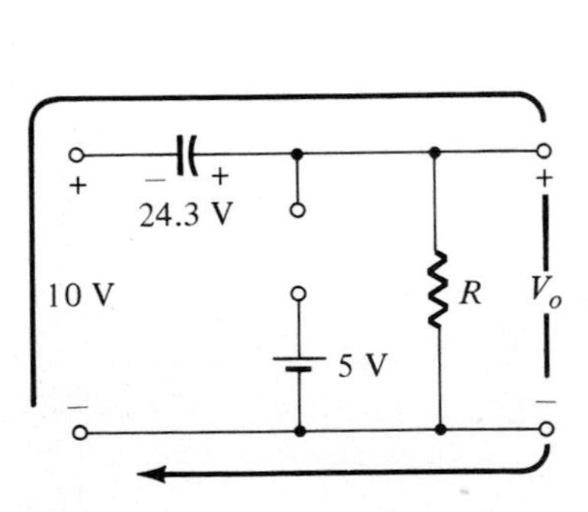

Figure 2.90

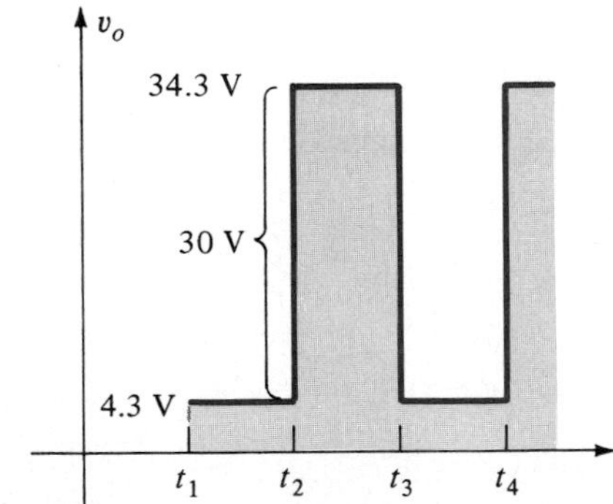

Figure 2.91

A number of clamping circuits and their effect on the input signal are shown in Fig. 2.92.

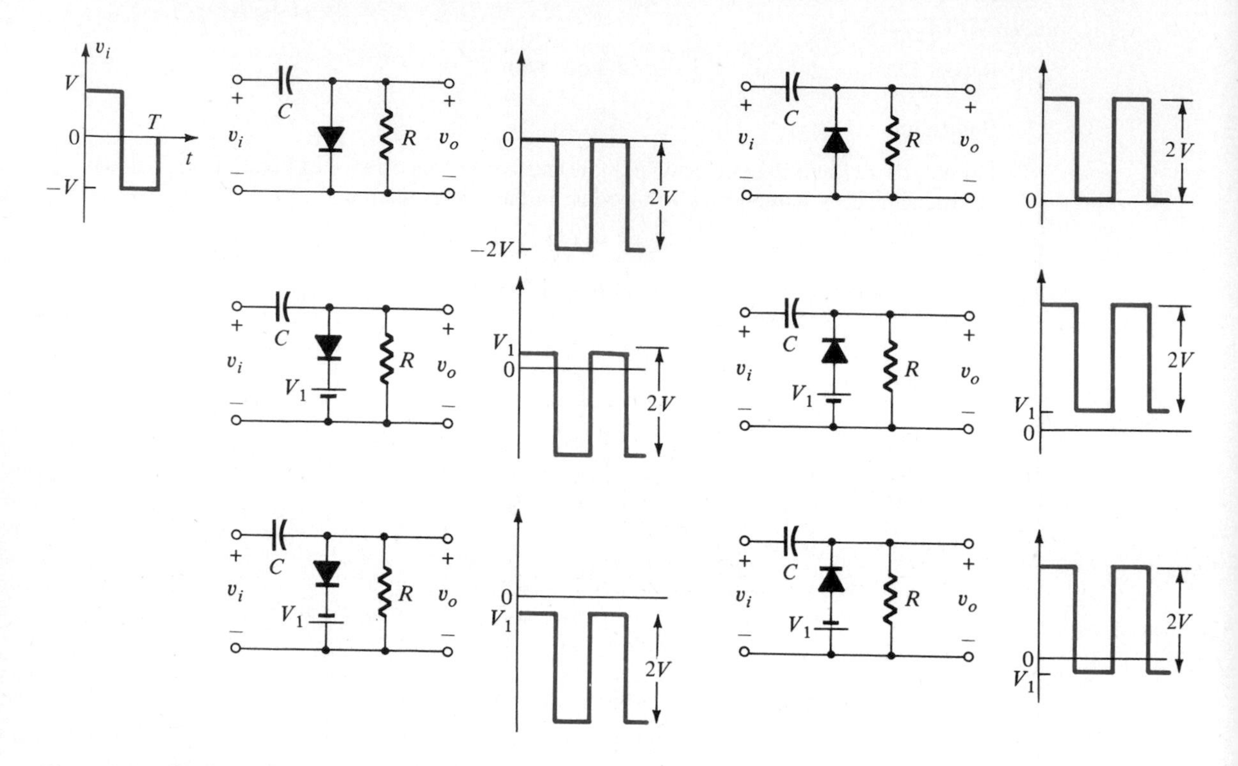

Figure 2.92 Clamping circuits ($5\tau = 5RC >> T/2$).

PROBLEMS

§ 2.2

1. Determine V_o and I_D for the networks of Fig. 2.93.

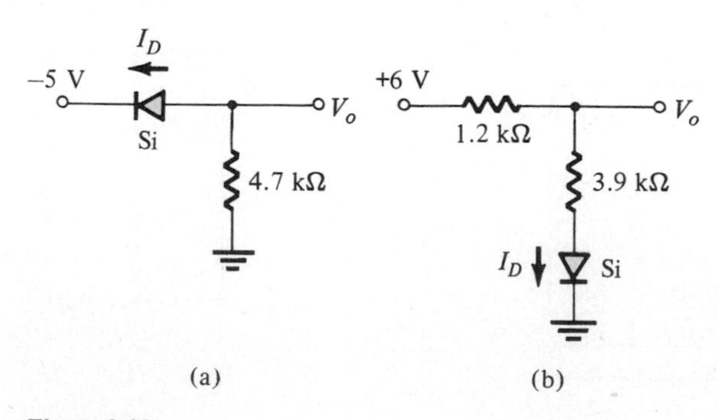

Figure 2.93

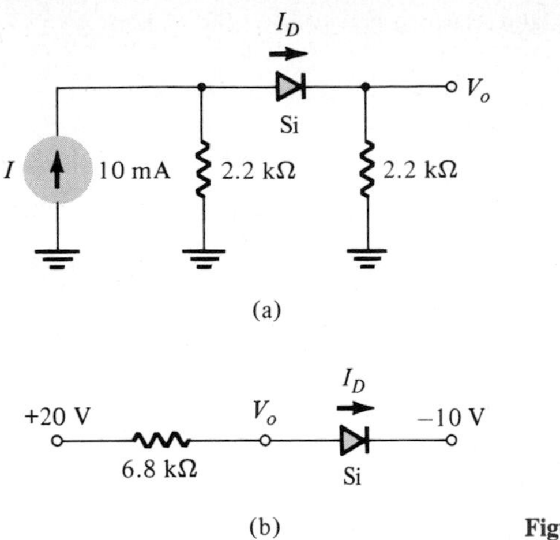

Figure 2.94

2. Determine V_o and I_D for the networks of Fig. 2.94.

3. Determine V_{o_1} and V_{o_2} for the networks of Fig. 2.95.

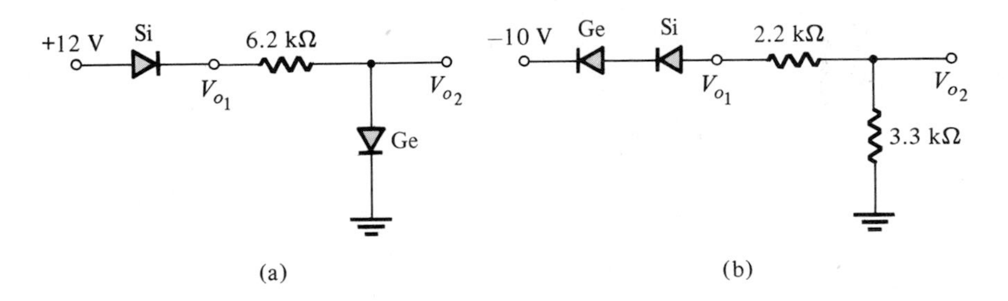

Figure 2.95

§ 2.3

4. Determine V_o and I_D for the networks of Fig. 2.96.

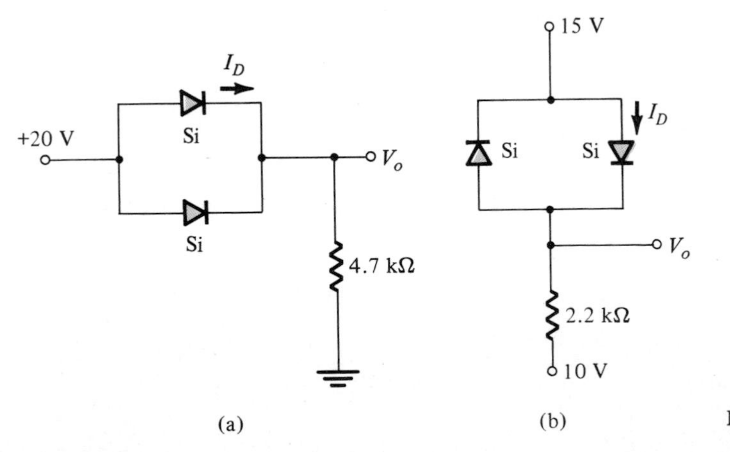

Figure 2.96

5. Determine V_o and I for the networks of Fig. 2.97.

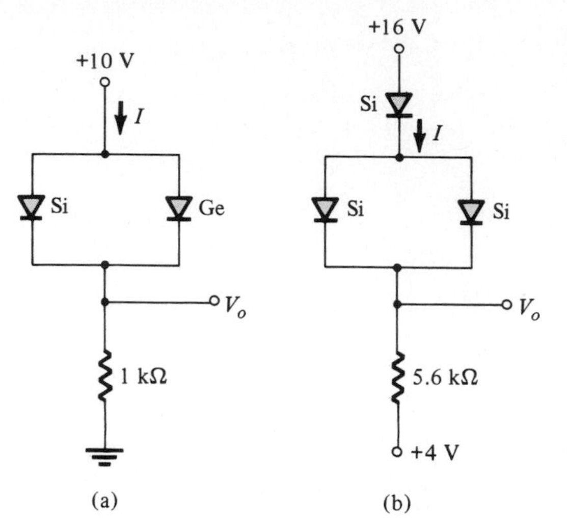

Figure 2.97

6. Determine V_{o_1} and V_{o_2} for the network of Fig. 2.98.

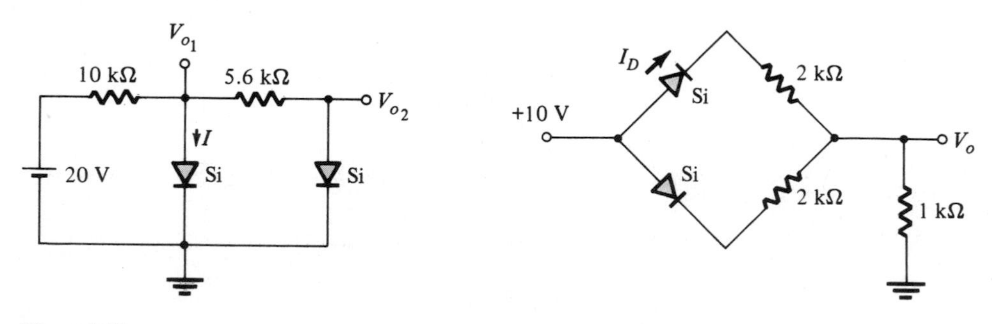

Figure 2.98

Figure 2.99

7. Determine V_o and I_D for the network of Fig. 2.99.

§ 2.4

8. Determine V_o for the network of Fig. 2.27 with 0 V on both inputs.
9. Determine V_o for the network of Fig. 2.27 with 10 V on both inputs.
10. Determine V_o for the network of Fig. 2.30 with 0 V on both inputs
11. Determine V_o for the network of Fig. 2.30 with 10 V on both inputs.
12. Determine V_o for the negative logic OR gate of Fig. 2.100.

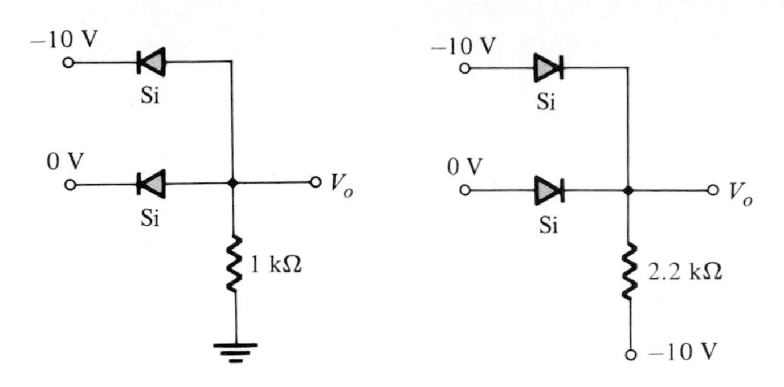

Figure 2.100 **Figure 2.101**

13. Determine V_o for the negative logic AND gate of Fig. 2.101.

§ 2.5

14. Assuming an ideal diode, sketch v_i, v_d, and i_d for the half-wave rectifier of Fig. 2.102. The input is a sinusoidal waveform with a frequency of 60 Hz.

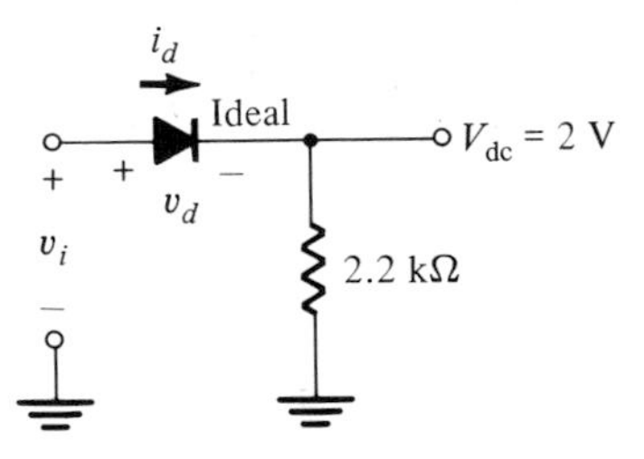

Figure 2.102

15. Repeat Problem 14 with a silicon diode ($V_T = 0.7$ V).

16. Repeat Problem 14 with a 6.8 kΩ load applied as shown in Fig. 2.103. Sketch v_L and i_L.

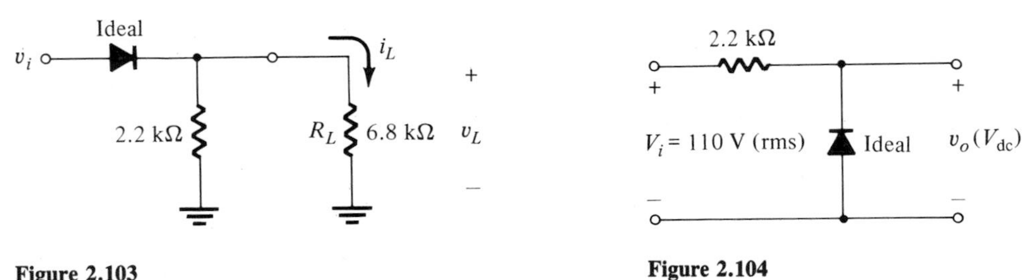

Figure 2.103 **Figure 2.104**

17. For the network of Fig. 2.104 sketch v_o and determine V_{dc}.

18. For the network of Fig. 2.105 sketch v_o and i_R.

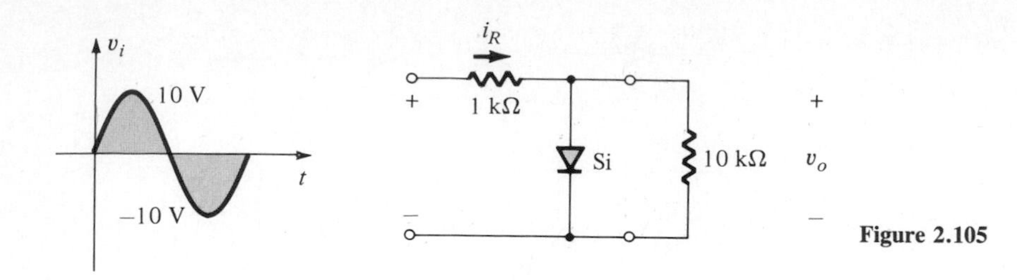

Figure 2.105

19. (a) Given $P_{max} = 14$ mW for each diode of Fig. 2.106, determine the maximum current rating of each diode.
(b) Determine I_{max} for $V_{i_{max}} = 160$ V.
(c) Determine the current through each diode for $V_m = 160$ V.
(d) Is the current determined in part (c) less than the maximum rating determined in part (a)?
(e) If only one diode were present, determine the diode current and compare it to the maximum rating.

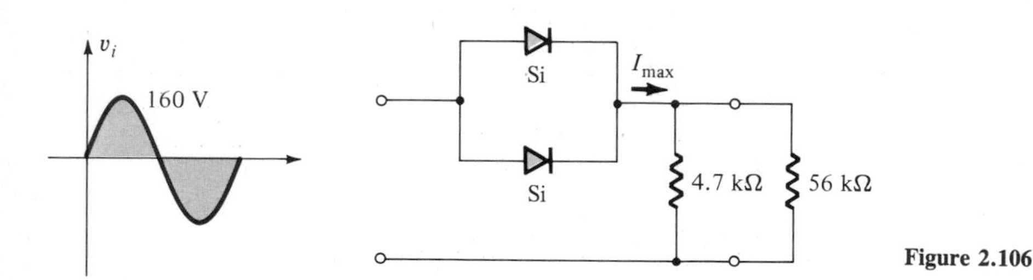

Figure 2.106

§ 2.6

20. A full-wave bridge rectifier with a 20 V rms sinusoidal input has a load resistor of 1 kΩ.
(a) If silicon diodes are employed, what is the dc voltage available at the load?
(b) Determine the required PIV rating of each diode.
(c) Find the maximum current through each diode during conduction.
(d) What is the required power rating of each diode?

21. Determine v_o and the required PIV rating of each diode for the configuration of Fig. 2.107.

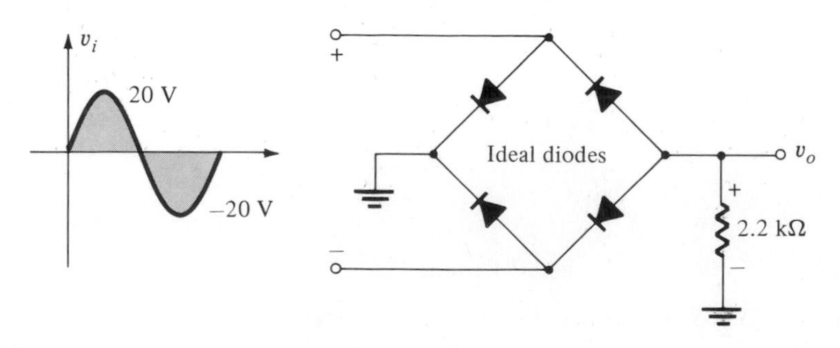

Figure 2.107

22. Sketch v_o for the network of Fig. 2.108 and determine the dc voltage available.

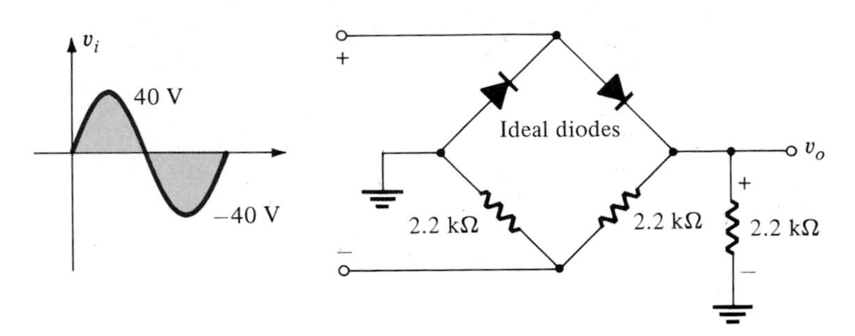

Figure 2.108

23. Sketch v_o for the network of Fig. 2.109 and determine the dc voltage available.

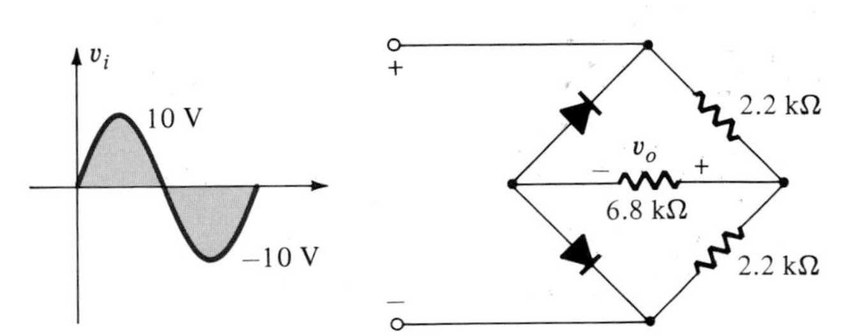

Figure 2.109

§ 2.7

24. Determine v_o for each network of Fig. 2.110 for the input shown.

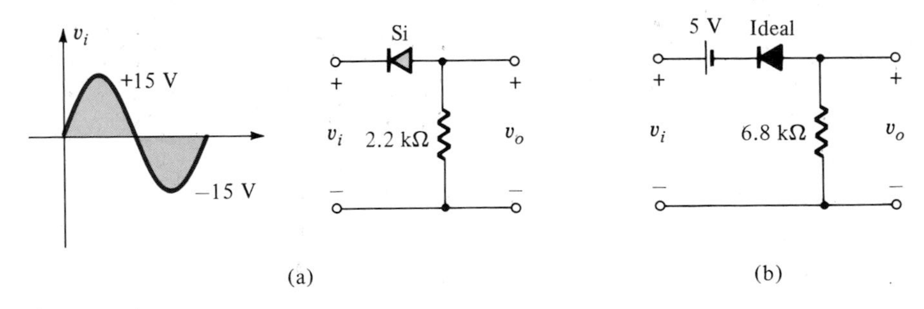

Figure 2.110

25. Determine v_o for each network of Fig. 2.111 for the input shown.

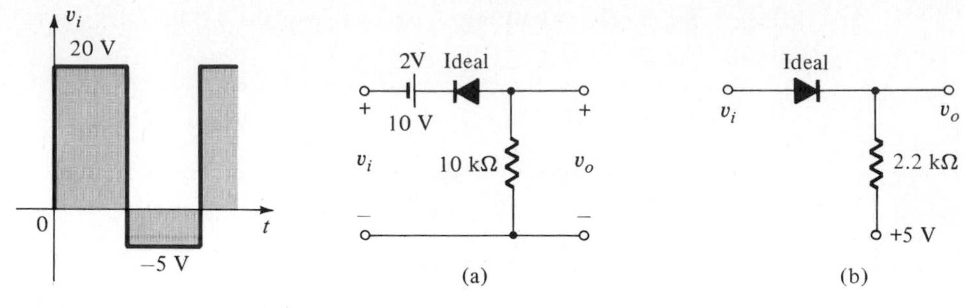

Figure 2.111

26. Determine v_o for each network of Fig. 2.112 for the input shown.

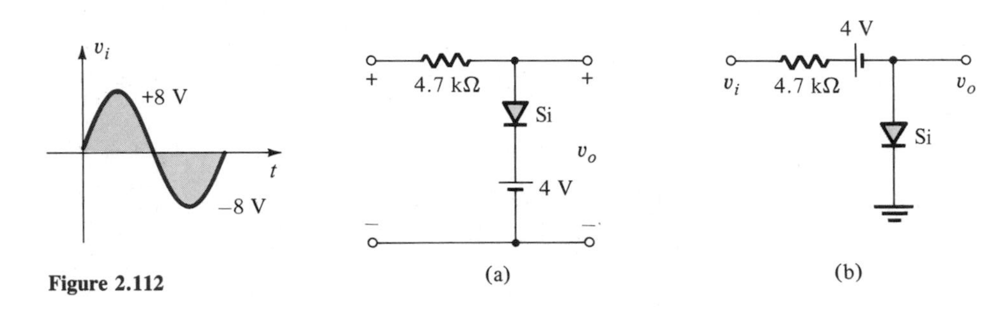

Figure 2.112

27. Sketch i_R for the network of Fig. 2.113 for the input shown.

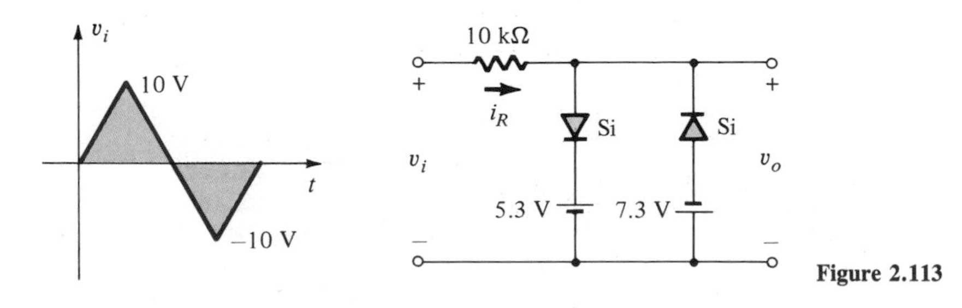

Figure 2.113

§ 2.8

28. Sketch v_o for each network of Fig. 2.114 for the input shown.

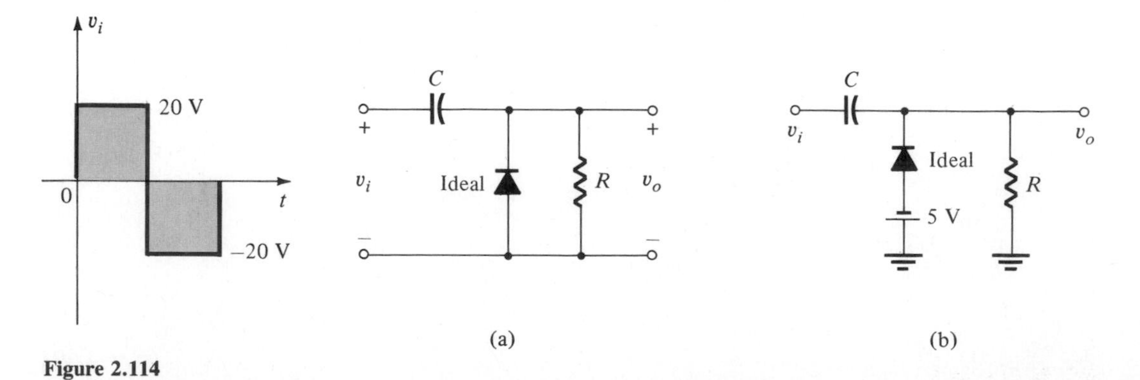

Figure 2.114

29. For the network of Fig. 2.115:
  (a) Calculate 5τ.
  (b) Compare 5τ to half the period of the applied signal.
  (c) Sketch v_o.

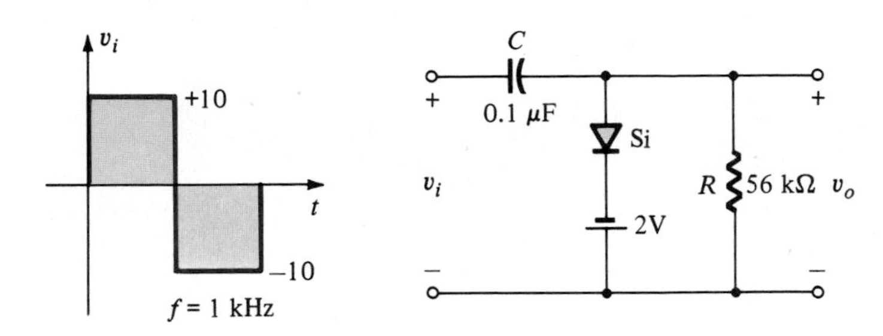

Figure 2.115

30. Using ideal diodes, design a clamper to perform the function indicated in Fig. 2.116.

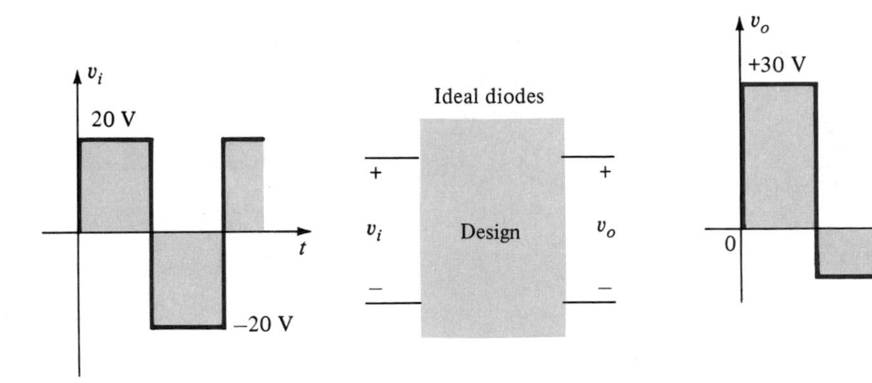

Figure 2.116

31. Using ideal diodes, design a clamper to perform the function indicated in Fig. 2.117.

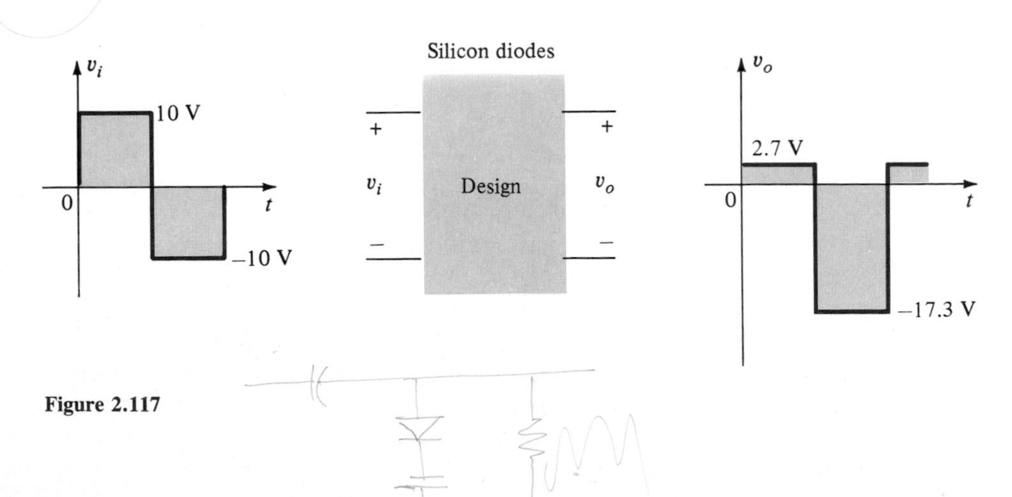

Figure 2.117

3

Zeners and Other Two-Terminal Devices

3.1 INTRODUCTION

There are a number of two-terminal devices having a single *p-n* junction like the semiconductor diode but with different modes of operation, terminal characteristics, and areas of application. A number, including the Zener, Schottky, tunnel, and varicap diodes, photodiodes, LEDs, and solar cells will be introduced in this chapter. In addition, two-terminal devices of a different construction, such as the photoconductive cell, LCD (liquid-crystal display), and thermistor will be examined.

3.2 ZENER DIODE CHARACTERISTICS AND NOTATION

The Zener and avalanche region of the semiconductor diode were discussed in detail in Section 1.6. It occurs at a reverse-bias potential of V_Z for the diode of Fig. 3.1a. The Zener diode is a device that is designed to make full use of this Zener region. If we present the characteristics as shown in Fig. 3.1b (the mirror image of Fig. 3.1a) to emphasize the region of interest by placing it in the first quadrant, a similarity appears between the characteristics and those of the silicon diode introduced in Section 1.10. The vertical rise approaches the ideal, although it is offset by a voltage V_Z. Any voltage from 0 to V_Z will result in an open-circuit equivalent as occurred below V_T for the silicon diode. There is, however, a significant difference between the characteristics of a silicon diode and the Zener diode in the reverse-bias region. Whereas the silicon diode maintains its open-circuit equivalent in the reverse-bias region the Zener

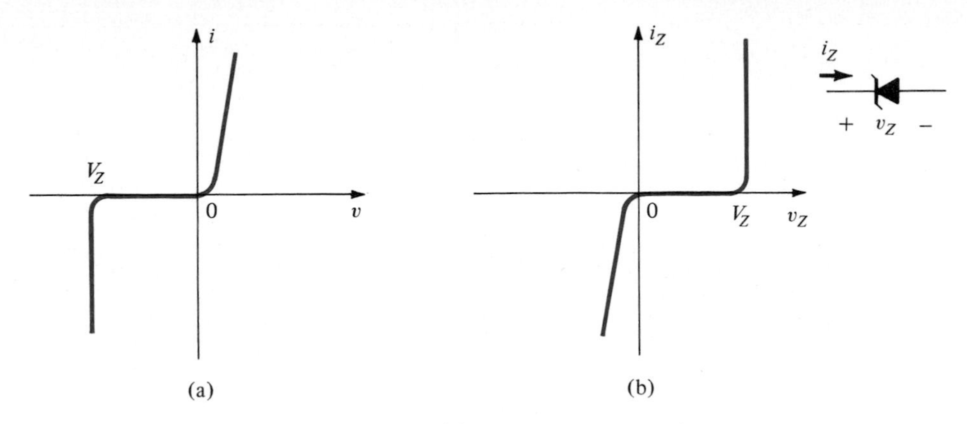

Figure 3.1 Zener diodes: (a) Zener potential; (b) characteristics and notation.

diode assumes a short-circuit state once the reverse offset voltage is reached. The first quadrant of Fig. 3.1b is defined by the polarities and current direction appearing next to the Zener diode symbol in the same figure. The application of a voltage V_Z with the polarity shown in Fig. 3.1b will cause the Zener device to switch from the open-circuit state to the "on" state in much the same manner as that described for the silicon diode in Chapter 2.

The location of the Zener region can be controlled by varying the doping levels. An increase in doping, producing an increase in the number of added impurities, will decrease the Zener potential. Zener diodes are available having Zener potentials of 2.4 to 200 V with power ratings from $\frac{1}{4}$ to 50 W. Because of its higher temperature and current capability, silicon is usually preferred in the manufacture of Zener diodes.

The complete equivalent circuit of the Zener diode in the Zener region includes a small dynamic resistance and dc battery equal to the Zener potential, as shown in Fig. 3.2. For all applications to follow, however, we shall assume as a first approximation that the external resistors are much larger in magnitude than the Zener-equivalent resistor and that the equivalent circuit is simply that indicated in Fig. 3.2b.

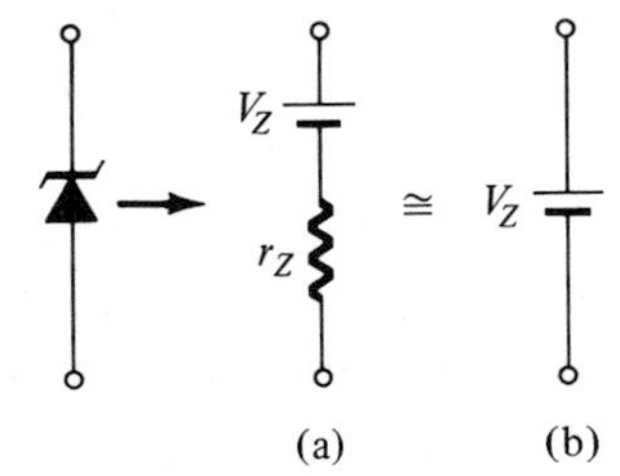

Figure 3.2 Zener equivalent circuit: (a) complete; (b) approximate.

A larger drawing of the Zener region is provided in Fig. 3.3 to permit a description of the Zener nameplate data appearing in Table 3.1 for a IN961, Fairchild, 500-m W, 20% diode.

The term "nominal" associated with V_Z indicates that it is a typical average value. Since this is a 20% diode, the Zener potential can be expected to vary as 10 V ± 20%

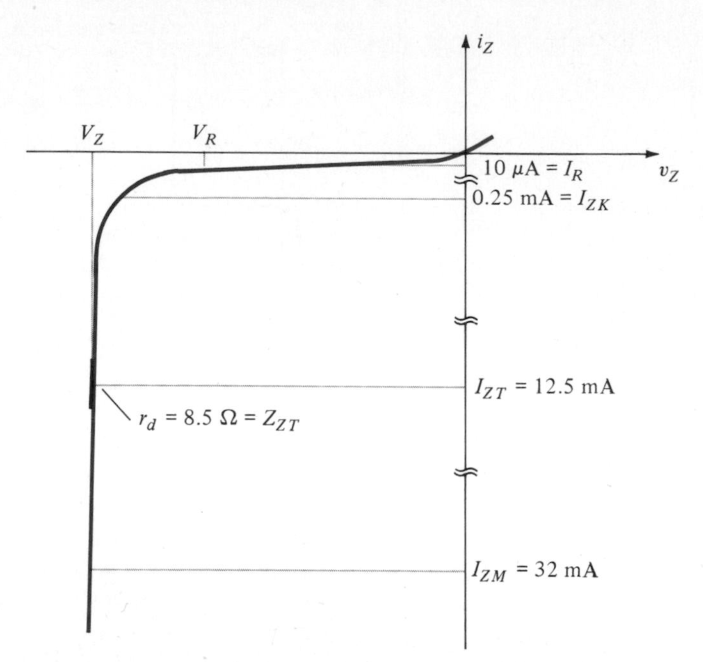

Figure 3.3 Zener test characteristics (Fairchild IN961).

TABLE 3.1 Electrical Characteristics (25°C Ambient Temperature Unless Otherwise Noted)

Jedec Type	*Zener Voltage Nominal,* V_Z *(V)*	*Test Current,* I_{ZT} *(mA)*	*Max Dynamic Impedance,* Z_{ZT} *at* I_{ZT} *(Ω)*	*Maximum Knee Impedance,* Z_{ZK} *at* I_{ZK} *(Ω) (mA)*	*Maximum Reverse Current,* I_R *at* V_R *(μA)*	*Test Voltage,* V_R *(V)*	*Maximum Regulator Current,* I_{ZM} *(mA)*	*Typical Temperature Coefficient (%/°C)*
1N961	10	12.5	8.5	700 0.25	10	7.2	32	+0.072

or from 8 to 12 V in its range of application. Also available are 10% and 5% diodes with the same specifications. The test current I_{ZT} is a typical operating level and Z_{ZT} is the dynamic impedance at this current level. The maximum knee impedance occurs at the knee current of I_{ZK}. The reverse saturation current is provided at a particular potential level and I_{ZM} is the maximum current for the 20% unit.

The temperature coefficient reflects the percent change in V_Z with temperature. It is defined by the equation

$$T_C = \frac{\Delta V_Z}{V_Z(T_1 - T_0)} \times 100\% \qquad \%/°C \tag{3.1}$$

where ΔV_Z is the resulting change in Zener potential due to the temperature variation. Note in Fig. 3.4a that the temperature coefficient can be positive, negative, or even zero for different Zener levels. A positive value would reflect an increase in V_Z with an increase in temperature, while a negative value would result in a decrease in value with increase in temperature. The 24-V, 6.8-V, and 3.6-V levels refer to three Zener

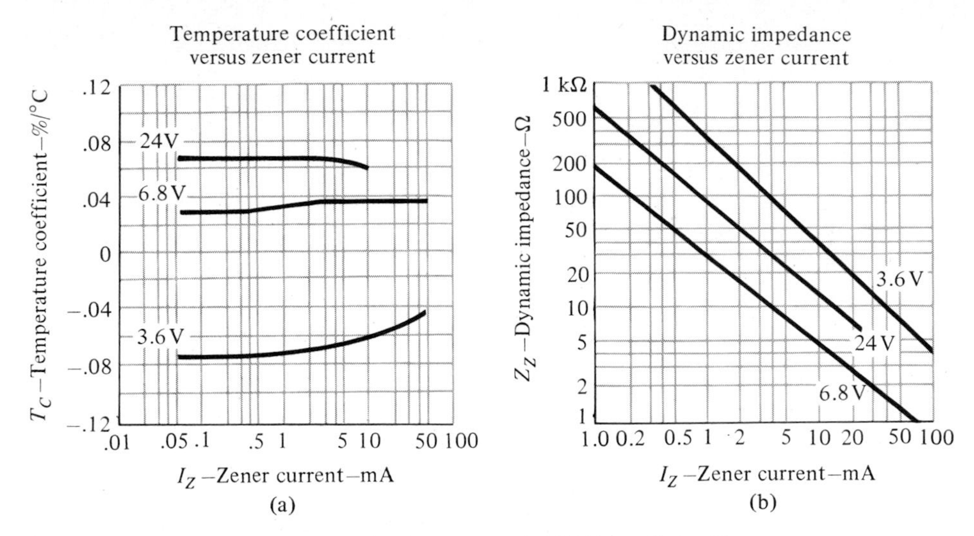

Figure 3.4 Electrical characteristics for a 500-mW Fairchild Zener diode. (Courtesy Fairchild Camera and Instrument Corporation.)

diodes having these nominal values within the same family of Zeners as the IN961. The curve for the 10-V IN961 Zener would naturally lie between the curves of the 6.8-V and 24-V devices. Note that it has a positive temperature coefficient for the entire region. Returning to Eq. (3.1), T_0 is the temperature at which V_Z is provided (normally room temperature—25°C) and T_1 is the new level. Example 3.1 will demonstrate the use of Eq. (3.1).

EXAMPLE 3.1

Determine the nominal voltage for a IN961 Fairchild Zener diode at a temperature of 100°C.

Solution:

From Eq. (3.1),

$$\Delta V_Z = \frac{T_C V_Z}{100}(T_1 - T_0)$$

Substituting yields

$$\Delta V_Z = \frac{(0.072)(10)}{100}(100 - 25)$$

$$= (0.0072)(75)$$

$$= 0.54 \text{ V}$$

and because of the positive temperature coefficient, the new Zener potential, defined by V_Z', is

$$V_Z = V_Z + 0.54$$

$$= \mathbf{10.54\ V}$$

The variation in dynamic impedance (fundamentally, its series resistance) with current appears in Fig. 3.4b. Again, the 10-V Zener appears between the 6.8-V and 24-V Zeners. Note that the heavier the current (or the farther up the vertical rise you are in Fig. 3.1b), the less the resistance value. Also note that as you approach the knee of the curve and beyond, the resistance increases to significant levels.

The terminal identification and the casing for a variety of Zener diodes appear in Fig. 3.5. Figure 3.6 is an actual photograph of a variety of Zener devices. Note that their appearance is very similar to the semiconductor diode. A few areas of application for the Zener diode will be examined in the next section.

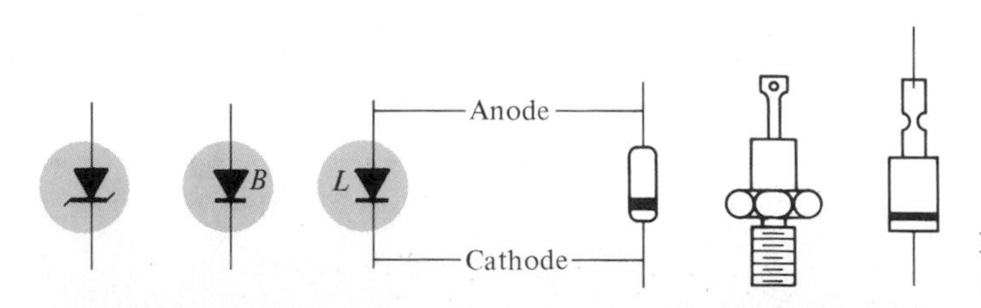

Figure 3.5 Zener terminal identification and symbols.

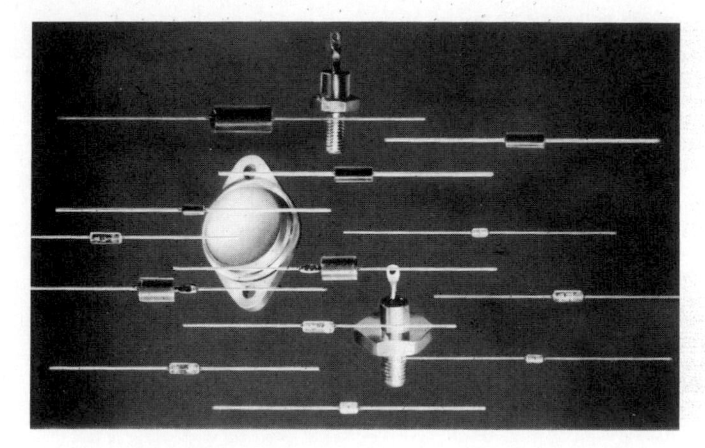

Figure 3.6 Zener diodes. (Courtesy Siemans Corporation.)

3.3 ZENER DIODE APPLICATIONS

The most common application of the Zener diode is to establish a fixed *reference* voltage for biasing and comparison purposes. Consider the network of Fig. 3.7, designed to maintain a fixed voltage V_Z across the load for variations in V_i or R_L. There are two situations to consider: one where the input voltage is fixed and R_L will change and the other where R_L is fixed and V_i will change. Each will be treated separately.

Fixed V_i, Variable R_L

Due to the offset voltage V_Z, there is a specific range of resistor values (and therefore load current) which will ensure that the Zener is in the "on" state. Too small a load resistance R_L will result in a voltage V_L across the load resistor less than V_Z and the Zener device will be in the "off" state.

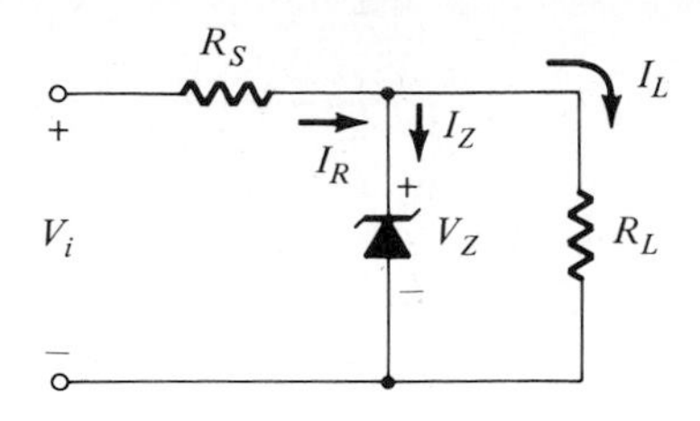

Figure 3.7

To determine the minimum load resistance (and therefore maximum load current) that will turn the Zener diode on, simply remove the Zener diode as shown in Fig. 3.8 and calculate the value of R_L that will result in a load voltage $V_L = V_Z$. That is,

$$V_L = V_Z = \frac{R_L V_i}{R_L + R_s}$$

From the voltage-divider rule and solving for R_L, we have

$$R_{L_{\min}} = \frac{R_s V_Z}{V_i - V_Z} \tag{3.2}$$

Any resistance value greater than the R_L obtained from Eq. (3.2) will ensure that the Zener diode is in the "on" state and the diode can be replaced by its V_Z source equivalent, as shown in Fig. 3.9.

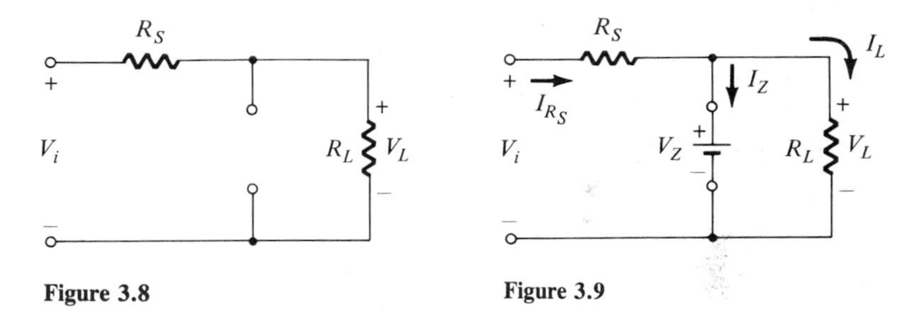

Figure 3.8

Figure 3.9

The condition defined by Eq. (3.2) establishes the minimum R_L, but the maximum I_L is

$$I_{L_{\max}} = \frac{V_L}{R_L} = \frac{V_Z}{R_{L_{\min}}} \tag{3.3}$$

Once the diode is in the "on" state, the voltage across R_S remains fixed at

$$V_{R_s} = V_i - V_Z \tag{3.4}$$

and I_R remains fixed at

$$I_{R_s} = \frac{V_{R_s}}{R_s} \tag{3.5}$$

The Zener current

$$I_Z = I_R - I_L \tag{3.6}$$

resulting in a minimum I_Z when I_L is a maximum and a maximum I_Z when I_L is a mimimum value since I_R is constant.

Since I_Z is limited to I_{ZM} as provided on the data sheet, it does affect the range of R_L and therefore I_L.

Substituting I_{ZM} for I_Z establishes the minimum I_L as

$$I_{L_{\min}} = I_{R_s} - I_{ZM} \tag{3.7}$$

and the maximum load resistance as

$$R_{L_{\max}} = \frac{V_Z}{I_{L_{\min}}} \tag{3.8}$$

EXAMPLE 3.2

(a) For the network of Fig. 3.10, determine the range of R_L and I_L that will result in V_{R_L} being maintained at 10 V.
(b) Determine the maximum wattage rating of the diode as a regulator.

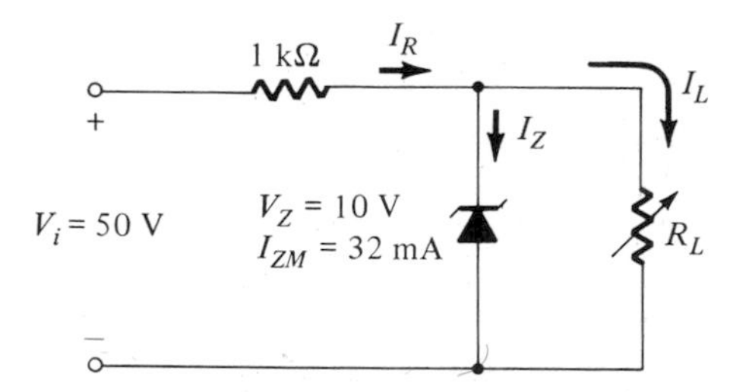

Figure 3.10

Solution:

(a) To determine the value of R_L that will turn the Zener diode on, apply Eq. (3.2):

$$R_{L_{\min}} = \frac{R_s V_Z}{V_i - V_Z} = \frac{(1\ \text{k}\Omega)(10)}{50 - 10} = \frac{10 \times 10^3}{40} = \mathbf{250\ \Omega}$$

The voltage across the resistor R_s is then determined by Eq. (3.4):

$$V_{R_s} = V_i - V_Z = 50 - 10 = \mathbf{40\ V}$$

and Eq. (3.5) provides the magnitude of I_{R_s}:

$$I_{R_s} = \frac{V_{R_s}}{R_s} = \frac{40}{1\ \text{k}\Omega} = \mathbf{40\ mA}$$

The minimum level of I_L is then determined by Eq. (3.7):

$$I_{L_{\min}} = I_{R_s} - I_{ZM} = 40 - 32 = \mathbf{8\ mA}$$

with Eq. (3.8) determining the maximum value of R_L:

$$R_{L_{\max}} = \frac{V_Z}{I_{L_{\min}}} = \frac{10}{8 \text{ mA}} = \mathbf{1.25\ k\Omega}$$

A plot of V_L versus R_L appears in Fig. 3.11a and for V_L versus I_L in Fig. 3.11b.

(b)
$$P_{\max} = V_Z I_{ZM}$$
$$= (10)(32 \text{ mA}) = \mathbf{320\ mW}$$

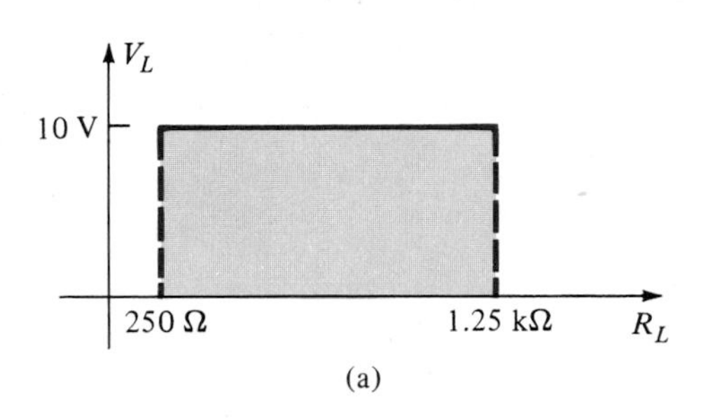

(a)

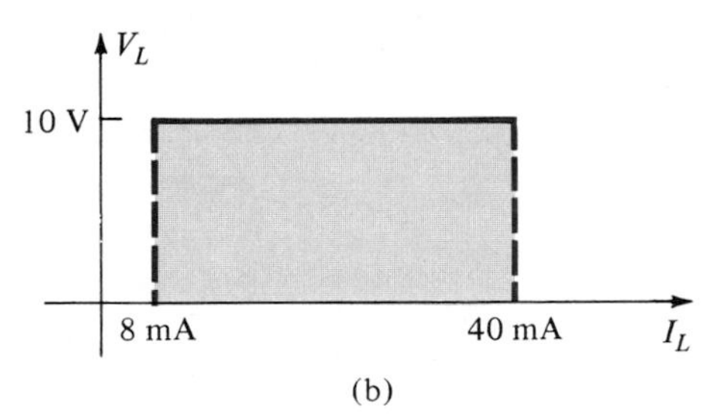

(b)

Figure 3.11

Fixed R_L, Variable V_i

For fixed values of R_L in Fig. 3.7, the voltage V_i must be sufficiently large to turn the Zener diode on. The turn-on voltage is determined by

$$V_L = V_Z = \frac{R_L V_i}{R_L + R_s}$$

and

$$\boxed{V_{i_{\min}} = \frac{(R_L + R_s)\, V_Z}{R_L}} \tag{3.9}$$

The maximum value of V_i is limited by the maximum Zener current I_{ZM}. Since $I_{ZM} = I_R - I_L$,

$$\boxed{I_{R_{\max}} = I_{ZM} + I_L} \tag{3.10}$$

Since I_L is fixed at V_Z/R_L and I_{ZM} is the maximum value of I_Z, the maximum V_i is defined by

$$V_{i_{\max}} = V_{R_{s_{\max}}} + V_Z$$

or

$$V_{i_{\max}} = I_{R_{\max}} R_s + V_Z \tag{3.11}$$

EXAMPLE 3.3

Determine the range of values of V_i that will maintain the Zener diode of Fig. 3.12 in the "on" state.

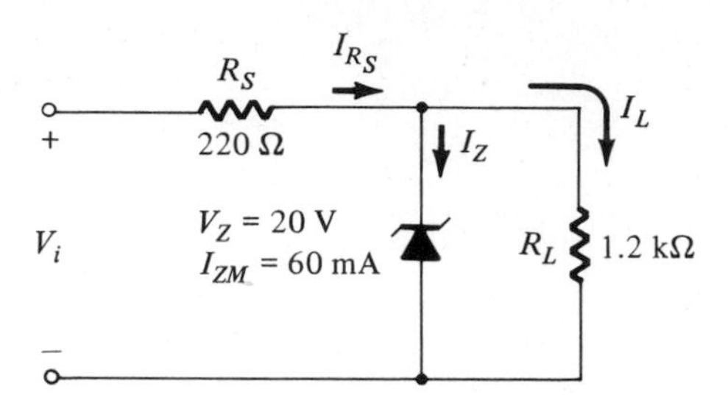

Figure 3.12

Solution:

[Eq. (3.9)] $$V_{i_{\min}} = \frac{(R_L + R_s)V_Z}{R_L} = \frac{(1200 + 220)(20)}{1200} = \mathbf{23.67\ V}$$

$$I_L = \frac{V_L}{R_L} = \frac{V_Z}{R_L} = \frac{20}{1.2\ \text{k}\Omega} = 16.67\ \text{mA}$$

[Eq. (3.10)] $$I_{R_{\max}} = I_{ZM} + I_L = (60 + 16.67)\ \text{mA}$$
$$= 76.67\ \text{mA}$$

[Eq. (3.11)] $$V_{i_{\max}} = I_{R_{\max}} R_s + V_Z$$
$$= (76.67\ \text{mA})(0.22\ \text{k}\Omega) + 20$$
$$= 16.87 + 20$$
$$= \mathbf{36.87\ V}$$

A plot of V_L versus V_i is provided in Fig. 3.13.

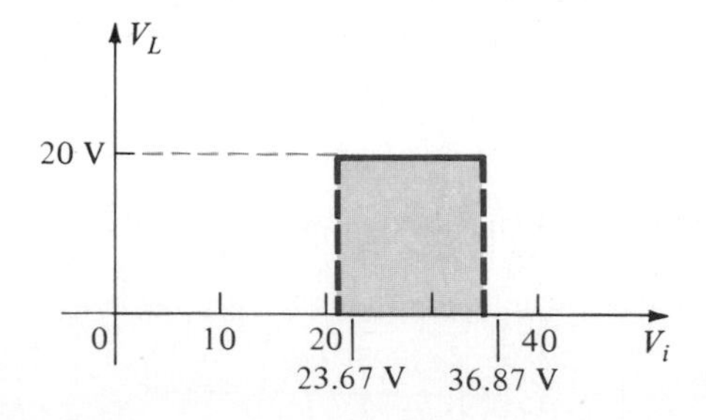

Figure 3.13

The results of Example 3.3 reveal that for the network of Fig. 3.12 with a fixed R_L, the output voltage will remain fixed at 20 V for a range of input voltage that extends from 23.67 to 36.87 V.

In fact, the input could appear as shown in Fig. 3.14 and the output would remain constant at 10 V, as shown in Fig. 3.13. The waveform appearing in Fig. 3.14 is obtained by *filtering* a half-wave- or full-wave-rectified output—a process described in detail in a later chapter. The net effect, however, is to establish a steady dc voltage such as that shown in Fig. 3.13 from a sinusoidal source with 0 average value.

Two or more reference levels can be established by placing Zener diodes in series as shown in Fig. 3.15. As long as E is greater than the sum of V_{Z_1} and V_{Z_2}, both diodes will be in the "on" state and the three reference voltages will be available.

Two back-to-back Zeners can also be used as an ac regulator as shown in Fig. 3.16. For the sinusoidal signal v_i the circuit will appear as shown in Fig. 3.16b at the instant $v_i = 10$ V. The region of operation for each diode is indicated in the adjoining

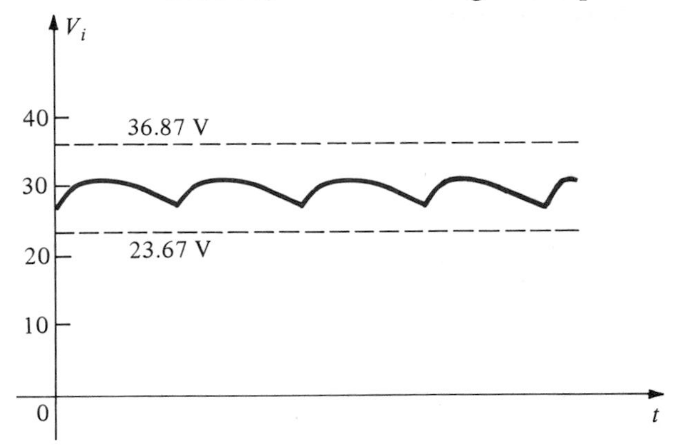

Figure 3.14

Figure 3.15 Three reference voltages.

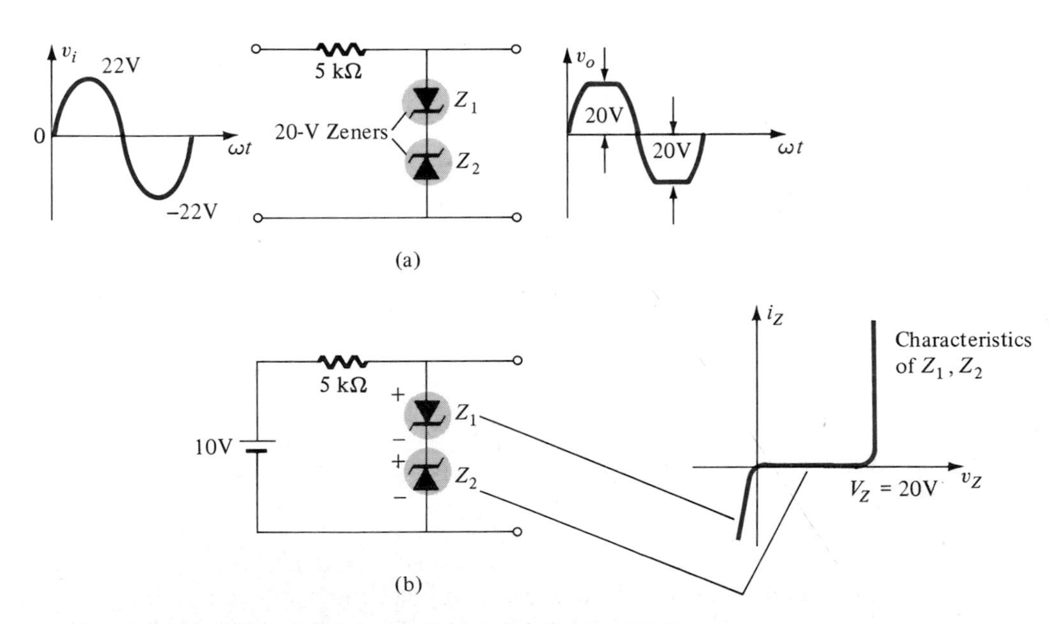

Figure 3.16 Sinusoidal ac regulation: (a) 40-V peak-to-peak sinusoidal ac regulator; (b) circuit operation at $V_i = 10$ V.

figure. Note that the impedance associated with Z_1 is very small, or essentially a short, since it is in series with 5 kΩ, while the impedance of Z_2 is very large corresponding to the open-circuit representation. Since Z_2 is an open circuit, $v_o = v_i = 10$ V. This will continue to be the case until v_i is slightly greater than 20 V. Then Z_2 will enter the low-resistance region (Zener region) and Z_1 will for all practical purposes be a short circuit and Z_2 will be replaced by $V_Z = 20$ V. The resultant output waveform is indicated in the same figure. Note that the waveform is not purely sinusoidal, but its rms value is closer to the desired 20-V peak sinusoidal waveform than the sinusoidal input having a peak value of 22 V (the rms value of a square wave is its peak value, while the rms value of a sinusoidal function is 0.707 times the peak value). The circuit of Fig. 3.16a can be extended to that of a simple square-wave generator (due to its clipping action) if the signal v_i is increased to perhaps a 50-V peak with 10-V Zeners. The resulting waveform appears in Fig. 3.17.

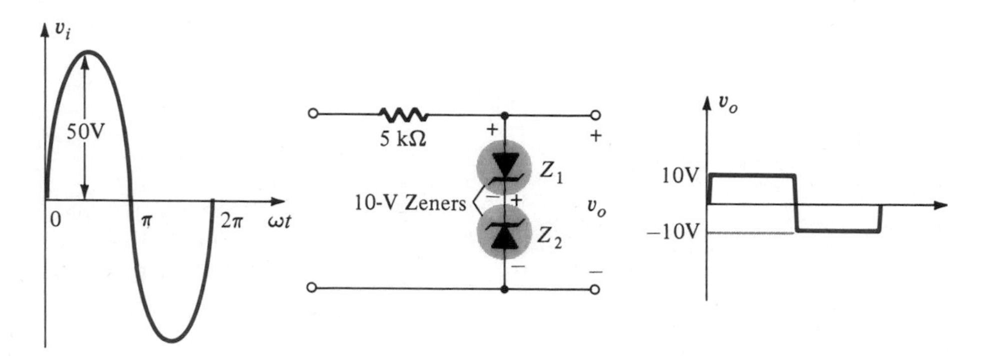

Figure 3.17 Simple square-wave generator.

3.4 SCHOTTKY BARRIER (HOT-CARRIER) DIODES

In recent years there has been increasing interest in a two-terminal device referred to as a *Schottky-barrier, surface-barrier,* or *hot-carrier* diode. Its areas of application were first limited to the very high frequency range as a competitor for the point-contact diode. It succeeded in this test because it was significantly more rugged and had a quicker response time (especially important at high frequencies) and a lower noise figure (a quantity of real importance in high-frequency applications). In recent years, however, it is appearing more and more in low-voltage/high-current power supplies and ac-to-dc converters. Other areas of application of the device include radar systems, Schottky TTL logic for computers, mixers and detectors in communication equipment, instrumentation, and analog-to-digital converters.

Its construction is very different from the conventional *p-n* junction in that a metal-semiconductor junction is created such as shown in Fig. 3.18. The semiconductor is normally *n*-type silicon (although *p*-type silicon is sometimes used), while a host of different metals, such as molybdenum, platinum, chrome, or tungsten, are used. Different construction techniques will result in a different set of characteristics for the device, such as increased frequency range, lower forward bias, and so on. Priorities do not permit an examination of each technique here, but information will

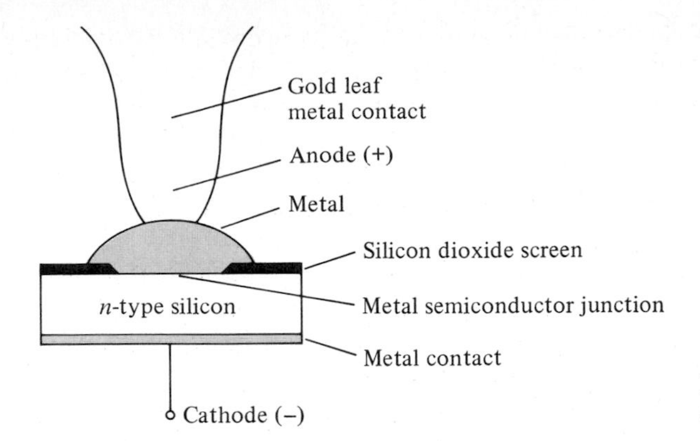

Figure 3.18 Passivated hot-carrier diode.

usually be provided by the manufacturer. In general, however, Schottky diode construction results in a more uniform junction region and increased ruggedness compared to the point-contact diode.

In both materials, the electron is the majority carrier. In the metal, the level of minority carriers (holes) is insignificant. When the materials are joined the electrons in the *n*-type silicon semiconductor material immediately flow into the adjoining metal, establishing a heavy flow of majority carriers. Since the injected carriers have a very high kinetic energy level compared to the electrons of the metal, they are commonly called "hot carriers." In the conventional *p-n* junction there was the injection of minority carriers into the adjoining region. Here the electrons are injected into a region of the same electron plurality. Schottky diodes are therefore unique, in that conduction is entirely by majority carriers. The heavy flow of electrons into the metal creates a region near the junction surface depleted of carriers in the silicon material—much like the depletion region in the *p-n* junction diode. The additional carriers in the metal establish a negative wall in the metal at the boundary between the two materials. The net result is a "surface barrier" between the two materials, preventing any further current. That is, any electrons (negatively charged) in the silicon material face a carrier-free region and a negative wall at the surface of the metal.

The application of a forward bias as shown in Fig. 3.18 will reduce the strength of the negative barrier through the attraction of the applied positive potential for electrons from this region. The result is a return to the heavy flow of electrons across the boundary, the magnitude of which is controlled by the level of the applied bias potential. The barrier at the junction for a Schottky diode is less than that of the *p-n* junction device in both the forward- and reverse-bias regions. The result is therefore a higher current at the same applied bias in the forward- and reverse-bias regions. This is a desirable effect in the forward-bias region but highly undesirable in the reverse-bias region.

The exponential rise in current with forward bias is described by Eq. (1.4) but with η dependent on the construction technique (1.05 for the metal whisker type of construction, which is somewhat similar to the germanium diode). In the reverse-bias region the current I_s is due primarily to those electrons in the metal passing into the semiconductor material. One of the areas of continuing research on the Schottky diode centers on reducing the high leakage currents that result with temperatures over 100°C. Through design, improvement units are now becoming available that have a temperature range from −65 to +150°C. At room temperature, I_s is typically in the microampere range for low-power units and milliampere range for high-power devices, although it is typically larger than that encountered using conventional *p-n* junction devices with the same current limits. In addition, even though Schottky diodes exhibit better characteristics than the point-contact diodes in the reverse-bias region as shown in Fig. 3.19, the PIV of these diodes is usually significantly less than that of a comparable *p-n* junction unit. Typically, for a 50-A unit, the PIV of the Schottky diode would be about 50 V as compared to 150 V for the *p-n* junction variety. Recent advances, however, have resulted in Schottky diodes with PIVs greater than 100 V at this current level. It is obvious from the characteristics of Fig. 3.19 that the Schottky diode is closer to the ideal set of characteristics than the point contact and has levels of V_T less than the typical silicon semiconductor *p-n* junction. The level of V_T for the "hot-carrier" diode is controlled to a large measure by the metal employed. There exists a required trade-off between temperature range and level of V_T. An increase in one appears to correspond to a resulting increase in the other. In addition, the lower the range of allowable current levels, the lower the value of V_T. For some low-level units, the value of V_T can be assumed to be essentially zero on an approximate basis. For the middle and high range, however, a value of 0.2 V would appear to be a good representative value.

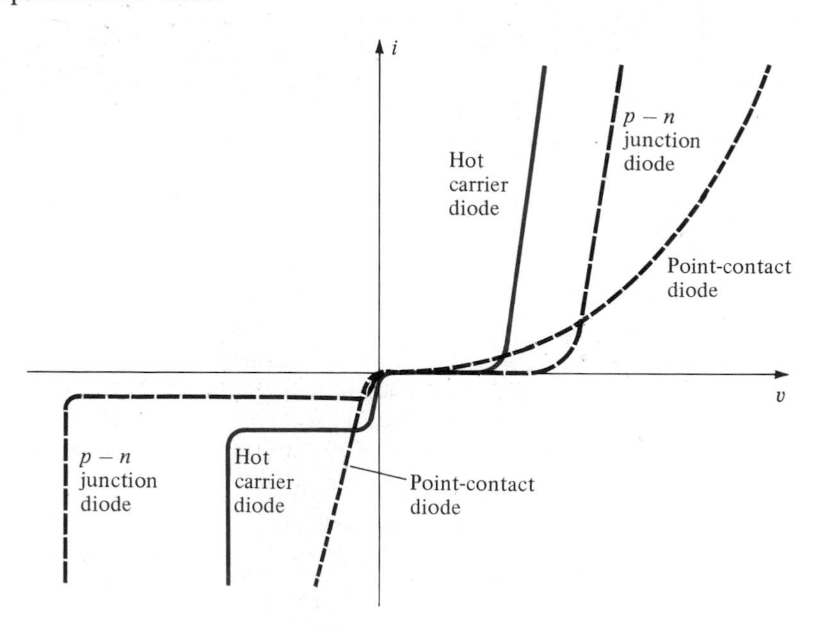

Figure 3.19 Comparison of characteristics of hot-carrier, point-contact, and *p-n* junction diodes.

The maximum current rating of the device is presently limited to about 75 A, although 100-A units appear to be on the horizon. One of the primary areas of application of this diode is in *switching power supplies* that operate in the frequency range of 20 kHz or more. A typical unit at 25°C may be rated at 50 A at a forward voltage of 0.6 V with a recovery time of 10 ns for use in one of these supplies. A *p-n* junction device with the same current limit of 50 A may have a forward voltage drop of 1.1 V and a recovery time of 30 to 50 ns. The difference in forward voltage may not appear significant, but consider the power dissipation difference: $P_{\text{hot carrier}} = (0.6)(50) = 30$ W compared to $P_{p\text{-}n} = (1.1)(50) = 55$ W, which is a measurable difference when efficiency criteria must be met. There will, of course, be a higher dissipation in the reverse-bias region for the Schottky diode due to the higher leakage current, but the total power loss in the forward- and reverse-bias regions is still significantly improved as compared to the *p-n* junction device.

Recall from our discussion of reverse recovery time for the semiconductor diode that the injected minority carriers accounted for the high level of t_{rr} (the reverse recovery time). The absence of minority carriers at any appreciable level in the Schottky diode results in a reverse recovery time of significantly lower levels, as indicated above. This is the primary reason Schottky diodes are so effective at frequencies approaching 20 GHz, where the device must switch states at a very high rate. For higher frequencies the point-contact diode, with its very small junction area, is still employed.

The equivalent circuit for the device (with typical values) and a commonly used symbol appear in Fig. 3.20. A number of manufacturers prefer to use the standard diode symbol for the device, since its function is essentially the same. The inductance L_P and capacitance C_P are package values, and r_B is the series resistance, which includes the contact and bulk resistance. The resistance r_d and capacitance C_J are values defined by equations introduced in earlier sections. For many applications, an excellent approximate equivalent circuit simply includes an ideal diode in parallel with the junction capacitance as shown in Fig. 3.21.

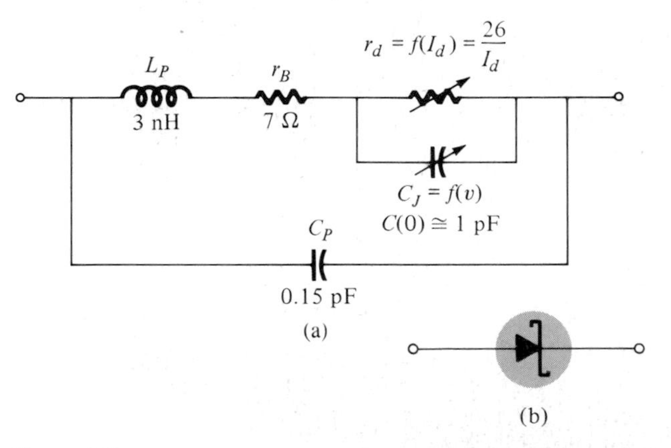

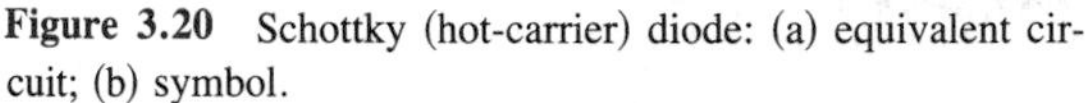

Figure 3.20 Schottky (hot-carrier) diode: (a) equivalent circuit; (b) symbol.

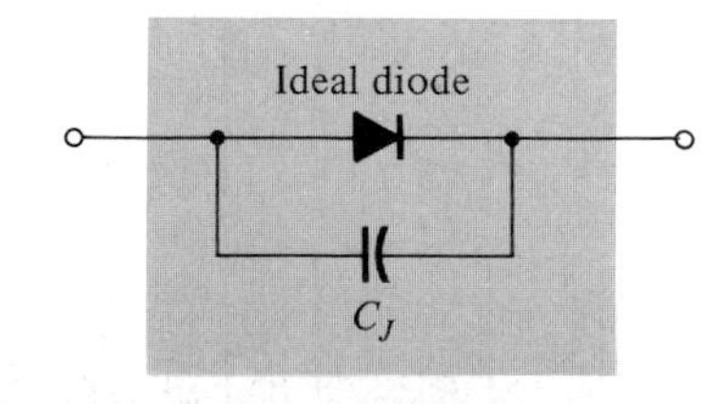

Figure 3.21 Approximate equivalent circuit for the Schottky diode.

A number of hot-carrier rectifiers manufactured by Motorola Semiconductor Products, Inc., appear in Fig. 3.22 with their specifications and terminal identification. Note that the maximum forward voltage drop V_F does not exceed 0.65 V for any of the devices, while this was essentially V_T for a silicon diode.

Three sets of curves for the Hewlett-Packard 5082-2300 series of general-purpose Schottky barrier diodes are provided in Fig. 3.23. Note at $T = 100°\text{C}$ in Fig. 3.23a that V_F is only 0.1 V at a current of 0.01 mA. Note also that the reverse current has been limited to nanoamperes in Fig. 3.23b and the capacitance to 1 pF in Fig. 3.23c to ensure a high switching rate.

I_O, Average rectified forward current (amperes)														
	0.5	1.0		3.0		3.0	5.0	15		25		40		
Case	51-02 (DO-7) Glass	59-04 Plastic		267 Plastic		60 Metal		245 (DO-4) Metal				257 (DO-5) Metal		430-2 (DO-21) Metal
V_{RRM} (Volts) — Anode: Cathode:														
20	MBR020	IN5817	MBR120P	1N5820	MBR320P	MBR320M	1N5823	1N5826	MBR1520	1N5829	MBR2520	1N5832	MBR4020	MBR4020PF
30	MBR030	IN5818	MBR130P	1N5821	MBR330P	MBR330M	1N5824	1N5827	MBR1530	1N5830	MBR2530	1N5833	MBR4030	MBR4030PF
35			MBR135P		MBR335P	MBR335M			MBR1535		MBR2535		MBR4035	MBR4035PF
40		IN5819	MBR140P	1N5822	MBR340P	MBR340M	1N5825	1N5828	MBR1540	1N5831	MBR2540	1N5834	MBR4040	
I_{FSM} (Amps)	5.0	100	50	250	200	500	500	500	500	800	800	800	800	800
T_C @ Rated I_O (°C)								85	80	85	80	75	70	50
T_J Max (°C)	125	125	125	125	125	125	125	125	125	125	125	125	125	125
Max V_F @ $I_{FM} = I_O$	0.50	*0.60	0.65	*0.525	0.60	0.45@5A	*0.38	*0.50	0.55	*0.48	0.55	*0.59	0.63	0.63

. . . Schottky barrier devices, ideal for use in low-voltage, high-frequency power supplies and as free-wheeling diodes. These units feature very low forward voltages and switching times estimated at less than 10 ns. They are offered in current ranges of 0.5 to 5.0 amperes and in voltages to 40.

V_{RRM} –respective peak reverse voltage
I_{FSM} –forward current, surge peak
I_{FM} –forward current, maximum

Figure 3.22 Motorola Schottky barrier devices. (Courtesy Motorola Semiconductor Products, Incorporated.)

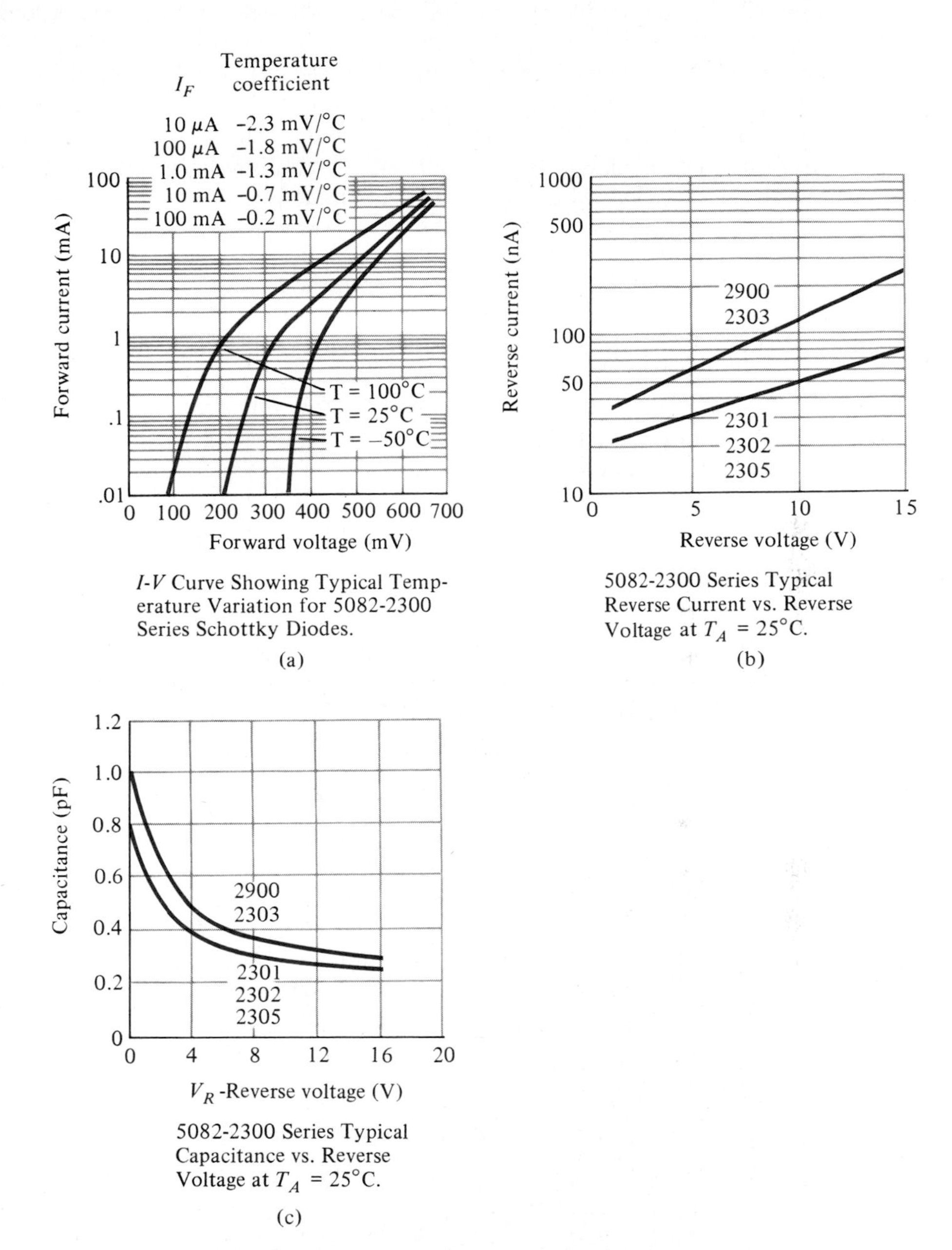

I-V Curve Showing Typical Temperature Variation for 5082-2300 Series Schottky Diodes.

(a)

5082-2300 Series Typical Reverse Current vs. Reverse Voltage at T_A = 25°C.

(b)

5082-2300 Series Typical Capacitance vs. Reverse Voltage at T_A = 25°C.

(c)

Figure 3.23 Characteristic curves for Hewlett-Packard 5082-2300 series of general-purpose Schottky barrier diodes. (Courtesy Hewlett-Packard Corporation.)

3.5 VARACTOR (VARICAP) DIODES

Varactor [also called varicap, VVC (voltage-variable capacitance), or tuning] diodes are semiconductor, voltage-dependent, variable capacitors. Their mode of operation depends on the capacitance that exists at the *p-n* junction when the element is reverse-biased. Under reverse-bias conditions, it was established that there is a region of uncovered charge on either side of the junction that together make up the depletion region and define the depletion width W_d. The transition capacitance (C_T) established by the isolated uncovered charges is determined by

$$\boxed{C_T = \epsilon \frac{A}{W_d}} \tag{3.12}$$

where ϵ is the permittivity of the semiconductor materials, A the *p-n* junction area, and W_d the depletion width.

As the reverse-bias potential increases, the width of the depletion region increases, which in turn reduces the transition capacitance. The characteristics of a typical commercially available varicap diode appear in Fig. 3.24. Note the initial sharp decline in C_T with increase in reverse bias. The normal range of V_r for VVC diodes is limited to about 20 V. In terms of the applied reverse bias, the transition capacitance is given approximately by

$$\boxed{C_T = \frac{K}{(V_T + V_r)^n}} \tag{3.13}$$

where K = constant determined by the semiconductor material and construction technique
V_T = knee potential as defined in Section 1.6
V_r = magnitude of the applied reverse-bias potential
n = $\frac{1}{2}$ for alloy junctions and $\frac{1}{3}$ for diffused junctions

In terms of the capacitance at the zero-bias condition $C(0)$, the capacitance as a

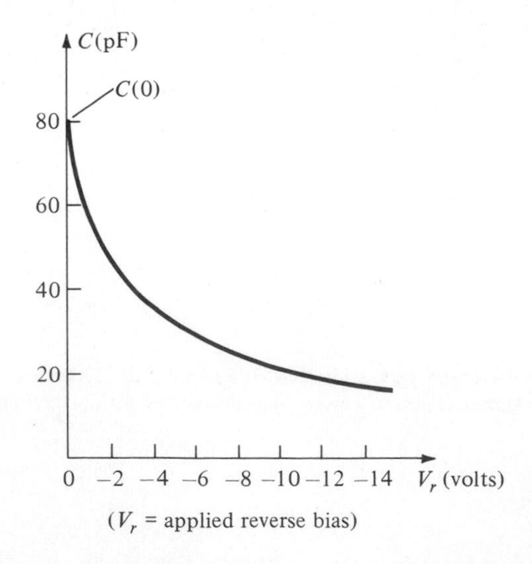

Figure 3.24 Varicap characteristics: C (pF) versus V_r.

function of V_r is given by

$$C_T(V_r) = \frac{C(0)}{\left(1 + \left|\frac{V_r}{V_T}\right|\right)^n} \tag{3.14}$$

The symbols most commonly used for the varicap diode and a first approximation for its equivalent circuit in the reverse-bias region are shown in Fig. 3.25. Since we are in the reverse-bias region, the resistance in the equivalent circuit is very large in magnitude—typically 1 MΩ or larger—while R_S, the geometric resistance of the diode, is, as indicated in Fig. 3.25, very small. The magnitude of C will vary from about 2 to 100 pF depending on the varicap considered. To ensure that R_r is as large (for minimum leakage current) as possible, silicon is normally used in varicap diodes. The fact that the device will be employed at very high frequencies requires that we include the inductance L_s even though it is measured in nanohenries. Recall that $X_L = 2\pi fL$ and a frequency of 10 GHz with $L_s = 1$ nH will result in an $X_{Ls} = 2\pi fL = (6.28)(10^{10})(10^{-9}) = 62.8\ \Omega$. There is obviously, therefore, a frequency limit associated with the use of each varicap diode.

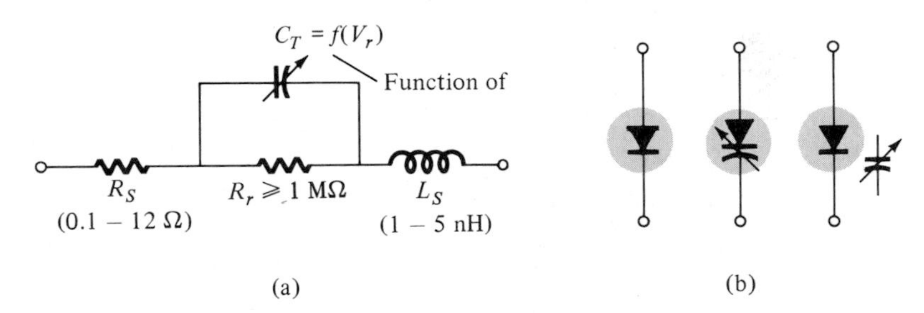

Figure 3.25 Varicap diode: (a) equivalent circuit in the reverse-bias region; (b) symbols.

Assuming the proper frequency range, a low value of R_S, and X_{L_s} compared to the other series elements, then the equivalent circuit for the varicap of Fig. 3.25a can be replaced by the variable capacitor alone. The complete data sheet and its characteristic curves appear in Figs. 3.26 and 3.27, respectively. The C_3/C_{25} ratio in Fig. 3.26 is the ratio of capacitance levels at reverse-bias potentials of 3 and 25 V. It provides a quick estimate of how much the capacitance will change with reverse-bias potential. The figure of merit is a quantity of consideration in the application of the device and is a measure of the ratio of energy stored by the capacitive device per cycle to the energy dissipated (or lost) per cycle. Since energy loss is seldom considered a positive attribute, the higher its relative value the better. The resonant frequency of the device is determined by $f_o = 1/2\pi\sqrt{LC}$ and affects the range of application of the device.

In Fig. 3.27, most quantities are self-explanatory. However, the *capacitance temperature coefficient* is defined by

$$TC_C = \frac{\Delta C}{C_0(T_1 - T_0)} \times 100\% \qquad \%/°\text{C} \tag{3.15}$$

BB139

VHF / FM VARACTOR DIODE

DIFFUSED SILICON PLANAR

- C_3/C_{25} . . . 5.0–6.5
- **MATCHED SETS** (Note 2)

ABSOLUTE MAXIMUM RATINGS (Note 1)

Temperatures

Storage Temperature Range	−55°C to +150°C
Maximum Junction Operating Temperature	+150°C
Lead Temperature	+260°C

Maximum Voltage

WIV	Working Inverse Voltage	30 V

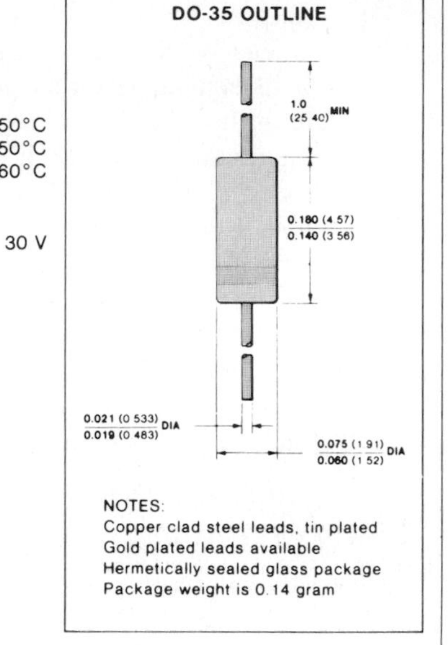

ELECTRICAL CHARACTERISTICS (25°C Ambient Temperature unless otherwise noted)

SYMBOL	CHARACTERISTIC	MIN	TYP	MAX	UNITS	TEST CONDITIONS
BV	Breakdown Voltage	30			V	I_R = 100 μA
I_R	Reverse Current		10	50	nA	V_R = 28 V
			0.1	0.5	μA	V_R = 28 V, T_A = 60°C
C	Capacitance		29		pF	V_R = 3.0 V, f = 1 MHz
		4.3	5.1	6.0	pF	V_R = 25 V, f = 1 MHz
C_3/C_{25}	Capacitance Ratio	5.0	5.7	6.5		V_R = 3 V/25 V, f = 1 MHz
Q	Figure of Merit		150			V_R = 3.0 V, f = 100 MHz
R_S	Series Resistance		0.35		Ω	C = 10 pF, f = 600 MHz
L_S	Series Inductance		2.5		nH	1.5 mm from case
f_o	Series Resonant Frequency		1.4		GHz	V_R = 25 V

NOTES:
1. These ratings are limiting values above which the serviceability of the diode may be impaired.
2. The capacitance difference between any two diodes in one set is less than 3% over the reverse voltage range of 0.5 V to 28 V.

Figure 3.26 Electrical characteristics for a VHF/FM Fairchild varactor diode. (Courtesy Fairchild Camera and Instrument Corporation.)

where ΔC is the change in capacitance due to the temperature change $T_1 - T_0$ and C_0 is the capacitance at T_0 for a particular reverse-bias potential. For example, Fig. 3.26 indicates that $C_0 = 29$ pF with $V_R = 3$ V and $T_0 = 25°C$. A change in capacitance ΔC could then be determined using Eq. (3.15) by simply substituting the new temperature T_1 and the TC_C as determined from the graph (= 0.013). At a new V_R the value of TC_C would change accordingly. Returning to Fig. 3.26, note that the maximum frequency

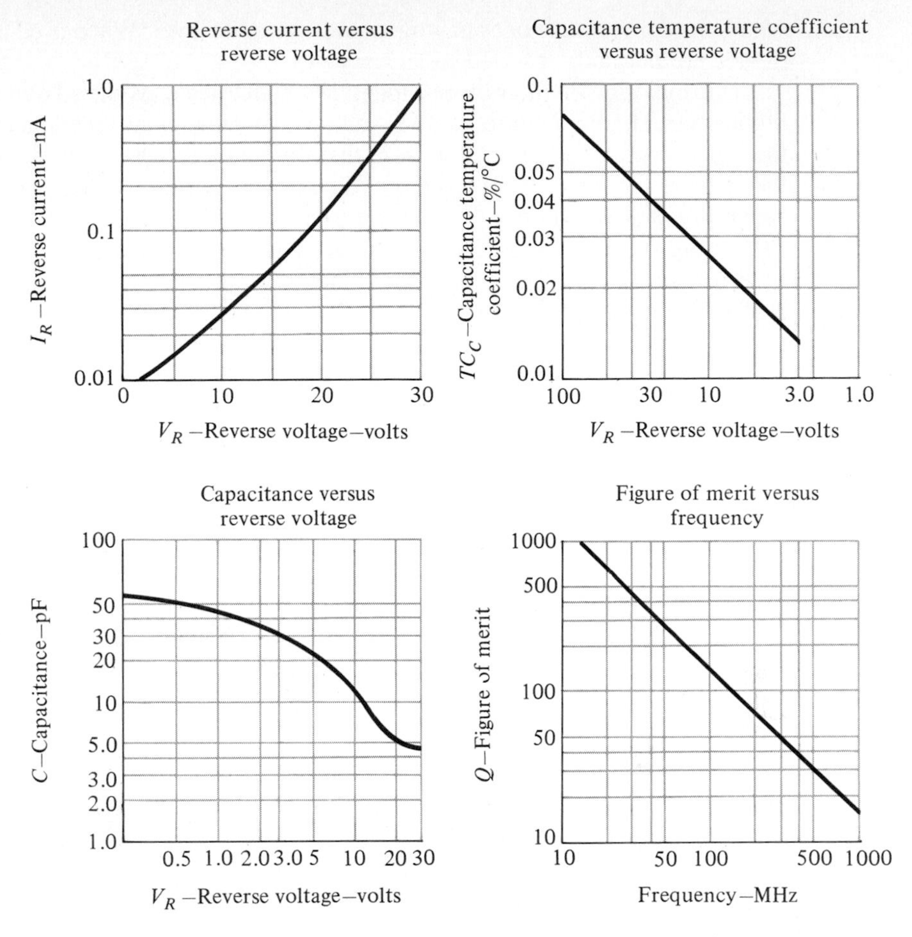

Figure 3.27 Characteristic curves for a VHF/FM Fairchild varactor diode. (Courtesy Fairchild Camera and Instrument Corporation.)

appearing is 600 MHz. At this frequency,

$$X_L = 2\pi fL = (6.28)(600 \times 10^6)(2.5 \times 10^{-9}) = 9.42\ \Omega$$

normally a quantity of sufficiently small magnitude to be ignored.

Some of the high-frequency (as defined by the small capacitance levels) areas of application include FM modulators, automatic-frequency-control devices, adjustable bandpass filters, and parametric amplifiers.

3.6 POWER DIODES

There are a number of diodes designed specifically to handle the high-power and high-temperature demands of some applications. The most frequent use of power diodes occurs in the rectification process, in which ac signals (having zero average value) are converted to ones having an average or dc level. When used in this capacity, diodes are normally referred to as *rectifiers*.

The majority of the power diodes are constructed using silicon because of its higher current, temperature, and PIV ratings. The higher current demands require that the junction area be larger, to ensure that there is a low forward diode resistance. If the forward resistance were too large, the I^2R losses would be excessive. The current capability of power diodes can be increased by placing two or more in parallel and the PIV rating can be increased by stacking the diodes in series.

Various types of power diodes and their current rating have been provided in Fig. 3.28a. The high temperatures resulting from the heavy current require, in many cases, that heat sinks be used to draw the heat away from the element. A few of the various types of heat sinks available are shown in Fig. 3.28b. If heat sinks are not employed, stud diodes are designed to be attached directly to the chassis, which in turn will act as the heat sink.

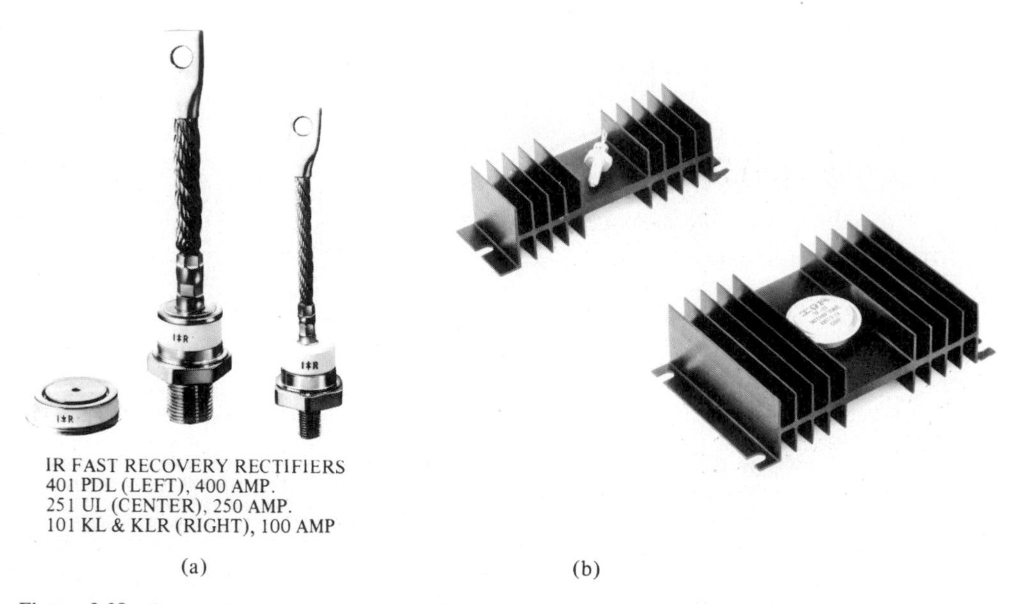

Figure 3.28 Power diodes and heat sinks. (Courtesy International Rectifier Corporation.)

3.7 TUNNEL DIODES

The tunnel diode was first introduced by Leo Esaki in 1958. Its characteristics, shown in Fig. 3.29, are different from any diode discussed thus far in that it has a negative-resistance region. In this region, an increase in terminal voltage results in a reduction in diode current.

The tunnel diode is fabricated by doping the semiconductor materials that will form the *p-n* junction at a level one hundred to several thousand times that of a typical semiconductor diode. This will result in a greatly reduced depletion region, of the order of magnitude of 10^{-6} cm, or typically about $\frac{1}{100}$ the width of this region for a

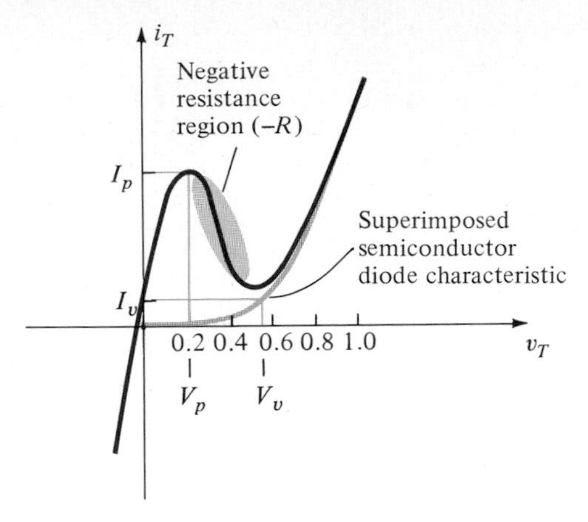

Figure 3.29 Tunnel diode characteristic.

typical semiconductor diode. It is this thin depletion region that many carriers can "tunnel" through, rather than attempt to surmount, at low forward-bias potentials that accounts for the peak in the curve of Fig. 3.29. For comparison purposes, a typical semiconductor diode characteristic has been superimposed on the tunnel-diode characteristic of Fig. 3.29.

This reduced depletion region results in carriers "punching through" at velocities that far exceed those available with conventional diodes. The tunnel diode can therefore be used in high-speed applications such as in computers, where switching times in the order of nanoseconds or picoseconds are desirable.

You will recall from Section 3.2 that an increase in the doping level will drop the Zener potential. Note the effect of a very high doping level on this region in Fig. 3.29. The semiconductor materials most frequently used in the manufacture of tunnel diodes are germanium and gallium arsenide. The ratio I_p/I_v is very important for computer applications. For germanium it is typically 10 : 1, while for gallium arsenide it is closer to 20 : 1.

The peak current, I_p, of a tunnel diode can vary from a few microamperes to several hundred amperes. The peak voltage, however, is limited to about 600 mV. For this reason, a simple VOM with an internal dc battery potential of 1.5 V can severely damage a tunnel diode if used improperly.

The tunnel diode equivalent circuit in the negative-resistance region is provided in Fig. 3.30, with the symbols most frequently employed for tunnel diodes. The values for each parameter are for the 1N2939 GE tunnel diode whose specifications appear in Table 3.2. The inductor L_S is due mainly to the terminal leads. The resistor R_S is due to the leads, ohmic contact at the lead-semiconductor junction, and the semiconductor materials themselves. The capacitance C is the junction diffusion capacitance and the R is the negative resistance of the region. The negative resistance finds application in oscillators to be described later.

Note the lead length of $\frac{1}{8}$ in. included in the specifications. An increase in this length will cause L_S to increase. In fact, it was given for this device that L_S will vary 1 to 12 nH, depending on lead length. At high frequencies ($X_{L_S} = 2\pi f L_S$) this factor can take its toll.

The fact that $V_{fp} = 500$ mV (typ.) and I_{forward} (max.) $= 5$ mA indicates that tunnel

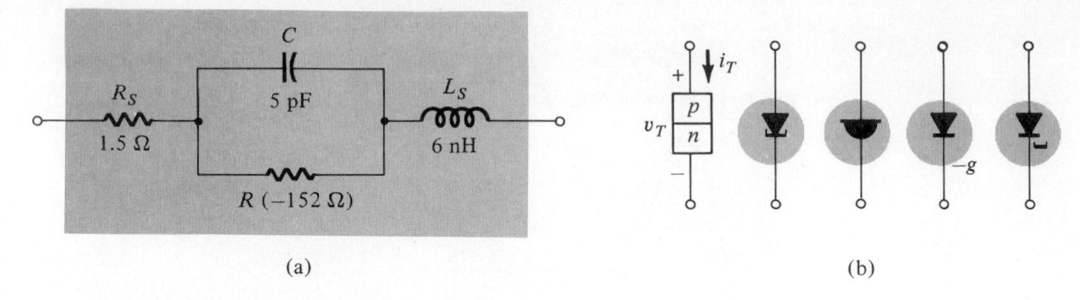

Figure 3.30 Tunnel diode: (a) equivalent circuit; (b) symbols.

TABLE 3.2 Specifications: Ge 1N2939

	Minimum	*Typical*	*Maximum*	
Absolute maximum ratings (25°C)				
Forward current (−55 to +100°C)		5 mA		
Reverse current (−55 to +100°C)		10 mA		
Electrical characteristics (25°C)($\frac{1}{8}$ in. leads)				
I_P	0.9	1.0	1.1	mA
I_V		0.1	0.14	mA
V_P	50	60	65	mV
V_V		350		mV
Reverse voltage (I_R = 1.0 mA)			30	mV
Forward peak point current voltage, V_{fp}	450	500	600	mV
I_p/I_v		10		
$-R$		−152		Ω
C		5	15	pF
L_S		6		nH
R_S		1.5	4.0	Ω

diodes are low-power devices [$P_D = (0.5)(5 \times 10^{-3}) = 2.5$ mW], which is also excellent for computer applications. A rendering of the device appears in Fig. 3.31.

Although the use of tunnel diodes in present-day high-frequency systems has been dramatically stalled by manufacturing techniques that suggest alternatives to the tunnel diode, its simplicity, linearity, low power drain, and reliability ensure its continued life and application. The basic construction of an advance design tunnel diode appears in Fig. 3.32 with a photograph of the actual junction.

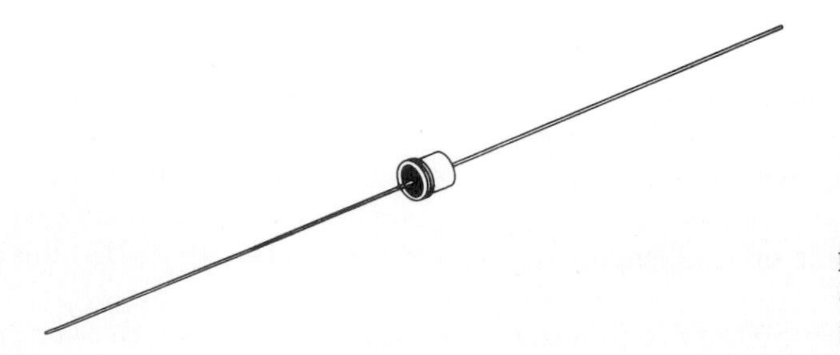

Figure 3.31 A Ge In2939 tunnel diode. (Courtesy General Electric Corporation.)

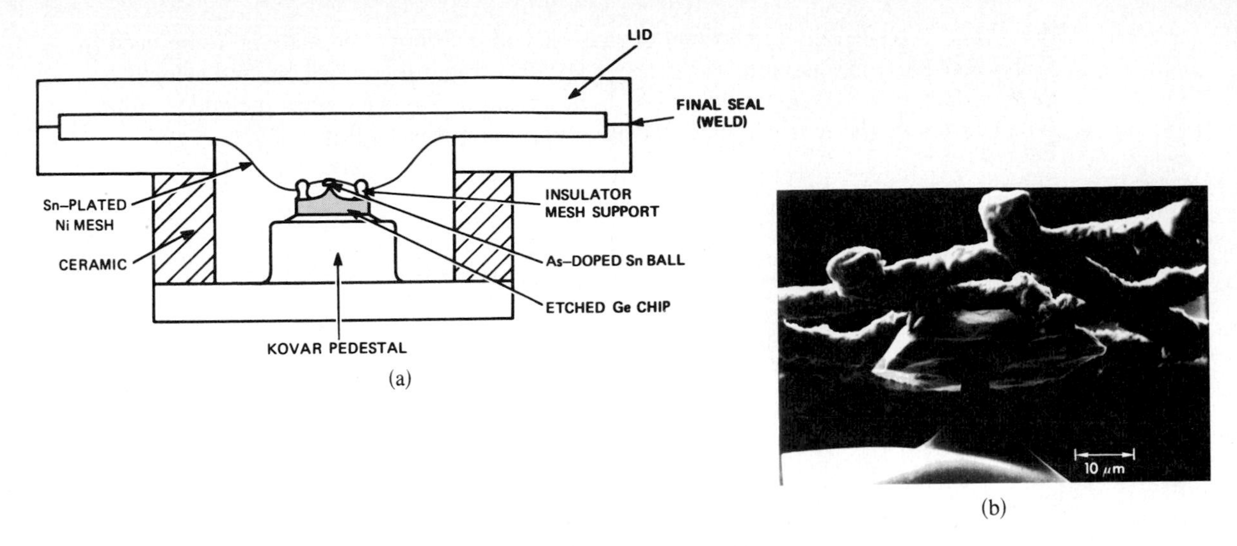

Figure 3.32 Tunnel diode: (a) construction; (b) photograph. (Courtesy *COM-SAT Technical Review*, P. F. Varadi and T. D. Kirkendall.)

3.8 PHOTODIODES

The interest in light-sensitive devices has been increasing at an almost exponential rate in recent years. The resulting field of *optoelectronics* will be receiving a great deal of research interest as efforts are made to improve efficiency levels.

Through the advertising media, the layperson has become quite aware that light sources offer a unique source of energy. This energy, transmitted as discrete packages called *photons*, has a level directly related to the frequency of the traveling light wave as determined by the following equation:

$$\boxed{W = \hbar f} \qquad \text{joules} \tag{3.16}$$

where $\hbar$ is called Planck's constant and is equal to 6.624×10^{-34} joule-second. It clearly states that since $\hbar$ is a constant, the energy associated with incident light waves is directly related to the frequency of the traveling wave.

The frequency is, in turn, directly related to the wavelength (distance between successive peaks) of the traveling wave by the following equation:

$$\boxed{\lambda = \frac{v}{f}} \tag{3.17}$$

where λ = wavelength, meters
v = velocity of light, 3×10^8 m/s
f = frequency of the traveling wave, hertz

The wavelength is usually measured in angstrom units (Å) or micrometers (μm), where

$$1\ \text{Å} = 10^{-10}\ \text{m} \quad \text{and} \quad 1\ \mu\text{m} = 10^{-6}\ \text{m}$$

The wavelength is important because it will determine the material to be used in the optoelectronic device. The relative spectral response for Ge, Si, and selenium is provided in Fig. 3.33. The visible light spectrum has also been included with an indication of the wavelength associated with the various colors.

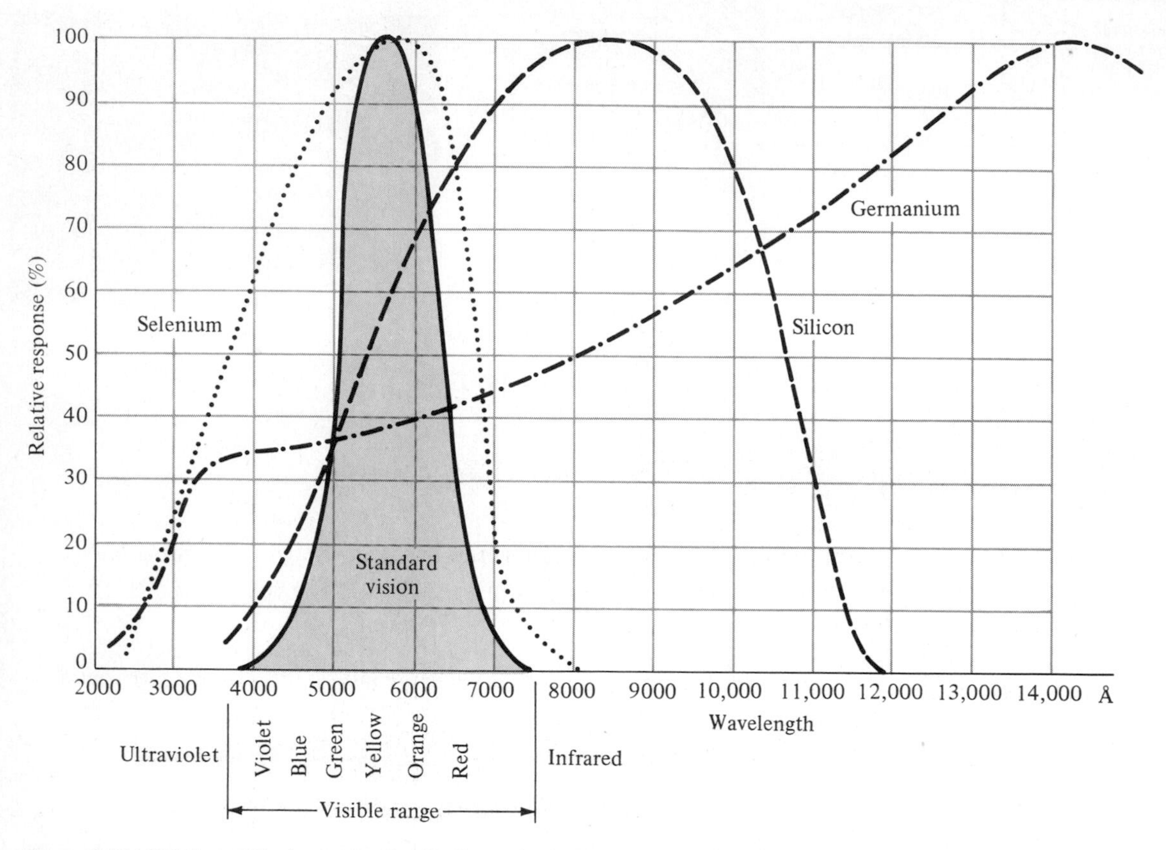

Figure 3.33 Relative spectral response for Si. Ge, and selenium as compared to the human eye.

The number of free electrons generated in each material is proportional to the *intensity* of the incident light. Light intensity is a measure of the amount of *luminous flux* falling in a particular surface area. Luminous flux is normally measured in *lumens* (lm) or watts. The two units are related by

$$1 \text{ lm} = 1.496 \times 10^{-10} \text{ W}$$

The light intensity is normally measured in lm/ft^2, foot-candles (fc), or W/m^2, where

$$1 \text{ lm/ft}^2 = 1 \text{ fc} = 1.609 \times 10^{-12} \text{ W/m}^2$$

The photodiode is a semiconductor *p-n* junction device whose region of operation is limited to the reverse-bias region. The basic biasing arrangement, construction, and symbol for the device appear in Fig. 3.34.

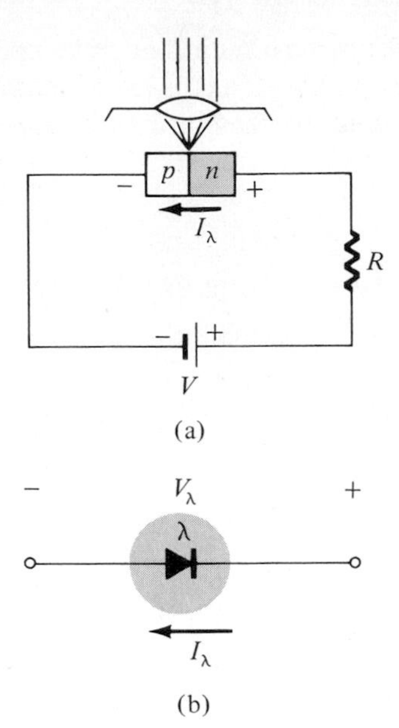

Figure 3.34 Photodiode: (a) basic-biasing arrangement and construction; (b) symbol.

Recall from Chapter 1 that the reverse saturation current is normally limited to a few microamperes. It is due solely to the thermally generated minority carriers in the *n*- and *p*-type materials. The application of light to the junction will result in a transfer of energy from the incident traveling light waves (in the form of photons) to the atomic structure, resulting in an increased number of minority carriers and an increased level of reverse current. This is clearly shown in Fig. 3.35 for different intensity levels. The *dark* current is that current that will exist with no applied illumination. Note that the

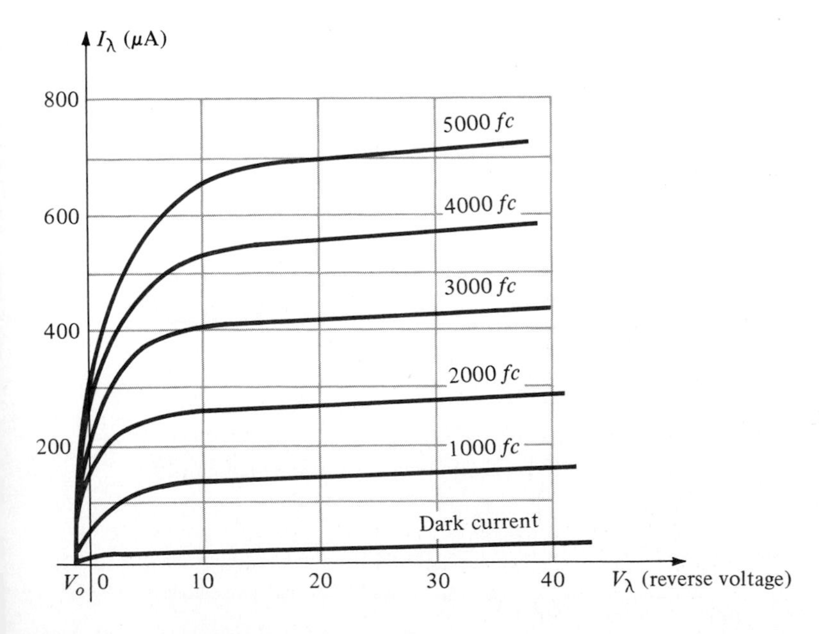

Figure 3.35 Typical set of photodiode characteristics.

current will only return to zero with a positive applied bias equal to V_o. In addition, Fig. 3.34 demonstrates the use of a lens to concentrate the light on the junction region. An actual device showing the lens in the cap appears in Fig. 3.36.

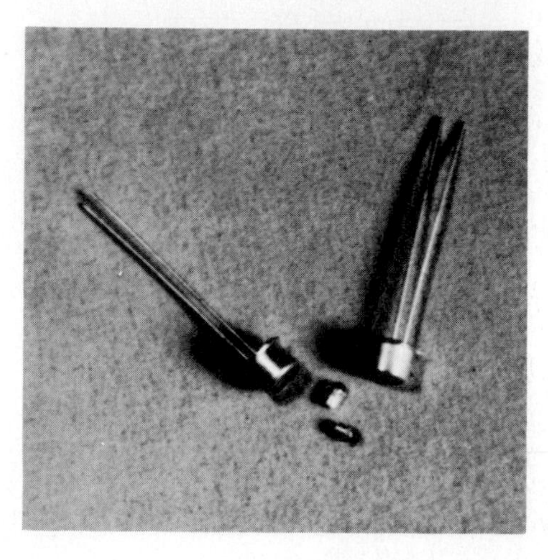

Figure 3.36 Photograph of Hewlett-Packard 5082-4200 S pin photodiodes. (Courtesy Hewlett-Packard Corporation.)

The almost equal spacing between the curves for the same increment in luminous flux reveals that the reverse current and luminous flux are almost linearly related. In other words, an increase in light intensity will result in a similar increase in reverse current. A plot of the two to show this linear relationship appears in Fig. 3.37 for a fixed voltage V_λ of 20 V. On the relative basis we can assume that the reverse current is essentially zero in the absence of incident light. Since the rise and fall times (change-of-state parameters) are very small for this device (in the nanosecond range), the device can be used for high-speed counting or switching applications. Returning to Fig. 3.33, we note that Ge encompasses a wider spectrum of wavelengths than Si. This would make it suitable for incident light in the infrared region as provided by lasers and IR (infrared) light sources to be described shortly. Of course, Ge has a

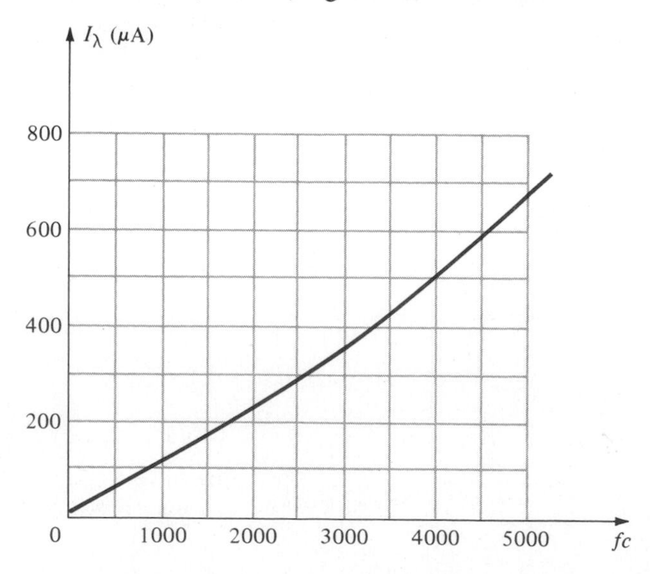

Figure 3.37 I_λ versus f_c (at V_A = 20 V) for the photodiode of Fig. 3.35.

higher dark current than silicon, but it also has a higher level of reverse current. The level of current generated by the incident light on a photodiode is not such that it could be used as a direct control, but it can be amplified for this purpose.

3.9 PHOTOCONDUCTIVE CELLS

The photoconductive cell is a two-terminal semiconductor device whose terminal resistance will vary (linearly) with the intensity of the incident light. For obvious reasons, it is frequently called a photoresistive device. A typical photoconductive cell and the most widely used graphic symbol for the device appear in Fig. 3.38.

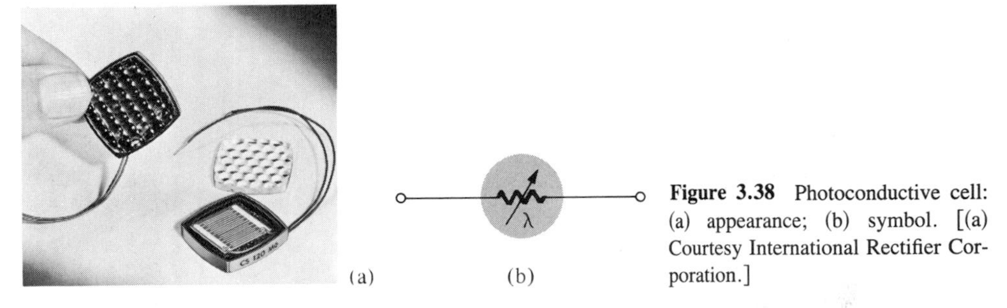

Figure 3.38 Photoconductive cell: (a) appearance; (b) symbol. [(a) Courtesy International Rectifier Corporation.]

The photoconductive materials most frequently used include cadmium sulfide (CdS) and cadmium selenide (CdSe). The peak spectral response of CdS occurs at approximately 5100 Å and for CdSe at 6150 Å, as shown in Fig. 3.33. The response time of CdS units is about 100 ms and 10 ms for CdSe cells.

The photoconductive cell does not have a junction like the photodiode. A thin layer of the material connected between terminals is simply exposed to the incident light energy.

As the illumination on the device increases in intensity, the energy state of a larger number of electrons in the structure will also increase, because of the increased availability of the photon packages of energy. The result is an increasing number of relatively "free" electrons in the structure and a decrease in the terminal resistance. The sensitivity curve for a typical photoconductive device appears in Fig. 3.39. Note

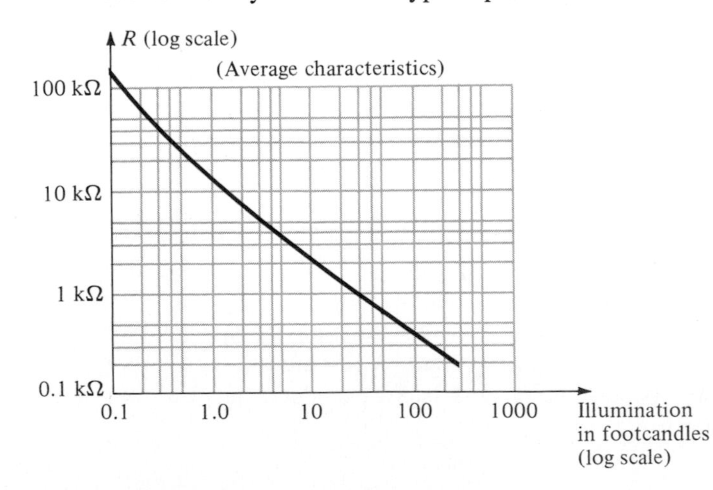

Figure 3.39 Photoconductive cell-terminal characteristics (GE type B425).

the linearity (when plotted using a log-log scale) of the resulting curve and the large change in resistance (100 kΩ → 100 Ω) for the indicated change in illumination.

One rather simple, but interesting, application of the device appears in Fig. 3.40. The purpose of the system is to maintain V_o at a fixed level even though V_i may fluctuate from its rated value. As indicated in the figure, the photoconductive cell, bulb, and resistor, all form part of this voltage-regulator system. If V_i should drop in magnitude for any number of reasons, the brightness of the bulb would also decrease. The decrease in illumination would result in an increase in the resistance (R_λ) of the photoconductive cell to maintain V_o at its rated level as determined by the voltage-divider rule; that is,

$$V_o = \frac{R_\lambda V_i}{R_\lambda + R_1} \tag{3.18}$$

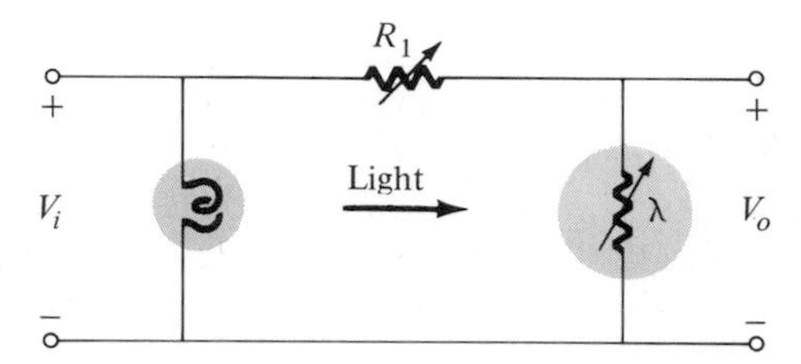

Figure 3.40 Voltage regulator employing a photoconductive cell.

In an effort to demonstrate the wealth of material available on each device from manufacturers, consider the CdS (cadmium sulfide) photoconductive cell described in Fig. 3.41. Note again the concern with temperature and response time.

3.10 IR EMITTERS

Infrared-emitting diodes are solid-state gallium arsenide devices that emit a beam of radiant flux when forward biased. The basic construction of the device is shown in Fig. 3.42. When the junction is forward biased, electrons from the *n*-region will recombine with excess holes of the *p*-material in a specially designed recombination region sandwiched between the *p*- and *n*-type materials. During this recombination process, energy is radiated away from the device in the form of photons. The generated photons will either be reabsorbed in the structure or leave the surface of the device as radiant energy, as shown in Fig. 3.42.

The radiant flux in mW versus the dc forward current for a typical device appears in Fig. 3.43. Note the almost linear relationship between the two. An interesting pattern for such devices is provided in Fig. 3.44. Note the very narrow pattern for devices with an internal collimating system. One such device appears in Fig. 3.45, with its internal construction and graphic symbol. A few areas of application for such devices include card and paper-tape readers, shaft encoders, data-transmission systems, and intrusion alarms.

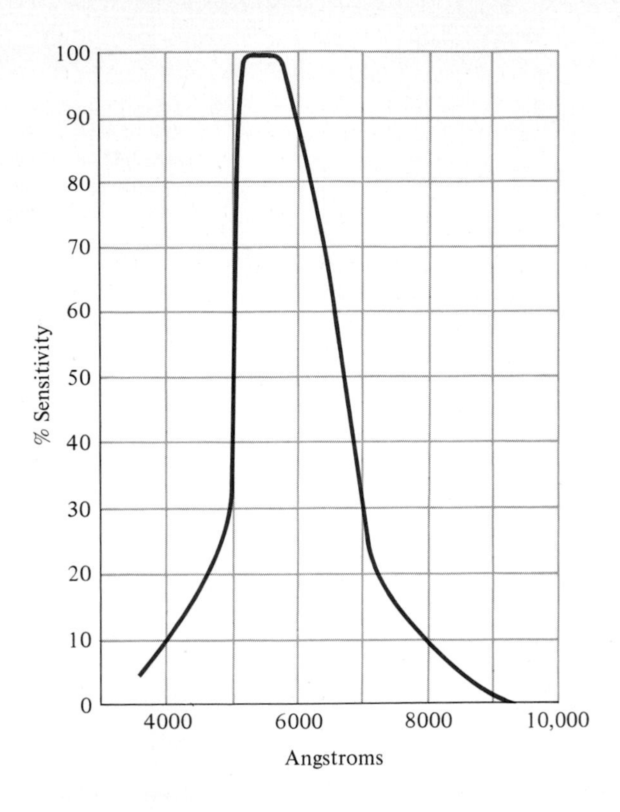

Variation of Conductance With Temperature and Light

Foot Candles	*.01*	*0.1*	*1.0*	*10*	*100*
Temperature			*% Conductance*		
−25°C	103	104	104	102	106
0	98	102	102	100	103
25°C	100	100	100	100	100
50°C	98	102	103	104	99
75°C	90	106	108	109	104

Response Time Versus Light

Foot Candles	*.01*	*0.1*	*1.0*	*10*	*100*
Rise (seconds)*	0.5	.095	.022	.005	.002
Decay (seconds)**	.125	.021	.005	.002	.001

* Time to (1 − 1/e) of final reading after 5 seconds Dark adaptation.
** Time to 1/e of initial reading.

Figure 3.41 Characteristics of a Clairex CdS photoconductive cell. (Courtesy Clairex Electronics.)

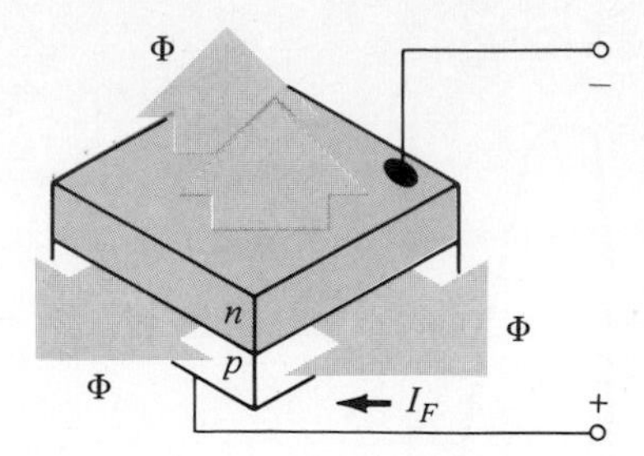

Figure 3.42 General structure of a semiconductor IR-emitting diode. (Courtesy RCA Solid State Division.)

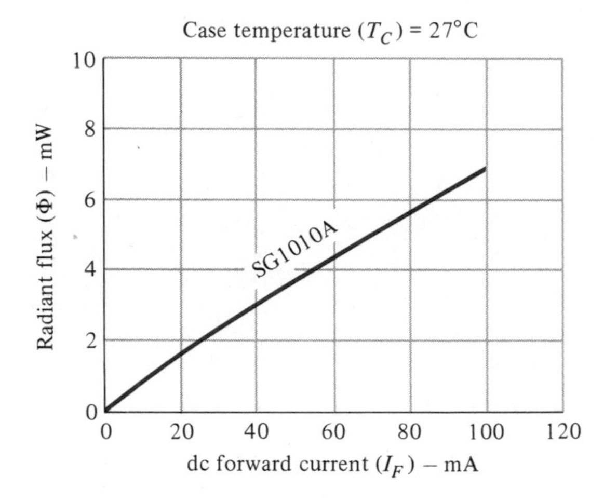

Figure 3.43 Typical radiant flux versus dc forward current for an IR-emitting diode. (Courtesy RCA Solid State Division.)

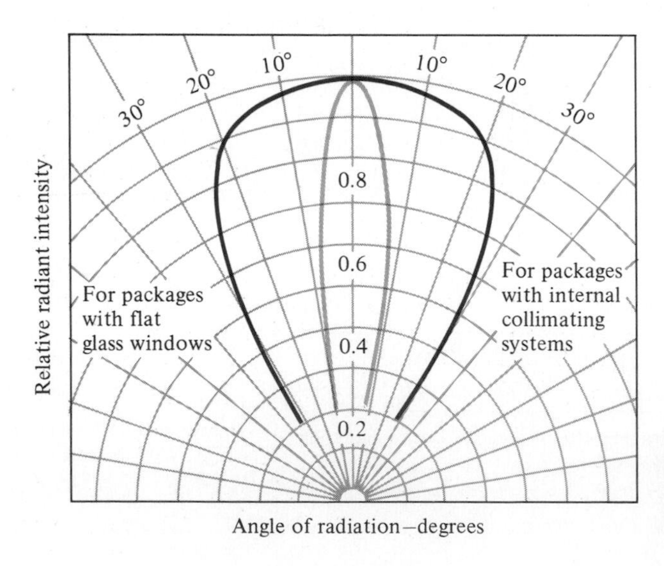

Figure 3.44 Typical radiant intensity patterns of RCA IR-emitting diodes. (Courtesy RCA Solid State Division.)

Figure 3.45 RCA IR-emitting diode: (a) construction; (b) photo; (c) symbol. (Courtesy RCA Solid State Division.)

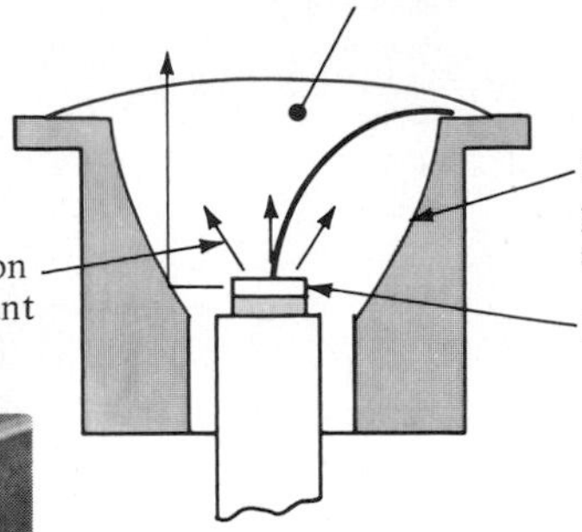

Epoxy resin

Reflecting parabolic surface

Direction of radiant flux

Pellet

(a)

Approx. 2X actual size

(b)

(c)

3.11 LIGHT-EMITTING DIODES

The light-emitting diode (LED) is, as the name implies, a diode that will give off visible light when it is energized. In any forward-biased *p-n* junction there is, within the structure and primarily close to the junction, a recombination of holes and electrons. This recombination requires that the energy possessed by the unbound free electron be transferred to another state. In all semiconductor *p-n* junctions some of this energy will be given off as heat and some in the form of photons. In silicon and germanium the greater percentage is given up in the form of heat and the emitted light is insignificant. In other materials, such as gallium arsenide phosphide (GaAsP) or gallium phosphide (GaP), the number of photons of light energy emitted is sufficient to create a very visible light source. The process of giving off light by applying an electrical source of energy is called *electroluminescence*. As shown in Fig. 3.46, the conducting surface connected to the *p*-material is much smaller, to permit the emergence of the maximum number of photons of light energy. Note in the figure that the recombination of the injected carriers due to the forward-biased junction is resulting in emitted light at the site of recombination. There may, of course, be some absorption of the packages of photon energy in the structure itself, but a very large percentage are able to leave, as shown in the figure.

The appearance and characteristics of a subminiature high-efficiency solid-state lamp manufactured by Hewlett-Packard appears in Fig. 3.47. Two quantities yet

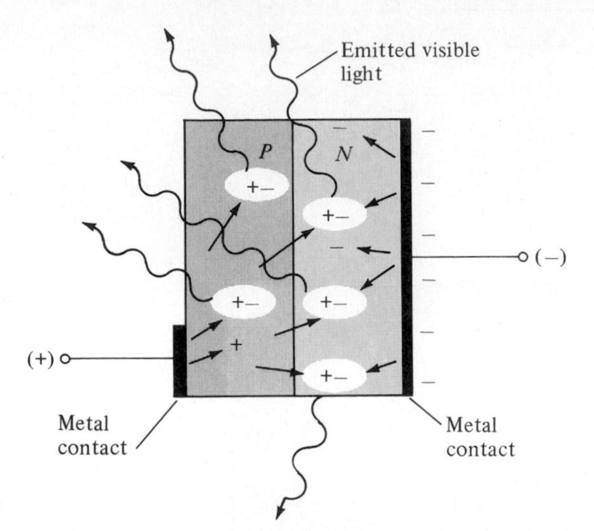

Figure 3.46 Process of electroluminescence in the LED.

undefined appear under the heading electrical/optical characteristics at T_A = 25°C. They are the *axial luminous intensity* (I_V) and the *luminous efficacy* (η_v). Light intensity is measured in *candela*. One candela emits a light flux of 4π lumens and establishes an illumination of 1 foot-candle on a 1-ft^2 area 1 ft from the light source. Even though this description may not provide a clear understanding of the candela as a unit of measure, its level can certainly be compared between similar devices. The term "efficacy" is, by definition, a measure of the ability of a device to produce a desired effect. For the LED this is the ratio of the number of lumens generated per applied watt of electrical energy. Note the high efficacy for the high-efficiency red LED. The relative efficiency is defined by the luminous intensity per unit current, as shown in Fig. 3.47h. Note also how close the peak wavelength of the light waves produced by each LED (red, yellow, green) approaches the defined wavelength (λ_d) for each color. The relative intensity of each color versus wavelength appears in Fig. 3.47e.

Since the LED is a *p-n* junction device, it will have a forward-biased characteristic (Fig. 3.47f) similar to the diode response curves introduced in Chapter 1. Note the almost linear increase in relative luminous intensity with forward current (Fig. 3.47g). Figure 3.47i reveals that the longer the pulse duration at a particular frequency, the lower the permitted peak current (after you pass the break value of t_p). Figure 3.47j simply reveals that the intensity is greater at 0° (or head on) and the least at 90° (when you view the device from the side).

LED displays are available today in many different sizes and shapes. The light-emitting region is available in lengths from 0.1 to 1 in. Numbers can be created by segments such as shown in Fig. 3.48. By applying a forward bias to the proper *p*-type material segment, any number from 0 to 9 can be displayed.

The display of Fig. 3.49 is used in calculators and will provide eight digits. There are also two-lead LED lamps that contain two LEDs, so that a reversal in biasing will change the color from green to red, or vice versa. LEDs are presently available in red, green, yellow, orange, and white. It would appear that the introduction of the color

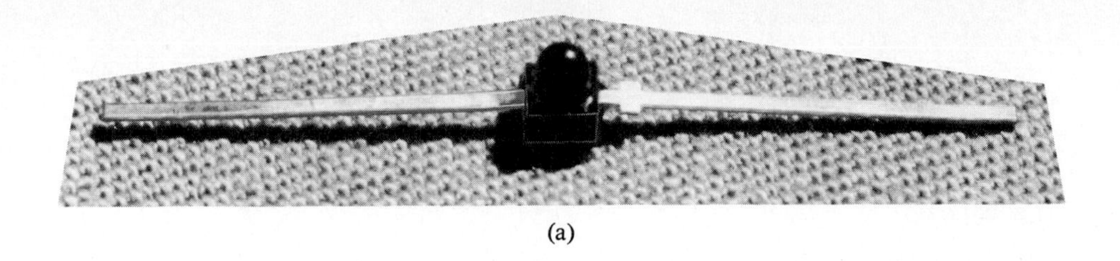

(a)

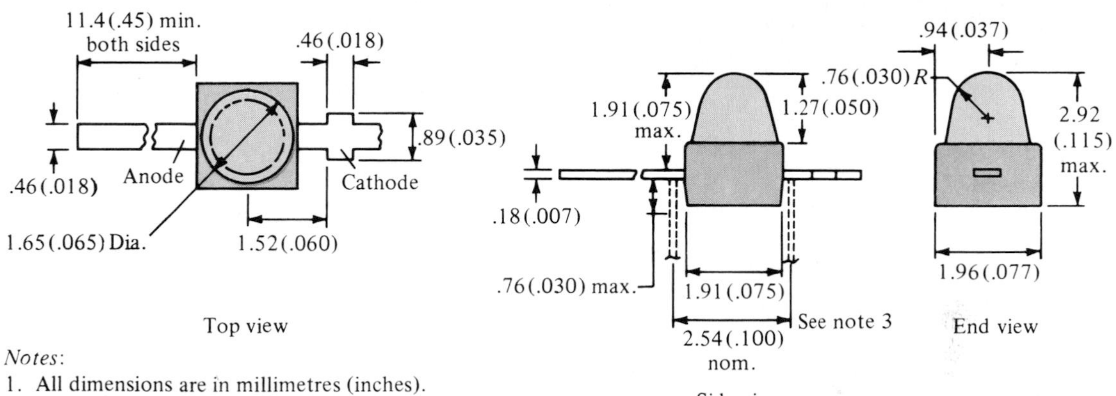

Notes:

1. All dimensions are in millimetres (inches).
2. Silver-plated leads. See application bulletin 3.
3. User may bend leads as shown.

(b)

Absolute Maximum Ratings at T_A = 25°C

Parameter	*Red 4100/4101*	*High Eff. Red 4160*	*Yellow 4150*	*Green 4190*	*Units*
Power dissipation	100	120	120	120	mW
Average forward current	50[1]	20[1]	20[1]	30[2]	mA
Peak forward current	1000	60	60	60	mA
Operating and storage temperature range		−55°C to 100°C			
Lead soldering temperature [1.6mm (0.063 in.) from body]		230°C for 3 seconds			

[1]. Derate from 50°C at 0.2mA/°C
[2]. Derate from 50°C at 0.4mA/°C

(c)

Figure 3.47 Hewlett-Packard subminiature high-efficiency red solid-state lamp: (a) appearance; (b) package dimensions; (c) absolute maximum ratings. (*Continued on pp.* 128 *and* 129.)

Electrical/Optical Characteristics at $T_A = 25°C$

Symbol	Description	5082–4100/4101			5082–4160			5082–4150			5082–4190			Units	Test Conditions
		Min.	Typ.	Max.	Min.	Typ.	Max.	Min.	Typ.	Max.	Min.	Typ.	Max.		
I_V	Axial luminous intensity	–/0.5	.7/1.0		1.0	3.0		1.0	2.0		0.8	1.5 At $I_F = 20mA$		mcd	$I_F = 10mA$,
$2\theta_{1/2}$	Included angle between half luminous intensity points		45			80			90			70		deg.	Note 1
λ_{peak}	Peak wavelength		655			635			583			565		nm	Measurement at peak
λ_d	Dominant wavelength		640			628			585			572		nm	Note 2
τ_S	Speed of response		15			90			90			200		ns	
C	Capacitance		100			11			15			13		pF	$V_F = 0$; $f = 1$ MHz
θ_{JC}	Thermal resistance		125			120			100			100		°C/W	Junction to cathode lead at 0.79mm (.031 in) from body
V_F	Forward voltage		1.6	2.0		2.2	3.0		2.2	3.0		2.4	3.0 At $I_F = 20mA$	V	$I_F = 10mA$,
BV_R	Reverse breakdown voltage	3.0	10		5.0			5.0			5.0			V	$I_R = 100\mu A$
η_v	Luminous efficacy		55			147			570			665		lm/W	Note 3

NOTES:
1. $\theta_{1/2}$ is the off-axis angle at which the luminous intensity is half the axial luminous intensity.
2. The dominant wavelength, λ_d, is derived from the CIE chromaticity diagram and represents the single wavelength which defines the color of the device.
3. Radiant intensity, I_e, in watts/steradian, may be found from the equation $I_e = I_v/\eta_v$, where I_v is the luminous intensity in candelas and η_v is the luminous efficacy in lumens/watt.

(d)

Figure 3.47 *(continued)* (d) Electrical/optical characteristics.

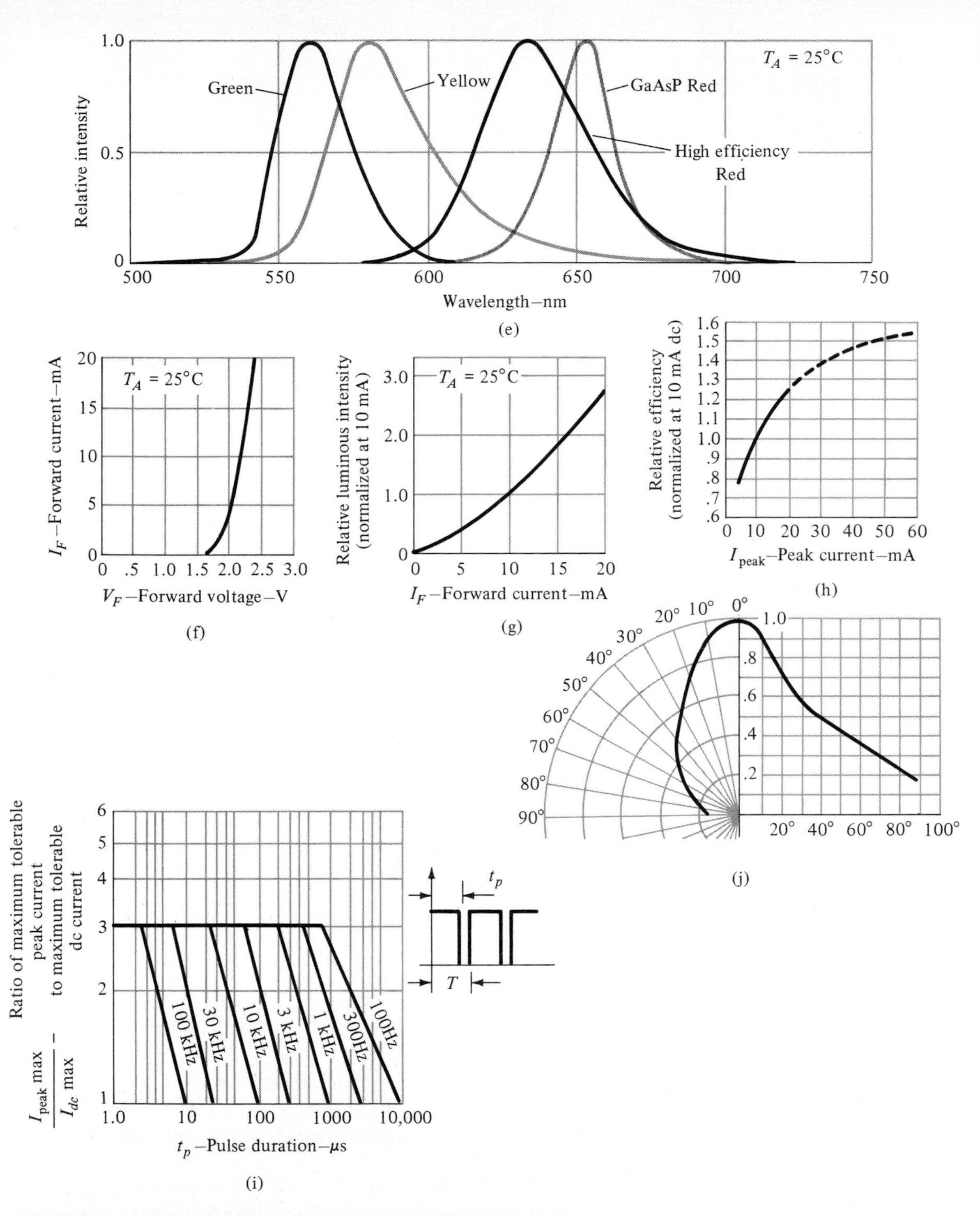

Figure 3.47 (*continued*) (e) Relative intensity versus wavelength; (f) forward current versus forward voltage; (g) relative luminous intensity versus forward current; (h) relative efficiency versus peak current; (i) maximum peak current versus pulse duration; (j) relative luminous intensity versus angular displacement. (Courtesy Hewlett-Packard Corporation.)

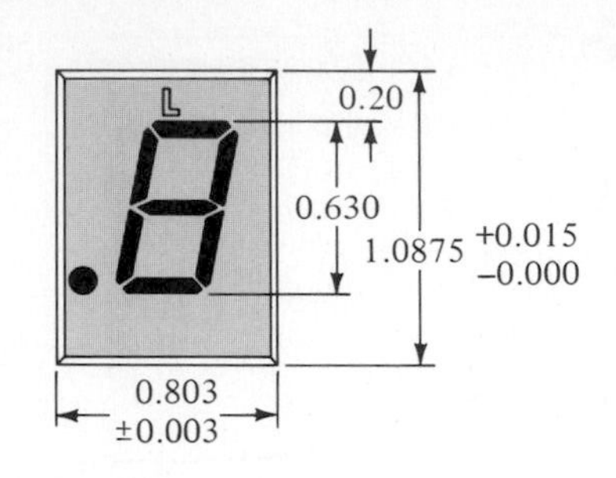

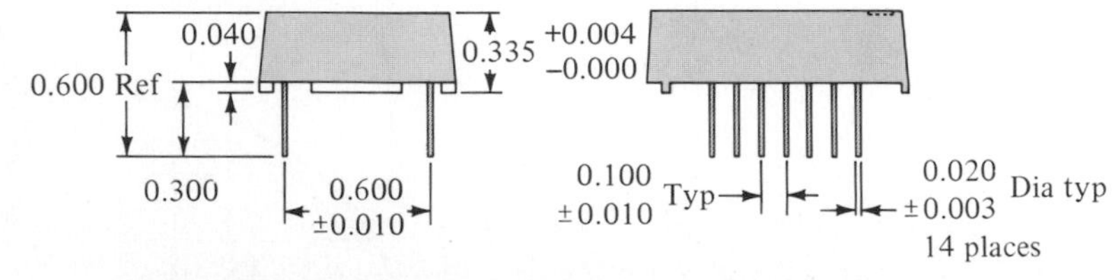

Figure 3.48 Litronix segment display.

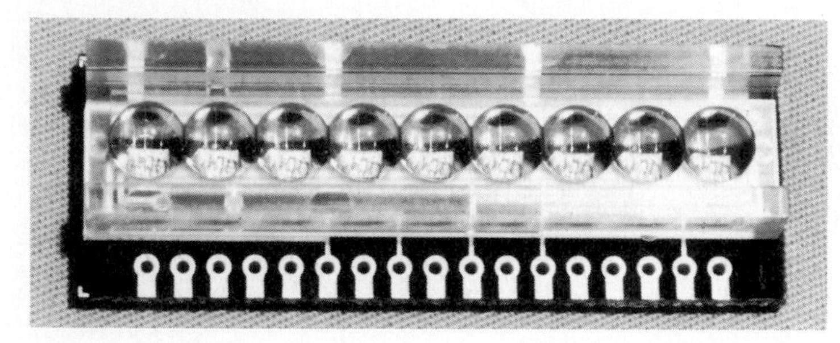

Figure 3.49 Eight-digit and a sign calculator display. (Courtesy Hewlett-Packard Corporation.)

blue is a possibility in the very near future. In general, LEDs operate at voltage levels from 1.7 to 3.3 V, which makes them completely compatible with solid-state circuits. They have a fast response time (nanoseconds) and offer good contrast ratios for visibility. The power requirement is typically from 10 to 150 mW with a lifetime of 100,000+ hours. Their semiconductor construction adds a significant ruggedness factor.

3.12 LIQUID-CRYSTAL DISPLAYS

The liquid-crystal display (LCD) has the distinct advantage of having a lower power requirement than the LED. It is typically in the order of microwatts for the display, as compared to the same order of milliwatts for LEDs. It does, however, require an external or internal light source, is limited to a temperature range of about 0° to 60°C, and lifetime is an area of concern because LCDs can chemically degrade. The types receiving the major interest today are the field-effect and dynamic-scattering units. Each will be covered in some detail in this section.

A liquid crystal is a material (normally organic for LCDs) that will flow like a liquid but whose molecular structure has some properties normally associated with solids. For the light-scattering units, the greatest interest is in the *nematic liquid crystal*, having the crystal structure shown in Fig. 3.50. The individual molecules

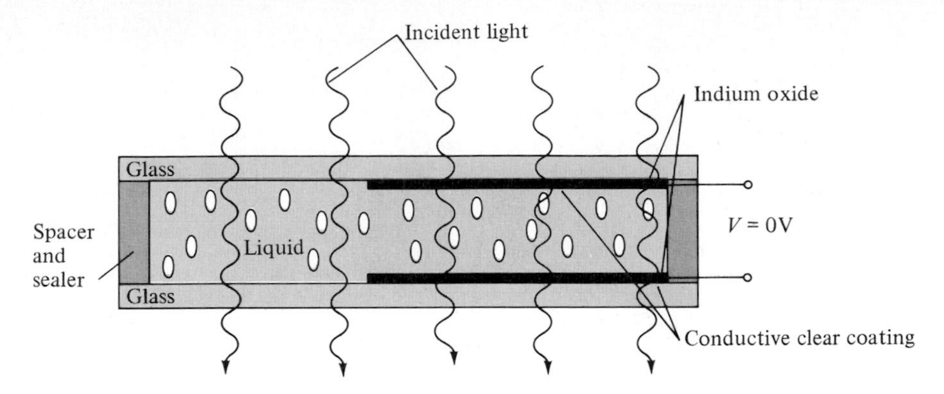

Figure 3.50 Nematic liquid crystal with no applied bias.

have a rodlike appearance as shown in the figure. The indium oxide conducting surface is transparent and, under the condition shown in the figure, the incident light will simply pass through and the liquid-crystal structure will appear clear. If a voltage (for commercial units the threshold level is usually between 6 and 20 V) is applied across the conducting surfaces, as shown in Fig. 3.51, the molecular arrangement is disturbed, with the result that regions will be established with different indices of refraction. The incident light is, therefore, reflected in different directions at the interface between regions of different indices of refraction (referred to as *dynamic scattering*—first studied by RCA in 1968) with the result that the scattered light has a frosted-glass appearance. Note in Fig. 3.51, however, that the frosted look occurs only where the conducting surfaces are opposite each other and that the remaining areas remain translucent.

A digit on an LCD display may have the segment appearance shown in Fig. 3.52. The black area is actually a clear conducting surface connected to the terminals below for external control. Two similar masks are placed on opposite sides of a sealed thick layer of liquid-crystal material. If the number 2 were required, the terminals 8, 7, 3, 4, and 5 would be energized and only those regions would be frosted while the other areas would remain clear.

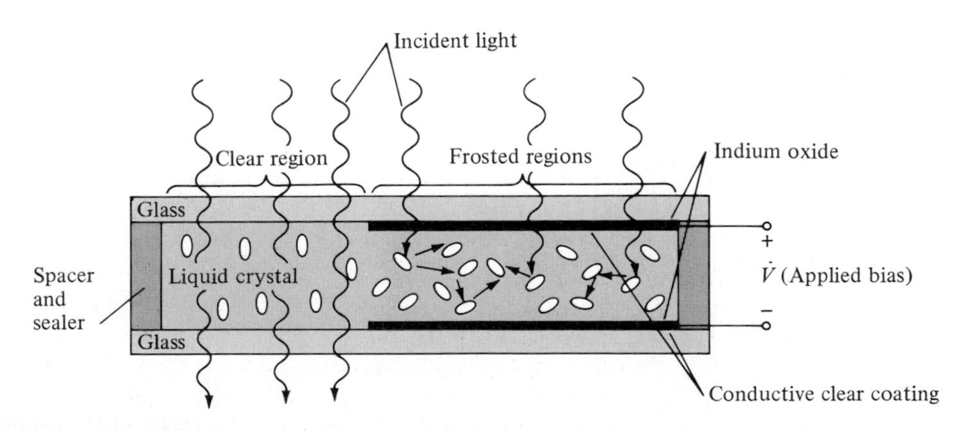

Figure 3.51 Nematic liquid crystal with applied bias.

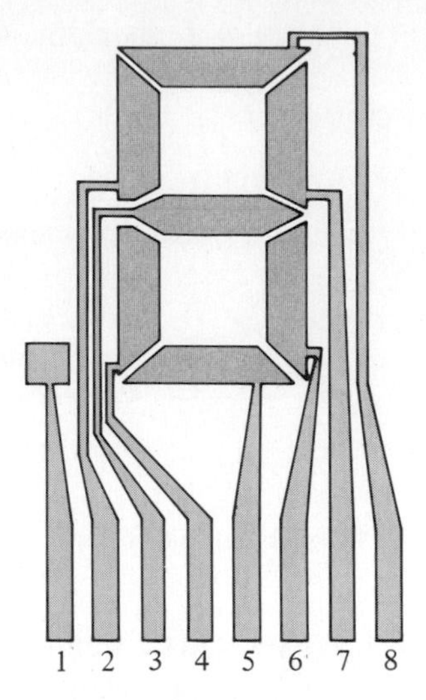

Figure 3.52 LCD eight-segment digit display.

As indicated earlier, the LCD does not generate its own light but depends on an external or internal source. Under dark conditions it would be necessary for the unit to have its own internal light source either behind or to the side of the LCD. During the day, or in lighted areas, a reflector can be put behind the LCD to reflect the light back through the display for maximum intensity. For optimum operation, current watch manufacturers are using a combination of the transmissive (own light source) and reflective modes called *transflective*.

The *field-effect* or *twisted nematic* LCD has the same segment appearance and thin layer of encapsulated liquid crystal, but its mode of operation is very different. Similar to the dynamic-scattering LCD, the field effect can be operated in the reflective or transmissive mode with an internal source. The transmissive display appears in Fig. 3.53. The internal light source is on the right and the viewer is on the left. This figure is most noticeably different from Fig. 3.50 in that there is an addition of a *light polarizer*. Only the vertical component of the entering light on the right can pass

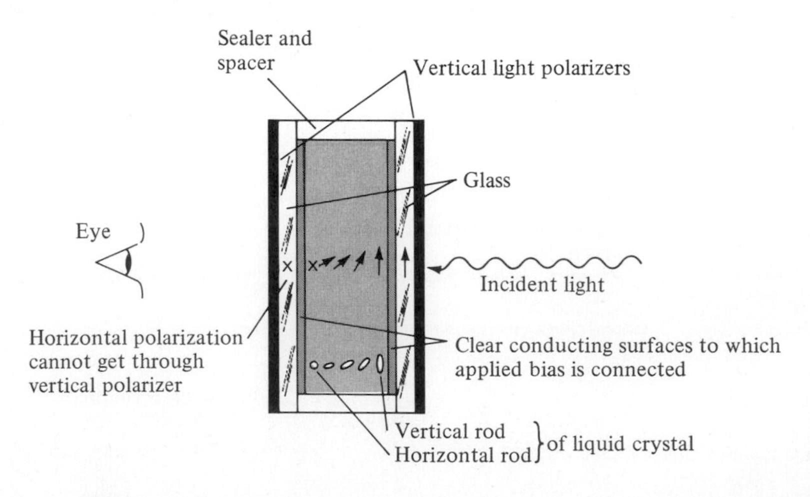

Figure 3.53 Transmissive field-effect LCD with no applied bias.

through the vertical-light polarizer on the right. In the field-effect LCD, either the clear conducting surface to the right is chemically etched or an organic film is applied to orient the molecules in the liquid crystal in the vertical plane, parallel to the cell wall. Note the rods to the far right in the liquid crystal. The opposite conducting surface is also treated to ensure that the molecules are 90° out of phase in the direction shown (horizontal) but still parallel to the cell wall. In between the two walls of the liquid crystal there is a general drift from one polarization to the other, as shown in the figure. The left-hand light polarizer is also such that it permits the passage of only the vertically polarized incident light. If there is no applied voltage to the conducting surfaces, the vertically polarized light enters the liquid-crystal region and follows the 90° bending of the molecular structure. Its horizontal polarization at the left-hand vertical light polarizer does not allow it to pass through, and the viewer sees a uniformly dark pattern across the entire display. When a threshold voltage is applied (for commercial units from 2 to 8 V), the rodlike molecules align themselves with the field (perpendicular to the wall) and the light passes directly through without the 90° shift. The vertically incident light can then pass directly through the second vertically polarized screen and a light area is seen by the viewer. Through proper excitation of the segments of each digit the pattern will appear as shown in Fig. 3.54. The reflective type field effect is shown in Fig. 3.55. In this case the horizontally polarized light at the far left encounters a horizontally polarized filter and passes through to the reflector, where it is reflected back into the liquid crystal, bent back to the other vertical polarization, and returned to the observer. If there is no applied voltage, there is a uniformly lit display. The application of a voltage results in a vertically incident light encountering a horizontally polarized filter at the left which will not be able to pass through and be reflected. A dark area results on the crystal, and the pattern as shown in Fig. 3.56 appears.

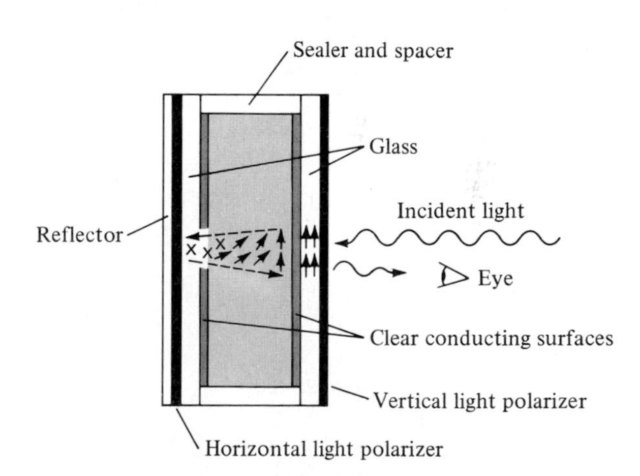

Figure 3.54 Reflective-type LCD. (Courtesy RCA Solid State Division.)

Figure 3.55 Reflective field-effect LCD with no applied bias.

Figure 3.56 Transmissive-type LCD. (Courtesy RCA Solid State Division.)

Field-effect LCDs are normally used when a source of energy is a prime factor (e.g., in watches, portable instrumentation, etc.) since they absorb considerably less power than the light-scattering types—the microwatt range compared to the low-milliwatt range. The cost is typically higher for field-effect units, and their height is limited to about 2 in., while light-scattering units are available up to 8 in. in height.

A further consideration in displays is turn-on and turn-off time. LCDs are characteristically much slower then LEDs. LCDs typically have response times in the range 100 to 300 ms, while LEDs are available with response times below 100 ns. However, there are numerous applications, such as in a watch, where the difference between 100 ns and 100 ms ($\frac{1}{10}$ of a second) is of little consequence. For such applications the lower power demand of LCDs is a very attractive characteristic. The lifetime of LCD units is steadily increasing beyond the 10,000+ hours limit. Since the color generated by LCD units is dependent on the source of illumination, there is a greater range of color choice.

3.13 SOLAR CELLS

In recent years there has been increasing interest in the solar cell as an alternative source of energy. When we consider that the power density received from the sun at sea level is about 100 mW/cm^2 (1 kW/m^2), it is certainly an energy source that requires further research and development to maximize the conversion efficiency from solar to electrical energy.

The basic construction of a silicon *p-n* junction solar cell appears in Fig. 3.57. As shown in the top view, every effort is made to ensure that the surface area perpendicular to the sun is a maximum. Also, note that the metallic conductor connected to the *p*-type material and the thickness of the *p*-type material are such that they ensure that a maximum number of photons of light energy will reach the junction. A photon of light energy in this region may collide with a valence electron and impart to it sufficient energy to leave the parent atom. The result is a generation of free electrons and holes. This phenomenon will occur on each side of the junction. In the *p*-type material the newly generated electrons are minority carriers and will move rather freely across the junction as explained for the basic *p-n* junction with no applied bias. A similar discussion is true for the holes generated in the *n*-type material. The result is an increase in the minority carrier flow which is opposite in direction to the

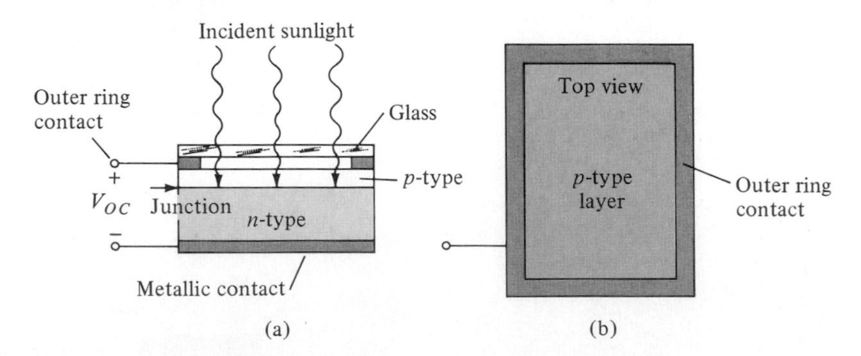

Figure 3.57 Solar cell: (a) cross section; (b) top view.

conventional forward current of a *p-n* junction. This increase in reverse current is shown in Fig. 3.58. Since $V = 0$ anywhere on the vertical axis and represents a short-circuit condition, the current at this intersection is called the *short-circuit current* and is represented by the notation I_{sc}. Under open-circuit conditions ($i_d = 0$) the *photovoltaic* voltage V_{oc} will result. This is a logarithmic function of the illumination, as shown in Fig. 3.59. V_{oc} is the terminal voltage of a battery under no-load (open-circuit) conditions. Note, however, in the same figure that the short-circuit current is a linear function of the illumination. That is, it will double for the same increase in illumination (f_{c_1} and $2f_{c_1}$ in Fig. 3.59) while the change in V_{oc} is less for this region. The major increase in V_{oc} occurs for lower-level increases in illumination. Eventually, a further increase in illumination will have very little effect on V_{oc}, although I_{sc} will increase, causing the power capabilities to increase.

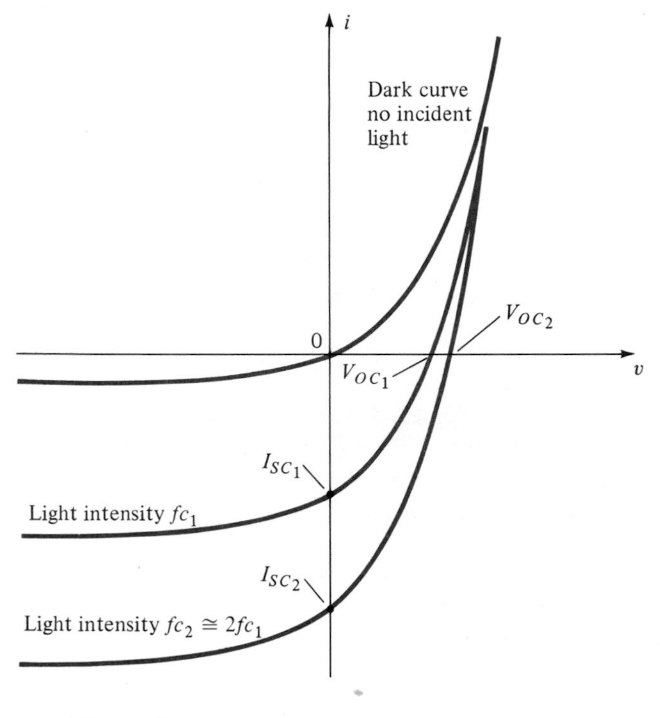

Figure 3.58 Short-circuit current and open-circuit voltage versus light intensity for a solar cell.

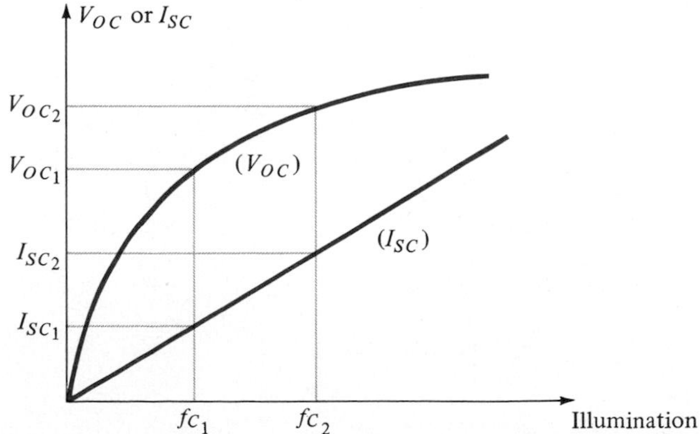

Figure 3.59 V_{oc} and I_{sc} versus illumination for a solar cell.

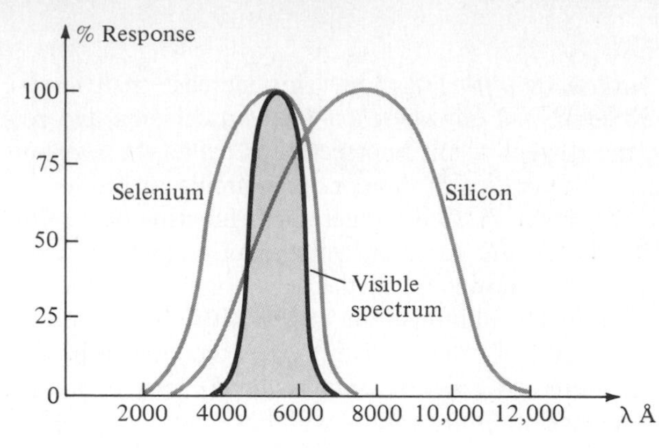

Figure 3.60 Spectral response of Se, Si, and the naked eye.

Selenium and silicon are the most widely used materials for solar cells, although gallium arsenide, indium arsenide, and cadmium sulfide, among other, are also used. The wavelength of the incident light will affect the response of the *p-n* junction to the incident photons. Note in Fig. 3.60 how closely the selenium cell response curve matches that of the eye. This fact has widespread application in photographic equipment such as exposure meters and automatic exposure diaphragms. Silicon also overlaps the visible spectrum but has its peak at the 0.8 μm (8000 Å) wavelength, which is in the infrared region. In general, silicon has a higher conversion efficiency and greater stability and is less subject to fatigue. Both materials have excellent temperature characteristics. That is, they can withstand extreme high or low temperatures without a significant drop-off in efficiency. A typical solar cell, with its electrical characteristics, appears in Fig. 3.61.

Figure 3.61 Typical solar cell and its electrical characteristics. (Courtesy International Rectifier Corporation.)

Electrical Characteristics*

IR Number	*Load Voltage (volts) (min.)*	*Load Current (milliamps) (min.)*	*Power (milliwatts) (min.)*
SP2A40B	1.6	36	58
SP2B48B	1.6	40	64
SP4C40B	3.2	36	115
SP2C80B	1.6	72	115
SP4D48B	3.2	40	128
SP2D96B	1.6	80	129
S2900E5M	.4	60	24
S2900E7M	.4	90	36
S2900E9.5M	.4	120	48

* Current Voltage characteristics are based on an illumination level of 100 mW/cm² (bright average sunlight).

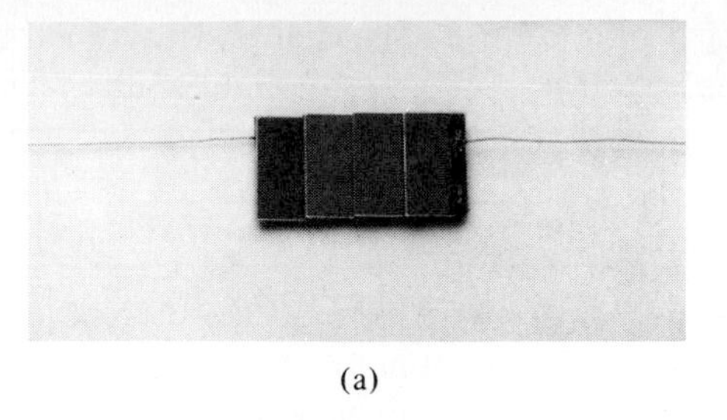

(a)

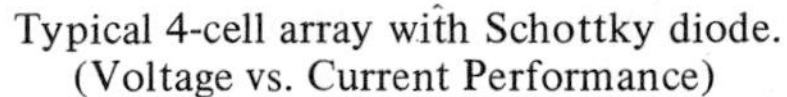

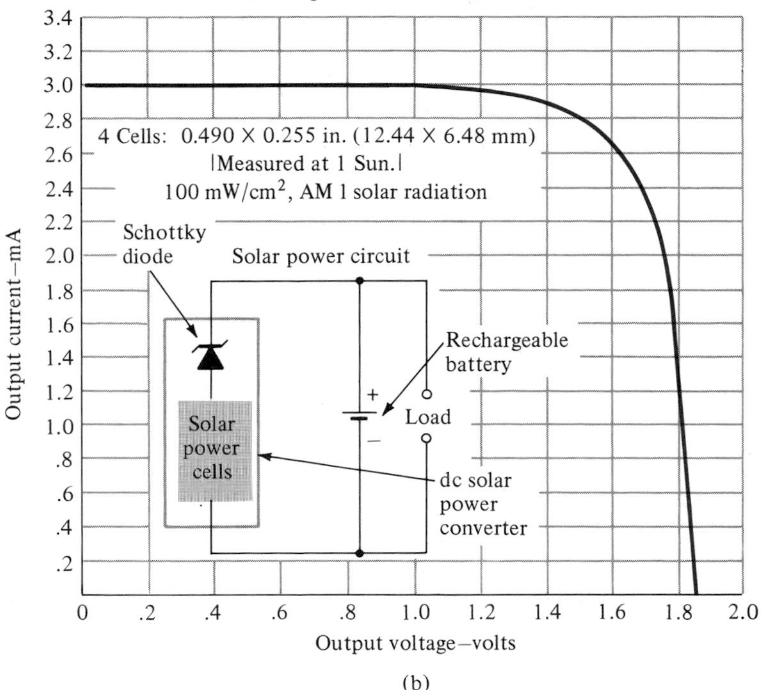

(b)

Figure 3.62 International Rectifier four-cell array: (a) appearance; (b) characteristics. (Courtesy International Rectifier Corporation.)

A very recent innovation in the use of solar cells appears in Fig. 3.62. The series arrangement of solar cells permits a voltage beyond that of a single element. The performance of a typical four-cell array appears in the same figure. At a current of approximately 2.6 mA the output voltage is about 1.6 V, resulting in an output power of 4.16 mW. The Schottky barrier diode is included to prevent battery current drain through the power converter. That is, the resistance of the Schottky diode is so high to charge flowing down through (+ to −) the power converter that it will appear as an open circuit to the rechargeable battery and not draw current from it.

It might be of interest to note that the Lockheed Missiles and Space Company has been awarded a grant from the National Aeronautics and Space Administration to develop a massive solar-array wing for the space shuttle. The wing will measure 13.5 ft by 105 ft when extended and will contain 41 panels, each carrying 3060 silicon solar cells. The wing can generate a total of 12.5 kW of electrical power.

The efficiency of operation of a solar cell is determined by the electrical power output divided by the power provided by the light source. That is,

$$\eta = \frac{P_{o(\text{electrical})}}{P_{i(\text{light energy})}} \times 100\% = \frac{P_{\max(\text{device})}}{(\text{area in cm}^2)(100\text{ mW/cm}^2)} \times 100\% \tag{3.19}$$

Typical levels of efficiency range from 10 to 40%—a level that should improve measurably if the present interest continues. A typical set of output characteristics for silicon solar cells of 10% efficiency with an active area of 1 cm^2 appears in Fig. 3.63. Note the optimum power locus and the almost linear increase in output current with luminous flux for a fixed voltage.

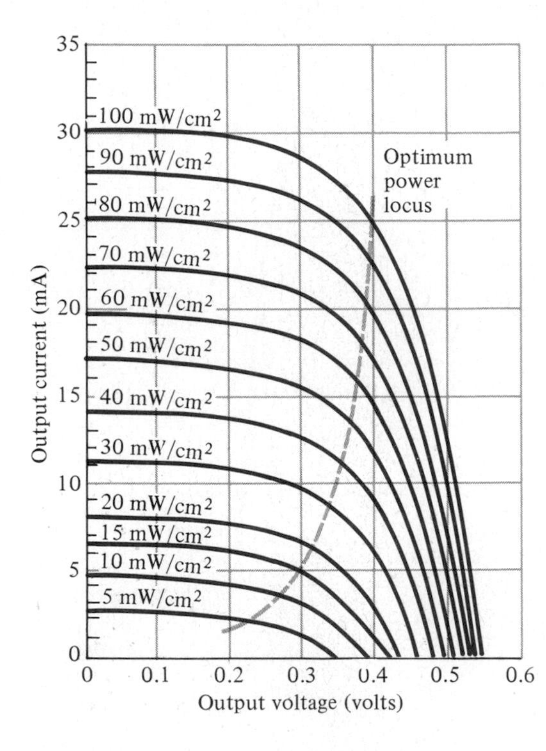

Figure 3.63 Typical output characteristics for silicon solar cells of 10% efficiency having an active area of 1 cm^2. Cell temperature is 30°C.

3.14 THERMISTORS

The thermistor is, as the name implies, a temperature-sensitive resistor; that is, its terminal resistance is related to its body temperature. It has a negative temperature coefficient, indicating that its resistance will decrease with an increase in its body temperature. It is not a junction device and is constructed of Ge, Si, or a mixture of oxides of cobalt, nickel, strontium, or manganese.

The characteristics of a representative thermistor are provided in Fig. 3.64, with the commonly used symbol for the device. Note, in particular, that at room temperature (20°C) the resistance of the thermistor is approximately 5000 Ω, while at 100°C (212°F) the resistance has decreased to 100 Ω. A temperature span of 80°C has therefore resulted in a 50 : 1 change in resistance. It is typically 3 to 5% per degree

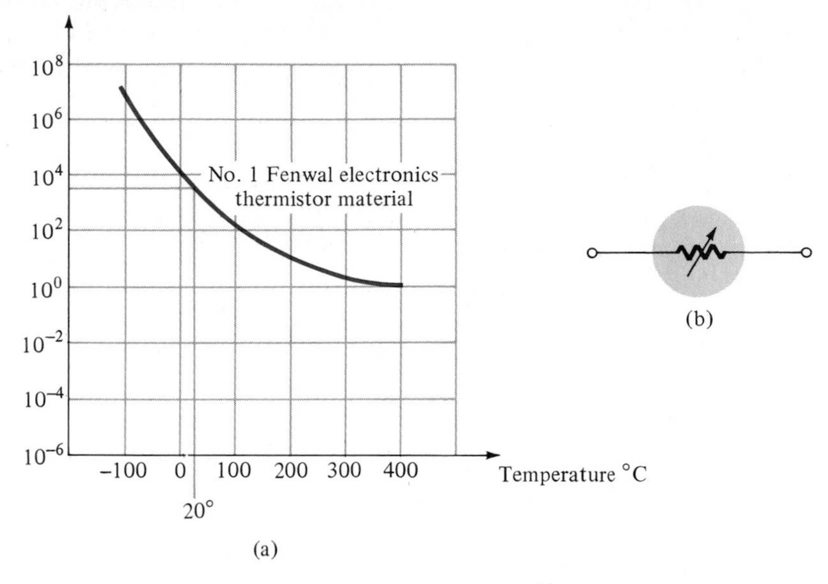

Figure 3.64 Termistor: (a) typical set of characteristics; (b) symbol.

change in temperature. There are, fundamentally, two ways to change the temperature of the device: internally and externally. A simple change in current through the device will result in an internal change in temperature. A small applied voltage will result in a current too small to raise the body temperature above that of the surroundings. In this region, as shown in Fig. 3.65, the thermistor will act like a resistor and have a positive temperature coefficient. However, as the current increases, the temperature will rise to the point where the negative temperature coefficient will appear as shown

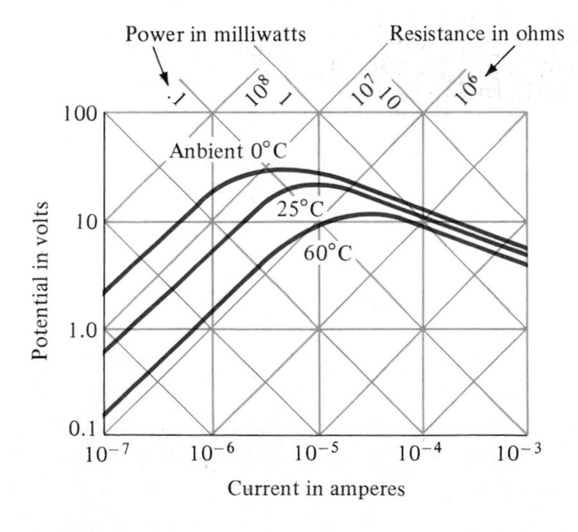

Figure 3.65 Steady-state voltage–current characteristics of Fenwal Electronics BK65V1 Thermistor. (Courtesy Fenwal Electronics, Incorporated.)

in Fig. 3.65. The fact that the rate of internal flow can have such an effect on the resistance of the device introduces a wide vista of applications in control, measuring techniques, and so on. An external change would require changing the temperature of the surrounding medium or immersing the device in a hot or cold solution.

A photograph of a number of commercially available thermistors is provided in Fig. 3.66.

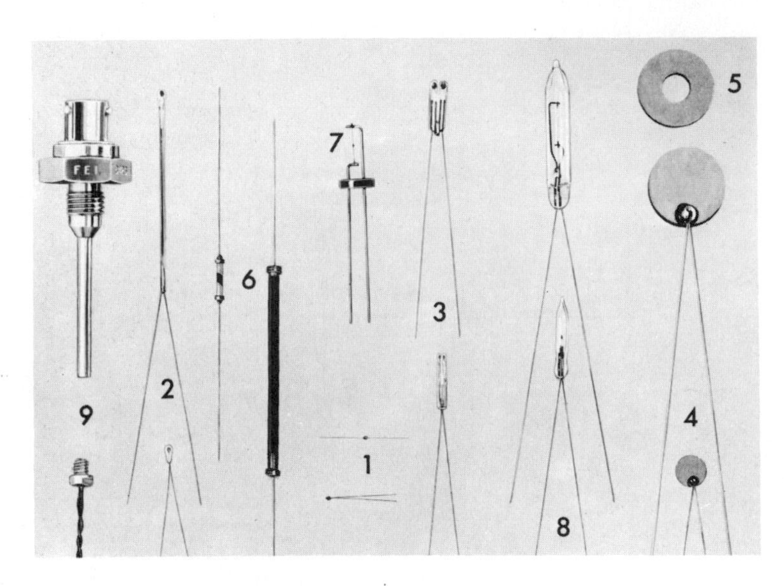

Figure 3.66 Various types of thermistors: (1) beads; (2) glass probes; (3) iso-curve interchangeable probes and beads; (4) disks; (5) washers; (6) rods; (7) specially mounted beads; (8) vacuum and gas-filled probes; (9) special probe assemblies. (Courtesy Fenwal Electronica, Incorporated.)

A simple temperature-indicating circuit appears in Fig. 3.67. Any increase in the temperature of the surrounding medium will result in a decrease in the resistance of the thermistor and an increase in the current I_T. An increase in I_T will produce an increased movement deflection, which when properly calibrated will accurately indicate the higher temperature. The variable resistance was added for calibration purposes.

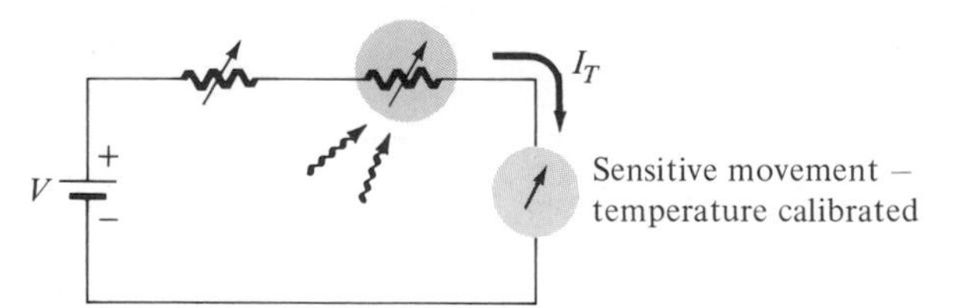

Figure 3.67 Temperature-indicating circuit.

PROBLEMS

§ 3.2

1. The following characteristics are specified for a particular Zener diode: $V_Z = 29$ V, $V_R = 16.8$ V, $I_{ZT} = 10$ mA, $I_R = 20$ μA, and $I_{ZM} = 40$ mA. Sketch the characteristic curve in the manner displayed in Fig. 3.3.

2. At what temperature will the IN961 10-V Fairchild Zener diode have a nominal voltage of 10.75 V? (*Hint:* Note data in Table 3.1.)

3. Determine the temperature coefficient of a 5-V Zener diode (rated 25°C value) if the nominal voltage drops to 4.8 V at a temperature of 100°C.

4. Using the curves of Fig. 3.4a, what level of temperature coefficient would you expect for a 20-V diode? Repeat for a 5-V diode. Assume a linear scale between nominal voltage levels and a current level of 0.1 mA.

5. Determine the dynamic impedance for the 24-V diode at I_Z = 10 mA from Fig. 3.4b. Note that it is a log scale.

6. Compare the levels of dynamic impedance for the 24-V diode of Fig. 3.4b at current levels of 0.2 mA, 1 mA, and 10 mA. How do the results relate to the shape of the characteristics in this region?

§ 3.3

7. (a) Determine V_L, I_L, I_Z, and I_R for the network of Fig. 3.68 if R_L = 180 Ω

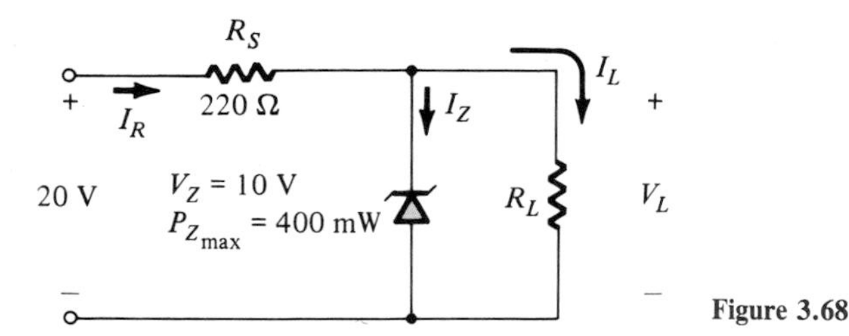

Figure 3.68

(b) Repeat part (a) if R_L = 470 Ω.
(c) Determine the value of R_L that will establish maximum power conditions for the Zener diode.
(d) Determine the minimum value of R_L to ensure that the Zener diode is in the "on" state.

8. (a) Design the network of Fig. 3.69 to maintain V_L at 12 V for a load variation (I_L) from 0 to 200 mA. That is, determine R_S and V_Z.
(b) Determine $P_{Z_{max}}$ for the Zener diode of part (a).

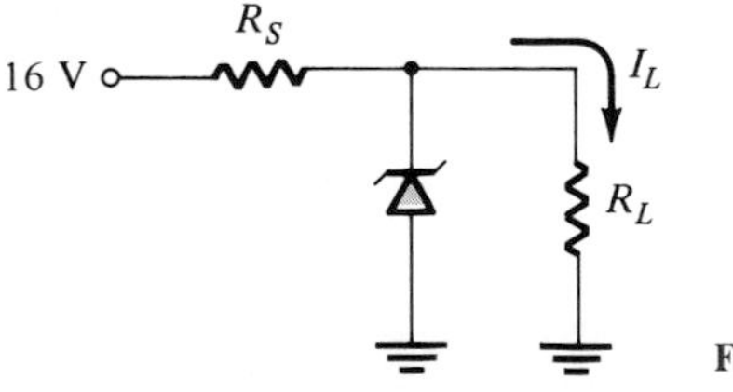

Figure 3.69

9. For the network of Fig. 3.70, determine the range of V_i that will maintain V_L at 8 V and not exceed the maximum power rating of the Zener diode.

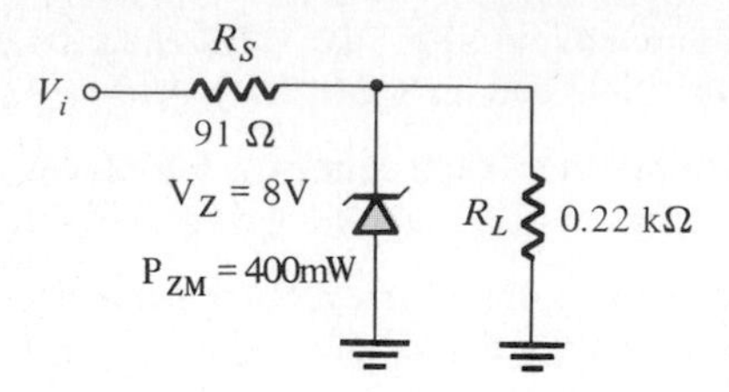

Figure 3.70

10. Design a voltage regulator that will maintain an output voltage of 20 V across a 1-kΩ load with an input that will vary between 30 and 50 V. That is, determine the proper value of R_S and the maximum current I_{ZM}.

11. Sketch the output of the network of Fig. 3.17 if the input is a 50-V square wave. Repeat for a 5-V square wave.

§ 3.4

12. (a) Describe in your own words how the construction of the hot-carrier diode is significantly different from the conventional semiconductor diode.
    (b) In addition, describe its mode of operation.

13. (a) Consult Fig. 3.19. How would you compare the dynamic resistances of the diodes in the forward-bias regions?
    (b) How do they compare at any level of reverse current more negative than I_S?

14. Referring to Fig. 3.22, how does the maximum surge current I_{FSM} relate to the average rectified forward current? Is it typically greater than 20 : 1? Why is it possible to have such high levels of current? What noticeable difference is there in construction as the current rating increases?

15. Referring to Fig. 3.23a, at what temperature is the forward voltage drop 300 mV at a current of 1 mA? Which current levels have the highest levels of temperature coefficients? Assume a linear progression between temperature levels.

16. For the curve of Fig. 3.23b denoted 2900/2303, determine the percent change in I_R for a change in reverse voltage from 5 to 10 V. At what reverse voltage would you expect to reach a reverse current of 1 μA? Note the log scale for I_R.

17. Determine the percent change in capacitance between 0 and 2 V for the 2900/2303 curve of Fig. 3.23c. How does this compare to the change between 8 and 10 V?

§ 3.5

18. (a) Determine the transition capacitance of a diffused junction varicap diode at a reverse potential of 4.2 V if $C(0) = 80$ pF and $V_o = 0.7$ V.
    (b) From the information of part (a), determine the constant K in Eq. (3.13).

19. (a) For a varicap diode having the characteristics of Fig. 3.24, determine the difference in capacitance between reverse-bias potentials of -3 and -12 V.
    (b) Determine the incremental rate of change $(\Delta C/\Delta V_r)$ at $V = -8$ V. How does this value compare with the incremental change determined at -2 V?

20. (a) The resonant frequency of a series RLC network is determined by $f_0 = 1/(2\pi\sqrt{LC})$. Using the value of f_0 and L_S provided in Fig. 3.26, determine the value of C.

(b) How does the value calculated in part (a) compare with that determined by the curve in Fig. 3.27 at $V_R = 25$ V?

21. Referring to Fig. 3.27, determine the ratio of capacitance at $V_R = 3$ V to $V_R = 25$ V and compare to the value of C_3/C_{25} given in Fig. 3.26 (maximum = 6.5).

22. Determine T_1 for a varactor diode if $C_0 = 22$ pF, $TC_C = 0.02\%/°\text{C}$, and $\Delta C = 0.11$ pF due to an increase in temperature above $T_0 = 25°\text{C}$.

23. What region of V_R would appear to have the greatest change in capacitance per change in reverse voltage for the BB139 varactor diode? Be aware that the scales are nonlinear.

24. If $Q = X_L/R = 2\pi fL/R$, determine the figure of merit (Q) at 600 MHz using the fact that $R_S = 0.35\ \Omega$ and $L_S = 2.5$ mH. Comment on the change in Q with frequency and the support or nonsupport of the curve in Fig. 3.27.

§ 3.6

25. Consult a manufacturer's data book and compare the general characteristics of a high-power device (> 10 A) to a low-power unit (< 100 mA). Is there a significant change in the data and characteristics provided? Why?

§ 3.7

26. What are the essential differences between a semiconductor junction diode and a tunnel diode?

27. Note in the equivalent circuit of Fig. 3.30 that the capacitor appears in parallel with the negative resistance. Determine the reactance of the capacitor at 1 MHz and 100 MHz if $C = 5$ pF, and determine the total impedance of the parallel combination (with $R = -152\ \Omega$) at each frequency. Is the magnitude of the inductive reactance anything to be overly concerned about at either of these frequencies if $L_S = 6$ nH?

28. Why do you believe the maximum reverse current rating for the tunnel diode can be greater than the forward current rating? (*Hint:* Note the characteristics and consider the power rating.)

§ 3.8

29. Determine the energy associated with the photons of green light if the wavelength is 5000 Å. Give your answer in joules and electron volts.

30. (a) Referring to Fig. 3.33, what would appear to be the frequencies associated with the upper and lower limits of the visible spectrum?
(b) What is the wavelength in microns associated with the peak relative response of silicon?
(c) If we define the bandwidth of the spectral response of each material to occur at 70% of its peak level, what is the bandwidth of silicon?

31. Referring to Fig. 3.35, determine I_λ if $V_\lambda = 30$ V and the light intensity is 4×10^{-9} W/m^2.

32. (a) Which material of Fig. 3.33 would appear to provide the best response to yellow, red, green, and infrared (less than 11,000 Å) light sources?
(b) At a frequency of 0.5×10^{15} Hz, which color has the maximum spectral response?

33. Determine the voltage drop across the resistor of Fig. 3.34 if the incident flux is 3000 fc, $V_\lambda = 25$ V, and $R = 100$ kΩ. Use the characteristics of Fig. 3.35.

§ 3.9

34. What is the approximate rate of change of resistance with illumination for a photoconductive cell with the characteristics of Fig. 3.39 for the ranges (a) 0.1→1 kΩ, (b) 1→10 kΩ, and (c) 10→kΩ? (Note that this is a log scale.) Which region has the greatest rate of change in resistance with illumination?

35. What is the "dark current" of a photodiode?

36. If the illumination on the photoconductive diode in Fig. 3.40 is 10 fc, determine the magnitude of V_i to establish 6 V across the cell if R_1 is equal to 5 kΩ. Use the characteristics of Fig. 3.39.

37. Using the data provided in Fig. 3.41, sketch a curve of percent conductance versus temperature for 0.01, 1.0, and 100 fc. Are there any noticeable effects?

38. (a) Sketch a curve of rise time versus illumination using the data from Fig. 3.41.
    (b) Repeat part (a) for the decay time.
    (c) Discuss any noticeable effects of illumination in parts (a) and (b).

39. Which colors is the CdS unit of Fig. 3.41 most sensitive to?

§ 3.10

40. (a) Determine the radiant flux at a dc forward current of 70 mA for the device of Fig. 3.43.
    (b) Determine the radiant flux in lumens at a dc forward current of 45 mA.

41. (a) Through the use of Fig. 3.44, determine the relative radiant intensity at an angle of 25° for a package with a flat glass window.
    (b) Plot a curve of relative radiant intensity versus degrees for the flat package.

42. If 60 mA of dc forward current is applied to an SG1010A IR emitter, what will be the incident radiant flux in lumens 5° off the center if the package has an internal collimating system? Refer to Figs. 3.43 and 3.44.

§ 3.11

43. (a) Convert the horizontal scale of Fig. 3.47e to angstrom units.
    (b) How do the peak values of the relative intensity fall into the color bands provided in Fig. 3.33?

44. Referring to Fig. 3.47f, what would appear to be an appropriate value of V_T for this device? How does it compare to the value of V_T for silicon and germanium?

45. Using the information provided in Fig. 3.47, determine the forward voltage across the diode if the relative luminous intensity is 1.5.

46. (a) What is the percent increase in relative efficiency of the device of Fig. 3.47 if the peak current is increased from 5 to 10 mA?
    (b) Repeat part (a) for 30 to 35 mA (the same increase in current).
    (c) Compare the percent increase from parts (a) and (b). At what point on the curve would you say there is little gained by further increasing the peak current?

47. (a) Referring to Fig. 3.47i, determine the maximum tolerable peak current if the period of the pulse duration is 1 ms, the frequency is 300 Hz, and the maximum tolerable dc current is 20 mA.
    (b) Repeat part (a) for a frequency of 100 Hz.

48. (a) If the luminous intensity at 0° angular displacement is 3.0 mcd for the device of Fig. 3.47, at what angle will it be 0.75 mcd?
    (b) At what angle does the loss of luminous intensity drop below the 50% level?

49. Sketch the current derating curve for the average forward current of the High Eff. Red LED of Fig. 3.47 as determined by temperature. (Note the absolute maximum ratings.)

§ 3.12

50. Referring to Fig. 3.52, which terminals must be energized to display number 7?

51. In your own words, describe the basic operation of an LCD.

52. Discuss the relative differences in mode of operation between an LED and an LCD display.

53. What are the relative advantages and disadvantages of an LCD display as compared to an LED display?

§ 3.13

54. A 1 cm by 2 cm solar cell has a conversion efficiency of 9%. Determine the maximum power rating of the device.

55. If the power rating of a solar cell is determined on a very rough scale by the product $V_{oc} I_{sc}$, is the greatest rate of increase obtained at lower or higher levels of illumination? Explain your reasoning.

56. (a) Referring to Fig. 3.63, what power density is required to establish a current of 24 mA at an output voltage of 0.25 V?
    (b) Why is 100 mW/cm^2 the maximum power density in Fig. 3.63?
    (c) Determine the output current if the power is 40 mW/cm^2 and the output voltage is 0.3 V.

57. (a) Sketch a curve of output current versus power density at an output voltage of 0.15 V using the characteristics of Fig. 3.63.
    (b) Sketch a curve of output voltage versus power density at a current of 19 mA.
    (c) Is either of the curves from parts (a) and (b) linear within the limits of the maximum power limitation?

§ 3.14

58. For the thermistor of Fig. 3.64, determine the dynamic rate of change in specific resistance with temperature at $T = 20°C$. How does this compare to the value determined at $T = 300°C$? From the results, determine whether the greatest change in resistance per unit change in temperature occurs at lower or higher levels of temperature. Note the vertical log scale.

59. Using the information provided in Fig. 3.64, determine the total resistance of a 2-cm length of the material having a perpendicular surface area of 1 cm^2 at a temperature of 0°C. Note the vertical log scale.

60. (a) Referring to Fig. 3.65, determine the current at which a 25°C sample of the material changes from a positive to negative temperature coefficient. (Figure 3.65 is a log scale.)

(b) Determine the power and resistance levels of the device (Fig. 3.65) at the peak of the 0°C curve.

(c) At a temperature of 25°C, determine the power rating if the resistance level is 1 MΩ.

61. In Fig. 3.67, $V = 0.2$ V and $R_{\text{variable}} = 10\ \Omega$. If the current through the sensitive movement is 2 mA and the voltage drop across the movement is 0 V, what is the resistance of the thermistor?

4

Bipolar Junction Transistors

4.1 INTRODUCTION

During the period 1904–1947, the vacuum tube was undoubtedly the electronic device of interest and development. In 1904, the vacuum-tube diode was introduced by J. A. Fleming. Shortly thereafter, in 1906, Lee De Forest added a third element, called the *control grid*, to the vacuum diode, resulting in the first amplifier, the *triode*. In the following years, radio and television provided great stimulation to the tube industry. Production rose from about 1 million tubes in 1922 to about 100 million in 1937. In the early 1930s the four-element tetrode and five-element pentode gained prominence in the electron-tube industry. In the years to follow, the industry became one of primary importance and rapid advances were made in design, manufacturing techniques, high-power and high-frequency applications, and miniaturization.

On December 23, 1947, however, the electronics industry was to experience the advent of a completely new direction of interest and development. It was on the afternoon of this day that Walter H. Brattain and John Bardeen demonstrated the amplifying action of the first transistor at the Bell Telephone Laboratories. The original transistor (a point-contact transistor) is shown in Fig. 4.1. The advantages of this three-terminal solid-state device over the tube were immediately obvious: it was smaller and lightweight; had no heater requirement or heater loss; had rugged construction; and was more efficient since less power was absorbed by the device itself; it was instantly available for use, requiring no warm-up period; and lower operating voltages were possible. Note in the discussion above that this chapter is our first discussion of devices with three or more terminals. You will find that all amplifiers

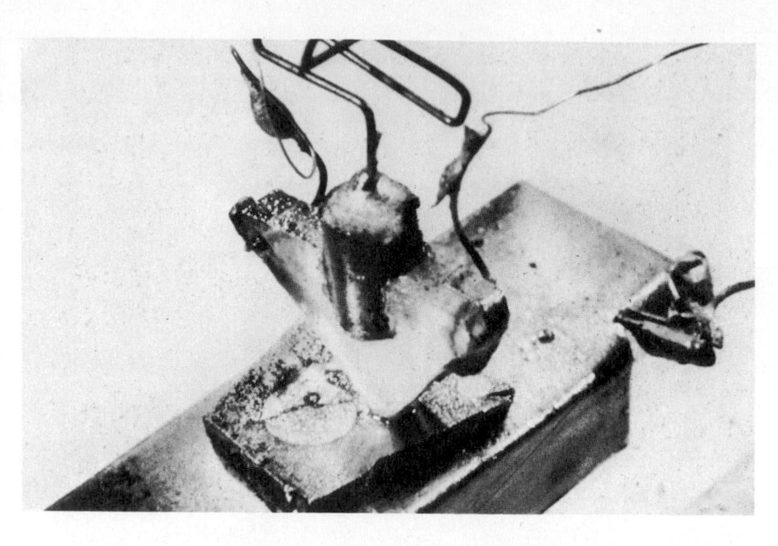

Figure 4.1 The first transistor. (Courtesy Bell Telephone Laboratories.)

(devices that increase the voltage, current, or power level) will have at least three terminals with one controlling the flow between two other terminals.

4.2 TRANSISTOR CONSTRUCTION

The transistor is a three-layer semiconductor device consisting of either two *n*- and one *p*-type layers of material or two *p*- and one *n*-type layers of material. The former is called an *npn transistor*, while the latter is called a *pnp transistor*. Both are shown in Fig. 4.2 with the proper dc biasing. We will find in Chapter 5 that the dc biasing is necessary to establish the proper region of operation for ac amplification. The outer layers of the transistor are heavily doped semiconductor materials having widths much greater than those of the sandwiched *p*- or *n*-type material. For the transistors shown in Fig. 4.2 the ratio of the total width to that of the center layer is 0.150/0.001 = 150:1. The doping of the sandwiched layer is also considerably less than that of the outer layers (typically 10:1 or less). This lower doping level decreases the conductivity (increases the resistance) of this material by limiting the number of "free" carriers.

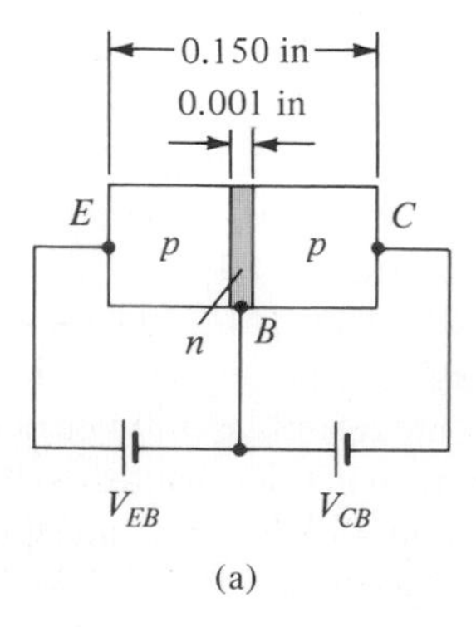

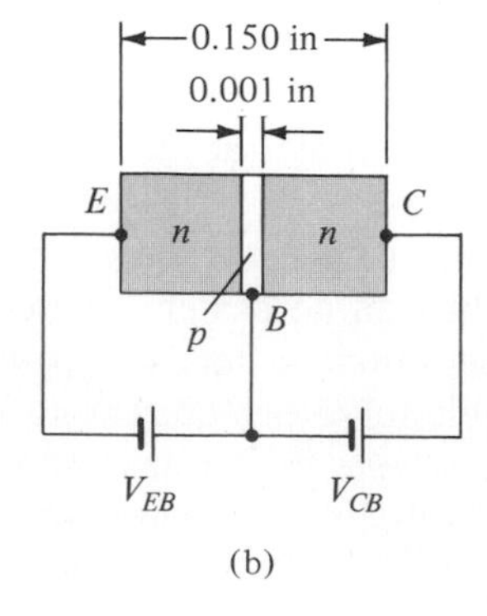

Figure 4.2 Types of transistors: (a) *pnp*; (b) *npn*.

For the biasing shown in Fig. 4.2 the terminals have been indicated by the capital letters E for *emitter*, C for *collector*, and B for *base*. An appreciation for this choice of notation will develop when we discuss the basic operation of the transistor.

The abbreviation BJT, from *bipolar junction transistor*, is often applied to this three-terminal device. The term *bipolar* reflects the fact that holes *and* electrons participate in the injection process into the oppositely polarized material. If only one carrier is employed (electron or hole), it is considered a *unipolar* device. Recall that the Schottky diode was such a device.

4.3 TRANSISTOR OPERATION

The basic operation of the transistor will now be described using the *pnp* transistor of Fig. 4.2a. The operation of the *npn* transistor is exactly the same if the roles played by the electron and hole are interchanged.

In Fig. 4.3 the *pnp* transistor has been redrawn without the base-to-collector bias. Note the similarities between this situation and that of the *forward-biased* diode in Chapter 1. The depletion region has been reduced in width due to the applied bias, resulting a heavy flow of majority carriers from the *p*- to the *n*-type material.

Let us now remove the base-to-emitter bias of the *pnp* transistor of Fig. 4.2a as shown in Fig. 4.4. Consider the similarities between this situation and that of the *reverse-biased* diode of Section 1.6. Recall that the flow of majority carriers is zero, resulting in only a minority-carrier flow, as indicated in Fig. 4.4.

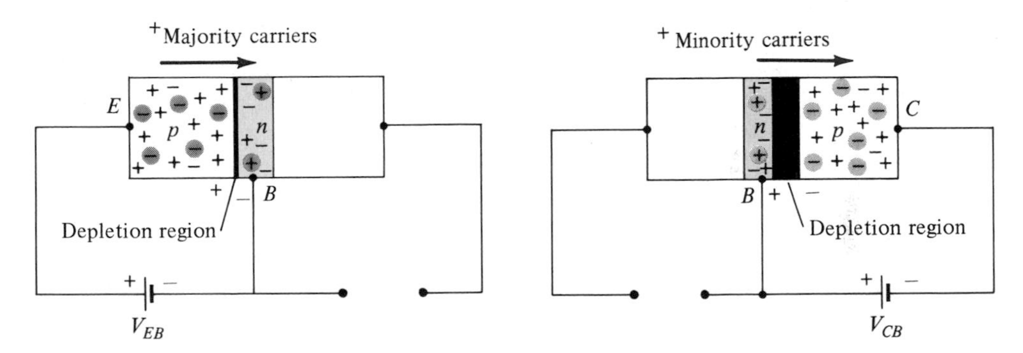

Figure 4.3 Forward-biased junction of a *pnp* transistor.

Figure 4.4 Reverse-biased junction of a *pnp* transistor.

In summary, therefore, one p-n junction of a transistor is reverse biased, while the other is forward biased.

In Fig. 4.5 both biasing potentials have been applied to a *pnp* transistor, with the resulting majority- and minority-carrier flow indicated. Note in Fig. 4.5 the widths of the depletion regions, indicating clearly which junction is forward-biased and which is reverse-biased. As indicated in Fig. 4.5, a large number of majority carriers will diffuse across the forward-biased *p-n* junction into the *n*-type material. The question then is whether these carriers will contribute directly to the base current I_B or pass

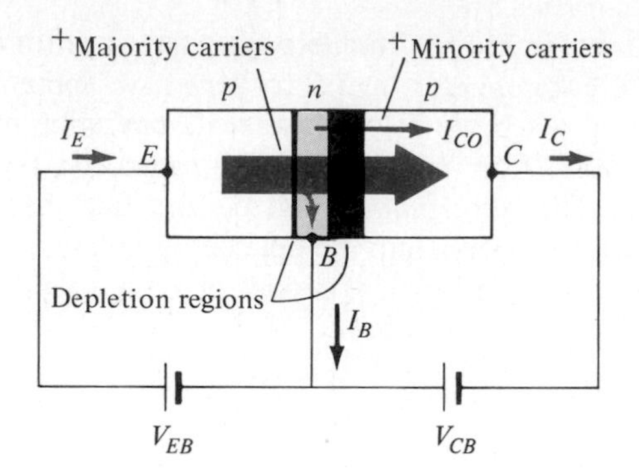

Figure 4.5 Majority and minority carrier flow of a *pnp* transistor.

directly into the *p*-type material. Since the sandwiched *n*-type material is very thin and has a low conductivity, a very small number of these carriers will take this path of high resistance to the base terminal. The magnitude of the base current is typically on the order of microamperes as compared to milliamperes for the emitter and collector currents. The larger number of these majority carriers will diffuse across the reverse-biased junction into the *p*-type material connected to the collector terminal as indicated in Fig. 4.5. The reason for the relative ease with which the majority carriers can cross the reverse-biased junction is easily understood if we consider that for the reverse-biased diode the injected majority carriers will appear as minority carriers in the *n*-type material. In other words, there has been an *injection* of minority carriers into the *n*-type base region material. Combining this with the fact that all the minority carriers in the depletion region will cross the reverse-biased junction of a diode accounts for the flow indicated in Fig. 4.5.

Applying Kirchhoff's current law to the transistor of Fig. 4.5 as if it were a single node, we obtain

$$\boxed{I_E = I_C + I_B} \tag{4.1}$$

and find that the emitter current is the sum of the collector and base currents. The collector current, however, is comprised of two components—the majority and minority carriers as indicated in Fig. 4.5. The minority-current component is called the *leakage current* and is given the symbol I_{CO} (I_C current with emitter terminal *O*pen). The collector current, therefore, is determined in total by Eq. (4.2.).

$$\boxed{I_C = I_{C_{\text{majority}}} + I_{CO_{\text{minority}}}} \tag{4.2}$$

For general-purpose transistors, I_C is measured in milliamperes, while I_{CO} is measured in microamperes or nanoamperes. I_{CO}, like I_S for a reverse-biased diode, is temperature sensitive and must be examined carefully when applications of wide temperature ranges are considered. It can severely affect the stability of a system at high temperatures if not considered properly.

Improvements in construction techniques have resulted in significantly lower levels of I_{CO}, to the point where its effect can often be ignored.

The configuration shown in Fig. 4.2 for the *pnp* and *npn* transistors is called the *common-base* configuration since the base is common to both input (emitter) and output (collector) terminals. For fixed values of V_{CB} in the common-base configuration the ratio of a small change in I_C to a small change in I_E is commonly called the *common-base, short-circuit amplification factor* and is given by the symbol α (alpha). In equation form, the magnitude of α is given by

$$\alpha = \left.\frac{\Delta I_C}{\Delta I_E}\right|_{V_{CB} = \text{constant}} \tag{4.3}$$

The term "short circuit" indicates that the load is short-circuited when α is determined. More will be said about the necessity for shorting the load and the operations involved with using equations of the type indicated by Eq. (4.3) when we consider equivalent circuits in Chapter 8. Typical values of α vary from 0.90 to 0.998. For most practical applications, a first approximation for the magnitude of α, usually correct to within a few percent, can be obtained using the following equation:

$$\alpha \cong \frac{I_C}{I_E} \tag{4.4}$$

where I_C and I_E are the magnitude of the collector and emitter currents, respectively, at a particular point on the transistor characteristics.

Equations (4.3) and (4.4) are employed to determine α from the device characteristics or network conditions. However, in the strictest sense, α is only a measure of the percentage of holes (majority carriers) originating in the emitter *p*-material of Fig. 4.5 that reach the collector terminal. As defined by Eq. (4.2), therefore,

$$I_C = \alpha I_{E_{\text{majority}}} + I_{CO_{\text{minority}}} \tag{4.5}$$

4.4 TRANSISTOR AMPLIFYING ACTION

The basic voltage-amplifying action of the common-base configuration can now be described using the network of Fig. 4.6. The dc biasing does not appear in the figure since our interest will be limited to the ac response. For the common-base configuration, the input resistance between the emitter and the base of a transistor will typically vary from 10 to 100 Ω, while the output resistance may vary from 100 kΩ

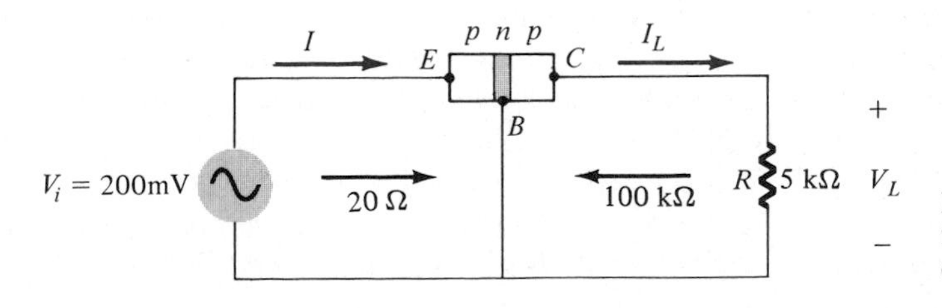

Figure 4.6 Basic voltage amplification action of the common-base configuration.

to 1 MΩ. The difference in resistance is due to the forward-biased junction at the input (base to emitter) and the reverse-biased junction at the output (base to collector). Using effective values and a common value of 20 Ω for the input resistance, we find

$$I = \frac{V_i}{R_i} = \frac{200 \times 10^{-3}}{20} = 10 \text{ mA}$$

If we assume for the moment that $\alpha = 1$ ($I_C = I_E$),

$$I_L = I = 10 \text{ mA}$$

and

$$\begin{aligned} V_L &= I_L R \\ &= (10 \times 10^{-3})(5 \times 10^{+3}) \\ &= 50 \text{ V} \end{aligned}$$

The voltage amplification is

$$A_v = \frac{V_L}{V_i} = \frac{50}{200 \times 10^{-3}} = 250$$

Typical values of voltage amplification for the common-base configuration vary from 50 to 300. The current amplification (I_C/I_E) is always less than 1 for the common-base configuration. This latter characteristic should be obvious since $I_C = \alpha I_E$ and α is always less than 1.

The basic amplifying action was produced by *transferring* a current I from a low- to a high-*resistance* circuit. The combination of the two terms in italics results in the label *transistor*; that is,

*tran*sfer + re*sistor* → *transistor*

4.5 COMMON-BASE CONFIGURATION

The notation and symbols used in conjunction with the transistor in the majority of texts and manuals published today are indicated in Fig. 4.7 for the common-base configuration with *pnp* and *npn* transistors. The common-base terminology is derived from the fact that the base is common to both the input and output sides of the configuration. Note the appearance of the letter B in both voltage sources and the fact that the base is the grounded or "supporting" terminal of the configuration. Throughout this text all current directions will refer to the conventional (hole flow) rather than the electron flow. This choice was based primarily on the fact that the vast majority of past and present publications in the field use conventional current.

Some texts prefer to show all the currents entering in Fig. 4.7 when they describe the basic operation of the transistor and simply include negative signs when appropriate. In other words, if the actual conventional flow direction is the opposite direction, a negative sign is included along with the magnitude. For clarity, all currents, as indicated in Fig. 4.7, will indicate the actual flow direction for the active region. *Note that the arrow in the symbol is the same as the direction of I_E (only true for conventional flow).* On specification sheets negative signs indicate that all currents are entering.

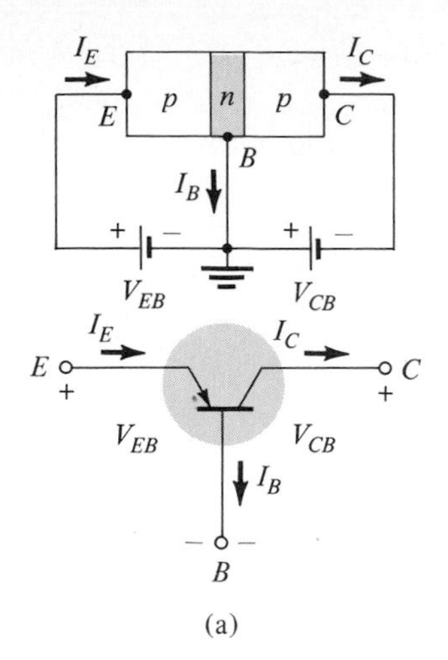

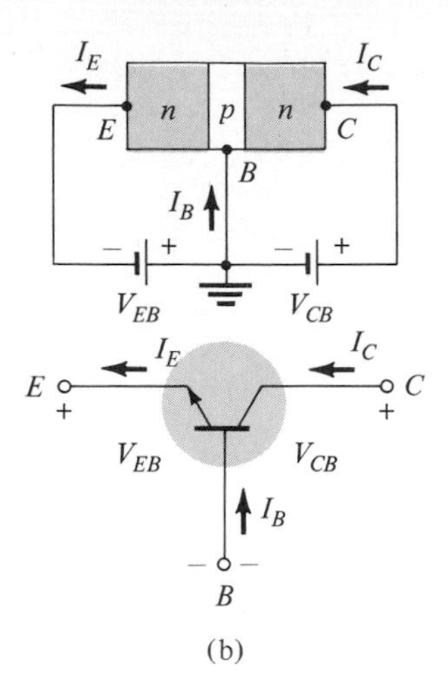

Figure 4.7 Notation and symbols used with the common-base configuration: (a) *pnp* transistor; (b) *npn* transistor.

For the common-base configuration the applied potentials are written with respect to the base potential resulting in V_{EB} and V_{CB}. In other words, the second subscript will always indicate the transistor configuration. In all cases the first subscript is defined to be the point of higher potential, as shown in Fig. 4.7. For the *pnp* transistor, therefore, V_{EB} is positive and V_{CB} is negative (since the battery V_{CB} sets the collector at the lower potential), as indicated on the characteristics of Fig. 4.8. For the *npn* transistor V_{EB} is negative and V_{CB} is positive.

In addition, note that two sets of characteristics are necessary to represent the behavior of the *pnp* common-base transistor of Fig. 4.7: the *driving point* (or input) and the *output* set.

The output or collector characteristics of Fig. 4.8a relate the collector (output) current to the collector-to-base (output) voltage and (input) emitter current. The collector characteristics have three basic regions of interest, as indicated in Fig. 4.8a: the *active*, *cutoff*, and *saturation* regions.

In the active region the collector junction is reverse-biased, while the emitter junction is forward-biased.

These conditions refer to the situation of Fig. 4.5. The active region is the only region employed for the amplification of signals with minimum distortion. When the emitter current (I_E) is zero, the collector current is simply that due to the reverse saturation current I_{CO}, as indicated in Fig. 4.8a. The current I_{CO} is so small (microamperes) in magnitude compared to the vertical scale of I_C (milliamperes) that it appears on virtually the same horizontal line as $I_C = 0$. The circuit conditions that exist when $I_E = 0$ for the common-base configuration are shown in Fig. 4.9. The notation most frequently used for I_{CO} on data and specification sheets is, as indicated in Fig. 4.9, I_{CBO}. Because of improved construction techniques, the level of I_{CBO} for general-purpose transistors (especially silicon) in the low- and midpower ranges is usually so low that its effect can be ignored. However, for higher power units I_{CBO} will still appear in the microampere range. In addition, keep in mind that I_{CBO} like I_S for the diode (both reverse leakage currents) is temperature sensitive. At higher temperatures the effect of I_{CBO} for any power level unit may become an important factor since it increases so rapidly with temperature.

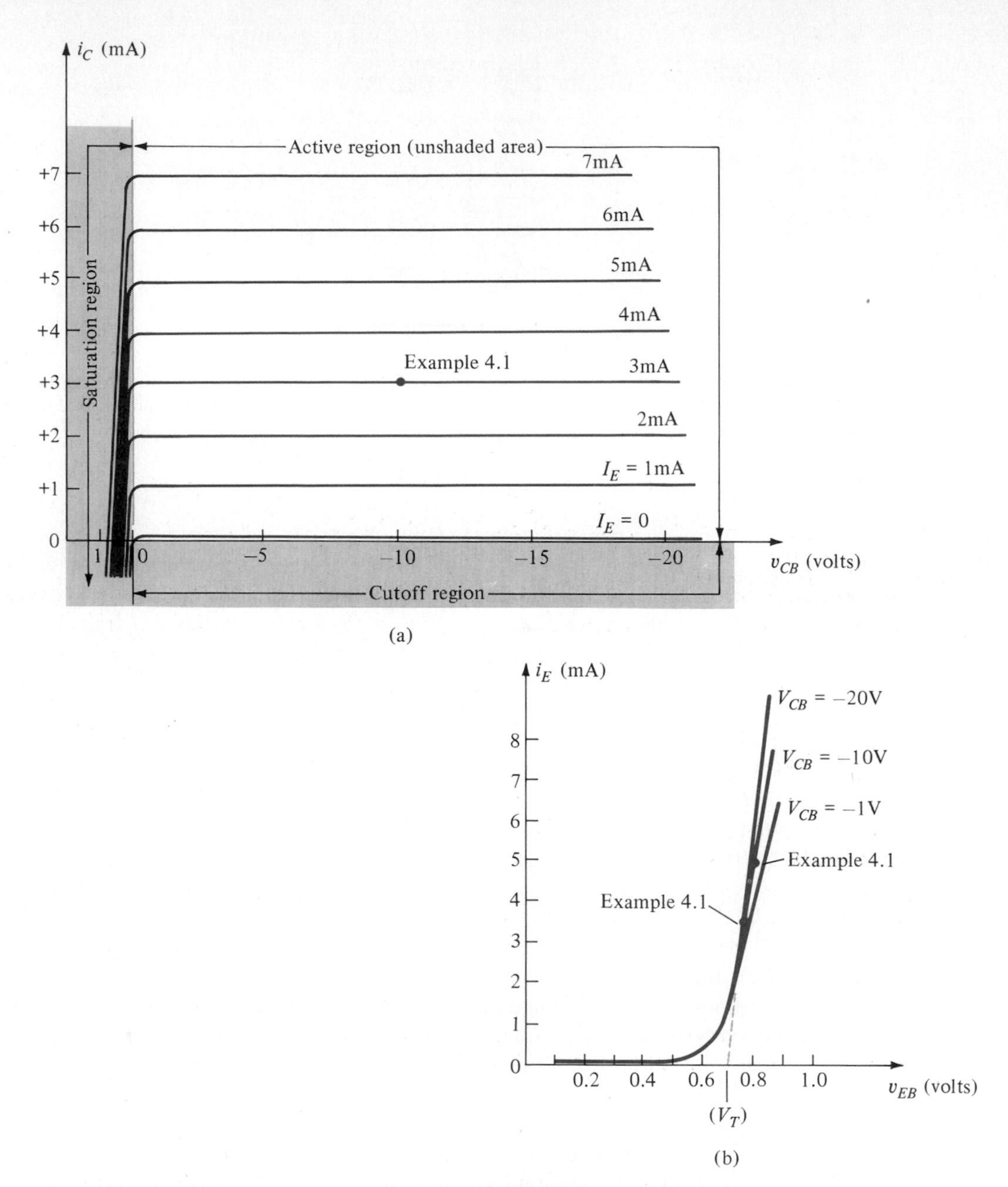

Figure 4.8 Characteristics of a *pnp* transistor in the common-base configuration: (a) collector or output characteristics; (b) emitter or input characteristics.

Note in Fig. 4.8a that as the emitter current increases above zero, the collector current increases to a magnitude essentially equal to that of the emitter current as determined by the basic transistor-current relations. Note also the almost negligible effect of V_{CB} on the collector current for the active region. The curves clearly indicate that *a first approximation to the relationship between I_E and I_C in the active region is*

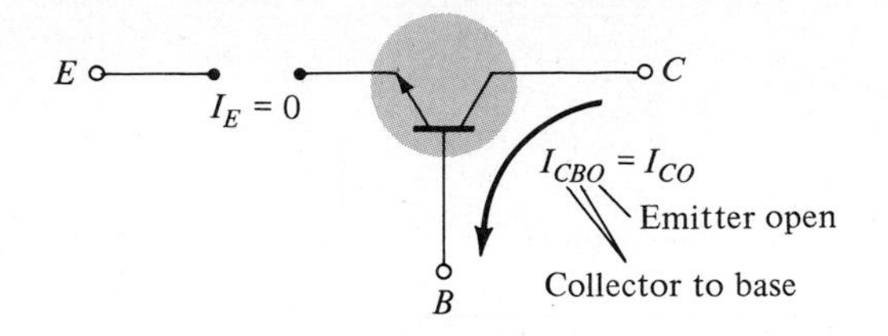

Figure 4.9 Reverse saturation current.

given by

$$\boxed{I_C \cong I_E} \tag{4.6}$$

In the cutoff region the collector and emitter junctions are both reverse-biased, resulting in negligible collector current as demonstrated in Fig. 4.8a.

The horizontal scale for V_{CB} has been expanded to the left of 0 V to clearly show the characteristics of this region. *In the region called the* saturation region, *the collector and emitter junctions are forward-biased,* resulting in the exponential change in collector current with small changes in collector-to-base potential.

The input or emitter characteristics have only one region of interest, as illustrated by Fig. 4.8b. For fixed values of collector voltage (V_{CB}), as the emitter-to-base potential increases, the emitter current increases, as shown. Increasing levels of V_{CB} result in a reduced level of V_{EB} to establish the same current. Note the tight grouping of the curves for the wide range of values for V_{CB}. In addition, consider how closely the average value of the curves appears to begin its rise or about $V_T = 0.7$ V for the silicon transistor.

As with the semiconductor silicon diode, *a first approximation for the forward-biased base-emitter junction in the dc mode would be that*

$$\boxed{V_{EB} \cong 0.7 \text{ V}} \tag{4.7}$$

EXAMPLE 4.1

Using the characteristics of Fig. 4.8:
(a) Find the resulting collector current if $I_E = 3$ mA and $V_{CB} = -10$ V.
(b) Determine I_C if $V_{EB} = 750$ mV and $V_{CB} = -10$ V.
(c) Find V_{EB} for the conditions $I_C = 5$ mA and $V_{CB} = -1$ V.

Solution:

(a) $I_C \cong I_E =$ **3 mA.**
(b) On the input characteristics $I_E = 3.5$ mA at the intersection of $V_{EB} = 750$ mV, $V_{CB} = -10$ V, and $I_C \cong I_E =$ **3.5 mA.**
(c) $I_E \cong I_C = 5$ mA. On the input characteristics the intersection of $I_E = 5$ mA and $V_{CB} = -1$ V results in $V_{EB} \cong 800$ mV $=$ **0.8 V.**

The proper biasing of the common-base can be quickly determined using the approximation $I_C \cong I_E$ and assuming for the moment that $I_B \cong 0$ μA. The result is the configuration of Fig. 4.10 for the *pnp* transistor. The arrow of the symbol defines the direction of conventional flow for $I_E \cong I_C$. The dc supplies are then inserted with

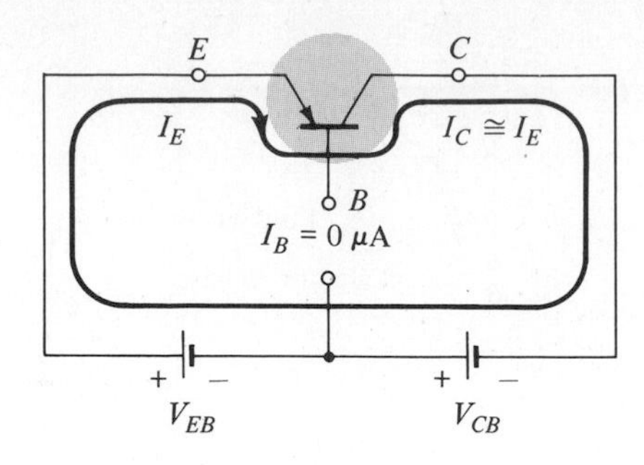

Figure 4.10

a polarity that will support the resulting current direction. For the *npn* transistor the polarities will be reversed.

Some students feel that they can remember whether the arrow of the device symbol in pointing in or out by matching the letters of the transistor type with the appropriate letters of the phrases "pointing in" or "not pointing in." For instance, there is a match between the letters *npn* and the italic letters of *n*ot *p*ointing i*n* and the letters *pnp* with *p*ointing i*n*.

4.6 COMMON-EMITTER CONFIGURATION

The most frequently encountered transistor configuration is shown in Fig. 4.11 for the *pnp* and *npn* transistors. It is called the *common-emitter configuration* since the emitter is common to both the input and output terminals (in this case, also common to both the base and collector terminals). Two sets of characteristics are again necessary to describe fully the behavior of the common-emitter configuration: one for the input or base circuit and one for the output or collector circuit. Both are shown in Fig. 4.12.

The emitter, collector, and base currents are shown in their actual conventional current direction, while the potentials have the capital letter *E* as the second subscript to indicate the configuration. Even though the transistor configuration has changed, the current relations developed earlier for the common-base configuration are still applicable.

For the common-emitter configuration the output characteristics will be a plot of the output current (I_C) versus output voltage (V_{CE}) for a range of values of input current (I_B). The input characteristics are a plot of the input current (I_B) versus the input voltage (V_{BE}) for a range of values of output voltage (V_{CE}).

Note that on the characteristics of Fig. 4.12 the magnitude of I_B is in microamperes as compared to milliamperes of I_C. Consider also that the curves of I_B are not as horizontal as those obtained for I_E in the common-base configuration, indicating that the collector-to-emitter voltage will influence the magnitude of the collector current.

The active region for the common-emitter configuration is that portion of the upper-right quadrant that has the greatest linearity, that is, that region in which the curves for I_B are nearly straight and equally spaced. In Fig. 4.12a this region exists to the right of the vertical dashed line at $V_{CE_{sat}}$ and above the curve for I_B equal to zero. The region to the left of $V_{CE_{sat}}$ is called the saturation region. *In the active region the collector junction is reverse biased, while the emitter junction is forward-biased.* You

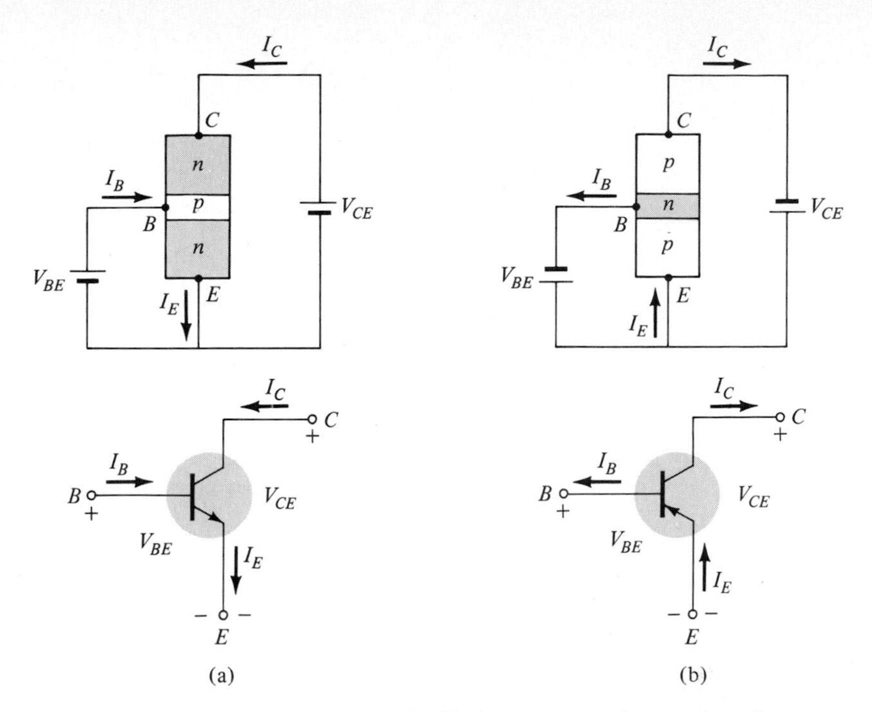

Figure 4.11 Notation and symbols used with the common emitter configuration: (a) *npn* transistor; (b) *pnp* transistor.

will recall that these were the same conditions that existed in the active region of the common-base configuration. The active region of the common-emitter configuration can be employed for voltage, current, or power amplification.

The cutoff region for the common-emitter configuration is not as well defined as for the common-base configuration. Note on the collector characteristics of Fig. 4.12 that I_C is not equal to zero when I_B is zero. For the common-base configuration, when the input current I_E was equal to zero, the collector current was equal only to the reverse saturation current I_{CO}, so that the curve $I_E = 0$ and the voltage axis were, for all practical purposes, one.

The reason for this difference in collector characteristics can be derived through the proper manipulation of Eqs. (4.1) and (4.5). That is,

$$I_C = \alpha I_E + I_{CO} \qquad \text{[Eq. (4.5)]}$$

but

$$I_E = I_C + I_B \qquad \text{[Eq. (4.1)]}$$

Therefore,

$$I_C = \alpha(I_C + I_B) + I_{CO} = \alpha I_C + \alpha I_B + I_{CO}$$

and

$$I_C(1 - \alpha) = \alpha I_B + I_{CO}$$

with

$$\boxed{I_C = \frac{\alpha I_B}{1 - \alpha} + \frac{I_{CO}}{1 - \alpha}} \qquad (4.8)$$

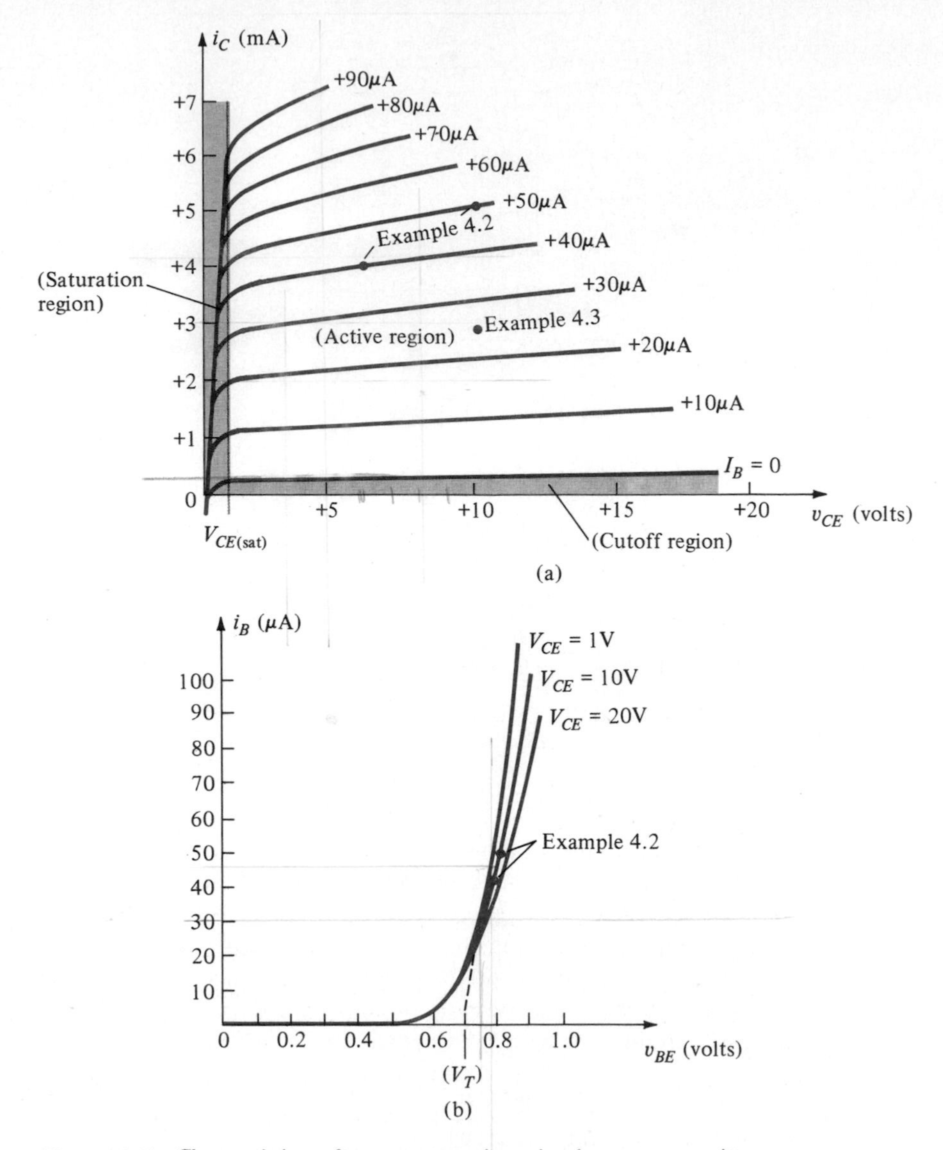

Figure 4.12 Characteristics of an *npn* transistor in the common-emitter configuration: (a) collector characteristics; (b) base characteristics.

If we consider the case discussed earlier, where $I_B = 0$, and substitute this value in Eq. (4.8), then

$$\boxed{I_C = \frac{I_{CO}}{1 - \alpha}\bigg|_{I_B=0}} \tag{4.9}$$

For $\alpha = 0.996$,

$$I_C = \frac{I_{CO}}{1 - 0.996} = \frac{I_{CO}}{0.004}$$

and $I_C = 250\, I_{CO}\big|_{I_B=0}$

which accounts for the vertical shift in the $I_B = 0$ curve from the horizontal voltage axis. For future reference, the collector current defined by Eq. (4.9) will be assigned the notation indicated by Eq. (4.10).

$$\boxed{I_{CEO} = \frac{I_{CO}}{1-\alpha}\bigg|_{I_B=0}} \tag{4.10}$$

In Fig. 4.13 the conditions surrounding this newly defined current are demonstrated with its assigned reference direction.

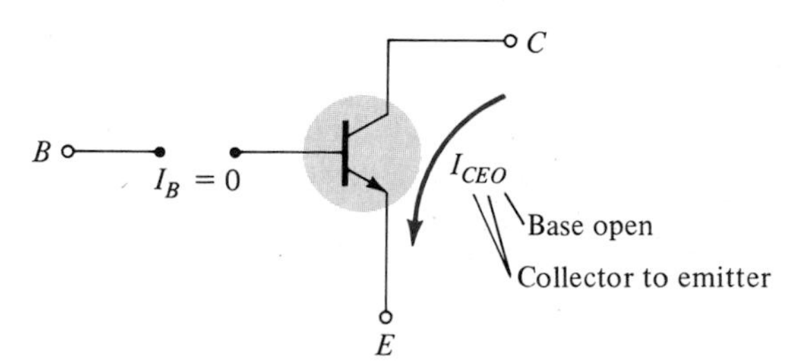

Figure 4.13 Circuit conditions related to I_{CEO}.

The magnitude of I_{CEO} is typically much smaller for silicon materials than for germanium materials. For transistors with similar ratings I_{CEO} would typically be a few microamperes for silicon but perhaps a few hundred microamperes for germanium.

For linear (least distortion) amplification purposes, cutoff for the common-emitter configuration will be (for this text) determined by $I_C = I_{CEO}$. In other words, the region below $I_B = 0$ is to be avoided if an undistorted output signal is required.

When employed as a switch in the logic circuitry of a computer, a transistor will have two points of operation of interest: one in the cutoff and one in the saturation region. The cutoff condition should ideally be $I_C = 0$ for the chosen V_{CE} voltage. Since I_{CEO} is typically low in magnitude for silicon materials, *cutoff will exist for switching purposes when* $I_B = 0$ *or* $I_C = I_{CEO}$ *for silicon transistors only. For germanium transistors, however, cutoff for switching purposes will be defined as those conditions that exist when* $I_C = I_{CBO} = I_{CO}$. This condition can normally be obtained for germanium transistors by reverse-biasing the normally forward-biased base-to-emitter junction a few tenths of a volt.

EXAMPLE 4.2

Using the characteristics of Fig. 4.12:
(a) Find the value of I_C corresponding to $V_{BE} = +800$ mV and $V_{CE} = +10$ V.
(b) Find the value of V_{CE} and V_{BE} corresponding to $I_C = +4$ mA and $I_B = +40$ μA.

Solution:

(a) On the input characteristics the intersection of $V_{BE} = +800$ mV and $V_{CE} = +10$ V results in

$$I_B \cong 50\ \mu\text{A}$$

On the output characteristics the intersection of $I_B = 50\ \mu A$ and $V_{CE} = 10$ V results in

$$I_C \cong \mathbf{5.1\ mA}$$

(b) On the output characteristics the intersection of $I_C = +4$ mA and $I_B = +40\ \mu A$ results in

$$V_{CE} = \mathbf{+6.2\ V}$$

On the input characteristics the intersection of $I_B = +40\ \mu A$ and $V_{CE} = +6.2$ V results in

$$V_{BE} \cong \mathbf{770\ mV}$$

In Section 4.3 the symbol alpha (α) was assigned to the forward current transfer ratio of the common-base configuration. For the common-emitter configuration, the ratio of a small change in collector current to the corresponding change in base current at a fixed collector-to-emitter voltage (V_{CE}) is assigned the Greek letter beta (β) and is commonly called the *common-emitter forward-current amplification factor*. In equation form, the magnitude of β is given by

$$\beta = \left.\frac{\Delta I_C}{\Delta I_B}\right|_{V_{CE}=\text{constant}} \tag{4.11}$$

As a first, but close, approximation, the magnitude of beta (β) can be determined by the following equation:

$$\beta \cong \frac{I_C}{I_B} \tag{4.12}$$

where I_C and I_B are collector and base currents of a particular operating point in the linear region (i.e., where the horizontal base current lines of the common-emitter characteristics are closest to being parallel and equally spaced). Since I_C and I_B in Eq. (4.12) are fixed or dc values, the value obtained for β from Eq. (4.12) is frequently called the *dc beta*, while that obtained by Eq. (4.11) is called the *ac* or *dynamic* value. Typical values of β vary from 20 to 600. Through the following manipulations of Eqs. (4.1), (4.4), (4.12):

[Eq. (4.12)] $\qquad \beta = \dfrac{I_C}{I_B}$ resulting in $I_B = \dfrac{I_C}{\beta}$

[Eq. (4.4)] $\qquad \alpha = \dfrac{I_C}{I_E}$ resulting in $I_E = \dfrac{I_C}{\alpha}$

[Eq. (4.1)] $\qquad I_E = I_C + I_B$

Substituting: $\qquad \dfrac{I_C}{\alpha} = I_C + \dfrac{I_C}{\beta}$

and dividing by I_C: $\qquad \dfrac{1}{\alpha} = 1 + \dfrac{1}{\beta}$

and $$\beta = \alpha\beta + \alpha$$

or $$\beta(1 - \alpha) = \alpha$$

we obtain $$\boxed{\beta = \frac{\alpha}{1 - \alpha}} \tag{4.13}$$

or $$\boxed{\alpha = \frac{\beta}{\beta + 1}} \tag{4.14}$$

In addition, since

$$I_{CEO} = \frac{I_{CO}}{1 - \alpha} = \frac{I_{CBO}}{1 - \alpha}$$

then $$\boxed{I_{CEO} = (\beta + 1)I_{CBO} \cong \beta I_{CBO}} \tag{4.15}$$

EXAMPLE 4.3

(a) Find the dc beta at an operating point of $V_{CE} = +10$ V and $I_C = +3$ mA on the characteristics of Fig. 4.12.
(b) Find the value of α corresonding with this operating point.
(c) At $V_{CE} = +10$ V find the corresponding value of I_{CEO}.
(d) Calculate the approximate value of I_{CBO} using the β_{dc} obtained in part (a).

Solution:

(a) At the intersection of $V_{CE} = +10$ V and $I_C = +3$ mA, $I_B = +25\ \mu$A, so that

$$\beta_{dc} = \frac{I_C}{I_B} = \frac{3 \times 10^{-3}}{25 \times 10^{-6}} = \mathbf{120}$$

(b) $\alpha = \dfrac{\beta}{\beta + 1} = \dfrac{120}{121} \cong \mathbf{0.992}$
(c) $I_{CEO} =$ **0.3 mA** at intersection of $V_{CE} = 10$ V and $I_B = 0\ \mu$A
(d) $I_{CBO} \cong \dfrac{I_{CEO}}{\beta} = \dfrac{0.3 \text{ mA}}{120} = \mathbf{2.5\ \mu A}$

The input characteristics for the common-emitter configuration are very similar to those obtained for the common-base configuration (Fig. 4.12). In both cases, the increase in input current is due to an increase in majority carriers crossing the base-to-emitter junction with increasing forward-bias potential. Note also that the variation in output voltage (V_{CE} for the CE configuration and V_{CB} for the CB configuration) does not result in a large relocation of the characteristics. In fact, for the dc voltage levels commonly encountered the variation in base-to-emitter voltage with change in output terminal voltage can, as a first approximation, be ignored. On this basis, if we use an average value, the curve of Fig. 4.14 for the CE configuration will result. Note the similarities with the silicon-diode characteristics. Recall also from the description of the semiconductor diode that for dc analysis we approximated the

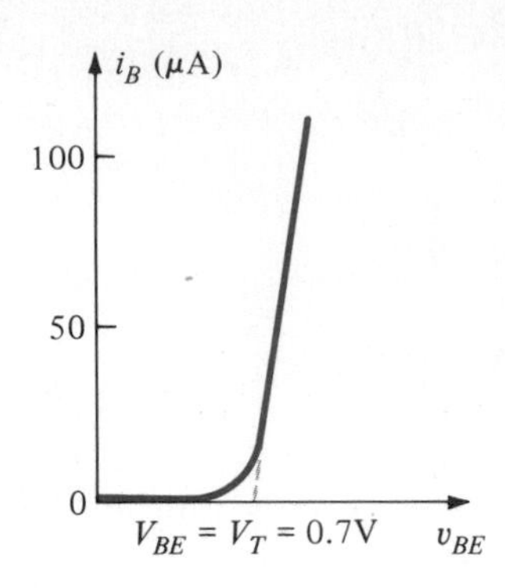

Figure 4.14 Reproduction of Fig. 4.12b ignoring the effects of V_{CE}.

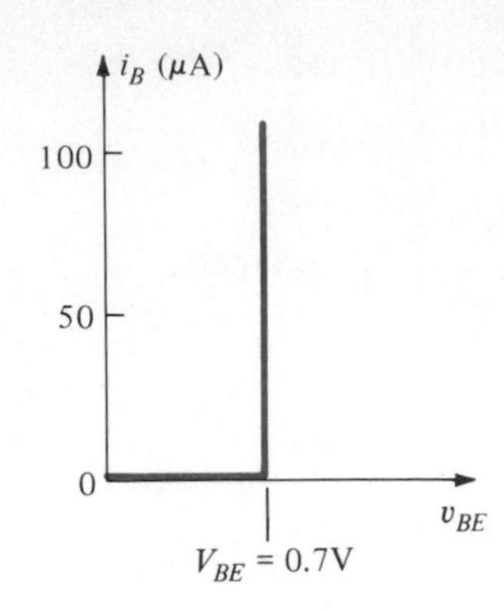

Figure 4.15 Approximate reproduction of Fig. 4.14 for dc analysis.

curve of Fig. 4.14 with that indicated in Fig. 4.15. Essentially, therefore, for dc analysis a first approximation to the base-to-emitter voltage of a transistor configuration is to assume that $V_{BE} \cong 0.7$ V for silicon and 0.3 V for germanium. If insufficient voltage is present to provide the 0.7 V bias (for silicon transistors) with the proper polarity, the transistor cannot be in the active region. A number of applications of this approximation will appear in Chapter 9. Since the CB characteristics had a similar set of input characteristics [also true for the common-collector (CC) configuration to be discussed], we can conclude *as a first approximation for dc analysis that the base-to-emitter voltage of a BJT is assumed to be V_T when biased in the active region of the characteristics*. Further, we found for the output characteristics of the CB configuration that $I_C \cong I_E$. For the CE configuration $I_C = \beta I_B$, where β is determined by the operating conditions.

In manuals, data sheets, and other transistor publications the common-emitter characteristics are the most frequently presented. The common-base characteristics can be obtained directly from the common-emitter characteristics using the basic current relations derived in the past few sections. In other words, for each point on the characteristics of the common-emitter configuration a sufficient number of variables can be obtained to substitute into the equations derived to come up with a point on the common-base characteristics. This process is, of course, time consuming, but it will result in the desired characteristics.

To determine the proper polarity for the applied dc potentials, it is only necessary first, to insert the direction of I_E in the same direction as the arrow of the symbol, as shown in Fig. 4.16 for an *npn* transistor. Since $I_E = I_C + I_B$ both I_C and I_B must enter

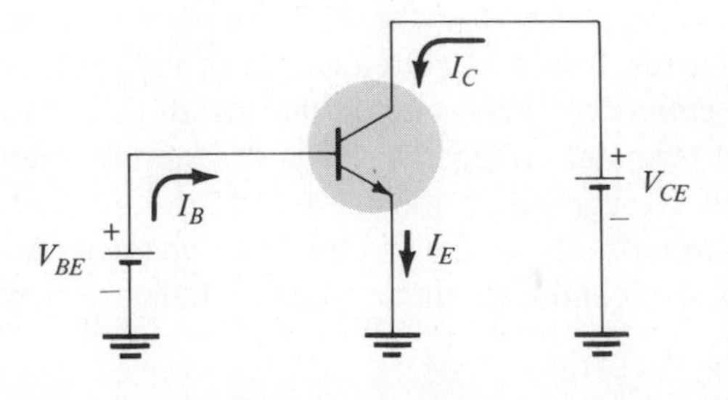

Figure 4.16 Determining the proper biasing for a common-emitter *npn* transistor.

the transistor, as shown in the same figure. It is then only necessary to insert the supplies V_{BE} and V_{CE} such that they "pressure" current in the direction indicated by I_B and I_C. For a *pnp* transistor all currents and therefore supplies will be reversed.

4.7 COMMON-COLLECTOR CONFIGURATION

The third and final transistor configuration is the *common-collector configuration*, shown in Fig. 4.17 with the proper current directions and voltage notation. The common-collector configuration is used primarily for impedance-matching purposes since it has a high input impedance and low output impedance, opposite to that which is true of the common-base and common-emitter configurations.

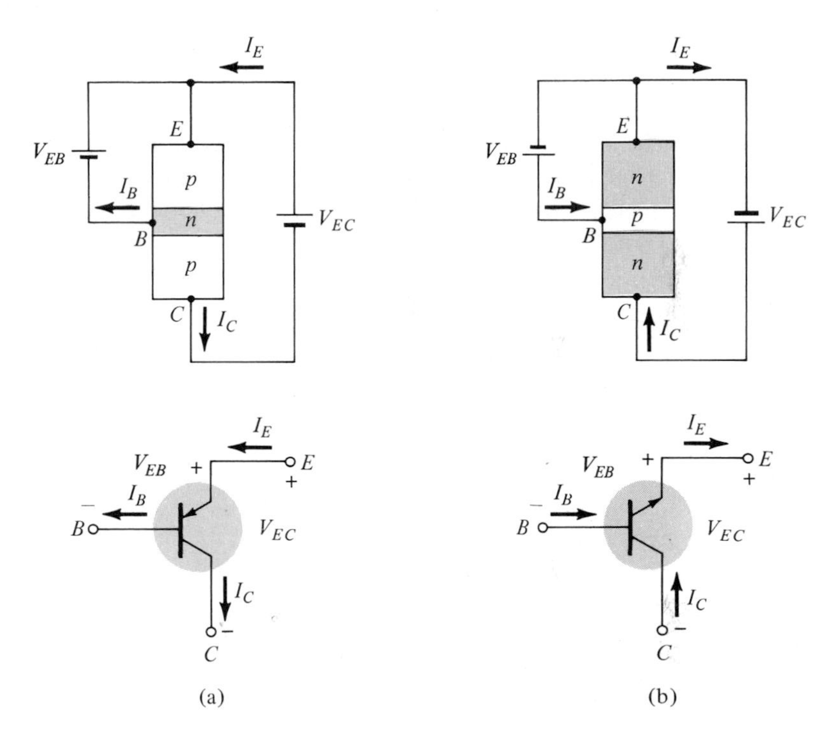

Figure 4.17 Notation and symbols used with the common-collector configuration: (a) *pnp* transistor; (b) *npn* transistor.

The common-collector circuit configuration is generally as shown in Fig. 4.18 with the load resistor from emitter to ground. Note that the collector is tied to ground even though the transistor is connected in a manner similar to the common-emitter configuration. From a design viewpoint, there is no need for a set of common-collector characteristics to choose the parameters of the circuit of Fig. 4.18. It can be designed using the common-emitter characteristics of Section 4.6. For all practical purposes, the output characteristics of the common-collector configuration are the same as for the common-emitter configuration. For the common-collector configuration the output characteristics are a plot of I_E versus V_{EC} for a range of values of I_B. The input current,

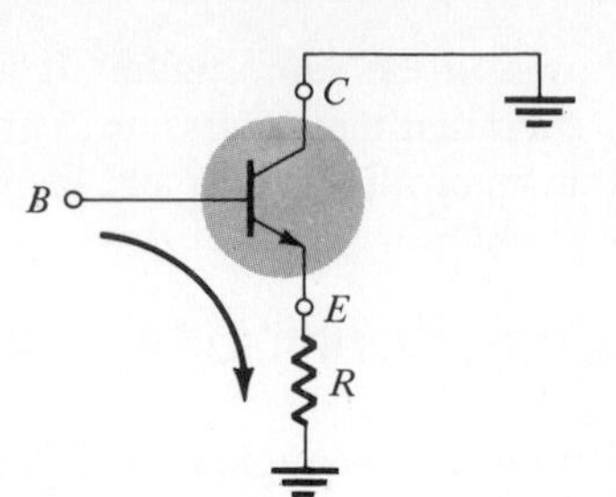

Figure 4.18 Common-collector configuration used for impedance matching purposes.

therefore, is the same for both the common-emitter and common-collector characteristics. The horizontal voltage axis for the common-collector configuration is obtained by simply changing the sign of the collector-to-emitter voltage of the common-emitter characteristics since $V_{EC} = -V_{CE}$. Finally, there is an almost unnoticeable change in the vertical scale of I_C of the common-emitter characteristics if I_C is replaced by I_E for the common-collector characteristics (since $\alpha \cong 1$). For the input circuit of the common-collector configuration the common-emitter base characteristics are sufficient for obtaining any required information by simply writing Kirchhoff's voltage law around the loop indicated in Fig. 4.18 and performing the proper mathematical manipulations.

4.8 TRANSISTOR MAXIMUM RATINGS

The standard transistor data sheet will include at least three maximum ratings: *collector dissipation, collector voltage, and collector current.* For the transistor whose characteristics were presented in Fig. 4.12, the following are the maximum ratings:

$$P_{C_{\text{max}}} = 30 \text{ mW}$$

$$I_{C_{\text{max}}} = 6 \text{ mA}$$

$$V_{CE_{\text{max}}} = 20 \text{ V}$$

The power or dissipation rating is the product of the collector voltage and current. For the common-emitter configuration.

$$\boxed{P_{C_{\text{max}}} = V_{CE} I_C} \tag{4.16}$$

The nonlinear curve determined by this equation is indicated in Fig. 4.19. The curve was obtained by choosing various values of V_{CE} (or I_C) and finding the other variable using Eq. (4.16). For example, at $V_{CE} = 10$ V,

$$I_C = \frac{P_{C_{\text{max}}}}{V_{CE}} = \frac{30 \times 10^{-3}}{10} = 3 \text{ mA}$$

as shown in Fig. 4.19. At $V_{CE} = 5$ V,

$$I_C = 6 \text{ mA}$$

and at $V_{CE} = 20$ V,

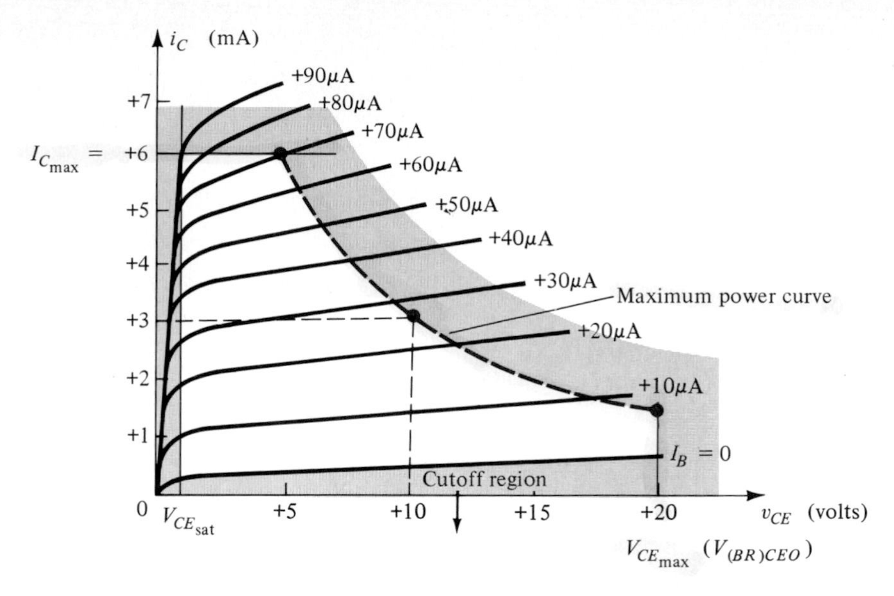

Figure 4.19 Region of operation for amplification purposes.

$$I_C = 1.5 \text{ mA}$$

Connecting the three points will result in the power curve. The region above this curve must be avoided in the design of systems using this particular transistor if the maximum power rating is not to be exceeded. The maximum collector voltage, in this case V_{CE}, is indicated as a vertical line in Fig. 4.19. The maximum collector current is also indicated as a horizontal line.

For the common-base configuration the collector dissipation is determined by the following equation. The maximum collector voltage would refer to V_{CB}.

$$P_{C_{max}} = V_{CB} I_C \tag{4.17}$$

For amplification purposes the nonlinear characteristics of the saturation and the cutoff regions are also avoided. The saturation region has been indicated by the vertical line at $V_{CE_{sat}}$ and the cutoff region by $I_B = 0$ in Fig. 4.19. The unshaded region remaining is the region employed for amplification purposes. Although it appears as though the area of operation has been drastically reduced, we must keep in mind that many signals are in the microvolt or millivolt range, while the horizontal axis of the characteristics is measured in volts. In addition to maximum ratings, data and specification sheets on transistors also include other important information about their operation. The discussion of these additional data will not be considered until each parameter is fully defined.

4.9 TRANSISTOR SPECIFICATION SHEET

The information provided in the RCA power transistors data book for the 2N1711 transistor appears in Figs. 4.20 through 4.27. As noted, this is a general-purpose small-signal/medium-power device.

Figure 4.20 Electrical characteristics of the RCA 2N1711 power transistor. (Courtesy RCA Solid State Division.)

Electrical Characteristics

Characteristic	Symbol	Case Temperature °C	Frequency kHz	Test Conditions: dc Collector-to-Base Voltage V V_{CB}	dc Collector-to-Emitter Voltage V V_{CE}	dc Emitter-to-Base Voltage V V_{EB}	dc Collector Current mA I_C	dc Emitter Current mA I_E	dc Base Current mA I_B	Limits: Min.	Max.	Units
Collector-cutoff current	I_{CBO}	25		60				0		—	0.01	
		150		60				0		—	10	μA
Emitter-cutoff current	I_{EBO}	25				5	0			—	0.005	μA
dc-pulse forward-current transfer ratio[a]	h_{FE}	25			10		10			75	—	
		25			10		150			100	300	
		25			10		500			40	—	
		25			10		0.01			20	—	
dc forward-current transfer ratio	h_{FE}	25			10		0.1			35	—	
		−55			10		10			35	—	
Collector-to-base breakdown voltage	$V_{(BR)CBO}$	25					0.1	0		75	—	V
Emitter-to-base breakdown voltage	$V_{(BR)EBO}$	25					0	0.1		7	—	V
Collector-to-emitter reach-through voltage	V_{RT}	25				1.5[b]	0.1			75	—	V
Collector-to-emitter sustaining voltage with external base-to-emitter resistance = 10 ohms	V_{CER} (sus)	25					100 (pulsed)			50	—	V

Electrical Characteristics (continued)

Characteristic	Symbol	Test Conditions								Limits		Units
		Case Temperature °C	Frequency kHz	dc Collector-to-Base Voltage V V_{CB}	dc Collector-to-Emitter Voltage V V_{CE}	dc Emitter-to-Base Voltage V V_{EB}	dc Collector Current mA I_C	dc Emitter Current mA I_E	dc Base Current mA I_B	Min.	Max.	
Collector-to-emitter saturation voltage	V_{CE} (sat)	25					150		15	—	1.5	V
Base-to-emitter saturation voltage	V_{BE} (sat)	25					150		15	—	1.3	V
Small-signal forward-current transfer ratio	h_{fe}	25	1		5		1			50	200	
		25	1		10		5			70	300	
		25	20 MHz		10		50			5	—	
Noise figure: Generator resistance (R_G) = 510 ohms, circuit bandwidth (BW) = 1 cycle	NF	25	1	10			0.3			—	8	dB
Output capacitance	C_{ob}	25		10				0		—	15	pF
Input capacitance	C_{ib}	25				0.5	0			—	80	pF
Input resistance	h_{ib}	25	1	5			1			24	34	Ω
		25	1	10			5			4	8	
Voltage-feedback ratio	h_{rb}	25	1	5			1			—	5×10^{-4}	
		25	1	10			5			—	5×10^{-4}	
Output conductance	h_{ob}	25	1	5			1			0.1	0.5	μS
		25	1	10			5			0.1	1	
Thermal resistance: Junction-to-case	$R_{\theta JC}$	—								—	58.3	°C/W
Junction-to-free air	$R_{\theta JA}$	—								—	219	

[A] Pulse duration = 300 μs; duty factor ≤ 2%.

[B] V_{EBF} = Emitter-to-base floating potential.

Solid State Division

Power Transistors

2N1711

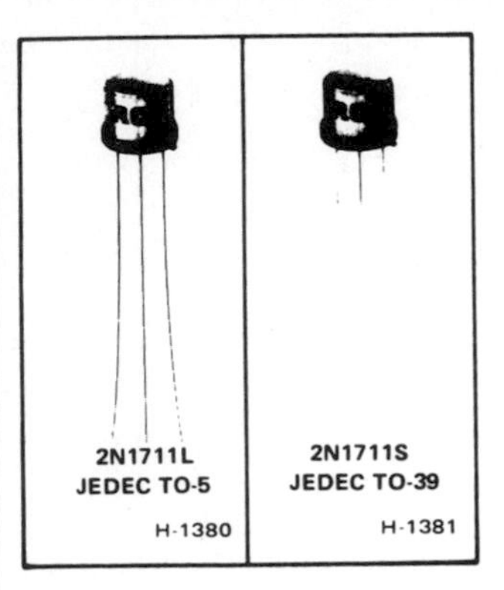

Silicon N-P-N Planar Transistors

General-Purpose Type for Small-Signal, Medium-Power Applications

Features:

- **Minimum gain-bandwidth product = 70 MHz; useful in applications from dc to 25 MHz**
- **Operation at high junction temperatures**
- **Planar construction for low-noise and low-leakage characteristics**
- **Low output capacitance**

These devices are available with either 1½-inch leads (TO-5 package) or ½-inch leads (TO-39 package). The longer-lead versions are specified by suffix "L" after the type number; the shorter-lead versions are specified by suffix "S" after the type number.

RCA-2N1711 is a silicon n-p-n planar transistor intended for a wide variety of small-signal and medium-power applications in military and industrial equipment. It features exceptionally low noise and leakage characteristics, high pulse beta (h_{FE}), high breakdown-voltage ratings, low saturation voltages, high sustaining voltages, and low output capacitance.

MAXIMUM RATINGS, *Absolute-Maximum Values:*

COLLECTOR-TO-BASE VOLTAGE	V_{CBO}	75	V
COLLECTOR-TO-EMITTER VOLTAGE:			
With external base-to-emitter resistance (R_{BE}) ≤ 10 Ω	V_{CER}	50	V
EMITTER-TO-BASE VOLTAGE	V_{EBO}	7	V
COLLECTOR CURRENT	I_C	1	A
TRANSISTOR DISSIPATION:	P_T		
At case temperatures up to 25°C		3	W
At case temperatures above 25°C		See Fig. 3.22	
At free-air temperatures up to 25°C		0.8	W
At free-air temperatures above 25°C		See Fig. 3.22	
TEMPERATURE RANGE:			
Storage and Operating (Junction)		−65 to +200	°C
LEAD TEMPERATURE (During soldering):			
At distance ≥ 1/16 in. (1.58 mm) from seating plane for 10 s max.		230	°C

Figure 4.21 RCA 2N1711 power transistor. (Courtesy RCA Solid State Division.)

The letter o at the end of a parameter indicates that the terminal not listed is left open. Note in Fig. 4.20 that I_{CBO} is only 0.01 μA at 25°C but 10 μA at 150°C. The quantity h_{FE}, which is synonymous with $\beta_{dc} = I_C/I_B$, has a minimum value of 20. Priorities do not permit an introduction to all the quantities listed in Fig. 4.20. However, most companies carefully define these quantities in the introductory pages of their catalogs. A number of the remaining quantities in the figure will be introduced

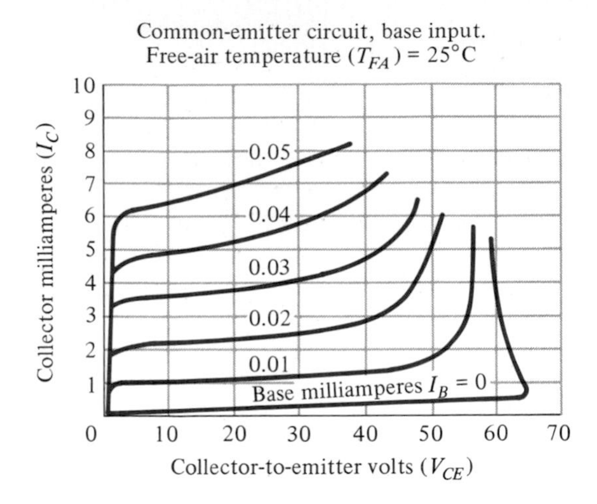

Figure 4.22 RCA 2N1711 output characteristics. (Courtesy RCA Solid State Division.)

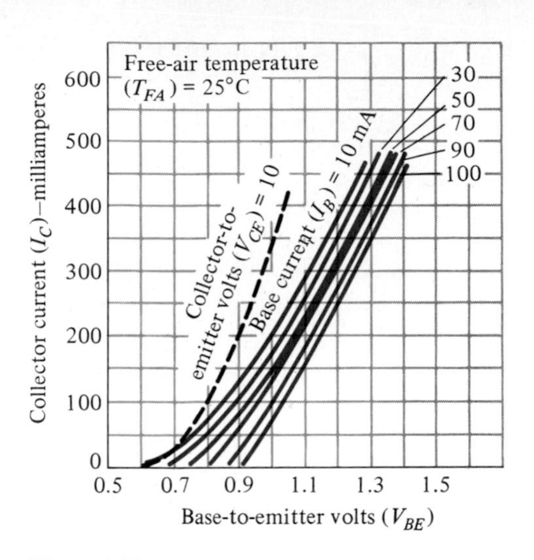

Figure 4.23 RCA 2N1711 transfer characteristics. (Courtesy RCA Solid State Division.)

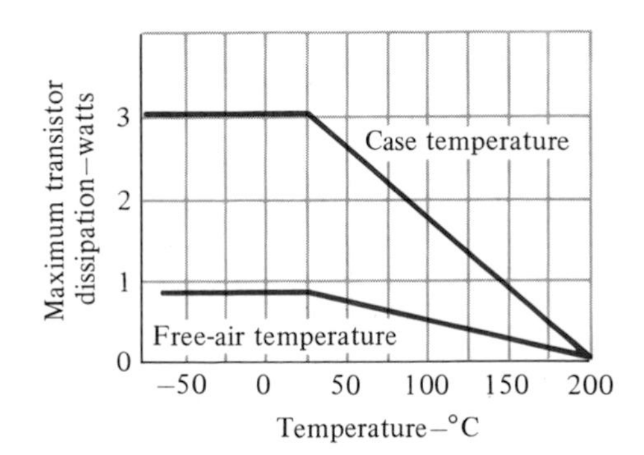

Figure 4.24 RCA 2N1711 derating curves. (Courtesy RCA Solid State Division.)

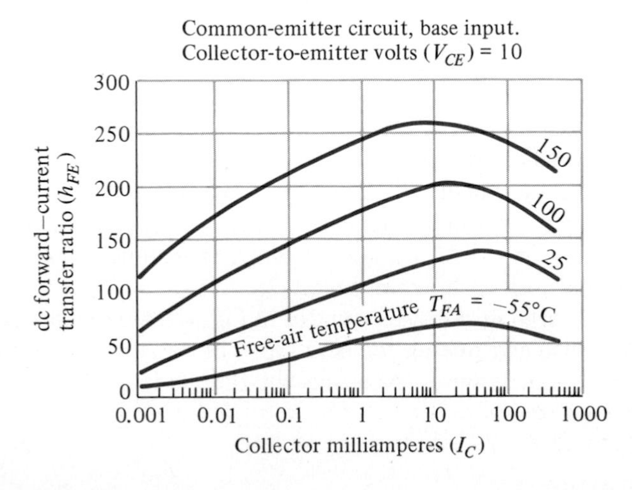

Figure 4.25 RCA 2N1711 dc beta characteristics. (Courtesy RCA Solid State Division.)

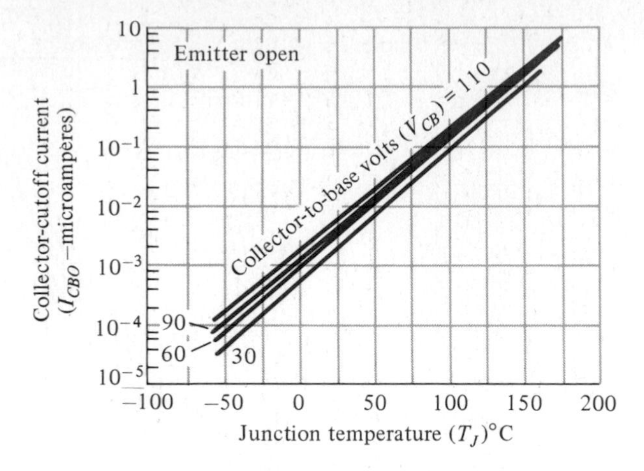

Figure 4.26 RCA 2N1711 collector-cutoff-current characteristics. (Courtesy RCA Solid State Division.)

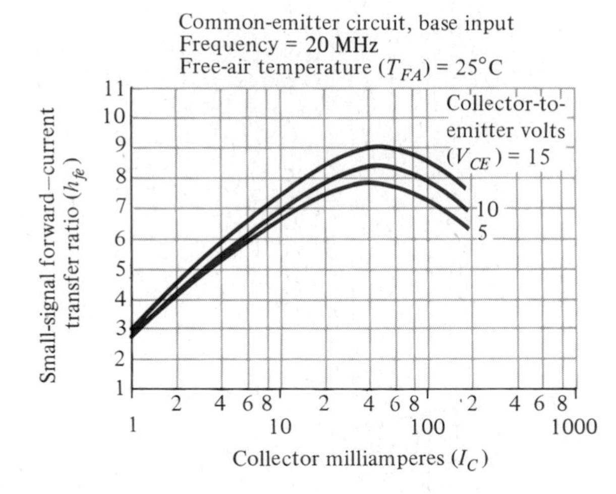

Figure 4.27 RCA 2N1711 small-signal beta characteristics. (Courtesy RCA Solid State Division.)

in a later chapter. Certainly, V_{CE}(sat), the capacitance levels and thermal resistance quantities are familiar. The *hybrid* parameters h_{fe}, h_{ib}, h_{rb}, and h_{ob} will be introduced in Chapter 9.

The output characteristics for the device appear in Fig. 4.22. Note the resulting distortion at high levels of voltage and current—a region that must be avoided for linear operation. Linear operation suggests that the output waveform has the same appearance as the input (but amplified) and is not distorted by the amplifying device.

Note the shift in the V_{BE} versus I_C curve for increasing levels of base current. Consider also that I_B is measured in milliamperes since it is a power device. It would appear that our use of V_{BE} = 0.7 V is a good average value since base currents will probably not exceed 25 mA, as shown in Fig. 4.28.

The familiar power derating curve is provided in Fig. 4.24. Note that a curve has been provided for the case and free-air temperature. The variation in h_{FE} is provided versus I_C for a range of temperatures. Note that the ratio is less at each temperature if the collector current is too high.

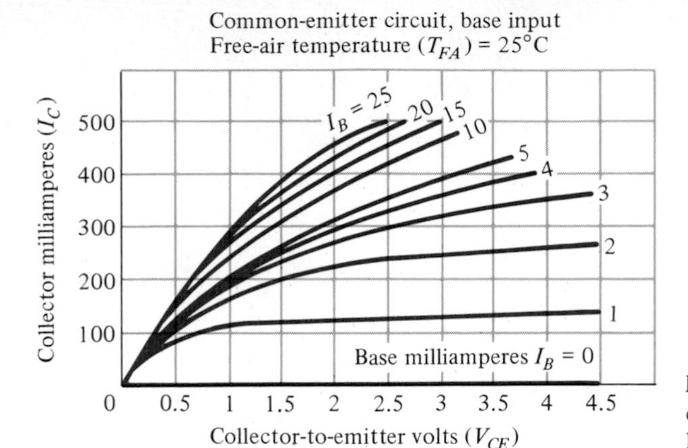

Figure 4.28 RCA 2N1711 output characteristics (high I_C). (Courtesy RCA Solid State Division.)

In Fig. 4.26 the level of I_{CBO} is provided versus junction temperature for different levels of collector-to-base voltage. It appears the I_{CBO} will not reach 1 μA until the junction temperature approaches 135°C. Even with a β of 100, I_{CEO} is still limited to $I_{CEO} \cong \beta I_{CBO} \cong 100(1) = 100\ \mu\text{A} = 0.1\ \text{mA}$ at this temperature. For a medium-power device, this is a fairly low level for this undesirable effect.

The small-signal forward-current transfer ratio (h_{fe}) will be defined in Chapter 9. Briefly, it is a measure of the small-signal ac gain of the device (the increase in the peak-to-peak level of a sinusoidal signal).

Figure 4.28 demonstrates the change in collector characteristics at the high levels of current. The nice even spacing between curves in Fig. 4.22 is no longer present, but the increasing density of the lines follows a fairly linear pattern.

We will refer to these figures as we introduce new quantities of importance in succeeding chapters.

4.10 TRANSISTOR FABRICATION

The majority of the methods used to fabricate transistors are simply extensions of the methods used to manufacture semiconductor diodes. The methods most frequently employed today include *point-contact*, *alloy junction*, *grown junction*, *and diffusion*. The following discussion of each method will be brief, but the fundamental steps included in each will be presented. A detailed discussion of each method would require a text in itself.

Point-Contact

The point-contact transistor is manufactured in a manner very similar to that used for point-contact semiconductor diodes. In this case two wires are placed next to an *n*-type wafer as shown in Fig. 4.29. Electrical pulses are then applied to each wire resulting in a *p-n* junction at the boundary of each wire and the semiconductor wafer. The result is a *pnp* transistor as shown in Fig. 4.29. This method of fabrication is today

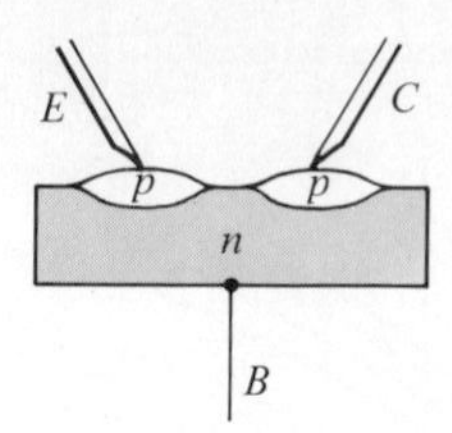

Figure 4.29 Point-contact transistor.

limited to high-frequency/low-power devices. It was the method used in the fabrication of the first transistor shown in Fig. 4.1.

Alloy Junction

The alloy junction technique is also an extension of the alloy method of manufacturing semiconductor diodes. For a transistor, however, two dots of the same impurity are deposited on each side of a semiconductor wafer having the oppostie impurity as shown in Fig. 4.30. The entire structure is then heated until melting occurs and each dot is alloyed to the base wafer resulting in the *p-n* junctions indicated in Fig. 4.30 as described for semiconductor diodes.

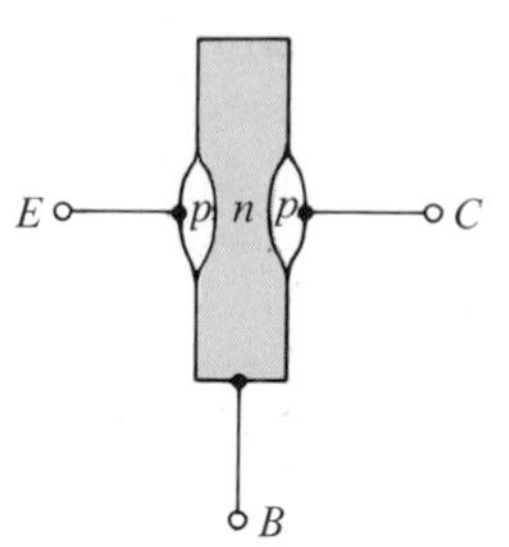

Figure 4.30 Alloy junction transistor.

The collector dot and resulting junction are larger, to withstand the heavy current and power dissipation at the collector-base junction. This method is not employed as much as the diffusion technique to be described shortly, but it is still used extensively in the manufacture of high-power diodes.

Grown Junction

The Czochralski technique (Section 1.18) is used to form the two *p-n* junctions of a grown-junction transistor. The process, as depicted in Fig. 4.31, requires that the impurity control and withdrawal rate be such as to ensure the proper base width and doping levels of the *n*- and *p*-type materials. Transistors of this type are, in general, limited to less than $\frac{1}{4}$-W rating.

Diffusion

The most frequently employed method of manufacturing transistors today is the diffusion technique. The basic process was introduced in the discussion of semi-

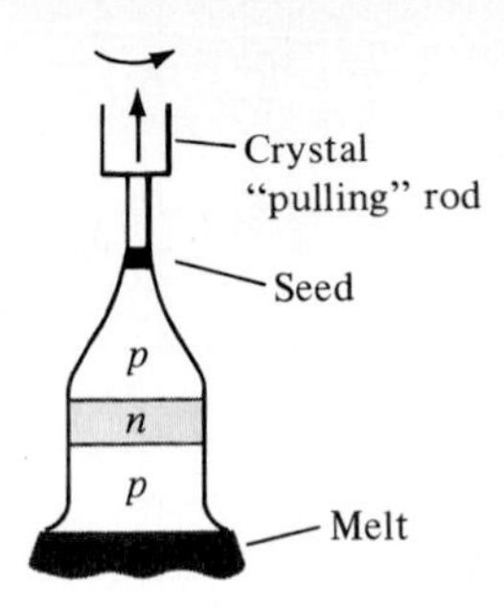

Figure 4.31 Grown junction transistor.

conductor diode fabrication. The diffusion technique is employed in the production of *mesa* and *planar* transistors, each of which can be of the *diffused* or *epitaxial* type.

In the *pnp*, diffusion-type mesa transistor the first process is an *n*-type diffusion into a *p*-type wafer, as shown in Fig. 4.32, to form the base region. Next, the *p*-type emitter is diffused or alloyed to the *n*-type base as shown in the figure. Etching is done to reduce the capacitance of the collector junction. The term "mesa" is derived from its similarities with the geographical formation. As mentioned earlier in the discussion of diode fabrication, the diffusion technique permits very tight control of the doping levels and thicknesses of the various regions.

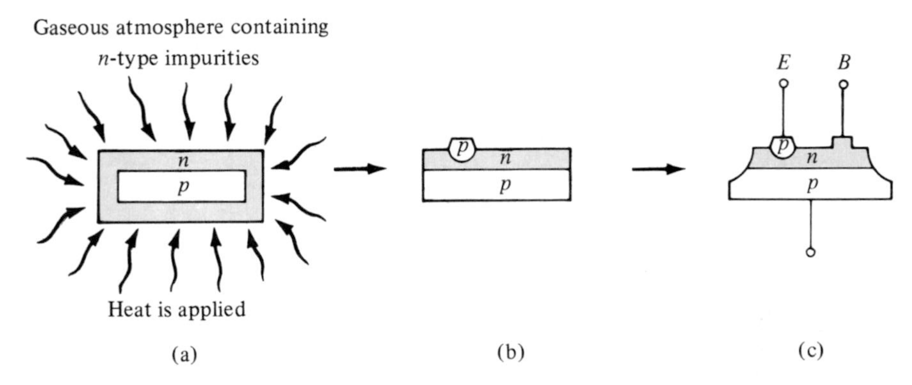

Figure 4.32 Mesa transistor: (a) diffusion process; (b) alloy process; (c) etching process.

The major difference between the epitaxial mesa transistor and the mesa transistor is the addition of an epitaxial layer on the original collector substrate. The term epitaxial is derived from the Greek words *epi*—upon, and *taxi*—arrange, which describe the process involved in forming this additional layer. The original *p*-type substrate (collector of Fig. 4.33) is placed in a closed container having a vapor of the same impurity. Through proper temperature control, the atoms of the vapor will *fall upon* and *arrange* themselves on the orignal *p*-type substrate resulting in the epitaxial layer indicated in Fig. 4.33. Once this layer is established, the process continues, as described above for the mesa transistor, to form the base and emitter regions. The original *p*-type substrate will have a higher doping level and correspondingly less

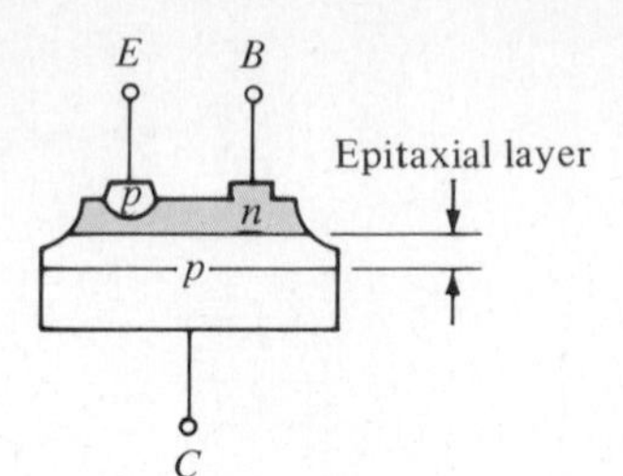

Figure 4.33 Epitaxial mesa transistor.

resistance than the epitaxial layer. The result is a low-resistance connection to the collector lead that will reduce the dissipation losses of the transistor.

The planar and epitaxial planar transistors are fabricated using two diffusion processes to form the base and emitter regions. The planar transistor, as shown in Fig. 4.34, has a flat surface, which accounts for the term "planar." An oxide layer is added as shown in Fig. 4.34 to eliminate exposed junctions, which will reduce substantially the surface leakage loss (leakage currents on the surface rather than through the junction).

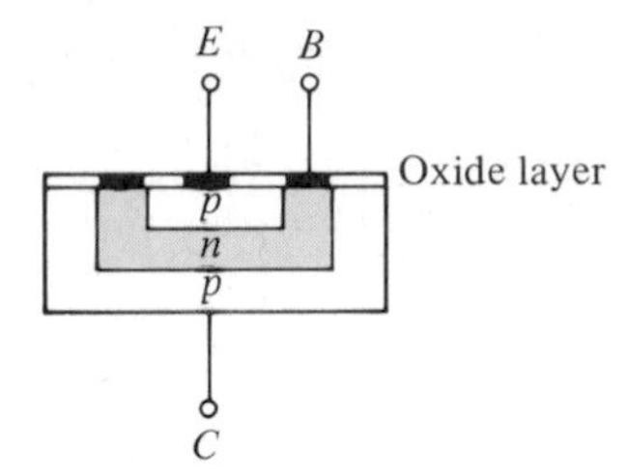

Figure 4.34 Planar transistor.

4.11 TRANSISTOR CASING AND TERMINAL IDENTIFICATION

After the transistor has been manufactured using one of the techniques indicated in Section 4.10, leads of, typically, gold, aluminum, or nickel are then attached and the entire structure is encapsulated in a container such as that shown in Fig. 4.35. Those with the studs and heat sinks are high-power devices, while those with the small can (top hat) or plastic body are low- to medium-power devices.

Whenever possible, the transistor casing will have some marking to indicate which leads are connected to the emitter, collector, or base of a transistor. A few of the methods commonly used are indicated in Fig. 4.36.

The internal construction of a TO-92 package in the Fairchild line appears in Fig. 4.37. Note the very small size of the actual semiconductor device. There are gold bond wires, a copper frame, and an epoxy encapsulation.

Four (quad) individual *pnp* silicon transistors can be housed in the 14-pin plastic dual-in-line package appearing in Fig. 4.38a. The internal pin connections appear in Fig. 4.38b. As with the diode IC package, the indentation in the top surface reveals the number 1 and 14 pins.

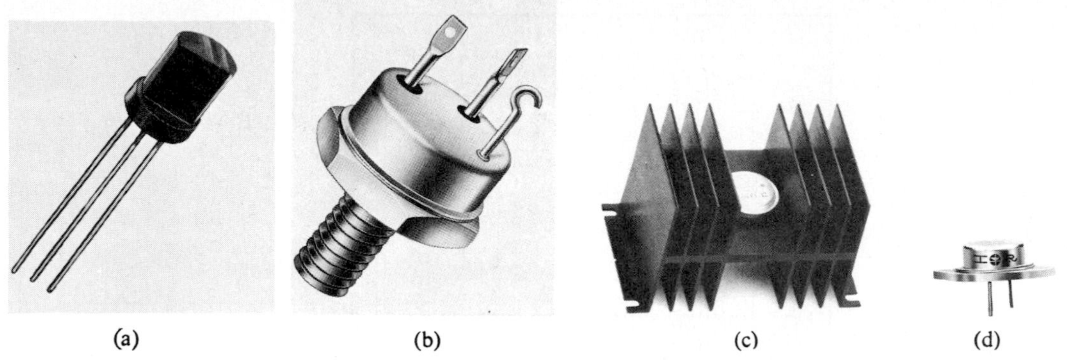

Figure 4.35 Various types of transistors. [(a) and (b) Courtesy General Electric Company; (c) and (d) Courtesy International Rectifier Corporation.]

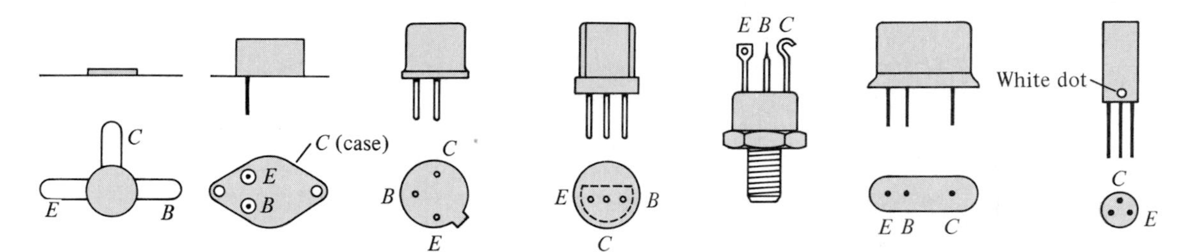

Figure 4.36 Transistor terminal identification.

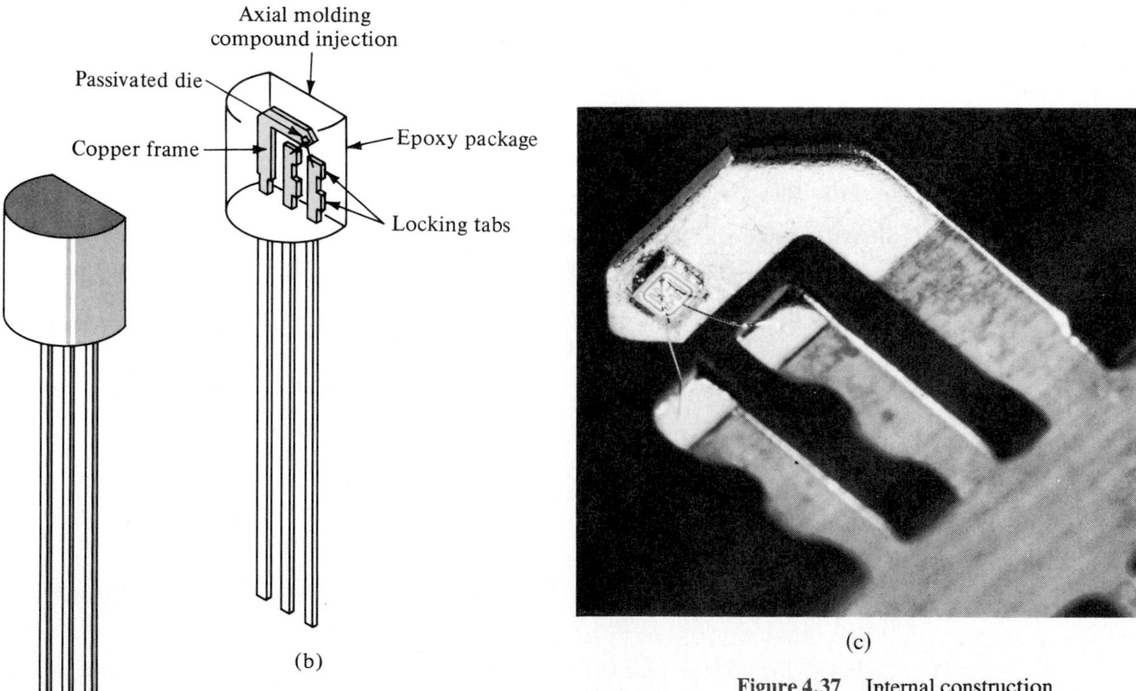

Figure 4.37 Internal construction of a Fairchild transistor in a TO-92 package. (Courtesy Fairchild Camera and Instrument Corporation.)

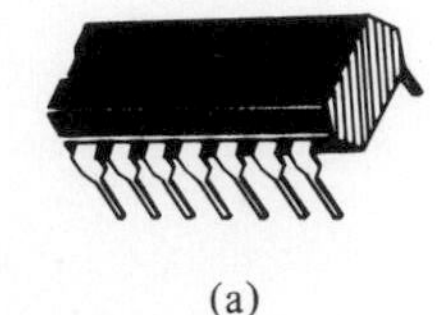

(a)

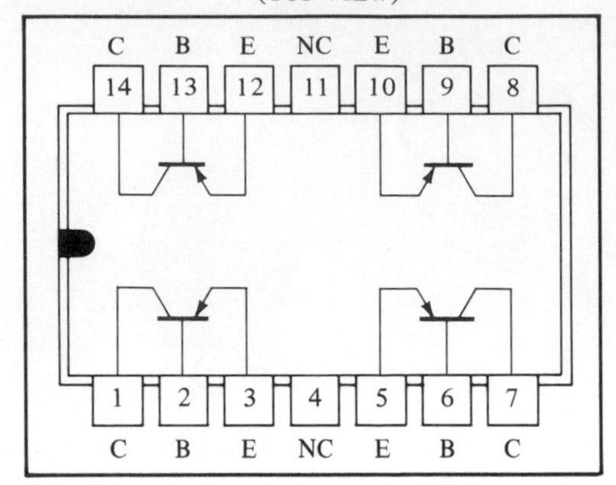

NC–No internal connection

(b)

Figure 4.38 Type Q2T2905 Texas Instruments quad *pnp* silicon transistors: (a) appearance; (b) pin connections. (Courtesy Texas Instruments Incorporated.)

PROBLEMS

§ 4.2

1. What names are applied to the two types of transistors? Sketch each transistor and indicate the type of majority and minority carrier in each layer. Is any of this information altered by changing from silicon to germanium material?

2. What is the major difference between a bipolar and a unipolar device?

§ 4.3

3. How must the two transistor junctions be biased for proper transistor operation?

4. What is the source of the leakage current in a transistor?

5. Sketch a figure similar to Fig. 4.3 for the forward-biased junction of an *npn* transistor. Describe the resulting carrier motion.

6. Sketch a figure similar to Fig. 4.4 for the reverse-biased junction of an *npn* transistor. Describe the resulting carrier motion.

7. Sketch a figure similar to Fig. 4.5 for the majority- and minority-carrier flow of an *npn* transistor. Describe the resulting carrier motion.

8. Determine the resulting change in emitter current for a change in collector current of 2 mA and an α of 0.98.

9. A transistor has an emitter current of 8 mA and an α of 0.99. How large is the collector current?

§ 4.4

10. Calculate the voltage gain ($A_v = V_o/V_i$) for the circuit of Fig. 4.6 if $V_i = 500$ mV and $R = 1\ \text{k}\Omega$. (The other circuit values remain the same.)

11. Calculate the voltage gain ($A_v = V_o/V_i$) for the circuit of Fig. 4.6 if the source has an internal resistance of 100 Ω in series with V_i.

§ 4.5

12. From memory, and memory only, sketch the common-base transistor configuration (for *npn* and *pnp*) and indicate the polarity of the applied bias and resulting current directions.

13. Using the characteristics of Fig. 4.8:
(a) Find the resulting collector current if $I_E = 5$ mA and $V_{CB} = -10$ V.
(b) Find the resulting collector current if $V_{EB} = 750$ mV and $V_{CB} = -10$ V.
(c) Find V_{EB} for the conditions $I_C = 4$ mA and $V_{CB} = -1$ V.

14. The characteristics of Fig. 4.8b are for a silicon transistor. What would be a first approximation for the base-to-emitter voltage of the forward-biased junction? How would you expect them to be different for a germanium transistor?

§ 4.6

15. Define I_{CO} and I_{CEO}. How are they different? How are they related? Are they typically close in magnitude?

16. Using the characteristics of Fig. 4.12:
(a) Find the value of I_C corresponding to $V_{BE} = +750$ mV and $V_{CE} = +5$ V.
(b) Find the value of V_{CE} and V_{BE} corresponding to $I_C = 3$ mA and $I_B = 30\ \mu$A.

17. (a) For the common-emitter characteristics of Fig. 4.12, find the dc beta at an operating point of $V_{CE} = +8$ V and $I_C = 2$ mA.
(b) Find the value of α corresponding to this operating point.
(c) At $V_{CE} = +8$ V, find the corresponding value of I_{CEO}.
(d) Calculate the approximate value of I_{CBO} using the dc beta value obtained in part (a).

§ 4.7

18. An input voltage of 2 V rms (measured from base to ground) is applied to the circuit of Fig. 4.18. Assuming that the emitter voltage follows the base voltage exactly and that V_{be} (rms) $= 0.1$ V, calculate the circuit voltage amplification ($A_v = V_o/V_i$) and emitter current for $R_E = 1$ kΩ.

19. For a transistor having the characteristics of Fig. 4.12, sketch the input and output characteristics of the common-collector configuration.

§ 4.8

20. Determine the region of operation for a transistor having the characteristics of Fig. 4.12 if $I_{C_{max}} = 5$ mA, $V_{CE_{max}} = 15$ V, and $P_{C_{max}} = 40$ mW.

21. Determine the region of operation for a transistor having the characteristics of Fig. 4.8 if $I_{C_{max}} = 6$ mA, $V_{CB_{max}} = -15$ V, and $P_{C_{max}} = 30$ mW.

§ 4.9

22. Referring to Fig. 4.21, determine the temperature range for the device in degrees Fahrenheit.

23. Determine the value of α when $I_C = 0.1$ mA from the provided value of dc forward-current transfer ratio (Fig. 4.20).

24. Using the results of Problem 23 and Eq. (4.9), calculate I_{CO} and note whether it falls within the limits of I_{CBO} in Fig. 4.20. Use $I_C = 0.1$ mA at $I_B = 0\ \mu$A with $T_{case} = 150°$C.

25. Using the information provided in Figs. 4.20 and 4.21, sketch the boundaries of the maximum power dissipation region for the *CE* configuration ($T_{case} = 25°C$).

26. Sketch the maximum power curve on the characteristics of Fig. 4.28 if the free-air temperature is 100°C.

27. Referring to Fig. 4.22, determine the following:
   (a) I_C if $V_{CE} = 30$ V, $I_B = 25$ μA.
   (b) V_{CE} if $I_C = 4$ mA, $I_B = 30$ μA.
   (c) I_B if $I_C = 3$ mA, $V_{CE} = 20$ V.

28. (a) At an operating point of $V_{CE} = 30$ V and $I_C = 7$ mA, determine the level of V_{BE} using Figs. 4.22 and 4.23. For base currents significantly less than 10 mA, use the curve of Fig. 3.23 denoted $V_{CE} = 10$ V.
   (b) At an operating point of $V_{CE} = 2$ V and $I_C = 300$ mA, determine the level of V_{BE} using Figs. 4.23 and 4.28.
   (c) What would be an appropriate level of V_T for the approximate equivalent circuit for each of the operating conditions of parts (a) and (b)?

29. (a) Calculate the power derating factor for each of the curves of Fig. 4.24.
   (b) Using the value obtained for the case-temperature curve, find the power rating at a temperature of 100°C.
   (c) Compare the results of part (b) with the value obtained from the graph.

30. (a) Determine the value of β_{dc} at $I_C = 5$ mA and a temperature of 100°C, using Fig. 4.25.
   (b) What is the level of α at this point?
   (c) What is the average difference in β level between room temperature and 100°C for the range $I_C = 0.01$ mA to $I_C = 10$ mA? Is it something we should carefully consider in a design problem? Why?

31. (a) Determine I_{CBO} from Fig. 4.26 for $V_{CB} = 30$ V and a junction temperature of 50°C. Note the log scale.
   (b) If $\beta = 200$, what is the level of I_{CEO}?
   (c) What is the rate of change of I_{CBO} per degree change in temperature near 50°C for $V_{CB} = 30$ V?

32. (a) Using Fig. 4.27, determine the rate of change of h_{fe} with change in collector current for the range $I_C = 2$ to 4 mA ($V_{CE} = 10$ V).
   (b) Why would you assume that the magnitude of h_{fe} in this particular figure is so small on the vertical scale when typical values of h_{fe} approach 100 or more?

§ 4.10

33. (a) Describe the basic differences among the various techniques of transistor construction.
   (b) Which would you categorize as acceptable for high-power applications?
   (c) Define the *diffusion* process.

§ 4.11

34. (a) Find three transistors with different casings, identify the terminals, and sketch the device.
   (b) Search through a manufacturer's data book for another IC structure limited totally to transistors. Sketch the internal schematic and identify the terminals.

5

DC Biasing: BJTs

5.1 INTRODUCTION

Transistors are used in a large variety of applications and in many different ways. It would be difficult if not impossible to learn each area and application. Instead, one studies the more fundamental circuit operation so that enough is known to carry over this knowledge to slightly different or even completely different applications. This chapter covers the basic concepts in the dc biasing of bipolar junction transistors (BJTs).

To use these devices for amplification of voltage or current, or as control (*on* or *off*) elements, it is necessary first to *bias* the device. The usual reason for this biasing is to turn the device on, and in particular, to place it in operation in the region of its characteristic where the device operates most linearly. Although the purpose of the bias network or biasing circuit is to cause the device to operate in this desired *linear* region of operation (which is best defined by the manufacturer for each device type), the bias components are still part of the overall application circuit—amplifier, waveform shaper, logic circuit, and so on. We could treat the overall circuit and consider all aspects of the operation at once, but that would be complex and more confusing. Each type of circuit application would have to be studied for all aspects of operation without a more basic understanding of those common features of operation. This chapter therefore provides basic concepts of dc biasing of the bipolar transistor, with the understood aim of getting the device operating in a desired region of the device characteristic. If these concepts are well understood, many different circuits, even new circuit applications, can be studied and analyzed more easily because a basic understanding of the circuit has been established. Amplifier gain and other factors affecting ac operation will be considered in Chapter 9. Basic dc bias concepts are presented in this chapter and then applied in Chapters 9, 11, and 12.

Dc biasing is a *static* operation since it deals with setting a fixed (steady) level of current (through the device) with a desired fixed voltage drop across the device. Necessary information about the device can be obtained from the device's static characteristics.

5.2 OPERATING POINT

Since the aim of biasing is to achieve a certain condition of current and voltage called the *operating point* (*quiescent* point or *Q*-point), some attention is given to the selection of this point in the device characteristic. Figure 5.1 shows a general device characteristic with four indicated operating points. The biasing circuit may be designed to set the device operation at any of these points or others within the *operating region*. The operating region is the area of current or voltage within the maximum limits for the particular device. These maximum ratings are indicated on the characteristic of Fig. 5.1 by a horizontal line for the maximum current, I_{max}, and a vertical line for the maximum voltage, V_{max}. An additional consideration of maximum power (product of voltage and current) must also be taken into consideration in defining the operating region of a particular device, as shown by the line marked P_{max} on Fig. 5.1.

The BJT device could be biased to operate outside these maximum limit points but the result of such operation would be either a considerable shortening of the lifetime of the device or destruction of the device. Confining ourselves to the safe operating region we may select many different operating areas or points. The exact point or area

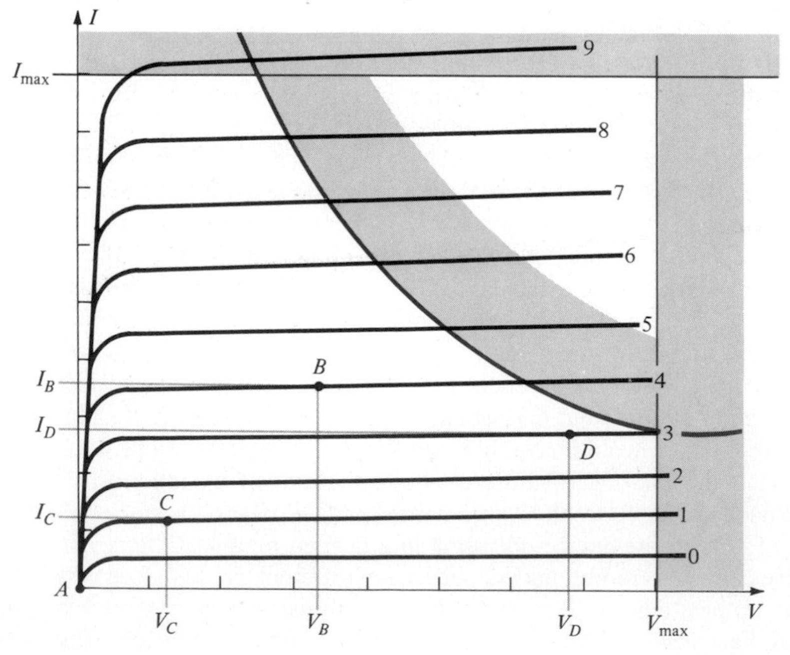

Figure 5.1 Various operating points on device static characteristics.

often depends on the intended use of the circuit. Still, we can consider some differences between operation at the various points shown in Fig. 5.1 to present some basic ideas about the operating point and, thereby, the bias circuit.

If no bias were used, the device would initially be completely off, which would result in the current of point *A*—namely, zero current through the device (and zero voltage across it). It is necessary to bias the device so that it can respond or change in current and voltage for the entire range of an input signal. While point *A* would not be suitable, point *B* provides this desired operation. If a signal is applied to the circuit, *in addition to the bias level*, the device will vary in current and voltage from operating point *B*, allowing the device to react to (and possibly amplify) both the positive and negative part of the input signal. If, as could be the case, the input signal is small, the voltage and current of the device will vary but not enough to drive the device into *cutoff* or *saturation*. Cutoff is the condition in which the device no longer conducts. Saturation is the condition in which voltage across the device is as small as possible with the current in the device path reaching a limiting or saturating value depending on the external circuit. The usual amplifier action desired occurs within the operating region of the device, that is, between saturation and cutoff.

Point *C* would also allow some positive and negative variation with the device still operating, but the output voltage could not decrease much because bias point *C* is lower in voltage than point *B*. Point *C* is in a region of operation in which the current level in the device is smaller and the device gain is *not* linear, that is, the spacing in going from one curve to the next is unequal. This nonlinearity shows that the amount of gain of the device is smaller when biased lower on the characteristic and larger when biased higher up. It is preferable to operate where the gain of the device is most constant (or linear) so that the amount of amplification over the entire swing of input signal is the same. Point *B* is a region of more linear spacing and, therefore, more linear operation, as shown in Fig. 5.1.

Point *D* sets the device operating point near the maximum voltage level. The output voltage swing in the positive direction is thus limited if the maximum voltage is not to be exceeded. Point *B*, therefore, seems the best operating point in terms of linear gain or largest possible voltage and current swing. This is usually the desired condition for small-signal amplifiers (Chapter 9) but not necessarily for power amplifiers which will be considered in Chapter 12. In this discussion, we will concentrate mainly on biasing the device for *small-signal* amplification operation.

One other very important biasing factor must be considered. Having selected and biased the BJT at a desired operating point, the effect of temperature must also be taken into account. Temperature causes device characteristics such as the transistor current gain and the transistor leakage current to change. Higher temperature results in more current in the device than at room temperature, thereby upsetting the operating condition set by the bias circuit. Because of this, the bias circuit must also provide a degree of *temperature stability* to the circuit so that temperature changes at the device produce minimum change in its operating point. This maintenance of operating point may be specified by a *stability factor*, *S*, indicating the amount of change in operating-point current due to temperature. A highly stable circuit is desirable and the stability of a few basic bias circuits will be compared.

Bipolar transistor operation may be specified sufficiently well by device parameters, and mathematical techniques can be used to determine its biasing. Nevertheless, the transistor characteristic still provides a convenient picture for understanding device operation and will be used on occasion.

For the BJT to be biased in its linear or active operating region the following must be true:

1. The base-emitter junction *must* be forward biased (*p*-region voltage more positive) with a resulting forward-bias voltage across the base-emitter of about 0.6 to 0.7 V.
2. The base-collector junction *must* be reverse biased (*n*-region more *p*ositive), with the reverse-bias voltage being any value within the maximum limits of the device.

[Notice that for forward bias the voltage across the *p-n* junction is *p*-*p*ositive, while for reverse bias it is opposite (reverse) with *n*-*p*ositive. This emphasis on the initial letter should provide a means of helping memorize the necessary voltage polarity.]

Operation in the cutoff, saturation, and linear regions of the BJT characteristic are provided as follows:

1. *Linear-region operation:*
   Base-emitter forward biased
   Base-collector reverse biased
2. *Cutoff-region operation:*
   Base-emitter reverse biased
3. *Saturation-region operation:*
   Base-emitter forward biased
   Base-collector forward biased

5.3 FIXED-BIAS CIRCUIT

The fixed-bias circuit shown in Fig. 5.2 provides a relatively straightforward and simple starting point in the dc bias considerations. While the present circuit analysis shows the calculations using an *npn* transistor, the equations and calculations apply

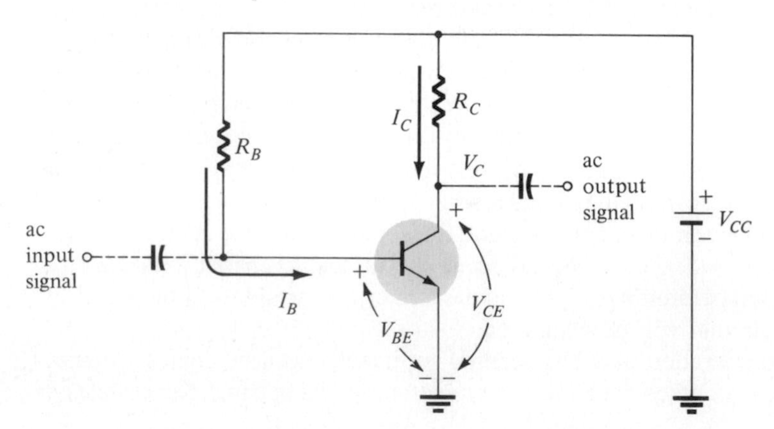

Figure 5.2 Fixed-bias circuit.

equally to a *pnp* transistor merely by changing all current directions and voltage polarities.

It is possible to consider the biasing of a BJT by separately analyzing the base-emitter and the base-collector dc bias loops. Recall that for operation of the BJT in its linear region the base-emitter must be forward biased (or else the transistor is biased *off*), and the base-collector reverse biased. Given the fixed-biased circuit of Fig. 5.2, how can we determine the dc bias currents and voltages for the base and collector of the transistor? In this section we develop the procedure that will determine the answers to this question.

Forward Bias of Base-Emitter

Consider first the base-emitter circuit loop shown in the partial circuit diagram of Fig. 5.3. Writing the Kirchhoff voltage equation for the loop, we get

$$+V_{CC} - I_B R_B - V_{BE} = 0$$

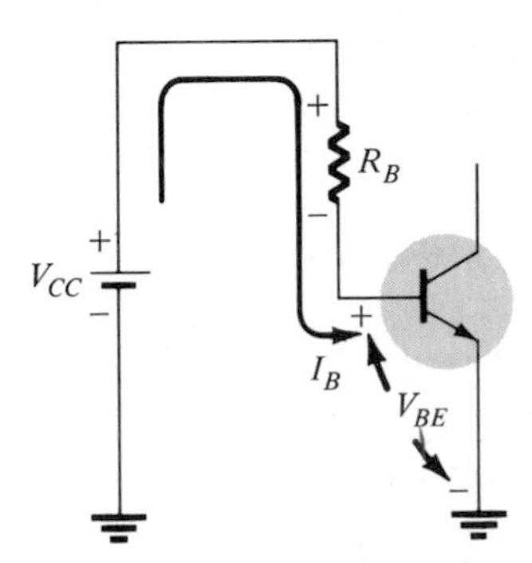

Figure 5.3 Base-emitter loop.

We can solve the foregoing equation for the base current I_B

$$\boxed{I_B = \frac{V_{CC} - V_{BE}}{R_B}} \tag{5.1}$$

Since the supply voltage V_{CC} and the base-emitter voltage V_{BE} are fixed values of voltage, the selection of a base-bias resistor fixes the value of the base current. As a good approximation we may even neglect the few tenths of a volt drop across the forward-biased base-emitter V_{BE}, obtaining the simplified form for calculating base current,

$$I_B \cong \frac{V_{CC}}{R_B} \tag{5.2}$$

Reverse Bias of Base-Collector

The collector-emitter section of the circuit (Fig. 5.4) consists of the supply battery, the collector resistor, and the transistor collector-emitter junction. The currents through the collector and emitter are about the same since I_B is small in comparison

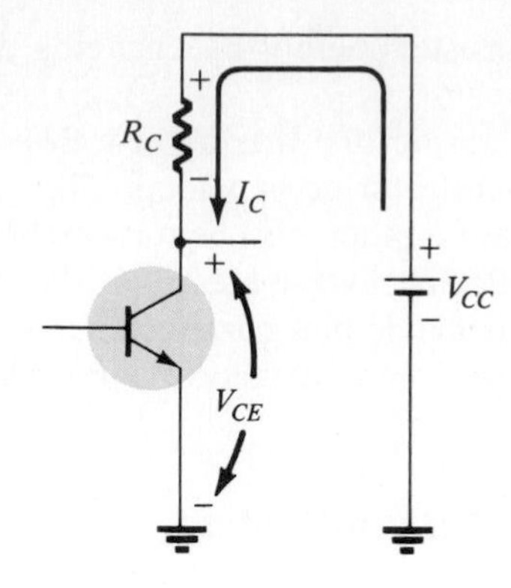

Figure 5.4 Collector-emitter loop.

to either. For linear amplifier operation the collector current is related to the base current by the transistor current gain, beta (β) or h_{FE}. Expressed mathematically,

$$\boxed{I_C = \beta I_B} \tag{5.3}$$

The base current is determined from the operation of the base-emitter section of the circuit as provided by Eq. (5.1) or (5.2). The collector current as shown by Eq. (5.3) is β times greater than the base current *and* not at all dependent on the resistance in the collector circuit. The collector current is therefore controlled by the base-emitter section of the circuit and not by the collector-base (or collector-emitter, in this case) section of the circuit.

Calculating voltage drops in the collector-emitter loop, we get

$$V_{CC} - I_C R_C - V_{CE} = 0$$

$$\boxed{V_{CE} = V_{CC} - I_C R_C} \tag{5.4}$$

EXAMPLE 5.1

Compute the dc bias voltages and currents for the circuit of Fig. 5.5.

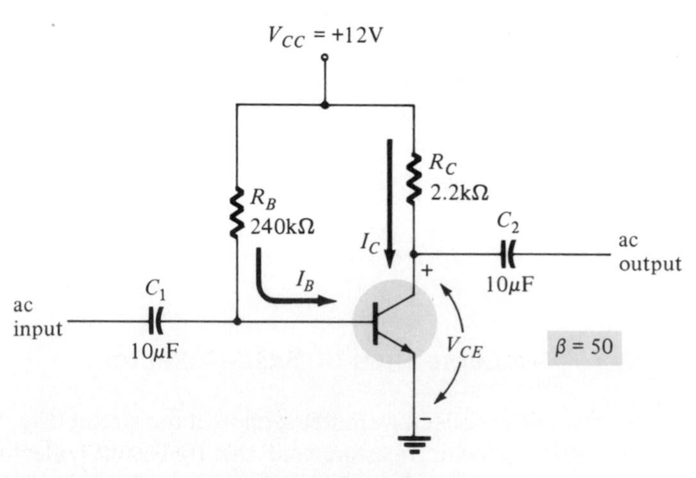

Figure 5.5 Dc fixed-bias circuit for Example 5.1.

Solution:

$$I_B = \frac{V_{CC} - V_{BE}}{R_B} = \frac{(12 - 0.7)\text{ V}}{240\text{ k}\Omega} = \mathbf{47.08\ \mu A}$$

$$I_C = \beta I_B = 50(47.08\ \mu\text{A}) = \mathbf{2.35\ mA}$$

$$V_{CE} = V_{CC} - I_C R_C = 12\text{ V} - (2.35\text{ mA})(2.2\text{ k}\Omega) = \mathbf{6.83\ V}$$

EXAMPLE 5.2

Compute the collector voltage and current for the circuit of Fig. 5.6.

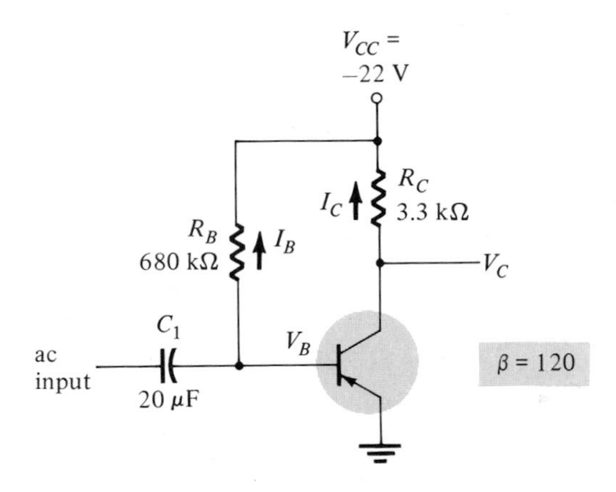

Figure 5.6 Circuit for Example 5.2.

Solution:

$$I_B = \frac{V_{CC} - V_{BE}}{R_B} = \frac{(22 - 0.7)\text{ V}}{680\text{ k}\Omega} = 31.32\ \mu\text{A}$$

$$I_C = \beta I_B = 120(31.32\ \mu\text{A}) = \mathbf{3.76\ mA}$$

$$V_C = -(V_{CC} - I_C R_C) = -[22\text{ V} - (3.76\text{ mA})(3.3\text{ k}\Omega)] = \mathbf{-9.6\ V}$$

Transistor Saturation

One additional consideration must be included in the above solution steps. The relation between collector and base current, namely, that $I_C = \beta I_B$, is true *only* if the transistor is properly biased in the linear region of the transistor's operation. If the transistor, for example, is biased in the *saturation* region, Eqs. (5.3) and (5.4) lead to incorrect results.

For the transistor to be biased in a region of linear amplifier operation (as opposed to regions of cutoff or saturation) the base-emitter junction must be forward biased *and* the base-collector junction reverse biased. Our concern here is with the second bias condition—that the collector-base be properly reverse biased. This is true only as long as the collector-emitter voltage V_{CE} is larger in value than the base-emitter forward-bias voltage V_{BE}. Since the collector-emitter voltage V_{CE} given by Eq. (5.4) is the difference between the supply voltage V_{CC} and the voltage drop across the collector resistor $(I_C R_C)$, the latter must be less than V_{CC} or in terms of the collector current, I_C must be less than V_{CC}/R_C. Stated mathematically,

$$I_C < \frac{V_{CC}}{R_C} \tag{5.5}$$

for the transistor to be biased in the active (linear) region of operation. A quick check would therefore be in order using Eq. (5.5) when performing the calculations of collector-emitter voltage to make sure that the condition just stated is correct in the circuit under consideration. If so, the three solution steps outlined above can be carried out, as representing the operation of the circuit. If, however, the above relation of maximum I_C allowable for operation in the transistor linear region is exceeded, the transistor is operating in the saturation region. In this case the collector current will be the maximum value set by the circuit:

$$I_{C_{\text{sat}}} \cong \frac{V_{CC}}{R_C} \tag{5.6}$$

and

$$V_{CE_{\text{sat}}} \cong 0 \text{ V} \qquad \text{(actually a few tenths volt)} \tag{5.7}$$

The base current calculated from Eq. (5.1) is correct in any case.

If the circuit to be analyzed is used as an amplifier, we shall not expect it to be biased in the saturation region. If, however, some value used is incorrect, or some wiring error occurs, the resulting operation might possibly bias the transistor into saturation and we must be aware of this condition. (Keep in mind that the saturation condition is undesirable only for amplifier operation. For operation in computer switching circuits the saturation region of operation is important.)

5.4 DC BIAS CIRCUIT WITH EMITTER RESISTOR

The dc bias circuit of Fig. 5.7 contains an emitter resistor to provide better bias stability than the fixed-bias circuit considered in Section 5.3. For the analysis of the circuit operation we shall deal separately with the base-emitter loop of the circuit and the collector-emitter loop of Fig. 5.7.

Base-Emitter Loop

A partial circuit diagram of the base-emitter loop is shown in Fig. 5.8. Writing Kirchhoff's voltage equation for the loop, we get

$$V_{CC} - I_B R_B - V_{BE} - I_E R_E = 0$$

We can replace I_E with $(\beta + 1)I_B$ so that the equation above can be written as[1]

---

[1] $I_E = I_C + I_B = \beta I_B + I_B = (\beta + 1)I_B$

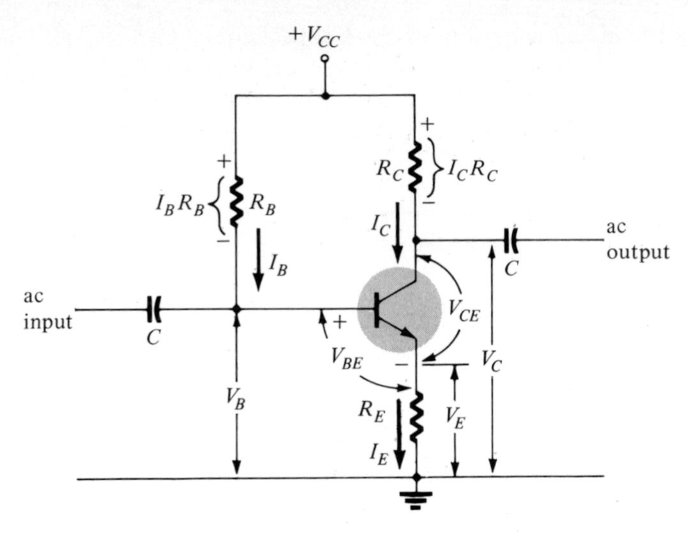

Figure 5.7 Dc bias circuit with emitter-stabilization resistor.

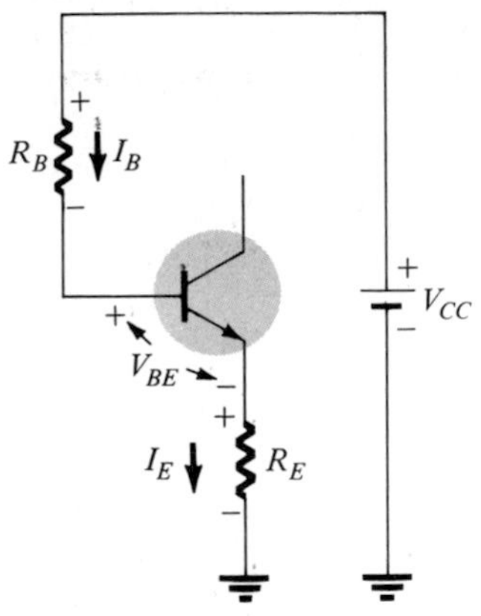

Figure 5.8 Base-emitter loop with emitter resistor.

$$V_{CC} - I_B R_B - V_{BE} - (\beta + 1)I_B R_E = 0$$

Solving for the base current, we get

$$\boxed{I_B = \frac{V_{CC} - V_{BE}}{R_B + (\beta + 1)R_E}} \qquad (5.8)$$

Note that the difference between the fixed-bias current calculation [Eq. (5.1)] and Eq. (5.8) is the additional term of $(\beta + 1)R_E$ in the denominator.

Collector-Emitter Loop

The collector-emitter loop is shown in Fig. 5.9. Writing the Kirchhoff voltage equation for this loop, we get

$$V_{CC} - I_C R_C - V_{CE} - I_E R_E = 0$$

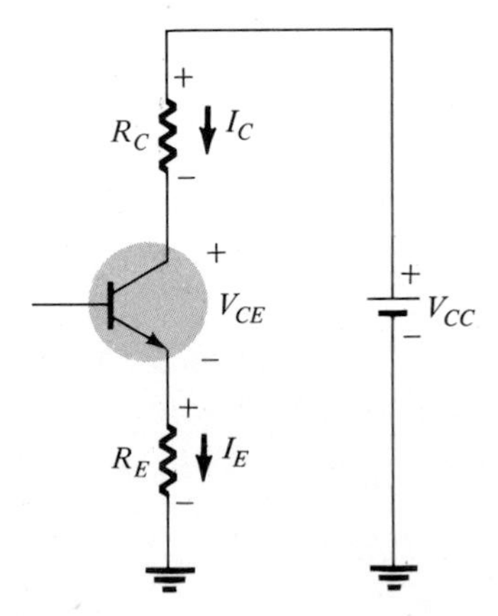

Figure 5.9 Collector-emitter loop with emitter resistor.

The collector current I_C is calculated using Eq. (5.3). Using the relation

$$I_E \cong I_C$$

we can solve for the voltage across the collector-emitter:

$$\boxed{V_{CE} \cong V_{CC} - I_C(R_C + R_E)} \tag{5.9}$$

The voltage measured from emitter to ground is

$$\boxed{V_E = I_E R_E \cong I_C R_E} \tag{5.10}$$

and the voltage measured from collector to ground is

$$\boxed{V_C = V_{CC} - I_C R_C} \tag{5.11}$$

The voltage at which the transistor is biased is measured from collector to emitter, V_{CE}, which is given by Eq. (5.9) and may also be calculated as

$$V_{CE} = V_C - V_E$$

EXAMPLE 5.3

Calculate the dc bias voltage V_{CE} and current I_C in the circuit of Fig. 5.10.

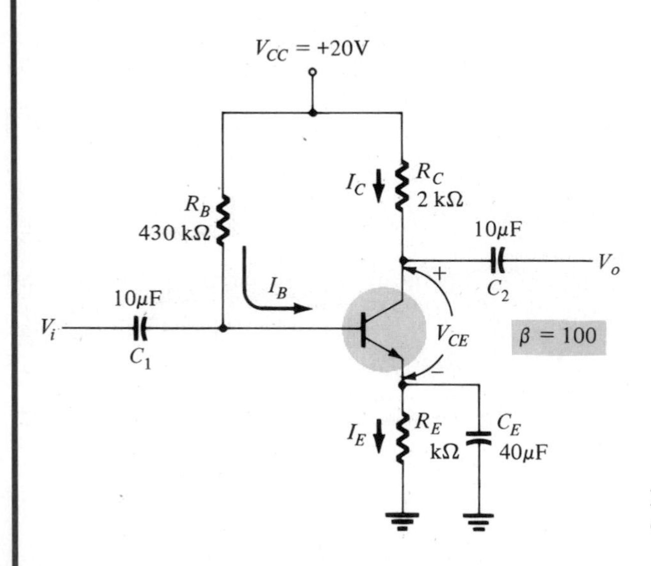

Figure 5.10 Emitter-stabilized bias circuit for Examples 5.3 and 5.4.

Solution:

$$I_B = \frac{V_{CC} - V_{BE}}{R_B + (\beta + 1)R_E} = \frac{20\text{ V} - 0.7\text{ V}}{430\text{ k}\Omega + 101(1\text{ k}\Omega)} = \frac{19.3\text{ V}}{531\text{ k}\Omega} = 36.35\ \mu\text{A}$$

$$I_C = \beta I_B = 100(36.35\ \mu\text{A}) = \mathbf{3.635\ mA} \cong I_E$$

$$V_{CE} = V_{CC} - I_C R_C - I_E R_E = 20\text{ V} - 3.635\text{ mA}(2\text{ k}\Omega) - 3.635\text{ mA}(1\text{ k}\Omega) \cong \mathbf{9.1\ V}$$

EXAMPLE 5.4

Calculate the value of collector resistor R_C needed to obtain $V_C = 10$ V, using the circuit of Fig. 5.10.

Solution:

$$I_B = \frac{V_{CC} - V_{BE}}{R_B + (\beta + 1)R_E} = \frac{20\text{ V} - 0.7\text{ V}}{430\text{ k}\Omega + 101(1\text{ k}\Omega)} = 36.35\ \mu\text{A}$$

$$I_C = \beta I_B = 100(36.35\ \mu\text{A}) = 3.635\text{ mA}$$

Note that I_B and I_C are still the same values as calculated in Example 5.3. Using Eq. (5.11)

$$V_C = V_{CC} - I_C R_C$$

$$10 = 20 - (3.635 \times 10^{-3})R_C$$

which can be solved for R_C as

$$R_C = \frac{20 - 10}{3.635 \times 10^{-3}} = \mathbf{2.75\ k\Omega} \quad (\text{use } 2.7\text{ k}\Omega)$$

IMPROVED BIAS STABILITY

The addition of the emitter resistor to the dc bias of the BJT provides improved stability; that is, the dc bias currents and voltages remain closer to where they were set by the circuit even when outside conditions, such as supply voltage, temperature, and even transistor beta, change. While a mathematical analysis is provided in Section 5.10, some comparison of the improvement can be obtained as demonstrated by Example 5.5.

EXAMPLE 5.5

Prepare a table comparing the bias voltage and currents of the circuit of Fig. 5.5 for the given value of $\beta = 50$ and for a new value of $\beta = 100$, and for the circuit in Fig. 5.10 for the given value of $\beta = 100$ and for a new value of $\beta = 50$.

Solution:

Using the results calculated in Example 5.1 and then repeating for a value of $\beta = 100$ yields the following:

β	I_B (μA)	I_C (mA)	V_{CE} (V)
50	47.08	2.35	6.83
100	47.08	4.71	1.64

The BJT collector current is seen to change by 100% due to the 100% change in the value of β (and no change in I_B).

Using the results calculated in Example 5.3 and then repeating for a value of $\beta = 50$, we have the following:

β	I_B (μA)	I_C (mA)	V_{CE} (V)
50	40.12	2.01	13.97
100	36.35	3.635	9.095

The BJT collector current changes by less than 50% due to the 100% change in β. Notice that I_B increased, helping maintain the value of I_C—or at least reducing the overall change in I_C due ω the change in β.

5.5 DC BIAS CIRCUIT INDEPENDENT OF BETA

Approximate Analysis

In the previous dc bias circuits the values of the bias current and voltage of the collector depended on the current gain (β) of the transistor. But the value of beta is temperature sensitive, especially for silicon transistors and since, also, the nominal value of beta is not well defined, it would be desirable for these as well as other reasons (transistor replacement and stability) to provide a dc bias circuit that is *independent* of the transistor beta. The circuit of Fig. 5.11 meets these conditions and is thus a very popular bias circuit.

Let us first analyze the base-emitter input circuit. If the resistance seen looking into the base (see Fig. 5.12) is much larger than that of resistor R_{B2}, then the base voltage is set by the voltage divider of R_{B1} and R_{B2}. If this is so, then the current through R_{B1} goes almost completely into R_{B2} and the two resistors may be considered effectively in series. The voltage at the junction of the resistors, which is also the voltage of the

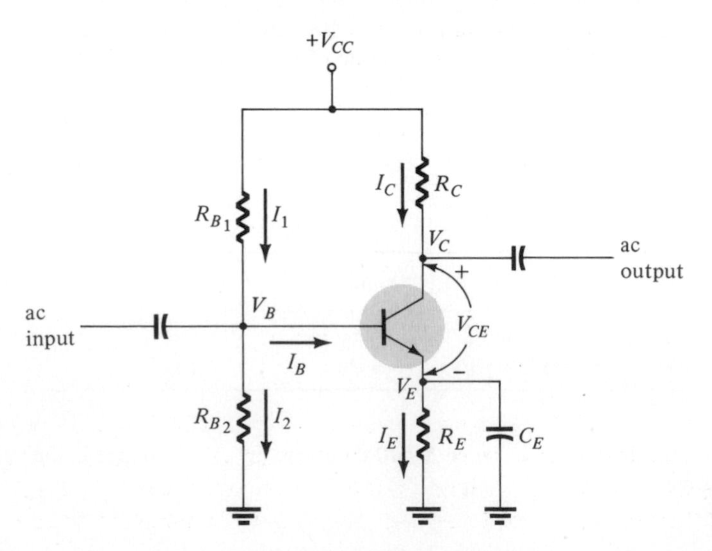

Figure 5.11 Beta-independent dc bias circuit.

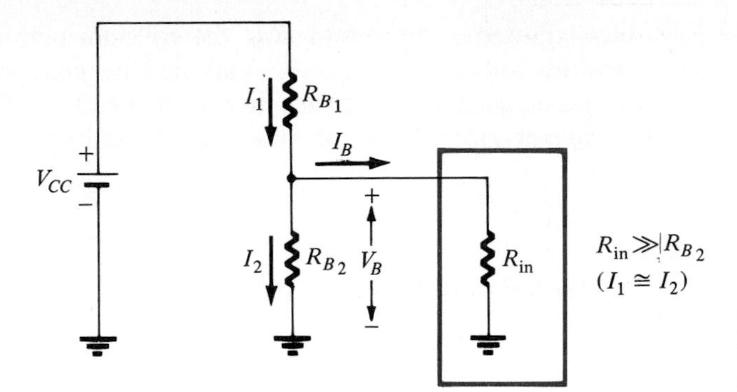

Figure 5.12 Partial bias circuit for calculating approximate base voltage V_B.

base of the transistor, is then determined simply by the voltage-divider network of R_{B1} and R_{B2} and the supply voltage. Calculating the voltage at the transistor base due to the voltage-divider network of resistors R_{B1} and R_{B2}, we get

$$V_B = \frac{R_{B2}}{R_{B1} + R_{B2}} V_{CC} \tag{5.12}$$

where V_B is the voltage measured from base to ground.

We then calculate the voltage at the emitter from

$$V_E = V_B - V_{BE} \tag{5.13}$$

The current in the emitter may then be calculated from

$$I_E = \frac{V_E}{R_E} \tag{5.14}$$

and the collector current is then

$$I_C \cong I_E \tag{5.15}$$

The voltage drop across the collector resistor is

$$V_{R_C} = I_C R_C$$

The voltage at the collector (measured with respect to ground) can then be obtained

$$V_C = V_{CC} - V_{R_C} = V_{CC} - I_C R_C \tag{5.16}$$

and finally, the voltage from collector to emitter is calculated from

$$V_{CE} = V_C - V_E$$

$$V_{CE} = V_{CC} - I_C R_C - I_E R_E \tag{5.17}$$

Look back at the procedure just outlined and notice that the value of beta was never used in Eqs. (5.12) through (5.17). The base voltage is set by resistors R_{B1} and R_{B2} and the supply voltage. The emitter voltage is fixed at approximately the same voltage value as the base. Resistor R_E then determines emitter and collector currents. Finally, R_C determines the collector voltage and, thereby, the collector-emitter bias voltage.

The base voltage V_B is best adjusted using resistor R_{B2}, the collector current by resistor R_E, and the collector-emitter voltage by resistor R_C. Varying other components will have less effect on the dc bias adjustments. The capacitor components are part of the ac amplifier operation but have no effect on the dc bias and will not be discussed at this time.

EXAMPLE 5.6

Calculate the dc bias voltage V_{CE} and current I_C for the circuit of Fig. 5.13.

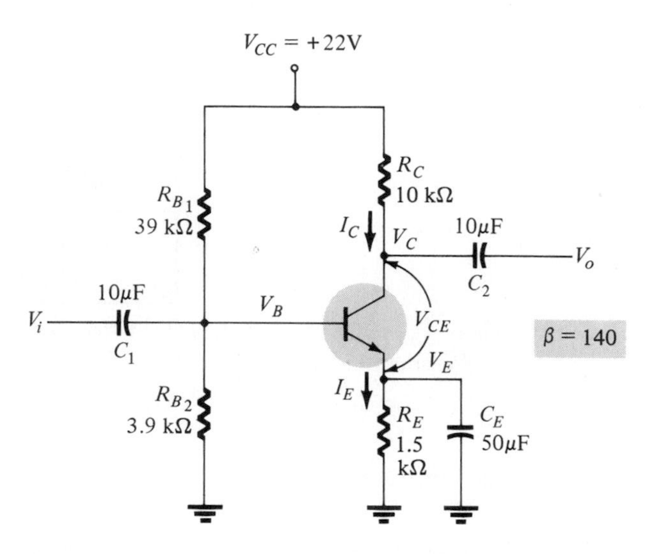

Figure 5.13 Beta-stabilized circuit for Example 5.6.

Solution:

$$V_B = \frac{R_{B2}}{R_{B1} + R_{B2}} V_{CC} = \frac{39}{39 + 3.9}(22) = 2 \text{ V}$$

$$V_E = V_B - V_{BE} = 2 - 0.7 = 1.3 \text{ V}$$

$$I_E = \frac{V_E}{R_E} \cong I_C = \frac{1.3 \text{ V}}{1.5 \text{ k}\Omega} = \mathbf{0.867\ mA}$$

$$V_C = V_{CC} - I_C R_C = 22 - (0.867 \text{ mA})(10 \text{ k}\Omega) = 13.33 \text{ V}$$

$$V_{CE} = V_C - V_E = 13.33 - 1.3 = \mathbf{12.03\ V}$$

Exact Analysis

The circuit of Fig. 5.11 can be analyzed as covered above only if the voltage divider is not loaded by the dc impedance looking into the transistor. A more exact analysis can be obtained by using the Thévenin equivalent of the voltage divider as described by the following analysis:

The Thévenin equivalent resistance of R_{B1} and R_{B2} is

$$R_{BB} = \frac{R_{B1}R_{B2}}{R_{B1} + R_{B2}} \tag{5.18}$$

The Thévenin equivalent voltage is

$$V_{BB} = \frac{R_{B2}}{R_{B1} + R_{B2}} V_{CC} \tag{5.19}$$

The dc circuit to be analyzed then can be redrawn as in Fig. 5.14. From Fig. 5.14, we can calculate I_B

$$I_B = \frac{V_{BB} - V_{BE}}{R_{BB} + (\beta + 1)R_E} \tag{5.20}$$

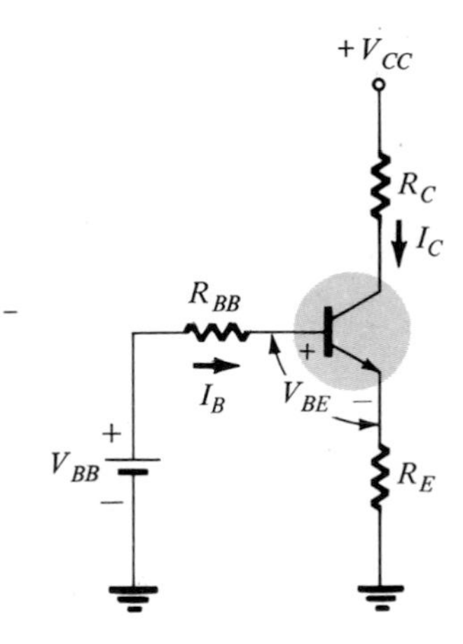

Figure 5.14 Dc circuit to analyze using Thévenin equivalent.

The collector current is then

$$I_C = \beta I_B$$

and the value of V_{CE} can be obtained using Eq. (5.9). The exact analysis is demonstrated in the following example.

EXAMPLE 5.7

Calculate the dc bias voltage V_{CE} and current I_C for the circuit of Fig. 5.13.

Solution:

$$V_{BB} = \frac{R_{B2}}{R_{B1} + R_{B2}} V_{CC} = \frac{3.9\text{ k}\Omega}{39\text{ k}\Omega + 3.9\text{ k}\Omega}(22\text{ V}) = 2\text{ V}$$

$$R_{BB} = \frac{R_{B1}R_{B2}}{R_{B1} + R_{B2}} = \frac{39\text{ k}\Omega \cdot 3.9\text{ k}\Omega}{39\text{ k}\Omega + 3.9\text{ k}\Omega} = 3.55\text{ k}\Omega$$

$$I_B = \frac{V_{BB} - V_{BE}}{R_{BB} + (\beta + 1)R_E} = \frac{2\text{ V} - 0.7\text{ V}}{3.55\text{ k}\Omega + 141(1.5\text{ k}\Omega)} = 6.05\ \mu\text{A}$$

$$I_C = \beta I_B = 140(6.05\ \mu\text{A}) = \mathbf{0.85\ mA} \cong I_E$$

$$V_{CE} = V_{CC} - I_C(R_C + R_E) = 22\text{ V} - 0.85\text{ mA}(10\text{ k}\Omega + 1.5\text{ k}\Omega)$$
$$= 22\text{ V} - 9.8\text{ V} = \mathbf{12.2\ V}$$

Comparing these values with the values calculated in Example 5.6, we note the difference is only about 2%.

EXAMPLE 5.8

Using an exact bias analysis of the circuit of Fig. 5.13, compare the collector current and the collector-emitter voltage for the given β of 140 and for a new β of 70.

Solution:

Using the results calculated in Example 5.7 and repeating for a value of $\beta = 70$, we have

β	I_C (mA)	V_{CE} (V)
140	0.85	12.2
70	0.83	12.46

Notice how well the circuit maintains the collector current and collector-emitter bias voltage; even with a change in β of 100%, the bias values only changed by less than 3% in this circuit.

The choice of using the exact or approximate analysis depends on whether βR_E is much greater than R_{BB}. Using 10% tolerance resistors, for example, engineering judgment would provide that any analysis giving a result within 10% of the exact value would be acceptable. In this consideration the approximate analysis would be satisfactory as long as

$$\beta R_E > 10R_{BB}$$

Transistor Saturation

If the transistor in the circuit of Fig. 5.7 is biased in saturation, the saturation voltage across the collector-emitter, $V_{CE_{sat}}$, is approximately zero, and the saturation current is

$$\boxed{I_{C_{sat}} = \frac{V_{CC}}{R_C + R_E}} \tag{5.21}$$

As long as I_C calculated using $I_C = \beta I_B$ is less than $I_{C_{sat}}$, the dc bias calculations of Eqs. (5.12) through (5.20) hold true; otherwise, $V_{CE} = V_{CE_{sat}}$ of the transistor at the collector current of $I_{C_{sat}}$ calculated using Eq. (5.21).

5.6 DC BIAS WITH VOLTAGE FEEDBACK

Apart from the use of an emitter resistor to provide improved bias stability, voltage feedback also provides improved dc bias stability. Figure 5.15 shows a dc bias circuit with voltage feedback. This section shows how to calculate the dc currents and voltages of this circuit.

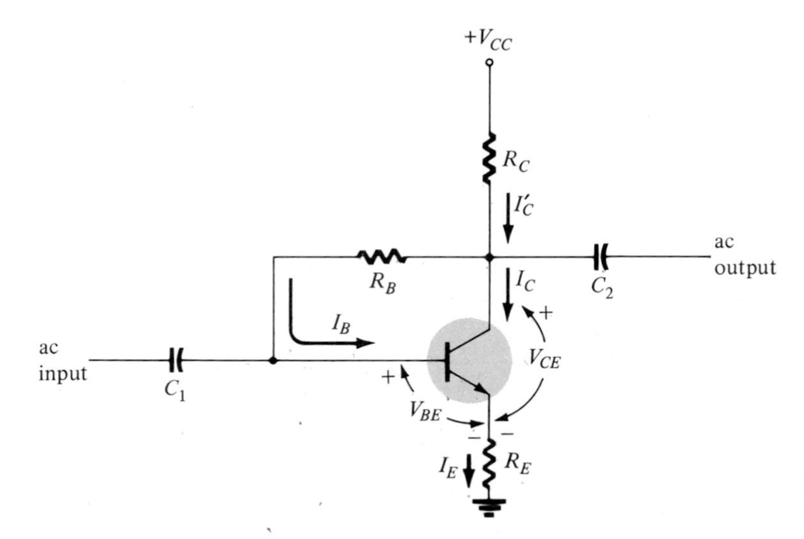

Figure 5.15 Dc bias circuit with voltage feedback.

Base-Emitter Loop

Figure 5.16 shows the base-emitter loop of the voltage feedback circuit. Writing the Kirchhoff voltage equation around the loop gives

$$+V_{CC} - I'_C R_C - I_B R_B - V_{BE} - I_E R_E = 0$$

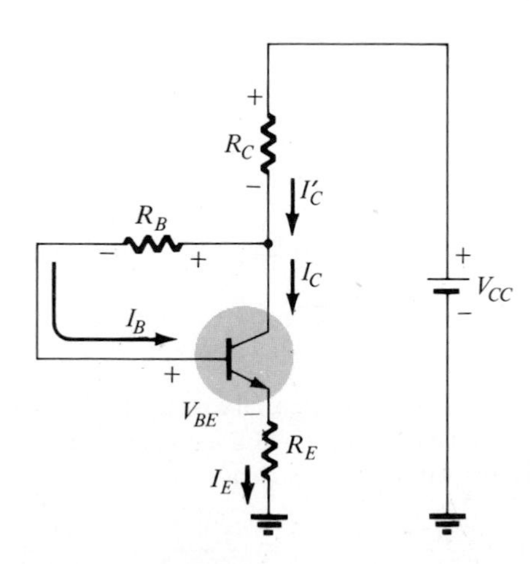

Figure 5.16 Partial circuit showing base-emitter loop.

The current I'_C is the sum of I_C and I_B

$$I'_C = I_C + I_B = I_E = (\beta + 1)I_B$$

which, substituted into the Kirchhoff voltage equation, yields

$$V_{CC} - (\beta + 1)I_B R_C - I_B R_B - V_{BE} - (\beta + 1)I_B R_E = 0$$

Solving for the base current I_B, we get

$$\boxed{I_B = \frac{V_{CC} - V_{BE}}{R_B + (\beta + 1)(R_C + R_E)}} \tag{5.22}$$

Collector-Emitter Loop

From the partial circuit diagram of the collector-emitter section show in Fig. 5.17 the Kirchhoff voltage equation is

$$+V_{CC} - I'_C R_C - V_{CE} - I_E R_E = 0$$

and using $I'_C = I_E$, we can solve for V_{CE}:

$$\boxed{V_{CE} = V_{CC} - I_E(R_C + R_E)} \tag{5.23}$$

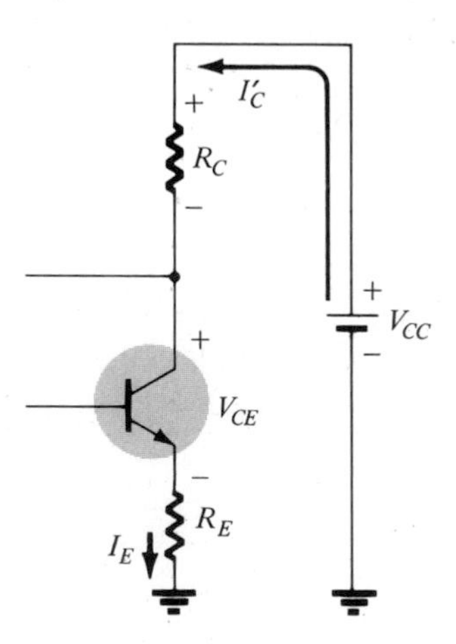

Figure 5.17 Partial circuit showing collector-emitter loop.

EXAMPLE 5.9

Calculate the dc bias current I_E and voltage V_{CE} for the circuit of Fig. 5.18 using voltage feedback.

Solution:

The feedback resistor R_B is the sum of the resistors between collector and base (the capacitor in the ac feedback path provides for attenuating or blocking the ac feedback signal and has no effect on the dc bias calculation).

$$I_B = \frac{V_{CC} - V_{BE}}{R_B + (\beta + 1)(R_C + R_E)} = \frac{(10 - 0.7)\text{ V}}{250\text{ k}\Omega + (51)(3\text{ k}\Omega + 1.2\text{ k}\Omega)} = 20.03\ \mu\text{A}$$

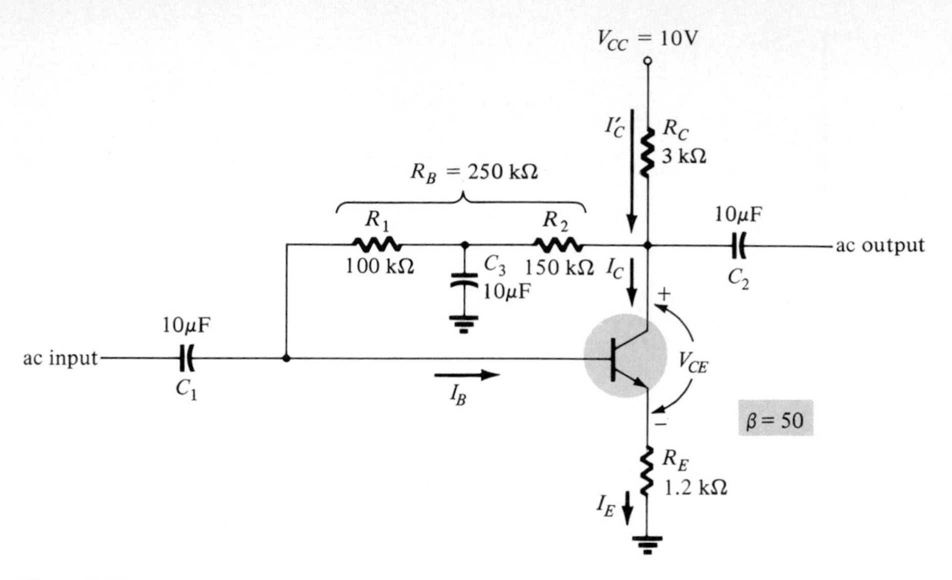

Figure 5.18 Voltage-feedback-stabilized bias circuit for Example 5.9.

$$I_E = (\beta + 1)I_B = (51)(20.03\ \mu\text{A}) = \mathbf{1.02\ mA}$$

$$V_{CE} = V_{CC} - I_E(R_C + R_E) = 10\text{ V} - (1.02\text{ mA})(3\text{ k}\Omega + 1.2\text{ k}\Omega)$$

$$= 10 - 4.28 = \mathbf{5.72\ V}$$

EXAMPLE 5.10

Calculate the dc collector current I_C and voltage V_C for the bias circuit of Fig. 5.19

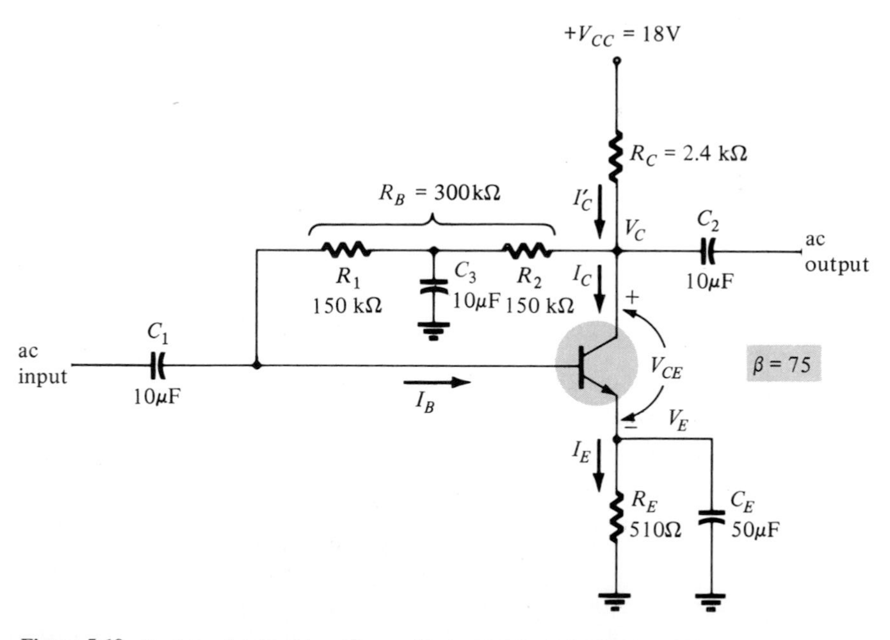

Figure 5.19 Dc bias circuit with emitter resistor and voltage feedback stabilization.

Solution:

$$I_B = \frac{V_{CC} - V_{BE}}{R_B + (\beta + 1)(R_C + R_E)} = \frac{(18 - 0.7)\text{ V}}{300\text{ k}\Omega + (76)(2.4\text{ k}\Omega + 510\ \Omega)} = 33.2\ \mu\text{A}$$

$$I_C = \beta I_B = 75(33.2\ \mu\text{A}) = \mathbf{2.49\ mA}$$

$$V_C = V_{CC} - I_C R_C = 18\text{ V} - (2.49\text{ mA})(2.4\text{ k}\Omega) = \mathbf{12.02\ V}$$

5.7 ANALYSIS OF VARIOUS DC BIAS CIRCUITS

While the preceding sections analyzed standard *npn* bias circuits, we find many various circuit configurations in real devices. This section provides dc bias calculations for a number of circuits that are not constructed in the standard form covered previously. The techniques used, however, still apply, as will be shown.

EXAMPLE 5.11

Calculate the collector current I_C and voltage V_{CE} for the circuit of Fig. 5.20.

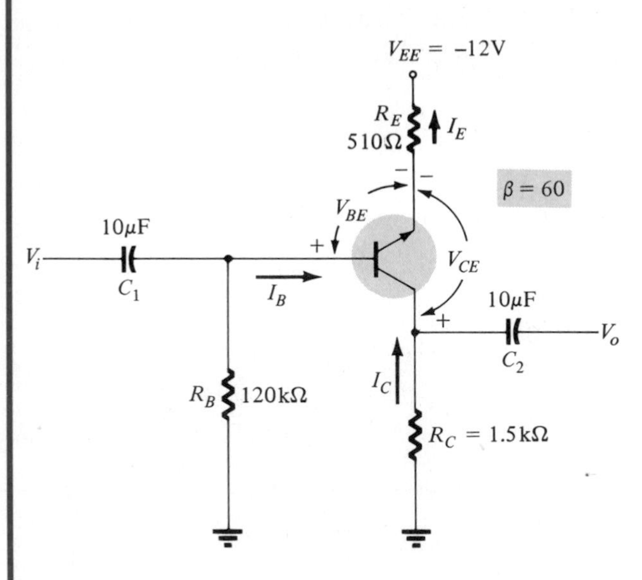

Figure 5.20 Bias circuit for Example 5.11.

Solution:

Base-Emitter Loop: $-I_B R_B - V_{BE} - I_E R_E + V_{EE} = 0$

$$I_B = \frac{V_{EE} - V_{BE}}{R_B + (\beta + 1)R_E} = \frac{12\text{ V} - 0.7\text{ V}}{120\text{ k}\Omega + 61(0.510\text{ k}\Omega)} = 74.78\ \mu\text{A}$$

$$I_C = \beta I_B = 60(74.78\ \mu\text{A}) = \mathbf{4.49\ mA}$$

Collector-Emitter Loop: $-V_{EE} + I_E R_E + V_{CE} + I_C R_C = 0$

$$V_{CE} \cong V_{EE} - I_C(R_C + R_E) = 12\text{ V} - (4.49\text{ mA})(1.5\text{ k}\Omega + 0.510\text{ k}\Omega) = \mathbf{2.975\ V}$$

EXAMPLE 5.12

Calculate the bias voltage V_E and current I_C for the circuit of Fig. 5.21.

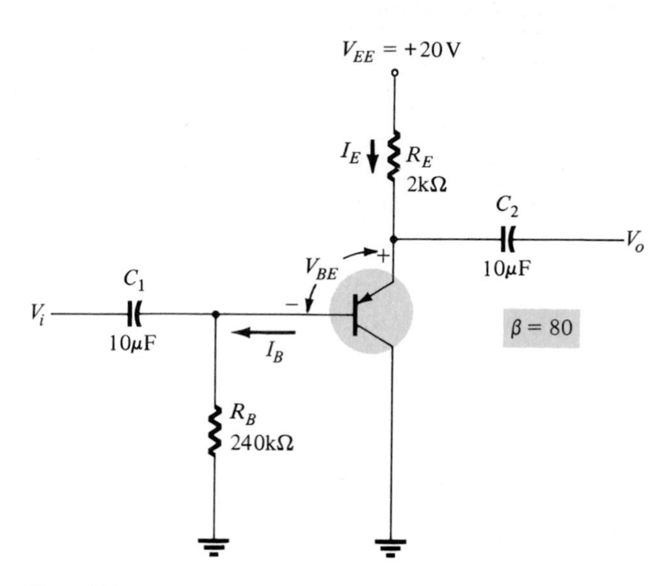

Figure 5.21 Bias circuit for Example 5.12.

Solution:

Writing the base-emitter loop equation

$$V_{EE} - I_E R_E - V_{BE} - I_B R_B = 0$$

$$I_B = \frac{V_{EE} - V_{BE}}{R_B + (\beta + 1)R_E} = \frac{20\text{ V} - 0.7\text{ V}}{240\text{ k}\Omega + 81(2\text{ k}\Omega)} = 48.01\ \mu\text{A}$$

$$I_C = \beta I_B = 80(48.01\ \mu\text{A}) = \mathbf{3.84\ mA} \cong I_E$$

$$V_E = V_{EE} - I_E R_E = 20\text{ V} - (3.84\text{ mA})(2\text{ k}\Omega) = \mathbf{12.32\ V}$$

EXAMPLE 5.13

Calculate the collector voltage V_C for the circuit of Fig. 5.22. (Use the approximate voltage-divider method.)

Solution:

$$V_B \cong \frac{R_{B1}}{R_{B1} + R_{B2}} V_{EE} = \frac{43\text{ k}\Omega}{43\text{ k}\Omega + 10\text{ k}\Omega}(10\text{ V}) = 8.11\text{ V}$$

$$V_E = V_B + V_{BE} = 8.11\text{ V} + 0.7\text{ V} = 8.81\text{ V}$$

$$I_E = \frac{V_{EE} - V_E}{R_E} = \frac{10\text{ V} - 8.81\text{ V}}{2\text{ k}\Omega} = \mathbf{0.595\ mA} \cong I_C$$

$$V_C = I_C R_C = (0.595\text{ mA})(6.2\ k\Omega) = \mathbf{3.69\ V}$$

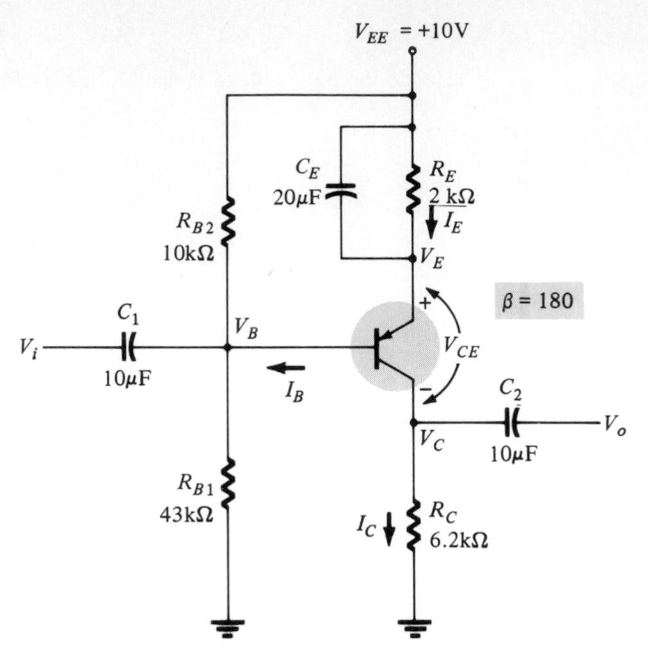

Figure 5.22 Bias circuit for Example 5.13.

EXAMPLE 5.14

Determine the collector voltage V_C and current I_C for the circuit of Fig. 5.23.

Solution:

$$-I_B R_B - V_{BE} + V_{EE} = 0$$

$$I_B = \frac{V_{EE} - V_{BE}}{R_B} = \frac{(9 - 0.7)\text{ V}}{100\text{ k}\Omega} = 83\ \mu\text{A}$$

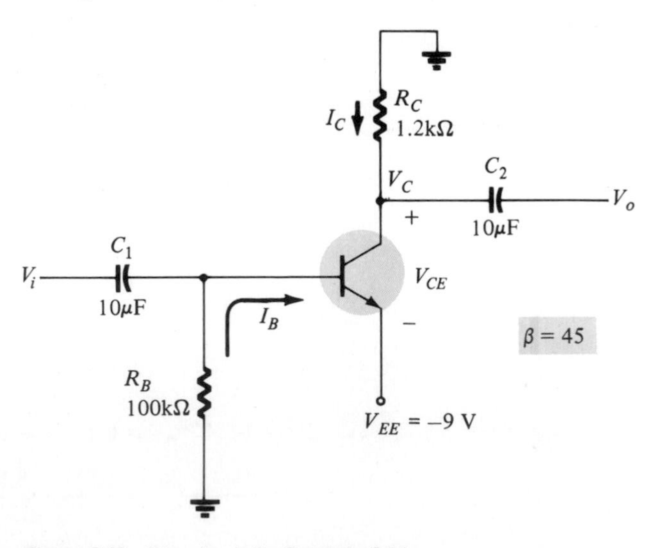

Figure 5.23 Bias circuit for Example 5.14.

$$I_C = \beta I_B = 45(83\ \mu\text{A}) = \mathbf{3.735\ mA}$$

$$V_C = -I_C R_C = -(3.735\ \text{mA})(1.2\ \text{k}\Omega) = \mathbf{-4.48\ V}$$

EXAMPLE 5.15

Calculate the emitter current I_E and collector voltage V_C for the circuit of Fig. 5.24.

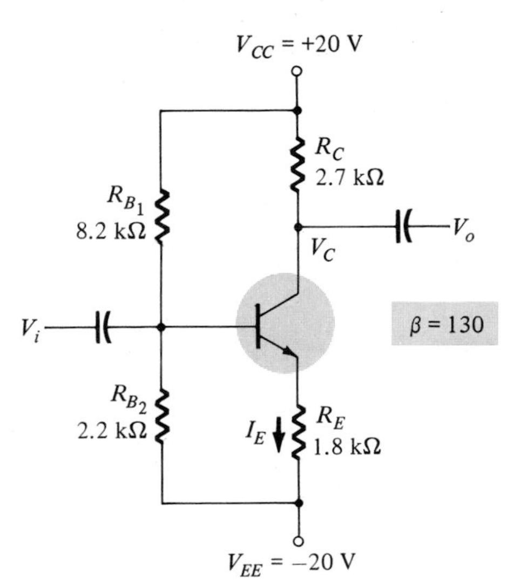

Figure 5.24 Circuit for Example 5.15.

Solution:

$$R_{BB} = \frac{(8.2\ \text{k}\Omega)(2.2\ \text{k}\Omega)}{8.2\ \text{k}\Omega + 2.2\ \text{k}\Omega} = 1.735\ \text{k}\Omega$$

$$V_{BB} = \frac{2.2\ \text{k}\Omega}{8.2\ \text{k}\Omega + 2.2\ \text{k}\Omega}(20\ \text{V}) + \frac{8.2\ \text{k}\Omega}{8.2\ \text{k}\Omega + 2.2\ \text{k}\Omega}(-20\ \text{V})$$

$$= 4.23\ \text{V} - 15.77\ \text{V} = -11.54\ \text{V}$$

$$-V_{BB} - I_B R_{BB} - V_{BE} - I_E R_E + V_{EE} = 0$$

$$I_B = \frac{V_{EE} - V_{BB} - V_{BE}}{R_{BB} + (\beta + 1)R_E} = \frac{20 - 11.54 - 0.7}{1.735\ \text{k}\Omega + 131(1.8\ \text{k}\Omega)}$$

$$= 32.67\ \mu\text{A}$$

$$I_E = (\beta + 1)I_B = 131(32.67\ \mu\text{A}) = \mathbf{4.28\ mA}$$

$$V_C = V_{CC} - I_C R_C = 20\ \text{V} - (4.28\ \text{mA})(2.7\ \text{k}\Omega) = \mathbf{8.4\ V}$$

5.8 GRAPHICAL DC BIAS ANALYSIS

The previous analysis of the dc bias currents and voltages was carried out mathematically for a number of transistor circuits. The only factors of interest used were the current gain (β) and base-emitter voltage (V_{BE}) when forward-biased. This section shows a graphical technique for finding the operating point of a transistor circuit. The graphical method demonstrated provides additional insight into the choice of operating point and leads into Section 5.9 on the design of a dc bias circuit.

The typical CE collector characteristic, shown in Fig. 5.25, only defines the overall operation of the transistor device. The circuit constraints must also be taken into account in obtaining the actual operating point (called the *quiescent operating point* or *Q*-point).

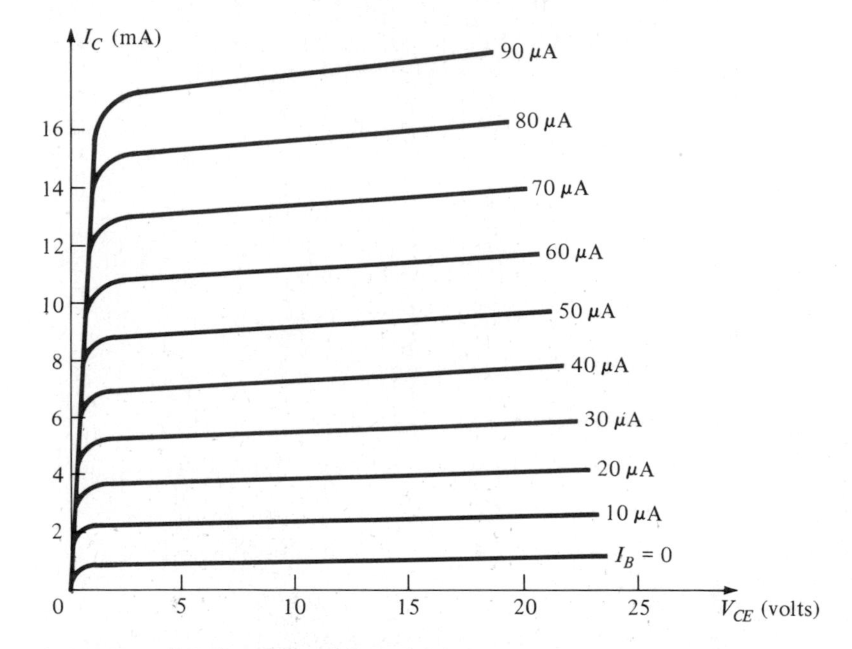

Figure 5.25 Transistor collector characteristic.

The equation of the collector-emitter loop may be graphically plotted onto the collector characteristic. Using Eq. (5.9) as typical of most circuits previously covered, we can rewrite that equation to solve for the collector current as follows:

$$V_{CE} = V_{CC} - I_C(R_C + R_E)$$

$$I_C = \frac{V_{CC}}{R_C + R_E} - \frac{V_{CE}}{R_C + R_E}$$

$$I_C = \underbrace{\frac{-1}{R_C + R_E}}_{m} \cdot V_{CE} + \underbrace{\frac{V_{CC}}{R_C + R_E}}_{b} \qquad (5.24)$$

$$y = m \cdot x + b$$

Equation (5.24) shows the circuit equation as that of a straight line with slope

$$m = \frac{-1}{R_C + R_E}$$

and y-intercept $$b = \frac{V_{CC}}{R_C + R_E}$$

with the variable I_C and V_{CE}. The straight line representing Eq. (5.24) can be drawn on the graph of Fig. 5.26 by obtaining the two extreme points of the straight line as follows:

1. For $I_C = 0$, $V_{CE} = V_{CC}$ in Eq. (5.24).
2. For $V_{CE} = 0$, $I_C = V_{CC}/(R_C + R_E)$ in Eq. (5.24).

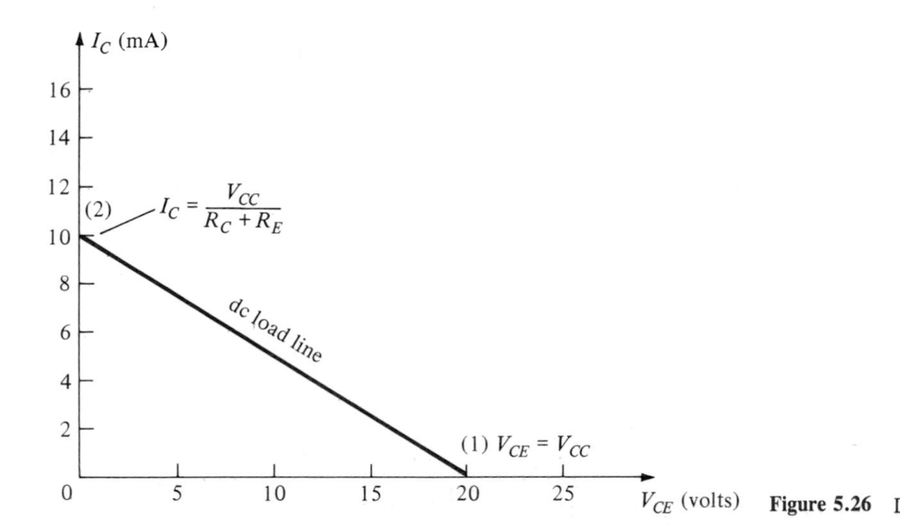

Figure 5.26 Dc load line.

These points are marked in Fig. 5.26 as (1) and (2), respectively, and the straight line connecting them is called the *dc load line*. Although the same voltage-current axis as that of the transistor collector characteristic is used, no characteristic is shown to reinforce the fact that the dc load line has nothing to do with the device itself. The load line drawn depends only on the supply voltage, V_{CC}, and the values of R_C and R_E.

The slope of the load line depends only on the values of R_C and R_E. Figure 5.27a shows the load line slopes for $(R_C + R_E)$ values smaller and larger than that of Fig. 5.26. Figure 5.27b shows that changing only the supply voltage will move the load line parallel to that of Fig. 5.26, and the slope remains the same since $(R_C + R_E)$ has not changed.

Since circuit operation depends on both the transistor characteristic and the circuit elements, plotting *both* curves (transistor characteristic and dc load line) on *one* graph allows determination of the circuit Q-point. This is shown in Fig. 5.28. The typical dc bias point shown in Fig. 5.28 is somewhat in the center of the voltage range (0 to V_{CC}) and the center of the current range [0 to $V_{CC}/(R_C + R_E)$]. A large-signal amplifier with output voltage swing near the voltage range set by the voltage supply value would require a centered operating point.

For circuits other than amplifiers different bias points may be desired. The dc load line describes all the possible values of voltage and current in the collector-emitter

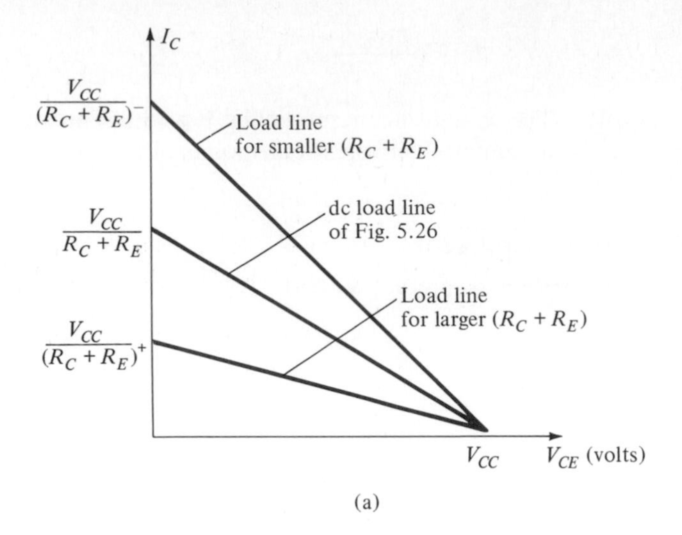

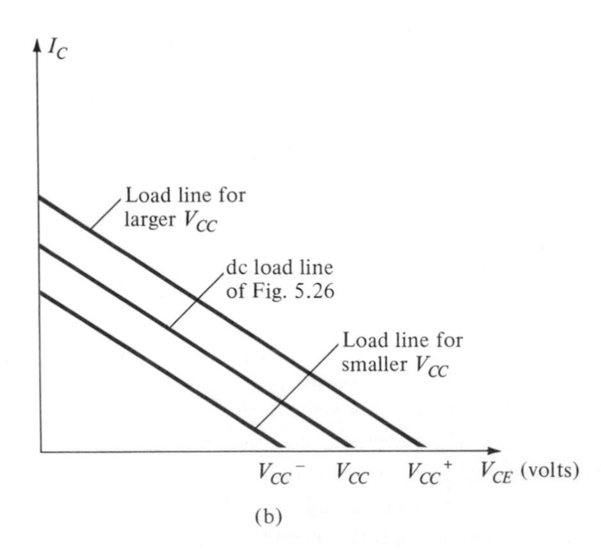

Figure 5.27 Effect of varying $(R_C + R_E)$ or V_{CC} on dc load line: (a) effect of resistor on dc load line; (b) effect of supply voltage on dc load line.

section of the circuit. Figure 5.28 shows a typical bias point set by the amount of base current and the dc load line. Adjusting the base current to higher values moves the operating point toward saturation along the load line, whereas reducing the base current moves the bias point toward transistor cutoff. For the characteristic and load line of Fig. 5.28 base currents in excess of 60 μA will drive the transistor into saturation. Note that for the load line and operating point indicated, an ac input, which adds to the dc base bias current, can go positive only by about 25 μA (from 30 to 55 μA) before the limiting condition (saturation) occurs. The variation of the ac base current, on the other hand, can go negative by 30 μA (30 to 0 μA) before cutoff is reached so that the particular bias point in Fig. 5.28 is not centered. For small-signal amplifiers with output voltage swings of less than a few volts, the exact centering of the Q-point is not essential—usually a region of largest transistor gain or most linear operation is sought.

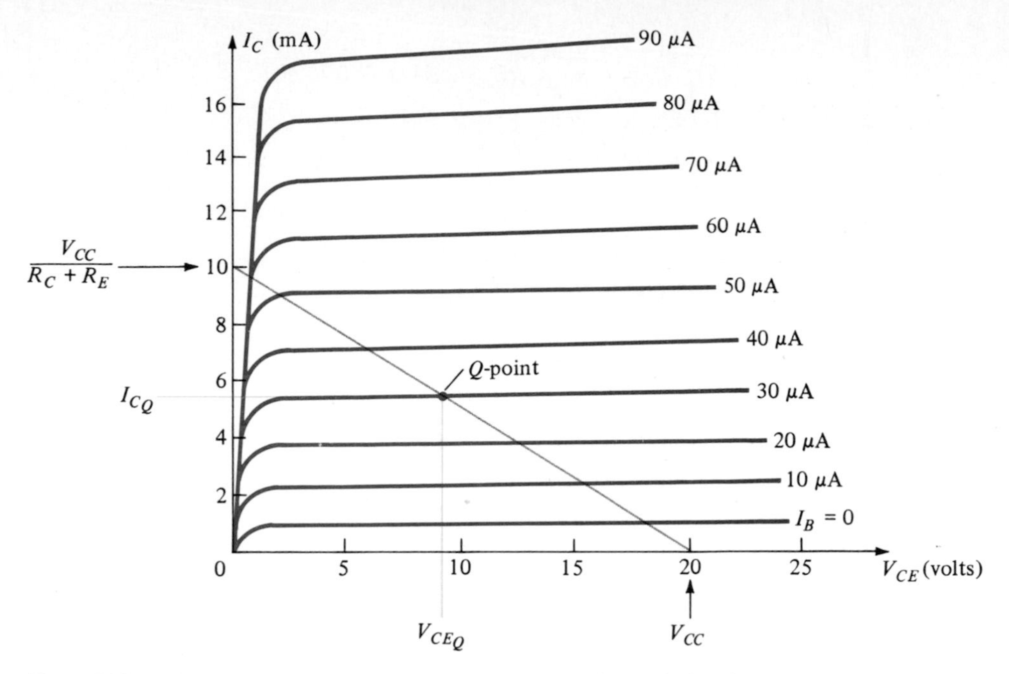

Figure 5.28 Using transistor collector characteristic and dc load line to obtain quiescent operating point (Q-point).

EXAMPLE 5.16

Determine the quiescent operating point for the circuit of Fig. 5.29 using the transistor collector characteristic of Fig. 5.25.

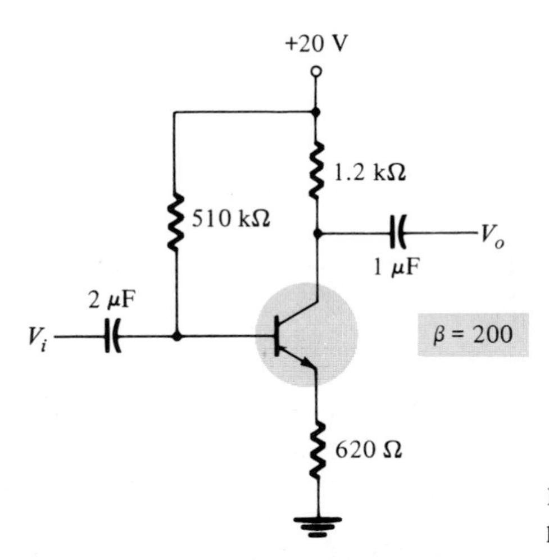

Figure 5.29 Bias circuit for Example 5.16.

Solution:

A dc load line should be plotted on the collector characteristic of Fig. 5.25. This dc load line is plotted by drawing a straight line from the point

$$\frac{V_{CC}}{R_C + R_E} = \frac{20 \text{ V}}{1.2 \text{ k}\Omega + 0.62 \text{ k}\Omega} \cong 11 \text{ mA}$$

along the I_C axis, to a point

$$V_{CC} = 20 \text{ V}$$

along the V_{CE} axis.

The base current is calculated using Eq. (5.8):

$$I_B = \frac{V_{CC} - V_{BE}}{R_B + (\beta + 1)R_E}$$

$$= \frac{(20 - 0.7) \text{ V}}{510 \text{ k}\Omega + 201(0.62 \text{ k}\Omega)} = 30.4 \ \mu\text{A}$$

Figure 5.30 shows the circuit dc load line and the transistor collector characteristic with the Q-point marked at the intersection of the dc load line and base current of $I_B = 30.4 \ \mu A$.

The transistor is seen to be biased at

$$V_{CE} = \mathbf{10 \ V} \quad \text{and} \quad I_C = \mathbf{5.5 \ mA}$$

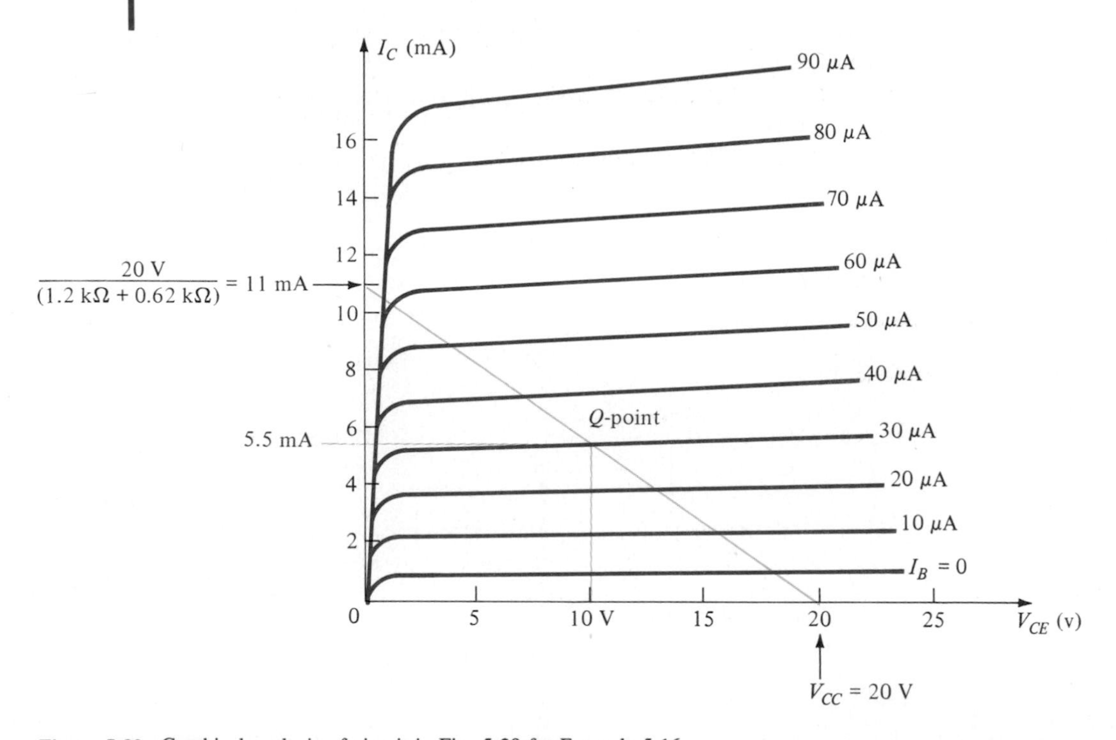

Figure 5.30 Graphical analysis of circuit in Fig. 5.29 for Example 5.16.

5.9 DESIGN OF DC BIAS CIRCUITS

Up to now the discussion has been directed to the techniques of analyzing a given transistor circuit to determine the dc operating point. Although it is often necessary to determine the Q-point of a given circuit, it is also important to be able to design a circuit to operate at a desired or specified bias point. Often the manufacturer's specification (spec) sheets provide information stating a suitable operating point (or operating region) for a particular transistor. In addition, other circuit factors connected with the given amplifier stage may also dictate some conditions of current swing, voltage swing, value of common supply voltage, and so on, which can be used in determining the Q-point in a design.

The techniques of synthesis (or design) readily follow from the previous discussions of circuit analysis. In almost all cases the calculation of the circuit elements proceeds in the reverse to those in the analysis consideration. Basically, the problem of concern in this section can be briefly stated as follows: Given a desired point or region of operation for a particular transistor, design the bias circuit (resistor and supply voltage values) to obtain the specified operating point.

In actual practice many other factors may have to be considered and may go into the selection of the desired operating point. For the moment we shall concentrate, however, on determining the component values to obtain a specified operating point. Since the basic relations and operation of a number of bias circuits have already been considered, no new theory has to be developed.

Design of Bias Circuit with Emitter Feedback Resistor

Consider first the design of the dc bias components of an amplifier circuit having emitter-resistor bias stabilization (see Fig. 5.31). The supply voltage and operating point will be selected from the manufacturer's information on the transistor used in the amplifier.

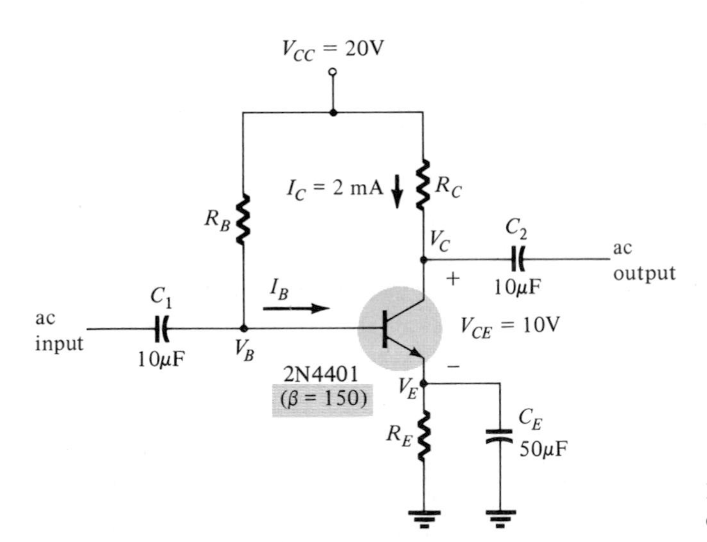

Figure 5.31 Emitter-stabilized bias circuit for design consideration.

The selection of collector and emitter resistors cannot proceed directly from the information just specified. In Eq. (5.9), which relates the voltages around the collector-emitter loop, there are two unknown quantities present—the values of the collector and emitter resistors, R_C and R_E, respectively. To make the solution to the problem easy (and meaningful), some engineering judgment may be used; that is, if some reasonable approximation of, say, the emitter voltage, can be made, the problem will then be straightforward.

Recall that the need for including a resistor from emitter to ground was to provide a means of dc bias stabilization so that the change of collector current due to leakage currents in the transistor and transistor β would not cause a large (if any) shift in the operating point. The emitter resistor cannot be unreasonably large because the voltage developed across it limits the range of voltage swing of the voltage from collector to emitter. The examples in Section 5.4 show that the voltage from emitter to ground, V_E, is typically around one-fifth to one-tenth of the supply voltage, V_{CC}. Selecting the emitter voltage in this way will permit calculating the emitter resistor, R_E, and then the collector resistor, R_C. Carrying out these calculations we get

$$V_{E_Q} \cong \frac{1}{10}V_{CC} = \frac{20\text{ V}}{10} = 2\text{ V}$$

$$R_E = \frac{V_{E_Q}}{I_{E_Q}} \cong \frac{V_{E_Q}}{I_{C_Q}} = \frac{2\text{ V}}{2\text{ mA}} = 1\text{ k}\Omega$$

$$R_C = \frac{V_{CC} - V_{CE_Q} - V_{E_Q}}{I_{C_Q}} = \frac{(20 - 10 - 2)\text{ V}}{2\text{ mA}} = 4\text{ k}\Omega$$

$$I_{B_Q} = \frac{I_{C_Q}}{\beta} = \frac{2\text{ mA}}{150} = 13.33\ \mu\text{A}$$

$$R_B = \frac{V_{CC} - V_{BE} - V_{E_Q}}{I_{B_Q}} = \frac{(20 - 0.7 - 2)\text{ V}}{13.33\ \mu\text{A}} \cong 1.3\text{ M}\Omega$$

EXAMPLE 5.17

Calculate the resistor values R_E, R_C, and R_B for a transistor amplifier circuit having emitter-resistor stabilization (Fig. 5.31). The current gain of an *npn* 2N4401 transistor is typically 90 at a collector current of 5 mA. Use a supply voltage of 20 V.

Solution:

The operating point selected from the information of supply voltage and transistor is $I_{C_Q} = 5$ mA and $V_{CE_Q} = 10$ V.

$$V_{E_Q} \cong \frac{1}{10}(V_{CC}) = \frac{1}{10}(20\text{ V}) = 2\text{ V}$$

The emitter resistor is then

$$R_E \cong \frac{V_E}{I_{C_Q}} = \frac{2\text{ V}}{5\text{ mA}} = \mathbf{400\ \Omega}$$

The collector resistor is calculated to be

$$R_C = \frac{V_{CC} - V_{CE_Q} - V_{E_Q}}{I_{C_Q}} = \frac{(20 - 10 - 2)\text{ V}}{5\text{ mA}} = \frac{8\text{ V}}{5\text{ mA}} = \mathbf{1.6\ k\Omega}$$

Calculating the base current using

$$I_{B_Q} = \frac{I_{C_Q}}{\beta} = \frac{5\text{ mA}}{90} \cong 55.56\ \mu\text{A}$$

we see that the base resistor is calculated to be

$$R_B = \frac{V_{CC} - V_{BE} - V_{E_Q}}{I_{B_Q}} = \frac{(20 - 0.7 - 2)\text{ V}}{55.56\ \mu\text{A}} = \frac{17.3\text{ V}}{55.56\ \mu\text{A}}$$

$$= \mathbf{311\ k\Omega} \quad (\text{use } R_B = 300\text{ k}\Omega)$$

Design of Current Gain Stabilized (Beta Independant) Circuit

The circuit of Fig. 5.32 provides stabilization both for leakage current and current gain changes. The values of the four resistors shown must be obtained for a specified operating point. Engineering judgment in selecting a value of emitter voltage, V_E, as in the previous design consideration, leads to a simple straightforward solution for all the resistor values. The design steps are as follows:

The emitter voltage will be selected to be approximately one-tenth of the supply voltage (V_{CC}).

$$V_{E_Q} \cong \frac{1}{10}V_{CC} = \frac{1}{10}(20\text{ V}) = 2\text{ V}$$

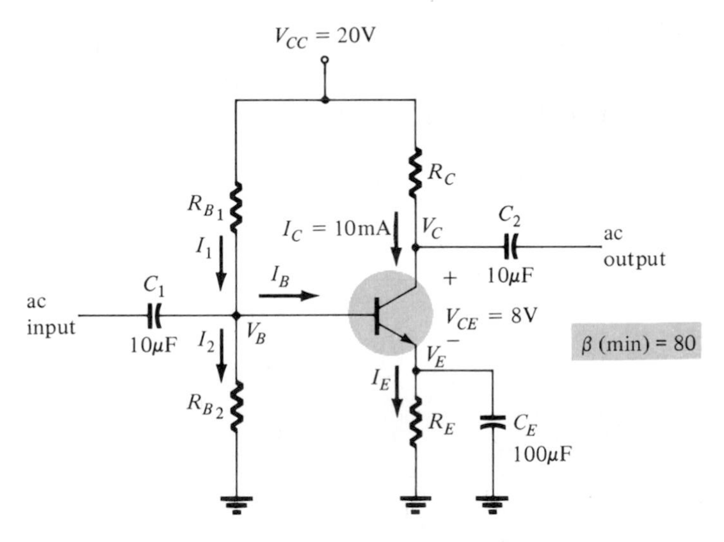

Figure 5.32 Current-gain-stabilized circuit for design considerations.

Using this value of V_E, we see that the emitter-resistor value is calculated to be

$$R_E \cong \frac{V_{E_Q}}{I_{C_Q}} = \frac{2\text{ V}}{10\text{ mA}} = 200\ \Omega$$

The collector resistance is then obtained using

$$R_C = \frac{V_{CC} - V_{CE_Q} - V_{E_Q}}{I_{C_Q}} = \frac{(20 - 8 - 2)\text{ V}}{10\text{ mA}} = 1\text{ k}\Omega$$

The base voltage is approximately equal to the emitter voltage or, more exactly,

$$V_B = V_E + V_{BE} = 2\text{ V} + 0.7\text{ V} = 2.7\text{ V}$$

The equation for calculation of the base resistors R_{B1} and R_{B2} will require a little thought. Using the values of base voltage calculated above and the value of the supply voltage will provide one equation—but there are two unknowns, R_{B1} and R_{B2}. An additional equation can be obtained from an understanding of the operation of these two resistors in providing the base voltage. For the circuit to operate properly the current through the two resistors should be approximately equal and therefore larger than the base current by an order of magnitude (at least 10 times larger). The two equations that will enable calculating the resistors R_{B1} and R_{B2} are

$$V_B = \frac{R_{B2}}{R_{B1} + R_{B2}}(V_{CC})$$

$$R_{B2} \leq \frac{1}{10}(\beta R_E)$$

Solving these equations results in

$$R_{B1} = 10.25\text{ k}\Omega\text{ (use 10 k}\Omega) \quad \text{and} \quad R_{B2} \cong 1.6\text{ k}\Omega$$

EXAMPLE 5.18

Design a dc bias circuit for an amplifier circuit as in Fig. 5.32. For this example, the manufacturer's spec states that the transistor has a current gain of 150, typical, at a collector current of 1 mA, and the supply voltage for the present circuit is 16 V. Provide design for $V_{C_Q} = V_{CC}/2$.

Solution:

Select $V_{E_Q} = \frac{1}{10}(V_{CC}) = \frac{1}{10}(16\text{ V}) = 1.6\text{ V}$ and calculate R_E:

$$R_E \cong \frac{V_{E_Q}}{I_{C_Q}} = \frac{1.6\text{ V}}{1\text{ mA}} = \mathbf{1.6\text{ k}\Omega}$$

For

$$V_{C_Q} = \frac{V_{CC}}{2} = \frac{16\text{ V}}{2} = 8\text{ V}$$

$$V_{CE_Q} = V_{C_Q} - V_{E_Q} = 8\text{ V} - 1.6\text{ V} = 6.4\text{ V}$$

Then calculate R_C,

$$R_C = \frac{V_{CC} - V_{CE_Q} - V_{E_Q}}{I_{C_Q}} = \frac{(16 - 6.4 - 1.6)\text{ V}}{1\text{ mA}} = \mathbf{8\text{ k}\Omega}\text{ (use 8.2 k}\Omega)$$

Calculate V_{B_Q}:

$$V_{B_Q} = V_{E_Q} + V_{BE} = 1.6\text{ V} + 0.7\text{ V} = 2.3\text{ V}$$

Finally, calculate R_{B1} and R_{B2}:

$$R_{B2} \leq \frac{1}{10}(\beta R_E) = \frac{150(1.6\text{ k}\Omega)}{10} = \mathbf{24\text{ k}\Omega}$$

and since $$\frac{R_{B2}}{R_{B1} + R_{B2}} V_{CC} = V_{BQ} = 2.3 \text{ V}$$

$$R_{B1} \cong \mathbf{143\ k\Omega} \quad (\text{use } 150 \text{ k}\Omega)$$

5.10 BIAS STABILIZATION

While the fixed-bias circuit provides suitable gain as an amplifier it has difficulty maintaining bias stability. In any amplifier circuit the collector current, I_C, will vary with change in temperature because of the three following main factors:

1. Reverse saturation current (leakage current), I_{CO}, which doubles for every 10° increase in temperature
2. Base-emitter voltage, V_{BE}, which decreases by 2.5 mV per °C
3. Transistor current gain, β, which increases with temperature

Any or all of these factors can cause the bias point to shift from the values originally set by the circuit because of a change in temperature. Table 5.1 lists typical parameter values for silicon transistors.

TABLE 5.1 Typical Silicon Transistor Parameters

T (°C)	I_{CO} (nA)	β	V_{BE} (V)
−65	0.2×10^{-3}	20	0.85
25	0.1	50	0.65
100	20	80	0.48
175	3.3×10^{3}	120	0.3

We first demonstrate the effect of leakage current and current gain change on the dc bias point initially set by the circuit. Consider the graphs of Figs. 5.33a and 5.33b, which show a transistor collector characteristic at room temperature (25°C) and the same transistor at some elevated temperature (100°C). Notice that the significant increase of leakage current not only causes the curves to rise but also that an increase in beta occurs as shown by the larger spacing between the curves at the higher temperature.

The operating point may be specified by drawing the circuit dc load line on the graph of the collector characteristic and noting the intersection of the load line and the dc base current set by the input circuit. An arbitrary point is marked as an example in Fig. 5.33a. Since the fixed-bias circuit provides a base current whose value depends approximately on the supply voltage and base resistor, neither of which is affected by temperature or the change in leakage current or beta, the same base current magnitude will exist at high temperatures as indicated on the graph of Fig. 5.33b. As the figure shows, this will result in the dc bias point's shifting to a higher collector current and a lower collector-emitter voltage operating point. In the extreme, the transistor could be driven into saturation. In any case the new operating point may not be at all

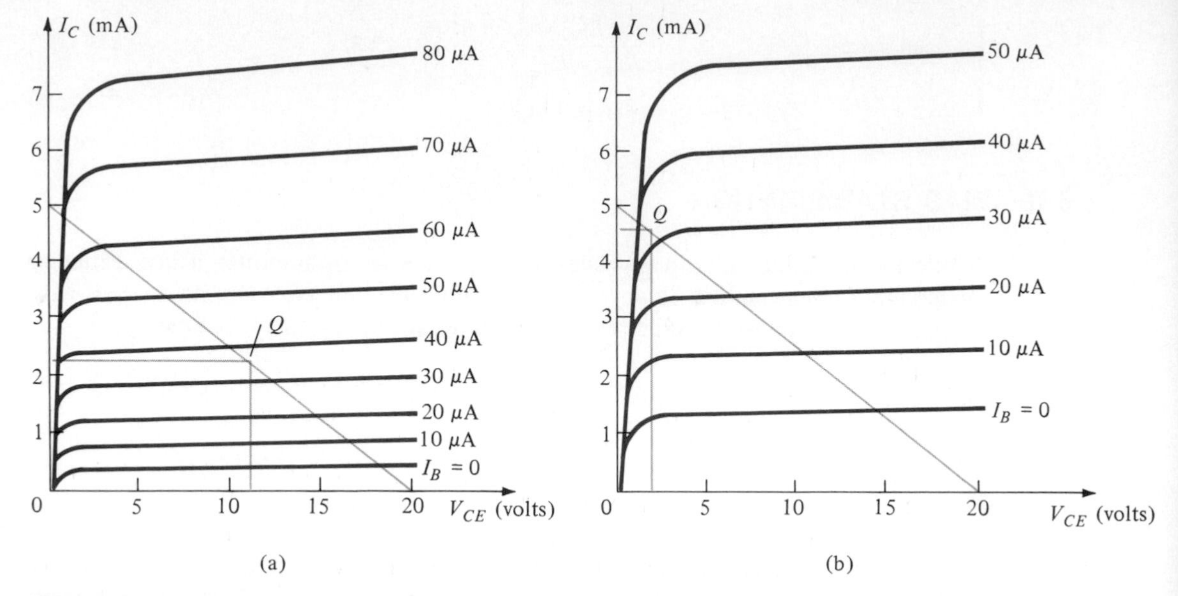

Figure 5.33 Shift in dc bias point (Q-point) due to change in temperature: (a) 25°C; (b) 100°C.

satisfactory and considerable distortion may result because of the bias-point shift. A better bias circuit is one that will stabilize or maintain the dc bias initially set, so that the amplifier can be used in a changing-temperature environment.

Stability Factor, *S*

A stability factor, S, can be defined for each of the parameters affecting bias stability. These are

$$S(I_{CO}) = \frac{\Delta I_C}{\Delta I_{CO}} \qquad S(V_{BE}) = \frac{\Delta I_C}{\Delta V_{BE}} \qquad S(\beta) = \frac{\Delta I_C}{\Delta \beta}$$

The stability factor is a numerical quantity representing the amount the collector current changes due to changes in each of the parameters because of temperature. The results of detailed mathematical analysis will be used to provide a comparison of how the device and circuit components affect bias stability to allow comparison of the effects on I_C by each of the transistor parameters.

S(I$_{CO}$)

Figure 5.34a shows a basic transistor circuit and the effect of I_{CO}. Figure 5.34b uses the result of analyzing the stability based on change due only to I_{CO} (β and V_{BE} considered held constant). Referring to Fig. 5.34b, we see that the stability factor varies from the ideal case (best condition) of $S = 1$ up to a maximum value of $S = \beta + 1$, which occurs for the fixed-bias circuit, or when the ratio R_B/R_E is greater than $\beta + 1$. Essentially, the stability factor is smallest for larger values of R_E so that inclusion of an emitter resistor improves the bias stability (makes S smaller).

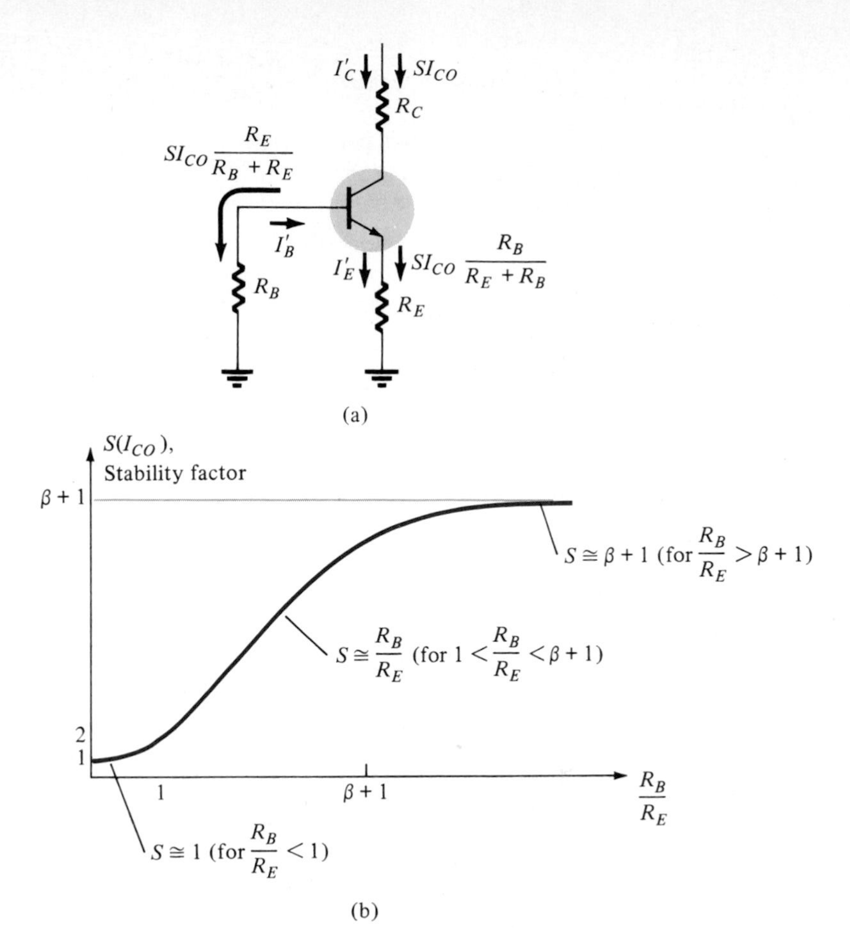

Figure 5.34 Effect of I_{CO} on bias point.

$$S(I_{CO}) = \frac{(\beta + 1)(1 + R_B/R_E)}{(\beta + 1) + (R_B/R_E)} \tag{5.25}$$

For values of $R_B/R_E \gg (\beta + 1)$ Eq. (5.25) approaches $(\beta + 1)$:

$$S(I_{C_O}) \to \frac{(\beta + 1)(R_B/R_E)}{(R_B/R_E)} = (\beta + 1) \qquad R_B/R_E \gg (\beta + 1)$$

In fact, for $R_E = 0$ (fixed bias operation)

$$S(I_{C_O}) = (\beta + 1)$$

provides the largest value of stability factor.

For values of R_B/R_E between 1 and $(\beta + 1)$

$$S(I_{C_O}) \to \frac{(\beta + 1)}{(\beta + 1) + (R_B/R_E)} (R_B/R_E) \cong R_B/R_E$$

Finally, for values of R_B/R_E less than 1 ($R_E > R_B$ however, provides poor circuit bias)

$$S(I_{C_O}) \rightarrow \frac{(\beta + 1)(1)}{(\beta + 1)} = 1$$

result in the best stability but poorest circuit bias.

The value of I_{CO} in modern transistors is so low (refer to Table 5.1), however, that the change in bias point in a circuit with even a large value of $S(I_{CO})$ would not be considerable, as the following example illustrates.

EXAMPLE 5.19

In a circuit using a transistor typified by the parameters in Table 5.1 calculate the change in I_C from 25°C to 100°C for (a) fixed bias ($R_B/R_E \rightarrow \infty$), (b) $R_B/R_E = 11$, and (c) $R_B/R_E = 0.01$.

Solution:

From 25°C to 100°C the change in I_{CO} is

$$\Delta I_{CO} = (20 - 0.1) \text{ nA} \cong 20 \text{ nA}$$

(a) For fixed bias, $S = \beta + 1 = 51$. Using the definition of stability, we get

$$\Delta I_C = S(\Delta I_{CO}) = 51(20 \text{ nA}) \cong \mathbf{1\ \mu A}$$

(b) For $R_B/R_E = 11$, $S = 51(1 + 11)/(51 + 11) \cong 10$

$$\Delta I_C = 10\ \Delta I_{CO} = 10(20 \text{ nA}) = \mathbf{0.2\ \mu A}$$

(c) For $R_B/R_E = 0.01$, $S = 51(1.01)/(51 + 0.1) \cong 1$

$$\Delta I_C = 1(20 \text{ nA}) = \mathbf{20\ nA}$$

While the change in I_C is considerably different in a circuit having ideal stability ($S = 1$) and one having the maximum stability factor ($S = 51$, in this example), the change in I_C is not significant. For example, the amount of change in I_C from a dc bias current set at, say, 2 mA, would be from 2 mA to 2.001 mA (only 0.05%) in the worst case, which is obviously small enough to be ignored. Some power transistors exhibit larger leakage current, but for most amplifier circuits and effect of I_{CO} change with temperature is slight.

$S(V_{BE})$

Analysis of the stability factor due to change in V_{BE} will result in

$$S(V_{BE}) = \frac{\Delta I_C}{\Delta V_{BE}} = \frac{-\beta}{R_B + R_E(\beta + 1)}$$

$$= \frac{-1}{R_E}, \quad \text{for } (\beta + 1) \gg \frac{R_B}{R_E} \quad \text{and} \quad \beta \gg 1 \tag{5.26}$$

Since smaller values of S indicate better stability, the larger the value of R_E the better the circuit stability due to changes in V_{BE} with temperature.

EXAMPLE 5.20

Determine the change in I_C for a transistor having parameters as listed in Table 5.1 over a temperature range from 25 to 100°C for a circuit having $R_E = 1\ \text{k}\Omega$ (and $\beta + 1 \gg R_B/R_E$).

Solution:

Using $S(V_{BE}) = -1/R_E = -1/1\ \text{k}\Omega = -10^{-3}$ and $V_{BE} = (0.65 - 0.48) = 0.17$ V from 25 to 100°C yields

$$I_C = S\ \Delta V_{BE} = -10^{-3}(0.17\ \text{V}) = -170\ \mu\text{A}$$

which is seen to be a reasonably large current change compared to that resulting from a change in I_{CO}. Using a typical collector current of 2 mA, we see that the collector current would change from 2 mA at 25°C to 1.830 mA at 100°C: an **8.5% change.**

The effect of V_{BE} changing with temperature can be somewhat compensated by the use of diode compensation in which the voltage change across a diode compensates for the change in V_{BE} to maintain the bias value of I_C. Thermistor and transistor compensation techniques are also used when needed.

S(β)

Analysis of the effect of β changing with temperature on the circuit bias stabilty results in

$$\frac{\Delta I_C}{I_C(T_1)} = \left(1 + \frac{R_B}{R_E}\right)\frac{\Delta\beta}{\beta(T_1)\,\beta(T_2)} = \left(1 + \frac{R_B}{R_E}\right)\frac{[\beta(T_2)/\beta(T_1)] - 1}{\beta(T_2)} \tag{5.27}$$

where $\beta(T_1)$ = beta at temperature T_1
$\beta(T_2)$ = beta at temperature T_2
$I_C(T_1)$ = collector current at temperature T_1

EXAMPLE 5.21

Calculate the change in collector current for the transistor having parameters as given in Table 5.1 from room temperature to 100°C. Assume that $R_B/R_E = 20$ for the circuit used and that I_C at room temperature is 2 mA.

Solution:

The change in I_C is calculated to be

$$\Delta I_C = I_C(T_1)\left[\left(1 + \frac{R_B}{R_E}\right)\frac{[\beta(T_2)/\beta(T_1)] - 1}{\beta(T_2)}\right]$$

$$= 2\ \text{mA}\left[(1 + 20)\frac{(80/50) - 1}{80}\right] = 0.315\ \text{mA} = \mathbf{315\ \mu A}$$

The collector current changing from 2 mA at room temperature to 2.315 mA at 100°C represents a change of about **16%.**

Comparison of the three examples shows that of the three parameters affecting bias stability the change due to β variation is probably greatest. These changes in parameter value need not be only due to temperature. Although the value of I_{CO} at room temperature varies negligibly between transistors, even of the same manufacture type, as does V_{BE}, the value of β varies considerably. For example, the same numbered transistor may have $\beta = 125$ for one device and $\beta = 300$ for another. In addition, the value of β for specific transistors will be different at different values of bias current.

For all these reasons the design of a good bias-stabilized circuit usually concentrates most on stabilizing the effect of changes in transistor beta.

5.11 COMPUTER SOLUTION OF DC BIAS

While a variety of bias circuits can be analyzed using the voltage loop equation of the base-emitter loop, then solving for the base current, there are many that can be represented by a relatively standard circuit form. If we are doing lots of work with a particular circuit, it is appropriate to develop a computer program to help carry out the many tedious calculations. A few examples should help organize an approach to using the computer in a meaningful way. To start, consider writing a program module or subroutine to do the dc bias analysis of the standard bias circuit of Fig. 5.11. This one circuit should cover bias calculations of the emitter-stabilized circuit of Fig. 5.7 by selecting R_{B2} as open (or infinite resistance), and also the fixed-bias circuit of Fig. 5.2 by selecting R_{B2} as open and R_E as a short (zero resistance). A summary of the equations is provided in Listing 5.1, together with a summary of equations variables and computer variables in Listing 5.2.

The program listing of Module 10000 will be discussed next. A main program must be included to obtain the circuit values needed by the module and to print the values of the resulting calculations.

Bias Module

A program module starting at line 10000 is written in the BASIC language to do the calculations for the dc bias of a circuit as in Fig. 5.11. Line 10010 calculates the Thévenin equivalent base resistance of R_{B_1} in parallel with R_{B_2}. If R_{B_2} is not present

LISTING 5.1 Equations and Computer Statements for DC Bias Calculations Module

Equation	*Computer Statement*
$R_{BB} = \dfrac{R_{B_1}R_{B_2}}{R_{B_1} + R_{B_2}}$	RT = (R1 * R2)/(R1 + R2)
$V_{BB} = \dfrac{R_{B_2}}{R_{B_1} + R_{B_2}} V_{CC}$	VT = (R2 * CC)/(R1 + R2)
$I_B = \dfrac{V_{BB} - 0.7}{R_{BB} + (\beta + 1)R_E}$	IB = (VT − 0.7)/(RT + (BETA + 1) * RE)
$I_C = \beta I_B$	IC = BETA * IB
$I_E = (\beta + 1)I_B$	IE = (BETA + 1) * IB
$V_E = I_E R_E$	VE = IE * RE
$V_B = V_E + 0.7$	VB = VE + 0.7
$V_C = V_{CC} - I_C R_C$	VC = CC − IC * RC
$V_{CE} = V_C - V_E$	CE = VC − VE

LISTING 5.2 Equation and Computer Variables for DC Bias Calculations Module

Equation Variable	*Computer Variable*
R_{B1}	R1
R_{B2}	R2
R_{BB}	RT
V_{CC}	CC
V_{BB}	VT
I_B	IB
V_{BE}	BE
β	BETA
R_E	RE
I_C	IC
I_E	IE
V_E	VE
V_B	VB
V_C	VC
V_{CE}	CE

(open circuit), the user will have to input a value which is as large as the computer can handle, so that the calculated value of R_T reduces to that of R_{B_1}. Line 10020 calculates the Thévenin equivalent voltage at the base. Again, if R_{B_2} is not present, the calculated value is that of V_{CC}. Line 10030 calculates I_B using a base-emitter voltage of $V_{BE} = 0.7$ V. Line 10040 then tests for a cutoff bias condition which occurs if the value V_T is less than $V_{BE} = 0.7$ V, in which case I_B is set to zero; otherwise, I_B remains as calculated in line 10030. Lines 10060 and 10070 then calculate I_C and I_E, respectively.

Line 10090 tests for the condition of circuit saturation, setting I_C (and I_E) to the saturation value in that case; otherwise, the values of I_C and I_E remain those previously calculated. Lines 10100 through 10120 then calculate V_E, V_B, and V_E, respectively. Line 10130 calculates V_{CE} and the program module then returns to the main program.

The main program, requesting input of all appropriate circuit data, Module 10000 to do the dc bias calculations, and main program steps to print out the results are provided in Listing 5.3. Some sample runs are shown in Listing 5.4 to provide program results for a variety of circuits using values from previous examples.

LISTING 5.3

```
10 REM ***********************************************************
20 REM
30 REM          DC BIAS CALCULATIONS OF STANDARD CIRCUIT
40 REM
50 REM ***********************************************************
60 REM
100 PRINT "This program calculates the dc bias"
110 PRINT "for a standard circuit as shown in Figure 5.11."
120 PRINT
130 PRINT "First, enter the following circuit data:"
140 INPUT "RB1=";R1
150 INPUT "RB2(use 1E30 if 'open')=";R2
```

LISTING 5.3 (cont.)

```
160 INPUT "RE=";RE
170 INPUT "RC=";RC
180 PRINT
190 INPUT "Vcc=";CC
200 PRINT
210 INPUT "Transistor beta=";BETA
220 PRINT
230 REM Now do circuit calculations
240 GOSUB 10000
250 PRINT "The results of dc bias calculations are:"
260 PRINT
270 PRINT "Circuit currents:"
280 PRINT "IB=";IB*1000000!;"uA"
290 PRINT "IC=";IC*1000;"mA"
300 PRINT "IE=";IE*1000;"mA"
310 PRINT
320 PRINT "Circuit voltages:"
330 PRINT "VB=";VB;"volts"
340 PRINT "VE=";VE;"volts"
350 PRINT "VC=";VC;"volts"
360 PRINT "Vce=";CE;"volts"
370 PRINT :PRINT
380 END
10000 REM Module to calculate dc bias of BJT circuit
10010 RT=R1*(R2/(R1+R2))
10020 VT=CC*(R2/(R1+R2))
10030 IB=(VT-.7)/(RT+(BETA+1)*RE)
10040 REM Test for cutoff condition
10050 IF VT<=.7 THEN IB=0
10060 IC=BETA*IB
10070 IE=(BETA+1)*IB
10080 REM Test for saturation condition
10090 IF IC*(RC+RE)>=CC THEN IC=CC/(RE+RC) :IE=IC
10100 VE=IE*RE
10110 VB=VE+.7
10120 VC=CC-IC*RC
10130 CE=VC-VE
10140 RETURN
```

LISTING 5.4

```
RUN
This program calculates the dc bias
for a standard circuit as shown in Figure 5.11.

First, enter the following circuit data:
RB1=? 240E3
RB2(use 1E30 if 'open')=? 1E30
RE=? 0
RC=? 2.2E3
```

```
Vcc=? 12

Transistor beta=? 50

The results of dc bias calculations are:

Circuit currents:
IB= 47.08333 uA
IC= 2.354167 mA
IE= 2.40125 mA

Circuit voltages:
VB= .7 volts
VE= 0 volts
VC= 6.820833 volts
Vce= 6.820833 volts


This program calculates the dc bias
for a standard circuit as shown in Figure 5.11.

First, enter the following circuit data:
RB1=? 430E3
RB2(use 1E30 if 'open')=? 1E30
RE=? 1E3
RC=? 2E3

Vcc=? 20

Transistor beta=? 100

The results of dc bias calculations are:

Circuit currents:
IB= 36.34652 uA
IC= 3.634652 mA
IE= 3.670998 mA

Circuit voltages:
VB= 4.370998 volts
VE= 3.670998 volts
VC= 12.7307 volts
Vce= 9.0597 volts


This program calculates the dc bias
for a standard circuit as shown in Figure 5.11.

First, enter the following circuit data:
RB1=? 39E3
RB2(use 1E30 if 'open')=? 3.9E3
RE=? 1.5E3
RC=? 10E3
```

LISTING 5.4 (cont.)

```
Vcc=? 22

Transistor beta=? 140

The results of dc bias calculations are:

Circuit currents:
IB= 6.045233 uA
IC= .8463327 mA
IE= .8523779 mA

Circuit voltages:
VB= 1.978567 volts
VE= 1.278567 volts
VC= 13.53667 volts
Vce= 12.25811 volts
```

PROBLEMS

§ 5.3

1. Calculate the collector voltage for the circuit of Fig. 5.5 with $R_B = 330\ \text{k}\Omega$, $R_C = 2.7\ \text{k}\Omega$, $V_{CC} = 12$ V, and $\beta = 50$.

2. Calculate the collector-base voltage for the circuit of Fig. 5.5 with $R_B = 150\ \text{k}\Omega$, $R_C = 2.1\ \text{k}\Omega$, $V_{CC} = 9$ V, and $\beta = 45$.

3. Calculate the collector current and collector-emitter voltage for the circuit of Fig. 5.5 with $R_B = 240\ \text{k}\Omega$, $R_C = 1.8\ \text{k}\Omega$, $V_{CC} = 12$ V, and $\beta = 70$.

4. What value of R_B should be used to set $V_C = 8$ V, for the circuit of Fig. 5.5 with $R_C = 2.4\ \text{k}\Omega$, $V_{CC} = 18$ V, and $\beta = 90$?

5. What value of R_C is needed to set $V_{CE} = 6$ V for the circuit of Fig. 5.5 with $R_B = 510\ \text{k}\Omega$, $V_{CC} = 22$ V, and $\beta = 120$?

6. What value of R_C is needed to obtain a collector voltage of -8.4 V in the circuit of Fig. 5.6 with $\beta = 85$?

§ 5.4

7. Calculate the dc bias voltage V_{CE} and current I_C for an emitter-stabilized bias circuit as in Fig. 5.7 for circuit values of $R_B = 220\ \text{k}\Omega$, $R_C = 2.7\ \text{k}\Omega$, $R_E = 1.5\ \text{k}\Omega$, $\beta = 55$, and $V_{CC} = 18$ V.

8. Calculate I_E for the circuit of Fig. 5.7 for circuit values of $R_B = 510\ \text{k}\Omega$, $R_E = 1.2\ \text{k}\Omega$, $R_C = 2.4\ \text{k}\Omega$, $V_{CC} = 20$ V, and $\beta = 100$.

9. Determine the value of transistor β for the circuit of Fig. 5.7 with values $R_B = 330\ \text{k}\Omega$, $R_E = 1\ \text{k}\Omega$, $R_C = 1.8\ \text{k}\Omega$, and $V_{CC} = 16$ V, resulting in an emitter voltage of $V_E = 3$ V.

10. Determine the value of R_B needed to provide a base voltage of $V_B = 4.4$ V for the circuit of Fig. 5.7 with values of $R_E = 2.2\ \text{k}\Omega$, $R_C = 2.7\ \text{k}\Omega$, $V_{CC} = 12$ V, and $\beta = 150$.

11. Determine the bias voltage V_{CE} for the circuit of Fig. 5.7 with values of $R_B = 1.5\ M\Omega$, $R_E = 1.1\ k\Omega$, $R_C = 4.3\ k\Omega$, $V_{CC} = 25$ V, and $\beta = 140$.

12. Determine the value of R_B to just bias the circuit of Fig. 5.7 into saturation for circuit values of $R_E = 820\ \Omega$, $R_C = 2.4\ k\Omega$, $V_{CC} = 18$ V, and $\beta = 85$.

13. Determine the value of R_B to bias the circuit of Fig. 5.7 for $I_C = 0.5\ I_{C_{sat}}$ for circuit values of $R_E = 620\ \Omega$, $R_C = 1.8\ k\Omega$, $V_{CC} = 20$ V, and $\beta = 110$.

14. Determine the percent change of V_C for the circuit of Fig. 5.7 with circuit values of $R_B = 680\ k\Omega$, $R_E = 910\ \Omega$, $R_C = 2.2\ k\Omega$, and $V_{CC} = 15$ V from $\beta = 90$ to $\beta = 180$.

15. Calculate all bias voltages, V_B, V_E, and V_C for the circuit of Fig. 5.7 with circuit values of $R_B = 750\ k\Omega$, $R_E = 0.82\ k\Omega$, $R_C = 3.3\ k\Omega$, $V_{CC} = 9$ V, and $\beta = 75$.

§ 5.5

16. Calculate the base voltage V_B in the circuit of Fig. 5.11 with circuit values of $R_{B_1} = 470\ k\Omega$, $R_{B_2} = 68\ k\Omega$, $R_E = 3.3\ k\Omega$, $R_C = 15\ k\Omega$, $V_{CC} = 18$ V, and $\beta = 120$.

17. Calculate the base and collector currents in the circuit of Fig. 5.11 for circuit values of $R_{B_1} = 91\ k\Omega$, $R_{B_2} = 11\ k\Omega$, $R_E = 1.2\ k\Omega$, $R_C = 4.7\ k\Omega$, $V_{CC} = 18$ V, and $\beta = 70$.

18. What value of R_E would result in $V_C = 6$ V for the circuit of Fig. 5.11 with values of $R_{B_1} = 82\ k\Omega$, $R_{B_2} = 24\ k\Omega$, $R_C = 5.6\ k\Omega$, $V_{CC} = 16$ V, and $\beta = 150$?

19. Calculate I_C and V_{CE} for the circuit of Fig. 5.11 with circuit values of $R_{B_1} = 100\ k\Omega$, $R_{B_2} = 22\ k\Omega$, $R_C = 8.2\ k\Omega$, $R_E = 2.2\ k\Omega$, $V_{CC} = 9$ V, and $\beta = 100$.

20. Determine the value of R_E that results in biasing the circuit of Fig. 5.11 at $0.5\ I_{C_{sat}}$ for circuit values of $R_{B_1} = 220\ k\Omega$, $R_{B_2} = 51\ k\Omega$, $R_C = 3.3\ k\Omega$, $V_{CC} = 18$ V, and $\beta = 130$.

21. Calculate the collector-base voltage, V_{CB}, for the circuit of Fig. 5.11 with values of $R_{B_1} = 62\ k\Omega$, $R_{B_2} = 9.1\ k\Omega$, $R_E = 0.68\ k\Omega$, $R_C = 3.9\ k\Omega$, $V_{CC} = 16$ V, and $\beta = 110$.

22. What value of R_C should be used to bias the circuit of Fig. 5.11 at $0.5V_{CC}$ for circuit values of $R_{B_1} = 220\ k\Omega$, $R_{B_2} = 33\ k\Omega$, $R_E = 1.8\ k\Omega$, $V_{CC} = 25$ V, and $\beta = 180$?

23. Determine the percent change of V_{CE} for the circuit of Fig. 5.11 with circuit values of $R_{B_1} = 75\ k\Omega$, $R_{B_2} = 24\ k\Omega$, $R_C = 2.4\ k\Omega$, $R_E = 1.2\ k\Omega$, and $V_{CC} = 16$ V from $\beta = 80$ to $\beta = 160$.

24. What value of R_E should be used to bias the circuit of Fig. 5.11 at $I_C = 0.5\ I_{C_{sat}}$ for circuit values of $R_{B_1} = 100\ k\Omega$, $R_{B_2} = 10\ k\Omega$, $R_C = 3.3\ k\Omega$, $V_{CC} = 30$ V, and $\beta = 200$?

25. What value of R_E should be used in the circuit of Fig. 5.11 to bias the collector voltage at $V_C = 12$ V for circuit values of $R_{B_1} = 91\ k\Omega$, $R_{B_2} = 11\ k\Omega$, $R_E = 1.1\ k\Omega$, $V_{CC} = 18$ V, and $\beta = 90$?

26. Determine the saturation current in the circuit of Fig. 5.11 for circuit values of $R_{B_1} = 12\ k\Omega$, $R_{B_2} = 2.2\ k\Omega$, $R_E = 1.1\ k\Omega$, $R_C = 2.7\ k\Omega$, $V_{CC} = 9$ V, and $\beta = 120$.

§ 5.6

27. Calculate V_C in the circuit of Fig. 5.35a.

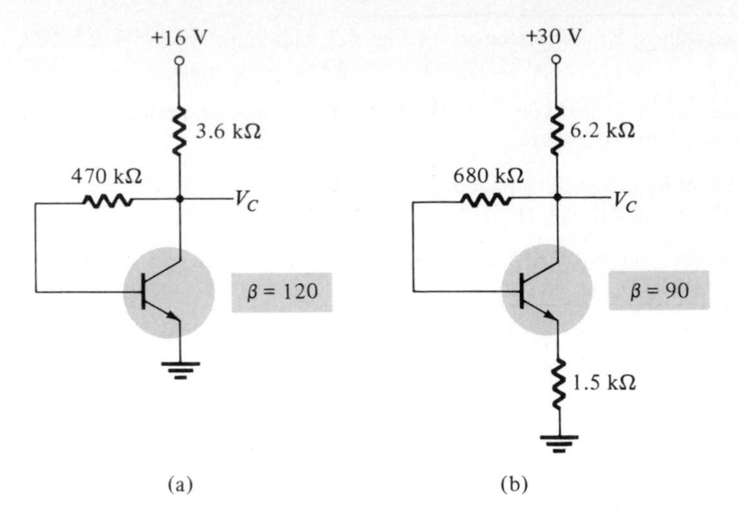

Figure 5.35 Bias circuits for Problems 27 through 31.

28. Determine the value of feedback resistor needed in Fig. 5.35a to result in $V_C = 0.5\,V_{CC} = 8$ V.

29. Calculate V_C in the circuit of Fig. 5.35b.

30. Determine the value of feedback resistor that will just bias the circuit in Fig. 5.35b at $I_C = 0.5\,I_{C_{sat}}$.

31. Determine the value of collector resistor needed in the circuit of Fig. 5.35b to result in $V_C = 15$ V.

32. Calculate V_{CE} and I_C for the circuit of Fig. 5.36a.

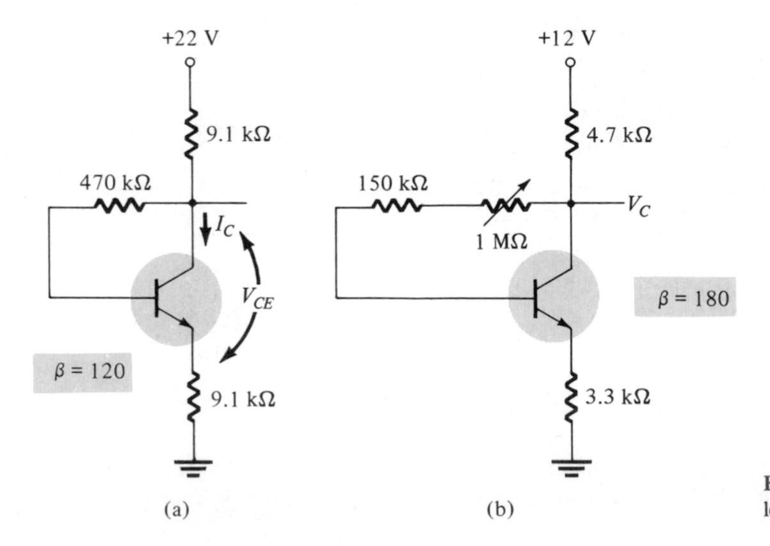

Figure 5.36 Bias circuits for Problems 32 through 37.

33. Calculate the percent change of V_C in the circuit of Fig. 5.36a if the transistor is replaced by one having $\beta = 60$.

34. Determine the smallest and largest values of I_C in the circuit of Fig. 5.36b for the 1-MΩ potentiometer set at 1 MΩ and at 0 Ω.

35. Calculate the voltage V_C in the circuit of Fig. 5.36b when the 1-MΩ potentiometer is set at its midvalue.

36. What is the largest value that V_C could reach in the circuit of Fig. 5.36b when the potentiometer is adjusted from 0 Ω to 1 MΩ?

37. What percent change in I_C results when the potentiometer in Fig. 5.36b is adjusted from 0 Ω to 1 MΩ?

§ 5.7

38. Calculate V_C for the bias circuit of Fig. 5.37a.

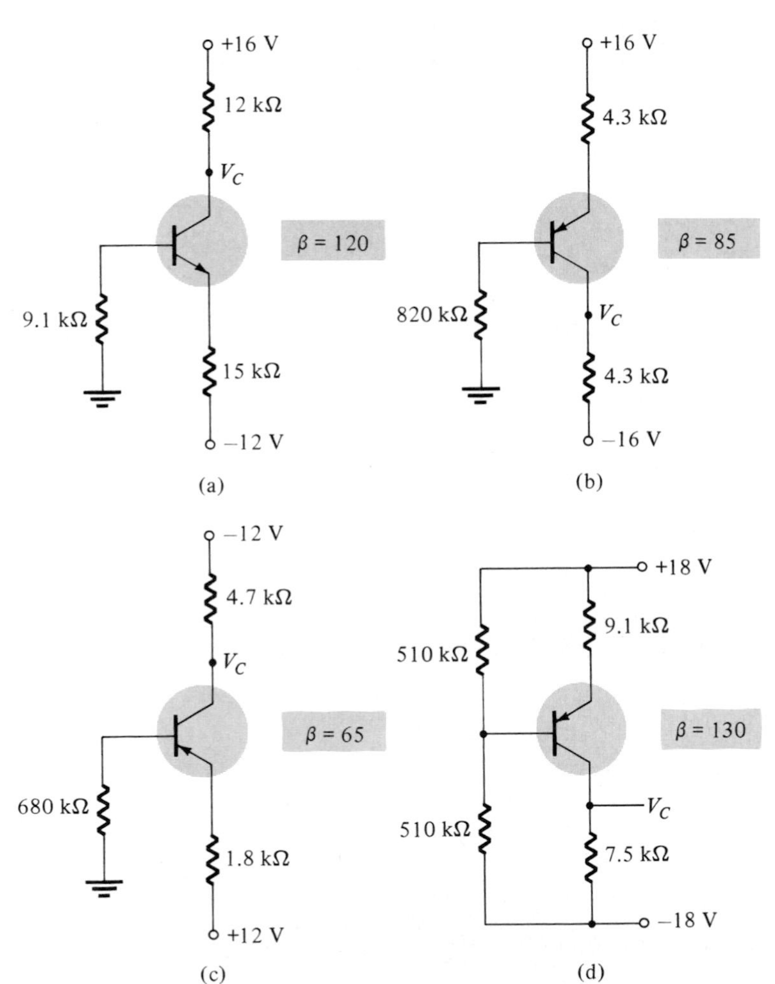

Figure 5.37 Bias circuits for Problems 38 through 44.

39. Calculate V_C for the circuit of Fig. 5.37b.

40. Calculate V_C for the circuit of Fig. 5.37c.

41. Calculate the voltage V_{CE} in the circuit of Fig. 5.37b if β increases to 120.

42. Calculate the voltage V_C in the circuit of Fig. 5.37c if the supply voltages are $\pm$ 9 V.

43. Calculate V_C for the circuit of Fig. 5.37d.

44. Calculate the voltage V_C for the circuit of Fig. 5.37d if the negative voltage is set to 0 V.

§ 5.8

45. For a fixed-bias circuit as in Fig. 5.7 with circuit values of R_B = 82 kΩ, R_C = 4.3 kΩ, $V_{CC} = 2\upsilon$ V, V_{BE} = 0.7 V, and transistor collector characteristic as shown in Fig. 5.38, do the following:
(a) Draw the dc load line.
(b) Obtain the quiescent operating point (Q-point).
(c) Find the operating point if R_C is changed to 8.2 kΩ.
(d) Find the operating point if V_{CC} is changed to 15 V (R_C is still 4.3 kΩ).

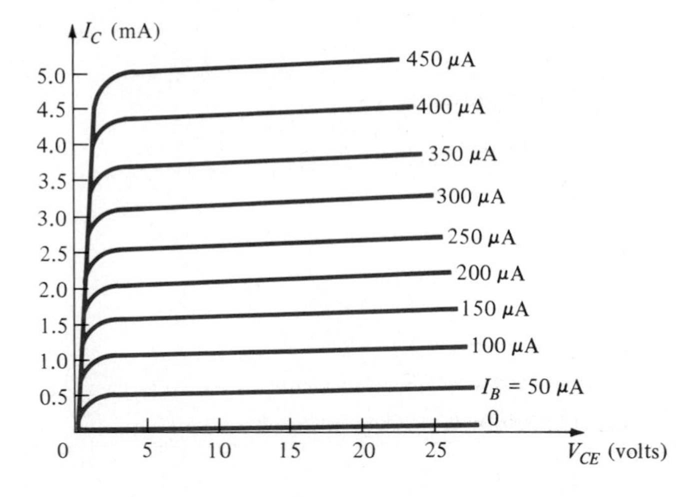

Figure 5.38 Transistor collector characteristics for Problem 45.

46. Determine graphically the operating point for a fixed-bias circuit using a *pnp* transistor having a collector characteristic as in Fig. 5.39. The circuit values are R_B = 150 kΩ, R_C = 2 kΩ, and V_{CC} = −20 V.

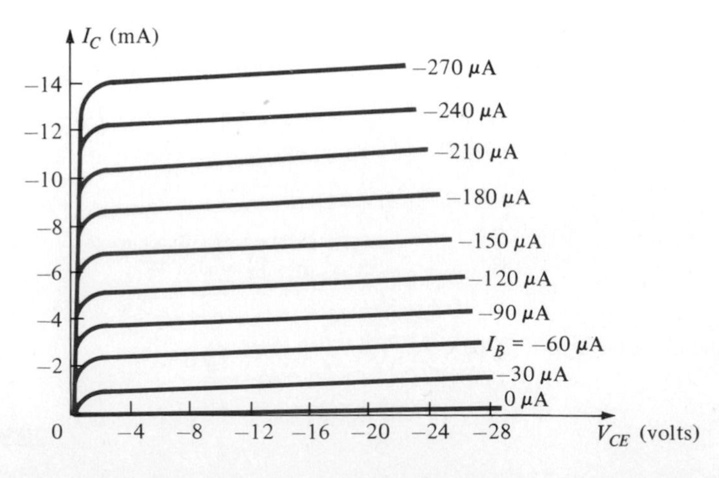

Figure 5.39 Transistor collector characteristics for Problem 46.

47. What value of R_C is needed in a bias circuit as in Fig. 5.7 with $V_{CC} = 22$ V, $R_B = 75$ kΩ, and $R_E = 3.3$ kΩ to bias the circuit at $V_{CE} = 12$ V using the transistor characteristic of Fig. 5.38?

48. What is the range of the collector voltage if the transistor of Fig. 5.38 is biased by a circuit such as that of Fig. 5.7 with $R_B = 3.3$ kΩ, $V_{CC} = 22$ V, and R_C a resistance that can be varied from 1 to 5 kΩ.

§ 5.9

49. Design an emitter-stabilized bias circuit to operate at $V_{CE} = 8$ V, $I_C = 5$ mA, using a supply of $V_{CC} = 18$ V and a transistor having $\beta = 100$.

50. Design an emitter-stabilized bias circuit to operate at 0.5 $I_{C_{sat}}$ for a supply voltage $V_{CC} = 22$ V and transistor with $\beta = 120$.

51. Design an emitter-stabilized bias circuit to operate at $V_{CE} = 0.5\,V_{CC}$ for a supply voltage of $V_{CC} = 16$ V, transistor with $\beta = 180$, and $R_C = 4.3$ kΩ.

52. Design a β-independent bias circuit to operate at $V_{CE} = 12$ V, $I_C = 5$ mA using a supply of $V_{CC} = 25$ V and a transistor with $\beta = 140$.

53. Design a β-independent bias circuit to operate at $V_{CE} = 0.5V_{CC}$ using a supply of $V_{CC} = 12$ V, a transistor with $\beta = 80$, and $R_E = 1.2$ kΩ.

54. Complete the design of a β-independent bias circuit having $R_{B_1} = 68$ kΩ, $R_{B_2} = 7.5$ kΩ, $V_{CC} = 16$ V, and transistor $\beta = 80$, to operate at $V_{CE} = 6$ V and $I_C = 2$ mA.

COMPUTER PROBLEMS

Write BASIC programs to:

1. Compute the dc bias values I_C and V_{CE} for a fixed-bias circuit.
2. Compute I_C for an emitter-stabilized bias circuit.
3. Compute V_B for a circuit using voltage-divider biasing using the approximate method.
4. Compute V_B for a voltage-divider biasing circuit using the exact method (Thévenin equivalent).
5. Compute I_C and V_{CE} for a voltage-divider biasing circuit using the exact method.
6. Compute I_C and V_{CE} using the approximate method in a voltage-divider biasing circuit.
7. Calculate the bias values I_C and V_{CE} for a collector-feedback-stabilized bias circuit.

PRACTICAL PROBLEMS

1. For the circuit of Fig. 5.40, answer the following:
   (a) Does V_C increase or decrease if R_B is increased?
   (b) Does I_C increase or decrease if β is reduced?
   (c) What happens to the saturation current if β is increased?

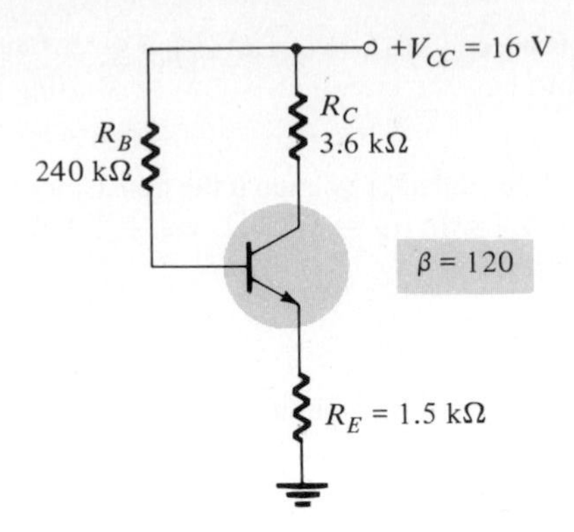

Figure 5.40 Circuit for Practical Problem 1.

(d) Does the collector current increase or decrease if V_{CC} is reduced?
(e) What happens to V_{CE} if the transistor is replaced by one with smaller β?

2. Answer the following questions about the circuit of Fig. 5.41:
(a) What happens to the voltage V_C if the transistor is replaced by one having a larger value of β?
(b) What happens to the voltage V_{CE} if the ground leg of resistor R_{B_2} opens (does not connect to ground)?
(c) What happens to I_C if the supply voltage is low?
(d) What voltage V_{CE} would occur if the transistor base-emitter junction fails by becoming open?
(e) What voltage V_{CE} would result if the transistor base-emitter junction fails by becoming a short?

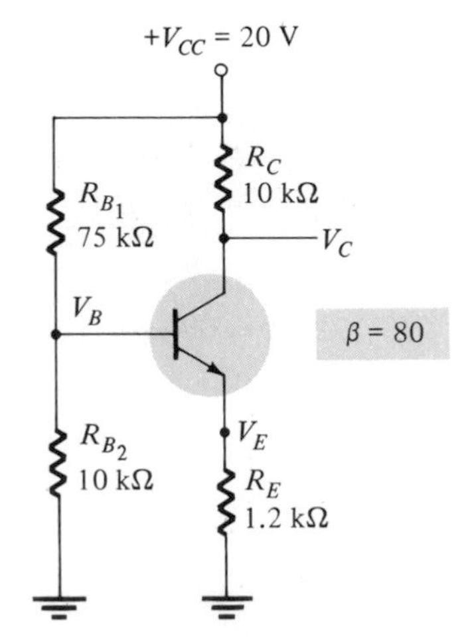

Figure 5.41 Circuit for Practical Problem 2.

3. Answer the following questions about the circuit of Fig. 5.42.
(a) What happens to the voltage V_C if the resistor R_B is open?
(b) What should happen to V_{CE} if β increases due to temperature?
(c) How will V_E be affected when replacing the collector resistor with one whose resistance is at the lower end of the tolerance range?

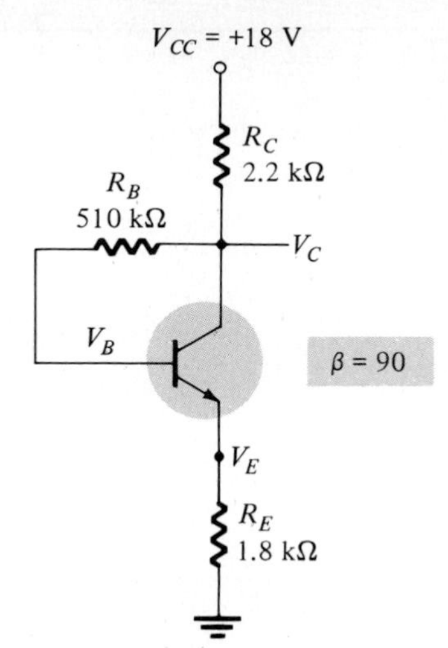

Figure 5.42 Circuit for Practical Problem 3.

(d) If the transistor collector connection becomes open, what will happen to V_{CE}?
(e) What might cause V_{CE} to become nearly 0 V?

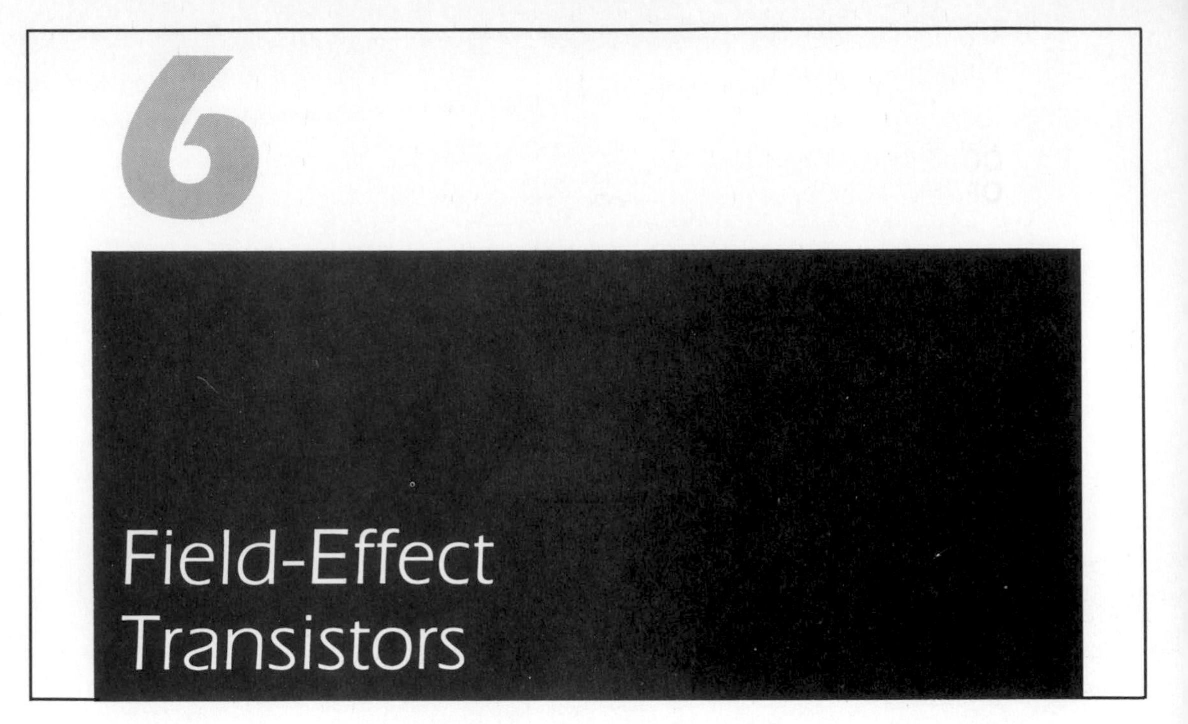

6.1 GENERAL DESCRIPTION OF FET

A bipolar junction transistor (BJT) made as *npn* or as *pnp* is a current-controlled device in which both electron current and hole current are involved. The field-effect transistor (FET) is a unipolar device. It operates as a voltage-controlled device with either electron current in an *n*-channel FET or hole current in a *p*-channel FET. Either BJT or FET devices can be used to operate in an amplifier circuit (or other similar electronic circuits), with different bias considerations.

A few general comparisons between FET and BJT devices and resulting circuits can be made.

1. The FET has an extremely high input resistance with about 100 MΩ typical (BJT input resistance typically 2 kΩ).
2. The FET has no offset voltage when used as a switch (or chopper).
3. The FET is relatively immune to radiation, but the BJT is very sensitive (beta is particularly affected).
4. The FET is less "noisy" than a BJT and thus more suitable for input stages of low-level amplifiers (it is used extensively in hi-fi FM receivers).
5. The FET can be operated to provide greater thermal stability than a BJT.
6. The FET is smaller than a BJT and is thus far more popular in ICs.

Some disadvantages of the FET are the relatively small gain-bandwidth of the device compared to the BJT and the greater susceptibility to damage in handling the FET.

6.2 CONSTRUCTION AND CHARACTERISTICS OF JFETS

The FET is a three-terminal device containing one basic *p-n* junction and can be built as either a Junction FET (JFET) or a Metal-Oxide Semiconductor FET (MOS-FET). Although the FET was one of the earliest solid-state devices proposed[1] for amplifier operation, until the mid-1960s the development of a commercially useful device lagged because of manufacturing construction limitations. Large and very large-scale integrated circuits are built primarily using MOSFET transistor devices.

JFET Operation

The physical structure of a JFET is shown in Fig. 6.1. The *n*-channel JFET shown in Fig. 6.1a is constructed using a bar of *n*-type material into which a pair of *p*-type regions are diffused. A *p*-channel JFET is made using a bar of *p*-type material with *n*-type diffused regions as shown in Fig. 6.1b.

The device symbol is shown in Fig. 6.1 for each type JFET. For the *p*-type device of Fig. 6.1*a* the arrow at the gate shows the gate to be *p*-type material with channel *n*-type. The symbol for the *p*-channel JFET of Fig. 6.1*b* includes an arrow at the gate showing the gate to be *n*-type material with channel *p*-type.

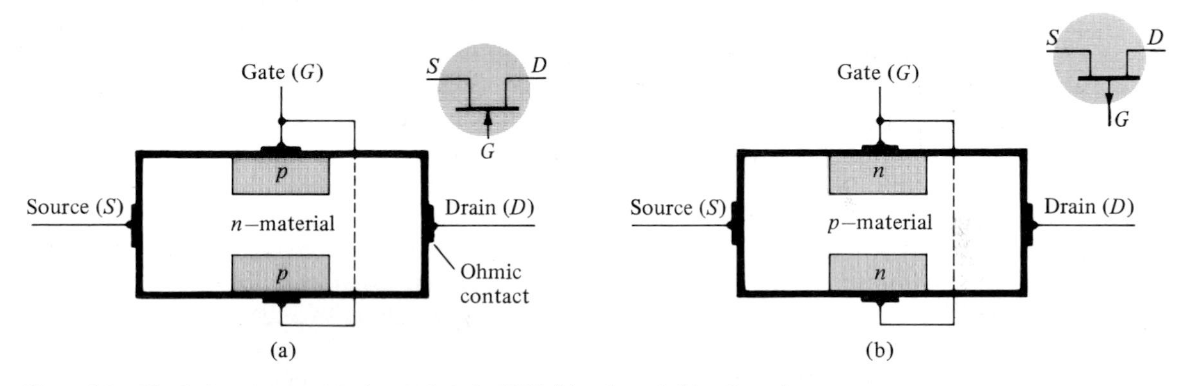

Figure 6.1 Physical structure and device symbol of a JFET: (a) *n*-channel; (b) *p*-channel.

To examine how the device is operated, consider the *n*-channel JFET of Fig. 6.2, shown with applied bias voltages to operate the device. The supply voltage, V_{DD}, provides a voltage across drain-source, $V_{DS,}$ which results in a current, I_D, from drain to source (electrons in an *n*-channel actually move from the source, hence name, to drain). This drain current passes through the *channel* surrounded by the *p*-type gate. A voltage between gate and source, V_{GS}, is shown here to be set by a voltage supply,

[1]W. Shockley, *Electrons and Holes in Semiconductors* (New York: Van Nostrand, 1953).

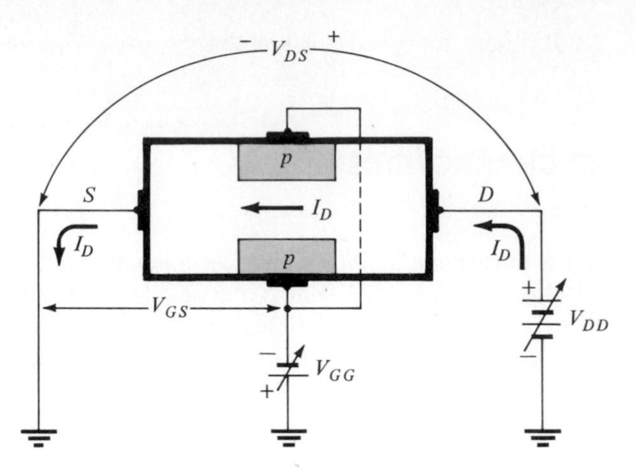

Figure 6.2 Basic operation of JFET.

V_{GG}. Since the polarity of this gate-source voltage will reverse bias the gate-source junction, no gate current will result. The effect of the gate-source voltage will be to create a depletion region in the channel and thereby reduce the channel width to increase the drain-source resistance resulting in less drain current.

We shall first consider the device operation with $V_{GS} = 0$ V and then with the reverse-bias voltage, V_{GS}, increased (made more negative for an n-channel device). Figure 6.3a shows that drain current through the n-material of the drain-source produces a voltage drop along the channel, which is more positive at the drain-gate junction than at the source-gate junction. This reverse-bias potential across the p-n junction causes a depletion region to form as shown in Fig. 6.3a. As the voltage, V_{DD}, is increased, the current, I_D, increases, resulting in a larger depletion region. As the voltage V_{DD} is increased, the depletion region forms fully across the channel as shown in Fig. 6.3b. Any further increase in V_{DD} will result in no increase in the drain current, the current I_D then remaining constant. This operation is depicted in the $V_{GS} = 0$ characteristic curve of Fig. 6.3c. As V_{DS} increases, the current I_D increases until the depletion region is fully formed across the channel, after which, the drain current saturates and remains a constant value even for increased voltage V_{DS}. This value of constant drain current at $V_{GS} = 0$ V is an important parameter used to specify the operation of the JFET and is referred to as I_{DSS}, the *D*rain-to-*S*ource current with the gate source *S*horted.

Figure 6.4 provides a summary of the operation of an n-channel JFET. When the gate-source voltage, V_{GS}, is made less than 0 V but greater than the pinch-off voltage (see Fig. 6.4a), a drain current I_D exists, this current being adjusted by the voltage V_{GS}. The gate current is then

$$I_G = 0 \tag{6.1}$$

since no current will pass through the reverse-biased gate-source junction.

When the gate-source voltage is set to exactly 0 V, the level of the drain current is a value of importance and is designated as I_{DSS} (see Fig. 6.4b). The gate current is still 0, as given by Eq. (6.1). If the gate-source voltage is increased beyond the pinch-off value (more negative than that needed to pinch-off the channel), the drain

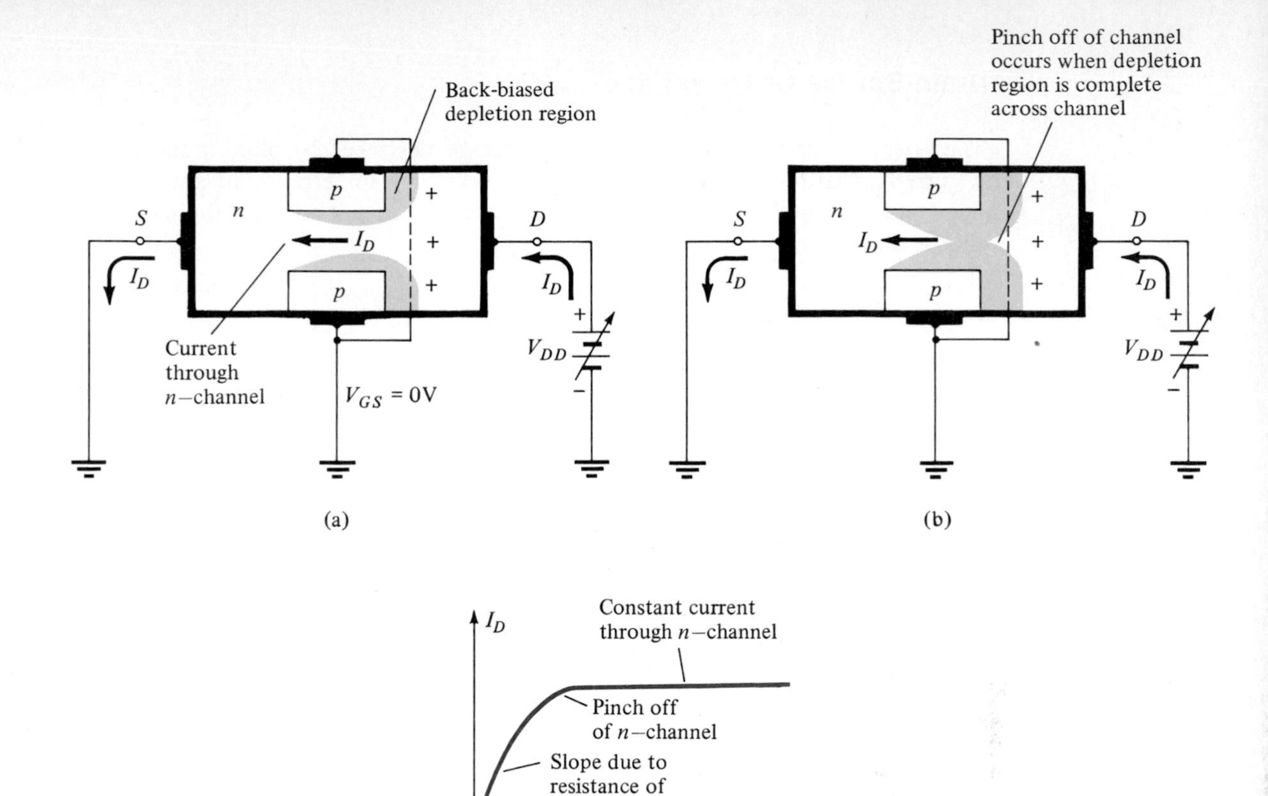

Figure 6.3 Pinch-off action due to channel current.

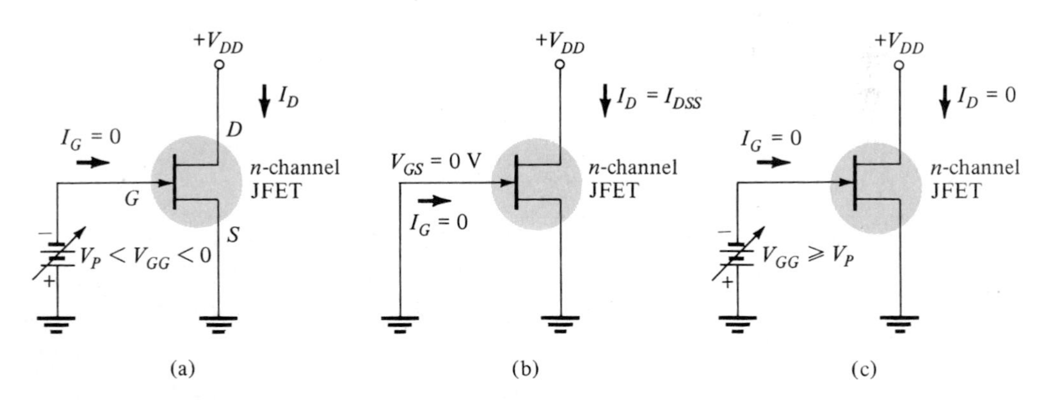

Figure 6.4 Operation of n-channel JFET using device symbol: (a) V_{GS} in range 0 V to V_p; (b) $V_{GS} = 0$ V; (c) $V_{GS} \geq V_P$.

current reduces to 0, with I_G also 0, and the JFET device is then completely turned off (see Fig. 6.4c).

Drain-Source Characteristic

The operation depicted in Fig. 6.4 can better be described by plotting the actual drain current at different values of drain-source voltage for a range of gate-source voltage values. This plot is called a drain characteristic since it plots the drain current versus the drain-source voltage.

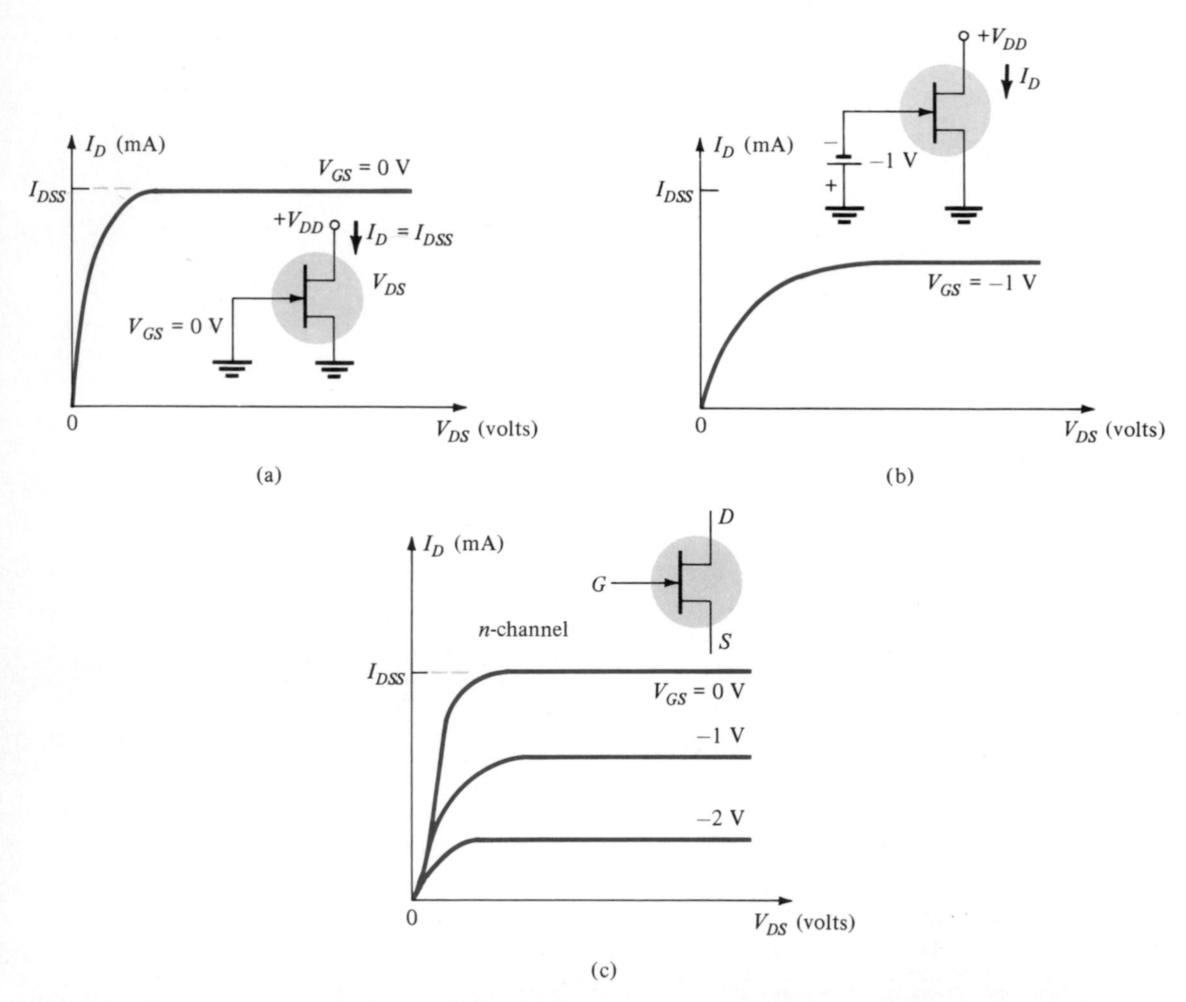

Figure 6.5 Drain-source characteristic: (a) V_{GS} = 0 V; (b) V_{GS} = − 1 V; (c) complete characteristic.

Figure 6.5a shows a typical n-channel JFET drain-source characteristic. For $V_{GS} = 0$ V the curve plotted shows that the drain current increases as V_{DS} is increased until a point at which the current levels off (or reaches saturation). From the previous discussion we know that the internal depletion region acts to limit the drain-source current. For the $V_{GS} = 0$ setting this saturation current is labeled I_{DSS}.

If the gate-source voltage is set at $V_{GS} = -1$ V (see Fig. 6.5b), the current increases as V_{DS} is increased until a saturation level is reached, this time at a lower level than for $V_{GS} = 0$ V, since the depletion region, starting partly formed due to $V_{GS} = -1$ V, fully forms at a lower level of drain-source current.

The drain-source characteristic is a set of curves for different values of V_{GS} from 0 V to the pinch-off voltage, the voltage at which the depletion region is formed without any drain-source current and at which no drain-source current can occur. The pinch-off voltage is usually specified as V_P or $V_{GS(\text{off})}$. Figure 6.6 shows a drain-source characteristic for a p-channel JFET with positive gate-source voltages reducing the drain-source current.

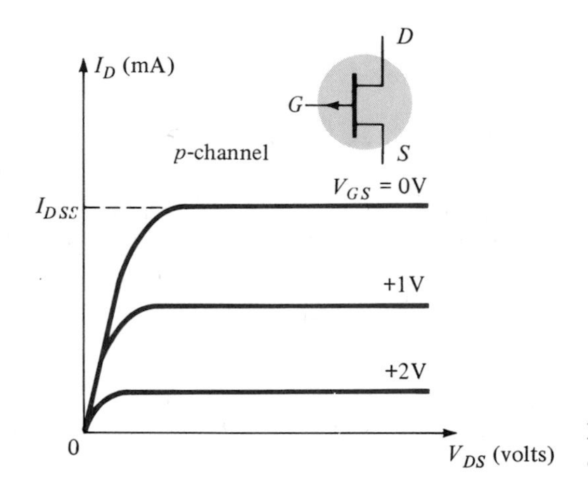

Figure 6.6 p-channel JFET drain-source characteristic.

Transfer Characteristic

Another form of device characteristic is the transfer characteristic which is a plot of drain current, I_D as a function of gate-source voltage, V_{GS}, for a constant value of drain-source voltage, V_{DS}. The transfer characteristic can be directly viewed on a curve tracer unit, obtained directly by measurement of device operation, or drawn from the drain characteristic as shown in Fig. 6.7. Two important points of the transfer curve

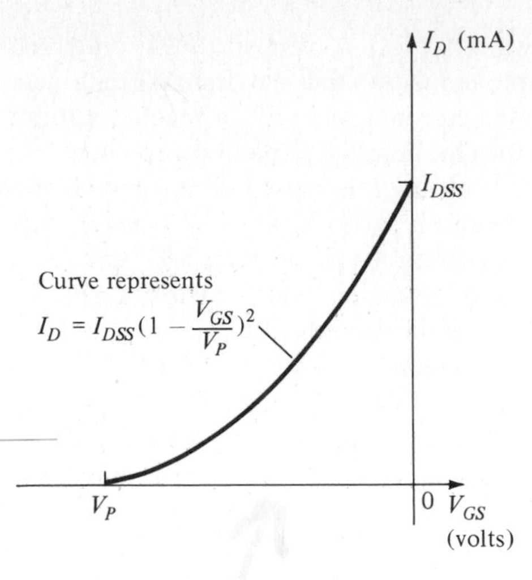

Figure 6.7 JFET transfer characteristic (n-channel).

shown are the values I_{DSS} and V_P. When these points are fixed, the rest of the curve can be seen on the transfer characteristic or obtained from theoretical consideration of the physical process occurring in the JFET, leading to the relation[2]

$$\boxed{I_D = I_{DSS}\left(1 - \frac{V_{GS}}{V_P}\right)^2} \tag{6.2}$$

which represents the transfer characteristic curve of Fig. 6.7. Notice that when $V_{GS} = 0$, $I_D = I_{DSS}$ and that when $I_D = 0$, $V_{GS} = V_P$ as seen on the transfer characteristic.

EXAMPLE 6.1

Determine the drain current of an n-channel JFET having pinch-off voltage $V_P = -4$ V and drain-source saturation current $I_{DSS} = 12$ mA at the following gate-source voltages: (a) $V_{GS} = 0$ V, (b) $V_{GS} = -1.2$ V, and (c) $V_{GS} = -2$ V.

Solution:

Using Eq. (6.2), we obtain

(a) $I_D = I_{DSS}\left(1 - \frac{V_{GS}}{V_P}\right)^2 = 12\text{ mA}\left(1 - \frac{0}{-4}\right)^2 = \mathbf{12\ mA}$

(b) $I_D = 12\text{ mA}\left(1 - \frac{-1.2}{-4}\right)^2 = \mathbf{5.88\ mA}$

(c) $I_D = 12\text{ mA}\left(1 - \frac{-2}{-4}\right)^2 = \mathbf{3\ mA}$

---

[2]Shockley's equation applies above pinch-off region for JFET device.

6.3 PLOTTING JFET TRANSFER CHARACTERISTIC

The two parameters that are used to specify the operation of a particular JFET device are its values of I_{DSS} and V_P. One can use the typical values given by the manufacturer's specifications sheet or measure these values for the particular JFET.

Figure 6.8 shows a circuit to measure I_{DSS}. When we set V_{GS} to 0 V, the circuit uses an ammeter to measure the value of drain-source current, which is I_{DSS}. The supply voltage, V_{DD}, need only be set large enough to raise the drain-source current to its saturation level (i.e., the supply voltage is increased until I_D no longer increases, the level reached being I_{DSS}).

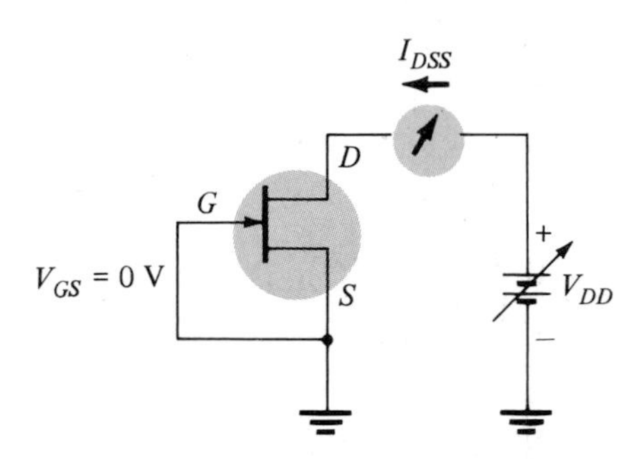

Figure 6.8 Circuit used to measure I_{DSS}.

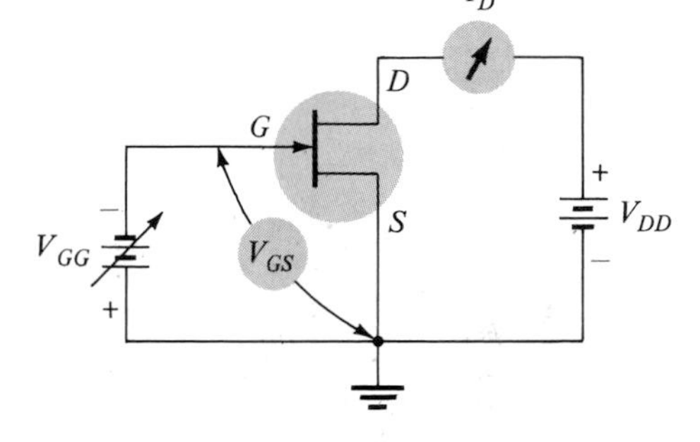

Figure 6.9 Circuit used to measure V_P.

The circuit of Fig. 6.9 can be used to measure V_P. The gate-source voltage is adjusted from 0 V to larger negative values until the drain current just reaches 0 (or falls below a minimum current level). The voltage V_{GS} to just cause the drain current to reach 0 is the measured value of $V_{GS(\text{off})}$ or V_P.

Having obtained values of I_{DSS} and V_P by measurement or from the device spec sheets, the user can now plot a transfer curve to use in later dc bias calculations or analysis of ac operation.

Plot of JFET Transfer Characteristic

The transfer characteristic of an n-channel JFET can be drawn on a set of axes with horizontal axis going from 0 V to negative voltages up to V_P, and a vertical axis of drain current, I_D, from 0 to I_{DSS}. Figure 6.10a shows such axes for values of $V_p = -5$ V and $I_{DSS} = 10$ mA. Two points of the transfer characteristic are shown:

1. $I_{DSS} = 10$ mA along the vertical axis ($V_{GS} = 0$ V).
2. $V_P = -5$ V along the horizontal axis ($I_D = 0$ mA).

These value are calculated using Eq. (6.2):

$$\text{For } V_{GS} = 0 \text{ V:} \quad I_D = I_{DSS}(1 - 0)^2 = I_{DSS} = 10 \text{ mA}$$

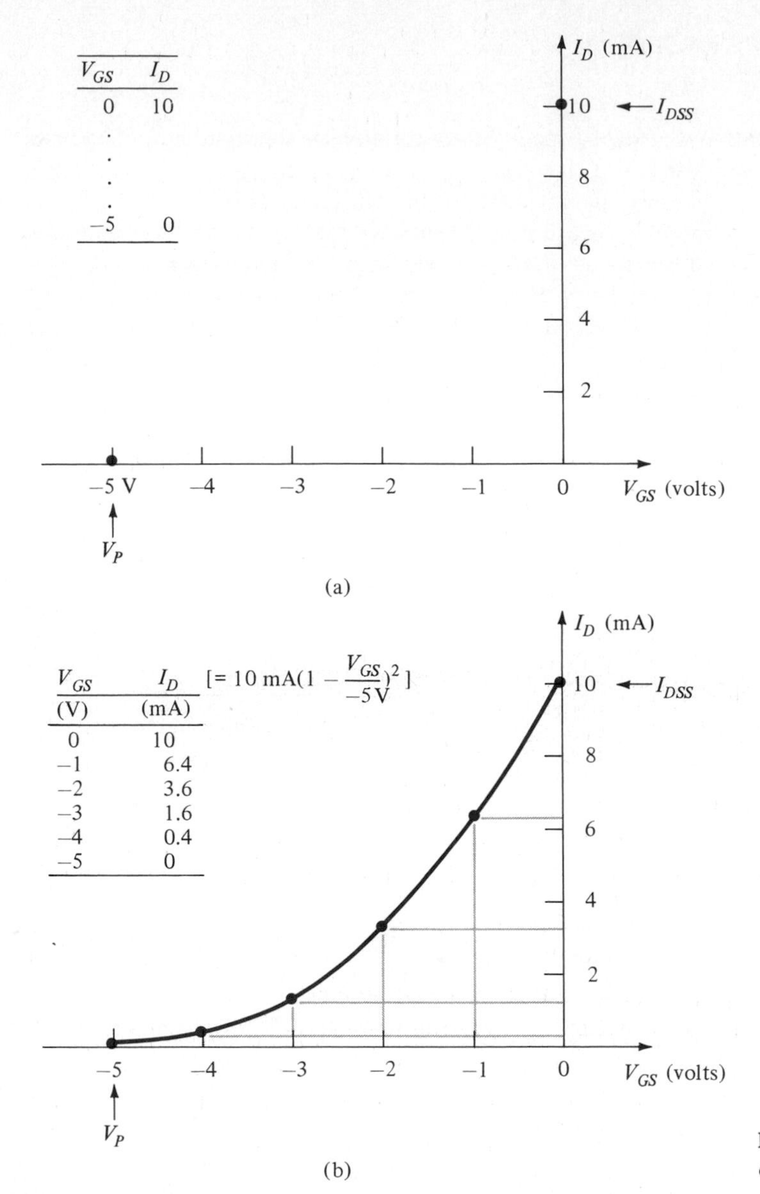

Figure 6.10 Plot of JFET transfer characteristic.

$$\text{For } I_D = 0 \text{ mA:} \quad 0 = 10 \text{ mA}\left(1 - \frac{V_{GS}}{-5 \text{ V}}\right)^2; \; V_{GS} = -5 \text{ V}$$

Additional points to plot the transfer curve can be selected using two or three values of voltage between 0 and V_P. Figure 6.10b shows the calculated value of I_D for each of a few voltages, each of these points then plotted on the transfer characteristic. The resulting transfer characteristic curve is then obtained by connecting these points as shown in Fig. 6.10b. The points to use in plotting the curve can be obtained in a relatively easy manner as described next.

1. Select $V_{GS} = 0$ V for which $I_D = I_{DSS}$:

$$V_{GS} = 0 \text{ V} \qquad I_{DSS} = 12 \text{ mA}$$

2. Select $V_{GS} = 0.3\ V_P$ for which

$$I_D = I_{DSS}\left(1 - \frac{0.3V_P}{V_P}\right)^2 = I_{DSS}(0.49) \cong I_{DSS}(0.5)$$

$$V_{GS} = 0.3(-4 \text{ V}) = -1.2 \text{ V} \qquad I_D = 0.5I_{DSS} = 0.5(12 \text{ mA}) = 6 \text{ mA}$$

3. Select $V_{GS} = 0.5V_P$ for which

$$I_D = I_{DSS}\left(1 - \frac{0.5V_P}{V_P}\right)^2 = 0.25I_{DSS}$$

$$V_{GS} = 0.5(-4 \text{ V}) = -2 \text{ V} \qquad I_D = 0.25I_{DSS} = 0.25(12 \text{ mA}) = 3 \text{ mA}$$

4. Select $V_{GS} = V_P$, for which

$$I_D = I_{DSS}\left(1 - \frac{V_P}{V_P}\right)^2 = 0$$

$$V_{GS} = V_P = -4 \text{ V} \qquad I_D = 0$$

Figure 6.11 shows the four data points above plotted on the transfer characteristic axes with a curve then connected between the four points. Although more points would provide a fuller plot, the use of four points is often sufficient for dc bias or ac

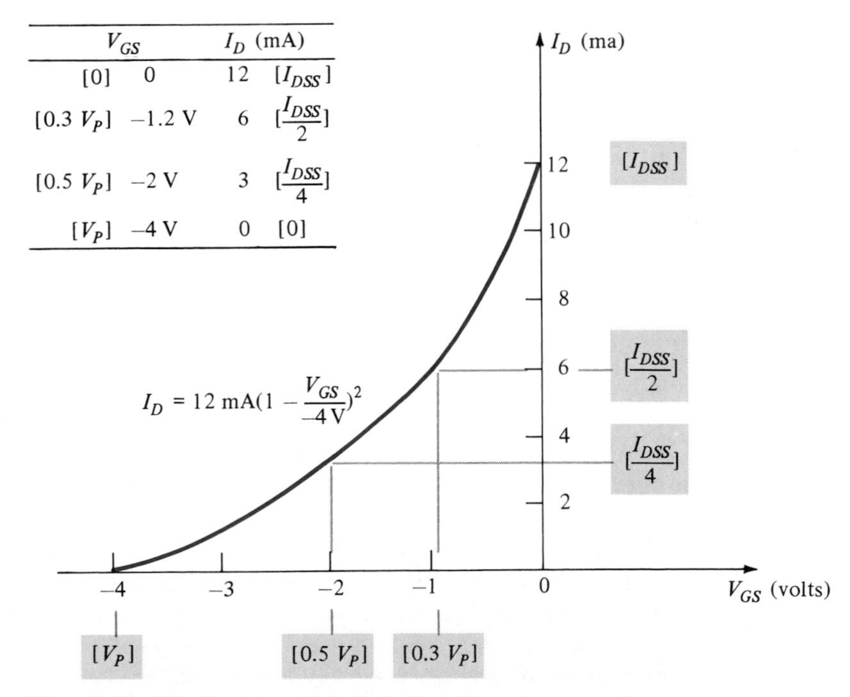

V_{GS}		I_D (mA)	
[0]	0	12	[I_{DSS}]
[0.3 V_P]	−1.2 V	6	[$\frac{I_{DSS}}{2}$]
[0.5 V_P]	−2 V	3	[$\frac{I_{DSS}}{4}$]
[V_P]	−4 V	0	[0]

Figure 6.11 Transfer characteristic for $I_{DSS} = 12$ mA, $V_P = -4$ V.

operation. The four points used for V_{GS} provide easily calculated values of I_D, most of which do not even require use of a calculator to obtain the results. A summary of the suggested values to use and results are summarized in Table 6.1

TABLE 6.1 Values to Use in Plotting JFET Transfer Characteristic

V_{GS}	I_D
0	I_{DSS}
$0.3V_P$	$\frac{I_{DSS}}{2}$
$0.5V_P$	$\frac{I_{DSS}}{4}$
V_P	0

6.4 JFET PARAMETERS

Manufacturers specify a number of parameters to describe the JFET device and to provide data necessary for selecting between various units. Some of the more useful parameters specified are:

1. I_{DSS}, the drain-source saturation current
2. $V_P = V_{GS(off)}$, the pinch-off or gate-source off voltage
3. BV_{GSS}, the device breakdown voltage with drain-source shorted
4. $g_m = g_{fs}$, the device transconductance
5. $r_{ds(on)}$, the drain-source resistance when the device is turned on

A number of other parameters relating to device capacitance, noise voltage, turn-on and turn-off times, and power handling are also usually provided in the manufacturer's spec sheets.

Drain-Source Saturation Current, I_{DSS}

The current at which the channel pinches off when gate-source is shorted ($V_{GS} = 0$) is a most important device parameter. The value of I_{DSS} can be easily measured using a circuit as in Fig. 6.8 and represents the largest drain current using the JFET reverse biased. For a small-signal device, this current is typically milliamperes (mA).

Gate-Source Cutoff (Pinch-Off) Voltage, $V_P = V_{GS(off)}$

The gate-source voltage at which the drain-source channel is cut off or pinched off, resulting in essentially no drain current, is $V_{GS(off)}$ or V_P in the manufacturer's specifications sheets. Practical measurements require determination of the cutoff voltage at a specified low value of drain current (some microamperes) since zero current or little drain current is too ambiguous a value to measure. A circuit to provide measurement of $V_{GS(off)}$ is shown in Fig. 6.9. For a specified value of V_{DS}, I_D and V_{GS} are measured as V_{GS} is varied until I_D is reduced to the sufficiently small current value indicating cutoff.

Source-Gate Breakdown Voltage, BV_{GSS}

The breakdown voltage of the source-gate junction, BV_{GSS}, is measured at a specified current with the drain-source shorted ($V_{DS} = 0$ V). The value of breakdown voltage indicates a limiting value of voltage across gate-source, above which the device current must be limited by the external circuit or else the device may be permanently damaged. The breakdown voltage provides a limiting voltage value for use in selecting the drain supply voltage.

Common-Source Forward Transconductance, $g_{fs} = g_m$

The measured device common-source forward transconductance, specified as g_{fs}, is also referred to in the text as g_m. It is measured with drain-source shorted; that is

$$g_{fs} = \left.\frac{\Delta I_D}{\Delta V_{GS}}\right|_{V_{DS}=0}$$

and is an indication of the JFET ac amplification. The value of g_{fs} (or g_m) tells how much the ac current will change due to an applied ac gate-source voltage.

The value of g_m is measured in *siemens* (S)—formerly mho (℧), with typical values from 1 mS to 10 mS or 1000 μS-10,000 μS. Mathematical derivation can be used to show that the relation given by Eq. (6.2) when mathematically differentiated results in [3]

$$\boxed{g_m = g_{m0}\left(1 - \frac{V_{GS}}{V_P}\right)} \tag{6.3}$$

where

$$\boxed{g_{m0} = \frac{2I_{DSS}}{|V_P|}} \tag{6.4}$$

The value of g_{m0} is the maximum ac gain parameter of the JFET occurring at a bias of $V_{GS} = 0$ V. For any other bias condition, the value of g_m is less, as calculated using Eqs. (6.3) and (6.4).

---

[3] Differentiating Eq. (6.2), $I_D = I_{DSS}\left(1 - \frac{V_{GS}}{V_P}\right)^2$ results in

$$g_m = \frac{\partial I_D}{\partial V_{GS}} = -\frac{2I_{DSS}}{V_P}\left(1 - \frac{V_{GS}}{V_P}\right) = \frac{2I_{DSS}}{|V_P|}\left(1 - \frac{V_{GS}}{V_P}\right)$$

EXAMPLE 6.2

Calculate the transconductance, g_m, of a JFET having specified values of $I_{DSS} = 12$ mA and $V_P = -4$ V at bias points (a) $V_{GS} = 0$ V and (b) $V_{GS} = -1.5$ V.

Solution:

Using Eqs. (6.3) and (6.4)

$$g_{m0} = \frac{2\,I_{DSS}}{|V_P|} = \frac{2(12\text{ mA})}{|-4\text{ V}|} = \frac{24 \times 10^{-3}}{|-4|} = 6 \times 10^{-3}\text{ S} = 6\text{ mS} = 6000\ \mu\text{S}$$

(a) $g_m = g_{m0}\left(1 - \frac{V_{GS}}{V_P}\right) = 6\text{ mS}\left(1 - \frac{0\text{ V}}{-4\text{ V}}\right) = 6\text{ mS} = \mathbf{6000\ \mu S}$

(b) $g_m = 6\text{ mS}\left(1 - \frac{-1.5\text{ V}}{-4\text{ V}}\right) = 3.75\text{ mS} = \mathbf{3750\ \mu S}$

Drain-Source *on* Resistance, $r_{ds(on)}$

The drain-source *on* resistance, measured at a specified gate-source voltage and drain current is important when using the JFET as a switch. When the JFET is biased in its saturation or ohmic region of operation, it exhibits a resistance between drain and source from ten to a few hundred ohms, as specified by the value of $r_{ds(\text{on})}$.

6.5 MOSFET CONSTRUCTION AND CHARACTERISTICS

A field-effect transistor can be constructed with the gate terminal insulated from the channel. The popular Metal-Oxide-Semiconductor FET (MOSFET) is constructed as either a *depletion* MOSFET (Fig. 6.12a) or *enhancement* MOSFET (Fig. 6.12b). In the depletion-mode construction a channel is physically constructed and current between drain and source will result from a voltage connected across the drain-source. The enhancement MOSFET structure has *no* channel formed when the device is constructed. Voltage must be applied at the gate to develop a channel of charge carriers so that a current results when a voltage is applied across the drain-source terminals.

Depletion MOSFET

The *n*-channel depletion MOSFET device of Fig. 6.12a is formed on a *p*-substrate (*p*-doped silicon material used as the starting material onto which the FET structure is formed). The source and drain are connected by metal (aluminum) to *n*-doped source and drain regions which are connected internally by an *n*-doped channel region. A metal layer is deposited above the *n*-channel on a layer of silicon dioxide (SiO_2) which is an insulating layer. This combination of a *metal* gate on an *oxide* layer over a *semi*conductor substrate forms the depletion MOSFET device. For the *n*-channel device of Fig 6.12a negative gate-source voltages push electrons out of the channel region to deplete the channel and a large enough negative gate-source voltage will

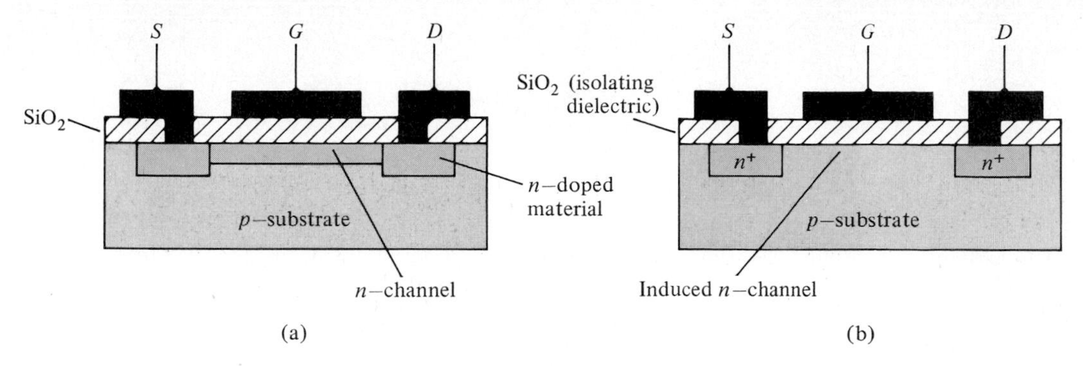

Figure 6.12 MOSFET construction: (a) depletion; (b) enhancement.

pinch off the channel. Positive gate-source voltage *on the other hand* will result in an increase in the channel size (pushing away p-type carriers), allowing more charge carriers and therefore greater channel current to result.

An n-channel depletion MOSFET device characteristic is shown in Fig. 6.13. The device is shown to operate with either positive or negative gate-source voltage, negative values of V_{GS} reducing the drain current until the pinch-off voltage, after which no drain current occurs. The transfer characteristic is the same for negative gate-source voltages, but it continues for positive values of V_{GS}. Since the gate is isolated from the channel for both negative and positive values of V_{GS}, the device can be operated with either polarity of V_{GS}—no gate current resulting in either case. The device schematic symbol in Fig. 6.13 shows the addition of a substrate terminal (in addition to gate, source, and drain leads) on which the device type is indicated, the arrow here indicating a p-substrate and thus n-channel device. A p-channel depletion MOSFET characteristic is shown in Fig. 6.14.

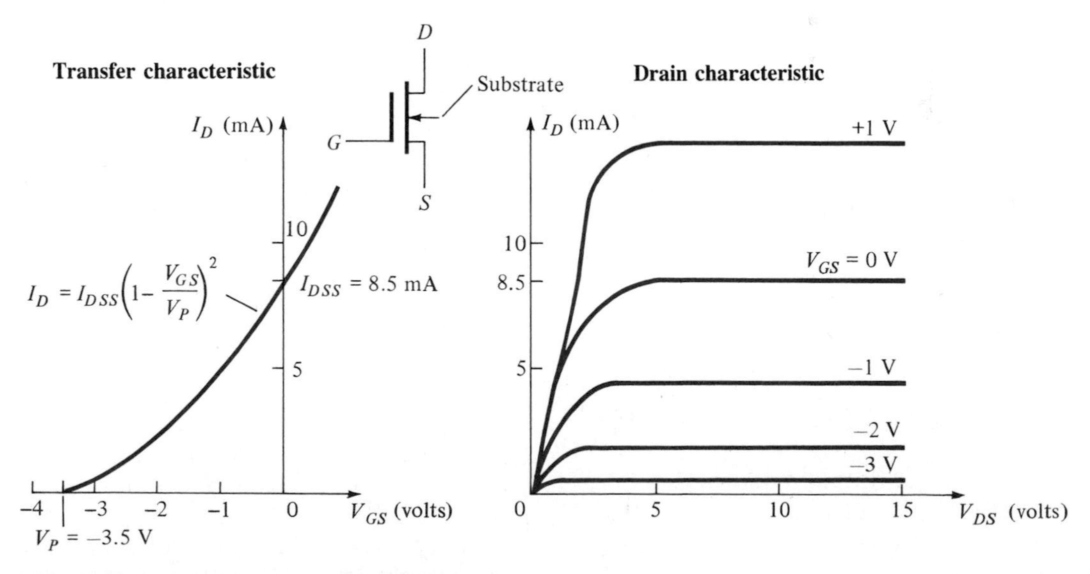

Figure 6.13 n-channel depletion MOSFET characteristics.

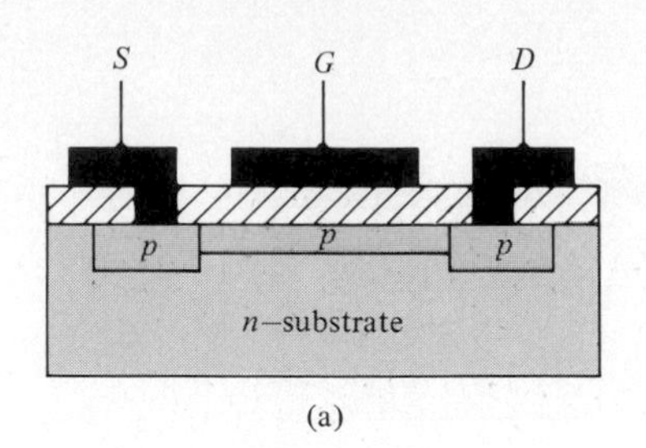

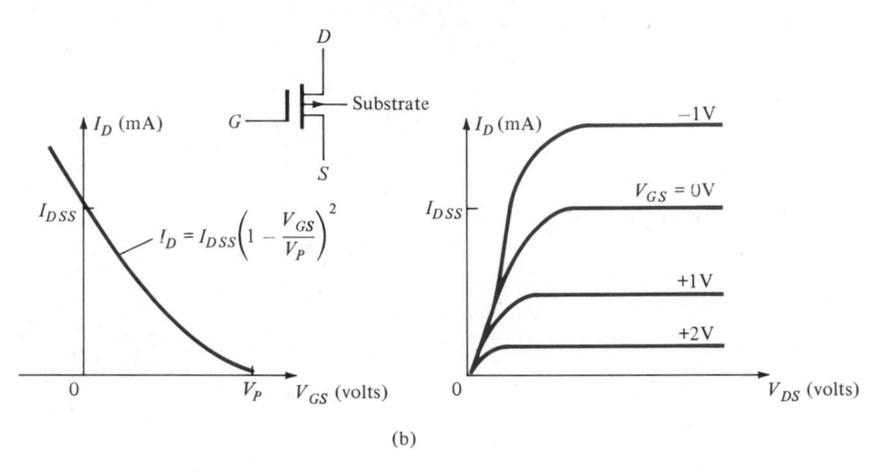

Figure 6.14 *p*-channel depletion MOSFET: (a) device structure; (b) device characteristic.

EXAMPLE 6.3

A depletion MOSFET has I_{DSS} = 12 mA and V_P = −4.5 V. Calculate the drain current at gate-source voltages of (a) 0 V, (b) −2 V, and (c) −3 V.

Solution:

Using Eq. (6.2) gives us

(a) $I_D = I_{DSS}\left(1 - \frac{V_{GS}}{V_P}\right)^2 = 12\text{ mA}\left(1 - \frac{0\text{ V}}{-4.5\text{ V}}\right)^2 = \mathbf{12\text{ mA}}$

(b) $I_D = 12\text{ mA}\left(1 - \frac{-2\text{ V}}{-4.5\text{ V}}\right)^2 = \mathbf{3.7\text{ mA}}$

(c) $I_D = 12\text{ mA}\left(1 - \frac{-3\text{ V}}{-4.5\text{ V}}\right)^2 = \mathbf{1.33\text{ mA}}$

Enhancement MOSFET

The enhancement MOSFET of Fig. 6.15 has no channel between drain and source as part of the basic device construction. Application of a positive gate-source voltage will repel holes in the substrate region under the gate leaving a depletion region. When

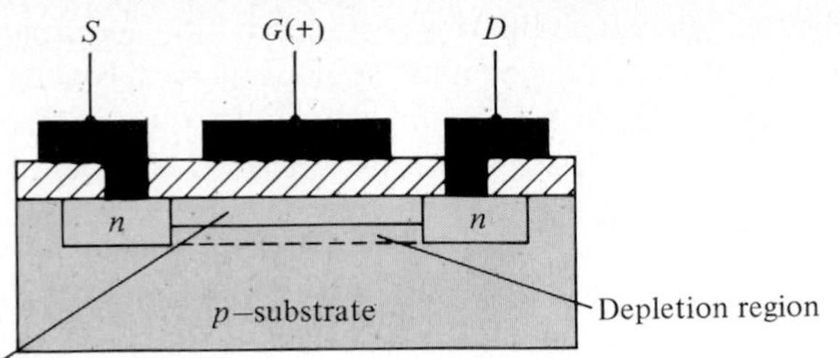

Figure 6.15 *n*-channel formed in enhancement MOSFET.

the gate voltage is sufficiently positive, electrons are attracted into this depletion region making it then act as an *n*-channel between drain and source. The resulting *n*-channel enhancement MOSFET characteristic is shown in Fig. 6.16. There is no drain current until the gate-source voltage exceeds the threshold value, V_T. Positive voltages above this threshold value result in increased drain current, the transfer characteristic being described by[4]

$$\boxed{I_D = K(V_{GS} - V_T)^2} \tag{6.5}$$

where K, typically 0.3 mA/V^2, is a property of the device construction. Note that no value I_{DSS} can be associated with an enhancement MOSFET because no drain current occurs with $V_{GS} = 0$ V. Although the enhancement MOSFET is more restricted in operating range than is the depletion device, the enhancement device is very useful in

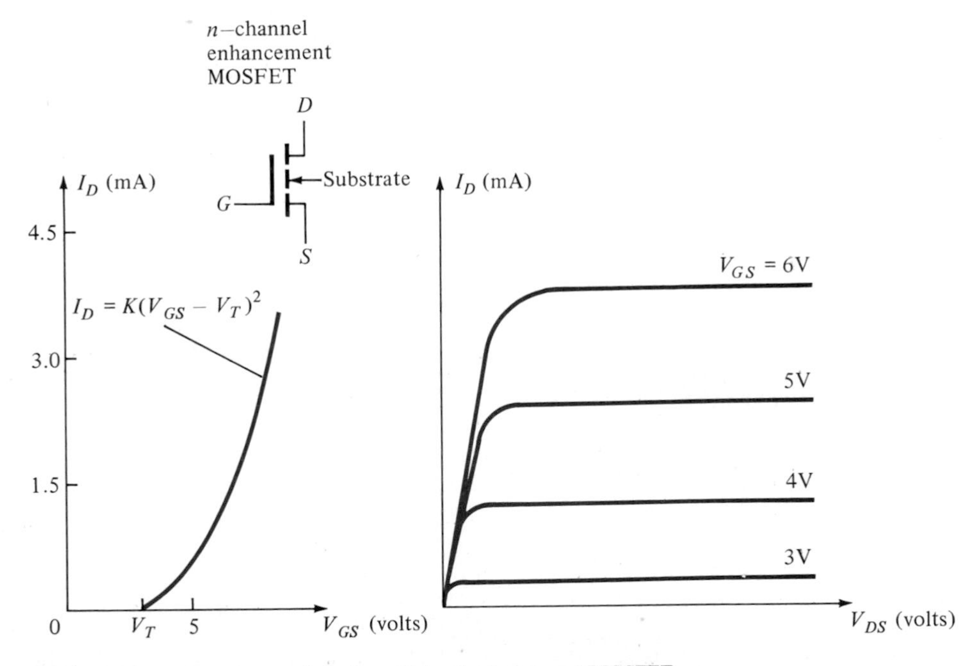

Figure 6.16 Device characteristic for *n*-channel enhancement MOSFET.

[4] Equation (6.5) is valid only for $|V_{GS}| > |V_T|$.

large-scale integrated circuits in which the simpler construction and smaller size make it a suitable device. The enhancement schematic symbol shows a broken line between drain and source indicating that there is no initial channel for the enhancement device. The substrate terminal arrow shows a p-substrate and an n-channel. P-channel enhancement MOSFETS can also be constructed, the device and characteristic being shown in Fig. 6.17.

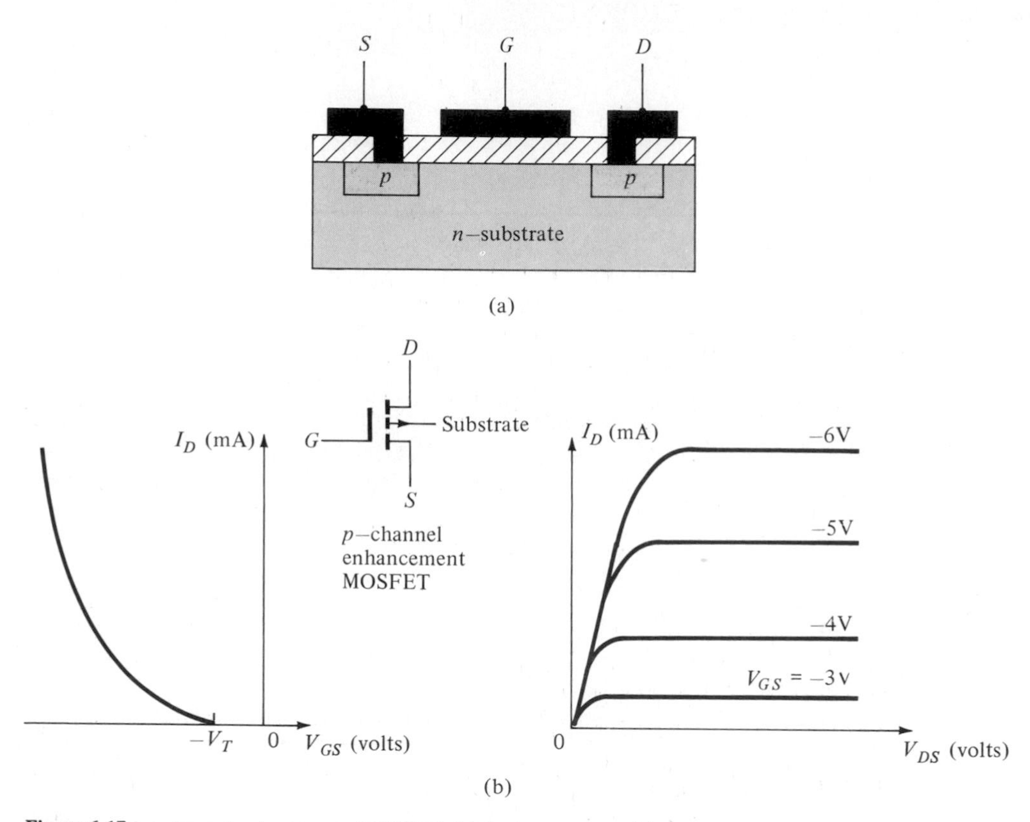

Figure 6.17 p-channel enhancement MOSFET: (a) device structure; (b) device characteristic.

EXAMPLE 6.4

For an n-channel enhancement MOSFET with threshold voltage of 2.5 V, determine the current at values of gate-source voltages (a) $V_{GS} = 2.5$ V, (b) $V_{GS} = 4$ V, and (c) $V_{GS} = 6$ V.

Solution:

Using $K = 0.3 \text{ mA/V}^2$ in Eq. (6.4) yields

(a) $I_D = 0.3\times10^{-3}(V_{GS} - V_T)^2 = 0.3\times10^{-3}(2.5 - 2.5)^2 = \mathbf{0\ mA}$

(b) $I_D = 0.3\times10^{-3}(4 - 2.5)^2 = \mathbf{0.675\ mA}$

(c) $I_D = 0.3\times10^{-3}(6 - 2.5)^2 = \mathbf{3.675\ mA}$

As with a JFET or depletion MOSFET, a value of transconductance can also be obtained for an enhancement MOSFET, the relation in this case being[5]

$$\boxed{g_m = 2K(V_{GS} - V_T)} \tag{6.6}$$

EXAMPLE 6.5

Determine the value of transconductance for an n-channel enhancement MOSFET having threshold voltage $V_T = 3$ V at the following operating points: (a) 6 V and (b) 8 V.

Solution:
(a) $g_m = 2(0.3 \times 10^{-3})(6 - 3) = \mathbf{1.8\ mS}$
(b) $g_m = 2(0.3 \times 10^{-3})(8 - 3) = \mathbf{3\ mS}$

6.6 SPECIFICATIONS SHEETS

The specifications sheet of a JFET provides information on a number of important electrical characteristics and ratings. Figure 6.18 provides typical data found in a spec sheet. The absolute maximum ratings provide such information as

$$\text{Maximum Drain-Source Voltage} = 30 \text{ V}$$

Any supply voltage at or below this value could permit reliable opertion of the JFET device.

$$\text{Total Device Dissipation at 25°C} = 360 \text{ mW}$$

The maximum power dissipated by the transistor at room temperature (25°C) should be at or below 360 mW. For example, at a bias condition of $V_{DS} = 20$ V and $I_D = 8$ mA, the power dissipated is

$$P_D = V_{DS} I_D = (20 \text{ V})(8 \text{ mA}) = 160 \text{ mW}$$

well below the maximum rating of 360 mW. In fact, the maximum current at 30 V would be

$$\text{maximum } I_D = \frac{\text{maximum } P_D}{\text{maximum } V_{DS}} = \frac{360 \text{ mW}}{30 \text{ V}} = 12 \text{ mA}$$

Among the published electrical characteristics are

$$BV_{GSS} = -30\ V$$

which specifies the typical voltage across the gate-source terminals at which the

---

[5] Obtained from differentiating Eq. (6.5),

$$g_m = \frac{\partial I_D}{\partial V_{GS}} = 2\,K(V_{GS} - V_T)$$

where $K = 0.3 \times 10^{-3}$ A/V^2(typical) $= 0.3$ mA/V^2.

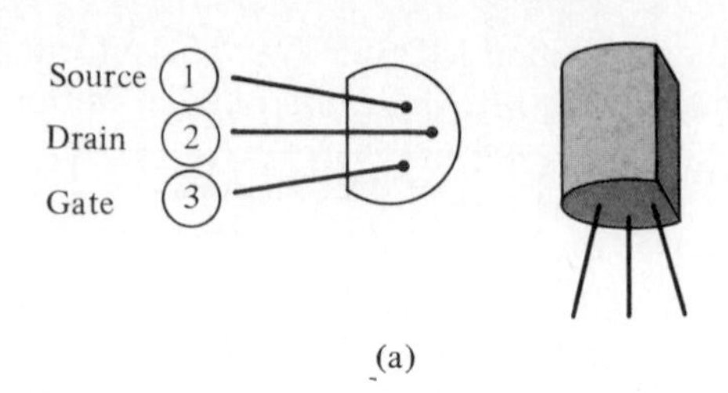

(a)

Absolute maximum ratings at 25°C free-air temperature

Drain-Gate Voltage	30 V
Reverse Gate-Source Voltage	−30 V
Continuous Forward Gate Current	10 mA
Continuous Device Dissipation at (or below) 25°C Free-Air-Temperature	360 mW
Continuous Device Dissipation at (or below) 25°C Lead Temperature	500 mW
Storage Temperature Range	−65°C to 150°C

(b)

Electrical characteristics at 25°C free-air temperature

PARAMETER		MIN	MAX
BV_{GSS}	Gate-Source Breakdown Voltage	−30 V	
$V_{GS(off)}$	Gate-Source Cutoff Voltage	−2.5 V	−6 V
I_{DSS}	Zero-Gate-Voltage Drain Current	10 mA	15 mA
$r_{ds(on)}$	Small-Signal Drain-Source On-State Resistance		210 Ω
y_{fs}	Small-Signal Common-Source Forward Transfer Admittance	3.5 μS	7.5 μS
y_{os}	Small-Signal Common-Source Output Admittance		75 μS

(c)

Figure 6.18 Specifications data for 2N5950 JFET. (a) Device leads; (b) absolute maximum ratings; (c) typical electrical characteristic data.

junction breaks down in the reverse-bias direction

$$V_{GS(\text{off})} = -2.5 \text{ V to } -6 \text{ V}$$

which specifies the gate-source cutoff voltage (V_P), measured at a drain-source voltage, $V_{DS} = 15$ V; and

$$I_{DSS} = 15 \text{ mA (maximum)}$$

which provides the value of saturation drain current measured at $V_{GS} = 0$ V and $V_{DS} = 15$ V.

Another parameter is

$$r_{ds(\text{on})} = 210\ \Omega \text{ (maximum)}$$

which is the resistance of the drain-source for small ac signals.

6.7 CMOS

A popular connection used primarily in digital circuits connects enhancement *p*MOS and *n*MOS transistors into a complementary or CMOS device. Figure 6.19 shows the basic CMOS connection. The input is connected in common to the gate of both *p*MOS and *n*MOS transistors. A positive input voltage drives the *p*MOS *off*, the *n*MOS *on*, with the output dropping to 0 V. A low-value input voltage will correspondingly drive the *p*MOS device *on*, the *n*MOS device *off*, with the output voltage rising to $+V_{DD}$. A plot showing the relation between input and output voltages is provided in Fig. 6.20.

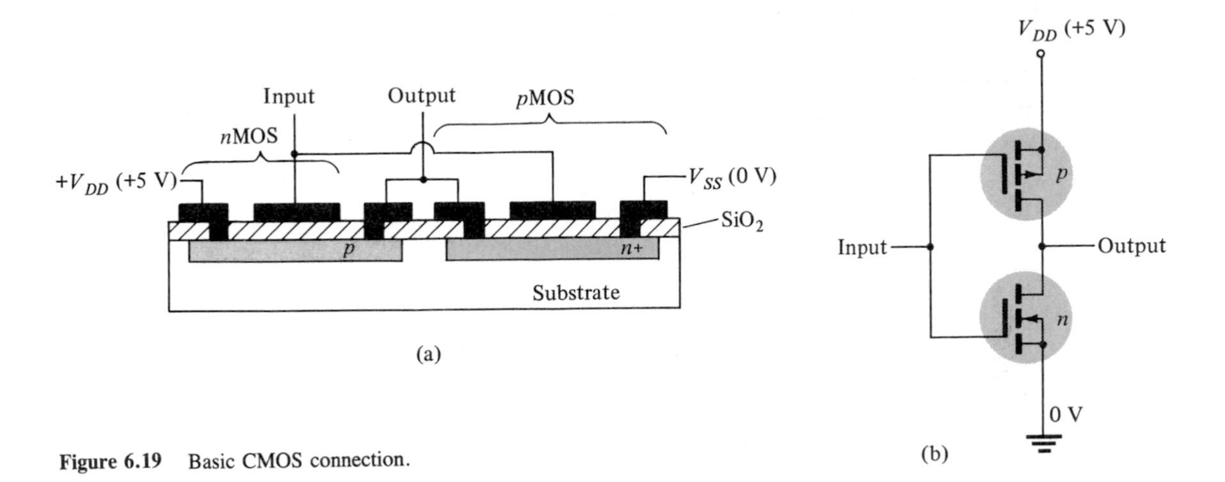

Figure 6.19 Basic CMOS connection.

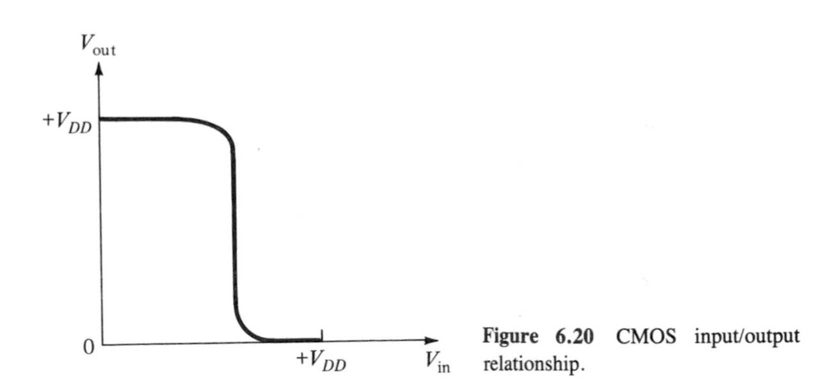

Figure 6.20 CMOS input/output relationship.

When the input voltage is low, the *n*MOS transistor remains *off* while the *p*MOS transistor is biased *on*, with the output then at the supply voltage level of $+V_{DD}$. As the input voltage is raised this condition remains as shown in Fig. 6.20 until the input goes high enough to start turning the *n*MOS transistor *on*. The output voltage then drops down as the *p*MOS device goes fully *on* and the *n*MOS device begins to turn *off*, the output voltage quickly dropping to 0 V.

Except for a short time as the voltage drops from $+V_{DD}$ to 0 V or rises from 0 V to $+V_{DD}$, the series connection of *p*MOS and *n*MOS transistors has one transistor *off*, with no current then drawn from the power supply. Thus the CMOS circuit operates with the output either high or low while drawing no power from the voltage supply, except during the brief time while switching between high and low output levels, when both transistors are *on* as one is turning *on* and the other is turning *off*. In fact, the power consumption of a CMOS circuit is 0 at dc conditions, increasing as the applied signal frequency increases since the circuit is switching more often.

The CMOS device is used primarily in digital circuits, operating to provide output of either 0 V or +5 V while drawing very little power from the supply. Most low-power integrated circuits are built using CMOS transistors.

PROBLEMS

§ 6.2

1. An *n*-channel JFET has device parameters $V_P = -3$ V, and $I_{DSS} = 8$ mA. Calculate the drain current at gate-source voltages of (a) −1 V, (b) 0 V, and (c) −2 V.

2. A *p*-channel JFET has device parameters of $I_{DSS} = 7.5$ mA and $V_P = +4$ V. Determine the drain current at gate-source voltages of (a) +2 V and (b) +3 V.

3. What gate-source voltage is required to obtain a drain current of 5 mA using an *n*-channel JFET having $V_P = -3$ V and $I_{DSS} = 8$ mA?

4. What gate-source voltage is required to obtain a drain current of 5 mA using a *p*-channel JFET having $V_P = +4$ V and $I_{DSS} = 7.5$ mA?

5. An *n*-channel JFET having value of $I_{DSS} = 8.5$ mA is operated with resulting measured values of $I_D = 2.125$ mA and $V_{GS} = -2.5$ V. What is the value of device V_P?

6. An *n*-channel JFET with $V_P = -6$ V is operated at $I_D = 6.75$ mA and $V_{GS} = -1.5$ V. Determine the value of I_{DSS} for the device.

7. An *n*-channel JFET with rated values of $I_{DSS} = 10$ mA and $V_P = -3.5$ V is operated with I_D measured to be 3.265 mA. What was the value of V_{GS}?

§ 6.3

8. Plot the transfer characteristic of an *n*-channel JFET having $I_{DSS} = 16$ mA and $V_P = -6$ V.

9. Plot the transfer characteristic of an *n*-channel JFET having $I_{DSS} = 10$ mA and $V_P = -5$ V.

10. Plot the transfer characteristic of a *p*-channel JFET having $I_{DSS} = 12$ mA and $V_P = 4.5$ V.

§ 6.4

11. Calculate the device transconductance, g_{m0}, for an *n*-channel JFET having $I_{DSS} = 8$ mA and $V_P = -4.5$ V.

12. What is the value of I_{DSS} for an *n*-channel JFET with $g_{m0} = 4.5$ mS and $V_P = -3$ V?

13. What is the value of V_P of a p-channel JFET having $I_{DSS} = 12$ mA and $g_{m0} = 6500\ \mu$S?

14. Determine the value of g_{m0} for a p-channel JFET having $V_P = -3.8$ V and $I_{DSS} =$ 6.8 mA.

15. What value of transconductance, g_m, results when operating an n-channel JFET ($I_{DSS} = 8$ mA, $V_P = -4$ V) at $V_{GS} = -1.5$ V?

16. What value of transconductance results when operating a p-channel JFET ($V_P = +3.5$ V, $I_{DSS} = 9$ mA) at $V_{GS} = +0.75$ V?

17. An n-channel JFET is biased at $V_{GS} = -1.5$ V and $I_D = 2.9$ mA. If $I_{DSS} = 7.5$ mA, what are the values of V_P and g_m?

18. An n-channel JFET ($I_{DSS} = 7.8$ mA) is operated at $I_D = 3.82$ mA and $V_{GS} = -1.2$ V. What is the value of g_m at this operating point?

19. A p-channel JFET ($I_{DSS} = 13.5$ mA, $V_P = +5$ V) is operated at $I_D = 9.5$ mA. What is the value of g_m at this operating point?

20. An n-channel JFET ($V_P = -4.5$ V, $I_{DSS} = 8$ mA) is operated at $V_{GS} = -1.2$ V. Determine the value of device transconductance, g_{m0}, and transconductance at the operating point g_m.

21. A JFET having $g_{m0} = 4200\ \mu$S is operated at $V_{GS} = -1$ V. What is the value of g_m at this operating point? ($V_P = -4$ V.)

22. A JFET ($I_{DSS} = 6$ mA, $V_P = -2.5$ V) is operated at $I_D = 5$ mA. What is the value of g_m at this operating point?

23. What is the maximum value of transconductance of a JFET ($V_P = -4$ V) if the transconductance is 4500 μS when operated at $V_{GS} = -1$ V.

§ 6.5

24. A depletion MOSFET ($I_{DSS} = 12$ mA, $V_P = -4$ V) is operated at $V_{GS} = -0.5$ V. What is the value of the transconductance at this operating point?

25. What is the value of transconductance of a depletion MOSFET ($I_{DSS} = 8$ mA, $V_P = -2$ V) when operated at $V_{GS} = 0$ V?

26. An enhancement MOSFET having threshold voltage of 3.5 V is operated at $V_{GS} = 5$ V. What current results (use $K = 0.3$ mA/V^2)?

27. What is the value of threshold voltage for an n-channel enhancement MOSFET that operates at $I_D = 4.8$ mA when biased at 7 V?

28. Determine the value of circuit transconductance for an n-channel enhancement MOSFET having $V_T = 2.8$ V when operated at 6 V.

29. An enhancement MOSFET operated at $V_{GS} = 7.5$ V has transconductance of 2.5 mS. What is the value of device threshold voltage?

30. An n-channel enhancement MOSFET ($V_T = 2.5$ V) when operated at $I_D = 6$ mA has what value of transconductance?

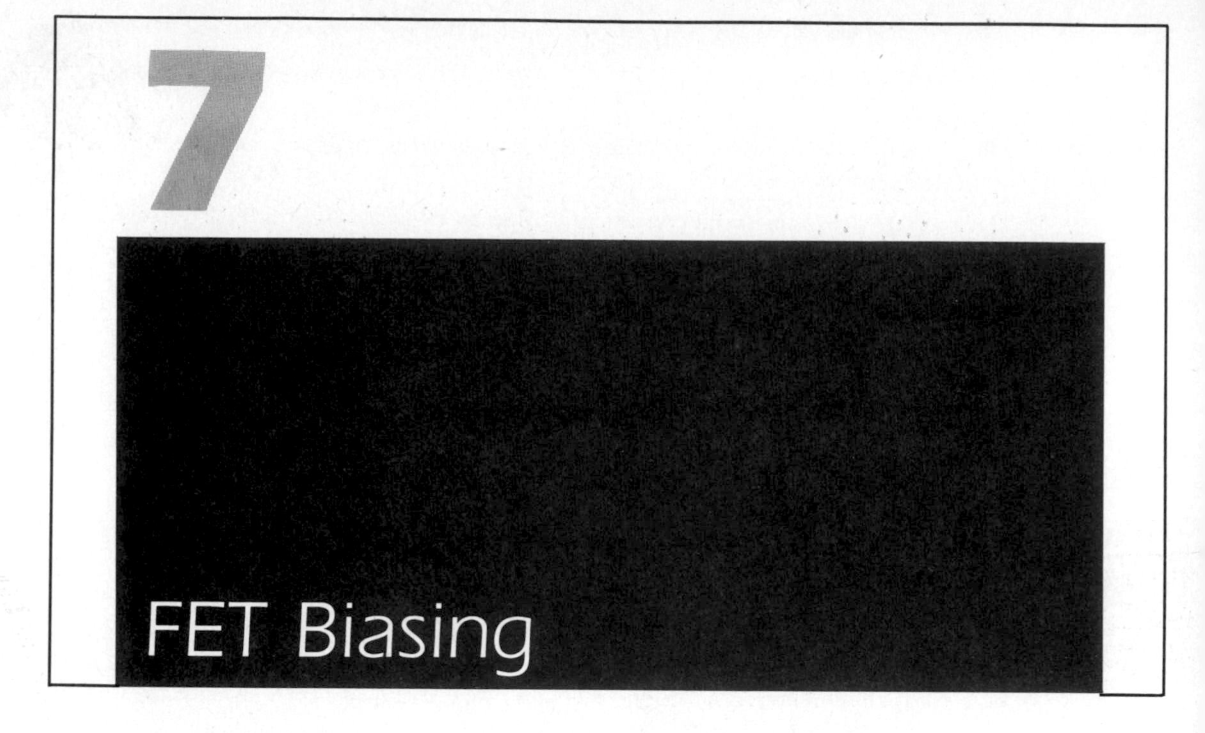

7.1 FIXED BIAS

Dc bias of a FET device requires setting the gate-source voltage, which results in a desired drain current. For a JFET the drain current is limited by the saturation current, I_{DSS}. A depletion MOSFET can be biased below, at, or above I_{DSS}. An enhancement MOSFET requires biasing at a gate-source voltage greater than the threshold value to turn on the device. Since the FET has such a high impedance seen looking into the gate (either reverse-biased *p-n* junction in a JFET or isolation by silicon-dioxide layer in an enhancement MOSFET) the dc voltage of the gate set by a voltage divider or a fixed battery voltage is not affected or loaded by the FET. Fixed dc bias is obtained using a battery to set the gate-source reverse bias voltage as in Fig. 7.1. Battery V_{GG} is used to set the reverse-bias voltage V_{GS} with no resulting current through R_G or the gate terminal.

$$I_G = 0$$

Since the gate-source is reverse biased, there is no current through that junction. No dc current passes through capacitor C so that no current results through resistor R_G. The battery provides a voltage V_{GS} to bias the *n*-channel JFET, but no resulting current is drawn from the battery, V_{GG}.

Resistor R_G is included to allow any ac signal applied through capacitor C to develop across R_G. While any ac signal will develop across R_G, the dc voltage drop across R_G is

$$V_{R_G} = I_G R_G = 0 \text{ V}$$

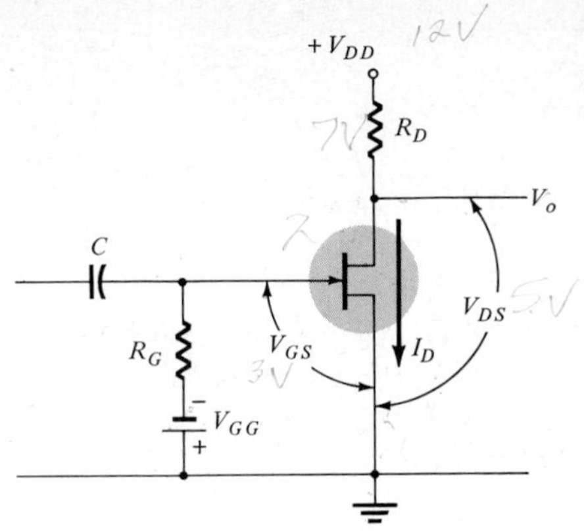

Figure 7.1 JFET circuit using fixed bias.

The gate-source voltage V_{GS} is then

$$V_{GS} = V_G - V_S = V_{GG} - 0 = V_{GG} \tag{7.1}$$

The drain-source current I_D is then set by the gate-source voltage V_{GS} as determined by Shockley's equation,

$$I_D = I_{DSS}\left(1 - \frac{V_{GS}}{V_P}\right)^2 \tag{7.2}$$

This current then results in a voltage drop across the drain resistor of

$$V_{R_D} = I_D R_D$$

with the voltage at the drain

$$V_D = V_{DD} - I_D R_D \tag{7.3}$$

EXAMPLE 7.1

Determine the drain current I_D and drain-source voltage V_{DS} for the fixed bias circuit of Fig. 7.2.

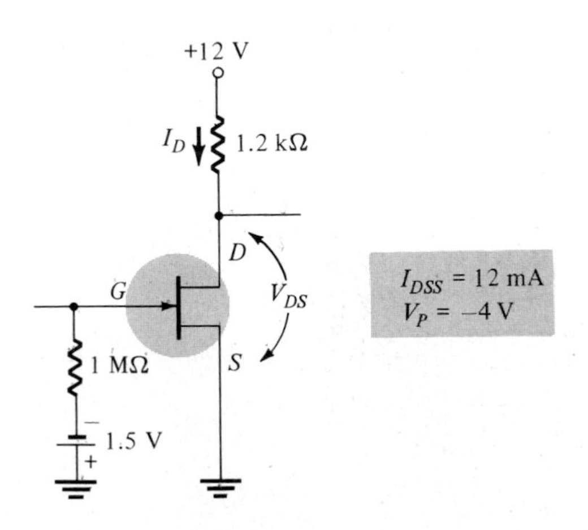

Figure 7.2 JFET fixed-bias circuit for Example 7.1

Solution:

$$V_{GS} = V_{GG} = -1.5 \text{ V}$$

$$I_D = I_{DSS}\left(1 - \frac{V_{GS}}{V_P}\right)^2 = 12 \text{ mA}\left(1 - \frac{-1.5 \text{ V}}{-4 \text{ V}}\right)^2 = \mathbf{4.69\ mA}$$

$$V_D = V_{DD} - I_D R_D = 12 \text{ V} - (4.69 \text{ mA})(1.2 \text{ k}\Omega) = 6.4 \text{ V}$$

$$V_{DS} = V_D - V_S = 6.4 \text{ V} - 0 \text{ V} = \mathbf{6.4\ V}$$

Graphical Analysis Using JFET Drain-Source Characteristic

A graphical analysis of the dc bias operation of a JFET circuit is provided next. Although the mathematical technique used above provides a clear and direct means for determining all the current and voltages in the circuit a graphical analysis adds another view of the bias operation and may further help in understanding why dc bias is needed. Figure 7.3 shows the drain-source characteristic of a JFET having $I_{DSS} = 12$ mA and $V_P = -4$ V.

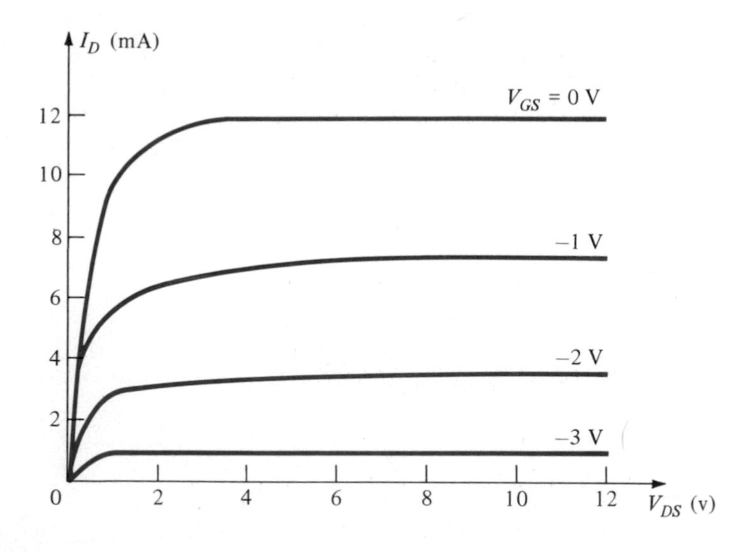

Figure 7.3 Drain-source characteristic of JFET with $I_{DSS} = 12$ mA and $V_P = -4$ V.

One can select the gate-source bias voltage by the value of battery V_{GG}. Using a 1.5-V battery would set the drain current and the choice of R_D would set the drain voltage. Since the gate-source characteristic line for $V_{GS} = -1.5$ V is not shown in Fig. 7.3, a line between $V_{GS} = -1$ V and $V_{GS} = -2$ V is approximated as shown by the dashed line $V_{GS} = -1.5$ V in Fig. 7.4.

A dc load line representing the operation provided by Eq. (7.3) is drawn as follows:

1. For $I_D = 0$ mA:

$$V_D = V_{DD} - (0)R_D = V_{DD}$$

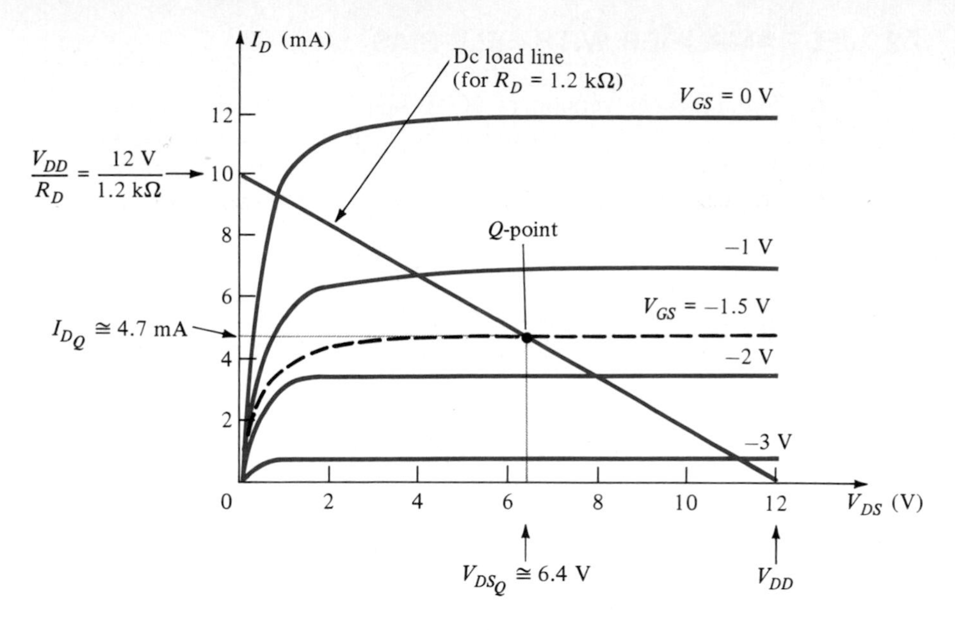

Figure 7.4 Drain-source characteristic and load line for Example 7.2.

(A point along $I_D = 0$, the horizontal axis, is marked at $V_D = V_{DD}$.)

2. For $V_{DS} = V_D = 0$ V:

$$0\text{ V} = V_{DD} - I_D R_D$$

$$I_D = \frac{V_{DD}}{R_D}$$

(A point along $V_{DS} = 0$, the vertical axis, is marked at $I_D = V_{DD}/R_D$.)

3. The dc load line is drawn between the points marked in steps 1 and 2.

EXAMPLE 7.2

Determine the quiescent point for the circuit of Fig. 7.2 using the JFET whose drain-source characteristic is shown in Fig. 7.3.

Solution:

The characteristic curve for $V_{GS} = -1.5$ V is sketched between those for -1 V and -2 V. The dc load line is drawn by connecting a line between points

$$\text{For } I_D = 0: \quad V_D = V_{DD} = 12\text{ V}$$

$$\text{For } V_{DS} = 0: \quad I_D = \frac{V_{DD}}{R_D} = \frac{12\text{ V}}{1.2\text{ k}\Omega} = 10\text{ mA}$$

The quiescent point shown on Fig. 7.4 is at

$$I_{D_Q} = 4.7\text{ mA} \quad \text{and} \quad V_{DS_Q} = 6.4\text{ V}$$

7.2 JFET AMPLIFIER WITH SELF-BIAS

A more practical version of JFET bias uses a source resistor R_S to provide the gate-source bias voltage without the need for a second supply voltage. Figure 7.5 shows the dc bias circuit using only the single voltage supply, V_{DD}. Since no gate current will pass through the reverse biased gate-source, the gate current is

$$I_G = 0$$

so that the gate voltage is

$$V_G = I_G R_G = 0 \text{ V} \tag{7.4}$$

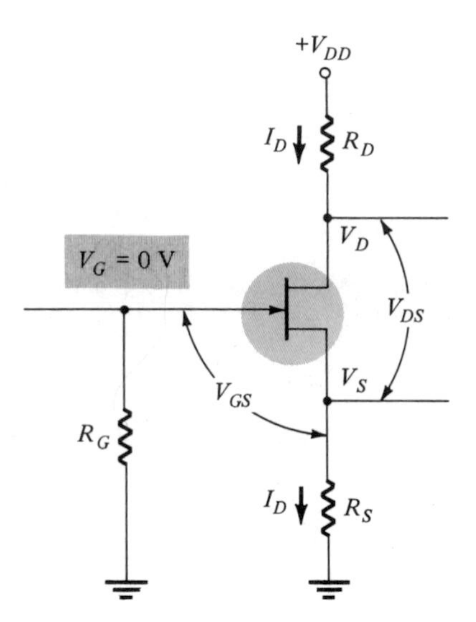

Figure 7.5 JFET self-bias circuit.

With a drain current set to I_D, the voltage at the source is

$$V_S = I_D R_S \tag{7.5}$$

The gate-source voltage is then

$$V_{GS} = V_G - V_S = 0 \text{ V} - I_D R_S$$

$$\boxed{V_{GS} = -I_D R_S} \tag{7.6}$$

The drain current must still satisfy the JFET as specified by Eq. (7.2).

Equations (7.2) and (7.6) are two equations with two unknown values, I_D and V_{GS}. One may mathematically solve by substituting either equation into the other and solving for one of the unknown variables. Although this may easily be carried out in a computer program, a more suitable method of solving the two equations uses a graphical technique, plotting both curves with their intersection being the desired solution.

First, plot the JFET transfer characteristic of Eq. (7.2). Second, plot the straight line of Eq. (7.6)—the self-bias line. This second line is plotted easily by selecting two points of the line

$$\text{For } I_D = 0: \quad V_{GS} = (0)R_S = 0 \text{ V}$$

$$\text{For } V_{GS} = V_P: \quad I_D = \frac{-V_P}{R_S}$$

The intersection of the self-bias line and the transfer characteristic provides the desired Q-point as demonstrated in Example 7.3.

EXAMPLE 7.3

Determine the values of V_{GS} and I_D for the circuit of Fig. 7.6.

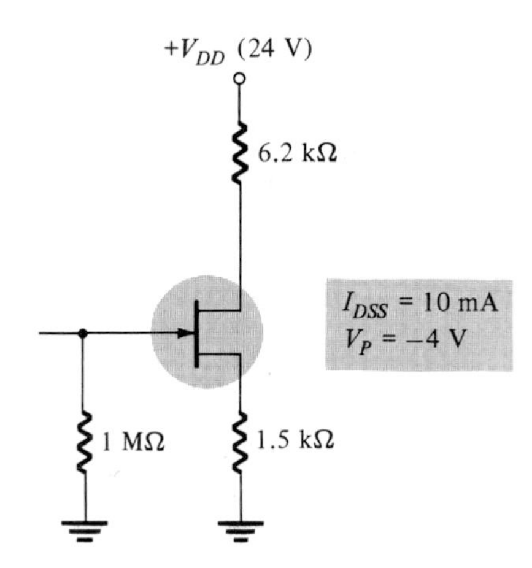

Figure 7.6 JFET bias circuit for Example 7.3.

Solution:

To plot the JFET transfer characteristic [Eq. (7.2)]:

	V_{GS} (V)	I_D (mA)	
	0	10	
$[0.3V_P]$	−1.2	5	$\left[\frac{I_{DSS}}{2}\right]$
$[0.5V_P]$	−2.0	2.5	$\left[\frac{I_{DSS}}{4}\right]$
$[V_P]$	−4.0	0	

To plot the self-bias line [Eq. (7.6)]:

	I_D (mA)	V_{GS} (V)
	0	0
$\left[-\frac{V_P}{R_S}\right]$	2.67	−4

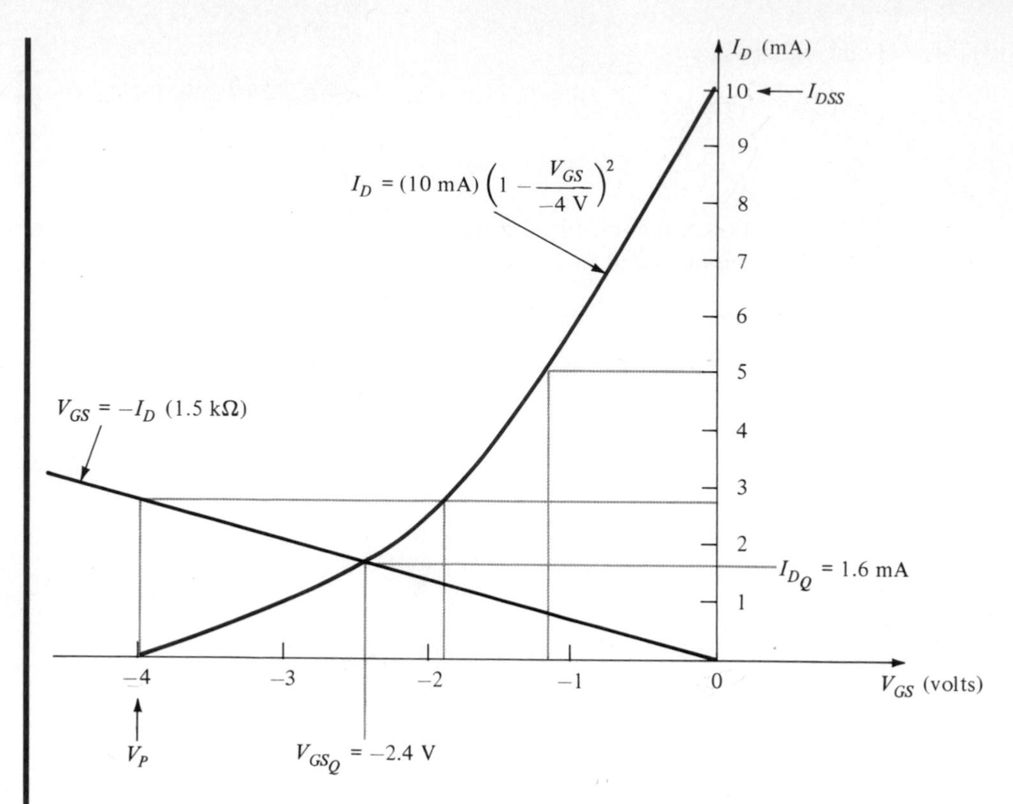

Figure 7.7 Plot of JFET transfer characteristic and self-bias line for Example 7.3.

Figure 7.7 shows the plot of the transfer characteristic and self-bias lines with resulting dc bias at

$$V_{GS_Q} = -2.4 \text{ V} \quad \text{and} \quad I_{D_Q} = 1.6 \text{ mA}$$

EXAMPLE 7.4

For the circuit of Fig. 7.6 calculate the dc bias value of V_{DS}.

Solution:

From Example 7.3, $I_D = 1.6$ mA

$$V_D = V_{DD} - I_D R_D = 24 \text{ V} - (1.6 \text{ mA})(6.2 \text{ k}\Omega) = 14.08 \text{ V}$$

$$V_S = I_D R_S = (1.6 \text{ mA})(1.5 \text{ k}\Omega) = 2.4 \text{ V}$$

$$V_{DS} = V_D - V_S = 14.08 \text{ V} - 2.4 \text{ V} = \mathbf{11.68 \text{ V}}$$

EXAMPLE 7.5

Determine the range of values of R_S to provide dc bias between $I_{DSS}/2$ and $I_{DSS}/4$ for the circuit of Fig. 7.6.

Solution:

Using the transfer characteristic plotted in Example 7.3, plot the self-bias line through the points

$$I_D = \frac{I_{DSS}}{2} = \frac{10 \text{ mA}}{2} = 5 \text{ mA}$$

and

$$I_D = \frac{I_{DSS}}{4} = \frac{10 \text{ mA}}{4} = 2.5 \text{ mA}$$

These points are marked on the transfer characteristic and self-bias lines, then connected from the (0, 0) axis point as shown in Fig. 7.8.

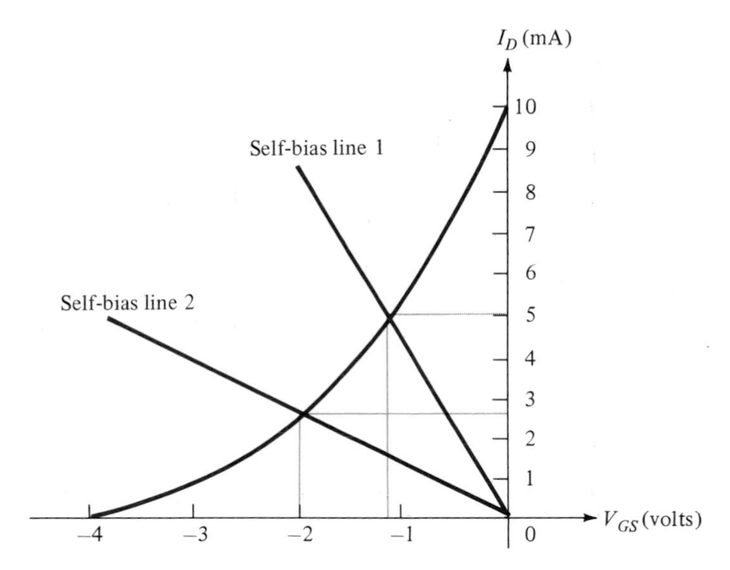

Figure 7.8 Plot for Example 7.5.

The values of R_S are then determined from the slope of each line:

$$R_S\left(\text{for } \frac{I_{DSS}}{2}\right) = \frac{|V_{GS}|}{I_D} = \frac{1.2 \text{ V}}{5 \text{ mA}} = \mathbf{240\ \Omega}$$

and

$$R_S\left(\text{for } \frac{I_{DSS}}{4}\right) = \frac{|V_{GS}|}{I_D} = \frac{2 \text{ V}}{2.5 \text{ mA}} = \mathbf{800\ \Omega}$$

Keeping values of R_S between 240 and 800 Ω will therefore bias the circuit for values of I_D between 2.5 and 5 mA, as desired.

EXAMPLE 7.6

Determine the values of I_D and V_{DS} for the circuit of Fig. 7.9a using the transfer characteristic of Fig. 7.9b. What are the values of I_{DSS} and V_P?

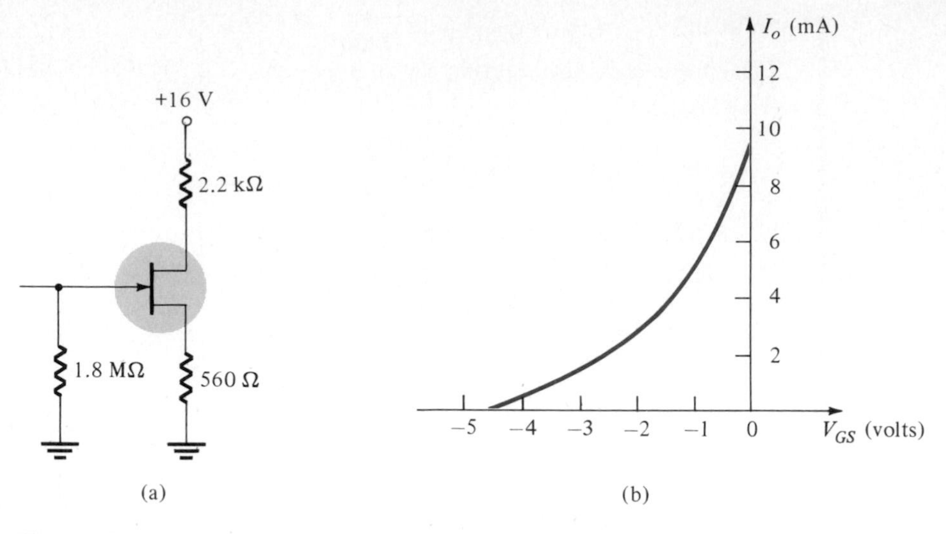

Figure 7.9 JFET circuit and device transfer characteristic for Example 7.6.

Solution:

From the transfer characteristic

$$\text{Along } I_D \text{ axis:} \quad I_{DSS} = \mathbf{9\ mA}$$

$$\text{Along } V_{GS} \text{ axis:} \quad V_P = \mathbf{-4.5\ V}$$

The R_S line is plotted from Eq. (7.6) (see Fig. 7.10):

$$V_{GS} = -I_D R_s$$

When $I_D = 0$: $V_{GS} = -0(0.56\ \text{k}\Omega) = 0\ \text{V}$

When $V_{GS} = V_P = -4.5$ V: $I_D = -\dfrac{V_P}{R_s} = \dfrac{-(-4.5\ \text{V})}{0.56\ \text{k}\Omega} = 8.04\ \text{mA}$

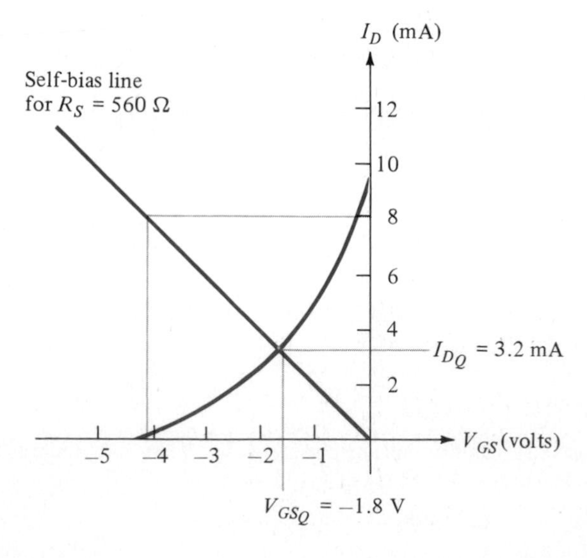

Figure 7.10 Plot of self-bias line for Example 7.6.

	I_D (mA)	V_{GS} (V)
	0	0
$\left[\frac{-V_P}{R_S}\right]$	8.04	−4.5

From the intersection of this line with the device transfer characteristic

$$I_{D_Q} = \mathbf{3.2\ mA} \text{ and } V_{GS_Q} = -1.8 \text{ V}$$

The drain voltage is then:

$$V_D = V_{DD} - I_D R_D = 16 \text{ V} - (3.2 \text{ mA})(2.1 \text{ k}\Omega) = 9.28 \text{ V}$$

and the source voltage

$$V_S = I_D R_S = (3.2 \text{ mA})(560\ \Omega) = 1.79 \text{ V}$$

the resulting bias being

$$V_{DS} = V_D - V_S = 9.28 \text{ V} - 1.79 \text{ V} = 7.49 \text{ V} \cong \mathbf{7.5\ V}$$

7.3 VOLTAGE-DIVIDER BIASING

A slightly modified form of dc bias is provided by the circuit of Fig. 7.11. The additional gate resistor R_G from gate to supply voltage results in greater adjustment of the dc bias point and permits larger values of R_S to be used. As will shortly be shown, the analysis of the dc bias is the same as that covered in Section 7.2 except that the self-bias line is shifted from the (0, 0) point.

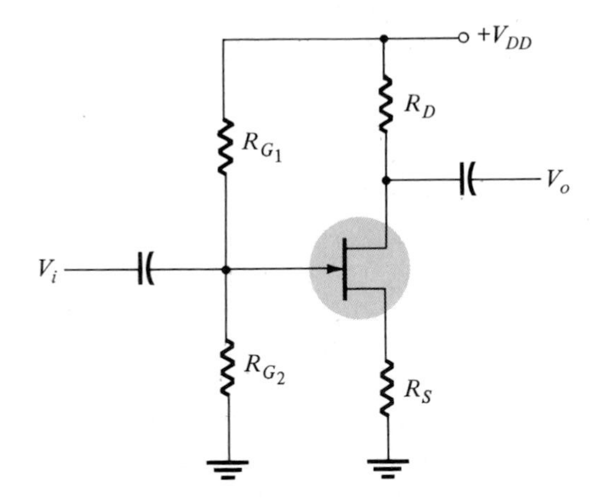

Figure 7.11 Voltage-divider bias circuit.

The gate is still reverse biased so that $I_G = 0$; the gate voltage, V_G, is

$$V_G = \frac{R_{G2}}{R_{G1} + R_{G2}} V_{DD} \tag{7.7}$$

and the JFET bias voltage is

$$V_{GS} = V_G - V_S = V_G - I_D R_S \tag{7.8}$$

EXAMPLE 7.7

Determine the bias current I_D in the circuit of Fig. 7.12.

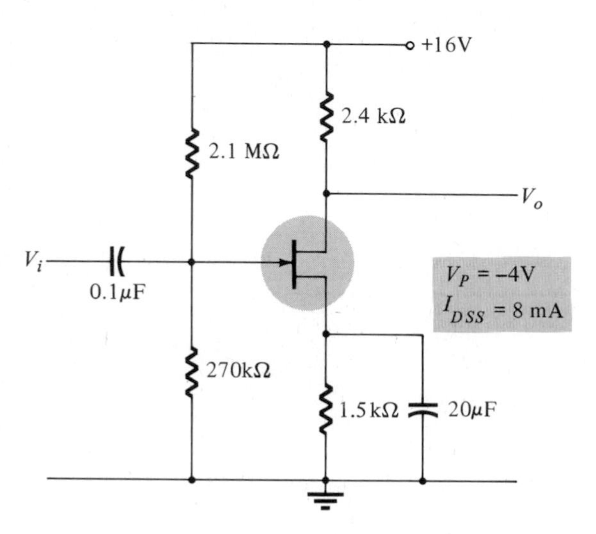

Figure 7.12 JFET bias circuit for Example 7.7.

Solution:

The transfer characteristic is drawn in Fig. 7.13 using

V_{GS} (V)	I_D (mA)
0	8
$[0.3V_P]$ −1.2	4 $\left[\frac{I_{DSS}}{2}\right]$
$[0.5V_P]$ −2	2 $\left[\frac{I_{DSS}}{4}\right]$
$[V_P]$ −4	0

The gate voltage V_G is calculated using Eq. (7.7):

$$V_G = \frac{270\text{ k}\Omega}{2.1\text{ M}\Omega + 270\text{ k}\Omega}(16\text{ V}) = +1.82\text{ V}$$

and the self-bias line is obtained using Eq. (7.8):

$$V_{GS} = 1.82\text{ V} - I_D(1.5\text{ k}\Omega)$$

$$\text{For } I_D = 0: \quad V_{GS} = -1.82\text{ V}$$

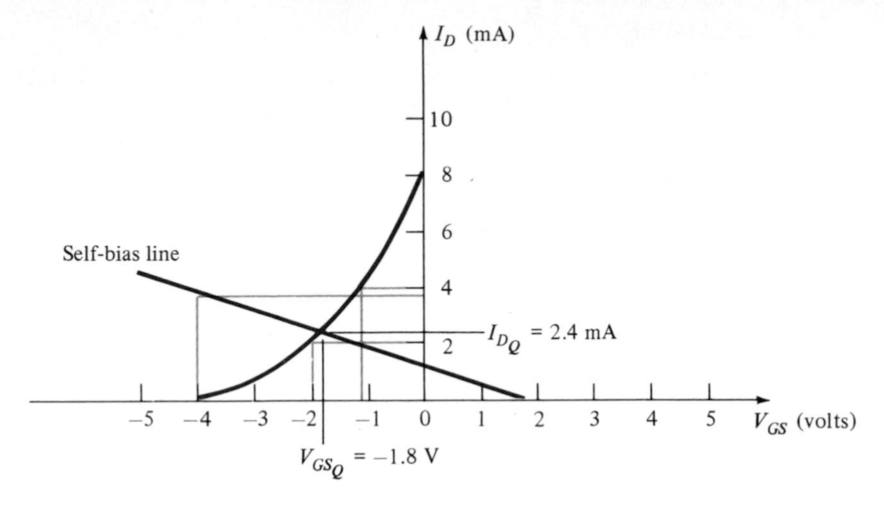

Figure 7.13 Plot of self-bias line for Example 7.7.

$$\text{For } V_{GS} = 0\text{:} \quad I_D = \frac{-V_{GS}}{R_S} = \frac{1.82 \text{ V}}{1.5 \text{ k}\Omega} = 1.21 \text{ mA}$$

I_D (mA)	V_{GS} (V)
0	1.82
1.21	0

This dc self-bias line intersects the device transfer characteristic at

$$V_{GSQ} = -1.8 \text{ V} \quad \text{and} \quad I_{DQ} = \mathbf{2.4\ mA}$$

EXAMPLE 7.8

Determine the bias voltages V_D, V_S, and V_{DS} for the circuit of Example 7.7.

Solution:

From Example 7.6, $I_D = 2.4$ mA:

$$V_D = V_{DD} - I_D R_D = 16 \text{ V} - (2.4 \text{ mA})(2.4 \text{ k}\Omega) = \mathbf{10.24\ V}$$

$$V_S = I_D R_S = (2.4 \text{ mA})(1.5 \text{ k}\Omega) = \mathbf{3.6\ V}$$

$$V_{DS} = V_D - V_S = 10.24 \text{ V} - 3.6 \text{ V} = \mathbf{6.64\ V}$$

A depletion MOSFET device can operate with either positive or negative gate-source voltage since the gate is isolated by the silicon dioxide dielectric and the gate dc current is *always* $I_G = 0$. The device transfer characteristic may go above I_{DSS} and the dc bias point may fall on either side of the $V_{GS} = 0$ V axis.

EXAMPLE 7.9

Draw the transfer characteristic for the n-channel depletion MOSFET of Fig. 7.14.

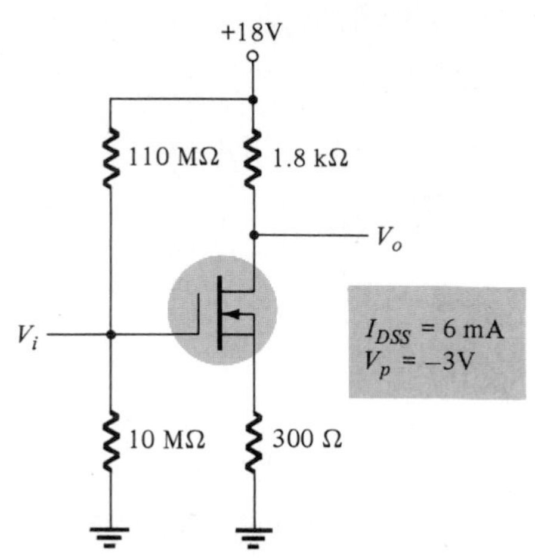

Figure 7.14 Depletion MOSFET bias circuit for Examples 7.9 and 7.10.

Solution:

The transfer characteristic plot may be obtained from Eq. (7.2) using $V_P = -3$ V and $I_{DSS} = 6$ mA.

$$I_D = 6 \text{ mA}\left(1 - \frac{V_{GS}}{-3 \text{ V}}\right)^2$$

In addition to the tabulation points shown previously one can also use $V_{GS} = 0.4\ V_P$ for which

$$I_D = I_{DSS}\left(1 - \frac{0.4\ V_P}{V_P}\right)^2 = 1.96\ I_{DSS} \cong 2I_{DSS}$$

	V_{GS} (V)	I_D (mA)	
$[-0.4V_P]$	+1.2	12	$[2I_{DSS}]$
	0	6	$[I_{DSS}]$
$[0.3V_P]$	−0.9	3	$\left[\frac{I_{DSS}}{2}\right]$
$[0.5V_P]$	−1.5	1.5	$\left[\frac{I_{DSS}}{4}\right]$
$[V_P]$	−3	0	

Figure 7.15 shows the resulting device transfer characteristic.

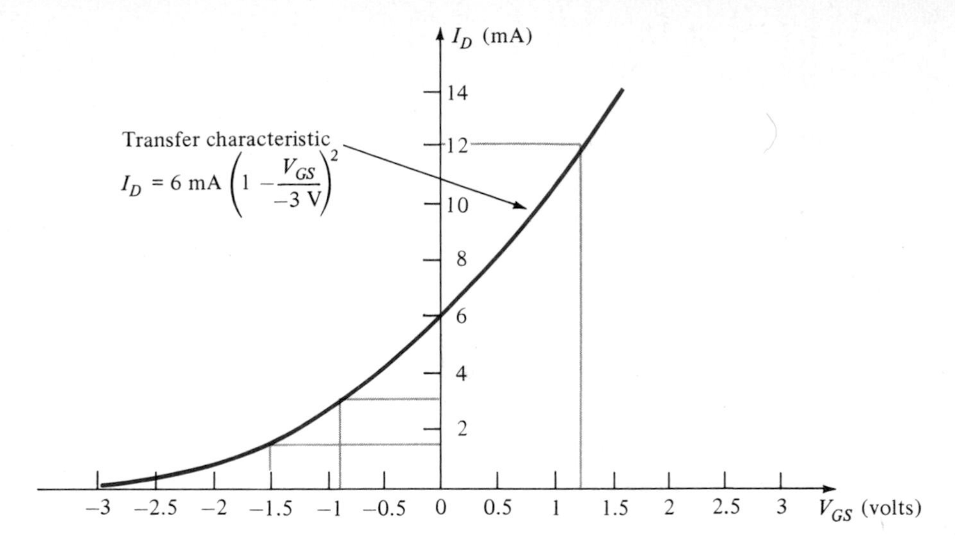

Figure 7.15 Transfer characteristic for depletion MOSFET of Example 7.14.

EXAMPLE 7.10

Using the transfer characteristic of Fig. 7.15 obtain the voltage V_{GS}, current I_D, and voltage V_{DS} for the circuit of Fig. 7.14.

Solution:

The gate voltage is calculated using Eq. (7.7):

$$V_G = \frac{R_{G2}}{R_{G1} + R_{G2}} V_{DD} = \frac{10 \text{ M}\Omega}{110 \text{ M}\Omega + 10 \text{ M}\Omega}(18 \text{ V}) = 1.5 \text{ V}$$

and the self-bias line is obtained from

$$V_{GS} = V_G - I_D R_S = 1.5 \text{ V} - I_D(300)$$

I_D (mA)	V_{GS} (V)
0	1.5
5	0

The self-bias line is shown in Fig. 7.16. The dc bias point obtained is

$$V_{GS} = \mathbf{-0.15 \text{ V}} \quad \text{and} \quad I_D = \mathbf{5.5 \text{ mA}}$$

The drain-source voltage is then

$$V_D = V_{DD} - I_D R_D = 18 \text{ V} - (5.5 \text{ mA})(1.8 \text{ k}\Omega) = 8.1 \text{ V}$$

$$V_S = I_D R_S = (5.5 \text{ mA})(300 \ \Omega) = 1.65 \text{ V}$$

$$V_{DS} = V_D - V_S = 8.1 \text{ V} - 1.65 \text{ V} = \mathbf{6.45 \text{ V}}$$

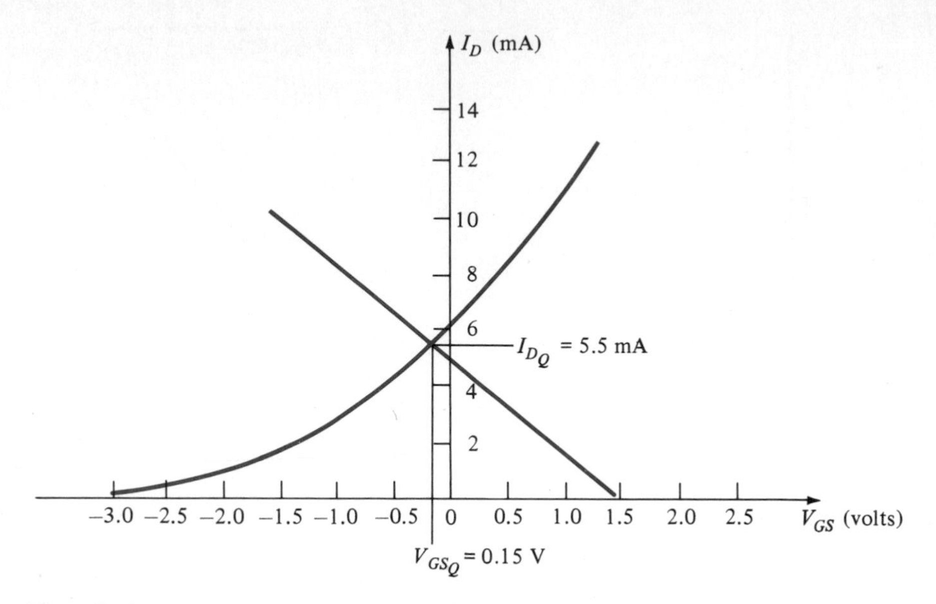

Figure 7.16 Plot showing self-bias line for Example 7.10.

EXAMPLE 7.11

Determine the bias condition I_D and V_{DS} for the circuit of Fig. 7.17.

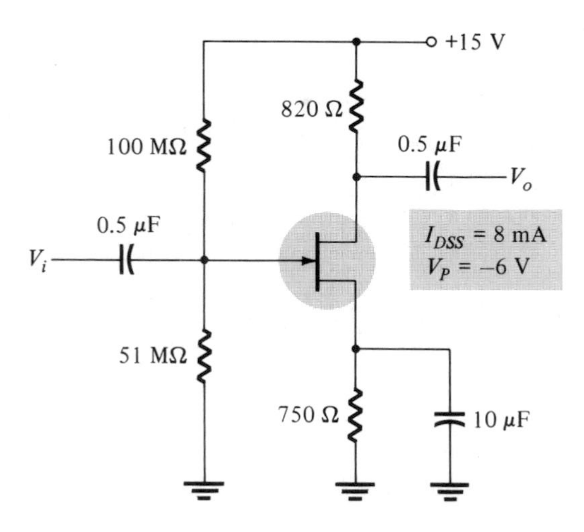

Figure 7.17 JFET circuit for Example 7.11.

Solution:

The *n*-channel JFET transfer characteristic is drawn using

$$I_D = 8 \text{ mA}\left(1 - \frac{V_{GS}}{-6}\right)^2$$

	V_{GS} (V)	I_D (mA)	
	0	8	$[I_{DSS}]$
$[0.3V_P]$	−1.8	4	$\left[\frac{I_{DSS}}{2}\right]$
$[0.5V_P]$	−3	2	$\left[\frac{I_{DSS}}{4}\right]$
$[V_P]$	−6	0	

The self-bias line is drawn using

$$V_G = \frac{R_{G2}}{R_{G1} + R_{G2}} V_{DD} = \frac{51\text{ M}\Omega}{100\text{ M}\Omega + 51\text{ M}\Omega}(15\text{ V}) = 5.07\text{ V}$$

and

$$V_{GS} = V_G - I_D R_S = 5.07\text{ V} - I_D(750\ \Omega)$$

I_D (mA)	V_{GS} (V)
0	5.07
6.76	0

From the intersection of the two curves drawn in Fig. 7.18, the bias condition is

$$I_{DQ} = \mathbf{7.2\text{ mA}} \quad \text{and} \quad V_{GS_Q} = -0.32\text{ V}$$

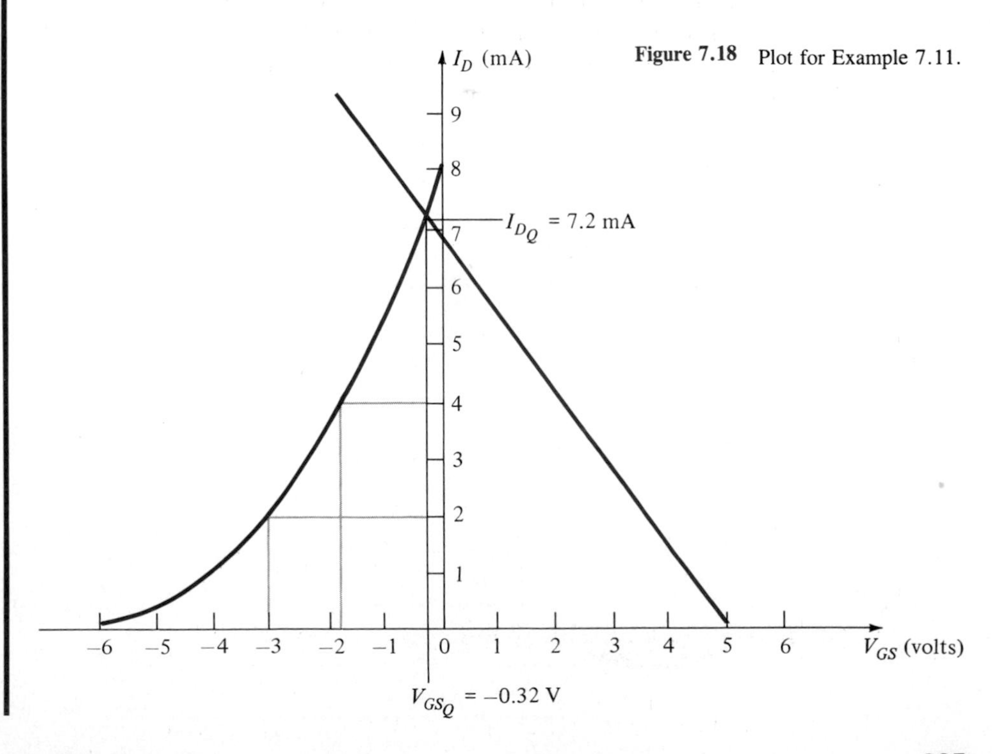

Figure 7.18 Plot for Example 7.11.

The voltage V_{DS} is then

$$V_D = V_{DD} - I_D R_D = 15\text{ V} - (7.2\text{ mA})(820\ \Omega) = 9.1\text{ V}$$

$$V_S = I_D R_S = (7.2\text{ mA})(0.75\text{ k}\Omega) = 5.4\text{ V}$$

$$V_{DS} = V_D - V_S = 9.1\text{ V} - 5.4\text{ V} = \mathbf{3.7\ V}$$

7.4 ENHANCEMENT MOSFET BIAS CIRCUITS

An enhancement MOSFET requires a gate-source voltage greater than the threshold voltage needed to drive the device on. A popular circuit arrangement to bias the enhancement MOSFET is shown in Fig. 7.19. Resistor R_G brings a suitably large voltage to the gate to drive the MOSFET on. The current then increases until an equilibrium condition of drain-source (or gate-source) voltage and drain current is established. The MOSFET drain current is set by the gate-source voltage as given by

$$\boxed{I_D = K(V_{GS} - V_T)^2} \qquad (7.9)$$

where V_T is the specified threshold voltage of the MOSFET. The current I_D also results in a voltage drop across R_D, so that

$$V_D = V_{DD} - I_D R_D \qquad (7.10)$$

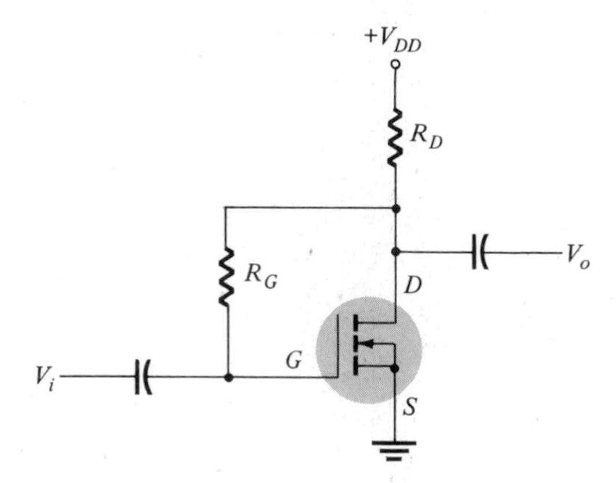

Figure 7.19 Enhancement MOSFET bias circuit.

Since the gate current $I_G = 0$, the voltage $V_{GS} = V_{DS}$, so that Eq. (7.9) may be written as

$$\boxed{I_D = K(V_{DS} - V_T)^2} \qquad (7.11)$$

With $V_S = 0$ V the value of V_{DS} is

$$\boxed{V_{DS} = V_D - V_S = V_{DD} - I_D R_D} \qquad (7.12)$$

The MOSFET drain characteristic can be plotted using Eq. (7.11). For given values of V_T and K, select values of V_{DS} calculating the resulting value of I_D to obtain a few points to plot. Figure 7.20 shows a typical plot of the MOSFET drain characteristic or dc bias calculation. The load line of Eq. (7.12) is also shown plotted on Fig. 7.20. Again, the intersection of the two curves in Fig. 7.20 provides the resulting quiescent operating values of the bias circuit.

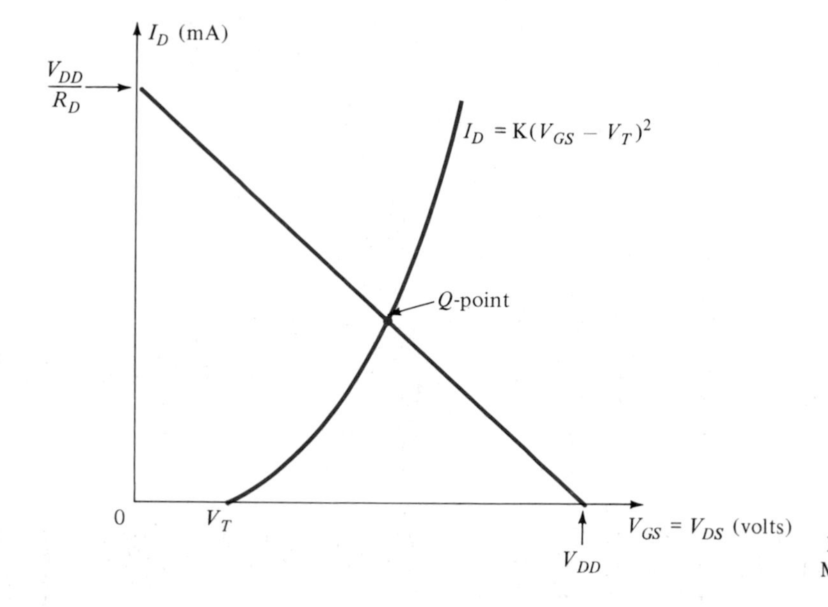

Figure 7.20 Plot of enhancement MOSFET drain characteristic.

EXAMPLE 7.12

Calculate the bias values of I_D and V_{DS} for the circuit of Fig. 7.21.

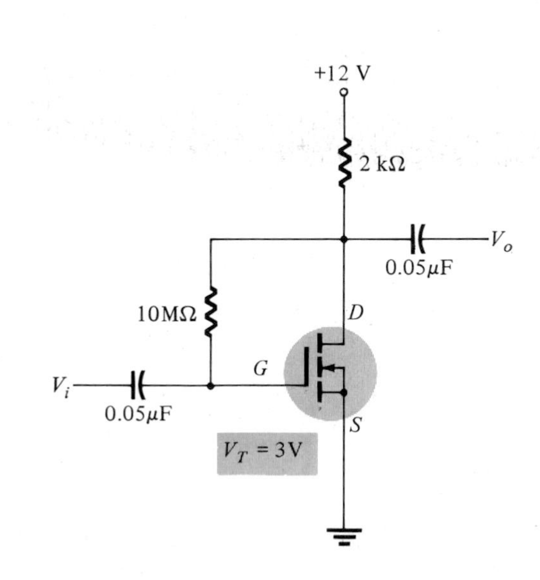

Figure 7.21 Enhancement MOSFET bias circuit.

Solution:

Using $V_T = 3$ V and $K = 0.3$ mA/V^2, a plot of the n-channel MOSFET drain characteristic can be made using Eq. (7.11) to calculate the data listed below.

$$I_D = 0.3(V_{DS} - 3)^2$$

	V_{DS} (V)	I_D (mA)
[V_T]	3	0
	5	1.2
	7	4.8
	9	10.8

The load line of Eq. (7.12) is plotted in Fig. 7.22 using the data below.

$$V_{DS} = 12 \text{ V} - I_D(2 \text{ k}\Omega)$$

I_D (mA)	V_{GS} (V)
0	12
6	0

From the intersection of the two curves in Fig. 7.22, the resulting bias values are

$$I_D = \mathbf{2.9\ mA} \quad \text{and} \quad V_{DS} = \mathbf{6.1\ V}$$

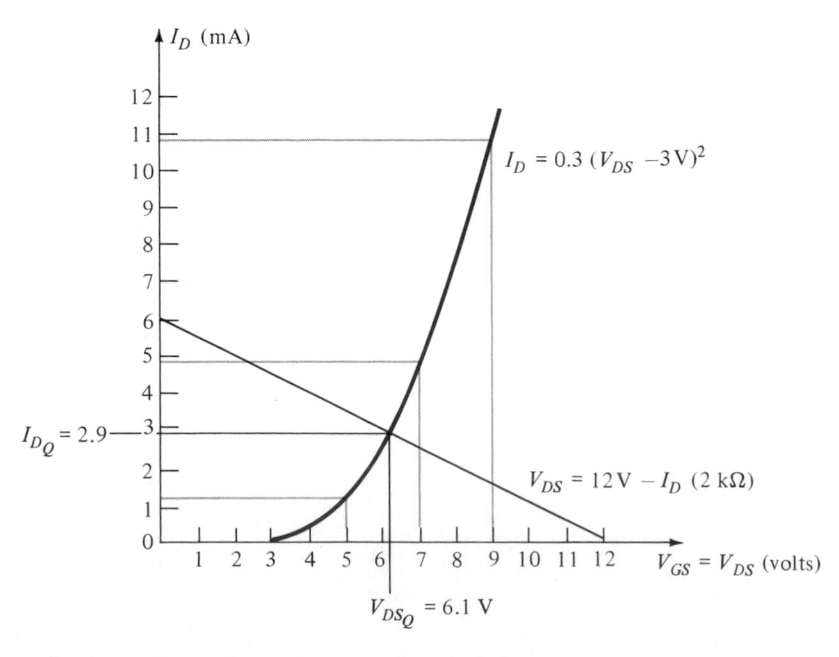

Figure 7.22 Plots of device and circuit curves for Example 7.12.

7.5 MISCELLANEOUS BIAS CIRCUITS

Although many circuits are built in the standard forms covered previously, some variations will occur. A number of miscellaneous circuit forms are analyzed by the examples in this section. No new theory is involved, as will be seen.

EXAMPLE 7.13

Calculate the bias values of I_D and V_{DS} for the circuit of Fig. 7.23.

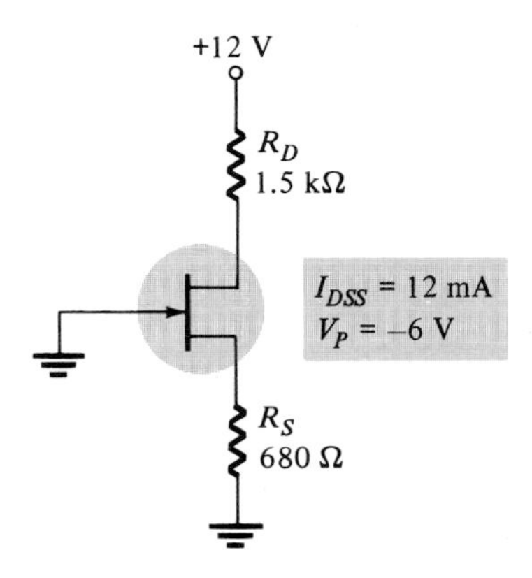

Figure 7.23 Bias circuit for Example 7.13.

Solution:

This is actually a standard circuit and can be analyzed by the graphical procedure covered previously. However, as an alternative approach the bias operating point can be obtained by successive calculations to quickly determine a pair of values I_D and V_{GS} which satisfy both device and circuit equations. The two equations to be satisfied are

$$V_{GS} = -(680\ \Omega)I_D \tag{1}$$

$$I_D = 12\ \text{mA}\left(1 - \frac{V_{GS}}{-6}\right)^2 \tag{2}$$

To determine values of I_D and V_{GS} that satisfy *both* equations select a value for I_D, calculate the values of V_{GS} using Eq. (1), and use this value in Eq. (2) to calculate I_D. If both values of I_D are the same (or reasonably close, say within 10%), the solution has been determined. If not, repeat with a new value of I_D *between* the two values. A tabulation of this procedure is demonstrated next.

I_D (mA)	*Use Eq. (1):* V_{GS} (V)	*Use Eq. (2):* I_D (mA)	*Determine Next Value of* I_D
6	−4.08	1.23	$\frac{6 + 1.23}{2} = 3.615$
3.615	−2.46	4.18	$\frac{3.615 + 4.18}{2} = 3.9$
3.9	−2.65	3.74	$\frac{3.9 + 3.74}{2} = 3.82$
3.82	−2.6	3.85	$\frac{3.82 + 3.85}{2} = 3.84$

While the procedure could be continued, the dc bias is approximately

$$I_D = \mathbf{3.84\ mA} \quad \text{and} \quad V_{GS} = -2.61\text{ V}$$

The drain and source voltages are then

$$V_D = V_{DD} - I_D R_D = 12\text{ V} - (3.84\text{ mA})(1.5\text{ k}\Omega) = 6.24\text{ V}$$

$$V_S = I_D R_S = (3.84\text{ mA})(680\ \Omega) = 2.61\text{ V}$$

so that

$$V_{DS} = V_D - V_S = 6.24\text{ V} - 2.61\text{ V} = \mathbf{3.63\ V}$$

EXAMPLE 7.14

Determine the drain current, drain voltage, and source voltage for the circuit of Fig. 7.24.

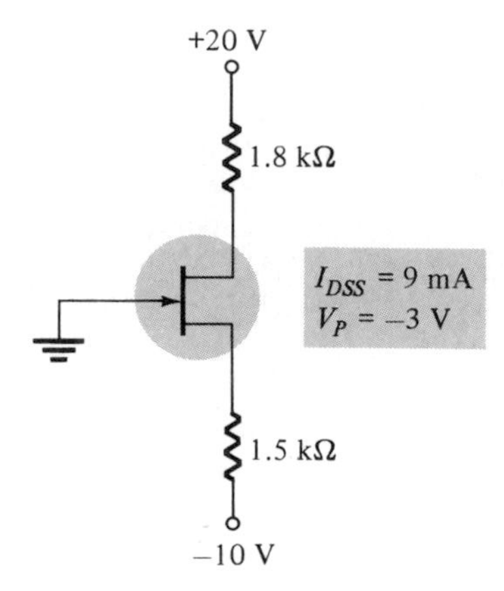

Figure 7.24 Bias circuit for Example 7.14.

Solution:

The circuit equation for the gate-source loop is

$$-V_{GS} - I_D R_S + 10\text{ V} = 0$$

$$V_{GS} = 10\text{ V} - I_D(1.5\text{ k}\Omega) \tag{1}$$

while the device equation is

$$I_D = I_{DSS}\left(1 - \frac{V_{GS}}{V_P}\right)^2 = 9\text{ mA}\left(1 - \frac{V_{GS}}{-3}\right)^2 \tag{2}$$

These curves are plotted in Fig. 7.25 using the following data:
From Eq. (1):

$$V_{GS} = 10\text{ V} - (1.5\text{ k}\Omega)I_D$$

I_D (mA)	V_{GS} (V)
0	10
6.67	0

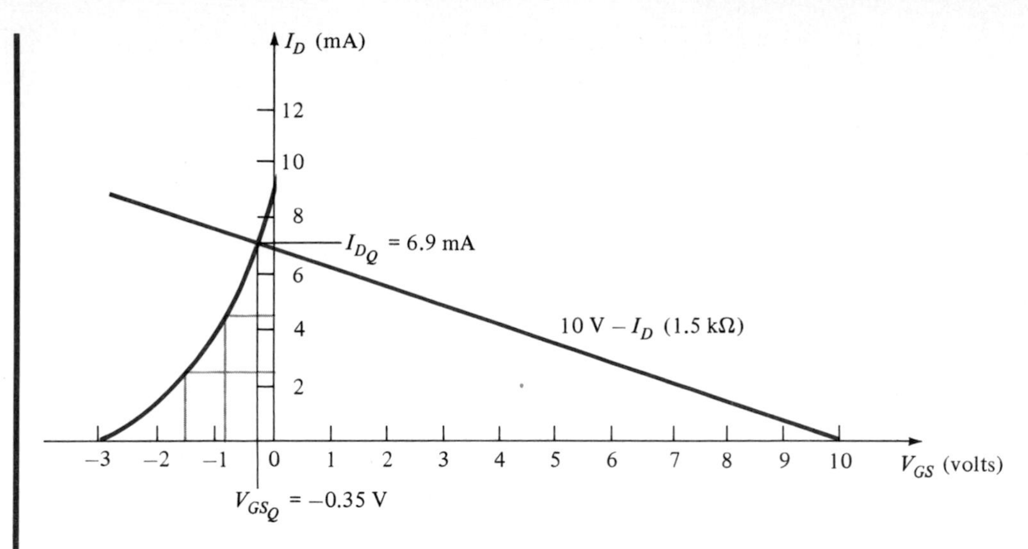

Figure 7.25 Plot of curves for Example 7.14.

From Eq. (2):

$$I_D = 9 \text{ mA}\left(1 - \frac{V_{GS}}{-3 \text{ V}}\right)^2$$

	V_{GS} (V)	I_D (mA)	
	0	9	$[I_{DSS}]$
$[0.3V_P]$	−0.9	4.5	$\left[\frac{I_{DSS}}{2}\right]$
$[0.5V_P]$	−1.5	2.25	$\left[\frac{I_{DSS}}{4}\right]$
$[V_P]$	−3	0	

From the intersection of the curves plotted in Fig. 7.25.

$$I_D = \mathbf{6.9 \text{ mA}} \quad \text{and} \quad V_{GS} = \mathbf{-0.35 \text{ V}}$$

The remaining circuit voltages are

$$V_D = 20 \text{ V} - (6.9 \text{ mA})(1.8 \text{ k}\Omega) = \mathbf{7.58 \text{ V}}$$

$$V_S = -10 \text{ V} + (6.9 \text{ mA})(1.5 \text{ k}\Omega) = +0.35 \text{ V}$$

EXAMPLE 7.15

Determine I_D and V_{DS} for the circuit of Fig. 7.26, using a *p*-channel JFET.

Solution:

For this *p*-channel JFET the equations to use are

$$V_{GS} = +I_D R_S = I_D(0.36 \text{ k}\Omega) \tag{1}$$

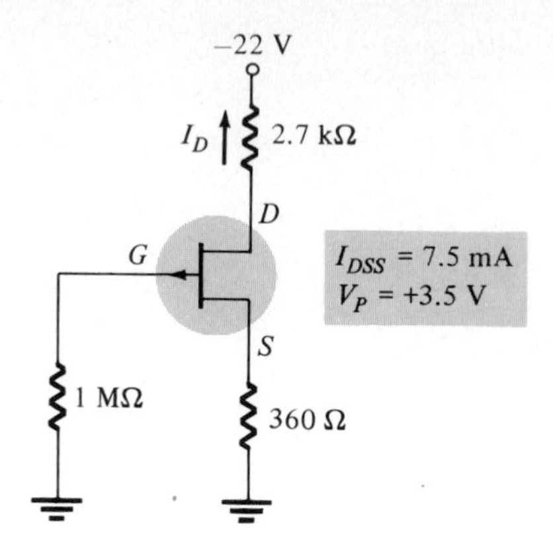

Figure 7.26 Circuit for Example 7.15.

and

$$I_D = I_{DSS}\left(1 - \frac{V_{GS}}{V_P}\right)^2 = 7.5 \text{ mA}\left(1 - \frac{V_{GS}}{3.5 \text{ V}}\right)^2 \qquad (2)$$

These curves are plotted in Fig. 7.27 using the data below.

I_D (mA)	*Use Eq. (1):* V_{GS} (V)			*Use Eq. (2):* V_{GS} (V)	I_D (mA)	
0	0			0	7.5	$[I_{DSS}]$
9.7	3.5	$[V_P]$	$[0.3V_P]$	+1.05	3.75	$\left[\frac{I_{DSS}}{2}\right]$
			$[0.5V_P]$	+1.75	1.875	$\left[\frac{I_{DSS}}{4}\right]$
				+3.5	0	

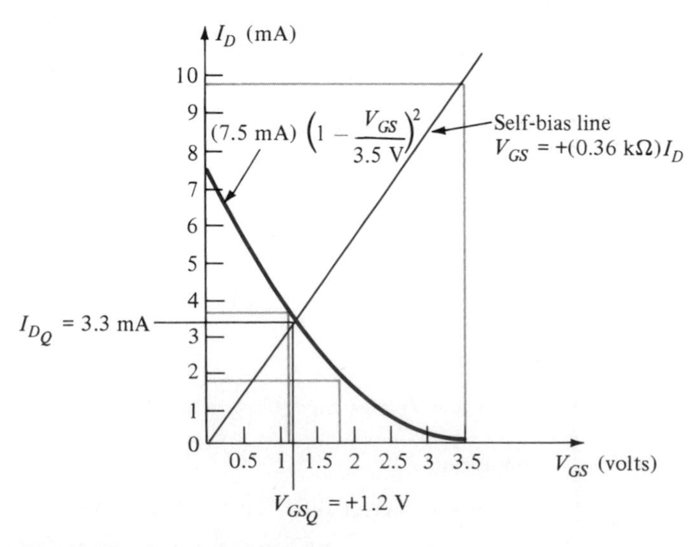

Figure 7.27 Plot for Example 7.15

From the intersection of these plots on Fig. 7.27,

$$I_D = \mathbf{3.3\ mA} \quad \text{and} \quad V_{GS} = +1.2\ \text{V}$$

The voltage V_{DS} is then

$$V_{DS} = V_{DD} + I_D R_D = -22\ \text{V} + (3.3\ \text{mA})(2.7\ \text{k}\Omega) = -13.09\ \text{V}$$

$$V_S = -I_D R_S = -(3.3\ \text{mA})(360\ \Omega) = -1.19\ \text{V}$$

$$V_{DS} = V_D - V_S = -13.09\ \text{V} - (-1.19\ \text{V}) = \mathbf{-11.9\ V}$$

EXAMPLE 7.16

Determine the values of I_D and V_D for the circuit of Fig. 7.28 (using *n*-channel depletion MOSFET).

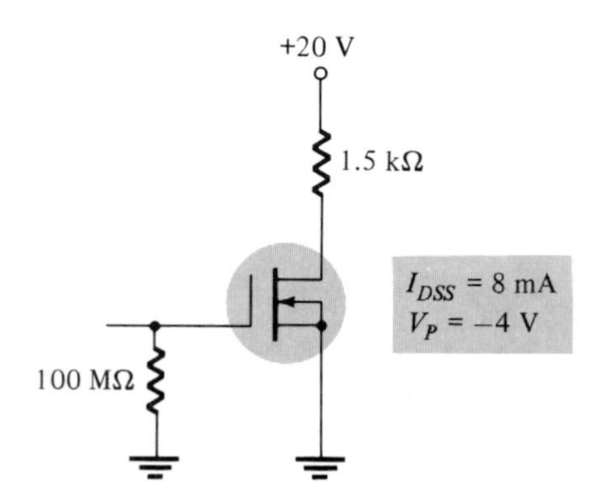

Figure 7.28 Circuit for Example 7.16.

Solution:

Since the gate-source voltage across a depletion MOSFET may be either polarity, the present circuit is biased at

$$V_{GS} = V_G - V_S = 0 - 0 = 0\ \text{V}$$

so that

$$I_D = I_{DSS} = \mathbf{8\ mA}$$

The drain voltage, V_D, is then

$$V_D = V_{DD} - I_D R_D = 20\ \text{V} - (8\ \text{mA})(1.5\ \text{k}\Omega) = \mathbf{8\ V}$$

EXAMPLE 7.17

Calculate I_D and V_S for the circuit of Fig. 7.29 (using an *n*-channel JFET).

Solution:

The circuit equation is

$$V_{GS} = 0 - I_D R_S = -(1.8\ \text{k}\Omega) I_D \qquad (1)$$

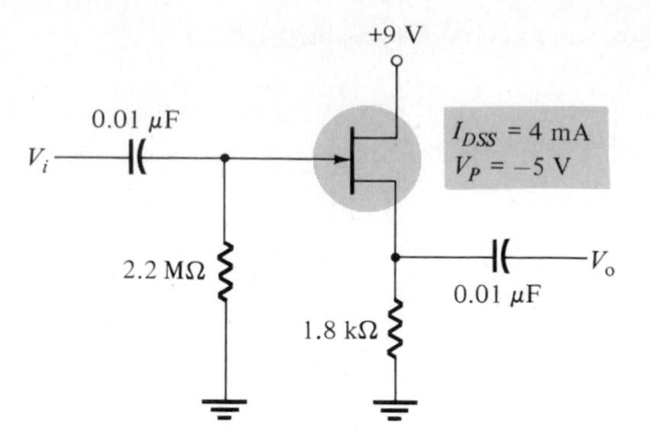

Figure 7.29 Circuit for Example 7.17.

and the device equation is

$$I_D = I_{DSS}\left(1 - \frac{V_{GS}}{V_P}\right)^2 = 4 \text{ mA}\left(1 - \frac{V_{GS}}{-5}\right)^2 \quad (2)$$

Using the tabulation method yields:

	I_D (mA)	*Use Eq. (1):* V_{GS} (V)	*Use Eq. (2):* I_D (mA)	*Next value of* I_D
$\left[\text{Choose } \frac{I_{DSS}}{2}\right]$	2	−3.6	0.31	$\frac{2 + 0.31}{2} = 1.16$
	1.16	−2.09	1.35	$\frac{1.16 + 1.35}{2} = 1.26$
	1.26	−2.27	1.19	$\frac{1.26 + 1.19}{2} = 1.23$

$$I_D \cong \mathbf{1.23\ mA} \quad \text{and} \quad V_{GS} \cong -2.2 \text{ V}$$

The source voltage is then

$$V_S = I_D R_S = (1.23 \text{ mA})(1.8 \text{ k}\Omega) = \mathbf{+2.21\ V}$$

7.6 DESIGN OF DC BIAS CIRCUITS

The design of a dc bias circuit involves a procedure which is a combination of doing the problems previously covered in a reverse manner and occasionally making some reasonable choice. Actually, one may approach design in a number of ways and those presented in this section reflect a selected but not unique way to obtain circuit values to achieve a desired bias condition.

Design of JFET Self-Bias Circuit

Consider the problem of selecting component values for a self-bias circuit such as that in Fig. 7.5. We start with the specification sheet information on the particular JFET device to be used.

1. Select V_{DD} at some value less than the device breakdown voltage BV_{GSS}. Use an available supply voltage (9 V, 12 V, etc.) if that meets the breakdown criteria.
2. Select R_G so that the voltage drop due to the device reverse current, I_{GSS}, is suitably small compared to other voltages in the circuit. The desired bias point is a value of I_D and V_{GS} within the range

$$0 < |V_{GS}| < |V_P|$$

   and

$$I_D < I_{DSS}$$

3. Select R_S to bias the gate-source voltage or the drain current at a specified value (possibly midway within the range of values listed above.)
4. Once I_D has been determined from the self-bias considerations, or if I_D is a desired bias value, the value of R_D may be obtained. Select R_D so that the drain voltage is either a specified value or one within the range

$$V_S < V_D < V_{DD}$$

EXAMPLE 7.18

Design a self-bias circuit such as that of Fig. 7.5 to operate a 2N5950 n-channel silicon JFET using a supply voltage of $V_{DD} = 20$ V.

Solution:

From the device spec sheet we obtain the following values:

$$BV_{GSS} = -30 \text{ V}$$

$$I_{GSS} = -200 \text{ nA} \quad (\text{at } 100°\text{C})$$

$$V_{GS(\text{off})} = -2.5 \text{ V to } -6 \text{ V} \quad (\text{use } V_P = -4 \text{ V})$$

$$I_{DSS} = 10 \text{ mA to } 15 \text{ mA} \quad (\text{use } I_{DSS} = 12 \text{ mA})$$

1. A supply voltage $V_{DD} = 20$ V is satisfactory with respect to the listed breakdown voltage of 30 V.
2. Selecting a voltage drop of 0.1 V as sufficiently negligible in the present circuit, we obtain

$$I_{GSS} R_G < 0.1 \text{ V}$$

$$R_G < 0.1 \text{ V}/200 \text{ nA} = 5 \times 10^5 = 500 \text{ k}\Omega \quad (\text{use } R_G = 470 \text{ k}\Omega)$$

3. Since $V_P = -4$ V and I_D is not specified, a selection of any voltage from, say,

-0.5 V to -3 V would fall well within the range. For the present example the choice is

$$V_{GS} = -1 \text{ V}$$

for which

$$I_D = 12 \text{ mA}\left(1 - \frac{-1}{-4}\right)^2 = 6.75 \text{ mA}$$

Since $V_G = 0$ V,

$$V_S = V_G - V_{GS} = 0 - (-1 \text{ V}) = +1 \text{ V}$$

From

$$V_S = I_D R_S$$

$$R_S = \frac{V_S}{I_D} = \frac{1 \text{ V}}{6.75 \text{ mA}} = 148 \ \Omega \quad (\text{use } R_S = 150 \ \Omega)$$

4. For the value of V_D within the range.

$$1 \text{ V} < V_D < 20 \text{ V}$$

select $V_D = 12$ V at the bias point. Since $V_D = V_{DD} - I_D R_D$,

$$12 \text{ V} = 20 \text{ V} - (6.75 \text{ mA})R_D$$

$$R_D = \frac{20 \text{ V} - 12 \text{ V}}{6.75 \text{ mA}} = 1.185 \text{ k}\Omega \quad (\text{use } R_D = 1.2 \text{ k}\Omega)$$

The resulting circuit is shown in Fig. 7.30.

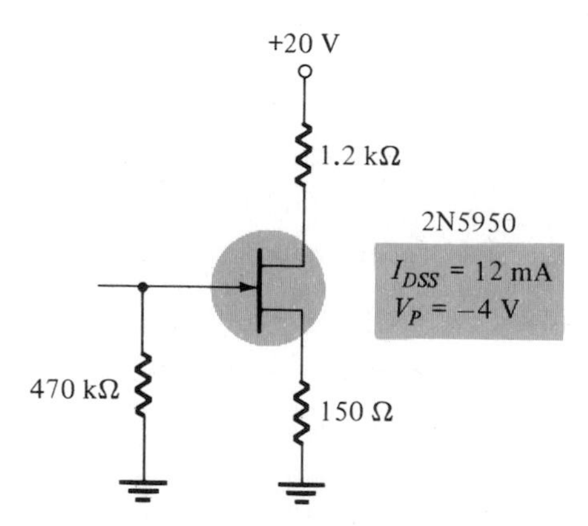

Figure 7.30 Resulting circuit for design in Example 7.18.

An analysis of the dc bias for the resulting circuit would be a good check that the values selected all provide acceptable bias operation.

EXAMPLE 7.19

Design a self-bias circuit such as that of Fig. 7.5 to operate a 2N5952. Use a 22-V supply and bias the device at $I_D = 5$ mA, $V_{DS} = 15$ V.

Solution:

The spec sheet for the 2N5952 shows

$$BV_{GSS} = -30\text{ V}$$
$$I_{GSS} = 200\text{ nA}$$
$$I_{DSS} = 4\text{ mA to }8\text{ mA}\quad(\text{use } I_{DSS} = 6\text{ mA})$$
$$V_{GS\,(\text{off})} = -1.3\text{ V to }-3.5\text{ V}\quad(\text{use } V_P = -2.5\text{ V})$$

Calculating V_{GS} for the desired drain current,

$$5\text{ mA} = 6\text{ mA}\left(1 - \frac{V_{GS}}{-2.5\text{ V}}\right)^2$$
$$V_{GS} = -0.22\text{ V}$$

Since $V_G = 0$ V,

$$V_S = V_G - V_{GS} = 0 - (-0.22\text{ V}) = 0.22\text{ V}$$

From $V_S = I_D R_S$

$$R_S = \frac{V_S}{I_D} = \frac{0.22\text{ V}}{5\text{ mA}} = 44\,\Omega\quad(\text{use } R_S = \mathbf{43\,\Omega})$$

To bias the device at $V_{DS} = 15$ V,

$$V_{DS} = V_D - V_S$$
$$V_D = V_{DS} + V_S = 15\text{ V} + 0.22\text{ V} = 15.22\text{ V}$$

Since $V_D = V_{DD} - I_D R_D$

$$15.22\text{ V} = 22\text{ V} - (5\text{ mA})R_D$$
$$R_D = \frac{22\text{ V} - 15.22\text{ V}}{5\text{ mA}} = 1.36\text{ k}\Omega\quad(\text{use } \mathbf{1.3\text{ k}\Omega})$$

Finally, for the voltage drop across R_G to be less than 0.1 V,

$$I_{GSS} R_G < 0.1\text{ V}$$
$$R_G < \frac{0.1\text{ V}}{200\text{ nA}} = 5 \times 10^5\quad(\text{use } R_G = \mathbf{470\text{ k}\Omega})$$

The resulting circuit is shown in Fig. 7.31.

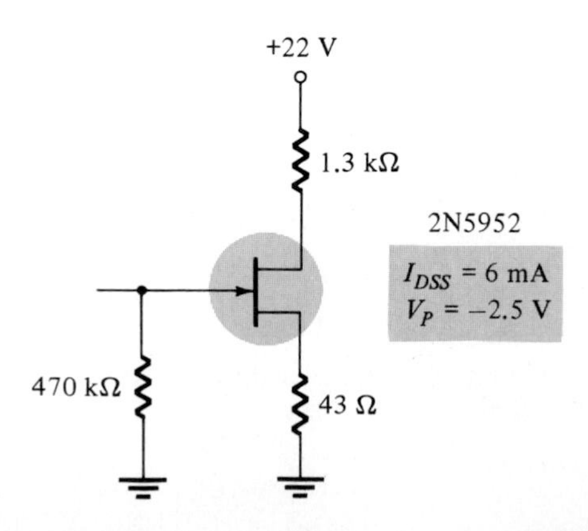

Figure 7.31 Resulting bias circuit for Example 7.19.

EXAMPLE 7.20

Design a self-bias circuit such as that of Fig. 7.5 using an n-channel JFET whose characteristic has been plotted as shown in Fig. 7.32. Use a supply of 18 V and bias the device at $I_D = 0.5I_{DSS}$ and $V_D = 0.5V_{DD}$.

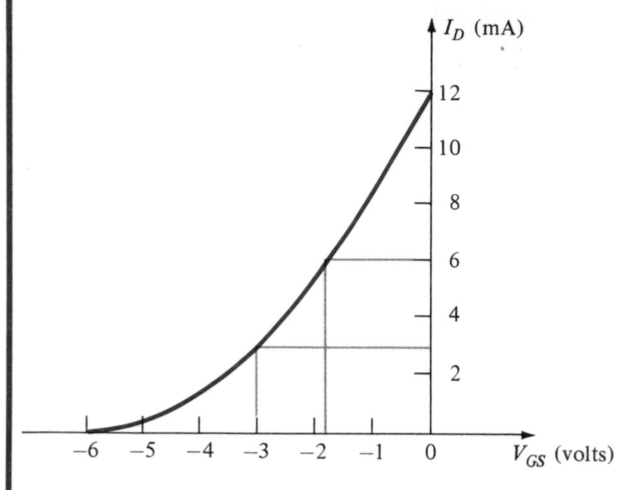

Figure 7.32 Transfer characteristic for device of Example 7.20.

Solution:

From the device characteristic,

$$I_{DSS} = 12 \text{ mA} \quad \text{and} \quad V_P = -6 \text{ V}$$

To bias the device at $I_D = I_{DSS}/2 = 6$ mA, the self-bias line shown in Fig. 7.33 was obtained by connecting a line from the origin through the desired bias point. At this bias condition

$$I_D = 6 \text{ mA}$$

$$V_{GS} = -1.75 \text{ V}$$

The value of R_S is then

$$V_S = I_D R_S$$

$$1.75 \text{ V} = (6 \text{ mA})R_S$$

$$R_S = \frac{1.75 \text{ V}}{6 \text{ mA}} = 291.7\ \Omega \quad (\text{use } R_S = \mathbf{300\ \Omega})$$

The value of R_D for bias at $V_D = V_{DD}/2$ is

$$V_D = V_{DD} - I_D R_D$$

$$9 \text{ V} = 18 \text{ V} - (6 \text{ mA})R_D$$

$$R_D = \frac{18 \text{ V} - 9 \text{ V}}{6 \text{ mA}} = \mathbf{1.5\ k\Omega}$$

Since no value of reverse gate current is available, choose R_G = **1 MΩ** as a suitable value. The resulting circuit is shown in Fig. 7.34.

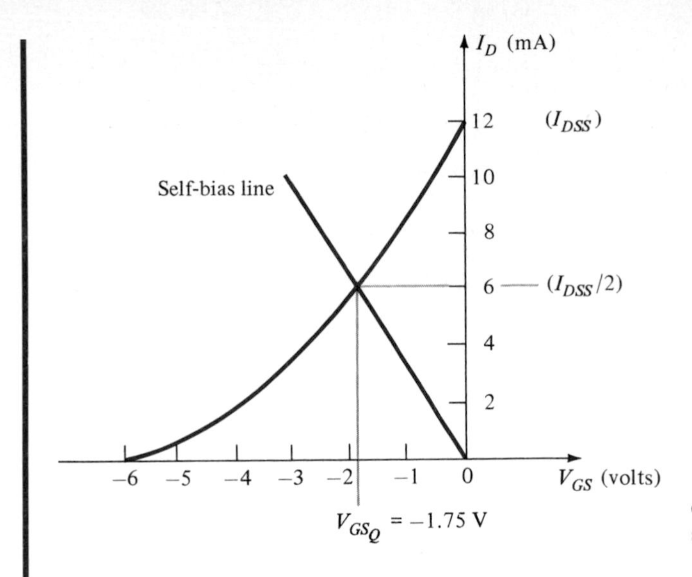

Figure 7.33 Transfer characteristic of Example 7.20 showing desired self-bias line.

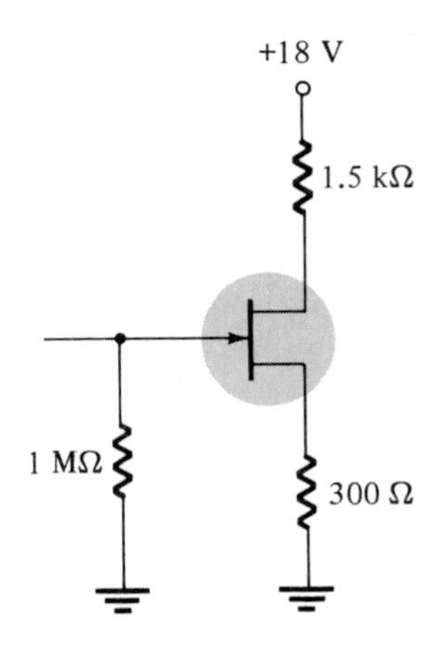

Figure 7.34 Resulting bias circuit for Example 7.20.

EXAMPLE 7.21

Design a voltage-divider bias circuit such as that of Fig. 7.14 using a 2N3797 n-channel depletion MOSFET transistor to operate at $I_D = \frac{I_{DSS}}{2}$. Use a 12-V supply.

Solution:

From the 2N3797 spec sheet

$$I_{DSS} = 6 \text{ mA} \quad \text{and} \quad V_{GS(\text{off})} = -6 \text{ V}$$

At $I_D = \frac{I_{DSS}}{2} = 3$ mA, the gate-source voltage is obtained as follows:

$$I_D = I_{DSS}\left(1 - \frac{V_{GS}}{V_P}\right)^2$$

$$3 \text{ mA} = 6 \text{ mA}\left(1 - \frac{V_{GS}}{-6 \text{ V}}\right)^2$$

$$V_{GS} = -1.76 \text{ V} \cong -1.8 \text{ V}$$

The desired dc bias point can be obtained using a self-bias line for R_S from 600 Ω to one as large as 4.6 kΩ, as demonstrated in Fig. 7.35. Arbitrary selection of $V_G = +2$ V for the self-bias line crossing the V_{GS}-axis requires a resistance of

$$V_G = V_{GS} + I_D R_S$$

$$2 \text{ V} = -1.8 \text{ V} + 3 \text{ mA}(R_S)$$

$$R_S = \frac{(2 \text{ V} + 1.8 \text{ V})}{3 \text{ mA}} = 1.27 \text{ k}\Omega \quad (\text{use } R_S = \mathbf{1.3 \text{ k}\Omega})$$

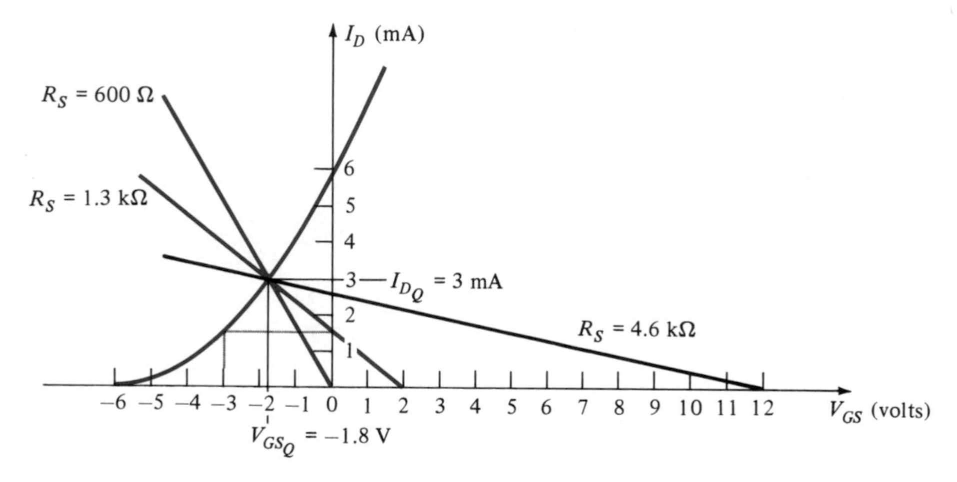

Figure 7.35 Transfer characteristic and self-bias line for Example 7.21.

Then $\qquad V_S = I_D R_S = (3 \text{ mA})(1.3 \text{ k}\Omega) = 3.9 \text{ V}$

Since, $\qquad V_{GS} = V_G - V_S$

$$V_G = V_{GS} + V_S = -1.8 \text{ V} + 3.9 \text{ V} = 2.1 \text{ V}$$

The voltage V_G is determined by

$$V_G = \frac{R_{G2}}{R_{G1} + R_{G2}} V_{DD}$$

for which

$$2.1 \text{ V} = \frac{R_{G2}}{R_{G1} + R_{G2}}(12 \text{ V})$$

$$\frac{R_{G1} + R_{G2}}{R_{G2}} = \frac{12 \text{ V}}{2.1 \text{ V}} = 5.7$$

$$\frac{R_{G1}}{R_{G2}} + 1 = 5.7$$

$$\frac{R_{G1}}{R_{G2}} = 5.7 - 1 = 4.7$$

Arbitrarily selecting R_{G2} = **10 MΩ** gives us

$$R_{G1} = 4.7(10\text{ M}\Omega) = \mathbf{47\ M\Omega}$$

With V_D limited between V_{DD} = 12 V and V_S = 3.9 V, selection of V_D = 8 V seems reasonable. Then

$$V_D = V_{DD} - I_D R_D$$

$$8\text{ V} = 12\text{ V} - (3\text{ mA})R_D$$

$$R_D = \frac{12\text{ V} - 8\text{ V}}{3\text{ mA}} = 1.33\text{ k}\Omega \quad (\text{use } R_D = \mathbf{1.3\ k\Omega})$$

The resulting bias circuit is that shown in Fig. 7.36.

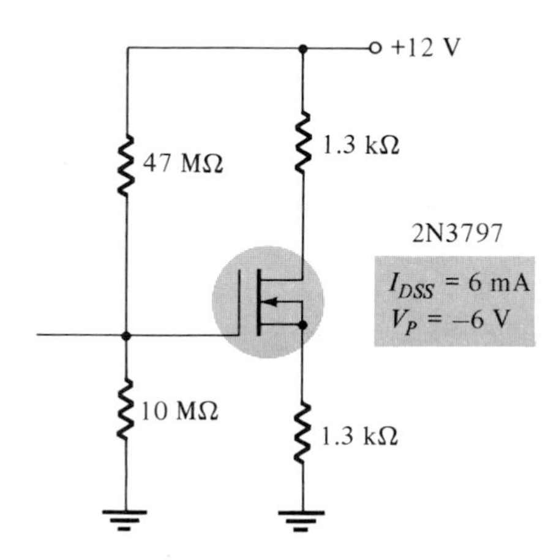

Figure 7.36 Bias circuit designed in Example 7.21.

EXAMPLE 7.22

Complete the design of a *p*-channel JFET voltage-divider bias circuit using a 2N5462 and supply voltage of $V_{DD} = -30$ V, biased as shown in Fig. 7.37.

Solution:

The device spec sheet provides

$$I_{DSS} = 4\text{ mA to }16\text{ mA} \quad (\text{use } I_{DSS} = 10\text{ mA})$$

$$V_{GS(\text{off})} = 1.8\text{ V to }9\text{ V} \quad (\text{use } V_P = 6\text{ V})$$

The desired bias point shown in Fig. 7.37 is

$$V_{GS} = +1\text{ V and } I_D = 6.9\text{ mA}$$

For

$$V_{GS} = V_G - V_S$$

$$+1\text{ V} = -8\text{ V} - V_S$$

$$V_S = -9\text{ V}$$

Since

$$V_S = -I_D R_S$$

$$R_S = \frac{-V_S}{I_D} = -\frac{(-9\text{ V})}{6.9\text{ mA}} = 1.3 \times 10^3 \quad (\text{use } R_S = \mathbf{1.3\ k\Omega})$$

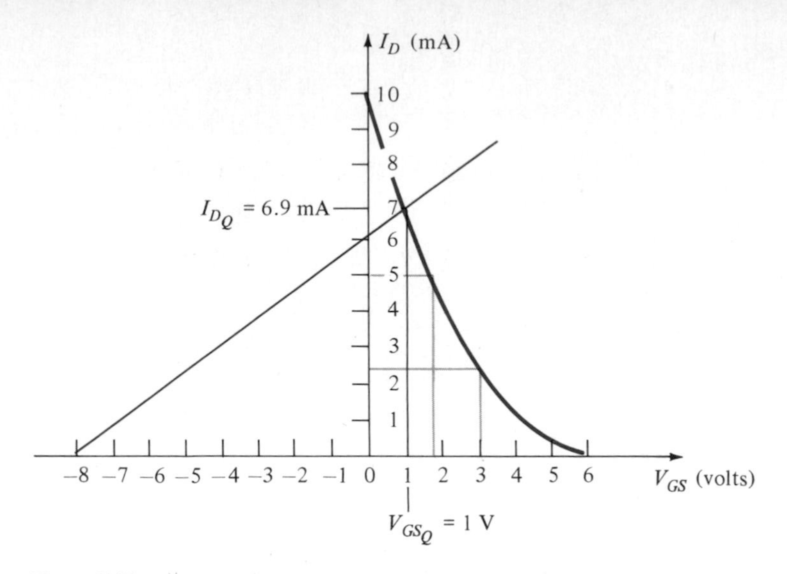

Figure 7.37 Bias line and transfer characteristic for Example 7.22.

For

$$V_G = \frac{R_{G2}}{R_{G1} + R_{G2}} V_{DD}$$

$$-8 \text{ V} = \frac{R_{G2}}{R_{G1} + R_{G2}}(-30 \text{ V})$$

$$1 + \frac{R_{G1}}{R_{G2}} = \frac{-30 \text{ V}}{-8 \text{ V}} = 3.75$$

$$\frac{R_{G1}}{R_{G2}} = 2.75$$

Selecting $R_{G_2} = \mathbf{10\ M\Omega}$

$$R_{G_1} = 2.75(10 \text{ M}\Omega) = 27.5 \text{ M}\Omega \quad (\text{use } R_{G_2} = \mathbf{27\ M\Omega})$$

The drain voltage, V_D, must fall between $V_{DD} = -30$ V and $V_S = -9$ V. Selecting $V_D = -12$ V, we obtain

$$V_D = V_{DD} + I_D R_D$$

$$-12 \text{ V} = -30 \text{ V} + (6.9 \text{ mA})R_D$$

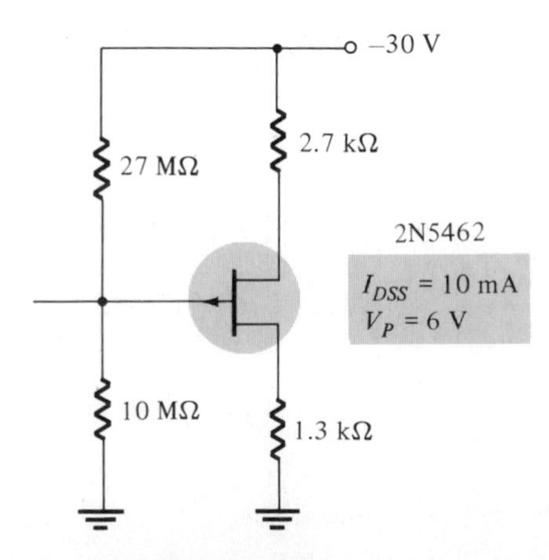

Figure 7.38 Design circuit for Example 7.22.

$$R_D = \frac{30\text{ V} - 12\text{ V}}{6.9\text{ mA}} = 2.61 \times 10^3 \quad (\text{use } R_D = \mathbf{2.7\ k\Omega})$$

The resulting circuit is shown in Fig. 7.38.

7.7 DC BIAS USING UNIVERSAL JFET BIAS CURVE

To reduce some of the effort in dc bias calculation with JFET (or depletion MOSFET) a normalized n-channel curve as shown in Fig. 7.39 may be used. The JFET transfer

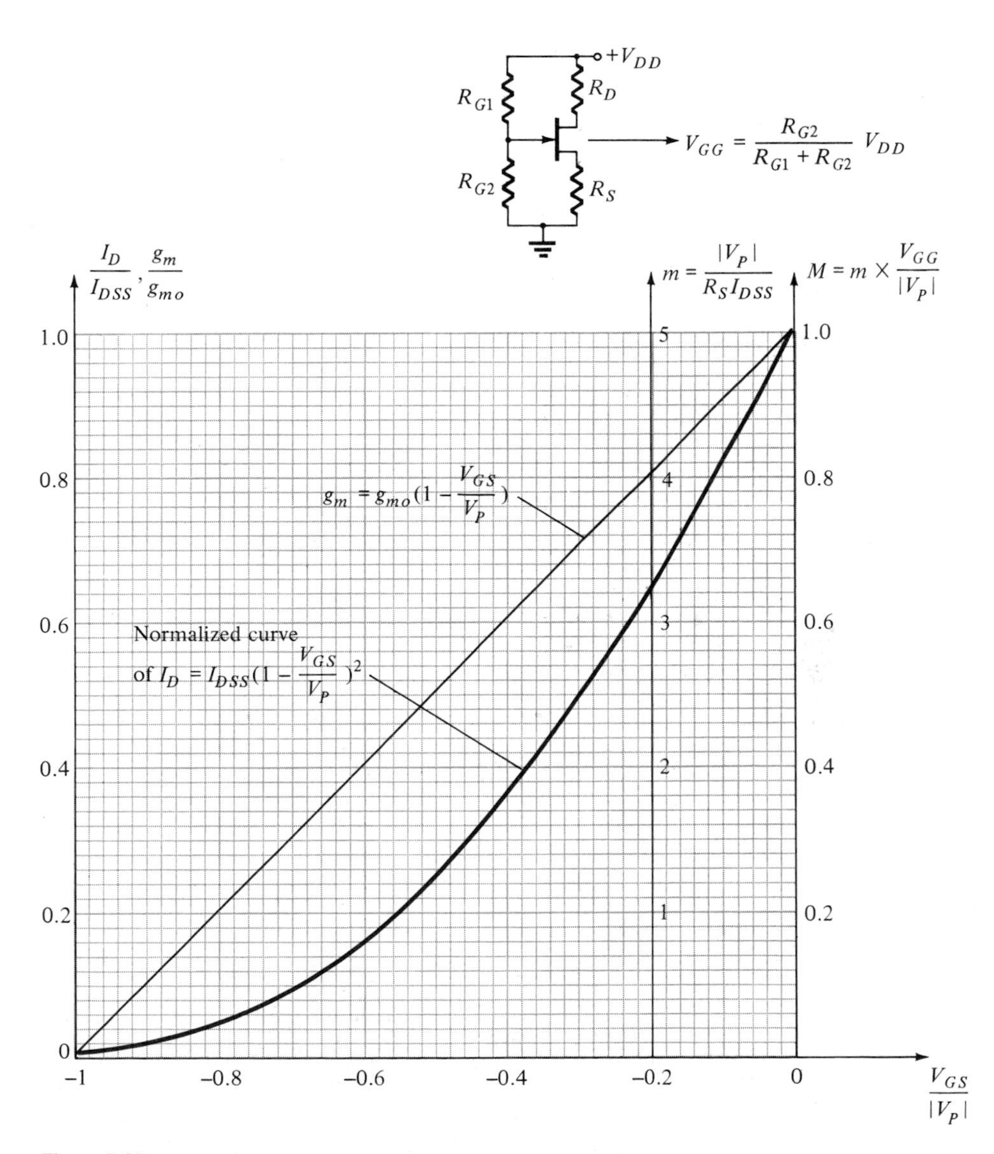

Figure 7.39 Universal JFET characteristic.

characteristic is plotted on normalized axes. To simplify the plotting of the R_S self-bias line an axis of R_S values is plotted as the value m, where

$$m = \frac{|V_P|}{I_{DSS} R_S} \tag{7.13}$$

with V_P in volts, I_{DSS} in milliamperes and R_S in kilohms. For the voltage-divider bias stabilized circuit the values of M and V_{GG} are used, as will be demonstrated shortly.

$$V_{GG} = \frac{R_{G2}}{R_{G1} + R_{G2}} \cdot V_{DD} \tag{7.14}$$

$$M = m \times \frac{V_{GG}}{|V_P|} \tag{7.15}$$

An example for each type of bias circuit will help show how this universal characteristic is used.

EXAMPLE 7.23

Determine the dc bias voltages and currents for the circuit of Fig. 7.40.

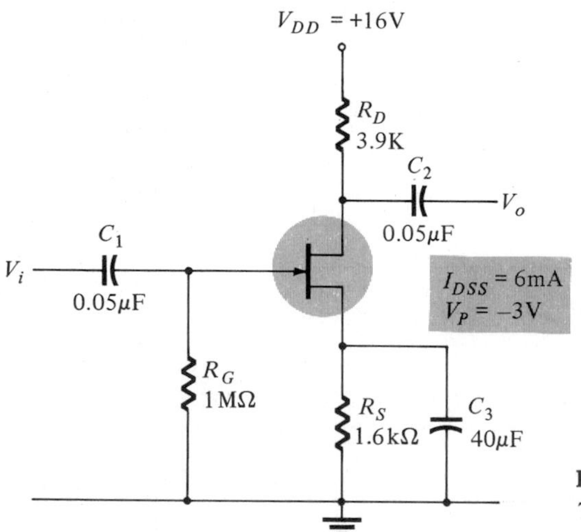

Figure 7.40 Circuit for Example 7.23.

Solution:

Calculating the value of m.

$$m = \frac{|V_P|}{I_{DSS} R_S} = \frac{|-3|}{6(1.6)} = 0.31$$

we plot the R_S bias line from the point 0 on the M-axis through the point $m = 0.31$ along the m-axis. The resulting bias point is seen in Fig. 7.41 to be

$$\frac{I_D}{I_{DSS}} = 0.18 \quad \text{and} \quad \frac{V_{GS}}{|V_P|} = -0.575$$

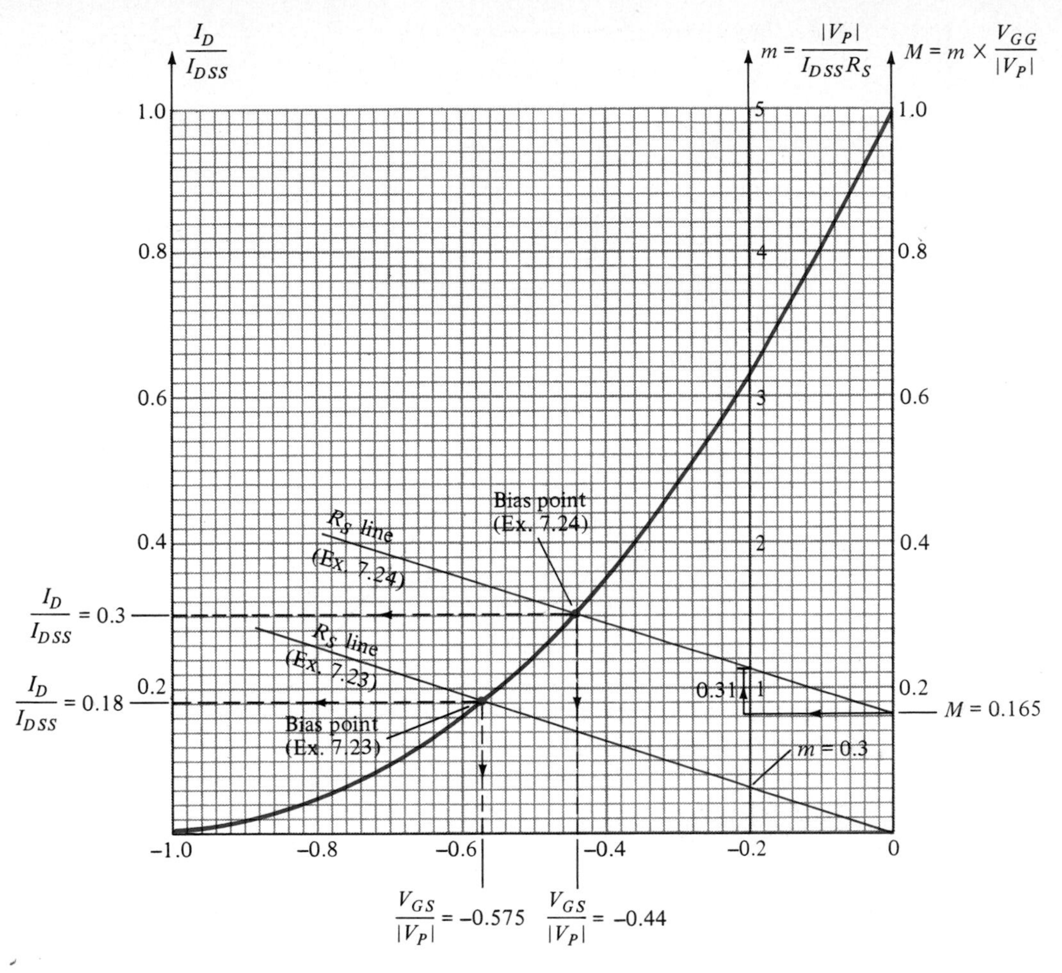

Figure 7.41 Universal curve for Examples 7.23 and 7.24.

from which we calculate

$$I_{D_Q} = 0.18(6 \text{ mA}) = \mathbf{1.08 \text{ mA}}$$

$$V_{GS_Q} = -0.575(|-3|) = \mathbf{-1.73 \text{ V}}$$

Using the value of I_{D_Q} obtained, we then calculate

$$V_{DS_Q} = V_{DD} - I_{D_Q}(R_D + R_S) = 16 - 1.08 \text{ mA}(3.9 \text{ k}\Omega + 1.6 \text{ k}\Omega) = \mathbf{10.06 \text{ V}}$$

EXAMPLE 7.24

Determine the dc bias condition for the circuit in Fig. 7.42.

Solution:

For the voltage-divider bias circuit of Fig. 7.42 we first calculate

$$m = \frac{|V_P|}{I_{DSS}R_S} = \frac{|-3|}{6(1.6)} = 0.31 \qquad \text{(same as in Fig. 7.40)}$$

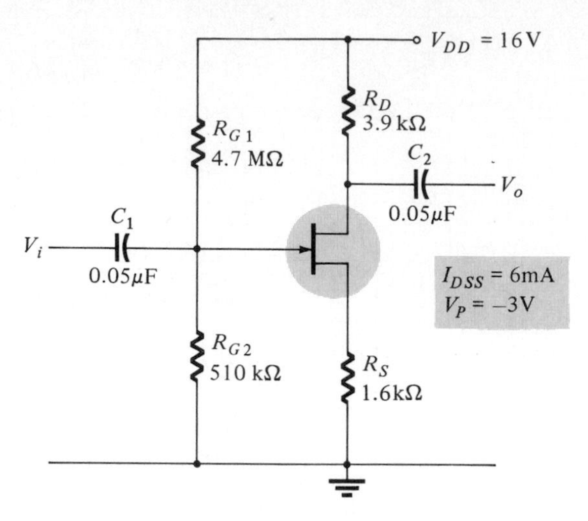

Figure 7.42 Circuit for Example 7.24.

$$V_{GG} = \frac{R_{G_2}}{R_{G_1} + R_{G_2}} V_{DD} = \frac{510 \text{ k}\Omega}{4.7 \text{ M}\Omega + 510 \text{ k}\Omega}(16) \cong 1.6 \text{ V}$$

$$M = m\frac{V_{GG}}{|V_P|} = 0.31\frac{1.6}{|-3|} = 0.165$$

To plot the bias line we must connect a line having the same slope (same R_S and JFET values) as in the previous example passing through the $M = 0.165$ point. This is directly accomplished by connecting a line between the point $M = 0.165$ along the M-axis and a point along the m-axis which is 0.31 higher. From Fig. 7.41 we see that this line gives a bias point

$$\frac{V_{GS}}{|V_P|} = -0.44 \quad \text{and} \quad \frac{I_D}{I_{DSS}} = 0.3$$

from which we calculate

$$V_{GS_Q} = -0.44(|-3 \text{ V}|) = \mathbf{-1.32 \text{ V}}$$

$$I_{D_Q} = 0.3(6 \text{ mA}) = \mathbf{1.8 \text{ mA}}$$

We can then calculate

$$V_{DS_Q} = V_{DD} - I_{D_Q}(R_D + R_S) = 16 - 1.8 \text{ mA}(3.9 \text{ k}\Omega + 1.6 \text{ k}\Omega) = \mathbf{6.1 \text{ V}}$$

7.8 COMPUTER ANALYSIS OF JFET BIAS CIRCUIT

Computer programs can be used to analyze the dc bias of a JFET circuit for a variety of circuit conditions. A single program module is developed next to determine the dc bias current and voltages for a standard JFET circuit such as that of Fig. 7.43a. A program module that uses the equations listed and depicted in Fig. 7.43b can determine the bias voltage and current. The program can provide solution for the case of R_{G1} = *open* or R_D = *short* and be suitable for either *n*-channel or *p*-channel JFET, depending on the voltage polarity used.

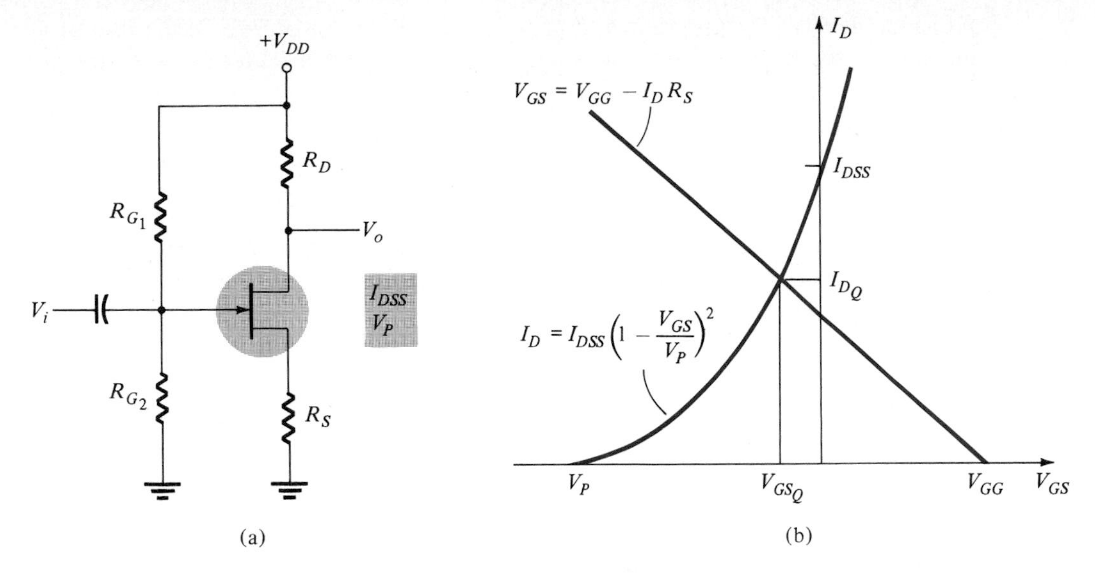

Figure 7.43 Circuit and equations to use in computer analysis of JFET dc bias.

The first part of developing the computer program is to express the equation to use in the solution. Although many of these equations have been specified previously in the chapter, they are used here with the final goal the computer solution for all bias voltages and currents.

$$V_G = V_{GG} = \frac{R_{G2}}{R_{G1} + R_{G2}} V_{DD} \tag{7.16}$$

$$V_S = I_D R_S \tag{7.17}$$

$$V_{GS} = V_G - V_S = V_{GG} - I_D R_S \tag{7.18}$$

$$I_D = I_{DSS}\left(1 - \frac{V_{GS}}{V_P}\right)^2 \tag{7.19}$$

Inserting Eq. (7.19) into Eq. (7.18) results in

$$V_{GS} = V_{GG} - I_{DSS} R_S \left(1 - \frac{V_{GS}}{V_P}\right)^2 \tag{7.20a}$$

which when expanded results in the quadratic equation

$$\underbrace{\left(\frac{I_{DSS} R_S}{V_P^2}\right)}_{\text{A}} V_{GS}^2 + \underbrace{\left(1 - \frac{2 I_{DSS} R_S}{V_P}\right)}_{\text{B}} V_{GS} + \underbrace{(I_{DSS} R_S - V_{GG})}_{\text{C}} = 0 \tag{7.20b}$$

A solution of the gate-source bias voltage is then

$$V_{1,2} = \frac{-B \pm \sqrt{B^2 - 4AC}}{2A} \tag{7.21}$$

with the solution being that value V_1 or V_2 which falls within the range 0 to $-V_P$ for

n-channel JFETs. The program will, of course, test the value of $B^2 - 4AC$, indicating no solution for the case that value is negative. The drain and source voltages are then

$$V_D = V_{DD} - I_D R_D \tag{7.22}$$

$$V_S = I_D R_S \tag{7.23}$$

$$V_{DS} = V_D - V_S \tag{7.24}$$

A summary of the variables and equations used in module 11000 are provided in Listings 7.1 and 7.2. The program Listing 7.3 is used to repeat some of the calculations in previous examples as shown in Listing 7.4.

LISTING 7.1 Equations and Computer Statements for Module 11000

Equation	*Computer Statement*
$V_{GG} = \dfrac{R_{G_2}}{R_{G_1} + R_{G_2}} V_{DD}$	GG = (R2/(R1 + R2)) * DD
$V_S = I_D R_S$	VS = ID * RS
$V_{GS} = V_G - V_S$	GS = VG − VS
$I_D = I_{DSS}\left(1 - \dfrac{V_{GS}}{V_P}\right)^2$	ID = SS * (1 − GS/VP) ↑ 2
$A = \dfrac{I_{DSS} R_S}{V_P^2}$	A = SS * RS/VP ↑ 2
$B = 1 - \dfrac{2I_{DSS} R_S}{V_P}$	B = 1 − 2 * SS * RS/VP
$C = I_{DSS} R_S - V_{GG}$	C = SS * RS − GG
$D = B^2 - 4AC$	D = B ↑ 2 − 4 * A * C
$V_1 = \dfrac{-B + \sqrt{D}}{2A}$	V1 = (−B + SQR(D))/(2 * A)
$V_2 = \dfrac{-B - \sqrt{D}}{2A}$	V2 = (−B − SQR(D))/(2 * A)
$V_D = V_{DD} - I_D R_D$	VD = DD − ID * RD
$V_S = I_D R_S$	VS = ID * RS
$V_{DS} = V_D - V_S$	DS = VD − VS

LISTING 7.2 Equation and Program Variables for Module 11000

Equation Variable	Program Variable
V_G	VG
V_S	VS
V_D	VD
V_{GG}	GG
V_{DD}	DD
V_{GS}	GS
V_{DS}	DS
V_P	VP
I_D	ID
I_{DSS}	SS
R_{G_1}	R1
R_{G_2}	R2
R_S	RS
R_D	RD

```
10 REM ***************************************************
20 REM
30 REM Module for FET dc Bias Calculations
40 REM
50 REM ***************************************************
60 REM
100 PRINT "This program provides dc bias calculations"
110 PRINT "for a JFET or depletion MOSFET circuit as"
120 PRINT "that of Figure 7.43."
130 PRINT
140 PRINT "Enter the following circuit data:"
150 PRINT
160 INPUT "RG1 (use 1E30 if open)=";R1
170 INPUT "RG2                   =";R2
180 INPUT "RS=";RS
190 INPUT "RD=";RD
200 PRINT
210 INPUT "Supply voltage, VDD=";DD
220 PRINT
230 PRINT "Enter the following device data:"
240 INPUT "Drain-source saturation current, IDSS=";SS
250 INPUT "Gate-source pinchoff voltage, VP=";VP
260 PRINT :PRINT
270 REM Now do bias calculations
280 GOSUB 11000
290 PRINT "Bias current is, ID=";ID*1000!;"mA"
300 PRINT "Bias voltages are:"
310 PRINT "VGS=";GS;"volts"
320 PRINT "VD=";VD;"volts"
330 PRINT "VS=";VS;"volts"
340 PRINT "VDS=";DS;"volts"
350 END
```

```
11000 REM Module for FET dc bias calculations
11010 GG=(R2/(R1+R2))*DD
11020 A=SS*RS/VP^2
11030 B=1-2*SS*RS/VP
11040 C=SS*RS-GG
11050 D=B^2-4*A*C
11060 IF D<0 THEN PRINT "No Solution!!!" :STOP
11070 V1=(-B+SQR(D))/(2*A)
11080 V2=(-B-SQR(D))/(2*A)
11090 IF ABS(V1)>ABS(VP) THEN GS=V2
11100 IF ABS(V2)>ABS(VP) THEN GS=V1
11110 ID=SS*(1-GS/VP)^2
11120 VS=ID*RS
11130 VG=GG
11140 VD=DD-ID*RD
11150 DS=VD-VS
11160 RETURN
```

```
Load resistance (1E30 if none), RL=? RUN
This program provides dc bias calculations
for a JFET or depletion MOSFET circuit as
that of Figure 7.43.

Enter the following circuit data:

RG1 (use 1E30 if open)=? 110E6
RG2                   =? 10E6
RS=? 300
RD=? 1.8E3

Supply voltage, VDD=? 18

Enter the following device data:
Drain-source saturation current, IDSS=? 6E-3
Gate-source pinchoff voltage, VP=? -3


Bias current is, ID= 5.460326 mA
Bias voltages are:
VGS=-.1380968 volts
VD= 8.171413 volts
VS= 1.638098 volts
VDS= 6.533316 volts
Ok

This program provides dc bias calculations
for a JFET or depletion MOSFET circuit as
that of Figure 7.43.

Enter the following circuit data:
```

```
RG1 (use 1E30 if open)=? 1E30
RG2                   =? 1.8E6
RS=? 560
RD=? 2.1E3

Supply voltage, VDD=? 16

Enter the following device data:
Drain-source saturation current, IDSS=? 9E-3
Gate-source pinchoff voltage, VP=? -4.5


Bias current is, ID= 3.225263 mA
Bias voltages are:
VGS=-1.806147 volts
VD= 9.226949 volts
VS= 1.806147 volts
VDS= 7.420802 volts
```

PROBLEMS

§ 7.1

1. Determine the drain current and drain-source voltage for the circuit of Fig. 7.44.

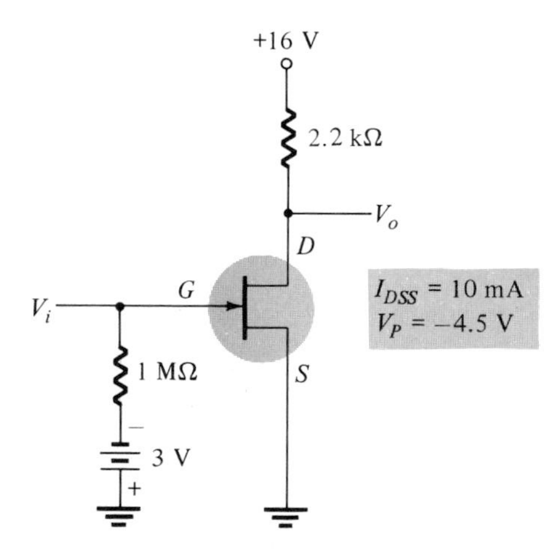

Figure 7.44 Circuit for Problems 1 through 5.

2. Determine the drain current and drain-source voltage for the circuit of Fig. 7.44 if the gate battery is changed from 3 V to 1.5 V.

3. What value of R_D must be used to result in a drain voltage of +8 V in the circuit of Fig. 7.44?

4. What value of V_{DS} results if the JFET of Fig. 7.44 is changed to one with I_{DSS} = 8 mA, V_P = −4 V?

5. What value of gate supply voltage is needed to obtain a bias current of $I_D = 5$ mA in the circuit of Fig. 7.44?

§ 7.2

6. Determine the dc bias voltage V_D for the circuit of Fig. 7.45.

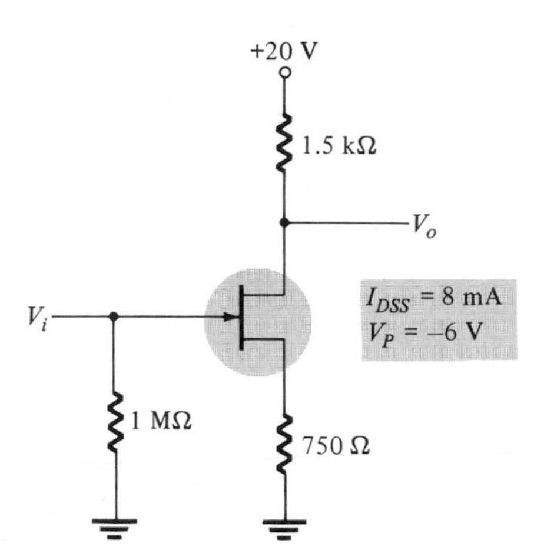

Figure 7.45 Circuit for Problems 6 through 9.

7. Determine the dc bias voltage V_S for the circuit of Fig. 7.45 with R_S changed to 1.2 kΩ.

8. What is the dc bias voltage V_{DS} for the circuit of Fig. 7.45 with R_D changed to 5.1 kΩ?

9. What value of R_S is needed for the circuit of Fig. 7.45 to modify the bias point to $V_{GS} = -2$ V?

10. For the circuit of Fig. 7.46, determine the drain current.

11. Determine the source voltage V_S for the circuit of Fig. 7.46.

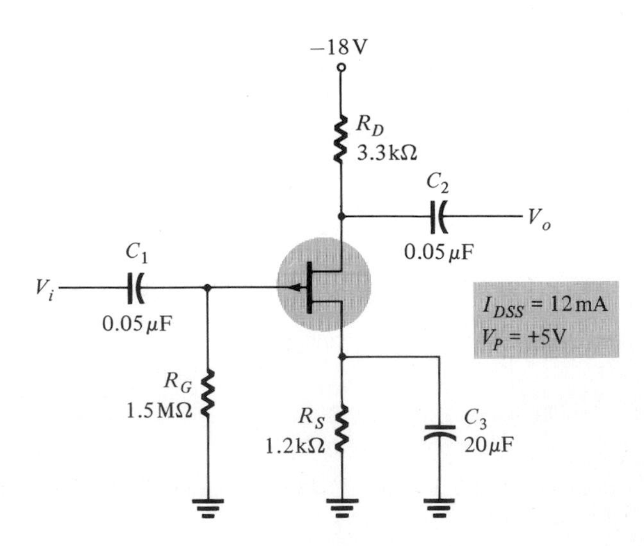

Figure 7.46 Circuit for Problems 10 through 12.

12. What is the value of V_{DS} for the circuit of Fig. 7.46?

13. What value of I_D results in the circuit of Fig. 7.47?

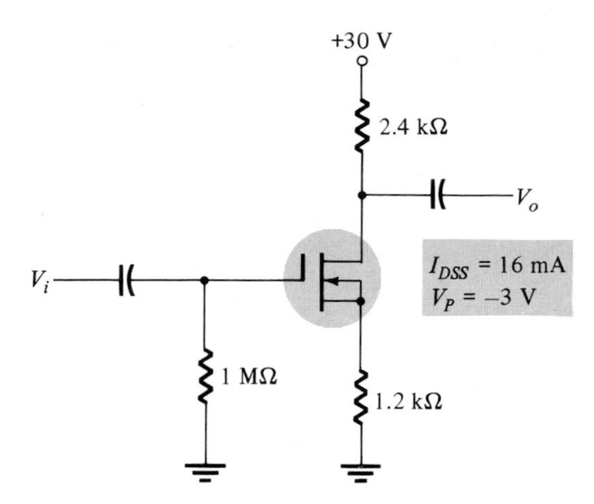

Figure 7.47 Circuit for Problems 13 through 17.

14. What is the value of I_D in the circuit of Fig. 7.47 if the source resistance is changed to 330 Ω?

15. What bias values of I_D and V_{DS} result in the circuit of Fig. 7.47 if the source resistance is changed to 51 Ω?

16. What value of V_D results if the JFET in FIG. 7.47 is changed to one having $I_{DSS} = 10$ mA, $V_P = -4$ V?

17. What value R_S is needed to bias the circuit of Fig. 7.47 at -1 V?

18. Determine I_D and V_{DS} for the circuit of Fig. 7.48.

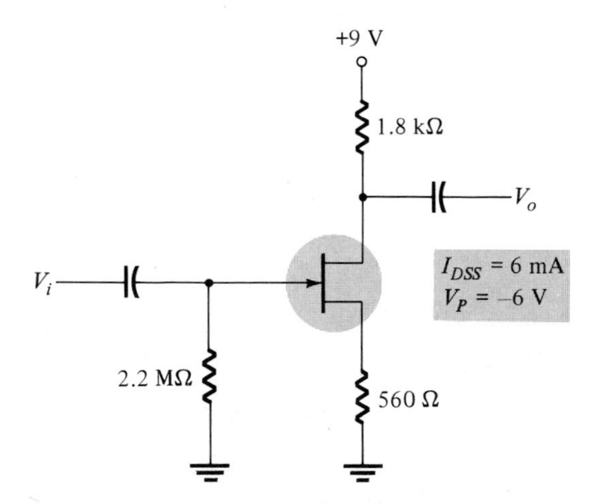

Figure 7.48 Circuit for Problems 18 through 21.

19. What values of I_D and V_{DS} result when the JFET of Fig. 7.48 is changed to one having $I_{DSS} = 8$ mA, $V_P = -6$ V?

20. What values of I_D and V_{DS} result when the JFET of Fig. 7.48 is changed to one having $I_{DSS} = 6$ mA, $V_P = -4$ V?

21. What values of I_D and V_{DS} result when the JFET of Fig. 7.48 is changed to one having $I_{DSS} = 8$ mA, $V_P = -4$ V?

§ 7.3

22. Determine V_{GS} and I_D for the circuit of Fig. 7.49.

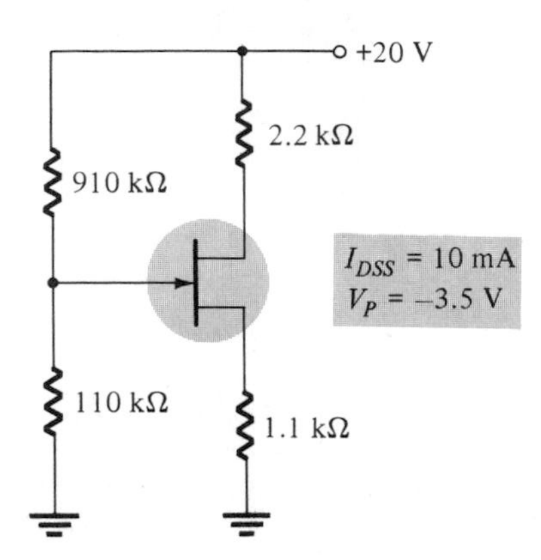

Figure 7.49 Circuit for Problems 22 through 27.

23. Determine V_{DS} for the circuit of Fig. 7.49.

24. What value of resistor R_S is needed to set the bias point of Fig. 7.49 to $V_{GS} = -2$ V?

25. What value of R_D is needed to set the drain voltage of Fig. 7.49 to 12 V?

26. Does the drain current in the circuit of Fig. 7.49 increase or decrease if R_S is changed to 750 Ω?

27. Does the source voltage in the circuit of Fig. 7.49 increase or decrease if the JFET is replaced with one having $I_{DSS} = 16$ mA, $V_P = -3.5$ V?

28. Determine the gate-source voltage for the circuit of Fig. 7.50.

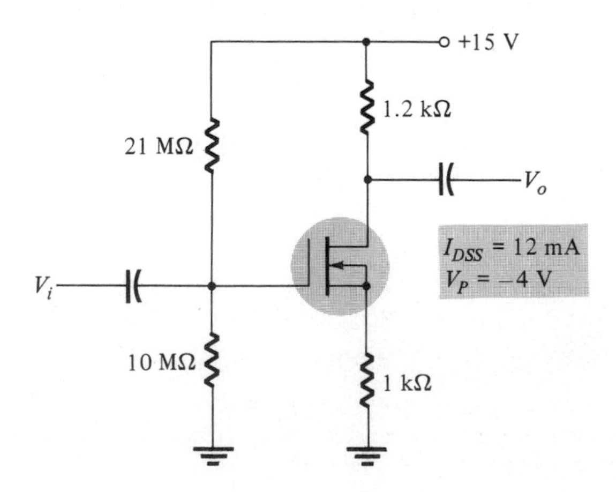

Figure 7.50 Circuit for Problems 28 through 34.

29. Determine the bias values of I_D and V_{DS} for the circuit of Fig. 7.50.

30. Determine the values of I_D and V_{DS} if the MOSFET of Fig. 7.50 is replaced by one having $I_{DSS} = 10$ mA, $V_P = -5$ V.

31. What values of bias I_D and V_{DS} result if the 7.5-MΩ resistor in Fig. 7.50 is removed?

32. What values of I_D and V_{DS} result if the 21-MΩ resistor in Fig. 7.50 is removed?

33. Does the value of I_D increase or decrease if the MOSFET of Fig. 7.50 is replaced by one having $V_P = -2.5$ V?

34. What value of V_D results if the supply of Fig. 7.50 is replaced by $V_{DD} = 12$ V?

§ 7.4

35. Determine the value of V_D for the circuit of Fig. 7.51.

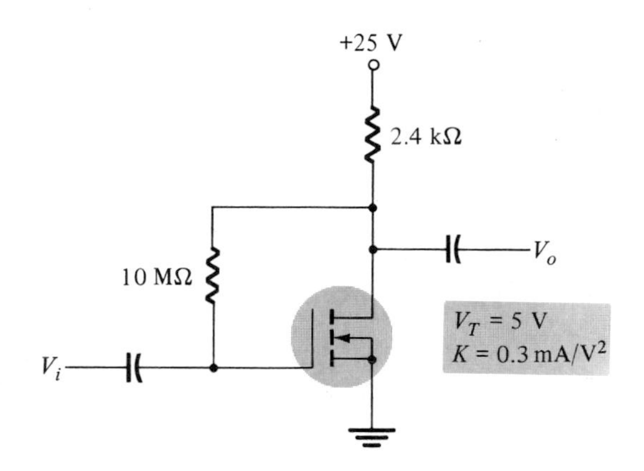

Figure 7.51 Circuit for Problems 35 through 40.

36. What value of V_D results if the drain resistor is replaced by 1.8 kΩ?

37. What is the value of K if the drain voltage in the circuit of Fig. 7.51 is 10 V?

38. What value of V_D results if the MOSFET in the circuit of Fig. 7.51 is replaced by one having $V_T = 3$ V ($K = 0.3$)?

39. What value of V_D results if the supply voltage in the circuit of Fig. 7.51 is reduced to 15 V?

40. Does the drain current increase or decrease if the MOSFET of Fig. 7.51 is replaced by one having $V_T = 7.5$ V?

§ 7.5

41. Determine the dc bias values of I_D and V_{DS} for the circuit of Fig. 7.52.

42. Determine the dc bias values of I_D and V_{DS} for the circuit of Fig. 7.53.

43. What is the value of I_{DSS} for the circuit of Fig. 7.54 with $V_P = -3$ V?

44. Determine the values of V_D, V_S, and I_D for the circuit of Fig. 7.54.

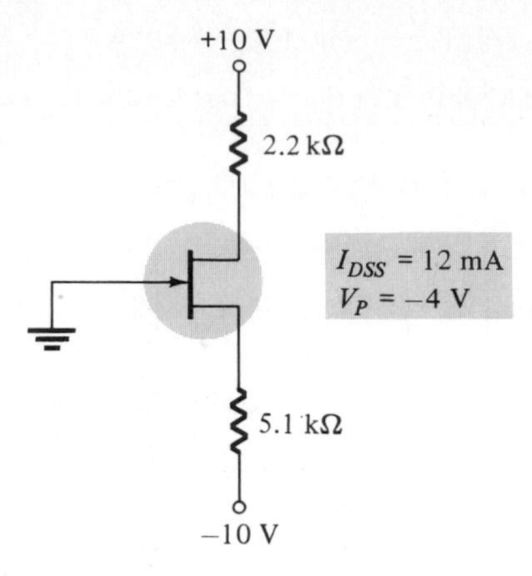

Figure 7.52 Circuit for Problem 41.

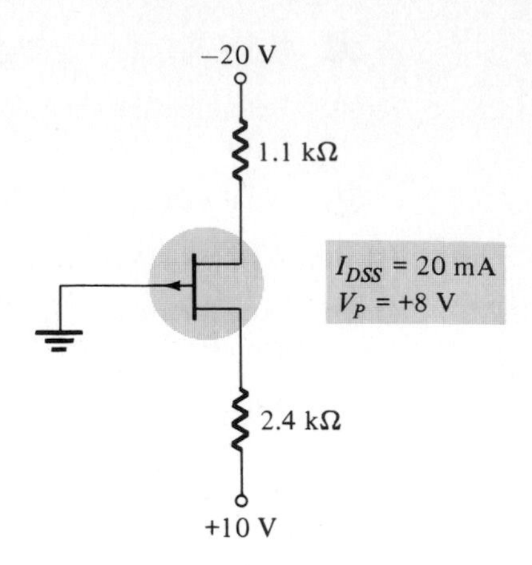

Figure 7.53 Circuit for Problem 42.

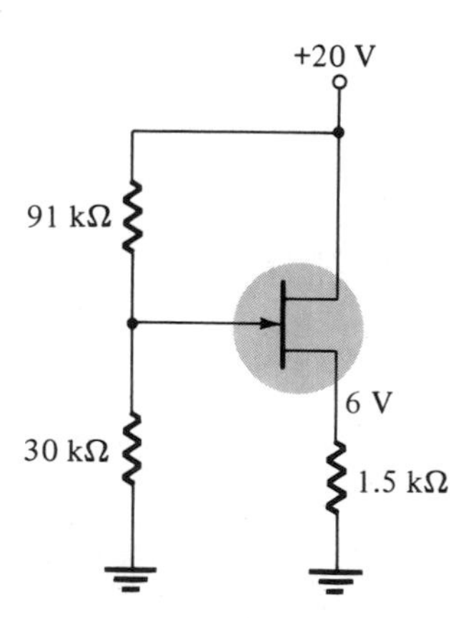

Figure 7.54 Circuit for Problem 43.

§ 7.6

45. Design a JFET bias circuit such as that of Fig. 7.55 for operation at I_D = 3 mA, V_{DS} = 10 V using a supply of 22 V and a JFET having I_{DSS} = 6 mA, and V_P = −4 V.

46. Design a depletion MOSFET circuit such as that of Fig. 7.11 for operation at I_D = 4 mA and V_D = 10 V using a supply of 30 V and a FET having I_{DSS} = 10 mA, V_P = −4.5 V.

47. Design an enhancement MOSFET bias circuit such as that of Fig. 7.19 for operation at I_D = 6 mA, V_D = 8 V using an 18-V supply and a FET having V_T = 3.5 V, K = 0.3 mA/V^2.

48. Complete the design of the circuit of Fig. 7.56.

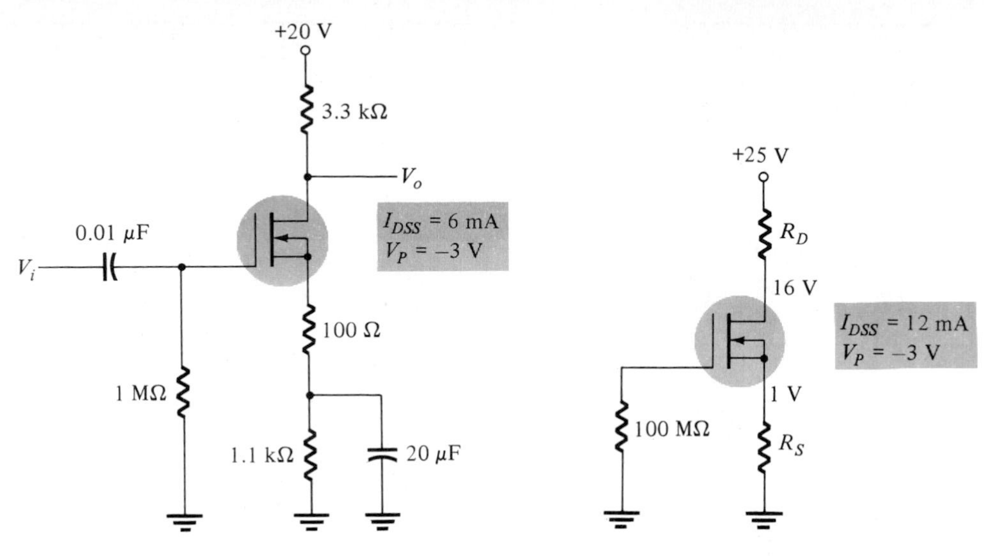

Figure 7.55 Circuit for Problem 44.

Figure 7.56 Circuit for Problem 48.

§ 7.7

49. Determine the dc bias voltage V_{GS} for the circuit of Fig. 7.49 using a JFET with $V_P = -4.5$ V and $I_{DSS} = 8.5$ mA using the universal JFET characteristic.

50. Using the universal JFET characteristic, determine the value of dc bias current I_D for the circuit of Fig. 7.49 with the JFET changed to one having $I_{DSS} = 10$ mA, $V_P = -5$ V.

51. Using the universal JFET bias curves, determine the dc bias current for the circuit of Fig. 7.49 with R_S changed to 1.2 kΩ.

52. Using the universal JFET curve, determine the dc bias voltage V_{DS} when R_{G2} is changed to 270 kΩ in Fig. 7.49.

53. What value of V_{GS} results in the circuit of Fig. 7.49 when the JFET used has $I_{DSS} = 12$ mA, $V_P = -4.5$ V?

COMPUTER PROBLEMS

Write BASIC programs to:

1. Compute a table of points of I_D and V_{DS} for a JFET circuit for given values of I_{DSS} and $V_{GS(OFF)}$.
2. Compute I_D and V_{DS} for a JFET circuit using fixed battery biasing.
3. Compute I_D and V_{DS} for a self-biased JFET circuit.
4. Compute I_D and V_{DS} for a JFET circuit with a voltage-divider biasing.
5. For a self-bias JFET circuit, tabulate the points to plot the transfer characteristic for a given device.
6. Compute and tabulate the values of I_D and V_{DS} for an enhancement MOSFET to plot the device drain characteristic.

8

Transistor Modeling

8.1 INTRODUCTION

The basic contruction, appearance, and characteristics of the transistor were introduced in Chapter 4. The dc biasing of the device was then examined in detail in Chapter 5. We now begin to examine the *small-signal* ac response of the BJT amplifier by reviewing the *models* most frequently used to represent the transistor in the sinusoidal ac domain.

One of our first concerns in the sinusoidal ac analysis of transistor networks is the magnitude of the input signal. It will determine whether *small-signal* or *large-signal* techniques should be applied. There is no set dividing line between the two, but the application, and the magnitude of the variables of interest relative to the scales of the device characteristics, will usually make it quite clear which method is appropriate. The small-signal technique is introduced in this chapter and large-signal applications are examined in Chapter 12.

There are two models commonly used in the small-signal ac analysis of transistor networks: the *hybrid equivalent* and the r_e *model*. This chapter not only introduces both models but defines the role each plays and how the two are related.

8.2 AMPLIFICATION IN THE AC DOMAIN

It was noted earlier that the transistor is an amplifying device. That is, the output sinusoidal signal is greater than the input signal or, stated another way, the output ac power is greater than the input ac power. The question often arises as to where this additional ac power has been generated. The sole purpose of the next few paragraphs will be to answer that question, since it is fundamental to the clear understanding of a number of important efficiency criteria to be defined later in the text.

Analogies are seldom perfect, but the following will be useful in describing the events leading to the foregoing conclusions. In Fig. 8.1a a steady heavy flow of a liquid has been established by the pump. The electrical analogy of this system appears in Fig. 8.1b. In each case there is some resistance to the flow, with the result that the magnitude of that flow is determined by an Ohm's law relationship. A graph of the flow versus time appears in each figure.

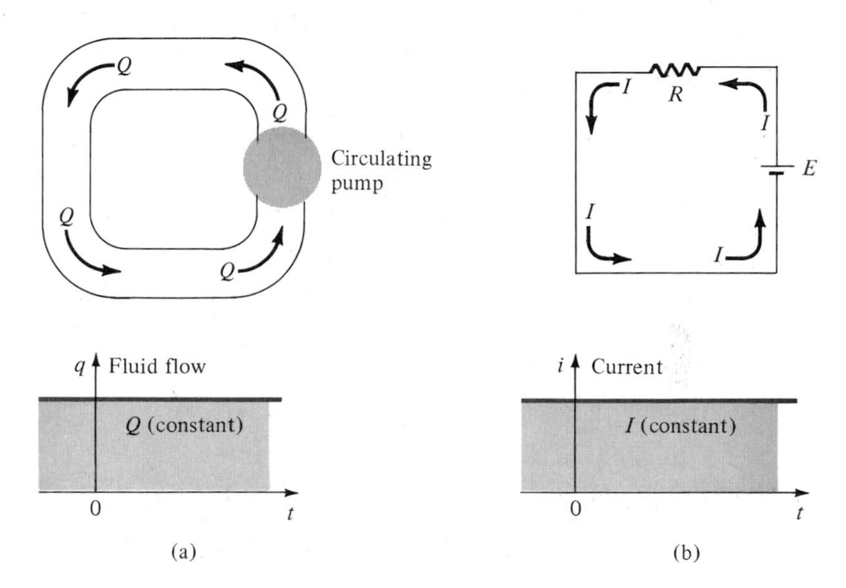

Figure 8.1 Fluid-flow analogy of a series electrical circuit with a dc input: (a) fluidic system; (b) electrical system.

Let us now install a *control mechanism* in each system, as shown in Fig. 8.2. A small signal at the input to each of these control elements can have a marked effect on the established steady-state (dc) flow of each system. For the fluid system it could be the oscillatory partial closing of the passage to limit the flow through the pipe. For the electrical system a mechanism is established for the control of the current i through the system. Recall that very small variations in I_B can have a pronounced effect on the collector current for the common-emitter transistor configuration.

In other words, a small input signal can have a pronounced effect on the steady-state flow of the system. Consider that the resulting output flow for the two systems may be as shown in Fig. 8.2. The sinusoidal swing of the output flow is certainly greater than the applied input—*amplification in the ac domain is therefore a reality!*

We can conclude, therefore, that most amplifiers are simply devices having a control point or terminal that can establish a heavy variation in flow between the other two terminals (normally part of the output circuit). The dc biasing circuits are necessary to establish the heavy flow of charge that will be very sensitive to the magnitude of the input signal. The increased ac power is the result of the conversion of some of

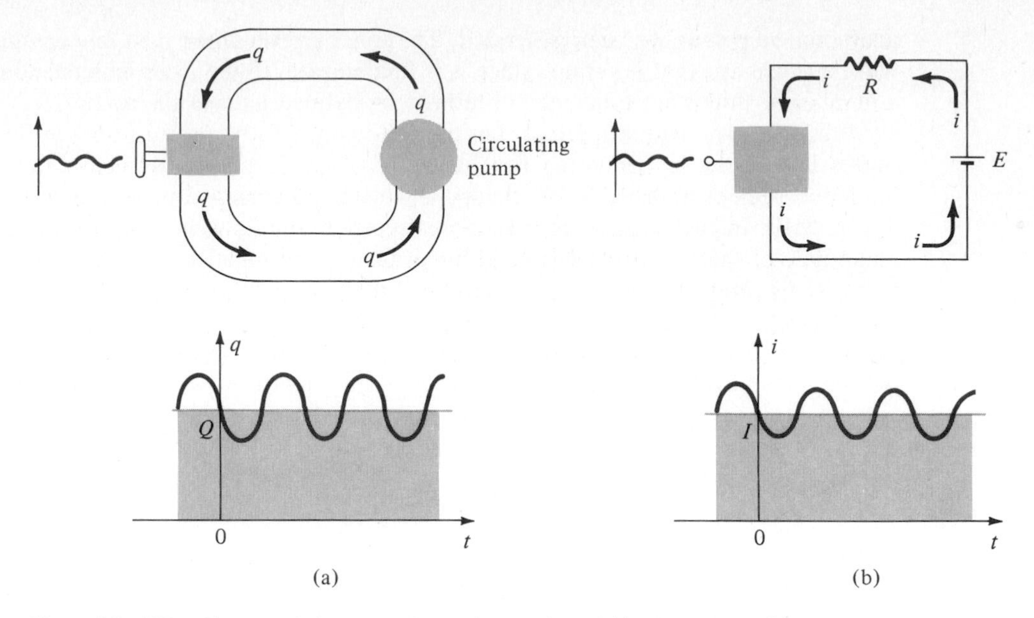

Figure 8.2 Effect of a control element on the steady-state flow of (a) a fluid system; (b) an electrical system.

the dc power to the sinusoidal domain. The efficiency of an electronic amplifier is typically the ratio of the ac power out to the dc power in.

8.3 TRANSISTOR MODELING

The key to the small-signal approach is the use of equivalent circuits (models) to be derived in this chapter. It is that combination of circuit elements, properly chosen, that will best approximate the actual semiconductor device in a particular operating region. Once the ac equivalent circuit has been determined, the graphic symbol of the device can be replaced in the schematic by this circuit and the basic methods of ac circuit analysis (branch-current analysis, mesh analysis, nodal analysis, and Thévenin's theorem) can be applied to determine the response of the circuit.

There are two schools of thought in prominence today regarding the equivalent circuit to be substituted for the transistor. For many years the industrial and educational institutions relied heavily on the *hybrid parameters* (to be introduced shortly). The hybrid-parameter equivalent circuit continues to be very popular, although it must now share the spotlight with an equivalent circuit derived directly from the operating conditions of the transistor-the r_e model. Manufacturers continue to specify the hybrid parameters for a particular operating region on their specification sheets. The parameters (or components) of the r_e model can be derived directly from the hybrid parameters in this region. However, the hybrid equivalent circuit suffers from being limited to a particular set of operating conditions if it is to be considered accurate. The parameters of the other equivalent circuit can be determined for any region of operation within the active region and are not limited by the single set of parameters provided by the specification sheet. In turn, however, the r_e model fails to have a parameter defining the output impedance level of the device and the feedback effect from output to input.

Since both models are used extensively today, they are both examined in detail in this text. In some analysis and examples the hybrid model will be employed, while in others the r_e model will be used exclusively. The text will make every effort, however, to show how closely related the two models are and how a proficiency with one leads to a natural proficiency with the other.

In an effort to demonstrate the effect that the ac equivalent circuit will have on the analysis to follow, consider the circuit of Fig. 8.3. Let us assume for the moment that the small-signal ac equivalent circuit for the transistor has already been determined. Since we are interested only in the ac response of the circuit, all the dc supplies can be replaced by a zero-potential equivalent (short circuit) since they determine only the dc or quiescent level of the output voltage and not the magnitude of the swing of the ac output. This is clearly demonstrated by Fig. 8.4. The dc levels were simply

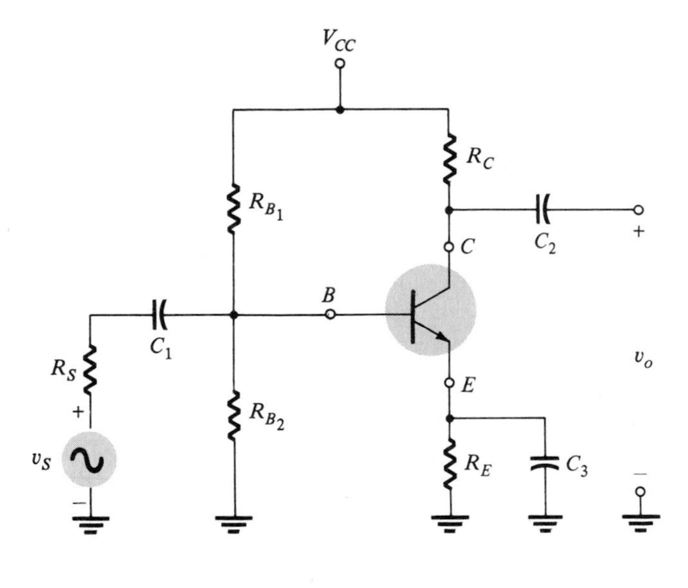

Figure 8.3 Transistor circuit under examination in this introductory discussion.

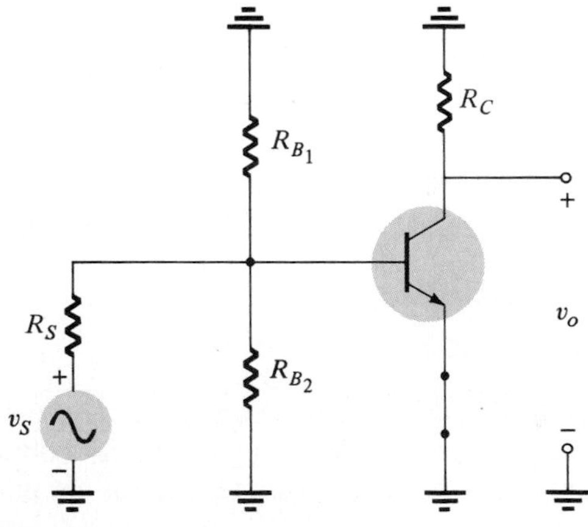

Figure 8.4 The network of Fig. 8.3 following the removal of the dc supply and inserting the short-circuit equivalent for the capacitors.

important for determining the proper Q-point of operation. Once determined, the dc levels can be ignored for the ac analysis of the network. In addition, the coupling capacitors C_1 and C_2 and bypass capacitor C_3 were chosen to have a very small reactance at the frequency of application. Therefore, they too may for all practical purposes be replaced by a low-resistance path (short circuit). Note that this will result in the "shorting out" of the dc biasing resistor R_E. Recall that capacitors assume an open-circuit equivalent under dc steady-state conditions, permitting an isolation between stages for the dc levels and quiescent conditions.

Connecting common grounds will result in a parallel combination for resistors R_{B_1} and R_{B_2} and R_C will appear from collector to emitter as shown in Fig. 8.5. Since the components of the transistor equivalent circuit inserted in Fig. 8.5 are those we are already familiar with (resistors, controlled sources, etc.), analysis techniques such as superposition, Thévenin's theorem, and so on, can be applied to determine the desired quantities.

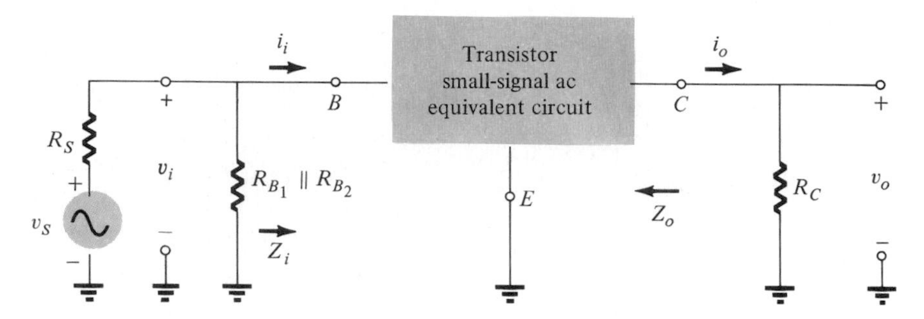

Figure 8.5 Circuit of Fig. 8.4 redrawn for small-signal ac analysis.

Let us further examine Fig. 8.5 and identify the important quantities to be determined for the system. Certainly, we would like to know the input and output impedance Z_i and Z_o as shown in Fig. 8.5. Since we know that the transistor is an amplifying device, we would expect some indication of how the output voltage v_o is related to the input voltage v_i—the *voltage gain*. Note in Fig. 8.5 for this configuration that $i_i = i_b$ and $i_o = i_c$, which define the *current gain* $A_i = i_o/i_i$.

In Chapter 4 we found that the collector-to-emitter voltage did have some effect (if even slight) on the input relationship between i_B and v_{BE}. We might, therefore, expect some "feedback" from the output to input circuit in the equivalent circuit. The following section, through its brief introduction to *two-port theory*, will introduce the hybrid equivalent circuit, which will have parameters that will permit a determination of each of the quantities discussd above.

8.4 TRANSISTOR HYBRID EQUIVALENT CIRCUIT

The development that follows a brief introduction to a subject called *two-port* theory. For the basic three-terminal electronic device or system it is obvious, from Fig. 8.6, that there are two ports (pairs of terminals) of interest. For our purposes, the set at the

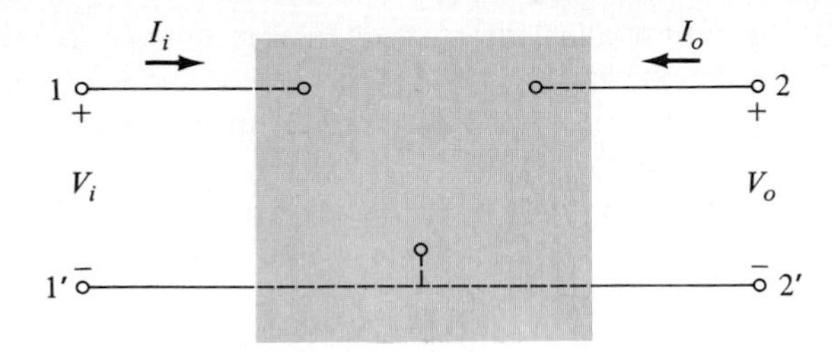

Figure 8.6 Two-port system.

left will represent the input terminals, and the set at the right, the output terminals. Note that, for each set of terminals, there are two variables of interest.

The following set of equations (8.1) is only one of a number of ways in which the four variables can be related. It is the most frequently employed in transistor circuit analysis, however, and, therefore, will be discussed in detail in this chapter.

$$\boxed{V_i = h_{11}I_i + h_{12}V_o} \tag{8.1a}$$

$$\boxed{I_o = h_{21}I_i + h_{22}V_o} \tag{8.1b}$$

The parameters relating the four variables are called *h-parameters* from the word "hybrid." The term "hybrid" was chosen because the mixture of variables (v and i) in each equation results in a "hybrid" set of units of measurement for the h-parameters. A clearer understanding of what the various h-parameters represent and how we can expect to treat them later can be developed by isolating each and examining the resulting relationship.

If we arbitrarily set $V_o = 0$ (short circuit the output terminals), and solve for h_{11} in Eq. (8.1a), the following will result:

$$\boxed{h_{11} = \left.\frac{V_i}{I_i}\right|_{V_o=0}} \quad \text{ohms} \tag{8.2}$$

The ratio indicates that the parameter h_{11} is an impedance parameter to be measured in ohms. Since it is the ratio of the *input* voltage to the *input* current with the output terminals *shorted*, it is called the *short-circuit input-impedance parameter*. The subscript 11 of h_{11} defines the facts that the parameter is determined by a ratio of quantities measured at the input terminals.

If I_i is set equal to zero by opening the input leads, the following will result for h_{12}:

$$\boxed{h_{12} = \left.\frac{V_i}{V_o}\right|_{I_i=0}} \quad \text{unitless} \tag{8.3}$$

The parameter h_{12}, therefore, is the ratio of the input voltage to the output voltage with the input current equal to zero. It has no units since it is a ratio of voltage levels. It is called the *open-circuit reverse transfer voltage ratio parameter*. The subscript 12 of h_{12} reveals that the parameter is a transfer quantity determined by a ratio of input to output measurements. The first integer of the subscript defines the measured quantity to appear in the numerator; the second integer defines the source of the quantity to appear in the denominator. The term *reverse* is included to indicate that the voltage ratio is an input quantity over an output quantity rather than the reverse, which is usually the ratio of interest.

If in Eq. (8.1b), V_o is equal to zero by again shorting the output terminals, the following will result for h_{21}:

$$h_{21} = \left.\frac{I_o}{I_i}\right|_{V_o=0} \qquad \text{unitless} \tag{8.4}$$

Note that we now have the ratio of an output quantity to an input quantity. The term *forward* will now be used rather than *reverse* as indicated for h_{12}. The parameter h_{21} is the ratio of the output current to the input current with the output terminals shorted. It is, for most applications, the parameter of greatest interest. This parameter, like h_{12}, has no units since it is the ratio of current levels. It is formally called the *short-circuit forward transfer current ratio parameter*. The subscript 21 again indicates that it is a transfer parameter with the output quantity in the numerator and the input quantity in the denominator.

The last parameter, h_{22}, can be found by again opening the input leads to set $I_1 = 0$ and solving for h_{22} in Eq. (8.1b):

$$h_{22} = \left.\frac{I_o}{V_o}\right|_{I_i=0} \qquad \text{siemens} \tag{8.5}$$

Since it is the ratio of the output current to the output voltage, it is the output conductance parameter and is measured in siemens (S) [formerly mhos (Ω)]. It is called the *open-circuit output admittance parameter*. The subscript 22 reveals that it is determined by a ratio of output quantities.

Since each term of Eq. (8.1a) has the unit volt, let us apply Kirchhoff's voltage law in reverse to find a circuit that "fits" the equation. Performing this operation will result in the circuit of Fig. 8.7. Since the parameter h_{11} has the unit ohm, it is represented by a resistor in Fig. 8.7. The quantity h_{12} is dimensionless and therefore simply appears as a multiplying factor of the "feedback" term in the input circuit.

Since each term of Eq. (8.1b) has the units of current, let us now apply Kirchhoff's current law in reverse to obtain the circuit of Fig. 8.8. Since h_{22} has the units of admittance, which for the transistor model is conductance, it is represented by the

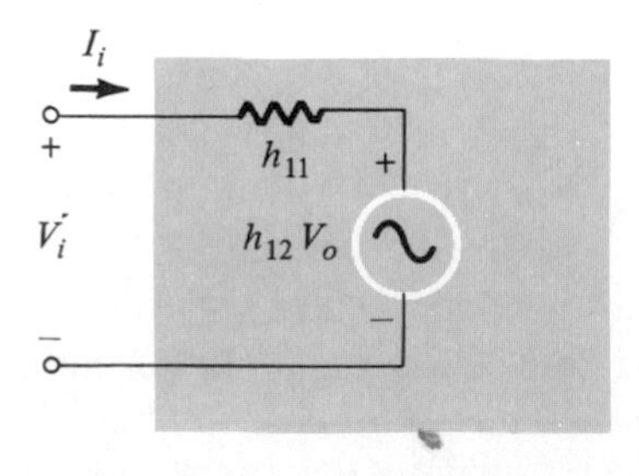

Figure 8.7 Hybrid input equivalent circuit.

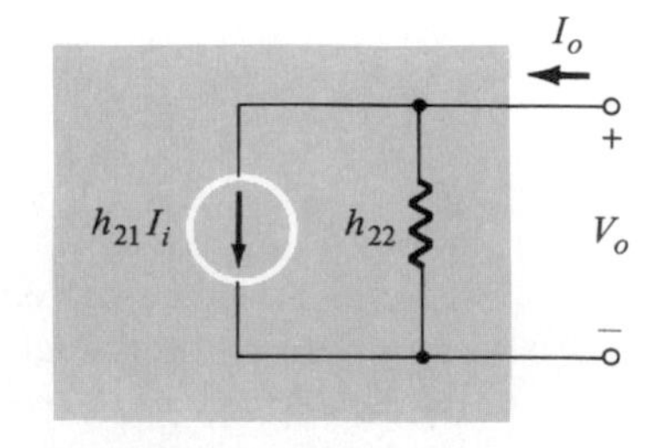

Figure 8.8 Hybrid output equivalent circuit.

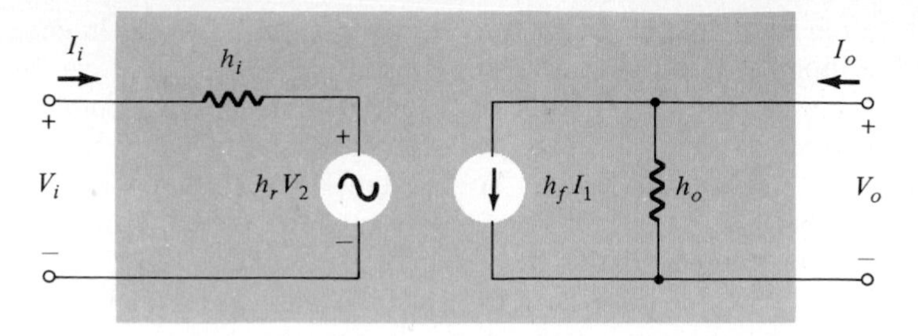

Figure 8.9 Complete hybrid equivalent circuit.

resistor symbol. Keep in mind, however, that the resistance in ohms of this resistor is equal to the reciprocal of conductance ($1/h_{22}$).

The complete "ac" equivalent circuit for the basic three-terminal linear device is indicated in Fig. 8.9 with a new set of subscripts for the h-parameters. The notation of Fig. 8.9 is of a more practical nature since it relates the h-parameters to the resulting ratio obtained in the last few paragraphs. The choice of letters is obvious from the following listing:

$$h_{11} \rightarrow \textit{i}\text{nput resistance} \rightarrow h_i$$

$$h_{12} \rightarrow \textit{r}\text{everse transfer voltage ratio} \rightarrow h_r$$

$$h_{21} \rightarrow \textit{f}\text{orward transfer current ratio} \rightarrow h_f$$

$$h_{22} \rightarrow \textit{o}\text{utput conductance} \rightarrow h_o$$

The circuit of Fig. 8.9 is applicable to any linear three-terminal electronic device or system with no internal independent sources. For the transistor, therefore, even though it has three basic configurations, *they are all three-terminal configurations*, so that the

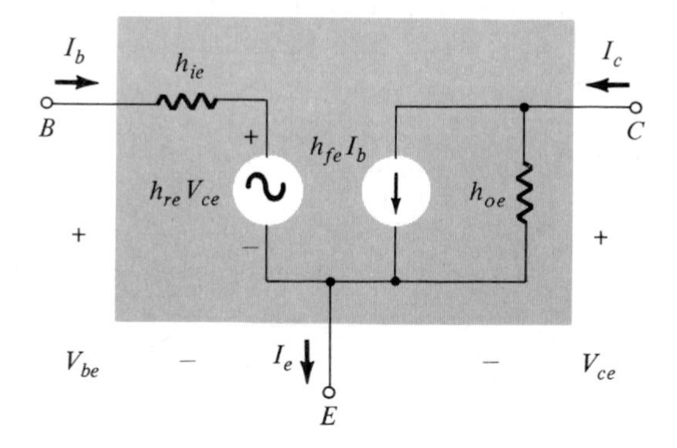

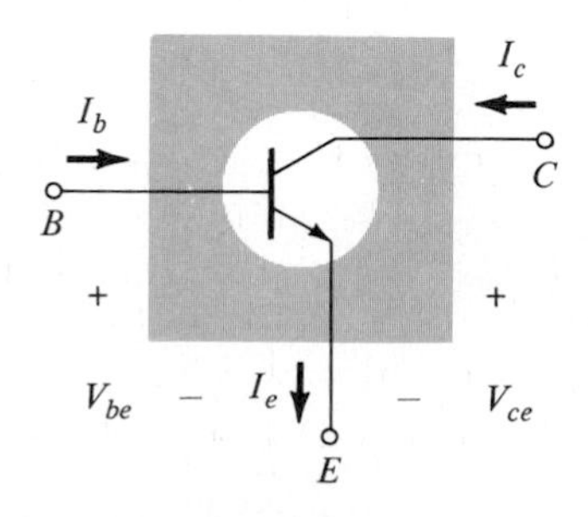

Figure 8.10 Common-emitter configuration.

resulting equivalent circuit will have the same format as shown in Fig. 8.9. In each case the bottom of the input and output sections of the network of Fig. 8.9 can be connected as shown in Fig. 8.10, since the potential level is the same. Essentially, therefore, the transistor model is a three-terminal two-port system. The h-parameters, however, will change with each configuration. To distinguish which parameter has been used or which is available, a second subscript has been added to the h-parameter notation. For the common-base configuration the lowercase letter b was added, while for the common-emitter and common-collector configurations the letters e and c were added, respectively. The hybrid equivalent network for the common-emitter configuration appears with the standard notation in Fig. 8.10. Note that $I_i = I_b$, $I_o = I_c$, and through an application of Kirchhoff's current law, $I_e = I_b + I_c$. The input voltage is now V_{be} with the output voltage V_{ce}. For the common-base configuration of Figure 8.11, $I_i = I_e$, $I_o = I_c$ with $V_{eb} = V_i$ and $V_{cb} = V_o$. The networks of Figs. 8.10 and 8.11 are applicable for *pnp* or *npn* transistors.

The hybrid equivalent circuit of Fig. 8.9 is an extremely important one in the area of electronics today. It will appear over and over again in the analysis to follow. It would be time well spent, at this point, for the reader to memorize and draw from memory its basic construction and define the significance of the various parameters [see Eqs. (8.2) through (8.5)]. The fact that both a Thévenin and Norton circuit appear

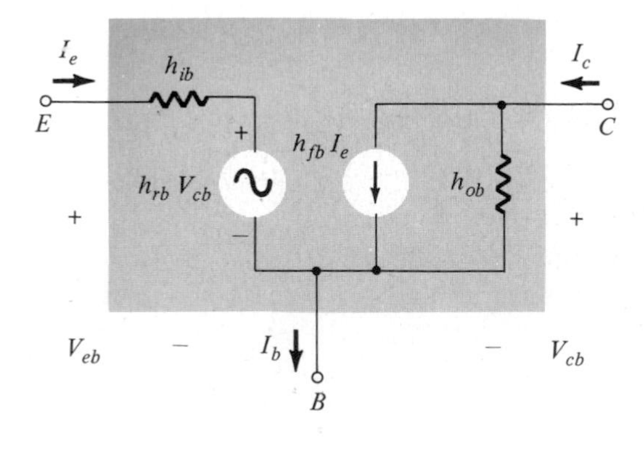

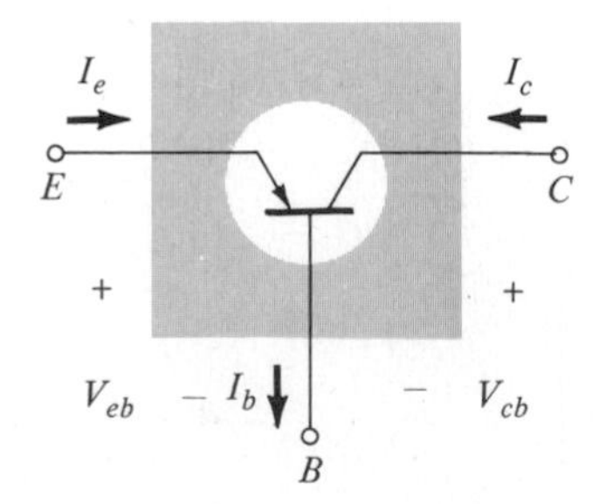

Figure 8.11 Common-base configuration.

in the circuit of Fig. 8.9 was further impetus for calling the resultant circuit a *hybrid* equivalent circuit. Two additional transistor equivalent circuits, not to be discussed in this text, called the **z**-parameter and **y**-parameter equivalent circuits, use either the voltage source or the current source but not both in the same equivalent circuit. In Section 8.7 the magnitude of the various parameters will be found from the transistor characteristics in the region of operation resulting in the desired *small-signal equivalent network* for the transistor.

This chapter is limited solely to an introduction to the models. Chapter 9 will apply each to a variety of standard configurations. When applied in Chapter 9, an approximate equivalent model will be employed in the early sections that is a reduced form of Fig. 8.9.

For the common-emitter and common-base configurations the magnitude of h_r and h_o is often such that the results obtained for the important parameters such as Z_i, Z_o, A_v and A_i are only slightly affected if they (h_r and h_o) are not included in the model.

Since h_r is normally a relatively small quantity, its removal is approximated by $h_r \cong 0$ and $h_r V_o = 0$, resulting in a short-circuit equivalent for the feedback element as shown in Fig. 8.12. The resistance determined by $1/h_o$ is normally large enough to be ignored, in comparison to a parallel load, which can be replaced by an open-circuit equivalent for the CE and CB models, as shown in Fig. 8.12.

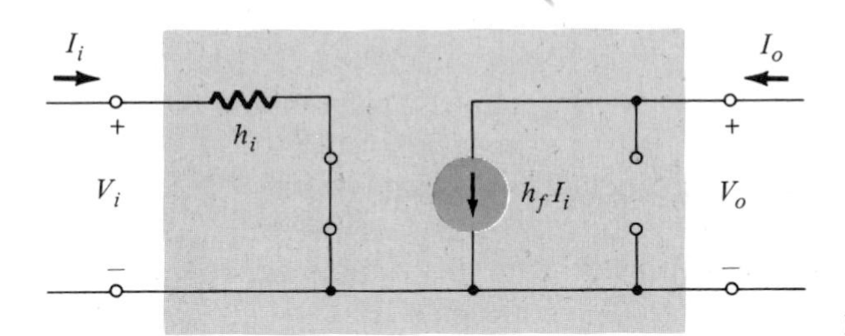

Figure 8.12

The resulting equivalent of Fig. 8.13 will be employed extensively in Chapter 9 to ensure that the approach is not clouded by the additional mathematical complexity introduced by h_r and h_o. In Section 9.10 we compare solutions obtained by the complete and approximate models to verify some of the general conclusions and further demonstrate the validity of the approximate approach for a wide range of applications.

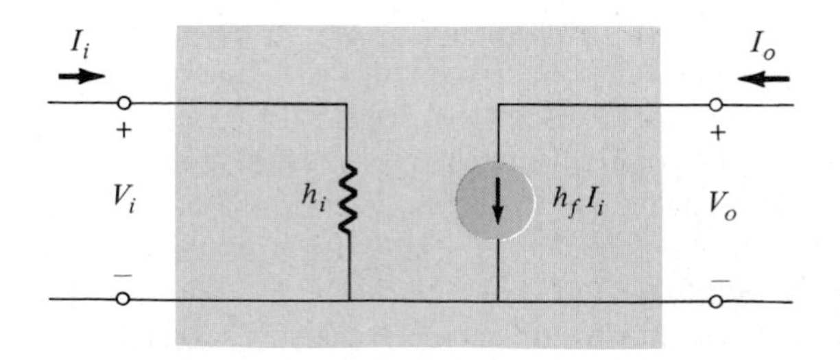

Figure 8.13 Approximate hybrid equivalent model.

For the common-emitter configuration the approximate equivalent model will appear as shown in Fig. 8.14, where it is fairly obvious that

$$I_i = I_b$$

$$Z_i = h_{ie}$$

$$I_o = I_c = h_{fe} I_b \quad \text{(a magnification of } I_b \text{ by the factor } h_{fe}\text{)}$$

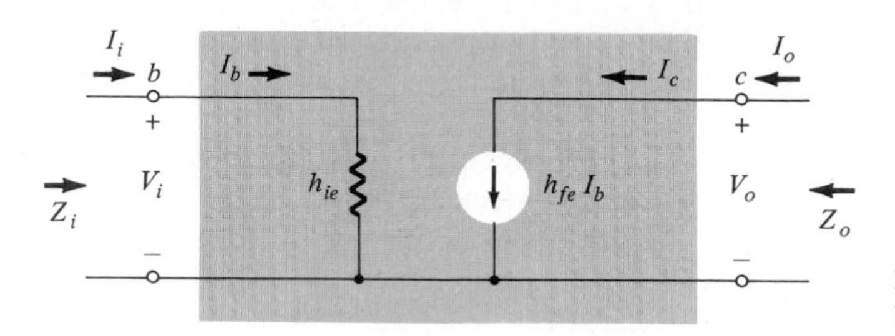

Figure 8.14 Approximate CE hybrid equivalent model.

For Z_o, V_i is set to zero, resulting in $I_b = 0$ and $h_{fe} I_b = 0$, which results in an open-circuit equivalent for the current source. The result,

$$Z_o \cong \infty\Omega \quad \text{(open-circuit)}$$

For future reference, the model of Fig. 8.14 will be referred to as the *approximate hybrid equivalent circuit* for the common-emitter configuration, and the model of Fig. 8.9 will be called the *complete hybrid equivalent circuit*.

The discussion above did not include the common-collector configuration since it is seldom employed in the analysis of small-signal ac transistor networks. Rather, the common-emitter configuration is substituted in a manner demonstrated in Chapter 9. Further, the magnitude of h_{rc} is such that it cannot be ignored, although the effects of h_{oc} can normally be ignored (on an approximate basis), as indicated above for the common-emitter and common-base configurations.

8.5 THE r_e MODEL

In recent years there has been an increasing interest in an approximate equivalent circuit for the transistor in which one of the parameters is determined by the dc operating conditions. You will possibly recall from the transistor specification sheet provided in Chapter 4 that the hybrid parameter h_{ie} was specified at a particular operating point. Figure 8.30 will reveal a significant variation in h_{ie} with I_C ($\cong I_E$). The question then arises of what one would do with the provided value of h_{ie} if the conditions of operation (level of $I_C \cong I_E$) were different from those indicated on the specification sheet. The equivalent circuit derived below will permit the determination of an equivalent h_{ie} using the dc operating conditions of the network, thereby not limiting itself to the data on the device as provided by the manufacturer.

The derivation of the alternate equivalent circuit begins with a close examination of the input and output characteristics of the CB transistor configuration, as redrawn in Fig. 8.15, on an approximate basis. Note that straight-line segments are used to

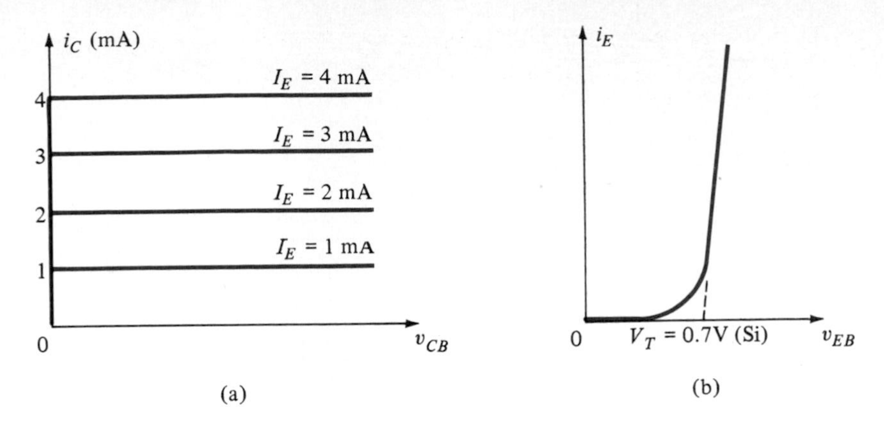

Figure 8.15 Approximate CB characteristics: (a) output; (b) input.

represent the collector characteristics and a single diode characteristic for the emitter circuit (neglecting the variation in the input charcteristics with the change in V_{CB}), resulting in an equivalent circuit such as shown in Fig. 8.16b. For ac conditions, therefore, the input impedance at the emitter of the CB transistor can be determined using Eq. (1.21) for the dynamic resistance of the diode. Since the diode current will be the emitter current, the notation for the diode resistance is r_e and its value is determined by

$$r_e = \frac{26 \text{ mV}}{I_E} \quad \text{ohms} \tag{8.6}$$

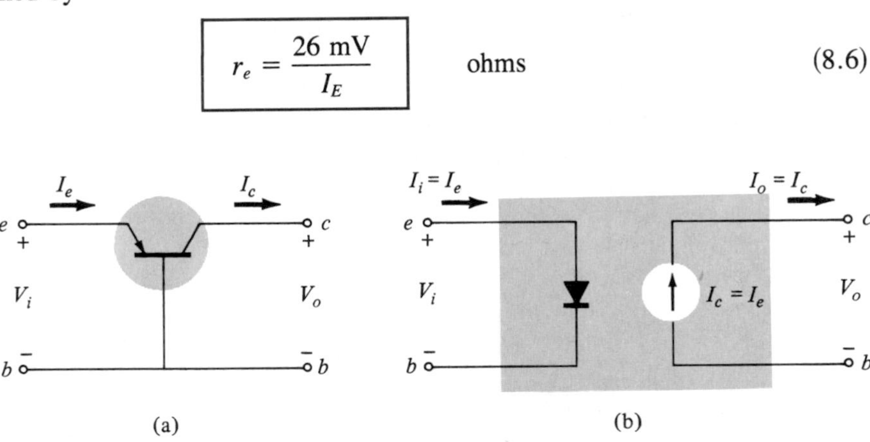

Figure 8.16 (a) CB configuration; (b) approximate CB equivalent circuit as defined by Fig. 8.15.

For the analysis to follow it is assumed that the operating conditions are such that any bulk or contact resistance [of Eq. (1.8)] can be ignored compared to the value calculated by Eq. (8.6).

Substituting r_e will result in the r_e *model* of the common-base configuration appearing in Fig. 8.17. Note in the figure that $I_c = I_e$, as verified by the curves of Fig. 8.15. Note the similarities of the r_e model of Fig. 8.17 with the approximate hybrid equivalent model of Fig. 8.18 for the common-base configuration. A comparison of

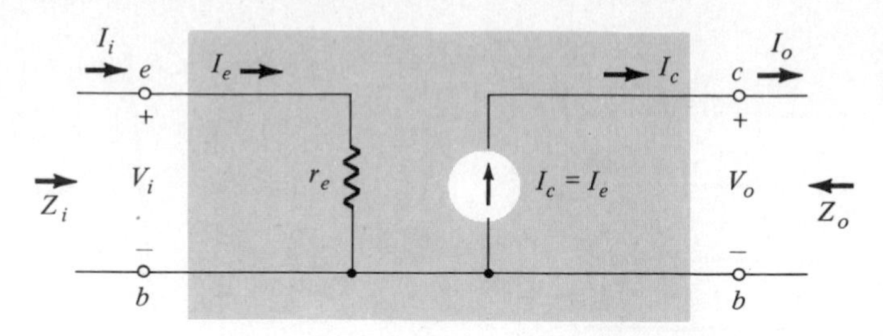

Figure 8.17 CB r_e equivalent circuit.

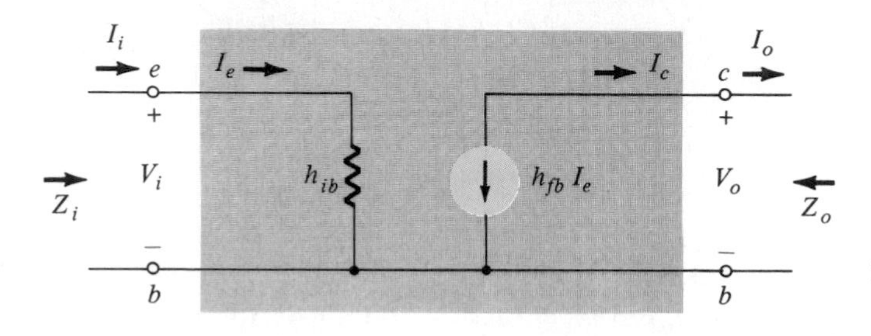

Figure 8.18 Approximate CB hybrid equivalent circuit.

the two clearly reveals that

$$h_{ib} = r_e \tag{8.7}$$

$$h_{fb} = -1 \tag{8.8}$$

The fact that the collctor characteristics have been approximated by horizontal lines requres that $1/h_{ob} \cong \infty\,\Omega$ or an open-circuit equivalent for the output section of the transistor as appearing in Fig. 8.17. In addition, the fact that the various curves at the input side were replaced by one ignores the effects of V_{cb} on the input quantities such as covered by h_{rb} in the hybrid equivalent network. As indicated earlier, the approximations $h_{rb} = h_{ob} = 0$ are always assumed for the r_e model. There is no mechanism for including these effects except to return to the characteristics or specification sheets and add the determined quantities to the equivalent circuit.

For the common-base model of Fig. 8.17, the following parameters are defined:

$$I_i = I_e$$

$$Z_i = r_e$$

$$Z_o = \infty\,\Omega \quad (I_i = I_e = 0;\ \text{therefore, } I_c = I_e = 0)$$

$$I_o = I_c = I_e$$

For the common-emitter configuration appearing in Fig. 8.19a, the input and output characteristics have been approximated by the set appearing in Fig. 8.19b and 8.19c, respectively. The base characteristics are again approximated to be those of a diode (the effect of V_{CE} on the characteristics is ignored) and

$$r_{ac} = \frac{26\text{ mV}}{I_B} \tag{8.9}$$

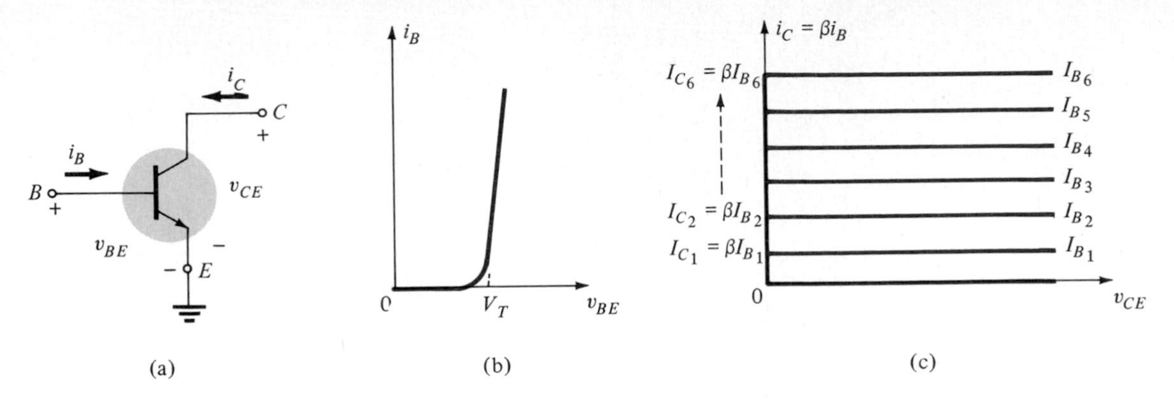

Figure 8.19 (a) CE configuration; (b) input characteristics; (c) output characteristics.

But
$$I_E \cong I_C = \beta I_B \quad \text{and} \quad I_B \cong \frac{I_E}{\beta}$$

so that
$$r_{ac} = \frac{26 \text{ mV}}{I_B} = \frac{26 \text{ mV}}{I_E/\beta} = \beta\left(\frac{26 \text{ mV}}{I_E}\right)$$

and
$$\boxed{r_{ac} = \beta r_e} \tag{8.10}$$

as appearing in Fig. 8.20. In words, the result states that the input resistance of a transistor in the common-emitter configuration with a grounded emitter is equal to β times the resistance determined by Eq. (8.6). For reasons to be more obvious in Chapter 9, the resistor r_e is included in the emitter leg, as shown in Fig. 8.20.

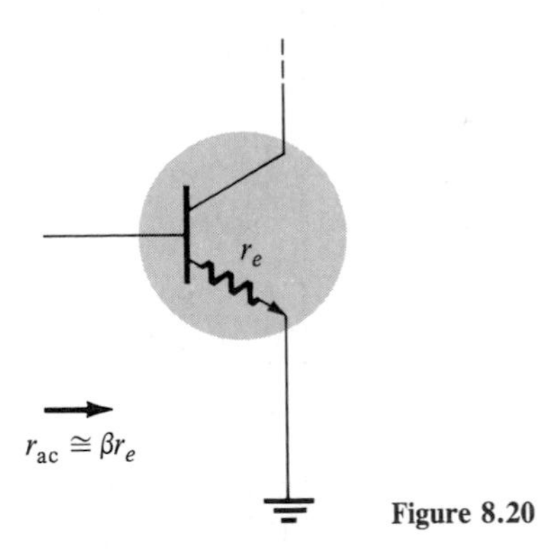

Figure 8.20

The input circuit for the CE configuration is approximated, for the reasons discussed above, by the diode circuit appearing in Fig. 8.21, but the input impedance appears as βr_e in Fig. 8.22 since r_e is determined by I_E and not by I_B. In Fig. 8.19c the approximation was employed that β is the same throughout the device characteristics. This we know is absolutely untrue. However, its variation about the provided value for the typical application in the active region is assumed to be minimal and a fixed value a valid first approximation. For our analysis, we will consider it to be a

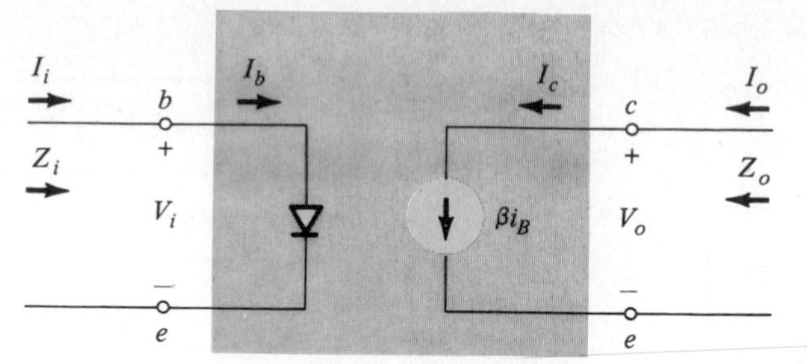

Figure 8.21

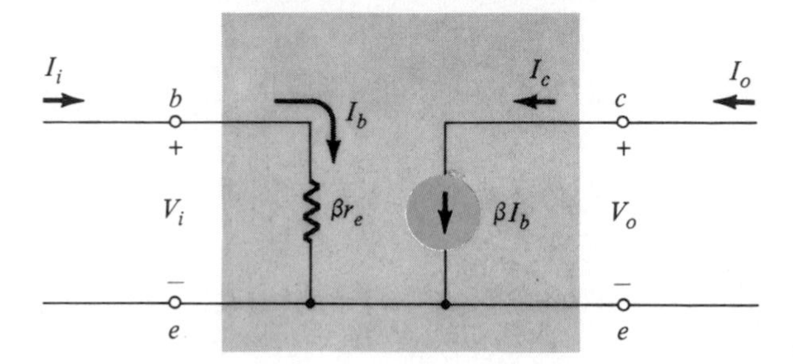

Figure 8.22 CE r_e model.

constant at the provided value resulting in the output equivalent circuits of Figs. 8.21 and 8.22. The approximate hybrid equivalent circuit for the common-emitter configuration appears in Fig. 8.23 for comparison purposes. Comparing the circuits, it is fairly obvious that

$$\boxed{\beta = h_{fe}} \tag{8.11}$$

$$\boxed{\beta r_e = h_{ie}} \tag{8.12}$$

From Fig. 8.22 we find

$$I_i = I_b \quad \text{and} \quad I_o = I_c = \beta I_b$$

$$Z_i = \beta r_e$$

$$Z_o = \infty\ \Omega\ (V_i = 0\ \text{V} \Rightarrow I_i = I_b = 0\ \text{A and } h_{fe} I_b = 0\ \text{A})$$

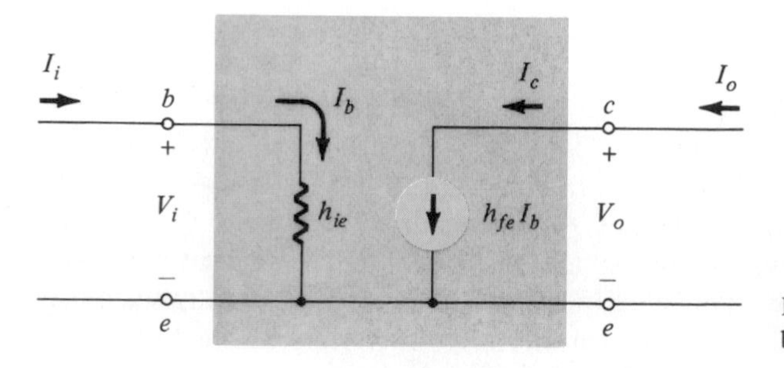

Figure 8.23 Approximate CE hybrid equivalent circuit.

Now that the r_e model has been defined, keep in mind for the chapters to follow that r_e is determined by the resulting dc emitter current of the dc bias configuration. In other words, before the ac analysis can begin, the emitter current of the configuration must first be determined using methods introduced in Chapter 5. The procedure is demonstrated in a number of examples in Chapter 9.

8.6 WHICH MODEL?

Now that more than one model has been introduced for the transistor, the question naturally arises as to which model should be used for a given situation.

The obvious advantage of the r_e model is that only β is required to perform an approximate analysis of an ac configuration. The value of β can be determined using instrumentation such as a curve tracer negating any need for the specification sheet or detailed information on the transistor beyond its maximum ratings. In fact, the specification sheet only provides a range of $\beta(=h_{fe})$ values suggesting that the measured value is the one that should be used in the calculations using either model if a reasonable level of accuracy is desired. The input impedance can then be determined following a calculation of r_e from a dc analysis of the system and an application of the equation βr_e for the common-emitter configuration.

As noted in an earlier section, however, the r_e model fails to include the effects of the feedback element h_{re} and output impedance $1/h_{oe}$-two quantities provided on a typical transistor specification sheet. For most applications in this text the approximation $h_{re} = h_{oe} \cong 0$ can be applied. However, there are situations where the effects of h_{re} or $1/h_{oe}$ may have to be considered to maintain the desired level of accuracy. For instance, if h_{re} and V_o are sufficiently large the feedback element $h_{re} V_o$ in the input equivalent circuit of the hybrid model can reduce the input base current by more than 10% and have a measurable effect on the gain of the system. In the majority of situations, however, the effect of h_{re} can be ignored in the analysis. The above comments were provided solely to ensure an awareness of the effect of the feedback element and when it may be a contributing element that cannot be ignored.

If the ratio of $1/h_{oe}$ to an applied load R_L across the output terminals is less than 10:1, the effect of $1/h_{oe}$ may have to be included. For example, if $R_L = 5\text{ k}\Omega$ and $1/h_{oe} = 50\text{ k}\Omega$ then $R_L' = R_L \| 1/h_{oe} = 5\text{ k}\Omega \| 50\text{ k}\Omega \cong 4.55\text{ k}\Omega$ and the 10:1 ratio between elements results in an almost 10% drop in the effective value. Certainly, any lower ratio would have even more impact. In summary, *if $1/h_{oe}$ is greater than or equal to ten times R_L then the effect of $1/h_{oe}$ can be ignored on an approximate basis.*

If conditions are such that the effects of h_{re} and $1/h_{oe}$ can be ignored the basic format of the two models is identical. Once $h_{fe}\,(=\beta)$ is known one-half the approximate equivalent model is known for each configuration. If the r_e model is used βr_e can be determined directly from the dc analysis. For the hybrid model the average value of the provided range for h_{ie} can be used although the range is normally provided for a particular operating collector current and a different level may significantly affect the value of h_{ie} since it is quite sensitive to the magnitude of the collector current. Of course, a *measured* value of h_{ie} under operating conditions is a better choice than the calculated βr_e value.

In any case, it is important that the student be aware of both models since they both have their followings in education and industry. Each has their advantages and disadvantages. The choice is primarily a matter of choice often influenced by personal experience. Keep in mind the similarities between the two models and note in the ac analysis to follow the similarities in procedure to determine the desired quantities.

The next section will demonstrate that the hybrid parameter h_{fe} is the least sensitive of the hybrid parameters to a change in collector current. Assuming, therefore, that $h_{fe} = \beta$ is a constant for the range of interest is a fairly good approximation. It is $h_{ie} = \beta r_e$ that will vary significantly with I_C and should be determined at operating levels, since it can have a real impact on the gain levels of a transistor amplifier.

8.7 GRAPHICAL DETERMINATION OF THE *h*-PARAMETERS

Using partial derivatives (calculus), we can show that the magnitude of the *h*-parameters for the small-signal transistor equivalent circuit in the region of operation for the common-emitter configuration can be found using the following equations:

$$h_{ie} = \frac{\partial v_i}{\partial i_i} = \frac{\partial v_{be}}{\partial i_b} \cong \left.\frac{\Delta v_{be}}{\Delta i_b}\right|_{V_{CE}=\text{constant}} \tag{8.13}$$

$$h_{re} = \frac{\partial v_i}{\partial v_o} = \frac{\partial v_{be}}{\partial v_{ce}} \cong \left.\frac{\Delta v_{be}}{\Delta v_{ce}}\right|_{I_B=\text{constant}} \tag{8.14}$$

$$h_{fe} = \frac{\partial i_o}{\partial i_i} = \frac{\partial i_c}{\partial i_b} \cong \left.\frac{\Delta i_c}{\Delta i_b}\right|_{V_{CE}=\text{constant}} \tag{8.15}$$

$$h_{oe} = \frac{\partial i_o}{\partial v_o} = \frac{\partial i_c}{\partial v_{ce}} \cong \left.\frac{\Delta i_c}{\Delta v_{ce}}\right|_{I_B=\text{constant}} \tag{8.16}$$

In each case the symbol Δ refers to a small change in that quantity around the quiescent point of operation. In other words, the *h*-parameters are determined in the region of operation for the applied signal, so that the equivalent circuit will be the most accurate available. The constant values of V_{CE} and I_B in each case refer to a condition that must be met when the various parameters are determined from the characteristics of the transistor. For the common-base and common-collector configurations the proper equation can be obtained by simply substituting the proper values of v_i, v_o, i_i, and i_o. *In Appendix A a list has been provided that relates the hybrid parameters of the three basic transistor configurations.* In other words, if the *h*-parameters for the common-emitter configuration are known, the *h*-parameters for the common-base or common-collector configurations can be found using these tables.

The parameters h_{ie} and h_{re} are determined from the input or base characteristics, while the parameters h_{fe} and h_{oe} are obtained from the output or collector characteristics. Since h_{fe} is usually the parameter of greatest interest, we shall discuss the operations involved with equations, such as Eqs. (8.13) through (8.16), for this parameter first. The first step in determining any of the four hybrid parameters is to find the quiescent point of operation as indicated in Fig. 8.24. In Eq. (8.15) the condition V_{CE} = constant requires that the changes in base current and collector current be taken along a vertical straight line drawn through the Q-point representing a fixed collector-to-emitter voltage. Equation (8.15) then requires that a small change in collector current be divided by the corresponding change in base current. For the greatest accuracy these changes should be made as small as possible.

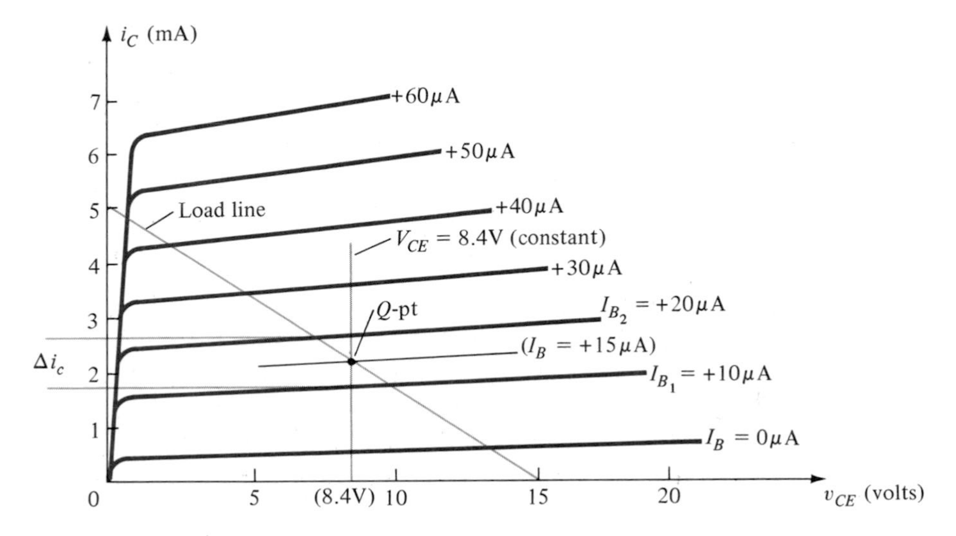

Figure 8.24 h_{fe} determination.

In Fig. 8.24 the change in i_B was chosen to extend from I_{B_1} to I_{B_2} along the perpendicular straight line at V_{CE}. The corresponding change in i_C is then found by drawing the horizontal lines from the intersections of I_{B_1} and I_{B_2} with V_{CE} = constant to the vertical axis. All that remains is to substitute the resultant changes of i_B and i_C into Eq. (8.15); that is,

$$|h_{fe}| = \left.\frac{\Delta i_c}{\Delta i_b}\right|_{V_{CE}=\text{constant}} = \left.\frac{(2.7 - 1.7) \times 10^{-3}}{(20 - 10) \times 10^{-6}}\right|_{V_{CE}=8.4\,\text{V}}$$

$$= \frac{10^{-3}}{10 \times 10^{-6}} = \mathbf{100}$$

In Fig. 8.25 a straight line is drawn tangent to the curve I_B through the Q-point to establish a line I_B = constant as required by Eq. (8.16) for h_{oe}. A change in v_{CE} was then chosen and the corresponding change in i_C is determined by drawing the horizontal lines to the vertical axis at the intersections on the I_B = constant line. Substituting

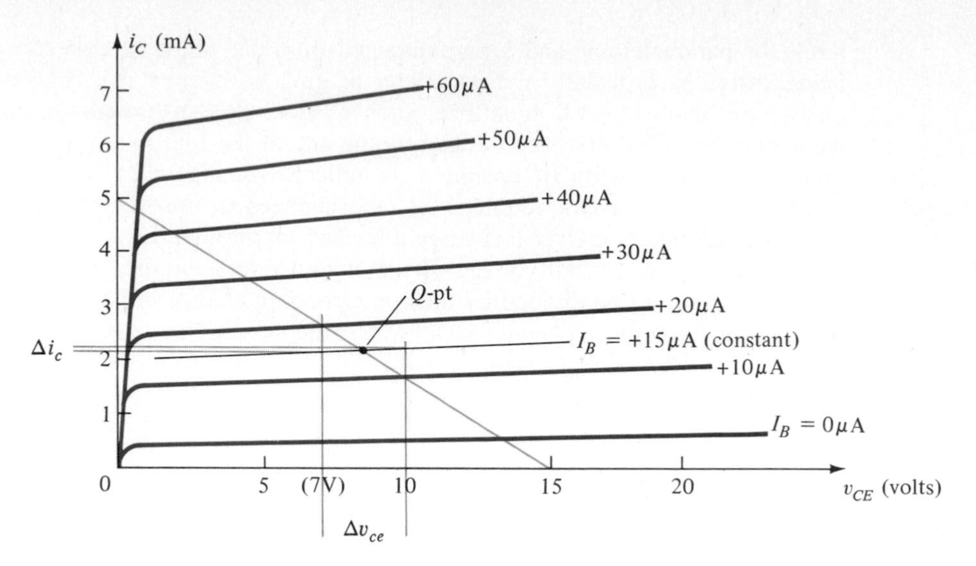

Figure 8.25 h_{oe} determination.

into Eq. (8.16), we get

$$|h_{oe}| = \left.\frac{\Delta i_c}{\Delta v_{ce}}\right|_{I_B=\text{constant}} = \left.\frac{(2.2 - 2.1) \times 10^{-3}}{10 - 7}\right|_{I_B=+15\ \mu\text{A}}$$

$$= \frac{0.1 \times 10^{-3}}{3} = \mathbf{33\ \mu A/V = 33 \times 10^{-6}\ S = 33\ \mu S}$$

To determine the parameters h_{ie} and h_{re} the Q-point must first be found on the input or base characteristics as indicated in Fig. 8.26. For h_{ie}, a line is drawn tangent to the curve $V_{CE} = 8.4$ V through the Q-point, to establish a line V_{CE} = constant as required by Eq. (8.13). A small change in v_{be} was then chosen, resulting in a corresponding change in i_B. Substituting into Eq. (8.13), we get

$$|h_{ie}| = \left.\frac{\Delta v_{be}}{\Delta i_b}\right|_{V_{CE}=\text{constant}} = \left.\frac{(733 - 718) \times 10^{-3}}{(20 - 10) \times 10^{-6}}\right|_{V_{CE}=8.4\ \text{V}}$$

$$= \frac{15 \times 10^{-3}}{10 \times 10^{-6}} = \mathbf{1.5\ k\Omega}$$

The last parameter, h_{re}, can be found by first drawing a horizontal line through the Q-point at $I_B = 15\mu$A. The natural choice then is to pick a change in v_{CE} and find the resulting change in v_{BE} as shown in Fig. 8.27.

Substituting into Eq. (8.14), we get

$$|h_{re}| = \left.\frac{\Delta v_{be}}{\Delta v_{ce}}\right|_{I_B=\text{constant}} = \frac{(733 - 725) \times 10^{-3}}{20 - 0} = \frac{8 \times 10^{-3}}{20} = \mathbf{4 \times 10^{-4}}$$

For the transistor whose characteristics have appeared in Figs. 8.25 through 8.27 the resulting hybrid small-signal equivalent circuit is shown in Fig. 8.28.

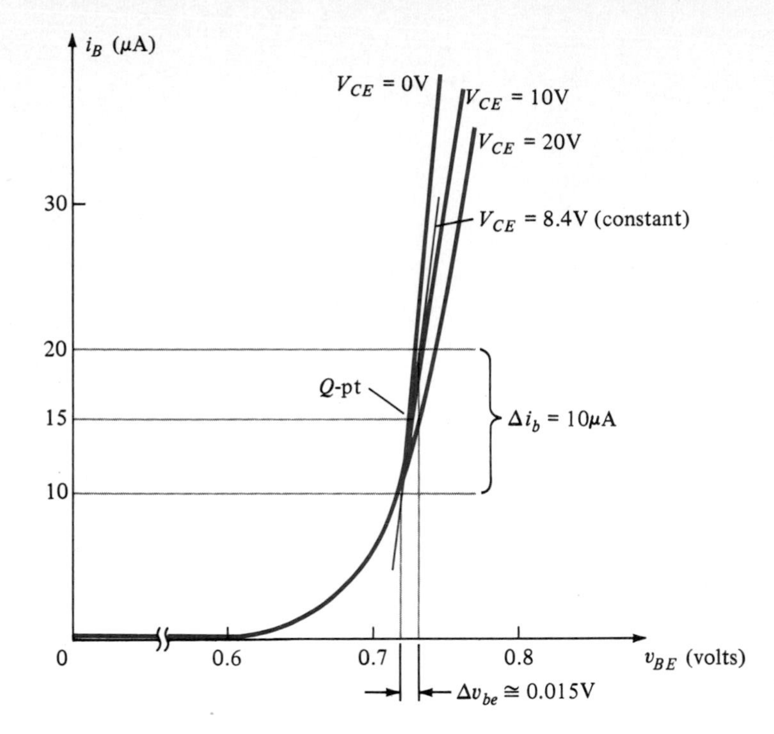

Figure 8.26 h_{ie} determination.

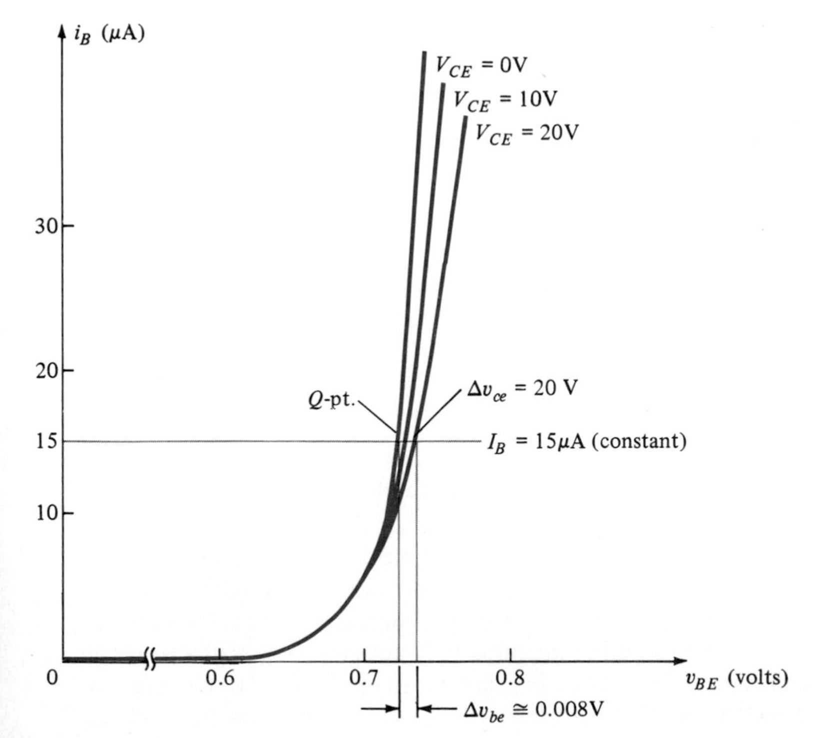

Figure 8.27 h_{re} determination.

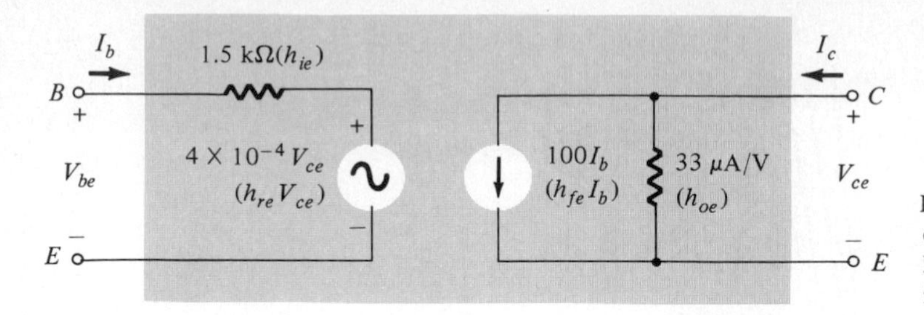

Figure 8.28 Complete hybrid equivalent circuit for a transistor having the characteristics that appear in Figs. 8.25 through 8.28.

As mentioned earlier, the hybrid parameters for the common-base and common-collector configurations can be found using the same basic equations with the proper variables and characteristics.

Typical values for each parameter for the broad range of transistors available today in each of its three configurations are provided in Table 8.1. The minus sign indicates that in Eq. (8.15) as one quantity increased in magnitude, within the change chosen, the other decreased in magnitude.

TABLE 8.1 Typical Parameter Values for the CE, CC, and CB Transistor Configurations

Parameter	*CE*	*CC*	*CB*
h_i	1 kΩ	1 kΩ	20 Ω
h_r	2.5×10^{-4}	$\cong 1$	3.0×10^{-4}
h_f	50	−50	−0.98
h_o	25 μA/V	25 μA/V	0.5 μA/V
$1/h_o$	40 kΩ	40 kΩ	2 MΩ

Note in retrospect (Section 4.4: Transistor Amplifying Action) that the input resistance of the common-base configuration is low, while the output resistance is high. Consider also that the short-circuit current gain is very close to 1. For the common-emitter and common-collector configurations note that the input resistance is much higher than that of the common-base configuration and that the ratio of output to input resistance is about 40:1. Consider also for the common-emitter and common-base configuration that h_r is very small in magnitude. Transistors are available today with values of h_{fe} that vary from 20 to 600. For any transistor the region of operation and conditions under which it is being used will have an effect on the various

h-parameters. The effect of temperature and collector current and voltage on the h-parameters is discussed in Section 8.8.

8.8 VARIATIONS OF TRANSISTOR PARAMETERS

There are a large number of curves that can be drawn to show the variations of the h-parameters with temperature, frequency, voltage, and current. The most interesting and useful at this stage of the development include the h-parameter variations with junction temperature and collector voltage and current.

In Fig. 8.29 the effect of the collector current on the h-parameter has been indicated. Take careful note of the logarithmic scale on the vertical and horizontal axes. Logarithmic scales will be examined in Chapter 11. The parameters have all been normalized to unity so that the relative change in magnitude with collector current can easily be determined. On every set of curves, such as in Fig. 8.30, the operating point at which the parameters were found is always indicated. For this particular situation the quiescent point is at the intersection of $V_{CE} = 5.0$ V and $I_C = 1.0$ mA. Since the frequency and temperature of operation will also affect the h-parameters, these quantities are also indicated on the curves. At 0.1 mA, h_{fe} is about 0.5 or 50% of its value at 1.0 mA, while at 3 mA, it is 1.5 or 150% of that value.

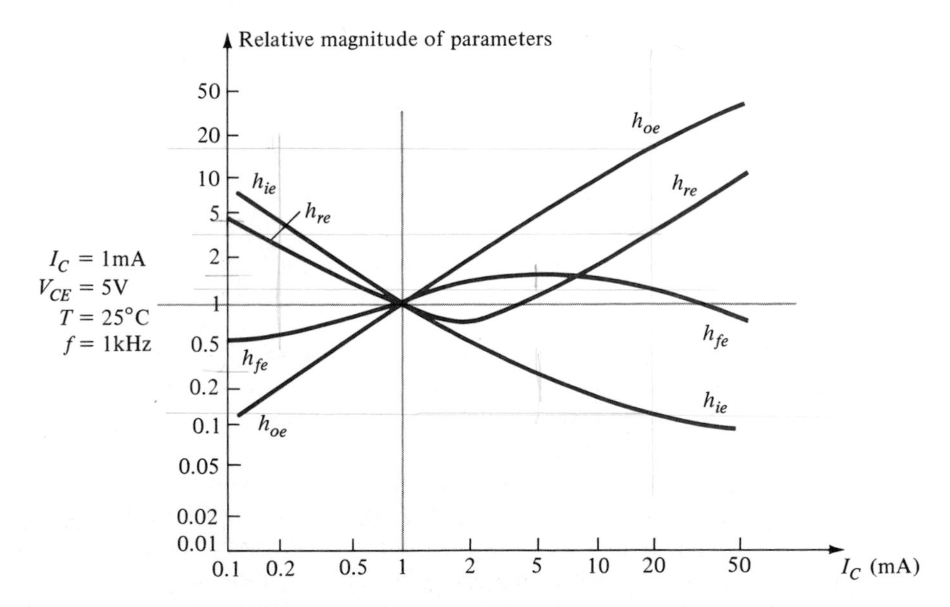

Figure 8.29 Hybrid parameter variations with collector current.

In other words, if $h_{fe} = 50$ at $I_C = 1.0$ mA, h_{fe} has changed from a value of $0.5(50) = 25$ to $1.5(50) = 75$ with a change of I_C from 0.1 mA to 3 mA. Consider, however, the point of operation at $I_C = 50$ mA. The magnitude of h_{re} is now approximately 11 times that at the defined Q-point, a magnitude that may not permit eliminating this parameter from the equivalent circuit. The parameter h_{oe} is approximately 35 times the normalized value. This increase in h_{oe} will decrease the magnitude of the output resistance of the transistor to a point where it may approach the magnitude of the load resistor. There would then be no justification in eliminating h_{oe} from the equivalent circuit on an approximate basis.

In Fig. 8.30 the variation in magnitude of the h-parameters on a normalized basis has been indicated with change in collector voltage. This set of curves was normalized at the same operating point of the transistor discussed in Fig. 8.29 so that a comparison between the two sets of curves can be made. Note that h_{ie} and h_{fe} are relatively steady in magnitude, while h_{oe} and h_{re} are much larger to the left and right of the chosen operating point. In other words, h_{oe} and h_{re} are much more sensitive to changes in collector voltage than are h_{ie} and h_{fe}.

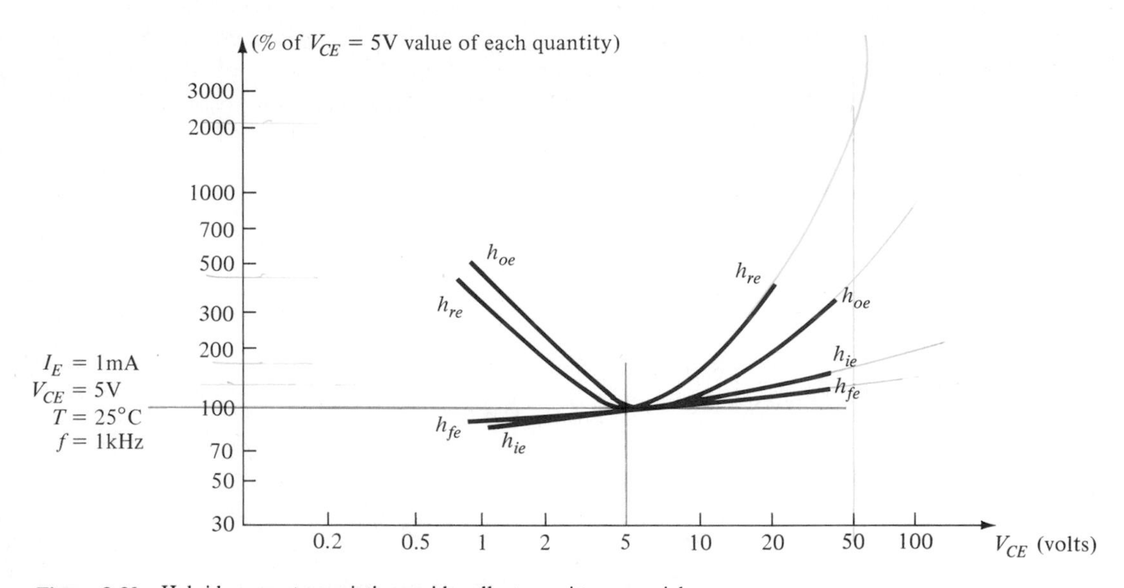

Figure 8.30 Hybrid parameter variations with collector-emitter potential.

It is interesting to note from Figs. 8.29 and 8.30 that the value of h_{fe} appears to change the least. Therefore, the specific value of current gain, whether h_{fe} or β, can, on an approximate and relative basis, be considered fairly constant for the range of collector current and voltage.

The value of $h_{ie} = \beta r_e$ does vary considerably with collector current as one might expect due to the sensitivity of r_e to emitter ($I_E \cong I_C$) current. It is therefore a quantity that should be determined as close to operating conditions as possible. For values below the specified V_{CE}, h_{re} is fairly constant, but it does increase measurably for higher values. It is indeed fortunate that for most applications the magnitude of h_{re} and

h_{oe} are such that they can usually be ignored. They are quite sensitive to collector current and collector-to-emitter voltage.

In Fig. 8.31 the variation in h-parameters has been plotted for changes in junction temperature. The normalization value is taken to be room temperature: $T = 25°C$. The horizontal scale is a linear scale rather than a logarithmic scale as was employed for Figs. 8.29 and 8.30. In general, all the parameters increase in magnitude with temperature. The parameter least affected, however, is h_{oe}, while the input impedance h_{ie} changes at the greatest rate. The fact that h_{fe} will change from 50% of its normalized value at $-50°C$ to 150% of its normalized value at $+150°C$ indicates clearly that the operating temperature must be carefully considered in the design of transistor circuits.

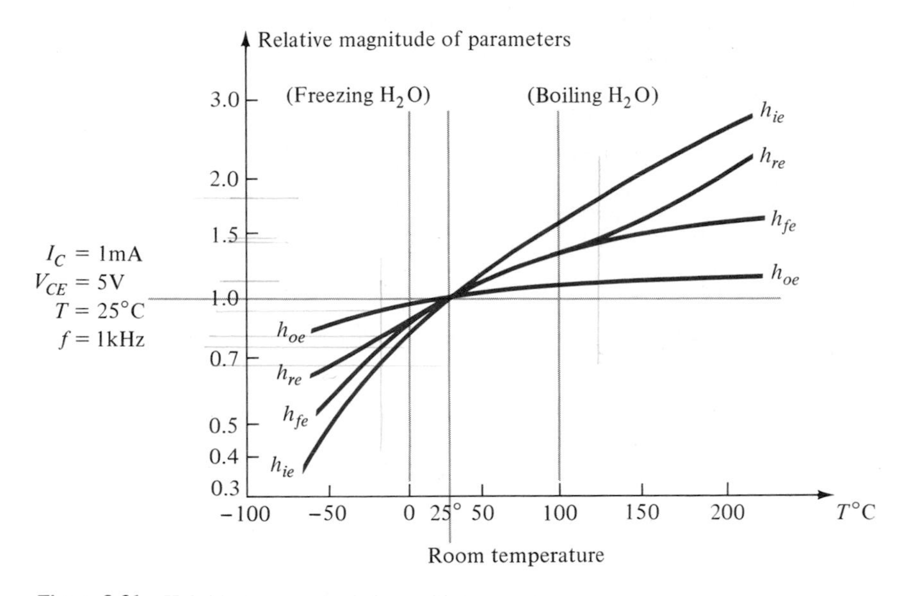

Figure 8.31 Hybrid parameter variations with temperature.

PROBLEMS

§ 8.2

1. Define, in general terms, the primary function of a "model" of a device or system.

§ 8.3

2. (a) Describe the difference between the hybrid and r_e models.
   (b) For each model, list the conditions under which it should be used.

§ 8.4

3. (a) Redraw Fig. 8.5 with the complete hybrid equivalent model inserted between the proper terminals.
   (b) Repeat part (a) with the approximate equivalent model for the common-emitter configuration.

4. Given the common-emitter configuration of Fig. 8.32,
   (a) Sketch the ac equivalent but do not substitute the transistor equivalent model. Leave it in the form of Fig. 8.5. Define Z_i, Z_o, v_i, v_o, i_i, and i_o.
   (b) Redraw the network of part (a) with the complete common-emitter hybrid equivalent circuit.
   (c) Redraw the network of part (a) with the approximate equivalent model.

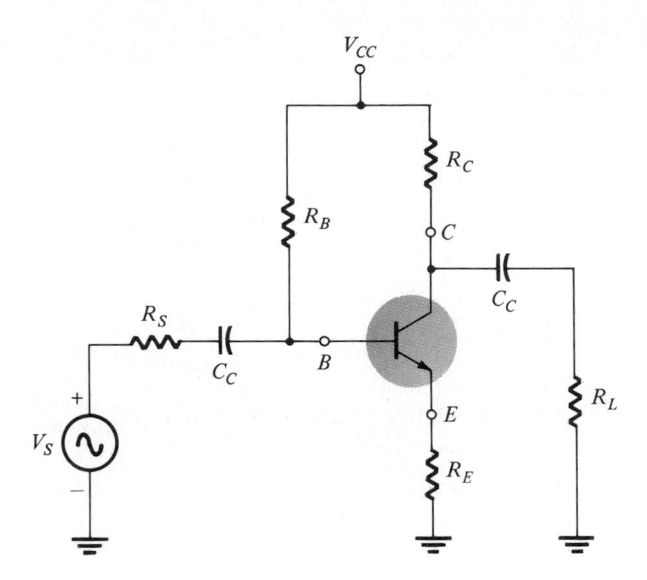

Figure 8.32 Hybrid parameter variations with temperature.

5. Given the common-base configuration of Fig. 8.33:
   (a) Sketch the ac equivalent but do not substitute the transistor equivalent model. Leave it in the form of Fig. 8.5. Define Z_i, Z_o, v_i, v_o, i_i, and i_o.
   (b) Redraw the network of part (a) with the complete common-base hybrid equivalent circuit.
   (c) Redraw the network of part (a) with the approximate equivalent model.

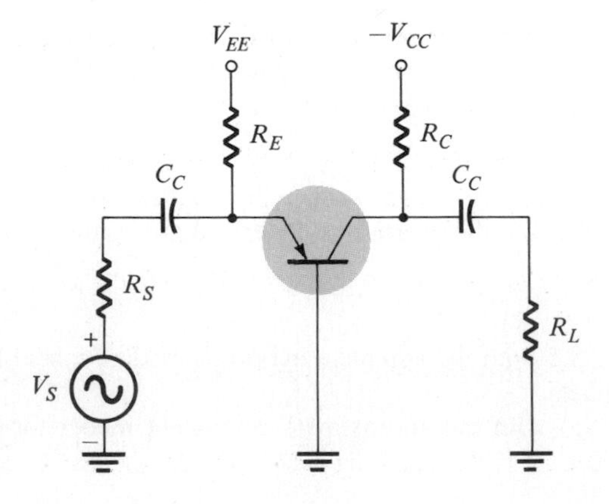

Figure 8.33

§ 8.5

6. For the network of Fig. 8.32:
(a) Repeat part (a) of Problem 4.
(b) Redraw the network of part (a) with the r_e model substituted between the appropriate terminals.

7. For the network of Fig. 8.33:
(a) Repeat part (a) of Problem 4.
(b) Redraw the network of part (a) with the r_e model substituted between the appropriate terminals.

§ 8.6

8. Given the typical values of $h_{re} = 2 \times 10^{-4}$ and $A_v = 160$, is it a good approximation to ignore the effects of h_{re} in the ac analysis of the system? What is the percent difference between the levels of I_b with and without the effects of h_{re}? Use

$$\% \text{ difference} = \frac{I_b(\text{w.o. } h_{re}) - I_b(\text{w. } h_{re})}{I_b(\text{w.o. } h_{re})} \times 100\%$$

9. Given the typical values of $R_L = 2.2\ k\Omega$ and $h_{oe} = 20\ \mu\text{S}$, is it a good approximation to ignore the effects of $1/h_{oe}$? What is the percent difference in load resistance using the following equation?

$$\% \text{ difference} = \frac{R_L(\text{w.o. } 1/h_{oe}) - R_L(\text{w.} 1/h_{oe})}{R_L(\text{w.o. } 1/h_{oe})} \times 100\%$$

10. Repeat Problem 8 for $h_{re} = 3.8 \times 10^{-4}$ and $A_v = 280$.

11. Repeat Problem 9 for $R_L = 6.8\ \text{k}\Omega$ and $h_{oe} = 25\ \mu\text{S}$.

§ 8.7

12. (a) Using the characteristics of Fig. 8.24, determine h_{fe} at $I_C = 6$ mA and $V_{CE} = 5$ V.
(b) Repeat part (a) at $I_C = 1$ mA and $V_{CE} = 15$ V.

13. (a) Using the characteristics of Fig. 8.25, determine h_{oe} at $I_C = 6$ mA and $V_{CE} = 5$ V.
(b) Repeat part (a) at $I_C = 1$ mA and $V_{CE} = 15$ V.

14. (a) Using the characteristics of Fig. 8.26, determine h_{ie} at $I_B = 20\ \mu$A and $V_{CE} = 20$ V.
(b) Repeat part (a) at $I_B = 5\ \mu$A and $V_{CE} = 10$ V.

15. (a) Using the characteristics of Fig. 8.27, determine h_{re} at $I_B = 20\ \mu$A and $V_{CE} = 20$ V.
(b) Repeat part (a) at $I_B = 30\ \mu$A and $V_{CE} = 10$ V.

16. Using the characteristics of Figs. 8.24 and 8.26, determine the approximate hybrid equivalent model at $I_B = 25\ \mu$A and $V_{CE} = 12.5$ V.

17. Determine the r_e model at $I_B = 25\ \mu$A and $V_{CE} = 12.5$ V using the characteristics of Figs. 8.24 and 8.26.

18. Using the results of Section 8.7, determine the r_e model at $I_B = 15\ \mu$A and $V_{CE} = 8.4$ V.

§ 8.8

19. (a) Using Fig. 8.29 determine the magnitude of the percent change in h_{fe} for an I_C change from 0.2 mA to 1 mA using the equation

$$\% \text{ change} = \left| \frac{h_{fe}(\text{low } I_C) - h_{fe}(\text{high } I_C)}{h_{fe}(\text{low } I_C)} \right| \times 100\%$$

(b) Repeat part (a) for an I_C change from 1 mA to 5 mA.
(c) Repeat part (a) for an I_C change from 0.2 mA to 5 mA.

20. Repeat problem 19 for h_{ie} (same changes in I_C).

21. (a) If $h_{oe} = 20\ \mu\text{S}$ at $I_C = 1$ mA on Fig. 8.29, what is the approximate value of h_{oe} at $I_C = 0.1$ mA?
(b) Determine its resistive value at 0.1 mA and compare to a resistive load of 6.8 kΩ. Is it a good approximation to ignore the effects of $1/h_{oe}$ in this case?

22. (a) If $h_{oe} = 20\ \mu\text{S}$ at $I_C = 1$ mA on Fig. 8.29, what is the approximate value of h_{oe} at $I_C = 10$ mA?
(b) Determine its resistive value at 10 mA and compare to a resistive load of 6.8 kΩ. Is it a good approximation to ignore the effects of $1/h_{oe}$ in this case?

23. (a) If $h_{re} = 2 \times 10^{-4}$ at $I_C = 1$ mA on Fig. 8.29, determine the approximate value of h_{re} at 0.1 mA.
(b) Using the value of h_{re} determined in part (a), can h_{re} be ignored as a good approximation if $A_v = 210$?

9

BJT Small-Signal Analysis

9.1 INTRODUCTION

The transistor models introduced in Chapter 8 will now be used to perform a small-signal ac analysis of a number of standard transistor network configurations. The networks to be analyzed represent the majority of those appearing in practice today. Modifications of the standard configurations will be relatively easy to examine once the content of this chapter is reviewed and understood.

Both the hybrid and r_e models will be employed to demonstrate the similarities in analysis and the common threads that exist between the models. Except for Section 9.10, the approximate hybrid equivalent circuit is employed throughout the chapter to ensure that the approach is not clouded by extensive mathematical manipulations and calculations. Section 9.10 demonstrates the effects of the complete model on the analysis and the results obtained.

The effects of an applied load R_L and a source resistance R_S are examined in Section 9.9, and a computer analysis of a few standard configurations is provided in Section 9.11. No attempt is made to introduce the BASIC computer language employed in the analysis. The presentation is included simply to demonstrate the versatility of the computer and how effective it can be if its capabilities are understood and properly applied.

9.2 COMMON-EMITTER FIXED-BIAS CONFIGURATION

The first configuration to be analyzed in detail is the common-emitter *fixed-bias* network of Fig. 9.1. Note that the input signal v_i is applied to the base of the transistor while the output v_o is off the collector. In addition, recognize that the input current I_i is not the base current but the source current, while the output current I_o is the collector current. The small-signal ac analysis begins by removing the dc effects of V_{CC} and replacing the dc blocking capacitors C_1 and C_2 by short-circuit equivalents, resulting in the network of Fig. 9.2.

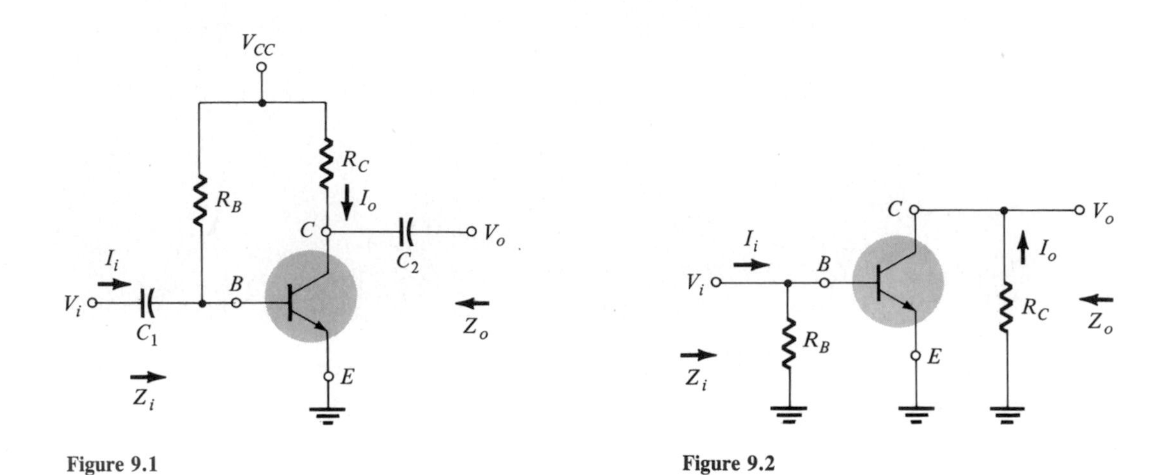

Figure 9.1

Figure 9.2

Note in Fig. 9.2 that the common ground of the dc supply and the emitter resistor permits the relocation of R_B and R_C in parallel with the input and output sections of the transistor, respectively. In addition, note the placement of the important network parameters Z_i, Z_o, I_i, and I_o on the redrawn network.

Substituting the approximate hybrid small-signal equivalent circuit for the transistor of Fig. 9.2 will result in the network of Fig. 9.3. Performing the analysis yields the following results.

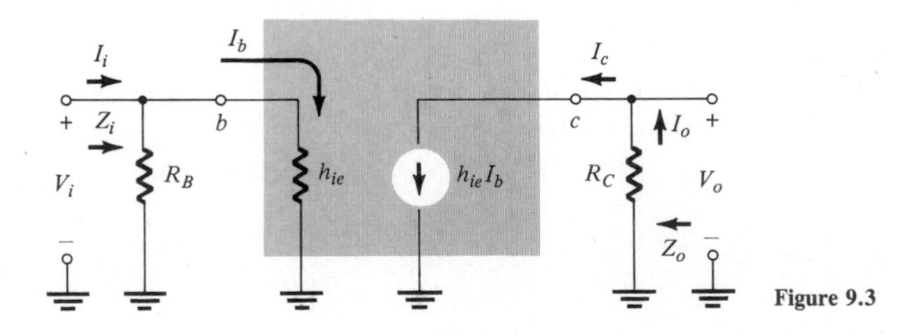

Figure 9.3

Z_i: Figure 9.3 clearly reveals that

$$\boxed{Z_i = R_B \| h_{ie}} \quad \text{ohms} \tag{9.1}$$

For situations where R_B is greater than h_{ie} by more than a factor of 10 (a typical situation), the following approximation is frequently applied:

$$\boxed{Z_i \cong h_{ie}} \quad \text{ohms} \tag{9.2}$$

Using the r_e model equivalence will result in the following equation for Z_i:

$$\boxed{Z_i \cong \beta r_e} \quad \text{ohms} \tag{9.3}$$

Z_o: The output impedance of any network is defined as the impedance Z_o determined when $V_i = 0$. For Fig. 9.3, when $V_i = 0$, I_i and therefore $I_b = 0$ and $h_{fe} I_b = 0$, resulting in an open-circuit equivalence for the current source. The result is

$$\boxed{Z_o = R_C} \quad \text{ohms} \tag{9.4}$$

for both the hybrid and r_e methods.

A_v: The voltage gain $A_v = V_o/V_i$ is determined by first assuming that $R_B \gg h_{ie}$ to permit the approximation $I_b \cong I_i$ and then solving for V_o.

$$V_o = -I_o R_C$$

The minus sign specifies that the polarity of V_o is opposite to that defined by the indicated direction of I_o. Substituting $I_o = h_{fe} I_b$ followed by $I_b = I_i$ will result in

$$V_o = -h_{fe} I_b R_C = -h_{fe} I_i R_C$$

but

$$I_i = \frac{V_i}{h_{ie}}$$

and

$$V_o = -h_{fe}\left(\frac{V_i}{h_{ie}}\right) R_C$$

or

$$\boxed{A_v = \frac{V_o}{V_i} = -\frac{h_{fe} R_C}{h_{ie}}} \tag{9.5}$$

The negative sign in the resulting equation reveals that a 180° phase shift occurs between the input and output signals, as shown in Fig. 9.4.

Substituting $h_{fe} = \beta$ and $h_{ie} = \beta r_e$ for the r_e model will result in

$$A_v = -\frac{h_{fe} R_C}{h_{ie}} = -\frac{\beta R_C}{\beta r_e}$$

and

$$\boxed{A_v = -\frac{R_C}{r_e}} \tag{9.6}$$

a very convenient form for the gain. Note the absence of β in Eq. (9.6), although we recognize that β must be utilized to determine r_e.

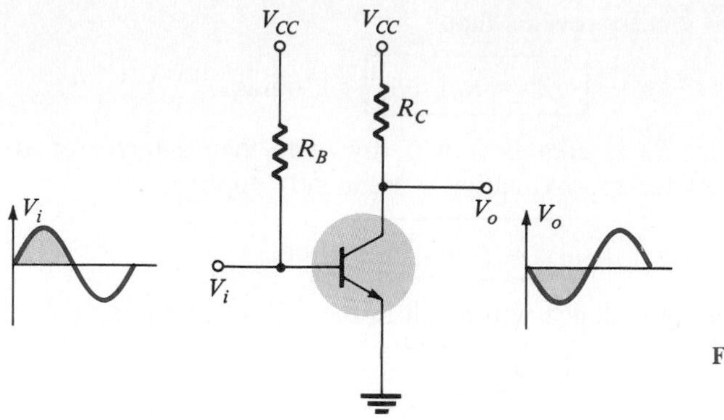

Figure 9.4

A_i: The current gain $A_i = I_o/I_i$ is determined in the following manner:

$$I_o = h_{fe} I_b \cong h_{fe} I_i$$

and

$$\boxed{A_i = \frac{I_o}{I_i} \cong h_{fe}} \tag{9.7}$$

or for the r_e model,

$$\boxed{A_i \cong \beta} \tag{9.8}$$

All the parameters of current importance for the CE fixed-bias configuration have now been determined. Note the relative simplicity of moving from one model to the other if the need arises. Simply recall that $h_{fe} = \beta$ and $h_{ie} = \beta r_e$ and both forms are immediately available. The examples that follow further demonstrate the interrelationship between the two models.

EXAMPLE 9.1

Determine Z_i, Z_o, A_v, and A_i for the network of Fig. 9.5.

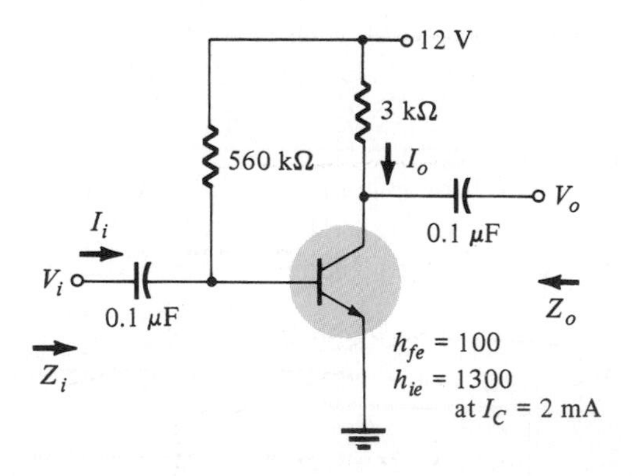

Figure 9.5

Solution:

Hybrid Parameters (Approximate):

Z_i:

$$R_B = 560\text{ k}\Omega \gg h_{ie} = 1300\ \Omega$$

From Eq. (9.2),

$$Z_i \cong h_{ie} = \mathbf{1300\ \Omega}$$

Z_o [Eq. (9.4)]:

$$Z_o = R_C = \mathbf{3\ k\Omega}$$

A_v [Eq. (9.5)]:

$$A_v = \frac{-h_{fe}R_C}{h_{ie}} = \frac{-(100)(3\text{ k}\Omega)}{1.3\text{ k}\Omega}$$

$$= \mathbf{-230.77}$$

A_i [Eq. (9.7)]:

$$A_i \cong h_{fe} = \mathbf{100}$$

r_e Model:

DC:

$$I_B = \frac{V_{CC} - V_{BE}}{R_B} = \frac{12 - 0.7}{560\text{ k}\Omega} \cong 20\ \mu\text{A}$$

$$I_C = \beta I_B = (100)(20\ \mu\text{A}) = 2\text{ mA} = I_E$$

$$r_e = \frac{26\text{ mV}}{I_E} = \frac{26}{2} = 13\ \Omega$$

Z_i [Eq. (9.3)]:

$$Z_i \cong \beta r_e = (100)(13) = \mathbf{1300\ \Omega}$$

Z_o [Eq. (9.4)]:

$$Z_o = R_C = \mathbf{3\ k\Omega}$$

A_v [Eq. (9.6)]:

$$A_v = -\frac{R_C}{r_e} = -\frac{3\text{ k}\Omega}{13} = \mathbf{-230.77}$$

A_i [Eq. (9.8)]:

$$A_i \cong \beta = \mathbf{100}$$

Effect of h_{oe}: If h_{oe} were included in the model of Fig. 9.3, the output network would appear as shown in Fig. 9.6. Z_i would remain unaffected but Z_o would become

$$\boxed{Z_o = R_C \left\| \frac{1}{h_{oe}}\right.} \qquad (9.9)$$

and

$$V_o = -h_{fe}I_b\left(R_C \left\| \frac{1}{h_{oe}}\right.\right)$$

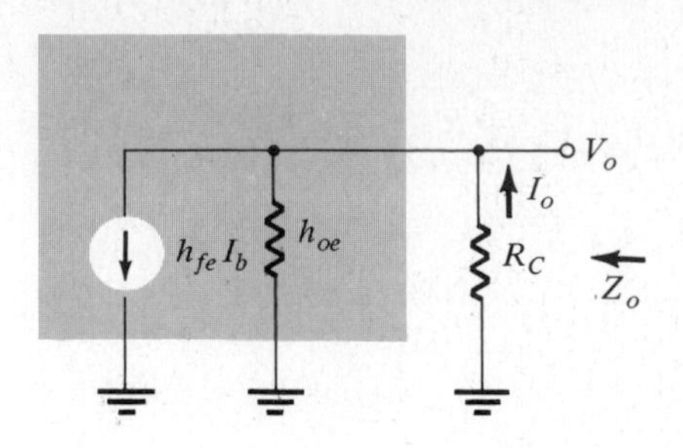

Figure 9.6

resulting in

$$A_v = \frac{-h_{fe}}{h_{ie}}\left(R_C \,\Big\|\, \frac{1}{h_{oe}}\right) \tag{9.10}$$

Note how the equation for each quantity is altered only by changing R_C to $R_C \| 1/h_{oe}$.

For A_i we must first apply the current-divider rule to the network of Fig. 9.6, to obtain

$$I_o = \frac{(1/h_{oe})h_{fe}I_b}{1/h_{oe} + R_C} = \frac{h_{fe}I_i}{1 + h_{oe}R_C}$$

and

$$A_i = \frac{I_o}{I_i} = \frac{h_{fe}}{1 + h_{oe}R_C} \tag{9.11}$$

In each case the effect of h_{oe} is to reduce the quantity of interest. However, since $1/h_{oe}$ is usually much greater than R_C, its effect is frequently ignored and the earlier equations applied. The modified equations for the r_e model can quickly be determined from the model relationships.

EXAMPLE 9.2

Determine the effect of h_{oe} on the quantities of Example 9.1 if $h_{oe} = 20\ \mu\text{S}$ and $1/h_{oe} = 50\ \text{k}\Omega$.

Solution:

$\boldsymbol{Z_i}$**:** The same.

$\boldsymbol{Z_o}$ [Eq. (9.9)]:

$$Z_o = R_C \,\Big\|\, \frac{1}{h_{oe}} = 3\ \text{k}\Omega \| 50\ \text{k}\Omega = \mathbf{2.83\ k\Omega} \text{ vs } 3\ \text{k}\Omega$$

$\boldsymbol{A_v}$ [Eq. (9.10)]:

$$A_v = -\frac{h_{fe}}{h_{ie}}\left(R_C \,\Big\|\, \frac{1}{h_{oe}}\right) = \frac{-100(2.83\ \text{k}\Omega)}{1.3\ \text{k}\Omega}$$

$$= \mathbf{-217.69} \text{ vs. } -230.77$$

$\boldsymbol{A_i}$ [Eq. (9.11)]:

$$A_i = \frac{h_{fe}}{1 + h_{oe}R_C} = \frac{100}{1 + (20 \times 10^{-6})(3 \times 10^3)} = \frac{100}{1.06}$$

$$= \mathbf{94.34} \text{ vs. } 100$$

9.3 VOLTAGE-DIVIDER BIAS

The next configuration to be analyzed is the *voltage-divider* bias network of Fig. 9.7. Recall that the name of the configuration is a result of the voltage-divider bias at the input side to determine the dc level of V_B.

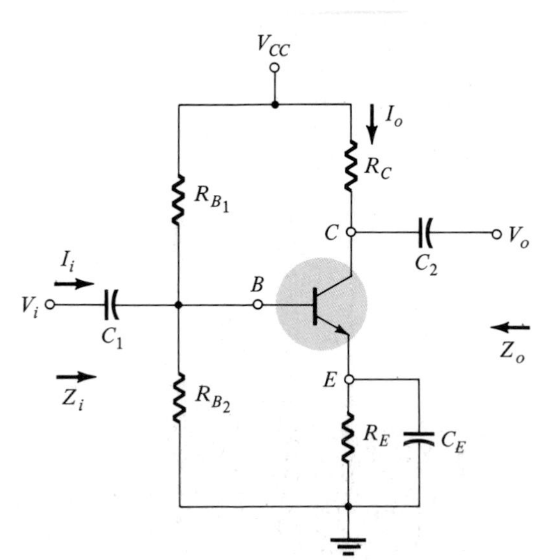

Figure 9.7

Substituting the approximate hybrid equivalent circuit will result in the network of Fig. 9.8. Note the absence of R_E due to the low-impedance shorting effect of C_E. That is, at the frequency (or frequencies) of operation, the reactance of the capacitor is so small compared to R_E and the remaining parameters of the network that it is treated as a short circuit across R_E.

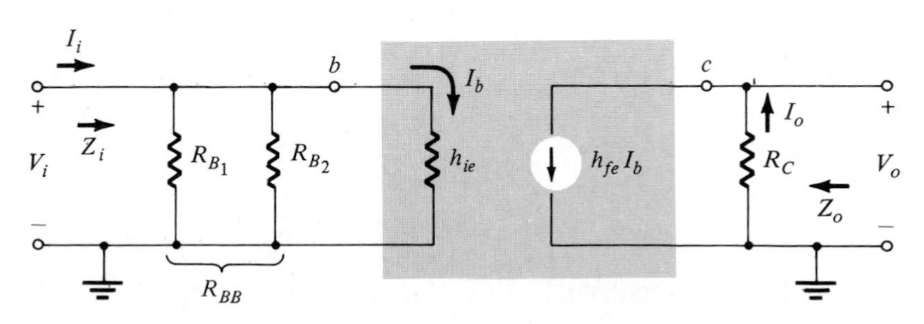

Figure 9.8

The parallel combination of R_{B_1} and R_{B_2} is defined by

$$R_{BB} = R_{B_1} \| R_{B_2} = \frac{R_{B_1} R_{B_2}}{R_{B_1} + R_{B_2}} \tag{9.12}$$

Z_i: From Fig. 9.8,

$$Z_i = R_{BB} \| h_{ie} \tag{9.13}$$

and for the r_e model,

$$Z_i = R_{BB} \| \beta r_e \tag{9.14}$$

Z_o: From Fig. 9.8,

$$Z_o = R_C \tag{9.15}$$

for both models.

A_v:

$$\begin{aligned} V_o &= -I_o R_C = -h_{fe} I_b R_C \\ &= -h_{fe} \left(\frac{V_i}{h_{ie}}\right) R_C \\ &= -\frac{h_{fe}}{h_{ie}} R_C V_i \end{aligned}$$

and

$$A_v = \frac{V_o}{V_i} = -\frac{h_{fe} R_C}{h_{ie}} \tag{9.16}$$

with the following for the r_e model:

$$A_v = -\frac{R_C}{r_e} \tag{9.17}$$

as obtained for the fixed-bias configuration.

A_i: Since the resistance R_{BB} is often too close in magnitude to h_{ie} to be ignored, the effect of R_{BB} should be included in the current gain equation. Referencing Fig. 9.8, we have

$$I_b = \frac{R_{BB} I_i}{R_{BB} + h_{ie}} \qquad \text{(current-divider rule)}$$

or

$$\frac{I_b}{I_i} = \frac{R_{BB}}{R_{BB} + h_{ie}}$$

For the output side,

$$I_o = h_{fe} I_b$$

or
$$\frac{I_o}{I_b} = h_{fe}$$

The current gain

$$A_i = \frac{I_o}{I_i} = \frac{I_o}{I_b}\frac{I_b}{I_i}$$

$$= h_{fe}\frac{R_{BB}}{R_{BB} + h_{ie}}$$

and
$$\boxed{A_i = \frac{R_{BB}h_{fe}}{R_{BB} + h_{ie}}} \qquad (9.18)$$

If $R_{BB} \gg h_{ie}$,

$$A_i \cong \frac{R_{BB}h_{fe}}{R_{BB}} = h_{fe}$$

as obtained earlier.

For the r_e model,

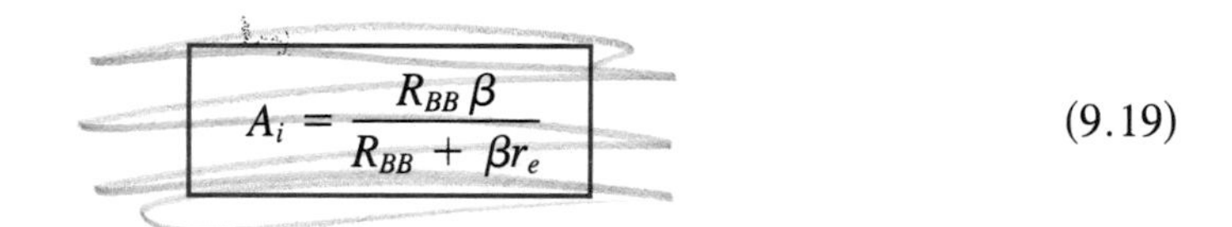

$$\boxed{A_i = \frac{R_{BB}\,\beta}{R_{BB} + \beta r_e}} \qquad (9.19)$$

EXAMPLE 9.3

Determine Z_i, Z_o, A_v, and A_i for the network of Fig. 9.9 using the r_e model approach.

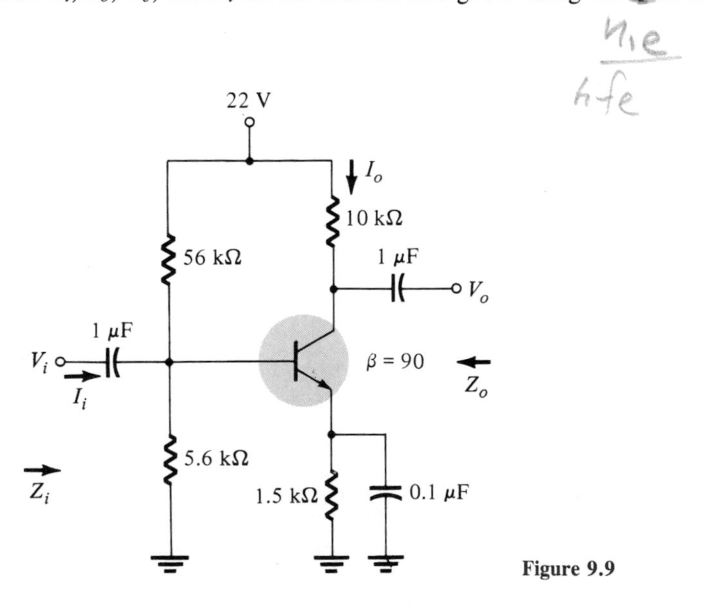

Figure 9.9

Solution:

DC:

$$V_{BB} = \frac{R_{B2}}{R_{B1} + R_{B2}} V_{CC} = \frac{5.6\text{ k}\Omega(22)}{56\text{ k}\Omega + 5.6\text{ k}\Omega} = 2\text{ V}$$

$$R_{BB} = R_{B1} \| R_{B2} = 56\text{ k}\Omega \| 5.6\text{ k}\Omega = 5.09\text{ k}\Omega$$

$$I_B = \frac{V_{BB} - V_{BE}}{R_{BB} + (1 + \beta)R_E} = \frac{2 - 0.7}{5.09\text{ k}\Omega + (1 + 90)1.5\text{ k}\Omega}$$

$$= 9.18\ \mu\text{A}$$

$$I_C = \beta I_B = (90)(9.18\ \mu\text{A}) = 0.826\text{ mA} = I_E$$

$$r_e = \frac{26\text{ mV}}{I_E} = \frac{26}{0.826} \cong 31.5\ \Omega$$

Z_i**:** From Eq. (9.12):

$$R_{BB} = R_{B1} \| R_{B2} = 56\text{ k}\Omega \| 5.6\text{ k}\Omega = 5.09\text{ k}\Omega$$

From Eq. (9.14):

$$Z_i = R_{BB} \| \beta r_e = 5.09\text{ k}\Omega \| (90)(31.5) = 5.09\text{ k}\Omega \| 2835\ \Omega$$

$$= \mathbf{1.821\text{ k}\Omega}$$

Z_o [Eq. (9.15)]:

$$Z_o = R_C = \mathbf{10\text{ k}\Omega}$$

A_v [Eq. (9.17)]:

$$A_v = -\frac{R_C}{r_e} = -\frac{10\text{ k}\Omega}{31.5} = \mathbf{-317.5}$$

A_i [Eq. (9.19)]:

$$A_i = \frac{R_{BB}\,\beta}{R_{BB} + \beta r_e} = \frac{5.09\text{ k}\Omega(90)}{5.09\text{ k}\Omega + (90)(31.5)} = \mathbf{57.8}$$

The same results would have been obtained using

$$h_{fe} = \beta = 90$$

$$h_{ie} = \beta r_e = 2.835\text{ k}\Omega$$

and the appropriate equations.

Effect of h_{oe}: Since h_{oe} appears in parallel with R_C in the same manner as in Fig. 9.6, the equations for Z_o and A_v are modified in the same manner.

$$\boxed{Z_o = R_C \left\| \frac{1}{h_{oe}} \right.} \tag{9.20}$$

and

$$\boxed{A_v = -\frac{h_{fe}}{h_{ie}}\left(R_C \left\| \frac{1}{h_{oe}} \right.\right)} \tag{9.21}$$

For A_i: $$I_o = \frac{(1/h_{oe})h_{fe}I_b}{1/h_{oe} + R_C} \quad \text{(current-divider rule)}$$

as obtained earlier, but the effects of R_{BB} result in the following equation:

$$A_i = \frac{I_o}{I_i} = \frac{I_o}{I_b}\frac{I_b}{I_i} = \frac{(1/h_{oe})h_{fe}}{1/h_{oe} + R_C}\frac{R_{BB}}{R_{BB} + h_{ie}}$$

and $$\boxed{A_i = \frac{R_{BB}h_{fe}}{(1 + h_{oe}R_C)(R_{BB} + h_{ie})}} \tag{9.22}$$

There is no general mechanism for including the effects of the output resistance $(1/h_{oe})$ using the r_e model—it is simply assumed that it is too large a quantity to be of any consequence to the analysis. However, substituting $h_{oe} = 20\ \mu\text{S}$ or $1/h_{oe} = 50\ \text{k}\Omega$ into the equations of Ex. 9.3 will have a definite impact since $1/h_{oe}$ and R_C are related by a 5:1 ratio.

9.4 CE UNBYPASSED EMITTER-BIAS CONFIGURATION

The networks examined in this section include an emitter resistor unbypassed in the ac domain. The most fundamental of such configurations appears in Fig. 9.10. Substituting the approximate hybrid equivalent model will result in the configuration of Fig. 9.11.

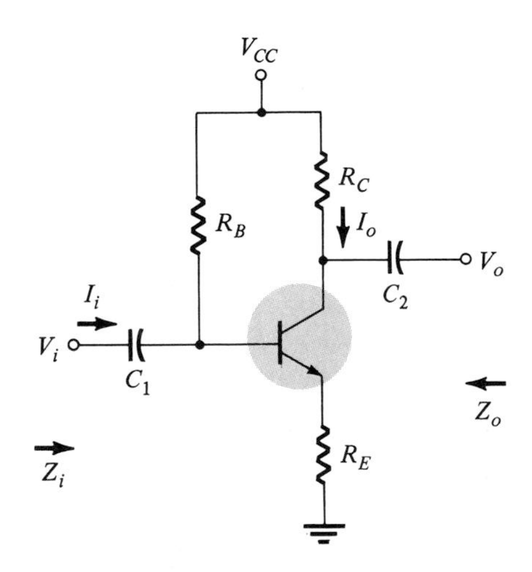

Figure 9.10

Applying Kirchhoff's voltage law to the input side of Fig. 9.11 will result in

$$V_i = I_b h_{ie} + (1 + h_{fe})I_b R_E$$

and $$Z_b = \frac{V_i}{I_b} = h_{ie} + (1 + h_{fe})R_E$$

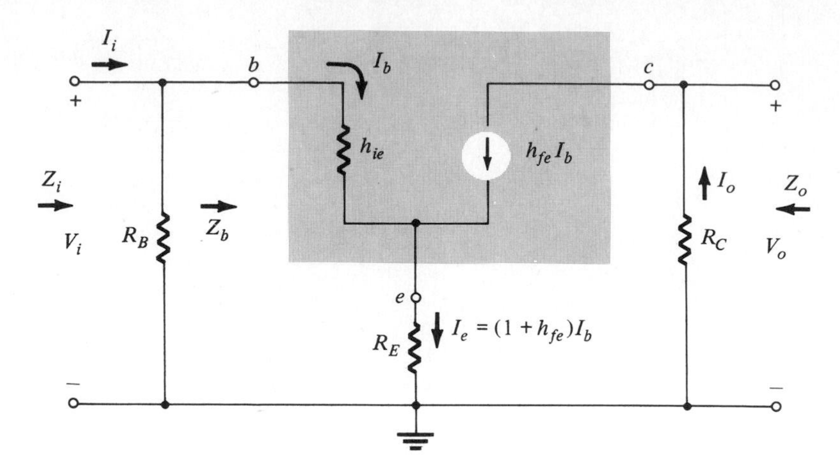

Figure 9.11

The result as displayed in Fig. 9.12 reveals that the input impedance to a transistor with an unbypassed resistor R_E is determined by

$$\boxed{Z_b = h_{ie} + (1 + h_{fe})R_E} \tag{9.23}$$

Since h_{fe} is normally much greater than 1, the equation reduces to

$$\boxed{Z_b = h_{ie} + h_{fe} R_E} \tag{9.24}$$

In most instances $h_{fe} R_E$ is also much greater than h_{ie}, resulting in the following approximation for the majority of situations:

$$\boxed{Z_b \cong h_{fe} R_E} \tag{9.25}$$

For the r_e model of Fig. 9.13, the following format is normally applied:

$$\boxed{Z_b = \beta(r_e + R_E)} \tag{9.26}$$

as derived from Eq. (9.23).

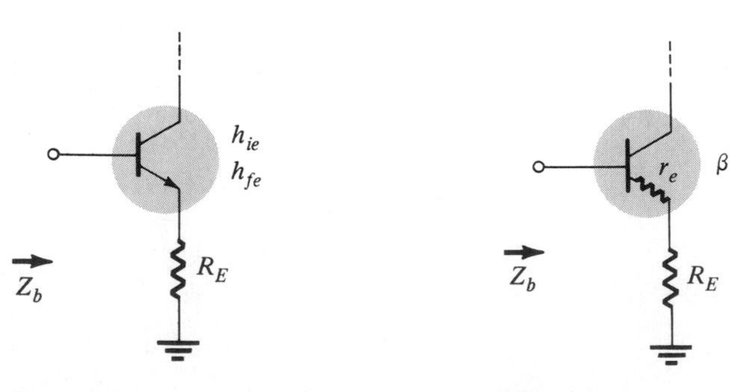

Figure 9.12

Figure 9.13

Assuming that $\beta \gg 1$ equation (9.26) then becomes

$$\boxed{Z_b \cong \beta R_E} \tag{9.27}$$

Z_i: Returning to Fig. 9.11, we have

$$\boxed{Z_i = R_B \| Z_b} \tag{9.28}$$

with Z_b defined as above for both forms.

Z_o: With V_i set to zero, $I_b = 0$ and $h_{fe}I_b$ can be replaced by an open-circuit equivalent. The result is

$$\boxed{Z_o = R_C} \tag{9.29}$$

for both forms.

A_v:

$$I_b = \frac{V_i}{Z_b}$$

and

$$V_o = -I_o R_C = -h_{fe} I_b R_C$$

$$= -h_{fe}\left(\frac{V_i}{Z_b}\right)R_C$$

and

$$\boxed{A_v = \frac{V_o}{V_i} = -\frac{h_{fe}R_C}{Z_b}} \tag{9.30}$$

or

$$\boxed{A_v = \frac{V_o}{V_i} = \frac{-R_C}{r_e + R_E}} \tag{9.31}$$

For the approximation $Z_b \cong h_{fe}R_E$,

$$A_v = \frac{-h_{fe}R_C}{h_{fe}R_E}$$

and

$$\boxed{A_v = \frac{V_o}{V_i} = -\frac{R_C}{R_E}} \tag{9.32}$$

which also applies to both models, due to the absence of any device parameters.

A_i: The magnitude of R_B is usually too close to Z_b to be ignored, requiring an application of the current-divider rule to the input circuit. That is,

$$I_b = \frac{R_B I_i}{R_B + Z_b}$$

and

$$\frac{I_b}{I_i} = \frac{R_B}{R_B + Z_b}$$

but

$$I_o = h_{fe} I_b$$

and $$\frac{I_o}{I_b} = h_{fe}$$

with $$A_i = \frac{I_o}{I_i} = \frac{I_o}{I_b}\frac{I_b}{I_i}$$

$$= h_{fe}\frac{R_B}{R_B + Z_b}$$

and $$\boxed{A_i = \frac{I_o}{I_i} = \frac{R_B h_{fe}}{R_B + Z_b}} \tag{9.33}$$

For the r_e model, simply substitute $\beta = h_{fe}$ and the proper value of Z_b (Eq. 9.26).

EXAMPLE 9.4

Determine Z_i, Z_o, A_v, and A_i for the network of Fig. 9.14.

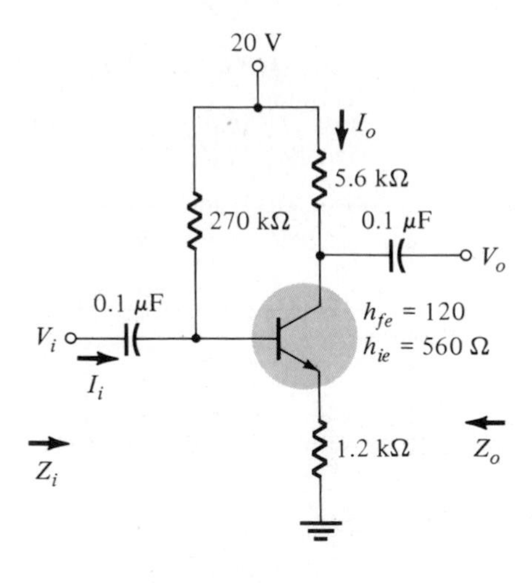

Figure 9.14

Solution:

Z_i From Eq. (9.23),

$$Z_b = h_{ie} + (1 + h_{fe})R_E$$
$$= 0.56\text{ k}\Omega + (1 + 120)1.2\text{ k}\Omega$$
$$= 145.76\text{ k}\Omega$$

From Eq. (9.28),

$$Z_i = R_B \| Z_b$$
$$= 270\text{ k}\Omega \| 145.76\text{ k}\Omega$$
$$= \mathbf{94.66\text{ k}\Omega}$$

Z_o [Eq. (9.29)]:

$$Z_o \cong \mathbf{5.6\ k\Omega}$$

$\mathbf{A_v}$ [Eq. (9.30)]:

$$A_v = -\frac{h_{fe}R_C}{Z_b}$$

$$= -\frac{(120)(5.6\ \text{k}\Omega)}{145.76\ \text{k}\Omega}$$

$$= \mathbf{-4.61}$$

$\mathbf{A_i}$ [Eq. (9.33)]:

$$A_i = \frac{R_B h_{fe}}{R_B + Z_b}$$

$$= \frac{(270\ \text{k}\Omega)(120)}{270\ \text{k}\Omega + 145.76\ \text{k}\Omega}$$

$$= \mathbf{77.93}$$

Applying the approximate equations for Z_b, we find

$$Z_b \cong h_{fe}R_E = (120)(1.2\ \text{k}\Omega) = 144\ \text{k}\Omega$$

and

$$Z_i = R_B \| Z_b = 270\ \text{k}\Omega \| 144\ \text{k}\Omega$$

$$= \mathbf{93.91\ k\Omega}$$

which is very close to the result (94.66 kΩ) obtained above; and

$$A_v \cong -\frac{R_C}{R_E}$$

$$= -\frac{5.6\ \text{k}\Omega}{1.2\ \text{k}\Omega}$$

$$= \mathbf{-4.67}$$

which is also very close to the −4.61 obtained above.

A dc analysis of the network of Fig. 9.14 would have resulted in

$$I_B = 46.5\ \mu\text{A} \qquad I_E = 5.578\ \text{mA}$$

and

$$r_e = 4.66\ \Omega \qquad \text{with } \beta r_e = 559\ \Omega$$

which could be used to determine the same results obtained above.

A second variation of an unbypassed emitter bias configuration appears in Fig. 9.15. For the dc analysis the emitter resistance is $R_{E_1} + R_{E_2}$, while for the ac analysis the resistor R_E in Eqs. (9.23) through (9.33) is simply R_{E_1}. R_{E_2} is bypassed by C_E for the ac analysis.

A third variation appears in Fig. 9.16. In this case the resistance R_B of Eqs. (9.23) through (9.33) is the parallel combination of R_{B_1} and R_{B_2}. The dc analysis to obtain r_e is the same as that performed in Example 9.3.

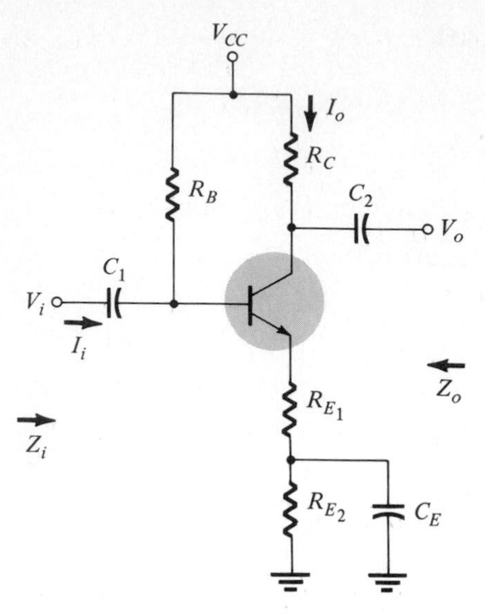

Figure 9.15

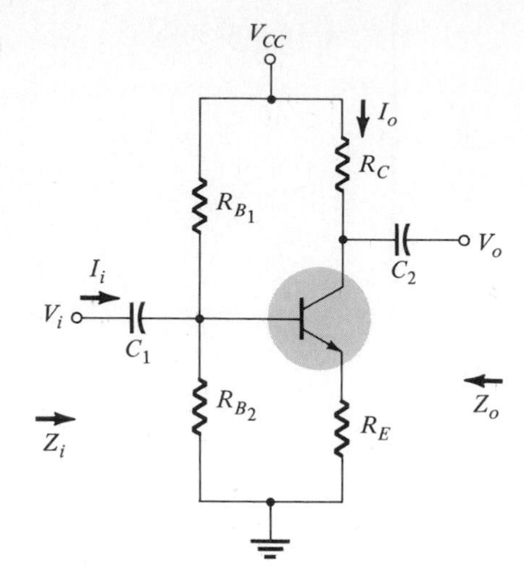

Figure 9.16

9.5 EMITTER-FOLLOWER CONFIGURATION

When the output is taken from the emitter terminal of the transistor as shown in Fig. 9.17 the network is referred to as an *emitter-follower*. The output voltage is always slightly less than the input signal, due to the drop from base to emitter, but the approximation $A_v \cong 1$ is usually a good one. Unlike the collector voltage, the emitter voltage is in phase with the signal v_i. That is, both v_o and v_i will attain their positive and negative peak values at the same time. The fact that v_o "follows" the magnitude of v_i with an in-phase relationship accounts for the terminology emitter-follower.

The most common emitter-follower configuration appears in Fig. 9.17. In fact, because the collector is grounded for ac analysis, it is actually a *common-collector* configuration. Other variations of Fig. 9.17 that draw the output off the emitter with $v_o \cong v_i$ will appear later in this section.

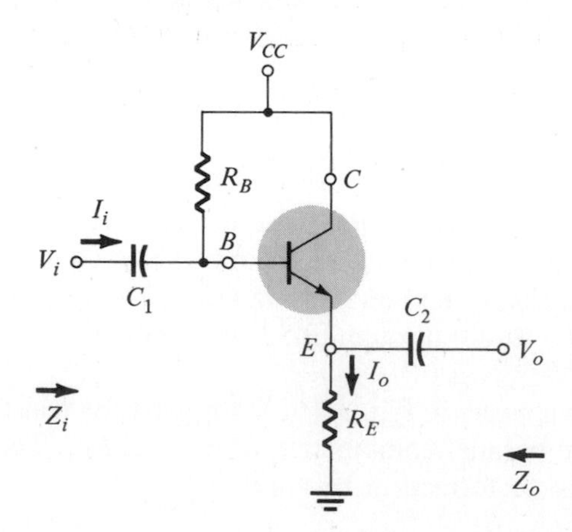

Figure 9.17

The emitter-follower configuration is frequently used for impedance-matching purposes. It presents a high impedance at the input and a low impedance at the output, which is the direct opposite of the standard fixes-bias configuration. The resulting effect is much the same as that obtained with a transformer, where a load is matched to the source impedance for maximum power transfer through the system.

Substituting the approximate equivalent circuit into the network of Fig. 9.17 will result in the network of Fig. 9.18.

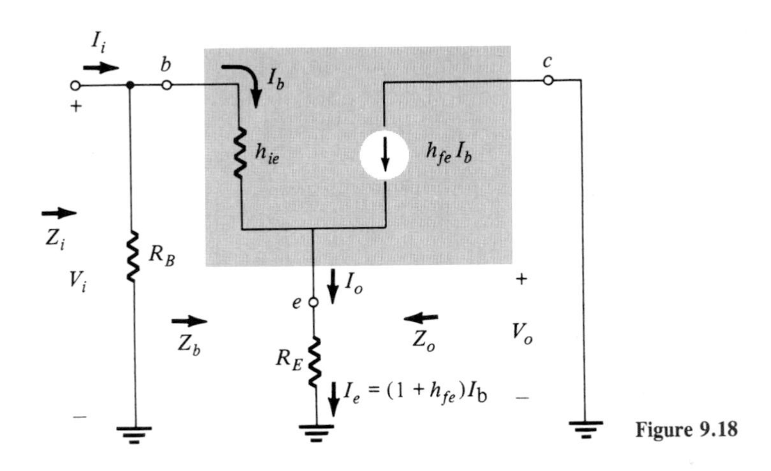

Figure 9.18

Z_i: The input impedance is determined in the same manner as described in the preceding section:

$$Z_i = R_B \| Z_b \tag{9.34}$$

with Z_b defined by Eqs. (9.23) through (9.27).

Z_o: The output impedance is best described by first writing the equation for the current I_b:

$$I_b = \frac{V_i}{Z_b}$$

and then multiplying by $(1 + h_{fe})$ to establish I_e. That is,

$$I_e = (1 + h_{fe})I_b = (1 + h_{fe})\frac{V_i}{Z_b}$$

$$= \frac{(1 + h_{fe})V_i}{h_{ie} + (1 + h_{fe})R_E}$$

or

$$I_e = \frac{V_i}{[h_{ie}/(1 + h_{fe})] + R_E} \tag{9.35}$$

If we now construct the network defined by Eq. (9.35), the configuration of Fig. 9.19 will result.

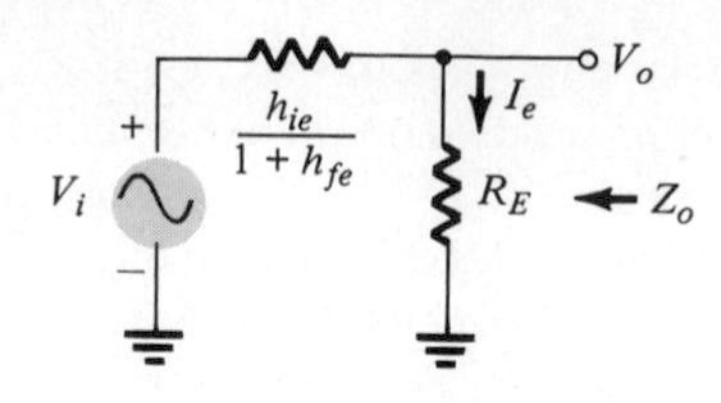

Figure 9.19

To determine Z_o, V_i is set to zero and

$$Z_o = R_E \left\| \frac{h_{ie}}{1 + h_{fe}} \right. \tag{9.36}$$

In other words, the network "seen" by the emitter branch is the input voltage V_i in series with a resistance determined by the hybrid parameters h_{ie} and h_{fe}. The resistance $h_{ie}/(1 + h_{fe})$ is usually quite small, dropping Z_o well below the R_E level.

A_v: Figure 9.19 can now be utilized to determine the voltage gain through an application of the voltage-divider rule:

$$V_o = \frac{R_E V_i}{R_E + [h_{ie}/(1 + h_{fe})]}$$

and

$$A_v = \frac{V_o}{V_i} = \frac{R_E}{R_E + [h_{ie}/(1 + h_{fe})]} \tag{9.37}$$

Note the absence of a negative sign to indicate that v_o and v_i are in phase and recognize that the factor $h_{ie}/(1 + h_{fe})$ is the only reason v_o does not equal v_i.

A_i: From Fig. 9.18,

$$I_b = \frac{R_B I_i}{R_B + Z_b}$$

and

$$\frac{I_b}{I_i} = \frac{R_B}{R_B + Z_b}$$

$$I_o = I_e = (1 + h_{fe})I_b$$

and

$$\frac{I_o}{I_b} = 1 + h_{fe}$$

$$A_i = \frac{I_o}{I_i} = \frac{I_o}{I_b}\frac{I_b}{I_i}$$

$$= (1 + h_{fe})\frac{R_B}{R_B + Z_b}$$

and

$$A_i = \frac{(1 + h_{fe})R_B}{R_B + Z_b} \tag{9.38}$$

The equations for the r_e model can be determined directly from the above simply by substituting $h_{ie} = \beta r_e$ and $h_{fe} = \beta$.

For $\beta >> 1$

$$Z_o = \frac{R_E}{r_e} \tag{9.39}$$

$$A_v = \frac{R_E}{R_E + r_e} \tag{9.40}$$

and

$$A_i = \frac{\beta R_B}{R_B + Z_b} \tag{9.41}$$

with

$$Z_b = \beta(r_e + R_E)$$

$$Z_b = \beta\left(\frac{h_{ie}}{h_{fe}} + R_E\right)$$

EXAMPLE 9.5

Determine Z_i, Z_o, A_v, and A_i for the emitter-follower network of Fig. 9.20.

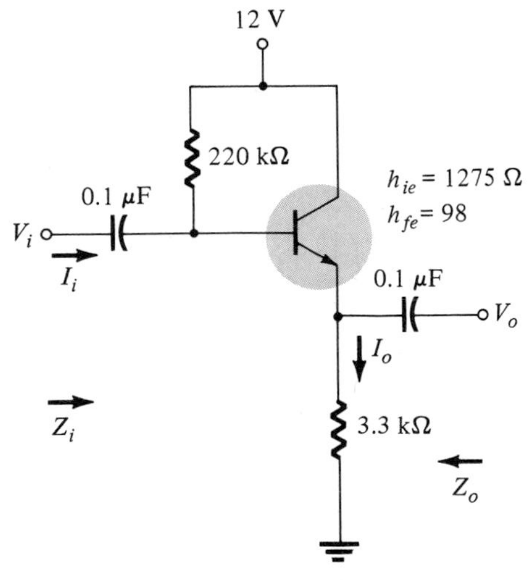

Figure 9.20

Solution:

Z_i:

$$Z_b = h_{ie} + (1 + h_{fe})R_E$$
$$= 1.275 \text{ k}\Omega + (1 + 98)3.3 \text{ k}\Omega$$
$$\cong 327.98 \text{ k}\Omega$$

and

$$Z_i = R_B \| Z_b$$
$$= 220 \text{ k}\Omega \| 327.98 \text{ k}\Omega$$
$$= \mathbf{131.68 \text{ k}\Omega}$$

Z_o:

$$Z_o = R_E \| \frac{h_{ie}}{1 + h_{fe}}$$

$$= 3.3\text{ k}\Omega \| \frac{1.275\text{ k}\Omega}{1 + 98}$$

$$= 3.3\text{ k}\Omega \| 12.9\ \Omega$$

$$\cong \mathbf{12.9\ \Omega}$$

$\boldsymbol{A_v}$**:**

$$A_v = \frac{R_E}{R_E + [h_{ie}/(1 + h_{fe})]}$$

$$= \frac{3300}{3300 + 12.9}$$

$$= \mathbf{0.996} \cong 1$$

$\boldsymbol{A_i}$**:**

$$A_i = \frac{(1 + h_{fe})R_B}{R_B + Z_b}$$

$$= \frac{(1 + 98)220\text{ k}\Omega}{220\text{ k}\Omega + 327.98\text{ k}\Omega}$$

$$= \mathbf{39.75}$$

A dc analysis of the system would result in

$$I_B = 20.7\ \mu\text{A}$$

$$I_C = \beta I_B \cong 2\text{ mA} \cong I_E$$

and $$r_e = \frac{26\text{ mV}}{I_E} = \frac{26}{2} = 13\ \Omega$$

with $$\beta r_e = 1274\ \Omega$$

Substituting $h_{ie} = \beta r_e = 1274\ \Omega$ ($\cong 1275\ \Omega$) and $h_{fe} = \beta = 98$ into the equations above would result in the same solutions.

The network of Fig. 9.21 is a variation of the network of Fig. 9.17 which employs a voltage-divider input section to set the bias conditions. Equations (9.34) through (9.38) are changed only by replacing R_B by $R_{BB} = R_{B_1} \| R_{B_2}$.

The network of Fig. 9.22 will also provide the input/output characteristics of an emitter-follower but includes a collector resistor R_C. In this case R_B is again replaced by the parallel combination of R_{B_1} and R_{B_2}. The input impedance Z_i and output impedance Z_o are unaffected by R_C since it is not reflected into the base or emitter equivalent networks. In fact, the only impact of R_C will be to determine the Q point of operation.

The effect of $1/h_{oe}$ is to divert some of the current $h_{fe}I_b$ from R_E, reducing the current gain and reducing the level of V_o. However, $1/h_{oe}$ is normally so much greater than R_E that the effect of $1/h_{oe}$ is usually ignored.

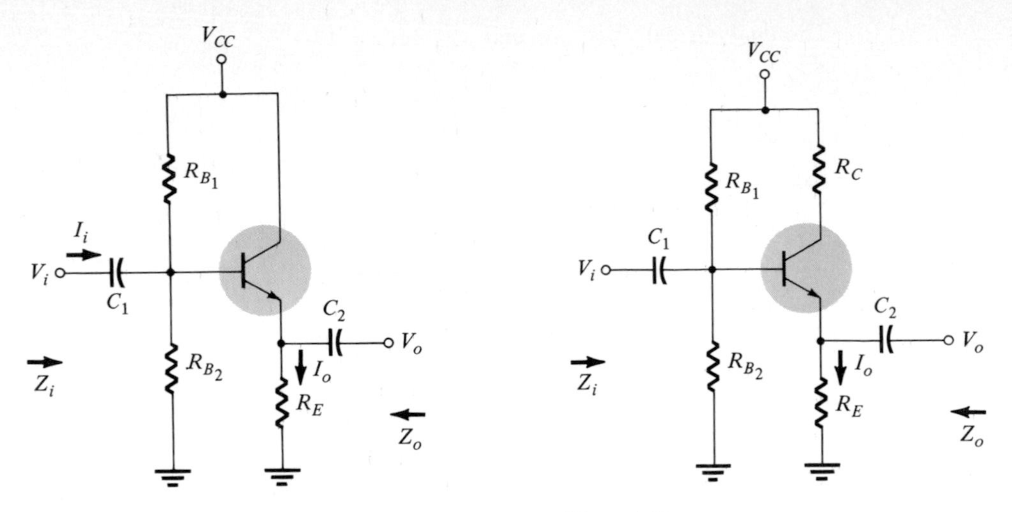

Figure 9.21

Figure 9.22

9.6 COMMON-BASE CONFIGURATION

The common-base configuration is characterized as having a low input and output impedance and a current gain less than 1. The voltage gain, however, can be quite respectable. The standard configuration appears in Fig. 9.23 with the approximate hybrid equivalent model substituted in Fig. 9.24. Note that the common-base model has the same layout as the common-emitter equivalent network except that now the parameters are the common-base parameters and I_b has been replaced by I_e.

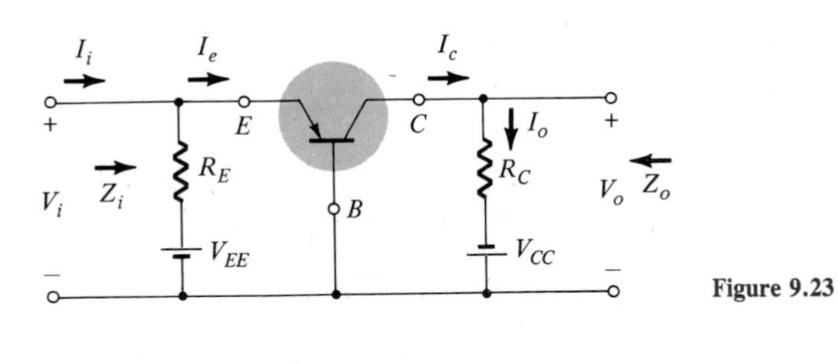

Figure 9.23

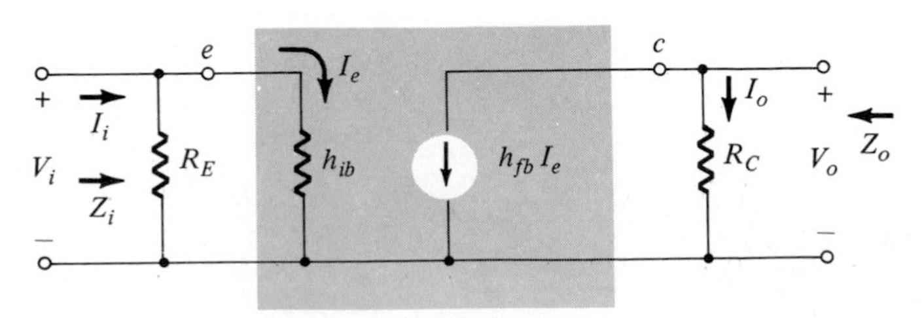

Figure 9.24

The analysis can then proceed as follows:

Z_i:

$$\boxed{Z_i = R_E \| h_{ib}} \tag{9.42}$$

Z_o:

$$\boxed{Z_o = R_C} \tag{9.43}$$

A_v:

$$V_o = I_o R_C = I_C R_C = -h_{fb} I_e R_C$$

with

$$I_e = \frac{V_i}{h_{ib}}$$

and

$$V_o = -h_{fb}\left(\frac{V_i}{h_{ib}}\right) R_C$$

so that

$$\boxed{A_v = \frac{V_o}{V_i} = -\frac{h_{fb}}{h_{ib}} R_C} \tag{9.44}$$

Since h_{fb} is a negative quantity, v_o and v_i are *in phase* for the common-base configuration.

A_i: Assume that $R_E \gg h_{ib}$. Then $I_e = I_i$

and

$$I_o = -h_{fb} I_e$$
$$= -h_{fb} I_i$$

and

$$\boxed{A_i = \frac{I_o}{I_i} = -h_{fb}} \tag{9.45}$$

For the r_e model, $h_{ib} = r_e$ and $h_{fb} = -1$, resulting in the approximate equivalent circuit of Fig. 9.25.

Z_i:

$$\boxed{Z_i = R_E \| r_e} \tag{9.46}$$

Z_o:

$$\boxed{Z_o = R_C} \tag{9.47}$$

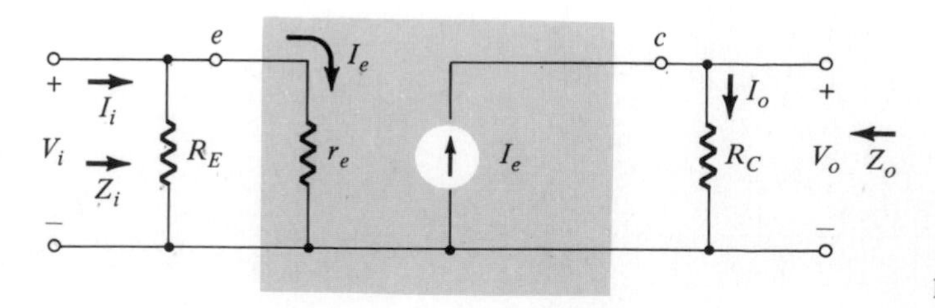

Figure 9.25

A_v:

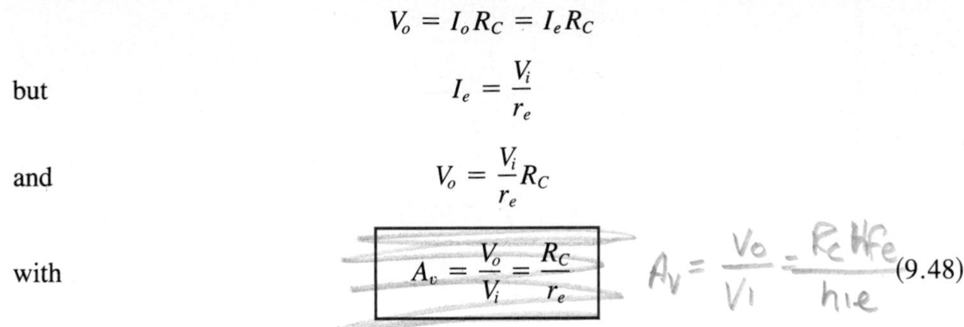

$$V_o = I_o R_C = I_e R_C$$

but
$$I_e = \frac{V_i}{r_e}$$

and
$$V_o = \frac{V_i}{r_e} R_C$$

with
$$\boxed{A_v = \frac{V_o}{V_i} = \frac{R_C}{r_e}} \tag{9.48}$$

A_i: Assuming that $R_E \gg r_e$ yields

$$I_e = I_i$$

and
$$I_o = I_e = I_i$$

with
$$\boxed{A_i = \frac{I_o}{I_i} = 1} \tag{9.49}$$

EXAMPLE 9.6

Determine Z_i, Z_o, A_v, and A_i for the network of Fig. 9.26. Note that the hybrid parameters have not been provided.

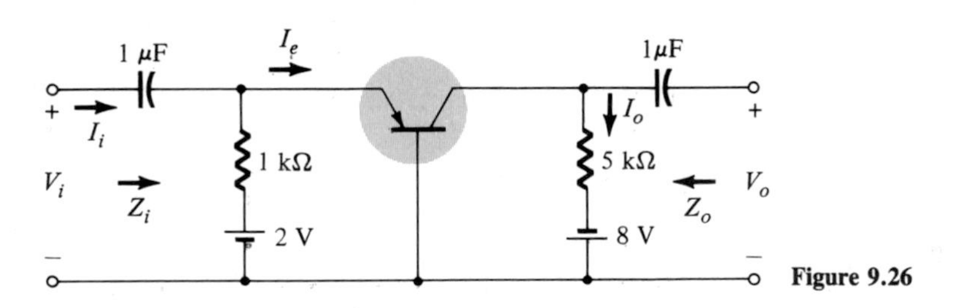

Figure 9.26

Solution:

$$I_E = \frac{V_{EE} - V_{BE}}{R_E} = \frac{2 - 0.7}{1\text{ k}\Omega} = \frac{1.3}{1\text{ k}\Omega} = 1.3\text{ mA}$$

$$r_e = \frac{26\text{ mV}}{I_E} = \frac{26}{1.3} = 20\ \Omega$$

Z_i:

$$Z_i = R_E \| r_e = 1\text{ k}\Omega \| 20$$
$$= \mathbf{19.6\ \Omega}$$

Z_o:

$$Z_o = R_C$$
$$= \mathbf{5\ k\Omega}$$

A_v:

$$A_v = \frac{R_C}{r_e} = \frac{5\ \text{k}\Omega}{20}$$
$$= \mathbf{250}$$

A_i:

$$A_i = \mathbf{1}$$

For the hybrid equivalent circuit, simply substitute $h_{ib} = r_e = 20\ \Omega$ and $h_{fb} = -1$ in Eqs. (9.42) through (9.45).

9.7 COLLECTOR DC FEEDBACK AND DIFFERENCE AMPLIFIER

The network of Fig. 9.27 has a dc feedback resistor for increased stability, yet the capacitor C_3 will shift portions of the feedback resistance to the input and output sections of the network in the ac domain. The portion of R_F shifted to the input or output side will be determined by the desired ac input and output resistance levels.

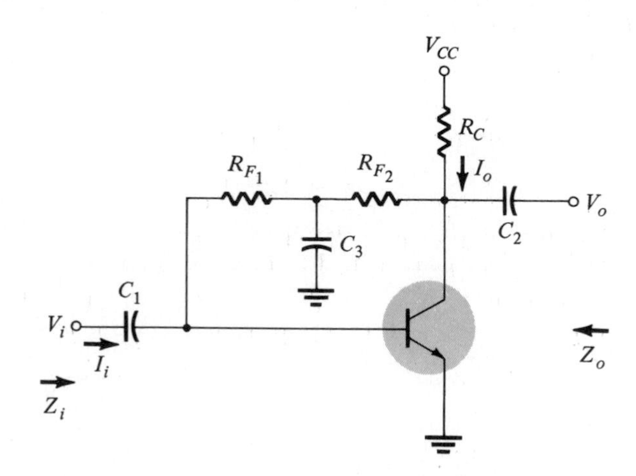

Figure 9.27

At the frequency or frequencies of operation, the capacitor will assume a short-circuit equivalent to ground due to its low impedance level compared to the other elements of the network. The small-signal ac equivalent circuit will then appear as shown in Fig. 9.28.

Z_i:

$$\boxed{Z_i = R_{F_1} \| h_{ie}} \tag{9.50}$$

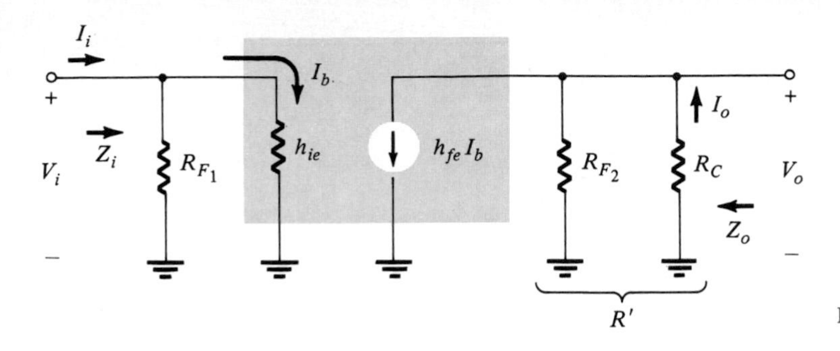

Figure 9.28

Z_o:

$$\boxed{Z_o = R_C \| R_{F_2}} \tag{9.51}$$

A_v:

$$R' = R_{F_2} \| R_C$$

and

$$V_o = -h_{fe} I_b R'$$

but

$$I_b = \frac{V_i}{h_{ie}}$$

and

$$V_o = -h_{fe}\left(\frac{V_i}{h_{ie}}\right)R'$$

so that

$$\boxed{A_v = \frac{V_o}{V_i} = -\frac{h_{fe}R'}{h_{ie}}} \tag{9.52}$$

A_i: For the input section,

$$I_b = \frac{R_{F_1} I_i}{R_{F_1} + h_{ie}} \quad \text{or} \quad \frac{I_b}{I_i} = \frac{R_{F_1}}{R_{F_1} + h_{ie}}$$

and for the output section,

$$I_o = \frac{R_{F_2} h_{fe} I_b}{R_{F_2} + R_C} \quad \text{or} \quad \frac{I_o}{I_b} = \frac{R_{F_2} h_{fe}}{R_{F_2} + R_C}$$

The current gain

$$A_i = \frac{I_o}{I_i} = \frac{I_o}{I_b}\frac{I_b}{I_i}$$

$$= \frac{R_{F_2} h_{fe}}{R_{F_2} + R_C}\frac{R_{F_1}}{R_{F_1} + h_{ie}}$$

and

$$\boxed{A_i = \frac{h_{fe} R_{F_1} R_{F_2}}{(R_{F_1} + h_{ie})(R_{F_2} + R_C)}} \tag{9.53}$$

EXAMPLE 9.7

Determine Z_i, Z_o, A_v, and A_i for the network of Fig. 9.29 using the r_e model.

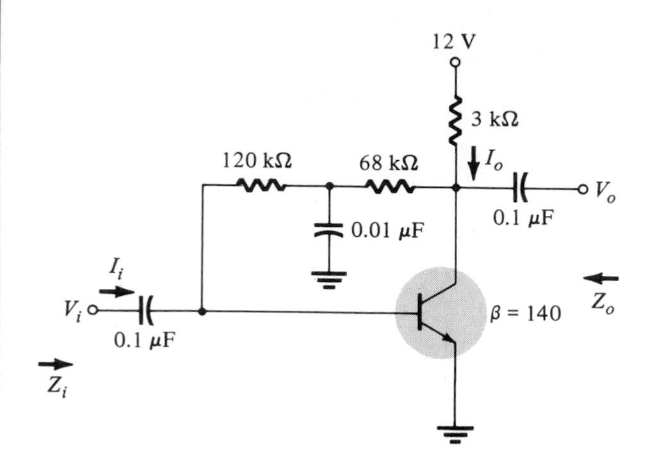

Figure 9.29

Solution:

Dc analysis (capacitor C_3 an open circuit):

$$I_B = \frac{V_{CC} - V_{BE}}{R_F + (\beta + 1)R_C}$$

$$= \frac{12 - 0.7}{(120\text{ k}\Omega + 68\text{ k}\Omega) + (140 + 1)3\text{ k}\Omega}$$

$$= \frac{11.3}{611\text{ k}\Omega} \cong 18.5\ \mu\text{A}$$

$$I_C = \beta I_B = (140)(18.5\ \mu\text{A}) = 2.59\text{ mA} = I_E$$

and

$$r_e = \frac{26\text{ mV}}{I_E} = \frac{26}{2.59} \cong 10.04\ \Omega$$

with

$$\beta r_e = (140)(10.04) \cong 1.4\text{ k}\Omega$$

The ac equivalent network appears in Fig. 9.30.

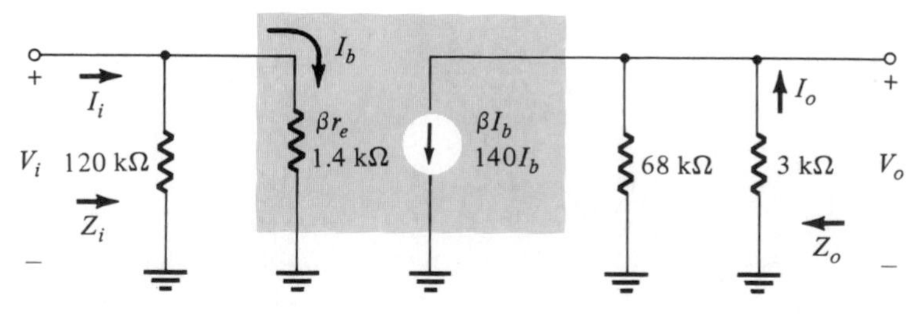

Figure 9.30

Z_i:

$$Z_i = R_{F_1} \| \beta r_e = 120 \text{ k}\Omega \| 1.4 \text{ k}\Omega$$

$$\cong \mathbf{1.38 \text{ k}\Omega}$$

Z_o:

$$Z_o = R_C \| R_{F_2} = 3 \text{ k}\Omega \| 68 \text{ k}\Omega$$

$$\cong \mathbf{2.87 \text{ k}\Omega}$$

A_v:

$$R' = R_C \| R_{F_2} = 2.87 \text{ k}\Omega$$

$$A_v = \frac{-h_{fe}R'}{h_{ie}} = \frac{-\beta R'}{\beta r_e} = -\frac{R'}{r_e}$$

$$= \frac{-2.87 \text{ k}\Omega}{10.04}$$

$$= \mathbf{-285.86}$$

A_i:

$$A_i = \frac{\beta R_{F_1} R_{F_2}}{(R_{F_1} + \beta r_e)(R_{F_2} + R_C)}$$

$$= \frac{(140)(120 \text{ k}\Omega)(68 \text{ k}\Omega)}{(120 \text{ k}\Omega + 1.4 \text{ k}\Omega)(68 \text{ k}\Omega + 3 \text{ k}\Omega)}$$

$$= \mathbf{132.54}$$

The difference amplifier of Fig. 9.31 is similar to the emitter-stabilized networks examined in Section 9.4 with the addition of a second signal in the emitter leg. The equation for the voltage gain can best be determined using the ac equivalent of Fig. 9.32.

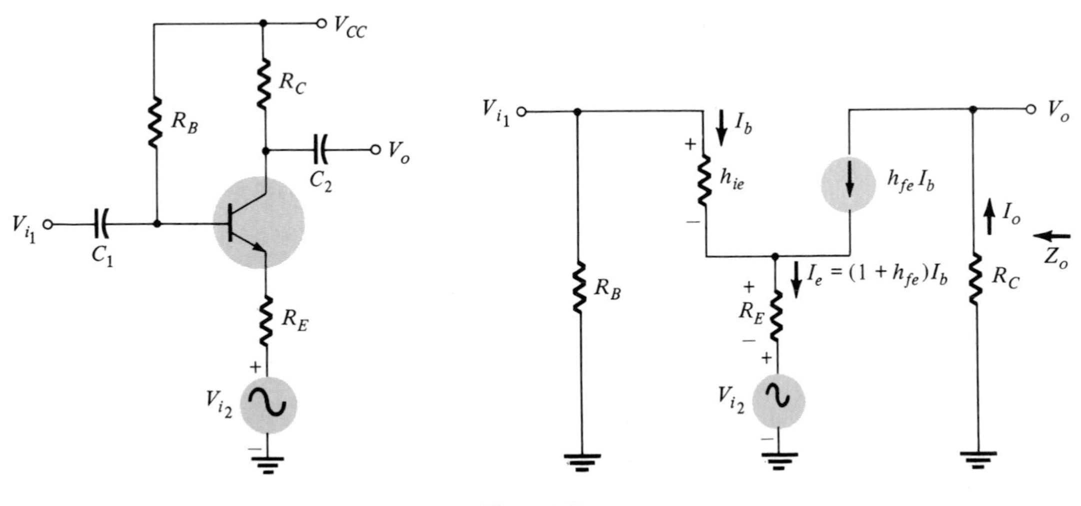

Figure 9.31

Figure 9.32

Applying Kirchhoff's voltage law to the input section will result in

$$V_{i_1} - I_b h_{ie} - (1 + h_{fe})I_b R_E - V_{i_2} = 0$$

Solving for I_b yields

$$I_b(h_{ie} + (h_{fe} + 1)R_E) = V_{i_1} - V_{i_2}$$

and

$$I_b = \frac{V_{i_1} - V_{i_2}}{h_{ie} + (h_{fe} + 1)R_E}$$

and the output voltage

$$V_o = -I_o R_C = -I_c R_C = -h_{fe} I_b R_C$$

and

$$V_o = -h_{fe}\left[\frac{V_{i_1} - V_{i_2}}{h_{ie} + (h_{fe} + 1)R_E}\right]R_C$$

or

$$\boxed{V_o = \frac{-h_{fe} R_C}{h_{ie} + (h_{fe} + 1)R_E}(V_{i_1} - V_{i_2})} \qquad (9.54)$$

The result is therefore an output voltage equal to the difference of the two signals multiplied by a factor sensitive to the network parameters and transistor equivalent circuit.

Since $h_{fe} \gg 1$ and $h_{fe} R_E$ is normally $\gg h_{ie}$,

$$V_o \cong -\frac{h_{fe} R_C}{h_{fe} R_E}(V_{i_1} - V_{i_2})$$

and

$$\boxed{V_o \cong -\frac{R_C}{R_E}(V_{i_1} - V_{i_2})} \qquad (9.55)$$

EXAMPLE 9.8

Given $R_C = 4.7\ \text{k}\Omega$, $R_E = 0.56\ \text{k}\Omega$, $h_{fe} = 200$, and $h_{ie} = 1.5\ \text{k}\Omega$ for the network of Fig. 9.31, determine V_o.

Solution:

From Eq. (9.54),

$$\begin{aligned} V_o &= \frac{-h_{fe} R_C}{h_{ie} + (h_{fe} + 1)R_E}(V_{i_1} - V_{i_2}) \\ &= \frac{-(200)(4.7\ \text{k}\Omega)}{1.5\ \text{k}\Omega + (200 + 1)0.56\ \text{k}\Omega}(V_{i_1} - V_{i_2}) \\ &= \mathbf{-8.24(V_{i_1} - V_{i_2})} \end{aligned}$$

Using Eq. (9.55),

$$V_o \cong -\frac{R_C}{R_E}(V_{i_1} - V_{i_2})$$

$$= \frac{4.7 \text{ k}\Omega}{0.56 \text{ k}\Omega}(V_{i_1} - V_{i_2})$$

$$= \mathbf{-8.39(V_{i_1} - V_{i_2})}$$

which compares very favorably with the result using Eq. (9.54).

9.8 COLLECTOR FEEDBACK CONFIGURATION

The collector feedback network of Fig. 9.33 employs a feedback path from collector to base to increase the stability of the system. The simple maneuver of connecting a resistor from base to collector rather than base to dc supply has a significant impact on the level of difficulty encountered when analyzing the network.

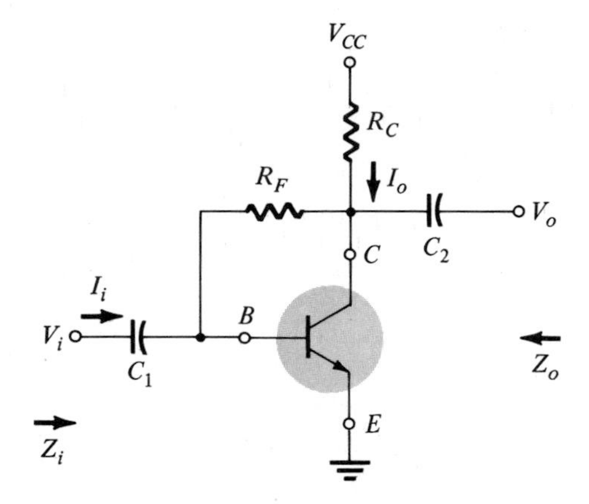

Figure 9.33

Some of the maneuvers to be performed below are the result of experience working with such configurations. It is not expected that a new student of the subject would choose the sequence of steps described below without taking a false path or two. Substituting the approximate equivalent circuit and redrawing the network will result in the configuration of Fig. 9.34.

The voltage gain will be determined first, followed by the current gain and the impedance levels.

$\mathbf{A_v}$**:** At node C:

$$I_o = h_{fe}I_b + I'$$

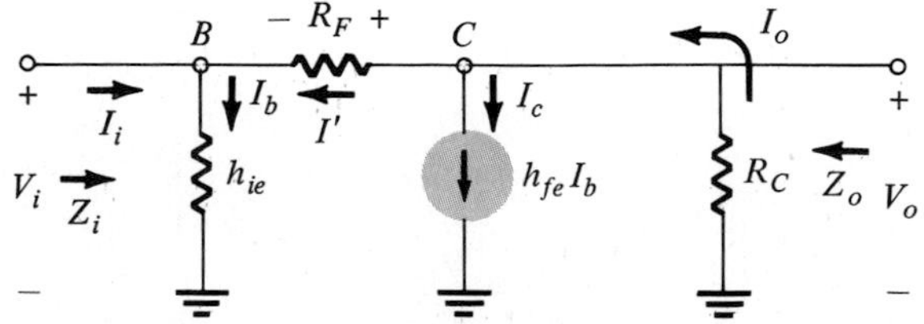

Figure 9.34

For typical values, $h_{fe}I_b \gg I'$ and $I_o \cong h_{fe}I_b$.

The output voltage

$$V_o = -I_oR_C = -(h_{fe}I_b)R_C$$

Substituting $I_b = V_i/h_{ie}$ gives us

$$V_o = -h_{fe}\left(\frac{V_i}{h_{ie}}\right)R_C$$

and

$$\boxed{A_v = \frac{V_o}{V_i} \cong -\frac{h_{fe}}{h_{ie}}R_C} \tag{9.56}$$

A_i: Applying Kirchhoff's voltage law around the outside network loop yields

$$V_i + V_{R_F} - V_o = 0$$

and

$$I_b h_{ie} + (I_b - I_i)R_F + I_oR_C = 0$$

Using $I_o \cong h_{fe}I_b$, we have

$$I_b h_{ie} + I_bR_F - I_iR_F + h_{fe}I_bR_C = 0$$

and

$$I_b(h_{ie} + R_F + h_{fe}R_C) = I_iR_F$$

Substituting $I_b = I_o/h_{fe}$ from $I_o \cong h_{fe}I_b$ yields

$$\frac{I_o}{h_{fe}}(h_{ie} + R_F + h_{fe}R_C) = I_iR_F$$

and

$$I_o = \frac{h_{fe}R_FI_i}{h_{ie} + R_F + h_{fe}R_C}$$

Ignoring h_{ie} compared to R_F and $h_{fe}R_C$ gives us

$$\boxed{A_i = \frac{I_o}{I_i} = \frac{h_{fe}R_F}{R_F + h_{fe}R_C}} \tag{9.57}$$

For $h_{fe}R_C \gg R_F$,

$$A_i = \frac{I_o}{I_i} = \frac{h_{fe}R_F}{h_{fe}R_C}$$

and

$$\boxed{A_i = \frac{I_o}{I_i} \cong \frac{R_F}{R_C}} \tag{9.58}$$

Z_i: From Fig. 9.34,

$$I_b = I_i + \frac{V_o - V_i}{R_F}$$

Since $V_o \gg V_i$,

$$I_b \cong I_i + \frac{V_o}{R_F}$$

and
$$V_i = I_b h_{ie}$$
$$= \left(I_i + \frac{V_o}{R_F}\right) h_{ie}$$
$$= I_i h_{ie} + \frac{h_{ie}}{R_F} V_o$$

Substituting $V_o = A_v V_i$ from $A_v = V_o/V_i$, we have
$$V_i = I_i h_{ie} + \frac{h_{ie} A_v V_i}{R_F}$$

Then
$$V_i\left(1 - \frac{h_{ie} A_v}{R_F}\right) = I_i h_{ie}$$

with
$$\frac{V_i}{I_i} = \frac{h_{ie}}{1 - h_{ie}(A_v/R_F)}$$

For parallel elements,
$$x \| y = \frac{xy}{x + y} = \frac{y}{1 + \frac{y}{x}}$$

which has the same format as the equation above with $y = h_{ie}$ and $x = R_F/A_v$. Therefore,

$$\boxed{Z_i = \frac{V_i}{I_i} = h_{ie} \left\| \frac{R_F}{|A_v|} \right.} \tag{9.59}$$

and we note that the magnitude of the voltage gain must be determined before the input impedance can be calculated.

Z_o: If we set V_i to zero as required to define Z_o, the network will appear as shown in Fig. 9.35. The effect of h_{ie} is removed and R_F appears in parallel with R_C and

$$\boxed{Z_o \cong R_C \| R_F} \tag{9.60}$$

for both models.

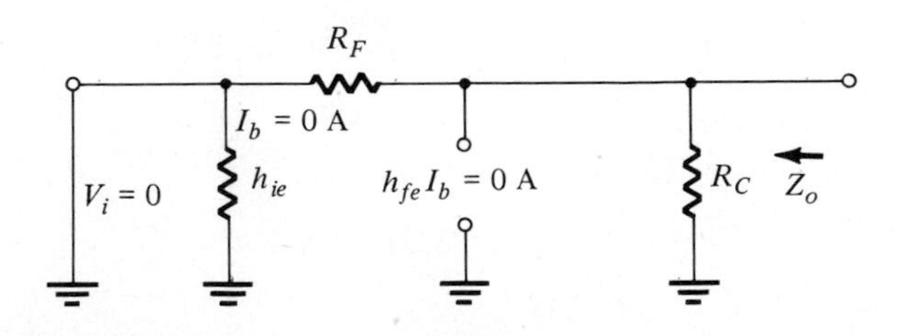

Figure 9.35

EXAMPLE 9.9

Determine A_v, A_i, Z_i, and Z_o for the network of Fig. 9.36 using the r_e model.

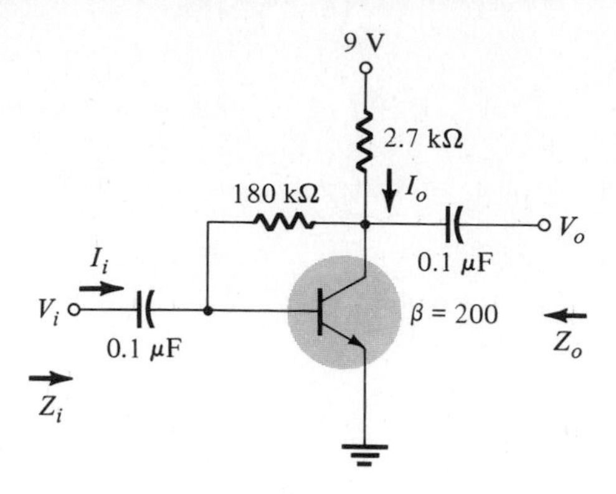

Figure 9.36

Solution:

$$I_B = \frac{V_{CC} - V_{BE}}{R_F + (\beta + 1)R_C} = \frac{9 - 0.7}{180\text{ k}\Omega + (200 + 1)2.7\text{ k}\Omega}$$

$$= 11.5\ \mu\text{A}$$

$$I_C = \beta I_B = (200)(11.5\ \mu\text{A}) = 2.3\text{ mA} = I_E$$

$$r_e = \frac{26\text{ mV}}{I_E} = \frac{26}{2.3} = 11.3\ \Omega$$

and $\beta r_e = (200)(11.3) = 2260\ \Omega$

A_v [Eq. (9.56)]:

$$A_v \cong -\frac{h_{fe}R_C}{h_{ie}} = -\frac{\beta R_C}{\beta r_e} = -\frac{R_C}{r_e}$$

$$= -\frac{2.7\text{ k}\Omega}{11.3} = \mathbf{-238.94}$$

A_i [Eq. (9.57)]:

$$A_i = \frac{h_{fe}R_F}{R_F + h_{fe}R_C} = \frac{\beta R_F}{R_F + \beta R_C} = \frac{(200)(180\text{ k}\Omega)}{180\text{ k}\Omega + (200)(2.7\text{ k}\Omega)}$$

$$= \mathbf{50}$$

Z_i [Eq. (9.59)]:

$$Z_i = h_{ie} \left\| \frac{R_F}{|A_v|} = \beta r_e \right\| \frac{R_F}{|A_v|} = (200)(11.3) \left\| \frac{180\text{ k}\Omega}{238.94} \right.$$

$$= 2.26\text{ k}\Omega \| 0.753\text{ k}\Omega = \mathbf{0.565\text{ k}\Omega}$$

Z_o [Eq. (9.60)]:

$$Z_o = R_C \| R_F = 2.7\text{ k}\Omega \| 180\text{ k}\Omega$$

$$= \mathbf{2.66\text{ k}\Omega}$$

For the configuration of Fig. 9.37, Eqs. (9.61) through (9.64) will determine the parameters of interest. The derivations are left as an exercise at the end of the chapter.

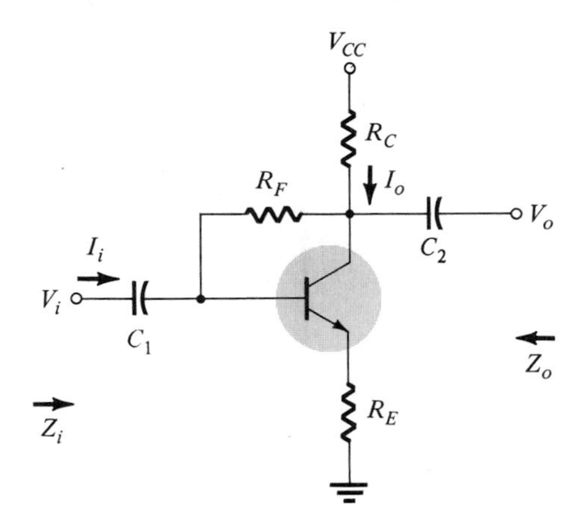

Figure 9.37

A_v:

$$A_v \cong -\frac{R_C}{R_E} \tag{9.61}$$

A_i:

$$A_i \cong \frac{R_F}{R_E + R_C + R_F/h_{fe}} \tag{9.62}$$

Z_i:

$$Z_i \cong h_{fe}R_E \left\| \frac{R_F}{|A_v|} \right. \tag{9.63}$$

Z_o:

$$Z_o \cong R_C \| R_F \tag{9.64}$$

9.9 LOADING EFFECTS OF R_S AND R_L

In practice, all sources have some internal resistance R_s, and loads (R_L) are applied to the output terminals of the amplifier as depicted in Fig. 9.38. At the input side the

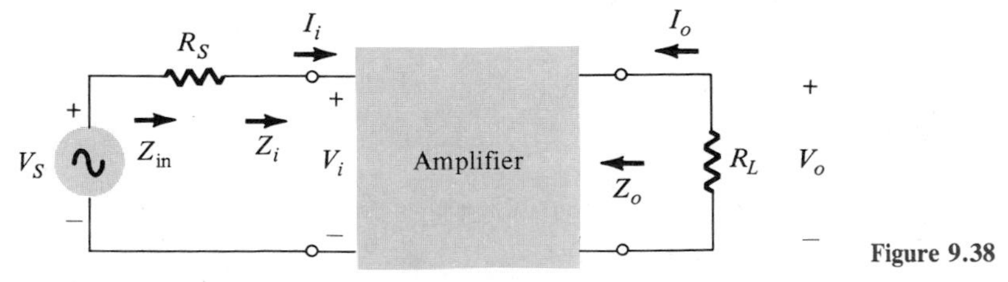

Figure 9.38

impedance Z_{in} is defined by

$$\boxed{Z_{in} = R_s + Z_i} \tag{9.65}$$

and the voltage V_i is related to V_s by an application of the voltage-divider rule

$$V_i = \frac{Z_i V_s}{Z_i + R_s}$$

or

$$\boxed{\frac{V_i}{V_s} = \frac{Z_i}{Z_i + R_s}} \tag{9.66}$$

The total voltage gain of the system can then be written as

$$\boxed{A_{v_s} = \frac{V_o}{V_s} = \frac{V_o}{V_i}\frac{V_i}{V_s}} \tag{9.67}$$

The input current is now

$$\boxed{I_i = \frac{V_s}{R_s + Z_i}} \tag{9.68}$$

For the output section the voltage V_o is now defined as across R_L with I_o the current through the load. Z_o is as defined in earlier sections.

For the variety of configurations the effect of R_s will be quite similar and the equations for the voltage gain will be only slightly affected by R_L. Defining I_o through the load resistor will alter the current gain equation somewhat, but the following approach will save both time and effort.

Once $A_v = V_o/V_i$ is determined (including the effects of R_L) the current gain for any system can be defined as follows using the defined variables of Fig. 9.38:

$$|A_i| = \left|\frac{I_o}{I_i}\right| = \left|\frac{V_o/R_L}{V_i/Z_i}\right| = \frac{Z_i}{R_L}\left|\frac{V_o}{V_i}\right|$$

and

$$\boxed{|A_i| = \left|\frac{I_o}{I_i}\right| = \frac{Z_i}{R_L}|A_v|} \tag{9.69}$$

Example 9.10 will demonstrate the use of Eq. (9.69) and the ease with which the current gain can be determined.

EXAMPLE 9.10

Determine Z_i, Z_{in}, Z_o, A_v, A_{v_s}, and A_i for the two-stage amplifier of Fig. 9.39.

Solution:

The coupling capacitors C_1 and C_2 isolate the dc biasing of each stage and the loading of the second stage is simply the input resistance $Z_{i_2} = 1.5\ \text{k}\Omega$. The ac equivalent appears in Fig. 9.40 with $R_{BB} = R_{B_1} \| R_{B_2} = 39\ \text{k}\Omega \| 9.1\ \text{k}\Omega = 7.378\ \text{k}\Omega$.

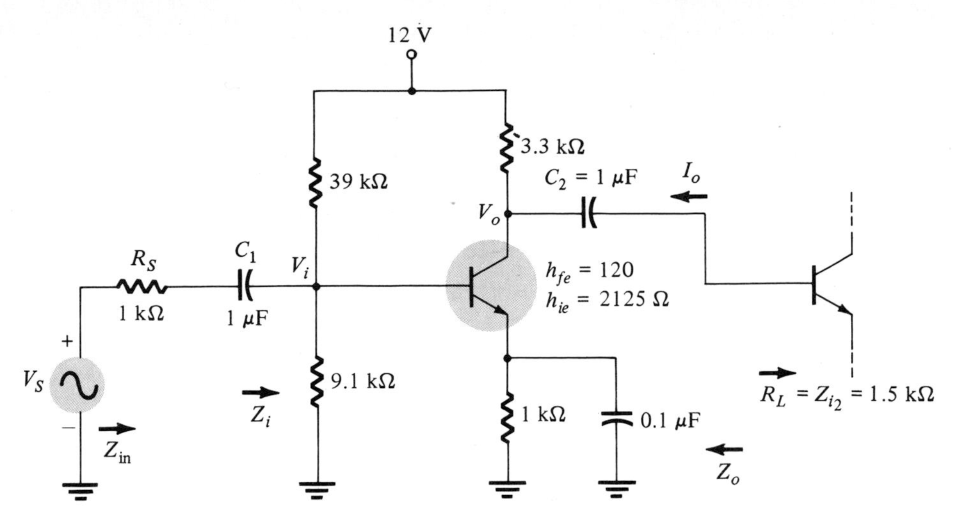

Figure 9.39

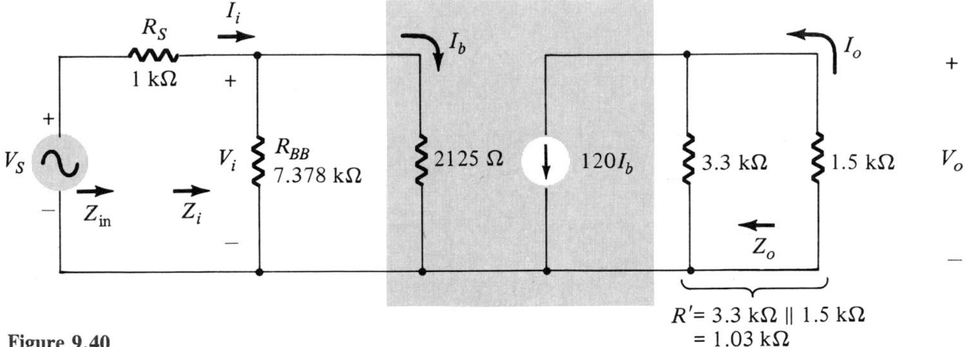

Figure 9.40

$\mathbf{Z_i}$: As before [Eq. (9.13)],

$$Z_i = R_{BB} \| h_{ie} = 7.378 \text{ k}\Omega \| 2.125 \text{ k}\Omega$$
$$= \mathbf{1.65 \text{ k}\Omega}$$

$\mathbf{Z_{in}}$ [Eq. (9.65)]:

$$Z_{\text{in}} = R_s + Z_i = 1 \text{ k}\Omega + 1.65 \text{ k}\Omega$$
$$= \mathbf{2.65 \text{ k}\Omega}$$

$\mathbf{Z_o}$: As before [Eq. (9.15)],

$$Z_o = R_C = \mathbf{3.3 \text{ k}\Omega}$$

$\mathbf{A_v}$: Equation (9.16) modified only by the fact that R_C is replaced by $R_C \| R_L$.

$$A_v = \frac{V_o}{V_i} = \frac{-h_{fe}(R_C \| R_L)}{h_{ie}} = \frac{-(120)(3.3 \text{ k}\Omega \| 1.5 \text{ k}\Omega)}{2.125 \text{ k}\Omega} = \frac{-(120)(1.03 \text{ k}\Omega)}{2.125 \text{ k}\Omega}$$
$$= \mathbf{-58.16}$$

Note that R_L had the effect of reducing the overall gain from a level of -186 with only R_C present.

A_{v_s}: From Eq. (9.67),

$$A_{v_s} = \frac{V_o}{V_s} = \frac{V_o}{V_i}\frac{V_i}{V_s}$$

with
$$\frac{V_i}{V_s} = \frac{Z_i}{Z_i + R_s} = \frac{1.65\text{ k}\Omega}{1.65\text{ k}\Omega + 1\text{ k}\Omega} = 0.623$$

and
$$A_{v_s} = \frac{V_o}{V_s} = (-58.16)(0.623)$$

$$= \mathbf{-36.23}$$

reducing the gain even further.

A_i [Eq. (9.69)]:

$$|A_i| = \frac{Z_i}{R_L}|A_v|$$

$$= \frac{1.65\text{ k}\Omega}{1.5\text{ k}\Omega}(58.16)$$

$$= \mathbf{63.98}$$

To check the result for A_i, let us now determine A_i through a detailed network analysis. For the input section,

$$I_b = \frac{R_{BB}I_i}{R_{BB} + h_{ie}}$$

and
$$\frac{I_b}{I_i} = \frac{R_{BB}}{R_{BB} + h_{ie}} = \frac{7.378\text{ k}\Omega}{7.378\text{ k}\Omega + 2.125\text{ k}\Omega} = 0.776$$

For the output section,

$$I_o = \frac{R_C h_{fe} I_b}{R_C + R_L}$$

and
$$\frac{I_o}{I_b} = \frac{R_C h_{fe}}{R_C + R_L} = \frac{(3.3\text{ k}\Omega)(120)}{3.3\text{ k}\Omega + 1.5\text{ k}\Omega} = 82.5$$

and
$$A_i = \frac{I_o}{I_i} = \frac{I_o}{I_b}\frac{I_b}{I_i} = (82.5)(0.776) = \mathbf{64.02}$$

as obtained above (the slight difference is due to the level of accuracy carried through the calculations).

For the r_e model, the dc analysis will result in $I_B = 12.23\ \mu\text{A}$, $I_C = 1.468\text{ mA}$, $r_e = 17.71\ \Omega$, and $\beta r_e = 2125\ \Omega$.

Effect of R_s on the Emitter Branch

The input section of a network with an emitter branch will appear as shown in Fig. 9.41 when the approximate equivalent circuit is substituted. Applying Kirchhoff's voltage law around the closed path gives us

$$V_s - I_b R_s - I_b h_{ie} - (1 + h_{fe}) I_b R_E = 0$$

and

$$V_s - I_b (R_s + h_{ie} + (1 + h_{fe}) R_E) = 0$$

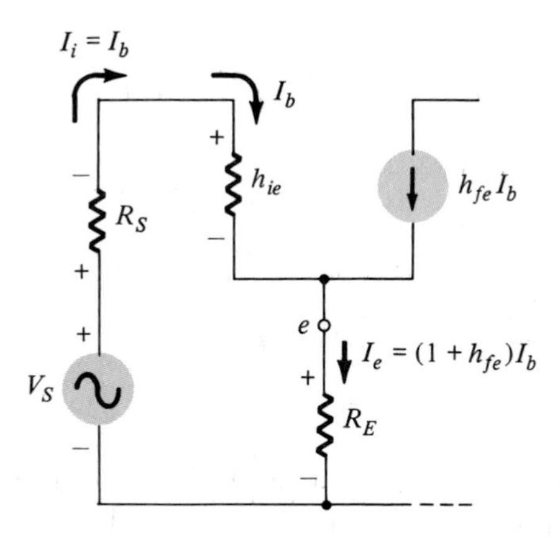

Figure 9.41

Solving for I_b yields

$$I_b = \frac{V_s}{R_s + h_{ie} + (1 + h_{fe}) R_E}$$

Establishing I_e, we have

$$I_e = (1 + h_{fe}) I_b = \frac{(1 + h_{fe}) V_s}{R_s + h_{ie} + (1 + h_{fe}) R_E}$$

and

$$\boxed{I_e = \frac{V_s}{[(R_s + h_{ie})/(1 + h_{fe})] + R_E}} \qquad (9.70)$$

Drawing the network to "fit" Eq. (9.70) will result in the configuration of Fig. 9.42, which is very similar to the network obtained earlier for an emitter branch except that now R_s is added to h_{ie} to increase the magnitude of the resistance in series with R_E.

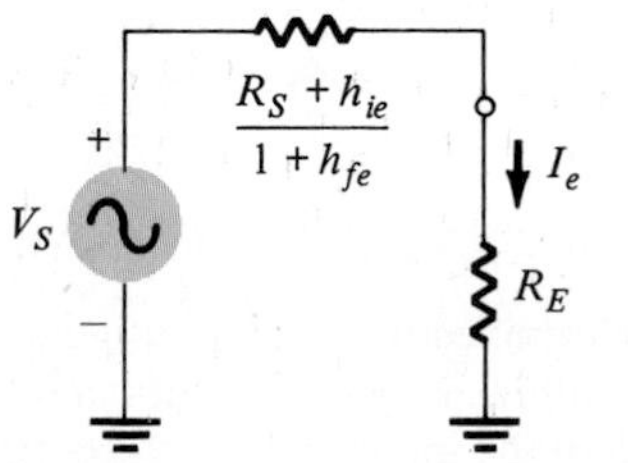

Figure 9.42

EXAMPLE 9.11

Determine Z_i, A_{v_s}, Z_o, and A_i for the network of Fig. 9.43 using the r_e model.

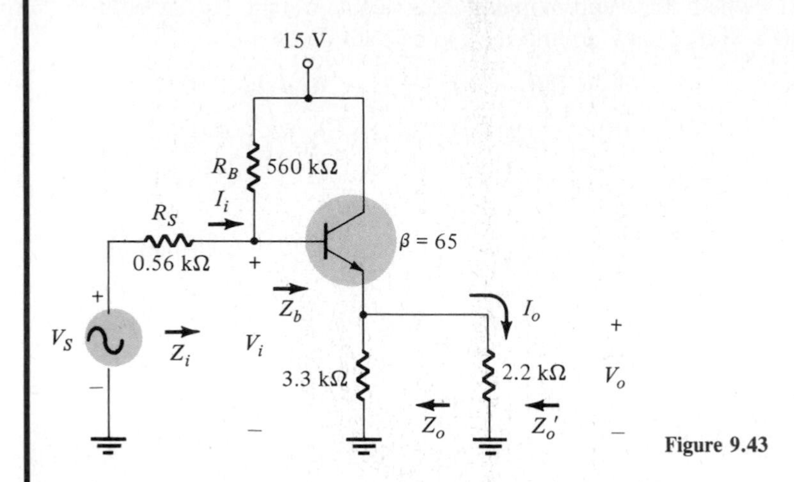

Figure 9.43

Solution:

DC analysis:

$$I_B = \frac{V_{CC} - V_{BE}}{R_B + (\beta + 1)R_E} = \frac{15 - 0.7}{560\text{ k}\Omega + (65 + 1)3.3\text{ k}\Omega}$$

$$= \frac{14.3}{560\text{ k}\Omega + 217.8\text{ k}\Omega} = 18.4\ \mu\text{A}$$

$$I_C = \beta I_B = 65(18.4\ \mu\text{A}) = 1.196\text{ mA} = I_E$$

$$r_e = \frac{26\text{ mV}}{I_E} = \frac{26}{1.196} = 21.74\ \Omega$$

$$\beta r_e = 65(21.74) = 1413\ \Omega \cong 1.4\text{ k}\Omega$$

Z_i: From Eq. (9.26),

$$Z_b = \beta(r_e + R_E)$$
$$= 65(21.74 + 3300)$$
$$= 215.9\text{ k}\Omega$$

From Eq. (9.28),

$$Z_i = R_B \| Z_b$$
$$= 560\text{ k}\Omega \| 215.9\text{ k}\Omega$$
$$= \mathbf{155.82\text{ k}\Omega}$$

A_{v_s}: A_{v_s} is best determined using the configuration of Fig. 9.42. However, V_s and R_s are in parallel with R_B in Fig. 9.43 and do not stand alone as shown in Fig. 9.41. If Thévenin's theorem is applied to the input section, the desired configuration can be obtained and the results surrounding Fig. 9.42 employed.

Determining the Thévenin equivalent circuit for the network of Fig. 9.44, for R_{Th} the network of Fig. 9.45 will result and

$$R_{Th} = R_B \| R_s = 560 \text{ k}\Omega \| 0.56 \text{ k}\Omega$$
$$= 0.559 \text{ k}\Omega$$

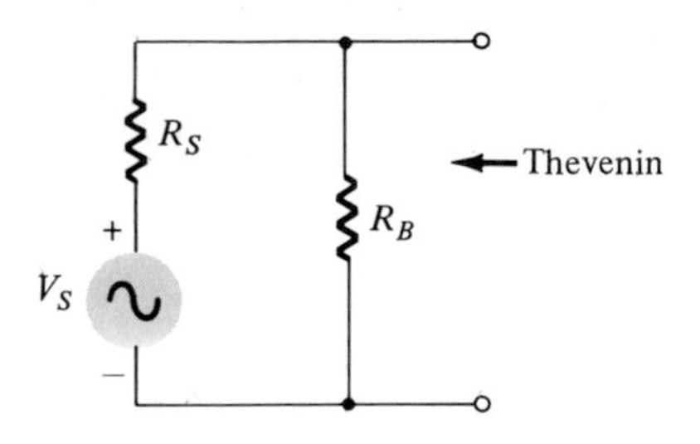

Figure 9.44

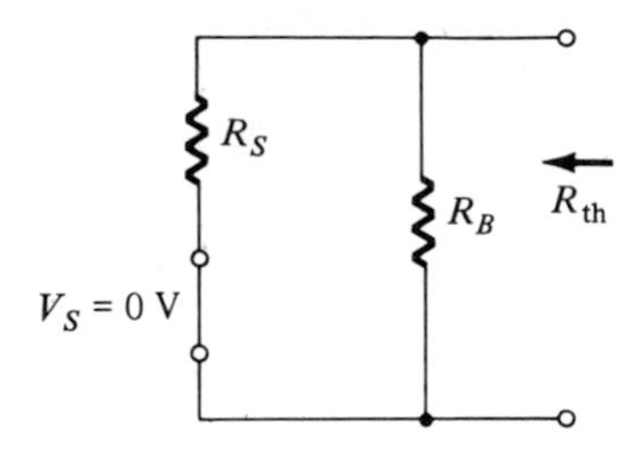

Figure 9.45

For E_{Th} the network of Fig. 9.46 is employed with the desired Thévenin voltage determined by the voltage-divider rule:

$$E_{Th} = \frac{R_B V_s}{R_B + R_s} = \frac{(560 \text{ k}\Omega)V_s}{560 \text{ k}\Omega + 0.56 \text{ k}\Omega}$$
$$= 0.999\ V_s$$

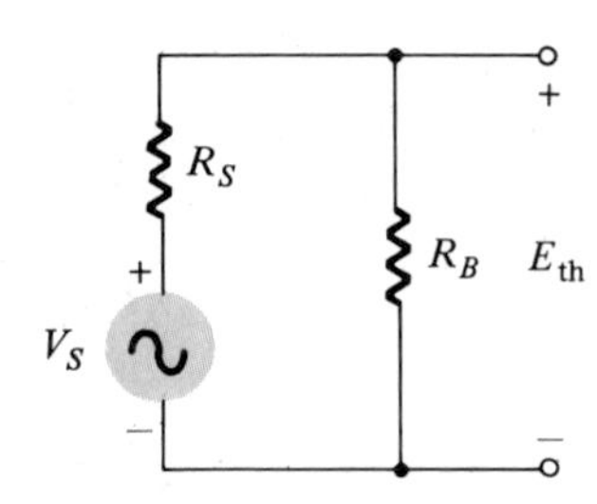

Figure 9.46

Note that $R_{Th} \cong R_s$ and $E_{Th} \cong V_s$, due to the large value of R_B—a typical result and frequently applied approximation. Defining

$$R'_E = R_E \| R_L = 3.3 \text{ k}\Omega \| 2.2 \text{ k}\Omega$$
$$= 1.32 \text{ k}\Omega$$

and referencing Fig. 9.42, we have

$$V_o = \frac{R'_E E_{Th}}{R'_E + (R_{Th} + h_{ie})/(1 + h_{fe})]}$$
$$= \frac{R'_E(0.999\ V_s)}{R'_E + [(0.559 \text{ k}\Omega + \beta r_e)/(1 + \beta)]}$$

and $$A_{v_s} = \frac{V_o}{V_s} = \frac{0.999\ R'_E}{R'_E + [(0.559 \text{ k}\Omega + \beta r_e)/(1 + \beta)]}$$

$$= \frac{0.999(1.32\text{ k}\Omega)}{1.32\text{ k}\Omega + [(0.559\text{ k}\Omega + 1.4\text{ k}\Omega)/(1 + 65)]}$$

$$= \frac{1.319\text{ k}\Omega}{1.32\text{ k}\Omega + .0297\text{ k}\Omega} = \frac{1.319\text{ k}\Omega}{1.3497\text{ k}\Omega}$$

$$= \mathbf{0.977}$$

revealing that the approximation of $A_{v_s} = 1$ is normally a good one for the emitter-follower configuration.

$\boldsymbol{Z_o'}$**:** From Fig. 9.42,

$$Z_o = R_E \| \frac{R_{\text{Th}} + h_{ie}}{1 + h_{fe}} = R_E \| \frac{R_{\text{Th}} + \beta r_e}{1 + \beta}$$

$$= 3.3\text{ k}\Omega \| \frac{0.559\text{ k}\Omega + 1.4\text{ k}\Omega}{1 + 65} = 3.3\text{ k}\Omega \| 0.0297\text{ k}\Omega$$

$$= 29.44\ \Omega$$

$$Z_o' = R_L \| Z_o$$

$$= 2.2\text{ k}\Omega \| 29.44\ \Omega$$

$$= \mathbf{29.05\ \Omega}$$

$\boldsymbol{A_i}$**:**

$$|A_i| = \frac{I_o}{I_i} = \frac{V_o/R_L}{V_s/Z_{\text{in}}} = \frac{Z_{\text{in}}}{R_L}|A_{v_s}|$$

Note that since we will be using the voltage gain $A_{v_s} = V_o/V_s$ rather than $A_v = V_o/V_i$, we must change Z_i to Z_{in} in the equation above.

$$Z_{\text{in}} = R_s + Z_i = 0.56\text{ k}\Omega + 155.82\text{ k}\Omega$$

$$= 156.38\text{ k}\Omega$$

and

$$|A_i| = \frac{Z_{\text{in}}}{R_L}|A_{v_s}| = \frac{156.38\text{ k}\Omega}{2.2\text{ k}\Omega}(0.977)$$

$$= \mathbf{69.45}$$

9.10 COMPLETE HYBRID EQUIVALENT CIRCUIT

Throughout the analysis of this chapter, the approximate hybrid equivalent model has been employed. The effects of h_{oe} were treated only for some introductory networks. In this section we derive the general equations for the current gain, voltage gain, input and output impedance, and power gain using the complete hybrid equivalent model. An example will demonstrate its use and compare the results with those obtained using the approximate model.

Consider the general configuration of Fig. 9.47 with the parameters of particular interest. The complete hybrid equivalent model is substituted in Fig. 9.48. The second

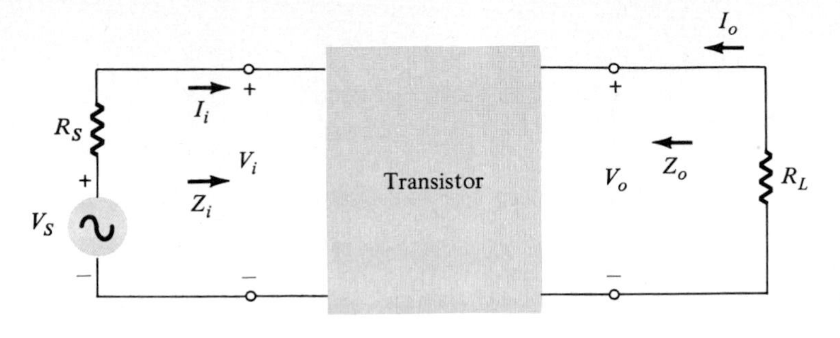

Figure 9.47

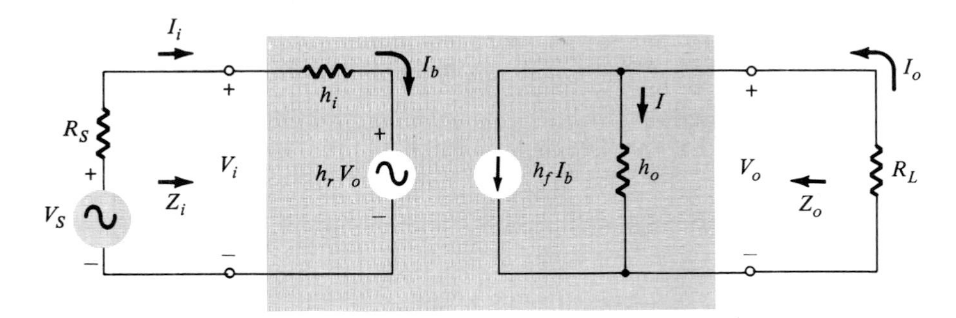

Figure 9.48

subscript for the hybrid parameters was not included, permitting the use of the resulting equations for any configuration (CE, CB, or CC) simply by substituting the appropriate parameters (h_{fe}, h_{fb}, or h_{fc} for h_f, etc.)

Current Gain, $A_i = I_o/I_i$

Applying Kirchhoff's current law to the output circuit yields

$$I_o = h_f I_b + I = h_f I_i + \frac{V_o}{1/h_o} = h_f I_i + h_o V_o$$

Substituting $V_o = -I_o R_L$ gives us

$$I_o = h_f I_i - h_o R_L I_o$$

Rewriting the equation above, we have

$$I_o + h_o R_L I_o = h_f I_i$$

and

$$I_o(1 + h_o R_L) = h_f I_i$$

so that

$$\boxed{A_i = \frac{I_o}{I_i} = \frac{h_f}{1 + h_o R_L}} \qquad (9.71)$$

Voltage Gain, $A_v = V_o/V_i$

Applying Kirchhoff's voltage law to the input circuit results in

$$V_i = I_i h_i + h_r V_o$$

Substituting $I_i = (1 + h_o R_L)I_o/h_f$ from Eq. (9.71) and $I_o = -V_o/R_L$ from above results in

$$V_i = \frac{-(1 + h_o R_L)h_i}{h_f R_L} V_o + h_r V_o$$

Solving for the ratio V_o/V_i yields

$$\boxed{A_v = \frac{V_o}{V_i} = \frac{-h_f R_L}{h_i + (h_i h_o - h_f h_r)R_L}} \tag{9.72}$$

Input Impedance, $Z_i = V_i/I_i$

For the input circuit,

$$V_i = h_i I_i + h_r V_o$$

substituting $\quad V_o = -I_o R_L$

we have $\quad V_i = h_i I_i - h_r R_L I_o$

Since $\quad A_i = \dfrac{I_o}{I_i}$

$$I_o = A_i I_i$$

so that the equation above becomes

$$V_i = h_i I_i - h_r R_L A_i I_i$$

Solving for the ratio V_i/I_i, we obtain

$$Z_i = \frac{V_i}{I_i} = h_i - h_r R_L A_i$$

and substituting

$$A_i = \frac{h_f}{1 + h_o R_L}$$

yields

$$\boxed{Z_i = \frac{V_i}{I_i} = h_i - \frac{h_f h_r R_L}{1 + h_o R_L}} \tag{9.73}$$

Output Impedance, $Z_o = V_o/I_o$

The output impedance of an amplifier is defined to be the ratio of the output voltage to the output current with the signal V_s set to zero. For the input circuit with $V_s = 0$,

$$I_i = \frac{-h_r V_o}{R_s + h_i}$$

Substituting this relationship into the following equation obtained from the output circuit yields

$$I_o = h_f I_i + h_o V_o$$
$$= \frac{-h_f h_r V_o}{R_s + h_i} + h_o V_o$$

and

$$Z_o = \frac{V_o}{I_o} = \frac{1}{h_o - [h_f h_r/(h_i + R_s)]} \tag{9.74}$$

Power Gain, $A_p = P_L/P_i$

The average power to a load is $V_L I_L \cos \theta$, which for the situation being considered is $V_o I_o \cos \theta$. If we limit our discussion to purely resistive loads, then $\cos \theta = 1$ and $P_L = P_o = V_o I_o$. The input power is $V_i I_i$, so that

$$A_p = \frac{P_L}{P_i} = \frac{V_o I_o}{V_i I_i}$$

but

$$A_v = \frac{V_o}{V_i} \quad \text{and} \quad A_i = \frac{I_o}{I_i}$$

and

$$A_p = A_v A_i \tag{9.75}$$

In terms of the h-parameters,

$$A_p = \frac{h_f^2 R_L}{(1 + h_o R_L)[h_i + (h_i h_o - h_f h_r) R_L]} \tag{9.76}$$

EXAMPLE 9.12

Determine A_i, A_v, Z_i, A_{v_s}, Z_o, and A_p for the network of Fig. 9.49. Compare the results to those obtained with the approximate equivalent circuit.

Solution:

The complete hybrid equivalent model is substituted in Fig. 9.50. A Thévenin equivalent circuit for the input section as shown in Fig. 9.50 will permit removing R_B from further analysis since $E_{\text{Th}} \cong V_s$ and $R_{\text{Th}} \cong R_s = 1\ \text{k}\Omega$. Defining I_o' and $R_L' = R_C \| R_L$ will then result in the configuration of Fig. 9.51, which matches Fig. 9.48 directly. The equations derived above can therefore be applied for all the quantities of interest by direct substitution.

$\boldsymbol{A_i}$: [Eq. (9.71)]:

$$A_i' = \frac{I_o'}{I_i} = \frac{h_{fe}}{1 + h_{oe} R_L'} = \frac{110}{1 + (20 \times 10^{-6})(2.56 \times 10^{+3})}$$

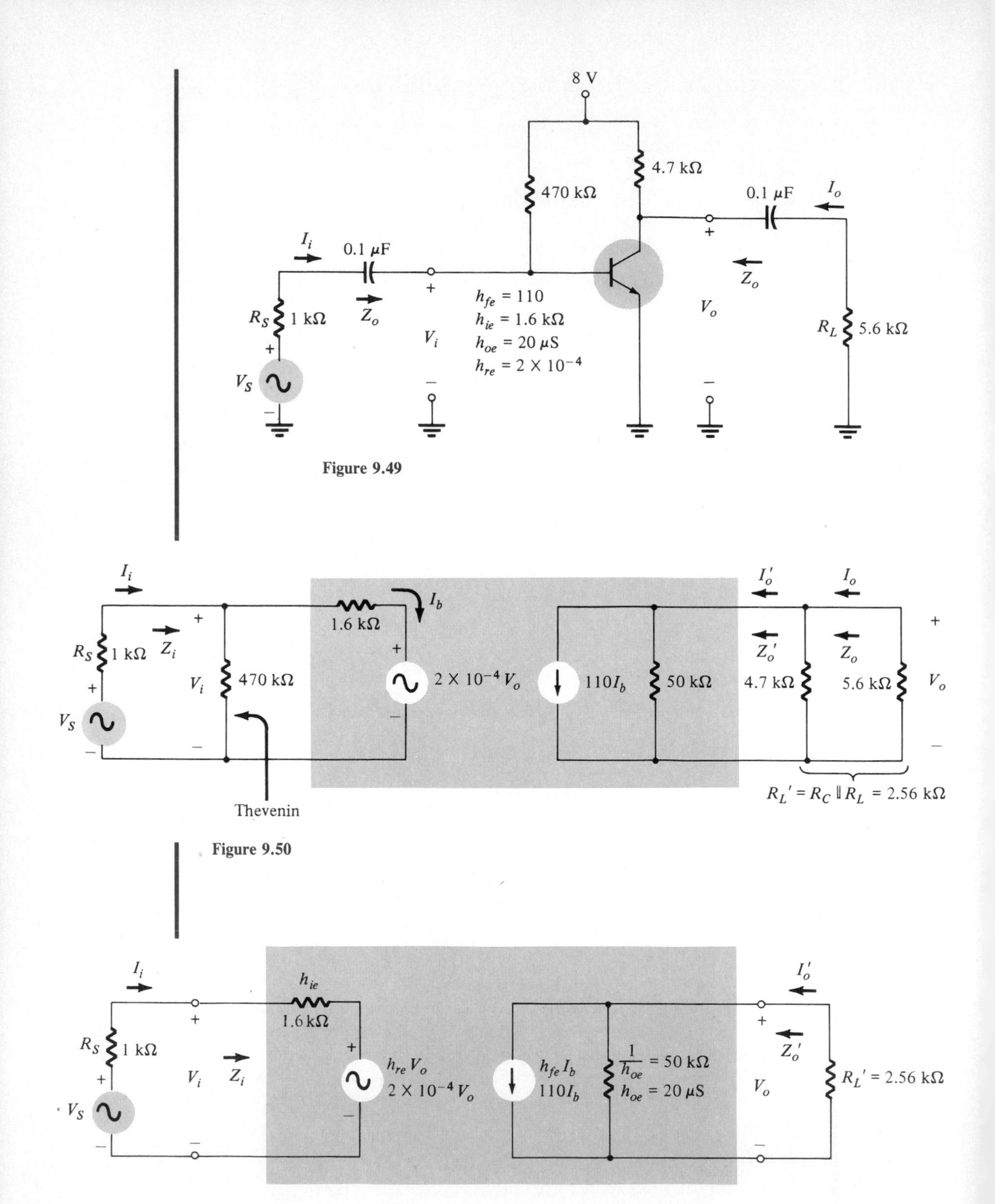

Figure 9.49

Figure 9.50

Figure 9.51

$$= \frac{110}{1 + 51.2 \times 10^{-3}} = \frac{110}{1 + 0.0512} = \frac{110}{1.0512}$$

$$= 104.64$$

I_o and I_o' are related by the current-divider rule:

$$I_o = \frac{(4.7\ \text{k}\Omega)I_o'}{4.7\ \text{k}\Omega + 5.6\ \text{k}\Omega}$$

and

$$\frac{I_o}{I_o'} = \frac{4.7\ \text{k}\Omega}{10.3\ \text{k}\Omega} = 0.456$$

with

$$A_i = \frac{I_o}{I_i} = \frac{I_o}{I_o'}\frac{I_o'}{I_i} = (0.456)(104.64)$$

$$\cong \mathbf{47.72}$$

A_v [Eq. (9.72)]:

$$A_v = \frac{V_o}{V_i} = \frac{-h_{fe}R_L'}{h_{ie} + (h_{ie}h_{oe} - h_{fe}h_{re})R_L'}$$

$$h_{ie}h_{oe} - h_{fe}h_{re} = (1.6 \times 10^3)(20 \times 10^{-6}) - (110)(2 \times 10^{-4})$$

$$= 32 \times 10^{-3} - 22 \times 10^{-3}$$

$$= 10 \times 10^{-3}$$

and

$$A_v = \frac{-(110)(2.56 \times 10^3)}{1.6 \times 10^3 + (10 \times 10^{-3})(2.56 \times 10^3)}$$

$$= \frac{-281.6 \times 10^3}{1600 + 25.6} = \frac{-281.6 \times 10^3}{1625.6}$$

$$= \mathbf{-173.23}$$

Z_i [Eq. (9.73)]:

$$Z_i = h_{ie} - \frac{h_{fe}h_{re}R_L'}{1 + h_{oe}R_L'}$$

$$= 1.6 \times 10^3 - \frac{(110)(2 \times 10^{-4})(2.56 \times 10^3)}{1.0512}$$

$$= 1600 - \frac{56.32}{1.0512} = 1600 - 53.58$$

$$\cong \mathbf{1546.4\ \Omega}$$

A_{v_s}**:** Referencing Fig. 9.51, we have

$$V_i = \frac{Z_iV_s}{Z_i + R_s}$$

and

$$\frac{V_i}{V_s} = \frac{Z_i}{Z_i + R_s} = \frac{1546.4}{1546.4 + 1000}$$

$$= 0.607$$

and
$$A_{v_s} = \frac{V_o}{V_s} = \frac{V_o}{V_i}\frac{V_i}{V_s} = (-173.23)(0.607)$$
$$= \mathbf{-105.15}$$

Z_o [Eq. (9.74)]:

$$Z'_o = \frac{1}{h_{oe} - [h_{fe}h_{re}/(h_{ie} + R_s)]}$$
$$= \frac{1}{20 \times 10^{-6} - [(110)(2 \times 10^{-4})/(1.6 \times 10^3 + 1 \times 10^3)]}$$
$$= \frac{1}{20 \times 10^{-6} - 8.46 \times 10^{-6}}$$
$$Z'_o = \frac{1}{11.54 \times 10^{-6}} = 86.66 \text{ k}\Omega$$
$$Z_o = Z'_o \| 4.7 \text{ k}\Omega = 86.66 \text{ k}\Omega \| 4.7 \text{ k}\Omega = \mathbf{4.46 \text{ k}\Omega}$$

A_p [Eq. (9.75)]:

$$|A_p| = |A_v| \cdot |A_i|$$
$$= (173.23)(47.72)$$
$$\cong \mathbf{8266.5}$$

Approximate Solution ($h_{re} \cong 0$, $1/h_{oe} \cong \infty\ \Omega$):

A_i:

$$A_i = \frac{I_o}{I_i} = \frac{I_o}{I'_o}\frac{I'_o}{I_i} = (0.456)(h_{fe})$$
$$= (0.456)(110)$$
$$= \mathbf{50.16} \text{ vs. } 47.72 \text{ above}$$

A_v:

$$A_v = \frac{-h_{fe}R'_L}{h_{ie}} = \frac{-(110)(2.56 \times 10^3)}{1.6 \times 10^3}$$
$$= \mathbf{-176} \text{ vs. } -173.23 \text{ above}$$

Z_i:

$$Z_i \cong h_{ie}$$
$$= \mathbf{1.6 \times 10^3} \text{ vs. } 1.546 \times 10^3 \text{ above}$$

A_{v_s}:

$$\frac{V_i}{V_s} = \frac{Z_i}{Z_i + R_s} = \frac{1.6 \times 10^3}{1.6 \times 10^3 + 1 \times 10^3}$$
$$= 0.615$$

and $$A_{v_s} = \frac{V_o}{V_s} = \frac{V_o}{V_i}\frac{V_i}{V_s} = (-176)(0.615)$$

$$= \mathbf{-108.24} \text{ vs. } 105.15 \text{ above}$$

***Z_o*:**

$$Z_o \cong R_C = \mathbf{4.7\ k\Omega} \text{ vs. } 4.46\ \text{k}\Omega \text{ above}$$

***A_p*:**

$$|A_p| = |A_v| \cdot |A_i| = (176)(50.16)$$

$$= \mathbf{8828.16} \text{ vs. } 8266.5 \text{ above}$$

The largest percent difference between the results obtained for the complete approximate models is 6.4% (from the power results), indicating that the approximate approach is a valid method of analysis when you further consider that the elements of the network will often be ±5% of the stated value. Be aware, however, that there are instances where the effects of $1/h_{oe}$ and h_{re} can be important to the general solution. Computer methods permit an analysis using the complete model for many large complex systems in virtually the same time as that required using the approximate approach. A sampling of computer programs applied to this area appears in the next section.

9.11 SYSTEMS APPROACH

In recent years the introduction of a wide variety of packaged networks and systems such as those in Chapter 14 has generated an increasing interest in the systems approach to design and analysis. Fundamentally, this approach works with the terminal characteristics of a package and treats each as a building block in the formation of the total package. In Fig. 9.52, for example, the important terminal characteristics for the "packaged" amplifier are indicated.

If we take a "Thévenin look" at the output terminals, we obtain the following for

$$R_{\text{Th}} = Z_o = R_o \quad (\text{with } V_i \text{ set to zero})$$

For E_{Th} the open-circuit voltage

$$E_{\text{Th}} = V_o$$

$$= A_v V_i \quad \text{from} \quad A_v = \frac{V_o}{V_i}$$

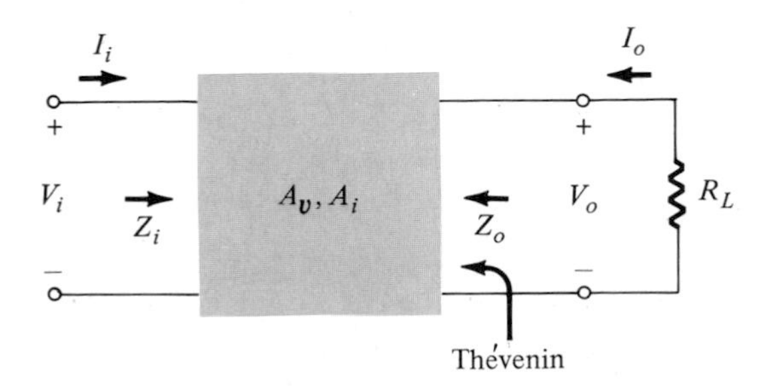

Figure 9.52

Substituting the Thévenin equivalent circuit between the output terminals will result in the configuration of Fig. 9.53. For the polarity shown for $A_v V_i$, A_v is a positive quantity. When using the format of Fig. 9.53, it is important to remember that A_v is determined without the applied load. For many of the networks analyzed in this chapter, R_C is included when A_v was determined but not R_L. A second format for Fig. 9.53, particularly popular with op-amps (operational amplifiers), appears in Fig. 9.54. The only change is in the appearance of the model.

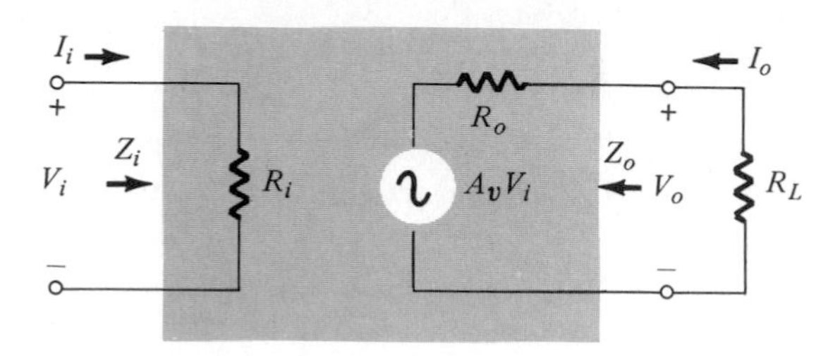

Figure 9.53

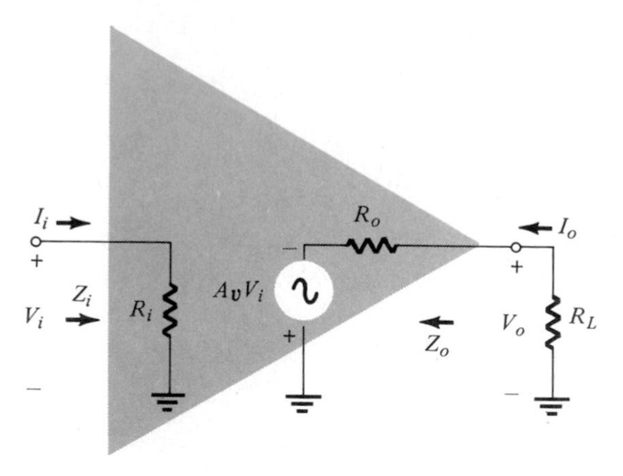

Figure 9.54

The output voltage can be determined from Fig. 9.53 or 9.54 by application of the voltage-divider rule:

$$V_{o_L} = \frac{R_L A_v V_i}{R_L + R_o}$$

and

$$\boxed{A_{v_L} = \frac{V_{o_L}}{V_i} = \frac{R_L A_v}{R_L + R_o}} \tag{9.77}$$

In other words, the gain with a load applied is $R_L/(R_L + R_o)$ times the gain without a load.

For the current gain,

$$|A_i| = \frac{I_o}{I_i} = \frac{V_{o_L}/R_L}{V_i/Z_i} = \frac{Z_i}{R_L}\frac{V_{o_L}}{V_i}$$

and

$$|A_i| = \frac{Z_i}{R_L}|A_{v_L}| \tag{9.78}$$

EXAMPLE 9.13

(a) Determine the parameters of Fig. 9.55 for the configuration of Fig. 9.53.
(b) Calculate Z_i, Z_o, A_v, A_{v_L}, A_i, and A_{v_s}.

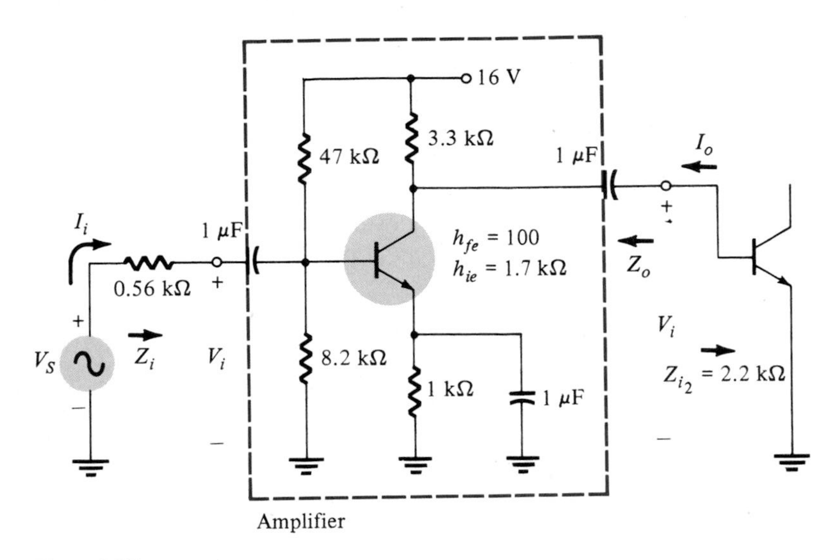

Figure 9.55

Solution:

(a) $\boldsymbol{R_i = Z_i}$:

$$R_{BB} = R_{B_1} \| R_{B_2} = 47\text{ k}\Omega \| 8.2\text{ k}\Omega = 6.98\text{ k}\Omega$$

$$Z_b = h_{ie} = 1.7\text{ k}\Omega$$

(bypassed emitter resistor)

and

$$R_i = R_{BB} \| Z_b = 6.98\text{ k}\Omega \| 1.7\text{ k}\Omega$$

$$R_i = \mathbf{1.367\text{ k}\Omega}$$

$\boldsymbol{R_o = Z_o}$:

$$R_o = R_C \cong \mathbf{3.3\text{ k}\Omega}$$

$\boldsymbol{A_v}$:

$$A_v = \frac{V_o}{V_i}\text{(no-load)}$$

$$= \frac{-h_{fe}R_C}{h_{ie}} = \frac{-(100)(3.3\text{ k}\Omega)}{1.7\text{ k}\Omega}$$

$$= \mathbf{-194.12}$$

The results appear in Fig. 9.56.

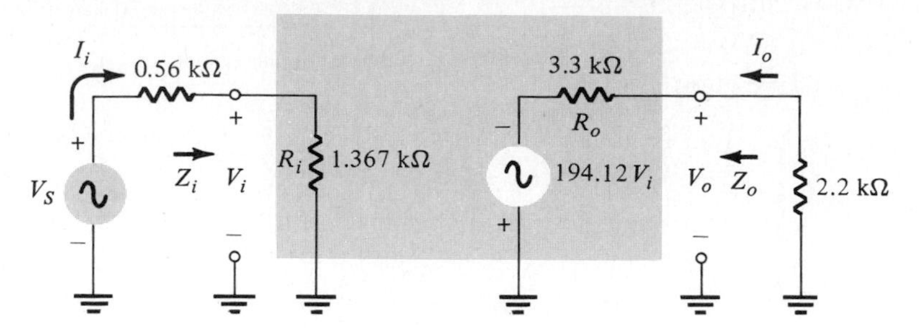

Figure 9.56

(b) Z_i, Z_o, and A_v are provided above.

A_{v_L} [Eq. (9.77)]:

$$A_{v_L} = \frac{R_L A_v}{R_L + R_o} = \frac{(2.2\text{ k}\Omega)(-194.12)}{2.2\text{ k}\Omega + 3.3\text{ k}\Omega}$$

$$= \mathbf{-77.65}$$

As a check using earlier methods,

$$A_{v_L} = \frac{-h_{fe}(R_C \| R_L)}{h_{ie}}$$

$$= \frac{-(100)(1.32\text{ k}\Omega)}{1700}$$

$$= \mathbf{-77.65} \quad \text{as above}$$

A_i [Eq. (9.78)]:

$$A_i = \frac{Z_i}{R_L}|A_{v_L}|$$

$$= \frac{1.367\text{ k}\Omega}{2.2\text{ k}\Omega}(77.65)$$

$$= \mathbf{48.2}$$

A_{v_s}:

$$V_i = \frac{R_i V_s}{R_i + R_s}$$

and

$$\frac{V_i}{V_s} = \frac{R_i}{R_i + R_s} = \frac{1.367\text{ k}\Omega}{1.367\text{ k}\Omega + 0.56\text{ k}\Omega} = 0.71$$

with

$$A_{v_s} = \frac{V_{OL}}{V_s} = \frac{V_{OL}}{V_i}\frac{V_i}{V_s} = (-77.65)(0.71)$$

$$= \mathbf{-55.13}$$

9.12 COMPUTER ANALYSIS

The relatively complex equations that result when the complete hybrid equivalent circuit is employed can easily be solved with a high degree of accuracy using the computer. Figure 9.57 shows a computer program that will return the important parameters of the network of Fig. 9.58 when the network elements are entered on lines 160 through 280. As indicated on the printout, lines 310 through 400 will print out the results.

```
10 REM ---------------------------------------------------------------
20 REM                        PROGRAM 11-1
30 REM ---------------------------------------------------------------
40 REM                      BJT AC ANALYSIS
50 REM                  USING HYBRID PARAMETERS
60 REM ---------------------------------------------------------------
70 REM
100 CLS
110 PRINT "This program performs ac analysis of a BJT circuit"
120 PRINT "using hybrid parameters."
130 PRINT
140 PRINT "Enter the following circuit data:"
150 PRINT
160 INPUT "RB1=";R1
170 INPUT "RB2=";R2
180 INPUT "RC=";RC
190 PRINT
200 INPUT "Load resistance, RL=";RL
210 INPUT "Source resistance, Rs=";RS
220 INPUT "Source voltage, Vs=";VS
230 PRINT :PRINT "Enter the BJT hybrid parameter values:"
240 PRINT
250 INPUT "hie=";HI
260 INPUT "hfe=";HF
270 INPUT "hoe=";HO
280 INPUT "hre=";HR
290 PRINT :PRINT
300 GOSUB 11000 :REM Perform ac analysis of BJT circuit
310 PRINT "The results of the ac analysis are:"
320 PRINT
330 PRINT "Input impedance, Ri=";RI/1000!;"kilohms"
340 PRINT "Output impedance, Ro=";RO/1000!;"kilohms"
350 PRINT "Voltage gain(no-load), Av=";AV
360 PRINT "Current gain(IL/Ii), Ai=";AI
370 PRINT
380 PRINT "Output voltage(without load), Vo=";VO;"volts"
390 PRINT
400 PRINT "Output voltage(with load), VL=";VL;"volts"
410 END
```

Figure 9.57 (From *BASIC for Electrical and Computer Technology*, Nashelsky, Boylestad, Prentice-Hall, Inc., Englewood Cliffs, N.J., 1986.)

```
11000 REM Module to perform ac calculations using hybrid parameters
11010 RP=RC*RL/(RC+RL)
11020 RZ=HI-HF*HR*RP/(1+HO*RP)
11030 RB=R1*R2/(R1+R2)
11040 RI=RZ*RB/(RZ+RB)
11050 IF HO<>0 AND HR<>0 THEN RT=1/(HO-HF*HR/(HI+RS)) ELSE RT=1E+30
11060 RO=RT*RC/(RT+RC)
11070 AI=(RB/(RB+RZ))*(HF/(1+HO*RP))*(RC/(RC+RL))
11080 IF AI<.000001 THEN AI=0
11090 AV=-HF*RC/(HI+(HI*HO-HF*HR)*RC)
11100 VI=RI*VS/(RI+RS)
11110 VO=AV*VI
11120 VL=VO*RL/(RO+RL)
11130 RETURN
```

```
This program performs ac analysis of a BJT circuit
using hybrid parameters.

Enter the following circuit data:

RB1=? 47E3
RB2=? 8.2E3
RC=? 3.3E3

Load resistance, RL=? 2.2E3
Source resistance, Rs=? 0
Source voltage, Vs=? 1E-3

Enter the BJT hybrid parameter values:

hie=? 1.7E3
hfe=? 100
hoe=? 20E-6
hre=? 3E-4


The results of the ac analysis are:

Input impedance, Ri= 1.34206 kilohms
Output impedance, Ro= 3.274574 kilohms
Voltage gain(no-load), Av=-192.622
Current gain(IL/Ii), Ai= 47.22017

Output voltage(without load), Vo=-.192622 volts

Output voltage(with load), VL=-7.740665E-02 volts
Ok
```

Figure 9.57 (cont.)

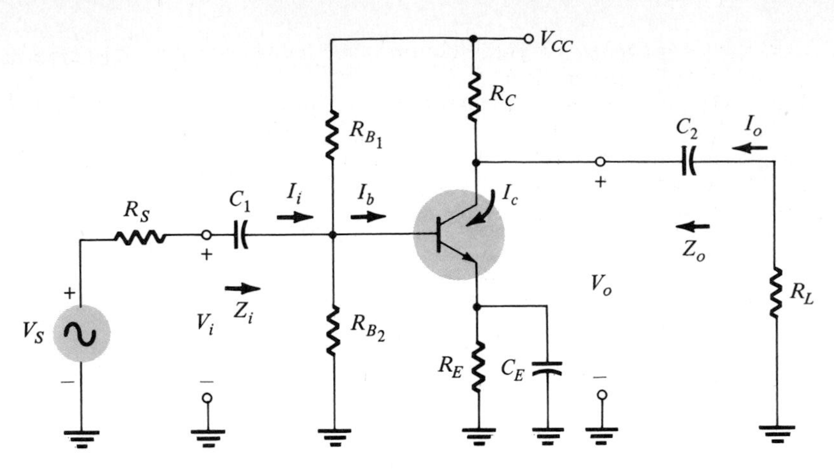

Figure 9.58

Line 11010 determines $R' = R_C \| R_L$ and line 11020 determines Z_b from

$$Z_b = h_{ie} - \frac{h_{fe}h_{re}R'}{1 + h_{oe}R'}$$

Line 11030 calculates R_{BB} from $R_{BB} = R_{B_1} \| R_{B_2}$ and line 11040 determines Z_i from

$$Z_i = R_{BB} \| Z_b$$

If h_{oe} and $h_{re} \neq 0$, line 11050 determines the output impedance of the transistor from

$$Z_o' = \frac{1}{h_{oe} - [h_{fe}h_{re}/(h_{ie} + R_s)]}$$

Otherwise, Z_o' is assumed to be an open circuit as approximated by $Z_o' = 1 \times 10^{30}\ \Omega$.

The output impedance is then determined by line 11060:

$$Z_o = R_C \| Z_o'$$

The current gain is determined by

$$A_i = \frac{I_o}{I_i} = \frac{I_b}{I_i}\frac{I_c}{I_b}\frac{I_o}{I_c}$$

where the current-divider rule provides

$$I_b = \frac{R_B I_i}{R_B + Z_i} \quad \text{and} \quad I_o = \frac{R_C I_c}{R_C + R_L}$$

$$\frac{I_c}{I_b} \text{ is the standard format of } A_i = \frac{h_{fe}}{1 + h_{oe}R'}$$

The voltage gain (unloaded) is determined from

$$A_v = \frac{-h_{fe}R_C}{h_{ie} + (h_{ie}h_{oe} - h_{fe}h_{re})R_C}$$

where A_v is the voltage gain in Fig. 9.53. At the input side the voltage-divider rule provides that

$$V_i = \frac{R_i V_s}{R_i + R_s}$$

and line 11110 determines $V_o = A_v V_i$ for the configuration of Fig. 9.53. Finally,

$$V_{o_L} = \frac{R_L V_o}{R_L + R_o} \qquad \text{where } V_o = A_v V_i$$

The run of the program used the same values of h_{fe} and h_{ie} employed in Example 9.13 for the same network configuration. h_{re} and h_{oe} were given the common values of 4×10^{-4} and 20 μA/V, respectively.

Note how closely the results match those obtained in Example 9.13 using the approximate approach. Recognize also that the magnitude of the output voltage is determined in the computer run, not the ratio. To match values the input voltage was set at 1 mV for the run provided. With the program in storage the important parameters of the configuration can quickly be determined for any combination of network elements.

Program 11-2 provides a detailed analysis of a BJT transistor configuration with voltage-divider bias and unbypassed and bypassed resistors R_{E_1} and R_{E_2}, respectively. The r_e model is employed, requiring a determination of r_e and $\beta(r_e + R_{E_1})$ as covered by lines 11210 through 11260. The ac equivalent circuit of Fig. 9.60 is then employed to determine the list of important characteristics of the system by lines 11270 through 11330. A number of the important equations employed appear in Fig. 9.60.

```
10 REM -------------------------------------------------------------------
20 REM                        PROGRAM 11-2
30 REM -------------------------------------------------------------------
40 REM                      BJT AC ANALYSIS
50 REM                USING re AND BETA PARAMETERS
60 REM -------------------------------------------------------------------
70 REM
100 CLS
110 PRINT :PRINT "This program performs the ac calculations"
120 PRINT "for a BJT circuit using the re and beta parameters."
130 PRINT
140 PRINT "Enter the following circuit data:
150 PRINT
160 INPUT "RB1=";R1
170 INPUT "RB2=";R2
180 INPUT "RC=";RC
190 INPUT "Unbypassed emitter resistance, RE1=";E1
200 INPUT "Bypassed emitter resistance, RE2=";E2
210 PRINT
220 INPUT "Beta=";BETA
230 INPUT "Supply voltage, Vcc=";CC
240 INPUT "Load resistance, RL=";RL
```

Figure 9.59 (From *BASIC for Electrical and Computer Technology*, Nashelsky, Boylestad, Prentice-Hall, Inc., Englewood Cliffs, N.J., 1986.)

```
250 INPUT "Source resistance, RS=";RS
260 INPUT "Source voltage, VS=";VS
270 PRINT :PRINT
280 GOSUB 11200 :REM Perform ac analysis
290 PRINT "The results of the ac analysis are:"
300 PRINT
310 PRINT "Transistor dynamic resistance, re=";RE;"ohms"
320 PRINT
330 IF CC-IE*(RC+E1+E2)<=0 THEN PRINT "Circuit in saturation." :GOTO 420
340 PRINT "Input impedance, Ri=";RI;"ohms"
350 PRINT "Output impedance, Ro=";RO;"ohms"
360 PRINT "Voltage gain(no-load), Av=";AV
370 PRINT "Current gain, Ai=";AI
380 PRINT
390 PRINT "Output voltage(no-load), Vo=";VO;"volts"
400 PRINT
410 PRINT "Output voltage(under load), VL=";VL;"volts"
420 PRINT
430 VM=CC-IE*(BETA/BETA+1)*(RC+E1+E2) :REM Maximum signal swing
440 IF ABS(VL)>VM THEN PRINT "but maximum undistorted output is";VM;"volts"
450 END

11200 REM Module to perform BJT ac analysis using re model.
11210 RB=R1*(R2/(R1+R2))
11220 RP=RC*(RL/(RC+RL))
11230 BB=R2*CC/(R1+R2)
11240 IE=(BB-.7)*(BETA+1)/(RB+BETA*(E1+E2))
11250 RE=.026/IE
11260 R3=BETA*(RE+E1)
11270 RI=RB*(R3/(RB+R3))
11280 RO=RC
11290 AI=(RC/(RC+RL))*BETA*(RB/(RB+R3))
11300 AV=-RC/(E1+RE)
11310 VI=VS*(RI/(RI+RS))
11320 VO=AV*VI
11330 VL=VO*(RL/(RO+RL))
11340 RETURN


This program performs ac analysis of a BJT circuit
using hybrid parameters.

Enter the following circuit data:

RB1=? 47E3
RB2=? 8.2E3
RC=? 3.3E3

Load resistance, RL=? 2.2E3
Source resistance, Rs=? 0
Source voltage, Vs=? 1E-3

Enter the BJT hybrid parameter values:
```

Figure 9.59 (cont.)

```
hie=? 1.7E3
hfe=? 100
hoe=? 20E-6
hre=? 3E-4


The results of the ac analysis are:

Input impedance, Ri= 1.34206 kilohms
Output impedance, Ro= 3.274574 kilohms
Voltage gain(no-load), Av=-192.622
Current gain(IL/Ii), Ai= 47.22017

Output voltage(without load), Vo=-.192622 volts

Output voltage(with load), VL=-7.740665E-02 volts
```

Figure 9.59 (cont.)

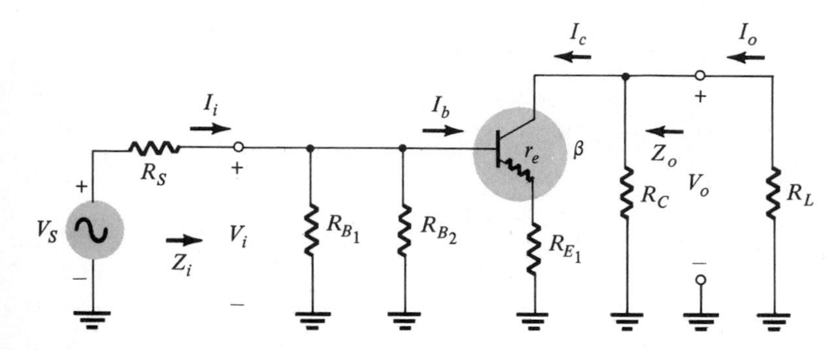

Figure 9.60

PROBLEMS

For all problems assume silicon-based transistors ($V_{BE} = 0.7$ V)

§ 9.2

1. (a) For the network of Fig. 9.61, determine Z_i, Z_o, A_v, and A_i.
   (b) For $h_{oe} = 20\ \mu$S, determine the quantities of part (a).

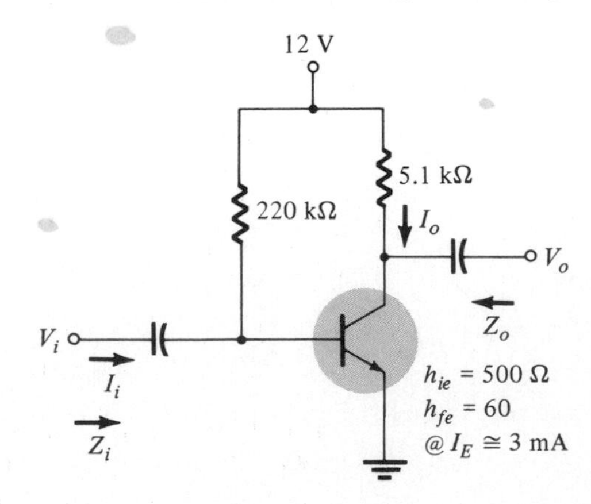

Figure 9.61

2. For the network of Fig. 9.62, determine V_{CC} for $A_v = -200$.

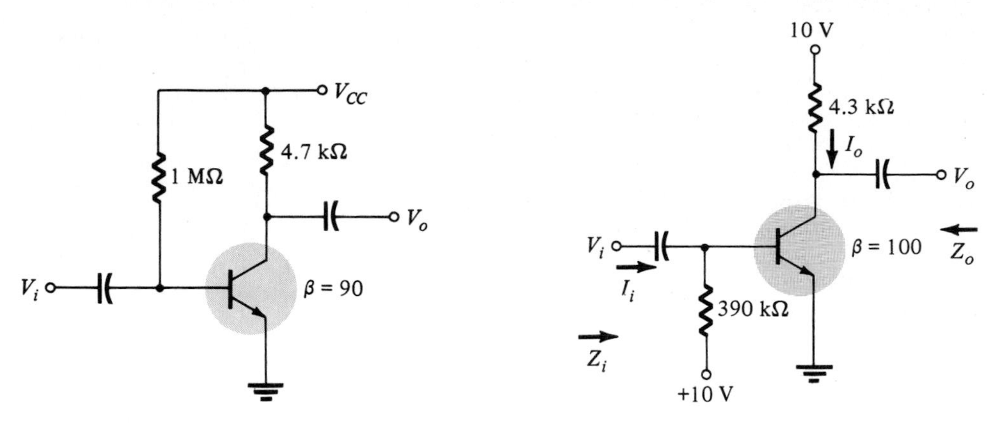

Figure 9.62

Figure 9.63

3. For the network of Fig. 9.63:
(a) Calculate I_B, I_C, and r_e.
(b) Determine h_{fe} and h_{ie}.
(c) Calculate Z_i, Z_o, A_v, and A_i.
(d) Determine the effect of $h_{oe} = 25\ \mu S$ on A_v and A_i.

§ 9.3

4. (a) Determine Z_i, Z_o, A_v, and A_i for the network of Fig. 9.64.
(b) For $h_{oe} = 20\ \mu S$, determine the effect on the quantities of part (a).
(c) Calculate r_e and compare to $r_e = h_{ie}/\beta$.

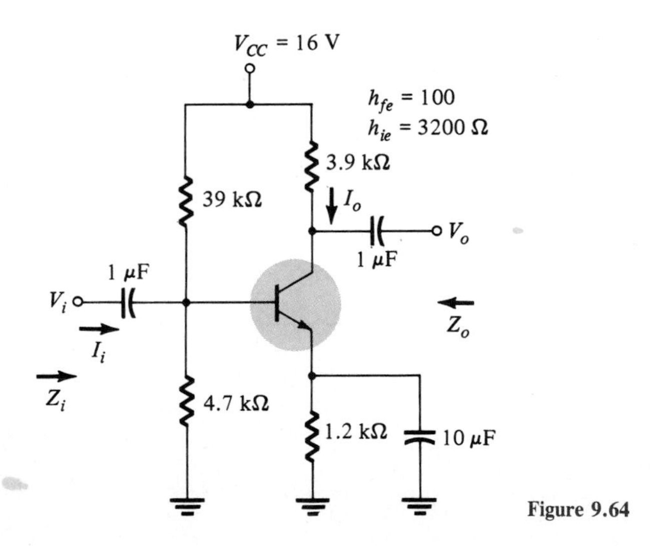

Figure 9.64

5. Determine V_{CC} for the network of Fig. 9.65 if $A_v = -160$.

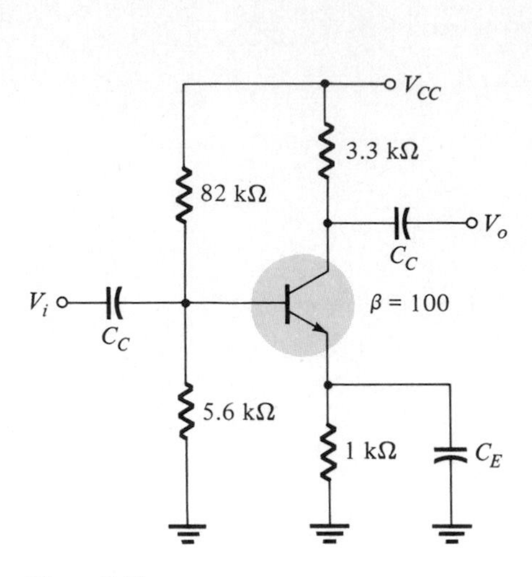

Figure 9.65

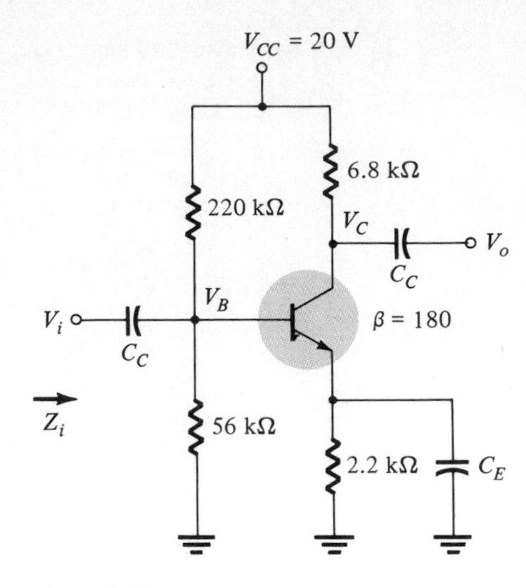

Figure 9.66

6. For the network of Fig. 9.66:
   (a) Determine r_e.
   (b) Calculate V_B and V_C.
   (c) Determine Z_i and $A_v = V_o/V_i$.

§ 9.4

7. For the network of Fig. 9.67:
   (a) Determine Z_i, Z_o, A_v, and A_i.
   (b) Calculate the effect of $h_{oe} = 20\ \mu S$ on the quantities of part (a).

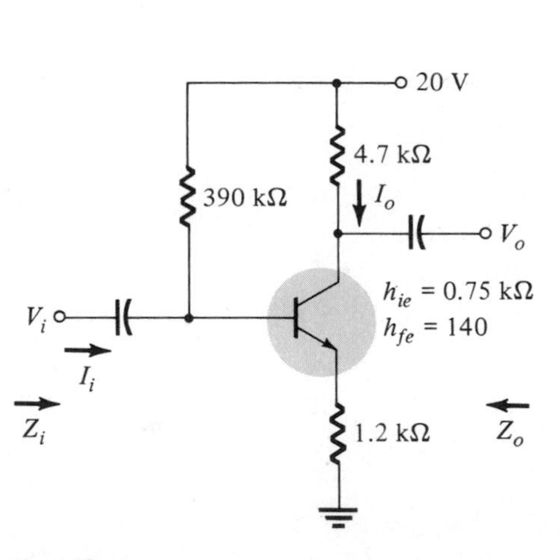

Figure 9.67

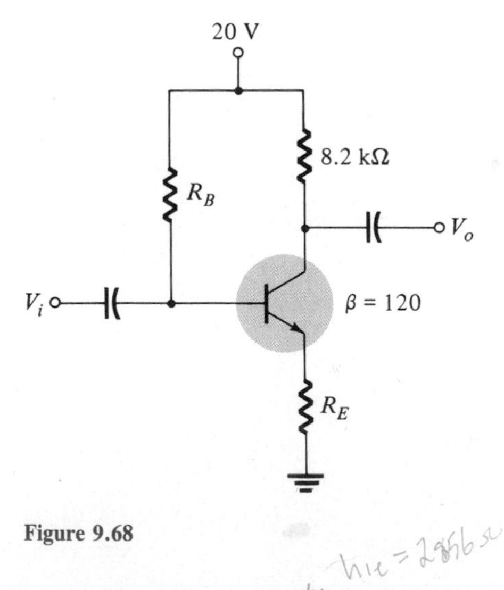

Figure 9.68

8. For the network of Fig. 9.68, determine R_E and R_B if $A_v = -10$ and $r_e = 3.8\ \Omega$. Assume that $Z_b \cong \beta R_E$.

9. For the network of Fig. 9.69:
(a) Determine r_e.
(b) Find Z_i and A_v.
(c) Calculate A_i.

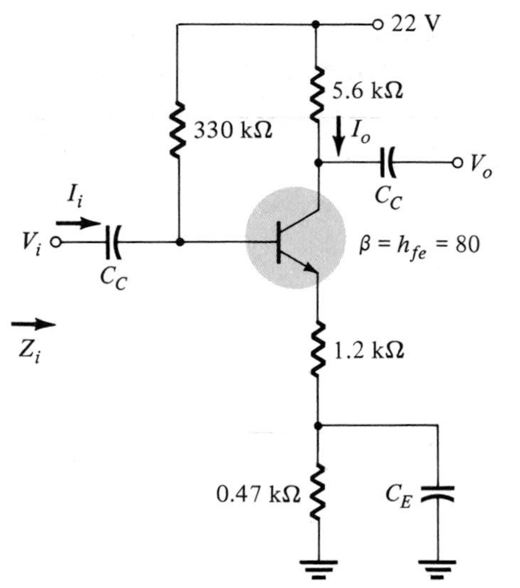

Figure 9.69

§ 9.5

10. For the network of Fig. 9.70:
(a) Calculate r_e and βr_e. How does βr_e compare to h_{ie}?
(b) Determine Z_i, Z_o, A_v, and A_i.

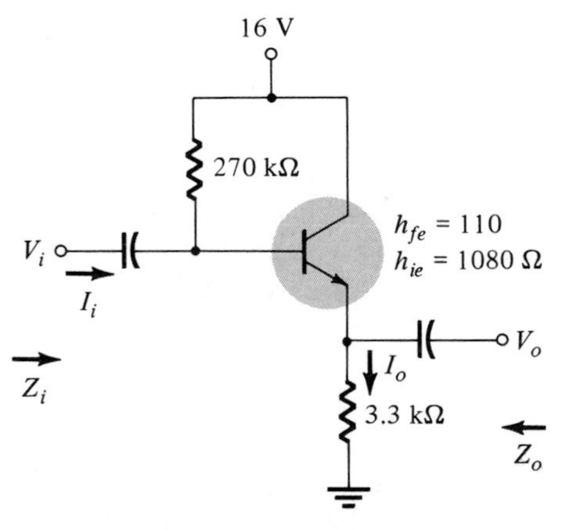

Figure 9.70

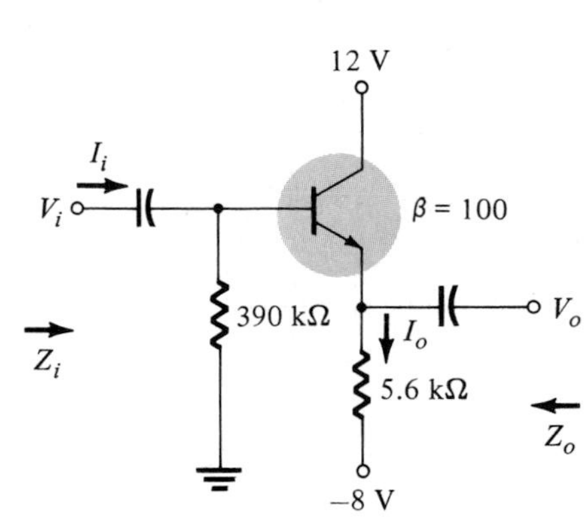

Figure 9.71

11. For the network of Fig. 9.71:
(a) Determine Z_i, Z_o, A_v, and A_i.
(b) Calculate V_o if V_i = 1 mV.

12. For the network of Fig. 9.72:
(a) Determine r_e.
(b) Calculate I_B and I_C.
(c) Determine A_v and A_i.

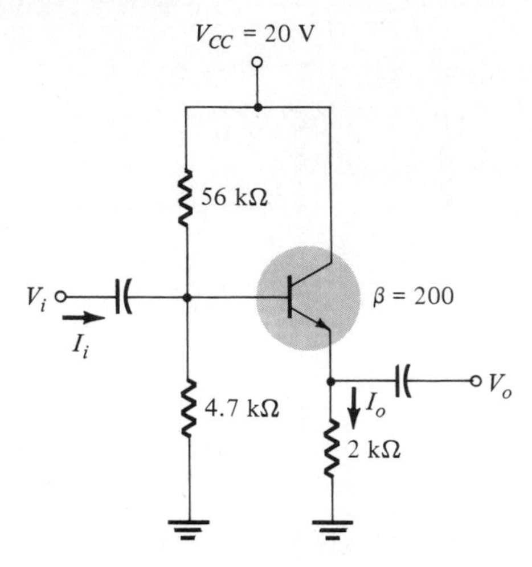

Figure 9.72

§ 9.6

13. For the common-base configuration of Fig. 9.73:
(a) Determine r_e.
(b) Calculate Z_i, Z_o, A_v, and A_i.

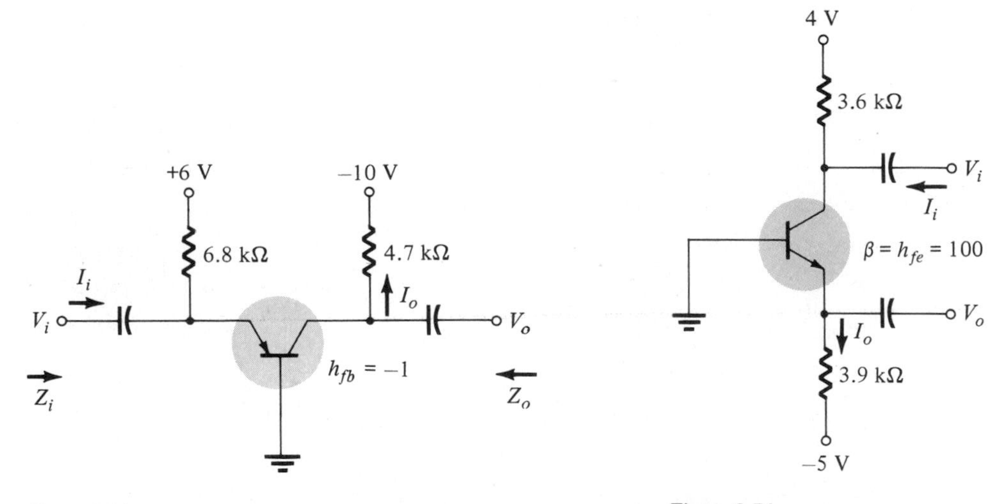

Figure 9.73

Figure 9.74

14. For the network of Fig. 9.74, determine A_v and A_i.

§ 9.7

15. For the network of Fig. 9.75, determine Z_i, Z_o, A_v, and A_i.

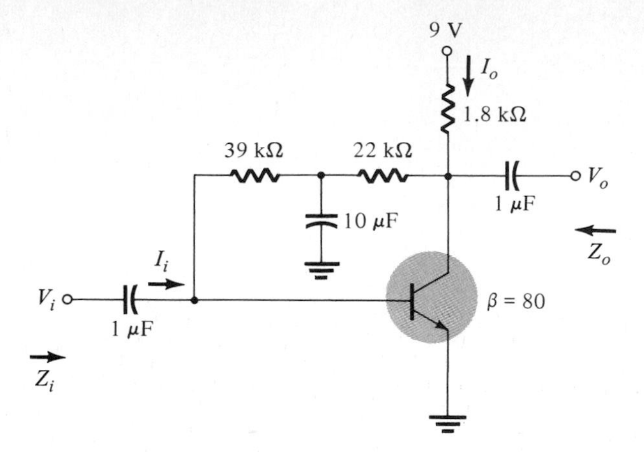

Figure 9.75

16. For the difference amplifier of Fig. 9.31, if $R_C = 5.6\ \text{k}\Omega$, $R_E = 1.1\ \text{k}\Omega$, $h_{fe} = 100$, $h_{ie} = 2\ \text{k}\Omega$, $V_{i_1} = 10$ mV, and $V_{i_2} = 4$ mV calculate the magnitude of:
(a) V_o.
(b) I_b.
(c) $I_c = I_e$.
(d) V_e.
Compare the magnitude of $V_b = V_i$ to V_e.

§ 9.8

17. For the collector FB configuration of Fig. 9.76:
(a) Determine r_e.
(b) Calculate Z_i, Z_o, A_v, and A_i.

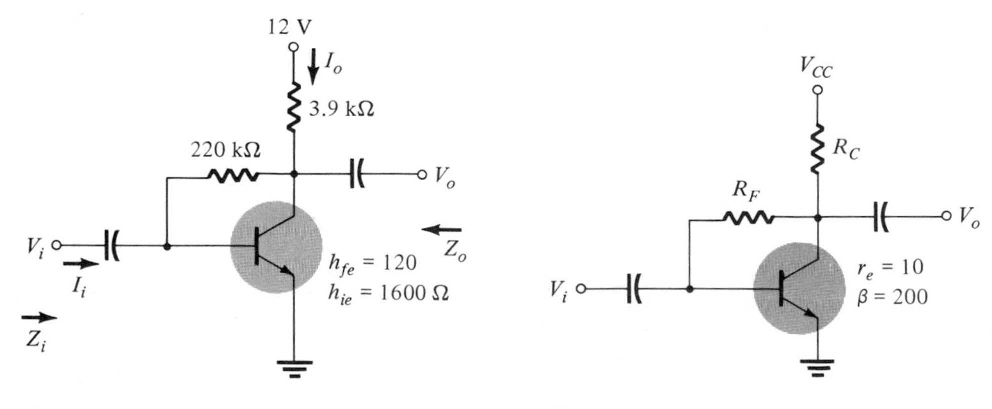

Figure 9.76

Figure 9.77

18. Given $r_e = 10$, $\beta = 200$, $A_v = -160$, and $A_i = +19$ for the network of Fig. 9.77, determine R_C, R_F, and V_{CC}.

19. For the network of Fig. 9.37:
(a) Derive the approximate equation for A_v.
(b) Derive the approximate equation for A_i.
(c) Derive the approximate equations for Z_i and Z_o.
(d) Given $R_C = 2.2\ \text{k}\Omega$, $R_F = 120\ \text{k}\Omega$, $R_E = 1.2\ \text{k}\Omega$, $\beta = 90$, $V_{CC} = 10$ V calculate the magnitude of A_v, A_i, Z_i, and Z_o using the equations of parts (a) through (c).

§ 9.9

20. (a) Determine $A_{v_s} = V_o/V_s$ and $A_i = I_o/I_i$ for the network of Fig. 9.61 with $R_s = 220\ \Omega$ and $R_L = 6.8\ \text{k}\Omega$. R_L is connected from collector to ground and I_o is the current through R_L.
(b) Compare with the results of Problem 1.

21. (a) Determine $A_{v_s} = V_o/V_s$ and $A_i = I_o/I_i$ for the network of Fig. 9.64 with $R_s = 2.2\ \text{k}\Omega$ and $R_L = 4.7\ \text{k}\Omega$. R_L is connected from collector to ground and I_o is the current through R_L.
(b) Compare with the results of Problem 4.

22. (a) Determine $A_{v_s} = V_o/V_s$ and $A_i = I_o/I_i$ for the network of Fig. 9.67 with $R_s = 1.2\ \text{k}\Omega$ and $R_L = 22\ \text{k}\Omega$. R_L is connected from collector to ground and I_o is the current through R_L.
(b) Compare to the results of Problem 7.

23. (a) Determine $A_{v_s} = V_o/V_c$ and $A_i = I_o/I_i$ for the network of Fig. 9.70 with $R_s = 6.8\ \text{k}\Omega$ and $R_L = 3.3\ \text{k}\Omega$. R_L is connected from emitter to ground and I_o is the current through R_L.
(b) Compare to the results of Problem 10.

24. (a) Determine $A_{v_s} = V_o/V_s$ and $A_i = I_o/I_i$ for the network of Fig. 9.73 with $R_s = 0.2\ \text{k}\Omega$ and $R_L = 6.8\ \text{k}\Omega$. R_L is connected from collector to ground and I_o is the current through R_L.
(b) Compare to the results of Problem 13.

§ 9.10

25. (a) Determine A_v, A_i, Z_i, Z_o, and A_p for the network of Fig. 9.61 if $h_{ie} = 0.5\ \text{k}\Omega$, $h_{fe} = 60$, $h_{re} = 1.5 \times 10^{-4}$, and $h_{oe} = 25\ \mu\text{S}$.
(b) Compare the results of part (a) with the results of Problem 1.
(c) Determine A_{v_s} and the same quantities as part (a) with $R_s = 220\ \Omega$ and $R_L = 6.8\ \text{k}\Omega$.
(d) Compare the results of part (c) with the results of Problem 20.

26. (a) Determine A_v, A_i, Z_i, Z_o, and A_p for the network of Fig. 9.64 if $h_{ie} = 3.2\ \text{k}\Omega$, $h_{fe} = 100$, $h_{re} = 2 \times 10^{-4}$, and $h_{oe} = 20\ \mu\text{S}$.
(b) Compare the results of part (a) with the results of Problem 4.
(c) Determine A_{v_s} and the same quantities as part (a) with $R_s = 2.2\ \text{k}\Omega$ and $R_L = 4.7\ \text{k}\Omega$.
(d) Compare the results of part (c) with the results of Problem 21.

27. (a) Determine A_v, A_i, Z_i, Z_o, and A_p for the network of Fig. 9.73 if $h_{ib} = 33.4\ \Omega$, $h_{fb} = -1$, $h_{rb} = 2 \times 10^{-4}$, and $h_{ob} = 0.5\ \mu\text{S}$.
(b) Compare the results of part (a) with the results of Problem 13.
(c) Determine the same quantities as part (a) with $R_s = 0.2\ \text{k}\Omega$ and $R_L = 6.8\ \text{k}\Omega$.
(d) Compare the results of part (c) with the results of Problem 24.

§ 9.11

28. (a) Determine the parameters of Fig. 9.53 for the network of Fig. 9.61 with $R_s = 220\ \Omega$ and $R_L = 6.8\ \text{k}\Omega$.

(b) Determine $A_{v_s} = V_o/V_s$ and compare with the solution of Problem 20.

29. (a) Determine the parameters of Fig. 9.53 for the network of Fig. 9.64 with $R_s = 2.2\ \text{k}\Omega$ and $R_L = 4.7\ \text{k}\Omega$.

(b) Determine $A_{v_s} = V_o/V_s$ and compare with the solution of Problem 21.

30. (a) Determine the parameters of Fig. 9.53 for the network of Fig. 9.73 with $R_s = 0.2\ \text{k}\Omega$ and $R_L = 6.8\ \text{k}\Omega$.

(b) Determine $A_{v_s} = V_o/V_s$ and compare with the solution of Problem 24.

COMPUTER PROBLEMS

31. Write a computer program to perform a complete analysis of the common-emitter fixed-bias network with both R_s and R_L. Determine r_e, h_{ie}, $A_v = V_o/V_i$, $A_{v_s} = V_o/V_s$, A_i, Z_i, and Z_o. All the resistor values, dc source voltage, and $h_{fe} = \beta$ are provided.

32. Repeat Problem 31 for the unbypassed emitter-bias configuration.

33. Repeat Problem 31 for the common-base configuration.

10

FET Small-Signal Analysis

10.1 GENERAL INTRODUCTION

FET devices can be used to build small-signal amplifier circuits providing voltage gain at very high input resistance. Both JFET and depletion MOSFET can operate with similar dc bias, providing the same magnitude of voltage gain. The MOSFET device however provides much higher input resistance.

The common-source amplifier configuration provides the best voltage-gain operation. An input signal is applied to the gate, and the output signal is taken from the drain, the source terminal being the reference or common. A common-drain amplifier provides a noninverted output with near-unity gain. A common-gate amplifier connection is used less frequently, providing voltage gain with no polarity inversion.

The FET ac equivalent circuit is even simpler than that for a BJT, having only an output current source with value dependent on the device transconductance, g_m, the principal ac parameter. Values of FET transconductance vary from about 1 mS to 20 mS, providing larger voltage gain for larger values of g_m.

The FET can be used as a linear device or as a digital device. In linear circuits, JFET and depletion MOSFET devices are most often used. Digital FET circuits use the enhancement MOSFET especially in large-scale integrated (LSI) circuits, very-large-scale integrated (VLSI) circuits, and ultra-large-scale integrated (ULSI) circuits.

10.2 JFET/DEPLETION MOSFET SMALL-SIGNAL MODEL

To aid in an analysis of ac circuits that use FET devices, we first need to obtain an ac equivalent circuit for the device. Figure 10.1 shows a simple equivalent circuit of the FET device. The ac voltage applied to the gate-source, V_{gs}, results in a drain current, I_d, of value $g_m V_{gs}$. The device transconductance, g_m, relates the amount of current resulting from an applied voltage across gate-source. The value of g_m can be obtained from the Shockley equation[1]

$$g_m = g_{mo}\left(1 - \frac{V_{GS}}{V_P}\right) \tag{10.1}$$

where

$$g_{mo} = \frac{2I_{DSS}}{|V_P|} \tag{10.2}$$

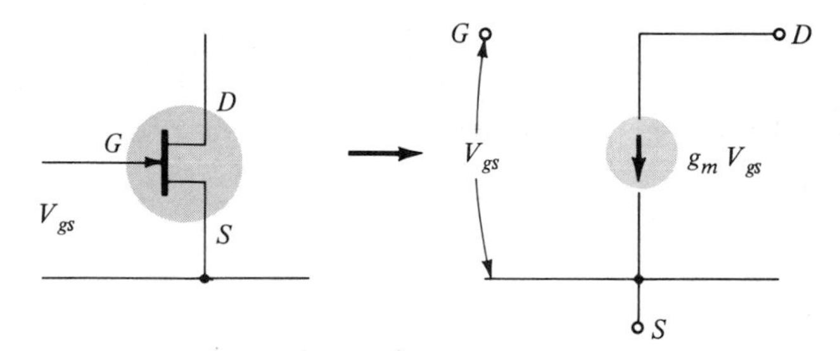

Figure 10.1 FET ac equivalent circuit.

The value g_{mo} is the value of the transconductance at the $V_{GS} = 0$ V bias point and represents a fixed value of the maximum gain of the JFET device. The value of g_{mo} is a constant for a particular FET and is not affected by the choice of dc bias point. At any bias point in the reverse-bias operating region, a value of g_m smaller than the value of g_{mo} is obtained.

---

[1] Since

$$i_D = I_{DSS}\left(1 - \frac{v_{GS}}{V_P}\right)^2$$

and

$$g_m = \left.\frac{\partial i_D}{\partial v_{GS}}\right|_{v_{DS}=\text{constant}}$$

we obtain

$$g_m = -\frac{2I_{DSS}}{V_P}\left(1 - \frac{v_{GS}}{V_P}\right) = g_{mo}\left(1 - \frac{v_{GS}}{V_P}\right)$$

where

$$g_{mo} = \frac{2I_{DSS}}{|V_P|}$$

Determination of g_m and g_{mo}

The value of g_{mo} is the transconductance at a bias value of $V_{GS} = 0$ V. That is, at $V_{GS} = 0$ V,

$$g_m = g_{mo}\left(1 - \frac{V_{GS}}{V_P}\right) = g_{mo}\left(1 - \frac{0}{V_P}\right) = g_{mo}$$

EXAMPLE 10.1

Calculate the value of g_{mo} for a JFET having $I_{DSS} = 12$ mA and $V_P = -3$ V.

Solution:

$$g_{mo} = \frac{2I_{DSS}}{|V_P|} = 2\left(\frac{12 \times 10^{-3}}{|-3\text{ V}|}\right) = \mathbf{8\ mS} = \mathbf{8{,}000\ \mu S}$$

At any bias voltage other than $V_{GS} = 0$ V, the value of V_{GS} is less than g_{mo} as given by Eq. (10.1).

EXAMPLE 10.2

Calculate V_{GS} at a voltage $V_{GS} = -1$ V for the JFET of Example 10.1 with $I_{DSS} = 12$ mA, $V_P = -3$ V.

Solution:

Using Eq. (10.1) yields

$$g_m = g_{mo}\left(1 - \frac{V_{GS}}{V_P}\right) = 8\text{ mS}\left(1 - \frac{-1}{-3}\right) = \mathbf{5.333\ mS} = \mathbf{5333\ \mu S}$$

AC Equivalent Circuit

Figure 10.2a shows a typical JFET circuit using self-bias resistor R_S for setting the dc bias. The ac equivalent circuit drawn in Fig. 10.2b shows the resistor R_S bypassed by capacitor C_S replaced by a short (capacitor's ac impedance = 0) and the resistor R_D connection to $+V_{DD}$ is connected to ac ground since the ac impedance of the voltage supply is replaced by an ac impedance of 0. The JFET device is replaced by the simple model for which an ac signal applied across gate-source V_{gs} results in a drain-source (channel) current of value $g_m V_{gs}$.

The value of g_m in Fig. 10.2b is determined from the dc bias voltage V_{GS} and the device parameters I_{DSS} and V_P. For an applied input ac voltage, V_i, there is a resulting source-drain current, $g_m V_{gs}$.

The ac model could also include the output resistance of the JFET, as shown in Fig. 10.3. This output resistance is usually listed on specification sheets as

$$y_{os} = \text{small-signal output conductance}$$

with the output ac resistance then being

$$r_d = \frac{1}{y_{os}}$$

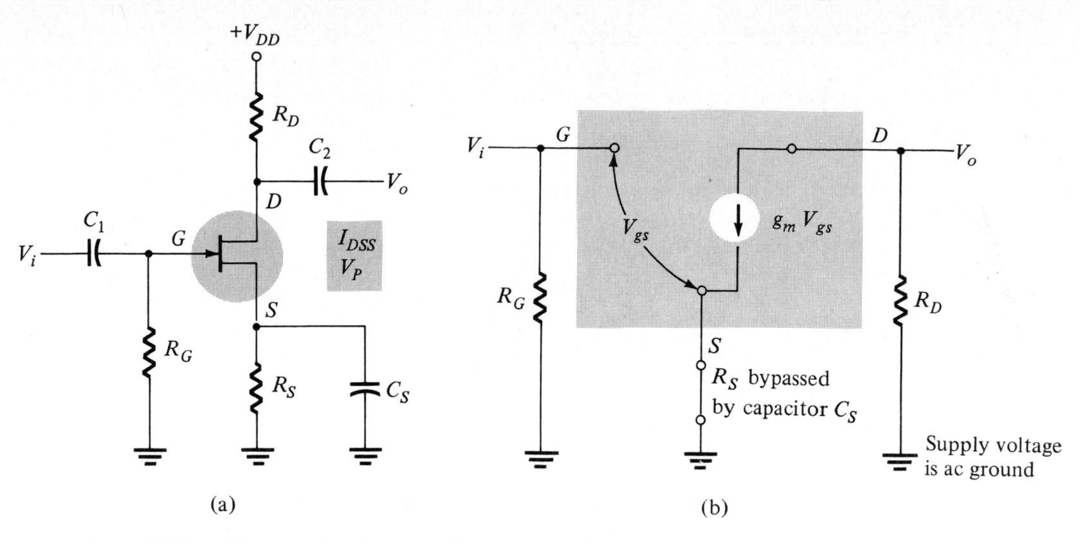

Figure 10.2 JFET amplifier circuit: (a) circuit; (b) ac equivalent circuit.

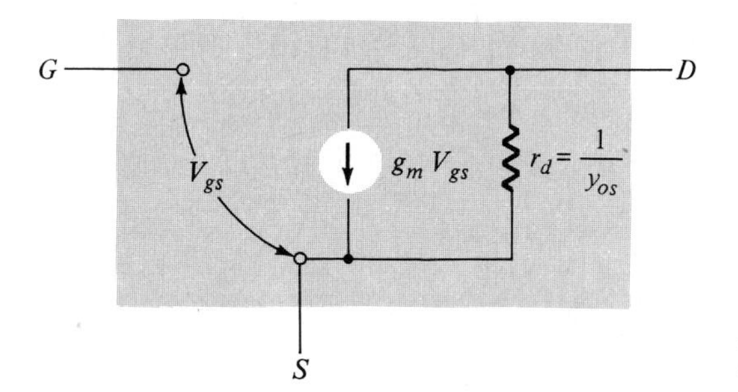

Figure 10.3 JFET ac model including drain-source resistance.

EXAMPLE 10.3

What are the values of g_{mo} and r_d for a device whose specifications are $I_{DSS} = 15$ mA, $V_P = -6$ V, and $y_{os} = 0.05$ mS?

Solution:

$$g_{mo} = \frac{2I_{DSS}}{|V_P|} = \frac{2(15 \times 10^{-3})}{|-6|} = \mathbf{5\ mS}$$

The output resistance is

$$r_d = \frac{1}{y_{os}} = \frac{1}{0.05 \times 10^{-3}} = \mathbf{20\ k\Omega}$$

10.3 AC SMALL-SIGNAL OPERATION

The FET ac equivalent circuit introduced in Section 10.2 can now be used in analyzing various FET amplifier configurations for voltage gain and input and output resistances. To demonstrate use of the ac equivalent circuit, consider the FET amplifier circuit of

Fig. 10.2a. The ac equivalent circuit is redrawn in Fig. 10.2b with capacitors replaced by a short for ac operation and with the FET device replaced by its simple equivalent circuit (r_d assumed infinite or open-circuit). The output ac voltage is

$$V_o = -I_d R_D = -g_m V_{gs} R_D$$

Since $V_i = V_{gs}$, the circuit voltage gain is

$$A_v = \frac{V_o}{V_i} = -g_m R_D \tag{10.3}$$

The ac impedance looking into the amplifier is

$$R_i = R_G \tag{10.4}$$

and the ac impedance seen looking from the load into the amplifier output terminal is

$$R_o = R_D \tag{10.5}$$

EXAMPLE 10.4

Calculate the circuit factors A_v, R_i and R_o for the JFET amplifier of Fig. 10.4 (assume r_d can be neglected).

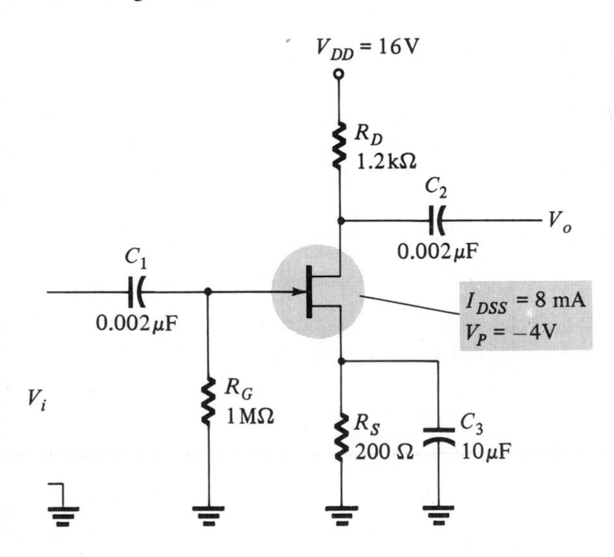

Figure 10.4 FET amplifier circuit for Example 10.4.

Solution:

Dc bias as covered in Chapter 7 results in $V_{GS_Q} = -0.94$ V. At this bias point, the value of g_m, using Eqs. (10.1) and (10.2) is

$$g_{mo} = \frac{2I_{DSS}}{|V_P|} = \frac{2(8\text{ mA})}{|-4\text{V}|} = 4\text{ mS}$$

and

$$g_m = g_{mo}\left(1 - \frac{V_{GS_Q}}{V_P}\right) = 4\text{ mS}\left(1 - \frac{-0.94\text{ V}}{-4\text{ V}}\right) = 3.06\text{ mS}$$

Using Eqs. (10.3) through (10.5) gives us

$$A_v = -g_m R_D = -(3.06\text{ mS})(1.2\text{ k}\Omega) = \mathbf{-3.67}$$

$$R_i = R_G = \mathbf{1\ M\Omega}$$

$$R_o = R_D = \mathbf{1.2\ k\Omega}$$

The voltage gain of magnitude 3.67 is typical of the lower values obtained using an FET circuit as opposed to a BJT circuit using discrete circuits.

The voltage-gain equation can be written in a modified form by identifying a resistance

$$\boxed{r_m = \frac{1}{g_m}} \tag{10.6}$$

There is a similarity between the JFET ac resistance, r_m, at the dc bias voltage V_{GS} and the BJT ac resistance, r_e, at the bias current I_E. The voltage gain given by Eq. (10.3) can then be written as

$$\boxed{A_v = -g_m R_D = \frac{-R_D}{r_m}} \tag{10.7}$$

Amplifier with Source Resistance

If the amplifier is built with part of the source resistance unbypassed (see Fig. 10.5), the relation for voltage gain can be determined using the ac equivalent circuit of Fig. 10.6 (with device output resistance, r_d, considered negligible).

$$V_{gs} = V_g - V_s = V_i - I_d R_{S1} = V_i - g_m V_{gs} R_{S1}$$

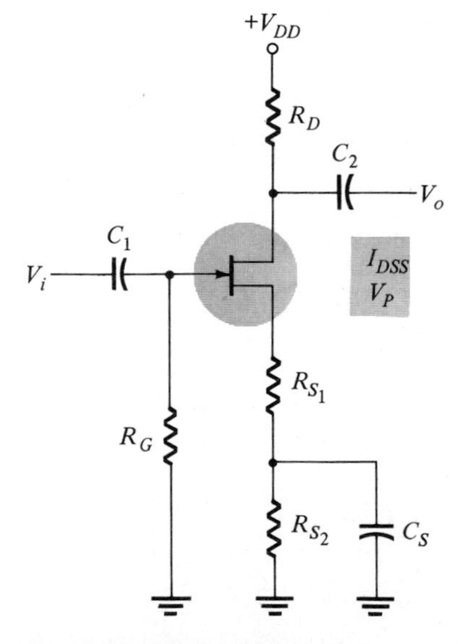

Figure 10.5 FET amplifier circuit with unbypassed source resistance.

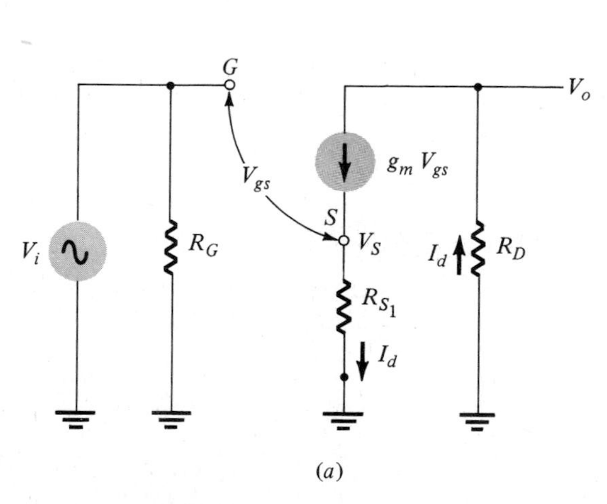

Figure 10.6 (a) Ac equivalent circuit for amplifier of Fig. 10.5.

Solving for V_i yields $\qquad V_i = (1 + g_m R_{S1})V_{gs}$

The output voltage is $\qquad V_o = -I_d R_D = -g_m V_{gs} R_D$

$$= -g_m R_D\left(\frac{V_i}{1 + g_m R_{S1}}\right)$$

so that

$$\boxed{A_v = \frac{V_o}{V_i} = \frac{-g_m R_D}{1 + g_m R_{S1}}} \tag{10.8a}$$

Using Eq. (10.6), the voltage gain of Eq. (10.8a) can be expressed as

$$\boxed{A_v = \frac{V_o}{V_i} = \frac{-\left(\frac{1}{r_m}\right)R_D}{1 + \left(\frac{1}{r_m}\right)R_{S1}} = -\frac{R_D}{r_m + R_{S1}}} \tag{10.8b}$$

If the FET ac equivalent model includes the FET output impedance, r_d, as shown in Fig. 10.6b the expression for the voltage gain A_v can be obtained as follows. After converting the FET current source into a voltage source as shown in the partial circuit of Fig. 10.6c the expression for V_o is

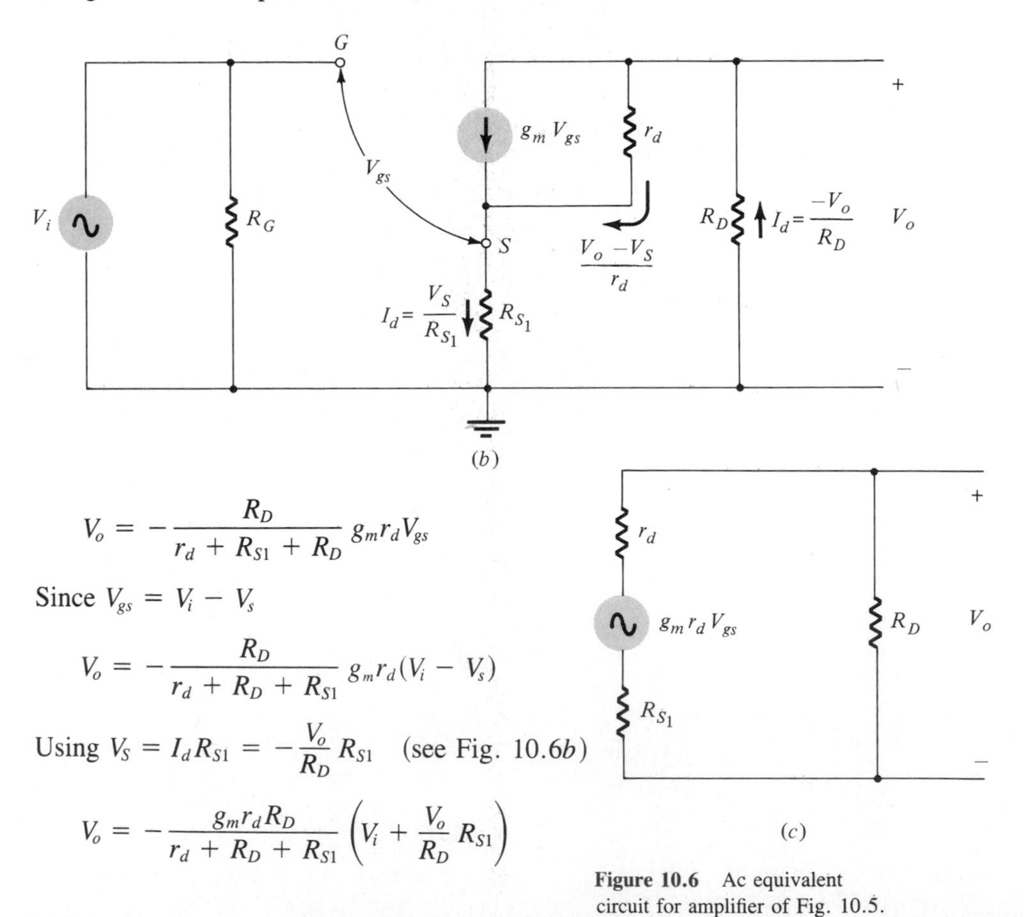

$$V_o = -\frac{R_D}{r_d + R_{S1} + R_D} g_m r_d V_{gs}$$

Since $V_{gs} = V_i - V_s$

$$V_o = -\frac{R_D}{r_d + R_D + R_{S1}} g_m r_d (V_i - V_s)$$

Using $V_S = I_d R_{S1} = -\frac{V_o}{R_D} R_{S1}$ (see Fig. 10.6*b*)

$$V_o = -\frac{g_m r_d R_D}{r_d + R_D + R_{S1}}\left(V_i + \frac{V_o}{R_D} R_{S1}\right)$$

Figure 10.6 Ac equivalent circuit for amplifier of Fig. 10.5.

which can be reworked to provide an expression for the voltage gain

$$A_v = \frac{V_o}{V_i} = \frac{-g_m R_D}{1 + g_m R_{S1} + \dfrac{R_D + R_{S1}}{r_d}} \tag{10.9a}$$

Notice that if $r_d > (R_D + R_{S1})$ the expression reduces to that of Eq. (10.8a).

One can also use $r_m = \dfrac{1}{g_m}$ and express the voltage gain as

$$A_v = \frac{V_o}{V_i} = -\frac{R_D}{r_m + R_{S1} + \dfrac{r_m}{r_d}(R_D + R_{S1})} \tag{10.9b}$$

Notice that Eq. (10.9b) reduces to that given by Eq. (10.8b) if the term $\dfrac{r_m}{r_d}(R_D + R_{S1})$ is negligible—that is, if r_d is large compared to the values of r_m, R_D and R_{S1}.

EXAMPLE 10.5

Calculate the voltage gain of the circuit in Fig. 10.7 (ignore r_d).

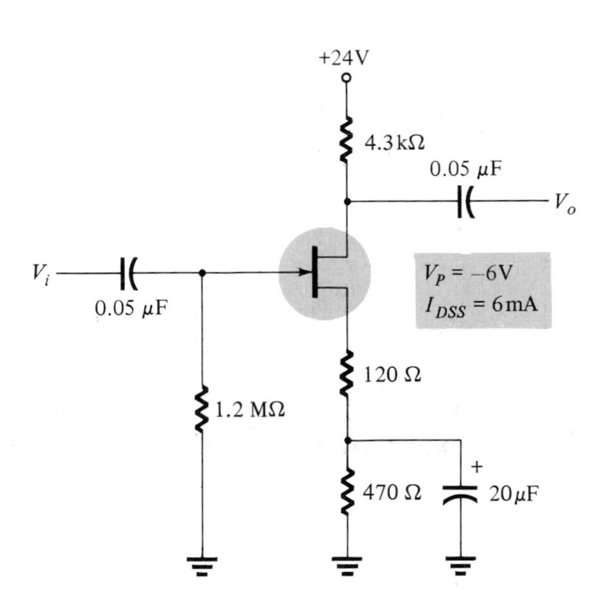

Figure 10.7 JFET amplifier for Example 10.5.

Solution:

Dc bias provides a gate-source voltage $V_{GS_Q} = -1.76$ V

The device transconductance is

$$g_{mo} = \frac{2I_{DSS}}{|V_P|} = \frac{2(6 \text{ mA})}{|-6 \text{ V}|} = 2 \text{ mS}$$

At the bias point established,

$$g_m = g_{mo}\left(1 - \frac{V_{GS_Q}}{V_P}\right) = 2\text{ mS}\left(1 - \frac{-1.76\text{ V}}{-6\text{ V}}\right) = 1.413\text{ mS}$$

The device resistance is then

$$r_m = \frac{1}{g_m} = 707.7\ \Omega$$

The circuit voltage gain is then [using Eq. (10.8*b*)]

$$A_v = \frac{-R_D}{r_m + R_{s1}} = \frac{-4.3 \times 10^3}{707.7 + 120} = \mathbf{-5.2}$$

EXAMPLE 10.6

Calculate the voltage gain and input and output resistance of the circuit in Fig. 10.8 (ignore r_d).

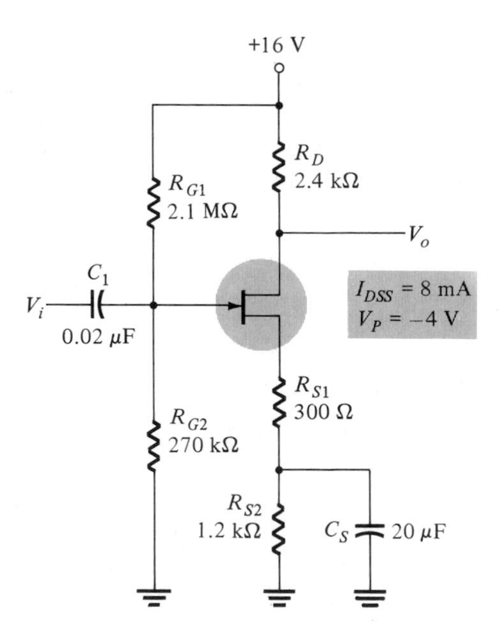

Figure 10.8 Circuit for Example 10.6.

Solution:

The dc bias was determined in Example 7.7 to be $V_{GS_Q} = -1.8$ V

At this bias point

$$g_m = \frac{2I_{DSS}}{|V_P|}\left(1 - \frac{V_{GS_Q}}{V_P}\right) = \frac{2(8\text{ mA})}{|-4\text{ V}|}\left(1 - \frac{-1.8\text{ V}}{-4\text{ V}}\right)$$

$$= 2.2 \times 10^{-3} = 2.2\text{ mS}$$

The device resistance is then

$$r_m = \frac{1}{g_m} = \frac{1}{2.2 \times 10^{-3}} = 454.5\ \Omega$$

The ac voltage gain is

$$A_v = \frac{-R_D}{r_m + R_{s1}} = \frac{-2.4 \times 10^3}{454.5 + 300} = \mathbf{-3.18}$$

The input resistance is

$$R_1 = R_{G_1} \| R_{G_2} = 2.1\ \text{M}\Omega \| 270\ \text{k}\Omega = \mathbf{239\ k\Omega}$$

The output resistance is

$$R_o = R_D = \mathbf{2.4\ k\Omega}$$

EXAMPLE 10.7

Calculate the voltage gain and input and output resistances for the circuit of Fig. 10.9. The FET output conductance is $y_{os} = 0.05$ mS.

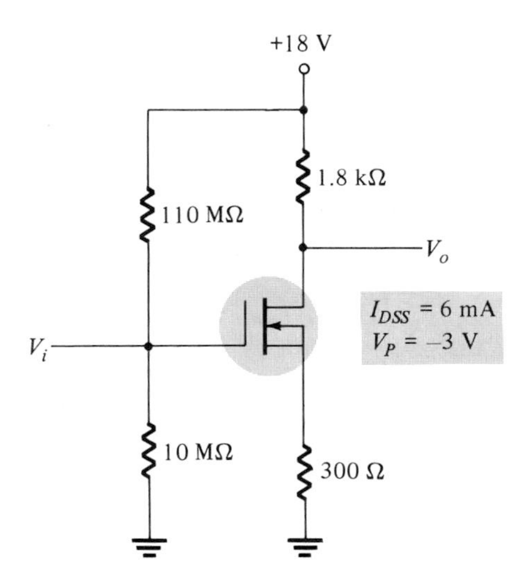

Figure 10.9 Circuit for Example 10.7.

Solution:

From dc bias calculations in Example 7.10, $V_{GS_Q} = -0.15$ V

Calculating g_{mo}:

$$g_{mo} = \frac{2I_{DSS}}{|V_P|} = \frac{2(6 \times 10^{-3})}{|-3|} = 4\ \text{mS}$$

The device g_m at the bias condition is

$$g_m = g_{mo}\left(1 - \frac{V_{GS_Q}}{V_P}\right) = 4\ \text{mS}\left(1 - \frac{-0.15\ \text{V}}{-3\ \text{V}}\right) = 3.8\ \text{mS}$$

for which

$$r_m = \frac{1}{g_m} = \frac{1}{3.8 \times 10^{-3}} = 263.2\ \Omega$$

The output resistance of the MOSFET is

$$r_d = \frac{1}{y_{os}} = \frac{1}{0.05 \times 10^{-3}} = 20\ \text{k}\Omega$$

The ac voltage gain is then

$$A_v = \frac{-R_D}{r_m + R_s + \dfrac{r_m}{r_d}(R_D + R_S)} = \frac{-1.8 \times 10^3}{263.2 + 300 + \dfrac{263.2}{20 \times 10^3}(1.8 \times 10^3 + 300)}$$

$$= \mathbf{-3.05}$$

The input impedance is

$$R_i = R_{G_1} \| R_{G_2} = 110\ \text{M}\Omega \| 10\ \text{M}\Omega = \mathbf{9.17\ M\Omega}$$

The output impedance of the circuit is

$$R_o = R_D \| r_d = 1.8\ \text{k}\Omega \| 20\ \text{k}\Omega = \mathbf{1.65\ k\Omega}$$

10.4 LOADING EFFECTS

When the output of an amplifier stage is connected to any other electronic circuit, the loading by that circuit will reduce the amplifier gain. Any source resistance will also cause a reduction in the output voltage, due to the loading effect of the amplifier stage input resistance.

Effect of Output Load

A load resistance at the output of a FET amplifier is shown in Fig. 10.10a. The ac equivalent of that circuit is shown in Fig. 10.10b. The load resistance is seen to be in parallel with the drain bias resistor, so that the voltage gain of the amplifier under

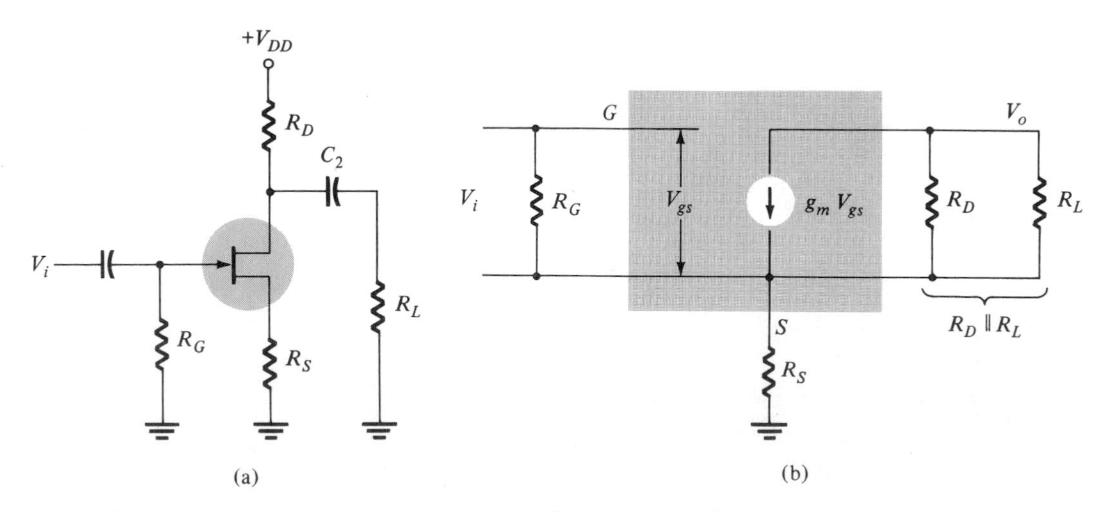

Figure 10.10 FET amplifier with output load: (a) circuit; (b) ac equivalent circuit.

load (neglecting r_d) is

$$A_v = \frac{V_o}{V_i} = \frac{-(R_D \| R_L)}{r_m + R_S} \tag{10.10a}$$

If r_d is considered

$$A_v = \frac{V_o}{V_i} = \frac{-(R_D \| R_L)}{r_m + R_s + \frac{r_m}{r_d}(R_D \| R_L + R_S)} \tag{10.10b}$$

EXAMPLE 10.8

Calculate the voltage gain and output voltage of the circuit of Fig. 10.11 (a) without the load connected and (b) with the load connected. Assume r_d can be ignored.

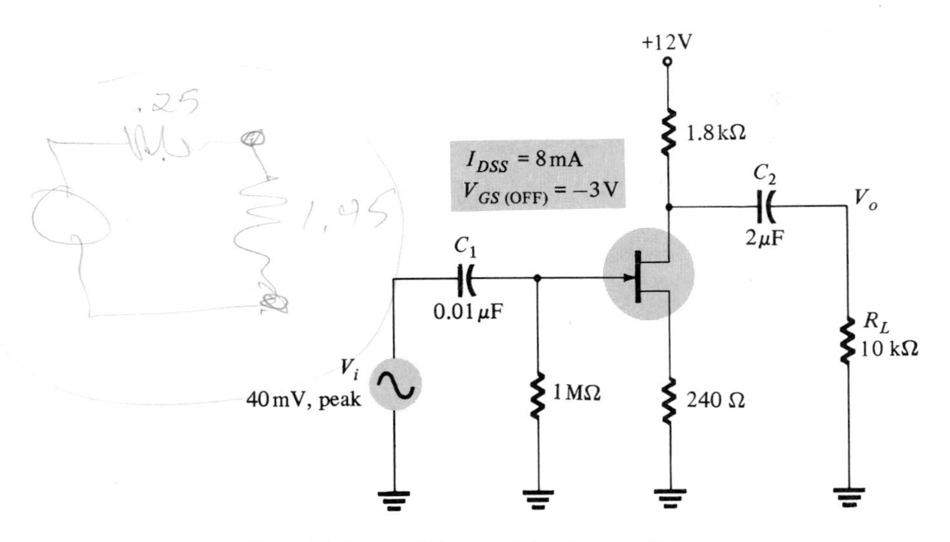

Figure 10.11 Amplifier circuit for Example 10.8.

Solution:

From the dc bias analysis $V_{GS_Q} = -0.92$ V

$$g_{mo} = \frac{2I_{DSS}}{|V_P|} = \frac{2(8 \times 10^{-3})}{|-3|} = 5.33 \times 10^{-3}\ \text{S} = 5.33\ \text{mS}$$

At the bias value of $V_{GS_Q} = -0.92$ V,

$$g_m = g_{mo}\left(1 - \frac{V_{GS_Q}}{V_P}\right) = 5.33 \times 10^{-3}\left(1 - \frac{-0.92}{-3}\right)$$

$$= 3.7 \times 10^{-3} = 3.7\ \text{mS}$$

and the FET ac resistance is

$$r_m = \frac{1}{g_m} = \frac{1}{3.7 \times 10^{-3}} = 270.3\ \Omega$$

(a) The voltage gain without the load connected is

$$A_v = \frac{-R_D}{r_m + R_S} = \frac{-1.8 \times 10^3}{270.3 + 240} = \mathbf{-3.53}$$

for which

$$V_o = A_v V_i = -3.53(40\text{ mV peak}) = \mathbf{-141.2\ mV\ peak}$$

(b) The voltage gain with load connected is

$$A_v = \frac{-R_D \| R_L}{r_m + R_S} = \frac{-(1.8\text{ k}\Omega) \| (10\text{ k}\Omega)}{270.3\ \Omega + 240\ \Omega}$$

$$= \mathbf{-2.99}$$

for which

$$V_o = A_v V_i = -2.99(40\text{ mV peak}) = \mathbf{-119.6\ mV\ peak}$$

Effect of Input Source Resistance

When the input signal is provided by a source having resistance, the input signal to the amplifier, V_i, is reduced from the unloaded value of the signal source V_S, by the loading between source resistance, R_S, and amplifier input resistance, R_i. Figure 10.12 details this loading. The input voltage to the amplifier is

$$\boxed{V_i = \frac{R_i}{R_S + R_i} V_S} \qquad (10.11)$$

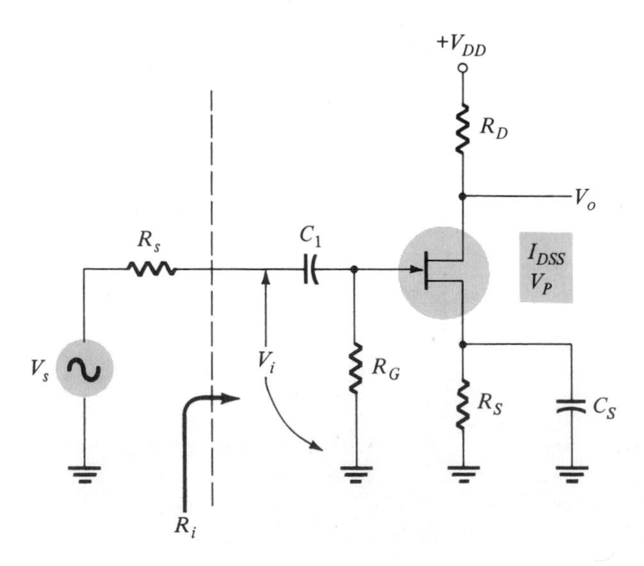

Figure 10.12 Amplifier stage showing loading of input source.

Once the input voltage to the amplifier is determined by Eq. (10.11), the remaining calculations are carried out as described previously.

EXAMPLE 10.9

For the circuit of Fig. 10.13 calculate (a) the voltage gain of only the amplifier stage, without any load connected; (b) the output voltage, V_o, with load connected; and (c) the overall voltage gain of the circuit, V_o/V_i, under load.

Solution:

The dc bias calculations for the circuit of Fig. 10.13 provides $V_{GS_Q} = -0.72$ V

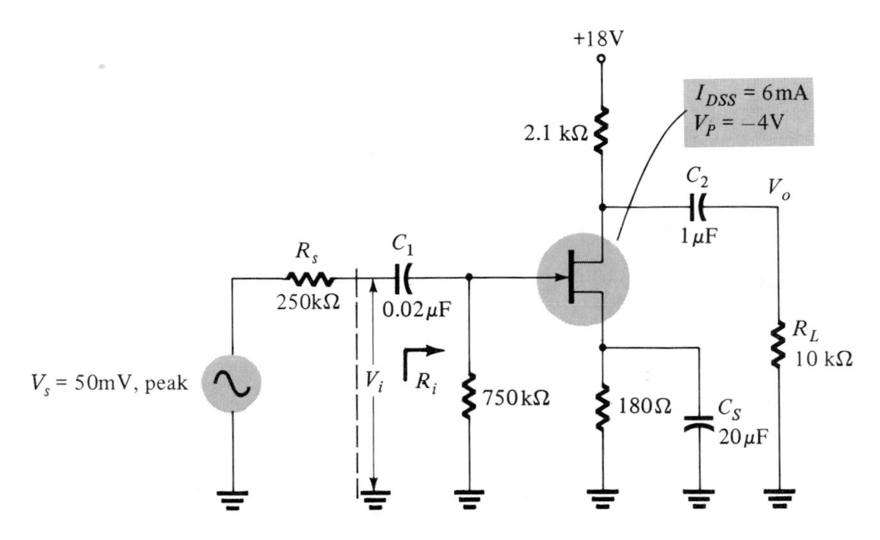

Figure 10.13 FET amplifier circuit for Example 10.9.

The device g_{mo} is

$$g_{mo} = \frac{2I_{DSS}}{|V_P|} = \frac{2(6 \times 10^{-3})}{|-4|} = 3 \times 10^{-3} = 3 \text{ mS}$$

At the dc bias value,

$$g_m = g_{mo}\left(1 - \frac{V_{GS_Q}}{V_P}\right) = 3 \times 10^{-3}\left(1 - \frac{-0.72}{-4}\right)$$

$$= 2.46 \times 10^{-3} = 2.46 \text{ mS}$$

for which the device ac resistance is

$$r_m = \frac{1}{g_m} = \frac{1}{2.46 \times 10^{-3}} = 406.5\ \Omega$$

(a) ac amplifier gain under load is ($R_S = 0\ \Omega$ due to capacitor bypass)

$$A_v = \frac{V_o}{V_i} = \frac{-R_D}{r_m + R_S} = \frac{-2.1 \times 10^3}{406.5} = \mathbf{-5.17}$$

(b) The stage input resistance, R_i, is

$$R_i = R_G = 750\ \text{k}\Omega$$

so that the input voltage, V_i, is,

$$V_i = \frac{R_i}{R_S + R_i} V_S$$

$$= \frac{750\ \text{k}\Omega}{250\ \text{k}\Omega + 750\ \text{k}\Omega} 50\ \text{mV peak} = 37.5\ \text{mV peak}$$

The output voltage (under load) is then

$$V_o = A_v V_i = -\left(\frac{2.1 \times 10^3 \| 10 \times 10^3}{406.5}\right) 37.5\ \text{mV peak} = \mathbf{-160.1\ mV\ peak}$$

(c) The overall circuit gain (under load) is

$$\frac{V_o}{V_S} = \frac{-160.1\ \text{mV peak}}{50\ \text{mV peak}} = \mathbf{-3.2}$$

So while the amplifier stage gain alone provides a gain of magnitude 5.17, the loading of the source and load resistances reduces the overall circuit voltage gain to a magnitude of 3.2.

10.5 SOURCE-FOLLOWER (COMMON-DRAIN) CIRCUIT

A second ac circuit configuration is the common-drain or source follower shown in Fig. 10.14a. The circuit is seen as a JFET version of a bipolar emitter-follower configuration. In fact, voltage gain is also less than unity with no polarity inversion, and the circuit provides high input resistance and lower output resistance than the common-source configuration.

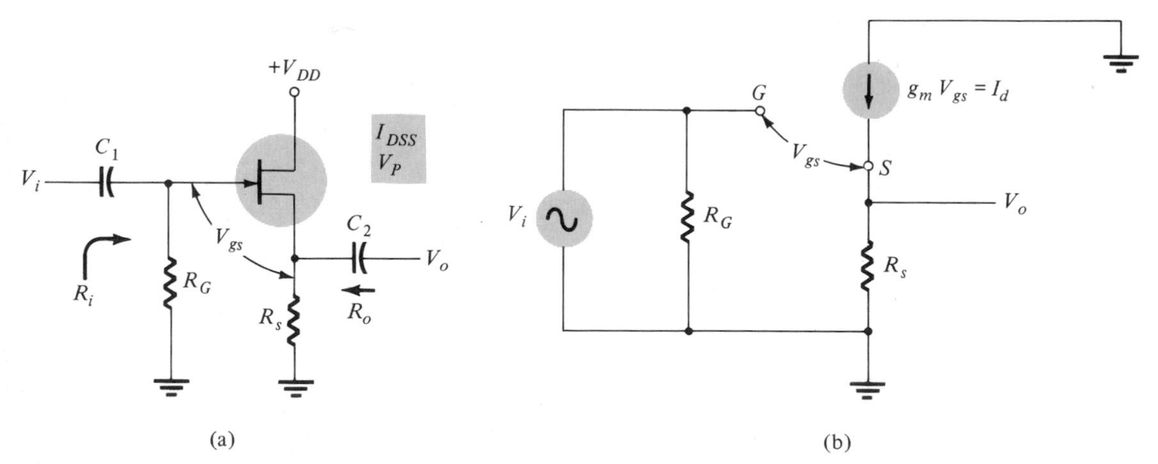

Figure 10.14 Source-follower amplifier circuit.

If the output is taken from the source terminal (see Fig. 10.14b), there is no polarity inversion between output and input, and the voltage amplitude is reduced from the input value. The ac voltage gain can be determined as follows. The gate-source voltage, V_{gs}, is

$$V_{gs} = V_i - V_o$$

Since $V_o = I_d R_s$ and $I_d = g_m V_{gs}$, the equation above can be expressed as

$$V_{gs} = V_i - I_d R_S = V_i - (g_m V_{gs}) R_S$$

for which

$$V_i = (1 + g_m R_S) V_{gs}$$

The amplifier voltage gain is then

$$A_v = \frac{V_o}{V_i} = \frac{(g_m V_{gs}) R_S}{(1 + g_m R_S) V_{gs}}$$

$$= \frac{g_m R_S}{1 + g_m R_S} \qquad (10.12)$$

Using $r_m = 1/g_m$,

$$\boxed{A_v = \frac{V_o}{V_i} = \frac{(1/r_m) R_S}{1 + (1/r_m) R_S} = \frac{R_S}{r_m + R_S}} \qquad (10.13)$$

The voltage gain is seen to be noninverting and less than 1, approaching unity as R_S is made larger compared to r_m. The amplifier input resistance is

$$\boxed{R_i = R_G} \qquad (10.14)$$

while the output resistance is the source bias resistor, R_S, in parallel with the device ac resistance, r_m,

$$\boxed{R_o = R_S \| r_m} \qquad (10.15)$$

EXAMPLE 10.10

Calculate the voltage gain and input and output impedance for the cicuit of Fig. 10.15.

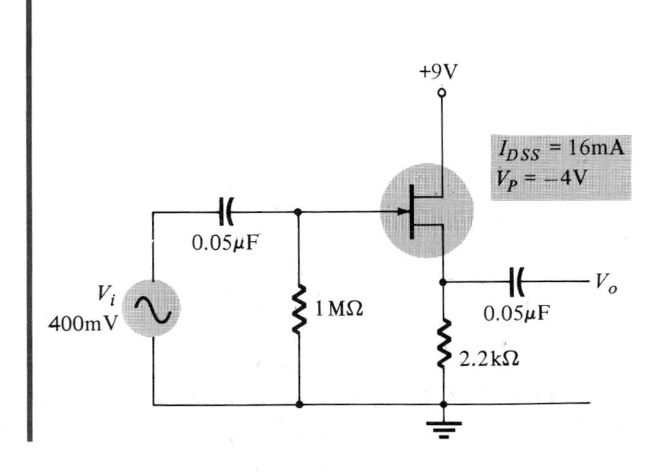

Figure 10.15 Source-follower amplifier circuit for Example 10.10.

Solution:

From the dc bias calculations $V_{GS_Q} = -2.86$ V. The device transconductance is

$$g_{mo} = \frac{2I_{DSS}}{|V_P|} = \frac{2(16 \times 10^{-3})}{|-4|} = 8 \times 10^{-3} = 8 \text{ mS}$$

At the bias condition,

$$g_m = g_{mo}\left(1 - \frac{V_{GS_Q}}{V_P}\right) = 8 \text{ mS}\left(1 - \frac{-2.86 \text{ V}}{-4 \text{ V}}\right) = 2.3 \text{ mS}$$

and the JFET ac resistance is then

$$r_m = 1/g_m = 1/2.3 \times 10^{-3} = 434.8 \ \Omega$$

The ac voltage gain is

$$A_v = \frac{V_o}{V_i} = \frac{R_S}{r_m + R_S} = \frac{2.2 \times 10^3}{434.8 + 2.2 \times 10^3} = \mathbf{0.835}$$

The amplifier input resistance is

$$R_i = R_G = \mathbf{1 \ M\Omega}$$

and the output impedance is

$$R_o = r_m \| R_S = 434.8 \| 2.2 \times 10^3 = \mathbf{363.05 \ \Omega}$$

EXAMPLE 10.11

Calculate the output voltage V_{o1} and V_{o2} for the circuit of Fig. 10.16.

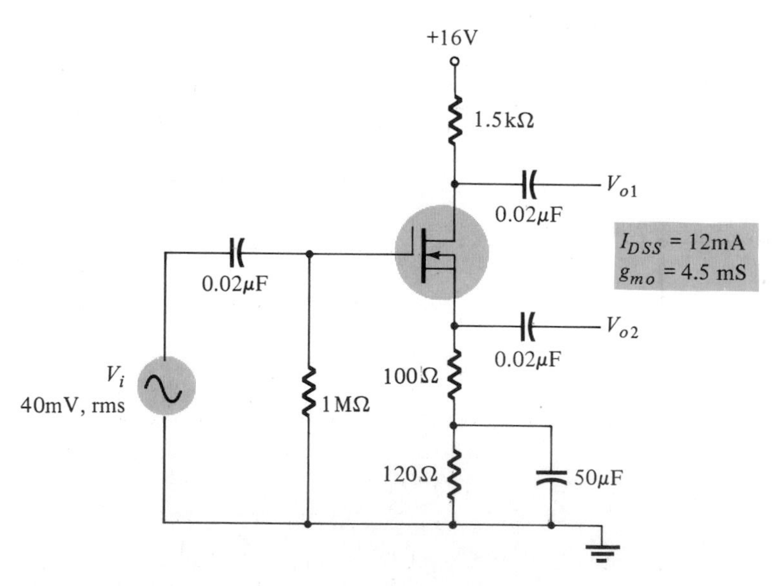

Figure 10.16 JFET amplifier circuit for Example 10.11.

Solution:

From the given data

$$|V_P| = \frac{2I_{DSS}}{g_{mo}} = \frac{2(12 \times 10^{-3})}{4.5 \times 10^{-3}} = 5.33 \text{ V}$$

From dc bias calculations $V_{GS} = -1.4$ V. At this bias value

$$g_m = g_{mo}\left(1 - \frac{V_{GS}}{V_P}\right) = 4.5 \text{ mS}\left(1 - \frac{-1.4 \text{ V}}{-5.33 \text{ V}}\right) = 3.32 \times 10^{-3} \text{ S} = 3.32 \text{ mS}$$

for which

$$r_m = \frac{1}{g_m} = \frac{1}{3.32 \times 10^{-3}} = 301.2 \ \Omega$$

The ac voltage at the drain output is

$$V_{O1} = A_v V_i = \frac{-R_D}{r_m + R_{S1}} V_i$$

$$= \frac{1.5 \times 10^3}{301.2 + 100}(40 \text{ mV rms}) = \mathbf{-\ 149.55\ mV\ rms}$$

The ac voltage at the source output is

$$V_{o2} = A_v V_i = \frac{R_{S1}}{r_m + R_{S1}} V_i$$

$$\mathbf{A}_v = \frac{100}{301.2 + 100}(40 \text{ mV rms}) = \mathbf{9.97\ mV\ rms}$$

Effect of Load and Source Resistance

When a load is connected across the output of a source follower, as shown in Fig. 10.17, the voltage gain is modified Eq. (10.13) by the load resistance R_L in parallel with the source resistance,

$$A_v = \frac{V_o}{V_i} = \frac{R_S \| R_L}{r_m + R_S \| R_L} \tag{10.16}$$

The amplifier stage input resistance is the parallel resistance of bias resistors R_{G1} and R_{G2}:

$$R_i = R_{G1} \| R_{G2} \tag{10.17}$$

Looking from the load back into the amplifier, the output resistance is the source resistance, R_s, in parallel with the device resistance, r_m:

$$R_o = R_S \| r_m \tag{10.18}$$

The loading of the source resistance results in a reduced input voltage due to the

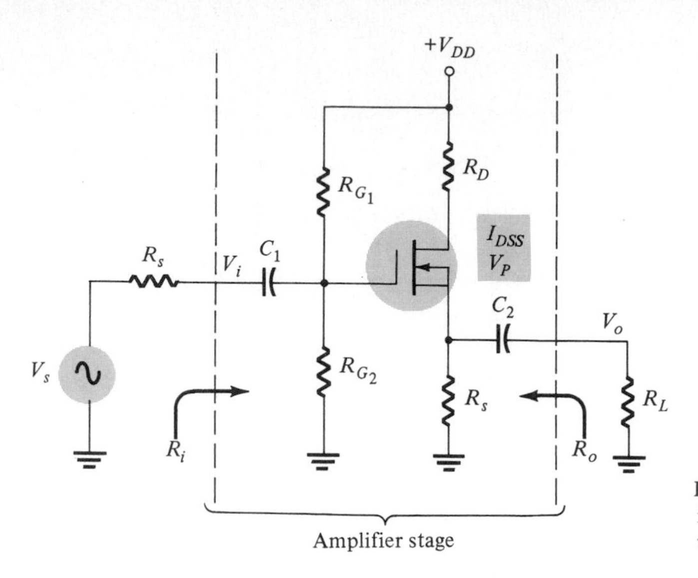

Figure 10.17 FET amplifier circuit including source and load resistances.

loading of the source by the amplifier stage input resistance

$$V_i = \frac{R_i}{R_S + R_i} V_S \tag{10.19}$$

EXAMPLE 10.12

Calculate the output voltage V_o and output resistance R_o for the circuit of Fig. 10.18.

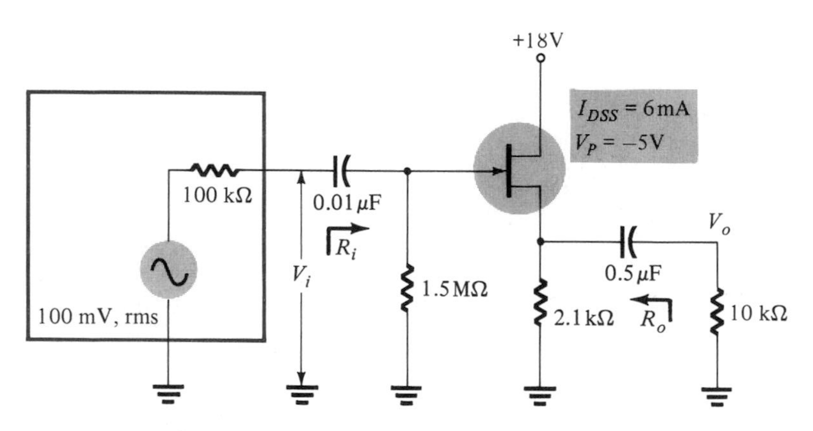

Figure 10.18 Source-follower circuit for Example 10.12.

Solution:

From dc bias calculations $V_{GS} = -2.69$ V. With device g_{mo}

$$g_{mo} = \frac{2I_{DSS}}{|V_P|} = \frac{2(6 \times 10^{-3})}{|-5|} = 2.4 \text{ mS}$$

the value of g_m at the bias condition is

$$g_m = g_{mo}\left(1 - \frac{V_{GS}}{V_P}\right) = 2.4 \text{ mS}\left(1 - \frac{-2.69 \text{ V}}{-5 \text{ V}}\right) = 1.11 \text{ mS}$$

for which

$$r_m = \frac{1}{g_m} = 900.9\ \Omega$$

The input voltage is

$$V_i = \frac{R_i}{R_S + R_i} V_S = \frac{1.5 \times 10^6}{100 \times 10^3 + 1.5 \times 10^6}(100 \text{ mV rms})$$

$$= 93.75 \text{ mV rms}$$

The amplifier gain is

$$A_V = \frac{R_S \parallel R_L}{r_m + R_S \parallel R_L}$$

$$= \frac{(2.1 \times 10^3) \parallel (10 \times 10^3)}{900.9 + (2.1 \times 10^3 \parallel 10 \times 10^3)} = 0.66$$

so that the output voltage is

$$V_o = A_v V_i = (0.66)(93.75 \text{ mV rms}) = \mathbf{61.9 \text{ mV rms}}$$

The output resistance is

$$R_o = r_m \parallel R_S = (900.9) \parallel (2.1 \times 10^3) = \mathbf{630.4\ \Omega}$$

10.6 COMMON-GATE CIRCUIT

A third circuit configuration is shown in Fig. 10.19a with ac input to source, ac output from drain—this circuit being the common-gate amplifier configuration. As will be shown, this amplifier form has low input resistance, noninverting voltage gain (similar in magnitude to a common drain), and output resistance the same as the common drain.

The ac equivalent for the circuit of Fig. 10.19a is drawn in Fig. 10.19b. The voltage gain is determined to be

$$V_o = -I_d R_D = -g_m V_{gs} R_D = -g_m(-V_i) R_D$$

$$\boxed{A_v = \frac{V_o}{V_i} = g_m R_D = \frac{R_D}{r_m}} \qquad (10.20)$$

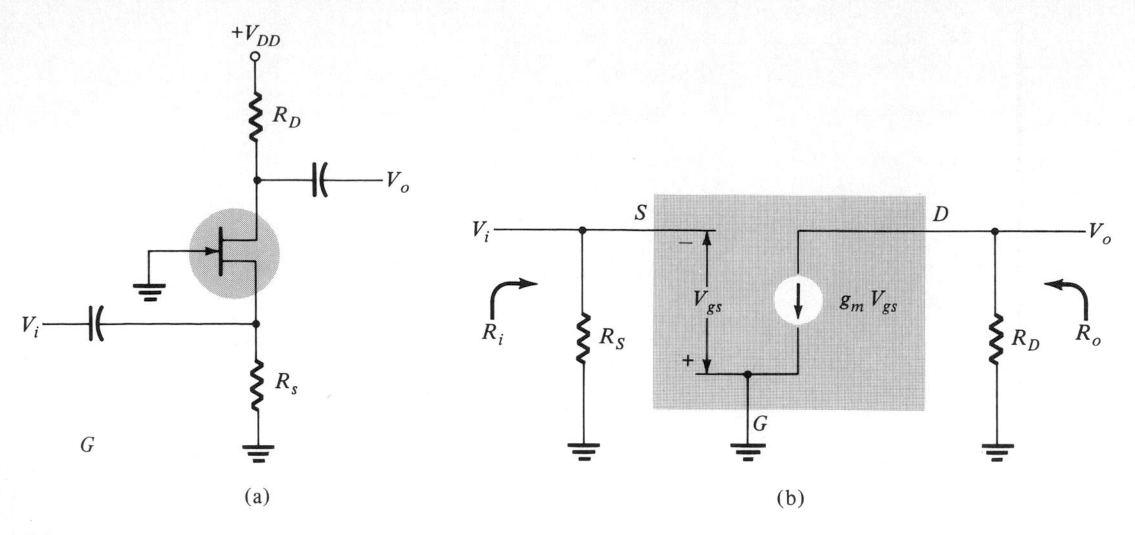

Figure 10.19 Common-gate amplifier: (a) circuit; (b) ac equivalent circuit.

The input resistance is

$$R_i = R_s \tag{10.21}$$

and the output resistance is

$$R_o = R_D \tag{10.22}$$

EXAMPLE 10.13

Calculate V_o for the circuit of Fig. 10.20.

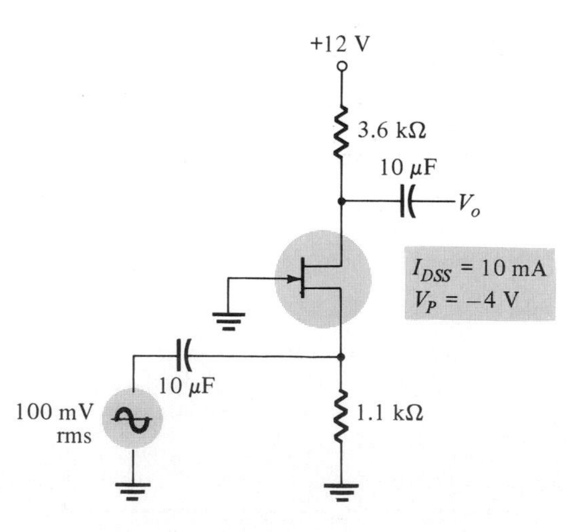

Figure 10.20 Circuit for Example 10.13.

Solution:

Dc bias calculations result in $V_{GS_Q} = -2.2$ V. With $g_{mo} = 5$ mS, the value of g_m is

$$g_m = g_{mo}\left(1 - \frac{V_{GS_Q}}{V_P}\right) = 5 \text{ mS}\left(1 - \frac{-2.2 \text{ V}}{-4 \text{ V}}\right) = 2.25 \text{ mS}$$

for which

$$r_m = \frac{1}{g_m} = \frac{1}{2.25 \times 10^{-3}} = 444.4 \ \Omega$$

The amplifier voltage gain is then

$$A_v = \frac{V_o}{V_i} = \frac{R_D}{r_m} = \frac{3.6 \times 10^3}{444.4} = 8.1$$

The output voltage is then

$$V_o = A_v V_i = 8.1(100 \text{ mV rms}) = \mathbf{0.81 \ V \ rms}$$

10.7 DESIGN OF FET AMPLIFIER CIRCUITS

The design of an amplifier circuit involves obtaining a desired voltage gain, with reasonable values of input and output resistance. For a common-source amplifier such as that in Fig. 10.2a, the voltage gain is given by Eq. (10.7). The selection of a JFET device is then quite important to achieve dc bias and the desired ac gain. Design is not necessarily carried out by a cookbook or step-by-step procedure. It depends on the information given, the information available, and the order in which the various circuit values of gain, resistance, and dc bias are considered. Some examples will demonstrate how design of a JFET circuit can be carried out for a particular set of desired circuit values. Other design approaches could also be used to achieve the desired voltage gain.

EXAMPLE 10.14

Design a JFET amplifier circuit as that of Fig. 10.2a to provide a voltage gain of magnitude 20. Use a 2N4416 *n*-channel JFET with a supply voltage of $V_{DD} = 20$ V.

Solution:

The manufacturer's specs for the 2N4416 show

$$I_{DSS} \text{ from 5 to 15 mA} \quad (\text{use } I_{DSS} = 10 \text{ mA})$$

$$V_P = -6 \text{ V}$$

Device transconductance for $I_{DSS} = 10$ mA is

$$g_{mo} = \frac{2I_{DSS}}{|V_P|} = \frac{2(10 \times 10^{-3})}{|-6|} = 3.33 \text{ mS}$$

One approach to starting the design is to develop a tabulation of a set of values to see whether some range of values appears more appropriate.

A table is provided below for which a range of bias voltages from 0 to -5 V are used.

1. Select a value of V_{GS}.
2. $I_D = I_{DSS}(1 - V_{GS}/V_P)^2$.
3. $g_m = g_{mo}(1 - V_{GS}/V_P)$, $r_m = 1/g_m$.
4. From $A_v = -R_D/r_m$, $R_D = |A_v|\, r_m$.
5. Calculate $I_D R_D$, the bias voltage drop across the drain resistor.

(1) V_{GS} (V)	(2) I_D (mA)	(3) r_m (Ω)	(4) R_D (kΩ)	(5) $I_D R_D$ (V)	
0	10	300	6	60	With V_{DD} = 20 V these voltages drops are too large
−1.75	5	424	8.5	42.5	
−3	2.5	600	12	30	
−4	1.1	900	18	20	
−4.5	0.625	1200	24	15	Any value in this range is acceptable
−5	0.278	1800	36	10	
−5.5	0.07	3600	72	5	

From the table data it is clear that selecting any dc bias in the range 0 to −4 V results in too large a voltage drop from the 20-V supply. Bias values of V_{GS} in the range of −4.5 to −5.5 V result in acceptable operation. Selecting $V_{GS} = -5$ V results in a drain resistor of value

$$R_D = 36\text{ k}\Omega$$

Since $V_{GS} = -I_D R_S$, the design value of R_S is

$$R_S = \frac{-V_{GS}}{I_D} = -\frac{-5}{0.278 \times 10^{-3}} = 17.98\text{ k}\Omega \quad \textbf{(use 18 k}\boldsymbol{\Omega}\textbf{)}$$

For a JFET circuit a value of $R_G = 1$ MΩ is suitable. The resulting circuit is shown in Fig. 10.21.

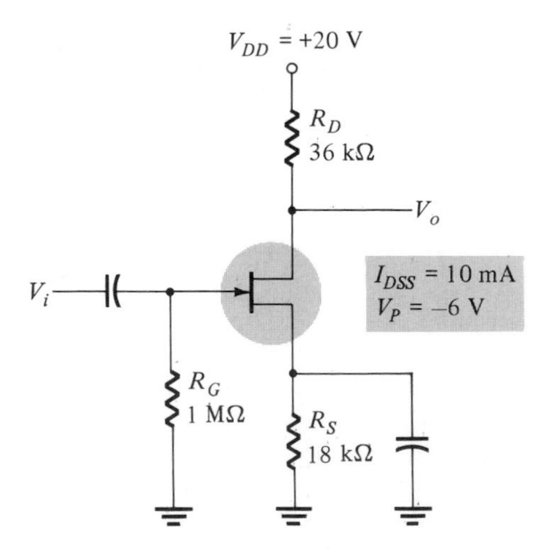

Figure 10.21 Circuit designed in Example 10.14.

EXAMPLE 10.15

Complete the design of an amplifier circuit as shown in Fig. 10.22 for a gain of magnitude 8.

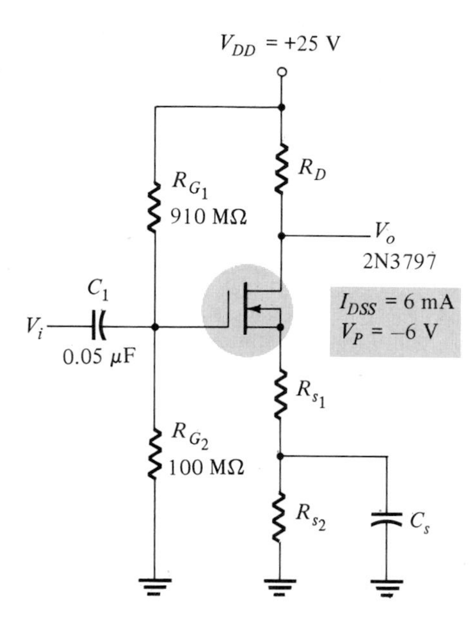

Figure 10.22 Circuit for design in Example 10.15.

Solution:

A tabulation similar to that of Example 10.14 shows that acceptable bias can be obtained over the complete range of V_{GS}. Selecting $V_{GS} = -3$ V results in

$$V_{GS} = -3 \text{ V} \quad \text{(selected value)}$$

$$I_D = I_{DSS}\left(1 - \frac{V_{GS}}{V_P}\right)^2 = 6 \text{ mA}\left(1 - \frac{-3 \text{ V}}{-6 \text{ V}}\right)^2 = 1.5 \text{ mA}$$

$$g_m = g_{mo}\left(1 - \frac{V_{GS}}{V_P}\right) = \frac{2(6 \times 10^{-3})}{|-6|}\left(1 - \frac{-3}{-6}\right) = 1 \text{ mS}$$

$$r_m = \frac{1}{g_m} = \frac{1}{1 \times 10^{-3}} = 1000\ \Omega$$

for a gain of

$$A_v = -8 = \frac{R_D}{r_m + R_{S_1}}$$

Selecting R_D so that the dc bias voltage drop is one-half of V_{DD},

$$I_D R_D = \frac{V_{DD}}{2} = 12.5 \text{ V}$$

$$R_D = \frac{I_D R_D}{I_D} = \frac{12.5 \text{ V}}{1.5 \text{ mA}} = 8.33 \text{ k}\Omega \text{ (use } 9.1 \text{ k}\Omega)$$

for a gain of

$$A_v = -8 = \frac{-R_D}{r_m + R_{S1}} = -\frac{9.1 \times 10^3}{1000 + R_{S1}}$$

$$R_{S1} = \frac{9.1 \times 10^3}{8} - 1000 = 137.5 \ \Omega \quad (\text{use } 130 \ \Omega)$$

Since $V_{GS} = V_G - V_S$

$$= \frac{100 \text{ M}\Omega}{910 \text{ M}\Omega + 100 \text{ M}\Omega}(25 \text{ V}) - 1.5 \times 10^{-3}(R_{S_1} + R_{S_2}) = -3 \text{ V}$$

$$R_{S1} + R_{S2} = \frac{2.475 + 3}{1.5 \times 10^{-3}} = 3.65 \times 10^3 \ \Omega$$

$$R_{S2} = 3650 - 130 = 3520 \quad (\text{use } 3.6 \text{ k}\Omega)$$

The present design results in resistor values of

$$R_D = \mathbf{9.1 \ k\Omega}$$

$$R_{S1} = \mathbf{130 \ \Omega}$$

$$R_{S2} = \mathbf{3.6 \ k\Omega}$$

Design of an amplifier circuit does not involve simply determining the component values to provide a desired gain but also consideration of the dc bias of the FET. A usual problem is that if the design selects the component values for a desired ac gain, these values may not be suitable for dc bias operation. If, on the other hand, the component values are selected to provide proper dc bias, the circuit may not provide a desired ac gain. Then, to complete the circuit, the limits of the FET as to values of V_{GS}, I_D must be included in the dc and ac considerations. With all these factors to consider, it should be clear that there could be numerous suitable design solutions, depending mainly on which circuit considerations are selected or calculated first and whether any particular circuit features are taken as being more important.

As one example, consider starting with the dc bias features that the bias circuit be set at $I_{DSS}/2$. If, in addition, we select the drain voltage to be $V_D = V_{DD}/2$, the remainder of the design can be examined to determine what range of values (if any) are possible with these starting restrictions. Figure 10.23 shows the circuit and equations to use in the design.

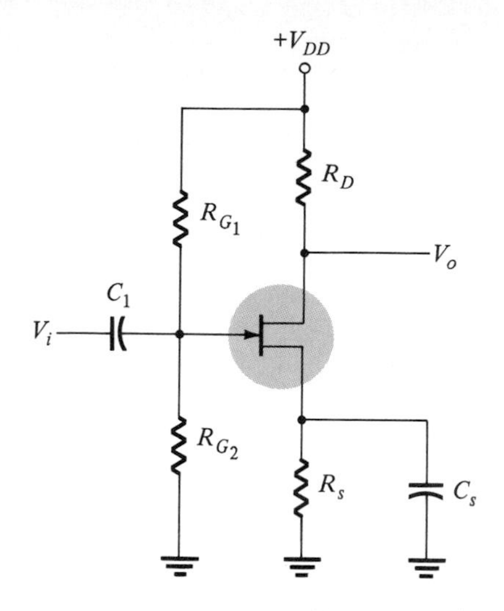

DC Bias	AC Calculations
$I_D = I_{DSS}\left(1 - \frac{V_{GS}}{V_P}\right)^2$	$A_v = -\frac{R_D}{r_m}$
$V_D = V_{DD} - I_D R_D$	$g_m = g_{mo}\left(1 - \frac{V_{GS}}{V_P}\right)$
$V_S = I_D R_S$; $V_G = \frac{R_{G_2}}{R_{G_1} + R_{G_2}} V_{DD}$ } $V_{GS} = V_G - V_S$	$g_{mo} = \frac{2I_{DSS}}{\lvert V_P \rvert}$
	$r_m = \frac{1}{g_m}$

Figure 10.23 Circuit and design equations.

EXAMPLE 10.16

Design a FET circuit to operate at dc bias levels of $I_D = I_{DSS}/2$ and $V_D = V_{DD}/2$ using a MOSFET having typical values of $I_{DSS} = 12$ mA, $V_P = -4$ V. (Use a supply voltage of $V_{DD} = 25$ V to provide the largest voltage gain.)

Solution:

From the desired dc bias conditions,

$$I_D = \frac{I_{DSS}}{2} = \frac{12\text{ mA}}{2} = 6\text{ mA}$$

for which

$$V_{GS} = \left(1 - \sqrt{\frac{I_D}{I_{DSS}}}\right)V_P = \left(1 - \sqrt{\frac{6\text{ mA}}{12\text{ mA}}}\right)(-4\text{ V}) = -1.17\text{ V}$$

With $V_D \doteq V_{DD}/2 = 25\text{ V}/2 = 12.5$ V,

$$I_D R_D = V_{DD} - V_D = 12.5\text{ V}$$

$$R_D = \frac{12.5\text{ V}}{6\text{ mA}} = 2.08\text{ k}\Omega \quad (\text{use } 2\text{ k}\Omega)$$

With

$$g_{mo} = \frac{2\,I_{DSS}}{\lvert V_P \rvert} = \frac{2(12 \times 10^{-3})}{\lvert -4 \rvert} = 6\text{ mS}$$

the value of g_m at the bias voltage $V_{GS} = -1.17$ V is

$$g_m = g_{mo}\left(1 - \frac{V_{GS}}{V_P}\right) = 6\text{ mS}\left(1 - \frac{-1.17\text{ V}}{-4\text{ V}}\right) = 4.25\text{ mS}$$

and $$r_m = \frac{1}{g_m} = \frac{1}{4.25 \times 10^{-3}} \cong 235\ \Omega$$

With R_S completely bypassed, the largest amplifier gain is

$$A_v = \frac{-R_D}{r_m} = \frac{-2 \times 10^3}{235} = -8.5$$

To select R_S, consider that a self-bias line from the origin through the bias point of 6 mA would result from

$$R_S = \frac{V_{GS}}{I_D} = \frac{1.17\ \text{V}}{6\ \text{mA}} = 195\ \Omega$$

while, in the extreme, a self-bias line from +25 V through the dc bias point at $I_D = 6$ mA and $V_{GS} = -1.17$ V would result from

$$R_S = \frac{V_G - V_{GS}}{I_D} = \frac{25\ \text{V} - (-1.17\ \text{V})}{6\ \text{mA}} = \frac{26.17\ \text{V}}{6\ \text{mA}} = 4.36\ \text{k}\Omega$$

However, with the latter value of R_S, the voltage drop across R_S would be

$$I_D R_S = (6\ \text{mA})(4.36\ \text{k}\Omega) = 26.16\ \text{V}$$

which is obviously too large.

Choosing dc bias at the lowest value of V_S,

$$R_S = 195\ \Omega \quad (\text{use } R_S = 200\ \Omega)$$

for which R_{G1} = *open* and R_{G2} would be 100 MΩ (or any other large value desired). The resulting circuit values are

$$R_D = \mathbf{2\ k\Omega}$$
$$R_S = \mathbf{200\ \Omega}$$
$$R_{G1} = \textit{open}$$
$$R_{G2} = \mathbf{100\ M\Omega}$$

10.8 HIGH-FREQUENCY EFFECTS—MILLER CAPACITANCE

Ac analysis so far has been applied only to the mid-frequency operation of the circuit. At higher frequencies, device capacitances between terminals cause reduction in the amplifier gain due to decreased capacitive impedance with increased frequency. These circuit capacitances resulting from device construction (or stray wiring) are connected with dashed lines in the circuit of Fig. 10.24a to indicate they are not capacitances that are connected into the circuit but arise as a result of the circuit and device construction. While the device capacitance between each set of terminals affects the overall amplifier gain, the capacitance between input and output has the greatest effect because of the *Miller effect*, which results in the effective capacitance being multiplied by the gain of the amplifier as will now be described.

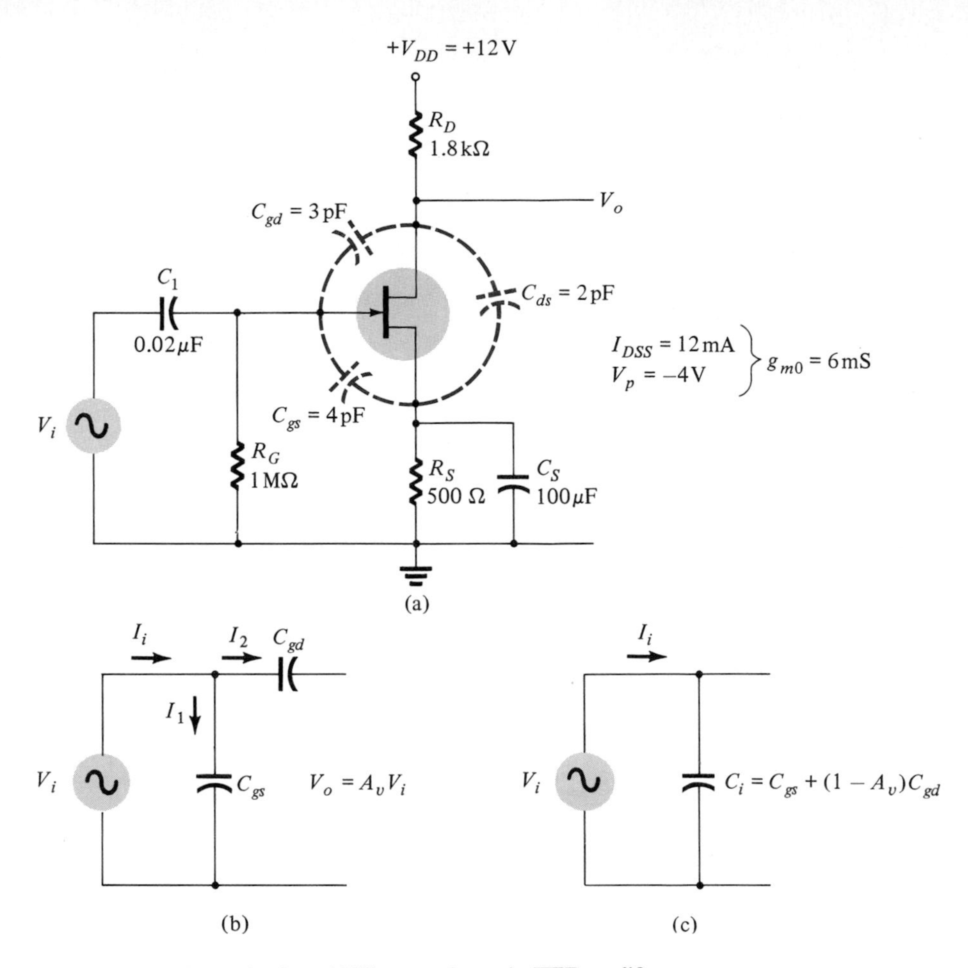

Figure 10.24 Determination of Miller capacitance in JFET amplifier.

Miller Effect (Miller Capacitance)

The most pronounced effect due to these device capacitances is due to the capacitance between the input and output terminals, C_{gd}, in the circuit of Fig. 10.24a. While each individual capacitive impedance provides a loading at higher frequencies, the magnitude of the effective loading due to the input/output capacitance, C_{gd}, is magnified by the amplifier gain. Figure 10.24b shows an ac equivalent circuit of the JFET amplifier of Fig. 10.24b, emphasizing the effect of the interterminal capacitances. In Fig. 10.24b, the input current is

$$I_i = V_i Y_i = I_1 + I_2 = V_i Y_{gs} + (V_i - A_v V_i)Y_{gd}$$

from which we get

$$Y_i = Y_{gs} + (1 - A_v)Y_{gd}$$

which can be expressed in terms of the capacitance values

$$\boxed{C_i = C_{gs} + (1 - A_v)C_{gd}} \tag{10.23a}$$

where

$$\boxed{C_M = (1 - A_v)C_{gd}} \tag{10.23b}$$

C_M being the Miller capacitance.

As shown in Fig. 10.24c, the effective input capacitance due to both C_{gs} and C_{gd} is the effective capacitance given by Eq. (10.23), which is mostly that due to the Miller effect.

EXAMPLE 10.17

Calculate the value of Miller capacitance in the circuit of Fig. 10.24.

Solution:

Calculating the dc bias to be $V_{GS_Q} = -1.8$ V, the device transconductance at the bias point is

$$g_{mo} = \frac{2I_{DSS}}{|V_p|} = \frac{2(12\text{ mA})}{|-4\text{ V}|} = 6\text{ mS}$$

$$g_m = 6 \times 10^{-3}\left(1 - \frac{-1.8}{-4}\right) = 3.3\text{ mS}$$

The value of r_s is then

$$r_m = \frac{1}{g_m} = \frac{1}{3.3 \times 10^{-3}} = 303\ \Omega$$

and the circuit midfrequency gain is

$$A_v = \frac{-R_D}{r_m} = \frac{-1.8 \times 10^3}{303} = -5.94$$

The Miller capacitance is then

$$C_M = (1 - A_v)C_{gd} = [1 - (-5.94)](3\text{ pF}) = \mathbf{20.82\text{ pF}}$$

10.9 COMPUTER ANALYSIS OF FET AMPLIFIER CIRCUITS

The computer can be used quite effectively to do dc bias or ac gain and impedance calculations for a variety of circuits. The dc bias calculations are needed before calculating ac gain. The dc bias calculations can be carried out using a module such as that written for Chapter 7. The equations and computer statements are provided in Listing 10.1. The following program (Listing 10.2) uses the previously developed dc bias module and includes a new module, module 13000, to do the ac calculations for the circuit of Fig. 10.25. Some typical results running the progam are provided in Listing 10.3.

LISTING 10.1 Equations and Computer Statements for Module 13000

Equation	Computer Statement
$g_{mo} = \dfrac{2\, I_{DSS}}{\lvert V_P \rvert}$	G0 = 2 * SS/ABS(VP)
$g_m = g_{mo}\left(1 - \dfrac{V_{GS}}{V_P}\right)$	GM = G0 * (1 − GS/VP)
$r_m = \dfrac{1}{g_m}$	RM = 1/GM
$R_P = R_D \Vert R_L$	RP = RD * RL/(RD + RL)
$A_v = -\dfrac{R_D}{r_m + R_{S1}}$	AV = −RP/(RM + S1)
$R_i = R_{G1} \Vert R_{G2}$	RI = R1 * R2/(R1 + R2)
$R_o = R_D$	RO = RD
$V_i = \dfrac{R_i}{R_S + R_i} V_S$	VI = RI * VA/(RA + RI)
$V_o = A_v V_i$	VO = AV * VI

```
10 REM ************************************************
20 REM
30 REM Module for FET ac Amplifier Calculations
40 REM
50 REM ************************************************
60 REM
100 PRINT "This program provides dc bias calculations"
110 PRINT "for a JFET or depletion MOSFET circuit then"
120 PRINT "ac amplifier calculations for a circuit"
130 PRINT "such as that of Fig. 10.23."
140 PRINT
150 PRINT "Enter the following circuit data:"
160 PRINT
170 INPUT "RG1 (use 1E30 if open)=";R1
180 INPUT "RG2                   =";R2
190 INPUT "Total source resistance, RS=";RS
200 INPUT "                         RD=";RD
210 PRINT
220 INPUT "Supply voltage, VDD=";DD
230 PRINT
240 PRINT "Enter the following device data:"
250 INPUT "Drain-source saturation current, IDSS=";SS
260 INPUT "Gate-source pinchoff voltage, VP=";VP
270 PRINT :PRINT
280 REM Now do bias calculations
290 GOSUB 11000
300 PRINT "Bias current is, ID=";ID*1000!;"mA"
310 PRINT "Bias voltages are:"
320 PRINT "VGS=";GS;"volts"
330 PRINT "VD=";VD;"volts"
340 PRINT "VS=";VS;"volts"
350 PRINT "VDS=";DS;"volts"
360 PRINT :PRINT
370 PRINT "Now accept data for ac amplifier calculations:"
380 PRINT
```

```
390 INPUT "Load resistance (1E30 if none), RL=";RL
400 INPUT "Source voltage, Vs=";VA
410 INPUT "Source resistance, Rs=";RA
420 INPUT "Unbypassed source resistance, RS1=";S1
430 REM Now do FET ac calculations
440 GOSUB 13000
450 PRINT
460 PRINT "The amplifier voltage gain is Av=";AV
470 PRINT "Output voltage across the load is";VL*1000 ;"mV"
480 PRINT
490 PRINT "Amplifier stage input resistance is, Ri=";RI/1000;"kilohms"
500 PRINT "Amplifier stage output resistance is, Ro=";RO/1000;"kilohms"
510 END
11000 REM Module for FET dc bias calculations
11010 GG=(R2/(R1+R2))*DD
11020 A=SS*RS/VP^2
11030 B=1-2*SS*RS/VP
11040 C=SS*RS-GG
11050 D=B^2-4*A*C
11060 IF D<0 THEN PRINT "No Solution!!!" :STOP
11070 V1=(-B+SQR(D))/(2*A)
11080 V2=(-B-SQR(D))/(2*A)
11090 IF ABS(V1)>ABS(VP) THEN GS=V2
11100 IF ABS(V2)>ABS(VP) THEN GS=V1
11110 ID=SS*(1-GS/VP)^2
11120 VS=ID*RS
11130 VG=GG
11140 VD=DD-ID*RD
11150 DS=VD-VS
11160 RETURN

13000 REM Module to do FET amplifier ac calculations
13010 G0=2*SS/ABS(VP)
13020 GM=G0*(1-GS/VP)
13030 RM=1/GM
13040 AV=-RD/(RM+S1)
13050 RI=R1*R2/(R1+R2)
13060 RO=RD
13070 VI=RI*VA/(RA+RI)
13080 VL=AV*VI*(RL/(RO+RL))
13090 RETURN
```

```
This program provides dc bias calculations
for a JFET or depletion MOSFET circuit then
ac amplifier calculations for a circuit
such as that of Fig. 10.23.

Enter the following circuit data:

RG1 (use 1E30 if open)=? 110E6
RG2                   =? 10E6
Total source resistance, RS=? 300
                         RD=? 1800
```

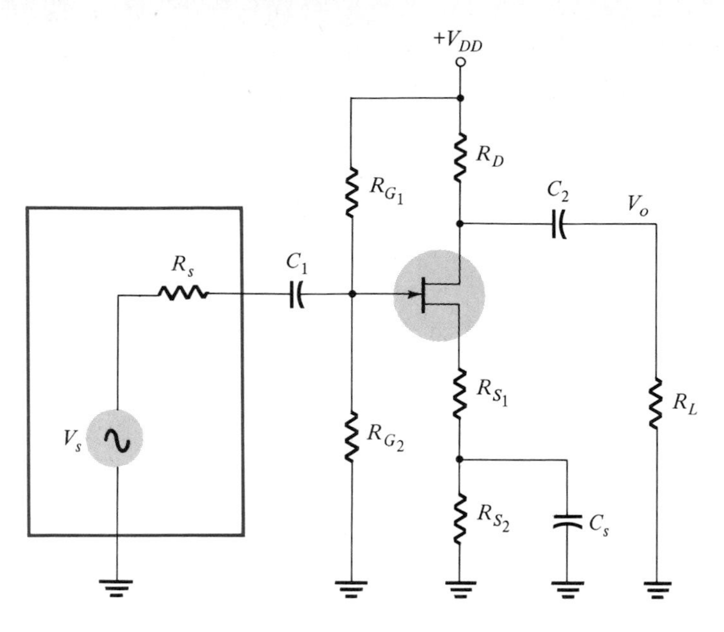

Figure 10.25 Amplifier circuit to use in dc and ac calculations.

```
Supply voltage, VDD=? 18

Enter the following device data:
Drain-source saturation current, IDSS=? 6E-3
Gate-source pinchoff voltage, VP=? -3


Bias current is, ID= 5.460326 mA
Bias voltages are:
VGS=-.1380968 volts
VD= 8.171413 volts
VS= 1.638098 volts
VDS= 6.533316 volts


Now accept data for ac amplifier calculations:

Load resistance (1E30 if none), RL=? 1E30
Source voltage, Vs=? 10E-3
Source resistance, Rs=? 0
Unoypassed source resistance, RS1=? 0

The amplifier voltage gain is Av=-6.868568
Output voltage across the load is-68.68568 mV

Amplifier stage input resistance is, Ri= 9166.666 kilohms
Amplifier stage output resistance is, Ro= 1.8 kilohms
This program provides dc bias calculations
for a JFET or depletion MOSFET circuit then
ac amplifier calculations for a circuit
such as that of Fig. 10.23.
```

```
Enter the following circuit data:

RG1 (use 1E30 if open)=? 2.1E6
RG2                   =? 270E3
Total source resistance, RS=? 1.5E3
                         RD=? 2.4E3

Supply voltage, VDD=? 16

Enter the following device data:
Drain-source saturation current, IDSS=? 8E-3
Gate-source pinchoff voltage, VP=? -4

Bias current is, ID= 2.416309 mA
Bias voltages are:
VGS=-1.801678 volts
VD= 10.20086 volts
VS= 3.624464 volts
VDS= 6.576393 volts

Now accept data for ac amplifier calculations:

Load resistance (1E30 if none), RL=? 10E3
Source voltage, Vs=? 100E-3
Source resistance, Rs=? 100E3
Unbypassed source resistance, RS1=? 300

The amplifier voltage gain is Av=-3.179261
Output voltage across the load is-180.8138 mV

Amplifier stage input resistance is, Ri= 239.2405 kilohms
Amplifier stage output resistance is, Ro= 2.4 kilohms
```

PROBLEMS

§ 10.3

1. Calculate the voltage gain for the circuit of Fig. 10.4 using a JFET having $I_{DSS} = 10$ mA and $V_P = -4$ V.
2. Calculate the voltage gain of the amplifier of Fig. 10.4 with R_D replaced by $R_D = 1.8$ kΩ.
3. Calculate the voltage gain of the circuit of Fig. 10.4 with R_S replaced by 330 Ω.
4. Determine the voltage gain of the circuit of Fig. 10.7 using a JFET having $V_P = -5$ V and $I_{DSS} = 8$ mA.
5. What is the voltage gain of the circuit in Fig. 10.7 with R_D replaced by 5.6 kΩ?
6. Calculate the output voltage for the circuit of Fig. 10.7 with input of 100 mV peak.
7. Calculate the output voltage of the circuit of Fig. 10.7 with $R_D = 3.6$ kΩ, $R_{S1} = 150$ Ω, and $V_i = 80$ mV, rms.

8. Determine the values of A_v, R_i, and R_o for the circuit of Fig. 10.8 with JFET having $I_{DSS} = 10$ mA and $V_P = -4$ V.

9. What output voltage results if the load resistance in Fig. 10.8 is changed to 3 kΩ for an input voltage $V_i = 75$ mV peak.

10. Calculate the voltage gain of the amplifier circuit of Fig. 10.9 with $R_S = 180$ Ω.

§ 10.4

11. Determine the output voltage for the circuit of Fig. 10.11 for input of $V_i = 80$ mV peak and load $R_L = 22$ kΩ.

12. Calculate the voltage gain of the circuit of Fig. 10.11 with $R_D = 2.4$ kΩ and $R_L = 50$ kΩ.

13. What is the output voltage of the circuit of Fig. 10.13 with $V_s = 120$ mV peak, $R_s = 100$ kΩ, and $R_L = 20$ kΩ?

14. Determine the value of signal input voltage needed in the circuit of Fig. 10.13 to obtain an output of 300 mV peak with load $R_L = 50$ kΩ.

15. Calculate the values of A_v, R_i, R_o, and V_o for the circuit of Fig. 10.13 with $R_D = 3.3$ kΩ, $R_L = 20$ kΩ, and JFET having $I_{DSS} = 8$ mA and $V_P = -5$ V.

16. What value of R_S is needed for JFET dynamic resistance of $r_m = 250$ Ω in the circuit of Fig. 10.26?

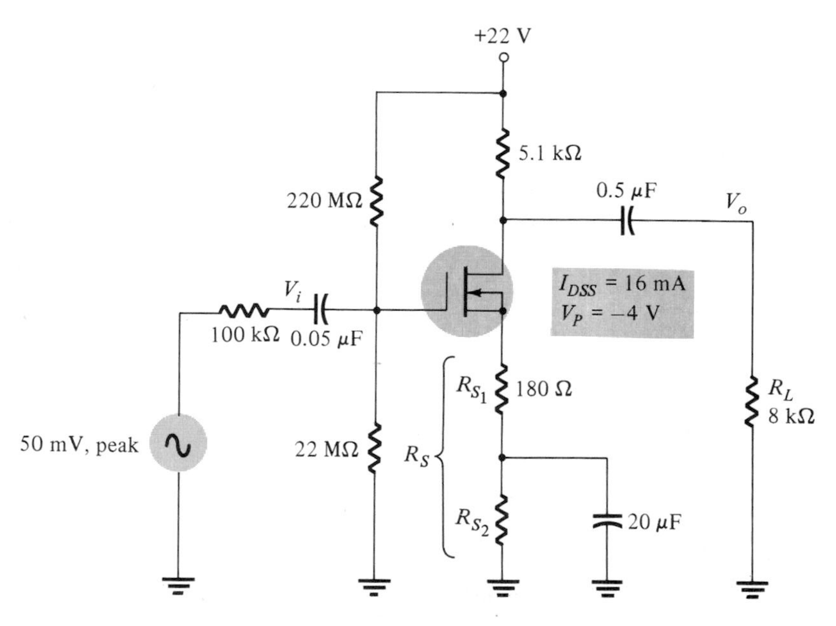

Figure 10.26 Circuit for Problems 16 through 20.

17. What is the circuit gain V_o/V_i for bias as for Problem 16?

18. What output voltage is developed across the load for the circuit of Fig. 10.26?

19. If R_{S2} is set to 220 Ω, calculate the circuit voltage gain, V_o/V_i, for the circuit of Fig. 10.26.

20. What output voltage results in the circuit of Fig. 10.26 if the load is changed to $R_L = 20\ \text{k}\Omega$?

§ 10.5

21. Calculate the output voltage for the circuit of Fig. 10.15 with JFET replaced by one having $I_{DSS} = 12\ \text{mA}$, $V_P = -3\ \text{V}$.

22. What output voltage results in the circuit of Fig. 10.15 if a load $R_L = 10\ \text{k}\Omega$ is connected to the output?

23. What output voltage results if the 2.2-kΩ resistor in Fig. 10.15 is replaced by one of value 1.2 kΩ?

24. What output voltages V_{O1} and V_{O2} result if the JFET of Fig. 10.16 is replaced by one having $I_{DSS} = 8\ \text{mA}$, $V_P = -4\ \text{V}$?

25. Calculate the output resistances looking back into the drain and source for the circuit of Fig. 10.16.

26. Calculate the output voltages V_{O1} and V_{O2} in the circuit of Fig. 10.16 if each output is connected to a load of 5 kΩ.

27. Calculate the output voltage in the circuit of Fig. 10.18 if the JFET is changed to one having $I_{DSS} = 12\ \text{mA}$, $V_P = -4\ \text{V}$.

28. Calculate the output voltage and output resistance if the source resistor of 2.1 kΩ in the circuit of Fig. 10.18 is replaced with one of value 1.5 kΩ.

§ 10.6

29. Calculate the gain of the circuit of Fig. 10.20 with JFET replaced by one having $I_{DSS} = 12\ \text{mA}$, $V_P = -4\ \text{V}$.

30. Calculate the output voltage of the circuit of Fig. 10.20 when a load of $R_L = 10\ \text{k}\Omega$ is connected to V_o.

31. Calculate the voltage gain of the circuit of Fig. 10.20 with 3.6-kΩ drain resistor replaced by one of value 2.7 kΩ.

32. What input voltage would be applied to the circuit of Fig. 10.20 for an output of 350 mV rms?

33. Calculate the output voltage in the circuit of Fig. 10.20 if the 1.1-kΩ source resistance is replaced by one of value 820 Ω.

§ 10.7

34. Design a circuit such as that in Fig. 10.4 using a 2N4416 having $I_{DSS} = 10\ \text{mA}$, $V_P = -6\ \text{V}$, for bias of $I_D = \frac{1}{2}I_{DSS}$, to provide an ac gain of magnitude at least 5 (use a supply of $V_{DD} = 20\ \text{V}$).

35. Complete the design of the circuit of Fig. 10.27 to provide an ac voltage gain of a least 10.

36. Design an amplifier circuit such as that in Fig. 10.23 for operation at $I_D = \frac{1}{2}I_{DSS}$ and $V_{DS} = \frac{1}{2}V_{DD}$ to provide a gain of at least 10 using a 2N5362 having $I_{DSS} = 6\ \text{mA}$, $V_P = 2.5\ \text{V}$, with a supply $V_{DD} = 30\ \text{V}$.

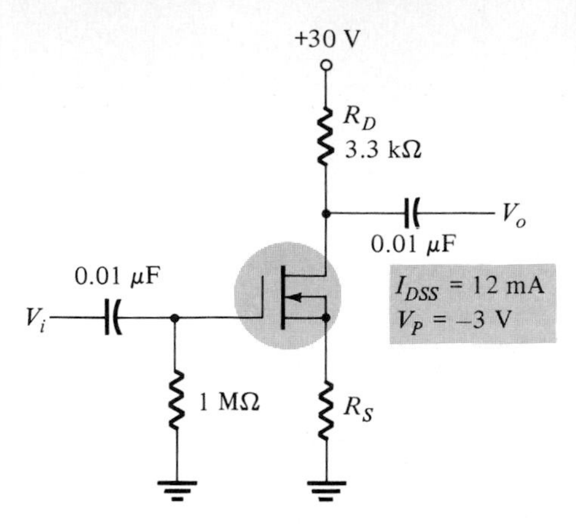

Figure 10.27 Circuit for Problem 35.

37. Complete the design of the MOSFET amplifier circuit shown in Fig. 10.28 for operation at $I_D = \frac{1}{2}I_{DSS}$ to provide a gain of magnitude greater than 3.5.

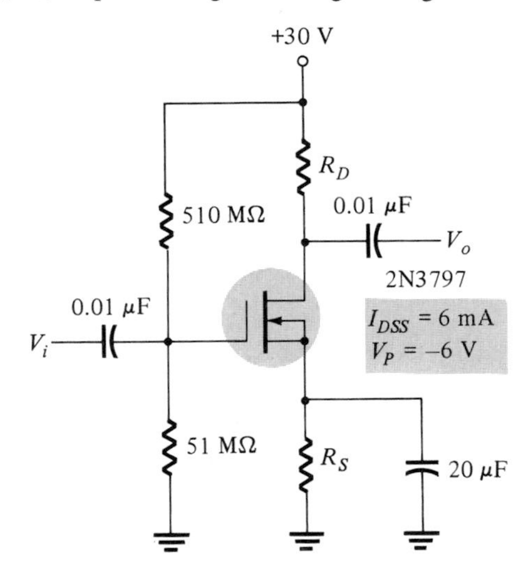

Figure 10.28 Circuit for Problem 37.

§ 10.8

38. Calculate the Miller capacitance of a circuit as in Fig. 10.24 with $R_D = 2.1$ kΩ and $C_{gd} = 4.5$ pF.

39. Calculate the input capacitance of a circuit as in Fig. 10.24 with $C_{gs} = 4.5$ pF and $R_D = 2.1$ kΩ.

COMPUTER PROBLEMS

Write BASIC programs to:

1. Calculate the voltage gain of a common-source circuit as in Fig. 10.5.
2. Calculate the input and output resistances of a JFET circuit as in Fig. 10.5.
3. Calculate the input and output resistances of a JFET source follower as in Fig. 10.14.
4. Calculate the voltage gain of a source follower as in Fig. 10.14.
5. Calculate the output voltage for a source follower as in Fig. 10.17.
6. Calculate the output voltage of a common-gate amplifier as in Fig. 10.19a.

11

Multistage Systems and Frequency Considerations

11.1 INTRODUCTION

This chapter will include, under the heading of multistage systems, both the *cascaded* and *compound* configurations. The *cascaded* system, for the purposes of this text, is one in which the stages are well defined and the connections between stages are very similar or identical. The *compound* system includes all other possible *multiple* active-device configurations, with a variety of interconnections.

The first few sections of this chapter examine multistage systems employing the technique of analysis developed in earlier chapters. This is followed by a detailed discussion of decibels (dB) and the effect of frequency on the response of single and multistage systems.

11.2 GENERAL CASCADED SYSTEMS

A discussion of cascaded systems is best initiated by considering the block-diagram representation of Fig. 11.1. The quantities of interest are indicated in the figure. The indicated A_v (voltage amplification) and A_i (current amplification) of each stage were determined with all stages connected as indicated in Fig. 11.1. In other words, A_v and A_i of each stage *do not* represent the gain of each stage on an independent basis. The loading effect of one stage on another was considered when these quantities were determined. All levels of gain, voltage, current, and impedance are magnitudes only and are not complex.

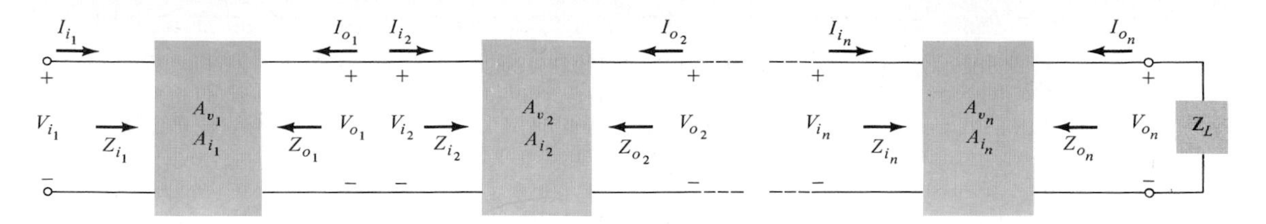

Figure 11.1 General cascaded system.

Rather than simply state the result for the overall gain of the system (voltage or current) a simple numerical example will clearly indicate the solution. If $A_{v_1} = -40$ and $A_{v_2} = -50$ with $V_{i_1} = 1$ mV, then $V_{o_1} = A_{v_1} \times V_{i_1} = -40(1 \text{ mV}) = -40$ mV. Since $V_{o_1} = V_{i_2}$,

$$V_{o_2} = A_{v_2}V_{i_2} = -50(-40 \text{ mV}) = 2000 \text{ mV} = 2 \text{ V}$$

The overall gain is $A_{v_T} = 2000 \text{ mV}/1 \text{ mV} = 2000$.

Obviously, the total gain of the two stages is simply the product of the individual gains A_{v_1} and A_{v_2}. In general, for n stages,

$$\boxed{A_{v_T} = \pm A_{v_1}A_{v_2}A_{v_3} \cdots A_{v_n}} \tag{11.1}$$

The same is true for the net current gain

$$\boxed{A_{i_T} = \pm A_{i_1}A_{i_2}A_{i_3} \cdots A_{i_n}} \tag{11.2}$$

The input and output impedance of each stage as indicated in Fig. 11.1 are also those values obtained by considering the effects of each and every stage of the system. There is no generally employed equation, such as Eq. (11.2), for the input or output impedances of the system in terms of the individual values. However, in a number of situations (BJT transistor, FET, or tube) the input (or output) impedance can normally be determined to an acceptable degree of accuracy by considering only one, or perhaps two, stages of the system.

The magnitude of the overall voltage gain of the representative system of Fig. 11.1 can be written as

$$|A_{v_T}| = \left|\frac{V_{o_n}}{V_{i_1}}\right| = \left|\frac{-I_{o_n}Z_L}{I_{i_1}Z_{i_1}}\right|$$

so that

$$\boxed{|A_{v_T}| = |A_{i_T}| \cdot \left|\frac{Z_L}{Z_{i_1}}\right|} \tag{11.3}$$

Equation (11.3) will prove useful in the analysis to follow. To go a step further, if the product of the voltage and current gain is formed for resistive loads

$$|A_{v_T}A_{i_T}| = \left|\frac{I_{o_n}R_L}{I_{i_1}R_{i_1}}\right| \cdot \left|\frac{I_{o_n}}{I_{i_1}}\right| = \left|\frac{I_{o_n}^2R_L}{I_{i_1}^2R_{i_1}}\right| = \frac{P_o}{P_i}$$

and

$$\boxed{|A_{p_T}| = |A_{v_T}| \cdot |A_{i_T}|} \tag{11.4}$$

this being the overall power gain of the system.

There are three types of coupling between stages of a system, such as in Fig. 11.1, that will be considered. The first to be described is the *RC-coupled* amplifier system, which is the most frequently applied of the three. This will be followed by the *transformer* and *direct-coupled* amplifier systems.

11.3 *RC*-COUPLED AMPLIFIERS

A cascaded *RC*-coupled transistor amplifier (two-stage) showing typical values and biasing techniques appears in Fig. 11.2. The terminology "*RC*-coupled" is derived from the biasing resistors and coupling capacitors employed between stages.

The primary function of the approximate technique is to obtain a "ball park" solution with a minimum of time and effort. A reduced time element obviously requires that the network be redrawn a minimum number of times. In fact, let us optimistically say that we can find the solutions for the network of Fig. 11.2 using only the original artwork.

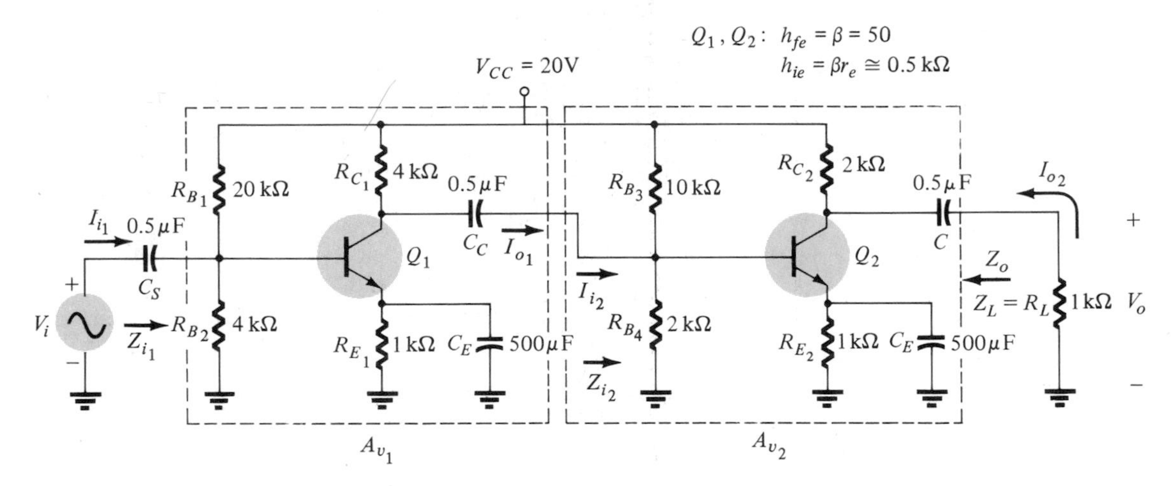

Figure 11.2 Two-stage *RC*-coupled amplifier.

Z_i

From the past experience with single amplifiers and the analysis just completed, it should be clear that for the ac response, both the 4-kΩ and 20-kΩ resistors will appear in parallel if the network is redrawn. They are also in parallel with the input impedance of Q_1, which is approximately $h_{ie} = \beta r_e = 0.5$ kΩ since the emitter resistor is bypassed by C_E.

The parallel combination:

$$Z_{i_1} = 20 \text{ k}\Omega \| 4 \text{ k}\Omega \| 0.5 \text{ k}\Omega = \mathbf{0.435\ k\Omega}$$

Z_o

Recall that the approximate collector-to-emitter equivalent circuit of a transistor is simply a current source $h_{fe}I_b$. This being the case, when $V_i = 0$, then $I_{b_1} = 0$, and

$I_{b_2} = 0$, resulting in $h_{fe} I_{b_2} = 0$, so that Z_o is simply R_{C_2} in parallel with the open-circuit representation of the controlled current source. That is,

$$Z_o \big|_{V_i=0} = R_{C_2} = \mathbf{2\ k\Omega}$$

A_i

Applying the current-divider rule (Fig. 11.3):

$$I_{b_1} = \frac{R' I_{i_1}}{R' + h_{ie}}$$

$$= \frac{3.333\ \text{k}\Omega\ I_{i_1}}{3.333\ \text{k}\Omega + 0.5\ \text{k}\Omega}$$

and $$I_{b_1} \cong 0.87\ I_{i_1}$$

The collector current of the first stage $I_{c_1} \cong h_{fe} I_{b_1}$. However, I_{c_1} will divide between the 4-kΩ resistor and the *loading* of the second stage (Fig. 11.3). The parallel combination of the R'' and 0.5-kΩ resistors as the loading of the next stage will result in an impedance of 0.385 kΩ.

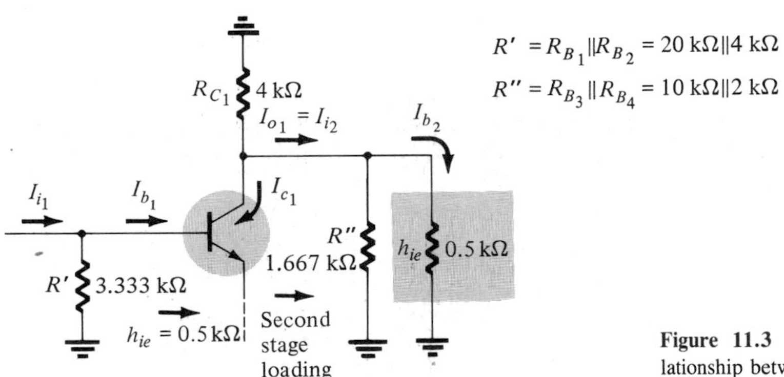

Figure 11.3 Determining the relationship between I_{i_1} and I_{b_2}.

Applying the current-divider rule:

$$I_{o_1} = \frac{-R_{C_1}(I_{c_1})}{R_{C_1} + 0.385\ \text{k}\Omega} = \frac{-4\ \text{k}\Omega\ (I_{c_1})}{4\ \text{k}\Omega + 0.385\ \text{k}\Omega}$$

$$= \frac{-4\ \text{k}\Omega\ (h_{fe} I_{b_1})}{4.385\ \text{k}\Omega} = \frac{-4\ \text{k}\Omega\ (50)(0.87\ I_{i_1})}{4.385\ \text{k}\Omega}$$

and $$A_{i_1} = \frac{I_{o_1}}{I_{i_1}} \cong \mathbf{-39.68}$$

For the second stage:

$$I_{b_2} = \frac{R'' I_{i_2}}{R'' + h_{ie}} = \frac{1.667\ \text{k}\Omega\ (I_{i_2})}{1.667\ \text{k}\Omega + 0.5\ \text{k}\Omega} = 0.769\ I_{i_2}$$

and $$I_{c_2} = h_{fe} I_{b_2} = 50(0.769\ I_{i_2}) = 38.45\ I_{i_2}$$

Applying the current-divider rule to the output circuit (see Fig. 11.2):

$$I_{o_2} = \frac{R_{C_2} I_{c_2}}{R_{C_2} + R_L} = \frac{2\text{ k}\Omega\ (h_{fe} I_{b_2})}{2\text{ k}\Omega + 1\text{ k}\Omega} = \frac{2\text{ k}\Omega(38.45\ I_{i_2})}{3\text{ k}\Omega} = 25.63\ I_{i_2}$$

and

$$A_{i_2} = \frac{I_{o_2}}{I_{i_2}} = 25.63$$

with

$$A_{i_T} = A_{i_1} A_{i_2} = (-39.68)(25.63) \cong \mathbf{-1017.0}$$

A_v

The direct connection under ac conditions clearly indicates in Fig. 11.2 that V_i appears directly at the base of the transistor of the first stage. Since the transistor has a grounded emitter terminal, the ac voltage gain can be obtained (on an approximate basis) using the following equation:

$$A_v \cong \frac{-h_{fe} R_L}{h_{ie}} = \frac{-R_L}{r_e}$$

R_L, the loading on the first stage, is the parallel combination of R_{C_1}, R_{B_3}, R_{B_4}, and $h_{ie}(=\beta r_e)$ which is $= 0.3509$ kΩ. The voltage amplification, A_{v_1}, therefore, equals $[-(50)(0.3509\text{ k}\Omega)]/0.5\text{ k}\Omega = -35.09$. For the second stage,

$$A_{v_2} = \frac{-(50)(R_{C_2} \| R_L)}{0.5\text{ k}\Omega} = \frac{-(50)(2\text{ k}\Omega \| 1\text{ k}\Omega)}{0.5\text{ k}\Omega} = \frac{-(50)(0.667\text{ k}\Omega)}{0.5\text{ k}\Omega} = -66.70$$

The net gain is, therefore,

$$A_{v_T} = A_{v_1} A_{v_2} = (-35.09)(-66.70)$$

$$A_{v_T} \cong \mathbf{2340.50}$$

Using Eq. (11.3)

$$|A_{v_T}| = |A_{i_T}| \cdot \left|\frac{Z_L}{Z_{i_1}}\right| = \frac{(1017.0)(1\text{ k}\Omega)}{0.435\text{ k}\Omega} = \mathbf{2337.93}$$

The very slight difference between the obtained values of A_{v_T} is due only to the decimal carry over in the individual solutions for A_{i_T} and A_{v_T}. In this case, the determination of A_{v_T} was markedly easier than that for A_{i_T}. In the future, therefore, there may be some savings in time if A_{v_T} is first determined, and A_{i_T} calculated from Eq. (11.3) in the following form:

$$|A_{i_T}| = |A_{v_T}| \cdot \left|\frac{Z_{i_1}}{Z_L}\right|$$

EXAMPLE 11.1

Calculate the input and output impedance, voltage gain, and current gain of the two-stage amplifier of Fig. 11.4. Note that the second stage is an emitter-follower configuration.

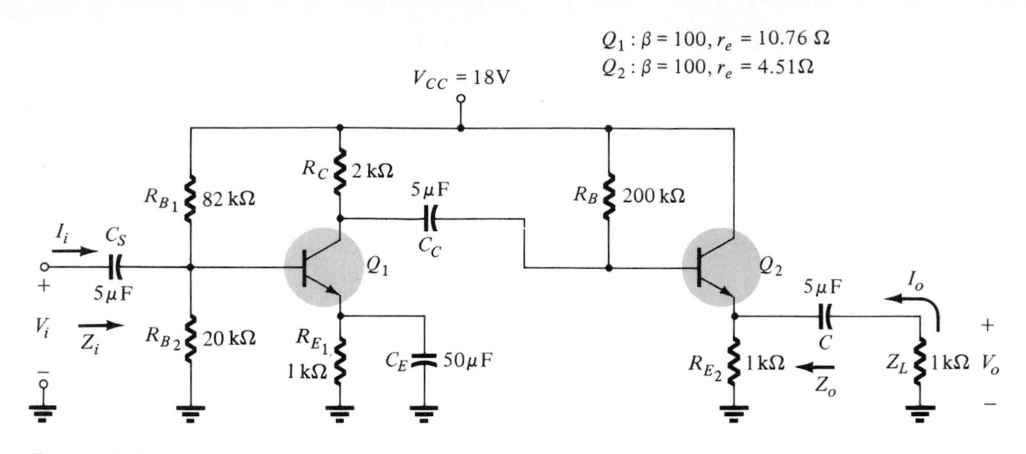

Figure 11.4 Two-stage transistor network to be examined in detail.

Solution:

Z_i: For ac conditions, R_E of the first stage is bypassed by C_E and the input impedance to Q_1 is $\cong \beta r_e = (100)(10.74) = 1.074\ \text{k}\Omega$.
Then

$$Z_i = R_{B_1} \| R_{B_2} \| \beta r_e = 82\ \text{k}\Omega \| 20\ \text{k}\Omega \| 1.074\ \text{k}\Omega \cong \beta r_e = \mathbf{1.074\ k\Omega}$$

Z_o: For ac conditions the network is redrawn as shown in Fig. 11.5. $Z_e = (R_s/\beta) + r_e$ where R_s is the source resistance connected to the base of the transistor. In this case $R_s = 2\ \text{k}\Omega \| 200\ \text{k}\Omega \cong 2\ \text{k}\Omega$ and

$$Z_e = \frac{2\ \text{k}\Omega}{100} + 4.51 = 20 + 4.51 = 24.51\ \Omega$$

$$Z_o = Z_e \| R_{E_2} = 24.51 \| 1\ \text{k}\Omega \cong \mathbf{24.51\ \Omega}$$

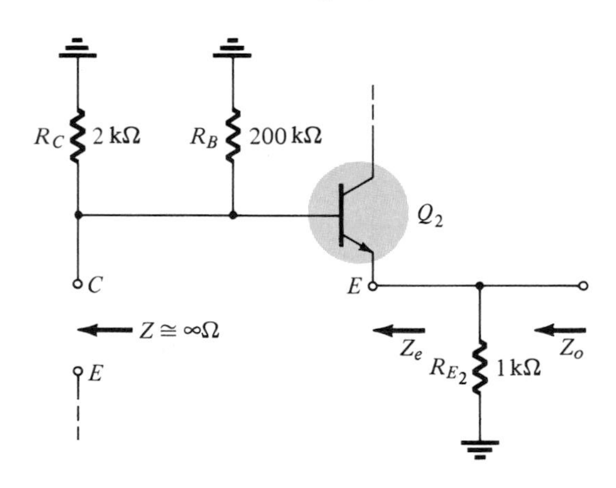

Figure 11.5 Determining Z_o for the network of Fig. 11.4.

A_v:

$$V_i = V_{b_1}$$

and

$$A_{v1} \cong \frac{-R_L}{r_{e1}} = \frac{-[R_C \| R_B \| \beta(R_{E_2} \| Z_L)]}{r_{e1}} = \frac{-[2\text{ k}\Omega \| 200\text{ k}\Omega \| 100(1\text{ k}\Omega \| 1\text{ k}\Omega)]}{r_{e1}}$$

$$= -\frac{2\text{ k}\Omega \| 200\text{ k}\Omega \| 50\text{ k}\Omega}{10.74} \cong -\frac{2\text{ k}\Omega}{10.74} = -186.22$$

$V_{be_2} \cong 0$ V and $V_{b_2} \cong V_o$ with $A_{v_2} = (V_{o_2}/V_{i_2}) = 1$. Then

$$A_{vT} = A_{v1}A_{v2} = (-172.27)(1) = \mathbf{-172.27}$$

A_i:

$$|A_{i_T}| = |A_{vT}| \cdot \left|\frac{Z_{i1}}{Z_L}\right|$$

$$= 172.27\left(\frac{1.074\text{ k}\Omega}{1\text{ k}\Omega}\right)$$

$$\cong \mathbf{185}$$

Note above how rapidly the solutions to fairly complex configurations are developing using the approximations developed in Chapter 9. In the following example the values of r_e will have to be determined.

EXAMPLE 11.2

Determine Z_i, Z_o, A_v, A_i, and A_p for the network of Fig. 11.6.

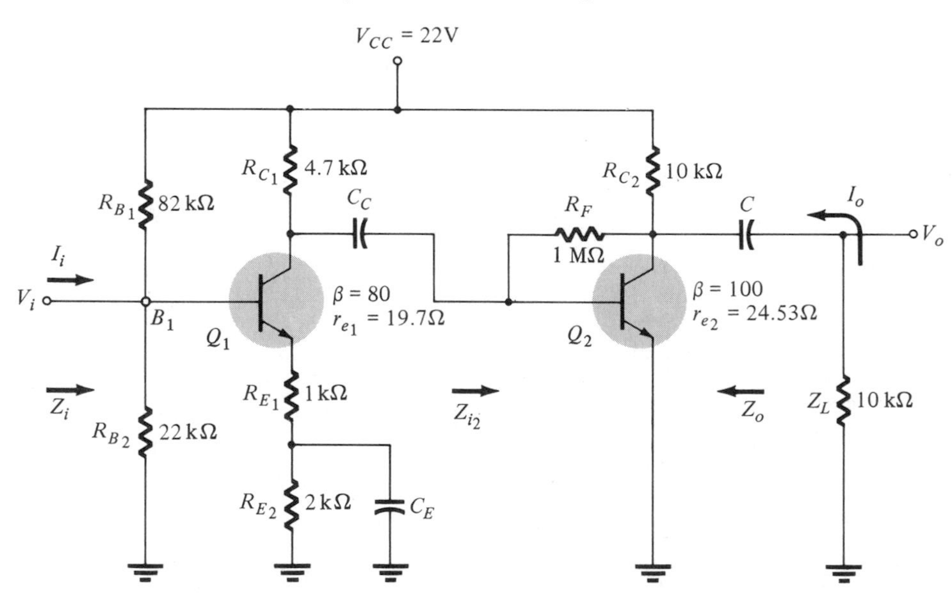

Figure 11.6

Solution:

The values of r_e must be determined. For Q_1:
Defining $R' = R_{B2} \| \beta(R_{E_1} + R_{E_2})$

$$= 22\text{ k}\Omega \| 80(3\text{ k}\Omega) = 22\text{ k}\Omega \| 240\text{ k}\Omega \cong R_{B2}$$

and $V_{B_1} = \dfrac{R' V_{CC}}{R' + R_{B_1}} \cong \dfrac{R_{B_2} V_{CC}}{R_{B_2} + R_{B_1}} = \dfrac{22 \text{ k}\Omega(22)}{22 \text{ k}\Omega + 82 \text{ k}\Omega}$

$$= \frac{484}{104} = 4.65 \text{ V}$$

and $V_{E_1} = V_{B_1} - V_{BE} = 4.65 - 0.7 = 3.95 \text{ V}$

with
$$I_{E_1} = \frac{V_{E_1}}{R_{E_1} + R_{E_2}} = \frac{3.95}{3 \text{ k}\Omega} = 1.32 \text{ mA}$$

and
$$\boldsymbol{r_{e_1}} = \frac{26 \text{ mV}}{I_{E_1}} = \frac{26}{1.32} = \mathbf{19.70\ \Omega}$$

For Q_2:

$$V_{CC} - (\beta + 1)I_B R_{C_2} - R_F I_B - V_{BE} = 0$$
$$22 - (101)I_B\ 10 \text{ k}\Omega - 10^6\ I_B - 0.7 = 0$$
$$21.3 = 2.01 \times 10^6\ I_B$$

and
$$I_B = 10.6\ \mu\text{A}$$

with
$$I_E \cong I_C = \beta I_B = (100)(10.6\ \mu\text{A}) = 1.06 \text{ mA}$$

and
$$\boldsymbol{r_{e_2}} = \frac{26 \text{ mV}}{I_E} = \frac{26}{1.06} = \mathbf{24.53\ \Omega}$$

$\boldsymbol{Z_i}$**:**

$$Z_{i_1} = R_{B_1} \| R_{B_2} \| \beta R_{E_1} = 82 \text{ k}\Omega \| 22 \text{ k}\Omega \| 80 \text{ k}\Omega \cong \mathbf{14.26\ k\Omega}$$

$\boldsymbol{Z_o}$**:**

$$Z_o \big|_{V_i=0} \cong R_{C_2} = \mathbf{10\ k\Omega}$$

$\boldsymbol{A_v}$**:**

$$V_{b_1} = V_i \text{ and } \quad A_{v_1} = \frac{-R_L}{R_{E_1} + r_{e_1}} = \frac{-(R_{C_1} \| Z_{i_2})}{R_{E_1} + r_{e_1}}$$

From Chapter 9 (Eq. 9.59):
$$Z_{i_2} = \frac{R_F}{|A_{v_2}|} \Big\| \beta r_{e_2}$$

and
$$A_{v_2} = \frac{-R_L}{r_{e_2}} = \frac{-R_{C_2} \| Z_L}{r_{e_2}} = -\frac{5 \text{ k}\Omega}{24.53} = -203.83$$

$$Z_{i_2} = \frac{10^6}{203.83} \Bigg\| 100(24.53) = 4.906 \text{ k}\Omega \| 2.453 \text{ k}\Omega = 1.6353 \text{ k}\Omega$$

so that
$$A_{v_1} = \frac{-(4.7 \text{ k}\Omega \| 1.6353 \text{ k}\Omega)}{1 \text{ k}\Omega + 0.0197 \text{ k}\Omega} = -\frac{1.213}{1.0197} \cong \mathbf{-1.19}$$

and
$$A_{v_T} = A_{v_1} A_{v_2} = (-1.19)(-203.83) \cong \mathbf{242.56}$$

A_i:

$$|A_{iT}| = |A_{vT}| \cdot \left|\frac{Z_{i_1}}{Z_L}\right| = \frac{(242.56)(14.26\text{ k}\Omega)}{10\text{ k}\Omega} \cong \mathbf{345.89}$$

A_p:

$$|A_p| = |A_{iT}| \cdot |A_{vT}| = (345.89)(242.56) \cong \mathbf{83.9 \times 10^3}$$

EXAMPLE 11.3 ***FET RC-Coupled Amplifier***

RC coupling is not limited to BJT transistor stages, as indicated by the two stage FET amplifier of Fig. 11.7. Determine the total voltage gain.

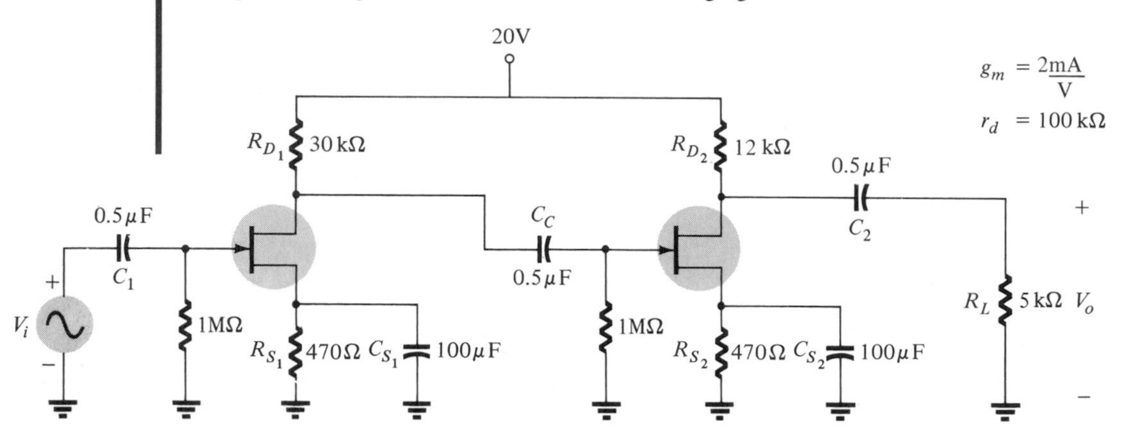

Figure 11.7 Two-stage FET amplifier.

Solution:

Substituting the small-signal equivalent circuit results in the configuration of Fig. 11.8.

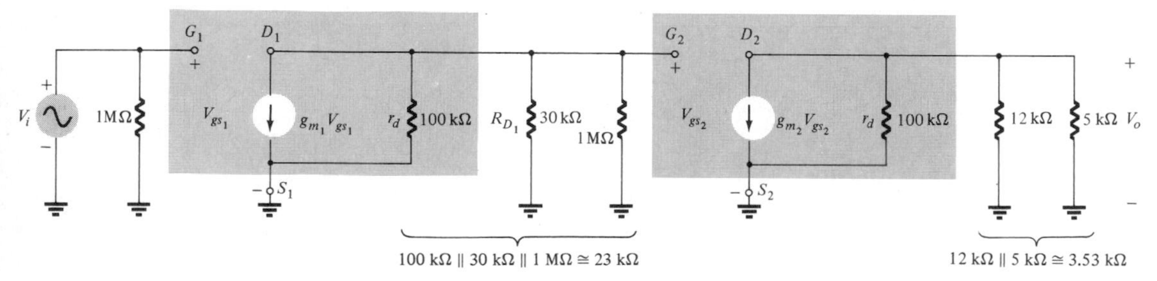

Figure 11.8 Network of Fig. 11.7 following the substitution of the small-signal ac equivalent circuits.

Combining parallel elements and eliminating those having no effect on the desired overall voltage gain will result in Fig. 11.9.

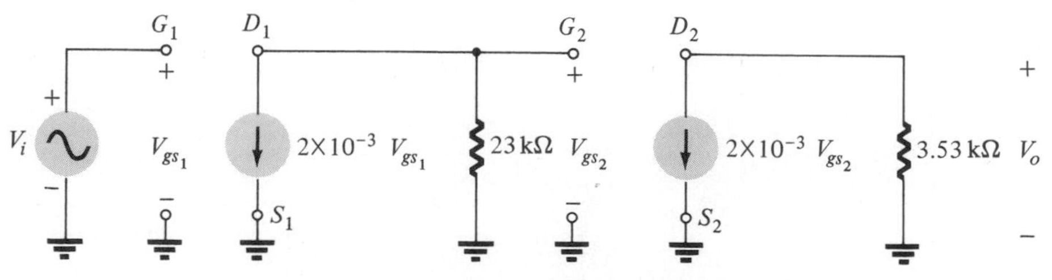

Figure 11.9 Network of Fig. 11.8 following the combination of parallel elements.

Obviously, $$V_{gs\,1} = V_i$$

and $$V_{gs\,2} = -(2 \times 10^{-3}\, V_{gs\,1})(23\text{ k}\Omega)$$

so that $$V_{gs\,2} = -46\, V_{gs\,1}$$

The minus sign indicates that the polarity of the voltage acros the 23-kΩ resistor due to the current source is the reverse of the defined polarities for $V_{gs\,2}$.

In conclusion:

$$V_o = -(2 \times 10^{-3}\, V_{gs\,2})(3.53\text{ k}\Omega) = -7.06\, V_{gs\,2}$$

so that $$V_o = -7.06\, V_{gs\,2} = -7.06(-46\, V_{gs\,1}) = 324.8\, V_{gs\,1} = 324.8\, V_i$$

and $$A_v = \frac{V_o}{V_i} = \mathbf{324.8}$$

11.4 TRANSFORMER-COUPLED TRANSISTOR AMPLIFIERS

A two-stage transformer-coupled transistor amplifier is shown in Fig. 11.10. Note that step-down transformers are employed between stages while a step-up transformer is connected to the source V_i. The step-up transformer increases the signal level while the step-down transformer matches, as closely as possible, the loading of each stage to the output impedance of the preceding stage. This is done in an effort to be as close to maximum-power-transfer conditions as possible. The effect of this matching tech-

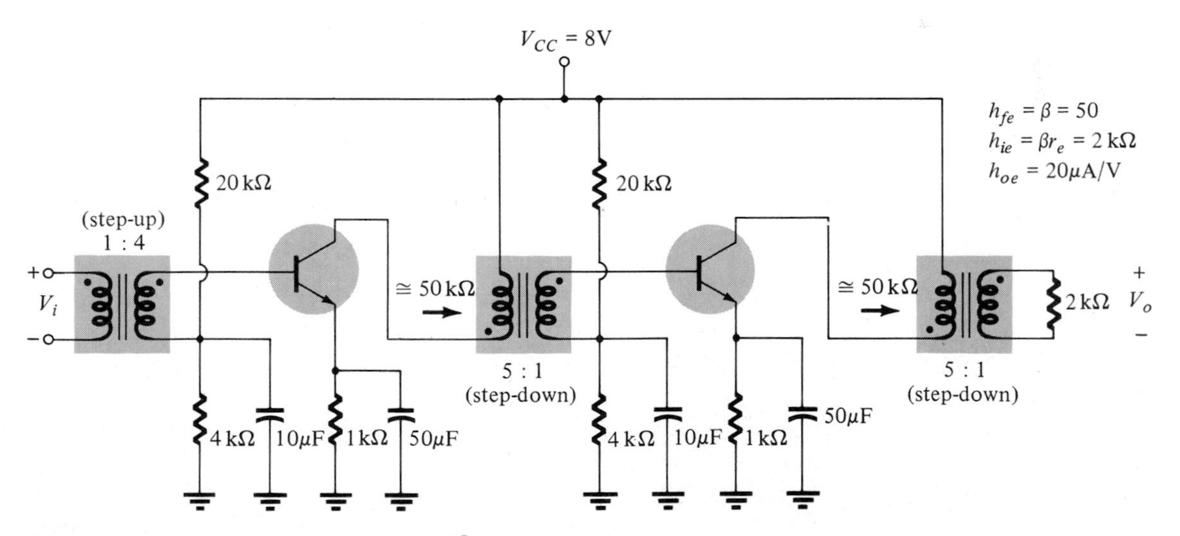

Figure 11.10 Two-stage transformer-coupled transistor amplifier.

nique through the use of transformer coupling will be clearly demonstrated in the following analysis.

Recall that a coupling capacitor was inserted to prevent any dc levels of one stage from affecting the bias conditions of another stage. The transformer provides this dc isolation very nicely.

The basic operation of this circuit is somewhat more efficient than the *RC*-coupled transistors due to the low dc resistance of the collector circuit of the transformer-coupled system. The primary resistance of the transformer is seldom more than a few ohms as compared to the large collector resistance R_C of the *RC*-coupled system. This lower dc resistance results in a lower dc power loss under operating conditions. The efficiency, as determined by the ratio of the ac power out to the dc power in, is therefore somewhat improved.

There are some decided disadvantages, however, to the transformer-coupled system. The most obvious is the increased size of such a system (due to the transformers) compared to *RC*-coupled stages. The second is a poorer frequency response due to the newly introduced reactive elements (inductance of coils and capacitance between turns). A third consideration, frequently an important one, is the increased cost of the transformer-coupled (as compared to the *RC*-coupled) system.

Before we consider the ac response of the system, the fundamental equations related to transformer action must be reviewed. For the configuration of Fig. 11.11,

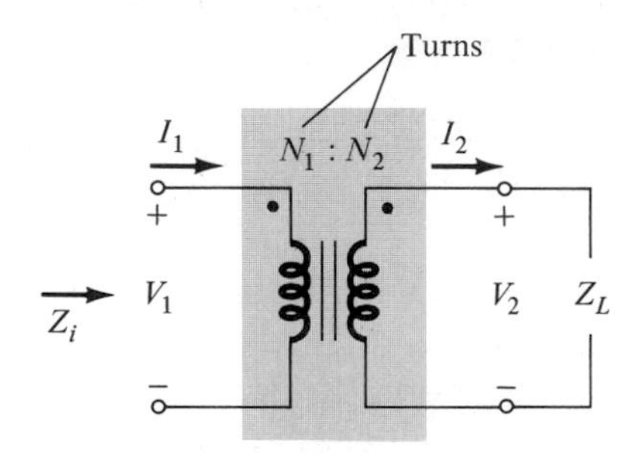

Figure 11.11 Basic transformer configuration.

$$\frac{V_1}{V_2} = \frac{N_1}{N_2} = a \quad \text{(transformation ratio)} \tag{11.5}$$

$$\frac{I_1}{I_2} = \frac{N_2}{N_1} = \frac{1}{a} \tag{11.6}$$

and

$$Z_i = a^2 Z_L \tag{11.7}$$

which states, in words, that the input impedance of a transformer is equal to the turns ratio squared times the load impedance.

For the ac response, the circuit of Fig. 11.10 will appear as shown in Fig. 11.12. For maximum power transfer the impedances Z_2 and Z_4 should be equal to the output impedance of each transistor: $Z_o \cong 1/h_{oe} = 1/20\ \mu\text{S} = 50\ \text{k}\Omega$. This is one system where the effect of $1/h_{oe}$ must be considered. The hybrid parameters will therefore be employed in its solution. Applying $Z_i = a^2 Z_L$, $Z_4 = a^2 R_L = (5)^2 2\ \text{k}\Omega = 50\ \text{k}\Omega$. Z_2

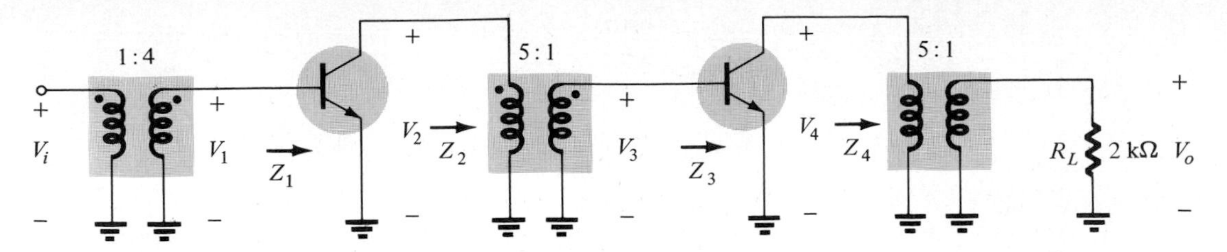

Figure 11.12 Cascaded transformer-coupled amplifiers of Fig. 11.10 redrawn to determine the small-signal ac response.

is also 50 kΩ since the input resistance to each stage (Z_1 and Z_3) is $\cong h_{ie} = 2\text{ k}\Omega$. Frequency considerations may not always permit Z_2 or Z_4 to be equal to $1/h_{oe}$. For situations of this type Z_2 and Z_4 are usually made as close as possible to $1/h_{oe}$ in magnitude.

Further analysis of the circuit of Fig. 11.12 results in

$$V_1 = \frac{N_2}{N_1}V_i = 4\ V_i$$

and

$$A_{v_1} = \frac{V_2}{V_1} = \frac{-h_{fe}Z_L}{h_{ie}} = \frac{-h_{fe}(\cong 1/h_{oe}\|Z_2)}{h_{ie}} = \frac{-50(50\text{ k}\Omega\|50\text{ k}\Omega)}{2\text{ k}\Omega} = -625$$

so that
$$V_2 = -625\ V_1 = -625(4\ V_i) = -2500\ V_i$$

but
$$V_3 = \frac{N_2}{N_1}V_2 = \frac{1}{5}V_2 = \frac{1}{5}(-2500\ V_i) = -500\ V_i$$

and
$$A_{v_2} = \frac{V_4}{V_3} = \frac{-h_{fe}Z_L}{h_{ie}} = \frac{-(50)(25\text{ k}\Omega)}{2\text{ k}\Omega} = -625$$

so
$$V_4 = -625\ V_3 = -625(-500\ V_i)$$
$$= 312.50 \times 10^3\ V_i$$

with
$$V_L = \frac{1}{5}V_4 = \frac{1}{5}(312.50 \times 10^3\ V_i)$$

and
$$A_{v_T} = \frac{V_L}{V_i} = \mathbf{62.50 \times 10^3}$$

11.5 DIRECT-COUPLED TRANSISTOR AMPLIFIERS

The third type of coupling between stages to be introduced in this chapter is *direct coupling*. The circuit of Fig. 11.13 is an example of a two-stage direct-coupled transistor system. Coupling of this type is necessary for very low-frequency applications. For a configuration of this type the dc levels of one stage are obviously related to the dc levels of the other stages of the system. For this reason the biasing arrange-

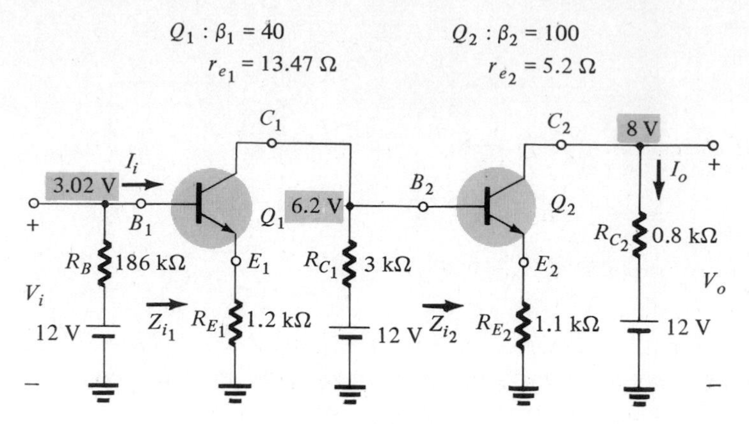

Figure 11.13 Direct-coupled transistor stages.

ment must be designed for the entire network rather than for each stage independently. Although three separate 12-V supplies are indicated, only one is required if the three terminals of higher (positive) potential for each supply are paralleled.

One of the biggest problems associated with direct-coupled networks is stability. Any variation in dc level in one stage is transmitted on an amplified basis to the other stages. The addition of the emitter resistor aids as a stabilizing element in each stage.

(DC) Bias Conditions

For an output voltage $V_{C_2} = 8$ V as indicated in Fig. 11.13,

***Q*₂:**

$$I_{0.8\text{k}\Omega} = \frac{12 - 8}{0.8\ \text{k}\Omega} = 5\ \text{mA}$$

therefore

$$I_{C_2} \cong I_{E_2} \cong 5\ \text{mA}$$

and

$$V_{E_2} = (5\ \text{mA})(1.1\ \text{k}\Omega) = 5.5\ \text{V}$$

For

$$V_{BE_2} = 0.7\ \text{V}$$

$$V_{B_2} = V_{C_1} = 5.5 + 0.7 = 6.2\ \text{V}$$

as indicated. Applying

$$I_{C_2} \cong \beta_2 I_{B_2}$$

$$I_{B_2} \cong \frac{I_{C_2}}{\beta_2} = \frac{5\ \text{mA}}{100} = 50\ \mu\text{A}$$

***Q*₁:**

$$I_{3\text{k}\Omega} = \frac{12 - 6.2}{3\ \text{k}\Omega} = \frac{5.8}{3\ \text{k}\Omega} = 1.93\ \text{mA}$$

and since

$$I_{3\text{k}\Omega} \gg I_{B_2}$$

assume

$$I_{C_1} \cong I_{3\text{k}\Omega} = 1.93\ \text{mA}$$

and

$$I_{E_1} = 1.93\ \text{mA}$$

so $$V_{E_1} = (1.93 \text{ mA})(1.2 \text{ k}\Omega) = 2.32 \text{ V}$$

and $$V_{B_1} = V_{E_1} + V_{BE_1} = 2.32 + 0.7 = 3.02 \text{ V}$$

as indicated. The verification above of the potential levels appearing in Fig. 11.13 demonstrates clearly the close tie-in required between bias levels of a direct-coupled amplifier.

Now for the ac response. The approach uses the approximate technique introduced earlier in this chapter. The input impedance to each emitter-follower configuration is $\cong \beta R_E$. Therefore,

$$Z_{i_1} = \beta_1 R_{E_1} = 40(1.2 \text{ k}\Omega) = 48 \text{ k}\Omega$$

and $$Z_{i_2} \cong \beta_2 R_{E_2} = 100(1.1 \text{ k}\Omega) = 110 \text{ k}\Omega$$

$$A_{v_1} = \frac{-R_{L_1}}{R_{E_1}} = \frac{-R_{C_1} \| \beta_2 R_{E_2}}{R_{E_1}} = -\frac{3 \text{ k}\Omega \| 110 \text{ k}\Omega}{1.2 \text{ k}\Omega} \cong \frac{-3 \text{ k}\Omega}{1.2 \text{ k}\Omega} = -2.5$$

$$A_{v_2} = \frac{-R_{L_2}}{R_{E_2}} = \frac{-R_{C_2}}{R_{E_2}} = \frac{-0.8 \text{ k}\Omega}{1.1 \text{ k}\Omega} = -0.7273$$

and $$A_{v_T} = A_{v_1} A_{v_2} = (-2.5)(-0.7273) = \mathbf{1.818}$$

$$|A_i| = |A_v| \cdot \left| \frac{Z_{i_1}}{Z_L} \right| = \frac{(1.818)(48 \text{ k}\Omega)}{0.8 \text{ k}\Omega} = \mathbf{109.08}$$

with $$|A_{p_T}| = |A_v| \cdot |A_i| = (1.818)(109.08) = \mathbf{198.3}$$

11.6 CASCODE AMPLIFIER

For high-frequency applications the CB configuration has the most desirable characteristics of the three configurations. However, it suffers from a very low input impedance ($Z_i \cong h_{ib} = r_e$). The *cascode* configuration in Fig. 11.14 is designed to improve the input impedance level for the CB configuration through the use of a typical CE

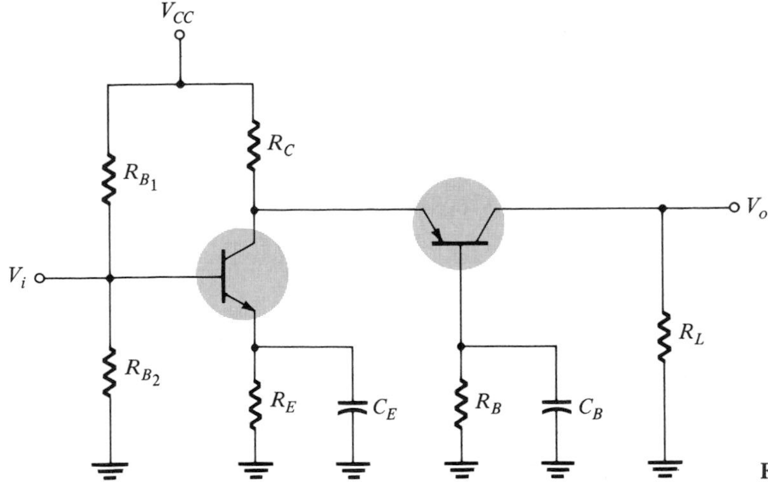

Figure 11.14 Cascode configuration.

network. The gain of the CE configuration is low to ensure that the input Miller capacitance level is a minimum (recall the discussion of Miller capacitance in Chapter 10 for the FET) for high-frequency applications.

A practical version of a cascode amplifier appears in Fig. 11.15. Note that the collector of the CE configuration is still tied directly to the emitter of the CB configuration.

DC:

$$I_{E_2} \cong I_{E_1} \quad \text{or} \quad I_{C_2} \cong I_{C_1}$$

or dividing each side by β since $\beta_1 = \beta_2 = \beta$

$$\frac{I_{C_2}}{\beta} \cong \frac{I_{C_1}}{\beta} \quad \text{or} \quad I_{B_2} \cong I_{B_1}$$

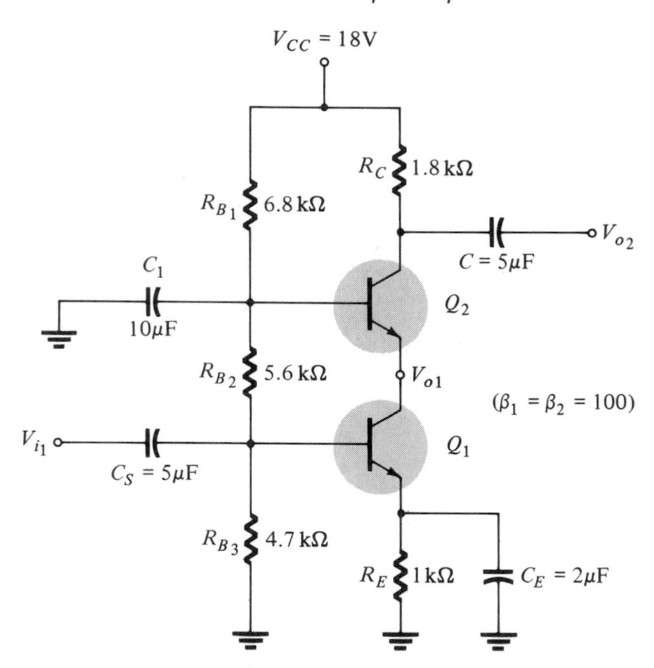

Figure 11.15 Practical cascode arrangement.

The current I_{B_1} will pass through βR_E of the parallel combination of R_{B_3} and βR_E. Since $\beta R_E = (100)(1\text{ k}\Omega) = 100\text{ k}\Omega$ and $R_{B_3} = 4.7\text{ k}\Omega$, we shall assume that I_{B_1} is significantly smaller than $I_{4.7\text{k}\Omega}$ to permit ignoring its effect. Since this approximation is applied to I_{B_1}, it can also be applied to I_{B_2} (since $I_{B_2} \cong I_{B_1}$) and

$$V_{B_1} = \frac{R_{B_3}(V_{CC})}{R_{B_3} + R_{B_2} + R_{B_1}} = \frac{4.7\text{ k}\Omega\ (18)}{4.7\text{ k}\Omega + 5.6\text{ k}\Omega + 6.8\text{ k}\Omega} = \frac{84.6}{17.1}$$

$$= 4.95\text{ V}$$

and

$$I_{E_1} = \frac{V_{E_1}}{R_E} = \frac{V_{B_1} - V_{BE}}{R_E} = \frac{4.95 - 0.7}{1\text{ k}\Omega} = 4.25\text{ mA}$$

with

$$r_{e_1} = \frac{26\text{ mV}}{I_{E_1}} = \frac{26}{4.25} = \mathbf{6.12\ \Omega}$$

and since

$$I_{E_1} \cong I_{E_2}$$

$$\boldsymbol{r_{e_2} = 6.12\ \Omega}$$

AC:

$$A_{v_1} = \frac{V_{o_1}}{V_{i_1}} \cong \frac{-R_L}{r_{e_1}}$$

with $R_L = r_{e_2} = h_{ib_2}$ of Q_2, the input impedance of that connected stage and

$$\boldsymbol{A_{v_1} = \frac{-r_{e_2}}{r_{e_1}} \cong -1} \qquad \text{(low as desired because of the Miller effect)}$$

with

$$A_{v_2} = \frac{R_L}{R_{e_2}} = \frac{R_C}{R_{e_2}} = \frac{1.8\ \text{k}\Omega}{6.12} \cong \mathbf{294.1}$$

and

$$A_{v_T} = \frac{V_{o_2}}{V_{i_1}} = A_{v_1}A_{v_2} = (-1)(294.1) = \mathbf{-294.1}$$

11.7 DARLINGTON COMPOUND CONFIGURATION

The Darlington circuit is a compound configuration that results in a set of improved amplifier characteristics. The configuration of Fig. 11.16 has a high input impedance with low output impedance and high current gain, all desirable characteristics for a current amplifier. We shall momentarily see, however, that the voltage gain will be less than one if the output is taken from the emitter terminal. A variation in the configuration can result in a trade-off between the output impedance and voltage gain.

The description of the biasing arrangement is similar to that of a single-stage emitter-follower configuration with current feedback. Note for the Darlington

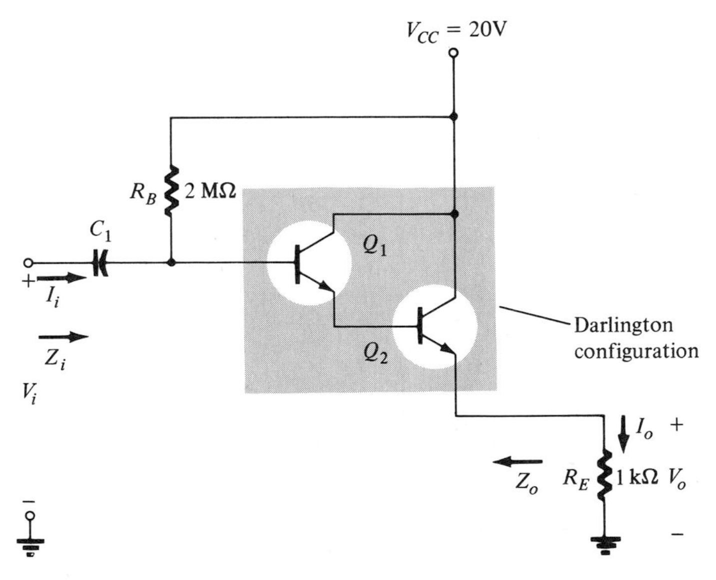

Figure 11.16 Darlington configuration.

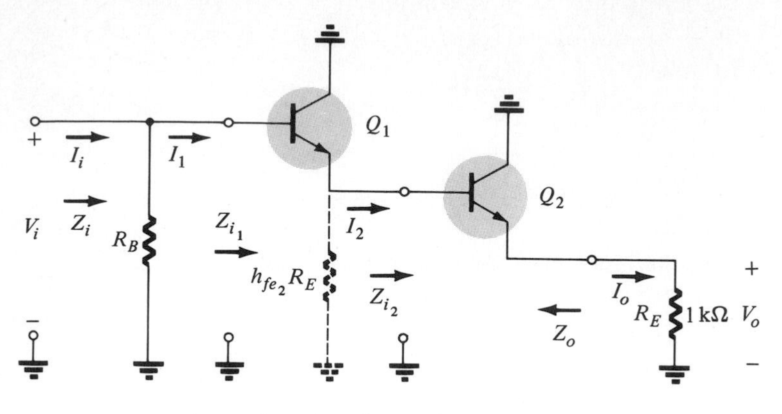

Figure 11.17 Darlington configuration of Fig. 11.16 redrawn to determine the small-signal ac response.

configuration that the emitter current of the first transistor is the base current for the second active device. In its small-signal ac form the circuit will appear as shown in Fig. 11.17.

For the second stage:

$$Z_{i_2} \cong h_{fe_2} R_E$$

and

$$A_{i_2} = \frac{I_o}{I_2} = \frac{I_{e_2}}{I_{b_2}} \cong h_{fe_2}$$

On a *good* approximate basis, these equations cannot be applied to the first stage. The "fly in the ointment" is the closeness with which Z_{i_2} compares with $1/h_{oe_1}$. You will recall that $1/h_{oe_1}$ could be eliminated in the majority of situations because the load impedance $Z_L \ll 1/h_{oe_1}$. For the Darlington configuration the input impedance Z_{i_2} is close enough in magnitude to $1/h_{oe_1}$ to necessitate considering the effects of h_{oe_1}. In Chapter 9 it was found that for the single-stage grounded-emitter transistor amplifier where $1/h_{oe}$ was considered (see Eq. 9.71),

$$A_i \cong \frac{h_{fe}}{1 + h_{oe} Z_L}$$

Applying the equation above to this situation, $Z_L = Z_{i_2} \cong h_{fe_2} R_E$ and

$$A_{i_1} = \frac{I_2}{I_1} = \frac{I_{e_1}}{I_{b_1}} \cong \frac{h_{fe_1}}{1 + h_{oe_1}(h_{fe_2} R_E)}$$

with

$$A_i = \frac{I_o}{I_1} = A_{i_1} A_{i_2} = \frac{h_{fe_1} h_{fe_2}}{1 + h_{oe_1}(h_{fe_2} R_E)} \tag{11.8}$$

For $h_{fe_1} = h_{fe_2} = h_{fe}$ and $h_{oe_1} = h_{oe_2} = h_{oe}$

$$A_i \cong \frac{h_{fe}^2}{1 + h_{oe} h_{fe} R_E} \tag{11.9}$$

For $h_{oe} h_{fe} R_E \leq 0.1$ a fairly good approximation (within 10%) is

$$A_i \cong h_{fe}^2 = \beta^2 \tag{11.10}$$

The current gain $A_{i_T} = I_o/I_i$, as defined by Fig. 11.17, can be determined through the use of the current-divider rule:

$$I_1 = \frac{R_B I_i}{R_B + Z_{i_1}}$$

Since $Z_{i_2} \cong h_{fe_2}R_E$ is the "emitter resistor" of the first stage (note Fig. 11.17), the input impedance to the first stages is $Z_{i_1} \cong h_{fe_1}(Z_{i_2} \| 1/h_{oe_1})$ since $Z_{i_2} = h_{fe_2}R_E$, and $1/h_{oe_1}$ will appear in parallel in the small-signal equivalent circuit. The result is

$$Z_{i_1} \cong h_{fe_1}\left(h_{fe_2}R_E \left\| \frac{1}{h_{oe_1}}\right.\right) = \frac{h_{fe_1}h_{fe_2}R_E(1/h_{oe_1})}{h_{fe_2}R_E + 1/h_{oe_1}}$$

and

$$\boxed{Z_{i_1} = \frac{h_{fe_1}h_{fe_2}R_E}{h_{oe_1}h_{fe_2}R_E + 1}} \tag{11.11}$$

which for $h_{fe_1} = h_{fe_2} = h_{fe}$ and $h_{oe_1} = h_{oe_2} = h_{oe}$,

$$\boxed{Z_{i_1} \cong \frac{h_{fe}^2 R_E}{1 + h_{oe}h_{fe}R_E}} \tag{11.12}$$

For $h_{oe}h_{fe}R_E \leq 0.1$,

$$\boxed{Z_{i_1} \cong h_{fe}^2 R_E = \beta^2 R_E} \tag{11.13}$$

For the following parameter values:

$$h_{fe_1} = h_{fe_2} = h_{fe} = 50$$

$$h_{ie_1} = 1\text{ k}\Omega, \qquad h_{ie_2} = 0.5\text{ k}\Omega$$

$$h_{oe_1} = h_{oe_2} = h_{oe} = 20\ \mu\text{A/V}$$

$$A_i = \frac{I_o}{I_1} \cong \frac{h_{fe}^2}{1 + h_{oe}h_{fe}R_E} = \frac{(50)^2}{1 + (20 \times 10^{-6})(50)(1\text{ k}\Omega)}$$

$$= \frac{2500}{1 + 1} = \mathbf{1250}$$

and

$$Z_{i_1} \cong \frac{h_{fe}^2 R_E}{1 + h_{oe}h_{fe}R_E} = \frac{(50)^2(1\text{ k}\Omega)}{2} = 1250\text{ k}\Omega = \mathbf{1.25\ M\Omega}$$

so that for $R_B = 2\text{ M}\Omega$,

$$\frac{I_1}{I_i} = \frac{R_B}{R_B + Z_{i_1}} = \frac{2\text{ M}\Omega}{2\text{ M}\Omega + 1.25\text{ M}\Omega} = \frac{2}{3.25} = 0.615$$

and

$$A_{i_T} = \frac{I_o}{I_i} = \frac{I_o}{I_1}\frac{I_1}{I_i} = A_i \times \frac{I_1}{I_i}$$

$$= (1250)(0.615) = \mathbf{769}$$

with

$$Z_i = 2\text{ M}\Omega \| Z_{i_1} = 2\text{ M}\Omega \| 1.25\text{ M}\Omega = \mathbf{769\ k\Omega}$$

Too frequently, the current gain of a Darlington circuit is assumed to be simply $A_i^2 \cong h_{fe}^2$ without any regard to the output impedance $1/h_{oe}$. In this case $A_i \cong (h_{fe})^2 = 2500$. Certainly, 2500 versus 1250 is *not* a good approximation. The effect of h_{oe_1} must therefore be considered when the current gain of the first stage is determined.

The output impedance Z_o can be determined directly from the emitter-equivalent circuits, as follows. For the first stage,

$$\boxed{Z_{o_1} \cong \frac{R_{s_1} + h_{ie_1}}{h_{fe_1}}} \tag{11.14}$$

$$= \frac{0 + 1\text{ k}\Omega}{50} \cong \mathbf{20.0\ \Omega}$$

and

$$\boxed{Z_{o_2} \cong \frac{(Z_{o_1} \| 1/h_{oe_1}) + h_{ie_2}}{h_{fe_2}}} \tag{11.15}$$

$$= \frac{(20\Omega \| 50\text{ k}\Omega) + 0.5\text{ k}\Omega}{50} \cong \frac{20\Omega + 0.5\text{ k}\Omega}{50} = \frac{520\Omega}{50}$$

$$= \mathbf{10.40\ \Omega}$$

Note, as indicated in the introductory discussion, that the input impedance is high, output impedance very low, and current gain high. We shall now examine the voltage gain of the system. Applying Kirchhoff's voltage law to the circuit of Fig. 11.16:

$$V_o = V_i - V_{be_1} - V_{be_2}$$

That the output potential is the input *less* the base-to-emitter potential of each transistor clearly indicates that $V_o < V_i$. It is closer in magnitude to one than to zero. On an approximate basis it is given by

$$\boxed{A_v \cong \frac{1}{1 + h_{ie_2}/(h_{fe_2} R_E)}} \tag{11.16}$$

as derived from the emitter equivalent circuit.

Substituting the numerical values of this general example:

$$A_v \cong \frac{1}{1 + 0.5\text{ k}\Omega/50\text{ k}\Omega} = \frac{1}{1 + 0.01} = \mathbf{0.99}$$

The ratings and characteristics for a 10-A RCA *npn* Darlington power transistor are provided in Fig. 11.18. Some of the characteristics appear in Figs. 11.19 through 11.24. The data provided are for the complete device—the individual β values are not provided.[1] Note in the characteristics that the level of V_{BE} is increased since it includes the drop across two transistors. Consider also that h_{fe} is over 3000 at 1 kHz but drops to only 20 at 2 MHz. Frequency will obviously have a pronounced effect on its performance. Note in Fig. 11.19 that the collector current is in amperes and that a pulsed operation results in an increased level of current—the longer the pulse, the

[1]That is, $h_{fe} = I_{C_2}/I_{b_1}$, thereby treating the transistor combination as a single device.

Solid State
Division

Power Transistors

2N6383 2N6384 2N6385

JEDEC TO-3

10-Ampere, N-P-N Darlington Power Transistors

40-60-80 Volts, 100 Watts
Gain of 1000 at 5 A

TERMINAL CONNECTIONS

Pin 1 - Base
Pin 2 - Emitter
Case - Collector
Mounting Flange - Collector

Features:

- Operates from IC without predriver
- Low leakage at high temperature
- High reverse second-breakdown capability

Applications:

- Power switching
- Audio amplifiers
- Hammer drivers
- Series and shunt regulators

The 2N6383, 2N6384, and 2N6385• are monolithic n-p-n silicon Darlington transistors designed for low- and medium-frequency power applications. The double epitaxial construction of these devices provides good forward and reverse second-breakdown capability; their high gain makes it possible for them to be driven directly from integrated circuits.

•Formerly RCA Dev. Nos. TA8349, TA8486, and TA8348.

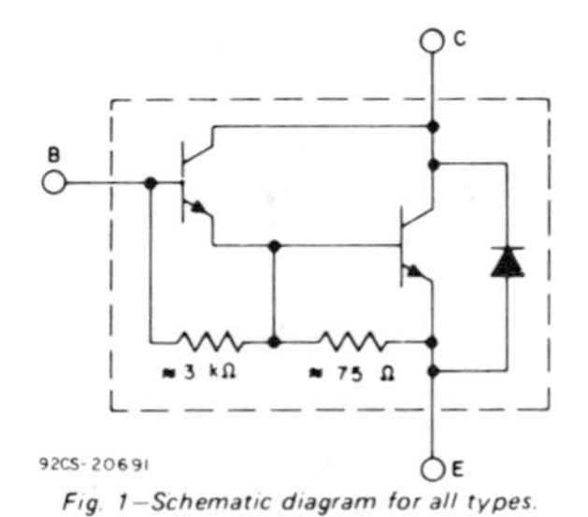

Fig. 1—Schematic diagram for all types.

MAXIMUM RATINGS, *Absolute-Maximum Values:*

		2N6385	2N6384	2N6383	
* COLLECTOR-TO-BASE VOLTAGE	V_{CBO}	80	60	40	V
COLLECTOR-TO-EMITTER VOLTAGE:					
With external base-to-emitter resistance (R_{BE}) = 100Ω, sustaining	$V_{CER(sus)}$	80	60	40	V
With base open, sustaining	$V_{CEO(sus)}$	80	60	40	V
* With base reverse-biased V_{BE} = −1.5 V, R_{BB} = 100Ω	V_{CEX}	80	60	40	V
* EMITTER-TO-BASE VOLTAGE	V_{EBO}	5	5	5	V
COLLECTOR CURRENT:	I_C				
* Continuous		10	10	10	A
Peak		15	15	15	A
* CONTINUOUS BASE CURRENT	I_B	0.25	0.25	0.25	A
* TRANSISTOR DISSIPATION:	P_T				
At case temperatures up to 25°C		100	100	100	W
At case temperatures above 25°C		← See Fig. 9.20 →			
* TEMPERATURE RANGE:					
Storage and Operating (Junction)		← −65 to +200 →			°C
* PIN TEMPERATURE (During Soldering):					
At distances ≥ 1/32 in. 0.8 mm from seating plane for 10 s max.		← 235 →			°C

*In accordance with JEDEC registration data format JS-6 RDF-2.

Figure 11.18 RCA NPN Darlington power transistors. (Courtesy RCA Solid State Division.)

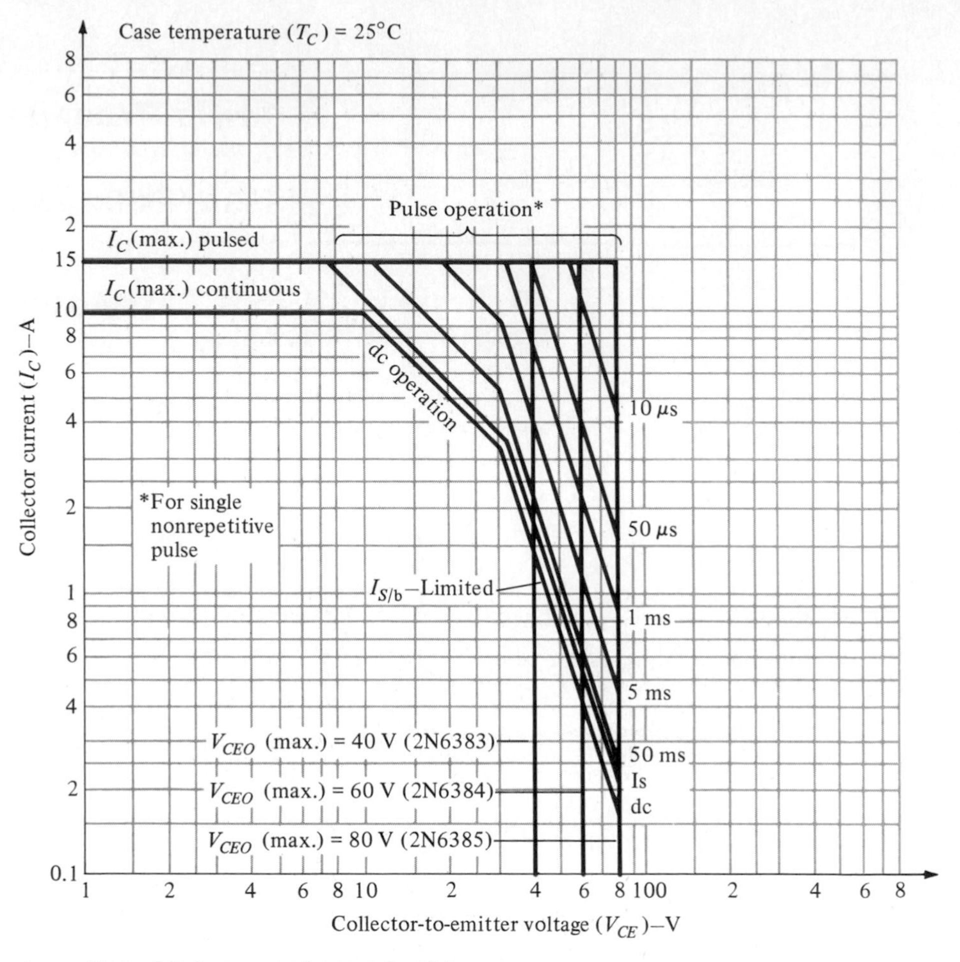

Figure 11.19 Maximum operating area for all types.

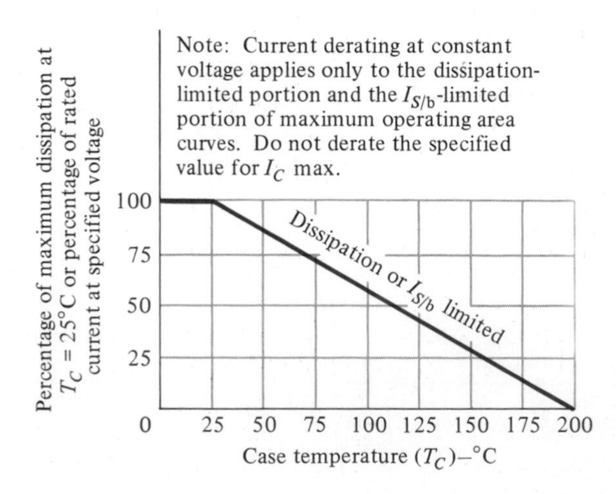

Figure 11.20 Derating curve for all types.

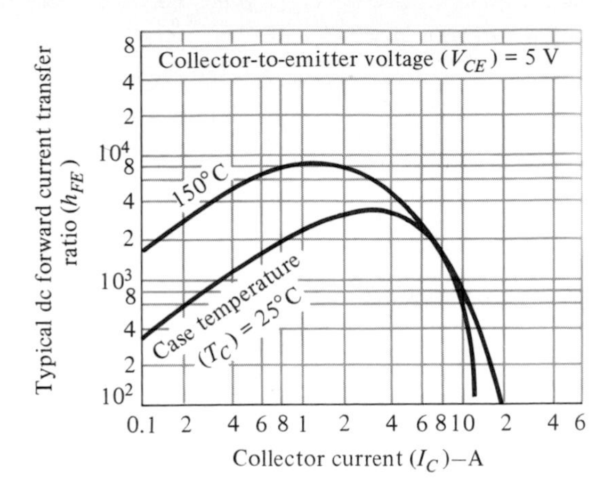

Figure 11.21 Typical dc beta characteristics for all types.

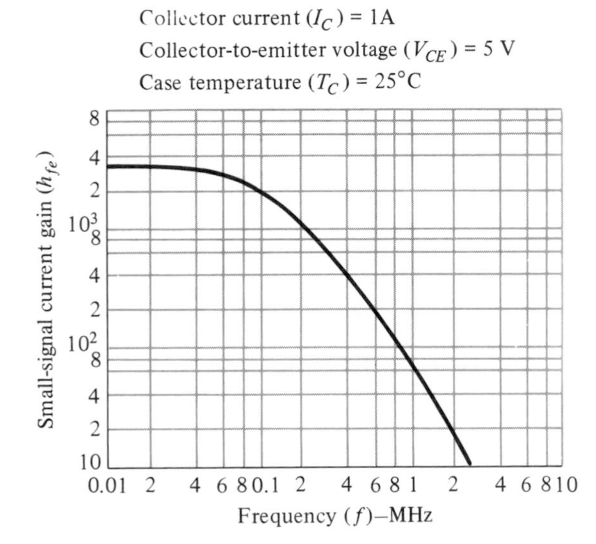

Figure 11.22 Typical small-signal gain for all types.

lower the permitted current. In Fig. 11.20 we note a drop in power rating starting with room temperature, and in Fig. 11.21 we find that the dc β is very sensitive to collector current. The effect of frequency on the ac gain is more carefully defined in Fig. 11.22. It begins a severe drop at about 0.1 MHz or 100 kHz. Note in Fig. 11.23 the increased level of V_T (cut-in voltage) due to the two transistors and the base current in mA in Fig. 11.24.

In this case, the hybrid parameters and admittance parameters are provided. The admittance parameters, not presented in this text, simply represent the parameters of another equivalent network for the transistor. Note also the high value of input impedance for the pairs and the low value of the equivalent $1/h_{oe} = 9.3\ \text{k}\Omega$. The dB gain in power is defined by equations to be derived in the next section.

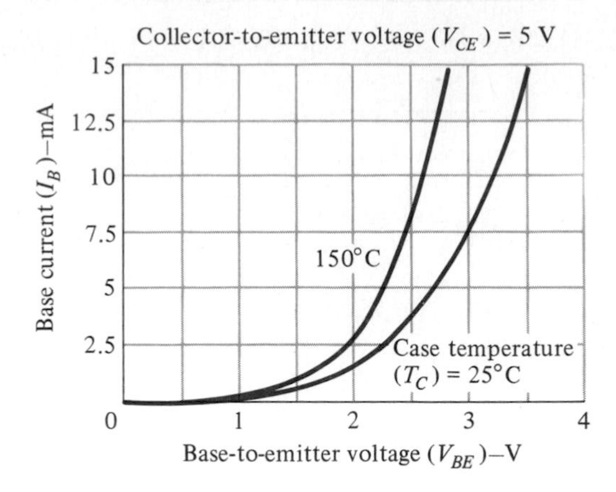

Figure 11.23 Typical input characteristics for all types.

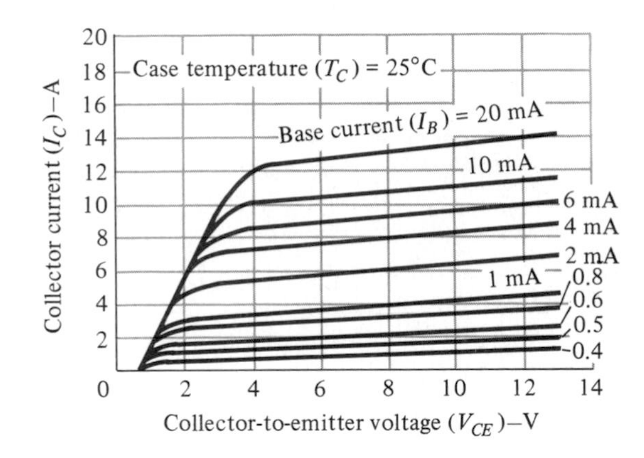

Figure 11.24 Typical output characteristics for all types.

11.8 DECIBELS

The concept of the *decibel* (dB) and the associated calculations will become increasingly important in the remaining sections of this chapter. The background surrounding the term *decibel* has its origin in the old established fact that power and audio levels are related on a logarithmic basis. That is, an increase in power level, say 4 to 16 W, for discussion purposes, does not mean that the audio level will increase by a factor of $16/4 = 4$. It will increase by a factor of 2 as derived from the power of 4 in the following manner: $(4)^2 = 16$. For a change of 4 to 64 W the audio level will increase by a factor of 3 since $(4)^3 = 64$. In logarithmic form, the relationship can be written as

$$\log_4 64 = 3$$

In words, the equation states that the logarithm of 64 to the base 4 is 3. In general: $\log_b a = x$ relates the variables in the same manner as $b^x = a$.

Because of pressures for standardization, the *bel* (B) was defined by the following equation to relate power levels P_1 and P_2:

$$G = \log_{10} \frac{P_2}{P_1} \qquad \text{bel} \tag{11.17}$$

Note that the common, or base 10, system was chosen to eliminate variability. Although the base is no longer the original power level, the equation will result in a basis for comparison of audio levels due to changes in power levels. The term *bel* was derived from the surname of Alexander Graham Bell.

It was found, however, that the bel was too large a unit of measurement for practical purposes, so the decibel (dB) was defined such that 10 decibels = 1 bel. Therefore,

$$G_{\text{dB}} = 10 \log_{10} \frac{P_2}{P_1} \qquad \text{dB} \tag{11.18}$$

The terminal rating of electronic communication equipment (amplifiers, microphones, etc.) is commonly rated in decibels. Equation (11.18) indicates clearly, however, that the decibel rating is a measure of the difference in magnitude between *two* power levels. For a specified terminal (output) power (P_2) there must be a reference power level (P_1). The reference level is generally accepted to be 1 mW although on occasion the 6-mW standard of earlier years is applied. The resistance to be associated with the 1-mW power level is 600 Ω, chosen because it is the characteristic impedance of audio transmission lines. When the 1-mW level is employed as the reference level, the decibel symbol frequently appears as dBm. In equation form,

$$G_{\text{dBm}} = 10 \log_{10} \frac{P_2}{1 \text{ mW}} \bigg|_{600\Omega} \qquad \text{dBm} \tag{11.19}$$

There exists a second equation for decibels that is applied frequently. It can be best described through the circuit of Fig. 11.25a. For V_i equal to some value V_1, $P_1 = V_1^2/R_i$ where R_i is the input resistance of the system of Fig. 11.25a. If V_i should be increased (or decreased) to some other level, V_2, then $P_2 = V_2^2/R_i$. If we substitute into Eq. (11.18) to determine the resulting difference in decibels between the power levels,

$$G_{\text{dB}} = 10 \log_{10} \frac{P_2}{P_1} = 10 \log_{10} \frac{V_2^2/R_i}{V_1^2/R_i} = 10 \log_{10} \left(\frac{V_2}{V_1}\right)^2$$

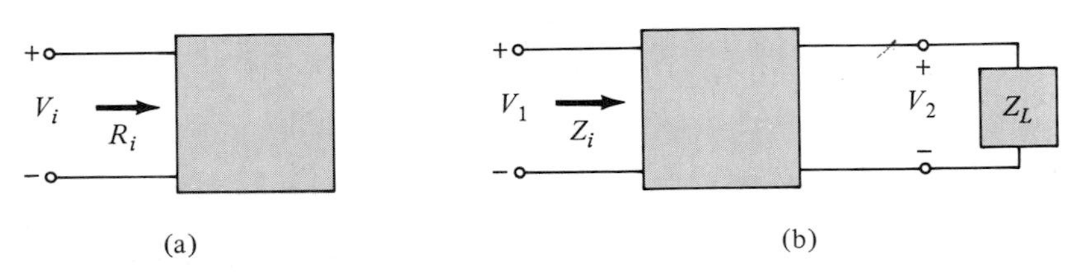

Figure 11.25 Configurations employed in the discussion of Eq. (11.18).

and $$G_{dB} = 20 \log_{10} \frac{V_2}{V_1} \quad \text{dB} \tag{11.20}$$

Keep in mind, however, that this equation is correct only if the associated resistance for each applied voltage is the same. For the system of Fig. 11.25b, where output and input levels are being compared, $Z_i \neq Z_L$ and Eq. (11.20) will not result. Equation (11.18) should therefore be employed.

If $Z_i = Z_i \cos \theta_i$ and $Z_L = Z_L \cos \theta_L$ are substituted into Eq. (11.18) such that $P_L = V_L^2/(Z_L \cos \theta_L)$, and so on, the following general equation will result:

$$G_{dB} = 20 \log_{10} \frac{V_L}{V_i} + 10 \log_{10} \frac{Z_i}{Z_L} + 10 \log_{10} \frac{\cos \theta_i}{\cos \theta_L} \tag{11.21}$$

For resistive elements, which are most commonly encountered, $\cos \theta_i = \cos \theta_L$, and the last term, $\log_{10}(1) = 0$. In addition, if $Z_i = Z_L$, the second term will also drop out, resulting in Eq. (11.20).

Frequently, the effect of different impedances ($Z_i \neq Z_L$) is ignored and Eq. (11.20) applied simply to establish a basis of comparison between levels—voltage or current. For situations of this type the decibel gain should more correctly be referred to as the *voltage or current gain in decibels* to differentiate it from the common usage of decibel as applied to power levels.

One of the advantages of the logarithmic relationship is the manner in which it can be applied to cascaded stages. For example, the overall voltage gain of a cascaded system is given by

$$A_{v_T} = A_{v_1} A_{v_2} A_{v_3} \cdots A_{v_n}$$

Applying the proper logarithmic relationship gives

$$G_v = 20 \log_{10} A_{v_T} = 20 \log_{10} A_{v_1} + 20 \log_{10} A_{v_2} + 20 \log_{10} A_{v_3} + \cdots + 20 \log_{10} A_{v_n} \quad \text{(dB)} \tag{11.22}$$

In words, the equation states that the decibel gain of a cascaded system is simply the sum of the decibel gains of each stage, that is,

$$G_v = G_{v_1} + G_{v_2} + G_{v_3} + \cdots + G_{v_n} \quad \text{dB} \tag{11.23}$$

The equations above can also be applied to current considerations. For $P_2 = I_2^2 R_o$ and $P_1 = I_1^2 R_o$,

$$G_{dB} = 20 \log_{10} \frac{I_2}{I_1} \quad \text{dB} \tag{11.24}$$

and $$G_i = G_{i_1} + G_{i_2} + G_{i_3} + \cdots + G_{i_n} \tag{11.25}$$

Before considering a few examples, the fundamental operations associated with logarithmic functions will be considered. For many it will simply be a review. For some, an extended amount of time may be required to fully understand the material to follow.

Each equation introduced in this section employs the common or base 10 logarithmic system. As indicated in the introductory discussion, the logarithms of numbers that are powers of the chosen base are easily determined. For example,

$$\log_{10} \overbrace{10{,}000}^{a} = x$$
$$\uparrow$$
$$b$$

$$\text{and } b^x = a$$

$$\text{or } (10)^x = 10{,}000$$

$$\text{resulting in } x = 4$$

Similarly,

$$\log_{10} 1000 = \log_{10} (10)^3 = 3$$
$$\log_{10} 100 = \log_{10} (10)^2 = 2$$
$$\log_{10} 10 = \log_{10} (10)^1 = 1$$
$$\log_{10} 1 = \log_{10} (10)^0 = 0$$

For the logarithm of a number such as 24.8,

$$\log_{10} 24.8 = x$$

or

$$10^x = 24.8$$

The unknown quantity x is obviously between 1 and 2, but a further determination would be purely a trial-and-error process if it were not for the logarithmic function. The procedure for determining the logarithm of a number requires that two components of the result be found separately. These two components are the *characteristic* and *mantissa*. The characteristic is simply the power of 10 associated with the number for which the logarithm is to be determined.

$$24.8 = 2.48 \times 10^1 \Rightarrow 1 = \text{characteristic}$$
$$4860.0 = 4.860 \times 10^3 \Rightarrow 3 = \text{characteristic}$$

The mantissa, or decimal portion of the logarithm, must be determined from a set of tables or a calculator.

$$\log_{10} 24.8 = 1.3945$$
$$\log_{10} 4860.0 = 3.6866$$

Of course, most calculators provide the characteristic and mantissa directly.

There will be many occasions in which the antilogarithm of a number must be determined; that is, for the example above, determine 24.8 and 4860.0 from the logarithm of these numbers. The process is simply the reverse of that applied to determine the logarithm. For example, find the antilogarithm of 2.140.

$$2.140\left\}\begin{matrix}\text{characteristic} \Rightarrow 10^2 \\ \text{mantissa} \quad \Rightarrow 138\end{matrix}\right\}1.38 \times 10^2 = 138$$

For the calculator, keep in mind that

$$\log_{10} x = 2.140$$

is equivalent to

$$10^{2.140} = x$$

and 10^y is a common calculator function.

For ratios less then 1, the logarithm can be determined by simply inverting the ratio and introducing a negative sign. Most calculators will include the insertion of the negative sign.

$$\log_{10} \frac{1.6}{24} = -\log_{10} \frac{24}{1.6} = -\log_{10} 15 = -1.1761$$

$$\log_{10} 0.788 = -\log_{10} \frac{1}{0.788} = -\log_{10} 1.269 = -0.1035$$

For power ratios, a negative decibel rating simply indicates a reduction in power level as compared to the initial or input power.

EXAMPLE 11.4

Find the magnitude gain corresponding to a decibel gain of 100.

Solution:

By Eq. (11.18),

$$G_{\text{dB}} = 10 \log_{10} \frac{P_2}{P_1} = 100 \text{ dB} \Rightarrow \log_{10} \frac{P_2}{P_1} = 10$$

so that

$$\frac{P_2}{P_1} = \mathbf{10^{10}} = \mathbf{10{,}000{,}000{,}000}$$

This example clearly demonstrates the range of decibel values to be expected from practical devices. Certainly a future calculation giving a decibel result in the neighborhood of 100 should be questioned immediately. In fact, a decibel gain of 50 dB corresponds with a magnitude gain of 100,000, which is still very large.

EXAMPLE 11.5

The input power to a device is 10,000 W at a voltage of 1000 V. The output power is 500 W while the output impedance is 20 Ω.

(a) Find the power gain in decibels.
(b) Find the voltage gain in decibels.
(c) Explain why parts (a) and (b) agree or disagree.

Solution:

(a) $$G_{\text{dB}} = 10 \log_{10} \frac{P_2}{P_1} = 10 \log_{10} \frac{0.5 \times 10^3}{10 \times 10^3} = 10 \log_{10} \frac{1}{20} = -10 \log_{10} 20$$

$$= -10(1.301) = \mathbf{-13.01\ dB}$$

(b) $G_v = 20 \log_{10} \frac{V_o}{V_i} = 20 \log_{10} \frac{\sqrt{PR}}{1000} = 20 \log_{10} \frac{\sqrt{500 \times 20}}{1000}$

$= 20 \log_{10} \frac{100}{1000} = 20 \log_{10} \frac{1}{10} = -20 \log_{10} 10 = \mathbf{-20\ dB}$

(c) $R_i = \frac{V^2}{P} = \frac{10^6}{10^4} = 10^2 \neq R_o = \mathbf{20\ \Omega}$

EXAMPLE 11.6

An amplifier rated at 40-W output is connected to a 10-Ω speaker.
(a) Calculate the input power required for full power output if the power gain is 25 dB.
(b) Calculate the input voltage for rated output if the amplifier voltage gain is 40 dB.

Solution:

(a) Eq. (11.18):

$$25 = 10 \log_{10} \frac{40}{P_i} \Rightarrow P_i = \frac{40}{\text{antilog}\ (2.5)} = \frac{40}{3.16 \times 10^2}$$

$$= \frac{40}{316} \cong \mathbf{126.5\ mW}$$

(b) $G_v = 20 \log_{10} \frac{V_o}{V_i} \Rightarrow 40 = 20 \log_{10} \frac{V_o}{V_i}$

$$\frac{V_o}{V_i} = \text{antilog}\ 2 = 100$$

$$V_o = \sqrt{PR} = \sqrt{40 \times 10} = 20\ \text{V}$$

$$V_i = \frac{V_o}{100} = \frac{20}{100} = 0.2\ V = \mathbf{200\ mV}$$

The use of log scales on a graph can significantly expand the range of variation of a particular variable—an effect to become obvious in the following section. Most plots provided, or graph paper available, is of the semilog or double-log (log-log) variety. The term *semi* (meaning one-half) indicates that only one of the two scales is a log scale, whereas double-log indicates that both scales are log scales. A semilog scale appears in Fig. 11.26. Note that the vertical scale is a linear scale with equal divisions. The source of the spacing between the lines of the log plot is shown on the graph. The log of 2 to the base 10 is approximately 0.3. The distance from $1(\log_{10}1 = 0)$ to 2 is therefore 30% of the span. Since $\log_{10}5 \cong 0.7$, it is marked off at a point 70% of the distance. Note that between any two digits the same compression of the lines appears as you progress from the left to the right. It is important to note the resulting numerical value and the spacing since plots will typically only have the tic marks indicated in Fig. 11.27 due to a lack of space. You must realize from past experience that the longer bars for this figure have the numerical values of 0.3, 3, and 30 associated with them, whereas the next shorter bars have values of 0.5, 5, and 50, and the shortest bars 0.7, 7, and 70.

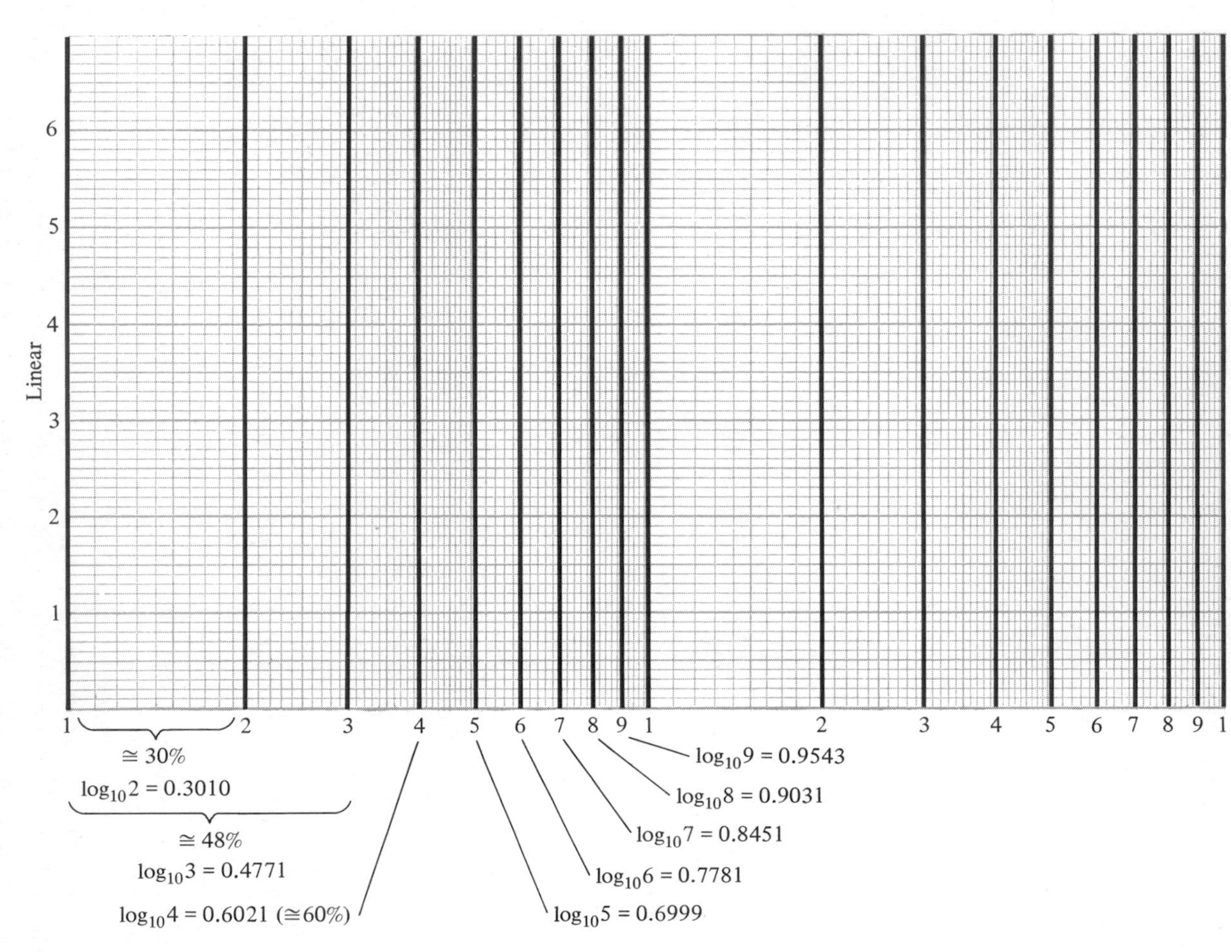

Figure 11.26 Semilog graph paper.

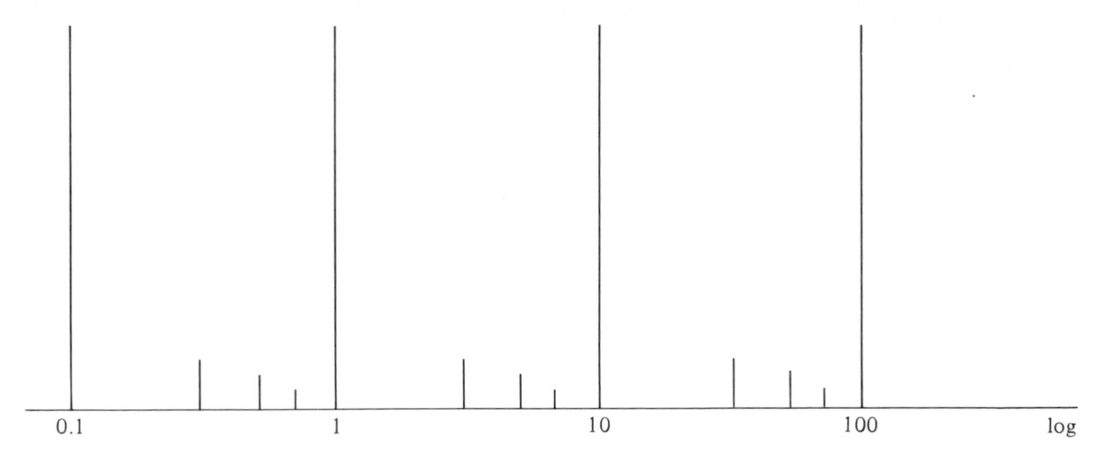

Figure 11.27 Identifying the numerical values of the tic marks on a log scale.

11.9 GENERAL FREQUENCY CONSIDERATIONS

The frequency of the applied signal can have a pronounced effect on the response of a single or multistage network. The analysis thus far has been for the midfrequency spectrum. At low frequencies we shall find that the coupling and bypass capacitors can no longer be replaced by the short-circuit approximation because of the resulting change in reactance of these elements. The frequency-dependent parameters of the small-signal equivalent circuits and the stray capacitive elements associated with the active device and the network will limit the high-frequency response of the system. An increase in the number of stages of a cascaded system will limit both the high- and low-frequency response.

The magnitude gain of an *RC*-coupled, direct-coupled, and transformer-coupled amplifier system are provided in Fig. 11.28. Note that the horizontal scale is a logarithmic scale to permit a plot extending from the low- to the high-frequency regions. For each plot, a low-, high-, and midfrequency region has been defined. In addition, the primary reasons for the drop in gain at low and high frequencies have also been indicated within the parentheses. For the *RC*-coupled amplifier the drop at low frequencies is due to the increasing reactance of C_C, C_S, or C_E, while its upper frequency limit is determined by either the parasitic capacitive elements of the network and active device or the frequency dependence of the gain of the active device. An explanation of the drop in gain for the transformer-coupled system requires a basic understanding of "transformer action" and the transformer equivalent circuit. For the moment let us say that it is simply due to the shorting effect (across the input terminals of the transformer) of a magnetizing inductive reactance at low frequencies ($X_L = 2\pi fL$). The gain must obviously be zero at $f = 0$ since at this point there is no longer a changing flux established through the core to induce a secondary or output voltage. As indicated in Fig. 11.28, the high-frequency response is controlled primarily by the stray capacitance between the turns of the primary and secondary windings. For the direct-coupled amplifier, there are no coupling or bypass capacitors to cause a drop in gain at low frequencies. As the figure indicates, it is a flat response to the

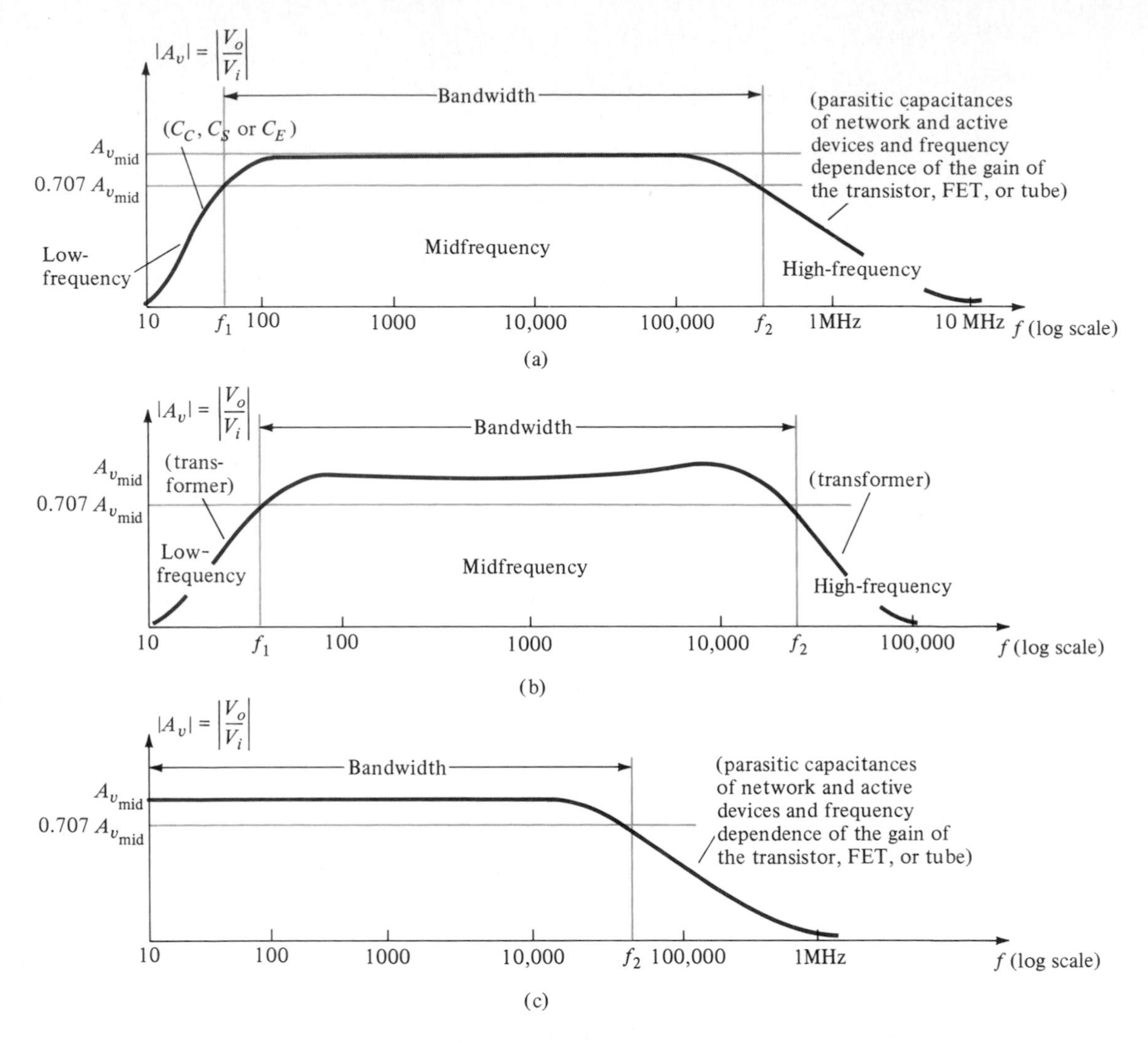

Figure 11.28 Gain versus frequency for (a) RC-coupled amplifiers; (b) transformer-coupled amplifiers; (c) direct-coupled amplifiers.

upper cutoff frequency which is determined by either the parasitic capacitances of the circuit and active device or the frequency dependence of the gain of the active device.

For each system of Fig. 11.28 there is a band of frequencies in which the magnitude of the gain is either equal or relatively close to the midband value. To fix the frequency boundaries of relatively high gain, $0.707\ A_{v_{mid}}$ was chosen to be the gain cutoff level. The corresponding frequencies f_1 and f_2 are generally called the corner, cutoff, band, break, or half-power frequencies. The multiplier 0.707 was chosen because at this level the output power is half the midband power output, that is, at midfrequencies,

$$P_{o_{mid}} = \frac{|V_o^2|}{R_o} = \frac{|A_{v_{mid}} V_i|^2}{R_o}$$

and at the half-power frequencies,

$$P_{o_{HPF}} = \frac{|0.707\, A_{v_{\text{mid}}} V_i|^2}{R_o} = 0.5\frac{|A_{v_{\text{mid}}} V_i|^2}{R_o}$$

and

$$\boxed{P_{o_{HPF}} = 0.5\, P_{o_{\text{mid}}}} \qquad (11.26)$$

The bandwidth (or passband) of each system is determined by f_1 and f_2, that is,

$$\boxed{\text{bandwidth (BW)} = f_2 - f_1} \qquad (11.27)$$

For applications of a communications nature (audio, video), a decibel plot of the voltage gain versus frequency is more useful than that appearing in Fig. 11.28. Before obtaining the logarithmic plot, however, the curve is generally normalized as shown in Fig. 11.29. In this figure the gain at each frequency is divided by the midband value. Obviously, the midband value is then 1 as indicated. At the half-power frequencies the resulting level is $0.707 = 1/\sqrt{2}$. A decibel plot can now be obtained by applying Eq. (11.20) in the following manner:

$$\boxed{\left|\frac{A_v}{A_{v_{\text{mid}}}}\right|_{\text{dB}} = 20 \log_{10} \left|\frac{A_v}{A_{v_{\text{mid}}}}\right|} \qquad (11.28)$$

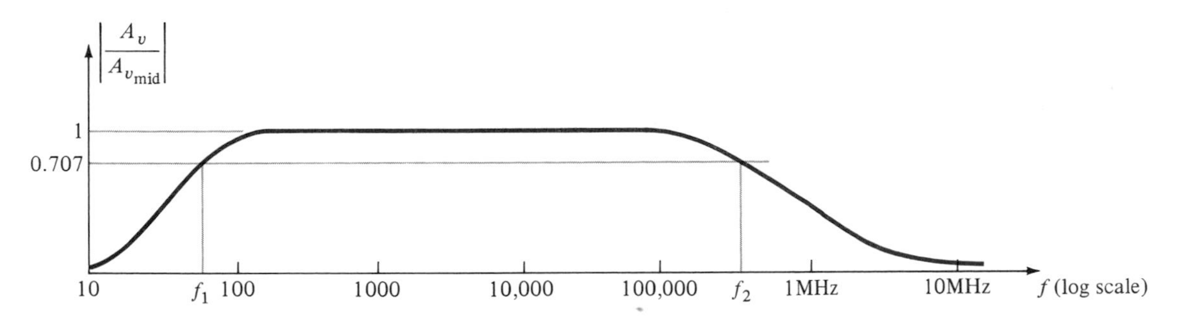

Figure 11.29 Normalized gain versus frequency plot.

At midband frequencies, $20 \log_{10} 1 = 0$, and at the cutoff frequencies, $20 \log_{10} 1/\sqrt{2} = -3$ dB. Both values are clearly indicated in the resulting decibel plot of Fig. 11.30. The smaller the fraction ratio, the more negative the decibel level due to the inversion process discussed in Section 11.8.

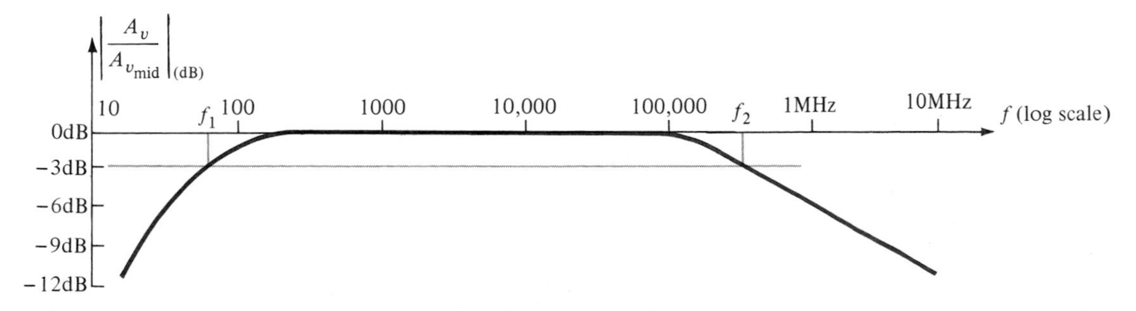

Figure 11.30 Decibal plot of the normalized gain versus frequency plot of Fig. 11.29.

For the greater part of the discussion to follow, a decibel plot will be made only for the low- and high-frequency regions. Keep Fig. 11.30 in mind, therefore, to permit a visualization of the broad system response.

It should be understood that an amplifier usually introduces an inversion between input and output signals. This fact must now be expanded to indicate that this is only the case in the midband region. At low frequencies there is a phase shift such that V_o lags V_i by an increased angle. At high frequencies the phase shift will drop below 180°. Figure 11.31 is a standard phase plot for an *RC*-coupled amplifier.

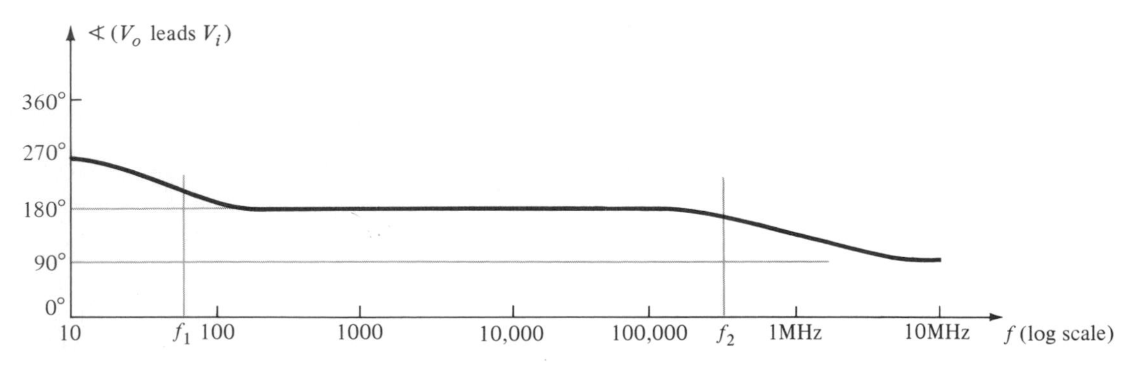

Figure 11.31 Phase plot for an *RC*-coupled amplifier system (for each stage).

11.10 SINGLE-STAGE TRANSISTOR AMPLIFIER—LOW-FREQUENCY CONSIDERATIONS

Before considering the basic BJT amplifier in the low-frequency region, let us first examine the series *RC* circuit appearing in Fig. 11.32 and note the effect the applied frequency will have on the ratio $A_v = V_o/V_i$. At very high frequencies

$$X_C = \frac{1}{2\pi f_1 C} \cong 0\ \Omega$$

and the short-circuit equivalent can be substituted for the capacitor as shown in Fig. 11.33. The result is that $V_o \cong V_i$ at high frequencies. At $f = 0$ Hz,

$$X_C = \frac{1}{2\pi f C} = \frac{1}{2\pi(0)C} = \infty\ \Omega$$

and the open-circuit approximation can be applied as shown in Fig. 11.34 with the result that $V_o = 0$ V.

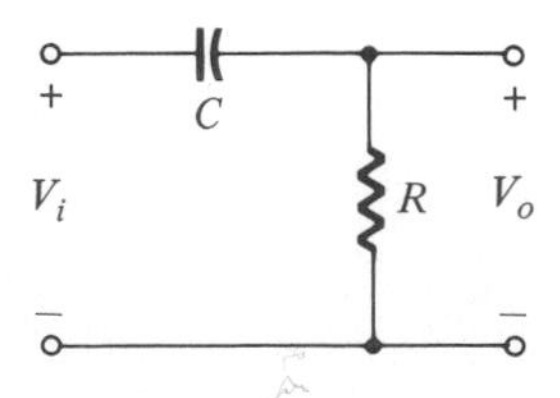

Figure 11.32

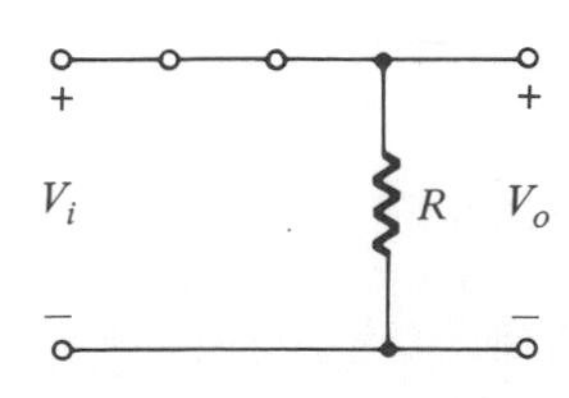

Figure 11.33

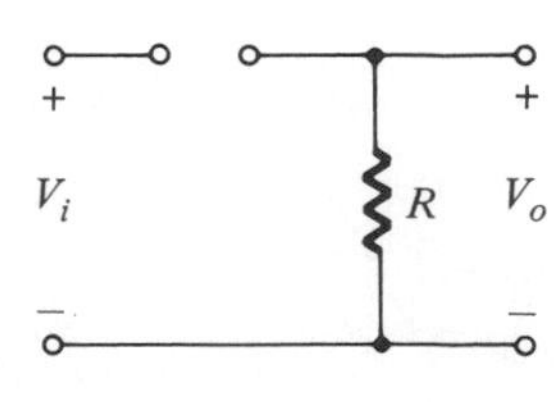

Figure 11.34

Between the two extremes the ratio $A_v = V_o/V_i$ will vary as shown in Fig. 11.35. As the frequency increases the capacitive reactance decreases and more of the input voltage appears across the output terminals.

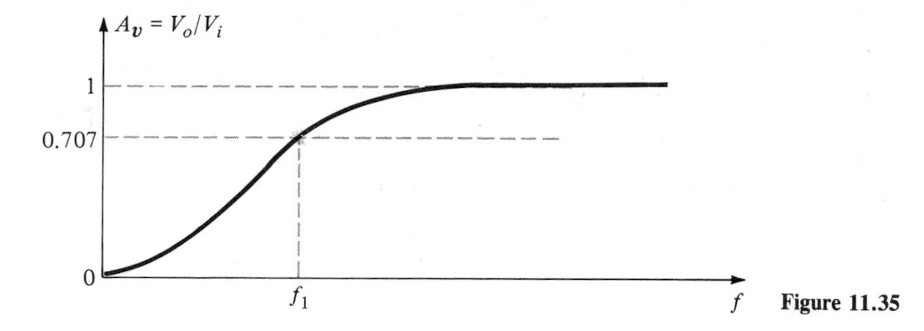

Figure 11.35

The output and input voltages are related by the voltage-divider rule in the following manner:

$$\mathbf{V}_o = \frac{\mathbf{R}\mathbf{V}_i}{\mathbf{R} + \mathbf{X}_C}$$

with the magnitude of V_o determined by

$$V_o = \frac{RV_i}{\sqrt{R^2 + X_C^2}}$$

For the special case where $X_C = R$,

$$V_o = \frac{RV_i}{\sqrt{R^2 + X_C^2}} = \frac{RV_i}{\sqrt{R^2 + R^2}} = \frac{RV_i}{\sqrt{2R^2}} = \frac{RV_i}{\sqrt{2}R} = \frac{1}{\sqrt{2}}V_i$$

and

$$\boxed{A_v = \frac{V_o}{V_i} = \frac{1}{\sqrt{2}} = 0.707\big|_{X_C = R}} \tag{11.29}$$

the level of which is indicated on Fig. 11.35. In other words, at the frequency where $X_C = R$, the output will be 70.7% of the input for the network of Fig. 11.32.

The frequency at which this occurs is determined from

$$X_C = \frac{1}{2\pi f_1 C} = R$$

and

$$\boxed{f_1 = \frac{1}{2\pi RC}} \tag{11.30}$$

In terms of logs,

$$G_v = 20 \log_{10} A_v = 20 \log_{10} \frac{1}{\sqrt{2}} = -3 \text{ dB}$$

while at $A_v = V_o/V_i = 1$ or $V_o = V_i$ (the maximum value),

$$G_v = 20 \log_{10} 1 = 20(0) = 0 \text{ dB}$$

In Fig. 11.30 we recognize that there is a 3-dB drop in gain from the midband level when $f = f_1$. In a moment we will find that an RC network will determine the low-frequency cutoff frequency for a BJT transistor and f_1 will be determined by Eq. (11.30).

If the gain equation is written as

$$A_v = \frac{V_o}{V_i} = \frac{R}{R - jX_C} = \frac{1}{1 - j(X_C/R)} = \frac{1}{1 - j(1/\omega CR)} = \frac{1}{1 - j(1/2\pi fCR)}$$

and using the defined frequency above,

$$\boxed{A_v = \frac{1}{1 - j(f_1/f)}} \tag{11.31}$$

In the magnitude and phase form,

$$A_v = \frac{V_o}{V_i} = \underbrace{\frac{1}{\sqrt{1 + (f_1/f)^2}}}_{\text{magnitude of } A_v} \underbrace{\angle \tan^{-1}(f_1/f)}_{\substack{\text{phase} \sphericalangle \text{ by which} \\ V_o \text{ leads } V_i}} \tag{11.32}$$

For the magnitude, when $f_1 = f$,

$$A_v = \frac{1}{\sqrt{1 + (1)^2}} = \frac{1}{\sqrt{2}} = 0.707 \Rightarrow -3 \text{ dB}$$

In the logarithmic form; the gain in dB is

$$A_v|_{\text{dB}} = 20 \log_{10} \frac{1}{\sqrt{1 + (f_1/f)^2}} = -20 \log_{10}\left[1 + \left(\frac{f_1}{f}\right)^2\right]^{1/2}$$

$$= -(\tfrac{1}{2})(20) \log_{10}\left[1 + \left(\frac{f_1}{f}\right)^2\right]$$

$$= -10 \log_{10}\left[1 + \left(\frac{f_1}{f}\right)^2\right]$$

For frequencies where $f \ll f_1$ and $(f_1/f)^2 \gg 1$, the equation above can be approximated by

$$= -10 \log_{10}\left(\frac{f_1}{f}\right)^2$$

and finally,

$$\boxed{A_v|_{\text{dB}} = -20 \log_{10} \frac{f_1}{f}}_{f \ll f_1} \tag{11.33}$$

Ignoring the condition $f \ll f_1$ for a moment, a plot of Eq. (11.33) on a frequency log scale will yield a result of a very useful nature for future decibel plots.

$$\text{At } f = f_1: \quad \frac{f_1}{f} = 1 \text{ and } -20 \log_{10} 1 = 0 \text{ dB}$$

$$\text{At } f = \tfrac{1}{2} f_1: \quad \frac{f_1}{f} = 2 \text{ and } -20 \log_{10} 2 \cong -6 \text{ dB}$$

$$\text{At } f = \tfrac{1}{4} f_1: \quad \frac{f_1}{f} = 4 \text{ and } -20 \log_{10} 4 \cong -12 \text{ dB}$$

$$\text{At } f = \tfrac{1}{10} f_1: \quad \frac{f_1}{f} = 10 \text{ and } -20 \log_{10} 10 = -20 \text{ dB}$$

A plot of these points is indicated in Fig. 11.36 from $0.1 f_1$ to f_1. Note that this results in a straight line when plotted against a log scale. In the same figure a straight line is also drawn for the condition of 0 dB for $f \gg f_1$. As stated earlier, the straight-line segments (asymptotes) are only accurate for 0 dB when $f \gg f_1$, and the sloped line when $f_1 \gg f$. We know, however, that when $f = f_1$, there is a 3-dB drop from the mid-band level. Employing this information in association with the straight-line segments permits a fairly accurate plot of the frequency response as indicated in the same figure. The piecewise linear plot of the asymptotes and associated breakpoints is called a *Bode plot*.

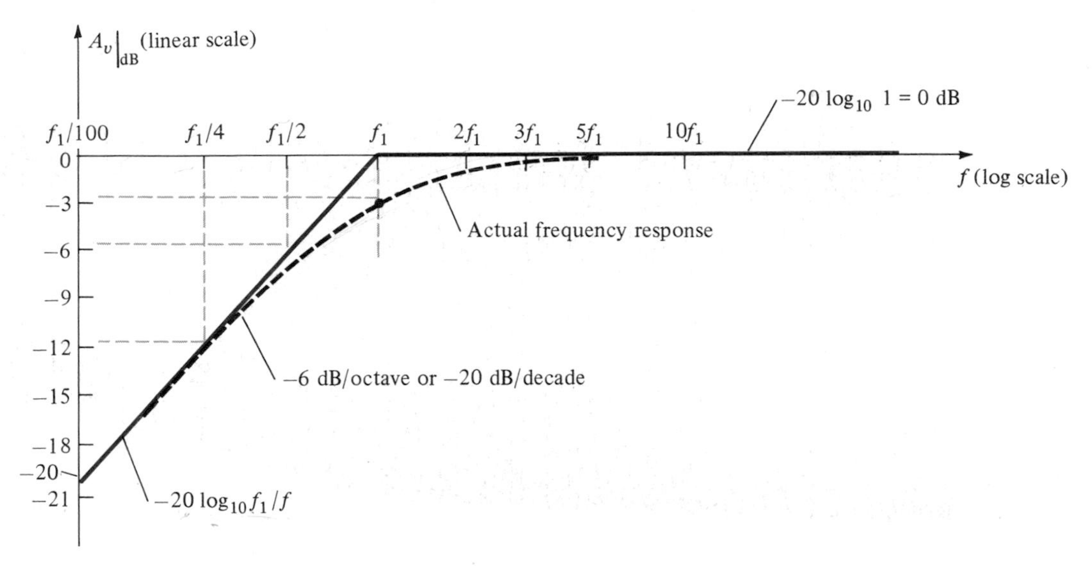

Figure 11.36 Bode plot for the low-frequency region.

The above calculations and the curve itself demonstrate clearly that a change in frequency by a factor of 2, equivalent to *1 octave*, results in a 6-dB change in the ratio. In particular, note the change in gain from $f_1/2$ to f_1. For a 10:1 change in frequency, equivalent to *1 decade*, there is a 20-dB change in the ratio, as demonstrated between the frequencies of $f_1/10$ and f_1. In the future, therefore, a decibel plot can be easily obtained for a function having the format of Eq. (11.33). First simply find f_1 from the circuit parameters and then sketch two asymptotes—one along the 0-dB line and the other drawn through f_1 sloped at 6 dB/octave or 20 dB/decade. Then find the 3-dB point corresponding to f_1 and sketch the curve.

EXAMPLE 11.7

For the network of Fig. 11.37:

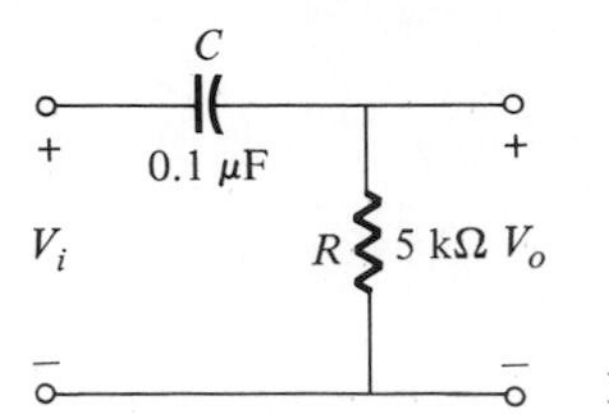

Figure 11.37

(a) Determine the break frequency.
(b) Sketch the asymptotes and locate the −3-dB point.
(c) Sketch the frequency response curve.

Solution:

(a) $f_1 = \dfrac{1}{2\pi RC} = \dfrac{1}{(6.28)(5 \times 10^3)(0.1 \times 10^{-6})}$

$\cong$ **318.5 Hz**

(b) and (c). See Fig. 11.38.

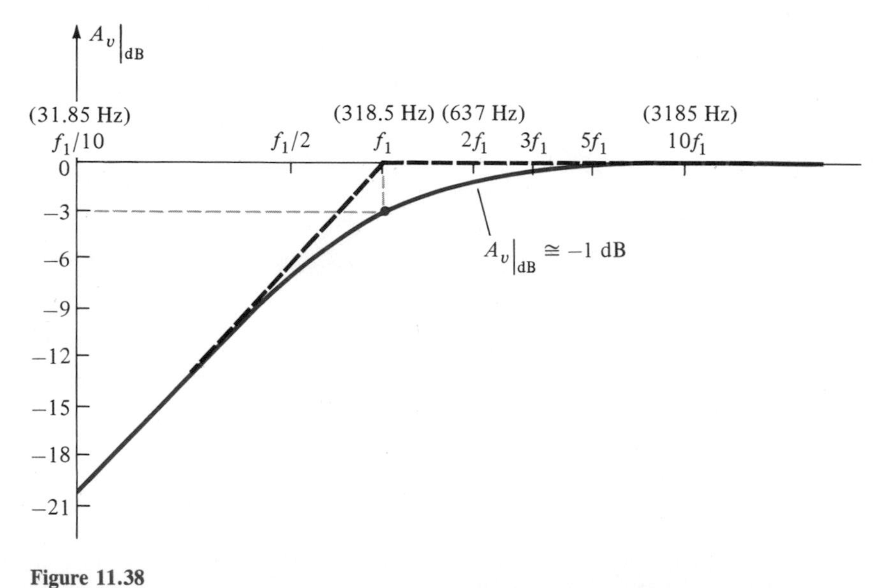

Figure 11.38

The gain at any frequency can then be determined from the Bode plot in the following manner:

$$A_v|_{dB} = 20 \log_{10} \frac{V_o}{V_i}$$

but

$$\frac{A_v|_{dB}}{20} = \log_{10} \frac{V_o}{V_i}$$

and

$$\boxed{A_v = \frac{V_o}{V_i} = 10^{\frac{A_v|_{dB}}{20}}} \tag{11.34}$$

For example, if $A_v|_{dB} = -3$ dB,

$$A_v = \frac{V_o}{V_i} = 10^{-3/20} = 10^{-0.15} = 0.707 \quad \text{as expected}$$

The quantity $10^{-0.15}$ is determined using the 10^y function found on most scientific calculators.

From Fig. 11.38, $A_v|_{dB} \cong -1$ dB at $f = 2f_1 = 637$ Hz. The gain at this point is

$$A_v = \frac{V_o}{V_i} = 10^{\frac{A_v|_{dB}}{20}} = 10^{-1/20} = 10^{-0.05} = 0.891$$

and

$$V_o = 0.891V_i$$

or V_o is 89.1% of V_i at $f = 637$ Hz. The phase angle of θ is determined from

$$\boxed{\theta = \tan^{-1}\left(\frac{f_1}{f}\right)} \tag{11.35}$$

from Eq. (11.32).

For frequencies $f \ll f_1$,

$$\theta = \tan^{-1}\frac{f_1}{f} \Rightarrow 90^\circ$$

For instance, if $f_1 = 100f$,

$$\theta = \tan^{-1}\frac{f_1}{f} = \tan^{-1}(100) = 89.4^\circ$$

For $f = f_1$,

$$\theta = \tan^{-1}\frac{f_1}{f} = \tan^{-1} 1 = 45^\circ$$

For $f \gg f_1$:

$$\theta = \tan^{-1}\frac{f_1}{f} \Rightarrow 0^\circ$$

For instance, if $f = 100f_1$,

$$\theta = \tan^{-1}\frac{f_1}{f} = \tan^{-1} 0.01 = 0.573^\circ$$

A plot of $\theta = \tan^{-1}(f_1/f)$ is provided in Fig. 11.39. If we add the additional 180° phase shift introduced by an amplifier, the phase plot of Fig. 11.31 will be obtained.

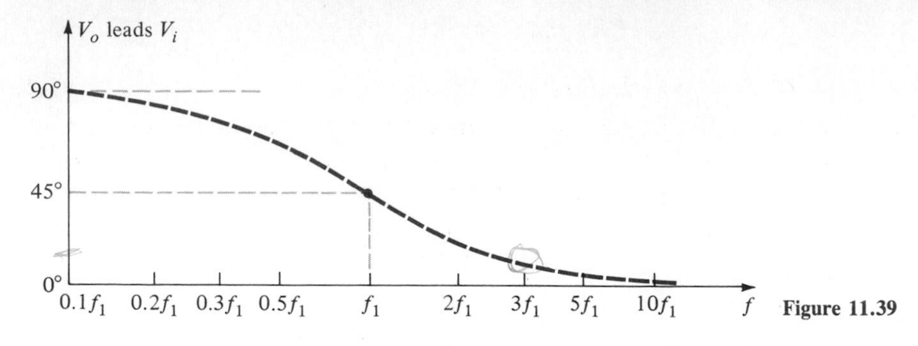

Figure 11.39

Let us now turn our attention to the basic BJT amplifier appearing in Fig. 11.40

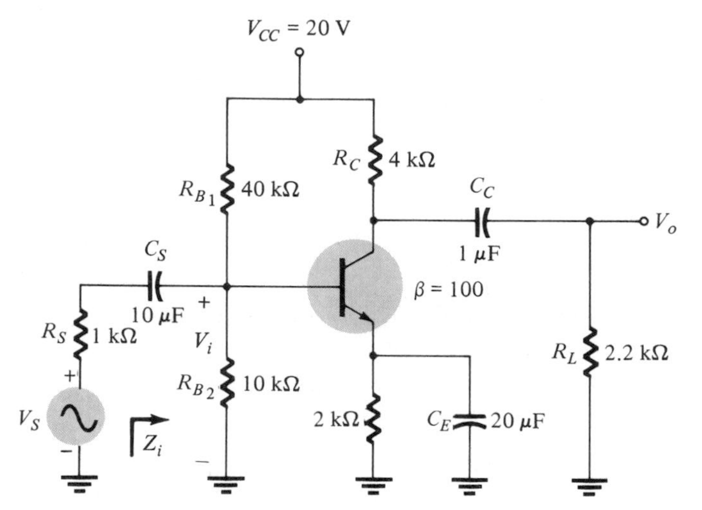

Figure 11.40

with the capacitors C_S, C_C, C_E that will affect the low-frequency response. We shall examine the effect of each independently.

C_S

The reduced ac equivalent circuit surrounding C_S appears in Fig. 11.41. The voltage V_i can then be determined using the voltage-divider rule:

$$\mathbf{V}_i = \frac{R_i \mathbf{V}_S}{R_S + R_i - jX_{C_S}} \tag{11.36}$$

and the cutoff frequency will be determined by

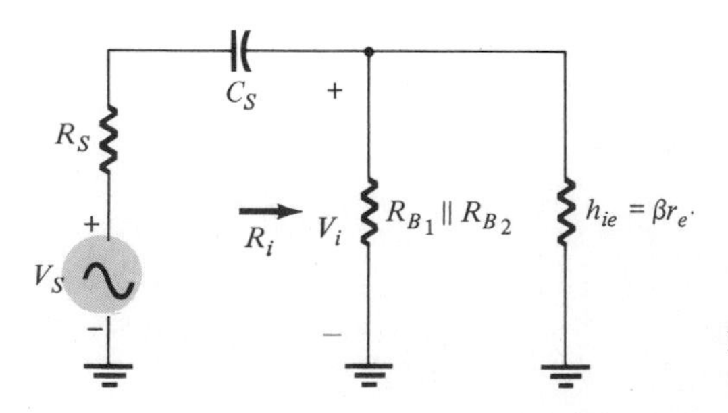

Figure 11.41 Localized ac equivalent for C_S.

$$R_S + R_i = X_{C_S}$$

as similarly derived for the network of Fig. 11.32. Using this result, the lower cutoff frequency determined by C_S is

$$f_{L_S} = \frac{1}{2\pi(R_S + R_i)C_S} \tag{11.37}$$

At mid- or high frequencies the reactance of the capacitor will be sufficiently small to permit a short-circuit approximation for the element. The voltage V_i will then be related to V_S by

$$V_i\big|_{\text{mid}} = \frac{R_i V_S}{R_i + R_s} \tag{11.38}$$

At f_{L_S} the voltage V_i will be 70.7% of the value determined by Eq. (11.38).

C_C

The effect of the coupling capacitor C_C can best be determined by referring to the output equivalent network of Fig. 11.42. At mid- or high frequencies the capacitor C_C can be replaced by a short-circuit equivalent and the output voltage V_o will equal the collector voltage of the transistor.

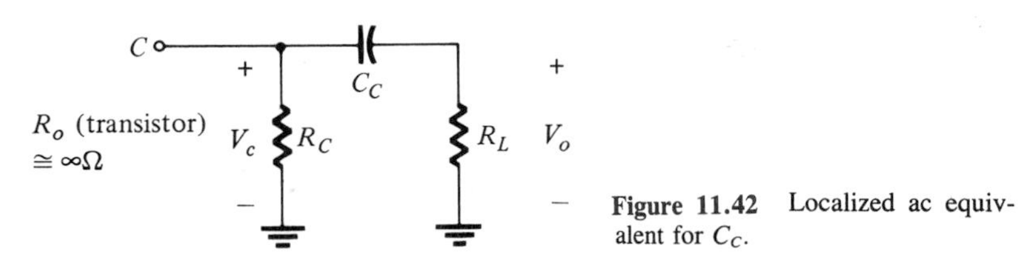

Figure 11.42 Localized ac equivalent for C_C.

The cutoff frequency is defined by the condition

$$R_C + R_L = X_{C_C}$$

and is equal to

$$f_{L_C} = \frac{1}{2\pi(R_C + R_L)C_C} \tag{11.39}$$

At f_{L_C} the output voltage V_o will be 70.7% of the mid-band value of V_c.

C_E

The equivalent circuit external to the capacitor C_E appears in Fig. 11.43. The condition

$$R_e = R_E \bigg\| \left[\frac{R_S'}{\beta} + r_e\right] = X_{C_E}$$

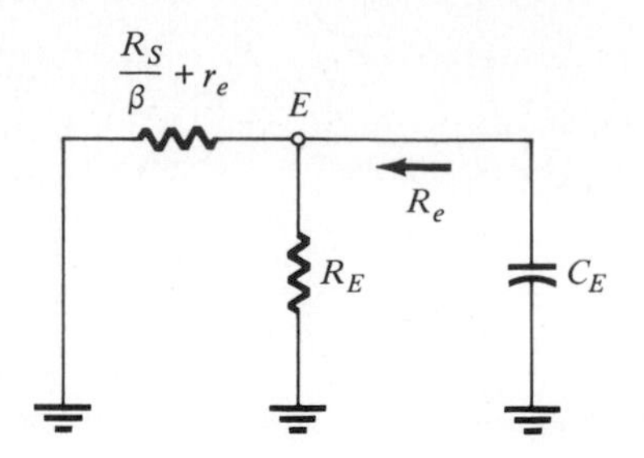

Figure 11.43 Localized ac equivalent for C_E.

defines the break frequency determined by the emitter bypass capacitor:

$$f_{L_E} = \frac{1}{2\pi R_e C_E} \tag{11.40}$$

The effect of C_E on the gain is best described in a quantitative manner by recalling that the gain for the configuration of Fig. 11.44 is given by

$$A_v = \frac{-R_C}{r_e + R_E}$$

The maximum gain is obviously available when R_E is zero ohms. At low frequencies, with the bypass capacitor C_E in its open-circuit equivalent state, all of R_E appears in the gain equation above, resulting in the minimum gain. As the frequency increases the reactance of the capacitor C_E will decrease, reducing the parallel impedance of R_E and C_E until the resistor R_E is effectively "shorted out" by C_E. The result is a maximum or midband gain determined by $A_v = -R_C/r_e$. At f_{L_E} the gain will be 3 dB below the midband value determined with R_E "shorted out."

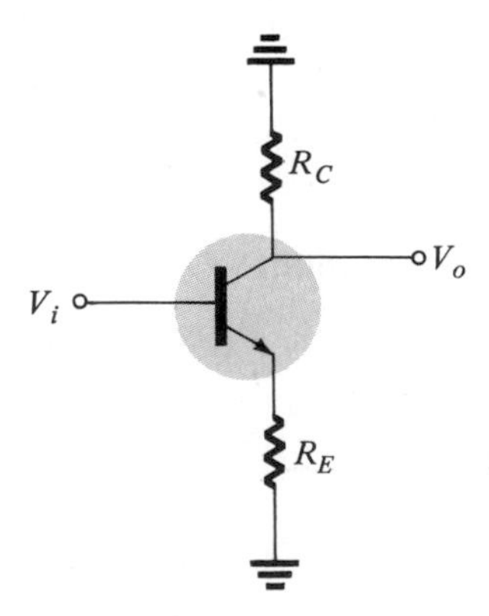

Figure 11.44

Before examining the next example, keep in mind that C_S, C_C, and C_E will affect only the low-frequency response. At the midband frequency level the short-circuit equivalents for the capacitors can be inserted. Although each will affect the gain $A_v = V_o/V_i$ in a similar frequency range, the highest cutoff frequency determined by C_S, C_C, or C_E will have the greatest impact since it will be the last encountered before the midband level. If the frequencies are relatively far apart, the highest cutoff frequency will essentially determine the lower cutoff frequency for the entire system. If there are two or more "high" cutoff frequencies, the effect will be to raise the lower cutoff frequency and reduce the resulting bandwidth of the system—a negative result.

EXAMPLE 11.8

Determine the lower cutoff frequency for the network of Fig. 11.40 and sketch the Bode plot. Include an estimate of the actual gain versus frequency curve.

Solution:

Determining r_e for dc conditions:

$$\beta R_E = (100)(2\text{ k}\Omega) = 200\text{ k}\Omega \gg 10\text{ k}\Omega$$

βR_E can therefore be ignored on an approximate basis and

$$V_B \cong \frac{R_{B_2}V_{CC}}{R_{B_2} + R_{B_1}} = \frac{10\text{ k}\Omega(20)}{10\text{ k}\Omega + 40\text{ k}\Omega} = \frac{200}{50} = 4\text{ V}$$

with
$$I_E = \frac{V_E}{R_E} = \frac{4 - 0.7}{2\text{ k}\Omega} = \frac{3.3}{2\text{ k}\Omega} = 1.65\text{ mA}$$

so that
$$r_e = \frac{26\text{ mV}}{1.65\text{ mA}} \cong \mathbf{15.76\ \Omega}$$

and
$$\beta r_e = 100(15.76) = 1576\Omega = \mathbf{1.576\ k\Omega}$$

The following analysis will include computer printouts from the HP-85 desk-top scientific computer. The software support for the system included a circuit analysis pac that permitted obtaining a plot of the ratio of the output to input voltage for the range of frequencies of interest. The mid-band gain will be required for comparison purposes:

$$A_{v_{\text{mid}}} = \frac{V_o}{V_S} = \frac{V_o}{V_i}\frac{V_i}{V_S}$$

The gain

$$A_v = \frac{V_o}{V_i} = \frac{-R_C \| R_L}{r_e} = -\frac{(4\text{ k}\Omega) \| (2.2\text{ k}\Omega)}{15.76} \cong -90$$

The input impedance

$$\begin{aligned} Z_i = R_i &= R_{B_1} \| R_{B_2} \| \beta r_e \\ &= 40\text{ k}\Omega \| 10\text{ k}\Omega \| 1.576\text{ k}\Omega \\ &\cong 1.32\text{ k}\Omega \end{aligned}$$

and from Fig. 11.45,

$$V_i = \frac{R_i V_S}{R_i + R_S}$$

or
$$\frac{V_i}{V_s} = \frac{R_i}{R_i + R_s} = \frac{1.32\text{ k}\Omega}{1.32\text{ k}\Omega + 1\text{ k}\Omega} = 0.569$$

so that
$$\begin{aligned} A_{v_{\text{mid}}} &= \frac{V_o}{V_i}\frac{V_i}{V_S} = (-90)(0.569) \\ &= \mathbf{-51.21} \end{aligned}$$

C_S:

$$R_i = R_{B_1} \| R_{B_2} \| \beta r_e = 40\text{ k}\Omega \| 10\text{ k}\Omega \| 1.576\text{ k}\Omega \cong 1.32\text{ k}\Omega$$

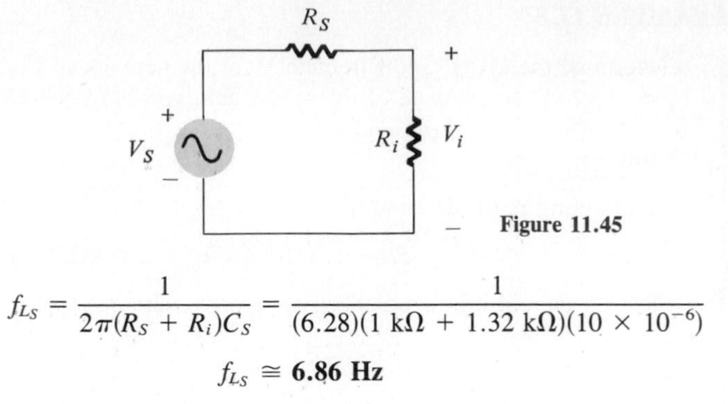

Figure 11.45

$$f_{L_S} = \frac{1}{2\pi(R_S + R_i)C_S} = \frac{1}{(6.28)(1\ \text{k}\Omega + 1.32\ \text{k}\Omega)(10 \times 10^{-6})}$$

$$f_{L_S} \cong \mathbf{6.86\ Hz}$$

The resulting HP-85 computer plot for the network with *just* C_S appears in Fig. 11.46. Recall that the 3-dB point corresponds with a gain equal to 0.707 of the midband value, or 36.21, as indicated in Fig. 11.46.

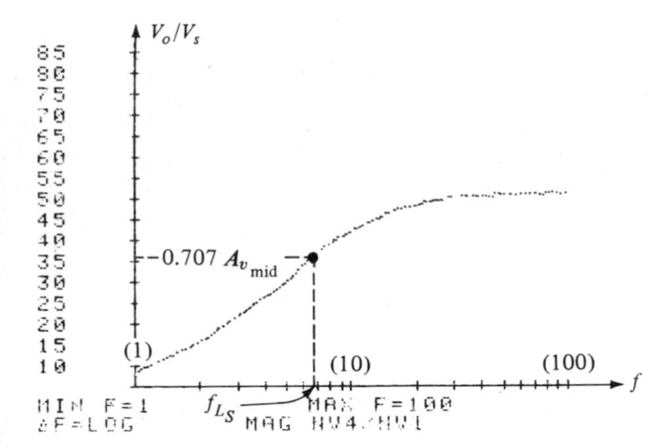

Figure 11.46

Dropping down to the log scale of Fig. 11.46, the cutoff frequency is as predicted ($\cong$ 6.86 Hz).

C_C:

$$f_{L_C} = \frac{1}{2\pi(R_C + R_L)C_C}$$

$$= \frac{1}{(6.28)(4\ \text{k}\Omega + 2.2\ \text{k}\Omega)(1 \times 10^{-6})}$$

$$\cong \mathbf{25.68\ Hz}$$

For C_C the computer plot appears in Fig. 11.47. Note again that the 0.707 level is achieved near the forecasted cutoff frequency of 25.68 Hz.

C_E:

$$R'_S = R_S \| R_{B_1} \| R_{B_2} = 1\ \text{k}\Omega \| 40\ \text{k}\Omega \| 10\ \text{k}\Omega \cong 1\ \text{k}\Omega$$

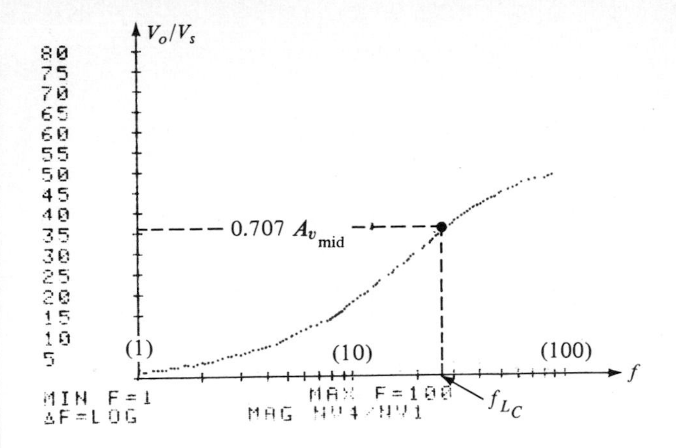

Figure 11.47

$$R_e = R_E \left\| \left(\frac{R_S'}{\beta} + r_e \right) = 2\text{ k}\Omega \right\| \left(\frac{1\text{ k}\Omega}{100} + 15.76 \right) = 2\text{ k}\Omega \| (10 + 15.76)$$

$$= 2\text{ k}\Omega \| 25.76\ \Omega \cong 25.76\ \Omega$$

$$f_{L_E} = \frac{1}{2\pi R_e C_E} = \frac{1}{(6.28)(25.76)(20 \times 10^{-6})} = \frac{10^6}{3235.46} \cong \mathbf{309.1\ Hz}$$

For C_E the computer plot appears in Fig. 11.48. The 0.707 level is again attained near the forecasted cutoff frequency of 309.1 Hz.

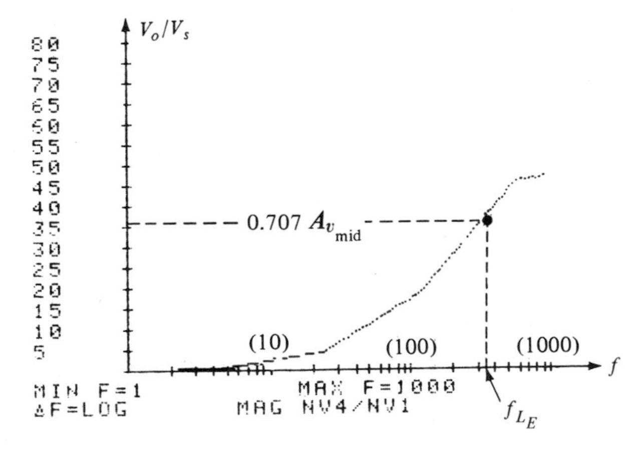

Figure 11.48

The fact that f_{L_E} is significantly higher than f_{L_S} or f_{L_C} suggests that it will be the predominant factor in determining the low-frequency response for the complete system. To test the accuracy of our hypothesis, the complete network was fed into the computer and the plot of Fig. 11.49 obtained. Note the very strong similarity with the plot of Fig. 11.48. For all practical purposes, C_C and C_S have only had an impact on the gain at frequencies below 100 Hz.

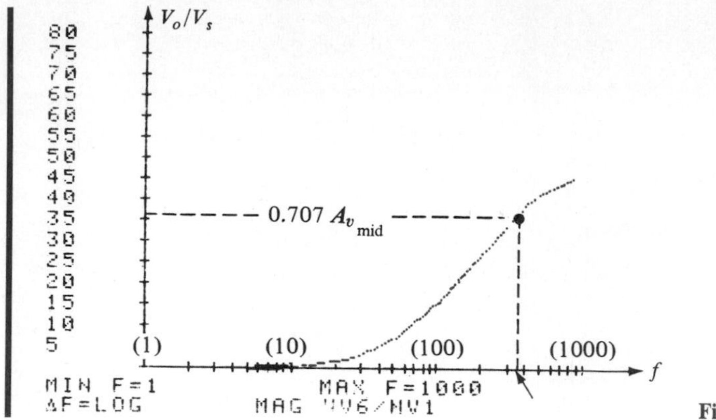

Figure 11.49

It was mentioned earlier that dB plots are usually normalized by dividing by the midband gain. For the network of Fig. 11.40 the midband gain is -51.21 and the ratio $|A_v/A_{v_{\text{mid}}}|$ becomes $|A_v/A_{v_{\text{mid}}}|_{\text{dB}} = 1$ in the midband region. The result is $20 \log_{10} 51.21/51.21 = 0$ dB, as shown in Fig. 11.50. At lower frequencies the gain A_v will drop but $A_{v_{\text{mid}}}$ remains fixed in magnitude at 51.21 and the ratio $|A_v/A_{v_{\text{mid}}}|$ will drop accordingly. For example, at the cutoff frequency $|A_v| = (0.707)(51.21) = 36.21$ and the ratio $|A_v/A_{v_{\text{mid}}}| = 36.21/51.21 = 0.707$ with $|A_v/A_{v_{\text{mid}}}|_{\text{dB}} = -3$ dB, as shown in Fig. 11.50.

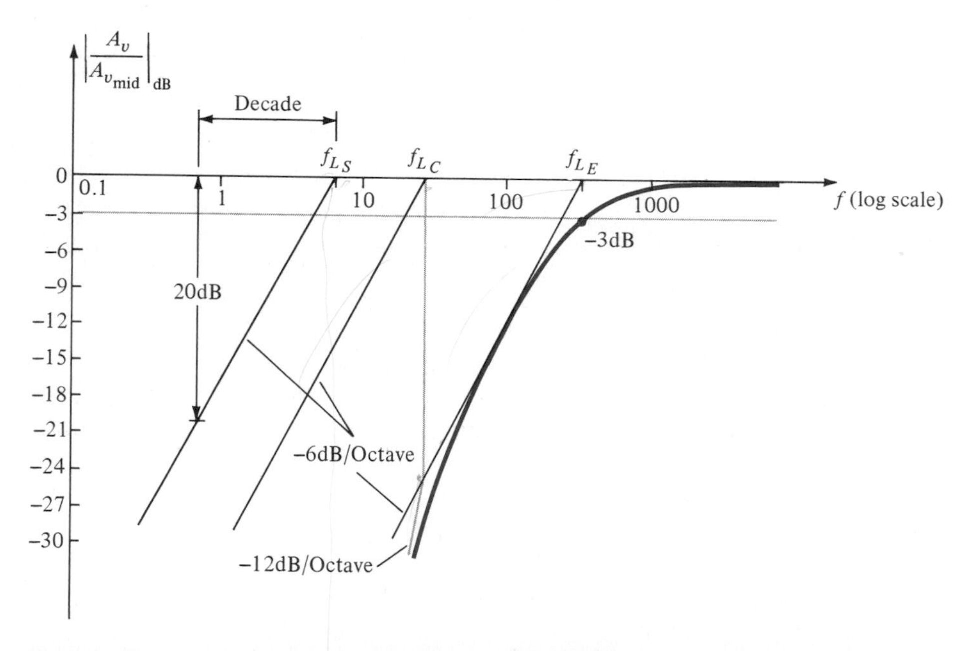

Figure 11.50 Low-frequency plot for the network of Example 11.8.

Figure 11.50 was drawn by sketching a -6-dB/octave asymptote at each cutoff frequency. For the element values it is quite clear from the figure that the cutoff defined by C_E will determine the -3-dB point usually employed to define the networks bandwidth. For frequencies within the bandwidth, at least one-half the maximum available power will reach the load. For $V_o = 0.707\ V_{max}$

$$P_L = \frac{V_L^2}{R_L} = \frac{(0.707V_{max})^2}{R_L} = \frac{0.5V_{max}^2}{R_L}$$

$$= 0.5P_{max}$$

Note in Fig. 11.50 that the slope of the asymptote used to sketch the actual frequency response dropped to -12 dB at f_{L_C}. For Bode plots the slope of the resultant asymptote is additive when an asymptote passes through a lower cutoff frequency. The same is true for the high-frequency region, as will be demonstrated in the next section.

For an unbypassed emitter resistor there will obviously be only two cutoff frequencies to be determined: that due to C_S and to C_C. The equation for f_{L_S} will be altered accordingly. There is an exercise on this very topic at the end of the chapter.

11.11 SINGLE-STAGE TRANSISTOR AMPLIFIER—HIGH-FREQUENCY CONSIDERATIONS

At the high-frequency end there are two factors that will define the -3 dB point: the network capacitance (parasitic and introduced) and the frequency dependence of $h_{fe}(\beta)$.

In the high-frequency region the *RC* network of concern has the configuration appearing in Fig. 11.51. At increasing frequencies the reactance X_C will decrease in magnitude, resulting in a shorting effect across the output and a decrease in gain. The derivation leading to the corner frequency for this *RC* configuration follows along similar lines to that encountered for the low-frequency region. The most significant difference is in the general form of A_v appearing below:

$$\boxed{A_v = \frac{1}{1 + j(f/f_2)}} \qquad (11.41)$$

which results in an asymptotic plot such as shown in Fig. 11.52 that drops off at 6 dB/octave with increasing frequency. Note that f_2 is in the denominator of the frequency ratio rather than the numerator as occurred for f_1 in Eq. (11.31).

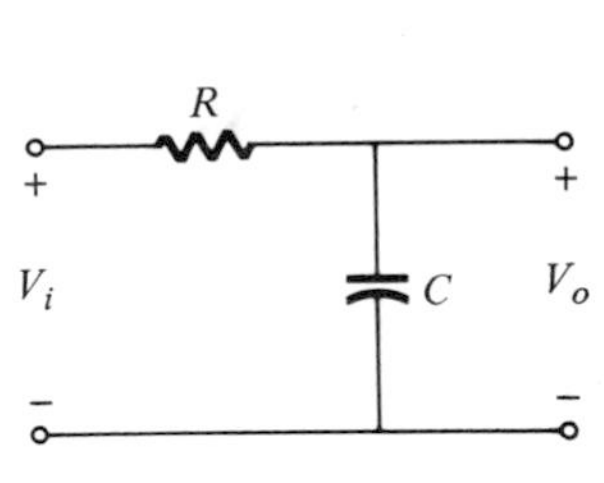

Figure 11.51

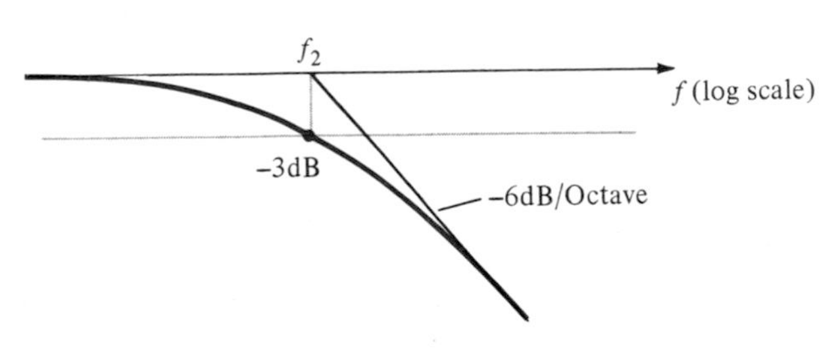

Figure 11.52 Asymptotic plot as defined by Eq. (11.41).

In Fig. 11.53 the various parasitic capacitances (C_{be}, C_{bc}, C_{ce}) of the transistor have been included with the wiring capacitance (C_{W_1}, C_{W_2}) introduced during construction. The high-frequency equivalent model for the network of Fig. 11.53 appears in Fig. 11.54. Note the absence of the capacitors C_S, C_C, and C_E, which are all assumed to be in the short-circuit state at these frequencies. Keep in mind that the capacitors that determined the low-frequency cutoff point appear between the input and output terminals of the *RC* network and not across the output as indicated in Fig. 11.53. The capacitance C_i includes the Miller capacitance C_M as introduced for the FET in Chapter 10, the capacitance C_{be}, and the input wiring capacitance C_{W_1}. The capacitance C_o includes the output wiring capacitance and the collector capacitance.

Determining the Thévenin equivalent circuit of the input and output networks of Fig. 11.53 will result in the configurations of Fig. 11.55. For the input network the

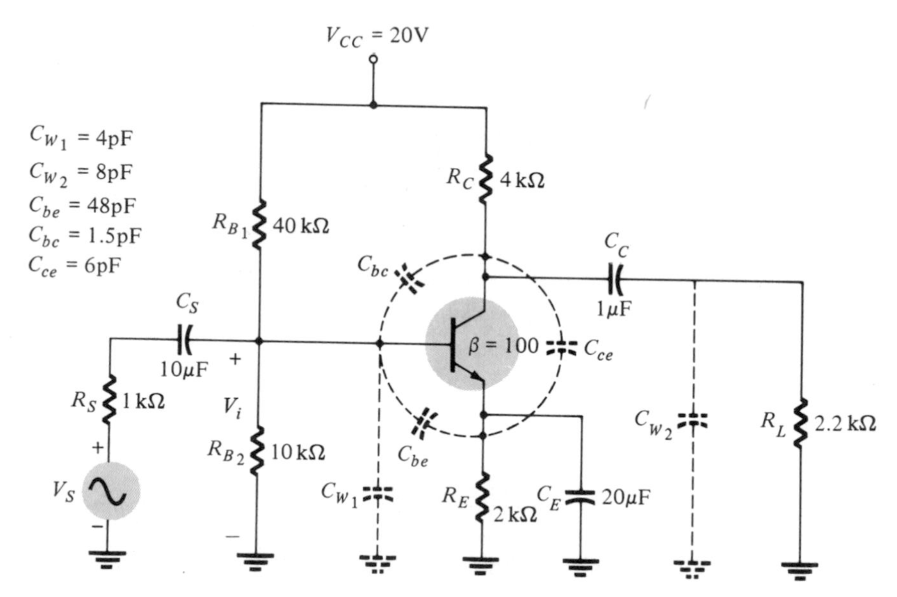

Figure 11.53 Network of Fig. 11.40 with the capacitors that affect the high-frequency response.

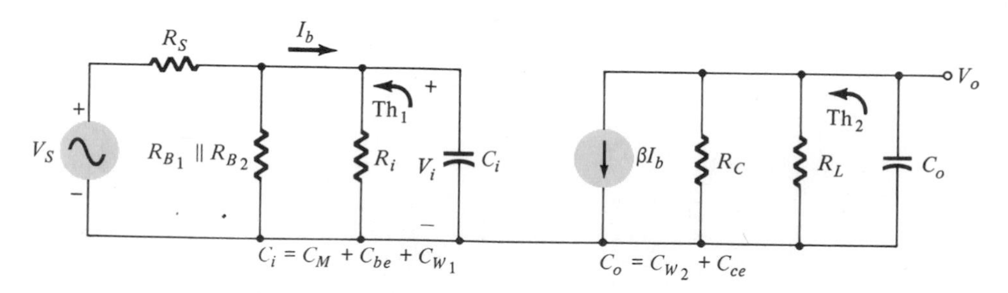

Figure 11.54 High-frequency model for the network of Fig. 11.53.

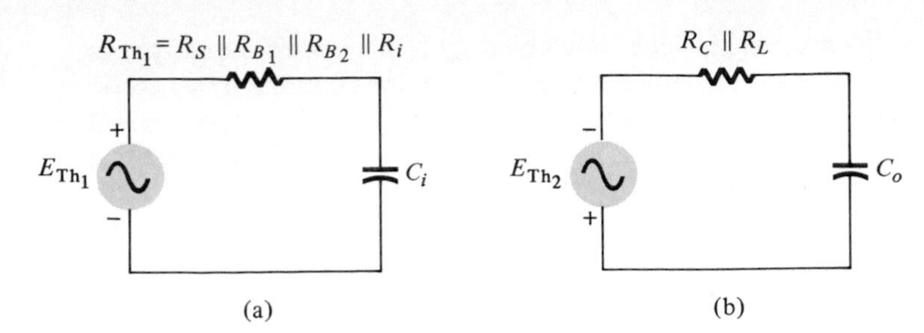

Figure 11.55 Thévenin circuits for the input and output networks of the network of Fig. 11.54.

−3-dB frequency is defined by

$$f_{H_i} = \frac{1}{2\pi R_{Th_1} C_i} \tag{11.42}$$

with $$R_{Th_1} = R_S \| R_{B_1} \| R_{B_2} \| R_i$$

and $$C_i = C_{W_1} + C_{be} + C_M = C_{W_1} + C_{be} + (1 + |A_v|)C_{bc}$$

At very high frequencies the effect of C_i is to reduce the total impedance of the parallel combination of R_{B_1}, R_{B_2}, R_i, and C_i in Fig. 11.54. The result is a reduced level of voltage across C_i and a reduction in I_b. The net result is a reduced level of gain for the complete system.

In the high-frequency range the voltage gain A_v will be a function of frequency due to the capacitive elements. It cannot be determined by a simple ratio of resistors as developed in Chapter 9. As a first approximation to the value of A_v, however, we shall use the mid-band value. As the frequency increases, the gain will obviously drop from the mid-band value due to the capacitive effects. If we therefore use the midband value, we have the maximum value of A_v and the maximum Miller capacitance. This will result in the maximum value for C_i and the minimum f_{H_i}. In essence, we have determined the lowest cutoff frequency due to C_i and defined a *worst-case* design situation. In other words, the actual cutoff due to C_i will always be higher than the level determined using the midband gain.

For the output network:

$$f_{H_o} = \frac{1}{2\pi R_{Th_2} C_o} \tag{11.43}$$

with $$R_{Th_2} = R_C \| R_L$$

and $$C_o = C_{W_2} + C_{ce}$$

At very high frequencies the capacitive reactance of C_o will decrease and consequently reduce the total impedance of the output parallel branches of Fig. 11.54. The net result is that V_o will also decline toward zero as the reactance X_C becomes smaller.

The h_{fe} or β variation with frequency must now be considered to ensure that it does not determine the high cutoff frequency of the amplifier. The variation of h_{fe} with frequency will approach, with some degree of accuracy, the following relationship:

$$h_{fe} = \frac{h_{fe_{\text{mid}}}}{1 + jf/f_\beta} \tag{11.44}$$

The only undefined quantity, f_β, is determined by a set of parameters employed in the *hybrid* π or *Giacoletto* model frequently applied to best represent the transistor in the high-frequency region. It appears in Fig. 11.56. The various parameters warrant a moment of explanation. The resistance $r_{bb'}$ includes the base contact, base bulk, and base spreading resistance. The first is due to the actual connection to the base. The second includes the resistance from the external terminal to the active region of the transistors, while the last is the actual resistance within the active base region. The resistances $r_{b'e}$, r_{ce}, and $r_{b'c}$ are the resistances between the indicated terminals when the device is in the active region. The same is true for the capacitance $C_{b'c}$ and $C_{b'e}$, although the former is a transition capacitance while the latter is a diffusion capacitance. A more detailed explanation of the frequency dependence of each can be found in a number of readily available texts.

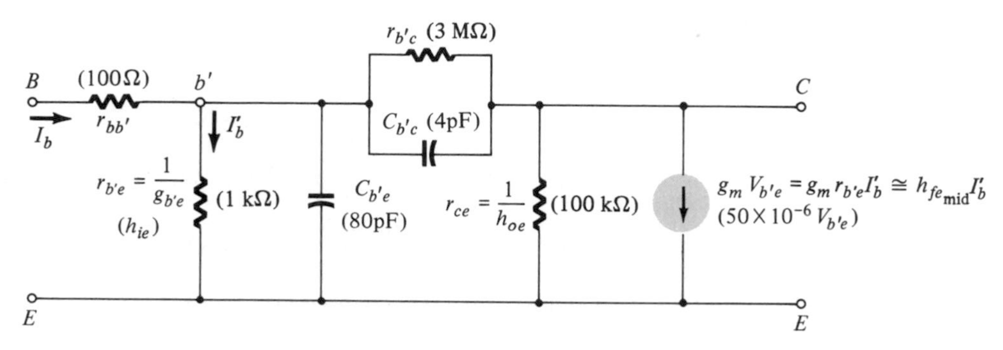

Figure 11.56 Giacoletto (or hybrid π) high-frequency transistor small-signal ac equivalent circuit.

In terms of these parameters:

$$f_\beta(\text{sometimes appearing as } f_{h_{fe}}) = \frac{g_{b'e}}{2\pi(C_{b'e} + C_{b'c})} \tag{11.45}$$

or since the hybrid parameter h_{fe} is related to the hybrid parameter $g_{b'e}$ through $g_m = h_{fe_{\text{mid}}} g_{b'e}$

$$f_\beta = \frac{1}{h_{fe_{\text{mid}}}} \frac{g_m}{2\pi(C_{b'e} + C_{b'c})} \tag{11.46}$$

The basic format of Eq. (11.44) should suggest some similarities between it and the curves obtained for the low-frequency response. The most noticeable difference is the fact that f_β appears in the denominator while f_1 appears in the numerator of the frequency ratio. This particular difference will have the effect depicted in Fig. 11.57;

the plot will drop off from the midband value rather than approach it with increase in frequency. The same figure has a plot of h_{fb} versus frequency. Note that it is almost constant for the frequency range. In general, the common-base configuration displays improved high-frequency characteristics over the common-emitter configuration. For this reason, common-base high-frequency parameters, rather than common-emitter parameters, are often specified for a transistor. The following equation permits a direct conversion for determining f_β if f_α and α are specified.

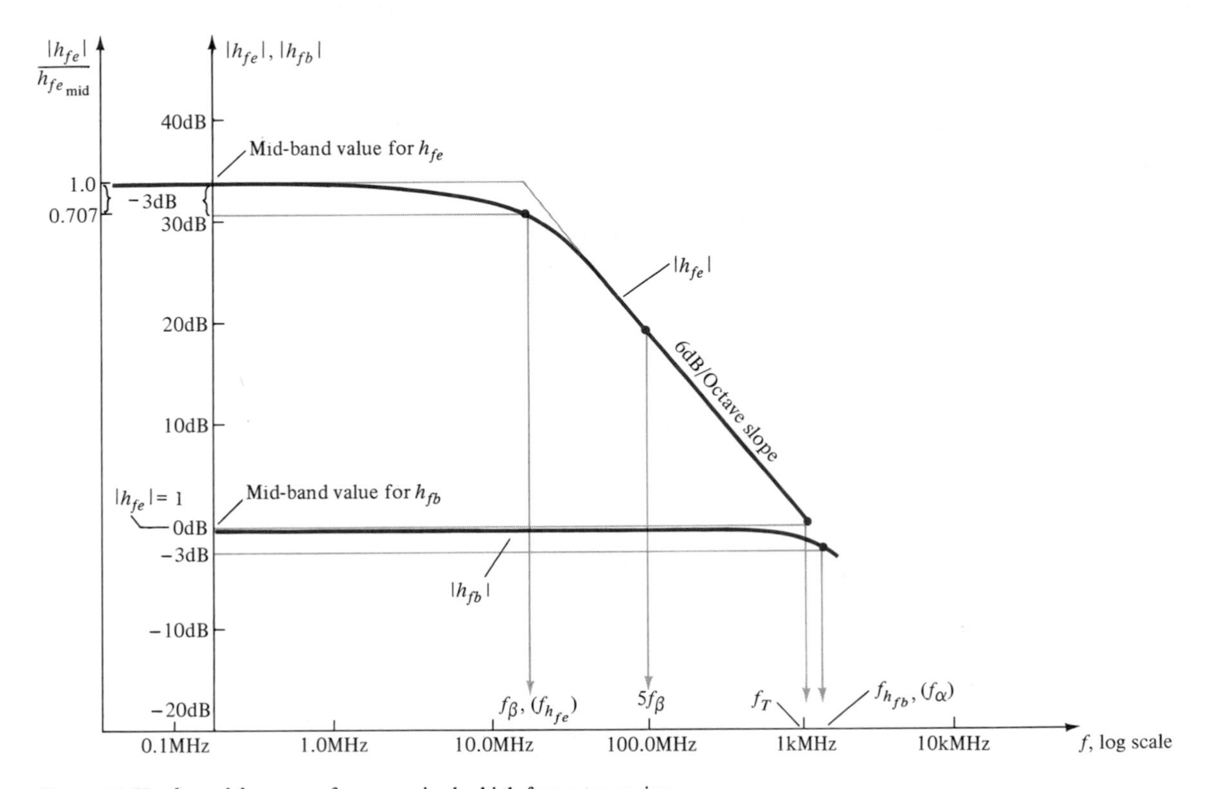

Figure 11.57 h_{fe} and h_{fb} versus frequency in the high-frequency region.

$$\boxed{f_\beta = f_\alpha(1 - \alpha)} \tag{11.47}$$

A quantity called the *gain-bandwidth product* is defined for the transistor by the condition

$$\left| \frac{h_{fe_{\text{mid}}}}{1 + jf/f_\beta} \right| = 1$$

so that $$|h_{fe}|_{dB} = 20 \log_{10} \left| \frac{h_{fe_{mid}}}{1 + jf/f_\beta} \right| = 20 \log_{10} 1 = 0 \text{ dB}$$

The frequency at which $|h_{fe}|_{dB} = 0$ dB is clearly indicated by f_T in Fig. 11.57. The magnitude of h_{fe} at the defined condition point ($f_T \gg f_\beta$) is given by

$$\frac{h_{fe_{mid}}}{\sqrt{1 + (f_T/f_\beta)^2}} \cong \frac{h_{fe_{mid}}}{f_T/f_\beta} = 1$$

so that $$\boxed{f_T \cong h_{fe_{mid}} \underbrace{f_\beta}_{(\cong \text{BW})} \qquad \text{(gain-bandwidth product)}} \tag{11.48}$$

or $$f_T \cong \beta f_\beta$$

with $$\boxed{f_\beta = \frac{f_T}{\beta}} \tag{11.49}$$

Substituting for f_β in Eq. (11.48) gives

$$f_T \cong h_{fe_{mid}} f_\beta = h_{fe_{mid}} \left[\frac{1}{h_{fe_{mid}}} \frac{g_m}{2\pi(C_{b'e} + C_{b'c})} \right]$$

Ignoring the effect of $r_{bb'}$ as a first approximation will result in the joining of terminals B and b' in Fig. 11.56 and $C_{b'e} = C_{be}$ and $C_{b'c} = C_{bc}$. The result is

$$\boxed{f_T \cong \frac{g_m}{2\pi(C_{be} + C_{bc})}} \tag{11.50}$$

EXAMPLE 11.9

(a) Determine f_{H_i}, f_{H_o}, f_β, and f_T for the network of Fig. 11.53 if $g_{b'e} = 1 \times 10^{-3}$ S.
(b) Sketch the Bode plot for the low- and high-frequency regions.
(c) Determine the resulting bandwidth of the system.

Solution:

(a) $R_{Th_1} = R_S \| R_{B_1} \| R_{B_2} \| R_i = 1 \text{ k}\Omega \| 40 \text{ k}\Omega \| 10 \text{ k}\Omega \| 1.576 \text{ k}\Omega \cong 0.568 \text{ k}\Omega$

$$\begin{aligned} C_i &= C_{W1} + C_{be} + (1 + |A_v|)C_{bc} \\ &= 4 \text{ pF} + 48 \text{ pF} + (1 + 90)(1.5 \text{ pF}) \\ &= 188.5 \text{ pF} \end{aligned}$$

Note that A_v defined by $A_v = V_o/V_i$ in Example 11.8 was employed since it provides the level of gain associated with the Miller capacitance.

$$f_{H_i} = \frac{1}{2\pi R_{\text{Th}_1} C_i} = \frac{1}{(6.28)(0.568 \times 10^3)(188.5 \times 10^{-12})}$$

$$= \frac{1000 \times 10^6}{672.4} \cong \mathbf{1.49\ MHz}$$

$$R_{\text{Th}_2} = R_C \| R_L = 4\ \text{k}\Omega \| 2.2\ \text{k}\Omega = 1.419\ \text{k}\Omega$$

$$C_o = C_{W2} + C_{ce} = 8\ \text{pF} + 6\ \text{pF} = 14\ \text{pF}$$

$$f_{H_o} = \frac{1}{2\pi R_{\text{Th}_2} C_o} = \frac{1}{(6.28)(1.419 \times 10^3)(14 \times 10^{-12})}$$

$$= \frac{1000 \times 10^6}{124.76} \cong \mathbf{8.02\ MHz}$$

Assuming that $C'_{be} = C_{be}$ and $C'_{b'c} = C_{bc}$ yields

$$f_\beta = \frac{g_{b'e}}{2\pi(C_{be} + C_{bc})}$$

$$= \frac{1 \times 10^{-3}}{(6.28)(49.5 \times 10^{-12})} = \frac{1000 \times 10^6}{310.86}$$

$$f_\beta = \mathbf{3.217\ MHz}$$

and

$$f_T \cong h_{fe} f_\beta$$

$$= 100\ (3.217 \times 10^6)$$

$$= \mathbf{321.7\ MHz}$$

(b) For the low-, mid-, and high-frequency regions a Bode plot for the network of Fig. 11.53 appears in Fig. 11.58. Note in the high- and low-frequency regions that each cutoff frequency defines a −6-dB/octave asymptote and the slope increases by −6 dB/octave (for the asymptote that will define the actual response) each time it passes a cutoff frequency. A curve approximating the actual response also appears in the figure. Note the −3-dB drop at the highest low-frequency cutoff frequency and at the lowest high-frequency cutoff frequency.

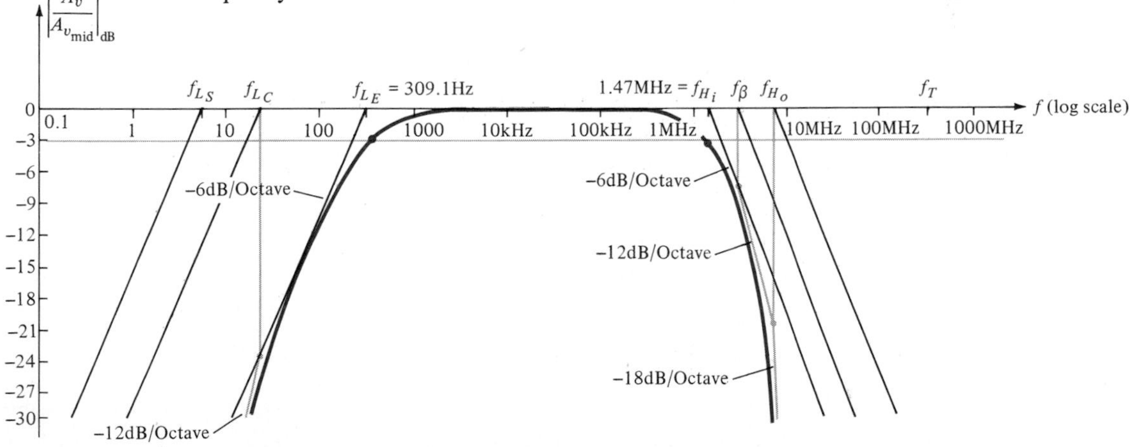

Figure 11.58 A_v dB versus frequency (log scale) for the network of Fig. 11.50.

(c) The bandwidth:

$$\text{BW} \cong f_{H_i} - f_{L_E} \cong f_{H_i} = \mathbf{1.49\ MHz}$$

11.12 MULTISTAGE FREQUENCY EFFECTS

For a second transistor stage connected directly to the output of a first stage there will be a significant change in the overall frequency response. In the high-frequency region the output capacitance C_o must now include the wiring capacitance (C_{W_1}), parasitic capacitance (C_{be}), and Miller capacitance (C_M), of the following stage. Further, there will be additional low-frequency cutoff levels due to the second stage that will further reduce the overall gain of the system in this region. For each additional stage the upper cutoff frequency will be determined primarily by that stage having the lowest cutoff frequency. The low-frequency cutoff is primarily determined by that stage having the highest low-frequency cutoff frequency. Obviously, therefore, one poorly designed stage can offset an otherwise well-designed cascaded system.

The effect of increasing the number of *identical* stages can be clearly demonstrated by considering the situations indicated in Fig. 11.59. In each case the upper and lower cutoff frequencies of each of the cascaded stages are identical. For a single stage the cutoff frequencies are f_1 and f_2 as indicated. For two identical stages in cascade the drop-off rate in the high- and low-frequency regions has increased to −12 dB/octave or −40 dB/decade. At f_1 and f_2, therefore, the decibel drop is now −6 dB rather than the defined band frequency gain level of −3 dB. The −3 dB point has shifted to f_1' and f_2' as indicated with a resulting drop in the bandwidth. A −18 dB/octave or −60 dB/decade slope will result for a three-stage system of identical stages with the indicated reduction in bandwidth (f_1'' and f_2'').

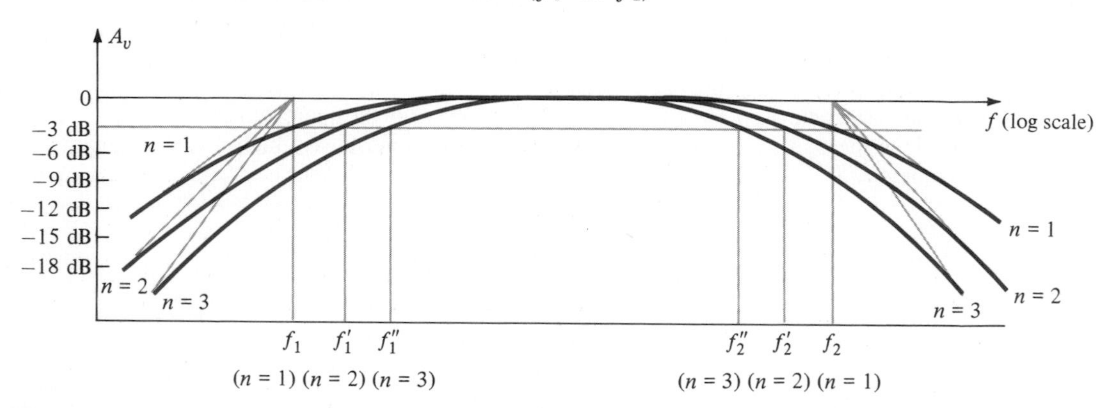

Figure 11.59 Effect of an increased number of stages on the cutoff frequencies and the bandwidth.

Assuming identical stages, an equation for each band frequency as a function of the number of stages (n) can be determined in the following manner:

For the low-frequency region,

$$A_{v_{\text{low, (overall)}}} = A_{v_{1_{\text{low}}}} A_{v_{2_{\text{low}}}} A_{v_{3_{\text{low}}}} \cdots A_{v_{n_{\text{low}}}}$$

but since each stage is identical, $A_{v_{1_{\text{low}}}} = A_{v_{2_{\text{low}}}} =$ etc. and

$$A_{v\,\text{low, (overall)}} = (A_{v\,1_{\text{low}}})^n$$

or
$$\frac{A_{v\,\text{low}}}{A_{v\,\text{mid}}}(\text{overall}) = \left(\frac{A_{v\,1_{\text{low}}}}{A_{v\,\text{mid}}}\right)^n = \frac{1}{(1 + jf_1/f)^n}$$

Setting the magnitude of this result equal to $1/\sqrt{2}$ (-3 dB level) results in

$$\frac{1}{[\sqrt{1 + (f_1/f_1')^2}]^n} = \frac{1}{\sqrt{2}}$$

or
$$\left\{\left[1 + \left(\frac{f_1}{f_1'}\right)^2\right]^{1/2}\right\}^n = \left\{\left[1 + \left(\frac{f_1}{f_1'}\right)^2\right]^n\right\}^{1/2} = (2)^{1/2}$$

so that
$$\left[1 + \left(\frac{f_1}{f_1'}\right)^2\right]^n = 2$$

and
$$1 + \left(\frac{f_1}{f_1'}\right)^2 = 2^{1/n}$$

with the result
$$\boxed{f_1' = \frac{f_1}{\sqrt{2^{1/n} - 1}}} \tag{11.51}$$

In a similar manner, it can be shown that for the high-frequency region,

$$\boxed{f_2' = \sqrt{2^{1/n} - 1}\, f_2} \tag{11.52}$$

Note the presence of the same factor $\sqrt{2^{1/n} - 1}$ in each equation. The magnitude of this factor for various values of n is listed below.

n	$\sqrt{2^{1/n} - 1}$
1	1
2	0.64
3	0.51
4	0.43
5	0.39

For $n = 2$, consider that the upper cutoff frequency $f_2' = 0.64f_2$ or 64% of the value obtained for a single stage, while $f_1' = (1/0.64)f_1 = 1.56f_1$. For $n = 3$, $f_2' = 0.51f_2$ or approximately $\frac{1}{2}$ the value of a single stage with $f_1' = (1/0.51)f_1 = 1.96f_1$ or approximately *twice* the single-stage value.

Consider the example of the past few sections, where $f_2 = 1.49$ MHz and $f_1 = 309.1$ Hz. For $n = 2$,

$$f_2' = 0.64f_2 = 0.64(1.49)\text{ MHz}) = 0.95\text{ MHz}$$

and
$$f_1' = 1.56f_1 = 1.56(309.1) \cong 482.2\text{ Hz}$$

The bandwidth is now 0.95 MHz − 482.2 Hz $\cong f_2' \Rightarrow$ 64.4% of its single-stage value, a drop of some significance.

For the *RC*-coupled transistor amplifier, if $f_2 = f_\beta$, or if they are close enough in magnitude for both to affect the upper 3-dB frequency, the number of stages must be increased by a factor of 2 when determining f_2' due to the increased number of factors $1/(1 + jf/f_x)$.

A decrease in bandwidth is not always associated with an increase in the number of stages if the midband gain can remain fixed independent of the number of stages. For instance, if a single-stage amplifier produces a gain of 100 with a bandwidth of 10,000 Hz, the resulting gain-bandwidth product is $10^2 \times 10^4 = 10^6$. For a two-stage system the same gain can be obtained by having two stages with a gain of 10 since $(10 \times 10 = 100)$. The bandwidth of each stage would then increase by a factor of 10 to 100,000 due to the lower gain requirement and fixed gain-bandwidth product of 10^6. Of course, the design must be such as to permit the increased bandwidth and establish the lower gain level.

This discussion of the effects of an increased number of stages on the frequency response was included here, rather than at the conclusion of this chapter, to add a note of completion to the analysis of cascaded transistor amplifier systems. The results, however, can be applied directly to the discussion of FET and vacuum-tube cascaded systems.

11.13 FREQUENCY RESPONSE OF CASCADED FET AMPLIFIERS

A representative cascaded system employing FET amplifiers appears in Fig. 11.60.

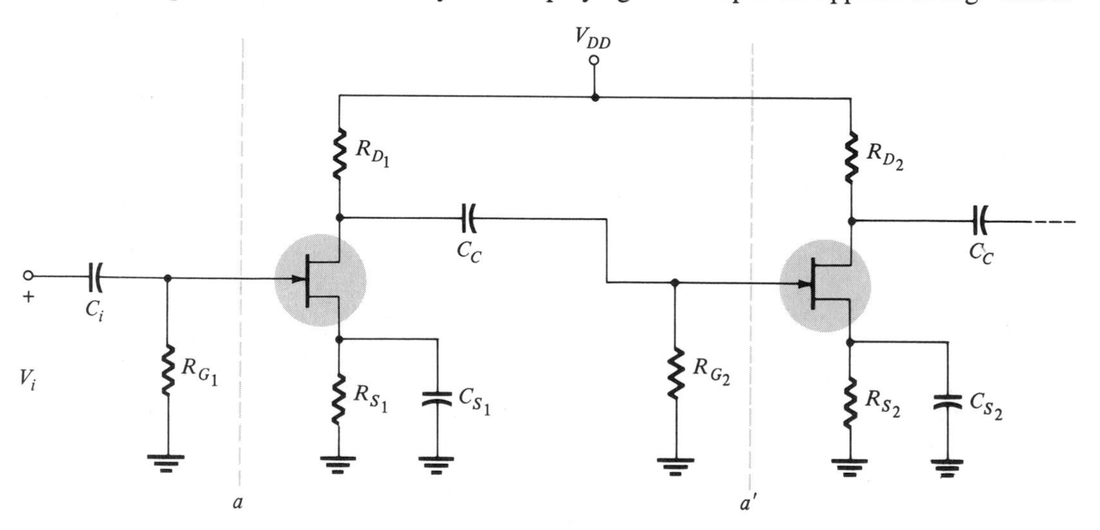

Figure 11.60 Cascaded FET amplifiers.

The equivalent circuit for the section a-a' indicated in Fig. 11.60 is presented in Fig. 11.61 for both the high- and low-frequency regions. It is assumed for the low-frequency response that the breakpoint frequency due to C_S is sufficiently less than that due to C_C, to result in C_C determining the lower-band frequency. For this reason C_S does not appear in the low-frequency model. For future reference, the breakpoint frequency determined by C_S is given by

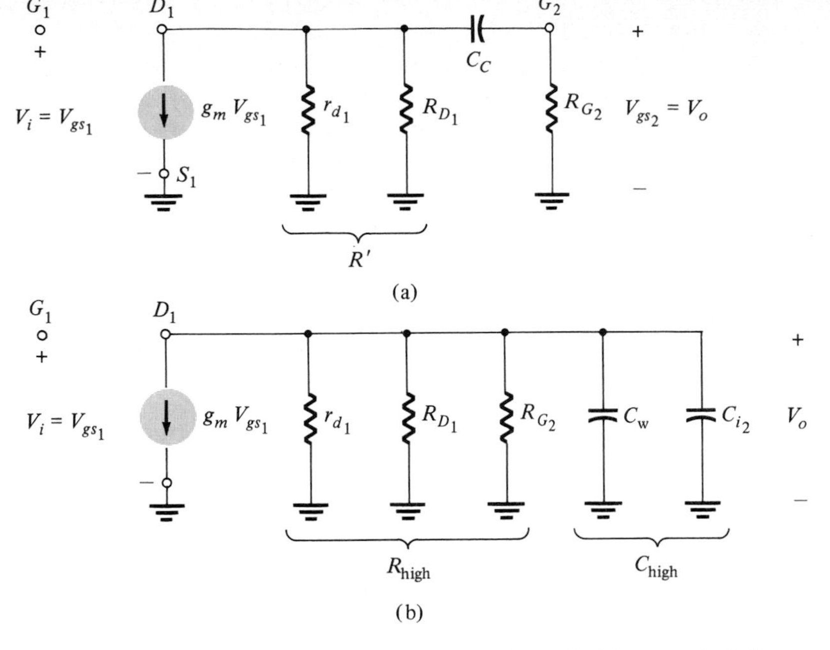

Figure 11.61 Small-signal ac equivalent circuit of section *a-a'* of the network of Fig. 11.60: (a) low frequency; (b) high frequency.

$$f_S = \frac{1 + R_{S_1}(1 + g_m r_d)/(r_d + R_{D_1} \| R_L)}{2\pi C_S R_S} \tag{11.53}$$

It was demonstrated in this chapter and in Chapter 10 that the midband voltage gain is given by

$$A_{v_{\text{mid}}} = \frac{V_o}{V_i} = -g_m R \qquad \text{where } R = r_{d_1} \| R_{D_1} \| R_{G_2}$$

The similarities between the circuits of Fig. 11.61 and those introduced in the last few sections on the transistor cascaded system are obvious. With this in mind, the steps leading to the following results should also be somewhat obvious.

$$\frac{A_{v_{\text{low}}}}{A_{v_{\text{mid}}}} = \frac{1}{1 - jf_1/f} \tag{11.54}$$

where

$$f_1 = \frac{1}{2\pi C_C(R' + R_{G_2})}$$

and

$$R' = r_{d_1} \| R_{D_1}$$

In addition,

$$\frac{A_{v_{\text{high}}}}{A_{v_{\text{mid}}}} = \frac{1}{1 + jf/f_2} \tag{11.55}$$

where $$f_2 = \frac{1}{2\pi C_{\text{high}} R_{\text{high}}}$$

and $$C_{\text{high}} = C_W + C_{i_2}$$

with $$R_{\text{high}} = R = r_{d_1} \| R_{D_1} \| R_{G_2}$$

For FET amplifiers, the input capacitance of the succeeding stage, as derived in Chapter 10 is given by

$$\boxed{C_i = C_{gs} + C_{gd}(1 + |A_v|)} \tag{11.56}$$

The gain-bandwidth product is

$$(\text{gain})(\text{BW}) = A_{v_{\text{mid}}}(f_2 - f_1 \cong f_2) = g_m R \frac{1}{2\pi R(C_W + C_i)}$$

and $$\boxed{\text{GBW} = \frac{g_m}{2\pi(C_W + C_i)} = f_T = A_{v_{\text{mid}}} f_2} \tag{11.57}$$

PROBLEMS

§ 11.2

1. (a) Determine A_{i_T} and A_{v_T} for a cascaded system if the power gain is 12.8×10^3 and $Z_L = 4\ \text{k}\Omega$ with the input impedance of the first stage $Z_{i_1} = 2\ \text{k}\Omega$.
(b) If the system of part (a) is comprised of two identical stages, determine the voltage and current gain of each stage.

§ 11.3

2. For the two-stage RC-coupled amplifier in Fig. 11.62:
(a) Determine Z_i and Z_o.
(b) Calculate the voltage gain $A_v = V_o/V_i$.
(c) Determine the current gain $A_i = I_o/I_i$.

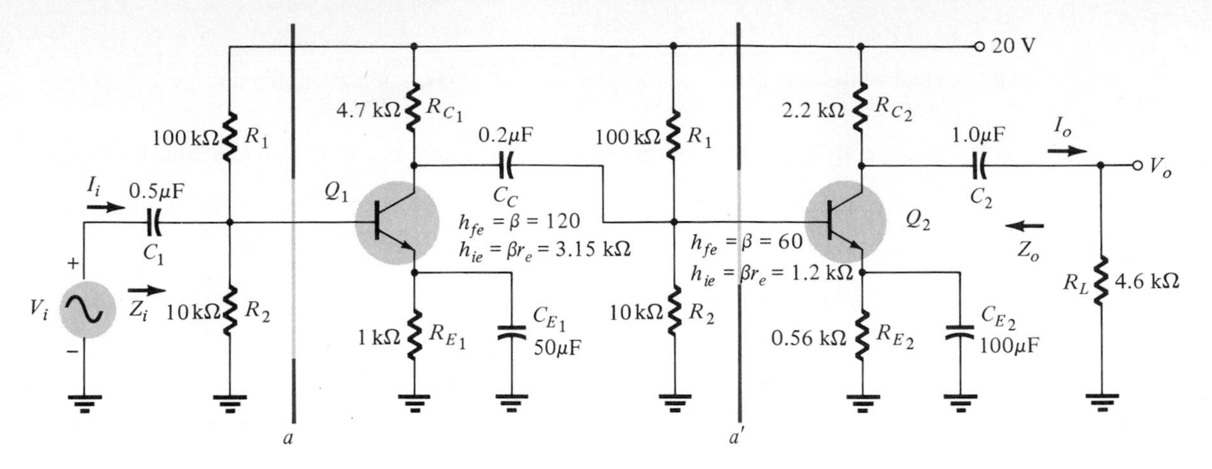

Figure 11.62 Two-stage *RC*-coupled amplifier.

3. Determine Z_i, Z_o, A_{vT}, A_{iT}, and A_{PT} for the *RC*-coupled amplifier of Fig. 00.0 if $R_{B_1} = 56$ kΩ and $R_{B_2} = 5.6$ kΩ for each transistor and $R_{C_1} = 6.8$ kΩ, $R_{C_2} = 3.3$ kΩ, $R_{E_1} = R_{E_2} = 0.56$ kΩ, $R_L = 2.2$ kΩ. All capacitor values are the same. For each transistor $\beta = 120$. Since h_{ie} is not provided, r_e will have to be determined for each transistor. Use all appropriate approximations.

4. Repeat Problem 3 if both emitter capacitors C_E are removed.

5. Repeat Example 11.1 if C_E is removed.

6. Repeat Example 11.1 if C_E is removed and the output load Z_L is connected through a capacitor to the collector of Q_2.

7. Repeat Example 11.2 if R_{E_1} is removed and Z_L is reduced to a significantly lower level of 0.56 kΩ.

8. Design a two-stage RC-coupled amplifier to provide an overall gain of ≅ 2000. The circuit is to operate into a load of 10 kΩ; the signal is supplied from a perfect voltage source ($R_S = 0$ Ω). Show typical (commercially available) component values for each element and calculate the voltage gain of the resulting circuit as a check. The list of commercially available resistors can be found in any electronic products publication.

9. Repeat Example 11.3 if C_{S_1} and C_{S_2} are removed and $R_{S_1} = 2.2$ kΩ with $R_{S_2} = 1$ kΩ.

§ 11.4

10. Calculate the impedance seen looking into the primary of a 5 : 1 step-down transformer connected to a load of 22 Ω.

11. Calculate the necessary transformer turns ratio to match a 50-Ω load to a 20-kΩ source impedance.

12. (a) Calculate the voltage gain (V_o/V_i) of the transformer-coupled amplifier of Fig. 11.63.

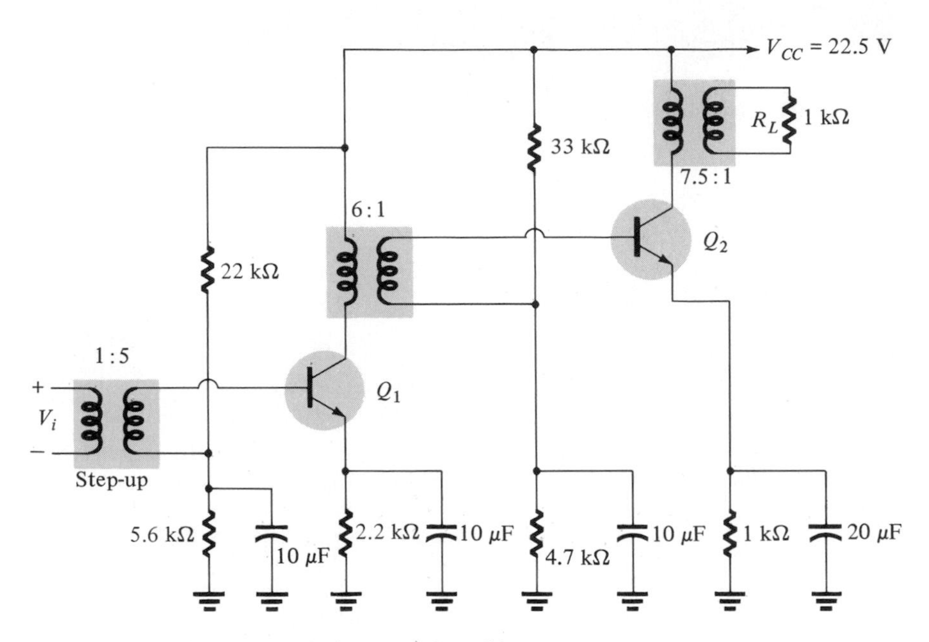

Figure 11.63 Two-stage transformer-coupled amplifier.

(b) What is the voltage gain of the circuit of Fig. 11.63 if the load is reduced to 0.5 kΩ?

§ 11.5

13. Determine the new dc levels in Fig. 11.13 if the 12-V batteries are replaced by 16-V supplies. In addition, determine the new value of r_e for each transistor. How are the ac voltage and current gain affected? What are their new levels if affected by this change?

§ 11.6

14. Determine the following for the cascode amplifier of Fig. 11.64:
(a) V_o.
(b) Z_i, Z_o.

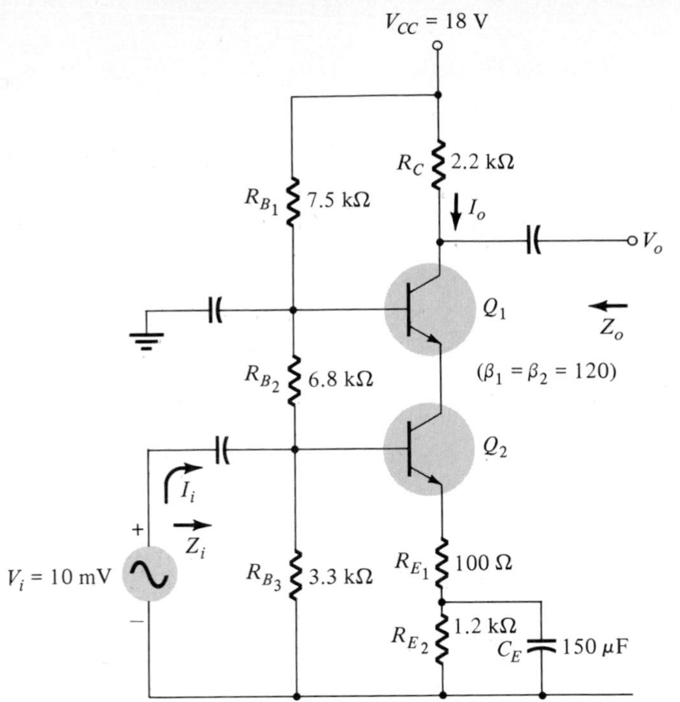

Figure 11.64

(c) I_o, I_i, and A_i.
(d) A_{PT}.

15. Determine the following for the cascode amplifier of Fig. 11.65:
(a) r_{e_1} and r_{e_2} for $\beta_1 = \beta_2 = 50$.
(b) A_{vT} and V_o if $V_i = 10$ mV.
(c) Z_i and Z_o.

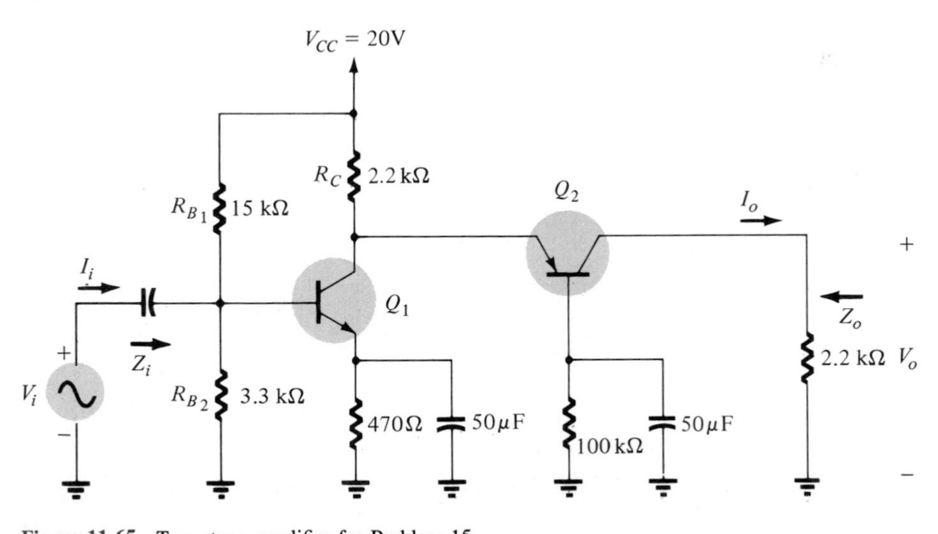

Figure 11.65 Two-stage amplifier for Problem 15.

§ 11.7

16. Determine A_i, Z_i, Z_o, and A_v for the Darlington configuration of Fig. 11.66.

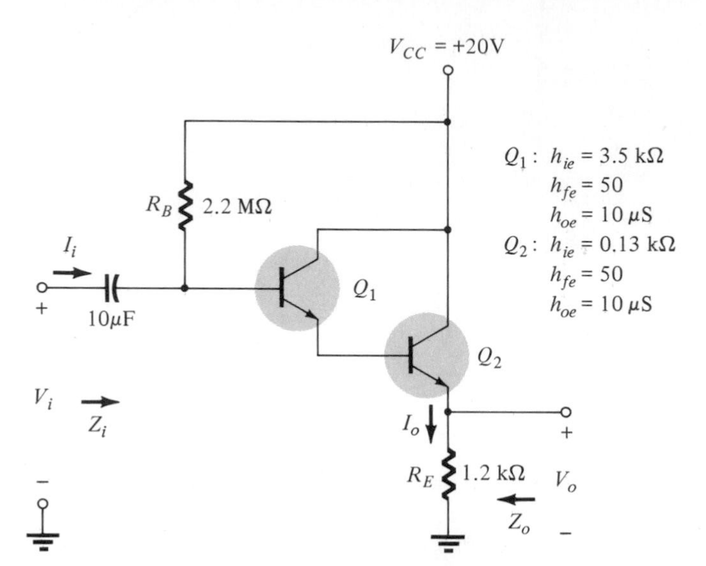

Figure 11.66 Amplifier circuit for Problem 16.

17. Repeat Problem 16 if a collector resistor of 2.2 kΩ is added between the collector of Q_1 and V_{CC} and the output is taken off the collector of the Darlington configuration. I_o is the current through the added 2.2-kΩ resistor.

18. In Fig. 11.67, h_{fe} and h_{ie} have been provided for the Darlington pair rather than for each transistor.

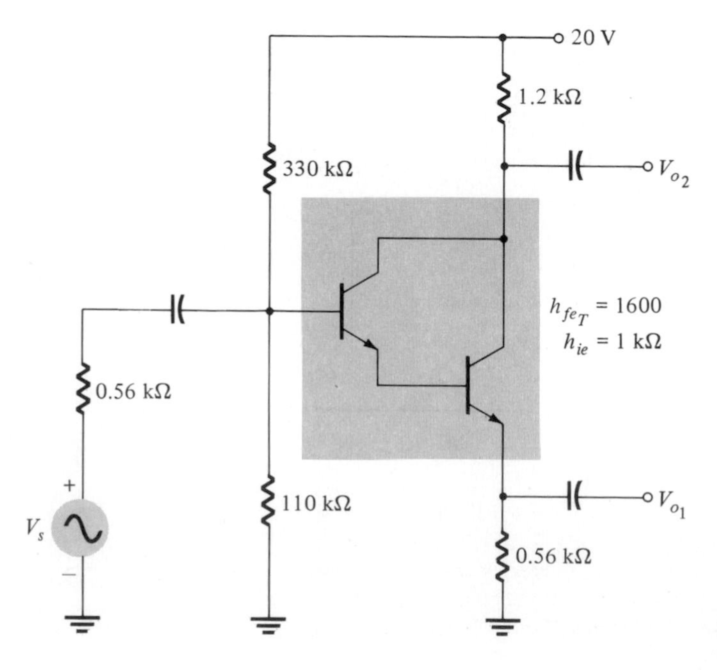

Figure 11.67

(a) Determine $A_{v_1} = V_{o1}/V_i$.
(b) Calculate $A_{v_2} = V_{o2}/V_i$.
(c) Determine $A_{v_s} = V_{o1}/V_s$.
(d) Find A_{v_1} if a 10-kΩ load is connected in parallel with the 0.56-kΩ load.

19. Repeat Problem 17 if R_E is changed to 150 Ω.

§ 11.8

20. Calculate the decibel power gain for the following:
(a) P_o = 100 W, P_i = 5 W.
(b) P_o = 100 mW, P_i = 5 mW.
(c) P_o = 100 μW, P_i = 20 μW.

21. Two voltage measurements made across the same resistance are V_1 = 25 V and V_2 = 100 V. Calculate the decibel power gain of the second reading over the first reading.

22. Input and output voltage measurements of V_i = 10 mV and V_o = 25 V are made. What is the voltage gain in decibels?

23. (a) The total decibel gain of a three-stage system is 120 dB. Determine the decibel gain of each stage if the second stage has twice the decibel gain of the first and the third has 2.7 times the decibel gain of the first.
(b) Determine the voltage gain of each stage.

§ 11.9–11.11

24. For the network of Fig. 11.68:

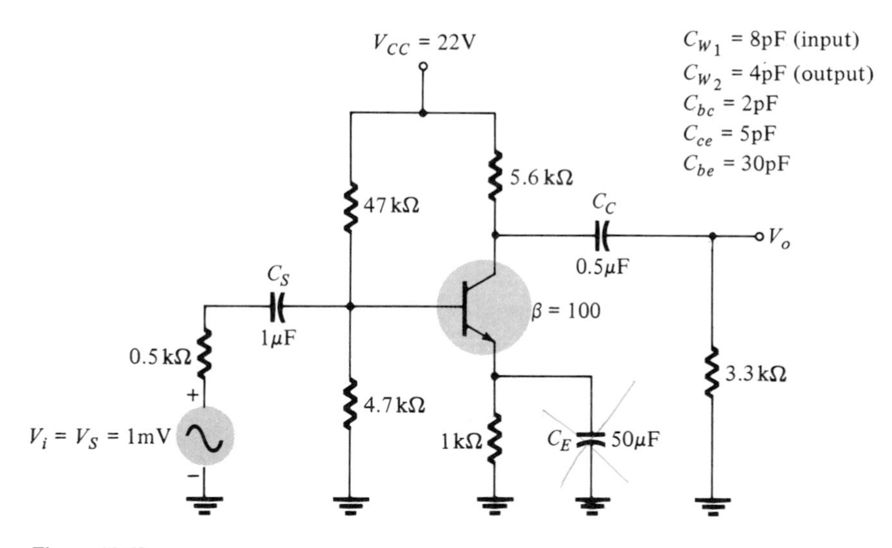

Figure 11.68

(a) Determine the low cutoff frequencies f_{L_S}, f_{L_C}, and f_{L_E}.
(b) Calculate the midband voltage gain.
(c) Determine the high cutoff frequencies f_{H_i} and f_{H_o}.
(d) Sketch a rough plot of $A_v = V_o/V_i$ on a log plane.

25. Repeat Problem 24 if the capacitor C_E is removed.

26. Repeat Problem 25 if C_E is reduced to 1 μF.

§ 11.12

27. Calculate the overall voltage gain of four identical stages of an amplifier, each having a gain of 20.

28. Calculate the overall upper 3-dB frequency for a four-stage amplifier having an individual stage value of $f_2 = 2.5$ MHz.

29. A four-stage amplifier has a lower 3-dB frequency for an individual stage of $f_1 = 40$ Hz. What is the value of f_1 for this full amplifier?

30. (a) Determine the low cutoff frequencies for the two stage amplifier of Fig. 11.69.
(b) Determine the high cutoff frequencies for the network of Fig. 11.69 ($f_\beta = 5$ MHz).
(c) Calculate the midband voltage gain and make a rough sketch of $A_v = V_o/V_i$ versus frequency (log plot).

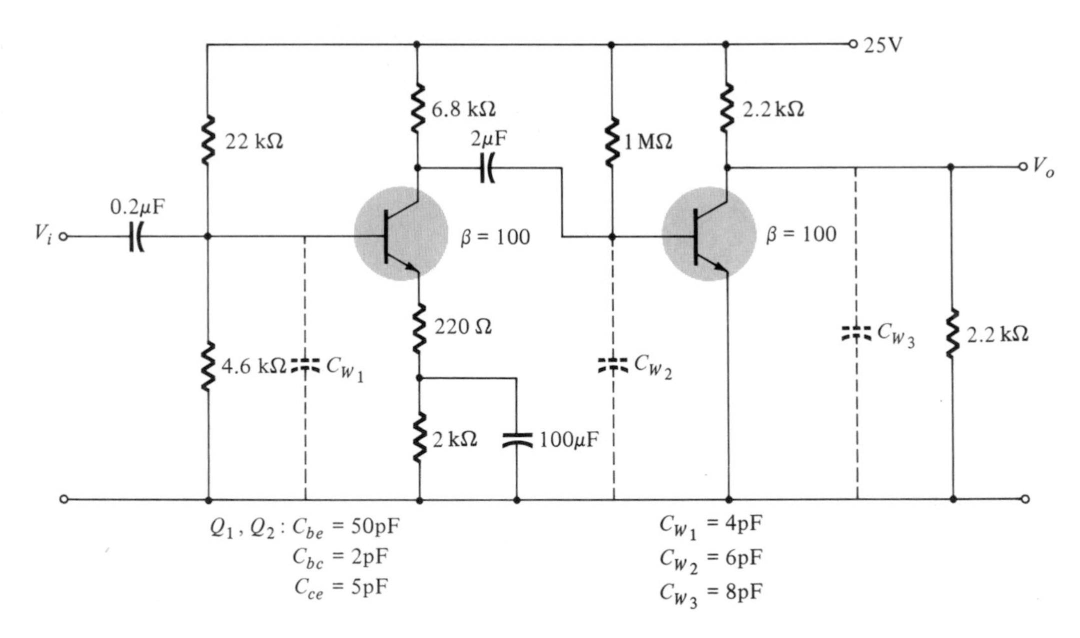

Figure 11.69

§ 11.13

31. Calculate the midfrequency gain for a stage of an FET amplifier as in Fig. 11.60 for circuit values: $g_m = 6000$ μmhos, $r_d = 50$ kΩ (Q_1 and Q_2); $R_{D_1} = R_{D_2} = 10$ kΩ; $R_{G_1} = R_{G_2} = 1$ MΩ; $C_S = 10$ μF, $R_S = 1$ kΩ, $C_C = 0.1$ μF, $C_{gs} = C_{gd} = 4$ pF, $C_W = 7$ pF.

32. For the circuit and values of Problem 31 calculate f_1 and f_2.

33. Calculate the gain-bandwidth product for the circuit and values of Problem 31.

34. For the network of Fig. 11.70:

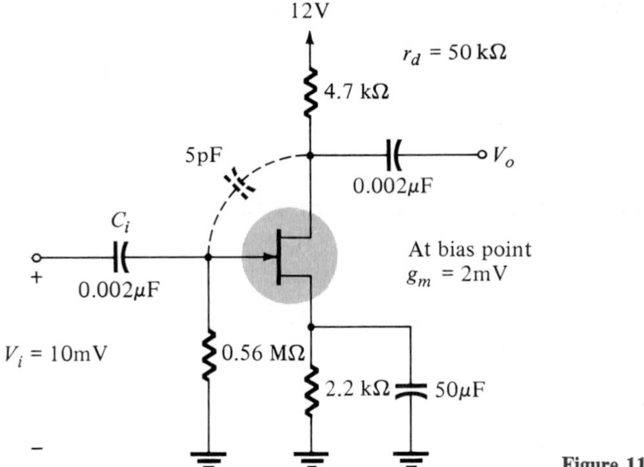

Figure 11.70

(a) Calculate the low and high cutoff frequencies as determined by the provided levels of capacitance. Use $V_{CC} = 20\ V$, $r_B = 0\ \Omega$.
(b) What value of C_i will make the low cutoff frequency 10 Hz?
(c) Determine V_o at the high cutoff frequency.

12

Large-Signal Amplifiers

12.1 INTRODUCTION

An amplifier system consists of a signal pickup transducer, followed by a small-signal amplifier, a large-signal amplifier, and an output transducer device. The input transducer signal is generally small and must be amplified sufficiently to be used to operate some output device. The factors of prime interest in small-signal amplifiers then are usually linearity and gain. Since the signal voltage and current from the input transducer is usually small, the amount of power-handling capacity and power efficiency are of slight concern. Voltage amplifiers provide a large enough voltage signal to the large-signal amplifier stages to operate such output devices as speakers and motors. A large-signal amplifier must operate efficiently and be capable of handling large amounts of power—typically, a few watts to hundreds of watts. This chapter concentrates on the amplifier stage used to handle large signals, typically a few volts to tens of volts. The amplifier factors of greatest concern are the power efficiency of the circuit, the maximum amount of power that the circuit is capable of handling, and impedance matching to the output device.

A class-A series-fed amplifier stage is considered first to show some of the limitations in using such a circuit connection. The single-ended transformer-coupled stage is then discussed to show one method of impedance matching between driver stage and load (output transducer). The push-pull connection, a very popular connection for low distortion and efficient coupling of the signal to a speaker or motor device is discussed. Finally, circuits using complementary transistors for push-pull operation without a transformer are presented.

12.2 SERIES-FED CLASS-A AMPLIFIER

The simple fixed-bias circuit connection can be used as a large-signal class-A amplifier as shown in Fig. 12.1. The only difference between this circuit and the small-signal version considered previously is that the signals handled by the large-signal circuit are in the range of volts and the transistor used is a power transistor capable of operating in the range of a few watts. As will be shown, this circuit is not the best to use for a large-signal amplifier. The dc bias is set by V_{CC} and R_B fixing a dc base-bias current of

$$I_B = \frac{V_{CC} - V_{BE}}{R_B} = \frac{V_{CC} - 0.7\text{ V}}{R_B} \tag{12.1}$$

with the collector current, then

$$I_C = \beta I_B = h_{FE} I_B \tag{12.2}$$

The collector-emitter voltage is

$$V_{CE} = V_{CC} - I_C R_C \tag{12.3}$$

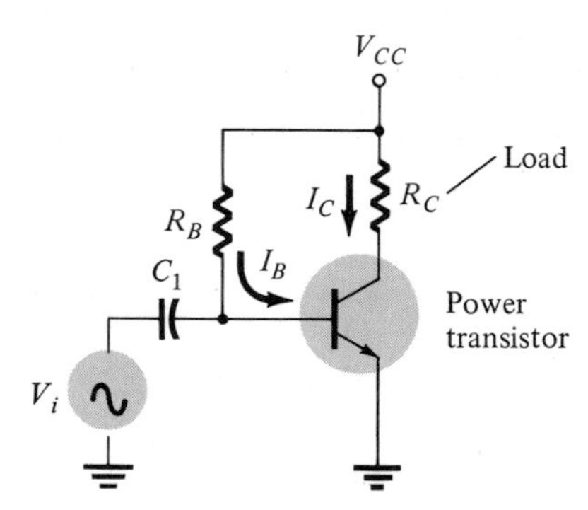

Figure 12.1 Series-fed class-A large-signal amplifier.

To better appreciate the importance of the dc bias on the operation of the amplifier, a collector characteristic is shown in Fig. 12.2. For the circuit values of V_{CC} and R_C, a dc load line may be drawn as shown in Fig. 12.2. The intersection of the bias value

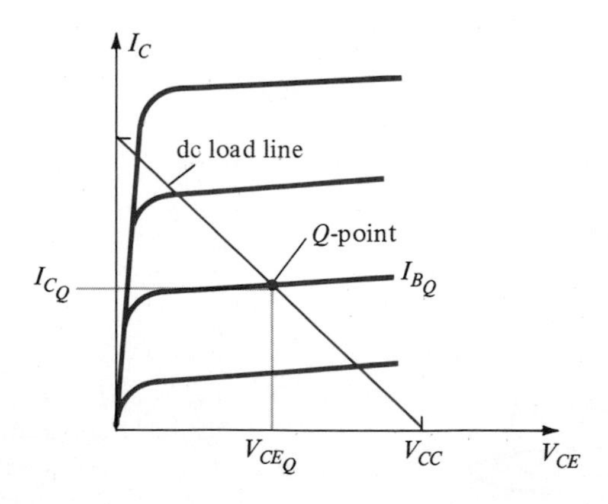

Figure 12.2 Transistor characteristic showing load line and Q-point.

of I_B with the dc load then determines the quiescent operating point (Q-point) for the circuit. The quiescent point values of I_C and V_{CE} are those calculated using Eqs. (12.1) through (12.3). If the dc bias collector current is set at one-half the possible signal swing (between 0 and V_{CC}/R_C), the largest ac collector current will be possible. Additionally, if the quiescent collector-emitter voltage is set at one-half the supply voltage, the largest collector voltage swing will be possible. If the Q-point is set at this optimum bias point, the power considerations for the circuit of Fig. 12.1 are determined as described next.

The input power from the voltage supply is

$$P_i(\text{dc}) = V_{CC} I_{C_Q} \tag{12.4}$$

The output power is that delivered to the load resistor, the collector resistor R_C in the present circuit. The ac signal V_i causes the base current to vary around the dc bias current and the collector current around its quiescent level, I_C. Figure 12.3 shows an ac base current input signal and the resulting ac collector current and collector-emitter voltage signals. If the input signal is small, the output signals only vary slightly around the quiescent point. As the input signal V_i is made larger and the ac base current signal becomes larger, the ac output signal also gets larger, the largest possible signal swings being limited by the values set by the circuit. The power developed across the load (R_C) by the ac signal can be expressed in a number of ways.

$$P_o(\text{ac}) = I_C^2(\text{rms})R_C = \frac{V_{CE}^2(\text{rms})}{R_C} \tag{12.5a}$$

$$= \left(\frac{I_C(\text{peak})}{\sqrt{2}}\right)^2 R_C = \frac{V_{CE}^2(\text{peak})}{2R_C} \tag{12.5b}$$

$$= \frac{I_C^2(\text{p-p})R_C}{8} = \frac{V_{CE}^2(\text{p-p})}{8R_C} \tag{12.5c}$$

It can also be expressed as

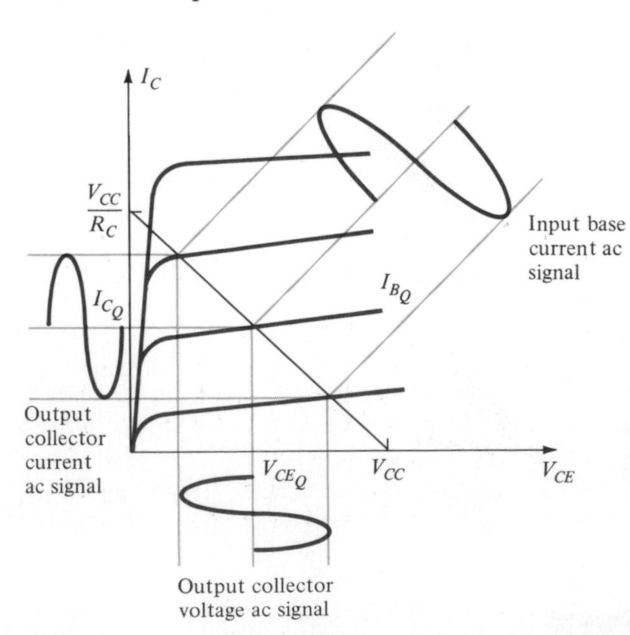

Figure 12.3 Transistor collector characteristic showing dc load line and ac signals.

$$P_o(\text{ac}) = I_C(\text{rms})V_{CE}(\text{rms}) \tag{12.6a}$$

$$= \frac{I_C(\text{peak})}{\sqrt{2}} \frac{V_{CE}(\text{peak})}{\sqrt{2}}$$

$$= \frac{I_C(\text{peak})V_{CE}(\text{peak})}{2} \tag{12.6b}$$

$$= \frac{I_C(\text{p-p})}{2\sqrt{2}} \frac{V_{CE}(\text{p-p})}{2\sqrt{2}}$$

$$\boxed{P_o(\text{ac}) = \frac{I_C(\text{p-p})V_{CE}(\text{p-p})}{8}} \tag{12.6c}$$

Any form of Eqs. (12.5) and (12.6) can be used—the resulting values being the same.

The efficiency of the amplifier is then calculated to be

$$\boxed{\%\eta = \frac{P_o(\text{ac})}{P_i(\text{dc})} \times 100\%} \tag{12.6c}$$

The efficiency is important not only because it indicates how much of the power drawn from the supply voltage gets to the load as ac signal but because it also indicates how much does not get to the load and must be dissipated as heat, primarily by the transistor. Assuming that all the power not delivered to the load must be dissipated by the transistor,

$$\boxed{P_{\text{transistor}} = P_Q = P_i(\text{dc}) - P_o(\text{ac})} \tag{12.8}$$

Maximum Power and Efficiency

If the Q-point is set at the midpoint of the maximum signal swing, the resulting maximum power condition may be achieved. For the circuit of Fig. 12.1 this would be determined using

$$\text{maximum } V_{CE}(\text{p-p}) = V_{CC}$$

$$\text{maximum } I_C(\text{p-p}) = \frac{V_{CC}}{R_C}$$

Using Eq. (12.6c), we have

$$\text{maximum } P_o(\text{ac}) = \frac{I_C(\text{p-p})V_{CE}(\text{p-p})}{8} = \frac{(V_{CC}/R_C)V_{CC}}{8} = \frac{V_{CC}^2}{8R_C}$$

For the quiescent point of

$$I_C = \frac{V_{CC}/R_C}{2}$$

the dc power from the supply voltage is

$$\text{maximum } P_i(\text{dc}) = V_{CC}I_C = V_{CC}\left(\frac{V_{CC}}{2R_C}\right) = \frac{V_{CC}^2}{2R_C}$$

The efficiency of an amplifier given by Eq. (12.7) can be calculated for the maximum conditions as

$$\text{maximum } \%\eta = \frac{\text{max } P_o(\text{ac})}{\text{max } P_i(\text{dc})} \times 100\%$$

$$= \frac{V_{CC}^2/8R_C}{V_{CC}^2/2R_C} \times 100\% = 25\%$$

This is the maximum percent efficiency for a series-fed class-A amplifier. Since this maximum efficiency will occur only under ideal conditions and for the maximum ac signal swings, most series-fed class-A amplifiers provide efficiencies much less than 25%.

EXAMPLE 12.1

Calculate the input power, output power, and efficiency of the amplifier in Fig. 12.4 for an input voltage resulting in a base current of 10 mA peak. Also calculate the power dissipated by the transistor.

Solution:

The Q-point for the circuit of Fig. 12.4a is determined as follows:

$$I_B = \frac{V_{CC} - 0.7\text{ V}}{R_B} = \frac{20\text{ V} - 0.7\text{ V}}{1\text{ k}\Omega} = 19.3\text{ mA}$$

$$I_C = \beta I_B = 25(19.3\text{ mA}) = 482.5\text{ mA}$$

$$V_{CE} = V_{CC} - I_C R_C = 20\text{ V} - (482.5\text{ mA})(20\Omega) = 10.35\text{ V}$$

This point is marked on the transistor characteristic of Fig. 12.4b. The ac variation of the output signal can be obtained graphically using the dc load line drawn on Fig. 12.4b by connecting $V_{CE} = V_{CC} = 20$ V with $I_C = V_{CC}/R_C = 1000$ mA, as shown. When the input ac base current increases the base current from its bias level, the collector current rises by

$$I_C(\text{peak}) = \beta I_B(\text{peak}) = 25(10\text{ mA peak}) = 250\text{ mA peak}$$

Using Eq. (12.5b) yields

$$P_o(\text{ac}) = \left(\frac{I_C(\text{peak})}{\sqrt{2}}\right)^2 R_C = \left(\frac{250 \times 10^{-3}}{\sqrt{2}}\right)^2(20) = 0.625\text{ W}$$

Using Eq. (12.4) gives us

$$P_i(\text{dc}) = V_{CC}I_C = (20\text{ V})(482.5 \times 10^{-3}) = \mathbf{9.65\text{ W}}$$

The amplifier's power efficiency is then calculated using Eq. (12.7):

$$\%\eta = \frac{P_o(\text{ac})}{P_i(\text{dc})} \times 100\% = \frac{0.625\text{ W}}{9.65\text{ W}} \times 100\% = \mathbf{6.48\%}$$

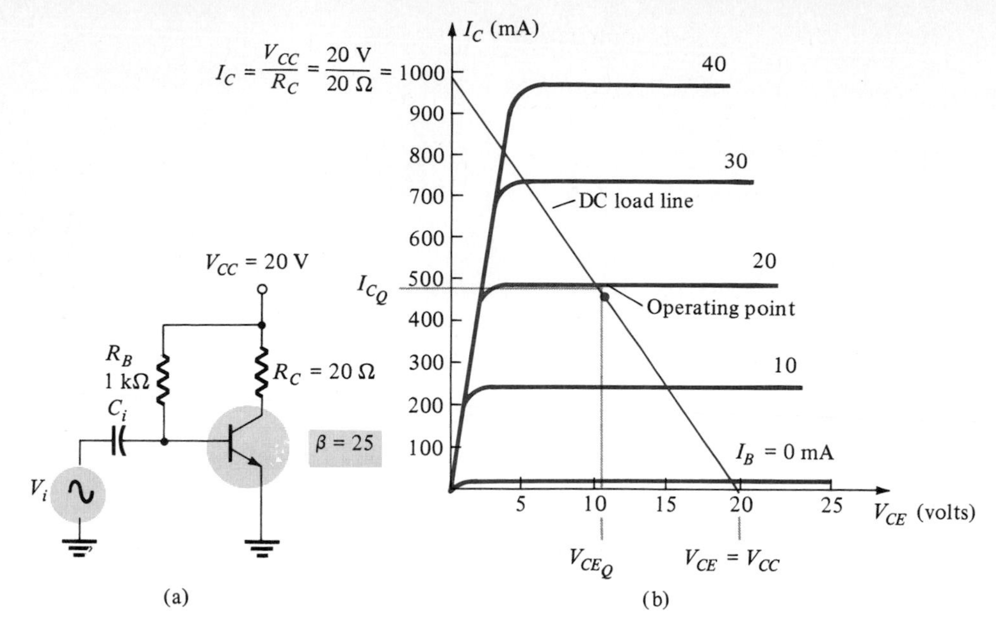

Figure 12.4 Operation of a series-fed circuit for Example 12.1.

The power dissipated by the transistor is then

$$P_Q = P_i - P_o = 9.65\text{ W} - 0.625\text{ W} = \mathbf{9.025\ W}$$

Example 12.1 shows the poor operation of a class-A series-fed circuit. While the signal swing was not the maximum possible, it still covered a good portion of the range around the operating point. Even so, the efficiency of the circuit was only 6.48%, well below the ideal maximum of 25%. More important, the transistor needed to dissipate 9.025 W to deliver only 0.625 W of signal power to the load. This circuit is obviously not the best to use.

12.3 TRANSFORMER-COUPLED POWER AMPLIFIER

A more reasonable class-A amplifier connection uses a transformer to couple the load to the amplifier stage as shown in Fig. 12.5a. This is a simple version of the circuit for the presentation of a few basic concepts. More practical circuit versions will be covered shortly. Figure 12.5b shows the output coupling transformer with voltage, current, and impedances indicated.

Transformer Impedance Matching

The resistance seen looking into the primary of the transformer is related to the resistance connected across the secondary. The ratio of secondary resistance to pri-

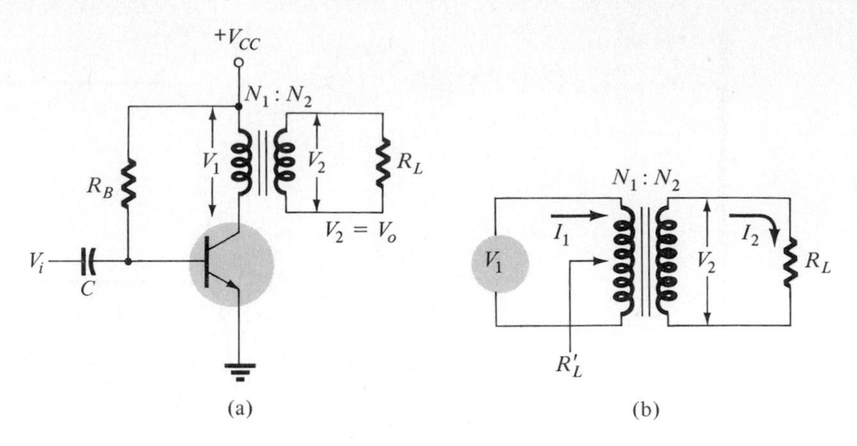

Figure 12.5 Transformer-coupled audio power amplifier.

mary resistance may be expressed as follows:

$$\frac{R_L'}{R_L} = \frac{V_1/I_1}{V_2/I_2} = \frac{V_1}{I_1}\frac{I_2}{V_2} = \frac{V_1}{V_2}\frac{I_2}{I_1} = \frac{N_1}{N_2}\frac{N_1}{N_2} = \left(\frac{N_1}{N_2}\right)^2$$

where $V_1/V_2 = N_1/N_2$ and $I_2/I_1 = N_1/N_2$. Hence the ratio of the transformer input and output resistance varies directly as the *square* of the transformer turns ratio:

$$\boxed{\frac{R_L'}{R_L} = \left(\frac{N_1}{N_2}\right)^2 = a^2} \tag{12.9}$$

and

$$\boxed{R_L' = a^2 R_L = \left(\frac{N_1}{N_2}\right)^2 R_L} \tag{12.10}$$

where R_L = resistance of load connected across the transformer secondary
R_L' = effective resistance seen looking into primary of transformer
a = N_1/N_2 is the step-down turns ratio needed to make the load resistance appear as a larger effective resistance seen from the transformer primary

EXAMPLE 12.2

Calculate the effective resistance (R_L') seen looking into the primary of a 15:1 transformer connected to an output load of 8 Ω.

Solution:

Using Eq. (12.10), we get

$$R_L' = \left(\frac{N_1}{N_2}\right)^2 R_L = (15)^2 8 = 1800\ \Omega = \mathbf{1.8\ k\Omega}$$

EXAMPLE 12.3

What transformer turns ratio is required to match a 16-Ω speaker load to an amplifier so that the effective load resistance is 10 kΩ?

Solution:

Using Eq. (12.9), we get

$$\left(\frac{N_1}{N_2}\right)^2 = \frac{R_L'}{R_L} = \frac{10{,}000}{16} = 625$$

$$\frac{N_1}{N_2} = \sqrt{625} = \mathbf{25:1}$$

DC Load Line

The transformer dc (winding) resistance determines the dc load line for the circuit. Typically, this dc resistance is small and is shown in Fig. 12.5a to be 0 Ω, providing a straight (vertical) load line. This is the ideal load line for the transformer. Practical transformer windings would provide a slight slope for the load line, but only the ideal case will be considered in this discussion. There is no dc voltage drop across the dc load resistance in the ideal case and the load line is drawn straight vertically from the voltage point, $V_{CE_Q} = V_{CC}$.

Quiescent Operating Point

The operating point is obtained graphically as the point of intersection of the dc load line and the transistor base current curve. From the operating point the quiescent collector current I_{C_Q} is read. The value of the base current is calculated separately from the circuit as considered in the dc bias calculations of Chapter 5.

AC Load Line

In order to obtain the ac signal operation it is necessary to first calculate the ac load resistance seen looking into the primary side of the transformer and then to draw the ac load line on the transistor characteristic. The effective load resistance is calculated using Eq. (12.10) from the values of secondary load resistance and transformer turns ratio. Having obtained the value of R_L' the ac load line must be drawn so that it passes through the operating point and has a slope equal to $-1/R_L'$, the load-line slope being the negative reciprocal of the ac load resistance. Since the collector signal passes through the operating point when no signal is applied, the load line must pass through the operating point.

In order to simplify drawing a load line of slope $-1/R_L'$ through the operating point the following technique may be used (see Fig. 12.6a):

If the ac signal were to vary from the quiescent level to 0 V, it would also vary from the quiescent current level, I_{C_Q}, by an amount

$$\Delta I_C = \frac{\Delta V_{CE}}{R_L'} \tag{12.11}$$

Mark a point on the y-axis of the transistor characteristic ΔI_C units above the quiescent level and connect this point through the operating point to draw the ac load line desired. Notice that the ac load line shows that the output signal swing can exceed the value of V_{CC}, the supply voltage. In fact, the voltage developed across the transformer

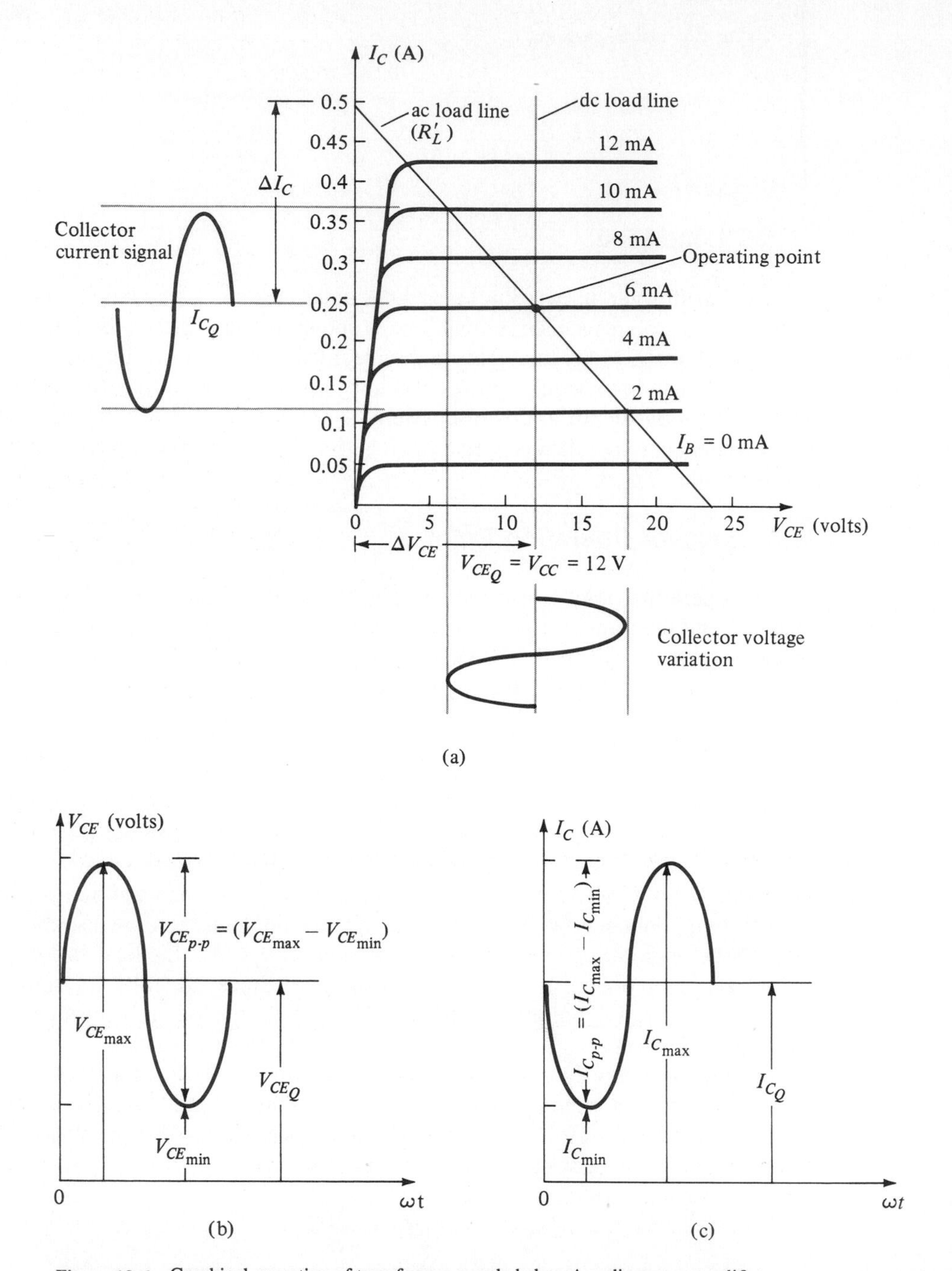

Figure 12.6 Graphical operation of transformer-coupled class-A audio power amplifier.

primary can be large. One of the maximum operating values that will have to be checked carefully is that of $V_{CE_{max}}$, as specified by the transistor manufacturer, to see that the value of maximum voltage obtained after drawing the ac load line does not exceed the transistor rated maximum value. Assuming that maximum power and voltage ratings are not exceeded, the ac signal swings of current and voltage are obtained as shown in Fig. 12.6a and redrawn in detail in Fig. 12.6b and 12.6c.

Signal Swing and Output AC Power

From the signal variations shown in Figs. 12.6b and 12.6c the values of the peak-to-peak signal swings are

$$V_{swing} = V_{CE}(\text{p-p}) = (V_{CE_{max}} - V_{CE_{min}}) \tag{12.12}$$

$$I_{swing} = I_C(\text{p-p}) = (I_{C_{max}} - I_{C_{min}}) \tag{12.13}$$

The ac power developed across the transformer primary can be calculated to be

$$\begin{aligned} P_o(\text{ac}) &= V_{CE}\ (\text{rms})\ I_C\ (\text{rms}) \\ &= \frac{V_{CE}\ (\text{peak})}{\sqrt{2}} \frac{I_C\ (\text{peak})}{\sqrt{2}} \\ &= \frac{V_{CE}\ (\text{p-p})/2}{\sqrt{2}} \times \frac{I_C\ (\text{p-p})/2}{\sqrt{2}} \end{aligned}$$

$$\boxed{P_o(\text{ac}) = \frac{(V_{CE_{max}} - V_{CE_{min}})(I_{C_{max}} - I_{C_{min}})}{8}} \tag{12.14}$$

The ac power calculated is that developed across the primary of the transformer. Assuming a highly efficient transformer the power across the speaker is approximately equal to that calculated by Eq. (12.14). For our purposes an ideal transformer will be assumed so that the ac power calculated using Eq. (12.14) is also the ac power delivered to the load.

For the ideal transformer considered, the voltage across the secondary of the transformer can be calculated from

$$V_2 = V_L = \left(\frac{N_2}{N_1}\right) V_1 \tag{12.15}$$

where the secondary voltage (V_2) equals the speaker or load voltage (V_L). The load voltage is related by the transformer turns ratio, N_2/N_1, to the voltage developed across the transformer primary (V_1). The voltage across the primary was previously labelled V_{CE} (rms), and for power considerations the rms values of voltage are usually used [unless otherwise stated, as in Eq. (12.14)].

The power across the load can be expressed as

$$P_L = \frac{V_L^2\ (\text{rms})}{R_L} \tag{12.16}$$

and equals the power calculated using Eq. (12.14). Thus the ac power can be calculated in a number of ways, including the following:

$$I_L(\text{rms}) = \frac{N_1}{N_2} I_C(\text{rms}) \tag{12.17}$$

where I_L is the rms value of the current through the load resistor (or speaker resistance) and the load current is related to the rms value of the ac component of collector current by the transformer turns ratio.

The ac power is then calculated from

$$P_L = I_L^2(\text{rms})\, R_L \tag{12.18}$$

EXAMPLE 12.4

The circuit of Fig. 12.7a shows a transformer-coupled class-A audio power amplifier driving an 8-Ω speaker. The coupling transformer has a 3:1 step-down turns ratio. If the circuit component values result in a dc base current of 6 mA and the input signal (V_i) results in a peak base current swing of 4 mA, calculate the following circuit values using the transistor characteristic of Fig. 12.7b: $V_{CE_{\max}}$, $V_{CE_{\min}}$, $I_{C_{\max}}$, $I_{C_{\min}}$, the rms values of load current and voltage, and the ac power developed across the load. Calculate the ac power using different equations as a check, that is, Eq. (12.14), Eq. (12.16), and Eq. (12.18).

Solution:

1. The dc load line can be drawn vertically from the voltage point $V_{CE_Q} = V_{CC} = 10$ V (see Fig. 12.7c).
2. For $I_B = 6$ mA the operating point on Fig. 12.7c is

$$V_{CE_Q} = 10 \text{ V} \quad \text{and} \quad I_{C_Q} = 140 \text{ mA}$$

3. The effective ac resistance R_L' is [use Eq. (12.10)]

$$R_L' = \left(\frac{N_1}{N_2}\right)^2 R_L = (3)^2 8 = 72\ \Omega.$$

4. Draw the ac load line as follows: Use Eq. (12.11) to calculate the current swing above the operating current.

$$\Delta I_C = \frac{\Delta V_{CE}}{R_L'} = \frac{10 \text{ V}}{72\ \Omega} = 139 \text{ mA}$$

Mark point A (Fig. 12.7c) $= I_{CE_Q} + \Delta I_C = 140 + 139 = 279$ mA along the y-axis. Connect point A to Q-point to draw ac load line.

5. For the given peak base current swing of 4 mA, the maximum and minimum values of collector current and voltage obtained from Fig. 12.7c are

$$V_{CE_{\min}} = 1.7 \text{ V} \qquad I_{C\min} = 25 \text{ mA}$$

$$V_{CE_{\max}} = 18.3 \text{ V} \qquad I_{C\max} = 255 \text{ mA}$$

6. Calculate the ac power across the transformer primary using Eq. (12.14).

$$P_o\,(\text{ac}) = \frac{(V_{CE_{\max}} - V_{CE_{\min}})(I_{C_{\max}} - I_{C_{\min}})}{8}$$

$$= \frac{(18.3 - 1.7)(255 - 25) \times 10^{-3}}{8} = \mathbf{0.477\ W}$$

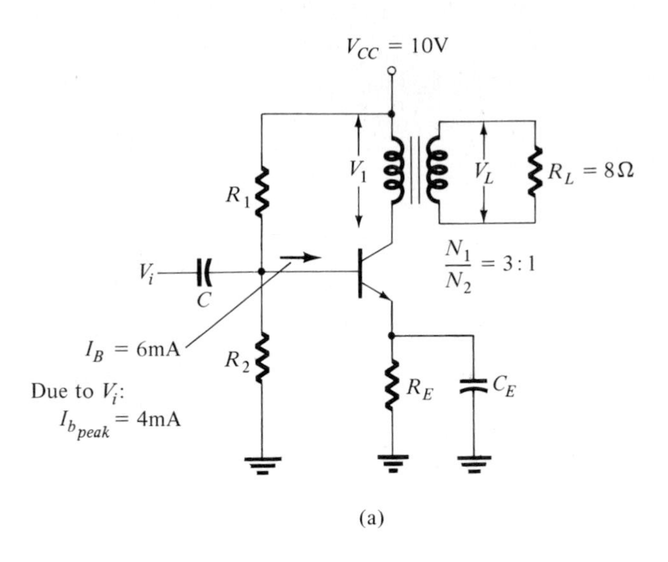

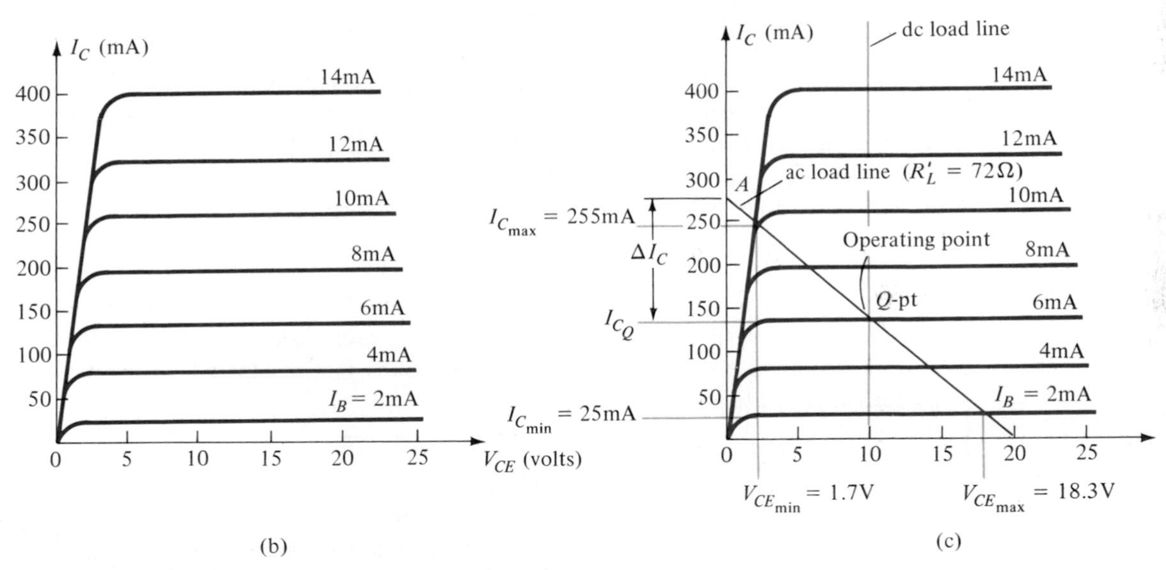

Figure 12.7 Transformer-coupled audio power amplifier and transistor characteristic for Example 12.4.

7. Calculate the rms voltage across the primary.

$$V_1\text{ (rms)} = \frac{V_1\text{(p-p)}}{2\sqrt{2}} = \frac{V_{CE_{\max}} - V_{CE_{\min}}}{2\sqrt{2}}$$

$$= \frac{16.6}{2.828} = 5.87\text{ V}$$

8. The rms value of the load voltage is [using Eq. (12.15)]

$$V_L\text{ (rms)} = \frac{N_2}{N_1}V_1\text{ (rms)} = \left(\frac{1}{3}\right)(5.87) = 1.96\text{ V}$$

9. Using Eq. (12.16) to calculate the ac power, we get

$$P_L\text{ (ac)} = \frac{V_L^2}{R_L} = \frac{(1.96)^2}{8} = \mathbf{0.480\ W}$$

10. Using Eq. (12.17) to calculate the rms component of the load current, we get

$$I_L\text{ (rms)} = \frac{N_1}{N_2}I_C\text{(rms)} = \frac{N_1}{N_2}\left[\frac{I_{C_{\max}} - I_{C_{\min}}}{2\sqrt{2}}\right] = (3)\,\frac{230\text{ mA}}{2.828} = 244\text{ mA}$$

11. Calculating the ac power using Eq. (12.18), we get

$$P_L\text{ (ac)} = I_L^2R_L = (244 \times 10^{-3})^2 8 = \mathbf{0.476\ W}$$

Power and Efficiency Calculations

So far we have considered calculating the ac power delivered to the load (the output ac power). We next consider the input power from the battery, power losses in the amplifier, and the overall power efficiency of the transformer-coupled class-A amplifier. The input dc power obtained from the battery is calculated from the values of dc battery voltage and average current from the battery as in Eq. (12.4).

$$P_i\text{ (dc)} = V_{CC}I_{C_Q}$$

For the transformer-coupled amplifier, as shown in Fig. 12.5, the power dissipated by the transformer is small and will be ignored in the present calculations. Thus for the transformer-coupled amplifier the only power lost is that dissipated by the power transistor as calculated by the following equation

$$\boxed{P_Q = P_i\text{ (dc)} - P_o\text{ (ac)}} \qquad (12.19)$$

when P_Q is the power dissipated as heat. The equation seems simple but is significant in operating a power amplifier. The amount of power dissipated by the transistor (which then sets the transistor power rating) is the difference between the average dc input power from the battery (which is a constant for a fixed battery and operating point) and the output ac power drawn by the load. If the output power is zero, then the transistor must handle the maximum amount, that set by the battery voltage and bias current for class-A operation. If the load does draw some of the power, then the transistor has to handle that much less (for the moment). In other words, the transistor has to work hardest (dissipate the most power) when the load is disconnected from the amplifier circuit in a class-A amplifier and the transistor dissipates least power when the load is drawing maximum power from the circuit. Obviously, the safest rating of the transistor used in a class-A amplifier is the maximum set when the load is disconnected. Since normal operation with the load connected requires the transistor to dissipate less power, it is always preferable to keep the load connected as long as the class-A amplifier unit is turned on.

EXAMPLE 12.5

Calculate the efficiency of the amplifier circuit of Example 12.4. Also calculate the power dissipated by the transistor.

Solution:

Using Eq. (12.4) to calculate the input power, we get

$$P_i\,(\text{dc}) = V_{CC}I_{C_Q} = (10)(140 \times 10^{-3}) = 1.4\text{ W}$$

Using Eq. (12.19), we see that the power dissipated by the transistor is

$$P_Q = P_i\,(\text{dc}) - P_o\,(\text{ac}) = 1.4\text{ W} - 0.48\text{ W} = \mathbf{0.92\ W}$$

The efficiency is

$$\eta = \frac{P_o\,(\text{ac})}{P_i\,(\text{dc})} \times 100 = \frac{0.48\text{ W}}{1.4\text{ W}} \times 100 = \mathbf{34.3\%}$$

Maximum Theoretical Efficiency

For a class-A amplifier the maximum theoretical efficiency for the series-fed circuit is 25% and for the transformer-coupled circuit it is 50%. From analysis of the operating range for a series-fed amplifier circuit the efficiency can be stated in the following form:

$$\boxed{\eta = 25\left[\frac{(V_{CE_{\max}} - V_{CE_{\min}})^2}{V_{CC}(V_{CE_{\max}} + V_{CE_{\min}})}\right]\ \%} \tag{12.20}$$

In practical operation the efficiency is less than 25%. In the circuit of Example 12.1 the efficiency was only 6.48%, indicating a poorly operating series-fed circuit.

The efficiency of a transformer-coupled class-A amplifier can be expressed by

$$\boxed{\eta = 50\left(\frac{V_{CE_{\max}} - V_{CE_{\min}}}{V_{CE_{\max}} + V_{CE_{\min}}}\right)^2\ \%} \tag{12.21}$$

The larger the value of $V_{CE_{\max}}$ and the smaller the value of $V_{CE_{\min}}$ the closer the efficiency approaches the theoretical limit of 50%. In the circuit of Example 12.4 the value obtained was 34.3%. Well-designed circuits can approach the limit of 50% closely so that the circuit of Fig. 12.7a would be considered average in operation. The larger the amount of power handled by the amplifier the more critical the efficiency becomes. For a few watts of power a simpler, cheaper circuit with less than maximum efficiency is acceptable (and sometimes desirable). For power levels in the tens to hundreds of watts, efficiency as close as possible to the theoretical maximum would be desired.

The value of 50% considered as the maximum for transformer-coupled amplifiers is only for class-A operation. There are additional classes for operating (biasing) the amplifier to obtain even higher efficiency, as will be considered.

EXAMPLE 12.6

Calculate the efficiency of a series-fed class-A amplifier as that of Fig. 12.1 for a supply voltage of $V_{CC} = 24$ V and output of
(a) $V_{peak} = 12$ V around a bias of $V_{CEQ} = 12$ V.
(b) $V_{peak} = 6$ V around a bias of $V_{CEQ} = 12$ V.
(c) $V_{peak} = 6$ V around a bias of $V_{CEQ} = 18$ V.

Solution:

Using Eq. (12.20), we obtain
(a) $V_{CE\,max} = V_{CE_Q} + V_{peak} = 12\text{ V} + 12\text{ V} = 24\text{ V}$
$V_{CE\,min} = V_{CE_Q} - V_{peak} = 12\text{ V} - 12\text{ V} = 0\text{ V}$

$$\eta = 25\left[\frac{(V_{CE\,max} - V_{CE\,min})^2}{V_{CC}(V_{CE\,max} + V_{CE\,min})}\right]\% = 25\left[\frac{(24\text{ V} - 0\text{ V})^2}{24\text{ V}(24\text{ V} + 0\text{ V})}\right]\% = \mathbf{25\ \%}$$

(b) $V_{CE\,max} = V_{CE_Q} + V_{peak} = 12\text{ V} + 6\text{ V} = 18\text{ V}$
$V_{CE\,min} = V_{CE_Q} - V_{peak} = 12\text{ V} - 6\text{ V} = 6\text{ V}$

$$\eta = 25\left[\frac{(18\text{ V} - 6\text{ V})^2}{24\text{ V}(18\text{ V} + 6\text{ V})}\right]\% = \mathbf{6.25\%}$$

(c) $V_{CE\,max} = V_{CE_Q} + V_{peak} = 18\text{ V} + 6\text{ V} = 24\text{ V}$
$V_{CE\,min} = V_{CE_Q} - V_{peak} = 18\text{ V} - 6\text{ V} = 12\text{ V}$

$$\eta = 25\left[\frac{(24\text{ V} - 12\text{ V})^2}{24\text{ V}(24\text{ V} + 12\text{ V})}\right]\% = \mathbf{4.17\%}$$

With bias at half the supply voltage the efficiency with maximum signal swing in part (a) is 25%. Reducing the signal swing to 6 V peak, around the same dc bias point, results in an efficiency of only 6.25% in part (b). The same 6 V signal swing around a bias point off the midpoint (12 V) is reduced to 4.17% in part (c).

EXAMPLE 12.7

Calculate the efficiency of a transformer-coupled class-A amplifier such as that of Fig. 12.5 for a supply voltage of $V_{CC} = 12$ V and outputs of (a) $V_{peak} = 12$ V; (b) $V_{peak} = 6$ V; (c) $V_{peak} = 2$ V.

Solution:

For a transformer-coupled class-A amplifier, $V_{CE_Q} = V_{CC} = 12$ V, using Eq. (12.21):
(a) $V_{CE\,max} = V_{CE_Q} + V_{peak} = 12\text{ V} + 12\text{ V} = 24\text{ V}$
$V_{CE\,min} = V_{CE_Q} - V_{peak} = 12\text{ V} - 12\text{ V} = 0\text{ V}$

$$\eta = 50\left(\frac{V_{CE\,max} - V_{CE\,min}}{V_{CE\,max} + V_{CE\,min}}\right)^2\% = 50\left(\frac{24\text{ V} - 0\text{ V}}{24\text{ V} + 0\text{ V}}\right)^2\% = \mathbf{50\%}$$

(b) $V_{CE\,max} = 12\text{ V} + 6\text{ V} = 18\text{ V}$
$V_{CE\,min} = 12\text{ V} - 6\text{ V} = 6\text{ V}$

$$\eta = 50\left(\frac{18\text{ V} - 6\text{ V}}{18\text{ V} + 6\text{ V}}\right)^2 \% = \mathbf{12.5\%}$$

(c) $V_{CE_{\max}} = 12 + 2 = 14\text{ V}$
$V_{CE_{\min}} = 12 - 2 = 10\text{ V}$

$$\eta = 50\left(\frac{14\text{ V} - 10\text{ V}}{14\text{ V} + 10\text{ V}}\right)^2 \% = \mathbf{1.39\%}$$

Notice how dramatically the amplifier efficiency drops from a maximum of 50% for $V_{\text{peak}} = V_{CC}$ to slightly over 1% for $V_{\text{peak}} = \frac{1}{6}V_{CC} = 2\text{ V}$.

12.4 CLASS-B AMPLIFIER OPERATION

Class-B operation is provided when the dc bias leaves the transistor biased just off, the transistor turning on when the ac signal is present. This is essentially *no* bias and the single transistor can conduct current for only one-half of the signal cycle. To obtain a desired output for the full signal cycle it is necessary to use two transistors and have *each* conduct on opposite half-cycles, the combined operation providing a full cycle of output signal. Since one part of the circuit *pushes* the signal high during one half-cycle and the other part *pulls* the signal low during the other half-cycle the circuits which operate class-B are also referred to as *push-pull* circuits. Figure 12.8 shows a diagram for push-pull operation. An input ac signal is applied to the push-pull circuit. Each half of the circuit operates on alternate half-cycles, the load then receiving a

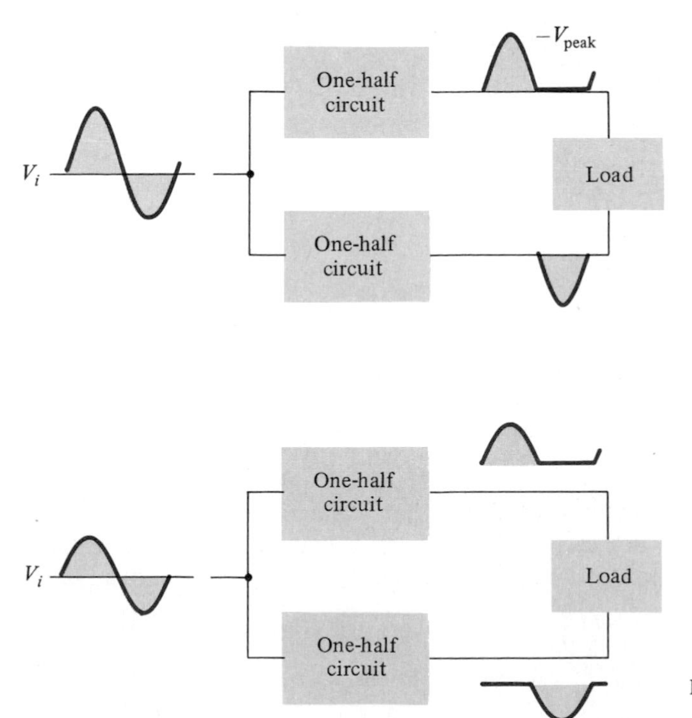

Figure 12.8 Block representation of push-pull operation.

signal for the full ac cycle. The power transistors used in the push-pull circuit are capable of delivering the desired power to the load, and the class-B operation of these transistors provide greater efficiency than was possible using a single transistor in class-A operation.

Power and Efficiency Calculations in Class-B Amplifiers

The power and efficiency calculations of a variety of class-B power amplifiers should help in understanding how these circuits operate and provide some comparison between important circuit values.

INPUT DC POWER

The power provided to the speaker of a power amplifier circuit is drawn from the power supply, (or power supplies; see Fig. 12.9), and is considered an input or dc power. The amount of this input power can be calculated from

$$P_i(\text{dc}) = V_{CC}I_{\text{dc}} \tag{12.22}$$

where I_{dc} is the average or dc current drawn from the power supplies.

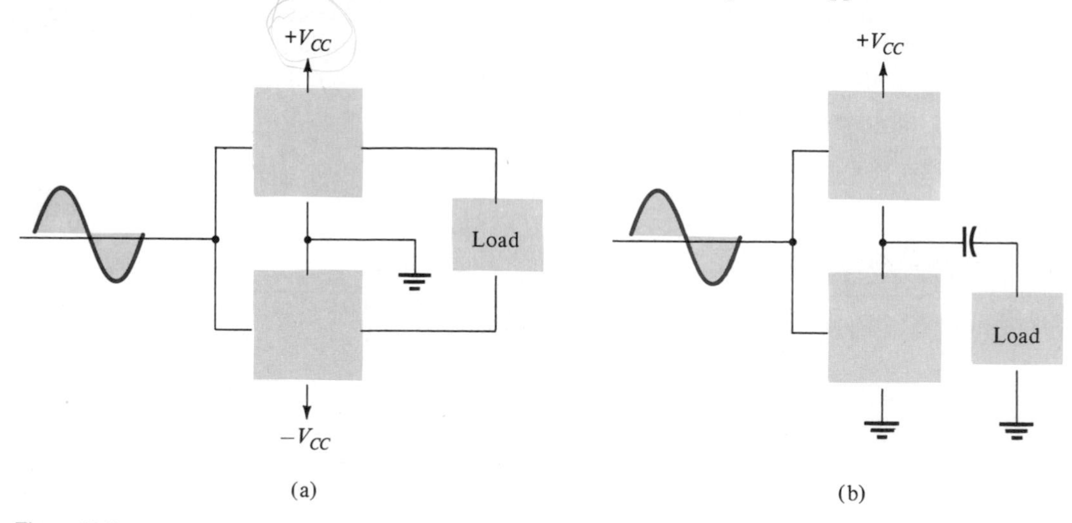

Figure 12.9 Connection of push-pull amplifier to load: (a) using two voltage supplies; (b) using one voltage supply.

In class-B operation the current drawn from a single power supply is a full-wave rectified signal, while that drawn from a circuit having two power supplies is a half-wave rectified waveform from each supply. In either case the value of average power can be expressed as

$$\boxed{I_{\text{dc}} = \frac{2}{\pi}I_{\text{peak}}} \tag{12.23}$$

where I_{peak} is the peak value of the output current waveform.

OUTPUT AC POWER

The power delivered to the load (usually referred to as a resistance, R_L) can be calculated from any one of a number of equal relations.

$$P_o\text{ (ac)} = \frac{V_L^2(p-p)}{8R_L} = \frac{V_L^2(p)}{2R_L} = \frac{V_L^2\text{ (rms)}}{R_L} \tag{12.24}$$

EFFICIENCY

The circuit's power efficiency is then calculated as

$$\eta = \frac{P_o}{P_i} \times 100\% \tag{12.25}$$

POWER DISSIPATED BY OUTPUT TRANSISTORS

The power dissipated (as heat) by the output power transistors is the difference between the input power from the supplies and the output power delivered to the load.

$$P_{2Q} = P_i - P_o \tag{12.26}$$

where P_{2Q} is the power dissipated by the *two* output power transistors. The power dissipated by each transistor is then

$$P_Q = \frac{P_{2Q}}{2} \tag{12.27}$$

EXAMPLE 12.8

Determine the input power, output power, and efficiency resulting in a class-B amplifier providing a signal of 20 V, peak to a 16-Ω load, using a single supply of V_{CC} = 30 V.

Solution:

For a 20-V, peak signal across a 16-Ω load,

$$I_{\text{peak}} = \frac{V_{\text{peak}}}{R_L} = \frac{20\text{ V}}{16\ \Omega} = 1.25\text{ A}$$

The dc value of current drawn from the voltage supply is

$$I_{\text{dc}} = \frac{2}{\pi} I_{\text{peak}} = \frac{2}{\pi}(1.25\text{ A}) = 0.796\text{ A}$$

and the input power from the voltage supply is

$$P_i\text{ (dc)} = V_{CC} I_{dc} = (30\text{ V})(0.796\text{ A}) = \mathbf{23.9\ W}$$

The output power delivered to the load is

$$P_o\text{ (ac)} = \frac{V_L^2(p)}{2R_L} = \frac{(20\text{ V})^2}{2(16\ \Omega)} = \mathbf{12.5\ W}$$

for a resulting circuit efficiency of

$$\%\eta = \frac{P_o\ (\text{ac})}{P_i\ (\text{dc})} \times 100\% = \frac{12.5\ \text{W}}{23.9\ \text{W}} \times 100\% = \mathbf{52.3\%}$$

Maximum-Power Considerations

For class-B operation the maximum output power to the load resulting when $V_L(p) = V_{CC}$ is

$$\text{maximum } P_o\ (\text{ac}) = \frac{V_{CC}^2}{2R_L} \tag{12.28}$$

A corresponding ac current signal developed through the load then varies to a peak value of

$$I_{\text{peak}} = \frac{V_{CC}}{R_L}$$

so that the average current from the power supply is

$$I_{\text{dc}} = \frac{2}{\pi} I_{\text{peak}} = \frac{2}{\pi}\frac{V_{CC}}{R_L} \tag{12.29}$$

The maximum input power drawn by the circuit is then

$$\text{maximum } P_i\ (\text{dc}) = V_{CC} I_{\text{dc}} = V_{CC}\left(\frac{2}{\pi}\frac{V_{CC}}{R_L}\right) = \frac{2\ V_{CC}^2}{\pi R_L}$$

The maximum circuit efficiency for class-B operation is then

$$\text{maximum } \eta = \frac{P_o}{P_i} \times 100 = \frac{V_{CC}^2/2R_L}{V_{CC}\left(\frac{2}{\pi}\frac{V_{CC}}{R_L}\right)} 100 = \frac{\pi}{4} \times 100 = 78.54\% \tag{12.31}$$

When the input signal results in less than the maximum output signal swing, the circuit efficiency is less than 78.5%. For class-B operation the maximum power dissipated by the output transistors *does not* occur at the maximum efficiency condition. The maximum power dissipated by the two output transistors occurs when the output voltage across the load is 0.636 $V_{CC}[\doteq (2/\pi)V_{CC}]$ and is

$$\text{maximum } P_{2Q} = \frac{2}{\pi^2}\frac{V_{CC}^2}{R_L} \tag{12.32}$$

EXAMPLE 12.9

Determine the maximum power values for a class-B amplifier using a power supply of $V_{CC} = 30$ V and driving a load of 16 Ω.

Solution:

The maximum output power is

$$\text{maximum } P_o \text{ (ac)} = \frac{V_{CC}^2}{2R_L} = \frac{(30 \text{ V})^2}{2(16\ \Omega)} = \mathbf{28.125\ W}$$

The maximum input power drawn from the voltage supply is

$$\text{maximum } P_i \text{ (dc)} = V_{CC}\frac{2}{\pi}\frac{V_{CC}}{R_L}$$

$$= (30 \text{ V})\frac{2}{\pi}\left(\frac{30 \text{ V}}{16\ \Omega}\right) = \mathbf{35.81\ W}$$

The circuit efficiency is

$$\text{maximum } \%\eta = \frac{P_o}{P_i} \times 100\% = \frac{28.125 \text{ W}}{35.81 \text{ W}} \times 100\% = 78.54\%$$

as expected.

The maximum power dissipated by each transistor is

$$\text{maximum } P_Q = \frac{\text{maximum } P_{2Q}}{2} = 0.5\left(\frac{2}{\pi^2}\right)\left(\frac{V_{CC}^2}{R_L}\right)$$

$$= 0.5\left(\frac{2}{\pi^2}\right)\frac{(30 \text{ V})^2}{16\ \Omega} = 5.7\ W$$

Under maximum conditions a pair of transistors, each handling 5.7 W at most, can deliver 28.125 W to a 16-Ω load.

The maximum efficiency of a class-B amplifier can also be expressed as follows:

$$P_o = \frac{V_{\text{peak}}^2}{2R_L}$$

$$P_i = V_{CC}I_{\text{dc}} = V_{CC}\frac{2}{\pi}\frac{V_{\text{peak}}}{R_L}$$

and

$$\eta = \frac{P_o}{P_i} \times 100\% = \frac{V_{\text{peak}}^2/2R_L}{V_{CC}\dfrac{2}{\pi}\dfrac{V_{\text{peak}}}{R_L}} \times 100\%$$

$$\boxed{\eta = 78.54\frac{V_{\text{peak}}}{V_{CC}}\ \%} \qquad (12.33)$$

EXAMPLE 12.10

Calculate the efficiency of a class-B amplifier for a supply voltage, $V_{CC} = 24$ V with peak-to-peak output of (a) $V_{\text{p-p}} = 22$ V; (b) $V_{\text{p-p}} = 6$ V.

Solution:

Using Eq. (12.33), gives us

$$\text{(a)}\ \eta = 78.54\frac{V_{\text{peak}}}{V_{CC}}\% = 78.54\left(\frac{22\text{ V}}{24\text{ V}}\right)\% = \mathbf{72\%}$$

$$\text{(b)}\ \eta = 78.54\left(\frac{6\text{ V}}{24\text{ V}}\right)\% = \mathbf{19.6\%}$$

Notice that a voltage near the maximum [22 V in part (a)] results in an efficiency near the maximum. Also note that even a small output voltage swing [6 V in part (b)] still provides operation at an efficiency near 20%. Similar power supply and signal swing values in a class-A amplifier result in significantly poorer efficiency.

12.5 CLASS-B AMPLIFIER CIRCUITS

A number of circuit arrangements for obtaining push-pull operation are possible. We shall consider a few of these and their various advantages and disadvantages. It is important to keep in mind the overall operation of the circuit in order to appreciate the different methods of obtaining the advantages of push-pull operation. For the push-pull circuit it is necessary to develop the output voltage across the load in such a manner that two stages operating in class-B will provide a full cycle of signal by conducting on alternate half-cycles.

Starting with an input signal obtained from a driver amplifier stage, it is necessary to operate the two-stage push-pull circuit on alternate half-cycles for class-B operation. The opposite-polarity input signals to the two stages of the push-pull circuit can be obtained in a number of ways. Figure 12.10a shows the use of an input transformer to provide the polarity inversion between the two push-pull input signals. With a center-tapped secondary the polarity of the voltage of the transformer ends with respect to the center tap are opposite. Other circuits used to obtain opposite-polarity input signals are shown in Fig. 12.10b and c. The input signal applied to the circuit of Fig. 12.10b appears of opposite polarity at the collector. The output from the emitter is the same polarity as the input so that the two output signals are of opposite polarity. Values of R_C, R_E, and h_{fe} can be chosen to make the voltage gain for the collector output signal to be 1. The gain for the signal taken from the emitter is 1 (emitter-follower operation). Thus the circuit results in opposite-polarity signals to drive the push-pull amplifier stage. The advantage of this driver connection is the saving on the use of a center-tapped transformer which is expensive and bulky and has a limited frequency operating range. A disadvantage is that the two signals do not come from similar impedance sources. The signal from the emitter provides a good driver connection since the source resistance viewed from the emitter is low. The collector circuit resistance, however, is relatively high and although the unloaded output signals are equal, they are different under load conditions. One possible improvement would be to add an additional emitter-follower stage to connect the output to the load since such stage would provide no additional voltage gain or polarity inversion but would drive the push-pull stage from a low-resistance source.

Another means of obtaining opposite-polarity signals to drive the push-pull stage is illustrated by the block diagram of Fig. 12.10c. The input signal is amplified and inverted by one amplifier stage and then attenuated for an overall gain of unity. The use of two emitter followers (possibly Darlington circuits) drives the push-pull stage from low-impedance sources.

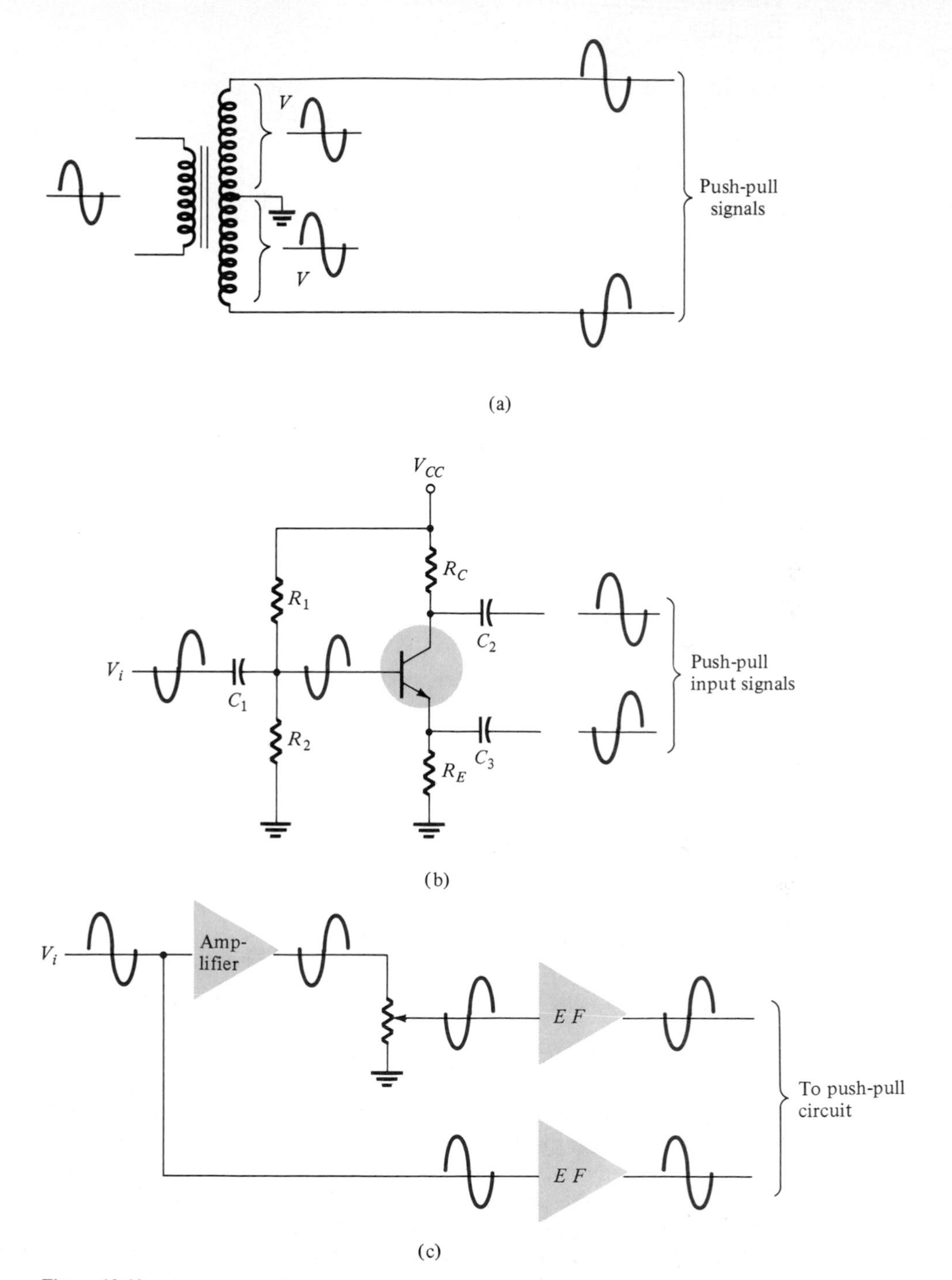

Figure 12.10 Phase-splitter circuits.

Transformer Coupled Push-Pull Circuits

The circuit of Fig. 12.11 uses an input transformer to produce opposite polarity signals to the two transistor inputs and an output transformer to drive the load in a push-pull mode of operation to be described.

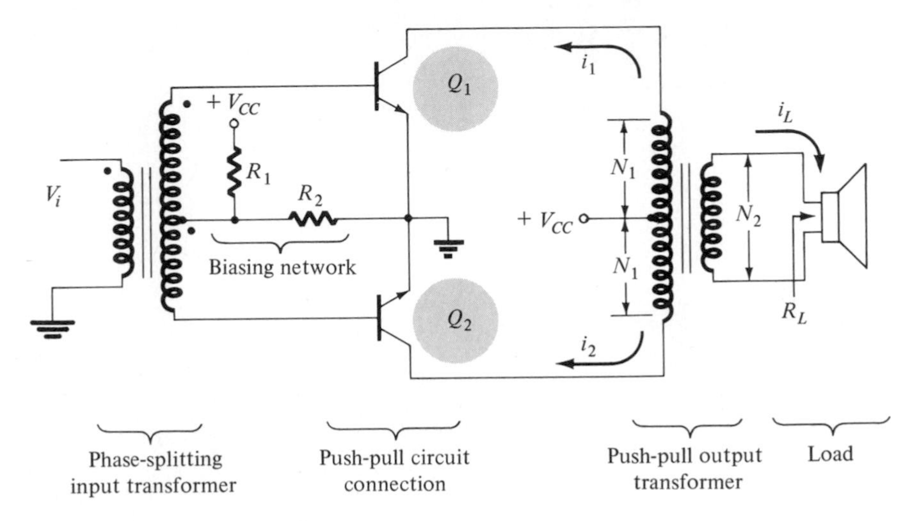

Figure 12.11 Push-pull circuit.

During the first half-cycle of operation, transistor Q_1 is driven into conduction, whereas transistor Q_2 is driven off. The current i_1 through the transformer results in the first half-cycle of signal to the load.

During the second half-cycle of the input signal, Q_2 conducts, whereas Q_1 stays off, the current i_2 through the transformer resulting in the second half-cycle to the load. The overall load signal varies over the full cycle of signal operation.

Complementary-Symmetry Circuits

A number of circuits go beyond eliminating only the input polarity-inverting transformer from the circuit. These circuits also remove the output transformer so that the circuit is completely transformerless. A simple version of a transformerless push-pull amplifier circuit is shown in Fig. 12.12. Complementary transistors are used, that is, *npn* and *pnp* transistors are used instead of using two of the same type. The single input signal required is applied to both base inputs. However, since the transistors are of opposite type, they will conduct on opposite half-cycles of the input. During the positive half-cycle of the input signal, for example, the *pnp* transistor will be reverse biased by the positive half-cycle signal and will not conduct. The *npn* transistor will be biased into conduction by the positive half-cycle signal with a resulting half-cycle of output across the load resistor (R_L) as shown in Fig. 12.12b. During the negative half-cycle of input signal the *npn* transistor is biased off and the output half-cycle

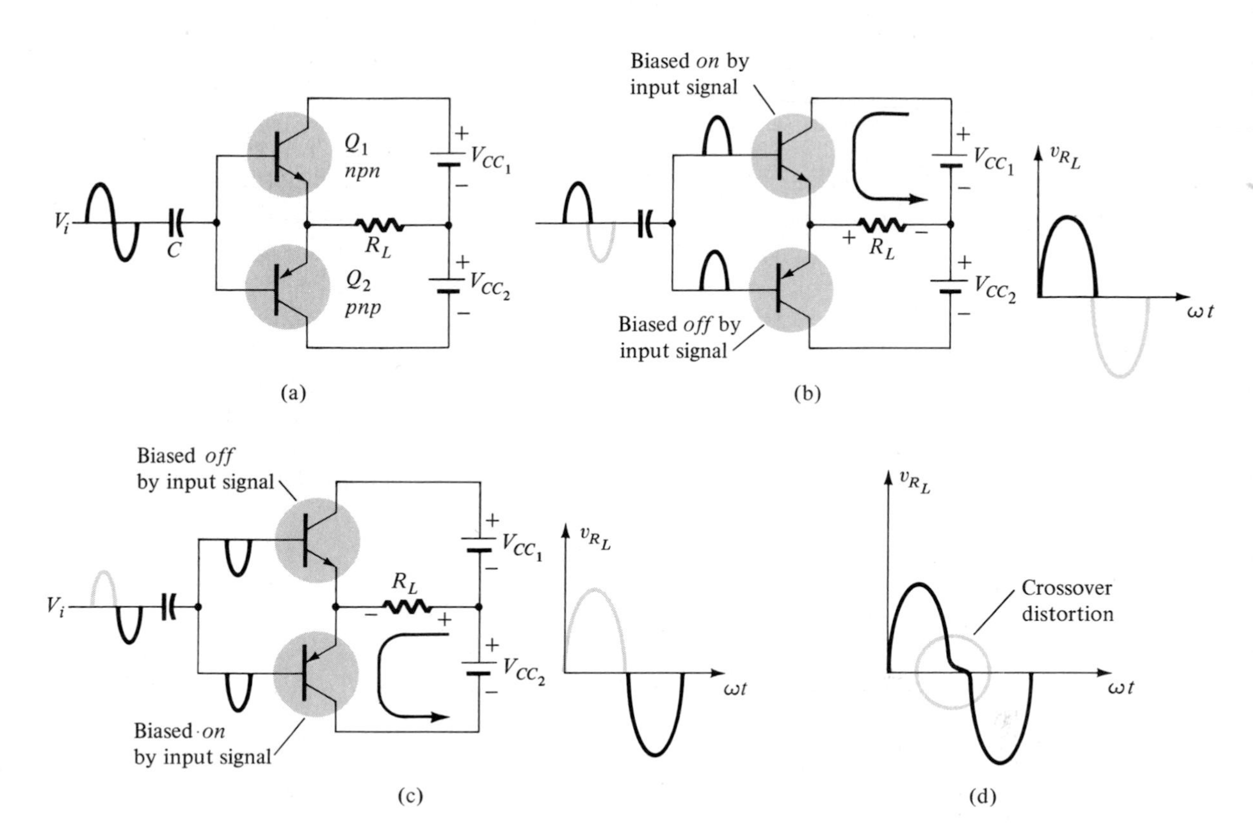

Figure 12.12 Complementary-symmetry push-pull circuit.

developed across the load is due to the operation of the *pnp* transistor at this time, as shown in Fig. 12.12c.

During a complete cycle of the input, a complete cycle of output signal is developed across the load. It should be obvious that one disadvantage of this circuit connection is the need for two supply voltages. Another, less obvious, but important, disadvantage with the complementary circuit as shown is the resulting *crossover* distortion in the output signal. Crossover distortion refers to the fact that during the signal crossover from positive to negative (or vice versa) there is some nonlinearity in the output signal as indicated in Fig. 12.12d. This results from the fact that for the simple circuit shown in Fig. 12.12a the operation of the circuit does not provide exact switching of one transistor *off* and the other *on* at the zero-voltage condition. Both may be off or partially conducting so that the output voltage is not exactly following the input and distortion occurs. This occurrence at the crossover point is of concern for the push-pull circuit of Fig. 12.11 as well, although not necessarily to the same degree. Bias of the transistors in class AB improves the operation by biasing the transistors so that each stays on for more than half of the cycle. For the circuit of Fig. 12.12a considerable effort must be made to reduce the crossover distortion. More practical circuit connections include additional biasing components in the base circuit to try to effect this improved operation.

Note that the load is driven as the output of an emitter-follower circuit so that the low resistance of the load is matched by low resistance from the driving source. Improved versions of the complementary circuit include the transistors, each connected in the Darlington arrangement, to provide even lower driver resistance than that with single transistors. The circuit of Fig. 12.13 shows a practical circuit connection

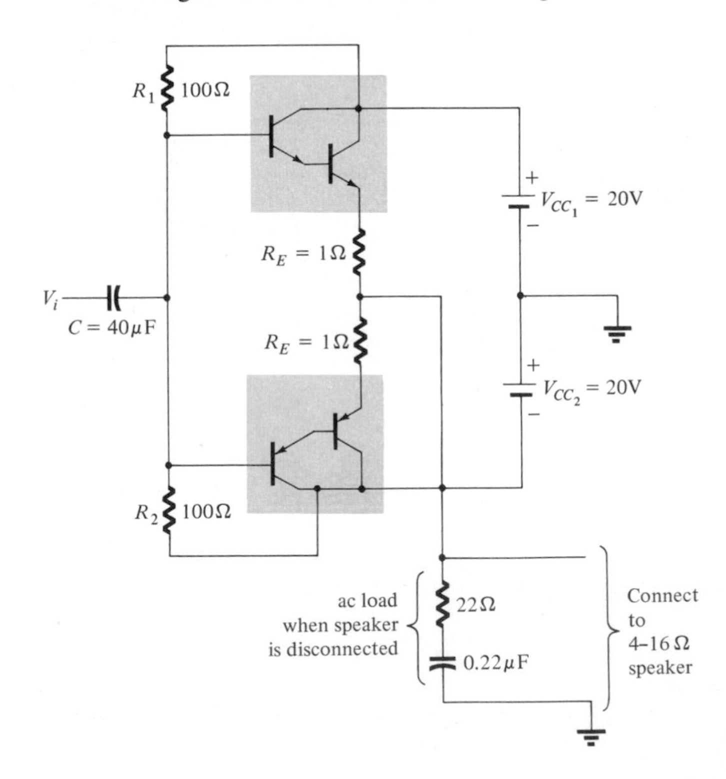

Figure 12.13 Complementary-symmetry push-pull circuit using Darlington transistors.

using the Darlington connection of the transistors and additional emitter resistors for temperature bias stabilization.

Quasi-complementary Push-Pull Amplifier

The push-pull circuit form is achieved in the circuit of Fig. 12.14 by using complementary transistors (Q_1 and Q_2) before the power output transistors (Q_3 and Q_4) so that both power output transistors can be *npn* types. This is a practically preferable arrangement, for the *npn* power transistors are presently the best available. Notice that transistors Q_1 and Q_3 form a *Darlington connection* that provides output at a low-impedance level from the emitter. The connection of transistors Q_2 and Q_4 forms a *feedback pair*, which similarly provides a low-impedance drive to the load. Resistor R_2 can be adjusted to minimize crossover distortion. The single signal applied as input to the push-pull stage then results in a full cycle output to the load R_L, each half of the circuit operating class-B for efficient power operation. This *quasi-complementary* push-pull power amplifier is presently a most popular circuit connection.

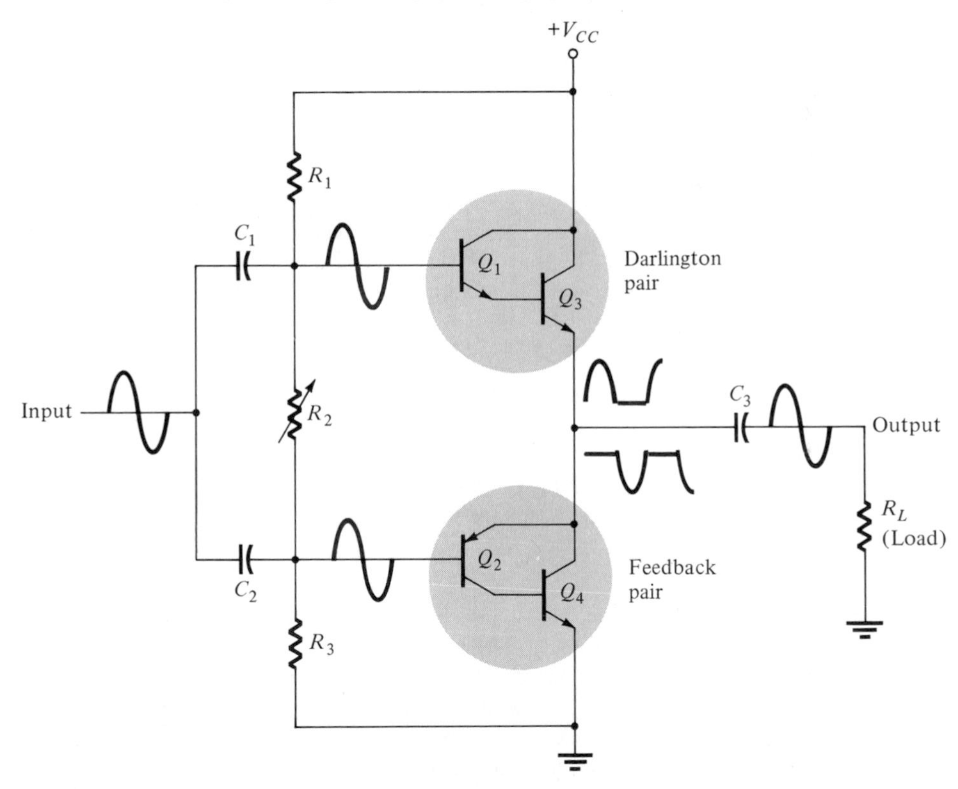

Figure 12.14 Quasi-complementary push-pull transformerless power amplifier.

EXAMPLE 12.11

For the circuit of Fig. 12.15:
(a) Calculate the input and output power handled by the circuit and the power dissipated by *each* output transistor for an input of 12 V rms.

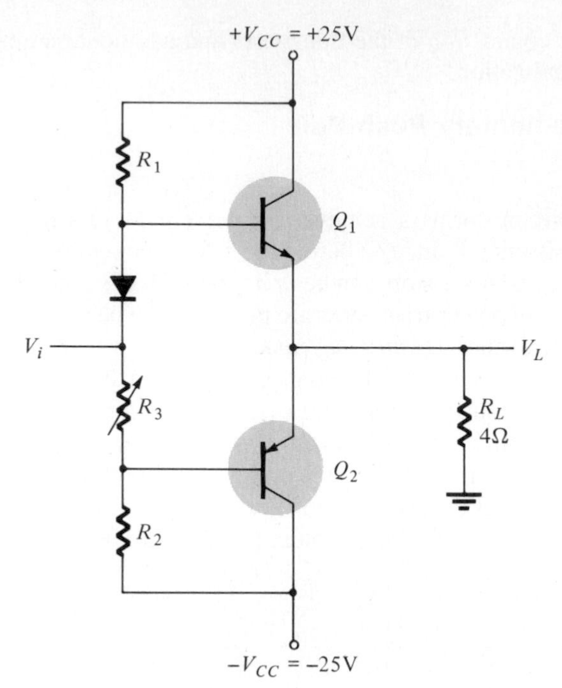

Figure 12.15 Class-B power amplifier for Example 12.8.

(b) If the input signal is increased to provide the maximum undistorted output, calculate the values of maximum input and output power and the power dissipated by *each* output transistor for this condition.
(c) Calculate the maximum power that *each* output transistor will have to handle.

Solution:

(a) The peak input voltage is

$$V_i\,(\text{p}) = \sqrt{2}\,V_i\,(\text{rms}) = \sqrt{2}\,(12\text{ V}) = 16.97\text{ V} \cong 17\text{ V}$$

Since the resulting voltage across the load is ideally the same as the input signal (the amplifier has, ideally, a voltage gain of unity),

$$V_L\,(\text{p}) = 17\text{ V}$$

$$P_o\,(\text{ac}) = \frac{V_L^2\,(\text{p})}{2R_L} = \frac{(17\text{ V})^2}{2(4\ \Omega)} = \mathbf{36.125\ W}$$

$$I_L\,(\text{p}) = \frac{V_L\,(\text{p})}{R_L} = \frac{17\text{ V}}{4\ \Omega} = 4.25\text{ A}$$

The dc current drawn from the two power supplies is then

$$I_{\text{dc}} = \frac{2}{\pi}I_L\,(\text{p}) = \frac{2(4.25\text{ A})}{\pi} = 2.71\text{ A}$$

so that the power supplied to the circuit is

$$P_i\,(\text{dc}) = V_{CC}I_{\text{dc}} = (25\text{ V})(2.71\text{ A}) = \mathbf{67.75\ W}$$

The circuit efficiency (for an input of $V_i = 12$ V rms) is

$$\eta = \frac{P_o}{P_i} \times 100\% = \frac{36.125 \text{ W}}{67.75 \text{ W}} \times 100\% = 53.3\%$$

and the power dissipated by each output transistor is

$$P_Q = \frac{P_{2Q}}{2} = \frac{P_i - P_o}{2} = \frac{67.75 \text{ W} - 36.125 \text{ W}}{2} = \mathbf{15.8 \text{ W}}$$

(b) If the input signal is increased to $V_i = 25$ V peak ($V_i = 17.68$ V rms) so that V_L (p) $= V_{CC} = 25$ V, the calculations are

$$\text{maximum } P_o = \frac{V_{CC}^2}{2R_L} = \frac{(25 \text{ V})^2}{2(4\ \Omega)} = \mathbf{78.125 \text{ W}}$$

$$\text{maximum } P_i = \frac{2}{\pi}\frac{V_{CC}^2}{R_L} = \frac{2}{\pi}\frac{(25 \text{ V})^2}{4\ \Omega} = \mathbf{99.47 \text{ W}}$$

and

$$\eta = \frac{P_o}{Pi} \times 100\% = \frac{78.125}{99.47} \times 100\% = 78.54\%$$

(the maximum circuit efficiency). At this maximum signal condition the power dissipated by *each* output transistor is

$$P_Q = \frac{P_{2Q}}{2} = \frac{P_i - P_o}{2} = \frac{99.47 \text{ W} - 78.125 \text{ W}}{2} = \mathbf{10.67 \text{ W}}$$

(c) The maximum power dissipation required of the output transistors is

$$\text{maximum } P_{2Q} = \frac{2}{\pi^2}\frac{V_{CC}^2}{R_L} = \frac{2}{\pi^2}\frac{(25 \text{ V})^2}{4\ \Omega} = 31.66 \text{ W}$$

so that

$$P_Q = \frac{P_{2Q}}{2} = \frac{31.66 \text{ W}}{2} = \mathbf{15.83 \text{ W}}$$

12.6 CLASSES OF AMPLIFIER OPERATION AND DISTORTION

Operating Classes

By definition class-A operation provides collector (output) current during the complete signal cycle (over a 360° interval). Figure 12.16a shows the output for class-A circuit operation. The bias level of current is I_{C_Q} and for the load line shown the output signal does not exceed values of $I_{C_{\max}}$ or $I_{C_{\min}}$, which would take the operation out of the linear region of device operation. Figure 12.16b shows class-B operation. The bias point is set at cutoff, the output current varying for only about 180° of the cycle, this being the definition of class-B operation. Note that the device is biased with no collector current and therefore no power dissipated by the transistor. Only when signal is applied does the transistor handle an average current which increases for larger input signals. Contrary to class-A operation, in which the worst condition occurs with no input signal and the least power is dissipated by the transistor

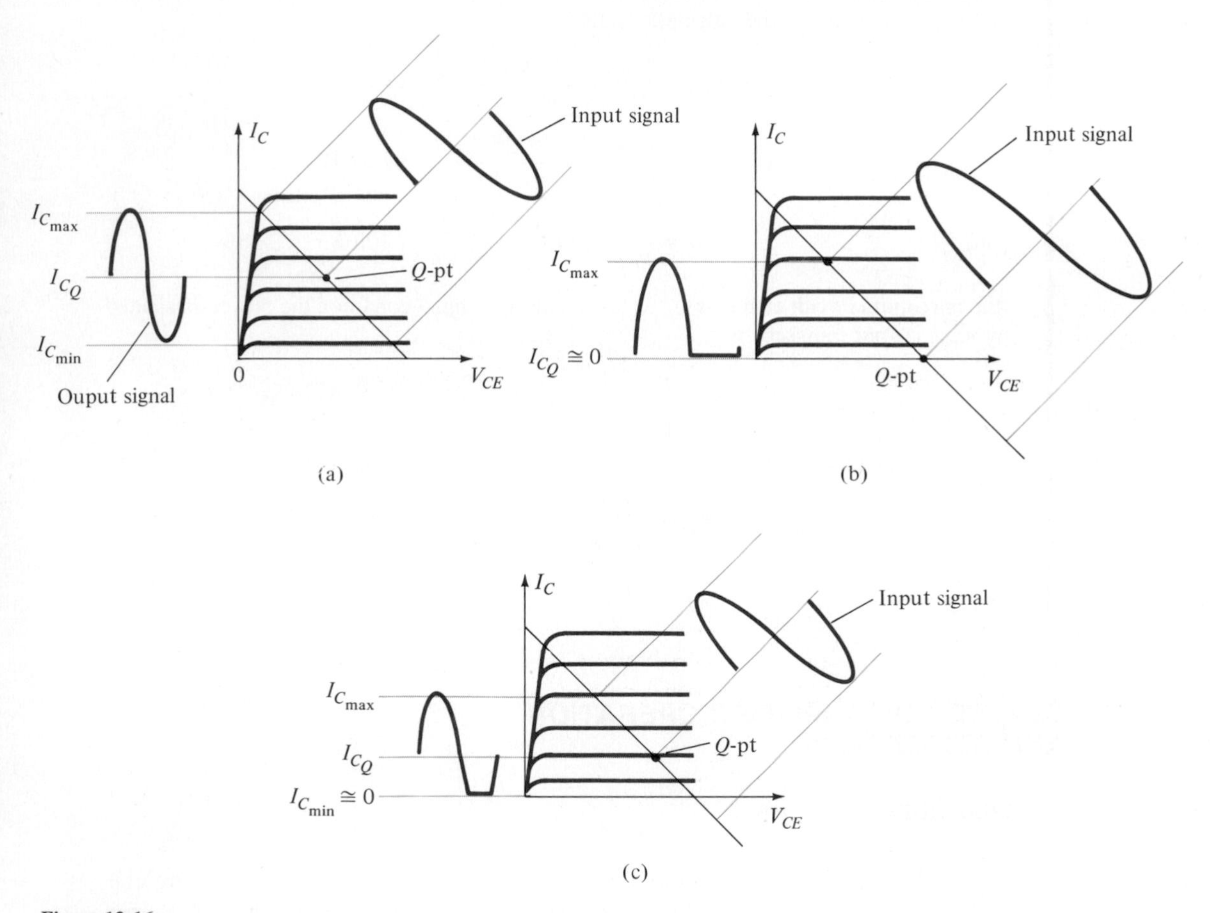

Figure 12.16 Various amplifier operating classes: (a) class-A; (b) class B; (c) class-AB.

for maximum input signal, the operation of a class-B circuit is to increase transistor dissipation for increased input signal. Since the average current in class-B operation is less than in class-A, the amount of power dissipated by the transistor is less in class-B.

In between class-A and class-B operation is class-AB operation, shown in Fig. 12.16c. The collector current occurs for more than 180° of the signal cycle but less than 360°. The maximum operating efficiency of class-AB is between that of class-A and class-B—that is, between 50% and 78.5%.

Operation with the output conducting for less than 180° is called class-C operation and is found in resonant or tuned amplifier circuits, as, for example, in radio or television. Operation on pulse-type signals is class-D.

Distortion

Output signal variations of less than 360° of the signal cycle are considered to have *distortion*. This means that the output signal is no longer just an amplified version of the input signal but in some ways is distorted or changed from that of the input. The poor quality of music coming from a radio or hi-fi system with the music or voice no longer sounding like that which was originally recorded or transmitted is the result of distortion. Distortion can come from a number of different places in any audio system.

Distortion can occur because the device characteristic is not linear: *nonlinear or amplitude distortion*. This can occur with all classes of operation. In addition, the circuit elements and the amplifying device can respond to the signal differently at various frequency ranges of operation: *frequency distortion*.

When distortion occurs the output signal no longer exactly represents the input signal. One technique of accounting for this change in the output signal is the method of *Fourier analysis*, which provides a means for describing a periodic signal in terms of its fundamental frequency component and frequency components at integer multiples—components called *harmonic components* or *harmonics*. For example, a signal which is originally 1000 Hz could result, after distortion, in a frequency component at 1000 Hz, and harmonic components at 2 kHz (2 × 1000 Hz), at 3 kHz (3 × 1000 Hz), 4 kHz (4 × 1000 Hz), and so on. The original frequency of 1000 Hz is called the fundamental frequency and those at integer multiples are the harmonics—that at 2 kHz is the second harmonic, the component at 3 kHz is the third harmonic, and so on. The fundamental signal is considered the first harmonic. (No harmonics at fractional amounts of the fundamental frequency exist using this technique.)

An instrument such as a spectrum analyzer would allow measurement of the harmonics present in the signal by providing a display of the fundamental component of the signal and a number of its harmonics on a CRT screen. Similarly, a wave analyzer instrument allows more precise measurement of the harmonic components of a distorted signal by filtering out each of these components and providing a reading of these components, one at a time.

In any case the technique of considering any distorted signal as containing a fundamental component and harmonic components is practical and useful. For a signal occurring in class-AB or class-B the distortion may be mainly even harmonics of which the second harmonic component is greatest. Thus, although the distorted signal contains all harmonic components from second harmonic on up, the most important in terms of the amount of distortion for the classes of operation we will consider is the second harmonic.

A collector current waveform is shown in Fig. 12.17 with the quiescent, minimum, and maximum signal levels, and the times they occur, marked on the waveform. The signal shown indicates that some distortion is present. An equation which approximately describes the distorted signal waveform is

$$i_C \cong I_{C_Q} + I_o + I_1 \cos(\omega t) + I_2 \cos(2\omega t) \tag{12.34}$$

The current waveform contains the original quiescent current I_{C_Q}, which occurs with zero input signal, an additional dc current I_o, due to the nonzero average of the distorted signal, the fundamental component of the distorted ac signal I_1, and a second harmonic component I_2, at twice the fundamental frequency. Although other harmonics are also present, only the second is considered here. Equating the resulting current from Eq. (12.34) at a few points in the cycle to that shown on the current waveform provides the following three relations:

At point 1 ($\omega t = 0$): $\quad i_C = I_{C_{\max}} = I_{C_Q} + I_o + I_1 \cos(0) + I_2 \cos(0)$

$$I_{C_{\max}} = I_{C_Q} + I_o + I_1 + I_2 \tag{12.35}$$

At point 2 ($\omega t = \pi/2$): $\quad i_C = I_{C_Q} = I_{C_Q} + I_o + I_1 \cos\left(\frac{\pi}{2}\right) + I_2 \cos\left(\frac{2\pi}{2}\right)$

$$I_{C_Q} = I_{C_Q} + I_o - I_2 \tag{12.36}$$

At point 3 ($\omega t = \pi$): $\quad i_C = I_{C_{\min}} = I_{C_Q} + I_o + I_1 \cos(\pi) + I_2 \cos(2\pi)$

$$I_{C_{\min}} = I_{C_Q} + I_o - I_1 + I_2 \tag{12.37}$$

Solving Eq. (12.35), (12.36), and (12.37) simultaneously gives the following results

$$I_o = I_2 = \frac{I_{C_{\max}} + I_{C_{\min}} - 2I_{C_Q}}{4}, \qquad I_1 = \frac{I_{C_{\max}} - I_{C_{\min}}}{2} \tag{12.38}$$

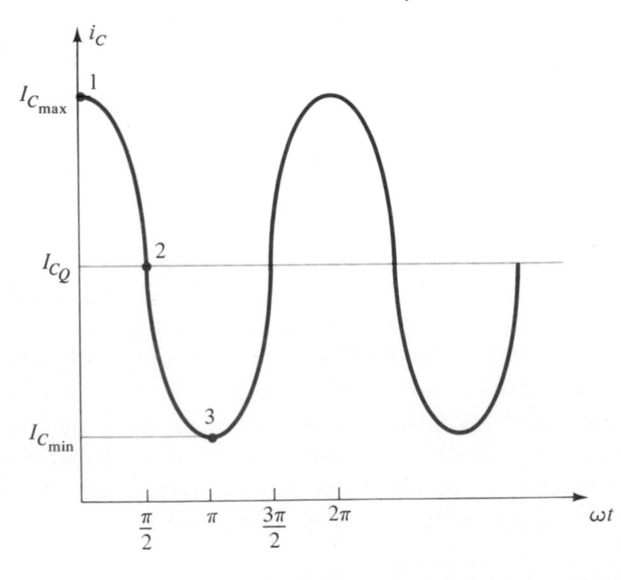

Figure 12.17 Waveform for obtaining second-harmonic distortion.

By definition the percent of the second harmonic distortion is given by

$$D_2 = \left| \frac{I_2}{I_1} \right| \times 100\% \tag{12.39}$$

The second harmonic distortion is the percent of the second harmonic component present in the output current waveform with respect to the amount of the fundamental component. Obviously, 0% distortion is the ideal condition of no distortion.

Using the results in Eq. (12.38) to express the second harmonic distortion defined by Eq. (12.39) gives

$$D_2 = \left| \frac{\frac{1}{2}(I_{C_{\max}} + I_{C_{\min}}) - I_{C_Q}}{I_{C_{\max}} - I_{C_{\min}}} \right| \times 100\% \tag{12.40}$$

In a similar manner, the amount of second harmonic distortion can be related to the measured values of the distorted output voltage waveform

$$D_2 = \left| \frac{\frac{1}{2}(V_{CE\max} + V_{CE\min}) - V_{CE_Q}}{V_{CE\max} - V_{CE\min}} \right| \times 100\% \tag{12.41}$$

EXAMPLE 12.12

An output waveform displayed on a scope provides the following measured values:
(a) $V_{CE\min} = 1$ V, $V_{CE\max} = 22$ V, $V_{CE_Q} = 12$ V.
(b) $V_{CE\min} = 4$ V, $V_{CE\max} = 20$ V, $V_{CE_Q} = 12$ V.
For each set of values calculate the amount of the second harmonic distortion.

Solution:

Using Eq. (12.41), we get

(a) $D_2 = \left| \frac{\frac{1}{2}(22 + 1) - 12}{22 - 1} \right| \times 100\% = \mathbf{2.38\%}$

(b) $D_2 = \left| \frac{\frac{1}{2}(20 + 4) - 12}{20 - 4} \right| \times 100\% = \mathbf{0\%}$ (no distortion)

The method used to obtain the amount of second harmonic distortion was called the three-point method since it involved equating the assumed form of the ouput voltage to the measured voltage at three points in the signal cycle. Using an assumed output signal equation containing more harmonic terms, along with choosing more points in the waveform, results in obtaining relations for the magnitude of the harmonic components at higher harmonic frequencies. Using a five-point method provides the dc component, first harmonic (fundamental), second harmonic, third harmonic, and fourth harmonic components. The harmonic distortion for each of these components is then defined as

$$D_2 = \left| \frac{I_2}{I_1} \right| \qquad D_3 = \left| \frac{I_3}{I_1} \right| \qquad D_4 = \left| \frac{I_4}{I_1} \right| \tag{12.42}$$

The total distortion may be defined, in general, using the individual distortion components

$$D = \sqrt{D_2^2 + D_3^2 + D_4^2 + \cdots} \tag{12.43}$$

When distortion does occur, the output power calculated by the undistorted case is no longer correct. Equation (12.14), for example, is true *only* for the nondistorted case. When distortion is present, the output power due to the fundamental component of the distorted signal is

$$P_1 = \frac{I_1^2 R_C}{2} \tag{12.44}$$

The total output power due to all the harmonic components of the distorted signal is

$$P = (I_1^2 + I_2^2 + I_3^2 + \cdots)\frac{R_C}{2} \tag{12.45}$$

The total power can also be expressed in terms of the total distortion

$$\boxed{P = (1 + D_2^2 + D_3^2 + \cdots)I_1^2\frac{R_C}{2} = (1 + D^2)P_1} \tag{12.46}$$

EXAMPLE 12.13

Using a five-point method to calculate harmonic components gives the following results: $D_2 = 0.1$, $D_3 = 0.02$, $D_4 = 0.01$, with $I_1 = 4$ A and $R_C = 8\ \Omega$. Calculate the total distortion, fundamental power component, and total power.

Solution:

Total distortion, using Eq. 12.43, gives

$$D = \sqrt{D_2^2 + D_3^2 + D_4^2} = \sqrt{(0.1)^2 + (0.02)^2 + (0.01)^2} \cong \mathbf{0.1}$$

Fundamental power, using Eq. (12.44), gives

$$P_1 = \frac{I_1^2 R_C}{2} = \frac{(4)^2 8}{2} = \mathbf{64\ W}$$

Total power, using Eq. (12.46), gives

$$P = (1 + D^2)P_1 = [1 + (0.1)^2]64 = (1.01)64 = \mathbf{64.64\ W}$$

(Total power mainly due to fundamental component even with 10% second harmonic distortion.)

GRAPHICAL DESCRIPTION OF HARMONIC COMPONENTS OF DISTORTED SIGNAL

A demonstration of the use of harmonic components to represent a distorted signal is provided to help clarify the concept. As an example, a distorted waveform such as that resulting from class-B operation is shown in Fig. 12.18a. The signal is clipped on the negative half-cycle so that only the positive sinusoidal half-cycle provides an output signal.

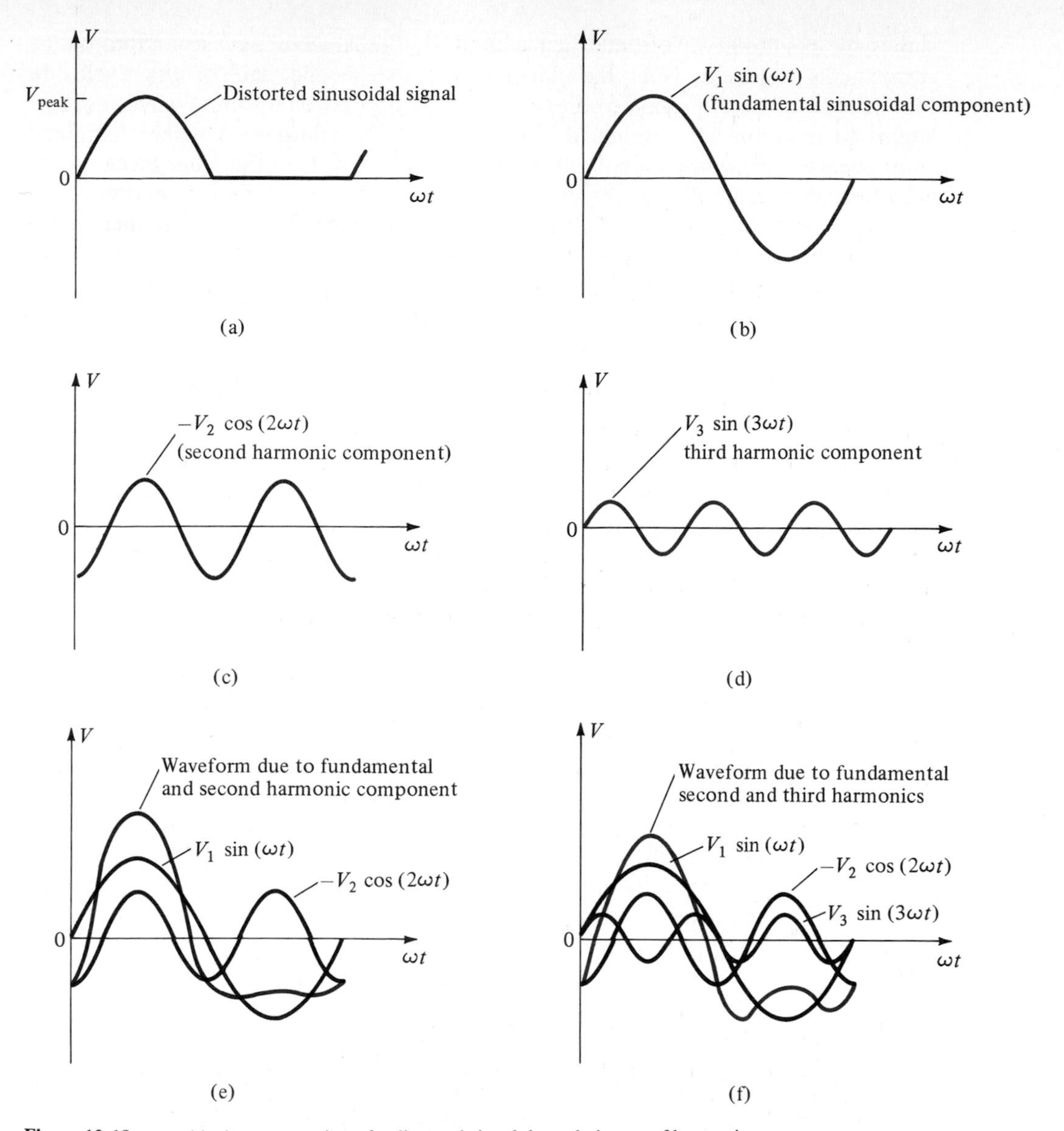

Figure 12.18 Graphical representation of a distorted signal through the use of harmonic components.

Using Fourier analysis techniques, we can calculate a fundamental component of the distorted signal as shown in Fig. 12.18b. Figure 12.18b does not show the distorted waveform, only the fundamental component (which is a perfectly sinusoidal signal itself). Similarly, the second and third harmonic components can be obtained and are shown in Figs. 12.18c and 12.18d, respectively.

We now wish to check whether these components, each a purely undistorted sinusoidal signal, add up approximately to the original distorted signal. Figure 12.18e shows the resulting waveform when adding the fundamental and second harmonic components together. Note the flattening of the second half of the cycle. In Fig. 12.18f the third harmonic component is added to give a resulting waveform that begins to resemble the original distorted signal. The addition of higher harmonic components of the correct amplitude and correct phase will further alter the resulting waveform to approximate the original distorted signal. In a relatively simple manner we can observe that addition of a fundamental component and harmonic components can result in the original distorted waveform. In general, any periodic waveform can be represented by a fundamental component and harmonic components, each of varying amplitudes and at various phase angles.

The concept of harmonics is useful in both analyzing distorted (nonsinusoidal) waveforms and in provided a means of working with such signals. Since all the harmonic components are sinusoidal signals, we can separately consider the effect of each component on the circuit and obtain the total effect using superposition—adding together the voltages or currents being considered.

12.7 POWER TRANSISTOR HEAT SINKING

While integrated circuits are used for small-signal and low-power applications, most high-power applications still require individual power transistors. Improvements in production techniques have provided higher power ratings in smaller-sized packaging cases, have increased the maximum transistor breakdown voltage, and have provided faster-switching power transistors.

The maximum power handled by a particular device and the temperature of the transistor junctions are related since the power dissipated by the device causes an increase in temperature at the junctions of the device. Obviously, a 100-W transistor will provide more power capability than a 10-W transistor. On the other hand, proper heat sinking techniques will allow operation of a device close to its maximum power rating.

We should note that of the two types of transistors—germanium and silicon—silicon transistors provide greater maximum temperature ratings. Typically, the maximum junction temperature of these types of power transistors is

Germanium: 100–110°C
Silicon: 150–200°C

For many applications the average power dissipated may be approximated by

$$P_D = V_{CE} I_C$$

This power dissipation, however, is only allowed up to some maximum temperature. Above this temperature the device power dissipation capacity must be reduced (or *derated*) so that at higher case temperatures the power-handling capacity is reduced—down at 0 W at the device maximum case temperature.

The greater the power handled by the transistor (dependent on power level set by the circuit) the higher the case temperature of the transistor. Actually, the limiting factor in power handled by a particular transistor is the temperature of the device collector junction. Power transistors are mounted in large metal cases to provide a large area from which the heat generated by the device may radiate. Even so, operating a transistor directly into air (mounting it on a plastic board, for example) severely limits the device power rating. If, instead (as is usual practice), the device is mounted on some form of *heat sink*, its power-handling capacity can approach the rated maximum value more closely. A few heat sinks are shown in Fig. 12.19. When the heat sink is used, the heat produced by the transistor dissipating power has a larger area from which to radiate the heat into the air, thereby holding the case temperature to a much lower value than would result without the heat sink. Even with an infinite heat sink (which, of course, is not available), for which the case temperature is held at the *ambient* (air) temperature, the junction will be heated above the case temperature and a maximum power rating must be considered.

Figure 12.19 Typical power heat sinks.

Since even a good heat sink cannot hold the transistor case temperature at ambient (which, by the way could be more than 25°C if the transistor circuit is in a confined area where other devices are also radiating a good deal of heat), it is necessary to derate the amount of *maximum power* allowed for a particular transistor as a function of increased case temperature.

Figure 12.20 shows a typical power derating curve for a silicon transistor. The curve shows that the manufacturer will specify an upper temperature point (not necessarily 25°C) after which a linear derating takes place. For silicon the maximum power that should be handled by the device does not reduce to 0 W until a case temperature of 200°C.

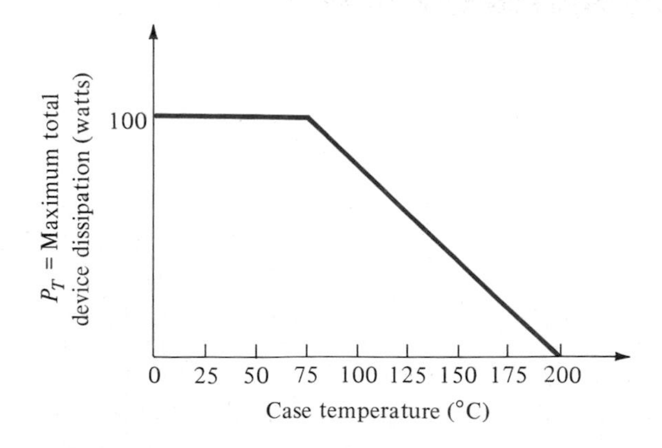

Figure 12.20 Typical power derating curve for silicon transistors.

It is not necessary to provide a derating curve since the same information could be given simply as a listed derating factor on the device specification sheet. Stated mathematically,

$$P_D(\text{temp}_1) = P_D(\text{temp}_0) - (\text{Temp}_1 - \text{Temp}_0)(\text{Derating factor}) \tag{12.47}$$

where the value of Temp_0 is the temperature at which derating should begin, the value of Temp_1 is the particular temperature of interest (above the value Temp_0), P_D (temp_0) and P_D (temp_1) are the maximum power dissipations at the temperatures specified, and the derating factor is the value given by the manufacturer in units of watts (or milliwatts) per degree of temperature.

EXAMPLE 12.14

Determine what maximum dissipation will be allowed for an 80-W silicon transistor (rated at 25°C) if derating is required above 25°C by a derating factor of 0.5 W/°C at a case temperature of 125°C.

Solution:

$$P_D(125°\text{C}) = P_D(25°\text{C}) - (125°\text{C} - 25°\text{C})(0.5\ \text{W}/°\text{C})$$
$$= 80\ \text{W} - 100°\text{C}(0.5\text{W}/°\text{C}) = \mathbf{30\ W}$$

It is interesting to note what power rating results using a power transistor without a heat sink. For example, a silicon transistor rated at 100 W at (or below) 100°C is rated only 4 W at (or below) 25°C, free-air temperature. Thus, operated without a heat sink the device can handle a maximum of only 4 W at the room temperature of 25°C. Using a heat sink large enough to hold the case temperature to 100°C at 100 W allows operation at the maximum power rating.

Thermal Analogy of Power Transistor

Selection of a suitable heat sink requires a considerable amount of detail which is not appropriate to our present basic considerations of the power transistor. However, more detail about the thermal characteristics of the transistor and its relation to the power dissipation of the transistor may help provide a clearer understanding of power as limited by temperature. The following discussion should provide some background information.

A picture of how the junction temperature (T_J), case temperature (T_C), and ambient (air) temperature (T_A) are related by the device heat-handling capacity—a temperature coefficient usually called *thermal resistance*—is presented in the thermal-electrical analogy shown in Fig. 12.21.

In providing a thermal-electrical analogy the term *thermal resistance* is used to describe heat effects by an electrical term. The terms in Fig. 12.21 are defined as follows:

θ_{JA} = total thermal resistance (junction to ambient)
θ_{JC} = transistor thermal resistance (junction to case)
θ_{CS} = insulator thermal resistance (case to heat sink)
θ_{SA} = heatsink thermal resistance (heat sink to ambient)

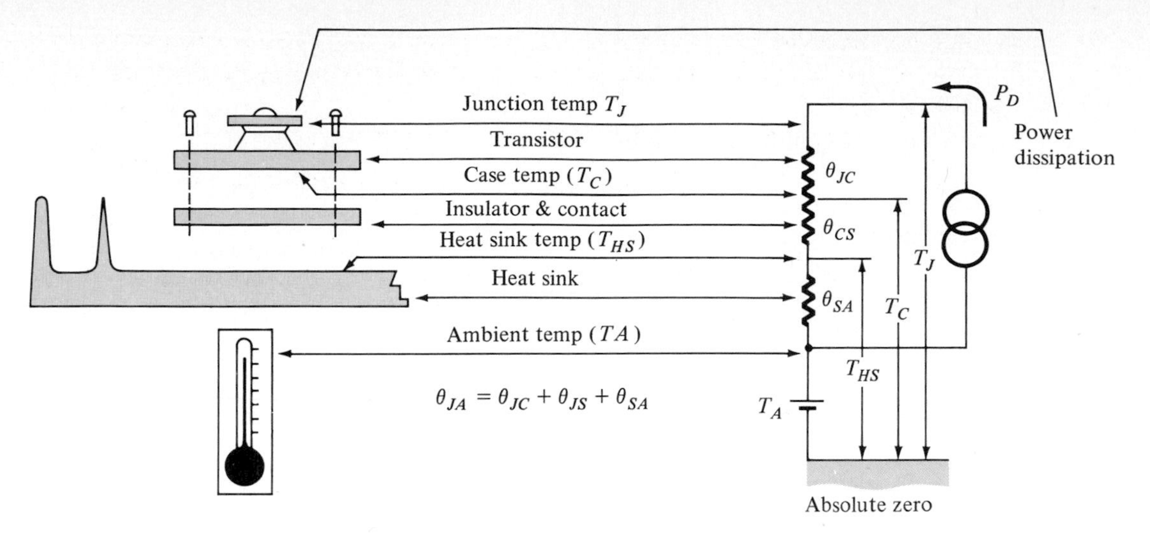

Figure 12.21 Thermal-to-electrical analogy.

Using the electrical analogy for thermal resistances, we can write

$$\boxed{\theta_{JA} = \theta_{JC} + \theta_{CS} + \theta_{SA}} \tag{12.48}$$

The analogy can also be used in applying Kirchhoff's law to obtain

$$\boxed{T_J = P_D\,\theta_{JA} + T_A} \tag{12.49}$$

The last relation shows that the junction temperature "floats" on the ambient temperature and that the higher the ambient temperature the lower the allowed value of device power dissipation.

The thermal factor θ provides information about how much temperature drop (or rise) results for a given amount of power dissipation. For example, the value of θ_{JC} is usually about 0.5°C/W. This means that for a power dissipation of 50 W the difference in temperature between case temperature (as measured by a thermocouple) and the inside junction temperature is only

$$T_J - T_C = \theta_{JC} P_D = (0.5°\text{C/W})(50\text{ W}) = 25°\text{C}$$

Thus, if the heat sink can hold the case at, say 50°C, the junction is then only at 75°C. This is a relatively small temperature difference, especially at lower power-dissipation levels.

The value of thermal resistance from junction to free air (using no heat sink) is, typically,

$$\theta_{JA} = 40°\text{C/W} \qquad \text{(into free air)}$$

For this thermal resistance only 1 W of power dissipation results in a junction temperature 40°C greater than the ambient.

A heat sink can now be seen to provide a low thermal resistance between case and air—much less than 40°C/W value of the transistor case alone. Using a heat sink having

$$\theta_{SA} = 2°C/W$$

and, with an insulating thermal resistance (from case to heat sink) of

$$\theta_{CS} = 0.8°C/W$$

and, finally for the transistor,

$$\theta_{CJ} = 0.5°C/W$$

we can obtain

$$\theta_{JA} = \theta_{SA} + \theta_{CS} + \theta_{CJ}$$

$$= 2.0°C/W + 0.8°C/W + 0.5°C/W = 3.3°C/W$$

So, with a heat sink, the thermal resistance between air and the junction is only 3.3°C/W as compared to 40°C/W for the transistor operating directly into free air. Using the value of θ_{JA} above for a transistor operated at, say, 2 W we calculate

$$(T_J - T_A) = \theta_{JA} P_D = (3.3°C/W)(2\ W) = 6.6°C$$

In other words, the use of a heat sink in this example provided only a 6.6°C increase in junction temperature as compared to an 80°C rise without a heat sink.

EXAMPLE 12.15

A silicon power transistor is operated with a heat sink ($\theta_{SA} = 1.5°C/W$). The transistor, rated at 150 W (25°C), has $\theta_{JC} = 0.5°C/W$ and the mounting insulation has $\theta_{CS} = 0.6°C/W$. What maximum power can be dissipated if the ambient temperature is 40°C and $T_{J\,max} = 200°C$?

Solution:

$$P_D = \frac{T_J - T_A}{\theta_{JC} + \theta_{CS} + \theta_{SA}} = \frac{200 - 40}{0.5 + 0.6 + 1.5} = \frac{160°C}{2.6°C/W} \cong \mathbf{61.5\ W}$$

PROBLEMS

§ 12.2

1. Calculate the input and output power for the circuit of Fig. 12.4 for a supply voltage of $V_{CC} = 25$ V. The input signal results in an ac base current of 5 mA rms.
2. Calculate the input power dissipated by the circuit of Fig. 12.4 if R_B is changed to 2 kΩ.
3. What maximum output power can be delivered by the circuit of Fig. 12.4 if R_B is changed to 2 kΩ?
4. If the circuit of Fig. 12.1 is biased at its center voltage and center collector current point, what is the input power for a maximum output of 1.5 W?

§ 12.3

5. A class-A transformer-coupled amplifier uses a 25:1 transformer to drive a 4-Ω load.

Calculate the effective ac load (seen by the transistor connected to the larger turns side of the transformer).

6. What turns ratio transformer is needed to couple to an 8-Ω load so that it appears as a 10-kΩ effective load?

7. Calculate the transformer turns ratio required to connect four parallel 16-Ω speakers so that they appear as an 8-kΩ effective load.

8. A transformer-coupled class-A amplifier drives a 16-Ω speaker through a $\sqrt{15}:1$ transformer. Using a power supply of 36 V (V_{CC}), the circuit delivers 2 W to the load. Calculate the following:
(a) The ac power across the transformer primary.
(b) The rms value of load voltage.
(c) The rms value of primary voltage.
(d) The rms value of load and primary current.

9. Calculate the efficiency of the circuit of Problem 8 if the bias current is $I_{C_Q} = 150$ mA.

10. Draw the circuit diagram of a class-A transformer-coupled amplifier using an *npn* transistor.

§ 12.4

11. Draw the circuit diagram of a class-B *npn* push-pull power amplifier using transformer-coupled input.

12. Sketch the input waveforms, both collector voltage waveforms, and both collector current waveforms for the circuit of Problem 11.

13. Draw the circuit diagram of an *npn* push-pull power amplifier operated class-AB. Show a phase-splitter driver stage before the push-pull stage.

§ 12.5

14. Sketch the circuit diagram of a quasi-complementary amplifier, showing voltage waveforms in circuit.

15. List any advantages of a transformerless circuit over a transformer-coupled circuit.

16. For a class-B power amplifier as in Fig. 12.14 using a 30-V power supply and $R_L = 8\ \Omega$, calculate (a) maximum P_o (ac), (b) maximum P_i (dc), (c) maximum $\%\eta$, and (d) maximum power dissipated by both output transistors. Use $R_L = 8\ \Omega$.

17. If the input voltage to the power amplifier in Fig. 12.14 is 8 V, rms using a 30-V power supply and $R_L = 8\ \Omega$, calculate (a) P_o (ac), (b) P_i (dc). (c) $\%\eta$, and (d) power dissipated by both output power transistors.

18. For power amplifier as in Fig. 12.15 using a 40-V power supply and signal input of 18 V, rms calculate (a) P_o (ac), (b) P_i (dc), (c) $\%\eta$, and (d) combined power dissipation in the output power transistors, when load is $R_L = 8\ \Omega$.

§ 12.6

19. Calculate the efficiency of the following amplifier classes and voltages:
(a) Class-A operation with $V_{CE_{max}} = 24$-V and $V_{CE_{min}} = 2$ V.
(b) Class-B transformer operation with $V_{CE_{max}} = 4$ V, and $V_{CC} = 22$ V.

20. For the following voltage values measured on a scope calculate the amount of the second-harmonic distortion: $V_{CE_{max}} = 27$ V, $V_{CE_{min}} = 14$ V, $V_{CE_Q} = 20$ V.

§ 12.7

21. Determine the maximum dissipation allowed for a 100-W silicon transistor (rated at 25°C) for a derating factor of 0.6 W/°C, at a case temperature of 150°C.

22. A 160-W silicon power transistor operated with a heat sink ($\theta_{SA} = 1.5$°C/W) has $\theta_{JC} = 0.5$°C/W and mounting insulation of $\theta_{CS} = 0.8$°C/W. What maximum power can be handled by the transistor at an ambient temperature of 80°C? (The junction temperature should not exceed 200°C.)

23. What maximum power can a silicon transistor ($T_{J_{max}} = 200$°C) dissipate into free air at an ambient of 80°C?

COMPUTER PROBLEMS

Write BASIC programs to:

1. Calculate I_{dc} and P_i for a class-A series-fed circuit as in Fig. 12.1.
2. Calculate P_o for a class-A series-fed circuit as in Fig. 12.1.
3. Tabulate the efficiency of a class-A series-fed circuit as in Fig. 12.1 over a range of output peak voltages from 0.1 V_{CC} to V_{CC}.
4. Calculate the efficiency of a class-A series-fed circuit as in Fig. 12.1 over a range of values for R_C.
5. Calculate the P_i for a transformer-coupled class-A amplifier as in Fig. 12.5.
6. Calculate the transformer turns ratio needed to optimally connect a given load, R_L, into a class-A transformer-coupled amplifier as in Fig. 12.5.
7. Tabulate the efficiency for a class-A transformer-coupled amplifier as in Fig. 12.5 for supply voltage values from 0.1 V_{CC} to V_{CC}.
8. Tabulate the value of P_i for a class-B amplifier as in Fig. 12.15 for supply voltage from 0.1 V_{CC} to V_{CC}.
9. Tabulate P_o for a class-B power amplifier as in Fig. 12.15 for values of supply voltage varying from 0.1 V_{CC} to V_{CC}.
10. Tabulate the power dissipated by both output transistors in the circuit of Fig. 12.15 over a range of supply voltages from 0.1 V_{CC} to V_{CC}.

13

pnpn and Other Devices

13.1 INTRODUCTION

In this chapter a number of important devices not discussed in detail in earlier chapters are introduced. The two-layer semiconductor diode has led to three-, four-, and even five-layer devices. A family of four-layer *pnpn* devices will first be considered: SCR (silicon controlled rectifier), SCS (silicon controlled switch), GTO (gate turn-off switch), LASCR (light activated SCR), followed by an increasingly important device—the UJT (unijunction transistor). Those four-layer devices with a control mechanism are commonly referred to as *thyristors*, although the term is most frequently applied to the SCR (silicon-controlled rectifier). The chapter closes with an introduction to the phototransistor, opto-isolators, and the PUT (programmable unijunction transistor).

pnpn DEVICES

13.2 SILICON-CONTROLLED RECTIFIER

Within the family of *pnpn* devices the silicon-controlled rectifier (SCR) is unquestionably of the greatest interest today. It was first introduced in 1956 by Bell Telephone Laboratories. A few of the more common areas of application for SCRs include relay controls, time-delay circuits, regulated power suppliers, static switches, motor controls, choppers, inverters, cycloconverters, battery chargers, protective circuits, heater controls, and phase controls.

In recent years, SCRs have been designed to *control* powers as high as 10 MW with individual ratings as high as 2000 A at 1800 V. Its frequency range of application has also been extended to about 50 kHz, permitting some high-frequency applications such as induction heating and ultrasonic cleaning.

13.3 BASIC SILICON-CONTROLLED RECTIFIER OPERATION

As the terminology indicates, the SCR is a rectifier constructed of silicon material with a third terminal for control purposes. Silicon was chosen because of its high temperature and power capabilities. The basic operation of the SCR is different from the fundamental two-layer semiconductor diode in that a third terminal, called a *gate*, determines when the rectifier switches from the open-circuit to short-circuit state. It is not enough simply to forward-bias the anode-to-cathode region of the device. In the conduction region the dynamic resistance of the SCR is typically 0.01 to 0.1 Ω. The reverse resistance is typically 100 kΩ or more.

The graphic symbol for the SCR is shown in Fig. 13.1 with the corresponding connections to the four-layer semiconductor structure. As indicated in Fig. 13.1a, if forward conduction is to be established, the anode must be positive with respect to the cathode. This is not, however, a sufficient criterion for turning the device on. A pulse of sufficient magnitude must also be applied to the gate to establish a turn-on gate current, represented symbolically by I_{GT}.

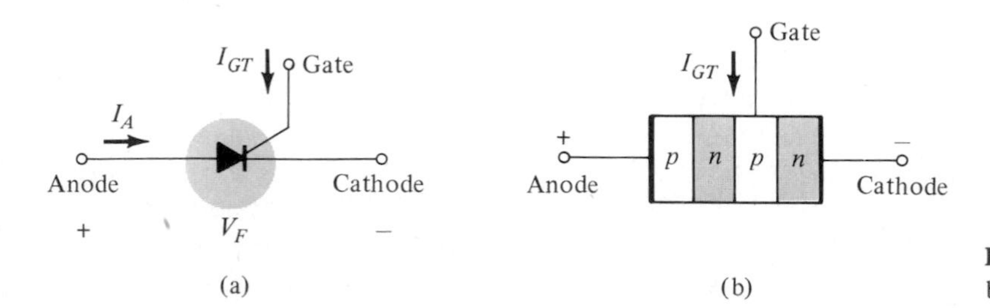

Figure 13.1 (a) SCR symbol; (b) basic construction.

A more detailed examination of the basic operation of an SCR is best effected by splitting the four-layer *pnpn* structure of Fig. 13.1b into two three-layer transistor structures as shown in Fig. 13.2a and then considering the resultant circuit of Fig. 13.2b.

Note that one transistor for Fig. 13.2 is an *npn* device while the other is a *pnp* transistor. For discussion purposes, the signal shown in Fig. 13.3a will be applied to the gate of the circuit of Fig. 13.2b. During the interval $0 \rightarrow t_1$, $V_{\text{gate}} = 0$ V, the circuit of Fig. 13.2b will appear as shown in Fig. 13.3b ($V_{\text{gate}} = 0$ V is equivalent to the gate terminal being grounded as shown in the figure). For $V_{BE_2} = V_{\text{gate}} = 0$ V, the base current $I_{B_2} = 0$ and I_{C_2} will be approximately I_{CO}. The base current of Q_1, $I_{B_1} = I_{C_2} = I_{CO}$, is too small to turn Q_1 on. Both transistors are therefore in the "off"

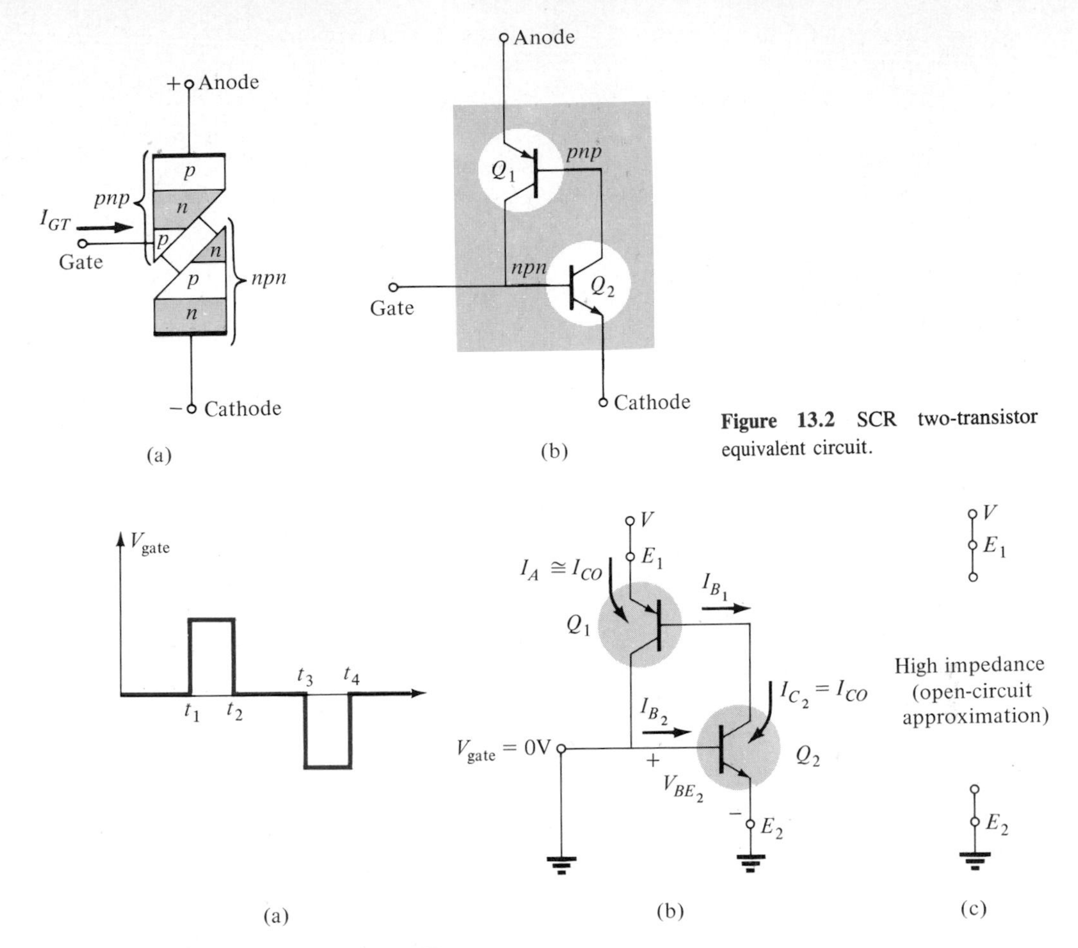

Figure 13.2 SCR two-transistor equivalent circuit.

Figure 13.3 "Off" state of the SCR.

state, resulting in a high impedance between the collector and emitter of each transistor and the open-circuit representation for the controlled rectifier as shown in Fig. 13.3c.

At $t = t_1$ a pulse of V_G volts will appear at the SCR gate. The circuit conditions established with this input are shown in Fig. 13.4a. The potential V_G was chosen sufficiently large to turn Q_2 on ($V_{BE_2} = V_G$). The collector current of Q_2 will then rise to a value sufficiently large to turn Q_1 on ($I_{B_1} = I_{C_2}$). As Q_1 turns on, I_{C_1} will increase, resulting in a corresponding increase in I_{B_2}. The increase in base current for Q_2 will result in a further increase in I_{C_2}. The net result is a regenerative increase in the collector current of each transistor. The resulting anode-to-cathode resistance [$R_{SCR} = V/(I_A - \text{large})$] is then very small, resulting in the short-circuit representation for the SCR as indicated in Fig. 13.4b. The regenerative action described above results in SCRs having typical turn-on times of 0.1 to 1 μs. However, high-power devices in the range 100 to 400 A may have 10- to 25-μs turn-on times.

In addition to gate triggering, SCRs can also be turned on by significantly raising the temperature of the device or raising the anode-to-cathode voltage to the breakover value shown on the characteristics of Fig. 13.7.

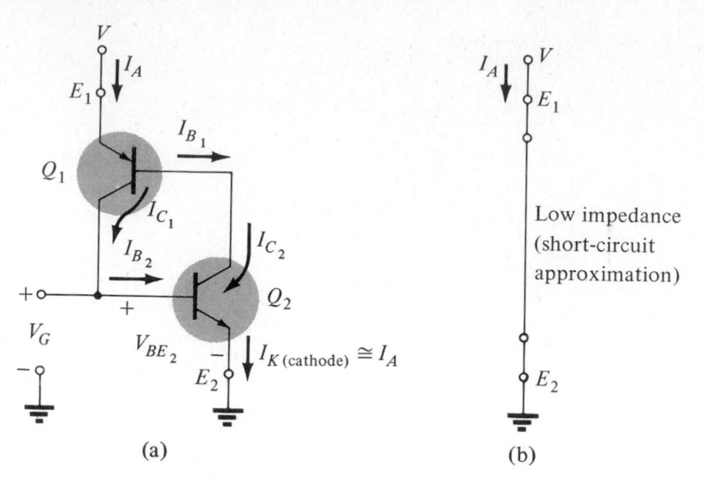

Figure 13.4 "On" state of the SCR.

The next question of concern is: How long is the turn-off time and how is turn-off accomplished? An SCR *cannot* be turned off by simply removing the gate signal, and only a special few can be turned off by applying a negative pulse to the gate terminal as shown in Fig. 13.3a at $t = t_3$. The two general methods for turning off an SCR are categorized as the *anode-current interruption* and the *forced-commutation technique*. The two possibilities for current interruption are shown in Fig. 13.5.

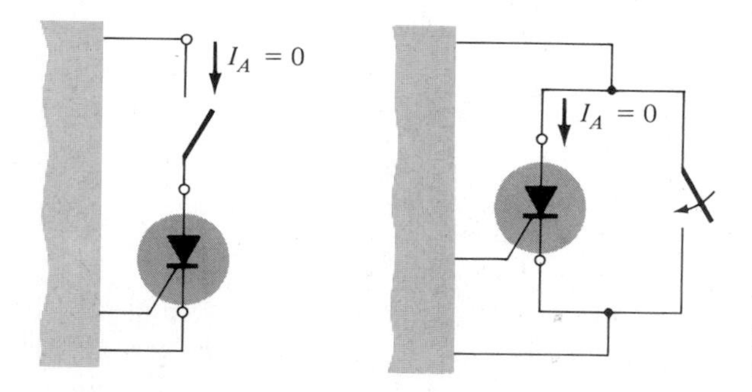

Figure 13.5 Anode current interruption.

In Fig. 13.5a, I_A is zero when the switch is opened (series interruption) while in Fig. 13.5b the same condition is established when the switch is closed (shunt interruption). Forced commutation is the "forcing" of current through the SCR in the direction opposite to forward conduction. There are a wide variety of circuits for performing this function, a number of which can be found in the manuals of major manufacturers in this area. One of the more basic types is shown in Fig. 13.6. As indicated in the figure, the turn-off circuit consists of an *npn* transistor, a dc battery V_B, and a pulse generator. During SCR conduction the transistor is in the "off" state; that is, $I_B = 0$ and the collector-to-emitter impedance is very high (for all practical purposes an open circuit). This high impedance will isolate the turn-off circuitry from affecting the operation of the SCR. For turn-off conditions, a positive pulse is applied to the base of the transistor, turning it heavily on, resulting in a very low impedance

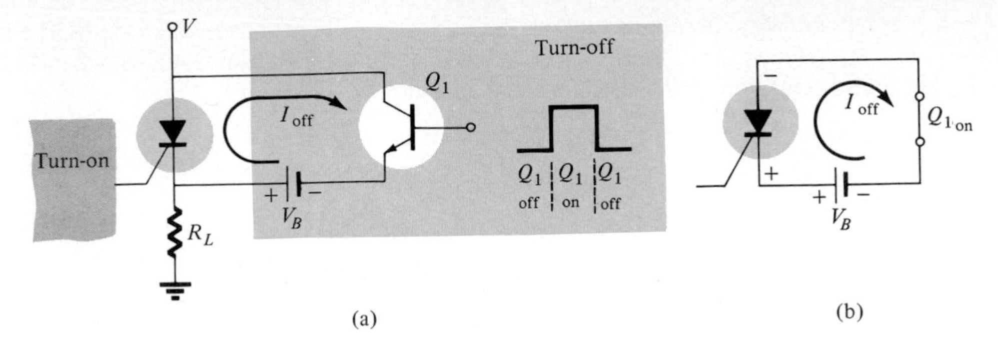

Figure 13.6 Forced-commutation technique.

from collector to emitter (short-circuit representation). The battery potential will then appear directly across the SCR as shown in Fig. 13.6b, forcing current through it in the reverse direction for turn-off. Turn-off times of SCRs are typically 5 to 30 μs.

13.4 SCR CHARACTERISTICS AND RATINGS

The characteristics of an SCR are provided in Fig. 13.7 for various values of gate current. The currents and voltages of usual interest are indicated on the characteristic. A brief description of each follows.

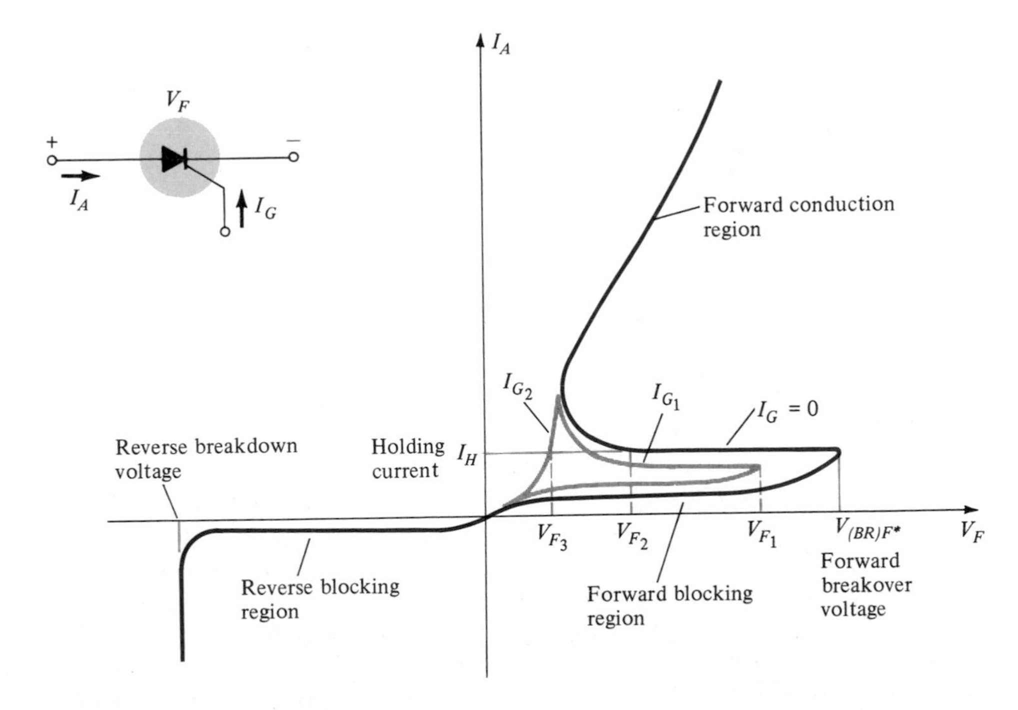

Figure 13.7 SCR characteristics.

1. *Forward breakover voltage* $V_{(BR)F^*}$ is that voltage above which the SCR enters the conduction region. The asterisk (*) is a letter to be added that is dependent on the condition of the gate terminal as follows:

$$O = \text{open circuit from } G \text{ to } K$$

$$S = \text{short circuit from } G \text{ to } K$$

$$R = \text{resistor from } G \text{ to } K$$

$$V = \text{fixed bias (voltage) from } G \text{ to } K$$

2. *Holding current* (I_H) is that value of current below which the SCR switches from the conduction state to the forward blocking region under stated conditions.
3. *Forward and reverse blocking regions* are the regions corresponding to the open-circuit condition for the controlled rectifier which *block* the flow of charge (current) from anode to cathode.
4. *Reverse breakdown voltage* is equivalent to the Zener or avalanche region of the fundamental two-layer semiconductor diode.

It should be immediately obvious that the SCR characteristics of Fig. 13.7 are very similar to those of the basic two-layer semiconductor diode except for the horizontal offshoot before entering the conduction region. It is this horizontal jutting region that gives the gate control over the response of the SCR. For the characteristic having the solid line in Fig. 13.7 ($I_G = 0$), V_F must reach the largest required breakover voltage ($V_{(BR)F^*}$) before the "collapsing" effect will result and the SCR can enter the conduction region corresponding to the *on* state. If the gate current is increased to I_{G_1}, as shown in the same figure by applying a bias voltage to the gate terminal, the value of V_F required for the conduction (V_{F_1}) is considerably less. Note also that I_H drops with increase in I_G. If increased to I_{G_2} the SCR will fire at very low values of voltage (V_{F_3}) and the characteristics begin to approach those of the basic *p-n* junction diode. Looking at the characteristics in a completely different sense, for a particular V_F voltage, say V_{F_2} (Fig. 13.7), if the gate current is increased from $I_G = 0$ to I_{G_1} or more, the SCR will fire.

The gate characteristics are provided in Fig. 13.8. The characteristics of Fig. 13.8b are an expanded version of the shaded region of Fig. 13.8a. In Fig. 13.8a the three gate ratings of greatest interest, P_{GFM}, I_{GFM}, and V_{GFM} are indicated. Each is included on the characteristics in the same manner employed for the transistor. Except for portions of the shaded region, any combination of gate current and voltage that falls within this region will fire any SCR in the series of components for which these characteristics are provided. Temperature will determine which sections of the shaded region must be avoided. At $-65°C$ the minimum current that will trigger the series of SCRs is 80 mA, while at $+150°C$ only 20 mA are required. The effect of temperature on the minimum gate voltage is usually not indicated on curves of this type since gate potentials of 3 V or more are usually obtained easily. As indicated on Fig. 13.8b, a minimum of 3 V is indicated for all units for the temperature range of interest.

Other parameters usually included on the specification sheet of an SCR are the turn-on time (t_{on}), turn-off time (t_{off}), junction temperature (T_J), and case temperature (T_C), all of which should by now be, to some extent, self-explanatory.

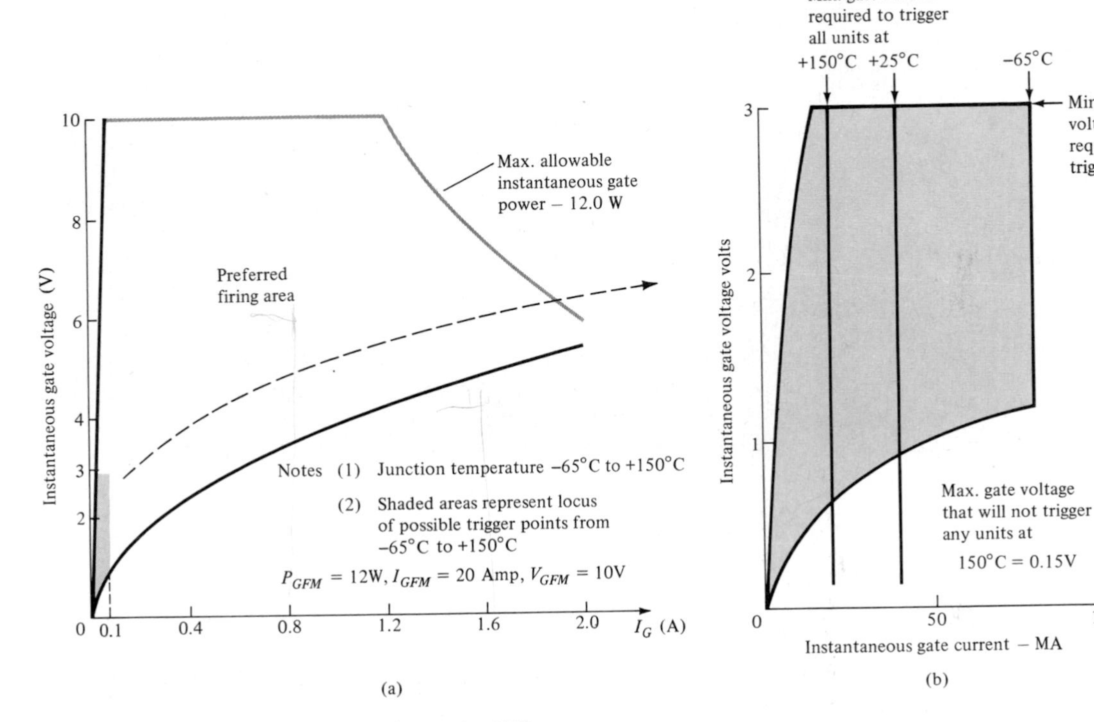

Figure 13.8 SCR gate characteristics (GE series C38).

13.5 SCR CONSTRUCTION AND TERMINAL IDENTIFICATION

The basic construction of the four-layer pellet of an SCR is shown in Fig. 13.9a. The complete construction of a thermal-fatigue-free, high-current SCR is shown in Fig.

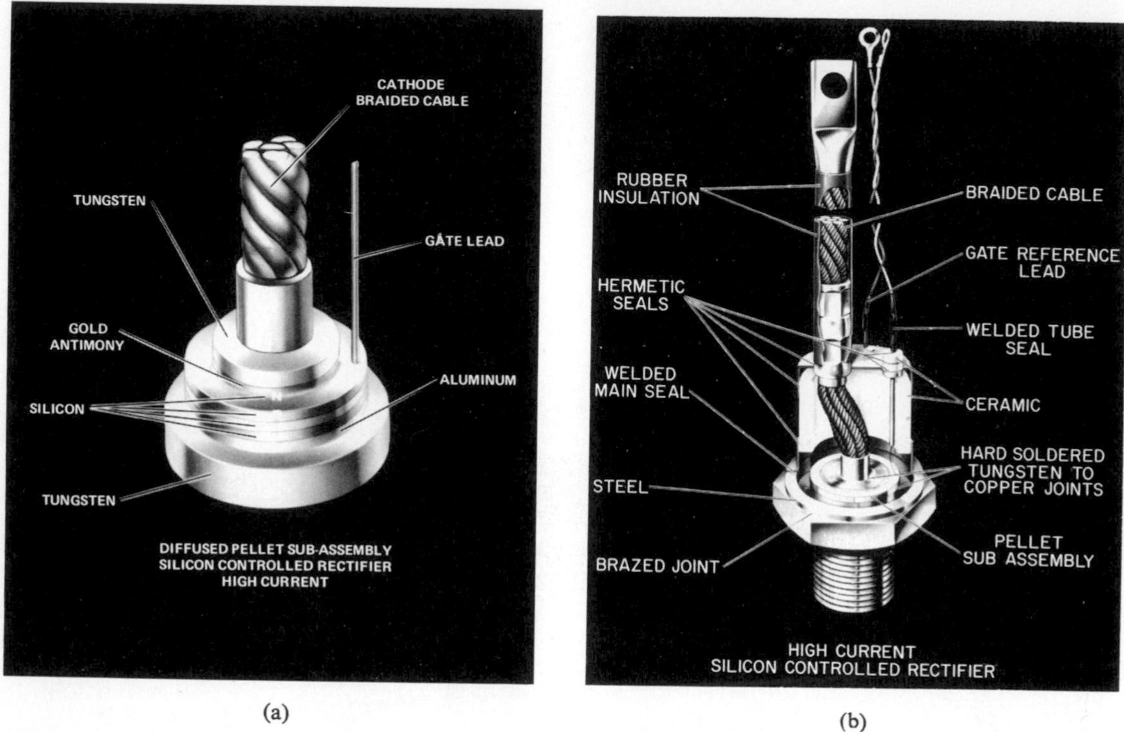

Figure 13.9 (a) Alloy-diffused SCR pellet; (b) thermal fatigue-free SCR construction. (Courtesy General Electric Company.)

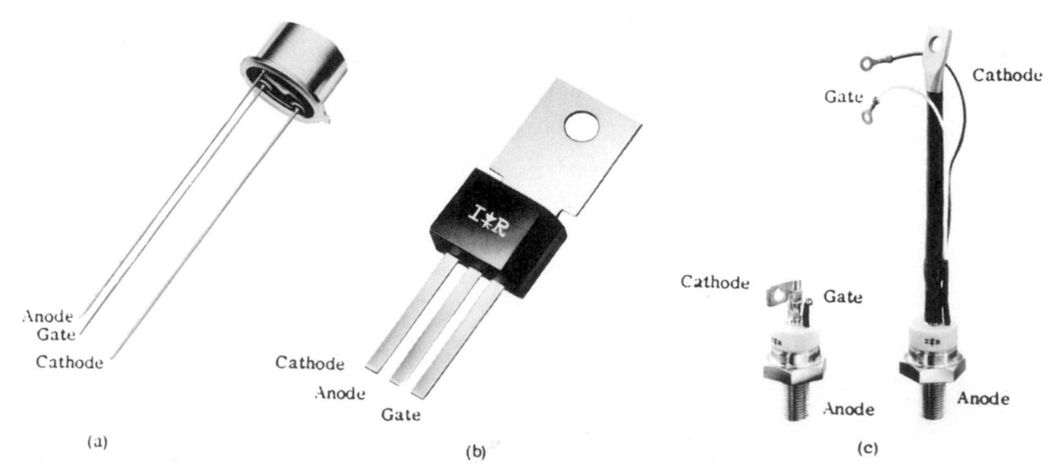

Figure 13.10 SCR case construction and terminal identification. [(a) Courtesy General Electric Company; (b) and (c) courtesy International Rectifier Corporation.]

13.9b. Note the position of the gate, cathode, and anode terminals. The pedestal acts as a heat sink by transferring the heat developed to the chassis on which the SCR is mounted. The case construction and terminal identification of SCRs will vary with the application. Other case-construction techniques and the terminal identification of each are indicated in Fig. 13.10.

13.6 SCR APPLICATIONS

A few of the possible applications for the SCR are listed in the introduction to the SCR (Section 13.2). In this section we consider five: a static switch, phase-control system, battery charger, temperature controller, and single-source emergency-lighting system.

A half-wave *series static switch* is shown in Fig. 13.11a. If the switch is closed as shown in Fig. 13.11b, a gate current will flow during the positive portion of the input signal, turning the SCR on. Resistor R_1 limits the magnitude of the gate current. When the SCR turns on, the anode-to cathode voltage (V_F) will drop to the conduction value, resulting in a greatly reduced gate current and very little loss in the gate circuitry. For the negative region of the input signal the SCR will turn off, since the anode is negative with respect to the cathode. The diode D_1 is included to prevent a reversal in gate current.

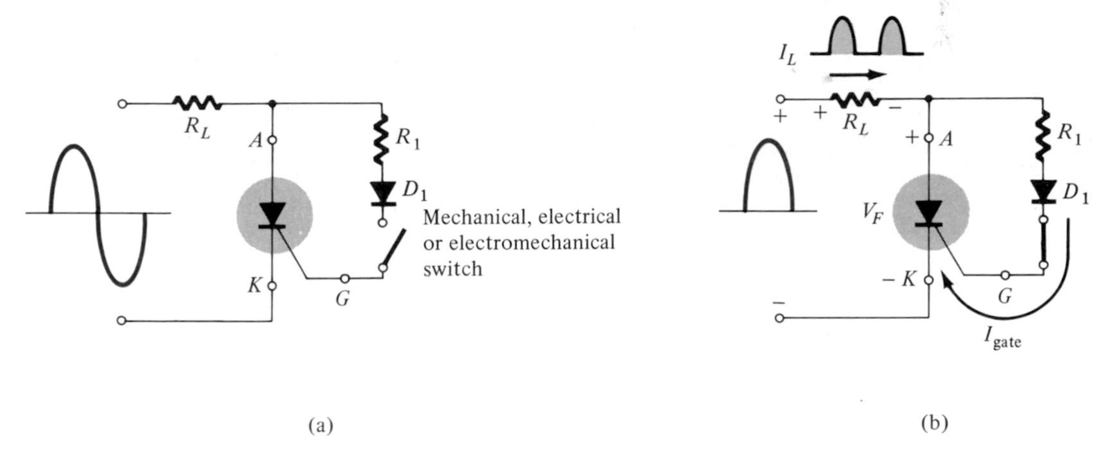

Figure 13.11 Half-wave series static switch.

The waveforms for the resulting load current and voltage are shown in Fig. 13.11b. The result is a half-wave-rectified signal through the load. If less than 180° conduction is desired, the switch can be closed at any phase displacement during the positive portion of the input signal. The switch can be electronic, electromagnetic, or mechanical, depending on the application.

A circuit capable of establishing a conduction angle between 90 and 180° is shown in Fig. 13.12a. The circuit is similar to that of Fig. 13.11a except for the addition of a variable resistor and the elimination of the switch. The combination of the resistors R and R_1 will limit the gate current during the positive portion of the input signal. If

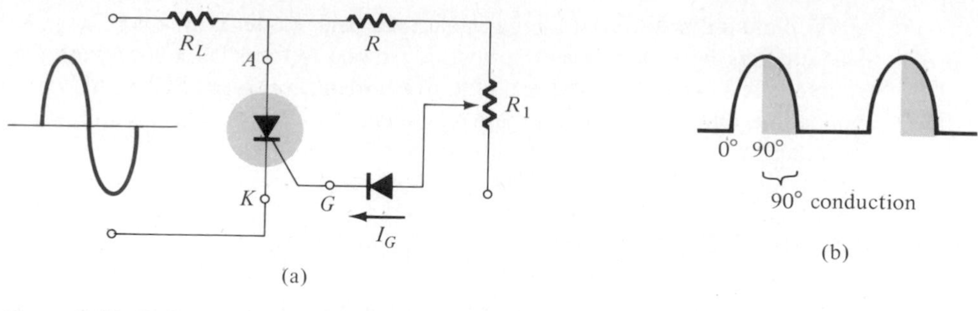

Figure 13.12 Half-wave variable-resistance phase control.

R_1 is set to its maximum value, the gate current may never reach turn-on magnitude. As R_1 is decreased from the maximum the gate current will increase from the same input voltage. In this way, the required turn-on gate current can be established in any point between 0 and 90° as shown in Fig. 13.12b. If R_1 is low, the SCR will fire almost immediately, resulting in the same action as that obtained from the circuit of Fig. 13.11a (180° conduction). However, as indicated above, if R_1 is increased, a larger input voltage (positive) will be required to fire the SCR. As shown in Fig. 13.12b, the control cannot be extended past a 90° phase displacement since the input is its maximum at this point. If it fails to fire at this and lesser values of input voltage on the positive slope of the input, the same response must be expected from the negatively sloped portion of the signal waveform. The operation here is normally referred to in technical terms as *half-wave variable-resistance phase control*. It is an effective method of controlling the rms current and therefore power to the load.

A third popular application of the SCR is in a *battery-charging regulator*. The fundamental components of the circuit are shown in Fig. 13.13. You will note that the control circuit has been blocked off for discussion purposes.

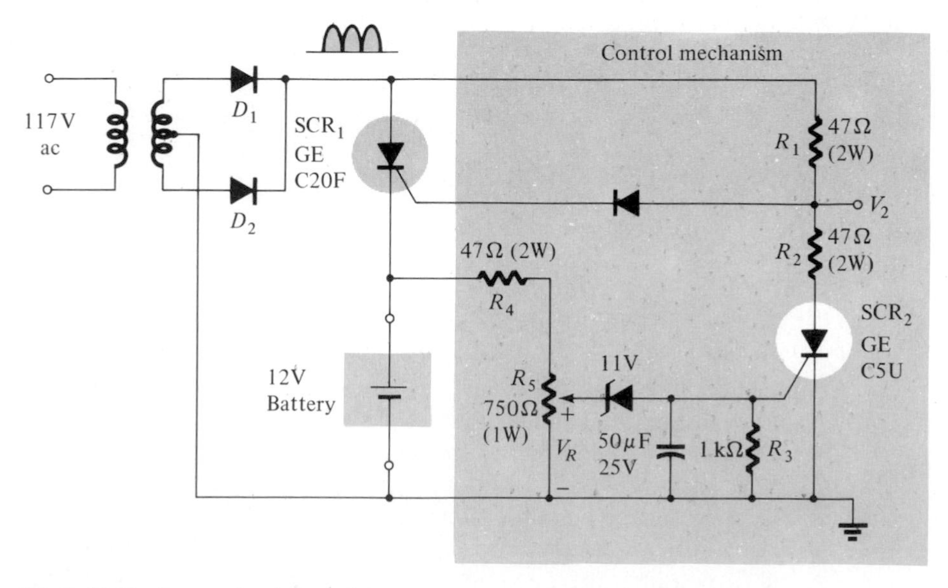

Figure 13.13 Battery-charging regulator.

As indicated in the figure, D_1 and D_2 establish a full-wave-rectified signal across SCR_1 and the 12-V battery to be charged. At low battery voltages SCR_2 is in the off state for reasons to be explained shortly. With SCR_2 open, the SCR_1 controlling circuit is exactly the same as the series static switch control discussed earlier in this section. When the full-wave-rectified input is sufficiently large to produce the required turn-on gate current (controlled by R_1), SCR_1 will turn on and charging of the battery will commence. At the start of charging, the low battery voltage will result in a low voltage V_R as determined by the simple voltage-divider circuit. Voltage V_R is in turn too small to cause 11.0-V Zener conduction. In the off state, the Zener is effectively an open circuit maintaining SCR_2 in the "off" state since the gate current is zero. The capacitor C_1 is included to prevent any voltage transients in the circuit from accidentally turning on SCR_2. Recall from your fundamental study of circuit analysis that the voltage cannot instantaneously change across a capacitor. In this way C_1 prevents transient effects from affecting the SCR.

As charging continues, the battery voltage rises to a point where V_R is sufficiently high to both turn on the 11.0-V Zener and fire SCR_2. Once SCR_2 has fired, the short-circuit representation for SCR_2 will result in a voltage-divider circuit determined by R_1 and R_2 that will maintain V_2 at a level too small to turn SCR_1 on. When this occurs, the battery is fully charged and the open-circuit state of SCR_1 will cut off the charging current. Thus the regulator recharges the battery whenever the voltage drops and prevents overcharging when fully charged.

The schematic diagram of a 100-W heater control using an SCR appears in Fig. 13.14. It is designed such that the 100-W heater will turn on and off as determined by thermostats. Mercury-in-glass thermostats are very sensitive to temperature change. In fact, they can sense changes as small as 0.1°C. It is limited in application, however, in that it can only handle very low levels of current—below 1 mA. In this application, the SCR serves as a current amplifier in a load-switching element. It is not an amplifier in the sense that it magnifies the current level of the thermostat. Rather it is a device whose higher current level is controlled by the behavior of the thermostat.

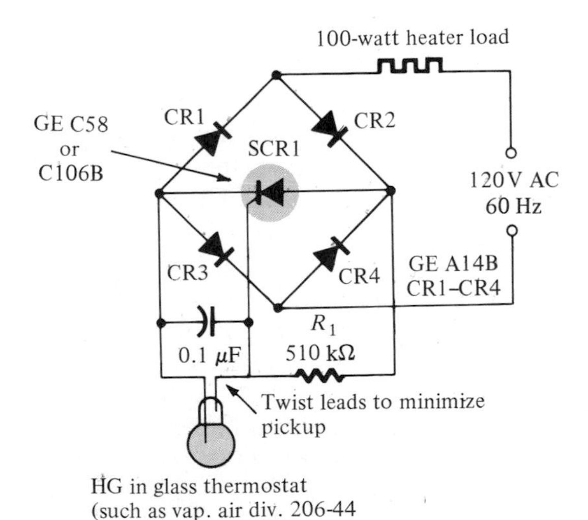

Figure 13.14 Temperature controller. (Courtesy General Electric Semiconductor Products Division.)

It should be clear that the bridge network is connected to the ac supply through the 100-W heater. This will result in a full-wave-rectified voltage across the SCR. When the thermostat is open, the voltage across the capacitor will charge to a gate-firing potential through each pulse of the rectified signal. The charging time constant is determined by the RC product. This will trigger the SCR during each half-cycle of the input signal, permitting a flow of charge (current) to the heater. As the temperature rises, the conductive thermostat will short-circuit the capacitor, eliminating the possibility of the capacitor charging to the firing potential and triggering the SCR. The 510-kΩ resistor will then contribute to maintaining a very low current (less than 250 μA) through the thermostat.

The last application for the SCR to be described is shown in Fig. 13.15. It is a single-source emergency-lighting system that will maintain the charge on a 6-V battery to ensure its availability and also provide dc energy to a bulb if there is a power shortage.

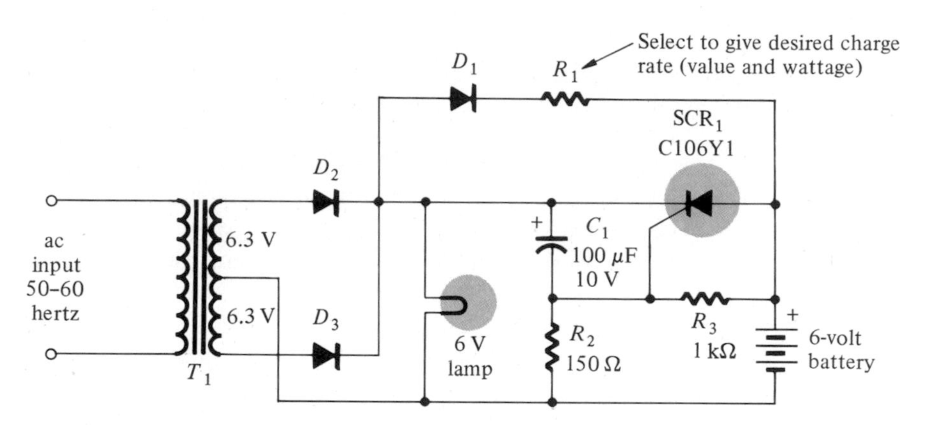

Figure 13.15 Single-source emergency lighting system. (Courtesy General Electric Semiconductor Products Division.)

A full-wave-rectified signal will appear across the 6-V lamp due to diodes D_2 and D_1. The capacitor C_1 will charge to a voltage slightly less than a difference between the peak value of the full-wave-rectified signal and the dc voltage across R_2 established by the 6-V battery. In any event, the cathode of SCR_1 is higher than the anode and the gate-to-cathode voltage is negative, ensuring that the SCR is nonconducting. The battery is being charged through R_1 and D_1 at a rate determined by R_1. Charging will only take place when the anode of D_1 is more positive than its cathode. The dc level of the full-wave-rectified signal will ensure that the bulb is lit when the power is on. If the power should fail, the capacitor C_1 will discharge through D_1, R_1, and R_3 until the cathode of SCR_1 is less positive than the anode. At the same time the junction of R_2 and R_3 will become positive and establish sufficient gate-to-cathode voltage to trigger the SCR. Once fired, the 6-V battery would discharge through the SCR_1 and energize the lamp and maintain its illumination. Once power is restored the capacitor C_1 will recharge and reestablish the nonconducting state of SCR_1 as described above.

13.7 SILICON-CONTROLLED SWITCH

The silicon-controlled switch (SCS), like the silicon-controlled rectifier, is a four-layer *pnpn* device. All four semiconductor layers of the SCS are available due to the addition of an anode gate, as shown in Fig. 13.16a. The graphic symbol and transistor equivalent circuit are shown in the same figure. The characteristics of the device are essentially the same as those for the SCR. The effect of an anode gate current is very similar to that demonstrated by the gate current in Fig. 13.7. The higher the anode gate current, the lower the required anode-to-cathode voltage to turn the device on.

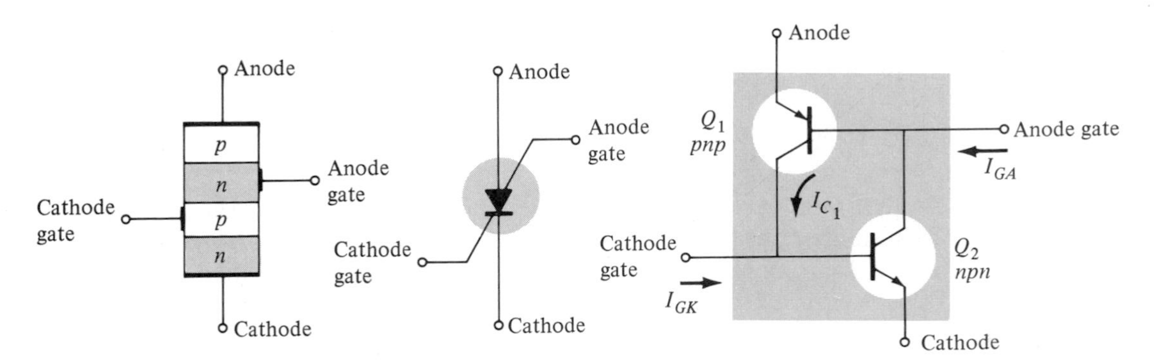

Figure 13.16 Silicon-controlled switch (SCS): (a) basic construction; (b) graphic symbol; (c) equivalent transistor circuit.

The anode gate connection can be used to turn the device either on or off. To turn on the device, a negative pulse must be applied to the anode gate terminal, while a positive pulse is required to turn off the device. The need for the type of pulse indicated above can be demonstrated using the circuit of Fig. 13.16c. A negative pulse at the anode gate will forward-bias the base-to-emitter junction of Q_1, turning it on. The resulting heavy collector current I_{C_1} will turn on Q_2, resulting in a regenerative action and the on state for the SCS device. A positive pulse at the anode gate will reverse-bias the base-to-emitter junction of Q_1, turning it off, resulting in the open-circuit off state of the device. In general, the triggering (turn-on) anode gate current is larger in magnitude than the required cathode gate current. For one representative SCS device, the triggering anode gate current is 1.5 mA while the required cathode gate current is 1 μA. The required turn-on gate current at either terminal is affected by many factors. A few include the operating temperature, anode-to-cathode voltage, load placement, and type of cathode, gate-to-cathode or anode gate-to-anode connection (short-circuit, open-circuit, bias, load, etc.). Tables, graphs, and curves are normally available for each device to provide the type of information indicated above.

Three of the more fundamental types of turn-off circuits for the SCS are shown in Fig. 13.17. When a pulse is applied to the circuit of Fig. 13.17a, the transistor conducts heavily, resulting in a low-impedance ($\cong$ short-circuit) characteristic between collector and emitter. This low-impedance branch diverts anode current away from the SCS, dropping it below the holding value and consequently turning it off.

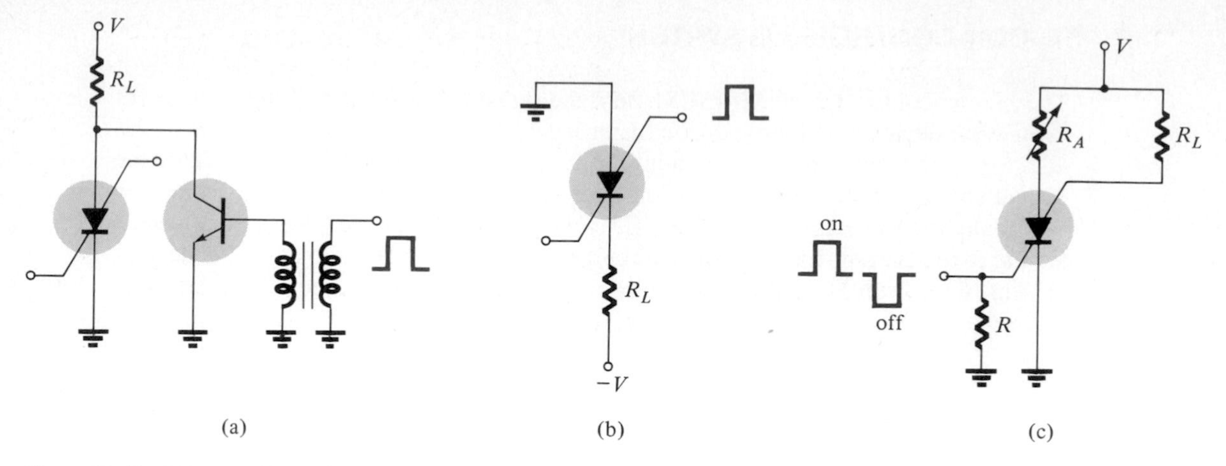

Figure 13.17 SCS turn-off techniques.

Similarly, the positive pulse at the anode gate of Fig. 13.17b will turn the SCS off by the mechanism described earlier in this section. The circuit of Fig. 13.17c can be turned either off *or* on by a pulse of the proper magnitude at the cathode gate. The turn-off characteristic is possible only if the correct value of R_A is employed. It will control the amount of regenerative feedback, the magnitude of which is critical for this type of operation. Note the variety of positions in which the load resistor R_L can be placed. There are a number of other possibilities that can be found in any comprehensive semiconductor handbook or manual.

An advantage of the SCS over a corresponding SCR is the reduced turn-off time, typically within the range 1 to 10 μs for the SCS and 5 to 30 μs for the SCR. Some of the remaining advantages of the SCS over an SCR include increased control and triggering sensitivity and a more predictable firing situation. At present, however, the SCS is limited to low power, current, and voltage ratings. Typical maximum anode currents range from 100 to 300 mA with dissipation (power) ratings of 100 to 500 mW.

A few of the more common areas of application include a wide variety of computer circuits (counters, registers, and timing circuits) pulse generators, voltage sensors, and oscillators. One simple application for an SCS as a voltage-sensing device is shown in Fig. 13.18. It is an alarm system with n inputs from various stations. Any single

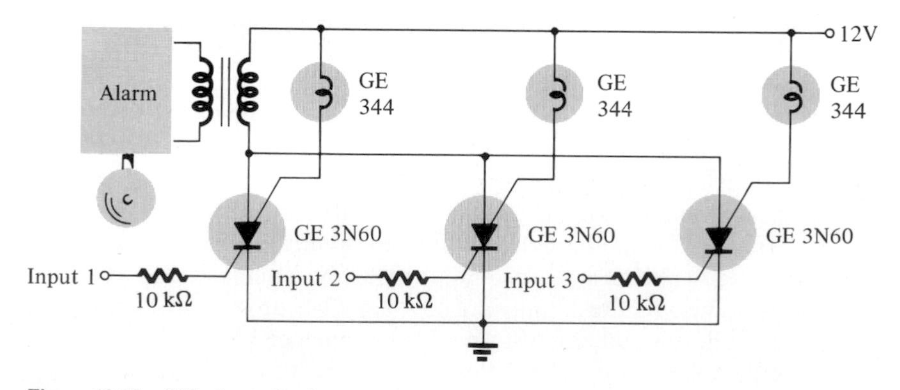

Figure 13.18 SCS alarm circuit.

input will turn that particular SCS on, resulting in an energized alarm relay and light in the anode gate circuit to indicate the location of the input (disturbance).

One additional application of the SCS is in the alarm circuit of Fig. 13.19. R_S represents a temperature-, light-, or radiation-sensitive resistor, that is, an element whose resistance will decrease with the application of any of the three energy sources listed above. The cathode gate potential is determined by the divider relationship established by R_S and the variable resistor. Note that the gate potential is at approximately 0 volts if R_S equals the value set by the variable resistor, since both resistors will have 12 V across them. However, if R_S decreases, the potential of the junction will increase until the SCS is forward biased, causing the SCS to turn on and energize the alarm relay.

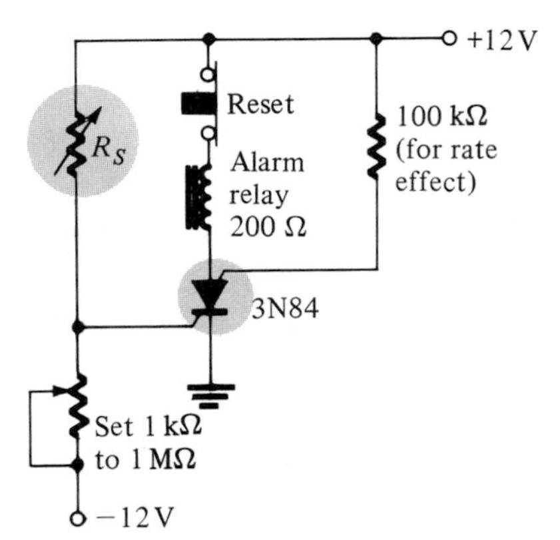

Figure 13.19 Alarm circuit. (Courtesy General Electric Semiconductor Products Division.)

The 100-kΩ resistor is included to reduce the possibility of accidental triggering of the device through a phenomena known as *rate effect*. It is caused by the stray capacitance levels between gates. A high-frequency transient can establish sufficient base current to turn the SCS on accidentally. The device is reset by pressing the reset button, which in turn opens the conduction path of the SCS and reduces the anode current to zero.

Sensitivity to resistors R_S that increase in resistance due to the application of any of the three energy sources described above can be accommodated by simply interchanging the location of R_S and the variable resistor. The terminal identification of an SCS is shown in Fig. 13.20 with a packaged SCS.

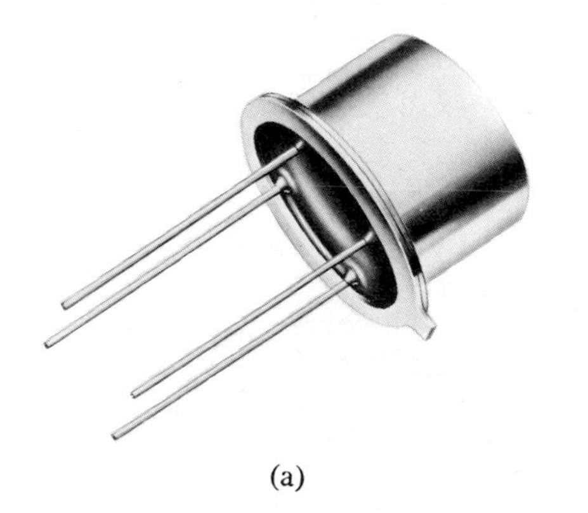

(a)

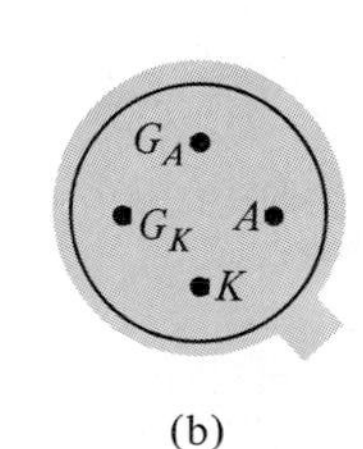

(b)

Figure 13.20 Silicon-controlled switch (SCS): (a) device; (b) terminal identification. (Courtesy General Electric Company.)

13.8 GATE TURN-OFF SWITCH

The gate turn-off switch (GTO) is the third *pnpn* device to be introduced in this chapter. Like the SCR, however, it has only three external terminals, as indicated in Fig. 13.21a. Its graphic symbol is also shown in Fig. 13.21b. Although the graphic symbol is different from either the SCR or the SCS, the transistor equivalent is exactly the same and the characteristics are similar.

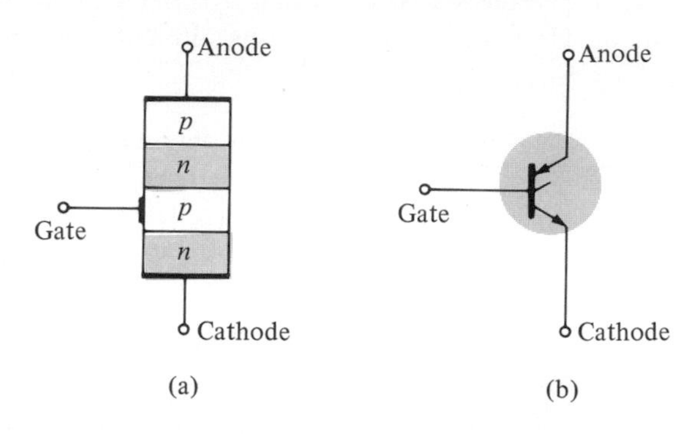

Figure 13.21 Gate turn-off switch (GTO): (a) basic construction; (b) symbol.

The most obvious advantage of the GTO over the SCR or SCS is the fact that it can be turned on *or* off by applying the proper pulse to the cathode gate (without the anode gate and associated circuitry required for the SCS). A consequence of this turn-off capability is an increase in the magnitude of the required gate current for triggering. For an SCR and GTO of similar maximum rms current ratings, the gate-triggering current of a particular SCR is 30 μA, while the triggering current of the GTO is 20 mA. The turn-off current of a GTO is slightly larger than the required triggering current. The maximum rms current and dissipation ratings of GTOs manufactured today are limited to about 3 A and 20 W, respectively.

A second very important characteristic of the GTO is improved switching characteristics. The turn-on time is similar to the SCR (typically 1 μs), but the turn-off time of about the *same* duration (1 μs) is much smaller than the typical turn-off time of an SCR (5 to 30 μs). The fact that the turn-off time is similar to the turn-on time rather than considerably larger permits the use of this device in high-speed applications.

A typical GTO and its terminal identification are shown in Fig. 13.22. The GTO gate input characteristics and turn-off circuits can be found in a comprehensive manual or specification sheet. The majority of the SCR turn-off circuits can also be used for GTOs.

Some of the areas of application for the GTO include counters, pulse generators, multivibrators, and voltage regulators. Figure 13.23 is an illustration of a simple sawtooth generator employing a GTO and a Zener diode.

When the supply is energized, the GTO will turn on, resulting in the short-circuit equivalent from anode to cathode. The capacitor C_1 will then begin to charge toward the supply voltage as shown in Fig. 13.23. As the voltage across the capacitor C_1 charges above the Zener potential, a reversal in gate-to-cathode voltage will result, establishing a reversal in gate current. Eventually, the negative gate current will be large enough to turn the GTO off. Once the GTO turns off, resulting in the open-circuit

Figure 13.22 Typical GTO and its terminal identification. (Courtesy General Electric Company.)

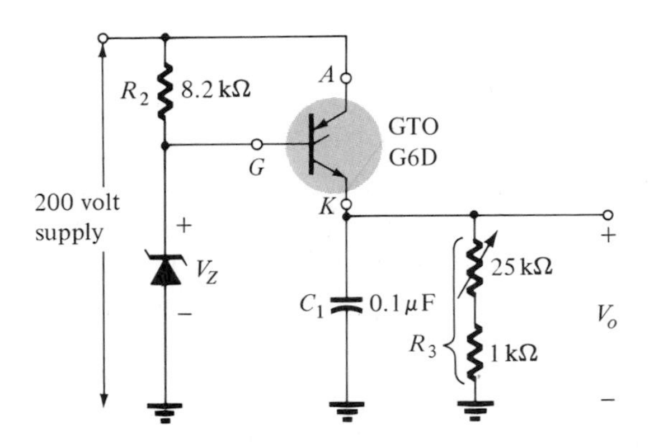

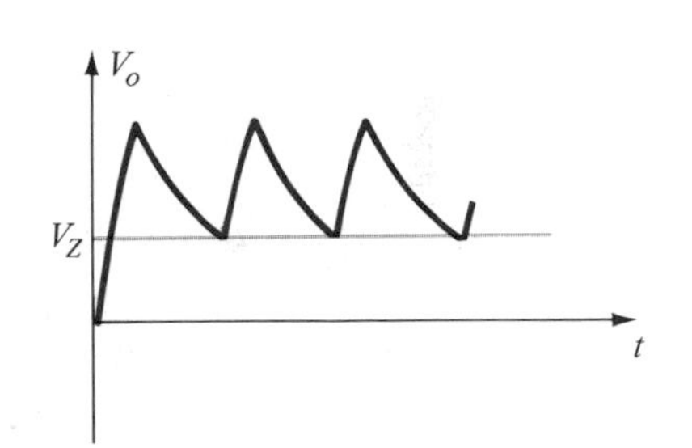

Figure 13.23 GTO sawtooth generator.

representation, the capacitor C_1 will discharge through the resistor R_3. The discharge time will be determined by the circuit time constant $\tau = R_3C_1$. The proper choice of R_3 and C_1 will result in the sawtooth waveform of Fig. 13.23. Once the output potential V_o drops below V_Z, the GTO will turn on and the process will repeat.

13.9 LIGHT-ACTIVATED SCR

The next in the series of *pnpn* devices is the light-activated SCR (LASCR). As indicated by the terminology, it is an SCR whose state is controlled by the light falling upon a silicon semiconductor layer of the device. The basic construction of an LASCR is shown in Fig. 13.24a. As indicated in Fig. 13.24a, a gate lead is also provided to permit triggering the device using typical SCR methods. Note also in the figure that the mounting surface for the silicon pellet is the anode connection for the device.

The graphic symbols most commonly employed for the LASCR are provided in Fig. 13.24b. The terminal identification and typical LASCRs are shown in Fig. 13.25a.

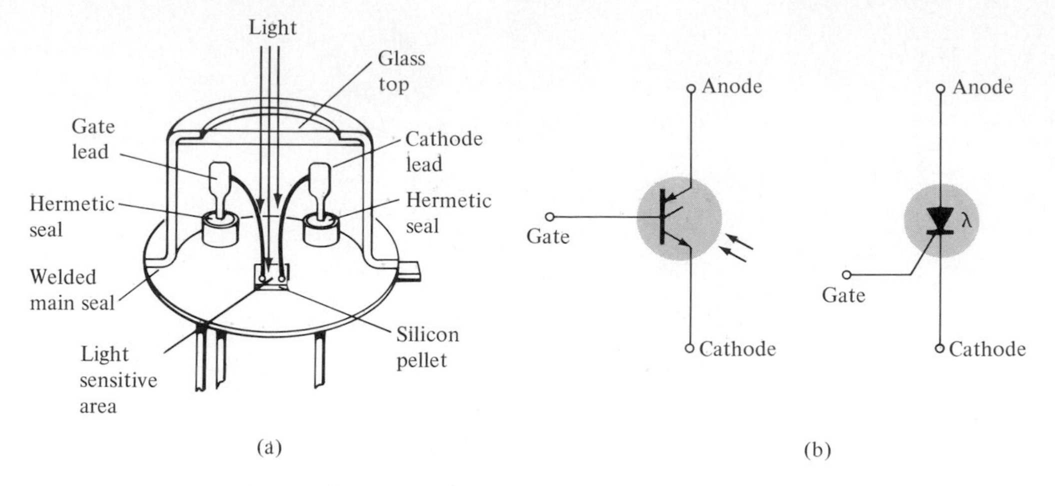

Figure 13.24 Light-activated SCR (LASCR): (a) basic construction; (b) symbols.

Some of the areas of application for the LASCR include optical light controls, relays, phase control, motor control, and a variety of computer applications. The maximum current (rms) and power (gate) ratings for LASCRs commercially available today are about 3 A and 0.1 W. The characteristics (light triggering) of a representative LASCR are provided in Fig. 13.25b. Note in this figure that an increase in junction temperature results in a reduction in light energy required to activate the device.

One interesting application of an LASCR is in the AND and OR circuits of Fig. 13.26. Only when light falls on $LASCR_1$ *and* $LASCR_2$ will the short-circuit representation for each be applicable and the supply voltage appear across the load. For the OR circuit, light energy applied to $LASCR_1$ *or* $LASCR_2$ will result in the supply voltage appearing across the load.

The LASCR is most sensitive to light when the gate terminal is open. Its sensitivity can be reduced and controlled somewhat by the insertion of a gate resistor, as shown in Fig. 13.26.

A second application of the LASCR appears in Fig. 13.27. It is the semiconductor analog of an electromechanical relay. Note that it offers complete isolation between the input and switching element. The energizing current can be passed through a light-emitting diode or a lamp, as shown in the figure. The incident light will cause the LASCR to turn on and permit a flow of charge (current) through the load as

Figure 13.25 LASCR: (a) appearance and terminal identification; (b) light-triggering characteristics. (Courtesy General Electric Company.)

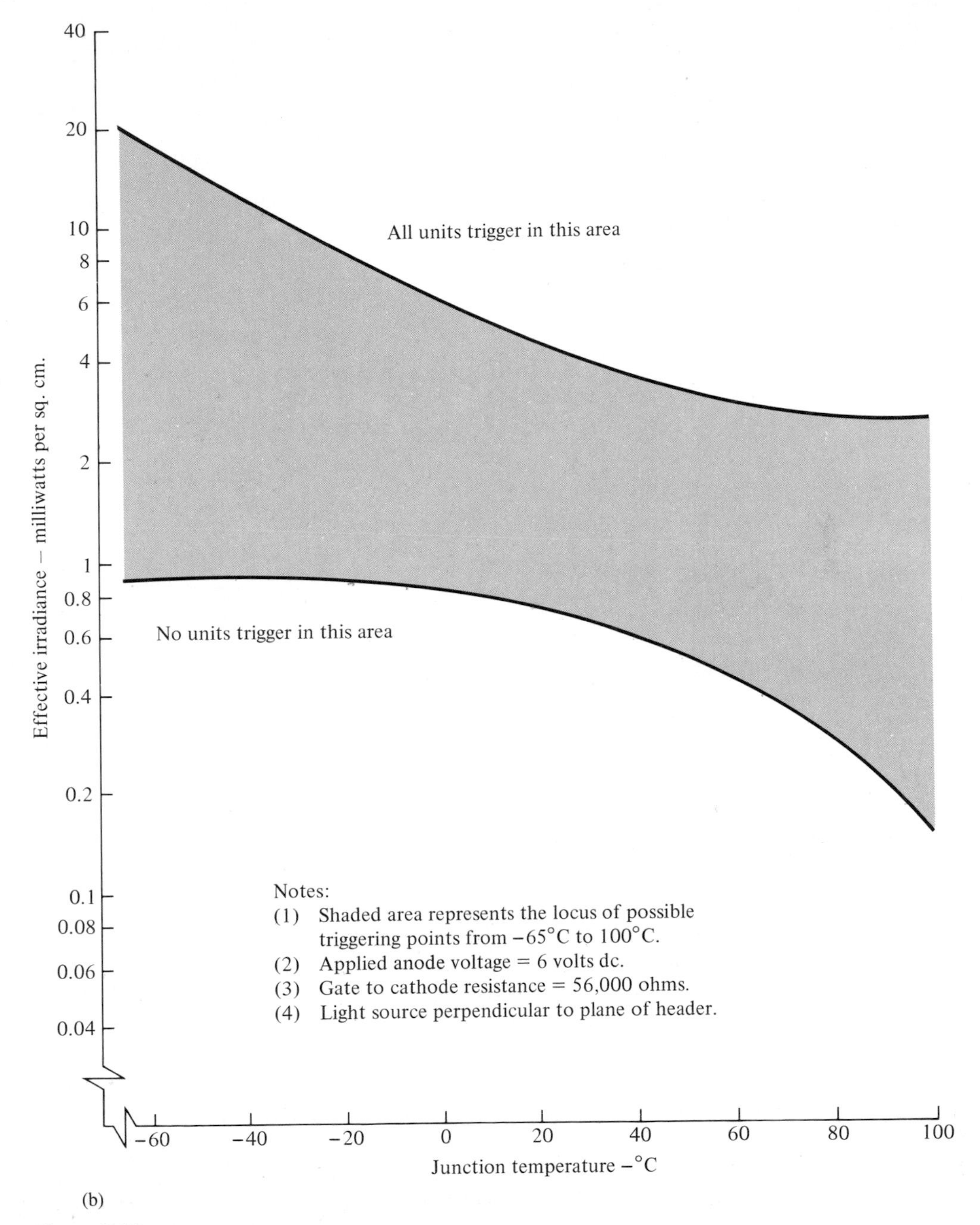

(b)

Figure 13.25 (*cont.*)

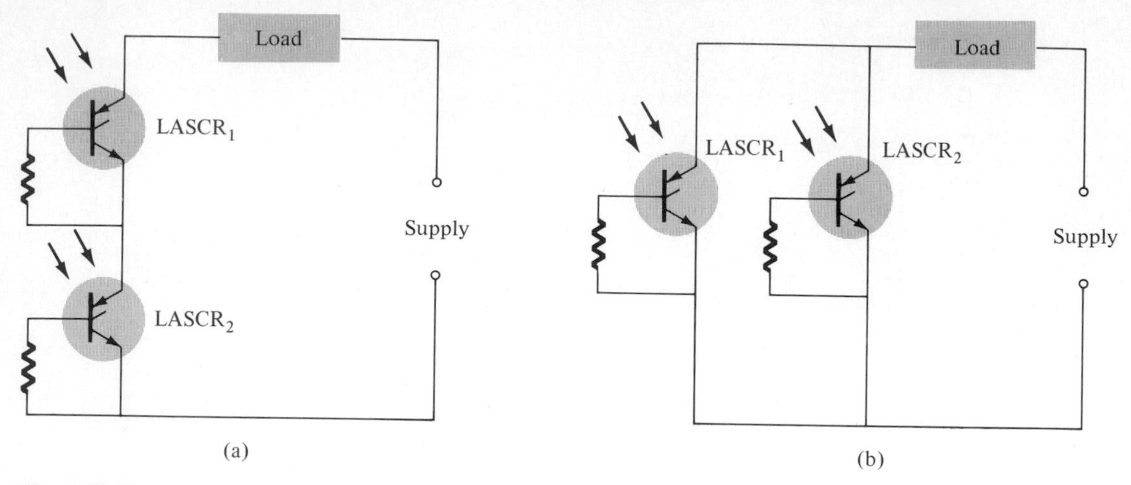

Figure 13.26 LASCR optoelectronic logic circuitry: (a) AND gate-input to $LASCR_1$ *and* $LASCR_2$ required for energization of the load; (b) OR gate—input to either $LASCR_1$ *or* $LASCR_2$ will energize the load.

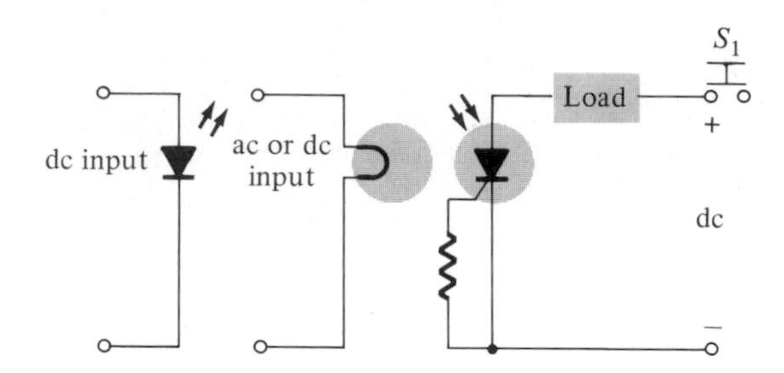

Figure 13.27 Latching relay. (Courtesy General Electric Semiconductor Products Division.)

established by the dc supply. The LASCR can be turned off using the reset switch S_1. This system offers the additional advantages over an electromechanical switch of long life, microsecond response, small size, and the elimination of contact bounce.

13.10 SHOCKLEY DIODE

The Shockley diode is a four-layer *pnpn* diode with only two external terminals, as shown in Fig. 13.28a with its graphic symbol. The characteristics (Fig. 13.28b) of the device are exactly the same as those encountered for the SCR with $I_G = 0$. As indicated by the characteristics, the device is in the off state (open-circuit representation) until the breakover voltage is reached, at which time avalanche conditions develop and the device turns on (short-circuit representation).

One common application of the Shockley diode is shown in Fig. 13.29, where it is employed as a trigger switch for an SCR. When the circuit is energized, the voltage across the capacitor will begin to change toward the supply voltage. Eventually, the voltage across the capacitor will be sufficiently high to first turn on the Shockley diode and then the SCR.

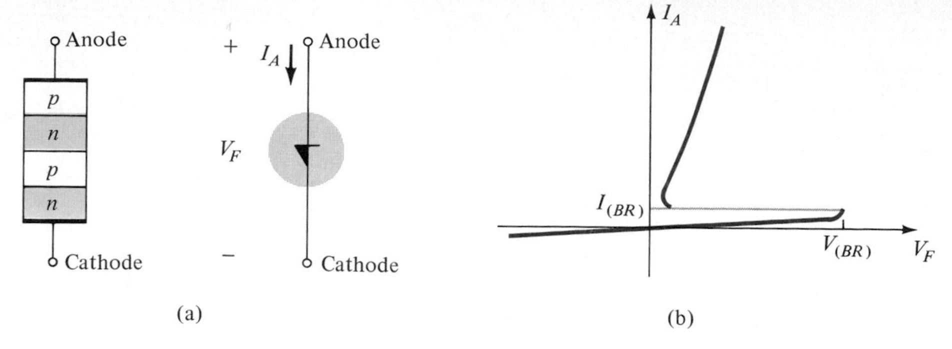

Figure 13.28 Shockley diode: (a) basic construction and symbol; (b) characteristics.

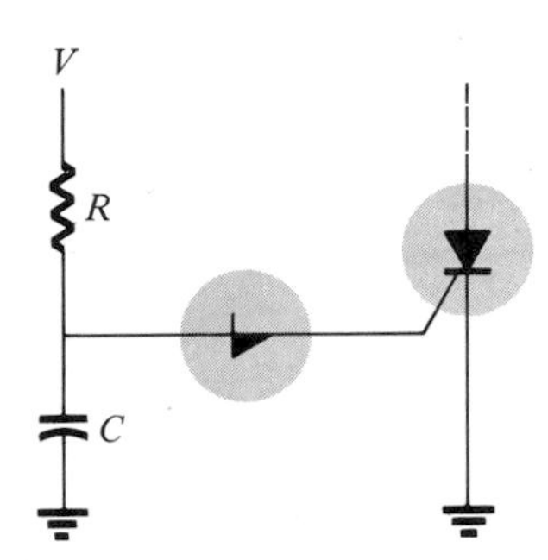

Figure 13.29 Shockley diode application—trigger switch for an SCR.

13.11 DIAC

The diac is basically a two-terminal parallel-inverse combination of semiconductor layers that permits triggering in either direction. The characteristics of the device, presented in Fig. 13.30a, clearly demonstrate that there is a breakover voltage in either direction. This possibility of an on condition in either direction can be used to its fullest advantage in *ac* applications.

The basic arrangement of the semiconductor layers of the diac is shown in Fig. 13.30b, along with its graphic symbol. Note that neither terminal is referred to as the cathode. Instead, there is an anode 1 (or electrode 1) and an anode 2 (or electrode 2). When anode 1 is positive with respect to anode 2, the semiconductor layers of particular interest are $p_1 n_2 p_2$ and n_3. For anode 2 positive with respect to anode 1 the applicable layers are $p_2 n_2 p_1$ and n_1.

For the unit appearing in Fig. 13.30, the breakdown voltages are very close in magnitude but may vary from a minimum of 28 V to a maximum of 42 V. They are related by the following equation provided in the specification sheet:

$$V_{BR_1} = V_{BR_2} \pm 10\% \ V_{BR_2} \tag{13.1}$$

The current levels (I_{BR_1} and I_{BR_2}) are also very close in magnitude for each device. For the unit of Fig. 13.30, both current levels are about 200 μA = 0.2 mA.

The use of the diac in a proximity detector appears in Fig. 13.31. Note the use of

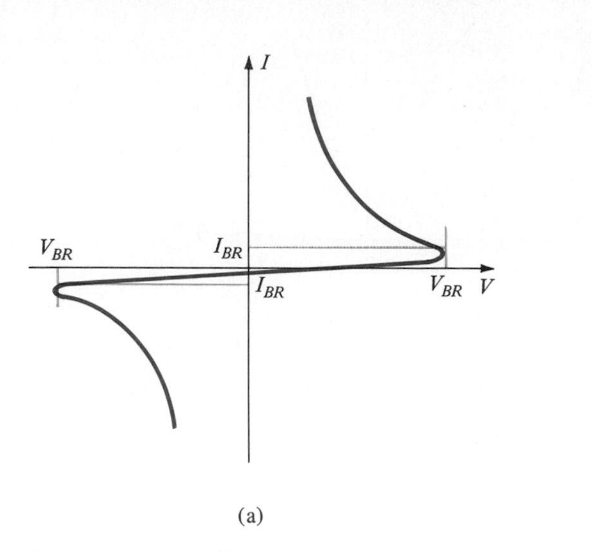

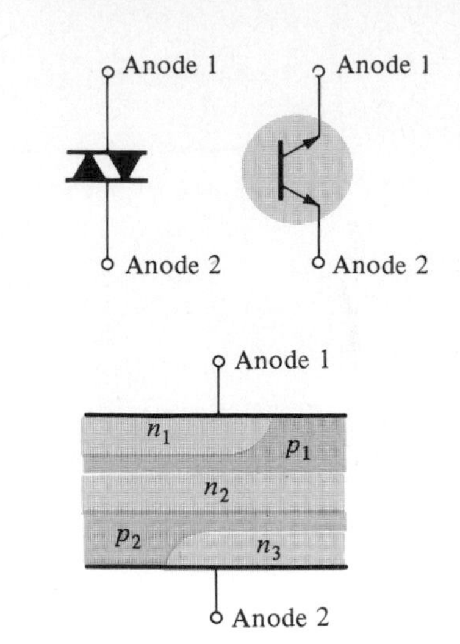

Figure 13.30 Diac: (a) characteristics; (b) symbols and basic construction. (Courtesy General Electric Company.)

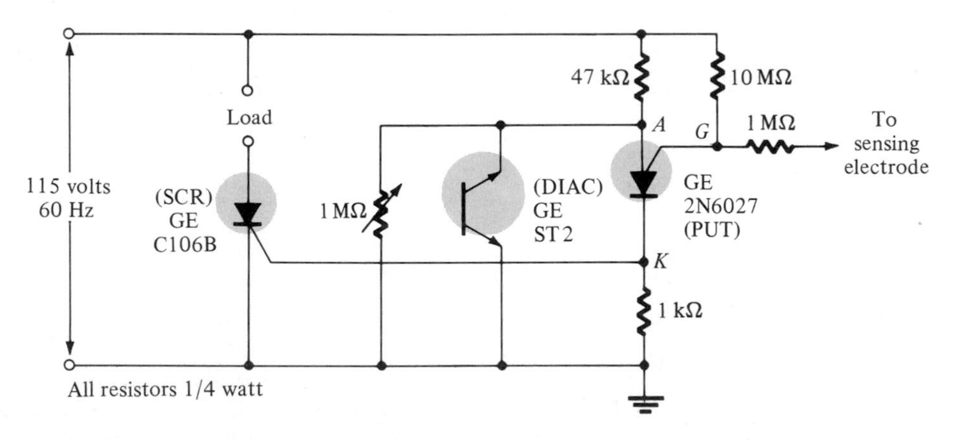

Figure 13.31 Proximity detector or touch switch. (Courtesy General Electric Semiconductor Products Division.)

an SCR in series with the load and the programmable unijunction transistor (to be described in Section 13.13) connected directly to the sensing electrode.

As the human body approaches the sensing electrode, the capacitance between the electrode and ground will increase. The programmable UJT (PUT) is a device that will fire (enter the short-circuit state) when the anode voltage (V_A) is at least 0.7 V (for silicon) greater than the gate voltage (V_G). Before the programmable device turns on, the system is essentially as shown in Fig. 13.32. As the input voltage rises, the diac voltage V_G will follow as shown in the figure until the firing potential is reached. It will then turn on and the diac voltage will drop substantially, as shown. Note that the diac

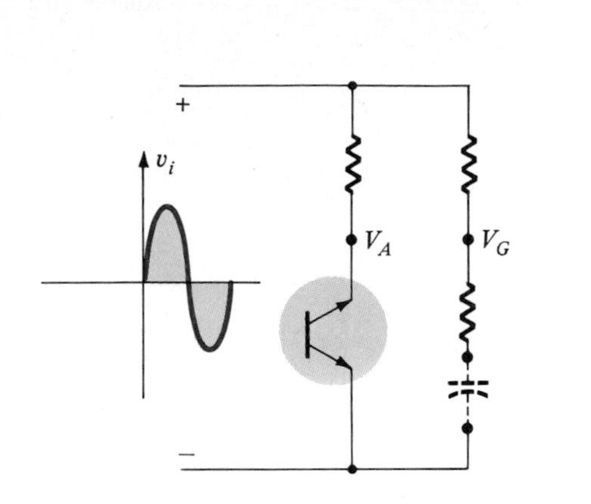

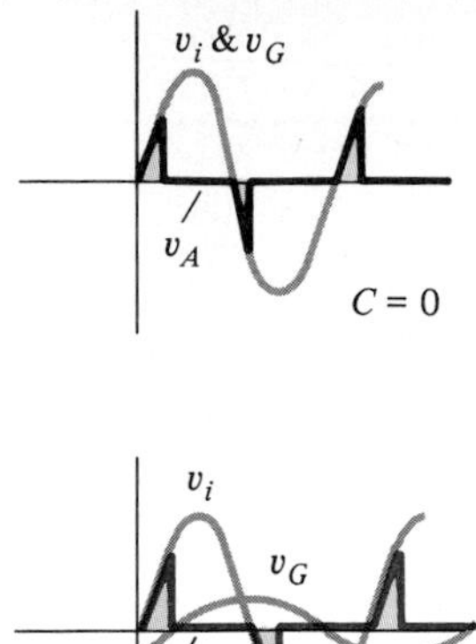

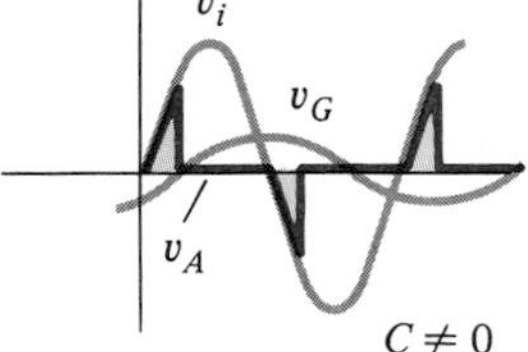

Figure 13.32 Effect of capacitive element on the behavior of the network of Fig. 13.31.

is in essentially an open-circuit state until it fires. Before the capacitive element is introduced, the voltage V_G will be the same as the input. As indicated in the figure, since both V_A and V_G follow the input, V_A can never be greater than V_G by 0.7 V and turn on the device. However, as the capacitive element is introduced, the voltage V_G will begin to lag the input voltage by an increasing angle, as indicated in the figure. There is therefore a point established where V_A can exceed V_G by 0.7 V and cause the programmable device to fire. A heavy current is established through the PUT at this point, raising the voltage V_K and turning on the SCR. A heavy SCR current will then exist through the load, reacting to the presence of the approaching person.

A second application of the diac appears in the next section (Fig. 13.34) as we consider an important power-control device: the triac.

13.12 TRIAC

The triac is fundamentally a diac with a gate terminal for controlling the turn-on conditions of the bilateral device in either direction. In other words, for either direction the gate current can control the action of the device in a manner very similar to that demonstrated for an SCR. The characteristics, however, of the triac in the first and third quadrants are somewhat different from those of the diac, as shown in Fig. 13.33c. Note the holding current in each direction not present in the characteristics of the diac.

The graphic symbol for the device and the distribution of the semiconductor layers are provided in Fig. 13.33 with photographs of the device. For each possible direction of conduction there is a combination of semiconductor layers whose state will be controlled by the signal applied to the gate terminal.

One fundamental application of the triac is presented in Fig. 13.34. In this capacity, it is controlling the ac power to the load by switching on and off during the positive and negative regions of input sinusoidal signal. The action of this circuit during the positive portion of the input signal is very similar to that encountered for

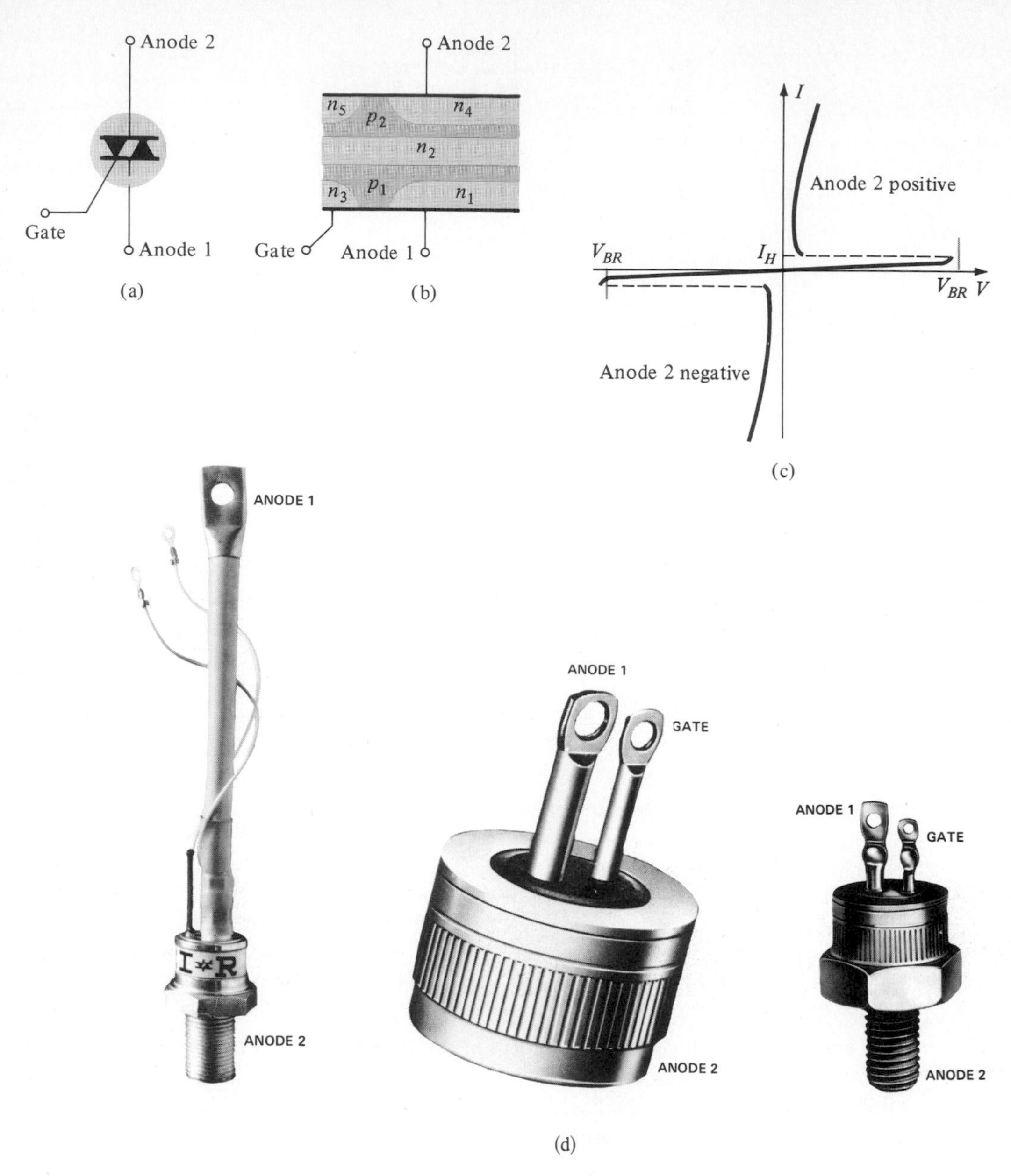

Figure 13.33 Triac: (a) symbol; (b) basic construction; (c) characteristics; (d) photographs.

the Shockley diode in Fig. 13.29. The advantage of this configuration is that during the negative portion of the input signal the same type of response will result, since both the diac and triac can fire in the reverse direction. The resulting waveform for the current through the load is provided in Fig. 13.34. By varying the resistor R the conduction angle can be controlled. There are units available today that can handle in excess of 10-kW loads.

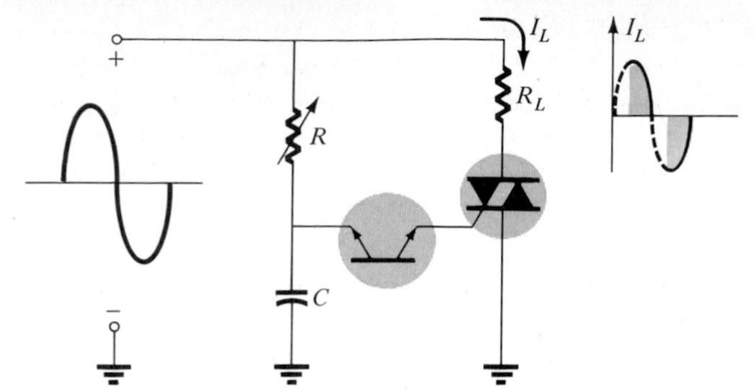

Figure 13.34 Triac application: phase (power) control.

OTHER DEVICES

13.13 UNIJUNCTION TRANSISTOR

Recent interest in the unijunction transistor (UJT) has, like that for the SCR, been increasing at an exponential rate. Although first introduced in 1948, the device did not become commercially available until 1952. The low cost per unit, combined with the excellent characteristics of the device, have warranted its use in a wide variety of applications. A few include oscillators, trigger circuits, sawtooth generators, phase control, timing circuits, bistable networks, and voltage- or current-regulated supplies. The fact that this device is, in general, a low-power-absorbing device under normal operating conditions is a tremendous aid in the continual effort to design relatively efficient systems.

The UJT is a three-terminal device having the basic construction of Fig. 13.35. A slab of lightly doped (increased resistance characteristic) *n*-type silicon material has two base contacts attached to both ends of one surface and an aluminum rod alloyed to the opposite surface. The *p-n* junction of the device is formed at the boundary of the aluminum rod and the *n*-type silicon slab. The single *p-n* junction accounts for the terminology unijunction. It was originally called a duo (double) base diode due to the presence of two base contacts. Note in Fig. 13.35 that the aluminum rod is alloyed to the silicon slab at a point closer to the base 2 contact than the base 1 contact and

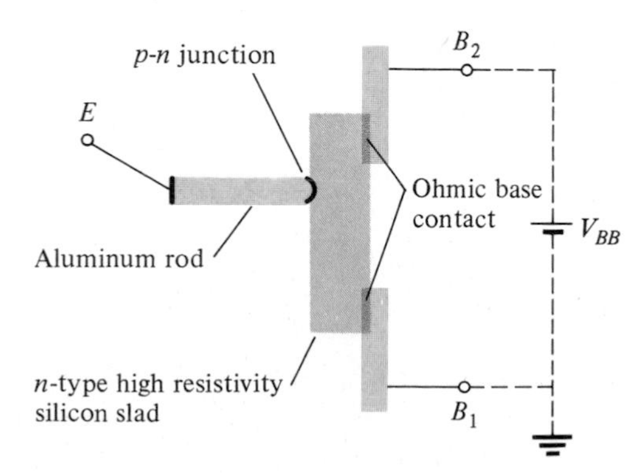

Figure 13.35 Unijunction transistor (UJT): basic construction.

that the base 2 terminal is made positive with respect to the base 1 terminal by V_{BB} volts. The effect of each will become evident in the paragraphs to follow.

The symbol for the unijunction transistor is provided in Fig. 13.36. Note that the emitter leg is drawn at an angle to the vertical line representing the slab of n-type material. The arrowhead is pointing in the direction of conventional current (hole) flow when the device is in the forward-biased, active, or conducting state.

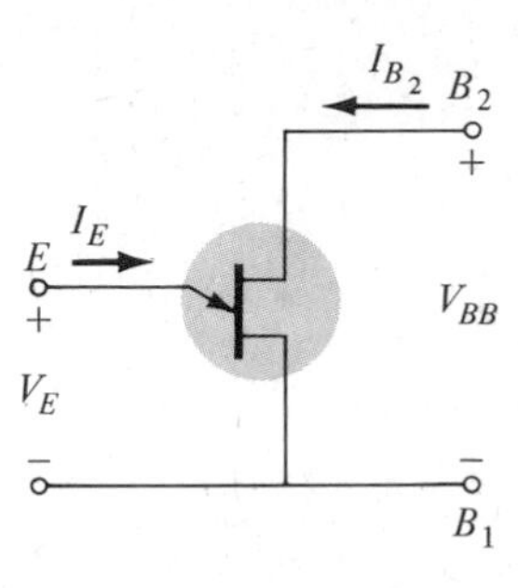

Figure 13.36 Symbol and basic biasing arrangement for the unijunction transistor.

The circuit equivalent of the UJT is shown in Fig. 13.37. Note the relative simplicity of this equivalent circuit: two resistors (one fixed, one variable) and a single diode. The resistance R_{B_1} is shown as a variable resistor since its magnitude will vary with the current I_E. In fact, for a representative unijunction transistor, R_{B_1} may vary from 5 kΩ down to 50 Ω for a corresponding change of I_E from 0 to 50 μA. The interbase resistance R_{BB} is the resistance of the device between terminals B_1 and B_2 when $I_E = 0$. In equation form,

$$\boxed{R_{BB} = (R_{B_1} + R_{B_2})\big|_{I_E=0}} \tag{13.2}$$

(R_{BB} is typically within the range of 4 to 10 kΩ.) The position of the aluminum rod of Fig. 13.35 will determine the relative values of R_{B_1} and R_{B_2} with $I_E = 0$. The magnitude of $V_{R_{B1}}$ (with $I_E = 0$) is determined by the voltage-divider rule in the following manner:

$$\boxed{V_{R_{B_1}} = \frac{R_{B_1} V_{BB}}{R_{B_1} + R_{B_2}} = \eta\, V_{BB}\bigg|_{I_E=0}} \tag{13.3}$$

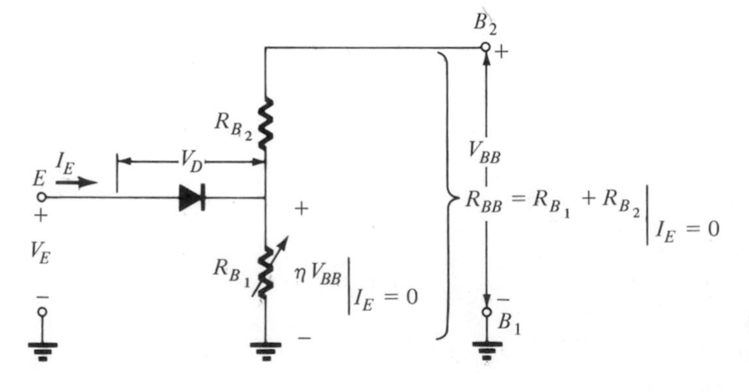

Figure 13.37 UJT equivalent circuit.

The Greek letter η (eta) is called the *intrinsic stand-off* ratio of the device and is defined by

$$\eta = \left.\frac{R_{B_1}}{R_{B_1} + R_{B_2}}\right|_{I_E=0} = \frac{R_{B_1}}{R_{BB}} \tag{13.4}$$

For applied emitter potentials (V_E) greater than $V_{R_{B1}} = \eta V_{BB}$ by the forward voltage drop of the diode, V_D (0.35 → 0.70 V) the diode will fire, assume the short-circuit representation (on an ideal basis), and I_E will begin to flow through R_{B_1}. In equation form the emitter firing potential is given by

$$V_P = \eta V_{BB} + V_D \tag{13.5}$$

The characteristics of a representative unijunction transistor are shown for $V_{BB} = 10$ V in Fig. 13.38. Note that for emitter potentials to the left of the peak point, the magnitude of I_E is never greater than I_{EO} (measured in microamperes). The current I_{EO} corresponds very closely with the reverse leakage current I_{CO} of the conventional bipolar transistor. This region, as indicated in the figure, is called the cutoff region. Once conduction is established at $V_E = V_P$, the emitter potential V_E will drop with increase in I_E. This corresponds exactly with the decreasing resistance R_{B_1} for increasing current I_E, as discussed earlier. This device, therefore, has a *negative resistance* region which is stable enough to be used with a great deal of reliability in the areas of application listed earlier. Eventually, the valley point will be reached, and any further increase in I_E will place the device in the saturation region. In this region the characteristics approach that of the semiconductor diode in the equivalent circuit of Fig. 13.37.

The decrease in resistance in the active region is due to the holes injected into the n-type slab from the aluminum p-type rod when conduction is established. The

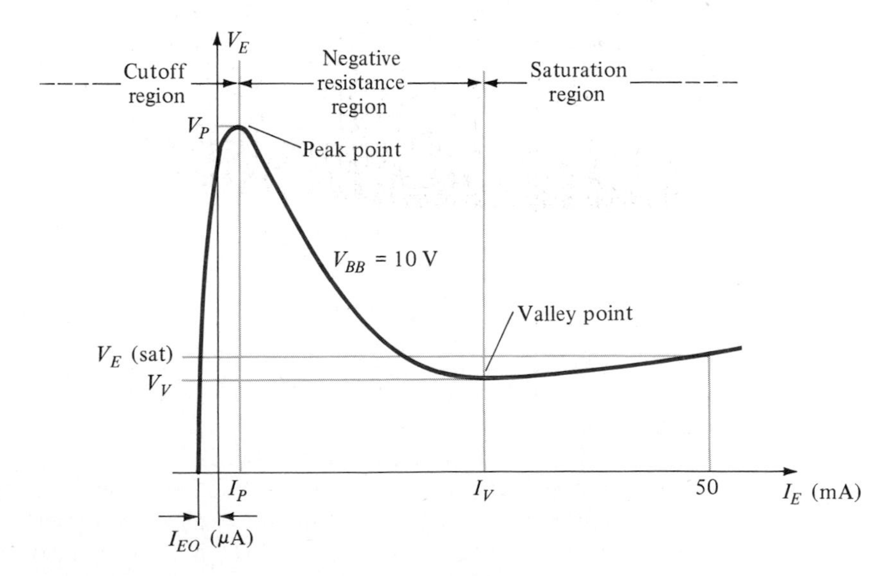

Figure 13.38 UJT static emitter-characteristic curve.

increased hole content in the n-type material will result in an increase in the number of free electrons in the slab, producing an increase in conductivity (G) and a corresponding drop in resistance ($R\downarrow = 1/G\uparrow$). Three other important parameters for the unijunction transistor are I_P, V_V, and I_V. Each is indicated on Fig. 13.38. They are all self-explanatory.

The emitter characteristics as they normally appear are provided in Fig. 13.39. Note that I_{EO} (μA) is not in evidence since the horizontal scale is in milliamperes. The intersection of each curve with the vertical axis is the corresponding value of V_P. For fixed values of η and V_D, the magnitude of V_P will vary as V_{BB}, that is,

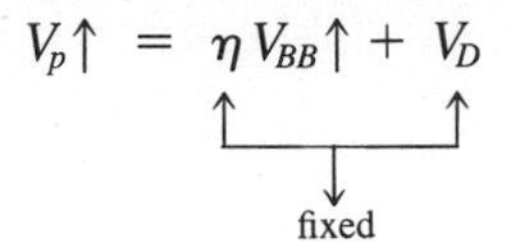

$$V_P\uparrow = \eta V_{BB}\uparrow + V_D$$

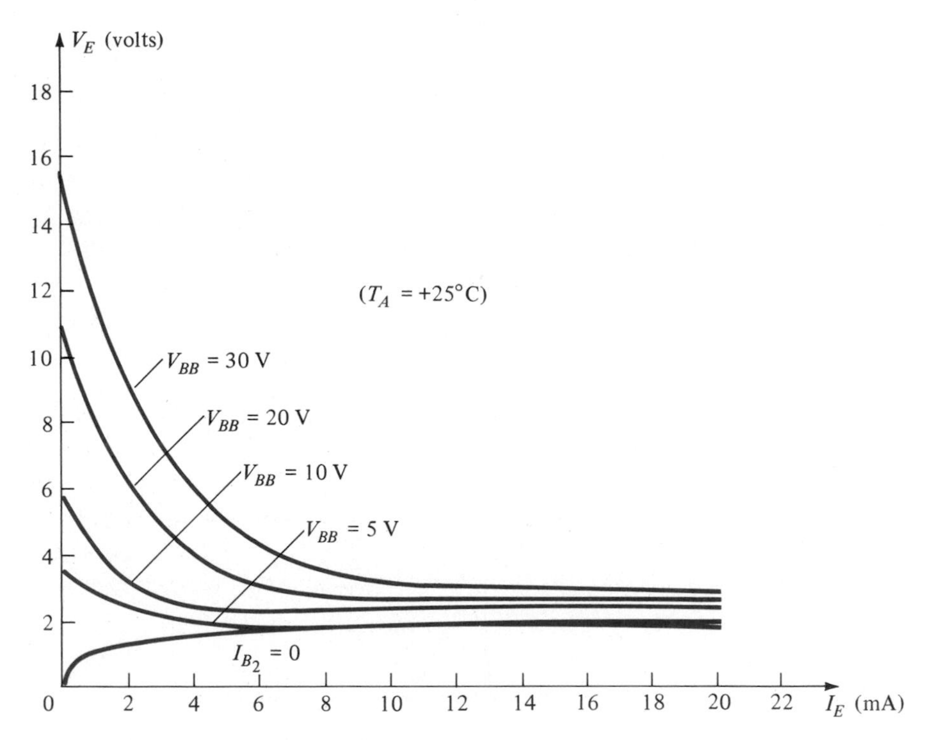

Figure 13.39 Typical static emitter-characteristic curves for a UJT.

A typical set of specifications for the UJT is provided in Fig. 13.40b. The discussion of the last few paragraphs should make each quantity readily recognizable. The terminal identification is provided in the same figure with a photograph of a representative UJT. Note that the base terminals are opposite each other while the emitter terminal is between the two. In addition, the base terminal to be tied to the higher potential is closer to the extension on the lip of the casing.

One rather common application of the UJT is in the triggering of other devices such as the SCR. The basic elements of such a triggering circuit are shown in Fig.

absolute maximum ratings: (25°C)

Power Dissipation	300 mw
RMS Emitter Current	50 ma
Peak Emitter Current	2 amperes
Emitter Reverse Voltage	30 volts
Interbase Voltage	35 volts
Operating Temperature Range	−65°C to +125°C
Storage Temperature Range	−65°C to +150°C

electrical characteristics: (25°C)

		Min.	Typ.	Max.
Intrinsic Standoff Ratio		0.56	0.65	
(V_{BB} = 10 V)	η	0.56	0.65	0.75
Interbase REsistance (kΩ)				
(V_{BB} = 3 V, I_E = 0)	R_{BB}	4.7	7	9.1
Emitter Saturation Voltage				
(V_{BB} = 10 V, I_E = 50 ma)	$V_{E(SAT)}$		2	
Emitter Reverse Current				
(V_{BB} = 30 V, I_{B1} = 0)	I_{EO}		0.05	12
Peak Point Emitter Current	I_P (μA)		0.04	5
(V_{BB} = 25 V)				
Valley Point Current				
(V_{BB} = 20 V)	I_V (mA)	4	6	

B1 B2 E

(a) (b) (c)

Figure 13.40 UJT: (a) appearance; (b) specification sheet; (c) terminal identification. (Courtesy General Electric Company.)

13.41. The resistor R_1 must be chosen to ensure that the load line determined by R_1 passes through the device characteristics in the negative resistance region, i.e., to the right of the peak point but to the left of the valley point as shown in Fig. 13.42. If the load line fails to pass to the right of the peak point, the device cannot turn on. An equation for R_1 that will ensure a turn-on condition can be established if we consider the peak point at which $I_{R_1} = I_P$ and $V_E = V_P$. (The equality $I_{R_1} = I_P$ is valid since the charging current of the capacitor, at this instant, is zero; that is, the capacitor is at this particular instant changing from a charging to a discharging state.) Then $V - I_{R_1}R_1 = V_E$ and $R_1 = (V - V_E)/I_{R_1} = (V - V_P)/I_P$ at the peak point. To ensure firing,

$$\boxed{R_1 < \frac{V - V_P}{I_P}} \tag{13.6}$$

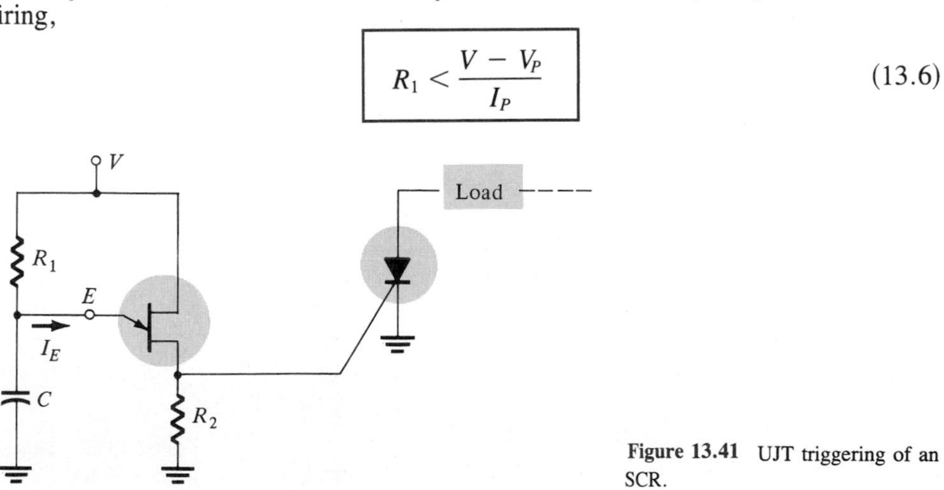

Figure 13.41 UJT triggering of an SCR.

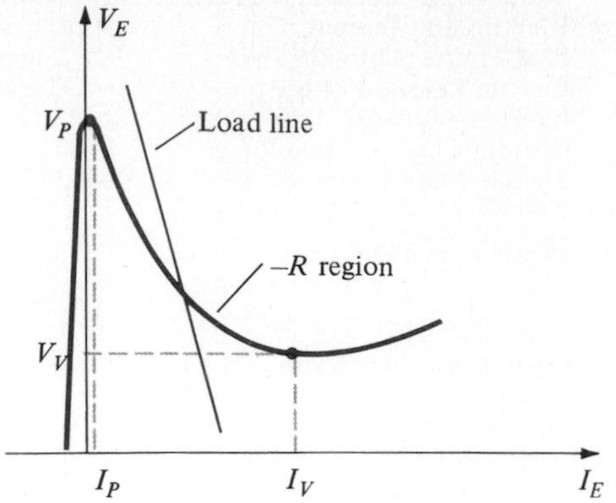

Figure 13.42

At the valley point $I_E = I_V$ and $V_E = V_V$, so that

$$V - I_{R_1}R_1 = V_E$$

becomes

$$V - I_V R_1 = V_V$$

and

$$R_1 = \frac{V - V_V}{I_V}$$

or to ensure turning off,

$$\boxed{R_1 > \frac{V - V_V}{I_V}} \tag{13.7}$$

The range of R_1 is therefore limited by

$$\boxed{\frac{V - V_V}{I_V} < R_1 < \frac{V - V_P}{I_P}} \tag{13.8}$$

The resistance R_2 must be chosen small enough to ensure that the SCR is not turned on by the voltage V_{R_2} of Fig. 13.43 when $I_E \cong 0$ A. The voltage

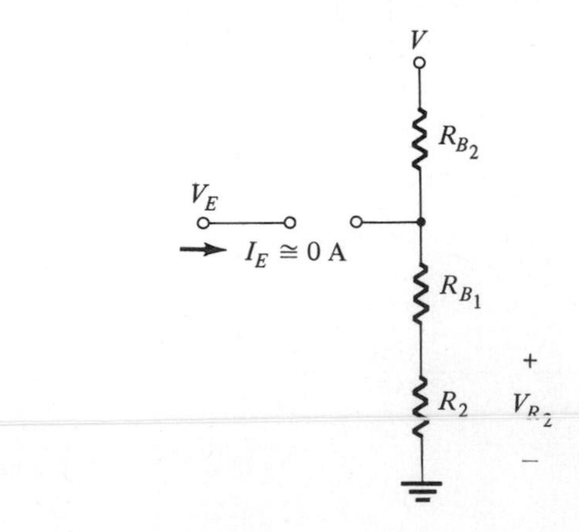

Figure 13.43 Triggering network when $I_E \cong 0$A.

$$V_{R_2} \cong \left. \frac{R_2 V}{R_2 + R_{BB}} \right|_{I_E \cong 0\,\text{A}} \tag{13.9}$$

The capacitor C will determine, as we shall see, the time interval between triggering pulses and the time span of each pulse.

At the instant the dc supply voltage V is applied, the voltage $v_E = v_C$ will charge toward V volts from V_V as shown in Fig. 13.44 with a time constant $\tau = R_1C$.

The general equation for the charging period is

$$v_C = V_V + (V - V_V)(1 - e^{-t/R_1C}) \tag{13.10}$$

As noted in Fig. 13.44, the voltage across R_2 is determined by Eq. (13.9) during this charging period. When $v_C = v_E = V_P$, the UJT will enter the conduction state and the capacitor will discharge through R_{B_1} and R_2 at a rate determined by the time constant $\tau = (R_{B_1} + R_2)C$.

The discharge equation for the voltage $v_C = v_E$ is the following:

$$v_C \cong V_p e^{-t/(R_{B_1}+R_2)C} \tag{13.11}$$

Equation (13.11) is complicated somewhat by the fact that R_{B_1} will decrease with increasing emitter current and the other elements of the network, such as R_1 and V, will affect the discharge rate and final level. However, the equivalent network appears as shown in Fig. 13.45 and the magnitude of R_1 and R_{B_2} are typically such that a Thévenin network for the network surrounding the capacitor C will be only slightly affected by these two resistors. Even though V is a reasonably high voltage, the voltage-divider contribution to the Thévenin voltage can be ignored on an approximate basis.

Using the reduced equivalent of Fig. 13.46 for the discharge phase will result in the following approximation for the peak value of V_{R_2}:

$$V_{R_2} \cong \frac{R_2(V_P - 0.7)}{R_2 + R_{B_1}} \tag{13.12}$$

The period t_1 of Fig. 13.44 can be determined in the following manner:

$$\begin{aligned} v_C \text{ (charging)} &= V_V + (V - V_V)(1 - e^{-t/R_1C}) \\ &= V_V + V - V_V - (V - V_V)e^{-t/R_1C} \\ &= V - (V - V_V)e^{-t/R_1C} \end{aligned}$$

when $v_C = V_P$, $t = t_1$ and $V_P = V - (V - V_V)e^{-t_1/R_1C}$, or

$$\frac{V_P - V}{V - V_V} = -e^{-t_1/R_1C}$$

and

$$e^{-t_1/R_1C} = \frac{V - V_P}{V - V_V}$$

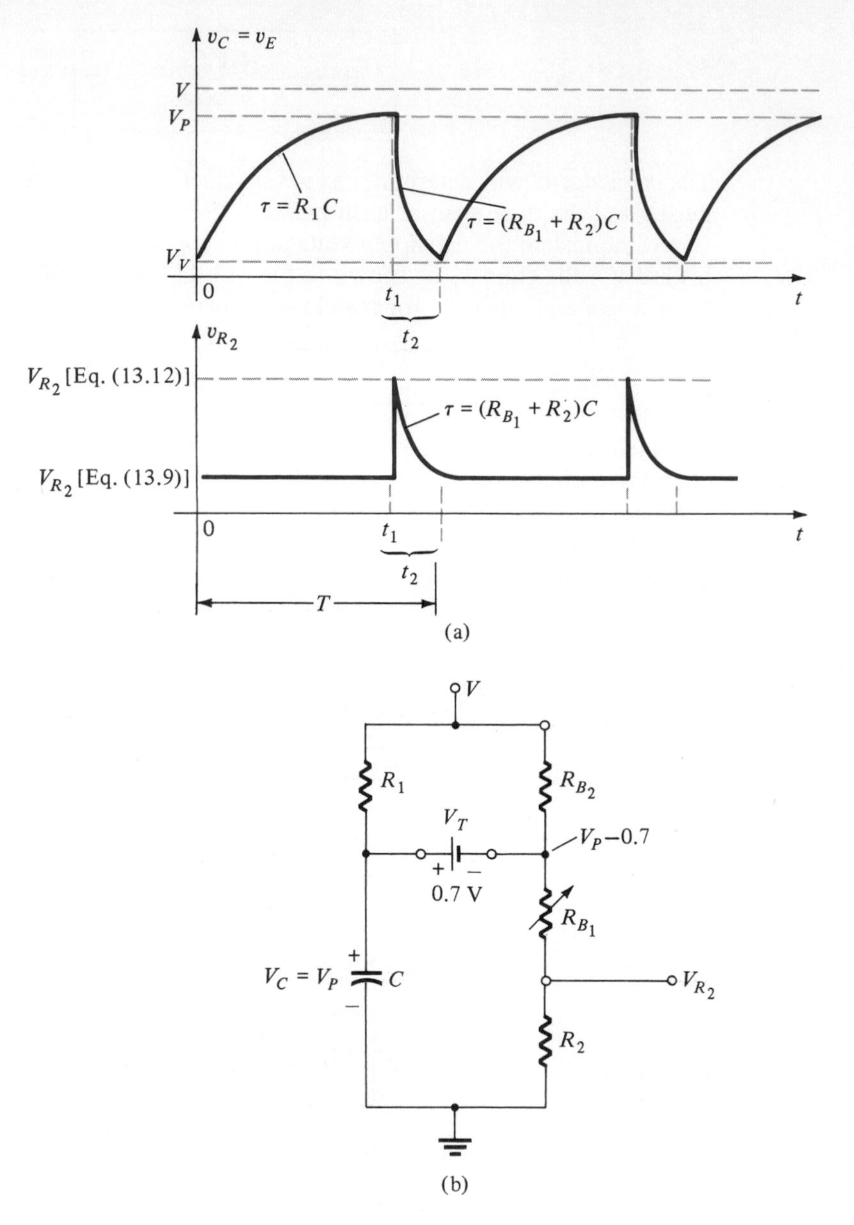

Figure 13.44 (a) Charging and discharging phases for trigger network of Fig. 13.41; (b) equivalent network when UJT turns on.

Using logs, we have

$$\log_e e^{-t_1/R_1C} = \log_e \frac{V - V_P}{V - V_V}$$

and

$$\frac{-t_1}{R_1C} = \log_e \frac{V - V_P}{V - V_V}$$

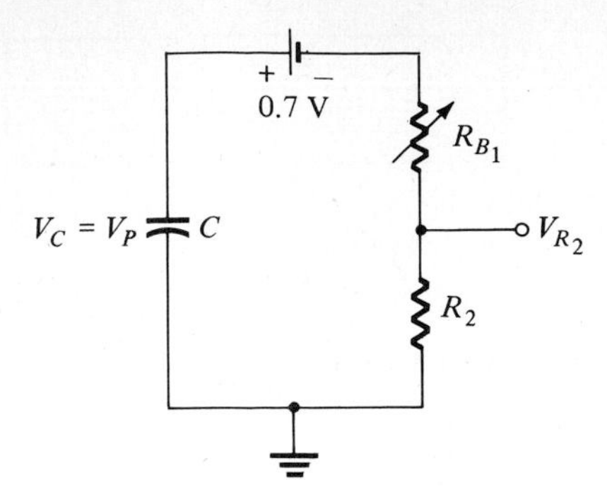

Figure 13.45 Reduced equivalent network when UJT turns on.

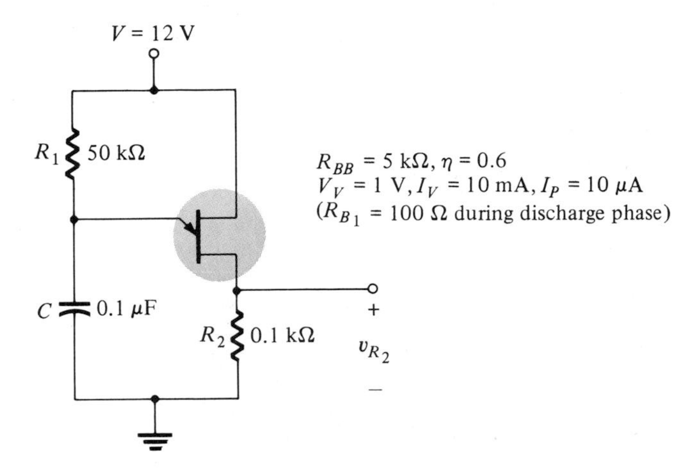

Figure 13.46

with

$$t_1 = R_1 C \log_e \frac{V - V_V}{V - V_P} \tag{13.13}$$

For the discharge period the time between t_1 and t_2 can be determined from Eq. (13.11) as follows:

$$v_C \text{ (discharging)} = V_P e^{-t/(R_{B_1}+R_2)C}$$

Establishing t_1 as $t = 0$ gives us

$$v_C = V_V \text{ at } t = t_2$$

and

$$V_V = V_P e^{-t_2/(R_{B_1}+R_2)C}$$

or

$$e^{-t_2/(R_{B_1}+R_2)C} = \frac{V_V}{V_P}$$

Using logs yields

$$\frac{-t_2}{(R_{B_1} + R_2)C} = \log_e \frac{V_V}{V_P}$$

and

$$t_2 = (R_{B_1} + R_2)C \log_e \frac{V_P}{V_V} \tag{13.14}$$

The period of time to complete one cycle is defined by T in Fig. 13.44. That is,

$$T = t_1 + t_2 \tag{13.15}$$

If the SCR were dropped from the configuration, the network would behave as a *relaxation oscillator*, generating the waveform of Fig. 13.44. The frequency of oscillation is determined by

$$f_{osc} = \frac{1}{T} \tag{13.16}$$

In many systems $t_1 \gg t_2$ and

$$T \cong t_1 = R_1 C \log_e \frac{V - V_V}{V - V_P}$$

Since $V \gg V_V$ in many instances,

$$T \cong t_1 = R_1 C \log_e \frac{V}{V - V_P}$$

$$= R_1 C \log_e \frac{1}{1 - V_P/V}$$

but $\eta = V_P/V$ if we ignore the effects of V_D in Eq. (13.5) and

$$T \cong R_1 C \log_e \frac{1}{1 - \eta}$$

or

$$f \cong \frac{1}{R_1 C \log_e [1/(1 - \eta)]} \tag{13.17}$$

EXAMPLE 13.1

Given the relaxation oscillator of Fig. 13.46:
(a) Determine R_{B_1} and R_{B_2} at $I_E = 0$ A.
(b) Calculate V_P, the voltage necessary to turn on the UJT.
(c) Determine whether R_1 is within the permissible range of values as determined by Eq. (13.8) to insure firing of the UJT.
(d) Determine the frequency of oscillation if $R_{B_1} = 100\ \Omega$ during the discharge phase.
(e) Sketch the waveform of v_C for a full cycle.
(f) Sketch the waveform of v_{R_2} for a full cycle.

Solution:

(a) $\eta = \dfrac{R_{B_1}}{R_{B_1} + R_{B_2}}$

$$0.6 = \frac{R_{B_1}}{R_{BB}}$$
$$R_{B_1} = 0.6R_{BB} = 0.6(5\ \text{k}\Omega) = \mathbf{3\ k\Omega}$$
$$R_{B_2} = R_{BB} - R_{B_1} = 5\ \text{k}\Omega - 3\ \text{k}\Omega = \mathbf{2\ k\Omega}$$

(b) At the point where $v_C = V_P$, if we continue with $I_E = 0$ A, the network of Fig. 13.47 will result where

$$V_P = 0.7 + \frac{(R_{B_1} + R_2)\,12}{\underbrace{R_{B_1} + R_{B_2}}_{R_{BB}} + R_2}$$

$$= 0.7 + \frac{(3\ \text{k}\Omega + 0.1\ \text{k}\Omega)\,12}{5\ \text{k}\Omega + 0.1\ \text{k}\Omega} = 0.7 + 7.294$$

$$\cong \mathbf{8\ V}$$

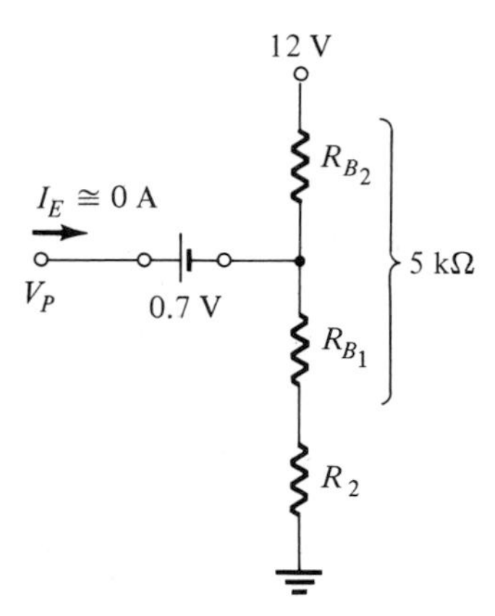

Figure 13.47

(c) $$\frac{V - V_V}{I_V} < R_1 < \frac{V - V_P}{I_P}$$
$$\frac{12 - 1}{10\ \text{mA}} < R_1 < \frac{12 - 8}{10\ \mu\text{A}}$$
$$1.1\ \text{k}\Omega < R_1 < 400\ \text{k}\Omega$$

The resistance $R_1 = 50\ \text{k}\Omega$ falls within this range.

(d) $$t_1 = R_1 C \log_e \frac{V - V_V}{V - V_P}$$
$$= (50 \times 10^3)(0.1 \times 10^{-6}) \log_e \frac{12 - 1}{12 - 8}$$
$$= 5 \times 10^{-3} \log_e \frac{11}{4} = 5 \times 10^{-3}(1.01)$$
$$= 5.05\ \text{ms}$$
$$t_2 = (R_{B_1} + R_2) C \log_e \frac{V_P}{V_V}$$

$$= (0.1\text{ k}\Omega + 0.1\text{ k}\Omega)(0.1 \times 10^{-6}) \log_e \frac{8}{1}$$
$$= (0.02 \times 10^{-6})(2.08)$$
$$= 41.6\ \mu\text{s}$$

and
$$T = t_1 + t_2 = 5.05\text{ ms} + 0.0416\text{ ms}$$
$$= 5.092\text{ ms}$$

with
$$f_{\text{OSC}} = \frac{1}{T} = \frac{1}{5.092 \times 10^{-3}} \cong \mathbf{196\ Hz}$$

Using Eq. (13.17) gives us

$$f \cong \frac{1}{R_1 C \log_e [1/(1 - \eta)]}$$
$$= \frac{1}{5 \times 10^{-3} \log_e 2.5}$$
$$= \mathbf{218\ Hz}$$

(e) See Fig. 13.48.

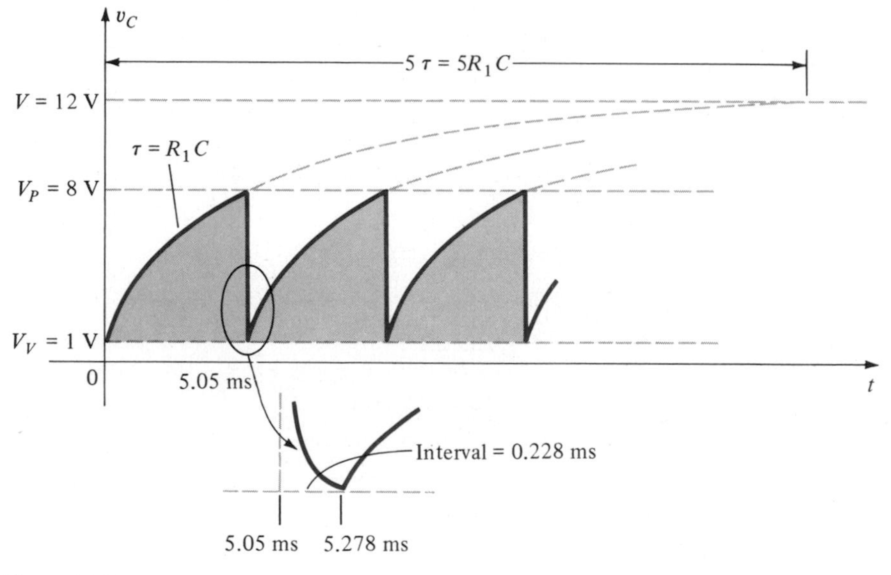

Figure 13.48

(f) During the charging phase, (Eq. 13.9)

$$V_{R_2} = \frac{R_2 V}{R_2 + R_{BB}} = \frac{0.1\text{ k}\Omega(12)}{0.1\text{ k}\Omega + 5\text{ k}\Omega} = \mathbf{0.235\ V}$$

When $v_C = V_P$ (Eq. 13.12)

$$V_{R_2} \cong \frac{R_2(V_P - 0.7)}{R_2 + R_{B_1}} = \frac{0.1\ \text{k}\Omega(8 - 0.7)}{0.1\ \text{k}\Omega + 0.1\ \text{k}\Omega}$$

$$= \mathbf{3.65\ V}$$

The plot of v_{R_2} appears in Fig. 13.49.

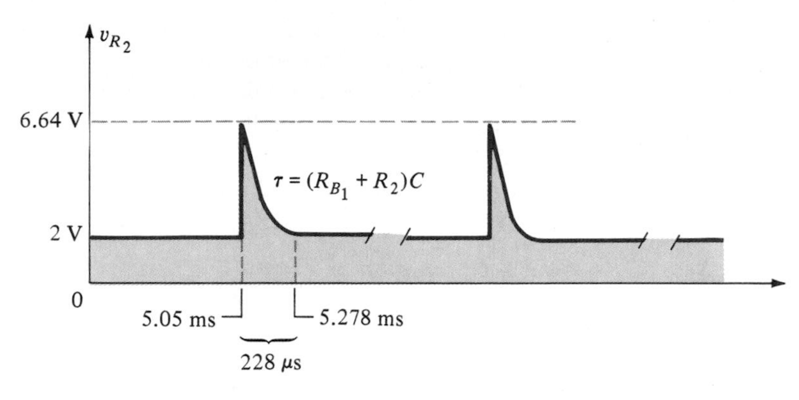

Figure 13.49

13.14 PHOTOTRANSISTORS

The fundamental behavior of photoelectric devices was introduced earlier with the description of the photodiode. This discussion will now be extended to include the phototransistor, which has a photosensitive collector-base *p-n* junction. The current induced by photoelectric effects is the base current of the transistor. If we assign the

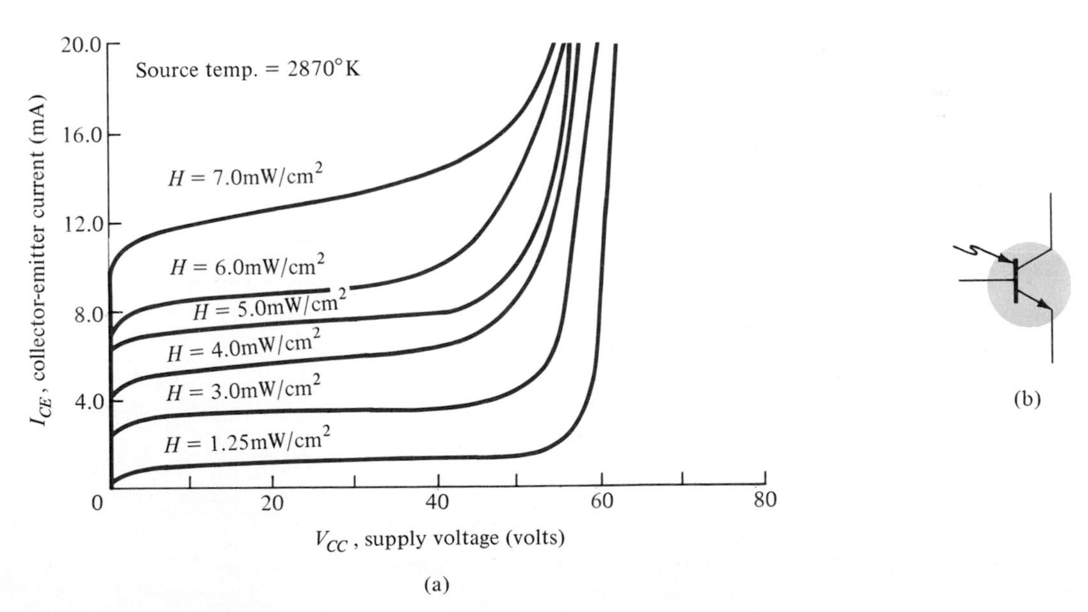

Figure 13.50 Phototransistor: (a) collector characteristics (MRD300); (b) symbol. (Courtesy Motorola, Inc.)

notation I_λ for the photoinduced base current, the resulting collector current, on an approximate basis, is

$$\boxed{I_c \cong h_{fe} I_\lambda} \tag{13.18}$$

A representative set of characteristics for a phototransistor is provided in Fig. 13.50 with the symbolic representation of the device. Note the similarities between these curves and those of a typical bipolar transistor. As expected, an increase in light intensity corresponds with an increase in collector current. To develop a greater degree of familiarity with the light-intensity unit of measurement, milliwatts per square

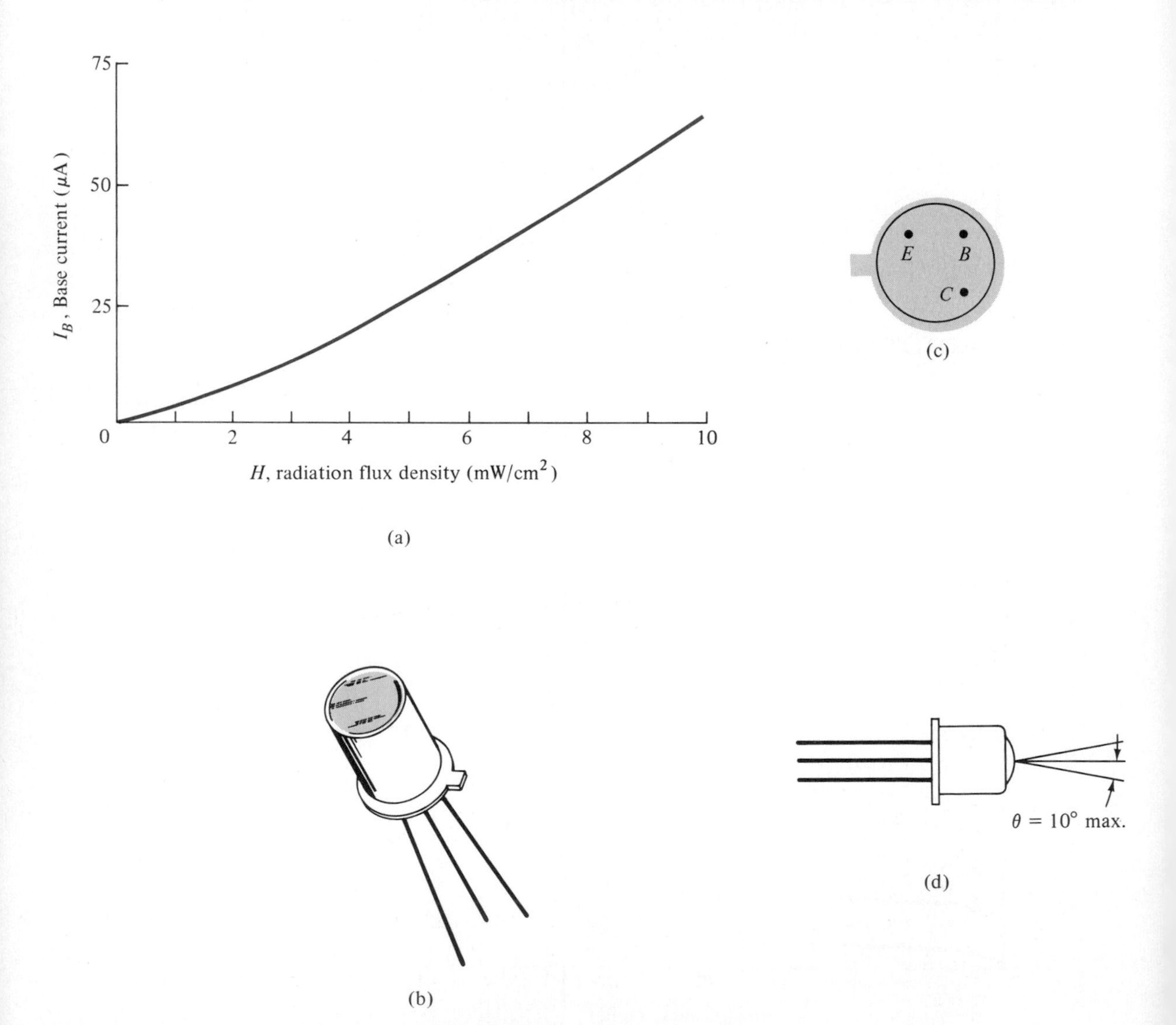

Figure 13.51 Phototransistor: (a) base current versus flux density; (b) device; (c) terminal identification; (d) angular alignment. (Courtesy Motorola, Inc.)

centimeter, a curve of base current versus flux density appears in Fig. 13.51a. Note the exponential increase in base current with increasing flux density. In the same figure a sketch of the phototransistor is provided with the terminal identification and the angular alignment.

Some of the areas of application for the phototransistor include punch-card readers, computer logic circuitry, lighting control (highways, etc.), level indication, relays, and counting systems.

A high-isolation AND gate is shown in Fig. 13.52 using three phototransistors and three LEDs (light-emitting diodes). The LEDs are semiconductor devices that emit light at an intensity determined by the forward current through the device. With the aid of discussions in Chapter 3, the circuit behavior should be relatively easy to understand. The terminology "high isolation" simply refers to the lack of an electrical connection between the input and output circuits.

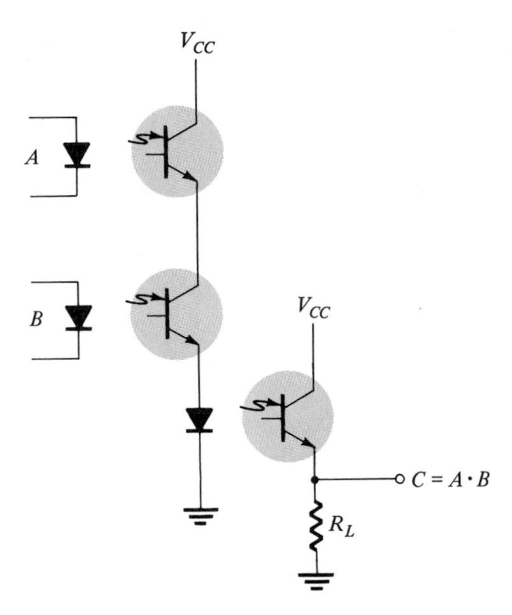

Figure 13.52 High-isolation AND gate employing phototransistors and light-emitting diodes (LEDS).

13.15 OPTO-ISOLATORS

The *opto-isolator* is a device that incorporates many of the characteristics described in the preceding section. It is simply a package that contains both an infrared LED and a photodetector such as a silicon diode, transistor Darlington pair, or SCR. The wavelength response of each device is tailored to be as identical as possible to permit the highest measure of coupling possible. In Fig. 13.53, two possible chip configurations are provided, with a photograph of each. There is a transparent insulating cap between each set of elements embedded in the structure (not visible) to permit the passage of light. They are designed with response times so small that they can be used to transmit data in the megahertz range.

ISO-LIT 1

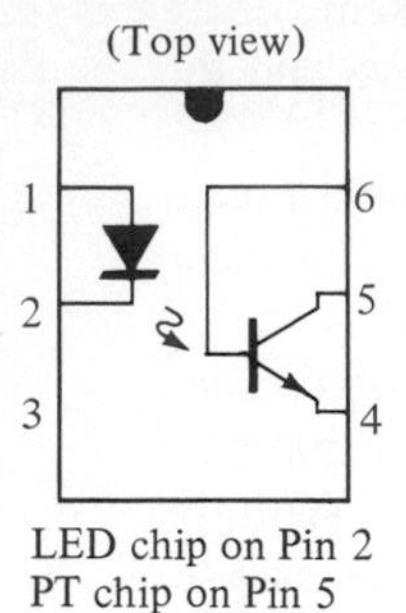

Pin No.	*Function*
1	anode
2	cathode
3	nc
4	emitter
5	collector
6	base

ISO-LIT Q1

(Top view)

1 2 3 4 5 6 7 8 16 15 14 13 12 11 10 9

Pin No.	*Function*
1	anode
2	cathode
3	cathode
4	anode
5	anode
6	cathode
7	cathode
8	anode
9	emitter
10	collector
11	collector
12	emitter
13	emitter
14	collector
15	collector
16	emitter

Figure 13.53 Two Litronix opto-isolators. (Courtesy Litronix, Inc.)

The maximum ratings and electrical characteristics for the IL-1 model are provided in Fig. 13.54. Note that I_{CEO} is measured in nanoamperes and that the power dissipation of the LED and transistor are about the same.

The typical optoelectronic characteristic curves for each channel are provided in Figs. 13.55 through 13.59. Note the very pronounced effect of temperature on the output current at low temperatures but the fairly level response at or above room temperature (25°C). As mentioned earlier, the level of I_{CEO} is improving steadily with

(a) Maximum Ratings

Gallium arsenide LED (each channel) IL-1	
Power dissipation @ 25°C	200 mW
Derate linearly from 25°C	2.6 mW/°C
Continuous forward current	150 mA
Detector silicon phototransistor (each channel) IL-1	
Power dissipation @ 25°C	200 mW
Derate linearly from 25°C	2.6 mW/°C
Collector-emitter breakdown voltage	30 V
Emitter-collector breakdown voltage	7 V
Collector-base breakdown voltage	70 V
Package IL-1	
Total package dissipation at 25°C ambient (LED plus detector)	250 mW
Derate linearly from 25°C	3.3 mW/°C
Storage temperature	−55°C to +150°C
Operating temperature	−55°C to +100°C

(b) Electrical Characteristics per Channel (at 25°C Ambient)

Parameter	*Min.*	*Typ.*	*Max.*	*Units*	*Test Conditions*
Gallium arsenide LED					
Forward voltage		1.3	1.5	V	$I_F = 60$ mA
Reverse current		0.1	10	μA	$V_R = 3.0$ V
Capacitance		100		pF	$V_R = 0$
Phototransistor detector					
BV_{CEO}	30			V	$I_C = 1$ mA
I_{CEO}		5.0	50	nA	$V_{CE} = 10$ V, $I_F = 0$
Collector-emitter capacitance		2.0		pF	$V_{CE} = 0$
BV_{ECO}	7			V	$I_E = 100$ μA
Coupled characteristics					
dc current transfer ratio	0.2	0.35			$I_F = 10$ mA, $V_{CE} = 10$ V
Capacitance, input to output		0.5		pF	
Breakdown voltage	2500			V	DC
Resistance, input to output		100		GΩ	
V_{SAT}			0.5	V	$I_C = 1.6$ mA, $I_F = 16$ mA
Propagation delay					
$t_{D\ ON}$		6.0		μs	$R_L = 2.4$ kΩ, $V_{CE} = 5$ V
$t_{D\ OFF}$		25		μs	$I_F = 16$ mA

Figure 13.54 The Litronix IL-1 opto-isolator.

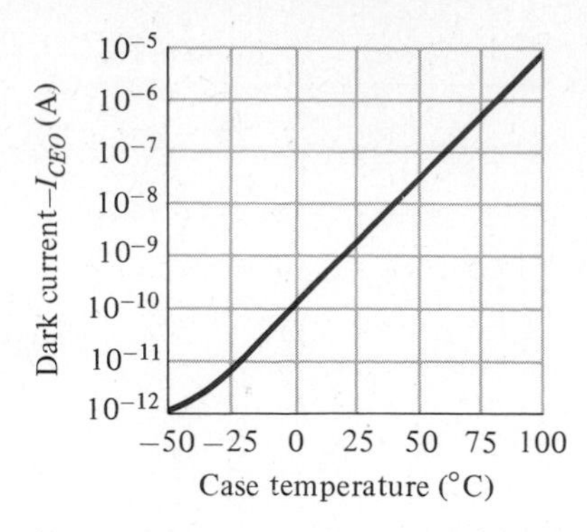

Figure 13.55 Dark current (I_{CEO}) versus temperature.

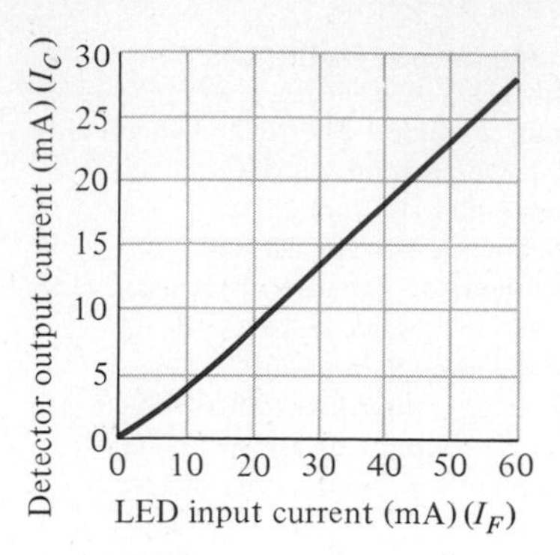

Figure 13.56 Transfer characteristics.

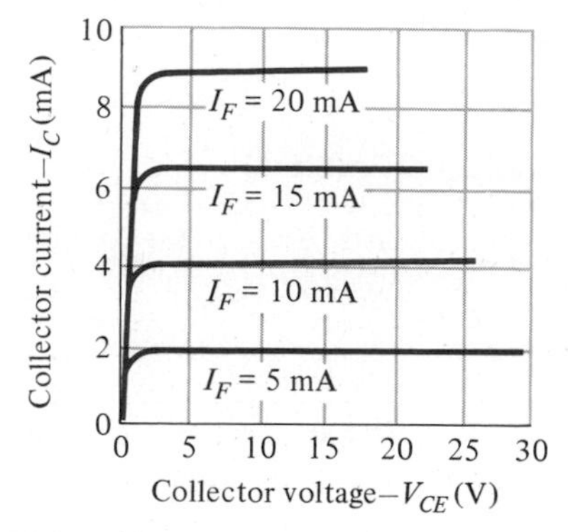

Figure 13.57 Detector output characteristics.

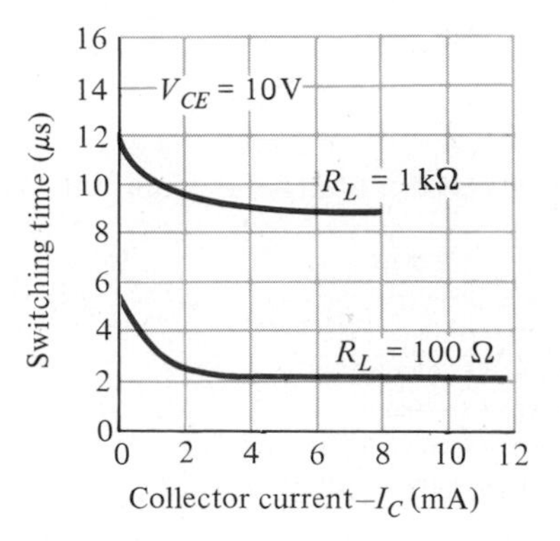

Figure 13.58 Switching time versus collector current.

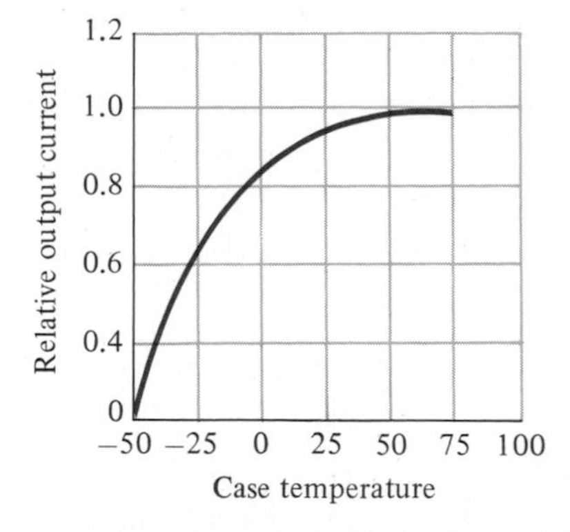

Figure 13.59 Relative output versus temperature.

improved design and construction techniques (the lower the better). In Fig. 13.55 we do not reach 1 μA until the temperature rises above 75°C. The transfer characteristics of Fig. 13.56 compare the input LED current (which establishes the luminous flux) to the resulting collector current of the output transistor (whose base current is determined by the incident flux). In fact, Fig. 13.57 demonstrates that the V_{CE} voltage affects the resulting collector current only very slightly. It is interesting to note in Fig. 13.58 that the switching time of an opto-isolator decreases with increased current, while for many devices it is exactly the reverse. Consider that it is only 2 μs for a collector current of 6 mA and a load R_L of 100 Ω. The relative output versus temperature appears in Fig. 13.59.

The schematic representation for a transistor coupler appears in Fig. 13.53. The schematic representations for a photodiode, photo-Darlington, and photo-SCR opto-isolator appear in Fig. 13.60.

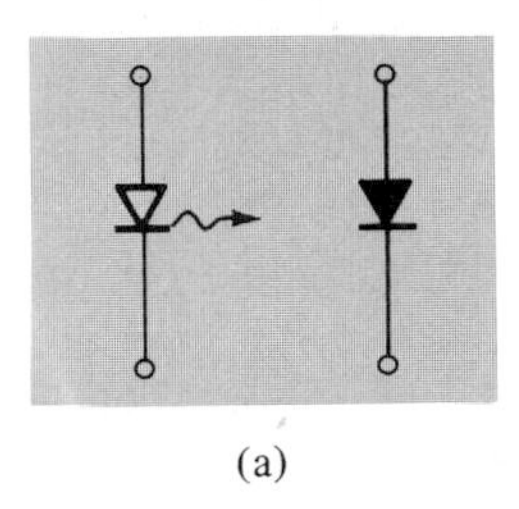

(a)

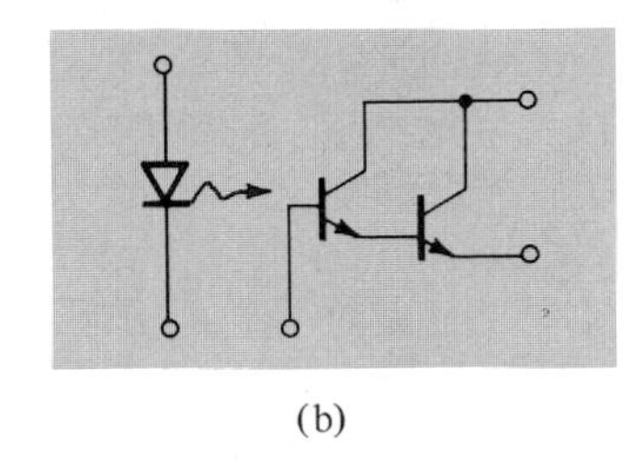

(b)

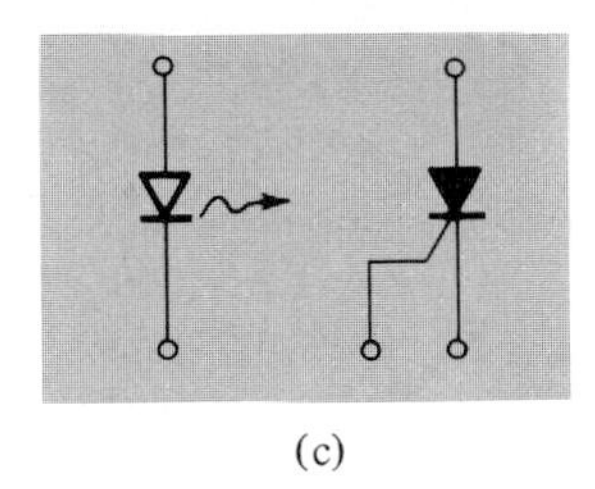

(c)

Figure 13.60 Opto-isolators: (a) photodiode; (b) photo-Darlington; (c) photo-SCR.

13.16 PROGRAMMABLE UNIJUNCTION TRANSISTOR

Although there is a similarity in name, the actual construction and mode of operation of the programmable unijunction transistor (PUT) is quite different from the unijunction transistor. The fact that the *I*—*V* characteristics and applications of each are similar prompted the choice of labels.

As indicated in Fig. 13.61, it is a four-layer *pnpn* device with a gate connected directly to the sandwiched *n*-type layer. The symbol for the device and the basic biasing arrangement appears in Fig. 13.62. As the symbol suggests, it is essentially

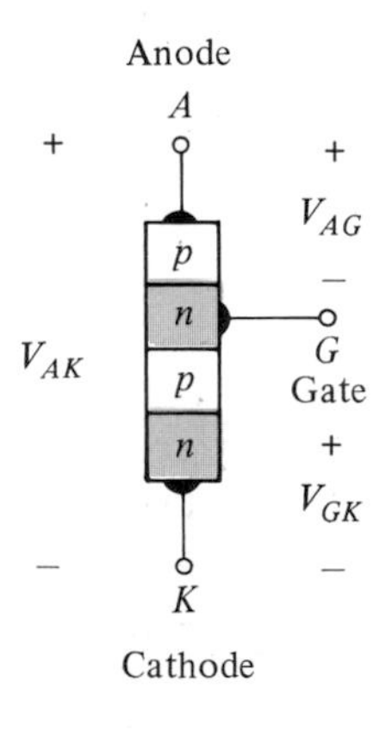

Figure 13.61 Programmable UJT (PUT).

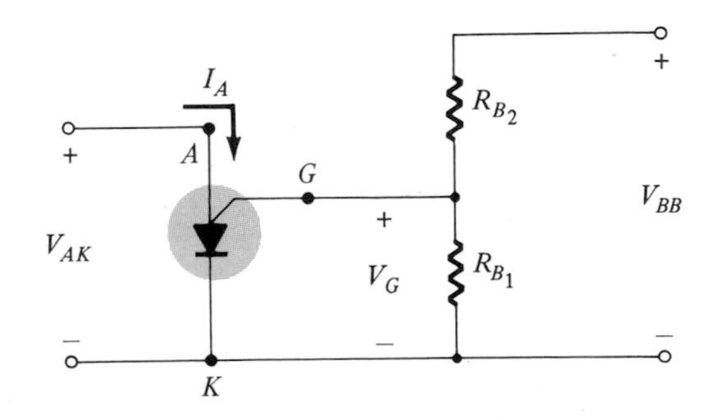

Figure 13.62 Basic biasing arrangement for the PUT.

an SCR with a control mechanism that permits a duplication of the characteristics of the typical SCR. The term "programmable" is applied because R_{BB}, η, and V_P as defined for the UJT can be controlled through the resistors, R_{B_1}, R_{B_2}, and the supply voltage V_{BB}. Note in Fig. 13.62 that through an application of the voltage-divider rule when $I_G = 0$:

$$V_G = \frac{R_{B_1}}{R_{B_1} + R_{B_2}} V_{BB} = \eta V_{BB} \tag{13.19}$$

where

$$\eta = \frac{R_{B_1}}{R_{B_1} + R_{B_2}}$$

as defined for the UJT.

The characteristics of the device appear in Fig. 13.63. As noted on the diagram, the off state (I low, V between 0 and V_P) and the on state ($I \geq I_V$, $V \geq V_V$) are separated by the unstable region as occurred for the UJT. That is, the device cannot stay in the unstable state—it will simply shift to either the off or on stable states.

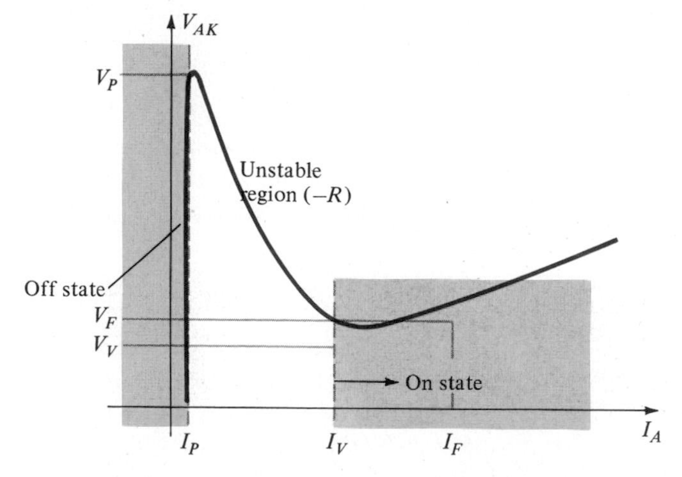

Figure 13.63 PUT characteristics.

The firing potential (V_P) or voltage necessary to "fire" the device is given by

$$V_P = \eta V_{BB} + V_D \tag{13.20}$$

as defined for the UJT. However, V_P represents the voltage drop V_{AK} in Fig. 13.61 (the forward voltage drop across the conducting diode). For silicon V_D is typically 0.7 V. Therefore,

$$V_{AK} = V_{AG} + V_{GK}$$

$$V_P = V_D + V_G$$

and

$$V_P = \eta V_{BB} + 0.7 \quad \text{silicon} \tag{13.21}$$

We noted above, however, that $V_G = \eta V_{BB}$ with the result that

$$\boxed{V_P = V_G + 0.7} \quad \text{silicon} \tag{13.22}$$

Recall that for the UJT both R_{B_1} and R_{B_2} represent the bulk resistance and ohmic base contacts of the device—both inaccessible. In the development above, we note that R_{B_1} and R_{B_2} are external to the device permitting an adjustment of η and hence V_G above. In other words, a measure of control on the level of V_P required to turn on the device.

Although the characteristics of the PUT and UJT are similar, the peak and valley currents of the PUT are typically lower than those of a similarly rated UJT. In addition, the minimum operating voltage is also less for a PUT.

If we take a Thévenin equivalent of the network to the right of gate terminal in Fig. 13.62, the network of Fig. 13.64 will result. The resulting resistance R_S is important because it is often included in specification sheets since it affects the level of I_V.

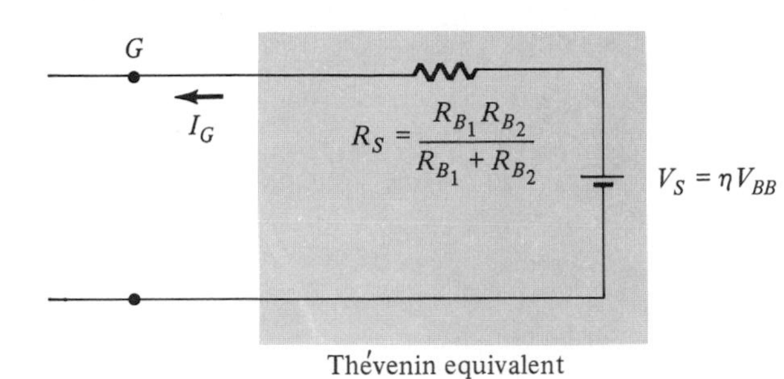

Figure 13.64 Thévenin equivalent for the network to the right of the gate terminal in Fig. 13.54.

The basic operation of the device can be reviewed through reference to Fig. 13.63. A device in the off state will not change state until the voltage V_P as defined by V_G and V_D is reached. The level of current until I_P is reached is very low, resulting in an open-circuit equivalent since $R = V$ (high)$/I$ (low) will result in a high resistance level. When V_P is reached the device will switch through the unstable region to the on state, where the voltage is lower but the current higher, resulting in a terminal resistance $R = V$ (low)$/I$ (high) which is quite small, representing short-circuit equivalent on an approximate basis. The device has therefore switched from essentially an open-circuit to a short-circuit state at a point determined by the choice of R_{B_1}, R_{B_2}, and V_{BB}. Once the device is in the on state, the removal of V_G will not turn the device off. The level of voltage V_{AK} must be dropped sufficiently to reduce the current below a holding level.

EXAMPLE 13.2

Determine R_{B_1} and V_{BB} for a silicon PUT if it is determined that $\eta = 0.8$, $V_P = 10.3$ V, and $R_{B_2} = 5$ kΩ.

Solution:

From Eq. (13.19),

$$\eta = \frac{R_{B_1}}{R_{B_1} + R_{B_2}} = 0.8$$

$$R_{B_1} = 0.8(R_{B_1} + R_{B_2})$$
$$0.2\,R_{B_1} = 0.8\,R_{B_2}$$
$$R_{B_1} = 4\,R_{B_2}$$
$$R_{B_1} = 4(5\text{ k}\Omega) = \mathbf{20\text{ k}\Omega}$$

From Eq. (13.20),

$$V_P = \eta V_{BB} + V_D$$
$$10.3 = (0.8)(V_{BB}) + 0.7$$
$$9.6 = 0.8V_{BB}$$
$$V_{BB} = \mathbf{12\text{ V}}$$

One popular application of the PUT is in the relaxation oscillator of Fig. 13.65. The instant the supply is connected, the capacitor will begin to charge toward V_{BB} volts, since there is no anode current at this point. The charging curve appears in Fig. 13.66.

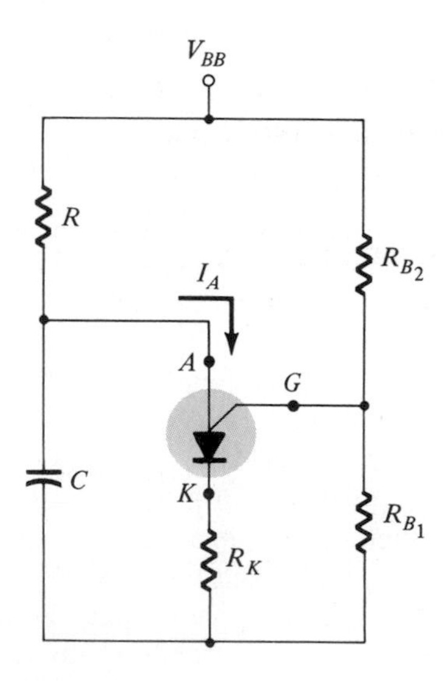

Figure 13.65 PUT relaxation oscillator.

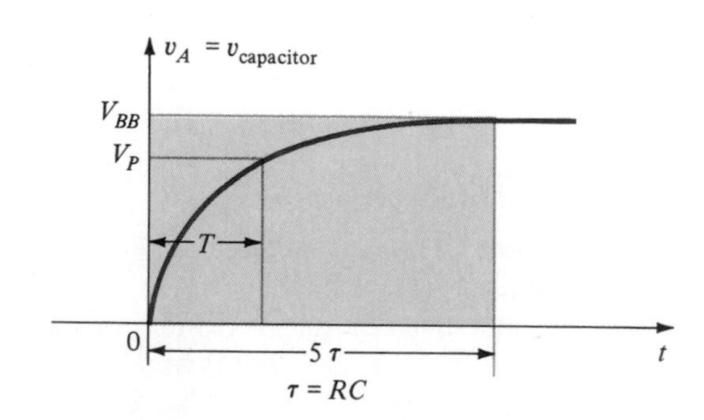

Figure 13.66 Changing wave for the capacitor C of Fig. 13.65.

The period T required to reach the firing potential V_P is given approximately by

$$T \cong RC \log_e \frac{V_{BB}}{V_{BB} - V_P} \tag{13.23}$$

or when $V_P \cong \eta V_{BB}$

$$T \cong RC \log_e \left(1 + \frac{R_{B_1}}{R_{B_2}}\right) \tag{13.24}$$

The instant the voltage across the capacitor equals V_P, the device will fire and a current $I_A = I_P$ established through the PUT. If R is too large, the current I_P cannot be established and the device will not fire. At the point of transition,

$$I_P R = V_{BB} - V_P$$

and

$$\boxed{R_{\max} = \frac{V_{BB} - V_P}{I_P}} \qquad (13.25)$$

The subscript is included to indicate that any R greater than $R_{\max}$ will result in a current less than I_P. The level of R must also be such to ensure it is less than I_V if oscillations are to occur. In other words, we want the device to enter the unstable region and then return to the off state. From reasoning similar to that above:

$$\boxed{R_{\min} = \frac{V_{BB} - V_V}{I_V}} \qquad (13.26)$$

The discussion above requires that R be limited to the following for an oscillatory system:

$$R_{\min} < R < R_{\max}$$

The waveforms of v_A, v_G, and v_K appear in Fig. 13.67. Note that T determines the maximum voltage v_A can charge to. Once the device fires, the capacitor will rapidly discharge through the PUT and R_K, producing the drop shown. Of course, v_K will peak at the same time due to the brief but heavy current. The voltage v_G will rapidly drop down from V_G to a level just greater than 0 volts. When the capacitor voltage drops to a low level, the PUT will once again turn off and the charging cycle repeated. The effect on V_G and V_K is shown in Fig. 13.67.

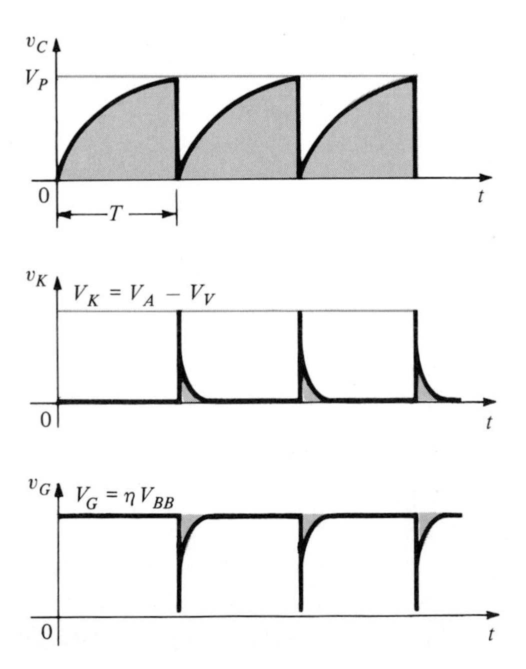

Figure 13.67 Waveforms for PUT oscillator of Fig. 13.65.

EXAMPLE 13.3

If $V_{BB} = 12$ V, $R = 20$ kΩ, $C = 1\ \mu$F, $R_K = 100\ \Omega$, $R_{B_1} = 10$ kΩ, $R_{B_2} = 5$ kΩ, $I_P = 100\ \mu$A, $V_V = 1$ V, and $I_V = 5.5$ mA, determine:

(a) V_P.
(b) $R_{\max}$ and $R_{\min}$.
(c) T and frequency of oscillation.
(d) The waveforms of v_A, v_G, and v_K.

Solution:

(a) From Eq. (13.20):

$$\begin{aligned} V_P &= \eta V_{BB} + V_D \\ &= \frac{R_{B_1}}{R_{B_1} + R_{B_2}} V_{BB} + 0.7 \\ &= \frac{10\text{ k}\Omega}{10\text{ k}\Omega + 5\text{ k}\Omega}(12) + 0.7 \\ &= (0.67)(12) + 0.7 = \mathbf{8.7\ V} \end{aligned}$$

(b) From Eq. (13.25):

$$\begin{aligned} R_{\max} &= \frac{V_{BB} - V_P}{I_P} \\ &= \frac{12 - 8.7}{100 \times 10^{-6}} = \mathbf{33\ k\Omega} \end{aligned}$$

From Eq. (13.26):

$$\begin{aligned} R_{\min} &= \frac{V_{BB} - V_V}{I_V} \\ &= \frac{12 - 1}{5.5 \times 10^{-3}} = \mathbf{2\ k\Omega} \end{aligned}$$

$$R\text{: } 2\text{ k}\Omega < 20\text{ k}\Omega < 33\text{ k}\Omega$$

(c) Eq. (13.23):

$$\begin{aligned} T &= RC \log_e \frac{V_{BB}}{V_{BB} - V_P} \\ &= (20 \times 10^3)(1 \times 10^{-6}) \log_e \frac{12}{12 - 8.7} \\ &= 20 \times 10^{-3} \log_e (3.64) \\ &= 20 \times 10^{-3}(1.29) \\ &= \mathbf{25.8\ ms} \end{aligned}$$

$$f = \frac{1}{T} = \frac{1}{25.8 \times 10^{-3}} = \mathbf{38.8\ Hz}$$

(d) as indicated in Fig. 13.68.

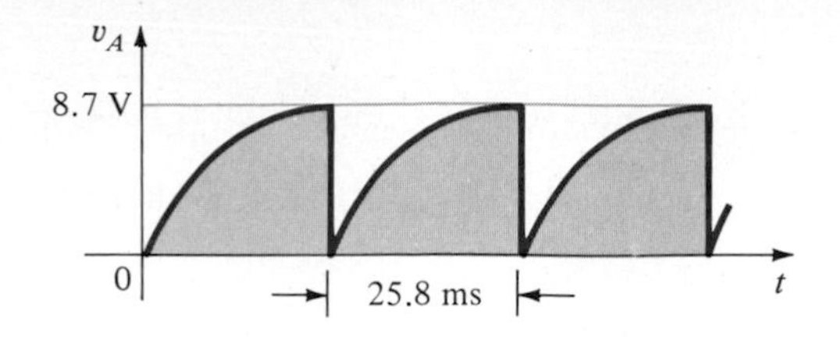

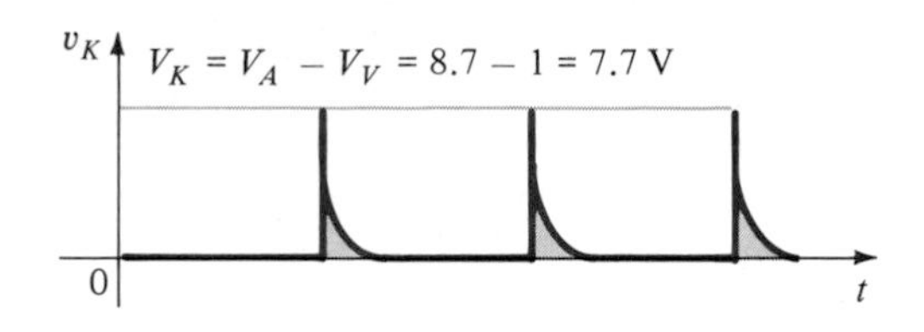

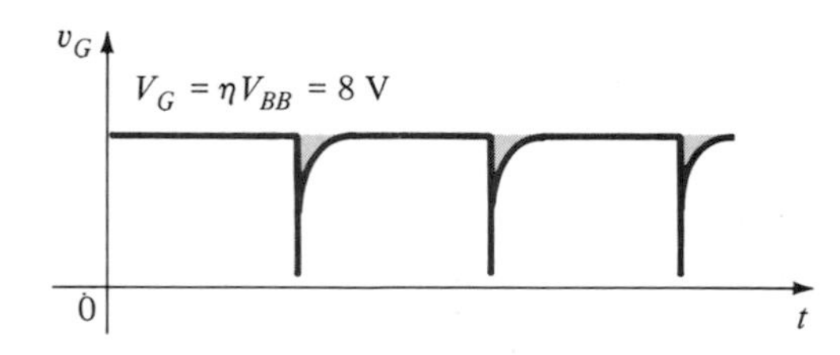

Figure 13.68 Waveforms for the oscillator of Example 13.3.

PROBLEMS

§ 13.3

1. Describe in your own words the basic behavior of the SCR using the two-transistor equivalent circuit.

2. Describe two techniques for turning an SCR off.

3. Consult a manufacturer's manual or specification sheet and obtain a turn-off network. If possible, describe the turn-off action of the design.

§ 13.4

4. (a) At high levels of gate current the characteristics of an SCR approach those of what two-terminal device?
(b) At a fixed anode-to-cathode voltage less than $V_{(BR)F^*}$, what is the effect on the firing of the SCR as the gate current is reduced from its maximum value to the zero level?
(c) At a fixed gate current greater than $I_G = 0$, what is the effect on the firing of the SCR as the gate voltage is reduced from $V_{(BR)F^*}$?
(d) For increasing levels of I_G, what is the effect on the holding current?

5. (a) In Fig. 13.8, will a gate current of 50 mA fire the device at room temperature (25C°)?
(b) Repeat part (a) for a gate current of 10 mA.
(c) Will a gate voltage of 2.6 V trigger the device at room temperature?
(d) Is $V_G = 6$ V, $I_G = 800$ mA a good choice for firing conditions? Would $V_G = 4$ V, $I_G = 1.6$ A be preferred? Explain.

§ 13.5

6. In Fig. 13.11b, why is there very little loss in potential across the SCR during conduction?

7. Fully explain why reduced values of R_1 in Fig. 13.12 will result in an increased angle of conduction.

8. Refer to the charging network of Fig. 13.13.
(a) Determine the dc level of the full-wave rectified signal if a 1 : 1 transformer were employed.
(b) If the battery in its uncharged state is sitting at 11 V, what is the anode-to-cathode voltage drop across SCR_1?
(c) What is the maximum possible value of V_R ($V_{GK} \cong 0.7$ V)?
(d) At the maximum value of part (c), what is the gate potential of SCR_2?
(e) Once SCR_2 has entered the short-circuit state, what is the level of V_2?

§ 13.7

9. Fully describe in your own words the behavior of the networks of Fig. 13.17.

§ 13.8

10. (a) In Fig. 13.23, if $V_Z = 50$ V, determine the maximum possible value the capacitor C_1 can charge to ($V_{GK} \cong 0.7$ V).
(b) Determine the approximate discharge time (5τ) for $R_3 = 20$ kΩ.
(c) Determine the internal resistance of the GTO if the rise time is one-half the decay period determined in part (b).

§ 13.9

11. (a) Using Fig. 13.25b, determine the minimum irradiance required to fire the device at room temperature (20°C).
(b) What percent reduction in irradiance is allowable if the junction temperature is increased from 0°C (32°F) to 100°C (212°F)?

§ 13.10

12. For the network of Fig. 13.29, if $V_{(BR)} = 6$ V, $V = 40$ V, $R = 10$ kΩ, $C = 0.2$ μF, and V_{GK} (firing potential) = 3 V, determine the time period between energizing the network and the turning on of the SCR.

§ 13.11

13. Using whatever reference you require, find an application of a diac and explain the network behavior.

14. If V_{BR_2} is 6.4 V, determine the range for V_{BR_1} using Eq. (13.1).

§ 13.12

15. Repeat Problem 13 for the triac.

§ 13.13

16. For the network of Fig. 13.41, in which $V = 40$ V, $\eta = 0.6$, $V_V = 1$ V, $I_V = 8$ mA, and $I_P = 10$ μA, determine the range of R_1 for the triggering network.

17. For a unijunction transistor with $V_{BB} = 20$ V, $\eta = 0.65$, $R_{B_1} = 2$ kΩ ($I_E = 0$), and $V_D = 0.7$ V, determine the following:

(a) R_{B_2}.
(b) R_{BB}.
(c) $V_{R_{B_1}}$.
(d) V_P.

18. Given the relaxation oscillator of Fig. 13.69:

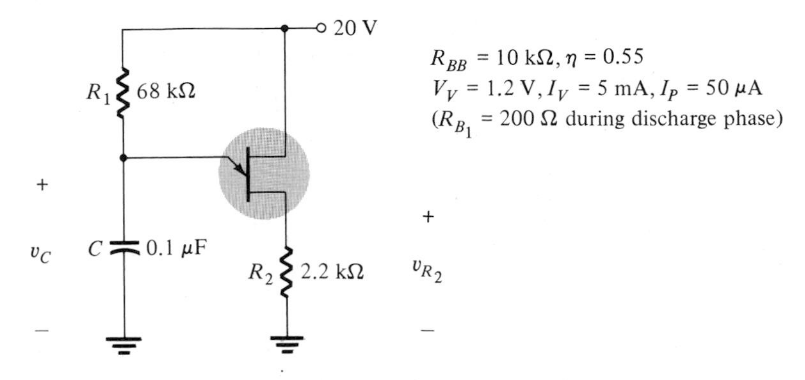

Figure 13.69

(a) Find R_{B_1} and R_{B_2} at $I_E = 0$ A.
(b) Determine V_P, the voltage necessary to turn on the UJT.
(c) Determine whether R_1 is within the permissible range of values defined by Eq. (13.8).
(d) Determine the frequency of oscillation if $R_{B_1} = 200\ \Omega$ during the discharge phase.
(e) Sketch the waveform of v_C for two full cycles.
(f) Sketch the waveform of v_{R_2} for two full cycles.
(g) Determine the frequency using Eq. (13.17) and compare to the value determined in part (d). Account for any major differences.

§ 13.14

19. For a phototransistor having the characteristics of Fig. 13.51, determine the photoinduced base current for a radiant flux density of 5 mW/cm^2. If $h_{fe} = 40$, find I_C.

20. Design a high-isolation OR-gate employing phototransistors and LEDs.

§ 13.15

21. (a) Determine an average derating factor from the curve of Fig. 13.59 for the region defined by temperatures less than 25°C.
(b) Is it fair to say that for temperature greater than room temperature (up to 100°C), the output current is somewhat unaffected by temperature?

22. (a) Determine, from Fig. 13.55, the average change in I_{CEO} per degree change in temperature for the range 25 to 50°C.
(b) Can the results of part (a) be used to determine the level of I_{CEO} at 35°C? Test your theory.

23. Determine, from Fig. 13.56, the ratio of LED input current to detector output current for an output current of 20 mA. Would you consider the device to be relatively efficient in its purpose?

24. (a) Sketch the maximum-power curve of P_D = 200 mW on the graph of Fig. 13.57. List any noteworthy conclusions.
(b) Determine β_{dc} (defined by I_C/I_F) for the system at V_{CE} = 15 V, I_F = 10 mA.
(c) Compare the results of part (b) with those obtained from Fig. 11.48 at I_F = 10 mA. Do they compare? Should they? Why?

25. (a) Referring to Fig. 13.58, determine the collector current above which the switching time does not change appreciably for R_L = 1 kΩ and R_L = 100 Ω.
(b) At I_C = 6 mA, how does the ratio of switching times for R_L = 1 kΩ and R_L = 100 Ω compare to the ratio of resistance levels?

§ 13.16

26. Determine η and V_G for a PUT with V_{BB} = 20 V and $R_{B_1} = 3\,R_{B_2}$.

27. Using the data provided in Example 13.3 determine the impedance of the PUT at the firing and valley points. Are the approximate open- and short-circuit states verified?

28. Can Eq. (13.24) be derived exactly as shown from Eq. (13.23)? If not, what element is missing in Eq. (13.24)?

29. (a) Will the network of Example 13.3 oscillate if V_{BB} is changed to 10 V? What minimum value of V_{BB} is required (V_V a constant)?
(b) Referring to the same example, what value of R would place the network in the stable on state and remove the oscillatory response of the system?
(c) What value of R would make the network a 2-ms time-delay network? That is, provide a pulse v_k 2 ms after the supply is turned on and then stay in the "on" state.

14

Integrated Circuits

14.1 INTRODUCTION

During the past decade the *integrated circuit* (IC) has through expanded usage and the various media of advertising, become a product whose basic function and purpose are now understood by the layperson. The most noticeable characteristic of an IC is its size. It is typically thousands of times smaller than a semiconductor structure built in the usual manner with discrete components. For example, the integrated circuit shown in Fig. 14.1 has 68,000 transistors in addition to a multitude of other elements

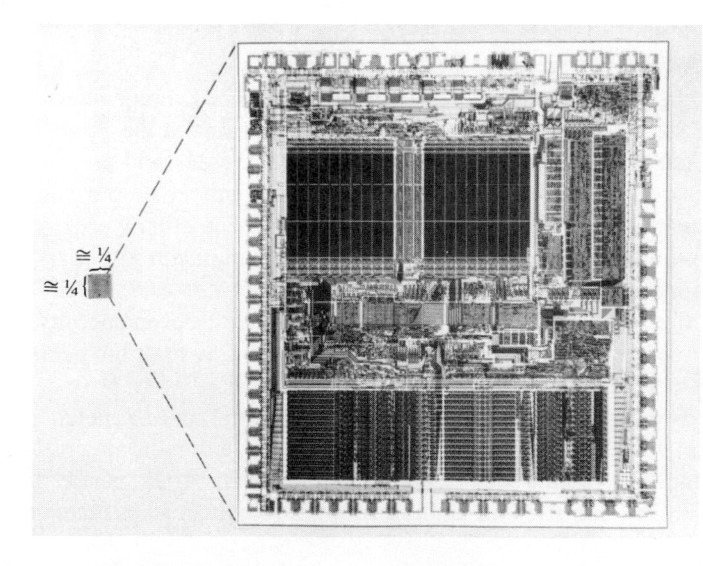

Figure 14.1 The MC6800 microprocessor and a discrete transistor package. The chip employed in the discrete device appears within the circled region. (Courtesy Motorola, Inc.)

although it is only 230 × 260 mils or about $\frac{1}{4}$ in. by $\frac{1}{4}$ in. The MC68000 is a microprocessor unit that is the heart of a microcomputer manufactured by Motorola.

Integrated circuits are seldom, if ever, repaired; that is, if a single component within an IC should fail, the entire structure (complete circuit) is replaced—a more economical approach. There are three types of ICs commercially available on a large scale today. They include the monolithic, thin (or thick) film, and hybrid integrated circuits. Each will be introduced in this chapter.

14.2 RECENT DEVELOPMENTS

Although the sequence of steps leading to the manufacture of an integrated circuit has not changed substantially during the past decade, the manner in which each step is performed has changed dramatically. In the early days the IC manufacturer designed, built, and maintained the equipment employed in the production cycle. Today, however, new industries have emerged that have assumed the responsibility of introducing the latest technological advances into the processing equipment. The result is that the manufacturer can concentrate on design, quality control, improved performance and reliability characteristics, and further miniaturization. The equipment available from the peripheral companies carries a high price tag (unit costs in excess of $\frac{1}{2}$ million are not unusual), however, and 24-hour operation is almost a necessity to ensure sound economic policy. In an effort to ensure continuous operation, the larger IC manufacturers have their own service staff rather than having to rely on immediate response from the equipment manufacturer.

Automation continues to be important in the production cycle. A great deal of microprocessor control introduced in the form of "cassette addressment" has significantly reduced the possibility of error owing to incorrect transfer of information to the processing unit. It also has a sensitivity to the process being performed that is unavailable through the human response curve. A complete set of instructions is placed on a magnetic tape cassette and identified for use with the preparation of a particular IC wafer. The increased level of automation also reduces the amount of "handling" and contact with the wafer, thereby reducing the number of sources of contaminants and increasing the yield factor.

One of the continuing areas of concerns is the yield level. The average number of "good" dies resulting from a wafer is improving but still remains at the 30 to 40% level. However, one must realize that as "feature" sizes decreases and density increases the yield level may not change significantly but the number of components produced in the same wafer area is increasing at a dramatic rate. In other words, if we utilized the improved production procedures of today on ICs manufactured 5 years ago the yield level would probably exceed 90%.

Developments of the last decade have resulted in a general acceptance by the industry that *IC density will double every two years*. Figure 14.2 reveals the decrease in size of BJTs in a period of one decade. At one time dimensions were provided in mils and square mils. Today, the micron or micrometer (1/millionth of a meter; μm) is the standard measure, with 1 mil = 25.4 μm. Note in Fig. 14.2 that the area of 1983 is about 10% of that of 1973, with a significant reduction in size in the period 1980 to 1983. The models of 1982 and 1983 are in the current production cycle. Statements

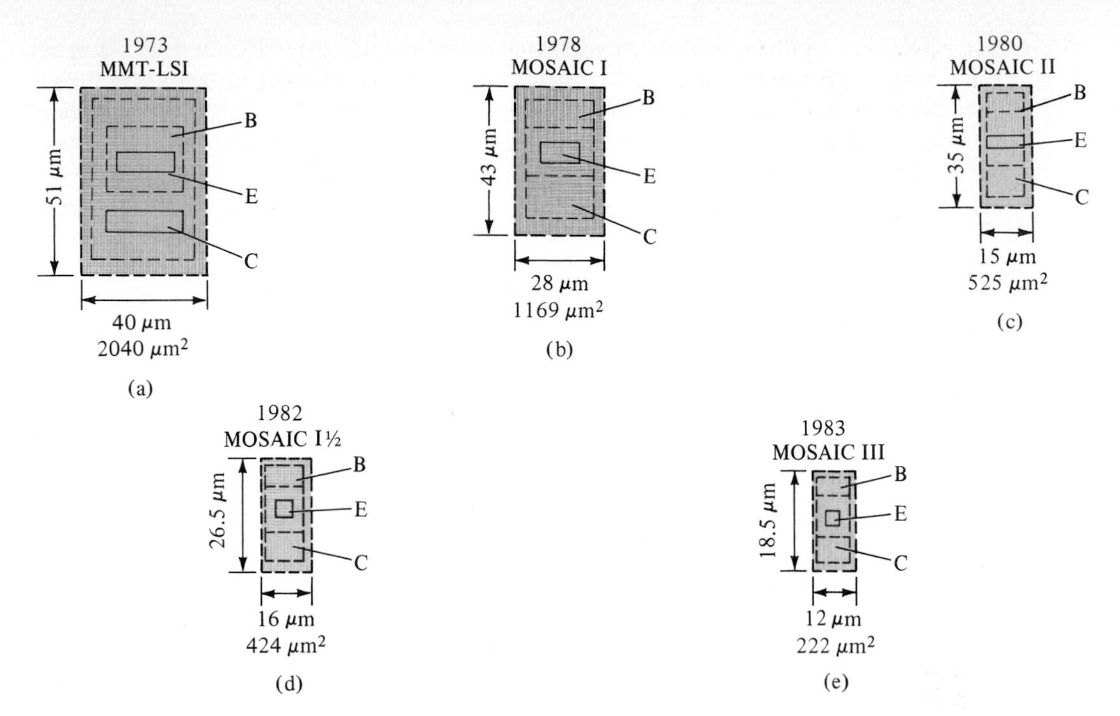

Figure 14.2 Bipolar process evolution. (Courtesy Motorola, Inc.)

regarding increasing density levels suggest that the current state of the art is approaching 100-μm^2 BJTs, which will probably be in the production cycle within the next few years.

The increased density and improved yield levels is due primarily to more sophisticated machinery in the production cycle, improved methods to detect and correct flaws, higher levels of cleanliness, increased purity levels of processing materials, improved manufacturing materials, and a reduction in the number of processing steps.

Whereas class 100 rooms were common 5 years ago, class 10 is the current industry standard. A class 10 room is tenfold cleaner than a typical hospital environment. The class number indicates the number of particles 1 μm or larger per cubic foot. The cost of establishing such an environment can only be described as staggering. A continual lamina flow of filtered air is established between the floor and ceiling to maintain the high level of cleanliness. The white robe, boots, and hat ("bunny suits") that appear in some of the photographs in this chapter are required in some production areas. Control is so tight that women working in many of these areas cannot wear makeup, to eliminate any possible introduction of foreign particles into the environment.

The water employed in the rinsing and cleaning operations is filtered at 0.2 μm and has an 18-MΩ resistivity level (recall the discussion of resistivity in Section 1.2). It is so free of organic contaminants that it will not support a culture growth. In addition, the purity of the processing materials, such as the chemicals, coatings, and other materials that "touch" the wafer, has improved to match the increased density levels.

The line widths of current manufacturing techniques extend from 1 to 2 μm for RF low power devices up to 10 to 15 μm for higher-power devices. The current state of the art is down to $\frac{1}{2}$-μm widths, but production is typically at the 1.5-μm level. It is expected that the production process will be down to 1 μm in the next 1 to 2 years.

The silicon wafer has been the mainstay for the industry from the birth of the industry to today's production cycle. As density levels continue to increase and line widths decrease there may be a need to turn to materials such as GaAs (gallium arsenide) with its range of improved performance characteristics.

Due to large investments, it is an absolute necessity that the product processing be tightly controlled through a strong management system. The computer is now playing a very important role in providing the data required for such continuous surveillance of the production cycle.

A number of improvements in the manufacturing process will be introduced in this chapter as each production step is described.

14.3 MONOLITHIC INTEGRATED CIRCUIT

The term *monolithic* is derived from a combination of the Greek words *monos*, meaning single, and *lithos* meaning stone, which in combination result in the literal translation, single-stone, or more appropriately, single-solid structure. As this descriptive term implies, the monolithic IC is constructed within a *single* wafer of semiconductor material. The greater portion of the wafer will simply act as a supporting structure for the very thin resulting IC. An overall view of the stages involved in the fabrication of monolithic ICs is provided in Fig. 14.3. The actual number of steps leading to a finished product is many times that appearing in Fig. 14.3. The figure does, however, point out the major production phases of forming a monolithic IC.

As indicated in the previous section, the processing equipment and not the steps have changed significantly in recent years. The initial preparation of the semiconductor wafer of Fig. 14.3 was discussed in Chapter 1 in association with the fabrication of diodes. As indicated in the figure, it is first necessary to design a circuit that will meet the specifications. The circuit must then be laid out in order to ensure optimum use of available space and a minimum of difficulty in performing the diffusion processes to follow. The appearance of the mask and its function in the sequence of stages indicated will be introduced in Section 14.5 For the moment, let it suffice to say that a mask has the appearance of a negative through which impurities may be diffused (through the light areas) into the silicon slice. The actual diffusion process for each phase is similar to that applied in the fabrication of diffused transistors in Chapter 4. The last mask of the series will control the placement of the interconnecting conducting pattern between the various elements. The wafer then goes through various testing procedures, is scribed and broken into individual chips, packaged and assembled as indicated. A processed silicon wafer appears in Fig. 14.4. The original wafer can be anywhere from $\frac{1}{2}$ to 5 in. in diameter. The size of each chip will, of course, determine the number of individual circuits resulting from a single wafer. The dimensions of each chip of the wafer in Fig. 14.4 are 50 $\times$ 50 mils. To point out the microminiature size of these chips, consider that 20 of them can be lined up along a 1-in. length. The average relative size of the elements of a monolithic IC appear in

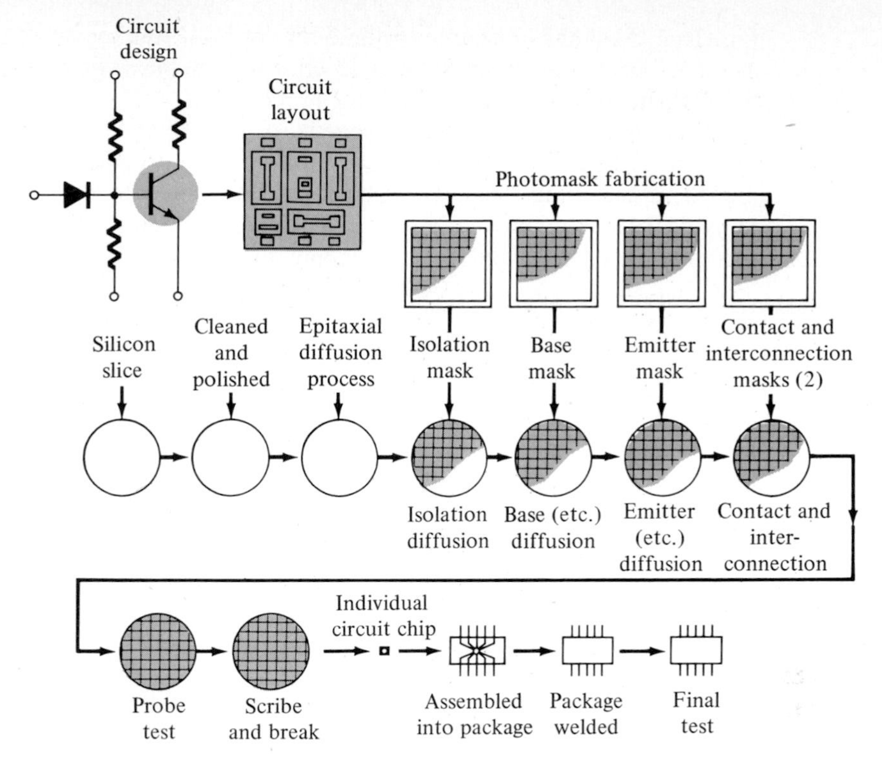

Figure 14.3 Monolithic integrated-circuit fabrication. (Courtesy Robert Hibberd.)

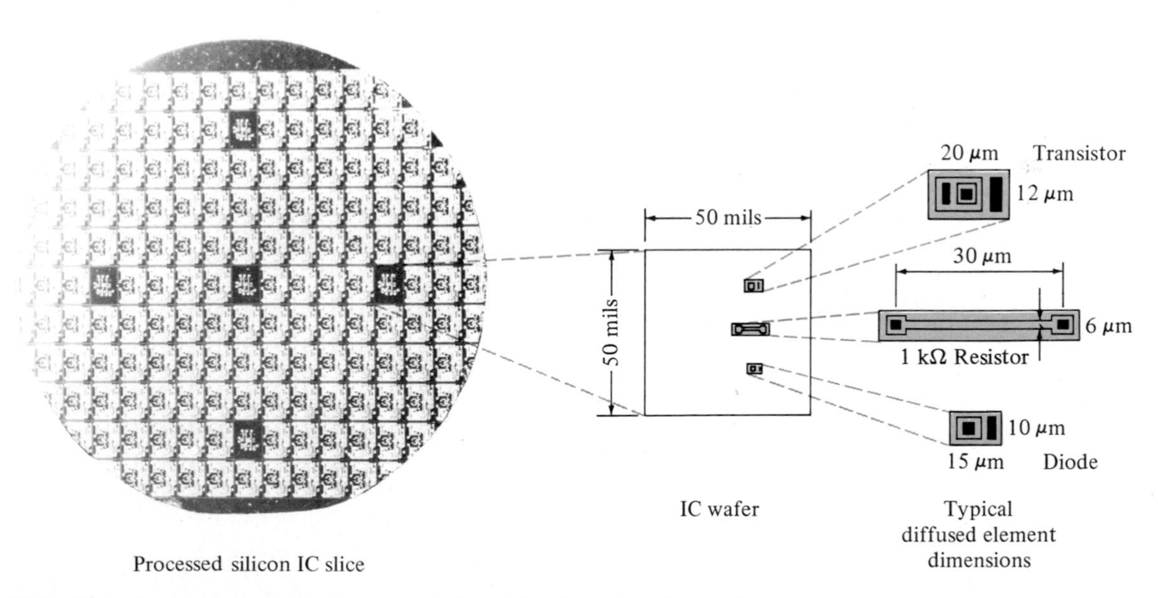

Figure 14.4 Processed monolithic IC wafer with the relative dimensions of the various elements. (Courtesy Robert Hibberd.)

Fig. 14.4. Note the large area required for the 1-kΩ resistor as compared to the other elements indicated. The next section will examine the basic construction of each of these elements.

A recent article indicated, by percentage, the relative costs of the various stages in the production of monolithic ICs as compared to discrete transistors. The resulting graphs appear in Fig. 14.5. The processing phase includes all stages leading up to the individual chips of Fig. 14.4. Note the difference in cost for the various phases of production as determined by the size and density of the chip.

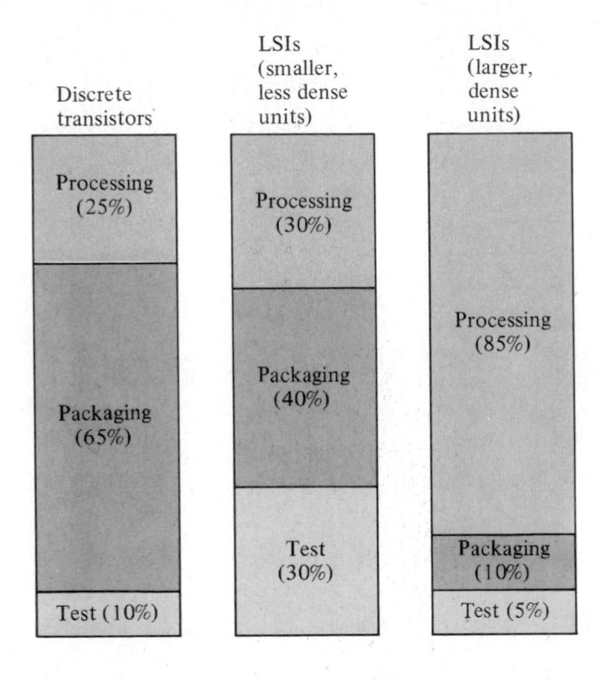

Figure 14.5 Cost breakdown for the manufacturing of discrete transistors and large-scale integrated circuits (LSIs).

14.4 MONOLITHIC CIRCUIT ELEMENTS

The surface appearance of the transistor, diode, and resistor appear in Fig. 14.4. We shall now examine the basic construction of each in more detail.

Resistor

You will recall that the resistance of a material is determined by the resistivity, length, area, and temperature of the material. For the integrated circuit, each necessary element is present in the sheet of semiconductor material appearing in Fig. 14.6.

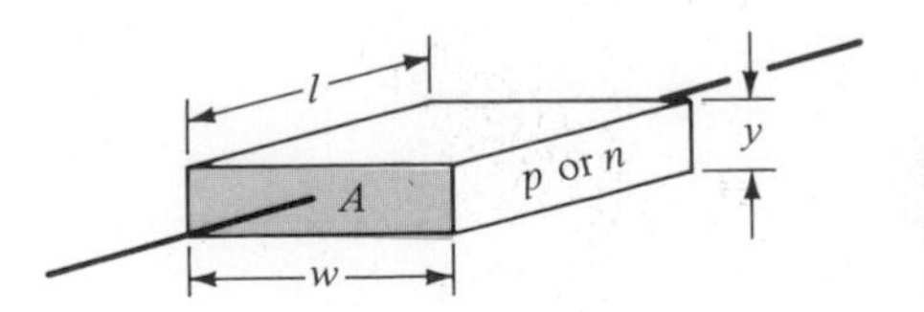

Figure 14.6 Parameters determining the resistance of a sheet of semiconductor material.

As indicated in the figure, the semiconductor material can be either p- or n-type, although the p-type is most frequently employed.

The resistance of any bulk material is determined by

$$R = \rho\frac{l}{A}$$

For $l = w$, resulting in a square sheet,

$$R = \frac{\rho l}{yw} = \frac{\rho l}{yl}$$

and

$$\boxed{R_S = \frac{\rho}{y}} \qquad \text{ohms} \qquad (14.1)$$

where ρ is in ohm-centimeters and y is in centimeters. R_S is called the sheet resistance and has the units ohms per square. The equation clearly reveals that the sheet resistance is independent of the size of the square.

In general, where $l \neq w$,

$$\boxed{R = R_S\frac{l}{w}} \qquad \text{ohms} \qquad (14.2)$$

For the resistor appearing in Fig. 14.4, $w = 6\ \mu\text{m}$, $l = 30\ \mu\text{m}$, and $R_S = 200\ \Omega/\text{square}$:

$$R = R_S\frac{l}{w} = 200 \times \frac{30}{6} = 1\ \text{k}\Omega$$

A cross-sectional view of a monolithic resistor appears in Fig. 14.7 along with the surface appearance of two monolithic resistors. In Fig. 14.7a the sheet resistive material (p) is indicated with its aluminum terminal connections. The n-isolation region performs exactly that function indicated by its name; that is, it isolates the monolithic resistive elements from the other elements of the chip. Note in Fig. 14.7b the method employed to obtain a maximum l in a limited area. The resistors of Fig. 14.7 are called base-diffusion resistors since the p-material is diffused into the p-type substrate during the base-diffusion process indicated in Fig. 14.3.

Capacitor

Monolithic capacitive elements are formed by making use of the transition capacitance of a reverse-biased p-n junction. At increasing reverse-bias potentials, there is an increasing distance at the junction between the p- and n-type impurities. The region between these oppositely doped layers is called the depletion region due to the absence of "free" carriers. The necessary elements of a capacitive element are therefore present—the depletion region has insulating characteristics that separate the two oppositely charged layers. The transition capacitance is related to the width (W) of the depletion region, the area (A) of the junction, and the permittivity (ϵ) of the material

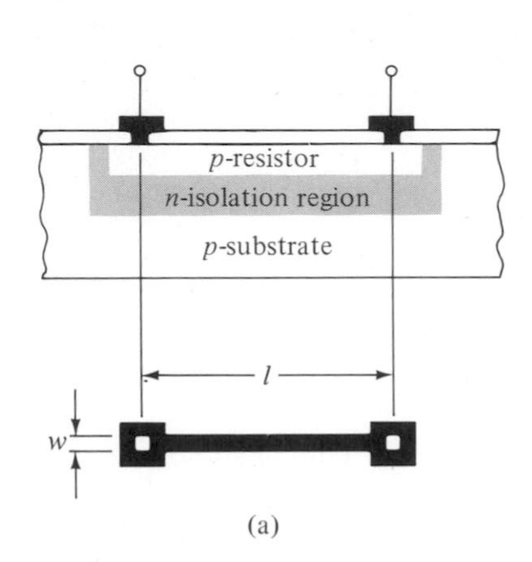

(a)

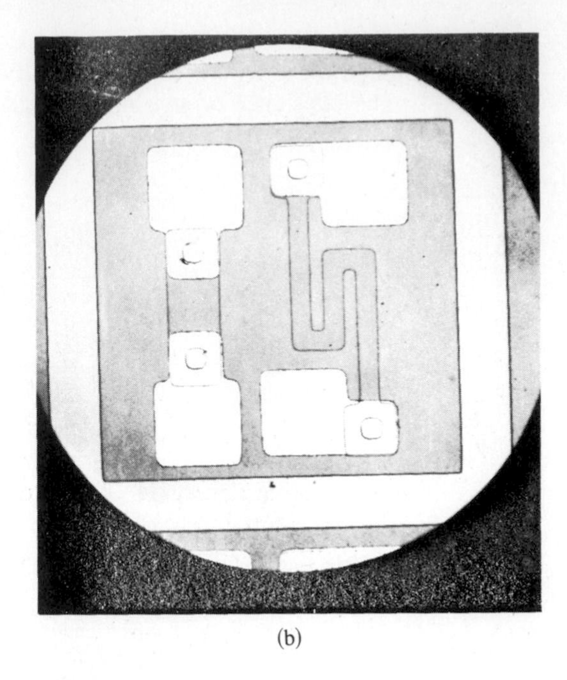
(b)

Figure 14.7 Monolithic resistors: (a) cross section and determining dimensions; (b) surface view of two monolithic resistances in a single die. (Courtesy Motorola, Inc.)

within the depletion region by

$$C_T = \frac{\epsilon A}{W} \tag{14.3}$$

The cross section and surface appearance of a monolithic capacitive element appear in Fig. 14.8. The reverse-biased junction of interest is J_2. The undesirable parasitic capacitance at junction J_1 is minimized through careful design. Due to the fact that aluminum is p-type impurity in silicon, a heavily doped n^+ region is diffused into the n-type region as shown to avoid the possibility of establishing an undesired p-n junction at the boundary between the aluminum contact and the n-type impurity region.

Inductor

Whenever possible, inductors are avoided in the design of integrated circuits. An effective technique for obtaining nominal values of inductances has so far not been devised for monolithic integrated circuits. In many instances, the need for inductive elements can be eliminated through the use of a technique known as *RC* synthesis. Thin (or thick) film or hybrid integrated circuits have an option open to them that cannot be employed in monolithic integrated circuits: the addition of discrete inductive elements to the surface of the structure. Even with this option, however, they are seldom employed due to their relatively bulky nature.

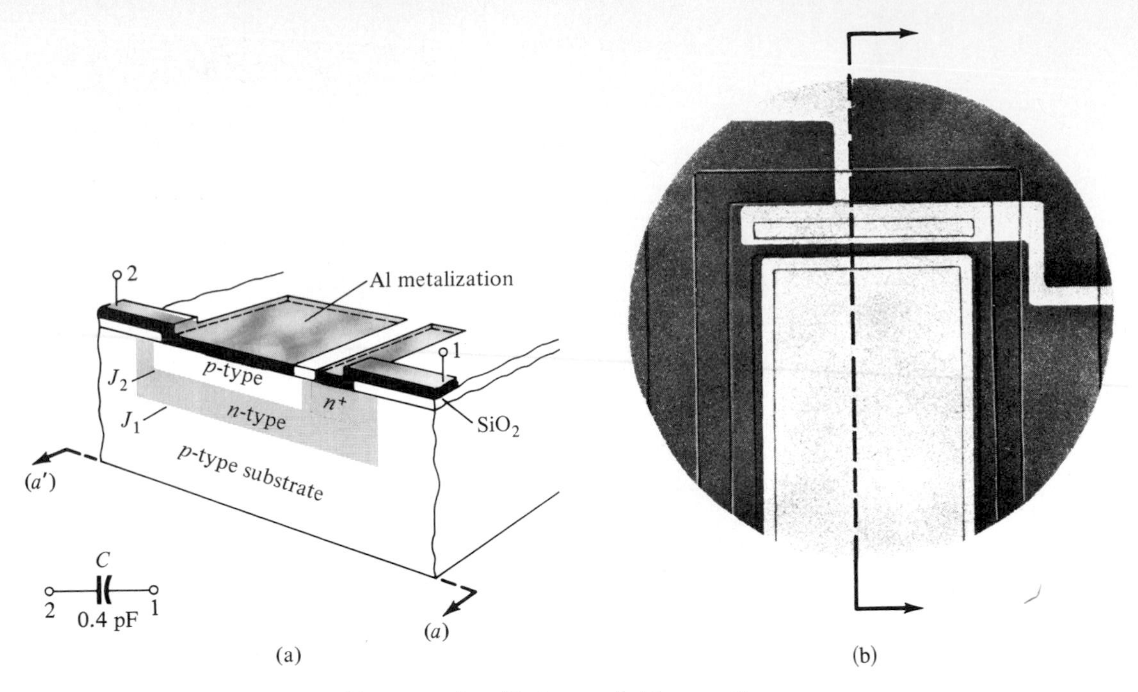

Figure 14.8 Monolithic capacitor: (a) cross section; (b) photograph. (Courtesy Motorola, Inc.)

Transistors

The cross section of a monolithic transistor appears in Fig. 14.9a. Note again the presence of the n^+ region in the n-type epitaxial collector region. The vast majority of monolithic IC transistors are *npn* rather than *pnp* for reasons to be found in more advanced texts on the subject. Keep in mind when examining Fig. 14.9 that the p-substrate is only a supporting and isolating structure forming no part of the active device itself. The base, emitter, and collector regions are formed during the corresponding diffusion processes of Fig. 14.3.

The top view of a typical monolithic transistor appears in Fig. 14.9b with the terminal diffusion areas. The order of diffusion is C, E, and B as defined by the depth and area of diffusion.

Diodes

The diodes of a monolithic integrated circuit are formed by first diffusing the required regions of a transistor and then masking the diode rather than transistor terminal connections. There is, however, more than one way of hooking up a transistor to perform a basic diode action. The two most common methods applied to monolithic integrated circuits appear in Fig. 14.10. The structure of a *BC-E* diode appears in Fig. 14.11.

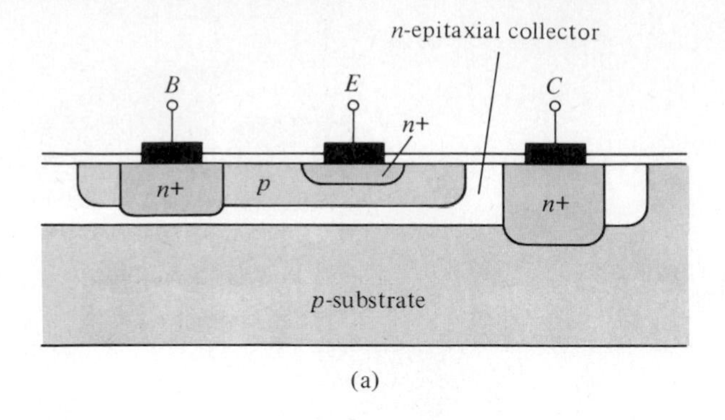

(a)

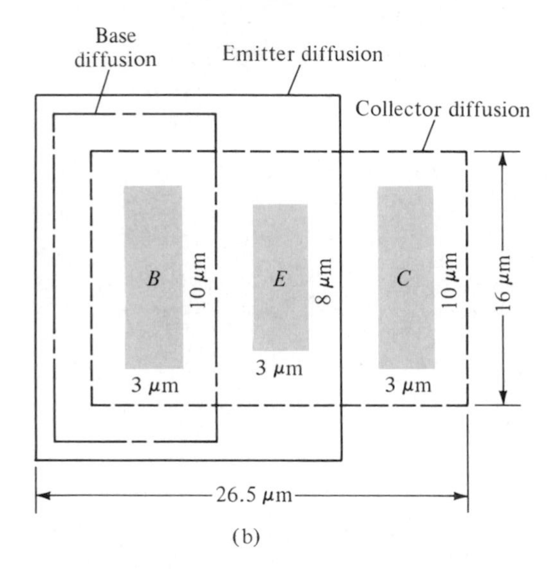

(b)

Figure 14.9 Monolithic transistor: (a) cross section; (b) surface appearance and dimension for a typical monolithic transistor. (Courtesy Motorola Monitor.)

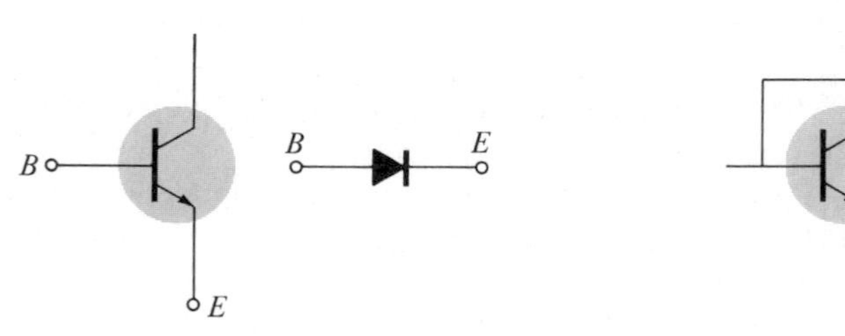

Figure 14.10 Transistor structure and two possible connections employed in the formation of monolithic diodes.

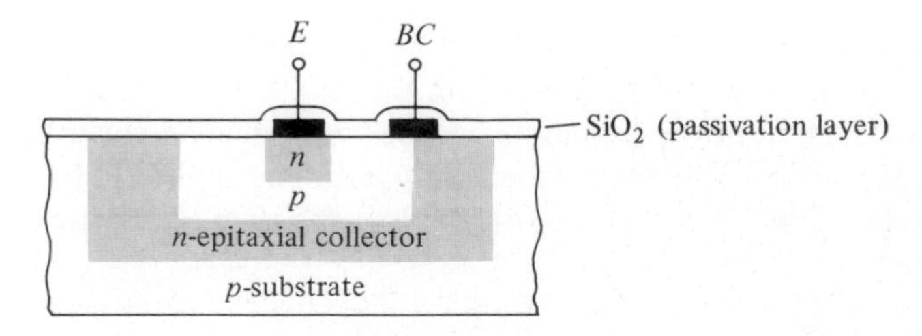

Figure 14.11 The cross-sectional view of a *BC–E* monolithic diode.

14.5 MASKS

The selective diffusion required in the formation of the various active and passive elements of an integrated circuit is accomplished through the use of masks such as that shown in Fig. 14.12. We shall find in Section 14.6 that the light areas are the only areas through which donor and acceptor impurities can pass. The dark areas will block the diffusion of impurities somewhat as a shade will prevent sunlight from changing the pigment of the skin. The next section will demonstrate the use of these marks in the formation of a computer logic circuit.

Figure 14.12 Mask. (Courtesy Motorola, Inc.)

The sequence of steps leading to the final mask is controlled by the micron width of the smallest features on the wafer. For the range 0.5 to 2 μm, electron beam litography must be employed, whereas for the range 3 to 5 μm, less sophisticated (and less expensive) methods can be employed.

At one time the making of a mask first required a large-scale stabilene drawing of all layers. The artwork was then transferred to a clear Mylar coated with red plastic called *Rubylith*. Very precise cuts were made in the red material and sections peeled off to reveal the regions through which the diffusion of impurities could take place. The resulting pattern was then photographed and reduced by 500X (500 times the desired size for production) in a series of steps until the desired master (redicle) was obtained.

Today, the same stabilene drawing is taken to a CAD (computer-aided-design) area called Calma (the manufacturer of the equipment employed) and is mounted on a digitizing board as shown in Fig. 14.13. The operator performs a function called *digitizing* through which the geometrical features from the stabilene drawing are transferred to the computer memory of the Calma system. The same procedure is repeated for each layer of the integrated circuit. Any section of the design can then be recalled and placed on the CRT of an editing station to modify the pattern, check continuity, etc., as shown in Fig. 14.14. One distinct advantage of the Calma system is that repetitive cells can be added from memory at the editing stage and not appear

Figure 14.13 Calma digitizing station. (Courtesy Motorola, Inc. and Calma, Inc.)

as part of the stabilene drawing. In addition, a library of cells could be established that could be introduced as required. The net result is that a skilled operator could design and develop a major portion of the complete reticle with a minimum of effort at the stabilene stage.

At this juncture there are two methods that can be applied to produce the final mask. One method employs the Xynetics plotter shown in Fig. 14.15 to generate a precise line drawing from the descriptive data provided by the Calma system. It is an amazing piece of equipment to watch in operation, with its multicolor option and very quick precise movements. It can enlarge a region to 10,000× if required. A photoreduction process can then be applied to reduce the pattern to 100× and then 10×. The 10× pattern can be obtained in one step utilizing a *pattern generator*. The magnetic tape on which the pattern data are stored is loaded into the console, and a 10× pattern is obtained through computer control of the exposure shapes and placement. For a complex LSI, 200,000 to 300,000 flashes may be required over a period of 20 to 36 hours. A programmed step-and-repeat machine will then generate the required format (multiple images on the same mark) from the 10× pattern for the product group.

Although the procedure described above is still employed today, many production cycles are changing over to a "direct write" approach using an electron beam (E-beam)

(a)

(b)

Figure 14.14 Circuit design/modification at a Calma (interactive computer graphics system) editing/tablet station. (Courtesy Motorola, Inc., and Calma, Inc.)

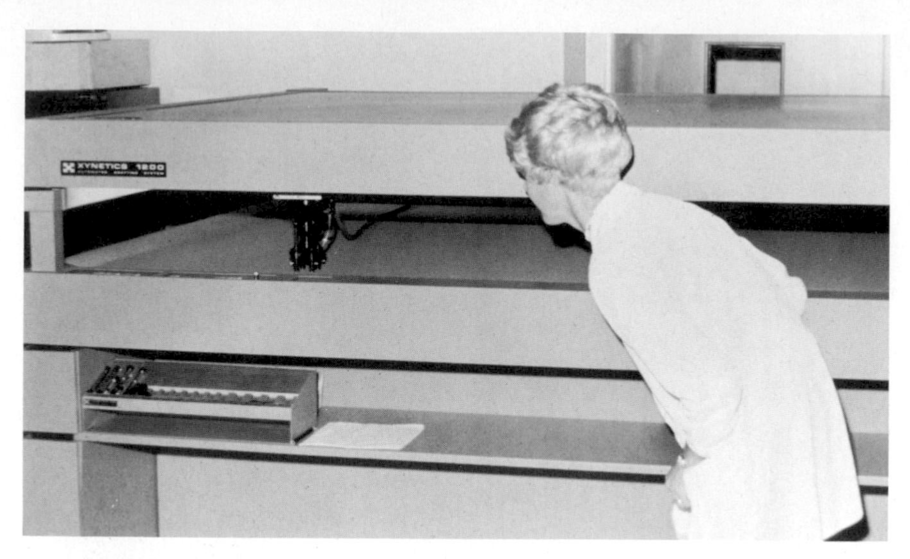

Figure 14.15 Xynetics 1200 Plotter (on line with the Calma system). (Courtesy Motorola, Inc., and Xynetics, Inc.)

system such as appearing in Fig. 14.16. It saves a number of intermediate steps by "cutting" the mask pattern directly from the Calma data using an electron beam as the cutting edge. The reduced number of steps and the direct cutting of the mark reduce the number of flaws and omissions that may appear in the final product. For LSI units, the time involved from initial design to mask availability may extend from a few months to 1 to 2 years.

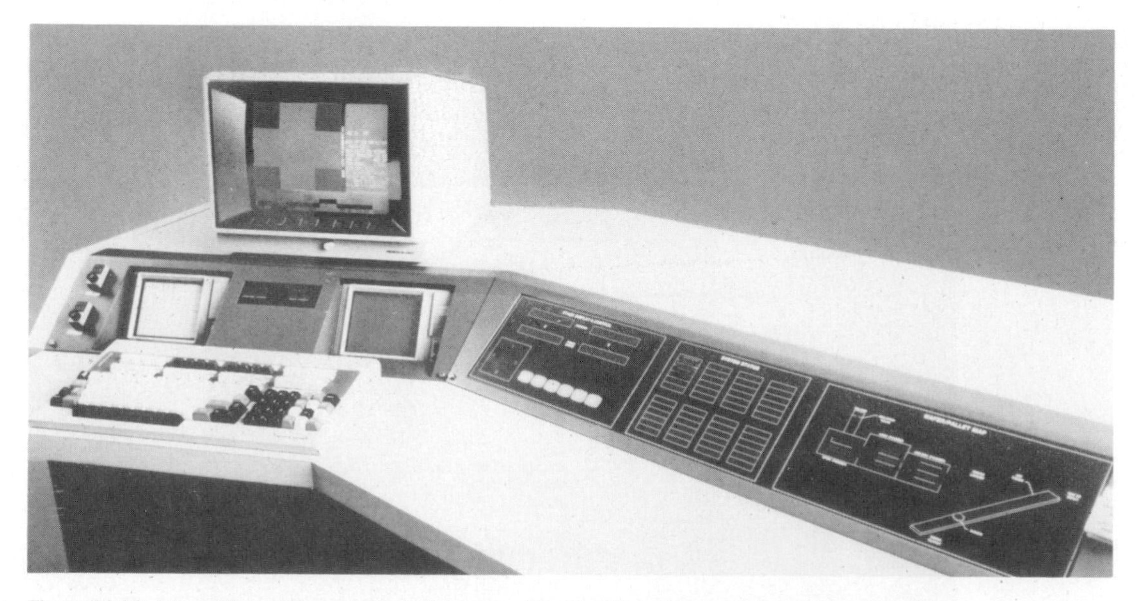

Figure 14.16 Direct-Write E-beam system. (Courtesy of Perkin–Elmer, Materials Research Corporation.)

14.6 MONOLITHIC INTEGRATED CIRCUIT—THE NAND GATE

This section is devoted to the sequence of production stages leading to a monolithic NAND-gate circuit. A detailed examination of each process would require many more pages that it is possible to include in this text. The description, however, should be sufficiently complete and informative to aid the reader in any future contact with this highly volatile area. The circuit to be prepared appears in Fig. 14.17a. The criteria of space allocation, placement of pin connection, and so on require that the elements be situated in the relative positions indicated in Fig. 14.17b. The regions to be isolated from one another appear within the solid heavy lines. A set of masks for the various diffusion processes must then be made up for the circuit as it appears in Fig. 14.17b.

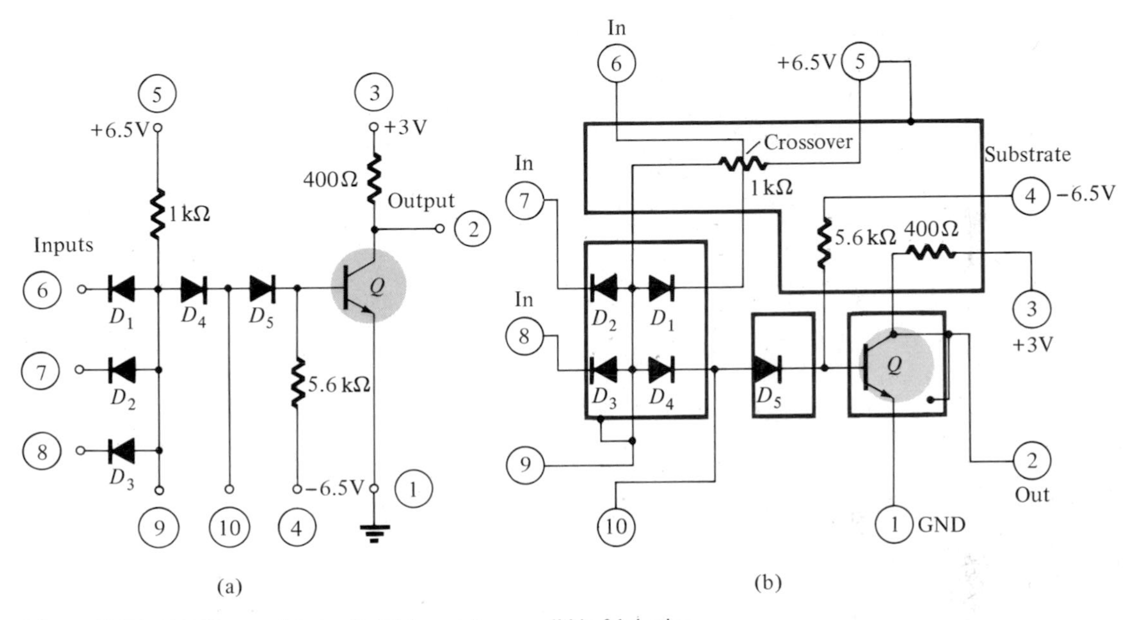

Figure 14.17 NAND gate: (a) circuit; (b) layout for monolithic fabrication.

Now we shall proceed slowly through the first diffusion process to demonstrate the natural sequence of steps that must be followed through each diffusion process indicated in Fig. 14.9.

p-Type Silicon Wafer Preparation

After being sliced from the grown ingot, a *p*-type silicon wafer is lapped, polished, and cleaned (Fig. 14.18a) to produce the structure of Fig. 14.17b. A chemical etch process is also applied to further smooth the surface and remove a layer of the wafer that may have been damaged during the lapping and polishing sequence.

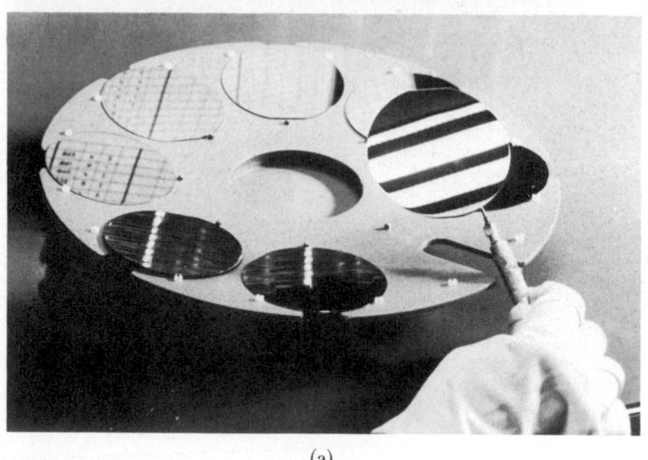

(a)

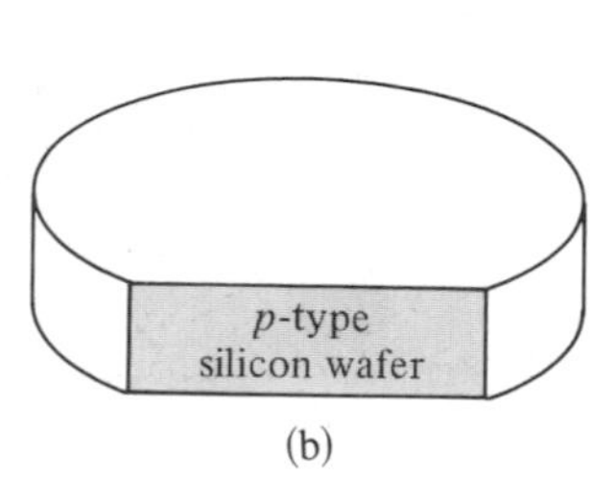

(b)

Figure 14.18 (a) Lapping and polishing stage of wafer preparation; (b) *p*-type silicon wafer. [(a) Courtesy Motorola, Inc.; (b) courtesy Texas Instruments, Inc.]

n-Type Epitaxial Region

An *n*-type epitaxial region is then diffused into the *p*-type substrate as shown in Fig. 14.19. It is deposited in a way that will result in a single-crystal structure having the same crystal structure and orientation as the substrate but with a different conductivity level. It is in this thin epitaxial layer that the active and passive elements will be diffused. The *p*-type area is essentially a supporting structure that adds some thickness to the structure to increase its strength and permit easier handling.

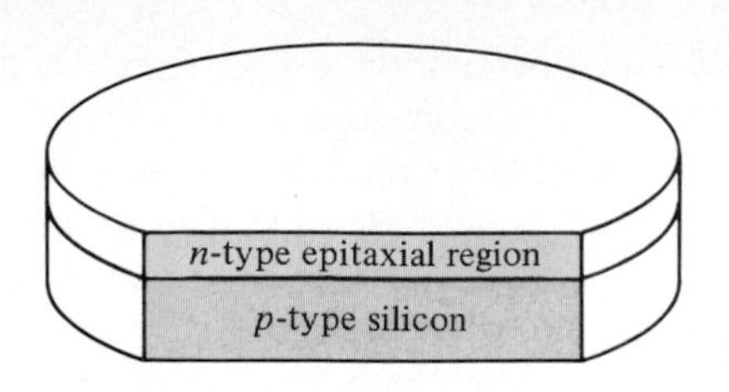

Figure 14.19 *p*-type silicon wafer after the *n*-type epitaxial diffusion process.

The apparatus employed in the deposition process includes a long quartz tube surrounded by a radio-frequency (RF) induction coil. The wafers are placed on a rectangular graphite structure called a *boat* and are inserted into the chamber as indicated in Fig. 14.20. The RF coils then heat the chamber to above 1100°C, while the impurities are introduced in a gaseous state. The entire process can be carefully monitored and controlled to ensure the proper epitaxial growth.

Figure 14.20 Deposition of the epitaxial layer.

Silicon Oxidation (SiO_2)

The resulting wafer is then subjected to an oxidation process resulting in a surface layer of SiO_2 (silicon dioxide) as shown in Fig. 14.21. This surface layer will prevent any impurities from entering the *n*-type epitaxial layer. However, selective etching of

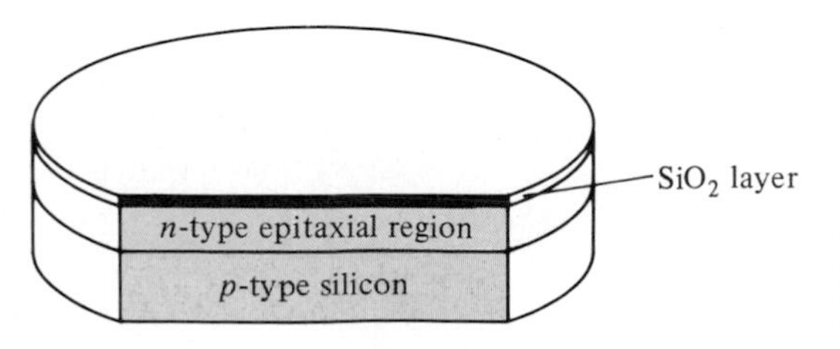

Figure 14.21 Wafer of Fig. 14.18 following the deposit of the SiO_2 layer.

this layer will permit the diffusion of the proper impurity into designated areas of the n-type epitaxial region of the silicon wafer.

The apparatus employed in the oxidation process is similar to that used to establish the epitaxial layer in that the wafers are placed in a boat (now made of quartz) and inserted into a quartz tube. Typically, about 20 wafers are introduced at the same time. In this case, however, a resistance heater is wound around the tube to raise the temperature to about 1100°C. Oxygen is introduced in the wet or dry form until the desired SiO_2 layer is established. Recent developments include raising the atmospheric pressure in the vessel to permit a measurable reduction in the processing temperature. For every 1 atm (atmosphere) increase in pressure, there is a 30°C reduction in required temperature. At 10 atm, the temperature can be reduced 300°C. Silicon is reasonably well behaved at 800°C but an entirely different animal at 1100°C. At lower processing temperatures there is also an improved quality of oxide, a reduction in introduced stresses, and reduction or elimination of a number of device design limitations.

The time involved for the oxidation process can extend from a few hours to as long as 24 hours depending on the thickness of the oxide and the quality desired.

Photolithographic Process

The selective etching of the SiO_2 layer is accomplished through the use of a photolithographic process. The wafer is first coated with a thin layer of photosensitive material, commonly called *photoresist*, by the system appearing in Fig. 14.22. The application of the photoresist is entirely microprocessor controlled. A deck of wafers is deposited into the receiving trays appearing in the left-hand region of Fig. 14.22. The equipment automatically applies a high-pressure scrub, a dehydration process, the resist coating, and a soft bake; then it develops and hard bakes the wafers.

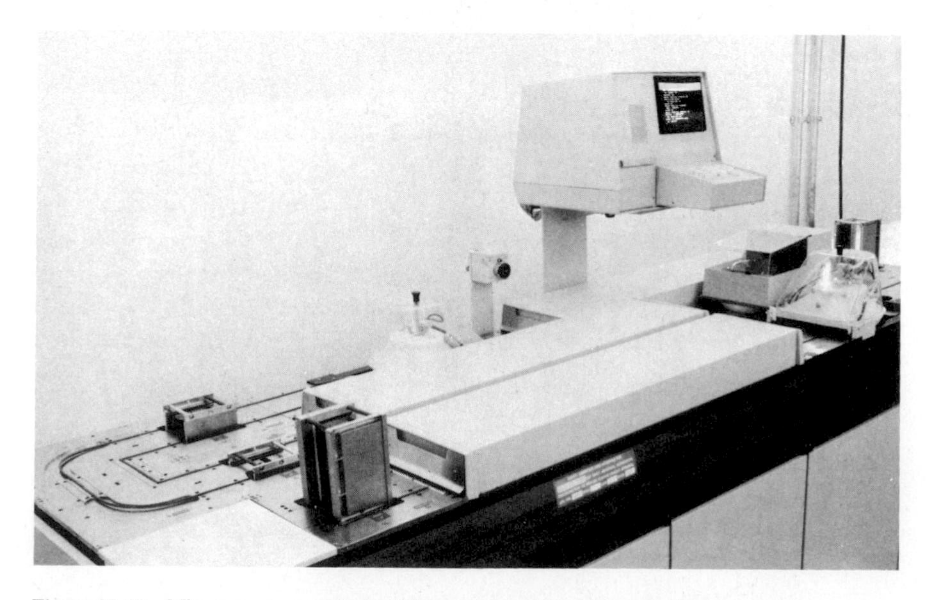

Figure 14.22 Microprocessor-controlled photoresist module. (Courtesy Motorola, Inc.)

The next step is to use one of the masks developed earlier to determine those areas of the SiO_2 layer to be removed in preparation for the isolation diffusion process. The mask can either be placed directly on the photoresist as shown in Fig. 14.23 or separated from the mask as shown in Fig. 14.24. When placed directly on the wafer an ultraviolet light is applied that will expose those regions of the photosensitive material not covered by the masking pattern (Fig. 14.23).

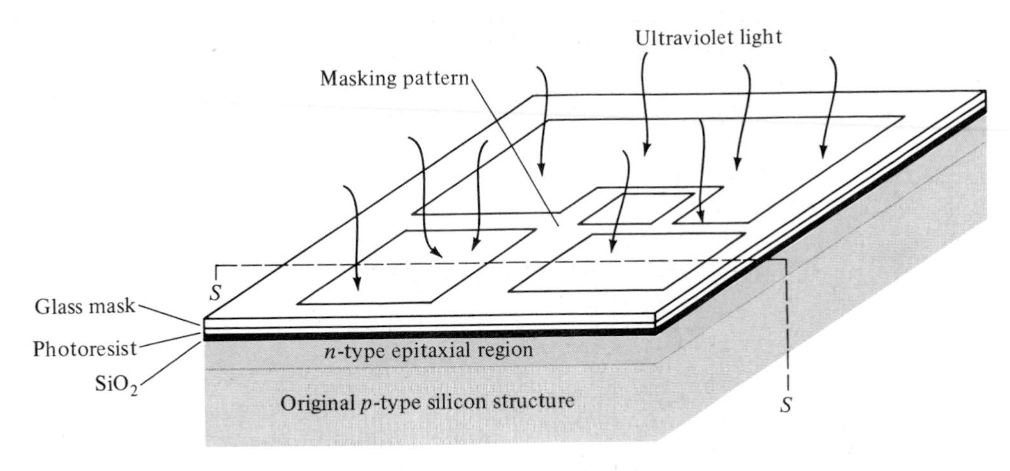

Figure 14.23 Photolithographic process: the application of ultraviolet light after the mask is properly set; the structure is only one of the 200,400, or even 500 individual NAND gate circuits being formed on the wafer of Figs. 14.18 through 14.21.

Figure 14.24 600 HT Micralign Projection Printing System (Courtesy Perkin-Elmer Inc.)

The method employed in Fig. 14.24 is called *projection printing*. It employs optics to expose the various regions. The primary advantage of this approach is that the mask cannot introduce contaminants to the wafer surface. It is an approach used more and more in recent years.

The resulting wafer is then subjected to a chemical solution that will remove the unexposed photosensitive material. A cross section of a chip (S-S of Fig. 14.23) will then appear as indicated in Fig. 14.25. A second solution will then etch away the SiO_2 layer from any region not covered by the photoresist material (Fig. 14.26). The final step before the diffusion process is the removal, by solution, of the remaining photosensitive material. The structure will then appear as shown in Fig. 14.27.

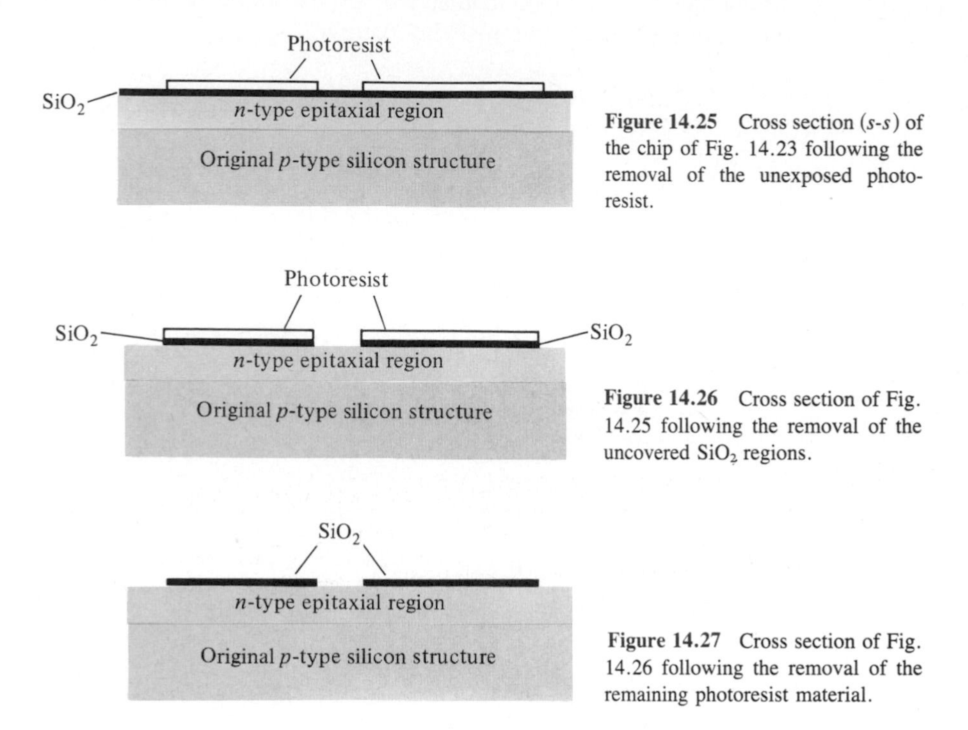

Figure 14.25 Cross section (*s-s*) of the chip of Fig. 14.23 following the removal of the unexposed photoresist.

Figure 14.26 Cross section of Fig. 14.25 following the removal of the uncovered SiO_2 regions.

Figure 14.27 Cross section of Fig. 14.26 following the removal of the remaining photoresist material.

Isolation Diffusion

The structure of Fig. 14.27 is then subjected to a p-type diffusion process resulting in the islands of n-type regions indicated in Fig. 14.28. The diffusion process ensures a heavily doped p-type region (indicated by p^+) between the n-type islands. The p^+ regions will result in improved *isolation* properties between the active and passive components to be formed in the n-type islands.

The apparatus employed includes a quartz boat and tube (to minimize the possibility of contaminating the processing environment) that is heated by a high-resistance

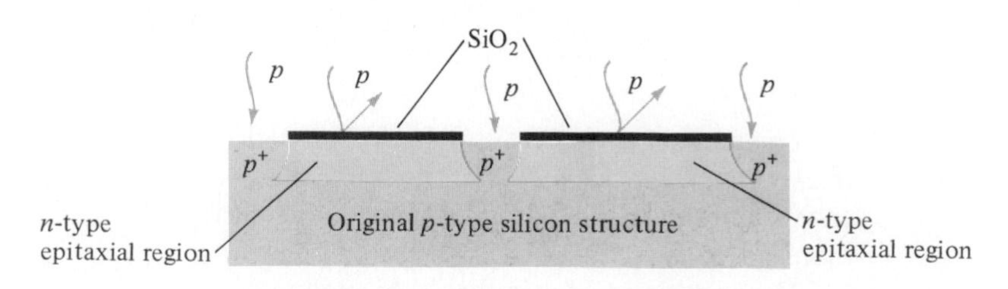

Figure 14.28 Cross section of Fig. 14.27 following the isolation diffusion process.

wire wrapped around the tube. The diffusion operation normally occurs at a temperature neighboring on 1200°C. The system as appearing in Fig. 14.29 is totally microprocessor controlled (through cassette addressment). Three or four people can operate 16 furnaces; and the entire operation, from pulling the boats in and out of the furnaces to monitoring the temperature and dopant level, is computer controlled.

Figure 14.29 Microprocessor-controlled diffusion operation (with 5-in. wafers). (Courtesy Motorola, Inc.)

An alternative to the high-temperature diffusion process is *ion implantation*. A beam of dopant ions (about the size of a pencil) is directed toward a wafer at a very high velocity by an ion-accelerator gun. The ions will penetrate the medium to a level that can be controlled to within $\frac{1}{2}$ micron as compared to $2\frac{1}{2}$ μm using other methods.

In addition to improved control, the processing temperature is lower, and a broader range of electrical parameters is now available. At present the primary use of this method is in establishing bases. In time, with the proper modifications, emitter diffusions will also be a possibility.

In preparation for the next masking and diffusion process, the entire surface of the wafer is coated with an SiO_2 layer as indicated in Fig. 14.30.

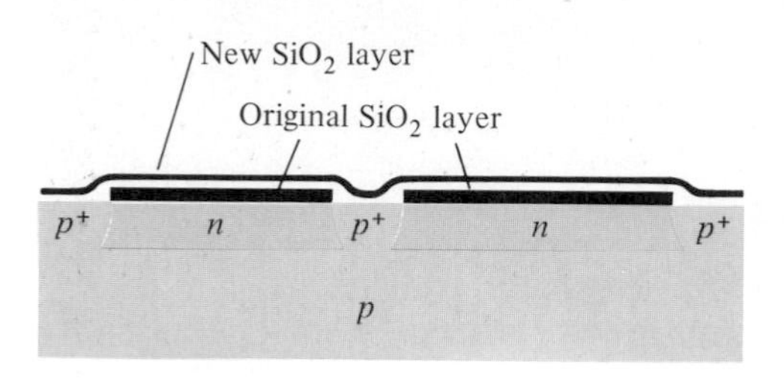

Figure 14.30 In preparation for the next diffusion process, the entire wafer is coated with an SiO_2 layer.

Base and Emitter Diffusion Processes

The isolation diffusion process is followed by the base and emitter diffusion cycles. The sequence of steps in either case is the same as that encountered in the description of the isolation diffusion process. Although "base" and "emitter" refer specifically to the transistor structure, necessary parts (layers) of each element (resistor, capacitor, and diodes) will be formed during each diffusion process. The surface appearance of the NAND gate after the isolation base and emitter diffusion processes appears in Fig. 14.31. The mask employed in each process is also provided above each photograph.

The cross section of the transistor will appear as shown in Fig. 14.32 after the base and emitter diffusion cycles.

Preohmic Etch

In preparation for a good ohmic contact, n^+ regions (see Section 14.4) are diffused into the structure as indicated by the light areas of Fig. 14.33. Note the correspondence between the light areas and the mask pattern.

Metalization

A final masking pattern exposes those regions of each element to which a metallic contact must be made. The entire wafer is then coated with a thin layer of aluminum, gold, molybdenum, or tantalum (all high-conductivity, low-boiling-point metals) that after being properly etched will result in the desired interconnecting conduction pattern. A photograph of the completed metalization process appears in Fig. 14.34.

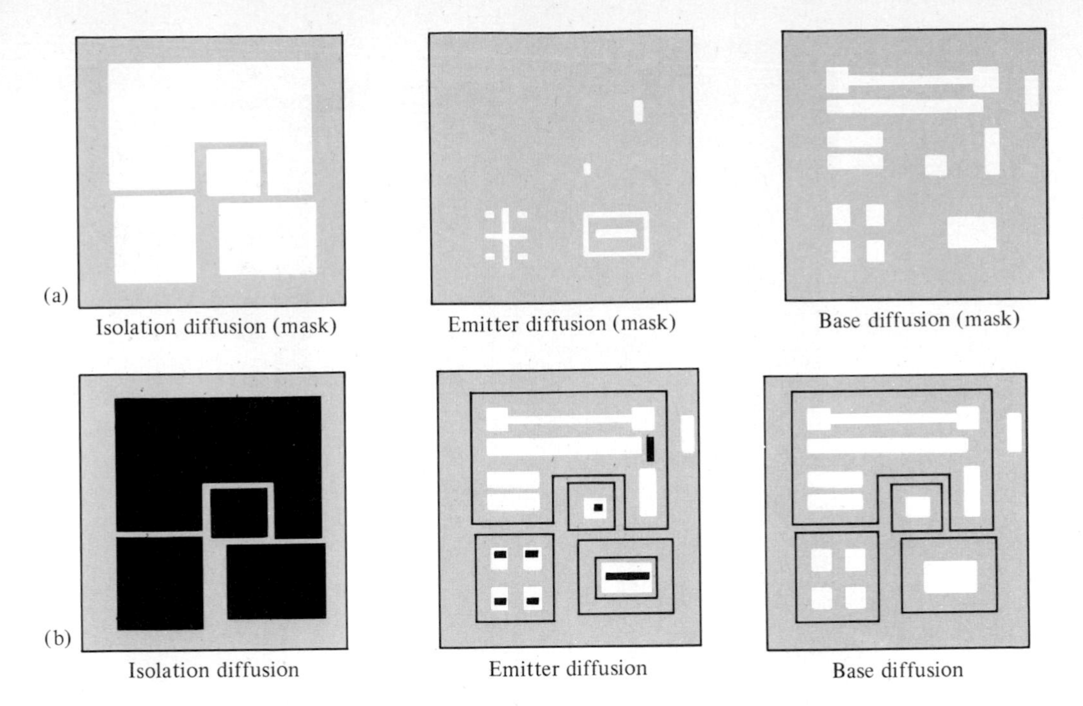

Figure 14.31 The masks (a) employed in each diffusion process (b). The surface appearance of the monolithic NAND gate after the isolation, base, and emitter diffusion processes. (Courtesy Motorola Monitor.)

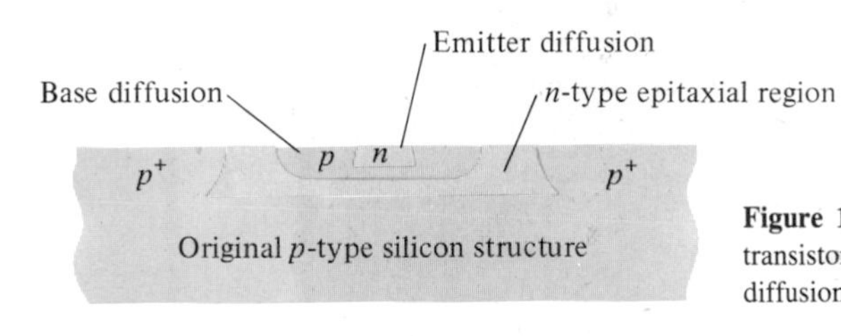

Figure 14.32 Cross section of the transistor after the base and emitter diffusion cycles.

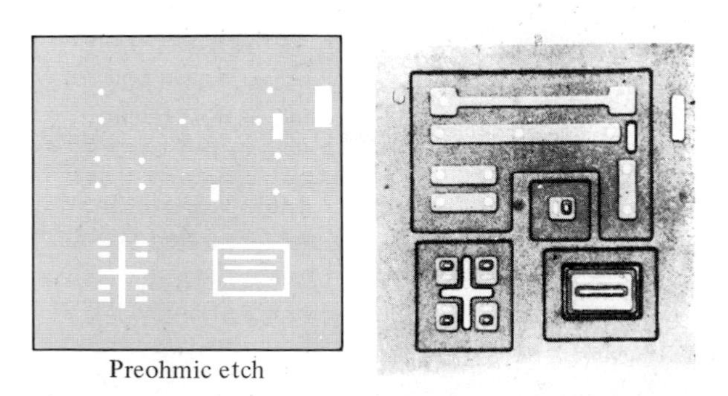

Figure 14.33 Surface appearance of the chip of Fig. 14.31 after the preohmic etch cycle. The mask employed is also included. (Courtesy Motorola Monitor.)

Figure 14.34 Completed metalization process. (Courtesy Motorola Monitor.)

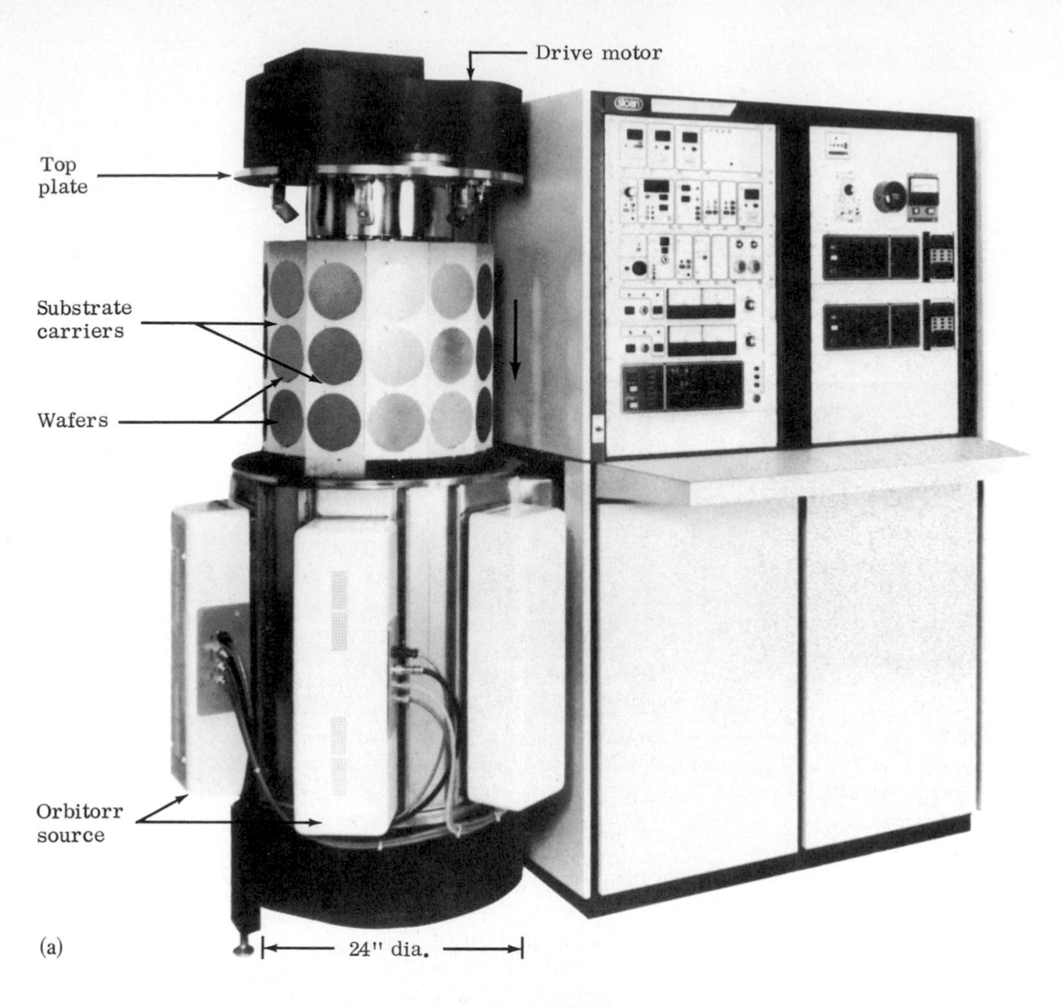

Figure 14.35 Deposition of the interconnect metal through metal evaporation. (Courtesy Motorola, Inc.)

The two methods most commonly applied to establish the uniform layer of conducting material are *evaporation* and *sputtering*. In the former, the metal is either melted through the use of heating coils or bombarded with an electron gun (E-gun) to result in the evaporation of the source metal. The metalization material is thereby sprayed over the wafers, which are held by clips in a drum or hemisphere structure such as is shown in Fig. 14.35.

An automated sputtering system that employs robotic units such as that shown in Fig. 14.36 places the source metal (at a very high negative potential) opposite, but not touching, an anode plate at a positive potential. An inert gas, such as argon, introduced between the plates will release positive ions that will bombard the negative plate and release some of the surface metal from the source. The "free" metal will then be deposited on the wafers on the surface of the anode.

Figure 14.36 Automated Sputter Coater (Courtesy Materials Research Corporation.)

The sputtering technique is often preferred over the evaporation method because the coverage is less line-of-sight. There is therefore a more uniform layer of deposition over abrupt junctions. For the decade to come it would appear that the sputtering and evaporation methods will share the metalization role in the production cycle.

The complete structure with each element indicated appears in Fig. 14.37. Try to relate the interconnecting metallic pattern to the original circuit of Fig. 14.17a.

Passivation

An SiO_2 layer deposited on the surface of the entire structure will be an effective protection layer for water vapor and some contaminants. However, certain metal ions can migrate through the SiO_2 layer and disturb the device characteristics. In an effort to improve the passivation process, a layer (2000 to 5000 Å) of glass (plasma silicon nitride) is applied to further reduce the degradation problem.

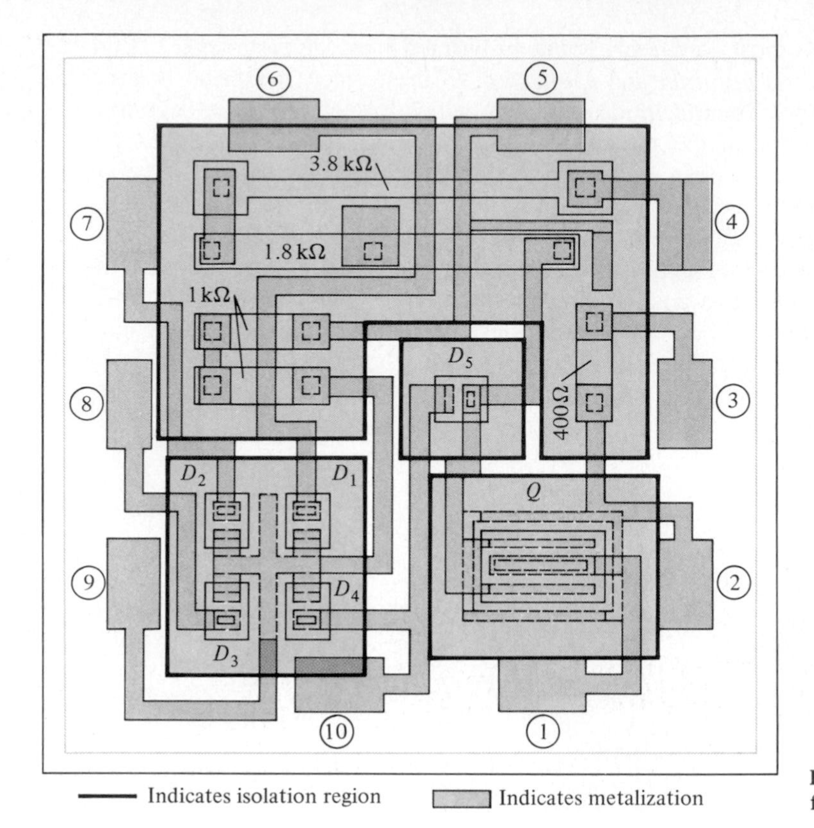

Figure 14.37 Monolithic structure for the NAND gate of Fig. 14.17.

Testing

Before breaking the wafer into the individual dies, an electrical test of each die is performed by the inspection system appearing in Fig. 14.38. The system automatically loads/unloads the wafers using carousels to further cut down on the degree of "handling." This process, as with most in the production cycle, is also microprocessor controlled. There is a probe card for each IC that will permit not only rejection but categorization of the type of faults (open, short, gain, etc). The bad die are identified by a red dot automatically deposited by the inspection system.

Packaging

Once the metalization and testing processes are complete, the wafer must be broken down into its individual chips. This is accomplished through the scribing and breaking processes depicted in Fig. 14.39. Each individual chip may then be packaged in one of the forms shown in Fig. 14.40. The name of each is provided in the figure.

(a) (b)

Figure 14.38 Electrical testing of the individual dies. [(a) Courtesy Electroglas, Inc.; (b) courtesy Texas Instruments, Inc.; (c) courtesy Autonetics, North American Rockwell Corporation.]

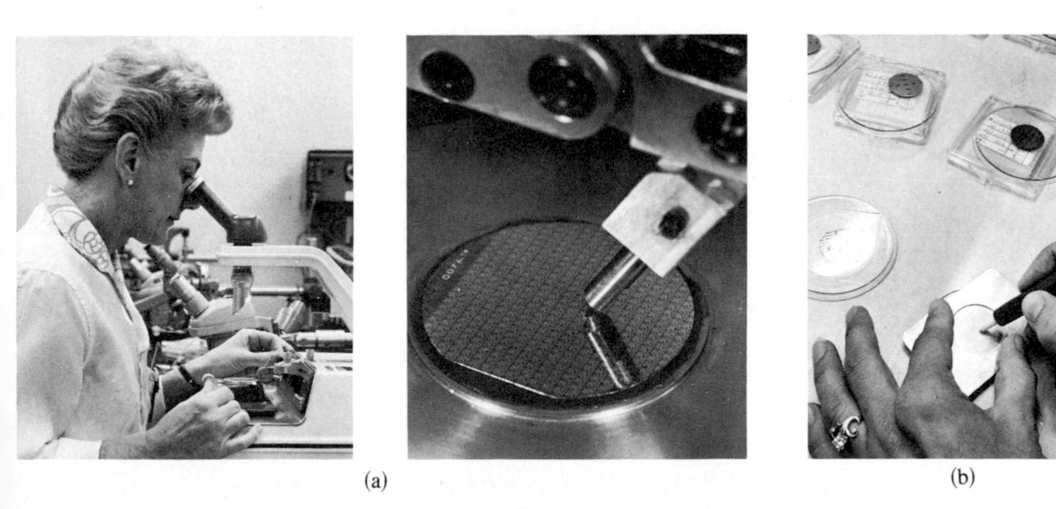

(a) (b)

Figure 14.39 Scribing (a) and breaking (b) of the monolithic wafer into individual chips. (*Left*, Courtesy Autonetics, North American Rockwell Corporation; *middle*, courtesy Texas Instruments, Inc., *right*, courtesy Motorola, Inc.)

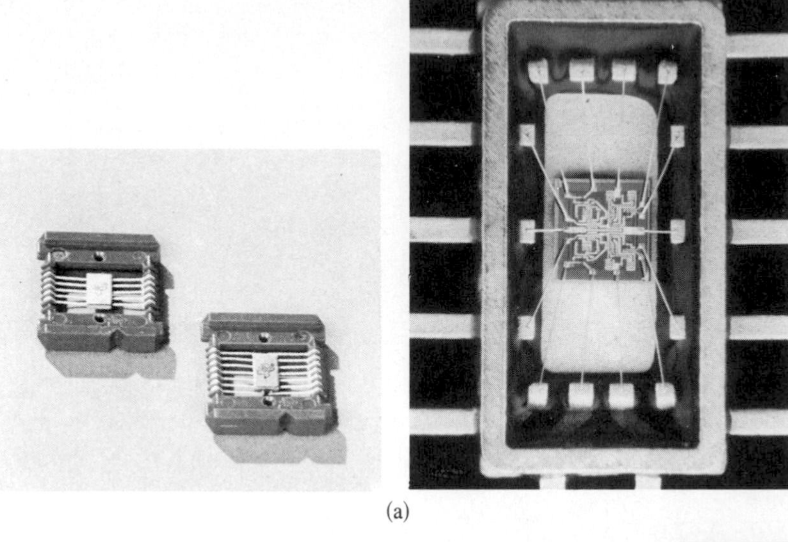

(a)

(b)

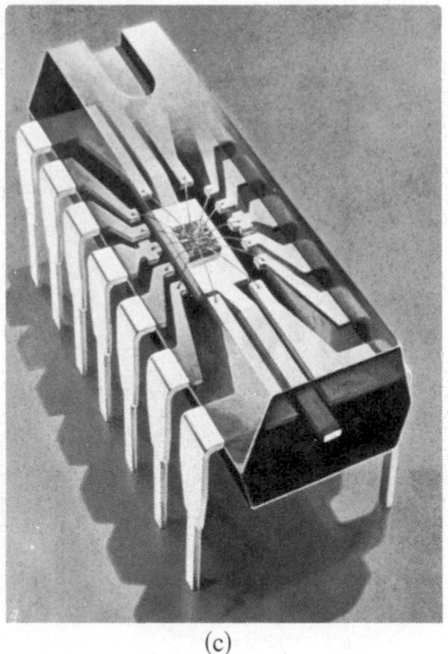

(c)

Figure 14.40 Monolithic packaging techniques: (a) flat package; (b) TO (top-hat)-type package; (c) dual-in-line plastic package. (Courtesy Texas Instruments, Inc.)

14.7 THIN- AND THICK-FILM INTEGRATED CIRCUITS

The general characteristics, properties, and appearance of thin- and thick-film integrated circuits are similar although they both differ in many respects from the monolithic integrated circuit. They are not formed within a semiconductor wafer but *on* the surface of an insulating substrate such as glass or an appropriate ceramic material. In addition, *only* passive elements (resistors, capacitors) are formed through thin or thick film techniques on the insulating surface. The active elements (transistors, diodes) are added as *discrete* elements to the surface of the structure after the passive elements have been formed. The discrete active devices are frequently produced using the monolithic process.

The primary difference between the thin and thick film techniques is the process employed for forming the passive components and the metallic conduction pattern. The thin-film circuit employs an evaporation or cathode-sputtering technique; the thick film employs silk-screen techniques. Priorities do not permit a detailed description of these processes here.

In general, the passive components of film circuits can be formed with a broader range of values and reduced tolerances as compared to the monolithic IC. The use of discrete elements also increases the flexibility of design of film circuits although, obviously, the resulting circuit will be that much larger. The cost of film circuits with a larger number of elements is also, in general, considerably higher than that of monolithic integrated circuits.

14.8 HYBRID INTEGRATED CIRCUITS

The term *hybrid integrated circuit* is applied to the wide variety of multichip integrated circuits and also those formed by a combination of the film and monolithic IC techniques. The multichip integrated circuit employs either the monolithic or film technique to form the various components, or set of individual circuits, which are then interconnected on an insulating substrate and packaged in the same container. Integrated circuits of this type appear in Fig. 14.41. In a more sophisticated type of hybrid integrated circuit the active devices are first formed within a semiconductor wafer which is subsequently covered with an insulating layer such as SiO_2. Film techniques are then employed to form the passive elements on the SiO_2 surface. Connections are made from the film to the monolithic structure through "windows" cut in the SiO_2 layer.

Figure 14.41 Hybrid integrated circuits. (Courtesy Texas Instruments, Inc.)

15

Linear ICs: Operational Amplifiers

15.1 BASIC DIFFERENTIAL AMPLIFIER

An amplifier is an electronic circuit containing BJT and FET devices, usually packaged in IC circuits, that provides voltage or current gain. It may also provide power gain, or allow impedance transformation. Since it is a basic part of practically every electronic application, the amplifier is an essential circuit. Amplifiers, as we have already discovered, may be classified in many ways. There are low-frequency amplifiers, audio amplifiers, ultrasonic amplifiers, radio-frequency (RF) amplifiers, wide-band amplifiers, video amplifiers, each type operating in a prescribed frequency range. We have considered small-signal and large-signal amplifiers and amplifiers that may be interconnected as either RC-coupled or transformer-coupled.

The *differential amplifier* is a special type of circuit that is used in a wide variety of applications. Let us first consider a number of basic properties of differential amplifiers. Figure 15.1 shows a block symbol of a differential amplifier unit. As shown, there are two separate input (1 and 2) and two separate output (3 and 4) terminals. We must first consider the relation between these terminals to obtain an

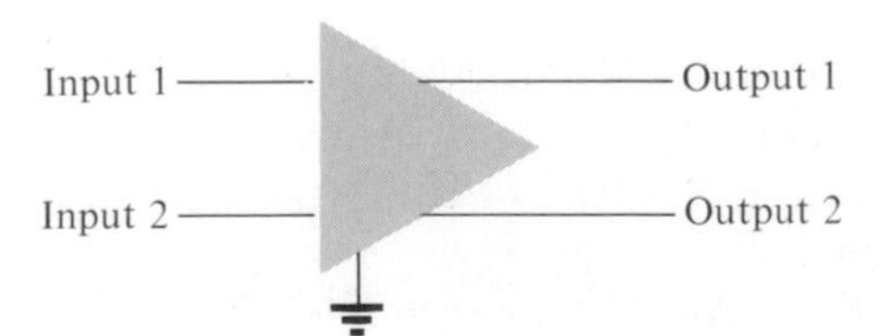

Figure 15.1 Block symbol of a differential amplifier.

understanding of how the differential amplifier may be applied. Notice that in Fig. 15.1 a ground connection is shown separately since both input or output terminals may be different from ground. Voltages may be applied to either or both input terminals and output voltages will appear at both output terminals. However, there are some very specific polarity relations between both input and both output terminals.

Figure 15.2 shows the block and circuit diagrams of a basic differential amplifier to be used in the following discussion. There are two inputs and two outputs shown in the block diagram. Inputs are applied essentially to each base of the two separate transistors. As shown, however, the transistor emitters are connected to a common-emitter resistor so that the two output terminals V_{o_1} and V_{o_2} are affected by *either* or *both* input signals. The outputs are taken from the collector terminals of each transistor. The input and output terminals are also numbered to facilitate reference. There are two supply voltages shown in the circuit diagram and it should be carefiully noted that no ground terminal is indicated within the circuit although the opposite points of both positive- and negative-voltage supplies are understood to be connected to ground. The amplifier could also operate using a single voltage supply.

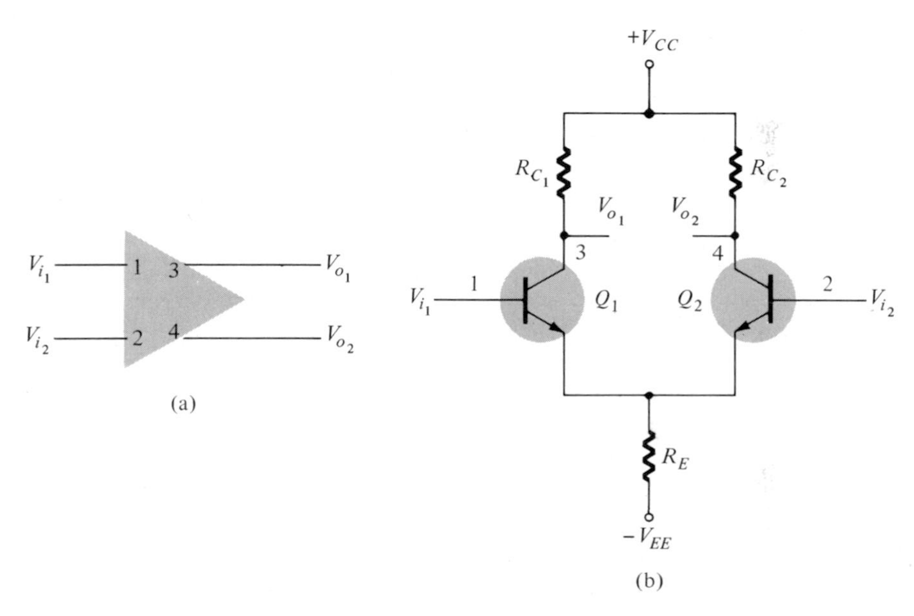

Figure 15.2 Basic differential amplifier: (a) block diagram; (b) circuit diagram.

Single-Ended Input Differential Amplifier

Consider first the operation of the differential amplifier with a single input signal applied to terminal 1, with terminal 2 connected to ground (0 V). Figure 15.3 shows the block and circuit diagrams for input signal V_{i_1} at terminal 1 and output V_{o_1} at terminal 3. The block diagram shows a sinusoidal input and an amplified, inverted output. The circuit diagram shows the sinusoidal input applied to the base of a

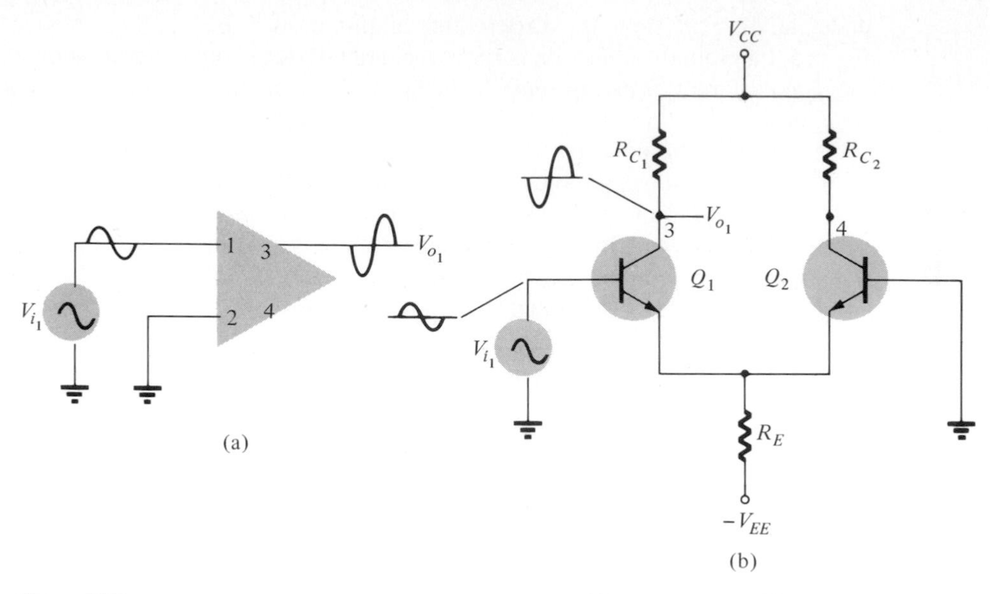

Figure 15.3 Single-ended operation of differential amplifier: (a) block diagram; (b) circuit diagram.

transistor with the amplified output at the collector inverted, as we would expect from past knowledge of a single-stage transistor amplifier.

With input 2 grounded it might seem that there is no output at terminal 4, but this is incorrect. The block diagram of Fig. 15.4 shows the operation of the differential amplifier with the V_{o_2} output at terminal 4 resulting from an input V_{i_1} at terminal 1. The input at terminal 1, V_{i_1}, is a small sinusodal voltage measured with respect to ground. Since the emitter resistor is connected in common with both emitters, a voltage due to V_{i_1} appears at the common-emitter point. This sinusoidal voltage, measured with respect to ground, is approximately one-half the magnitude and in phase with V_{i_1} because it results from emitter-follower action of the circuit.

The part of the circuit acting as an emitter follower is shown in Fig. 15.4c. An input applied to the base of Q_1 appears of the same polarity and about one-half the magnitude at the emitter of Q_1 for the emitter-follower part of the circuit shown. Recall that for an emitter follower the gain is less than unity (with no polarity reversal). This emitter signal is measured with respect to ground. Figure 15.4d shows the part of the circuit with the emitter voltage affecting the operation of transistor Q_2. The voltage at the emitter of Q_2 is the same as that of Q_1 (since the emitters are connected together) and appears from the emitter of Q_2 to ground or to the base of Q_2 (since that is connected to ground). If the voltage measured from emitter to base of Q_2 is in phase with input V_{i_1} as shown, the voltage measured from base to emitter of Q_2 is in the same signal with opposite polarity. Thus, by measuring from base to emitter of Q_2 a voltage of about one-half the magnitude of V_{i_1} is obtained, but the signal is opposite in polarity to that of V_{i_1}. The amplifier action of transistor Q_2 and load resistor R_{C_2} provides an output at the collector of Q_2 that is amplified and inverted from the signal developed across base to emitter of Q_2.

In summary, an input V_{i_1} is applied to input 1 and an amplified, in-phase signal V_{o_2}

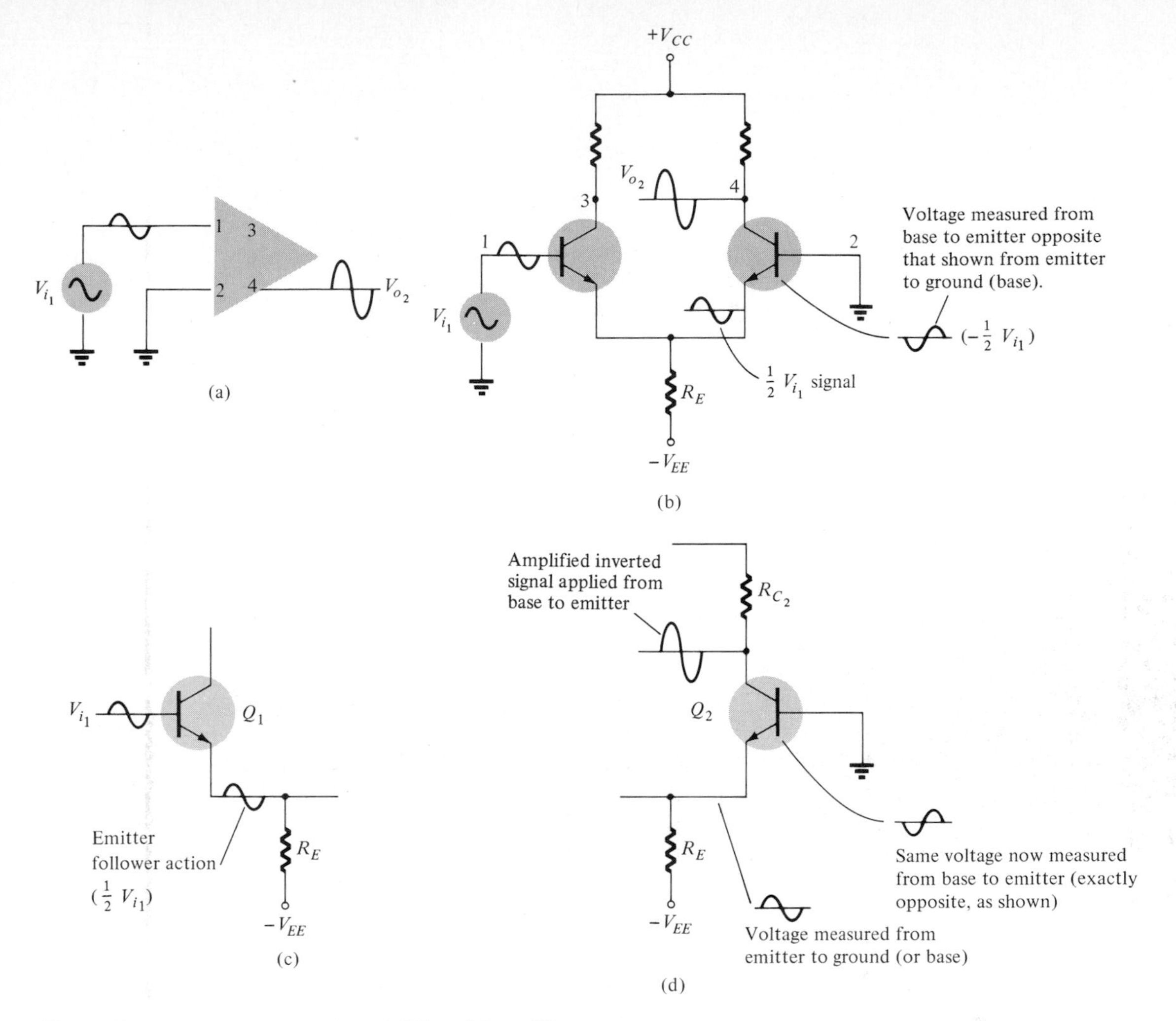

Figure 15.4 Single-ended operation of differential amplifier.

results at output terminal 4. Although the input at terminal 2 is grounded, an output still occurs at terminal 4. In fact, we can now see that the input at terminal 1 causes output signals at both terminals 3 and 4. In addition, these outputs are opposite in polarity and of about the same magnitude. Finally, we should see (as in Fig. 15.5a) that the output at terminal 4 is the same polarity as the input at terminal 1, while the output at terminal 3 is opposite in polarity to the input at terminal 1. It follows from the previous discussion that an input applied to terminal 2 (with terminal 1 grounded) will result in output voltages as shown in Fig. 15.5b.

Differential (Double-Ended) Input Operation

In addition to using only one input to operate the differential-amplifier circuitry, it is possible to apply signals to each input terminal, with opposite polarity outputs

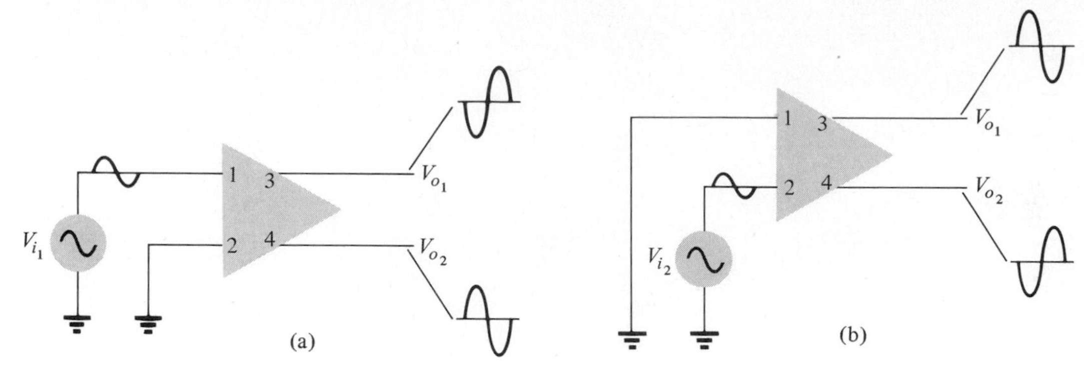

Figure 15.5 Single-input, opposite polarity outputs.

appearing at the two output terminals. The usual use of the *double-ended* or *differential* mode of input is when the two input signals are themselves opposite in polarity and about the same magnitude. Figure 15.6 shows such a situation.

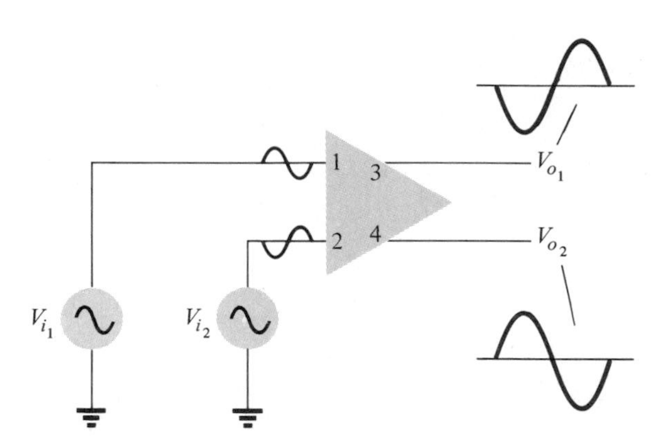

Figure 15.6 Operation with differential input signals.

We must now consider how each input affects the outputs and what the resulting output signal looks like. This can be done using the *superposition principle*, considering each input applied separately with the other at 0 V and summing the resulting output voltages at each terminal. Figure 15.7a and 15.7b show the result of each input acting alone and Fig. 15.7c shows the resulting overall operation. The input applied to terminal 1 results in an opposite-polarity amplified output at terminal 3 *and* a same-polarity amplified output at terminal 4. Assume that the inputs are about equal in magnitude and that the ouput magnitudes are about equal, of value V, for discussion purposes.

The input applied to terminal 2 results in an opposite-polarity amplified output at terminal 4 *and* a same polarity amplified ouput at terminal 3. The magnitudes of the ouputs will be V since the input magnitudes were assumed to be about the same, It is important to note that the outputs in each case are the same polarity at each output terminal. By superposition, the resulting signals at each output terminal are added, and we obtain the full operation of the circuit shown in Fig. 15.7c. The output at each

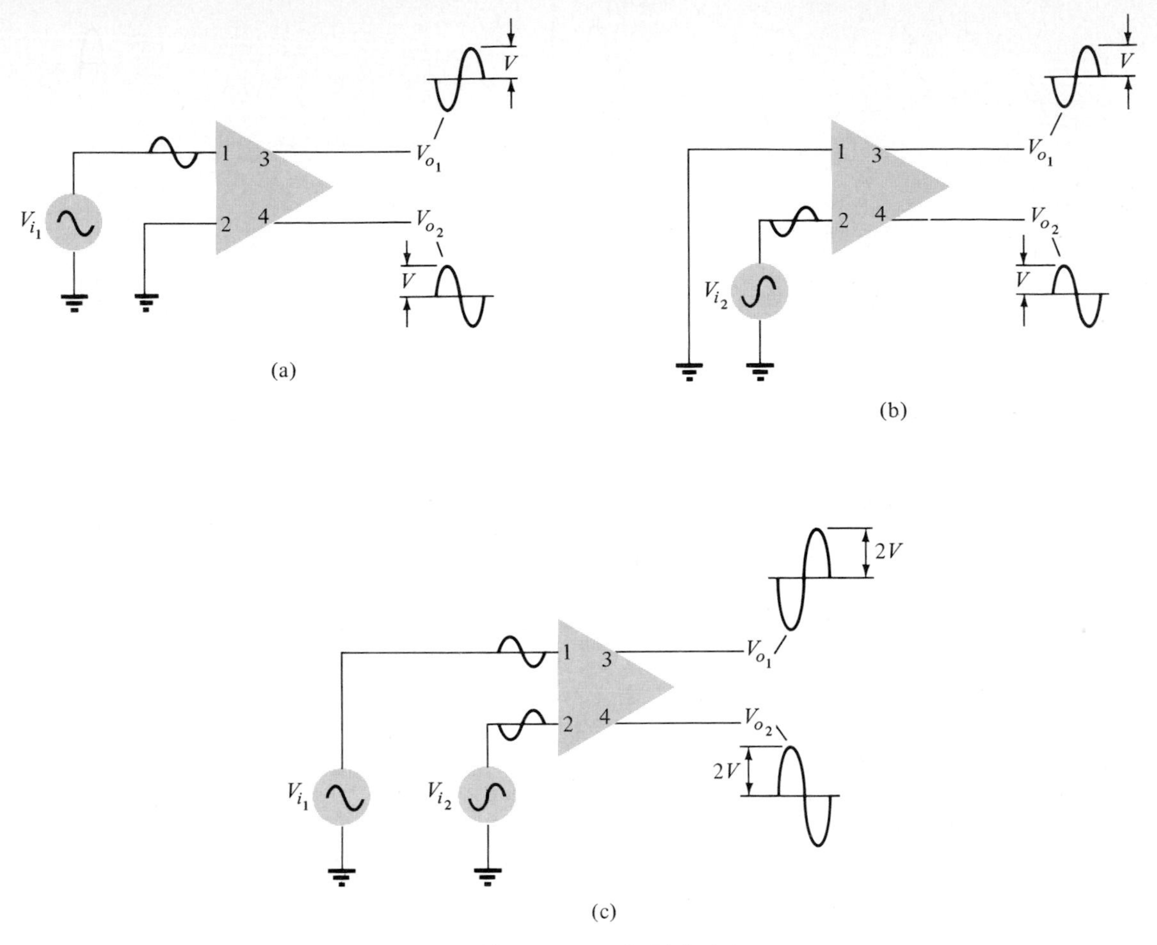

Figure 15.7 Differential operation of amplifier: (a) $V_{i_2} = V_{i_1} = 0$; (c) both inputs present.

terminal is twice that resulting from single-ended operation because the ouputs due to each input have the same polarity. If the inputs applied were both the same polarity (or if the same input were applied to both input terminals), the resulting signals due to each input acting alone would be opposite in polarity at each ouput and the resulting output would *ideally* be about 0 V, as shown in Fig. 15.8.

To bring the operation as single- and double-ended differential-amplifier states into full perspective consider the connection of two differential amplifiers shown in Fig. 15.9. From the previous discussion, if the amplifiers had identical single-ended gains, then the outputs of stage 1 would be larger than the inputs by the amount of amplifier gain, while the ouputs of stage 2 would be larger than the inputs to stage 2 by twice the amplifier gain. The initial signal, from an antenna of a radio, or a phonograph pickup cartridge, and so on, is single-ended and is used as such. The second differential-amplifier stage, however, could be operated double-ended to obtain twice the stage gain. Either output of stage 2 (or both) could then be used as amplified signals to the next section of the system. Although differential operation

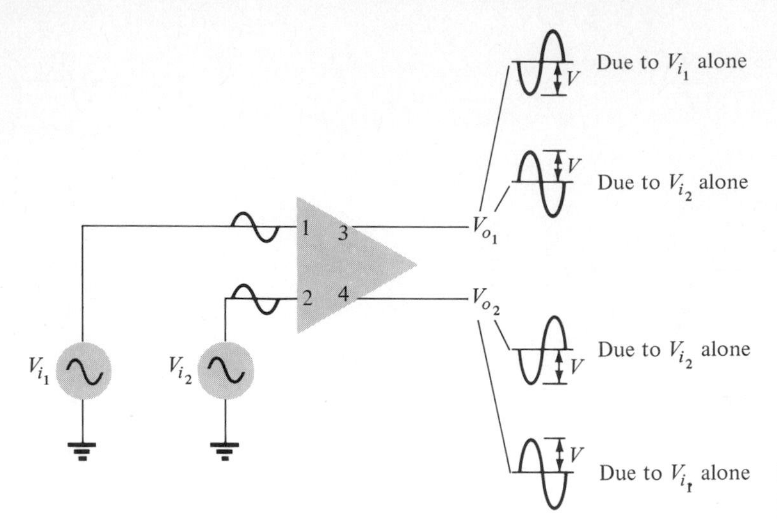

Figure 15.8 Operation with in-phase input signals.

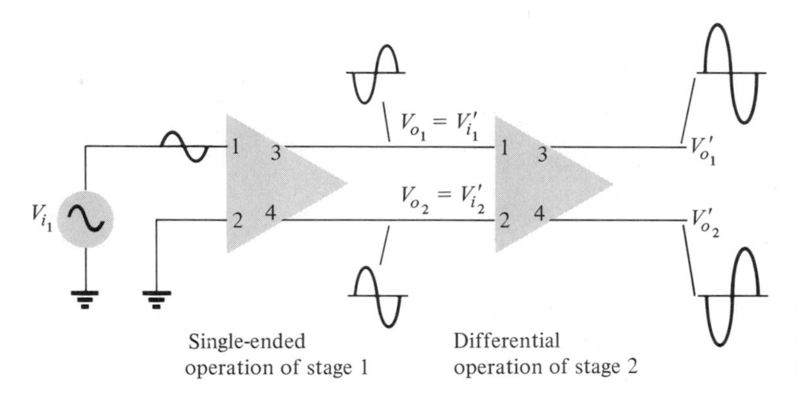

Figure 15.9 Single- and double-ended operation of differential-amplifier stages.

requires about equal and opposite polarity signals, this is often available, especially after one single-ended stage of gain.

15.2 DIFFERENTIAL-AMPLIFIER CIRCUITS

Having considered the basic features of a differential amplifier, we shall now look into some circuit details. In particular, we shall consider the voltage gain of the stage and its input and output impedance. We will first examine discrete-circuit versions to allow introducing circuit concepts. Then IC versions of various parts of the difference amplifier are presented, these being the typical circuits used in bulding IC operational amplifiers (op-amps). A basic circuit of a discrete differential amplifier is shown in Fig. 15.10. Input signals are shown as a voltage source with source resistance in the general case.

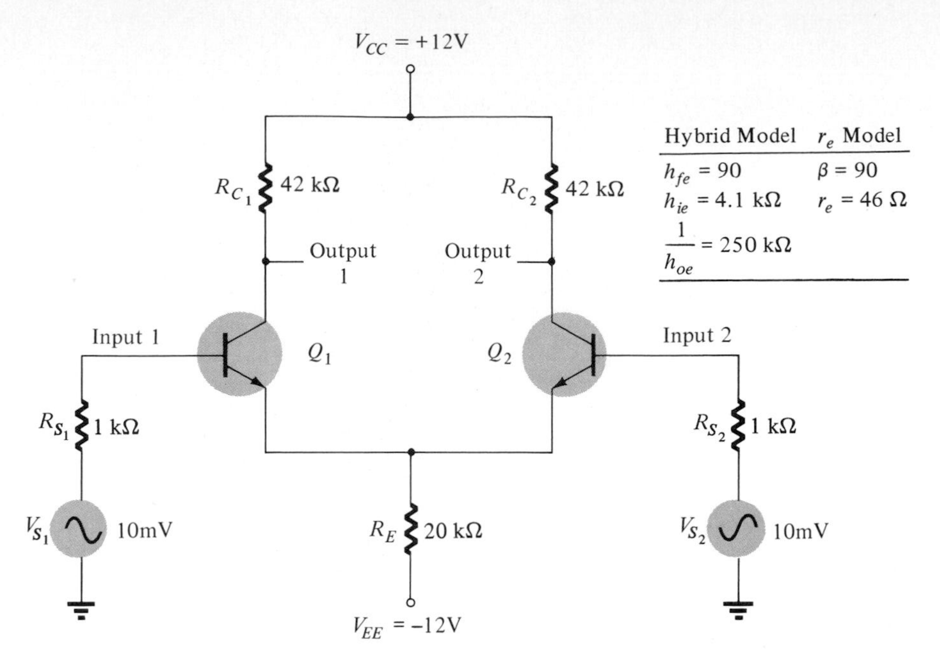

Figure 15.10 Basic differential-amplifier circuit.

DC Bias Action of Circuit

Before considering the operation of the circuit as a voltage amplifier, let us see how the circuit is biased. Figure 15.11 shows the main dc voltage and current values of the circuit. Ac signal sources have been set to 0 V with only the source resistances present. The base-emitter of Q_1 is forward biased by the V_{EE} battery from ground through resistor R_{S_1}, through the base emitter, through resistor R_E to $-V_{EE}$ (see Fig. 15.12a). We would have to write a number of equations to solve for the dc voltages and currents. However, it is possible to use good approximations to make the calculations more direct. For example, the dc voltage drop across source resistor R_{S_1} will be small as the following calculation indicates (assuming a typical base current in the order of microamperes)

$$I_{B_1} R_{S_1} = (100\ \mu\text{A})(1\ \text{k}\Omega) = 100\ \text{mV} = 0.1\ \text{V}$$

If the base current were only 10 μA, then the dc voltage drop across R_{S_1} would be 10 mV, which is negligible. On the other hand, a source resistance of 10 kΩ with base current of 100 μA would result in a voltage drop of 1 V, which is not negligible. For our purposes we shall assume the voltage drop to be small (which is often correct) and check later in the calculations to be sure we were able to make such an assumption.

If we assume that

$$V_{B_1} \cong 0\ \text{V}$$

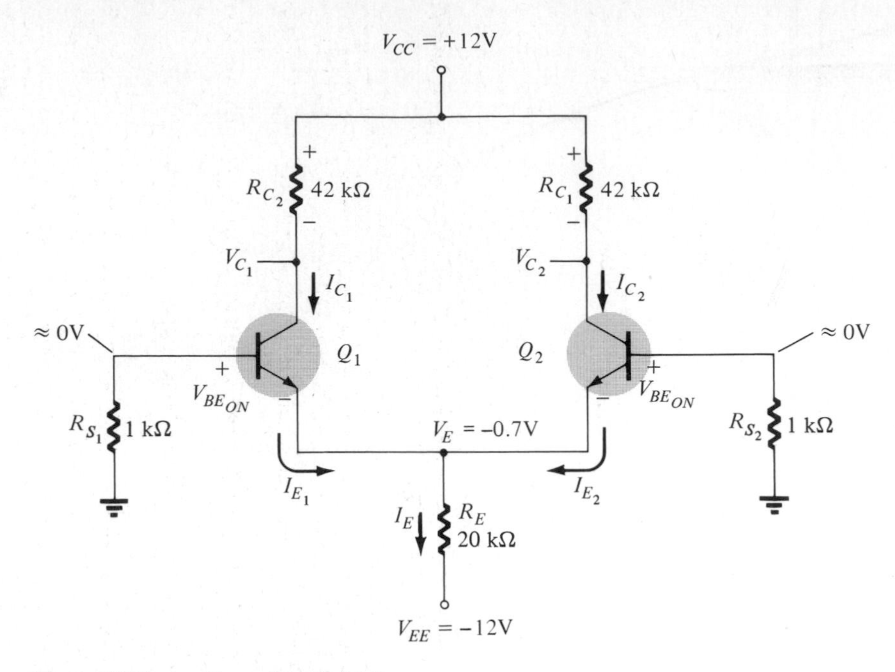

Figure 15.11 Dc bias action of circuit.

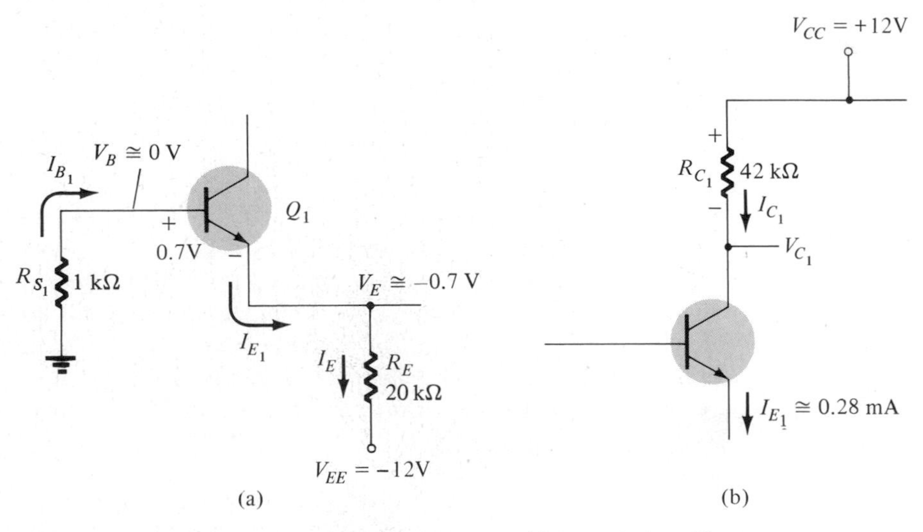

Figure 15.12 Partial circuits of differential amplifier: (a) input section; (b) output section.

then the emitter voltage is

$$V_E = V_{E_1} = V_{B_1} - V_{BE_1} \tag{15.1}$$
$$= 0 - 0.7 \text{ V} = -0.7 \text{ V}$$

The current through resistor R_E is then

$$I_E = \frac{V_E - V_{EE}}{R_E} \tag{15.2}$$

$$= \frac{-0.7 \text{ V} - (-12 \text{ V})}{20 \text{ k}\Omega} = \frac{11.3 \text{ V}}{20 \text{ k}\Omega} = 0.565 \text{ mA}$$

The current through resistor R_E is made up of the emitter currents from each transistor. If the transistors are matched, the emitter current of each transistor is one-half the total current through R_E.

$$I_{E_1} = I_{E_2} = \frac{I_E}{2} \tag{15.3}$$

$$= \frac{0.565 \text{ mA}}{2} = 0.2825 \text{ mA} \cong 0.28 \text{ mA}$$

We can now check our assumption of V_{B_1} by calculating I_{B_1} as follows:

$$I_{B_1} = \frac{I_{E_1}}{h_{fe_1} + 1} \tag{15.4}$$

$$= \frac{0.28 \text{ mA}}{90 + 1} = 3.08\ \mu\text{A}$$

$$V_{B_1} = I_{B_1} R_{S_1} = 3.08\ \mu\text{A} \times 1 \text{ k}\Omega = 3.08 \text{ mV}$$

which is negligible compared to the other voltage drops in the circuit. Figure 15.12b shows a partial circuit diagram of the output section of the circuit. The collector current is obtained from the calculation of emitter current.

$$I_{C_1} \cong I_{E_1} \tag{15.5}$$

$$= 0.28 \text{ mA}$$

and the collector voltage is

$$V_{C_1} = V_{CC} - I_{C_1} R_{C_1} \tag{15.6}$$

$$= 12 \text{ V} - (0.28 \text{ mA})(42 \text{ k}\Omega) = 0.24 \text{ V} \cong 0 \text{ V}$$

AC Operation of Differential-Amplifier Circuit

To consider the ac operation of the circuit all dc voltage supplies are set at zero and the transistors are replaced by small-signal ac equivalent circuits. Figure 15.13a shows the resulting ac equivalent circuit, with the transistors replaced by hybrid equivalent circuits. The circuit obviously appears complex, and analyzing the total circuit would become involved. Again, we can break up the calculations by using some simplifying approximations so that smaller parts of the circuit can be analyzed separately. We can assume that

$$h_{ie_1} = h_{ie_2} = h_{ie}, \qquad h_{fe_1} = h_{fe_2} = h_{fe}, \qquad h_{oe_1} = h_{oe_2} \cong 0$$

and

$$R_{C_1} = R_{C_2} = R_C, \qquad R_{S_1} = R_{S_2} = R_S$$

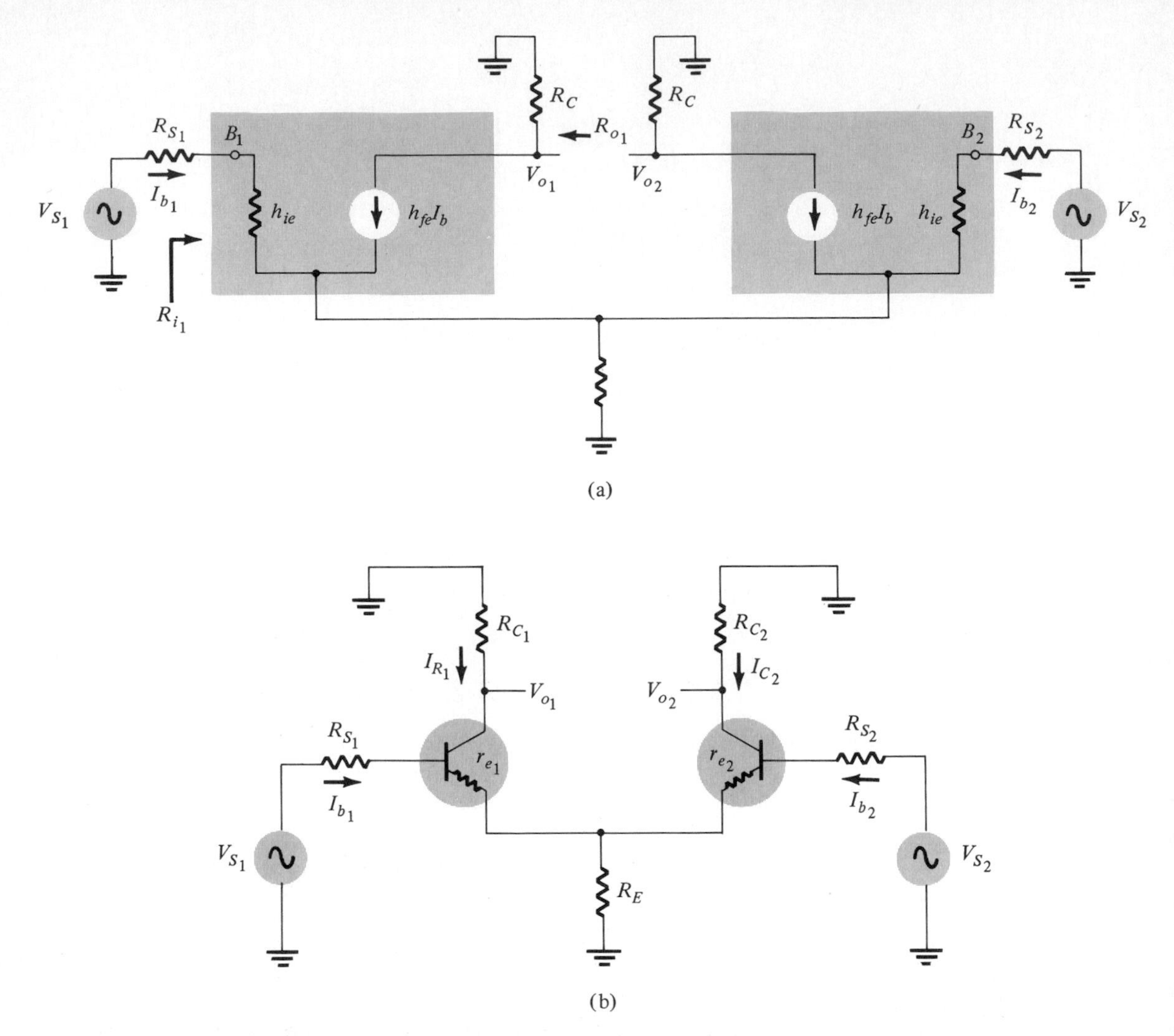

Figure 15.13 Ac equivalent circuit of differential amplifier: (a) hybrid model; (b) r_e model.

INPUT AC SECTION

Figure 15.14a shows the partial ac equivalent circuit of the input for transistor Q_1. Looking into the emitter of transistor Q_2 a small ac equivalent resistance is present, equal in value to

$$R_{e_2} = \frac{R_S + h_{ie}}{h_{fe} + 1} \tag{15.7}$$

For the values of Fig. 15.10,

$$R_{e_2} = \frac{1 \text{ k}\Omega + 4.1 \text{ k}\Omega}{90 + 1} = 56 \ \Omega$$

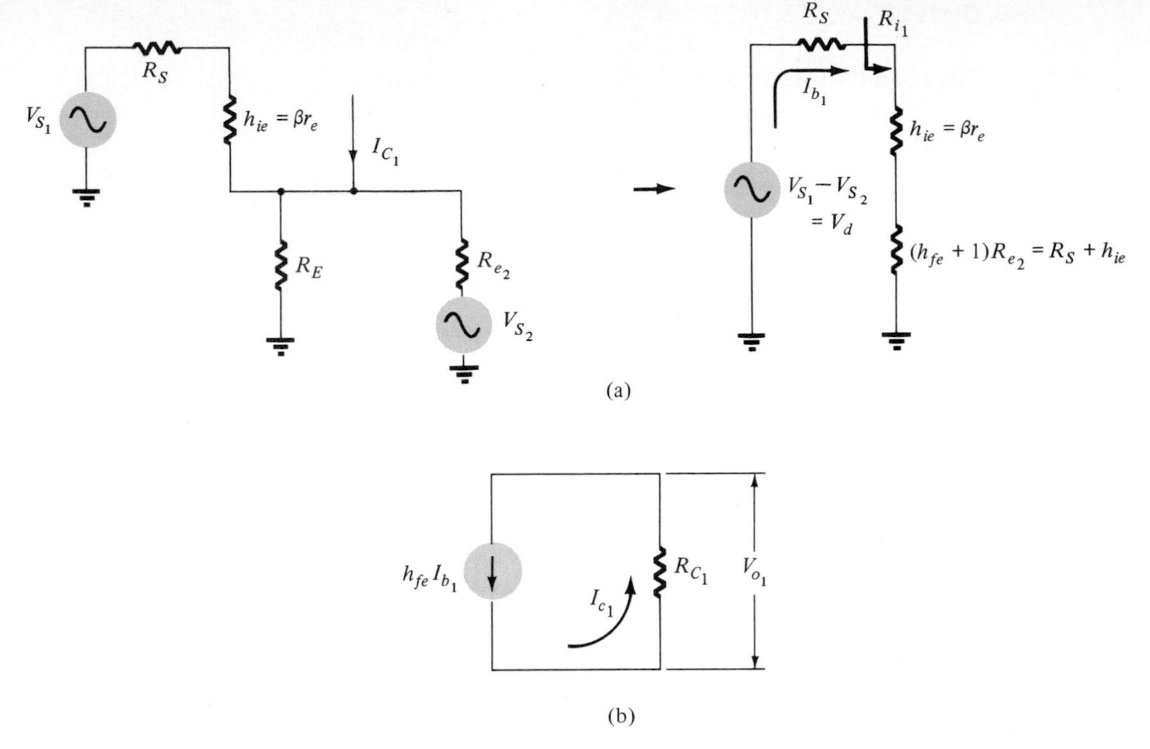

Figure 15.14 Partial ac equivalent circuit of differential amplifier.

The parallel combination of resistors R_E and R_{E_2} gives an equivalent ac resistance of

$$R_E \| R_{e_2} = \frac{R_{e_2} R_E}{R_{e_2} + R_E} = \frac{56 \times 20{,}000}{56 + 20{,}000} \cong 55.8\ \Omega$$

Since the differential-amplifier circuit generally has an R_E of large value, we can make the approximate statement that if

$$R_E \gg R_{e_2}$$

the parallel combination is approximately R_{e_2} in value, as shown in Fig. 15.14a. Using the resulting ac equivalent circuit, we see that the value of the ac base current is calculated to be

$$I_{b_1} = \frac{V_{S_1} - V_{S_2}}{R_S + h_{ie} + (h_{fe} + 1)R_{e_2}} = \frac{V_{S_1} - V_{S_2}}{2(R_S + h_{ie})} \tag{15.8a}$$

and defining $V_d \equiv V_{S_1} - V_{S_2}$ as the difference input voltage

$$I_{b_1} = \frac{V_d}{2(R_S + h_{ie})} \tag{15.8b}$$

OUTPUT AC SECTION

The output voltage can be written as

$$V_{o_1} = -I_{c_1} R_{C_1}$$

Using $I_{c_1} = h_{fe} I_{b_1}$ and I_{b_1} as expressed in Eq. (15.8b) results in

$$V_{o_1} = \frac{-h_{fe} R_{C_1}}{2(R_S + h_{ie})} V_d$$

The circuit ac difference gain is then

$$A_d = \left|\frac{V_o}{V_d}\right| = \left|-\frac{h_{fe} R_C}{2(R_S + h_{ie})}\right| = \left|-\frac{\beta R_C}{2(R_S + \beta r_e)}\right| \tag{15.9}$$

If $\beta r_e \gg R_S$ then

$$A_d = \left|\frac{-\beta R_c}{2(\beta r_e)}\right| = \left|\frac{-R_c}{2r_e}\right|$$

Using the values of the circuit of Fig. 15.10 results in

$$A_d \cong \left|-\frac{90 \times 42\text{ k}\Omega}{2(1\text{ k}\Omega + 4.1\text{ k}\Omega)}\right| = 370.6$$

INPUT RESISTANCE

From the ac equivalent circuit of Fig. 15.14a the input resistance of the circuit seen from the source is

$$R_{i_1} = h_{ie} + (h_{fe} + 1)R_{e_2}$$

which can be expressed after substituting for R_{e_2} from Eq. (15.7) as

$$R_i = R_{i_1} = R_{i_2} = 2h_{ie} + R_S = 2(\beta r_e + R_S) \tag{15.10}$$

For the circuit of Fig. 15.10,

$$R_i = 2(4.1\text{ k}\Omega) + 1\text{ k}\Omega = 9.2\text{ k}\Omega$$

OUTPUT RESISTANCE

From the ac equivalent circuit of Fig. 15.14b the resulting approximate output resistance (assuming that $h_{oe} \cong 0$) is

$$R_o = R_{o_1} = R_{o_2} = R_C \tag{15.11}$$

which is $R_o = 42\text{ k}\Omega$.

EXAMPLE 15.1

Calculate the difference gain (A_d), the input resistance (R_i), and the output resistance (R_o) of a differential amplifier circuit as in Fig. 15.10 for circuit values of ($R_{C_1} = R_{C_2} =$ 36 kΩ, $R_E = 18$ kΩ, $h_{fe} = \beta = 120$, $h_{ie} = 10$ kΩ, and $R_{S_1} = R_{S_2} = 1.2$ kΩ.

Solution:

$$A_d = \left| \frac{-h_{fe} R_C}{2(\text{R}_S + h_{ie})} \right| = \left| \frac{-120(36 \times 10^3)}{2(1.2 \times 10^3 + 10 \times 10^3)} \right| \cong \mathbf{193}$$

$$R_i = 2h_{ie} + R_S = 2(10 \times 10^3) + 1.2 \times 10^3 = \mathbf{21.2\ k\Omega}$$

$$R_o = R_C = \mathbf{36\ k\Omega}$$

EXAMPLE 15.2

Determine the value of collector resistor (R_C) needed to obtain a differential gain of 250 for a differential amplifier such as that in Fig. 15.10 for circuit values of $R_S = 0\ \Omega$, $R_E = 10\ \text{k}\Omega$, $h_{fe} = 80$, and $V_{CC} = 20$ V, $V_{EE} = -20$ V.

Solution:

$$I_E = \frac{V_E - V_{EE}}{R_E} = \frac{19.3\ \text{V}}{10\ \text{k}\Omega} = 1.93\ \text{mA}$$

$$I_{E1} = I_{E2} = \frac{I_E}{2} = \frac{1.93\ \text{mA}}{2} = 0.965\ \text{mA}$$

$$r_e = \frac{26}{I_E} = \frac{26}{0.965} \cong 27\ \Omega$$

Since

$$A_d = \left| \frac{-R_C}{2r_e} \right|$$

$$R_C = 2r_e A_d = 2(27)(250) = \mathbf{13.5\ k\Omega}$$

EXAMPLE 15.3

What output voltage, V_{o2}, is developed in the circuit of Fig. 15.10 for inputs $V_{i_1} = 0$ V, $V_{i_2} = 5$ mV rms, $R_{S_1} = R_{S_2} = 0\ \Omega$, $R_C = 24\ \text{k}\Omega$, and $r_{e_1} = r_{e_2} = r_e = 90\ \Omega$?

Solution:

$$A_d = \left| \frac{-R_C}{2r_e} \right| = \left| \frac{-24 \times 10^3}{2(90)} \right| = 133.3$$

$$V_o = A_d V_d = A_d(V_{i_1} - V_{i_2}) = 133.3(0 - 5\ \text{mV}) = \mathbf{-0.67\ V\ rms}$$

15.3 CONSTANT-CURRENT SOURCE

Differential-Amplifier Circuit with Constant-Current Source

One important thing to note in the previous circuit considerations was that with $R_{e_2} \ll R_E$, the value of R_E was very large and therefore negligible. In fact, the larger the value of R_E, the better certain desirable aspects of a differential-amplifier circuit. The main reason for R_E being very large is a circuit factor called *common-mode rejection*, which will be discussed in detail in Section 15.3.

DC bias calculations show that the emitter (and thus the collector) current is determined partly by the value of R_E. For a fixed negative-voltage supply of, say, $V_{EE} = -20$ V a value of R_E of 10 kΩ would limit the emitter-resistor current to about

$$I_E \cong \frac{V_{EE}}{R_E} = \frac{20\text{ V}}{10\text{ k}\Omega} = 2\text{ mA}$$

If a preferably larger value of $R_E = 100$ kΩ were used the value of dc emitter-resistor current would then be

$$I_E \cong \frac{V_{EE}}{R_E} = \frac{20\text{ V}}{100\text{ k}\Omega} = 0.2\text{ mA} = 200\ \mu\text{A}$$

and if a very large value of $R_E = 1$ MΩ were used,

$$I_E = \frac{20\text{ V}}{1\text{ M}\Omega} = 20\ \mu\text{A}$$

We see that as larger values of R_E are used the resulting dc emitter current becomes much too small for proper operation of the transistors since the emitter and collector current of each transistor is one-half the already very small emitter current.

One way to achieve high ac resistance while still allowing reasonable dc emitter currents is to use a constant-current source as shown in Fig. 15.15. The value of I_E could be set by the constant-current circuit to any desired value—1, 10, 20 mA, and so on. The ac resistance of a constant-current source is ideally infinite and practically from 100 kΩ to about 1 MΩ.

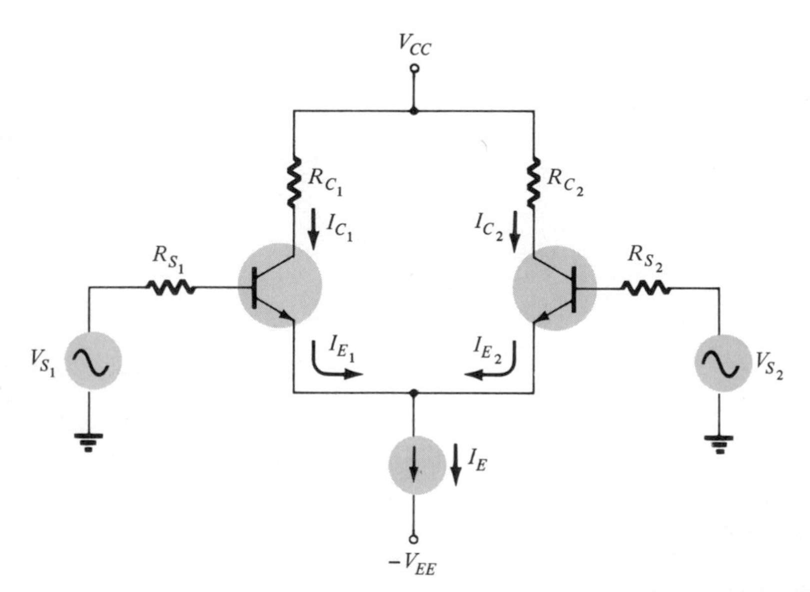

Figure 15.15 Differential amplifier with constant-current source.

The use of a constant-current source is critical to the construction of well-designed differential amplifiers. This is the case for discrete versions of the circuit and even more so for IC units. A few examples of both types of constant-current circuits are provided in this section.

Discrete Constant-Current Source

A constant-current source using resistors is shown in Fig. 15.16. The collector current I_C is the desired constant current and is set by the resistors R_1, R_2, and R_E and supply V_{EE}. The better the operation as a constant-current source, the less I_C is affected by the circuit to which it is connected.

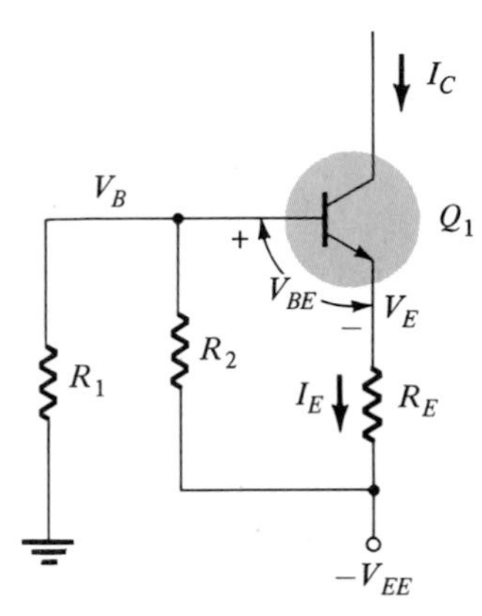

Figure 15.16 Discrete constant-current source.

The dc bias current I_E can be calculated as follows (assume that $h_{fe}R_E \gg R_2$):

$$V_B = \frac{R_1}{R_1 + R_2} V_{EE} \tag{15.12}$$

$$V_E = V_B - 0.7 \text{ V} \tag{15.13}$$

and

$$I_E = \frac{V_E - V_{EE}}{R_E} \cong I_C \tag{15.14}$$

Thus a collector or constant current is set by the circuit, for use by the rest of a differential amplifier circuit as considered previously.

An important feature for determining how well the constant-current source operates with ac signals is its ac output impedance. Ideally, the ac output impedance is infinite. For the circuit of Fig. 15.16 the actual output impedance can be determined using the ac equivalent circuit of Fig. 15.17. The output impedance for this ac equivalent circuit can be determined as

$$R_o \cong \frac{1}{h_{oe}}\left(1 + \frac{h_{fe}R_E}{R_E + h_{ie} + R_1 \parallel R_2}\right) \tag{15.15}$$

The larger this value, the better the operation as a constant current source.

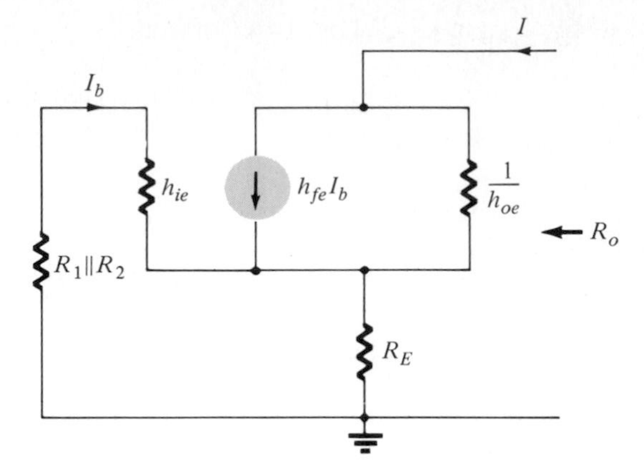

Figure 15.17 Ac considerations for a constant-current circuit.

EXAMPLE 15.4

Determine the value of the constant current and the output resistance for the circuit of Fig. 15.18.

Solution:

$$V_B = \frac{R_1}{R_1 + R_2} V_{EE} = \frac{5.1\text{ k}\Omega}{5.1\text{ k}\Omega + 5.1\text{ k}\Omega}(-20) = -10\text{ V}$$

$$V_E = V_B - 0.7\text{ V} = -10\text{ V} - 0.7\text{ V} = -10.7\text{ V}$$

$$I_E = \frac{V_E - V_{EE}}{R_E} = \frac{-10.7\text{ V} - (-20\text{ V})}{2\text{ k}\Omega} = \frac{9.3\text{ V}}{2\text{ k}\Omega}$$

$$= \mathbf{4.65\text{ mA}} = I$$

and the output impedance is

$$R_o = \frac{1}{h_{oe}}\left(1 + \frac{h_{fe}R_E}{R_E + h_{ie} + R_1 \| R_2}\right)$$

$$= 100\text{ k}\Omega\left(1 + \frac{100(2\text{ k}\Omega)}{2\text{ k}\Omega + 0.5\text{ k}\Omega + \dfrac{5.1\text{ k}\Omega}{2}}\right) = \mathbf{4.06\text{ M}\Omega}$$

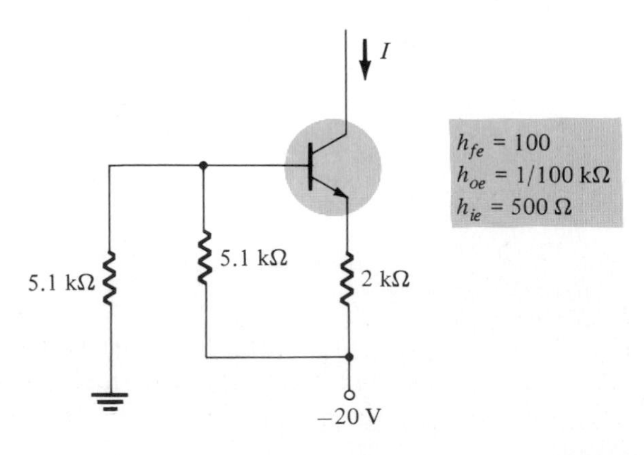

Figure 15.18 Constant-current source for Example 15.4.

Zener Constant-Current Source

Replacing resistor R_2 with a Zener diode, as shown in Fig. 15.19, provides an improved constant-current source over that of Fig. 15.16. The Zener diode keeps the constant current well regulated at a value calculated by

$$V_B = -V_{EE} + V_Z \tag{15.16}$$

$$V_E = V_B - 0.7\text{ V} \tag{15.17}$$

$$I_C \cong I_E = \frac{V_E - V_{EE}}{R_E} = \frac{V_B - 0.7\text{ V} - V_{EE}}{R_E} = \frac{V_Z + 0.7\text{ V}}{R_E} \tag{15.18}$$

Equation (15.18) shows that the effect of using the Zener diode is to set the constant current to a value dependent only on the Zener diode voltage, V_Z, and the emitter resistor, R_E.

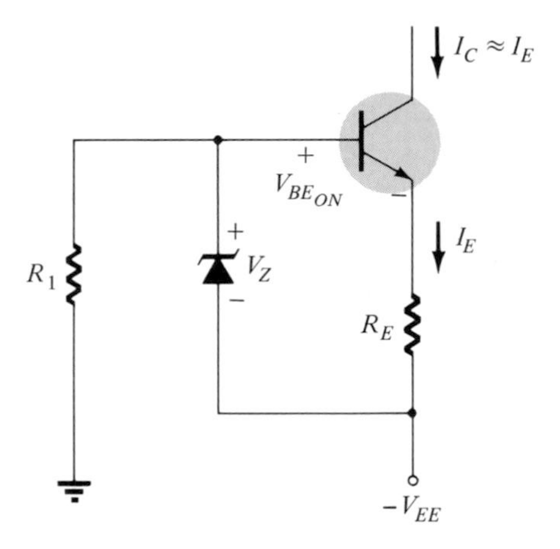

Figure 15.19 Constant-current circuit using Zener diode.

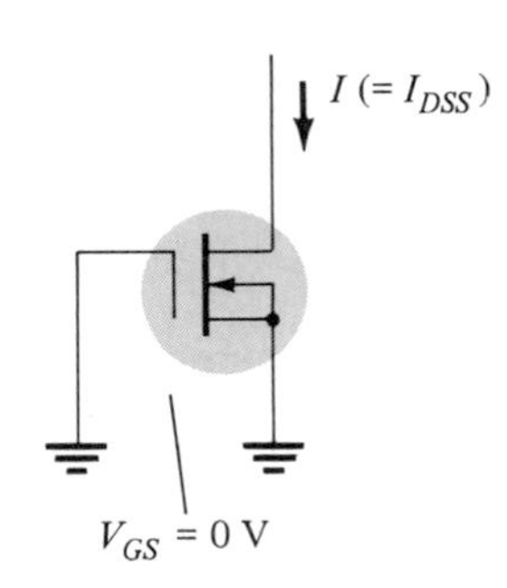

Figure 15.20 Depletion JFET constant-current source.

JFET Constant-Current Source

The JFET or depletion MOSFET provides an excellent constant-current source. If the FET device is biased at $V_{GS} = 0\ V$, the constant current is set at the I_{DSS} of the FET. Figure 15.20 shows how simple a constant-current source circuit can be.

IC CURRENT SOURCES:

The most popular IC circuit configuration is presently the current source. Figure 15.21 shows a few basic forms for constructing IC current sources. For low current values, the *current mirror* shown in Fig. 15.21a is quite popular. Transistor Q_1 is a diode-connected transistor providing temperature compensation for transistor Q_2 in

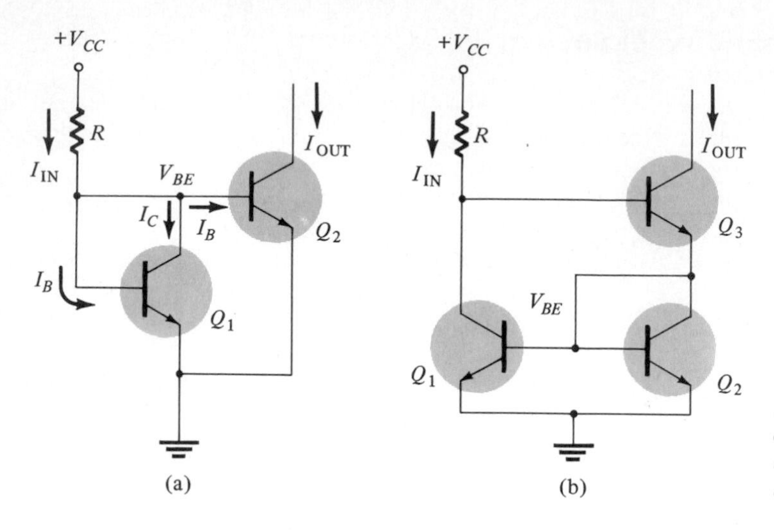

Figure 15.21 IC constant-current sources: (a) current mirror for low current values; (b) constant-current source with higher output impedance.

establishing a constant output current. If the two transistors are matched (typically the case when devices are formed on the same chip in very close proximity), then the output current will remain at the constant value set by supply V_{CC} and resistor R, regardless of the circuit connected to the current source, I_{out}. The circuit of Fig. 15.2lb provides an additional transistor, Q_3, in series with the output to provide a much higher value of output impedance from the current source. The larger the value of the current-source impedance, the more ideal is the circuit's operation.

For the circuit of Fig. 15.21a the input current is set at

$$I_{in} = \frac{V_{CC} - V_{BE}}{R} = I_C + 2I_B \qquad (15.19)$$

Assuming that the transistors Q_1 and Q_2 are well matched, the output is held constant at

$$I_C = \beta I_B = I_{in} - 2I_B \cong I_{in} \qquad (15.20)$$

for large values of β.

The input current for the circuit of Fig. 15.21b is set at

$$I_{in} = \frac{V_{CC} - 2V_{BE}}{R} = I_C + I_B \qquad (15.21)$$

and for matched transistors Q_1, Q_2, and Q_3 the output current is

$$I_{out} = I_C + I_B = I_{in} \qquad (15.22)$$

The output current is thus set by the voltage V_{CC}, the resistor R, and the transistor's base-emitter voltage drop.

IC Circuit Techniques

To construct the various parts of an op-amp on a single IC chip requires the use of a number of circuit techniques that achieve the desired function using mostly transistor components and keeping the number of resistors and their values low. The number of capacitors is also kept low as is their value. The following basic circuit parts are covered to aid in better understanding the overall IC circuit.

IC VOLTAGE SOURCES (BIAS CIRCUITS)

When a reference or dc bias voltage is required in an integrated circuit, connections such as those of Fig. 15.22 can be used. Figure 15.22a shows a JFET used to provide a fixed bias current through a Zener diode, which provides a fixed voltage. The resistor voltage divider then drops the Zener diode voltage to a desired bias voltage value, V_{BIAS}; this bias voltage is available for use in other parts of the op-amp circuit. Figure 15.22b adds a bipolar transistor in series with the Zener diode to provide temperature compensation to maintain the bias voltage over a range of temperature. The bipolar transistor base-emitter voltage drop varies with temperature opposite to that of the Zener diode so that the bias voltage is maintained even as temperature varies. The tracking of the Zener diode and bipolar transistor is quite good as both are formed near each other on the same IC chip.

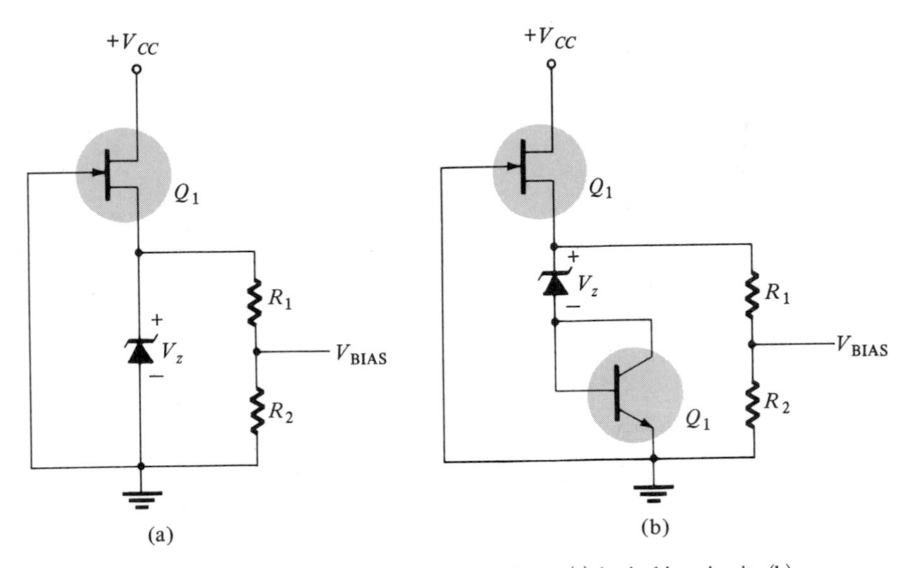

Figure 15.22 Developing a dc bias or reference voltage: (a) basic bias circuit; (b) temperature-compensated bias circuit.

DIFFERENTIAL AMPLIFIER STAGE WITH CONSTANT-CURRENT LOAD

To help appreciate how differential amplifier stages are built in ICs, Fig. 15.23a shows a basic stage with constant-current source, and Fig. 15.23b shows the load resistors replaced by *pnp* current-source loads for larger effective values of R_C and therefore larger stage voltage gain. When even larger gain is desired, the improved circuit of Fig. 15.23c can be used. The additional resistors and transistors provide larger effective load impedance and larger voltage gain for the single stage.

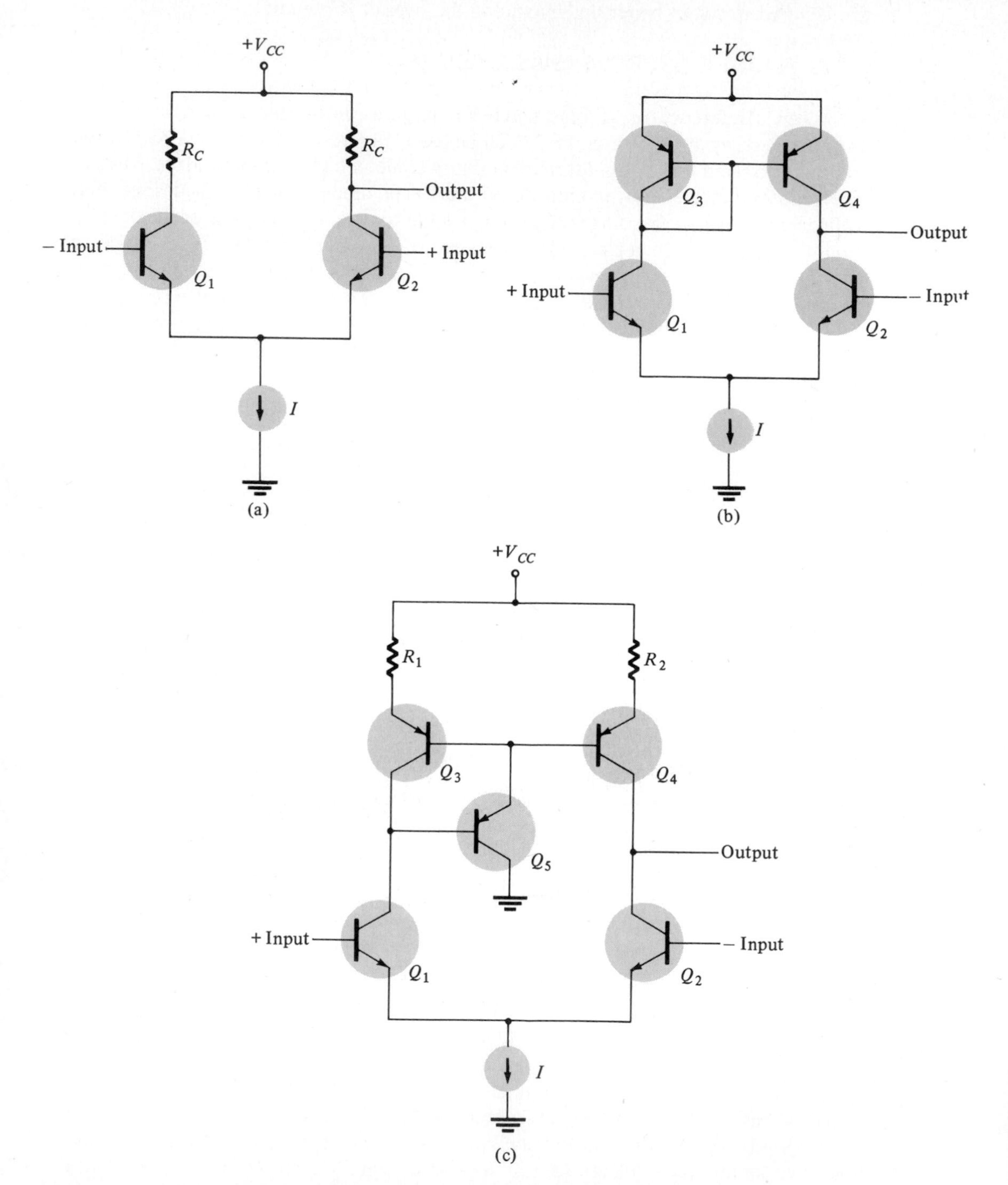

Figure 15.23 IC form of differential-amplifier stage using symmetrical constant-current sources.

LEVEL SHIFTING

To interface between input and output or couple stages together without the limitations imposed by using capacitor coupling, it is necessary to use a voltage-level-shifting circuit. Figure 15.24 shows a few typical circuits to provide a shift in dc level between input and output. In Fig. 15.24a, the output voltage follows the input voltage except for the lower dc voltage level of the output set by resistor R_1 and the current set through transistor Q_2. In the circuit of Fig. 15.24b, the voltage drop between input and output is essentially set by the Zener diode voltage (and the value of transistor Q_1 base-emitter voltage drop).

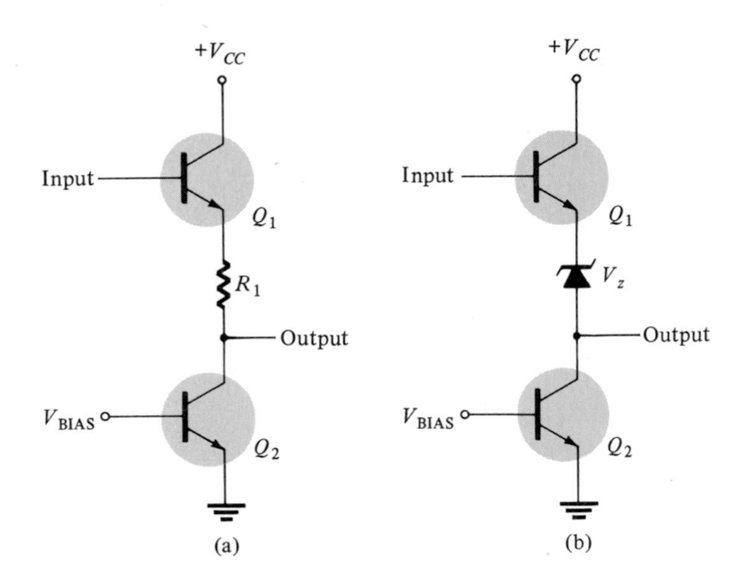

Figure 15.24 Level-shifting circuits: (a) resistive level shifting; (b) Zener-diode level shifting.

OUTPUT STAGE

After the inputs are amplified to the desired value of output voltage, an output stage is used to supply a signal capable of driving the load. Figure 15.25 shows a few output stage circuits. Figure 15.25a is simply a conventional emitter follower; Fig. 15.25b replaces R_E with a current source to provide large value of R_E using a small die area of the IC chip. The output circuit of Fig. 15.25c contains a driver with diode biasing that provides both current sourcing and sinking of the output. Figure 15.25d then shows a full-range output driver using transistor biasing. Finally, Fig. 15.25e shows the circuit of Fig. 15.25c modified by the addition of short-circuit protection for the output.

15.4 COMMON-MODE REJECTION

One of the more important features of a difference amplifier is its ability to cancel out or reject certain types of unwanted voltage signals. These unwanted signals are referred to as "noise" and can occur as voltages induced by stray magnetic fields in

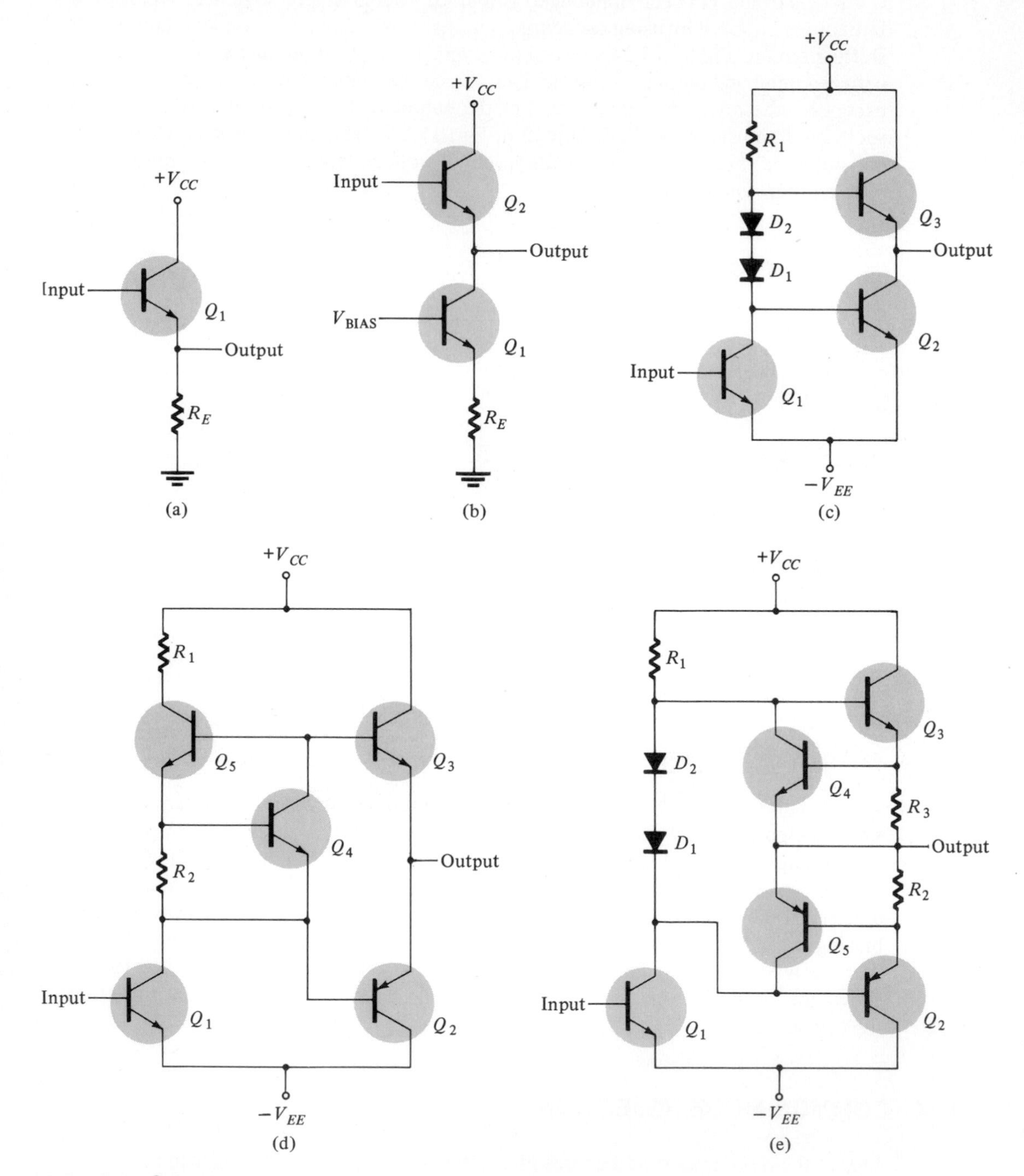

Figure 15.25 Output stages.

the ground or signal wires, as voltage variations in the voltage supply. What is important in this consideration is that these noise signals are not the signals that are desired to be amplified in the difference amplifier. Their distinguishing feature is that the noise signal appears *equally* at both inputs of the circuit.

We can say then that any unwanted (noise) signals that appear in polarity, or common to both input terminals, will be greatly rejected (canceled out) at the output of the difference amplifier. The signal that is to be amplified appears at only one input or opposite in polarity at both inputs. What we wish to consider in this section is, if undesirable noise does occur, how much does the amplifier reject or cancel out this noise? A measure of this rejection of signals common to both inputs is called the amplifier's *common-mode rejection* and a numerical value is assigned, which is called the *common-mode rejection ratio* (CMRR).

Figure 15.26a shows an amplifier with two input signals. These signals can, in general, be considered to contain components that are exactly opposite in polarity *and* components that are the same polarity. For ideal operation we would want the differential amplifier to provide high gain for the opposite-polarity components of the signals and zero gain for the in-polarity components of the signals.

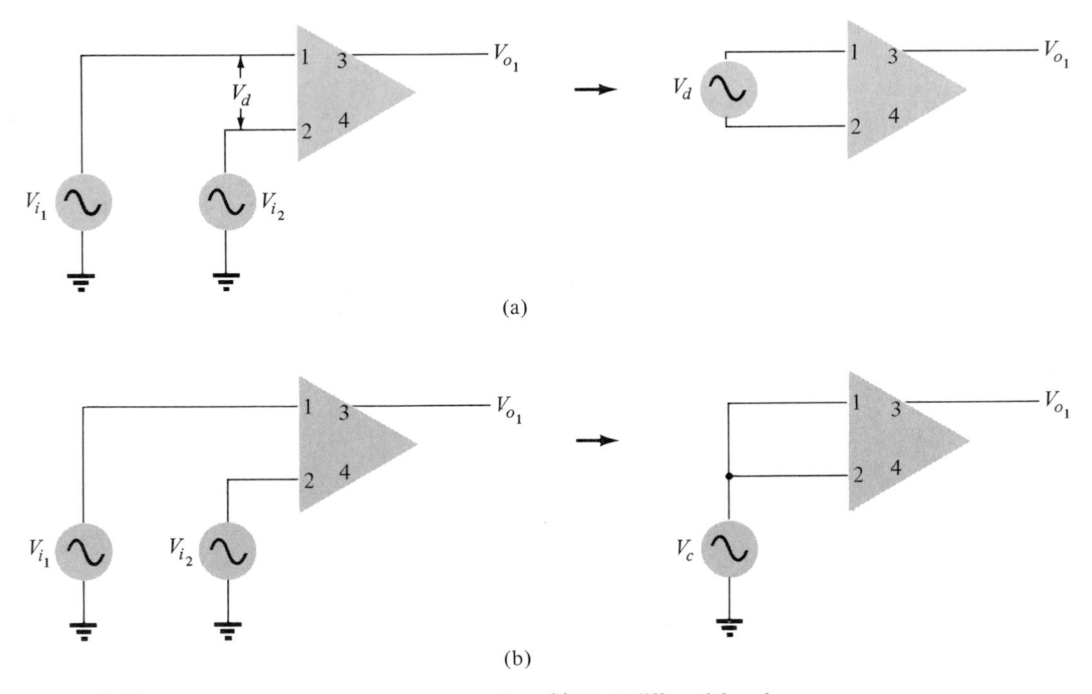

Figure 15.26 Differential and common-mode operation: (a) ideal differential-mode operation; (b) ideal common-mode operation.

The voltage measured *from* terminal 1 *to* terminal 2 can be considered as a differential voltage

$$\boxed{V_d = V_{i_1} - V_{i_2}} \tag{15.23}$$

If, as in the ideal case, $V_{i_1} = -V_{i_2}$, we note that

$$V_d = V_{i_1} - (-V_{i_1}) = 2V_{i_1} = -2V_{i_2}$$

In general, there may also exist common components to the input signals. We can define a common input as

$$\boxed{V_c = \tfrac{1}{2}(V_{i_1} + V_{i_2})} \tag{15.24}$$

The ideal case (shown in Fig. 15.26b) is $V_{i_1} = V_{i_2}$, for which

$$V_c = \tfrac{1}{2}(V_{i_1} + V_{i_2}) = V_{i_1} = V_{i_2}$$

From Eqs. (15.23) and (15.24) we can obtain expressions for V_{i_1} and V_{i_2} based on V_c and V_d:

$$V_{i_1} = V_c + \frac{V_d}{2} \tag{15.25a}$$

$$V_{i_2} = V_c - \frac{V_d}{2} \tag{15.25b}$$

The output voltages can then be expressed as

$$V_{o_1} = A_1 V_{i_1} + A_2 V_{i_2} \tag{15.26a}$$

$$V_{o_2} = A_2 V_{i_1} + A_1 V_{i_2} \tag{15.26b}$$

where A_1 = negative voltage gain from input terminal 1 to output terminal 3 (with input terminal 2 grounded)
A_2 = positive voltage gain from input terminal 2 to output terminal 3 (with input terminal 1 grounded)

It is more important to consider the differential and common-mode operation of the amplifier since this allows determining the common-mode rejection of the circuit. This second way of considering the operation of the amplifier provides an output voltage as

$$V_{o_1} = A_d V_d + A_c V_c \tag{15.27a}$$

$$V_{o_2} = -A_d V_d + A_c V_c \tag{15.27b}$$

where A_d = differential-mode gain of the amplifier
A_c = common-mode gain of the amplifier

V_d and V_c are defined in Eqs. (15.23) and (15.24), respectively.

OPPOSITE-POLARITY INPUTS

If the inputs are equal and opposite in polarity, $V_{i_1} = V_s$ and $V_{i_2} = -V_s$, then from Eq. (15.23),

$$V_d = V_{i_1} - V_{i_2} = V_s - (-V_s) = 2V_s$$

and from Eq. (15.24),

$$V_c = \tfrac{1}{2}(V_{i_1} + V_{i_2}) = \tfrac{1}{2}[V_s + (-V_s)] = 0$$

so that in Eq. (15.27a),

$$V_{o_1} = A_d V_d + A_c V_c = A_d(2V_s) + A_c(0)$$

$$V_{o_1} = 2A_d V_s$$

which shows that only differential-mode operation occurs (and that the overall gain is twice the value of A_d).

SAME-POLARITY INPUTS

If the inputs are equal and the same polarity, $V_{i_1} = V_s = V_{i_2}$, then from Eq. (15.23)

$$V_d = V_{i_1} - V_{i_2} = V_s - V_s = 0$$

and from Eq. (15.24),

$$V_c = \tfrac{1}{2}(V_{i_1} + V_{i_2}) = \tfrac{1}{2}(V_s + V_s) = V_s$$

so that in Eq. (15.27a),

$$V_{o_1} = A_d V_d + A_c V_c = A_d(0) + A_c V_s$$

$$= A_c V_s$$

which shows that only common-mode operation occurs.

Common-Mode Rejection Ratio

The solutions above indicate how A_d and A_c can be measured in differential-amplifier circuits.

1. *To measure A_d:* Set $V_{i_1} = -V_{i_2} = V_s = 0.5$ V so that $V_d = 1$ V and $V_c = 0$ V. Under these conditions the output voltage is $A_d \times (1 \text{ V})$ so that the output voltage equals A_d.
2. *To measure A_c:* Set $V_{i_1} = V_{i_2} = V_s = 1$ V so that $V_d = 0$ V and $V_c = 1$ V. Then the output voltage measured equals A_c.

Having measured A_d and A_c for the amplifier we can now calculate a common-mode rejection ratio, which is defined as

$$\boxed{\text{CMRR} = \frac{A_d}{A_c}} \qquad (15.28a)$$

The value of CMRR can also be expressed in logrithmic terms as

$$\boxed{\text{CMRR(log)} = 20 \log \frac{A_d}{A_c}} \qquad (15.28b)$$

It should be clear that the desired operation will have A_d very large with A_c very small. That is, the signals of opposite polarity will appear greatly amplified at the output terminal, whereas the same-polarity signals will mostly cancel out so that the common-mode gain A_c is very small. Ideally, A_d is very large and A_c is zero so that the value of CMRR is infinite. The larger the value of CMRR the better the common-mode rejection of the circuit.

It is possible to obtain an expression for the output voltage as follows:

$$V_{o_1} = A_d V_d \left(1 + \frac{1}{\text{CMRR}} \frac{V_c}{V_d}\right) \tag{15.29}$$

Even if both V_c and V_d components of voltage exist at the inputs, the value of $(1/\text{CMRR})(V_c/V_d)$ will be very small, for CMRR very large, and the output voltage will be approximately $A_d V_d$. In other words, the output will be almost completely due to the difference signal with the common-mode input signals rejected (or cancelled out). Some practical examples should help clarify these ideas.

EXAMPLE 15.5

Determine the output voltage of a differential amplifier for input voltages of $V_{i_1} = 150\ \mu\text{V}$, and $V_{i2} = 100\ \mu\text{V}$. The amplifier has a differential mode gain of $A_d = 1000$ and the value of CMRR is: (a) 100; (b) 10^5.

Solution:

$$V_d = V_{i_1} - V_{i_2} = (150 - 100)\ \mu\text{V} = 50\ \mu\text{V}$$

$$V_c = \frac{1}{2}(V_{i_1} + V_{i_2}) = \frac{(150 + 100)\ \mu\text{V}}{2} = 125\ \mu\text{V}$$

Note that the common signal is more than twice as large as the difference signal.

(a) $V_o = A_d V_d \left(1 + \frac{1}{\text{CMRR}} \frac{V_c}{V_d}\right) = A_d V_d \left(1 + \frac{1}{100} \times \frac{125}{50}\right) = A_d V_d (1.025)$

$= (1000)(50\ \mu\text{V})(1.025) = \mathbf{51.25\ mV}$

The output is only 0.025 or 2.5% more than was the output due only to a difference signal of 50 μV.

(b) $V_o = A_d V_d \left(1 + \frac{1}{10^5} \times \frac{125}{50}\right) = A_d V_d (1.000025) \cong 100 \times 50\ \mu\text{V} = \mathbf{50\ mV}$

Example 15.5 shows that the larger the value of CMRR the better the circuit rejects common-input signals. Thus, one of the important differential-amplifier factors to consider is the circuit's common-mode rejection ratio.

As summarized in Fig. 15.27 the differential gain between any input and output terminal is

$$|A_d| = \frac{h_{fe} R_C}{2h_{ie} + 2(h_{fe} + 1)r_E} \cong \frac{R_C}{2(r_e + r_E)} \tag{15.30}$$

the polarity relation between input and output depending on which set of terminals is used.

A common-mode gain can also be calculated as given by

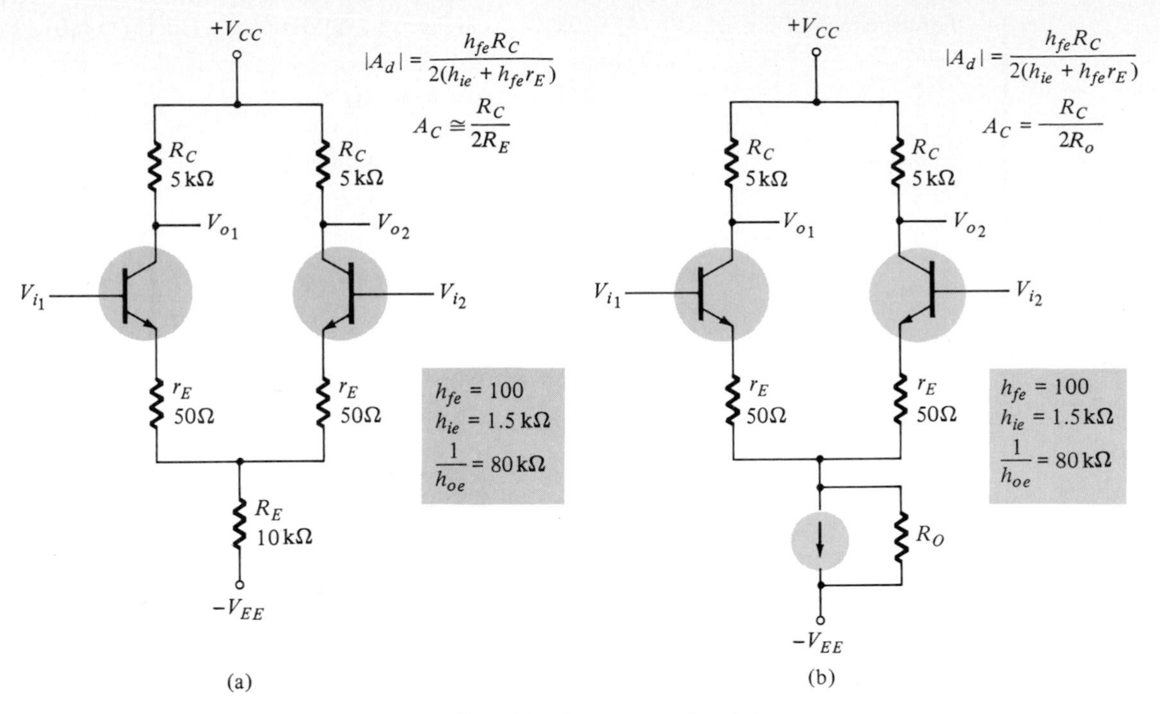

Figure 15.27 Differential amplifiers showing differential and common-mode gains.

$$A_c = \frac{R_C}{2R_E} \tag{15.31}$$

using a circuit with emitter resistor, or

$$A_c = \frac{R_C}{2R_o} \tag{15.32}$$

for a circuit with constant-current source having output resistance, R_o.

EXAMPLE 15.6

Calculate the differential and common-mode gains for the circuits of Fig. 15.27 and also the respective values of CMRR.

Solution:

For Fig. 15.27a

$$A_d = \frac{h_{fe}R_C}{2[h_{ie} + (h_{fe} + 1)r_E]} = \frac{100\ (5\ \text{k}\Omega)}{2[1.5\ k\Omega + 101\ (50\ \Omega)]} = 38.17$$

$$A_c = \frac{R_C}{2R_E} = \frac{5\ \text{k}\Omega}{2(10\ \text{k}\Omega)} = 0.25$$

$$\text{CMRR} = \frac{A_d}{A_c} = \frac{38.17}{0.25} = 152.68(=\mathbf{43.68\ dB})$$

For the circuit of Fig. 15.27b, $A_d = 38.17$ (as in Fig. 15.27a). Using Eq. (15.15) to calculate R_o, we obtain

$$R_o = 80\ \text{k}\Omega(101) = 8.08\ \text{M}\Omega$$

so that A_c is

$$A_c = \frac{R_C}{2R_o} = \frac{5\ \text{k}\Omega}{2(8.08\ \text{M}\Omega)} = 3.09 \times 10^{-4}$$

We then calculate

$$\text{CMRR} = \frac{A_d}{A_c} = \frac{38.17}{3.09 \times 10^{-4}} \cong 1.24 \times 10^5 (= \mathbf{101.9\ dB})$$

15.5 BASICS OF THE OPERATIONAL AMPLIFIER

An operational amplifier is a very high-gain differential amplifier that uses voltage feedback to provide a stabilized voltage gain. The basic amplifier used is essentially a difference amplifier having very high open-loop gain (no signal-feedback condition) as well as high input impedance and low output impedance. Typical uses of the operational amplifier are scale changing; analog computer operations, such as addition and integration; and a great variety of phase shift, oscillator, and instrumentation circuits.

Figure 15.28 shows an op-amp unit having two inputs and a single output. Recall how the two inputs affect an output in a difference amplifier. Here the inputs are marked with *plus* (+) and *minus* (−) to indicate noninverting and inverting inputs, respectively. A signal applied to the *plus* input will appear with the same polarity and amplified at the output, whereas an input applied to the *minus* (−) terminal will appear amplified but inverted at the output.

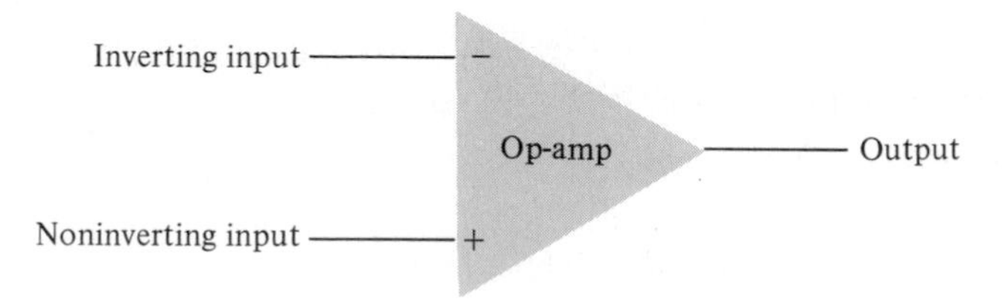

Figure 15.28 Basic op-amp.

The basic circuit connection of an operational amplifier is shown in Fig. 15.29a. As shown the circuit operates as a scale changer or constant-gain multiplier. An input signal V_1 is applied through a resistor R_1 to the minus input terminal. The output voltage is fed back through resistor R_f to the same input terminal. The plus input terminal is connected to ground. We now wish to determine the overall gain of the circuit (V_o/V_1). To do this we must consider some more details of the op–amp unit.

Figure 15.29 shows the op-amp replaced by an equivalent circuit of input resistance R_i and output voltage source and resistance. An ideal op-amp as shown in Fig. 15.29c has infinite resistance ($R_i = \infty$), zero output resistance ($R_o = 0$), and infinite

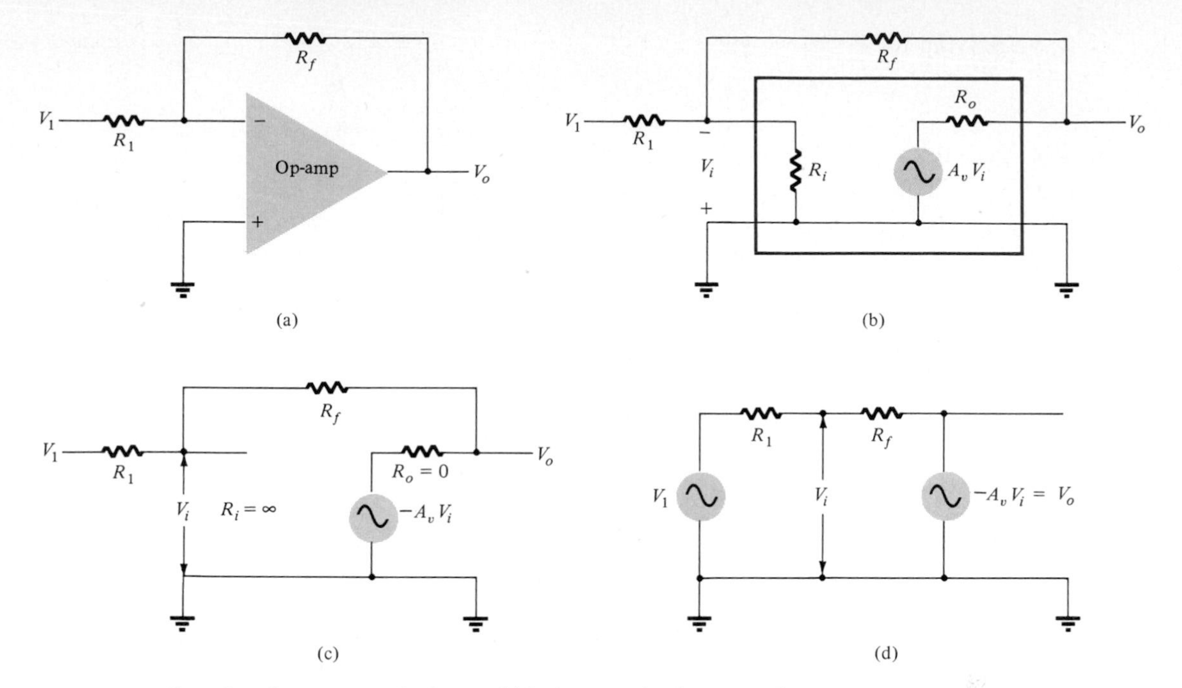

Figure 15.29 Operation of op-amp as scale changer: (a) basic connection (constant-gain multiplier); (b) effect of op-amp circuit; (c) ideal op-amp; (d) ideal equivalent circuit.

voltage gain ($A_v = \infty$). The connection for the ideal amplifier is shown redrawn in Fig. 15.29d.

Using superposition we can solve for the voltage V_i in terms of the components due to each of the sources. For source V_i only ($-A_v V_i$ set to zero),

$$V_{i_1} = \frac{R_f}{R_1 + R_f} V_1$$

For source $-A_v V_i$ only (V_1 set at zero),

$$V_{i_2} = \frac{R_1}{R_1 + R_f}(-A_v V_i)$$

The total voltage of V_i is then

$$V_i = V_{i_1} + V_{i_2} = \frac{R_f}{R_1 + R_f} V_1 + \frac{R_1}{R_1 + R_f}(-A_v V_i)$$

which can be solved for V_i as

$$V_i = \frac{R_f}{R_f + (1 + A_v)R_1} V_1 \qquad (15.33)$$

if $A_v \gg 1$ and $A_v R_1 \gg R_f$, as is usually true, then

$$V_i \cong \frac{R_f}{A_v R_1} V_1$$

Solving for V_o/V_1, we get

$$\frac{V_o}{V_1} = \frac{-A_v V_i}{V_i} = \frac{-A_v}{V_1}\left(\frac{R_f V_1}{A_v R_1}\right) = -\frac{R_f}{R_1}$$

$$\boxed{\frac{V_o}{V_1} = -\frac{R_f}{R_1}} \tag{15.34}$$

The result shows that the ratio of overall output to input voltage is dependent only on the values of resistors R_1, R_f—provided that A_v is very large.

If $R_f = R_1$, the gain is

$$A_v = -\frac{R_1}{R_1} = -1$$

and the circuit provides a sign change with no magnitude change.

If $R_f = 2R_1$, then

$$A_v = \frac{-2R_1}{R_1} = -2$$

and the circuit provides a gain of 2 along with polarity inversion of the input signal.

If we select precise resistor values for R_f and R_1, we can obtain a wide range of gains, the gain being as accurate as the resistors used and only slightly affected by temperature and other circuit factors.

Virtual Ground

The output voltage is limited by the supply voltage of, typically, a few volts. Voltage gains as stated before are very high. If, for example, $V_o = -10$ V and $A_v = 10{,}000$, the input voltage is

$$V_1 = -\frac{V_o}{A_v} = -\frac{-10}{10{,}000} = 1 \text{ mV}$$

If the circuit has an overall gain (V_o/V_1) of say, 1, the value of V_1 would be 10 V. The value of V_i, compared to all other voltages, is then small and may be considered 0 V. Note that although $V_i \cong 0$ V, it is not exactly 0 V since the output is the value of V_i times the gain of the amplifer.

The fact that $V_i \cong 0$ V leads to the concept that at the input to the amplifier there exists a virtual short circuit or *virtual ground*. The concept of a virtual short implies that although the voltage is nearly 0 V, there is no current through the amplifier input to ground. Figure 15.30 depicts the virtual-ground concept. The heavy line is used to indicate that we may consider that a short exists with $V_i \cong 0$ V, but that this is a virtual short in that no current goes through the short to ground. Current is through resistor R_1 and through R_f as shown.

Using the virtual-ground concept, we can write equations for the current I as follows:

$$I = \frac{V_1}{R_1} = -\frac{V_o}{R_f}$$

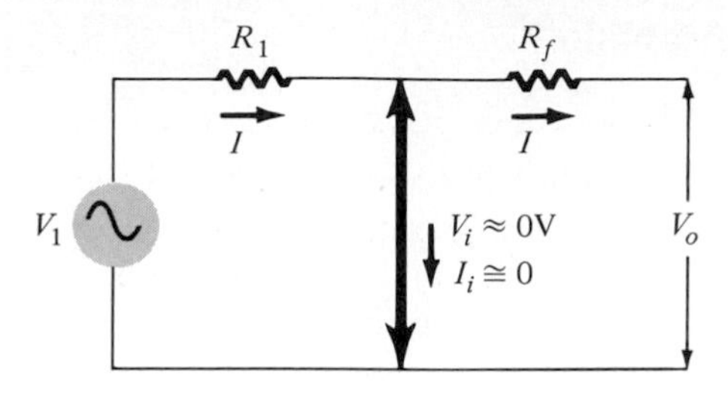

Figure 15.30 Virtual ground in an op-amp.

which can be solved for V_o/V_1:

$$\frac{V_o}{V_1} = -\frac{R_f}{R_1}$$

The virtual-ground concept, which depended on A_v being very large, allowed simple solution of overall voltage gain. It should be understood that although the circuit of Fig. 15.30 is not a physical circuit, it does allow an easy means for determining the overall circuit gain.

15.6 OP-AMP CIRCUITS

Constant-Gain Multiplier

An inverting constant-multiplier circuit has already been considered but is repeated here to provide a fuller listing of basic op-amp circuits. Figure 15.31 shows an inverting constant-gain-multiplier circuit.

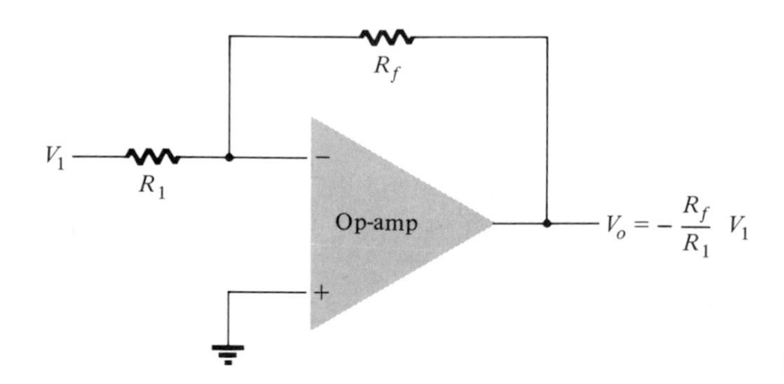

Figure 15.31 Inverting constant-gain multiplier.

EXAMPLE 15.7

The circuit of Fig. 15.31 has $R_1 = 100\text{ k}\Omega$ and $R_f = 500\text{ k}\Omega$. What is the output voltage for an input of $V_1 = -2\text{ V}$?

Solution:

Using Eq. (15.34) gives

$$V_o = -\frac{R_f}{R_1}V_1 = -\frac{500\text{ k}\Omega}{100\text{ k}\Omega}(-2\text{ V}) = \mathbf{+10\text{ V}}$$

Noninverting Amplifier

The connection of Fig. 15.32a shows an op-amp circuit that works as a noninverting constant-gain multiplier. To determine the voltage gain of the circuit we can use the equivalent virtual-ground representation in Fig. 15.32b. Note that the voltage across R_1 is V_1, since $V_i \cong 0$ V. This must be equal to the voltage due to the output, V_o, through a voltage divider of R_1 and R_f so that

$$V_1 = \frac{R_1}{R_1 + R_f} V_o$$

and

$$\boxed{\frac{V_o}{V_1} = \frac{R_1 + R_f}{R_1} = 1 + \frac{R_f}{R_1}} \qquad (15.35)$$

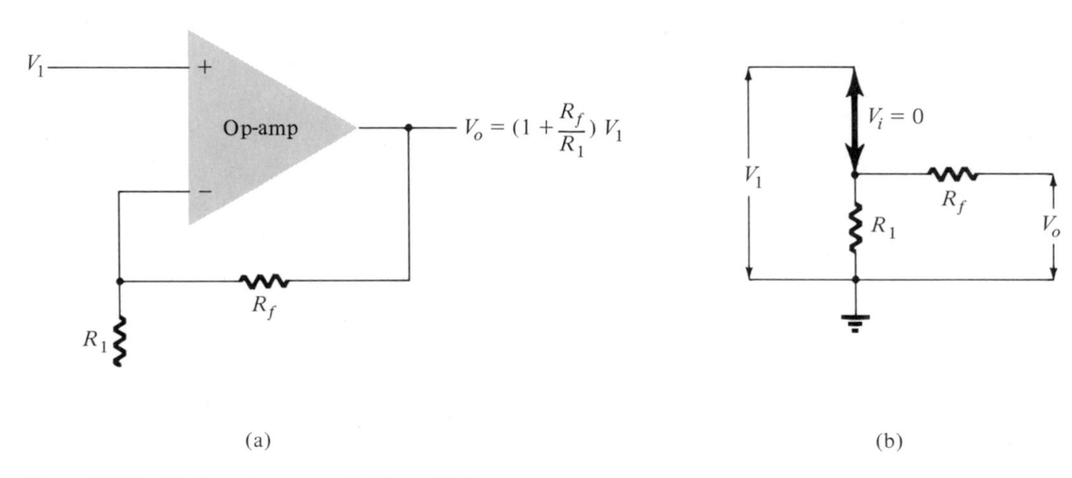

Figure 15.32 Noninverting constant-gain multiplier.

EXAMPLE 15.8

Calculate the output voltage of a noninverting constant-gain multiplier (as in Fig. 15.32) for value of $V_1 = 2$ V, $R_f = 500$ kΩ, and $R_1 = 100$ kΩ.

Solution:

Using Eq. (15.35), we get

$$V_o = \left(1 + \frac{R_f}{R_1}\right)V_1 = \left(1 + \frac{500\text{ k}\Omega}{100\text{ k}\Omega}\right)(2\text{ V}) = 6(2\text{ V}) = \mathbf{+12\ V}$$

Unity Follower

The unity follower, as in Fig. 15.33, provides a gain of 1 with no polarity reversal. From the equivalent circuit with virtual ground it is clear that

$$\boxed{V_o = V_1} \qquad (15.36)$$

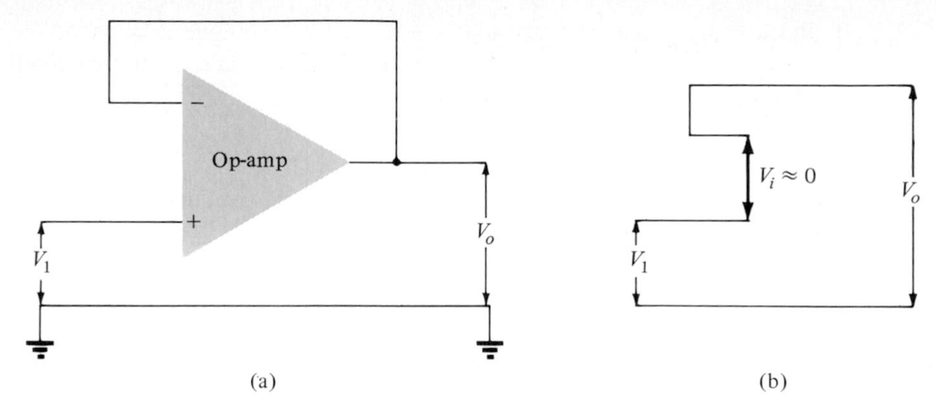

Figure 15.33 (a) Unity follower; (b) virtual-ground equivalent circuit.

and that the output is the same polarity and magnitude as the input. The circuit acts very much like an emitter follower except that the gain is very much closer to being exactly unity.

Summing Amplifier

Probably the most useful of the op-amp circuits used in analog computers is the summing-amplifier circuit. Figure 15.34 shows a three-input summing circuit, which provides a means of algebraically summing (adding) three-input voltages, each multiplied by a constant-gain factor.

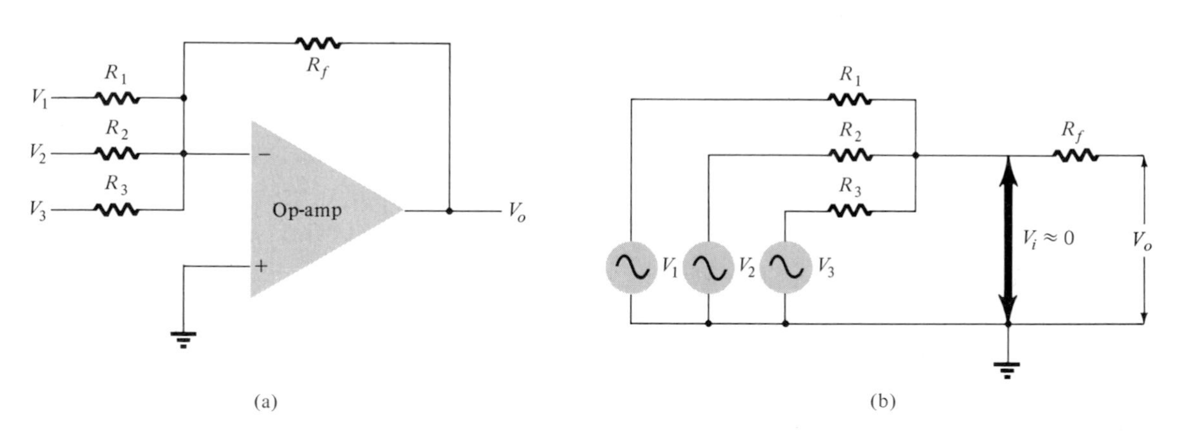

Figure 15.34 (a) Summing amplifier; (b) virtual-ground equivalent circuit.

If the virtual equivalent circuit is used, the output voltage can be expressed in terms of inputs as

$$V_o = -\left(\frac{R_f}{R_1}V_1 + \frac{R_f}{R_2}V_2 + \frac{R_f}{R_3}V_3\right) \tag{15.37}$$

In other words, each input adds a voltage to the output as obtained for an inverting constant-gain circuit. If more inputs are used, they add additional components to the output.

EXAMPLE 15.9

What is the output voltage of an op-amp summing amplifier for the following sets of input voltages and resistors ($R_f = 1\ \text{M}\Omega$ in all cases)?

(a) $V_1 = +1\ \text{V}, V_2 = +2\ \text{V}, V_3 = +3\ \text{V},$
$R_1 = 500\ \text{k}\Omega, R_2 = 1\ \text{M}\Omega, R_3 = 1\ \text{M}\Omega$

(b) $V_1 = -2\ \text{V}, V_2 = +3\ \text{V}, V_3 = +1\ \text{V},$
$R_1 = 200\ \text{k}\Omega, R_2 = 500\ \text{k}\Omega, R_3 = 1\ \text{M}\Omega$

Solution:

Using Eq. (15.37), we get

$$\text{(a)}\ V_o = -\left[\frac{1000\ \text{k}\Omega}{500\ \text{k}\Omega}(+1\ \text{V}) + \frac{1000\ \text{k}\Omega}{1000\ \text{k}\Omega}(+2\ \text{V}) + \frac{1000\ \text{k}\Omega}{1000\ \text{k}\Omega}(+3\ \text{V})\right]$$
$$= -[2(1\ \text{V}) + 1(2\ \text{V}) + 1(3\ \text{V})] = \mathbf{-7\ V}$$

$$\text{(b)}\ V_o = -\left[\frac{1000\ \text{k}\Omega}{200\ \text{k}\Omega}(-2\ \text{V}) + \frac{1000\ \text{k}\Omega}{500\ \text{k}\Omega}(+3\ \text{V}) + \frac{1\ \text{M}\Omega}{1\ \text{M}\Omega}(+1\ \text{V})\right]$$
$$= -[5(-2\ \text{V}) + 2(+3\ \text{V}) + 1(1\ \text{V})] = -(-10\ \text{V} + 6\ \text{V} + 1\ \text{V}) = \mathbf{+3\ V}$$

Integrator

So far the input and feedback components have been resistors. If the feedback component used is a capacitor, as in Fig. 15.35, the resulting circuit is an integrator.

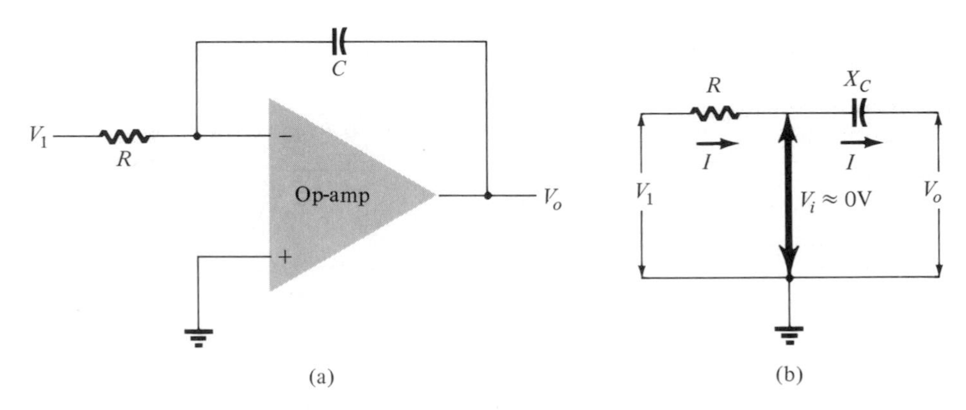

Figure 15.35 Integrator.

The virtual-ground equivalent circuit shows that an expression between input and output voltages can be derived from the current I, which flows from input to output. Recall that virtual ground means that we can consider the voltage at the junction point of R and X_C to be ground (since $V_i \cong 0\ \text{V}$) but that no current goes into ground at that point. The capacitive impedance can be expressed as

$$X_C = \frac{1}{j\omega C} = \frac{1}{sC}$$

where $s = j\omega$ is the Laplace notation. Solving for V_0/V_1 gives us

$$I = \frac{V_1}{R} = -\frac{V_0}{X_C} = \frac{-V_o}{1/sC} = -sCV_0$$

$$\frac{V_0}{V_1} = \frac{-1}{sCR} \tag{15.38a}$$

The last expression can be rewritten in the time domain as

$$v_0(t) = -\frac{1}{RC}\int v_1(t)\,dt \tag{15.38}$$

Equation (15.38b) shows that the output is the integral of the input, with an inversion and scale multiplier of $1/RC$. The ability to integrate a given signal provides the analog computer with the ability to solve differential equations and therefore allows setup of a wide variety of electrical circuit analogs of physical-system operations.

As an example, consider an input step voltage shown in Fig. 15.36. The integral of the step voltage is a ramp or linearly changing voltage. The circuit scale factor of $-1/RC$ is

$$-\frac{1}{RC} = -\frac{1}{10^6 \times 10^{-6}} = -1$$

so that

$$v_0(t) = -\int v_i(t)\,dt$$

and the output is a negative ramp as shown in Fig. 15.36b. If the scale factor is changed by making $R = 100\ \text{k}\Omega$, for example, then

$$-\frac{1}{RC} = -\frac{1}{10^5 \times 10^{-6}} = -10$$

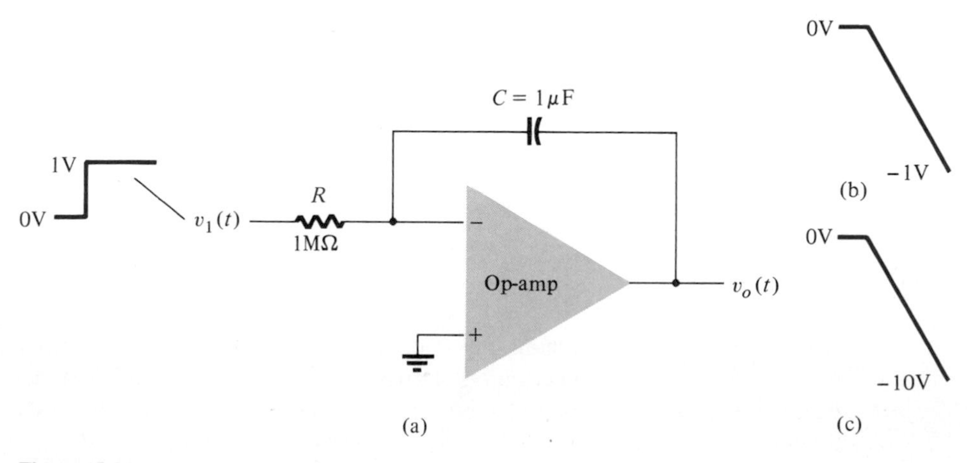

Figure 15.36 Operation of integrator with step input.

and the output is $$v_o(t) = -10 \int v_i(t)\, dt$$

which is shown in Fig. 15.36c.

More than one input may be applied to an integrator as shown in Fig. 15.37 with the resulting operation given by

$$v_o(t) = -\left[\frac{1}{R_1 C}\int v_1(t)\, dt + \frac{1}{R_2 C}\int v_2(t)\, dt + \frac{1}{R_3 C}\int v_3(t)\, dt\right] \qquad (15.39)$$

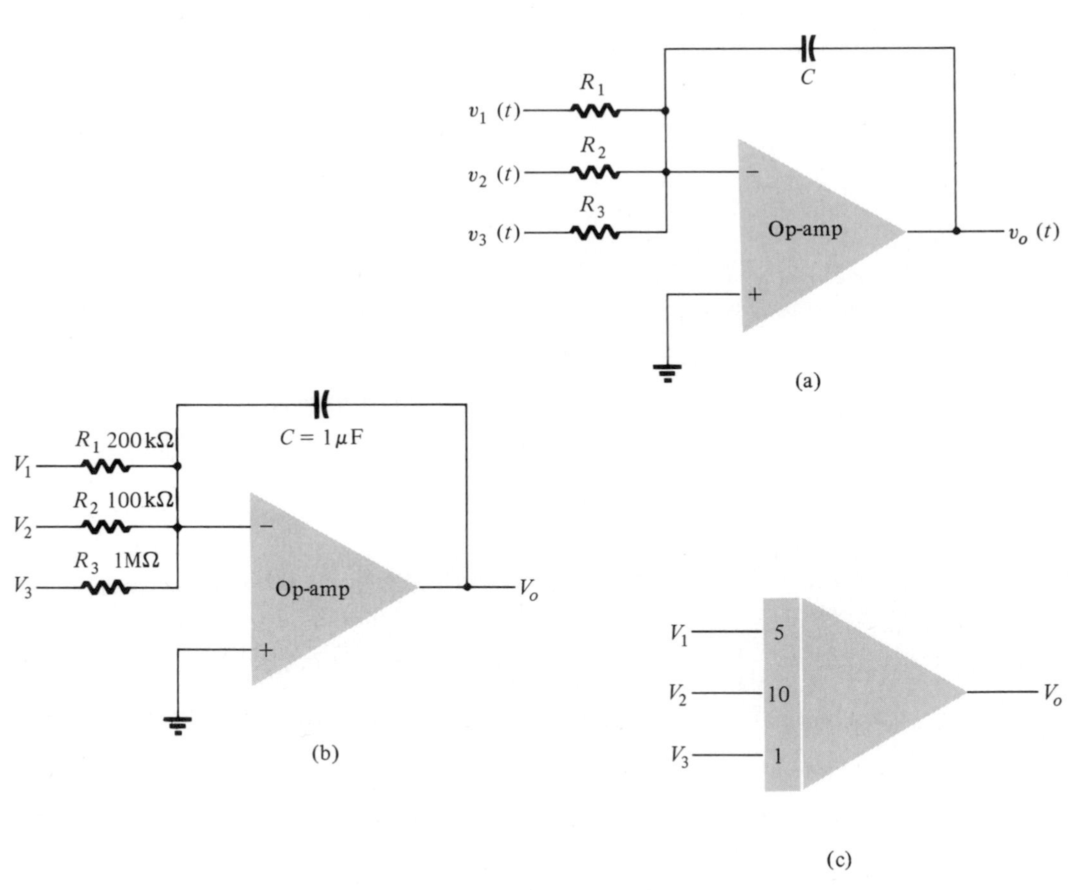

Figure 15.37 (a) Summing-integrator circuit; (b) op-amp; (c) analog-computer, integrator-circuit representation.

An example, showing a summing integrator as used in an analog computer, is given in Fig. 15.37. The actual circuit is shown with input resistors and feedback capacitor, whereas the analog-computer representation only indicates the scale factor for each input.

Differentiator

The differentiator circuit of Fig. 15.38 is not as useful a computer circuit as the integrator because of practical problems with noise. The resulting relation for the circuit is

$$v_o(t) = -RC\frac{dv_1(t)}{dt} \tag{15.40}$$

where the scale factor is $-RC$. Reference to any text on analog computers will show how differential equations are set up for solution using mainly summing and integrator circuits.

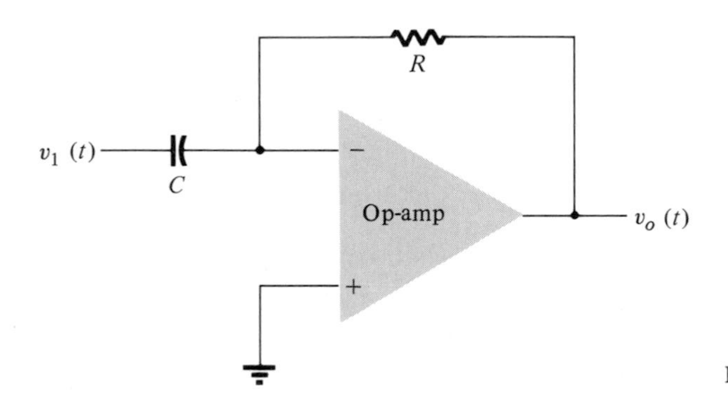

Figure 15.38 Differentiator circuit.

15.7 PRACTICAL OP-AMP CIRCUITS

Using multiple differential-amplifier stages in a single IC package results in an overall circuit called an *operational amplifier* or *op-amp*. Basic features of this circuit include extremely high voltage gain, high input resistance, and low output resistance. The differential-amplifier circuit considered earlier in this chapter is then a basic circuit used in building practical op-amp units. Integrated-circuit construction requires the use of the smallest sized components to form the many hundreds of components—mostly transistors—to build from one to four op-amps in a single IC chip. These circuits may be built using only BJT (bipolar), both bipolar and JFET (BiFET), or bipolar and MOSFET (BiMOS). At present, BiFET op-amps are most popular with high resistance provided by the JFET input transistor, high gain using bipolar difference-amplifier circuits, and low output resistance using an emitter-follower output stage as shown in Fig. 15.25.

In a BiFET op-amp, the JFET device is used in the input part of the circuit to obtain high input resistance. At present, mostly JFET transistors are used for input. For example, the schematic and connection diagram of a 347 op-amp are shown in Fig. 15.39. The 347 IC is a quad JFET input operational amplifier using BiFET technology. The circuit of one op-amp stage is shown in Fig. 15.39a; the pin connection diagram showing the four op-amp units is detailed in Fig. 15.39b. Some

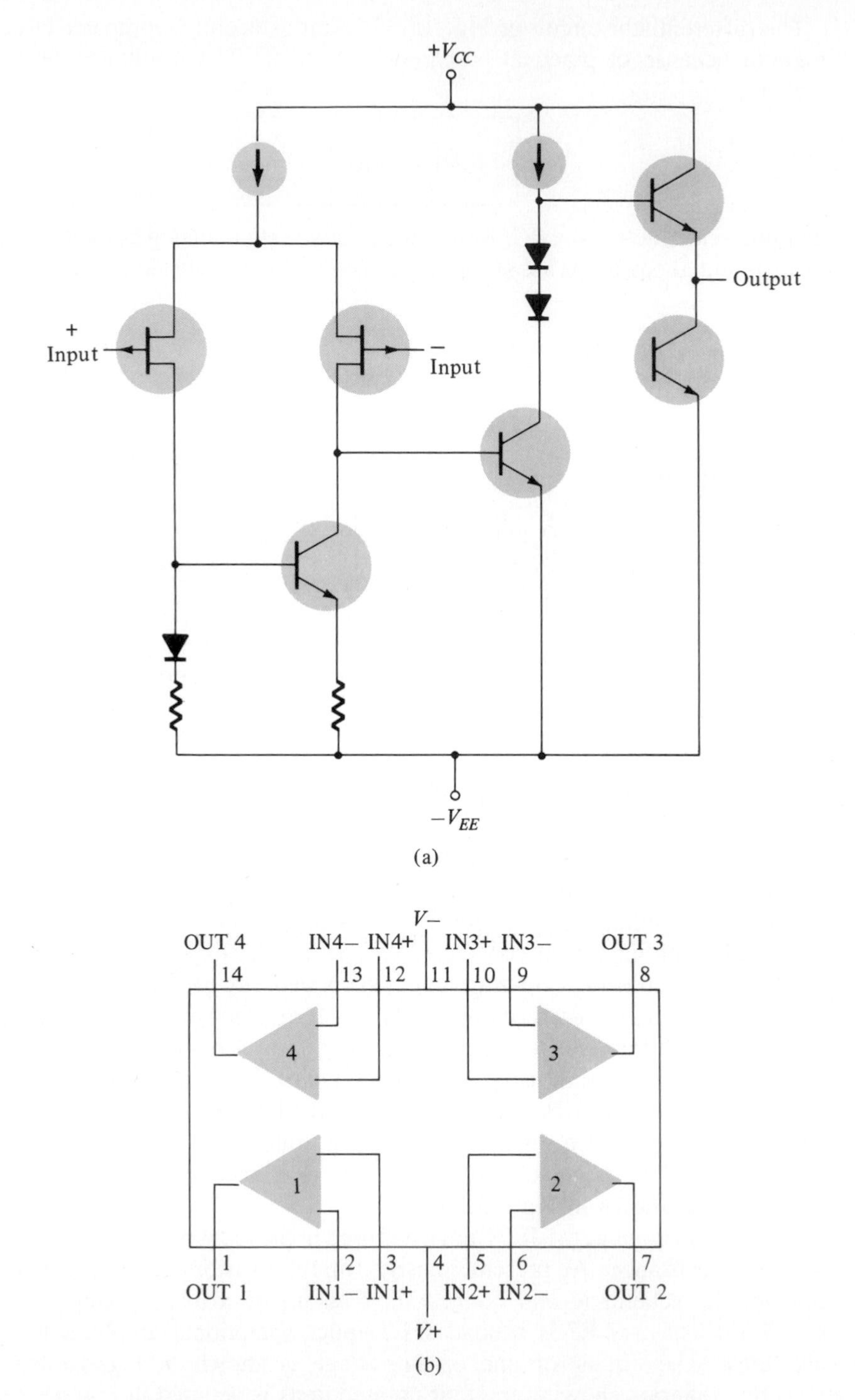

Figure 15.39 IC op-amp (347): (a) circuit diagram; (b) connection diagram.

of the manufacturer's listed unit specifications are:

R_{in}: Input resistance $= 10^{12}\ \Omega$

A_{vol}: Large-signal voltage gain $= 100$ V/mV $= 100{,}000\ (= 100$ dB)

CMRR: Common-mode rejection ratio $= 100$ dB

GB: Gain-bandwidth product $= 4$ MHz

SR: Slew rate $= 13\ \text{V}/\mu\text{s}$

Notice how the circuit of Fig. 15.39a has many of the parts covered in Section 15.3. However, we need not dwell on the details of the circuit in order to use it. The external features that will be considered next are sufficient to allow using the amplifier.

Table 15.1 is part of the manufacturer's listing of electrical characteristics and definitions of terms. The following discussion will elaborate on a number of the more important characteristics and give some examples.

TABLE 15.1 347 Op-Amp Electrical Characteristics at $T_A = 25°\text{C}$, $V_{CC} = +15$ V, $V_{EE} = -15$ V

		Limit			
Characteristics	*Symbols*	*Min.*	*Typ.*	*Max.*	*Units*
Dynamic characteristics					
Large-signal voltage gain	A_{vol}	25	100		V/mV
Input resistance	R_{in}		10^{12}		Ω
Common-mode rejection ratio	CMRR	70	100		dB
Output voltage swing	V_o(p-p)	±12	±13.5		V
Input common-mode voltage range	V_{CM}	±11	−12		V
Gain-bandwidth product	GB		4		MHz
Static characteristics					
Input offset voltage	V_{OS}		5	10	mV
Input offset current	I_{OS}		25	100	pA
Device dissipation	P_D			500	mW

15.8 OP-AMP SPECIFICATIONS

Definitions of Op-Amp Terms

Input Offset Voltage: The difference in the dc voltages that must be applied to the input terminals to obtain equal quiescent operating voltages (zero-output voltage) at the output terminals.

Input Offset Current: The difference in the currents at the two input terminals.

Quiescent Operating Voltage: The dc voltage at either output terminal, with respect to ground.

DC Device Dissipation: The total power drain of the device with no signal applied and no external load current.

Common-Mode Voltage Gain: The ratio of the signal voltages developed at either of the two output terminals to the common signal voltage applied to the two input terminals connected in parallel.

Differential Voltage Gain—Single-Ended Input–Output: The ratio of the change in output voltage at either output terminal with respect to ground, to difference in the input voltages.

Common-Mode Rejection Ratio: The ratio of the full differential voltage gain to the common-mode voltage gain.

Bandwidth at 3-dB Point (B)*:* The frequency at which the voltage gain of the device is 3 dB below the voltage gain at a specified lower frequency.

Maximum Output Voltage $V_o\,(p\text{-}p)$: The maximum peak-to-peak output voltage swing, measured with respect to ground, that can be achieved without clipping of the signal waveform.

Single-Ended Input Resistance (R_{in})*:* The ratio of the change in input voltage to the change in input current measured at either input terminal with respect to ground.

Single-Ended Output Resistance (R_o): The ratio of the change in output voltage to the change in output current measured at either output terminal with respect to ground.

Slew-Rate: Device parameter indicating how fast the output voltage changes with time.

DC Electrical Parameters

DIFFERENTIAL VOLTAGE GAIN—LARGE-SIGNAL VOLTAGE GAIN, A_{VOL}

The typical value of 106 dB is the gain from one input terminal to either output terminal. This was considered the gain A_v in Sections 15.1 through 15.3. The manufacturer lists the gain in units of decibels (dB). The relation of decibels and the gain as numerical ratio of output voltage (V_o) to input voltage (V_i) is

$$A_{dB} = 20 \log |A_v| = 20 \log \left| \frac{V_o}{V_i} \right| \qquad (15.41)$$

As an example, a gain of $A_v = 1000$ is the same as

$$A_{dB} = 20 \log 1000 = 20(3) = 60 \text{ dB}$$

and a gain of $A_v = 100{,}000$ is the same as

$$A_{dB} = 20 \log 100{,}000 = 20(5) = 100 \text{ dB}$$

A gain of 106 dB is then the same as a voltage gain above 100,000 and can be calculated exactly as

$$106 = 20 \log A_v$$

$$5.3 = \log A_v$$

$$A_v = \text{antilog } 5.3 \cong 2 \times 10^5 = 200{,}000$$

The large open-loop gain ($\cong$ 200,000) and limited output voltage swing ($\pm$13 V) requires that the input voltage for open-circuit operation be no greater than

$$V_d = \frac{V_o}{A_{vol}} = \frac{\pm 13\text{ V}}{200{,}000} = \pm 65\ \mu\text{V}$$

An op-amp with such large gain is typically operated closed loop as desired later in this chapter to allow operation with larger input voltages.

SINGLE-ENDED INPUT RESISTANCE, R_{in}

The input resistance is measured at either input terminal. A listed value of $10^{12}\ \Omega$ indicates a high value. It should be recalled from Chapter 11 how important input resistance values are when interconnecting amplifier stages or driving the amplifier with a practical voltage source. If the input resistance is not much larger than the source resistance, loading will cause the input voltage to be less than that of the unloaded source signal, resulting in less output voltage.

Bipolar op-amps typically provide input resistances of around 1 MΩ, BiFET op-amps are rated at $10^{12}\ \Omega$, and BiMOS typically are $10^{15}\ \Omega$.

OUTPUT RESISTANCE, R_o:

The output resistance, typically 100 Ω, depends on the output stage used to drive the signal to a load. Output stages can provide signal swings good in only one direction of voltage swing, in both directions of voltage swing, and may be short-circuit protected depending on which output stage circuit is used (see Fig. 15.25).

COMMON-MODE REJECTION RATIO

The common-mode rejection ratio, defined as

$$\text{CMRR} = \left|\frac{A_d}{A_c}\right|$$

may also be calculated in decibel units as

$$\boxed{\text{CMRR(dB)} = 20 \log \left|\frac{A_d}{A_c}\right| \quad \text{dB}} \tag{15.42}$$

A value of CMRR = 100 dB is a ratio of differential mode to common-mode gain of

$$\frac{A_d}{A_c} = \text{antilog}\frac{\text{CMRR(dB)}}{20} = \text{antilog}\frac{100\text{ dB}}{20} = \text{antilog } 5$$

$$= 10^5 = 100{,}000$$

The common or in-polarity signal is thus amplified by a gain of 100,000 less than the differential or opposite-polarity inputs.

The value of CMRR generally drops off at increasing frequency as shown by the plot of Fig. 15.40.

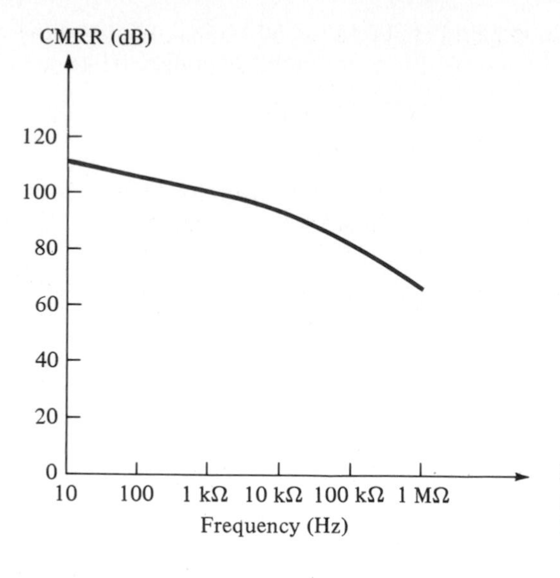

Figure 15.40 Plot of common-mode rejection ratio (CMRR) versus frequency.

OFFSET VOLTAGE, V_{OS}

The definition of input offset voltage, V_{OS}, can be stated as "that differential dc voltage required between inputs of an operational amplifier to force the output to 0 V." Ideally the value of V_{OS} should be 0 V and practically the value of V_{OS} is a few millivolts. When the op-amp is used primarily for its large-signal operation, the small offset voltage is acceptable. When used in applications where a small output voltage represents some measured quantity as in a converter, meter, or measuring device, any nonzero output voltage can result in substantial error. In such circuit application, an op-amp with very low offset voltage is used, or one having input terminals allowing offset voltage adjustment is used.

The 357 op-amp, for example, has balance input terminals for adjusting the offset voltage as shown in Fig. 15.41. A 25-kΩ potentiometer is suggested by the manufacture for connection between pins 1 and 5, the wiper arm being adjusted until

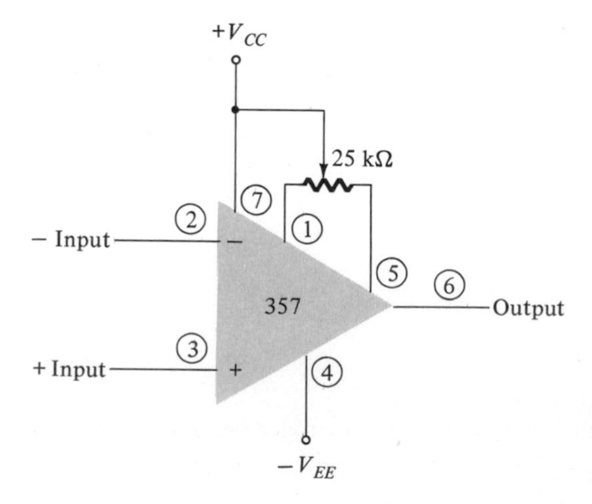

Figure 15.41 Operation of offset voltage (V_{OS}) adjustment using 357 op-amp.

the measured output voltage is adjusted to 0 V, when the inputs at pins 2 and 3 are grounded (V_d = 0 V).

INPUT BIAS CURRENT, I_{bias}

For the circuit inside the IC to operate properly, sufficient dc bias current must be provided as specified by the manufacturer's information. For BJT inputs, the current required is typically microamperes; for JFET input stages, the current required is some tens of picoamperes. Although the device value is typically rated at 25°C (room temperature), it increases greatly with temperature as shown in the plot of Fig. 15.42.

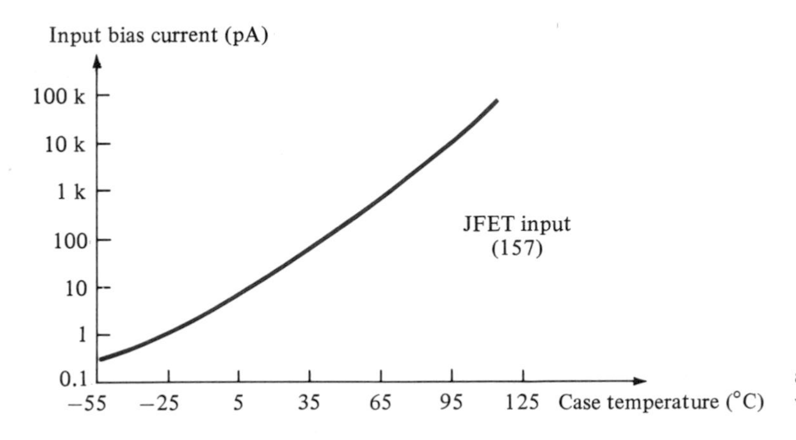

Figure 15.42 Plot showing variation of input bias current required versus device case temperature.

INPUT OFFSET CURRENT, I_{OS}

The small difference in bias currents at the inputs is magnified by the amplifier gain to provide an output offset voltage. Offset current for BJT input circuits is tens to hundreds of nanoamperes, whereas for JFET input stages the value is typically picoamperes.

DRIFT

Drift is the term describing the change in output voltage resulting from change in temperature. Even when the output voltage is adjusted to 0 V at room temperature, that value will change as temperature changes. Typically the offset-voltage drift, $\Delta V_{OS}/\Delta T$, is in the range of 5–40 μV/°C. As a rule of thumb, the drift is 3.3 μV/°C for each mV of initial offset voltage. The drift resulting from the offset current is typically $\Delta I_{OS}/\Delta T$ = 0.01 to 0.5 nA/°C.

AC Electrical Parameters

BANDWIDTH, B

The op-amp unity-gain bandwidth specifies the upper frequency at which the gain drops to unity (gain = 1) owing to the capacitances resulting from manufacture of the circuit. Typical values of unity gain bandwidth or gain bandwidth product are greater than 1 MHz. Figure 15.43 shows a plot of open-loop voltage gain versus frequency for a 357 IC. Notice that below about 100 Hz, the gain stays constant at the rated dc open-loop gain value, dropping to unity gain (0 dB) at a frequency around 10 MHz.

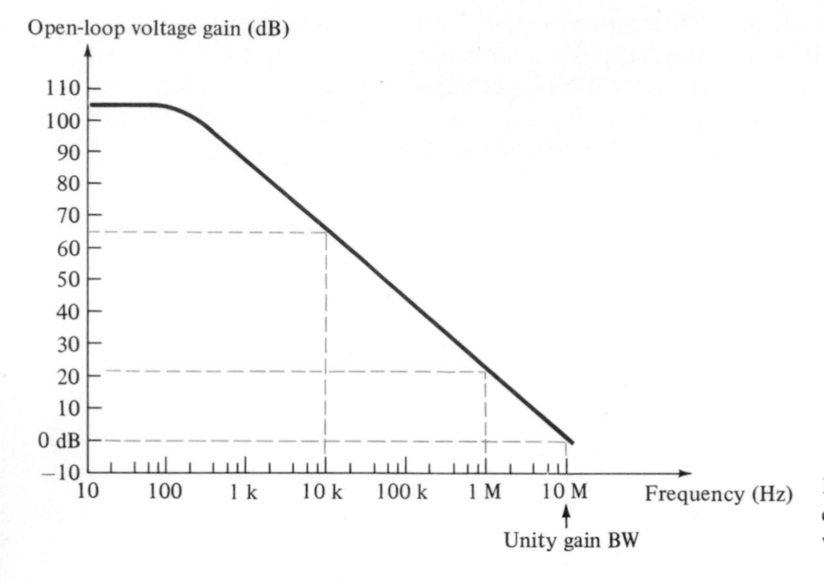

Figure 15.43 Gain versus frequency showing unity-gain bandwidth.

An alternate parameter used to specify the device bandwidth is the rise time, t_r. The value of rise time and bandwidth are related by

$$B = \frac{0.35}{t_r} \tag{15.43}$$

If the manufacturer lists the value of t_r at 0.3 μs, for example, the value of unity-gain bandwidth is then

$$B = \frac{0.35}{0.3\ \mu\text{s}} = 1.167\text{ MHz}$$

Various device spec sheets list unity-gain B, gain-bandwidth product, or rise time to specify the limited frequency range when using that op-amp. It should be clear that an op-amp having an open-loop gain rated at 100,000 (at dc) with gain-bandwidth of 1 MHz has a value of A_{vol} much smaller at, say, 10 kHz. Using the plot of Fig. 15.43, the op-amp gain is greater than 100 dB at dc but decreases to about 65 dB at 10 kHz. At 1 MHz the op-amp gain is around 20 dB ($A_{\text{vol}} = 10$) and obviously doesn't qualify as large gain for op-amp applications. In any case the larger value of bandwidth provides op-amp operation at higher frequency.

SLEW RATE, *SR*

Slew rate is a parameter indicating how fast the output voltage changes with time. Typical slew rate values are from 0.5 V/μs to 50 V/μs, with the larger values indicating the units that operate faster.

TABLE 15.2 Comparison of Op-Amp Parameters

	Parameter	*Bipolar*	*BiFET*	*Norton*	*Units*
Open-loop voltage gain	A_{vol}	200	200	2.8	V/mV
Input resistance	R_{in}	2	10^6	1	MΩ
Output resistance	R_{out}	75		8	Ω
Common-mode rejection ratio	CMRR	90	100		dB
Input offset voltage	V_{OS}	1	1		mV
Input bias current	I_{bias}	80 nA	30 pA	30 pA	nA or pA
Input offset current	I_{OS}	20 nA	3 pA		nA or pA
Drift	$\Delta V_{OS}/\Delta T$	15	3		μV/°C
Bandwidth	B	1	20	2.5	MHz
Slew rate	SR	0.5	50	0.5	V/μs

A comparison of the various device parameters for a number of ICs is listed in Table 15.2.

15.9 OP-AMP APPLICATIONS

To provide some indication of how useful op-amps can be in other than just analog-computer circuits (which is a very large op-amp application area) a few miscellaneous applications will be considered here, with additional oscillator applications in Chapter 18.

DC Millivoltmeter

Figure 15.44 shows a 741 op-amp used as the basic amplifier in a dc millivoltmeter. The amplifier provides a meter with high input impedance and scale factors

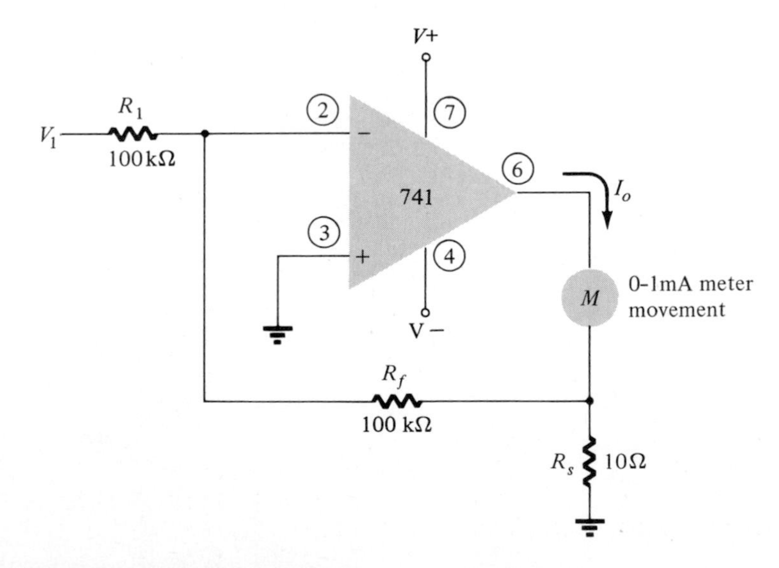

Figure 15.44 Op-amp dc millivoltmeter.

dependent only on resistor value and accuracy. Notice that the meter reading represents millivolts of signal *at* the circuit input. An analysis of the op-amp circuit yields the circuit transfer function.

$$\frac{I_0}{V_1} = \frac{R_f}{R_1}\left(\frac{1}{R_S}\right)$$

$$= \frac{100\ \text{k}\Omega}{100\ \text{k}\Omega} \times \frac{1}{10\ \Omega} = \frac{1\ \text{mA}}{10\ \text{mV}}$$

Thus an input of 10 mV will result in a current through the meter of 1 mA. If the input is 5 mV, the current through the meter will be 0.5 mA, which is half-scale deflection. Changing R_f to 200 kΩ, for example, would result in a circuit scale factor of

$$\frac{I_0}{V_1} = \frac{200\ \text{k}\Omega}{100\ \text{k}\Omega} \times \frac{1}{10\ \Omega} = \frac{1\ \text{mA}}{5\ \text{mV}}$$

showing that the meter now reads 5 mV, full scale. It should be kept in mind that building such a millivoltmeter requires purchasing an op-amp circuit, a few resistors, and a meter movement. The ability to obtain a completely operating, tested op-amp unit makes the overall meter unit easy to set up.

Constant-Current Source

Figure 15.45 shows op-amp circuits that provide constant current. The circuit of Fig. 15.45a provides an output current fixed at 1 mA by the three resistors. The voltage divider of R_2 and R_3 sets the input to the noninverting input to +3 V, resulting

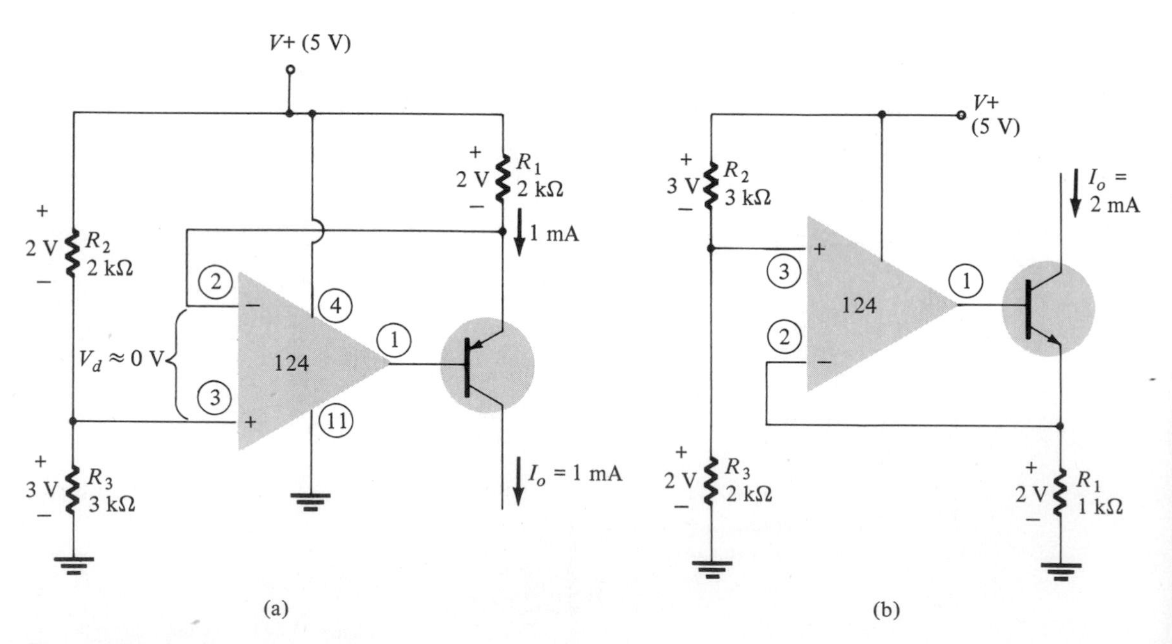

Figure 15.45 Constant-current circuits: (a) current source; (b) current sink.

in +2 V across R_1. The output current is then fixed at a value of $(2\text{ V})/R_1 = (2\text{ V})/2\text{ k}\Omega = 1\text{ mA}$. Low-current-source operation can be fixed over a range of values by selection of resistor R_1.

The circuit of Fig. 15.45b sets a fixed output sinking current (at 2 mA in this example). Setting the value of resistor R_1 fixes the output current at a desired current value.

Display Drivers

Figure 15.46 shows op-amp circuits that can be used to drive a lamp display or LED display. When the noninverting input to Fig. 15.46 goes above the inverting input, the output at terminal 1 goes to the positive saturation level (near +5 V in this example), and the lamp is driven on when transistor Q_1 goes on. As shown in the circuit, the output of the op-amp provides 30 mA of current to the base of transistor Q_1, which then drives 600 mA through a suitably selected transistor (with $\beta > 20$) capable of handling that amount of current.

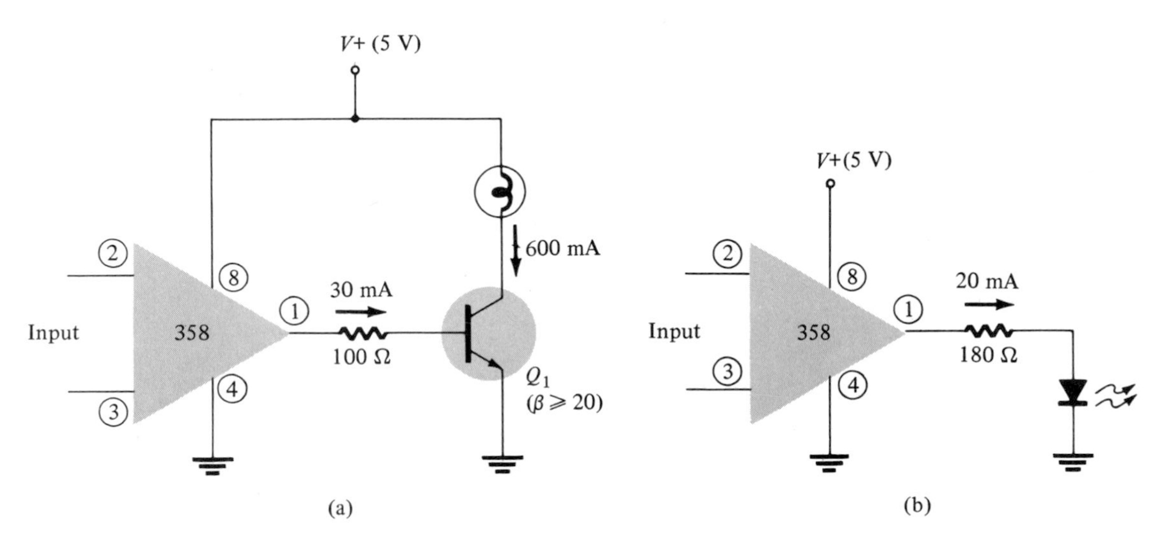

Figure 15.46 Display driver circuits: (a) lamp driver; (b) LED driver.

Figure 15.46b shows an op-amp circuit that can supply 20 mA to drive an LED display when the + input goes positive compared to the − input.

AC Millivoltmeter

As another example, an ac millivoltmeter circuit, is shown in Fig. 15.47. The resulting circuit transfer function is

$$\frac{I_0}{V_1} = \frac{R_f}{R_1}\left(\frac{1}{R_S}\right) = \frac{100\text{ k}\Omega}{100\text{ k}\Omega} \times \frac{1}{10\ \Omega} = \frac{1\text{ mA}}{10\text{ mV}}$$

which appears the same as the dc millivoltmeter, except that in this case it is for ac

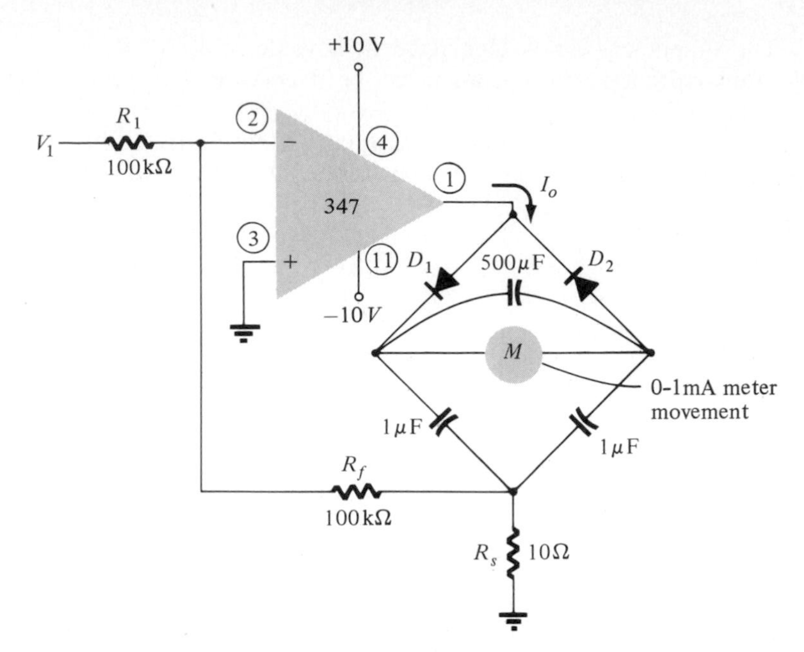

Figure 15.47 Ac millivoltmeter using op-amp.

signals. The meter indication provides a full-scale deflection, for an ac input voltage of 10 mV. An ac input signal of 10 mV will result in full-scale deflection, while an ac input of 5 mV will result in half-scale deflection, and the meter reading can be interpreted in millivolt units.

Active Filters

A popular application of op-amps is in building active filters. A filter circuit is built using passive components—resistors and capacitors. An active filter additionally uses an amplifier for voltage gain and signal isolation or buffering.

A filter that provides a constant output from dc up to a cutoff frequency, $f_{o,H}$ and then passes no signal is an ideal low-pass filter as described by the response plotted in Fig. 15.48a. A filter that passes signals only above a cutoff frequency is a high-pass filter, as idealized in Fig. 15.48b. When the filter circuit passes signals that are above one cutoff frequency and below a second cutoff frequency, it is a band-pass filter as idealized in Fig. 15.48c.

LOW-PASS FILTER

A first-order low-pass filter using a single resistor and capacitor as in Fig. 15.49a has a practical slope of 20 dB per decade, as shown in Fig. 15.49b (rather than the ideal response of Fig. 15.48a). The voltage gain below the cutoff frequency is constant at

$$A_v = 1 + \frac{R_{of}}{R_{o_1}} \tag{15.44}$$

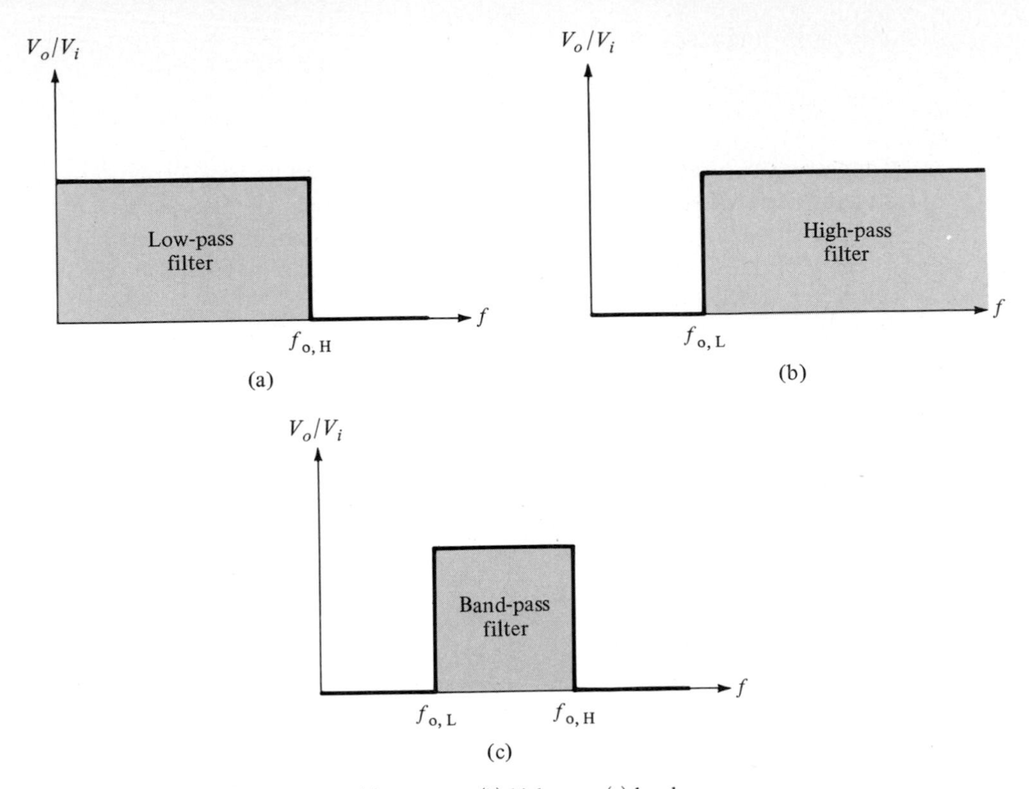

Figure 15.48 Ideal filter response: (a) low-pass; (b) high-pass; (c) bandpass.

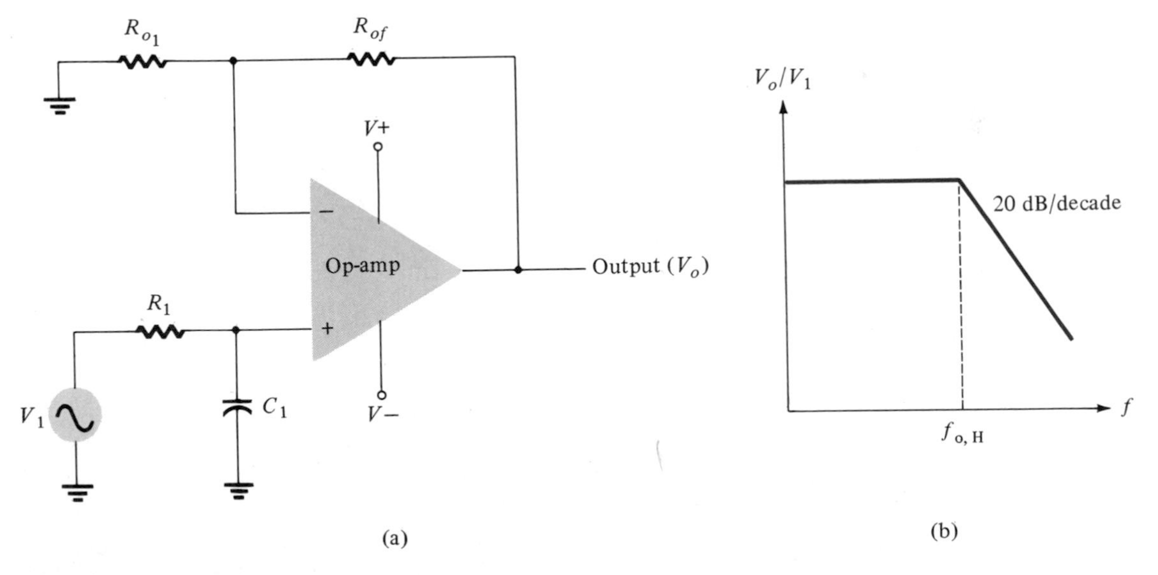

Figure 15.49 First-order low-pass active filter.

at a cutoff frequency of

$$f_{o,H} = \frac{1}{2\pi R_1 C_1} \tag{15.45}$$

Connecting two sections of filter as in Fig. 15.50 results in a second-order low-pass filter with cutoff at 40 dB/decade—closer to the ideal characteristic of Fig. 15.48a. The circuit voltage gain and cutoff frequency are the same for the second-order circuit as for the first-order filter circuit except that the filter response drops faster for a second-order filter circuit.

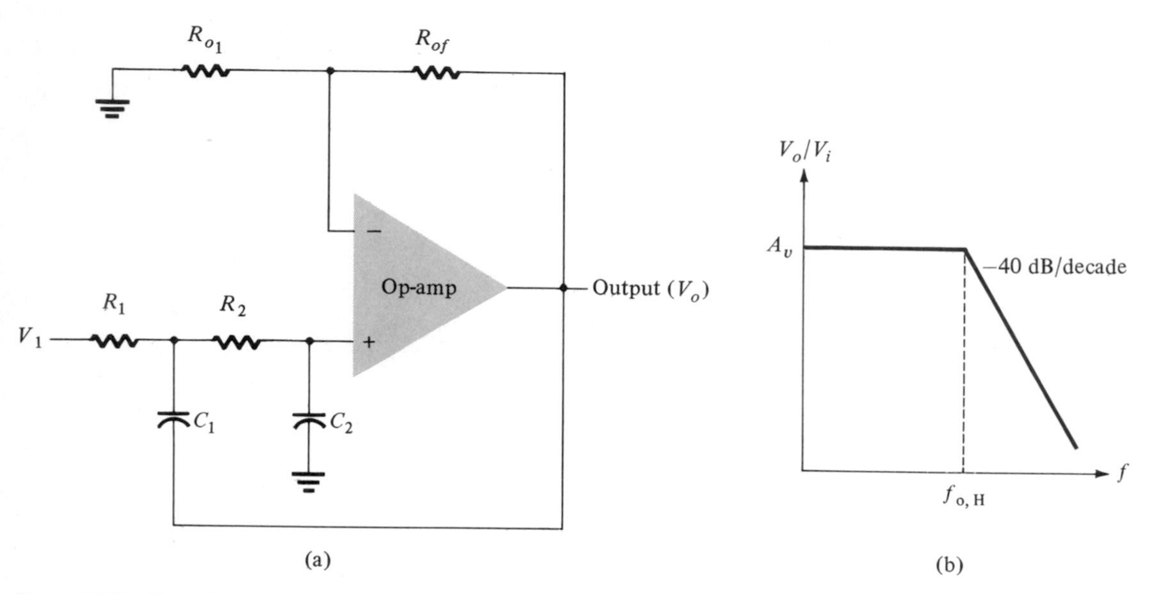

Figure 15.50 Second-order low-pass active filter.

EXAMPLE 15.10

Calculate the cutoff frequency of a first-order low-pass filter for $R_1 = 1.2\ \text{k}\Omega$ and $C_1 = 0.02\ \mu\text{F}$.

Solution:

$$f_{o,H} = \frac{1}{2\pi R_1 C_1} = \frac{1}{2\pi(1.2 \times 10^3)(0.02 \times 10^{-6})} = \mathbf{6.63\ kHz}$$

HIGH-PASS ACTIVE FILTER

First- and second-order high-pass active fiilters can be built as shown in Fig. 15.51. The amplifier gain is calculated using Eq. (15.44), with cutoff frequency

$$f_{o,L} = \frac{1}{2\pi R_1 C_1} \tag{15.46}$$

(with second-order filter $R_1 = R_2$, and $C_1 = C_2$ results in the same cutoff frequency).

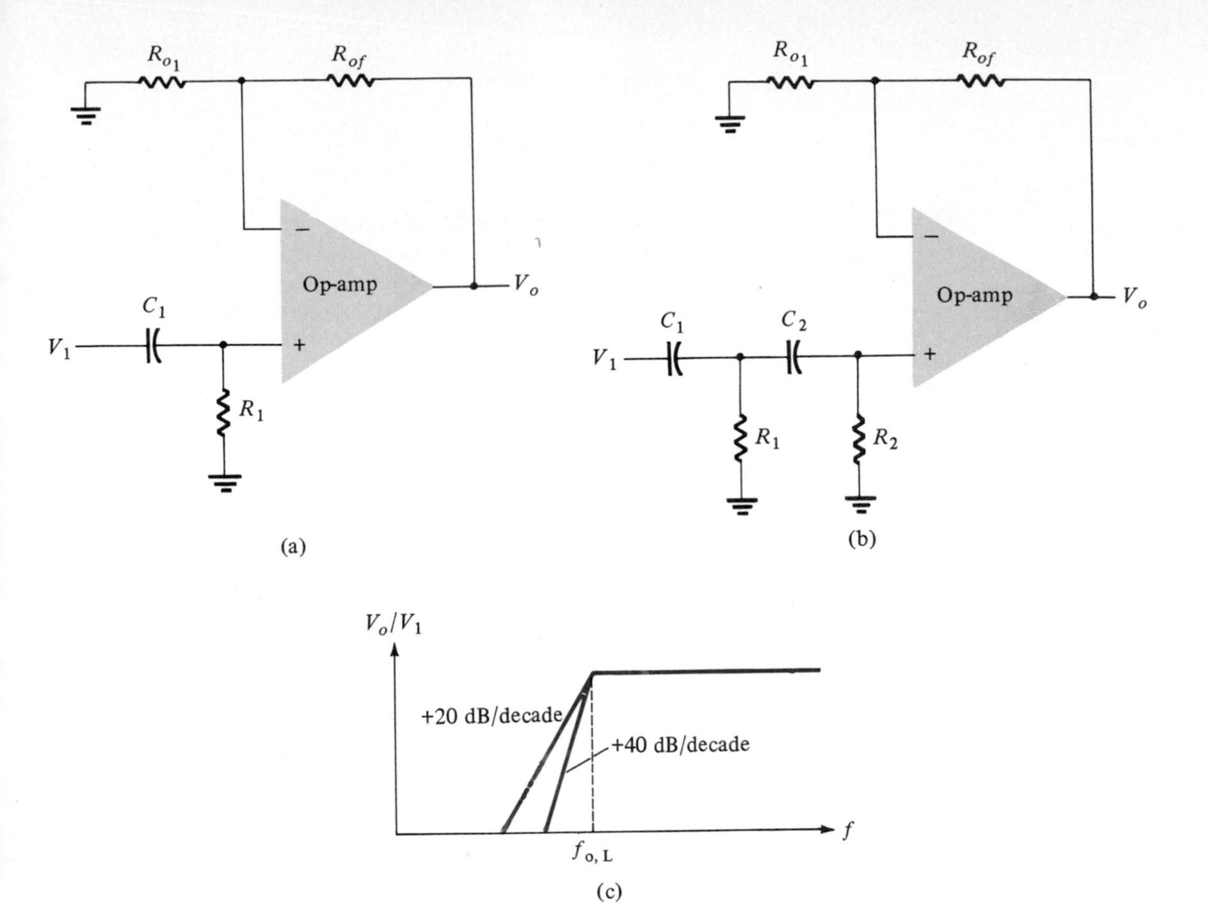

Figure 15.51 High-pass filter: (a) first order; (b) second order; (c) response plot.

EXAMPLE 15.11

Calculate the gain and cutoff frequency of a second-order high-pass filter as in Fig. 15.51b for $R_1 = R_2 = 2.1\ \text{k}\Omega$, $C_1 = C_2 = 0.05\ \mu\text{F}$, and $R_{o_1} = 10\ \text{k}\Omega$, $R_{of} = 50\ \text{k}\Omega$.

Solution:

$$A_v = 1 + \frac{R_{of}}{R_{o_1}} = 1 + \frac{50\ \text{k}\Omega}{10\ \text{k}\Omega} = 6$$

At cutoff frequency,

$$f_{o,L} = \frac{1}{2\pi R_1 C_1} = \frac{1}{2\pi(2.1 \times 10^3)(0.05 \times 10^{-6})} \cong \mathbf{1.5\ kHz}$$

BAND-PASS FILTER:

Figure 15.52 shows a band-pass filter using two stages, the first a high-pass filter and the second a low-pass, the combined operation being the desired band-pass response.

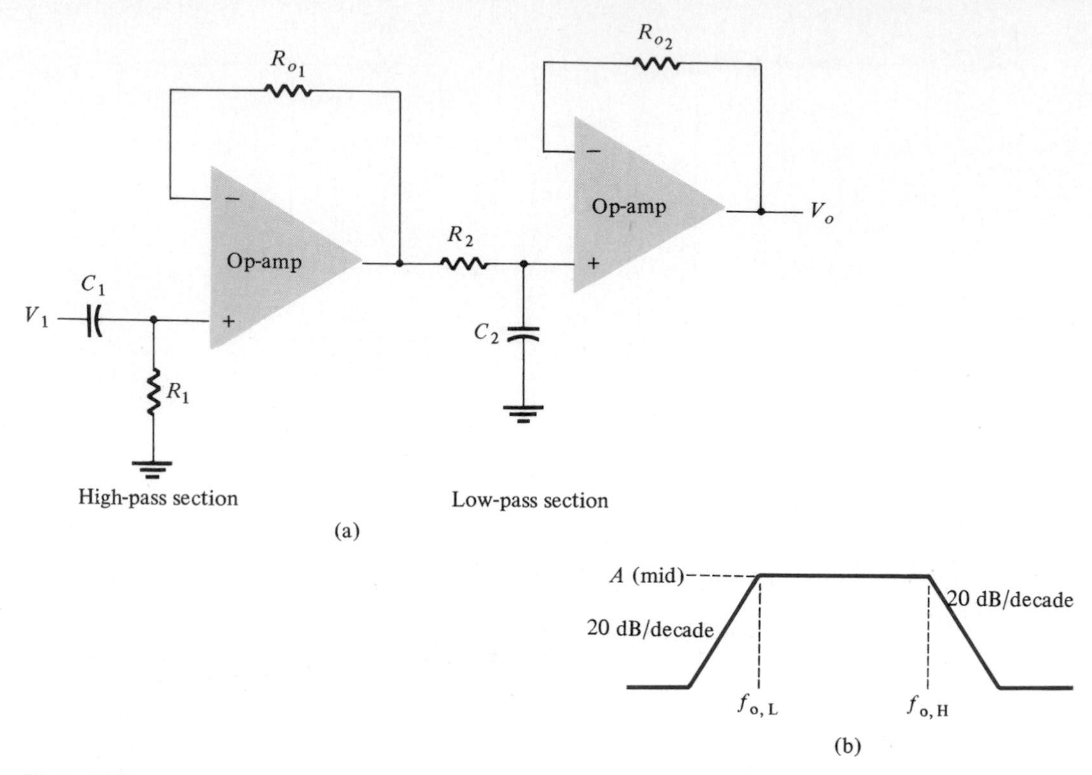

Figure 15.52 Bandpass active filter.

EXAMPLE 15.12

Calculate the cutoff frequencies of the band-pass filter circuit of Fig. 15.52 with $R_1 = R_2 = 10\ \text{k}\Omega$, $C_1 = 0.1\ \mu\text{F}$, and $C_2 = 0.002\ \mu\text{F}$.

Solution:

$$f_{o,L} = \frac{1}{2\pi R_1 C_1} = \frac{1}{2\pi(10 \times 10^3)(0.1 \times 10^{-6})} = \mathbf{159.15\ Hz}$$

$$f_{o,H} = \frac{1}{2\pi R_2 C_2} = \frac{1}{2\pi(10 \times 10^3)(0.002 \times 10^{-6})} = \mathbf{7.96\ kHz}$$

Voltage Buffer and Distributor

The circuit of Fig. 15.53 shows how a signal can be buffered and distributed to a number of outputs using op-amp circuits. The four op-amp circuits are all packaged in a single 347 quad op-amp IC unit. Supply voltages applied to pins 4 and 11 connect to all four op-amp circuits. The input stage, connected as a unity-gain amplifier provides output at pin 1, the same as the input signal. The other three stages are connected as unity-gain noninverting amplifiers and supply three identical signals to be distributed as three separate and isolated outputs.

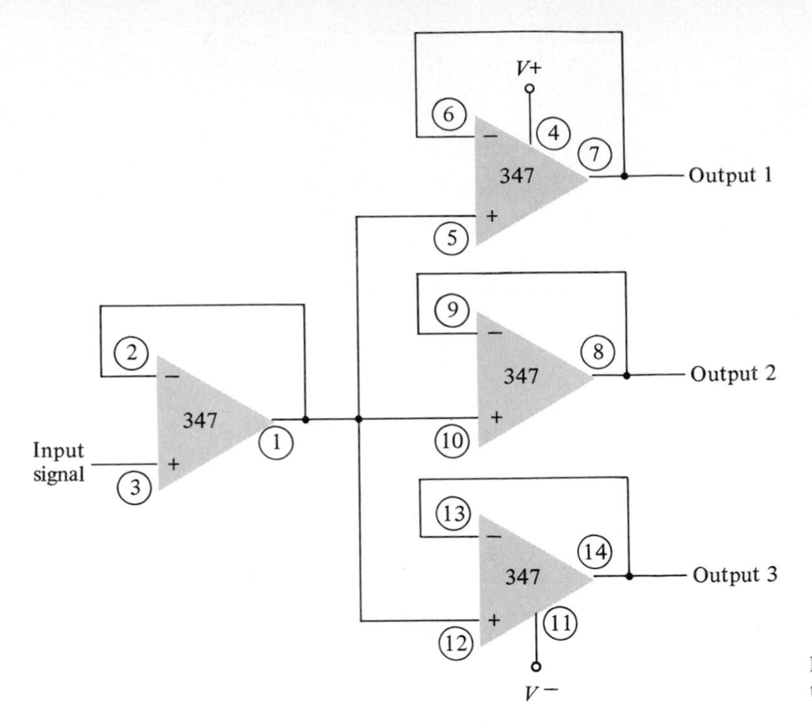

Figure 15.53 Use of buffer op-amp to provide signal distribution.

PROBLEMS

§ 15.1

1. Draw the output waveforms for the input signal and differential amplifier of Fig. 15.54. (Refer to details of Fig. 15.2.)

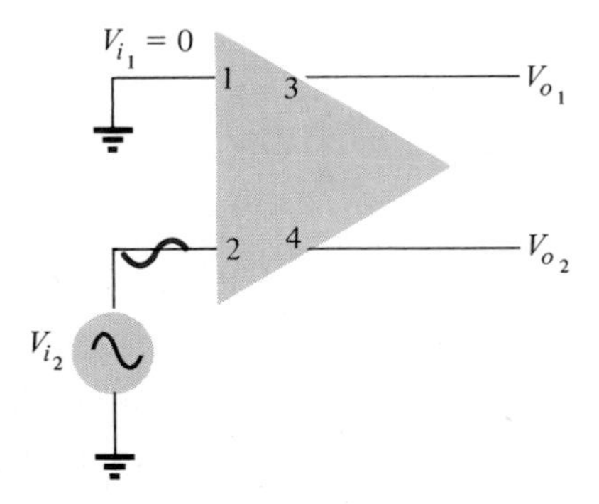

Figure 15.54 Differential amplifier and input waveform for Problems 1 and 2.

2. If the input signal V_{i_2} is an 8-mV, peak, 1000-Hz signal, sketch the output waveforms at V_{o_1} and V_{o_2} (the amplifier gain is 1000) in Fig. 15.54.

3. Draw the output waveform at V_{o_2} for the differential amplifier of Fig. 15.55.

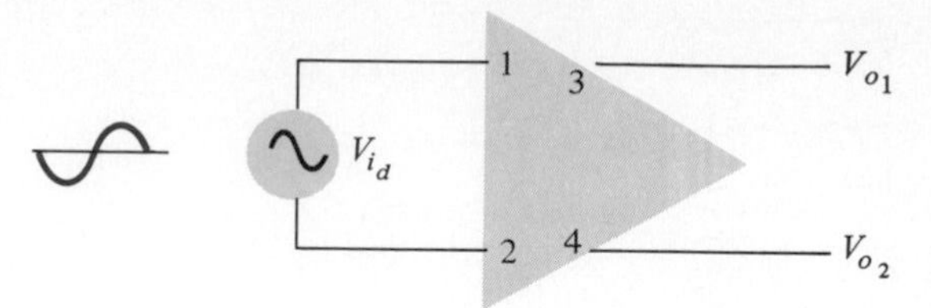

Figure 15.55 Differential amplifier and input for Problems 3 and 4.

4. Draw the output waveform at V_{o_1} from the circuit of Fig. 15.55 for an input of 5 mV peak if the amplifier gain is 1000.

§ 15.2

5. Determine the dc collector voltages in the circuit of Fig. 15.56.

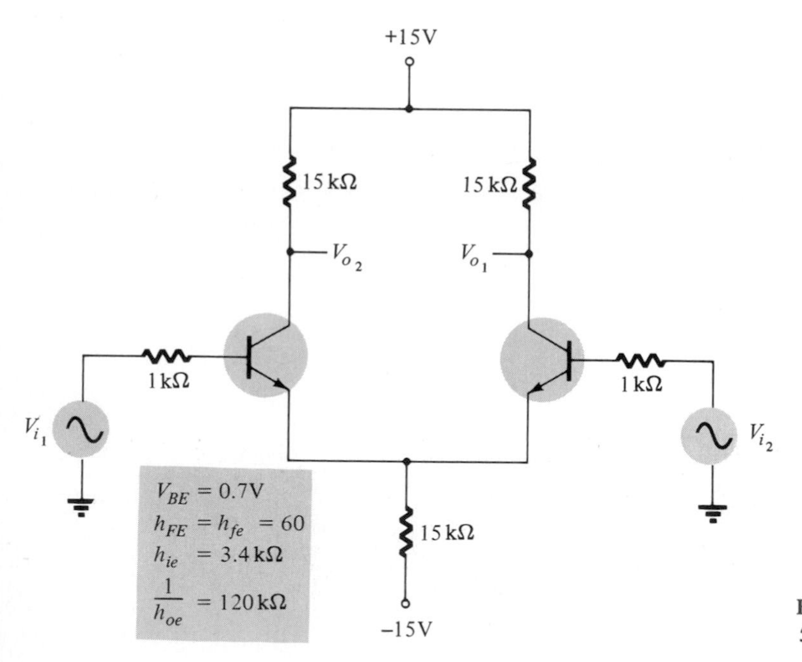

Figure 15.56 Circuit for Problems 5 through 7.

6. Calculate the ac voltage gain of the differential amplifier in Fig. 15.56.
7. Calculate the input and output resistance of the circuit in Fig. 15.56.
8. Calculate the value of the constant current, I_E, for the circuit of Fig. 15.57.
9. Calculate the dc collector voltage in the circuit of Fig. 15.57.
10. Calculate the constant current in a circuit such as that in Fig. 15.19 with values V_Z = 6.8 V, R_E = 1.8 kΩ, and R_1 = 5 kΩ.
11. Using the values of Problem 10, calculate the ac output resistance of circuit (h_{fe} = 150, $1/h_{oe}$ = 100 kΩ, h_{ie} = 1150 Ω).

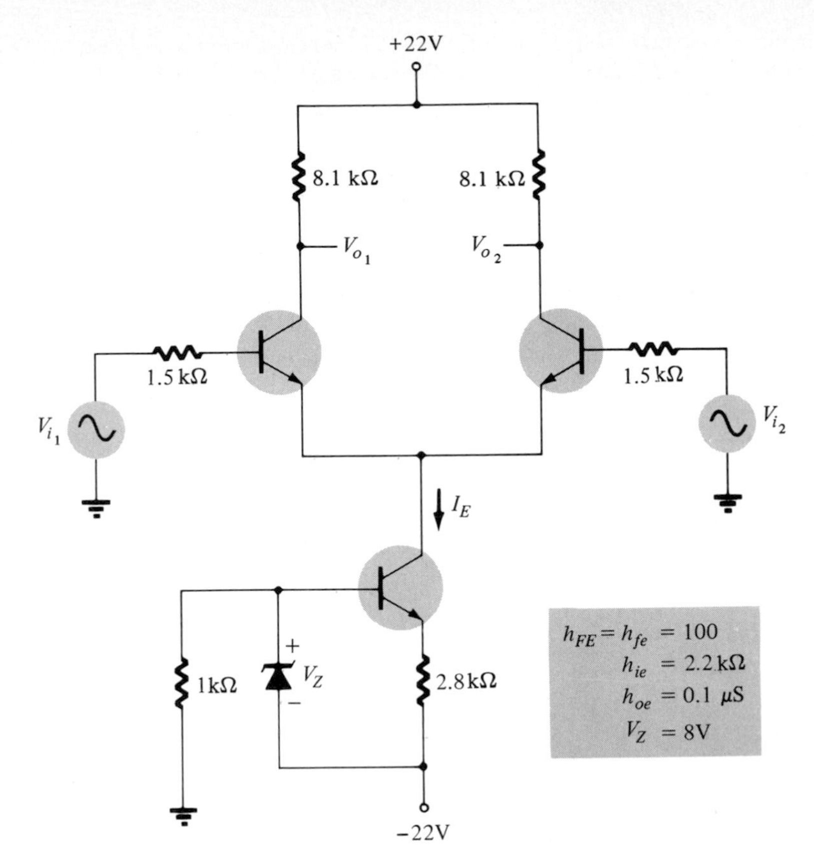

Figure 15.57 Circuit for Problems 8 and 9.

§ 15.4

12. An amplifier with differential voltage gain of 200 and common-mode rejection of 80 dB has what value common-mode gain?

13. What is the common-mode rejection ratio in dB of an amplifier having differential gain of 180 and common-mode gain of 0.5?

14. Calculate the differential- and common-mode gains for the circuit of Fig. 15.27a and the value of CMRR if $R_C = 4.3$ kΩ, $R_E = 11$ kΩ, and transistor $h_{fe} = 120$ and $h_{ie} = 1.5$ kΩ.

15. Calculate the differential- and common-mode gains and the value of CMRR for the circuit of Fig. 15.27b modified by $R_C = 4.3$ kΩ, $h_{fe} = 120$, $1/h_{oe} = 100$ kΩ, and $h_{ie} = 1.5$ kΩ.

16. Calculate the output voltage of a differential amplifier for inputs of $V_{i_1} = 0.5$ mV, $V_{i_2} = 0.45$ mV, $A_d = 450$, and CMRR $= 10^4$.

17. The input resistance of a differential amplifier is measured using a 25-kΩ resistor in series with an input voltage of 5 V. What is the value of R_i if the voltage into the amplifier is 1.5 V?

18. A differential amplifier has single-ended gain of $A_1 = 120$. When determining A_c the circuit measurements are $V_i = 2$ V and $V_o = 20$ mV. Calculate the value of CMRR in dB.

§ 15.5

19. What is meant by "virtual ground"?

§ 15.6

20. Calculate the output voltage of a noninverting op-amp circuit (such as in Fig. 15.31) for values of $V_1 = 4$ V, $R_f = 250$ kΩ, and $R_1 = 50$ kΩ.

21. Calculate the output voltage of a three-input summing amplifier (as in Fig. 15.34) for the following values: $R_1 = 200$ kΩ, $R_2 = 250$ kΩ, $R_3 = 500$ kΩ, $R_f = 1$ MΩ, $V_1 = -2$ V, $V_2 = +2$ V, and $V_3 = 1$ V.

22. Determine the output voltage of the circuits of Fig. 15.58.

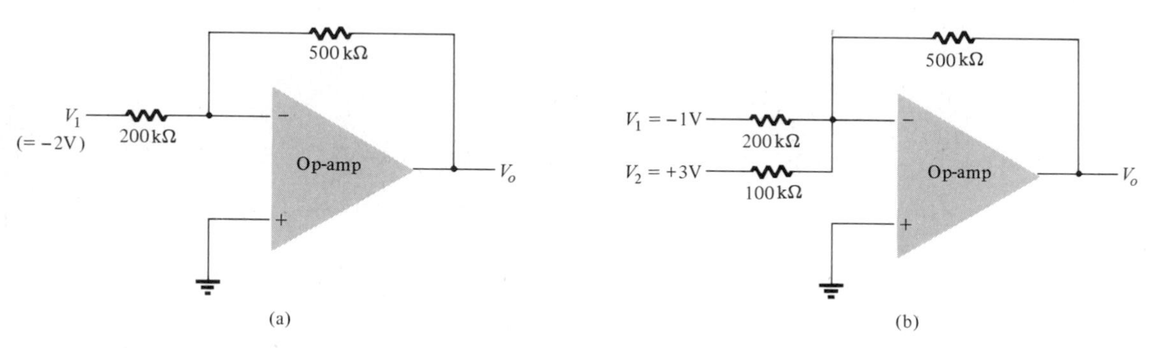

Figure 15.58 Circuits for Problem 22.

23. Repeat Problem 22 for the circuit of Fig. 15.59.

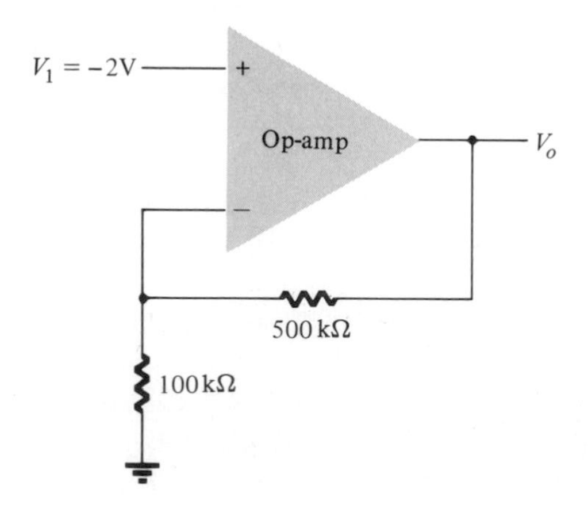

Figure 15.59 Circuit for Problem 23.

24. Draw the output waveform for the circuit and inputs of Fig. 15.60.

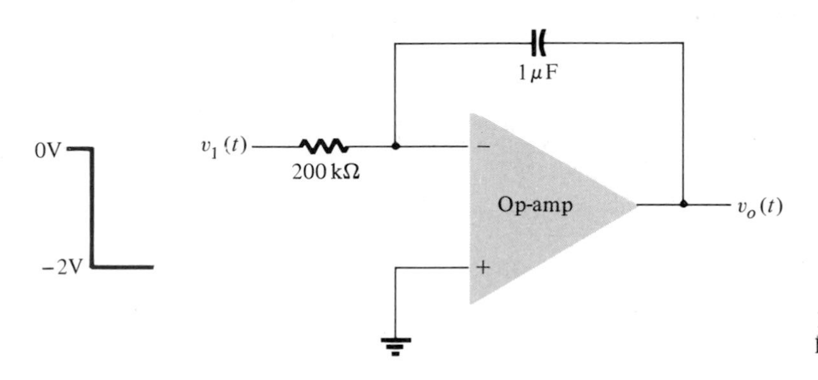

Figure 15.60 Circuit for Problem 24.

25. Repeat Problem 24 for Fig. 15.61.

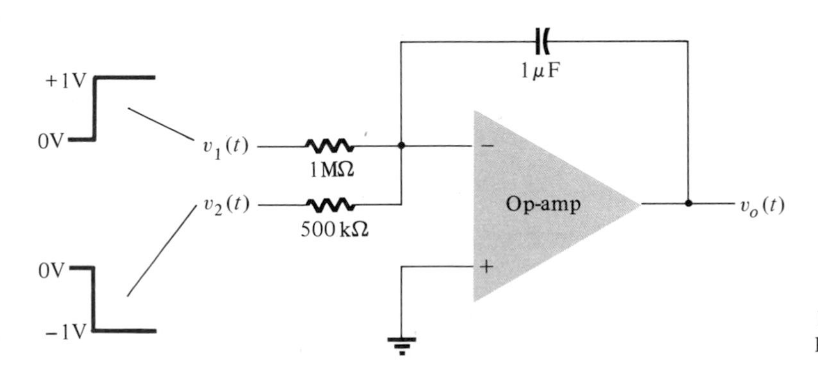

Figure 15.61 Circuit for Problem 25.

§ 15.7

26. What is the constant current output of the circuit of Fig. 15.45a with R_2 = 1.8 kΩ, R_3 = 3.3 kΩ, and R_1 = 1.8 $k\Omega$?

27. What is the value of output current, I_o, if R_2 = 2.7 kΩ, R_3 = 1.8 kΩ, and R_1 = 1.2 kΩ in the circuit of Fig. 15.45b?

28. Calculate the cutoff frequency of the filter-circuit in Fig. 15.62a

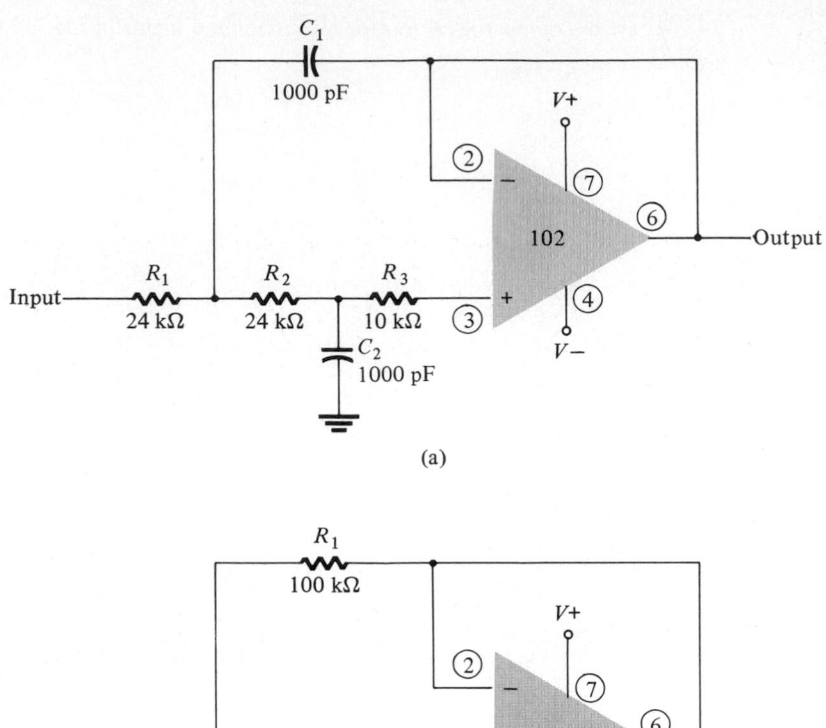

(b)

Figure 15.62 Active filter circuits for Problems 28 and 29.

29. Calculate the cutoff frequency of the filter circuit of Fig. 15.62b. Plot the ideal response curve of the filter.

16

Linear ICs: Regulators (Including Filters and Power Supplies)

16.1 INTRODUCTION

Voltage regulators are a popular group of linear ICs. A voltage regulator IC receives input of a fairly constant dc voltage and supplies as output a somewhat lower value of dc voltage, which the regulator maintains fixed or regulated over a wide range of load current, or input voltage, variation. Starting with an ac voltage supply, a steady dc voltage can be developed by rectifying the ac voltage, then filtering to a dc level, and finally regulating with an IC voltage-regulator circuit.

The present chapter introduces the operation of filter capacitors and the overall operation of converting an ac voltage into a dc voltage using transformer, rectifier, and filter. Refer back to Chapter 2 for coverage of diode rectifier circuits.

Voltage regulator ICs that provide a fixed output voltage are available in a range of output voltage values. Regulator ICs are selected to operate with either positive or negative voltages. Voltage regulators are also available to provide output at any set voltage over a range of values set by external resistor values.

A block diagram containing the parts of a typical power supply and the voltages at various points in the unit is shown in Fig. 16.1. The ac voltage, typically 120 V rms, is connected to a transformer which steps that voltage up, or more typically, down to the level for the desired dc output. A diode rectifier then provides a half-wave—or, more typically, full-wave—rectified voltage which is applied to a filter to smooth the varying signal. A simple capacitor filter is often sufficient to provide this smoothing action. The resulting dc voltage with some ripple or ac voltage variation is then provided as input to an IC regulator that provides as output a well-defined dc voltage level with extremely low ripple voltage over a range of load.

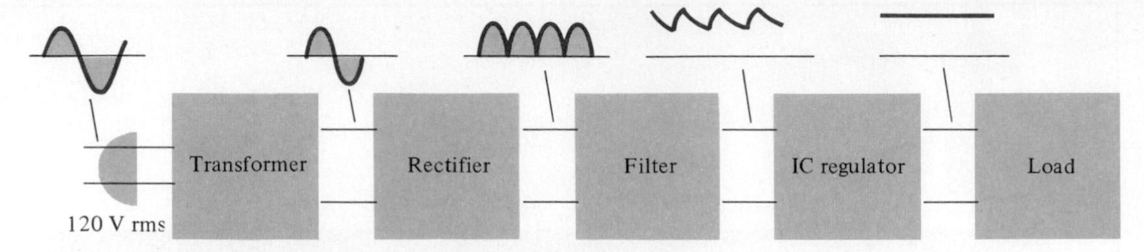

Figure 16.1 Block diagram showing parts of a power supply.

16.2 GENERAL FILTER CONSIDERATIONS

A rectifier circuit is necessary to convert a signal having zero average value to one that has a nonzero average. However, the resulting pulsating dc signal is not pure dc or even a good representation of it. Of course, for a circuit such as a battery charger the pulsating nature of the signal is no great detriment as long as the dc level provided will result in charging of the battery. On the other hand, for voltage supply circuits for a tape recorder or radio the pulsating dc will result in a 60-(or 120-)Hz signal appearing in the output, thereby making the operation of the overall circuit poor. For these applications, as well as for many more, the output dc developed will have to be much "smoother" than that of the pulsating dc obtained directly from half-wave or full-wave rectifier circuits.

Filter Voltage Regulation and Ripple Voltage

Before going into the details of the filter circuit it would be appropriate to consider the usual method of rating the circuits so that we are able to compare a circuit's effectiveness as a filter. Figure 16.2 shows a typical filter output voltage, which will be used to define some of the signal factors. The filtered output voltage of Fig. 16.2 has a dc value and some ac variation (*ripple*). Although a battery has essentially a constant or dc output voltage, the dc voltage derived from an ac source signal by rectifying and filtering will have some variation (ripple). The smaller the ac variation *with respect to* the dc level the better the filter circuit operation.

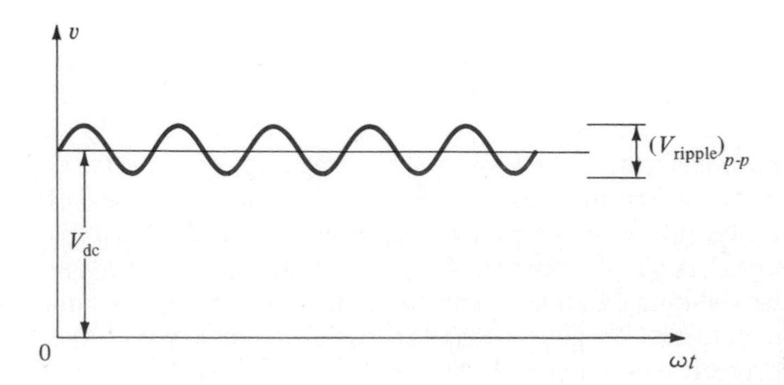

Figure 16.2 Filter voltage waveform showing dc and ripple voltages.

Consider measuring the output voltage of the filter circuit using a dc voltmeter and an ac (rms) voltmeter. The dc voltmeter will read only the average or dc level of the output voltage. The ac (rms) meter will read only the rms value of the ac component of the output voltage (assuming the signal is coupled to the meter through a capacitor to block out the dc level).

Definition: Ripple:

$$r = \text{ripple} = \frac{\text{ripple voltage (rms)}}{\text{dc voltage}} = \frac{V_r(\text{rms})}{V_{dc}} \times 100\% \qquad (16.1)$$

EXAMPLE 16.1

Using a dc and ac voltmeter to measure the output signal from a filter circuit, a dc voltage of 25 V and an ac ripple voltage of 1.5 V rms are obtained. Calculate the ripple of the filter output.

Solution:

$$r = \frac{V_r(\text{rms})}{V_{dc}} \times 100\% = \frac{1.5\text{ V}}{25\text{ V}} 100 = \mathbf{6\%}$$

VOLTAGE REGULATION

Another factor of importance in a voltage supply is the amount of change in the output dc voltage over the range of the circuit operation. The voltage provided at the output at no-load (no current drawn from the supply) is reduced when load current is drawn from the supply. How much this voltage changes with respect to either the loaded or unloaded voltage value is of considerable interest to anyone using the supply. This voltage change is described by a factor called *voltage regulation.*

Definition: Voltage regulation:

$$\text{voltage regulation} = \frac{\text{voltage at no-load} - \text{voltage at full-load}}{\text{voltage at full-load}}$$

$$\text{V.R.} = \frac{V_{NL} - V_{FL}}{V_{FL}} \times 100\% \qquad (16.2)$$

EXAMPLE 16.2

A dc voltage supply provides 60 V when the output is unloaded. When full-load current is drawn from the supply, the output voltage drops to 56 V. Calculate the value of voltage regulation.

Solution:

$$\text{V.R.} = \frac{V_{NL} - V_{FL}}{V_{FL}} \times 100\% = \frac{60\text{ V} - 56\text{ V}}{56\text{ V}} \times 100\% = \mathbf{7.14\%}$$

If the value of full-load voltage is the same as the no-load voltage, the *V.R.* calculated is 0%, which is the best to expect. This value means that the supply is a true voltage source for which the output voltage is independent of current drawn from the supply. The output voltage from most supplies decreases as the amount of current drawn from the voltage supply is increased. The smaller the voltage decreases, the smaller the percent of V.R. and the better the operation of the voltage supply circuit.

RIPPLE FACTOR OF RECTIFIED SIGNAL

Although the rectified voltage is not a filtered voltage, it nevertheless contains a dc component and a ripple component. We can calculate these values of dc voltage and ripple voltage (rms) and from them obtain the ripple factor for the half-wave and full-wave rectified voltages. The calculations will show that the full-wave rectified signal has less percent of ripple and is therefore a better rectified signal than the half-wave rectified signal, if lowest percent of ripple is desired. The percent of ripple is not always the most important concern. If circuit complexity or cost considerations are important (and the percent of ripple is secondary), then a half-wave rectifier may be satisfactory. Also, if the filtered output supplies only a small amount of current to the load and the filtering circuit is not critical, then a half-wave rectified signal may be acceptable. On the other hand, when the supply must have as low a ripple as possible, it is best to start with a full-wave-rectified signal since it has a smaller ripple factor, as will now be shown.

For a half-wave-rectified signal the output dc voltage is $V_{dc} = 0.318\,V_m$. The rms value of the ac component of output signal can be calculated (see Appendix B), and is V_r (rms) $= 0.385\,V_m$. Calculating the percent of ripple, we obtain

$$r = \frac{V_r(\text{rms})}{V_{dc}}(100) = \frac{0.385\,V_m}{0.318\,V_m}(100) = 1.21(100) = 121\% \qquad \text{(half-wave)} \qquad (16.3)$$

For the full-wave rectifier the value of V_{dc} is $V_{dc} = 0.636\,V_m$. From the results obtained in Appendix B the ripple voltage of a full-wave rectified signal is V_r (rms) $=$ $0.308\,V_m$. Calculating the percent of ripple yields

$$r = \frac{V_r(\text{rms})}{V_{dc}}(100) = \frac{0.308\,V_m}{0.636\,V_m}(100) = 48\% \qquad \text{(full-wave)} \qquad (16.4)$$

The amount of ripple factor of the full-wave-rectified signal is about 2.5 times smaller than that of the half-wave-rectified signal and provides a better filtered signal. Note that these values of ripple factor are absolute values and do not depend at all on the peak voltage. If the peak voltage is made larger, the dc value of the output increases but then so does the ripple voltage. The two increase in the same proportion so that the ripple factor stays the same.

16.3 SIMPLE-CAPACITOR FILTER

A popular filter circuit is the simple-capacitor filter circuit shown in Fig. 16.3. The capacitor is connected across the rectifier output and the dc output voltage is available across the capacitor. Figure 16.4a shows the rectifier output voltage of a full-wave

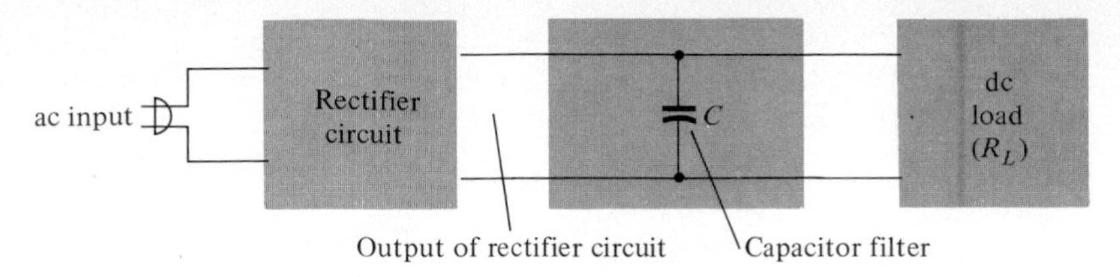

Figure 16.3 Simple capacitor filter.

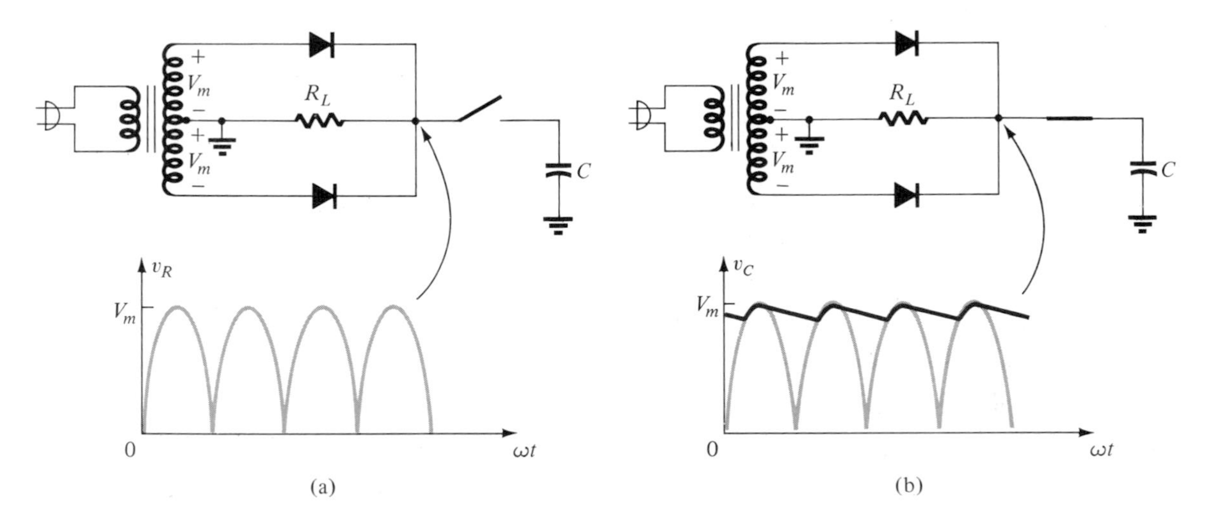

Figure 16.4 Capacitor filter operation: (a) full-wave rectifier voltage; (b) filtered output voltage.

rectifier circuit before the signal is filtered. Figure 16.4b shows the resulting waveform after the capacitor is connected across the rectifier output. As shown this filtered voltage has a dc level with some ripple voltage riding on it.

Figure 16.5a shows a full-wave rectifier and the output waveform obtained from the circuit when connected to an output load. If no load were connected to the filter, the output waveform would ideally be a constant dc level equal in value to the peak voltage (V_m) from the rectifier circuit. However, the purpose of obtaining a dc voltage is to provide this voltage for use by other electronic circuits, which then constitute a load on the voltage supply. Since there will always be some load on the filter, we must consider this practical case in our discussion. For the full-wave-rectified signal indicated in Fig. 16.5b there are two intervals of time indicated. T_1 is the time during which a diode of the full-wave rectifier conducts and charges the capacitor up to the peak rectifier output voltage (V_m). T_2 is the time during which the rectifier voltage drops below the peak voltage, and the capacitor discharges through the load.

If the capacitor were to discharge only slightly (due to a light load), the average voltage would be very close to the optimum value of V_m. The amount of ripple voltage would also be small for a light load. This shows that the capacitor filter circuit provides a large dc voltage with little ripple for light loads (and a smaller dc voltage with larger

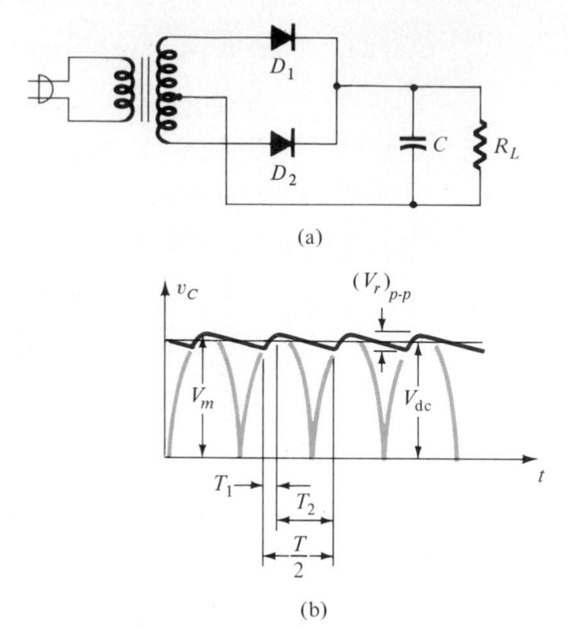

Figure 16.5 Capacitor filter: (a) capacitor filter circuit; (b) output voltage waveform.

ripple for heavy loads). To appreciate these quantities better we must further examine the output waveform and determine some relations between the input signal to be rectified, the capacitor value, the resistor (load) value, the ripple factor, and the regulation of the circuit.

Figure 16.6 shows the output waveform approximated by straight line charge and discharge. This is reasonable since the analysis with the nonlinear charge and discharge that actually takes place is complex to analyze and because the results obtained will yield values that agree well with actual measurements made on circuits. The waveform of Fig. 16.6 shows the approximate output voltage waveform for a full-wave-rectified signal. From an analysis of this voltage waveform the following re-

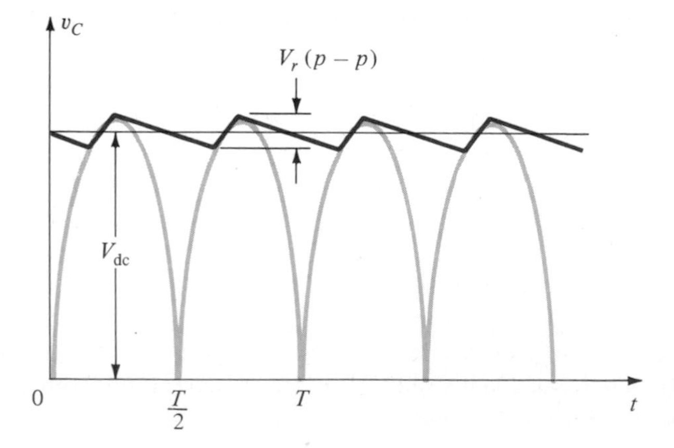

Figure 16.6 Approximate output voltage of capacitor filter circuit.

lations can be obtained:

$$\boxed{V_{dc} = V_m - \frac{V_r(\text{p-p})}{2}} \quad \text{(half-wave} \qquad (16.5)$$

and

$$\boxed{V_r(\text{rms}) = \frac{V_r(\text{p-p})}{2\sqrt{3}}} \quad \text{full-wave)} \qquad (16.6)$$

These relations, however, are only in terms of the waveform voltages and we must further relate them to the different components in the circuit. Since the form of the ripple waveform for half-wave is the same as for full-wave, Eqs. (16.5) and (16.6) apply to both rectifier-filter circuits.

Ripple Voltage, V_r (rms)

Appendix B provides the details for determining the value of the ripple voltage in terms of the other circuit parameters. The result obtained for V_r (rms) is the following:

$$V_r(\text{rms}) \cong \frac{I_{dc}}{4\sqrt{3}fC} \times \frac{V_{dc}}{V_m} \qquad \text{(full-wave)} \qquad (16.7a)$$

where f is the frequency of the sinusoidal ac power supply voltage (usually 60 Hz), I_{dc} is the average current drawn from the filter by the load, and C is the filter capacitor value.

Another simplifying approximation that can be made is to assume that when used typically for light loads[1] the value of V_{dc} is only slightly less than V_m so that $V_{dc} \cong V_m$, and the equation can be written as

$$V_r(\text{rms}) \cong \frac{I_{dc}}{4\sqrt{3}fC} \qquad \text{(full-wave, light load)} \qquad (16.7b)$$

Finally, we can include the typical value of line frequency (f = 60 Hz) and the other constants into the simpler equation

$$\boxed{V_r(\text{rms}) = \frac{2.4I_{dc}}{C} = \frac{2.4V_{dc}}{R_LC}} \qquad \text{(full-wave, light load)} \qquad (16.7c)$$

where I_{dc} is in milliamperes, C is in microfarads, and R_L is in kilohms.

EXAMPLE 16.3

Calculate the ripple voltage of a full-wave rectifier with a 100 μF filter capacitor connected to a load of 50 mA.

---

[1]Appendix B shows the relation of V_{dc} and V_m based on the amount of ripple. From Fig. B.3 we see that at ripple factors less than 6.5%, V_{dc} is within 10% of V_m. We can therefore define a *light load* as one resulting in a ripple less than 6.5%.

Solution:

Using Eq. (16.7c), we get

$$V_r(\text{rms}) = \frac{2.4(50)}{100} = \mathbf{1.2\ V}$$

DC Voltage, V_{dc}

Using Eqs. (16.5), (16.6), and (16.7a), we see that the dc voltage of the filter is

$$V_{dc} = V_m - \frac{V_r(\text{p-p})}{2} = V_m - \frac{I_{dc}}{4fC} \times \frac{V_{dc}}{V_m} \qquad \text{(full-wave)} \qquad (16.8a)$$

Again, using the simplifying assumption that V_{dc} is about the same as V_m for light loads, we get an approximate value of V_{dc} (which is less than V_m) of

$$V_{dc} = V_m - \frac{I_{dc}}{4fC} \qquad \text{(full-wave, light load)} \qquad (16.8b)$$

which can be written (using $f = 60$ Hz)

$$\boxed{V_{dc} = V_m - \frac{4.17 I_{dc}}{C}} \qquad \text{(full-wave, light load)} \qquad (16.8c)$$

where V_m is the peak rectified voltage, in volts, I_{dc} the load current in milliamperes, and C the filter capacitor in microfarads.

EXAMPLE 16.4

If the peak rectified voltage for the filter circuit of Example 16.3 is 30 V, calculate the filter dc voltage.

Solution:

Using Eq. (16.8c), we get

$$V_{dc} = V_m - \frac{4.17 I_{dc}}{C} = 30 - \frac{4.17(50)}{100} = \mathbf{27.9\ V}$$

The value of dc voltage is less than the peak rectified voltage. Note also from Eq. (16.8c) that the larger the value of average current drawn from the filter the less the value of output dc voltage, and the larger the value of the filter capacitor the closer the output dc voltage approaches the peak value of V_m.

Filter-Capacitor Ripple

Using the definition of ripple [Eq. (16.1)] and the equation for ripple voltage [Eq. 16.7c)], we obtain the expression for the ripple factor of a full-wave capacitor filter

$$\boxed{r = \frac{V_r(\text{rms})}{V_{dc}} \times 100\% \cong \frac{2.4 I_{dc}}{C V_{dc}} \times 100\%} \qquad \text{(full-wave, light load)} \qquad (16.9a)$$

Since V_{dc} and I_{dc} relate to the filter load R_L, we can also express the ripple as

$$\boxed{r = \frac{2.4}{R_L C} \times 100\%} \qquad \text{(full-wave, light load)} \qquad (16.9b)$$

where I_{dc} is in milliamperes, C is in microfarads, V_{dc} is in volts, and R_L is in kilohms.

This ripple factor is seen to vary directly with the load current (larger load current, larger ripple factor), and inversely with the capacitor size. This agrees with the previous discussion of the filter circuit operation.

EXAMPLE 16.5

A load current of 50 mA is drawn from a capacitor filter circuit ($C = 100\ \mu F$). If the peak rectified voltage is 30 V, calculate r.

Solution:

Using the results of Examples 16.3 and 16.4 in Eq. (16.9a), we get

$$r = \frac{2.4 I_{dc}}{C V_{dc}} \times 100\% = \frac{2.4(50)}{(100)(27.9)} \times 100\% = \mathbf{4.3\%}$$

From the basic definition of r we could also calculate

$$r = \frac{V_r(\text{rms})}{V_{dc}} \times 100\% = \frac{1.2}{27.9} \times 100\% = 4.3\%$$

Diode Conduction Period and Peak Diode Current

From the previous discussion it should be clear that larger values of capacitance provide less ripple and higher average voltages, thereby providing better filter action. From this one may conclude that to improve the performance of a capacitor filter it is only necessary to increase the size of the filter capacitor. However, the capacitor also affects the peak current through the rectifying diode and, as will now be shown, the larger the value of capacitance used the larger the peak current through the rectifying diode.

Referring back to the operation of the rectifier and capacitor filter circuit, we see that there are two periods of operation to consider. After the capacitor is charged to the peak rectified voltage (see Fig. 16.5b), a period of diode nonconduction elapses (time T_2) while the output voltage discharges through the load. After T_2 the input rectified voltage becomes greater than the capacitor voltage and for a time, T_1, the capacitor will charge back up to the peak rectified voltage. The average current supplied to the capacitor and load during this charge period must equal the average current drawn from the capacitor during the discharge period. Figure 16.7 shows the diode current waveform for half-wave-rectifier operation. Notice that the diode conducts for only a short period of the cycle. In fact, it should be seen that the larger the capacitor, the less the amount of voltage decay and the shorter the interval during which charging takes place. In this shorter charging interval the diode will have to pass the same amount of *average current*, and can do so only by passing larger peak

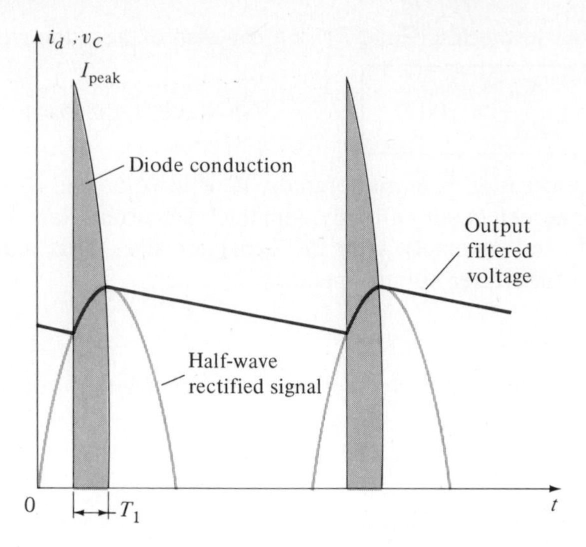

Figure 16.7 Diode conduction during charging part of cycle.

current. Figure 16.8 shows the output current and voltage waveforms for small and large capacitor values. The important factor to note is the increase in peak current through the diode for the larger values of capacitance. Since the average current drawn from the supply must equal the average of the current through the diode during the charging period, the following relation can be derived from Fig. 16.8:[2]

$$I_{dc} = \frac{T_1}{T} I_{peak} \qquad (16.10a)$$

from which we obtain

$$\boxed{I_{peak} = \frac{T}{T_1} I_{dc}} \qquad (16.10b)$$

where T_1 = diode conduction time
$T = 1/f = \frac{1}{60}$ for usual line 60-Hz voltage
I_{dc} = average current drawn from filter circuit
I_{peak} = peak current through the conducting diode

16.4 *RC* FILTER

It is possible to further reduce the amount of ripple across a filter capacitor while reducing the dc voltage by using an additional *RC* filter section as shown in Fig. 16.9.

---

[2]Assuming a rectangular-shaped pulse of duration T_1, peak value, I_{peak}, and a period of T, the area under the pulse divided by the period gives the average value, I_{dc}:

$$\frac{1}{T}(I_{peak}T_1) = I_{dc}$$

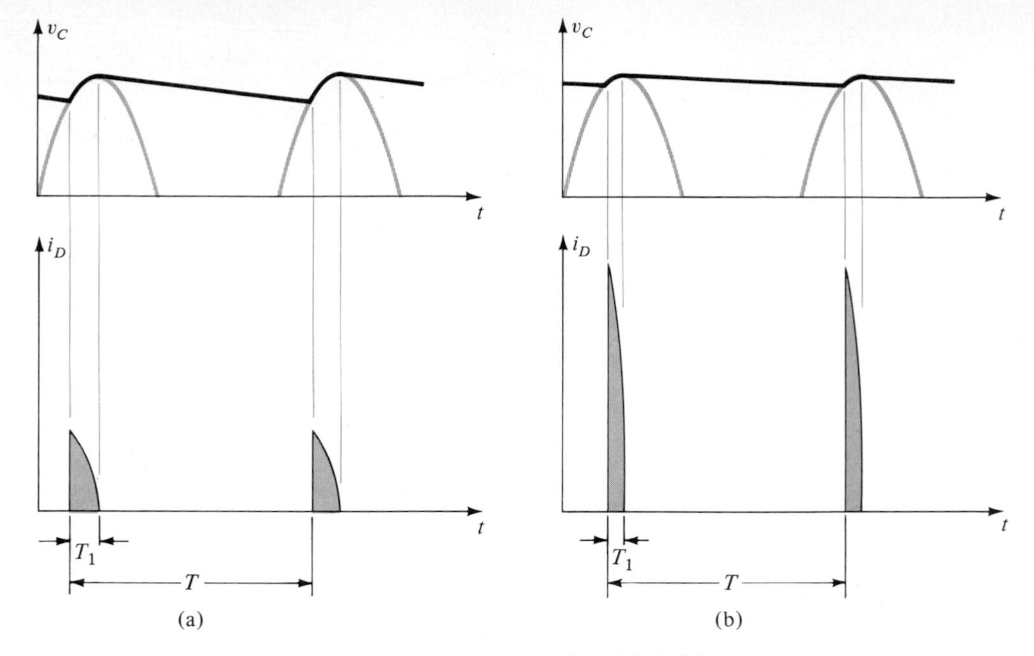

Figure 16.8 Output voltage and diode current waveforms: (a) small C; (b) large C.

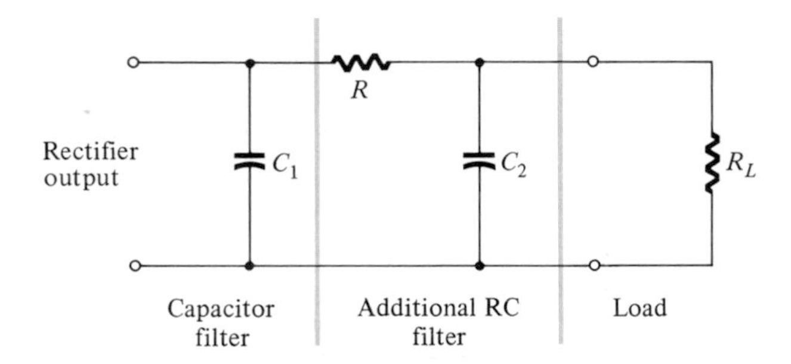

Figure 16.9 *RC* filter stage.

The purpose of the added network is to pass as much of the dc component of the voltage developed across the first filter capacitor C_1 and to attenuate as much of the ac component of the ripple voltage developed across C_1 as possible. This action would reduce the amount of ripple in relation to the dc level, providing better filter operation than for the simple-capacitor filter. There is a price to pay for this improvement, as will be shown; this includes a lower dc output voltage due to the dc voltage drop across the resistor and the cost of the two additional components in the circuit.

Figure 16.10 shows the rectifier filter circuit for full-wave operation. Since the rectifier feeds directly into a capacitor, the peak currents through the diodes are many times the average current drawn from the supply. The voltage developed across capacitor C_1 is then further filtered by the resistor-capacitor section (R, C_2) providing an output voltage having less percent of ripple than that across C_1. The load, represented by resistor R_L, draws dc current through resistor R with an output dc voltage

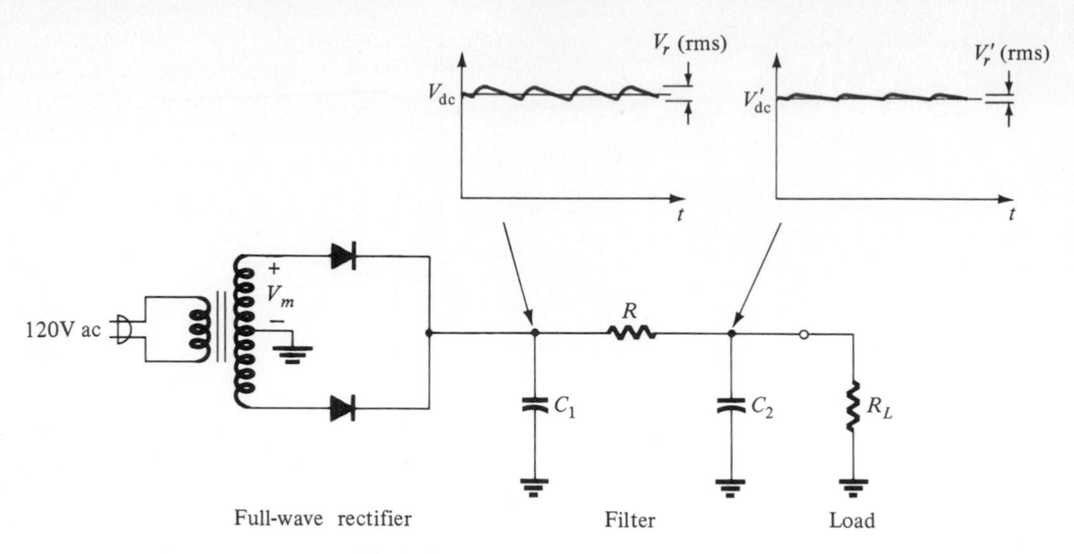

Figure 16.10 Full-wave rectifier and *RC* filter circuit.

across the load being somewhat less than that across C_1 due to the voltage drop across R. This filter circuit, like the simple-capacitor filter circuit, provides best operation at light loads, with considerably poorer voltage regulation and higher percent of ripple at heavy loads.

The analysis of the resulting ac and dc voltages at the output of the filter from that obtained across capacitor C_1 can be carried out by using superposition. We can separately consider the *RC* circuit acting on the dc level of the voltage across C_1 and then the *RC* circuit action on the ac (ripple) portion of the signal developed across C_1. The resulting values can then be used to calculate the overall circuit voltage regulation and ripple.

DC Operation of *RC* Filter Section

Figure 16.11a shows the equivalent circuit to use when considering the dc voltage and current in the filter and load. The two filter capacitors are open circuit for dc and

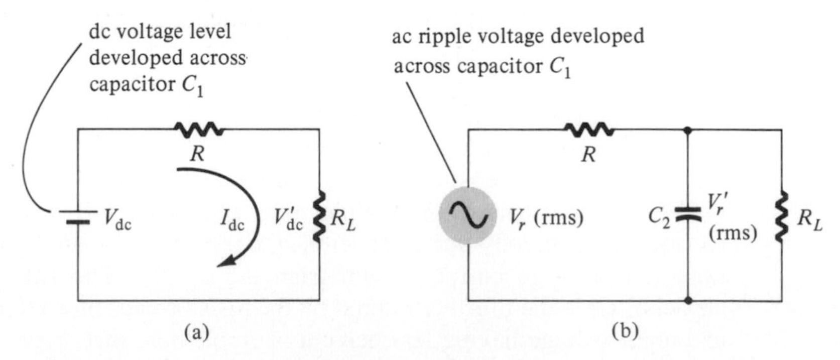

Figure 16.11 Dc and ac equivalent circuits of *RC* filter: (a) dc equivalent circuit; (b) ac equivalent circuit.

are thus removed from consideration at this time. Calculation of the dc voltage across filter capacitor C_1 was essentially covered in Section 16.3 and the treatment of the additional *RC* filter stage will proceed from there. Knowing the dc voltage across the first filter capacitor (C_1), we can calculate the dc voltage at the output of the additional *RC* filter section. From Fig. 16.11a we see that the voltage, V_{dc}, across capacitor C_1 is attenuated by a resistor-divider network of R and R_L (the equivalent load resistance); the resulting dc voltage across the load being V'_{dc}:

$$\boxed{V'_{dc} = \frac{R_L}{R + R_L} V_{dc}} \tag{16.11}$$

EXAMPLE 16.6

The addition of an *RC* filter section with $R = 120\ \Omega$, reduces the dc voltage across the initial filter capacitor from 60 V (V_{dc}). If the load resistance is 1 kΩ, calculate the value of the output dc voltage (V'_{dc}) from the filter circuit.

Solution:

Using Eq. (16.1), we get

$$V'_{dc} = \frac{R_L}{R + R_L} V_{dc} = \frac{1000}{120 + 1000} \times 60 = \mathbf{53.6\ V}$$

In addition, we may calculate the drop across the filter resistor and the load current drawn:

$$V_R = V_{dc} - V'_{dc} = 60 - 53.6 = 6.4\ \text{V}$$

$$I_{dc} = \frac{V'_{dc}}{R_L} = \frac{53.6}{1 \times 10^3} = 53.6\ \text{mA}$$

AC Operation of *RC* Filter Section

Figure 16.11b shows the equivalent circuit for analyzing the ac operation of the filter circuit. The input to the filter stage from the first filter capacitor (C_1) is the ripple or ac signal part of the voltage across C_1, V_r (rms), which is approximated now as a sinusoidal signal. Both the *RC* filter stage components and the load resistance affect the ac signal at the output of the filter.

For a filter capacitor (C_2) value of 10 μF at a ripple voltage frequency (f) of 60 Hz, the ac impedance of the capacitor is[3]

$$X_C = \frac{1}{\omega C} = \frac{1}{2\pi f C} = \frac{1}{6.28(60)(10 \times 10^{-6})} = 0.265\ \text{k}\Omega$$

Referring to Fig. 16.11b, we see that this capacitive impedance is in parallel with the load resistance. For a load resistance of 2 kΩ, for example, the parallel combination of the two components would yield an impedance of magnitude:

$$Z = \frac{R_L X_C}{\sqrt{R_L^2 + X_C^2}} = \frac{2(0.265)}{\sqrt{2^2 + (0.265)^2}} = \frac{2}{2.02}(0.265) = 0.263\ \text{k}\Omega$$

[3]X_C is understood here to represent only the *magnitude* of the capacitor's ac impedance.

This is close to the value of the capacitive impedance alone, as expected, since the capacitive impedance is much less than the load resistance and the parallel combination of the two would be smaller than the value of either. As a rule of thumb we can consider neglecting the loading by the load resistor on the capacitive impedance as long as the load resistance is at least five times as large as the capacitive impedance. Because of the limitation of light loads on the filter circuit the effective value of load resistance is usually large compared to the impedance of capacitors in the range of microfarads.

In the discussion above it was stated that the frequency of the ripple voltage was 60 Hz. Assuming that the line frequency was 60 Hz, the ripple frequency will also be 60 Hz for the ripple voltage from a half-wave rectifier. The ripple voltage from a full-wave rectifier, however, will be double since there are twice the number of half-cycles and the ripple frequency will then be 120 Hz. Referring to the relation for capacitive impedance $X_C = 1/\omega C$, we have value of $\omega = 377$ for 60 Hz and of $\omega = 754$ for 120 Hz. Using values of capacitance in μF, we can express the relation for capacitive impedance as

$$X_C = \frac{2.653}{C} \quad \text{(half-wave)} \tag{16.12a}$$

$$X_C = \frac{1.326}{C} \quad \text{(full-wave)} \tag{16.12b}$$

where C is in microfarads and X_C is in kilohms.[4]

EXAMPLE 16.7

Calculate the impedance of a 15-μF capacitor used in the filter section of a circuit using full-wave rectification.

Solution:

$$X_C = \frac{1.326}{C} = \frac{1.326}{15} = 0.0884\ \text{k}\Omega = \mathbf{88.4\ \Omega}$$

Using the simplified relation that the parallel combination of the load resistor and the capacitive impedance equals, approximately, the capacitive impedance, we can calculate the ac attenuation in the filter stage:

$$V_r'(\text{rms}) \cong \frac{X_C}{\sqrt{R^2 + X_C^2}} V_r(\text{rms}) \tag{16.13a}$$

---

[4]Equations (16.12a) and (16.12b) may also be expressed as

$$X_C = \frac{2653}{C} \quad \text{(half-wave)}$$

$$X_C = \frac{1326}{C} \quad \text{(full-wave)}$$

where C is in microfarads and X_C is in ohms.

The use of the square root of the sum of the squares in the denominator was necessary since the resistance and capacitive impedance must be added vectorially, not algebraically. If the value of the resistance is larger by a factor of 5 than that of the capacitive impedance, then a simplification of the denominator may be made yielding the following result:

$$\boxed{V_r'(\text{rms}) \cong \frac{X_C}{R} \cdot V_r(\text{rms})} \qquad (16.13\text{b})$$

EXAMPLE 16.8

The output of a full-wave rectifier and capacitor filter is further filtered by an RC filter section (see Fig. 16.12). The component values of the *RC* section are $R = 500\ \Omega$ and $C = 10\ \mu\text{F}$. If the initial capacitor filter develops 150 V dc with a 15 V ac ripple voltage, calculate the resulting dc and ripple voltage across a 5-kΩ load.

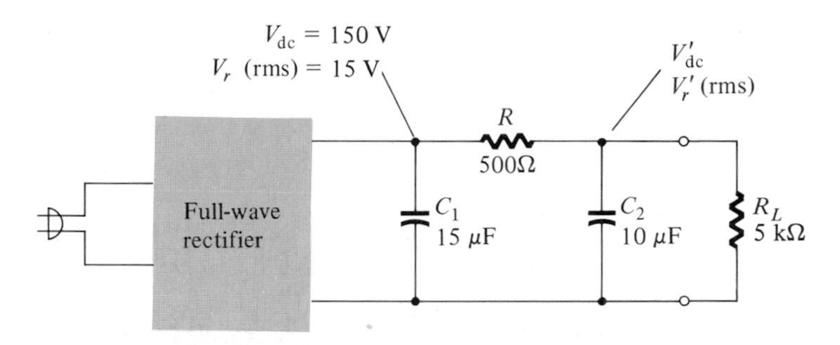

Figure 16.12 *RC* filter circuit for Example 16.8.

Solution:

DC Calculations: Calculating the value of V'_{dc} from Eq. (16.11):

$$V'_{\text{dc}} = \frac{R_L}{R + R_L} V_{\text{dc}} = \frac{5000}{500 + 5000}(150) = \frac{5000}{5500}(150) = \mathbf{136.4\ V}$$

AC Calculations: Calculating the value of the capacitive impedance first (for full-wave operation):

$$X_C = \frac{1.326}{C} = \frac{1.326}{10} = 0.133\ \text{k}\Omega = 133\Omega$$

Since this impedance is not quite 5 times smaller than that of the filter resistor ($R = 500\ \Omega$), we shall use Eq. (16.13a) for the calculation, and then repeat the calculation to show what the difference would have been using Eq. (16.13b) (since the components are almost 5 times different in size). Using Eq. (16.13a), we get

$$V_r'(\text{rms}) = \frac{X_C}{\sqrt{R^2 + X_C^2}} V_r(\text{rms}) = \frac{0.133}{\sqrt{(0.5)^2 + (0.133)^2}}(15)$$

$$= \frac{0.133}{0.517}(15) = \mathbf{3.86\ V}$$

Now using Eq. (16.13b), we get

$$V_r'(\text{rms}) = \frac{X_C}{R}V_r(\text{rms}) = \frac{0.133}{0.500}(15) = \mathbf{3.99\ V}$$

Comparing the results of 3.86 V and 3.99 V, use of Eq. (16.13b) would have yielded an answer within 3.5% of the more exact solution.

16.5 VOLTAGE-MULTIPLIER CIRCUITS

Voltage Doubler

A modification of the capacitor filter circuit allows building up a larger voltage than the peak rectified voltage (V_m). The use of this type of circuit allows keeping the transformer peak voltage rating low while stepping up the peak output voltage to two, three, four, or more times the peak rectified voltage.

Figure 16.13 shows a half-wave voltage doubler. During the positive-voltage half-cycle across the transformer, secondary diode D_1 conducts (and diode D_2 is cut off), charging capacitor C_1 up to the peak rectified voltage (V_m). Diode D_1 is ideally a short during this half-cycle and the input voltage charges capacitor C_1 to V_m with the polarity shown in Fig. 16.14a. During the negative half-cycle of the secondary voltage, diode D_1 is cut off and diode D_2 conducts charging capacitor C_2. Since diode D_2 acts as a short during the negative half-cycle (and diode D_1 is open), we can sum the voltages around the outside loop (see Fig. 16.14b):

$$-V_{C_2} + V_{C_1} + V_m = 0$$

$$-V_{C_2} + V_m + V_m = 0$$

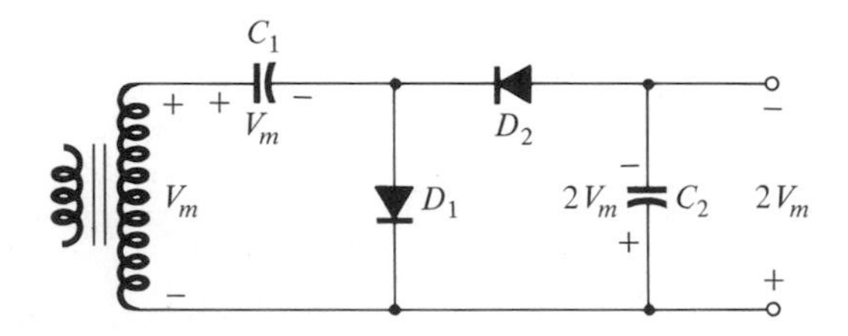

Figure 16.13 Half-wave voltage doubler.

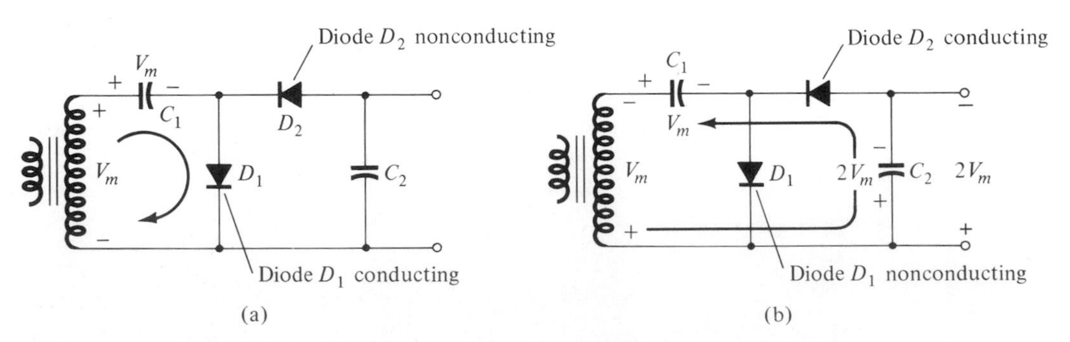

Figure 16.14 Double operation, showing each half-cycle of operation: (a) positive half-cycle; (b) negative half-cycle.

from which

$$V_{C_2} = 2V_m$$

On the next positive half-cycle, diode D_2 is nonconducting and capacitor C_2 will discharge through the load. If no load is connected across capacitor C_2, both capacitors stay charged—C_1 to V_m and C_2 to $2V_m$. If, as would be expected, there is a load connected to the output of the voltage doubler, the voltage across capacitor C_2 drops during the positive half-cycle (at the input) and the capacitor is recharged up to $2V_m$ during the negative half-cycle. The output waveform across capacitor C_2 is that of a half-wave signal filtered by a capacitor filter. The peak inverse voltage across each diode is $2V_m$.

Another doubler circuit is the full-wave doubler of Fig. 16.15. During the positive half-cycle of transformer secondary voltage (see Fig. 16.16a) diode D_1 conducts charging capacitor C_1 to a peak voltage V_m. Diode D_2 is nonconducting at this time.

During the negative half-cycle (see Fig. 16.16b) diode D_2 conducts charging capacitor C_2 while diode D_1 is nonconducting. If no load current is drawn from the circuit, the voltage across capacitors C_1 and C_2 is $2V_m$. If load current is drawn from

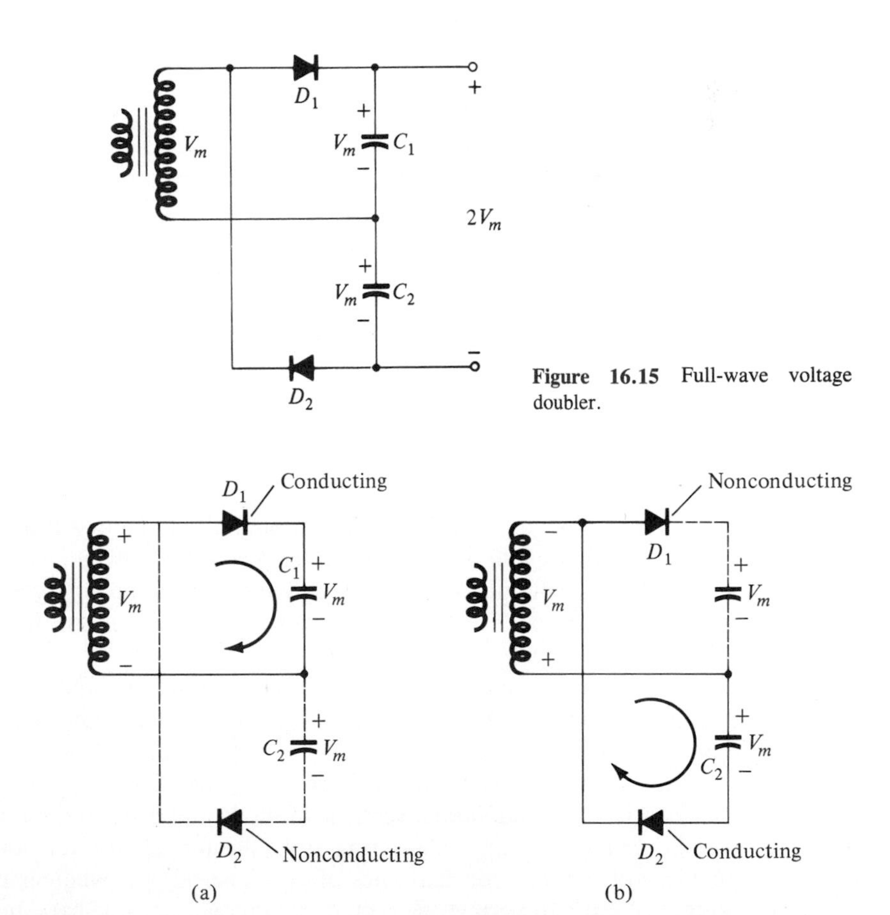

Figure 16.15 Full-wave voltage doubler.

Figure 16.16 Alternate half-cycles of operation for full-wave voltage doubler.

the circuit, the voltage across capacitors C_1 and C_2 is the same as that across a capacitor fed by a full-wave rectifier circuit. One difference is that the effective capacitance is that of C_1 and C_2 in series, which is less than the capacitance of either C_1 or C_2 alone. The lower capacitor value will provide poorer filtering action than the single-capacitor filter circuit.

The peak inverse voltage across each diode is $2V_m$ as it is for the filter capacitor circuit. In summary, the half-wave or full-wave voltage doubler circuits provide twice the peak voltage of the transformer secondary while requiring no center-tapped transformer and only $2V_m$ PIV rating for the diodes.

Voltage Tripler and Quadrupler

Figure 16.17 shows an extension of the half-wave voltage doubler, which develops three and four times the peak input voltage. It should be obvious from the pattern of the circuit connection how additional diodes and capacitors may be connected so that the output voltage may also be five, six, seven, etc., times the basic peak voltage (V_m).

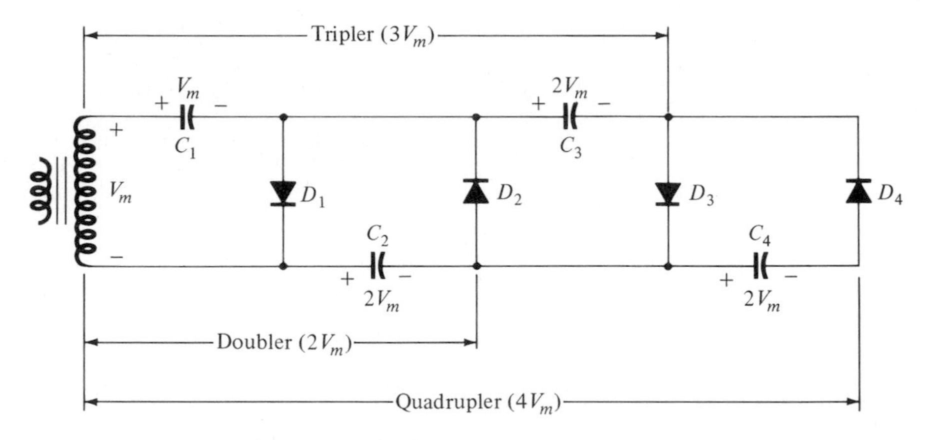

Figure 16.17 Voltage tripler and quadrupler.

In operation capacitor C_1 charges through diode D_1 to a peak voltage, V_m, during the positive half-cycle of the transformer secondary voltage. Capacitor C_2 charges to twice the peak voltage $2V_m$ developed by the sum of the voltages across capacitor C_1 and the transformer, during the negative half-cycle of the transformer secondary voltage.

During the positive half-cycle, diode D_3 conducts and the voltage across capacitor C_2 charges capacitor C_3 to the same $2V_m$ peak voltage. On the negative half-cycle, diodes D_2 and D_4 conduct with capacitor C_3, charging C_4 to $2V_m$.

The voltage across capacitor C_2 is $2V_m$, across C_1 and C_3 it is $3V_m$, and across C_2 and C_4 it is $4V_m$. If additional sections of diode and capacitor are used, each capacitor will be charged to $2V_m$. Measuring from the top of the transformer winding (Fig. 16.17) will provide odd multiples of V_m at the output, whereas measuring from the bottom of the transformer the output voltage will provide even multiples of the peak voltage, V_m.

The transformer rating is only V_m, maximum, and each diode in the circuit must be rated at $2V_m$ PIV. If the load is small and the capacitors have little leakage, extremely high dc voltages may be developed by this type of circuit, using many sections to step up the dc voltage.

16.6 DISCRETE VOLTAGE REGULATORS

Regulation Defined

A wide variety of circuit configurations are capable of performing voltage or current regulation. Only a few of the more commonly applied circuits will be considered in this text. Voltage and current regulation can best be defined through the use of the circuits of Figs. 16.18 and 16.19, respectively.

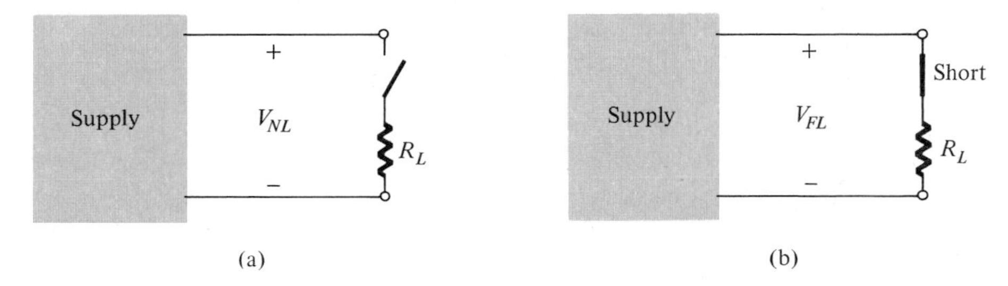

Figure 16.18 Voltage regulation: (a) no-load (NL) state; (b) full-load (FL) state.

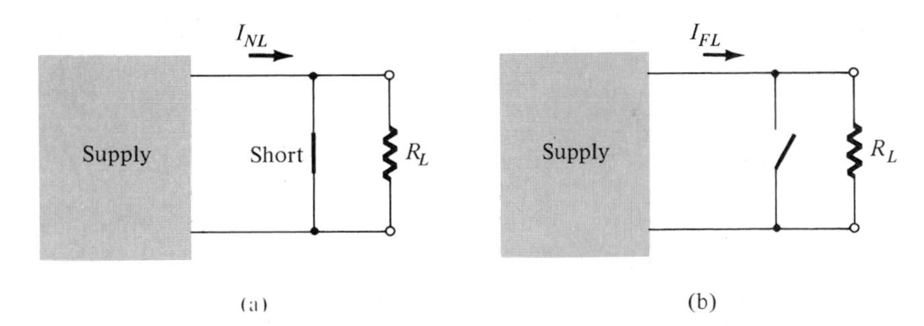

Figure 16.19 Current regulation: (a) no-load (NL) state; (b) full-load (FL) state.

In Fig. 16.18a the no-load (open-circuit) terminal voltage of the supply is designated by V_{NL}. The corresponding load current is $I_{NL} = 0$. The full-load conditions are shown in Fig. 16.18b. The ideal situation would require that $V_L = V_{FL} = V_{NL}$ for every value of R_L between no-load and full-load conditions. In other words, the terminal voltage V_L would be unaffected by variations in R_L. Unfortunately, there is no supply available today, whether it be semiconductor or electromechanical (generator), that can provide a terminal voltage completely independent of the load applied. Voltage regulation is defined formally as in Eq. (16.2). Current regulation is defined by

$$\text{current regulation} = \frac{I_{NL} - I_{FL}}{I_{FL}} \times 100\% \quad (16.14)$$

where I_{NL} and I_{FL} and the no-load full-load currents as defined by Fig. 16.19.

Zener and Thermistor Voltage Regulators

There are, fundamentally, two basic configurations for establishing voltage or current regulation. Each is shown in Fig. 16.20. The usual terminology for each is indicated in the figure. You will find as you progress through some of the more typical circuit configurations that the more sophisticated regulators will apply the benefits to be derived from both series and shunt regulation in the same system. Before continuing, let us examine the unregulated supply of Fig. 16.21. It clearly demonstrates the need for voltage and/or current regulators. Recall from your experimental work that when your dc supply indicates 10 V, you want it to be that value for any load you apply to its terminals. At any except infinite ohms (open circuit) in Fig. 16.21, would this be the case? Obviously not. As R_L increases, the voltage across R_L also increases, and V_L does not remain fixed. The function of the voltage regulator is to maintain V_L at 10 V for any value of R_L.

Our interest for the moment is in a voltage-regulating device. A simple shunt voltage regulating system is shown in Fig. 16.22. As indicated, it consists simply of

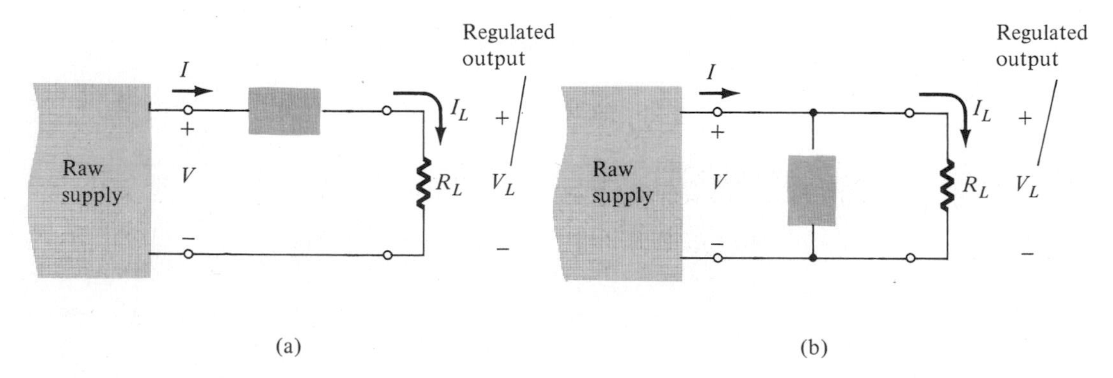

Figure 16.20 Regulators: (a) series; (b) parallel (shunt).

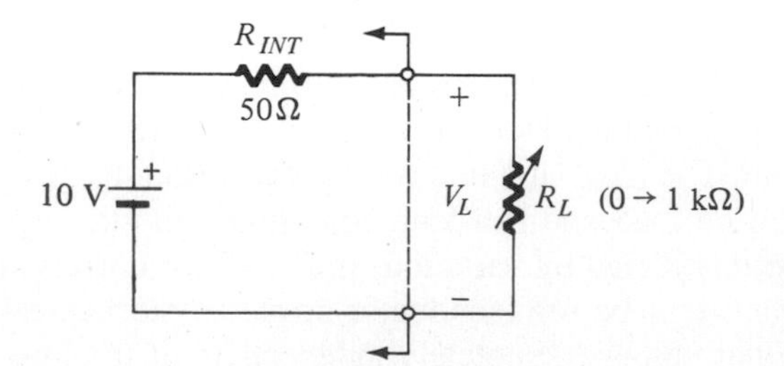

Figure 16.21 Circuit demonstrating the need for voltage and current regulators.

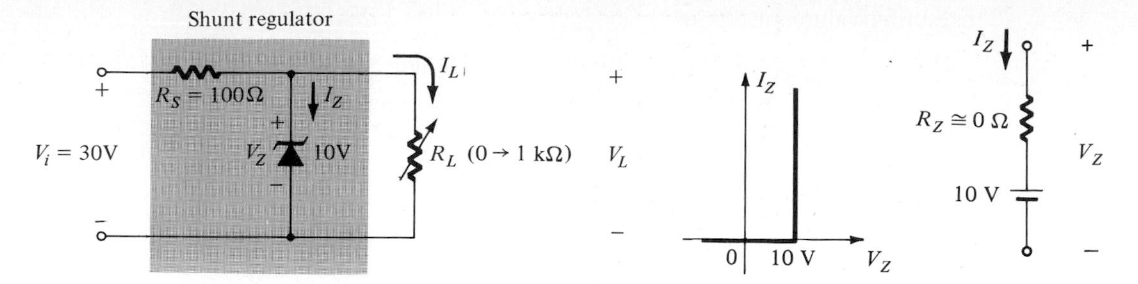

Figure 16.22 Zener diode shunt regulator.

a Zener diode and series resistance, R_S. Proper operation requires that the Zener diode be in the *on* state. The first requirement, therefore, is to find the minimum R_L (and corresponding I_L) to ensure that this condition is established. Before Zener conduction, the Zener diode is fundamentally an open circuit and the circuit of Fig. 16.22 can be replaced by that indicated in Fig. 16.23. As indicated in the figure, the load voltage will be determined by the voltage-divider rule. At Zener conduction, $V_L = V_Z = 10$ V. If these data are used, it is possible to find the minimum value of R_L that the Zener will tolerate in maintaining V_L constant. Applying the voltage-divider rule to the circuit of Fig. 16.24a results in

$$V_L = \frac{R_L}{R_L + R_s} V_i$$

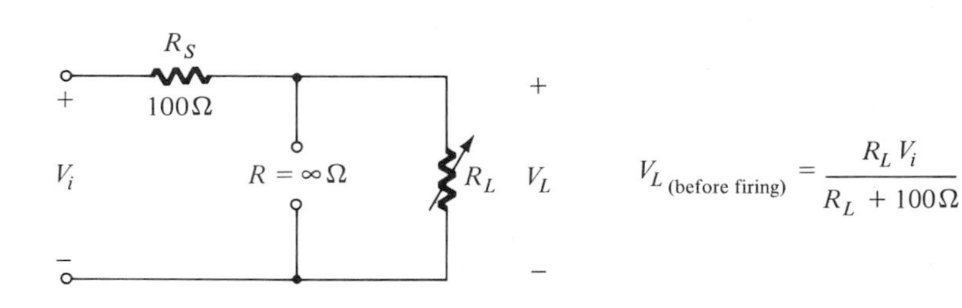

Figure 16.23 Regulator of Fig. 16.22 before firing.

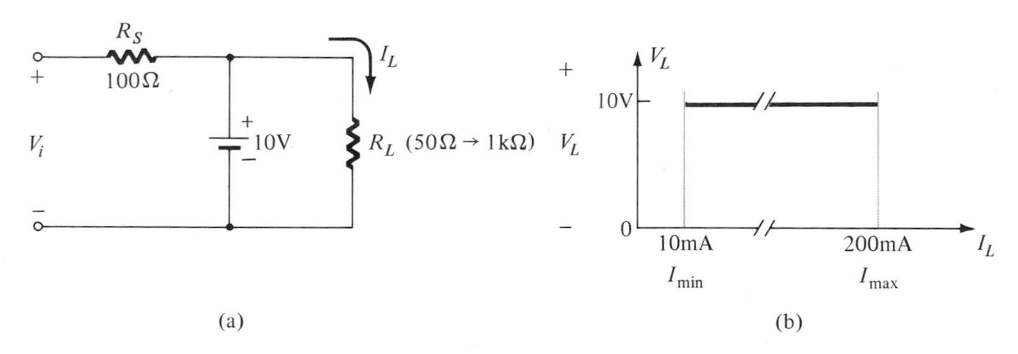

Figure 16.24 Zener diode shunt regulator: (a) after firing; (b) regulated output.

Substituting values results in

$$10 = \frac{R_L}{R_L + 0.1\ \text{k}\Omega} 30\ \text{V}$$

$$10R_L + 1\ \text{k}\Omega = 30R_L$$

and

$$R_L = 50\ \Omega$$

The *minimum* load R_L for this supply is, therefore, 50 Ω, corresponding to a *maximum* load current of

$$I_{\text{max}} = \frac{10\ \text{V}}{50\ \Omega} = 200\ \text{mA}$$

The maximum current is indicated on the plot of Fig. 16.24b. For any value of R_L from 50 Ω to 1 kΩ the Zener diode will be in the *on* state. An application of the voltage-divider rule to the circuit of Fig. 16.23 for any value of R_L less than 50 Ω will result in $V_L < V_Z$ and the diode will remain in the *off* state. At $R_L = 1\ \text{k}\Omega$, $I_L = (10/1\ \text{k}\Omega) = 10\ \text{mA}$ (indicated in Fig. 16.24b as I_{min}), and

$$I_{R_S} = \frac{30\ \text{V} - 10\ \text{V}}{0.1\ \text{k}\Omega} = 200\ \text{mA}$$

with

$$I_Z = I_{R_S} - I_L = 190\ \text{mA}$$

Our approximation, $R_Z \cong 0\ \Omega$, has resulted in the ideal characteristics of Fig. 16.24b between 10 and 200 mA. For this regulator the percent regulation would be determined between these two points of normal operation. The effect of R_Z on the regulation can be determined easily if we find the Thévenin equivalent circuit for the portion of the network of Fig. 16.25 at the left of points 1 and 2.

$$V_{\text{th}} = 10\ \text{V} + \frac{2\ \Omega(30\ \text{V} - 10\ \text{V})}{102\ \Omega} = 10 + \frac{40}{102} \cong 10.4\ \text{V}$$

$$R_{\text{th}} = 100\ \Omega \parallel 2\ \Omega \cong 2\ \Omega$$

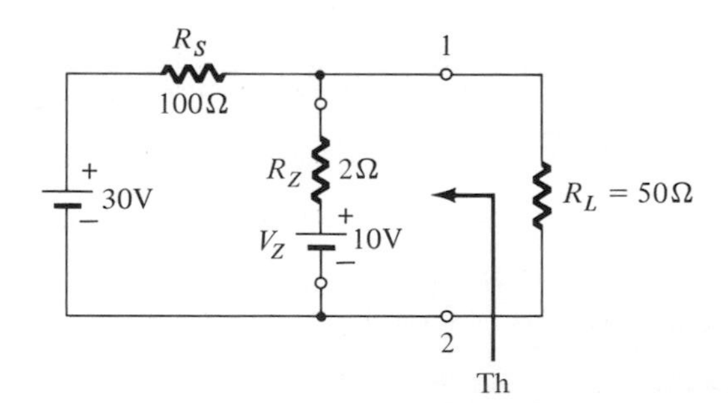

Figure 16.25 Determining the effect of R_z on the output of the Zener shunt regulator.

Substituting the Thévenin equivalent circuit (Fig 16.26) and

$$V_L = \frac{50\ \Omega(10.4\ \text{V})}{52\ \Omega} = 10\ \text{V}$$

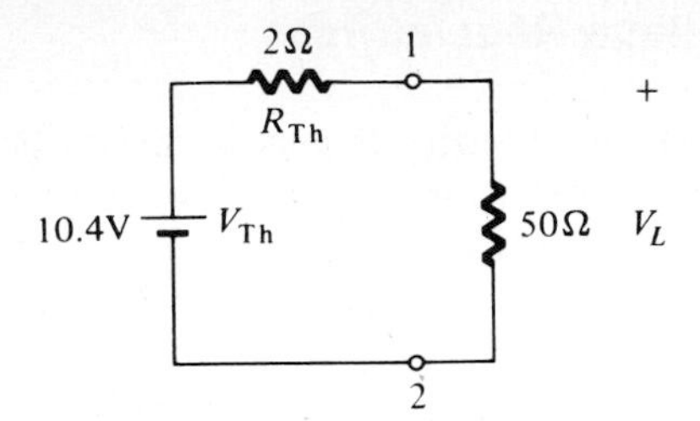

Figure 16.26 Thévenin equivalent circuit for the circuit of Fig. 16.25.

R_Z, therefore, has a negligible effect at minimum R_L (maximum I_L).

For $R_L = 1\ \text{k}\Omega$ (minimum I_L):

$$V_L = \frac{1\ \text{k}\Omega(10.4\ \text{V})}{1\ \text{k}\Omega + 2\ \Omega} \cong 10.4\ \text{V} > 10\ \text{V}$$

and

$$\text{V.R.} = \frac{V_{1\text{k}\Omega} - V_{50\Omega}}{V_{50\Omega}} \times 100\% = \frac{10.4\ \text{V} - 10.0\ \text{V}}{10.0\ \text{V}} \times 100\%$$

$$= \frac{0.4\ \text{V}}{10.0\ \text{V}} \times 100\% = 4\%$$

A shunt regulator employing a thermistor appears in Fig. 16.27. Any tendency for V_L to decrease due to a change in load will result in a decrease in current through the thermistor. The temperature of the thermistor element will thereby decrease, resulting in an increase in its resistance. The resulting resistance $R_T = R_L \parallel (R_{\text{th}} + 100\ \Omega)$ will increase somewhat and the load voltage $V_L = R_T V_i/(R_T + R_1)$ will tend to increase, offsetting the initial drop in V_L. The symbol $\uparrow$ is assigned to an increasing quantity and $\downarrow$ to a decreasing quantity in the following summary of the voltage-regulating action of this system (read from left to right):

$$V_L \downarrow,\ I_{\text{th}} \downarrow,\ R_{\text{th}} \uparrow,\ R_T \uparrow,\ V_L \uparrow$$

balance

An increasing V_L will have the opposite effect on each element and quantity of the summary above.

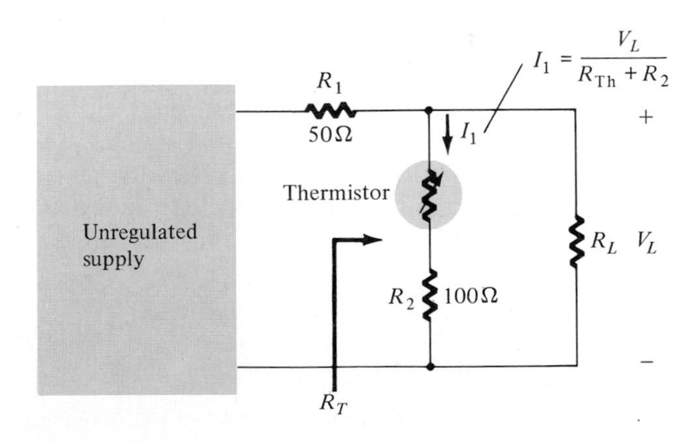

Figure 16.27 Thermistor shunt regulator.

Transistor Voltage Regulators

The characteristics of a voltage regulator can be markedly improved by using active devices such as the transistor. The simplest of the transistor-*series*-type voltage regulator appears in Fig. 16.28a. In this configuration the transistor behaves like a simple variable resistor whose resistance is determined by the operating conditions. The basic operation of the regulator is best described using the circuit of Fig. 16.28b, in which the transistor has been replaced by a variable resistor, R_T. For variations in R_L, if V_L is to remain constant, the ratio of R_L to R_T must remain fixed. Applying the voltage-divider rule results in

$$V_L = \overbrace{\frac{R_L}{R_L + R_T}}^{\text{[fixed for constant } V_L (V_i = \text{constant})]} V_i$$

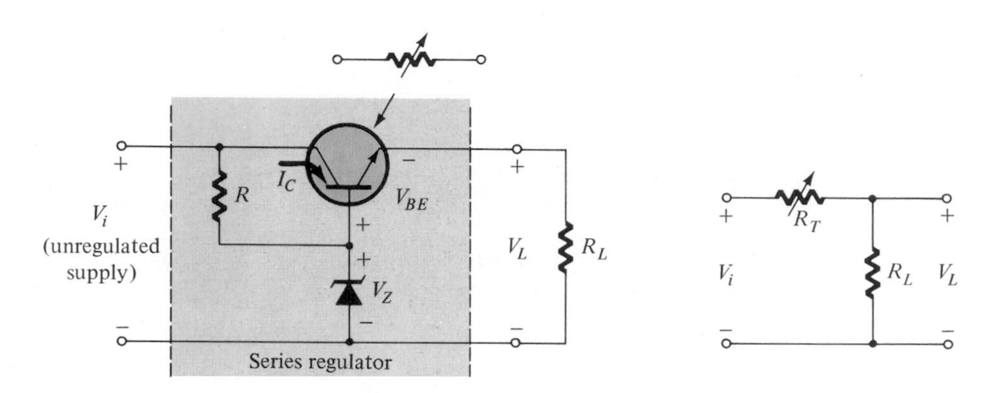

Figure 16.28 Transistor series voltage regulator.

For

$$\frac{R_L}{R_T} = k_1 \quad \text{or} \quad R_L = k_1 R_T$$

$$\frac{R_L}{R_L + R_T} = \frac{k_1 R_T}{k_1 R_T + R_T} = \frac{k_1}{k_1 + 1} = k \quad \text{(constant as required)}$$

In summary, for a decreasing or increasing load (R_L), R_T must change in the same manner and at the same rate to maintain the same voltage division.

Recall that the voltage regulation is determined by noting the variations in terminal voltage versus the load current demand. For this circuit an increasing current demand associated with a *decreasing* R_L will result in a tendency on the part of V_L to decrease in magnitude also. If, however, we apply Kirchoff's voltage law around the output loop of Fig. 16.28a.

$$V_{BE} = \overbrace{V_Z}^{\text{(fixed)}} - V_L$$

A decrease in V_L (since V_Z is fixed in magnitude) will result in an increase in V_{BE}.

This effect will, in turn, increase the level of conduction of the transistor, resulting in a *decrease* in its terminal (collector-to-emitter) resistance. This is, as described in the previous paragraphs of this section, the effect desired to maintain V_L at a fixed level.

A voltage regulator employing a transistor in the shunt configuration is provided in Fig. 16.29. Any tendency on the part of V_L to increase or decrease in magnitude will have the corresponding effect on V_{BE} since

$$V_{BE} = V_L - \overbrace{V_Z}^{\text{(fixed)}}$$

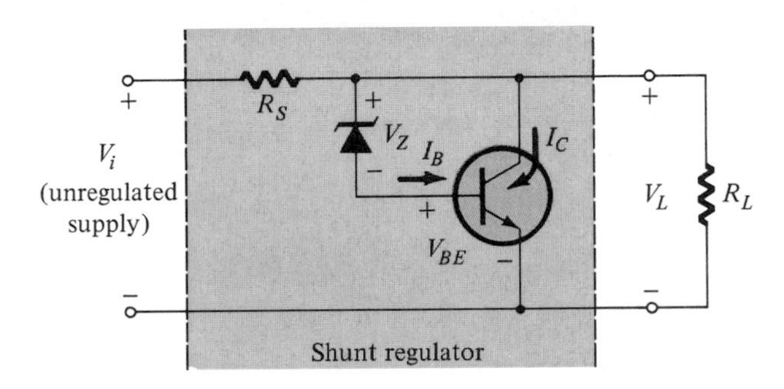

Figure 16.29 Transistor shunt voltage regulator.

For decreasing V_L, the current through the resistor R_S will decrease since the conduction level of the transistor has dropped ($V_{BE} \downarrow$). The reduced drop in potential across R_S will offset any tendency on the part of V_L to decrease in magnitude. In sequential logic

$$V_L \downarrow,\ V_{BE} \downarrow,\ I_B \downarrow,\ I_C \downarrow,\ I_{R_S} \downarrow,\ V_{R_S} \downarrow,\ V_L \uparrow$$

(balance between the first $V_L \downarrow$ and the final $V_L \uparrow$)

A similar discussion can be applied to increasing values of V_L.

A series voltage regulator employing a second transistor for control purposes can be found in Fig. 16.30. The base-to-emitter potential (V_{BE_2}) of the control transistor Q_2

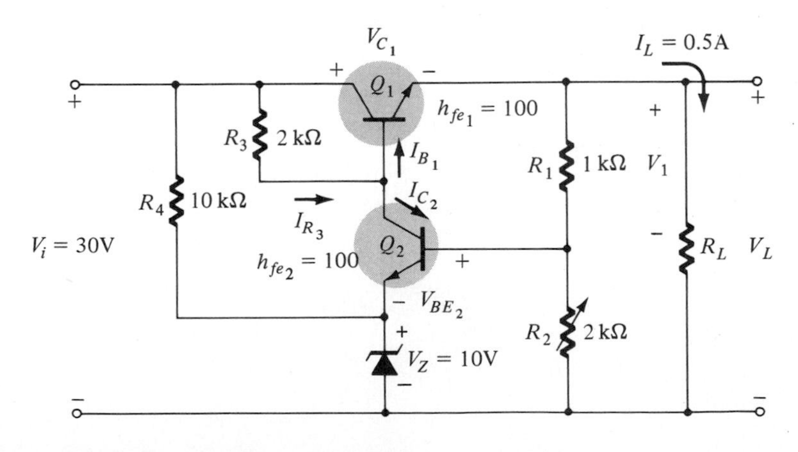

Figure 16.30 A series regulator employing two transistors.

is determined by the difference between V_1 and the reference voltage V_Z. The voltage level V_2 is sensitive to the changes in the terminal voltage V_L. Any tendency on the part of V_L to increase will result in an increase in V_2 and therefore in V_{BE_2} since $V_{BE_2} = V_2 - V_Z$. The difference in potential is amplified by the control transistor and carried to the variable series resistive element Q_1. An increase in V_{BE_2}, corresponding to an increase in I_{B_2} and I_{C_2} will result in a decreasing I_{B_1} (assuming I_{R_3} to be relatively constant or decreasing only slightly). The net result is a decrease in the conductivity of Q_1 corresponding to an increase in its terminal resistance and a stabilization of V_L. In sequential logic

$$V_L \uparrow, V_2 \uparrow, V_{BE_2} \uparrow, I_{C_2} \uparrow, I_{B_1} \downarrow, V_{C_1} \uparrow, V_L \downarrow$$

balance

Again, a similar discussion can be applied to decreasing values of V_L.

EXAMPLE 16.9

We shall now calculate various currents and voltages of the circuit of Fig. 16.30 for the input shown. Approximations, as introduced in previous chapters, will be the working tools of this analysis. Those of primary importance include $I_C \cong h_{fe} I_B$, $V_{BE} \cong 0$ V, and $I_C \cong I_E$.

Solution:

$$V_{R4} = V_i - V_Z = 30 \text{ V} - 10 \text{ V} = \mathbf{20 \text{ V}}$$

and

$$I_{R4} = \frac{20 \text{ V}}{10 \text{ k}\Omega} = \mathbf{2 \text{ mA}}$$

$$V_{R_2} \cong V_Z = \mathbf{10 \text{ V}} \qquad \text{since } V_{BE_2} \cong 0 \text{ V}$$

and

$$I_{R_2} = \frac{10 \text{ V}}{2 \text{ k}\Omega} = \mathbf{5 \text{ mA}}$$

Assuming $I_{B_2} \ll I_{R_1}, \; I_{R_2}$

then $I_{R_1} = I_{R_2} = \mathbf{5 \text{ mA}}$

and $V_L = 5 \text{ mA} \times 3 \text{ k}\Omega = \mathbf{15 \text{ V}}$

with

$$V_{R_3} = V_i - V_L \qquad (V_{BE_1} \cong 0 \text{ V})$$
$$= 30 \text{ V} - 15 \text{ V} = \mathbf{15 \text{ V}}$$

and

$$I_{R_3} = \frac{15 \text{ V}}{2 \text{ k}\Omega} = \mathbf{7.5 \text{ mA}}$$

Similarly,

$$V_{C_1} = V_i - V_L = 30 \text{ V} - 15 \text{ V} = \mathbf{15 \text{ V}}$$

$$I_{E_1} \cong h_{fe} I_{B_1} = 100 I_{B_1}$$

and

$$I_{B_1} = \frac{I_{E_1}}{100} = \frac{(500 + 5)\text{mA}}{100} = \mathbf{5.05 \text{ mA}}$$

$$I_{E_2} \cong I_{C_2} = I_{R_3} - I_{B_1}$$

$$= (7.5 - 5.05)\text{mA} = \mathbf{2.45\ mA}$$

$$I_{B_2} \cong \frac{I_{C_2}}{100} = \frac{2.45\text{ mA}}{100} = \mathbf{24.5\ \mu A}$$

(Certainly, $I_{B_2} \ll I_{R_2}$. I_{R_2} as employed above is an excellent approximation.) Finally,

$$I_Z = I_{R_4} + I_{E_2} = (2 + 2.45)\text{mA} = \mathbf{4.45\ mA}$$

The fact that $I_{B_2} \ll I_{R_1}, I_{R_2}$ permits the use of the circuit of Fig. 16.31 to derive a rather useful equation for the circuit of Fig. 16.30. Applying the voltage-divider rule results in

$$V_Z = \frac{R_2}{R_1 + R_2} V_L$$

or since V_Z is fixed,

$$V_L = V_Z\left(1 + \frac{R_1}{R_2}\right)$$

For the case above,

$$V_L = 10(1 + \tfrac{1}{2}) = 15\text{ V}$$

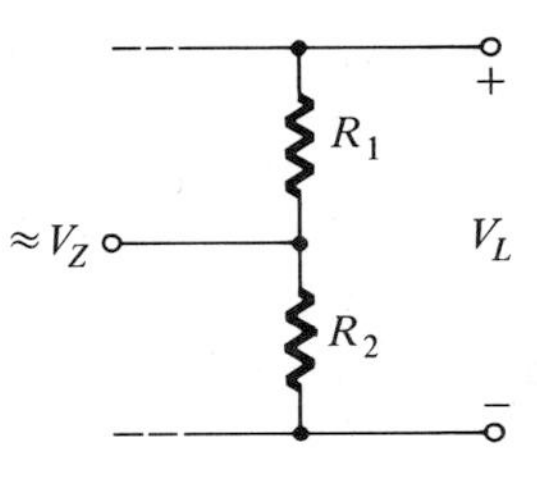

Figure 16.31 Circuit employed in Example 16.9.

You will find in Fig. 16.30 that R_2 is a variable resistor. Variations in this resistance will control V_L. The maximum voltage available is obviously 30 V (for V_i = 30 V) since at this point V_{C_1} = 0 V (saturation). The minimum is 10 V attainable with either $R_1 = 0$ or $R_2 = \infty$.

Complete Power Supply (Voltage Regulated)

A power supply employing a voltage regulator similar to the one described in Fig. 16.30 appears in Fig. 16.32. A *Darlington* circuit has replaced the single series transistor of Fig. 16.28 in an effort to increase the sensitivity of the regulator to changes in V_L. In the circuit of Fig. 16.30 changes in I_{C_2} will be reflected in changes in I_{R_3} causing a reduction in the sensitivity of I_{B_1} to changes in V_L. To minimize this undesirable effect R_3 should be as large as possible while still permitting the necessary R_3 for proper circuit behavior. Greater efficiency is achieved by employing a current source in place of R_3. The current source has, ideally, infinite terminal resistance along with the capability to supply the necessary current. This portion of a power supply as

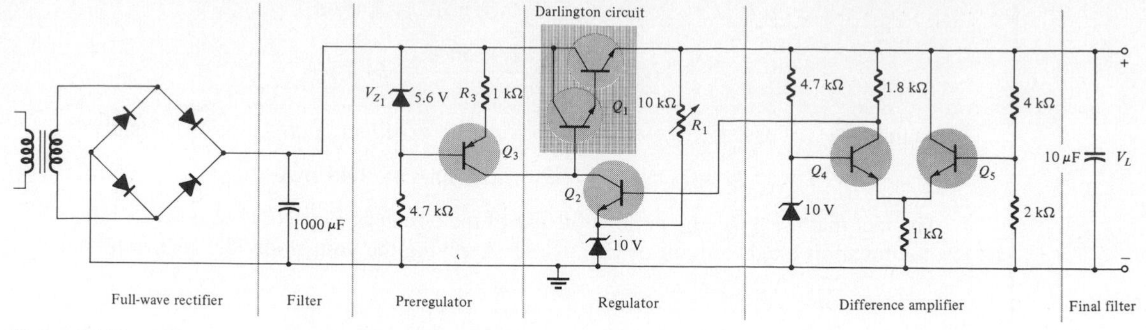

Figure 16.32 Complete voltage-regulated power supply.

indicated in Fig. 16.32 is sometimes referred to as a *preregulator*. For the circuit of Fig. 16.32,

$$I_{\text{current source}} = I_{C_3} \cong \frac{V_{Z1}}{R_3}$$

To improve further the sensitivity of the regulator to changes in V_L a *difference* amplifier has been introduced, the output of which is fed to the control transistor. The unregulated input is a full-wave-rectified signal to be passed through a capacitive filter. The 10-μF capacitor at the output is to reduce the possibility of oscillations and further filter the supply voltage. The supply voltage V_L can be varied by changing R_1 while still maintaining regulation.

Current Regulator

The analysis of current regulators will be limited to a brief discussion of the circuit of Fig. 16.33. Recall from the introductory discussion that a current regulator is designed to maintain a fixed current through a load for variations in terminal voltage. A decrease in $I_L = I_C$ due to a drop in V_L would result in a decrease in $I_E \cong I_C$ and, in turn, a drop in V_{R_E}. The base-to-emitter potential is

$$V_{BE} = \overbrace{V_Z}^{\text{(fixed)}} - V_{R_E}$$

A decrease in V_{R_E} will result in an increase in V_{BE} and the conductivity of the transistor, maintaining I_L at a fixed level.

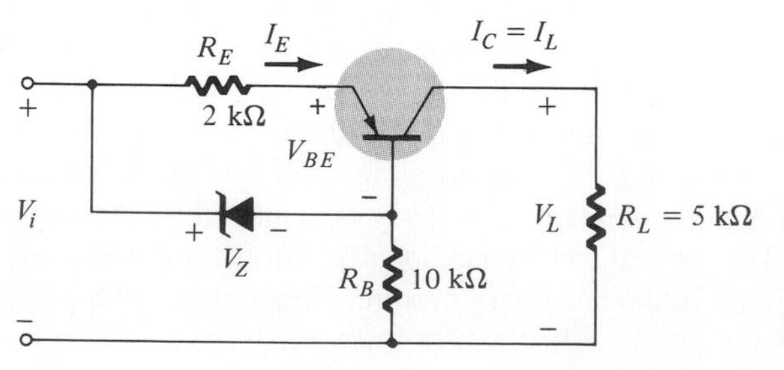

Figure 16.33 Series current regulator.

16.7 IC VOLTAGE REGULATORS

Voltage regulators comprise a class of widely used ICs. These units contain the circuitry for reference source, error amplifier, control device, and overload protection all in a single IC chip. Although the internal construction is somewhat different than that described for discrete voltage-regulator circuits, the external operation is much the same. We will examine operation using some of the popular 3-terminal fixed-voltage regulators (for both positive and negative voltages) and those allowing an adjustable output voltage.

A power supply can be built very simply using a transformer connected to the ac supply to step the voltage to a desired level, then rectifying with a half- or full-wave circuit, filtering the voltage using a simple capacitor filter, and finally regulating the dc voltage using an IC voltage regulator.

A basic category of voltage regulators includes those used with only positive voltages, those used with only negative voltages, and those further classified as having fixed or adjustable output voltages. These regulators can be selected for operation with load currents from hundreds of milliamperes to tens of amperes corresponding to power ratings from milliwatts to tens of watts. A representation of various types of voltage-regulator ICs is presented next. Various terms common to this area of electronics also will be introduced and defined.

Three-Terminal Voltage Regulators

Voltage regulators that provide a positive fixed regulated voltage over a range of load currents are schematically represented in Fig. 16.34. The fixed voltage regulator has an unregulated voltage, V_{in}, applied to one terminal, delivers a regulated output voltage, V_o, from a second terminal, with the third terminal connected to ground. For a particular IC unit, device specifications list a voltage range over which the input voltage can vary to maintain the regulated output voltage, V_o, over a range of load current, I_o. An output-input voltage differential must be maintained for the IC to operate, which means that the varying input voltage must always be kept large enough

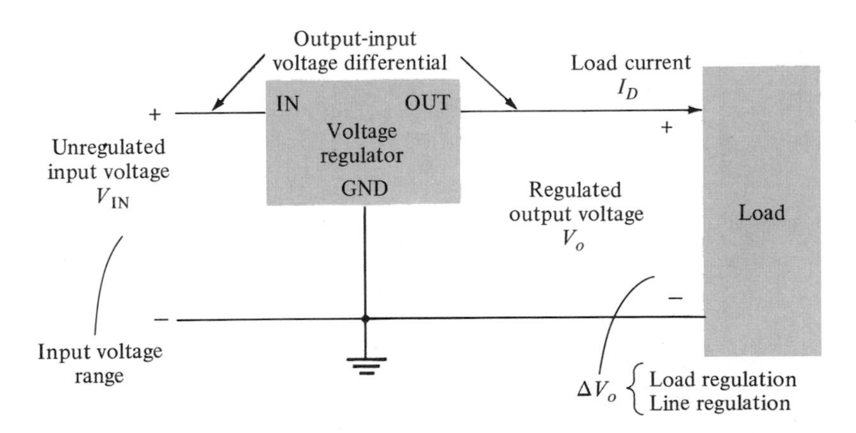

Figure 16.34 Block representation of three-terminal voltage regulator.

to maintain a voltage drop across the IC to permit proper operation of the internal circuit. The device specifications also list the amount of output voltage change, V_o, resulting from changes in load current (load regulation) and also from changes in input voltage (line regulation).

A group of fixed-positive-voltage regulators is the series 78, which provide fixed voltages from 5 V to 24 V. Figure 16.35a shows how many of these regulators are connected. A rectified and filtered unregulated dc voltage is the input, V_{in}, to pin 1 of the regulator IC. Capacitors connected from input or output to ground help to maintain the dc voltage and additionally to filter any high-frequency voltage variation. The output voltage from pin 2 is then available to connect to the load. Pin 3 is the IC circuit reference or ground. When selecting the desired fixed regulated output voltage, the two digits after the 78 prefix indicate the regulator output voltage. Table 16.1 lists some typical data.

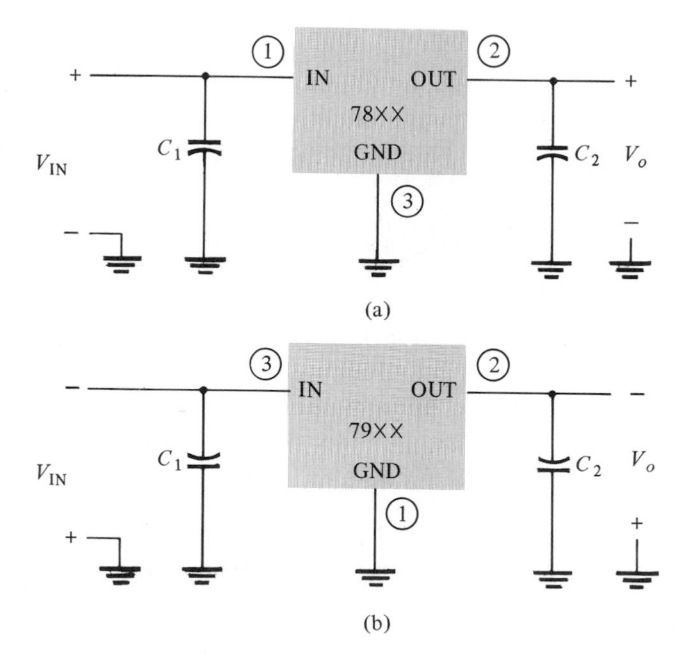

Figure 16.35 (a) Series 78XX positive voltage regulator; (b) series 79XX negative voltage regulator.

TABLE 16.1 Positive Series 78XX Voltage Regulator ICs

IC Part Number	*Regulated Positive Voltage (V)*	*Minimum V_{in} (V)*
7805	+ 5	7.3
7806	+ 6	8.35
7808	+ 8	10.5
7810	+ 10	12.5
7812	+ 12	14.6
7815	+ 15	17.7
7818	+ 18	21
7824	+ 24	27.1

Negative voltage regulator ICs are available in the 79 series (see Fig. 16.35b), which provide a series of ICs similar to the 78 series but operating on negative voltages, providing a regulated negative output voltage. Table 16.2 lists the 79XX series of fixed-negative voltage regulators and their corresponding regulated voltages.

TABLE 16.2 Fixed-Negative Voltage Regulators in the 79XX Series

IC Part Number	*Regulated Output Voltage (V)*	*Minimum V_{in} (V)*
7905	− 5	− 7.3
7906	− 6	− 8.4
7908	− 8	− 10.5
7909	− 9	− 11.5
7912	− 12	− 14.6
7915	− 15	− 17.7
7918	− 18	− 20.8
7924	− 24	− 27.1

Voltage regulators are also available in circuit configurations that allow the user to set the output voltage to a desired regulated value. The LM317, for example, can be operated with output voltage regulated at any setting over the range of voltage from 1.2 V to 37 V. Figure 16.36 shows a typical connection using the LM317 IC.

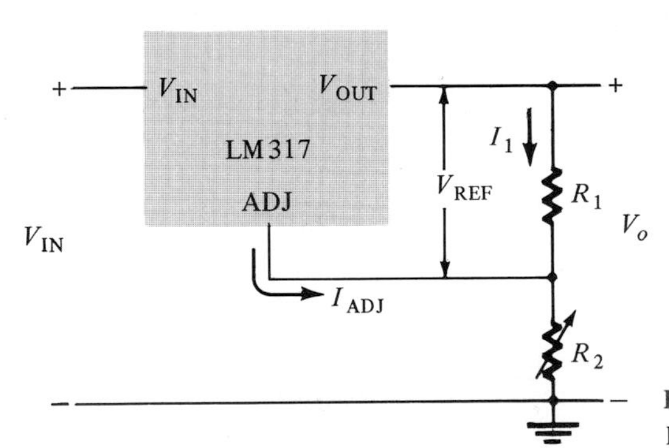

Figure 16.36 Connection of LM317 adjustable-voltage regulator.

Selection of resistors R_1 and R_2 allow the setting of the output to any desired voltage over the adjustment range (1.2 to 37 V). The output voltage desired can be calculated using

$$V_o = V_{\text{ref}}\left(1 + \frac{R_2}{R_1}\right) + I_{\text{adj}} R_2 \tag{16.15}$$

with typical values of

$$V_{\text{ref}} = 1.25 \text{ V} \qquad \text{and} \qquad I_{\text{adj}} = 100\ \mu\text{A}$$

EXAMPLE 16.10

Determine the regulated output voltage using an LM317, as in Fig. 16.36, with R_1 = 240 Ω and R_2 = 2.4 kΩ.

Solution:

Using Eq. (16.15), we obtain

$$V_O = 1.25\text{ V}\left(1 + \frac{2.4\text{ k}\Omega}{240\ \Omega}\right) + 100\ \mu\text{A}\ (2.4\text{ k}\Omega)$$

$$= 13.75\text{ V} + 0.24\text{ V} = \mathbf{13.99\ V}$$

16.8 PRACTICAL POWER SUPPLIES

A practical power supply can be built to convert the 120-V supply voltage into a desired regulated dc voltage. The standard circuit includes a transformer to step the voltage to a desired ac level, a diode rectifier to half-wave or full-wave-rectify the ac signal, and a capacitor filter to develop an unregulated dc voltage. The unregulated dc voltage is then connected as input to an IC voltage regulator, which provides the desired regulated output dc voltage. A few examples will show how a dc voltage supply can be built and how it operates.

EXAMPLE 16.11

Analyze the operation of the +12-V voltage supply shown in Fig. 16.37 connected to a load drawing 400 mA.

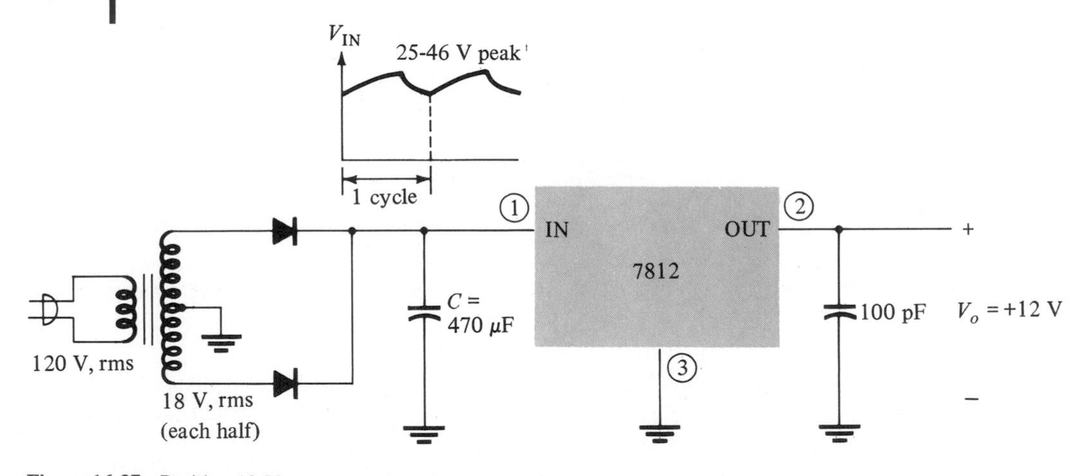

Figure 16.37 Positive 12-V power supply.

Solution:

The transformer steps down the line voltage from 120 V rms to a secondary voltage of 18 V rms across each transformer half. This results in a peak voltage across the transformer of

$$V_m = \sqrt{2}\, V_{\text{rms}} = \sqrt{2} \times 18 \text{ V} = 25.456 \text{ V}$$

The ripple voltage [using Eq. (16.7c)] is then

$$V_r \text{ (rms)} = \frac{2.4 I_{\text{dc}}}{C} = \frac{2.4(400)}{470} = 2.043 \text{ V}$$

and the peak ripple voltage is [using Eq. (16.6)]

$$V_r \text{ (peak)} = \sqrt{3}\, V_r \text{ (rms)} = \sqrt{3}\, (2.043 \text{ V}) = 3.539 \text{ V}$$

The dc level of the voltage across the 470-μF capacitor C is

$$V_{\text{dc}} = V_m - V_r \text{ (peak)} = 25.456 \text{ V} - 3.539 \text{ V} = 21.917 \text{ V}$$

The ripple factor of the filter capacitor when operating into a 400-mA load is then [Eq. (16.9a)]

$$r = \frac{2.4 I_{\text{dc}}}{C V_{\text{dc}}} \times 100\% = \frac{2.4(400)}{(470)(21.917)} \times 100\% \cong 9.3\%$$

The voltage across filter capacitor C has a ripple of about 9.3% and drops to a minimum voltage of

$$V_{\text{in}_{\text{min}}} = V_m - 2V_r \text{ (peak)} = 25.456 \text{ V} - 2(3.539 \text{ V}) = 18.378 \text{ V}$$

Device specifications list V_{in} as required to maintain line regulation at 14.6 V. The lowest voltage being maintained across the capacitor is somewhat greater at 18.378 V.

Lowering the value of the filter capacitor or increasing the load current will result in greater ripple voltage and lower minimum voltage across the capacitor. As long as this minimum voltage remains above 14.6 V, the 7812 will maintain the output voltage regulated at +12 V.

Device specifications for the 7812 list the maximum voltage change as 60 mV. This means that the output voltage regulation will be less than

$$\text{V. R.} = \frac{60 \text{ mV}}{12 \text{ V}} \times 100\% = \mathbf{0.5\%}$$

EXAMPLE 16.12

Analyze the operation of the 5-V supply of Fig. 16.38 operating at a load current of (a) 200 mA and (b) 400 mA.

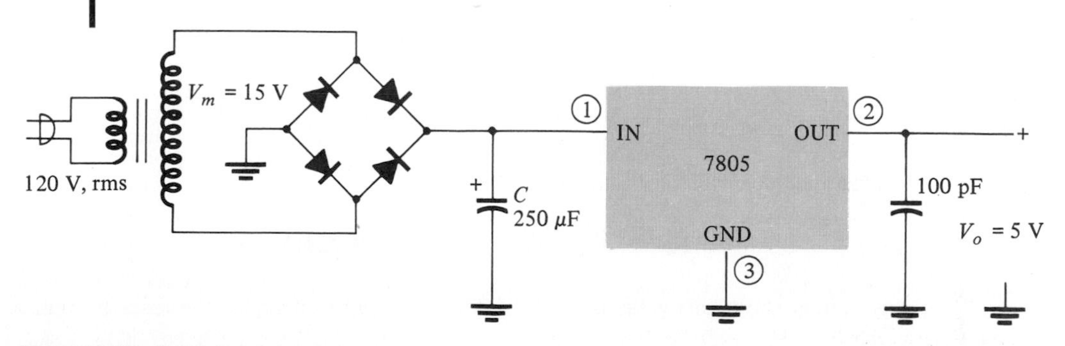

Figure 16.38 Positive 5-V power supply.

Solution:

The specifications for the 7805 list an input of 7.3 V as the minimum allowable to maintain line regulation.

(a) At a load of I_{dc} = 200 mA, the ripple voltage is

$$V_r \text{ (peak)} = \sqrt{3}\, V_r \text{ (rms)} = \sqrt{3} \times \frac{(2.4)I_{dc}}{C} = \sqrt{3} \times \frac{2.4(200)}{(250)} = \mathbf{3.326\ V}$$

and the dc voltage across the 250-μF filter capacitor is

$$V_{dc} = V_m - V_r \text{ (peak)} = 15\text{ V} - 3.326\text{ V} = \mathbf{11.674\ V}$$

The voltage across the filter capacitor will drop to a minimum value of

$$V_{in_{min}} = V_m - 2V_r \text{ (peak)} = 15\text{ V} - 2(3.326\text{ V}) = \mathbf{8.348\ V}$$

Since this is above the rated value of 7.3 V, the output will be maintained at the regulated +5 V.

(b) At a load of I_{dc} = 400 mA, the ripple voltage is

$$V_r \text{ (peak)} = \sqrt{3} \cdot \frac{(2.4)(400)}{250} = \mathbf{6.65\ V}$$

around a dc voltage of

$$V_{dc} = 15\text{ V} - 6.65\text{ V} = \mathbf{8.35\ V}$$

which is above the rated 7.3-V lower level. However, the input swings around this dc level by 6.65 V peak, dropping during part of the cycle to

$$V_{in_{min}} = 15\text{ V} - 2(6.65\text{ V}) = \mathbf{1.7\ V}$$

which is well below the minimum allowed input voltage of 7.3 V. Therefore, the output is not maintained at the regulated +5-V level over the entire input cycle. Regulation is maintained for load currents below 200 mA, but not at or above 400 mA.

EXAMPLE 16.13

Determine the maximum value of load current at which regulation is maintained for the circuit of Fig. 16.38.

Solution:

To maintain $V_{in} \geq 7.3$ V

$$V_r \text{ (p-p)} \leq V_m - V_{in_{min}} = 15\text{ V} - 7.3\text{ V} = 7.7\text{ V}$$

so that

$$V_r \text{ (rms)} = \frac{V_r \text{ (p-p)}/2}{\sqrt{3}} = \frac{7.7\text{ V}/2}{\sqrt{3}} = 2.2\text{ V}$$

We can then determine the value of I_{dc} (in mA)

$$I_{dc} = \frac{V_r \text{ (rms) } C}{2.4} = \frac{(2.2)(250)}{2.4} = \mathbf{229.2\ mA}$$

Any current above this value is too large for the circuit to maintain the regulator output at +5 V.

Using a positive adjustable voltage regulator IC it is possible to set the regulated output voltage to any desired voltage (within the device operating range).

EXAMPLE 16.14

Determine the regulated output voltage of the circuit in Fig. 16.39.

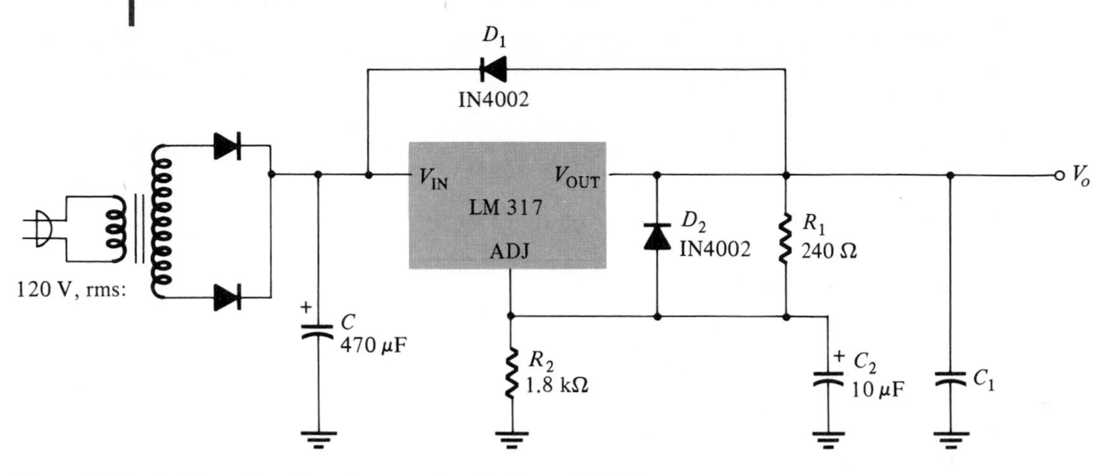

Figure 16.39 Positive adjustable voltage regulator for Example 16.14.

Solution:

The output voltage is

$$V_o = 1.25 \text{ V}\left(1 + \frac{1.8 \text{ k}\Omega}{240 \ \Omega}\right) + 100 \ \mu\text{A}(1.8 \text{ k}\Omega) \cong \mathbf{10.8 \ V}$$

A check of the filter-capacitor voltage shows that an input-output voltage differential of 2 V can be maintained up to at least 200 mA of load current.

PROBLEMS

§ 16.2

1. What is the ripple factor of a sinusoidal signal having a peak ripple of 2 V on an average of 50 V?
2. A filter circuit provides an output of 28 V unloaded and 25 V under full-load operation. Calculate the % voltage regulation.
3. A half-wave rectifier develops 20 V dc. What is the rms value of the ripple voltage?
4. What is the rms ripple voltage of a full-wave rectifier whose output voltage is 8 V dc?

§ 16.3

5. A simple capacitor filter fed by a full-wave rectifier develops 14.5 V dc at 8.5% ripple factor. What is the output ripple voltage (rms)?

6. A full-wave rectified signal of 18 V peak is fed into a capacitor filter. What is the voltage regulation of the filter circuit if the output dc is 17 V dc at full load?

7. A full-wave rectified voltage of 18 V peak is connected to a 400-μF filter capacitor. What is the dc voltage across the capacitor at a load of 100 mA?

8. A full-wave rectifier operating from the 60-Hz ac supply produces a 20-V peak rectified voltage. If a 200-μF filter capacitor is used, calculate the ripple at a load of 120 mA.

9. A capacitor filter circuit ($C = 100\ \mu$F) develops 12 V dc when connected to a load of 2.5 kΩ. Using a full-wave rectifier operating from the 60-Hz supply, calculate the ripple of the output voltage.

10. Calculate the size of the filter capacitor needed to obtain a filtered voltage with 15% ripple at a load of 150 mA. The full-wave-rectified voltage is 24 V dc, and supply is 60 Hz.

11. A 500-μF filter capacitor provides a load current of 200 mA at 8% ripple. Calculate the peak rectified voltage obtained from the 60-Hz supply and the dc voltage across the filter capacitor.

12. Calculate the size of the filter capacitor needed to obtain a filtered voltage with 7% ripple at a load of 200 mA. The full-wave-rectified voltage is 30 V dc, and supply is 60 Hz.

13. Calculate the percent ripple for the voltage developed across a 120-μF filter capacitor when providing a load current of 80 mA. The full-wave rectifier operating from the 60-Hz supply develops a peak rectified voltage of 25 V.

14. Calculate the amount of peak diode current through the rectifier diode of a full-wave rectifier feeding a capacitor filter when the average current drawn from the filter is 100 mA when diode conduction is for $\frac{1}{10}$ cycle.

§ 16.4

15. An *RC* filter stage is added after a capacitor filter to reduce the percent of ripple to 2%. Calculate the ripple voltage at the output of the *RC* filter stage providing 80 V dc.

16. An *RC* filter stage ($R = 33\ \Omega$, $C = 120\ \mu$F) is used to filter a signal of 24 V dc with 2 V rms operating from a full-wave rectifier. Calculate the percent ripple at the output of the *RC* section for a load of 100 mA. Calculate also the ripple of the filtered signal applied to the *RC* stage.

17. A simple-capacitor filter has an input of 40 V dc. If this voltage is fed through an *RC* filter section ($R = 50\ \Omega$, $C = 40\ \mu$F), what is the load current for a load resistance of 500 Ω?

18. Calculate the ripple voltage (rms) at the output of an *RC* filter section that feeds a 1-kΩ load when the filter input is 50 V dc with 2.5 V rms ripple from a full-wave rectifier and capacitor filter. The *RC* filter section components are $R = 100\ \Omega$ and $C = 100\ \mu$F.

19. If the output no-load voltage for the circuit of Problem 18 is 60 V, calculate the percent of voltage regulation with the 1-kΩ load.

§ 16.5

20. Draw the circuit diagram of a voltage doubler. Indicate the value of the diode PIV rating in terms of the transformer peak voltage, V_m.

21. Draw a voltage tripler circuit. Indicate diode PIV ratings and voltage across each circuit capacitor. Include polarity.

22. Repeat Problem 21 for a voltage quadrupler.

§ 16.6

23. Determine the percent voltage regulation of a dc supply providing 100 V under no-load conditions and 95 V at full-load.

24. Determine the voltage regulation of the network of Fig. 16.40 if R_L is limited to a minimum value of 1 kΩ before exceeding the current limits of the supply.

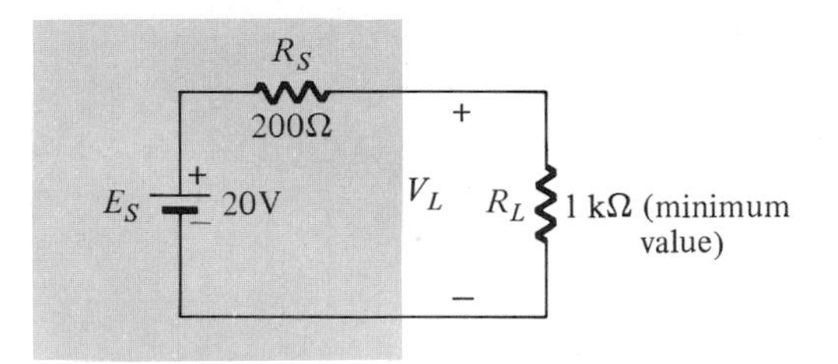

Figure 16.40 Circuit for problem 24.

25. Determine the current regulation of the network of Fig. 16.41 if R_L is limited to a minimum value of 1 kΩ before exceeding the voltage limits of the supply.

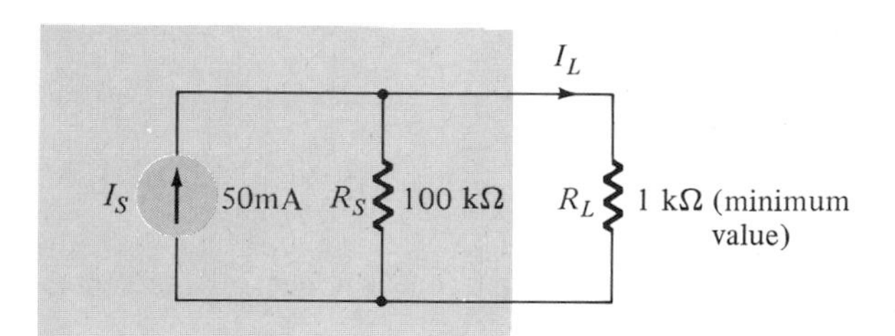

Figure 16.41 Circuit for problem 25.

26. (a) For the Zener diode shunt regulator of Fig. 16.22, determine the minimum value of R_L to ensure that the Zener diode is in the "on" state if V_i = 60 V, V_z = 12.5 V and R_S = 200 Ω.
 (b) For the conditions of part (a), determine the maximum load current I_L.
 (c) Find the minimum value of I_L if $R_{L_{max}}$ is 5 kΩ.
 (d) If R_Z is 1.5 Ω, determine the voltage regulation for the range indicated above.

27. A Zener regulator uses a diode with V_Z = 22 V and a maximum dissipation of 2 W. If the applied voltage is 50 V, find the minimum value of the resistance R_S.

28. For the network of Fig. 16.28, if R_L = 1 kΩ, R = 5 kΩ, V_Z = 10 V, V_{BE} = 0.7 V, and V_i = 20 V, determine the following:
 (a) The voltage V_L and the current I_L.
 (b) The collector current I_C.
 (c) The current through R.
 (d) The supply current.

29. For the network of Fig. 16.29, if R_L = 4 kΩ, R_S = 2 kΩ, V_Z = 10 V, V_{BE} = 0.7 V, and V_i = 20 V, determine the following:
 (a) The voltage V_L.

(b) The current I_L.
(c) The supply current through R_S.
(d) The Zener current if $\beta = 50$.

30. Determine the value of R_2 in the network of Fig. 16.30 to establish a load voltage of 20 V.

31. (a) Describe the complete behavior of the voltage-related supply of Fig. 16.32.
(b) Describe the resulting action to maintain a fixed V_L, if V_L begins to drop in level.

32. For the network of Fig. 16.33, calculate V_L if $V_{BE} = 0.7$ V, $V_Z = 10$ V, and $V_i = 20$ V. Resistor values remain the same.

§ 16.7

33. Determine the regulated output voltage using an LM317, as in Fig. 16.36, with $R_1 = 240\ \Omega$ and $R_2 = 1.8\ k\Omega$.

34. What output voltage results in a circuit as in Fig. 16.36 with $R_1 = 240\ \Omega$ and $R_2 = 3.3\ k\Omega$?

§ 16.8

35. Determine the % ripple across the filter capacitor of a voltage supply as in Fig. 16.37 operating into a load that draws 250 mA. Assume that $V_m = 25.5$ V.

36. If a voltage supply, as in Fig. 16.37, has a ripple of 12% at 20 V dc across capacitor C, to what lowest value does V_{in} drop during the cycle? Assume that $V_m = 25.5$ V.

37. For a 5-V supply as in Fig. 16.38 with $C = 500\ \mu F$ and load $I_{dc} = 150$ mA, determine:
(a) V_{dc} across capacitor C.
(b) V_r (peak) across capacitor C.

38. A +5-V supply as in Fig. 16.38 with $C = 330\ \mu F$ and load of 300 mA has what value of $V_{in_{min}}$? Will output be maintained at regulated +5-V level? (Assume that $V_m = 15$ V.)

17

Linear/Digital ICs

17.1 INTRODUCTION

A group of IC units are available containing both linear and digital circuits. These include comparator circuits, digital/analog converter circuits, interfacing circuits, and timer circuits, among the more popular units. The comparator circuit receives input of a linear voltage, comparing it to a reference input voltage to determine which is greater. The output of the unit is a digital signal that indicates whether or not the input exceeded the reference. The IC unit thus accepts input of a linear voltage providing as output a digital voltage. Digital/analog converter circuits are used to convert a digital value to a proportional analog or linear voltage.

A wide variety of circuits are available to interface or interconnect different types of signals, both linear and digital. Some interface circuits operate by matching impedance levels, some operate by adjusting to various voltage levels, some operate from specified transducers, and some into prescribed loads. A variety of interface circuits deal with the conversion of different digital signal levels.

A timer circuit also contains both linear and digital parts. A flexible arrangement of both linear comparator circuits and digital circuits allows use of the timer in a variety of applications, including generation of pulse signals that are triggered by an input signal, and generation of a clock signal that operates at a frequency set by external resistor and capacitor. The 555 IC timer unit, one of the more popular circuits, is covered in this chapter.

17.2 COMPARATOR UNITS AND OPERATION

A comparator circuit accepts input of linear voltages and provides a digital output that indicates when one input is less than or greater than the second. A basic comparator circuit can be represented as in Fig. 17.1a. The output is a digital signal that stays at a high level when the noninverting (+) input is greater than the inverting (−) input and switches to a lower voltage level when the noninverting input voltage goes below the inverting input reference voltage level. Figure 17.1b shows a typical connection with one input (the inverting input in this example) connected to a reference voltage, the other connected to the input signal voltage. As long as V_{in} is less than the reference voltage level of +2 V, the output remains at a low-voltage level (near −10 V). When the input rises just above +2 V, the output quickly switches to a high-voltage level (near +10 V). Thus, the high output indicates that the input signal is greater than +2 V.

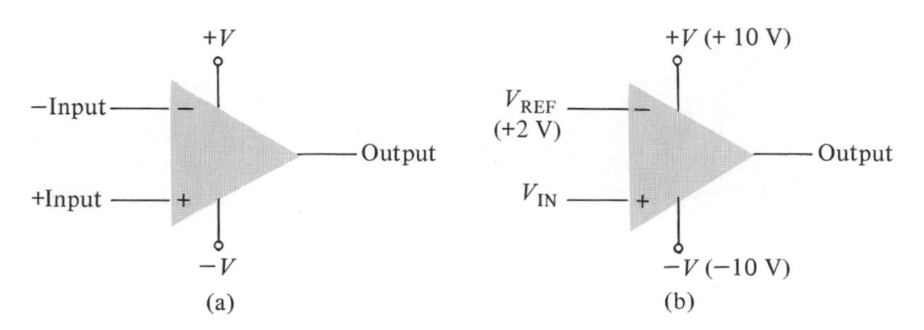

Figure 17.1 Comparator unit: (a) basic unit; (b) typical application.

Since the internal circuit used to build the comparator contains essentially an op-amp circuit with very high gain, we can first examine the operation of a comparator using a 741 op-amp, as shown in Fig. 17.2. With reference input (at pin 2) set to 0 V, a sinusoidal signal applied to the input terminal (at pin 3) will cause the output to switch between its two output states, as shown in Fig. 17.2b. The input, V_i, going even a fraction of a millivolt above the 0 V reference level will be amplified by the very high gain (typically over 100,000) so that the output rises to its positive output saturation level and remains there while the input stays above $V_{ref} = 0$ V. When the input drops just below the reference 0-V level, the output is driven to its lower saturation level and stays there while the input remains below $V_{ref} = 0$ V. Figure 17.2b clearly shows that the input signal is linear while the output is digital.

In general, the reference level need not be 0 V, but any desired value, either positive or negative. Either op-amp (or comparator) input may be used as the reference input, the other then connected to the input signal.

Figure 17.3a shows a circuit operating with a positive voltage reference level, with output driving an indicator LED. The reference level is set at

$$V_{ref} = \frac{10\text{ k}\Omega}{10\text{ k}\Omega + 10\text{ k}\Omega}(+12\text{ V}) = +6\text{ V}$$

Since the reference voltage is connected to the inverting input, the output will switch

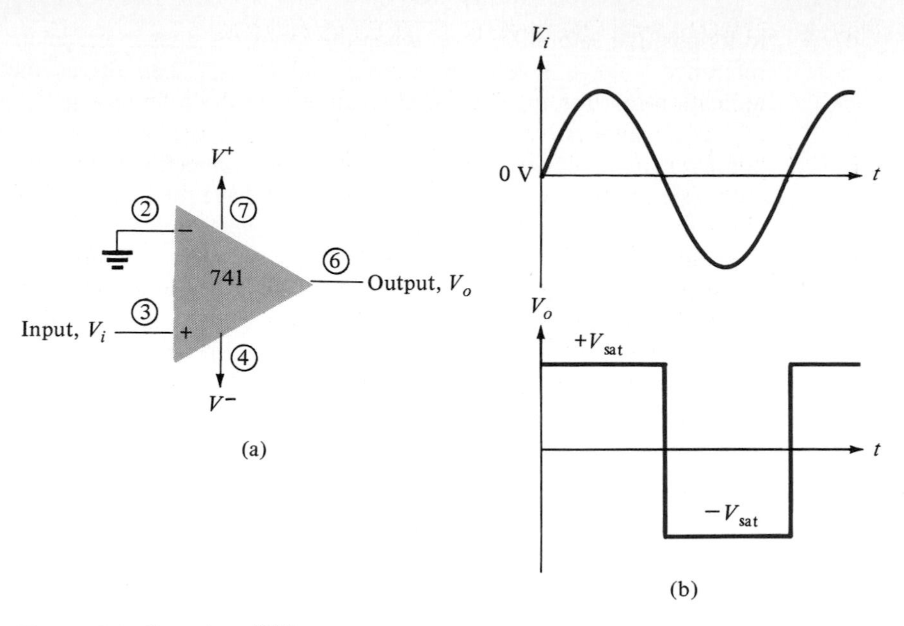

Figure 17.2 Operation of 741 op-amp as comparator.

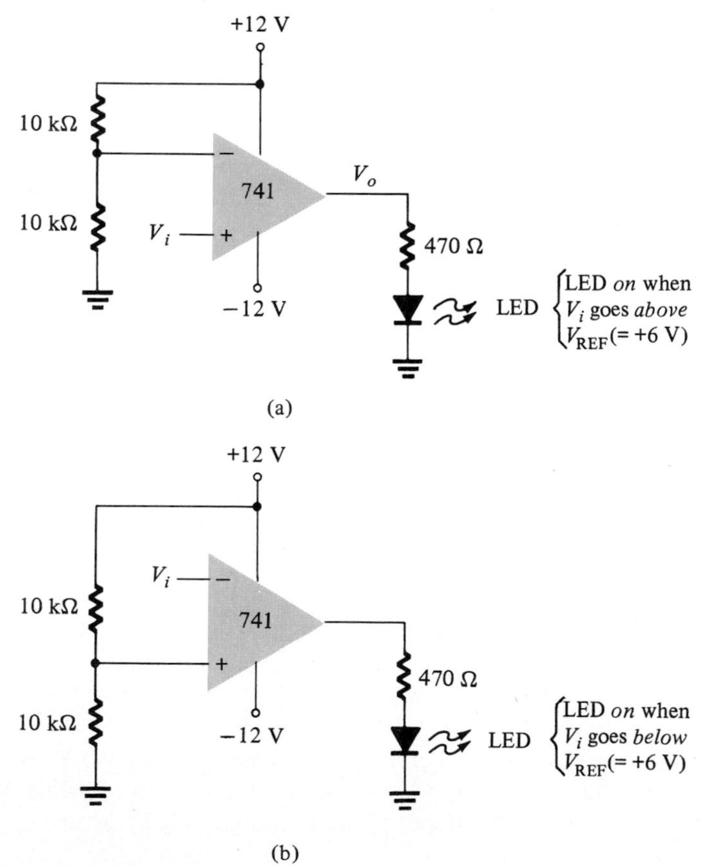

Figure 17.3 A 741 op-amp used as a comparator.

to its positive saturation level when the input, V_i, goes more positive than the +6-V reference voltage level. The output voltage, V_o, then drives the LED on, as an indication that the input is more positive than the reference level.

As an alternative connection, the reference voltage could be connected to the noninverting input (see Fig. 17.3b). With this connection the input signal going below the reference level would cause the output to drive the LED on. The LED can thus be made to go on when the input signal goes either above or below the reference level, depending on which input is connected as the input signal and which as the reference.

While op-amp ICs can be used as comparator circuits, separate comparator ICs are available for use in such applications. Some improvements built into comparator IC units are faster switching between the two output levels, built-in noise immunity to prevent the output from oscillating when the input passes by the reference level, and outputs capable of directly driving a variety of loads. A few popular IC comparators are covered next to show how they are defined and how they may be used.

The 311 voltage comparator shown in Fig. 17.4 contains a comparator circuit that can operate as well from dual power supplies of ±15 V as from a single +5-V supply (as used for digital logic circuits). The output can provide voltage at one of two distinct levels or can be used to drive lamps or relays. Notice that the output is taken from a bipolar transistor to allow driving a variety of loads. The unit also has balance and strobe inputs, the strobe input allowing gating of the output. A few examples will show how the comparator unit can be used for common applications.

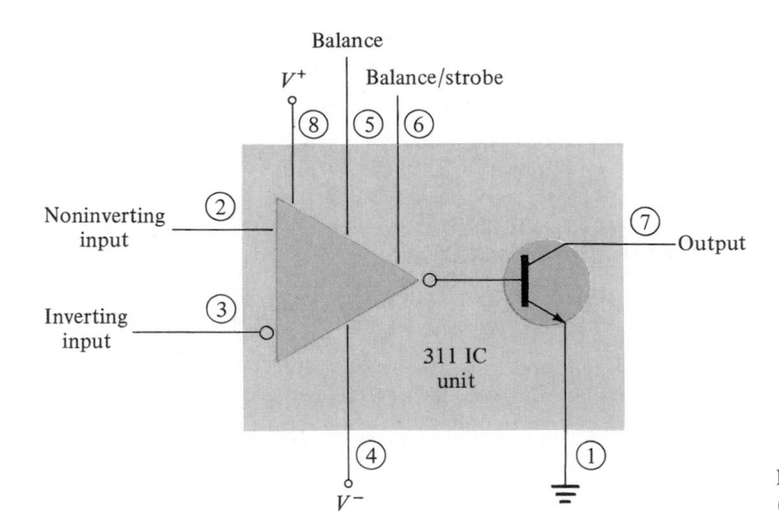

Figure 17.4 A 311 comparator (eight-pin DIP unit).

A zero-crossing detector can be built using the 311 as shown in Fig. 17.5. The input going positive (above 0 V) drives the output transistor on, the output going low, to −10 V in this connection. The input going below 0 V will drive the output transistor off, the output going to +10 V.

The output is thus an indication of whether the input is above or below 0 V. When the input is any positive voltage (above 0 V), the output is a low level, while any negative input voltage will result in the output going to a high-voltage level.

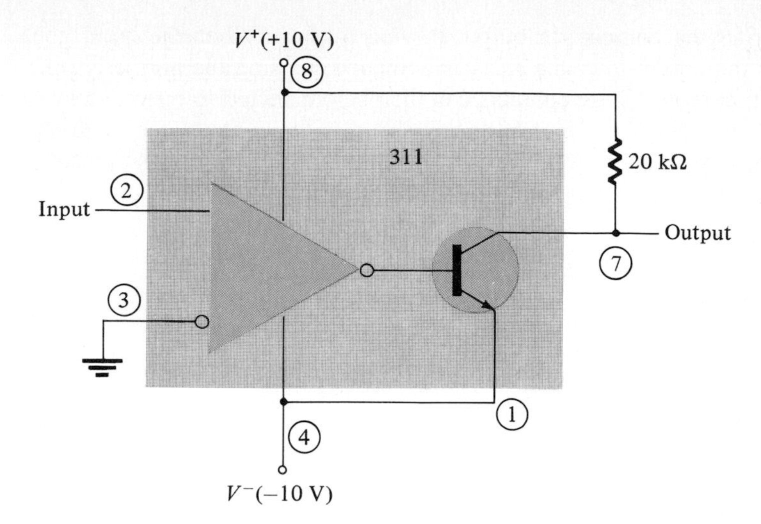

Figure 17.5 Zero-crossing detector using a 311 IC.

Figure 17.6 shows how a 311 comparator can be used with strobing. In this example the output will go high when the input goes above the reference level—but only if the TTL strobe input is off (or 0 V). If the TTL strobe input goes high, it drives the 311 strobe input at pin 6 low, causing the output to remain in the off state (with output high) regardless of the input signal. In effect, the output remains high unless strobed. If strobed, the output then acts normally, switching from high to low depending on the input signal level. In operation, the comparator output will respond to the input signal only during the time the strobe signal allows such operation.

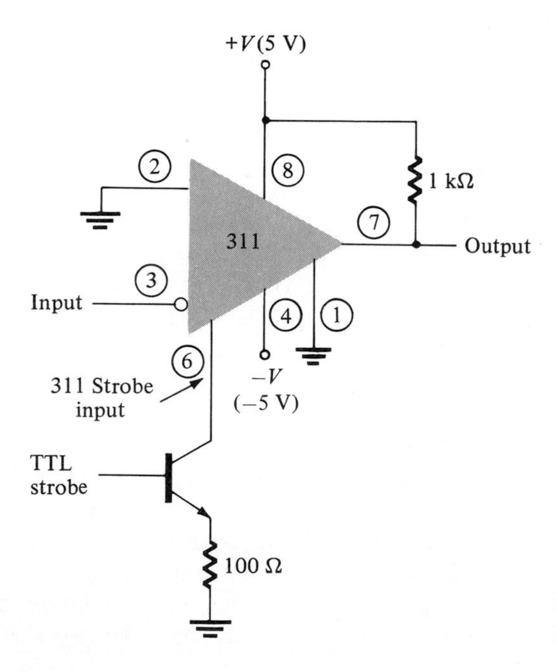

Figure 17.6 Operation of a 311 comparator with strobe input.

Figure 17.7 shows the comparator output driving a relay. When the input goes below 0 V, driving the output low, the relay is activated, closing the normally open (N.O.) contacts at that time. These contacts can then be connected to operate a large variety of devices. For example, a buzzer or bell wired to the contacts can be driven on whenever the input voltage drops below 0 V. As long as the voltage is present at the input terminal, the buzzer will remain off.

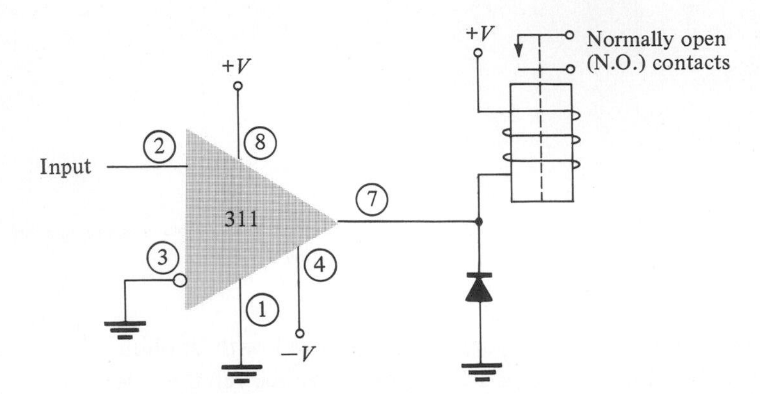

Figure 17.7 Operation of a 311 comparator with relay output.

Another popular comparator unit is packaged with four independent voltage comparator circuits in a single IC. The 339 is a quad comparator IC, all four comparator circuits connected to the external pins as shown in Fig. 17.8. Each comparator has inverting and noninverting input and an output. The supply voltage applied to a pair of pins connects to all four comparator circuits. Even if one wishes to use only some of the comparator circuits, all four are active, drawing power from the supply.

To see how these comparator circuits can be used, Fig. 17.9 shows one of the 339 comparator circuits connected as a zero-crossing detector. Whenever the input signal goes above 0 V, the output switches to V^+. The output switches to V^- only when the input goes below 0 V.

A reference level of other than 0 V can also be used and either input terminal could be used as reference, the other input then being the signal input. The operation of one of the comparator circuits is described next.

The differential input (voltage difference across input terminals) going positive drives the output transistor off (open circuit), while a negative differential input drives the output transistor on—the output then at the supply low level.

If the negative input is set at a reference level, V_{ref}, the positive input going above V_{ref} results in a positive differential input with output driven to the open-circuit state. When the noninverting input goes below V_{ref}, resulting in a negative differential input, the output will be driven to V^-.

If the positive input is set at the reference level, the inverting input going below V_{ref} results in the output open circuit, while the inverting input going above V_{ref} results in the output at V^-. This operation is summarized in Fig. 17.10.

Since the output of one of these comparator circuits is open-circuit collector, applications in which the outputs from more than one circuit can be wire-ORed are possible. Figure 17.11 shows two comparator circuits connected with common output,

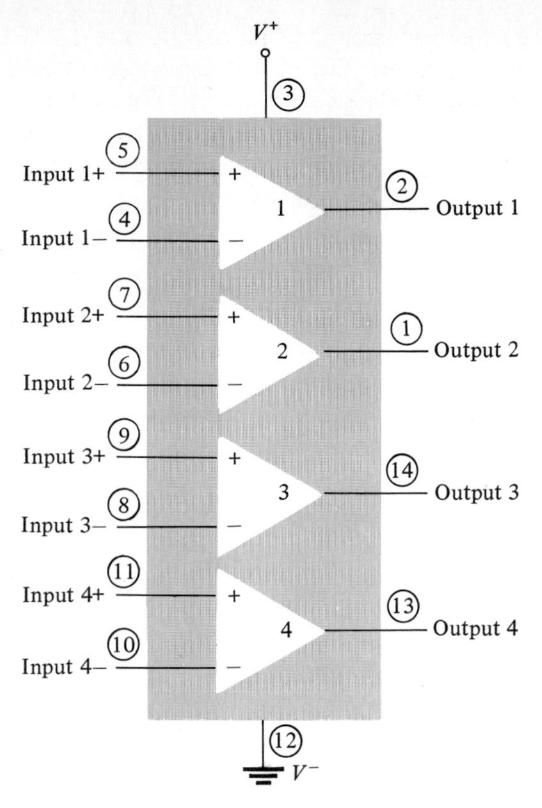

Figure 17.8 Quad comparator IC (339).

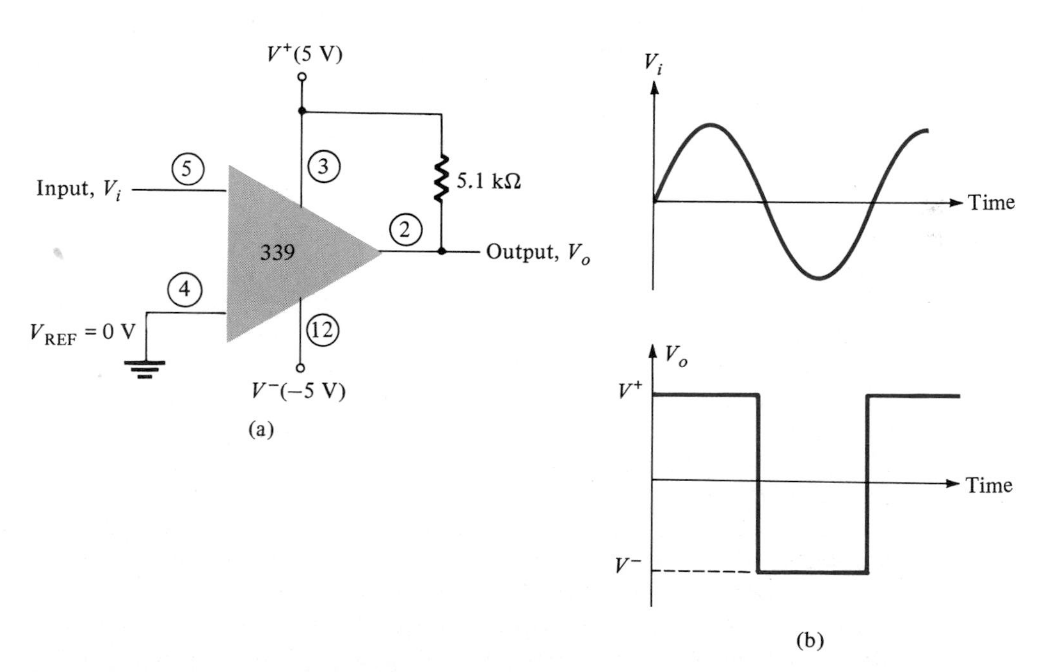

Figure 17.9 Operation of one 339 comparator circuit as a zero-crossing detector.

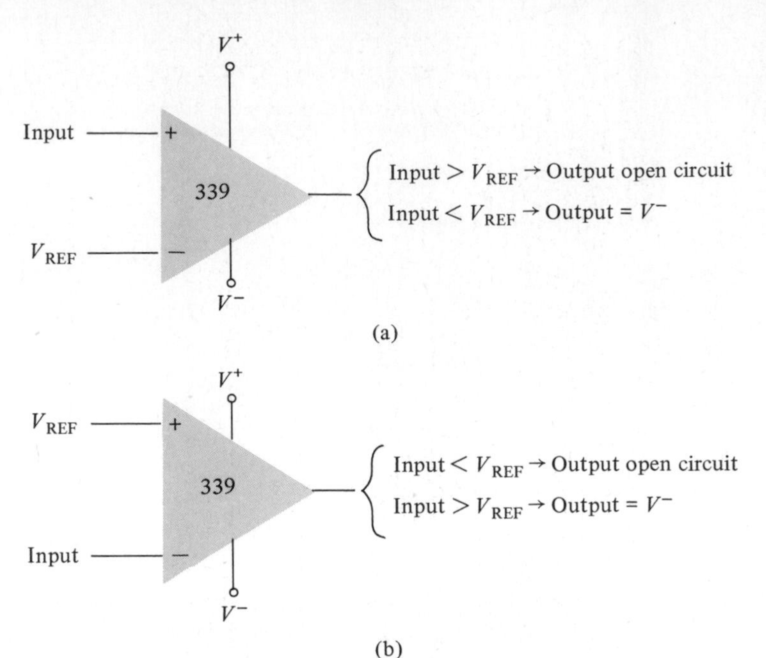

Figure 17.10 Operation of a 339 comparator circuit with reference input at (a) minus input; (b) plus input.

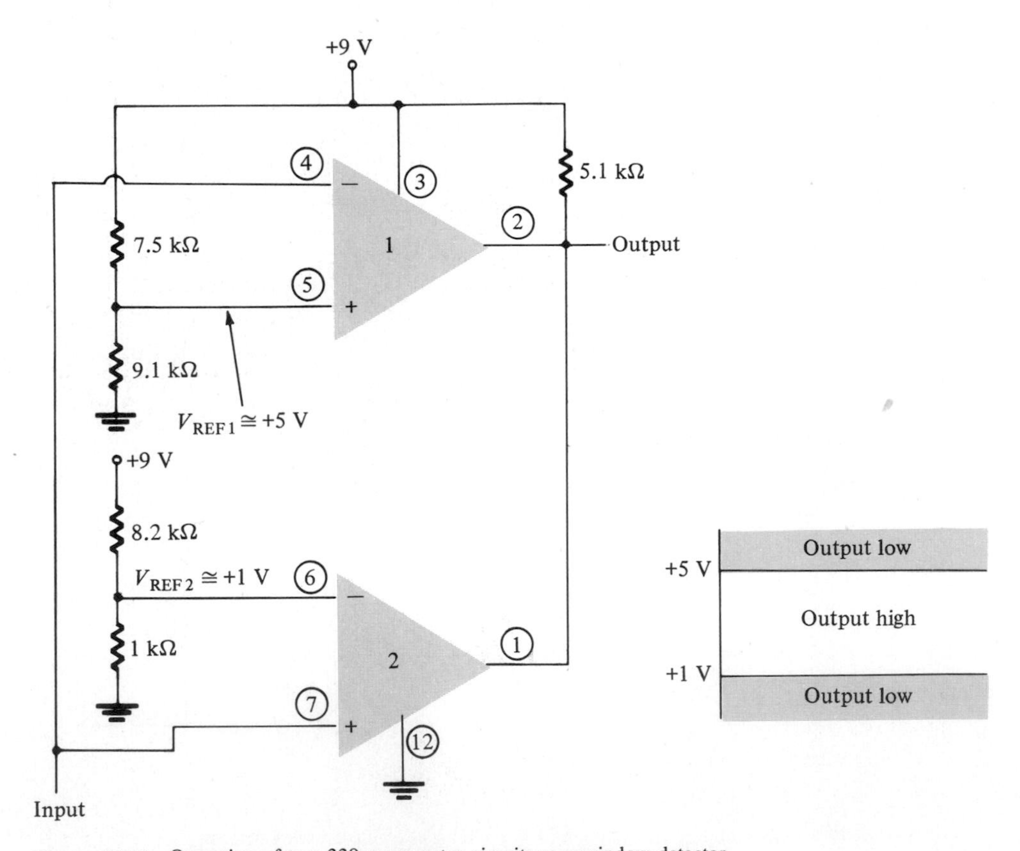

Figure 17.11 Operation of two 339 comparator circuits as a window detector.

and also with common input. Comparator 1 has a +5-V reference voltage input connected to the noninverting input. The output will be driven low by comparator 1 when the input signal goes above +5 V. Comparator 2 has a reference voltage of +1 V connected to the inverting input. The output of comparator 2 will be driven low when the input signal goes below +1 V. In total the output will go low whenever the input is below +1 V or above +5 V, as shown in Fig. 17.11, the overall operation being that of a voltage window detector. The output high indicates that the input is within a voltage window of +1 V to +5 V (these values being set by the reference voltage levels used).

17.3 DIGITAL/ANALOG CONVERTERS

Many of the voltage and current signals occurring in electronics are linear, in that they vary continuously over some range of values. In digital devices and computers the signals are digital, at one of two levels representing the binary values of one or zero.

If the signals to be used in some digital operations are linear (analog) voltages (e.g., dc voltages representing temperature or pressure, or position), a circuit must convert this analog voltage into a digital value—this conversion circuit being an analog-to-digital (A/D) converter. When a computer has a digital value to be output as an analog voltage, a digital-to-analog (D/A) converter circuit is used.

Digital-to-Analog Conversion

Digital-to-analog conversion can be achieved using a number of different methods. One popular scheme uses a network of resistors, called a ladder network. A ladder network accepts inputs of binary values at, typically, 0 V or V_{ref}, and provides an output voltage proportional to the binary input value. Figure 17.12a shows a ladder

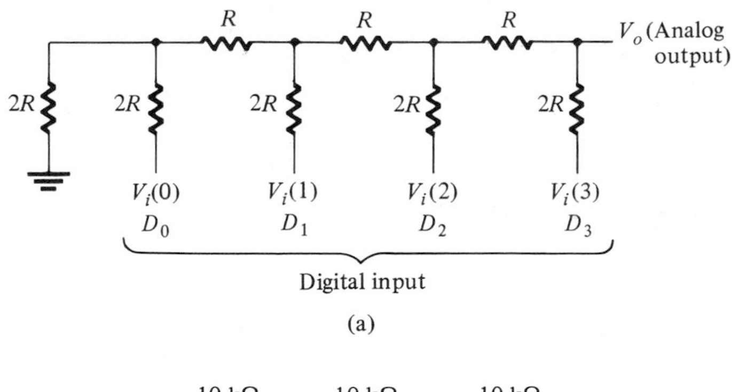

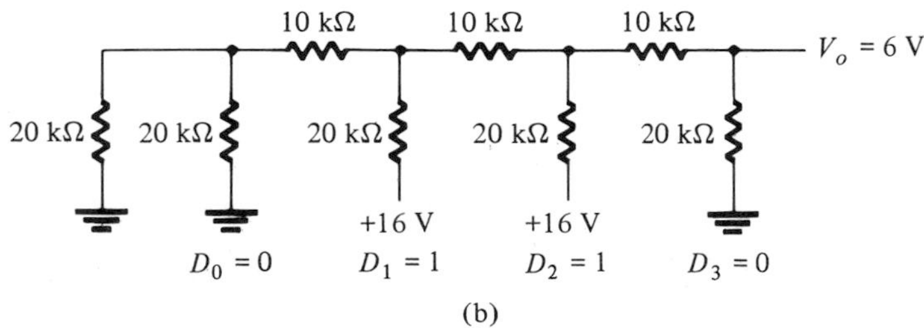

Figure 17.12 Four-stage ladder network used as D/A converter: (a) basic circuit; (b) circuit example with 0110 input.

network with four input voltages, representing 4 bits of digital data and a dc voltage output. The output voltage is proportional to the digital input value as given by the relation

$$V_o = \frac{D_0 \times 2^0 + D_1 \times 2^1 + D_2 \times 2^2 + D_3 \times 2^3}{2^4} \times V_{\text{ref}} \tag{17.1}$$

In the example shown in Fig. 17.12b, the output voltage resulting should be

$$V_o = \frac{0 \times 1 + 1 \times 2 + 1 \times 4 + 0 \times 8}{16} \times 16 \text{ V} = 6 \text{ V}$$

Therefore, 0110_2 converts to 6 V.

Use superposition to verify that the resulting value of V_o is indeed 6 V. The function of the ladder network is to convert the 16 possible binary values from 0000 to 1111 into one of 16 voltage levels in steps of $V_{\text{ref}}/16$. Using more sections of ladder allows having more binary inputs and greater quantization for each step. For example, a 10-stage ladder network could extend the number of voltage steps or voltage resolution to $V_{\text{ref}}/2^{10}$ or $V_{\text{ref}}/1024$. A reference voltage of 10 V would then provide output voltage steps of 10 V/1024 or approximately 10 mV. More ladder stages provide greater voltage resolution, in general the voltage resolution for n ladder stages being

$$\frac{V_{\text{ref}}}{2^n} \tag{17.2}$$

A block diagram of the main components of a typical IC D/A converter is shown in Fig. 17.13. The ladder network, referred to in the diagram as an R-$2R$ ladder is sandwiched between the reference current supply and current switches connected to each binary input with a resulting output current proportional to the input binary value. The binary inputs turn on selected legs of the ladder, the output current being a weighted summing of the reference current. Connecting the output across a resistor will produce an analog voltage, if desired.

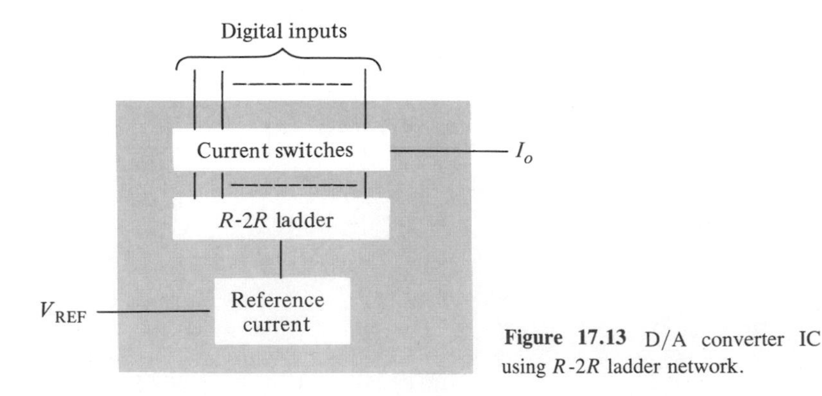

Figure 17.13 D/A converter IC using R-$2R$ ladder network.

Analog-to-Digital Conversion

DUAL-SLOPE CONVERSION

A popular method of converting analog voltage into a digital value is the dual-slope method. Figure 17.14a shows a block diagram of the basic dual-slope converter. The analog voltage to be converted is applied through an electronic switch to an integrator or ramp-generator circuit (essentially a constant current charging a capacitor to produce a linear-ramp voltage). The digital output is obtained from a counter operated during both positive and negative slope intervals of the integrator.

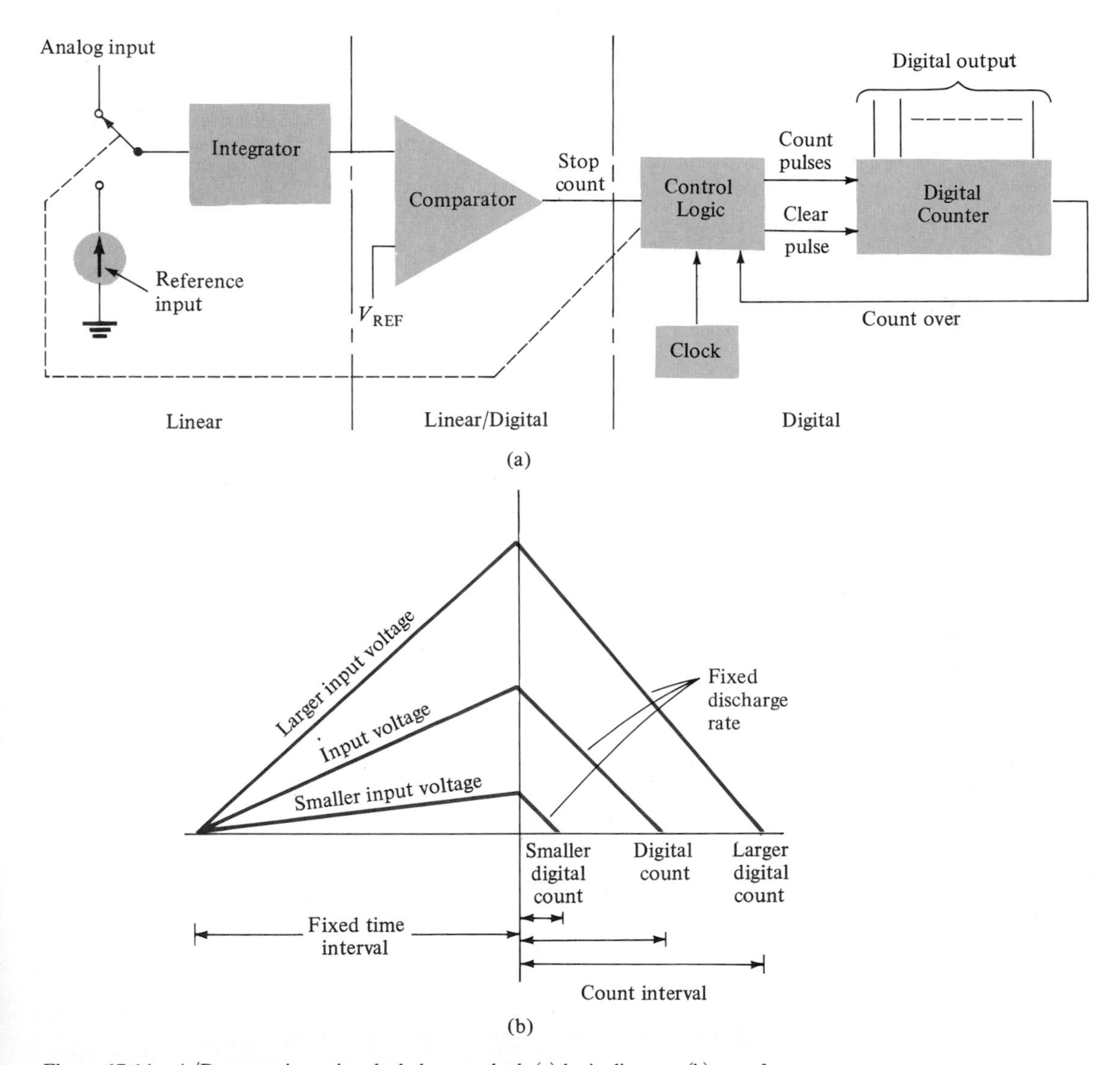

Figure 17.14 A/D conversion using dual-slope method: (a) logic diagram; (b) waveform.

The conversion method proceeds as follows. For a fixed time interval (usually the full count range of the counter), the analog input voltage, connected to the integrator, raises the voltage in the comparator to some positive level. Figure 17.14b shows that at the end of the fixed time interval the voltage from the integrator is greater for larger input voltages. At the end of the fixed count interval, the count is set at zero and the electronic switch connects the integrator to a reference or fixed input. The integrator output (or capacitor input) then decreases at a fixed rate. The counter advances during this time. The integrator output voltage decreases at a fixed rate until it drops below the comparator reference voltage, at which time the control logic receives a signal (the comparator output) to stop the count. The digital value stored in the counter is then the digital output of the converter.

Using the same clock and integrator to perform the conversion during positive and negative slope intervals tends to compensate for clock frequency drift and integrator accuracy limitations. Setting the reference input value and clock rate can scale the counter output as desired. The counter can be binary or BCD or other digital form, if desired.

LADDER-NETWORK CONVERSION

Another popular method of analog-to-digital conversion uses the ladder network along with counter and comparator circuits (see Fig. 17.15). A digital counter advances from a zero count while a ladder network driven by the counter outputs a staircase voltage as shown in Fig. 17.15b, which increases one voltage increment for each count step. A comparator circuit, receiving both staircase voltage and analog input voltage, provides a signal to stop the count when the staircase voltage rises above the input voltage. The counter value at that time is the digital output.

The amount of voltage change stepped by the staircase signal depends on the reference voltage applied to the ladder network and on the number of count bits used. A 12-stage counter operating a 12-stage ladder network using a reference voltage of 10 V would step each count by a voltage of

$$\frac{V_{\text{ref}}}{2^{12}} = \frac{10\text{ V}}{4096} = 2.4\text{ mV}$$

This would result in a conversion resolution of 2.4 mV. The clock rate of the counter would affect the time required to carry out a conversion. A clock rate of 1 MHz operating a 12-stage counter would need a maximum conversion time of

$$4096 \times 1\ \mu\text{s} = 4096\ \mu\text{s} \cong 4.1\text{ ms}$$

The minimum number of conversions that could be carried out each second would then be

$$\text{number of conversions} = 1/4.1\text{ ms} \cong 244\text{ conversion/second}$$

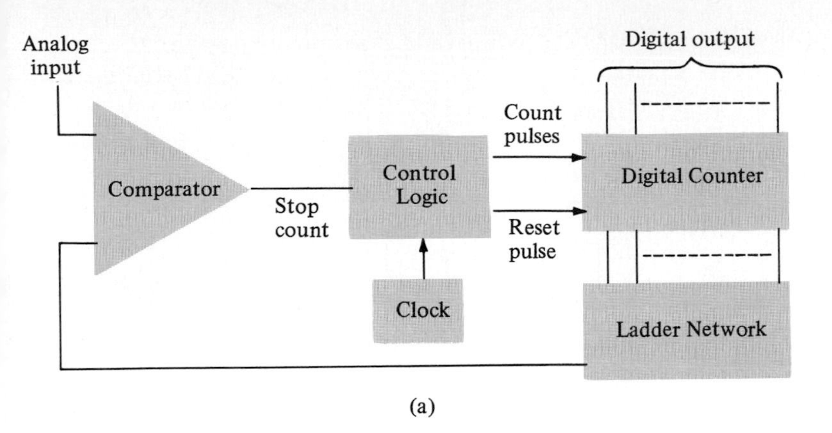

(a)

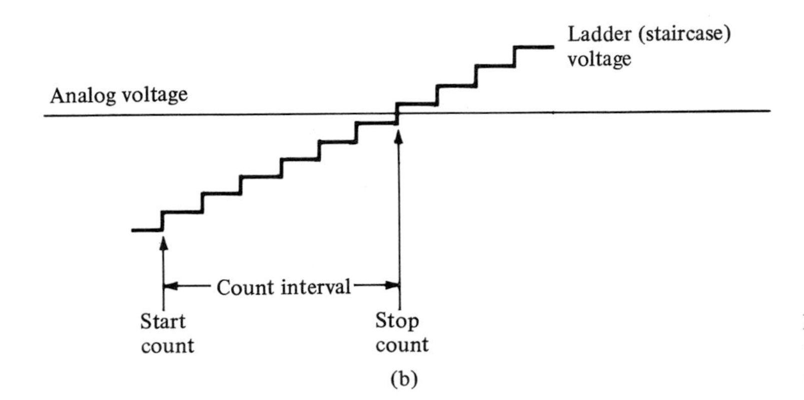

(b)

Figure 17.15 A/D conversion using ladder network: (a) logic diagram; (b) waveform.

Since on the average, with some conversions requiring little count time and others near maximum count time, a conversion time of (4.1 ms)/2 = 2.05 ms would be needed, the average number of conversions would be 2 × 244 = 488 conversions/second. A slower clock rate would result in fewer conversions per second. A converter using fewer count stages (and less conversion resolution) would carry out more conversions per second. The conversion accuracy depends on the accuracy of the comparator.

17.4 TIMER IC UNIT AND APPLICATIONS

Another popular analog-to-digital integrated circuit is the versatile 555 timer unit. The IC is made of a combination of linear comparators and digital flip-flop as described in Fig. 17.16. The entire circuit is usually housed in an eight-pin DIP package with pin numbers as specified in Fig. 17.16. A series connection of three resistors set the reference level inputs to the two comparators at $\frac{2}{3}V_{CC}$ and $\frac{1}{3}V_{CC}$, the outputs of these

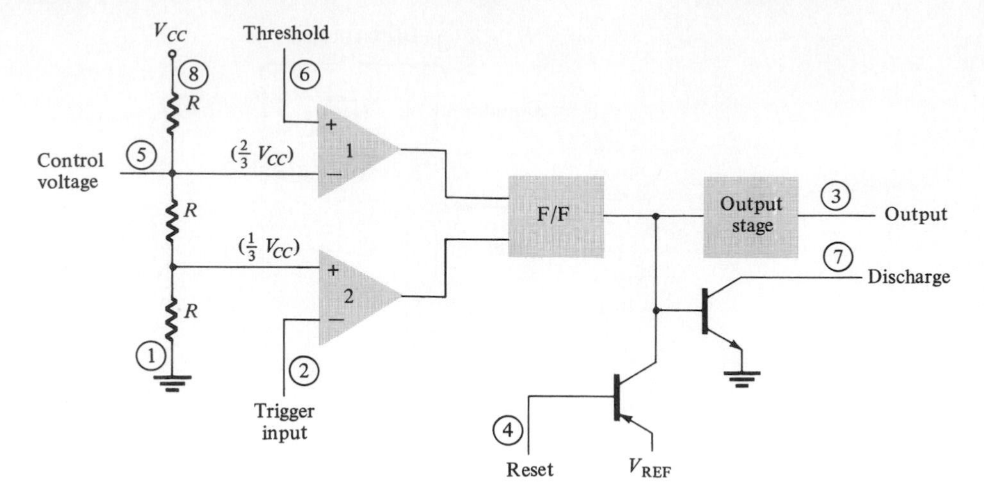

Figure 17.16 Details of 555 timer IC.

comparators setting or resetting the flip-flop unit. The flip-flop circuit output is then brought out through an output amplifier stage. The flip-flop circuit also operates a transistor inside the IC, the transistor collector usually being driven low to discharge a timing capacitor.

Astable

One popular application of the 555 timer IC is as an astable multivibrator or clock circuit. The following analysis of the operation of the 555 as an astable circuit will include details of the different parts of the unit and how the various inputs and outputs are utilized. Figure 17.17 shows an astable circuit using external resistor and capacitor to set the timing interval of the output signal.

Capacitor C charges toward V_{CC} through external resistors R_A and R_B. Referring to Fig. 17.17, the capacitor voltage rises until it goes above $\frac{2}{3}V_{CC}$. This voltage is the threshold voltage at pin 6, which drives comparator 1 to trigger the flip-flop so that the output at pin 3 goes low. In addition, the discharge transistor is driven on, causing the output at pin 7 to discharge the capacitor through resistor R_B. The capacitor voltage then decreases until it drops below the trigger level ($V_{CC}/3$). The flip-flop is triggered so that the output goes back high and the discharge transistor is turned off, so that the capacitor can again charge through resistors R_A and R_B toward V_{CC}.

Figure 17.18b shows the capacitor and output waveforms resulting from the astable circuit connection. Calculation of the time intervals during which the output

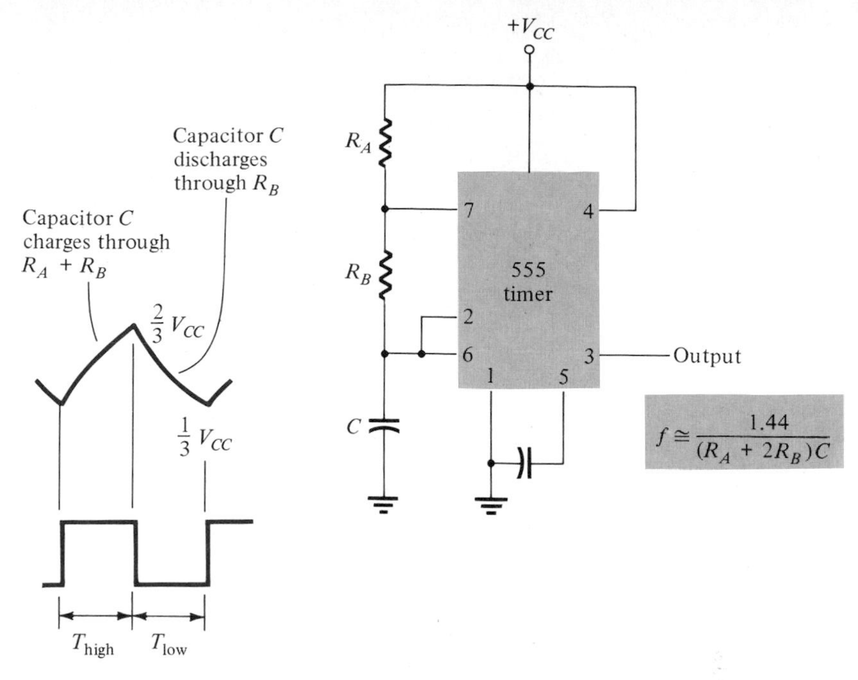

Figure 17.17 Astable multivibrator using 555 IC.

is high and low can be made using the relations:

$$T_{high} \cong 0.7\,(R_A + R_B)C \tag{17.3}$$

$$T_{low} \cong 0.7\,R_BC \tag{17.4}$$

The total period is

$$\text{period} = T = T_{high} + T_{low} \tag{17.5}$$

The frequency of the astable circuit is then calculated using[1]

$$f = \frac{1}{T} \cong \frac{1.44}{(R_A + 2R_B)C} \tag{17.6}$$

---

[1] The period can be directly calculated from

$$T = 0.693\,(R_A + 2R_B)C \cong 0.7(R_A + 2R_B)C$$

and frequency from

$$f \cong \frac{1.44}{(R_A + 2R_B)C}$$

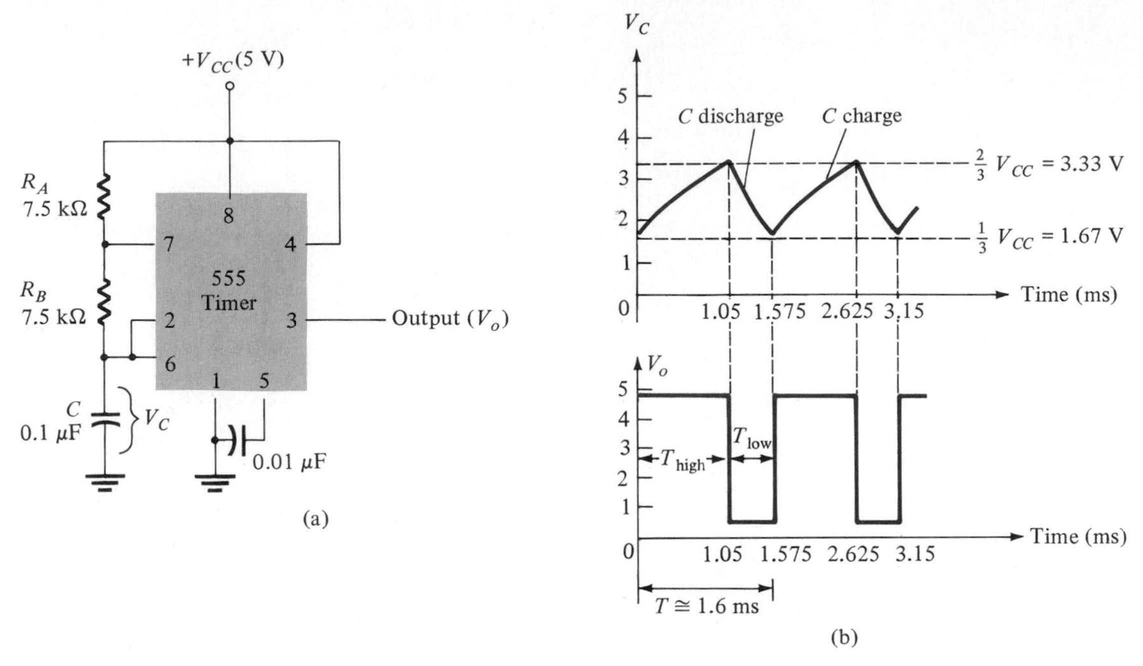

Figure 17.18 Astable multivibrator for Example 17.1: (a) circuit; (b) waveforms.

EXAMPLE 17.1

Determine the frequency and output waveforms for the circuit of Fig. 17.18a.

Solution:

Using Eqs. (17.3) through (17.6) yields

$$T_{\text{high}} = 0.7(R_A + R_B)C = 0.7(7.5 \times 10^3 + 7.5 \times 10^3)(0.1 \times 10^{-6}) = 1.05 \text{ ms}$$

$$T_{\text{low}} = 0.7R_BC = 0.7(7.5 \times 10^3)(0.1 \times 10^{-6}) = 0.525 \text{ ms}$$

$$T = T_{\text{high}} + T_{\text{low}} = 1.05 \text{ ms} + 0.525 \text{ ms} = 1.575 \text{ ms}$$

$$f = \frac{1}{T} = \frac{1}{1.575 \times 10^{-3}} \cong \mathbf{635\ Hz}$$

The waveforms are drawn in Fig. 17.18b.

Monostable

The 555 timer can also be used as a one-shot or monostable multivibrator circuit. Figure 17.19 shows such connection. When the trigger input signal goes negative, it

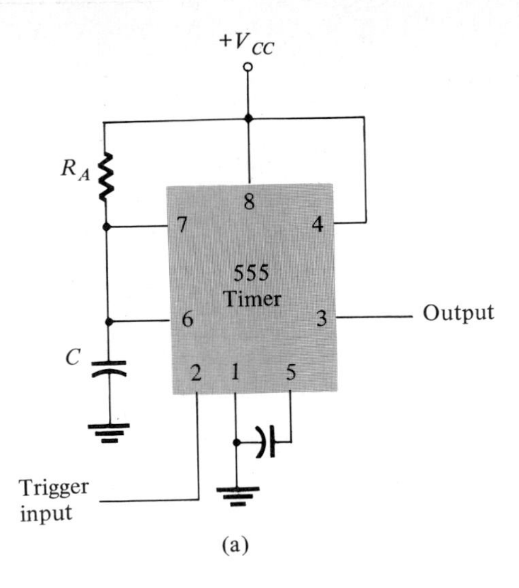

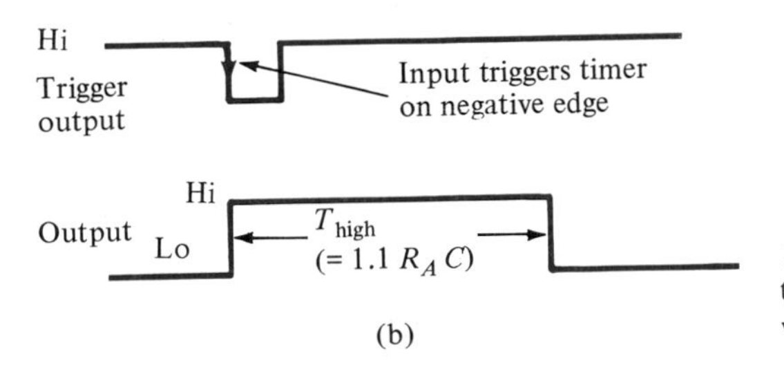

Figure 17.19 Operation of 555 timer as one-shot: (a) circuit; (b) waveforms.

triggers the one-shot with output at pin 3 then going high for a time period

$$\boxed{T_{\text{high}} = 1.1\, R_A C} \tag{17.7}$$

Referring back to Fig. 17.16, the negative edge of the trigger input causes comparator 2 to trigger the flip-flop with output at pin 3 going high. Capacitor C charges toward V_{CC} through resistor R_A. During the charge interval the output remains high. When the voltage across the capacitor reaches the threshold level of $\frac{2}{3}$ V_{CC} comparator 1 then triggers the flip-flop with output going low. The discharge transistor also goes low, causing the capacitor to remain at near 0 V until triggered again.

Figure 17.19b shows the input trigger signal and the resulting output waveform for the 555 timer IC operated as a one-shot. Time periods for this circuit can range from microseconds to many seconds, making this IC useful for a large range of applications.

EXAMPLE 17.2

Determine the period of the output waveform for the circuit of Fig. 17.20 when triggered by a negative pulse.

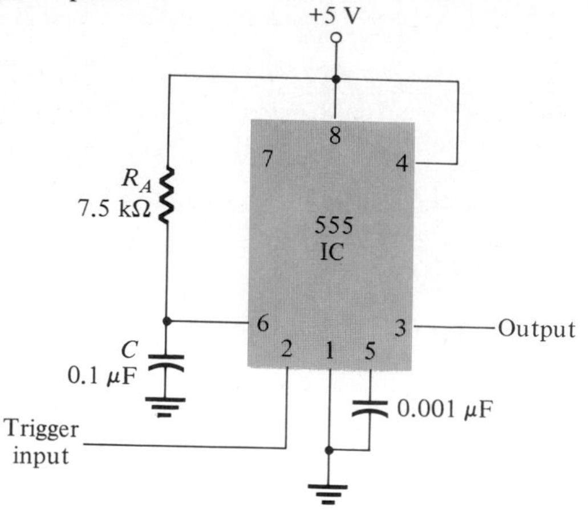

Figure 17.20 Monostable circuit for Example 17.2.

Solution:

Using Eq. (17.7), we obtain

$$T_{\text{high}} = 1.1R_AC = 1.1(7.5 \times 10^3)(0.1 \times 10^{-6}) = \mathbf{0.825\ ms}$$

17.5 VOLTAGE-CONTROLLED OSCILLATOR

A voltage-controlled oscillator (VCO) is a circuit that provides an oscillating output signal (typically of square-wave or triangular-wave form) whose frequency can be adjusted over a range controlled by a dc voltage. An example of a VCO is the 566 IC unit, which contains circuitry to generate both square-wave and triangular-wave signals whose frequency is set by an external resistor and capacitor and then varied by an applied dc voltage. Figure 17.21a shows that the 566 contains current sources to charge and discharge an external capacitor, C_1, at a rate set by external resistor, R_1, and the modulating dc input voltage. A Schmitt trigger circuit is used to switch the current sources between charging and discharging the capacitor, and the triangular voltage developed across the capacitor and square wave from the Schmitt trigger are provided as outputs through buffer amplifiers.

Figure 17.21b shows the pin connection of the 566 unit and a summary of formula and value limitations. The oscillator can be programmed over a 10-to-1 frequency range by proper selection of an external resistor and capacitor, and then modulated over a 10-to-1 frequency range by a control voltage, V_C.

A free-running or center-operating frequency, f_o, can be calculated from

$$f_o = \frac{2}{R_1C_1}\,\frac{V^+ - V_c}{V^+} \tag{17.8}$$

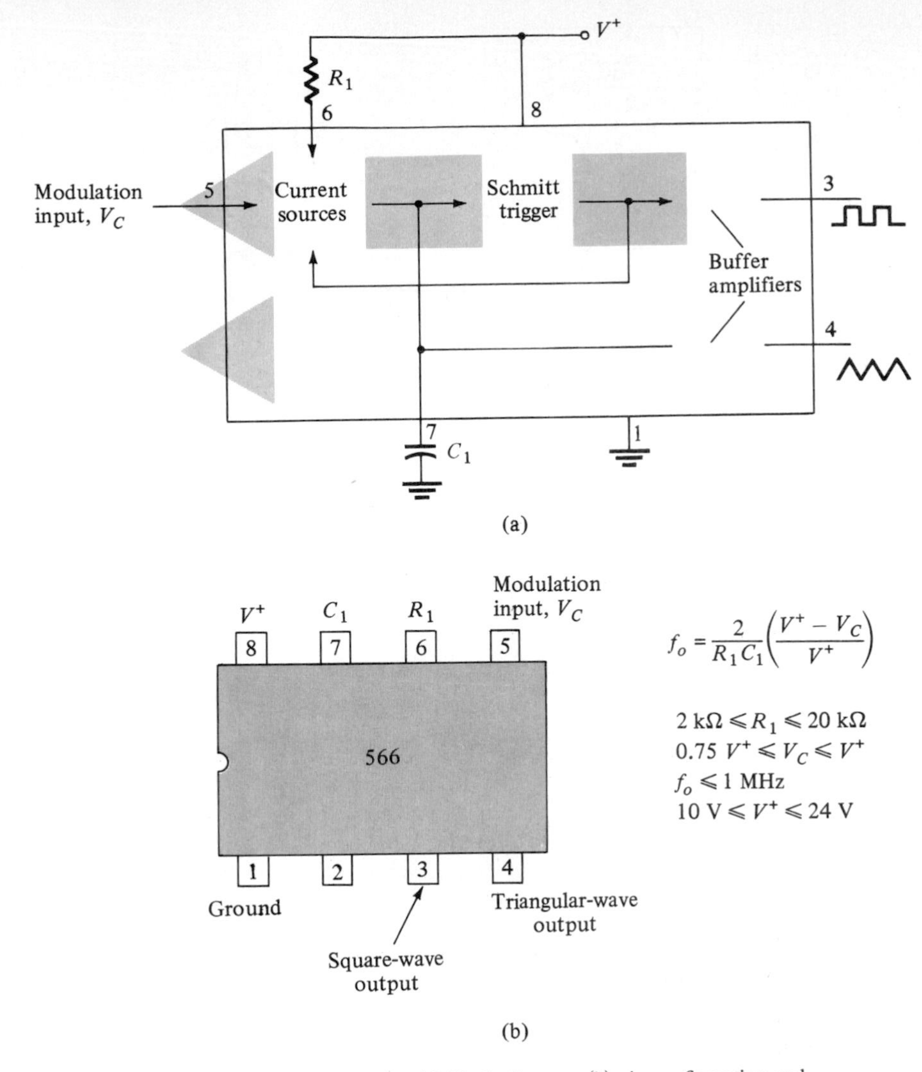

Figure 17.21 A 566 function generator: (a) block diagram; (b) pin configuration and summary of operating data.

with the following practical circuit value restrictions:

1. R_1 should be within range $2\ \text{k}\Omega \leq R_1 \leq 20\ \text{k}\Omega$.
2. V_c should be within range $\frac{3}{4}\text{V}^+ \leq V_c \leq V^+$.
3. f_o should be below 1 MHz.
4. V^+ should range between 10 V and 24 V.

Figure 17.22 shows an example in which the 566 function generator is used to provide both square-wave and triangular-wave signals at a fixed frequency set by R_1,

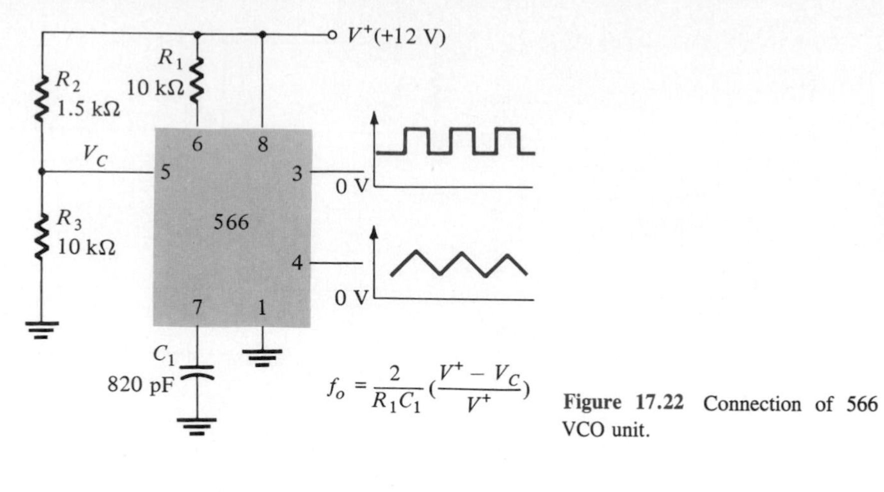

Figure 17.22 Connection of 566 VCO unit.

C_1, and V_c. A resistor divider R_2 and R_3 sets the dc modulating voltage at a fixed value

$$V_c = \frac{R_3}{R_2 + R_3}V^+ = \frac{10\ \text{k}\Omega}{1.5\ \text{k}\Omega + 10\ \text{k}\Omega}12\ \text{V} = 10.4\ \text{V}$$

(which falls properly in the voltage range $0.75\ V^+ = 9$ V and $V^+ = 12$ V). Using Eq. (17.8) yields

$$f_o = \frac{2}{(10 \times 10^3)(820 \times 10^{-12})}\left(\frac{12 - 10.4}{12}\right) \cong 32.5\ \text{kHz}$$

The circuit of Fig. 17.23 shows how the output square-wave frequency can be adjusted using the input voltage, V_c, to vary the signal frequency. Potentiometer R_3 allows varying V_c from about 9 V to near 12 V, over the full 10-to-1 frequency range. With the potentiometer wiper set at the top, the control voltage is

$$V_c = \frac{R_3 + R_4}{R_2 + R_3 + R_4}V^+ = \frac{5\ \text{k}\Omega + 18\ \text{k}\Omega}{510\ \Omega + 5\ \text{k}\Omega + 18\ \text{k}\Omega}(+12\ \text{V}) = 11.74\ \text{V}$$

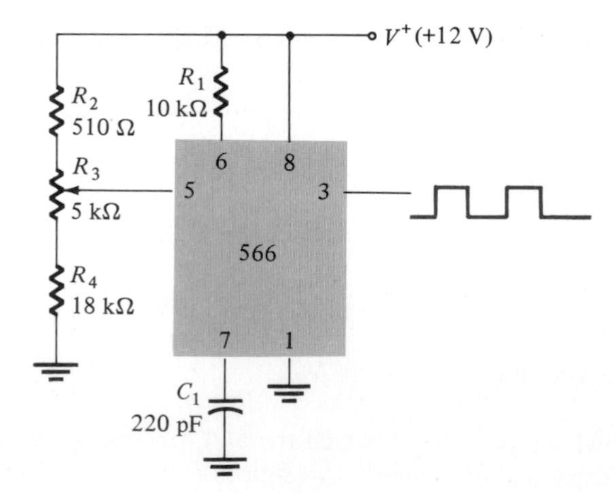

Figure 17.23 Connection of 566 as a VCO unit.

resulting in a lower output frequency of

$$f_o = \frac{2}{(10 \times 10^3)(220 \times 10^{-12})}\left(\frac{12 - 11.74}{12}\right) \cong 19.7 \text{ kHz}$$

With the wiper arm of R_3 set at the bottom, the control voltage is

$$V_c = \frac{R_4}{R_2 + R_3 + R_4}V^+ = \frac{18 \text{ k}\Omega}{510 \ \Omega + 5 \text{ k}\Omega + 18 \text{ k}\Omega}(+12 \text{ V}) = 9.19 \text{ V}$$

resulting in an upper output frequency of

$$f_o = \frac{2}{(10 \times 10^3)(220 \times 10^{-12})}\left(\frac{12 - 9.19}{12}\right) \cong 212.9 \text{ kHz}$$

The frequency of the output square wave can then be varied using potentiometer R_3 over a frequency range of at least 10 to 1.

Rather than varying a potentiometer setting to change the value of V_c, an input modulating voltage, V_{in}, can be applied as shown in Fig. 17.24. The voltage divider sets V_c at about 10.4 V. An input ac voltage of about 1.4 V, peak can drive V_c around the bias point between voltages of 9 V and 11.8 V, causing the output frequency to vary over about a 10-to-1 range. The input signal, V_{in}, thus frequency modulates the output voltage around the center frequency set by the bias value of V_c = 10.4 V (f_o = 121.2 kHz).

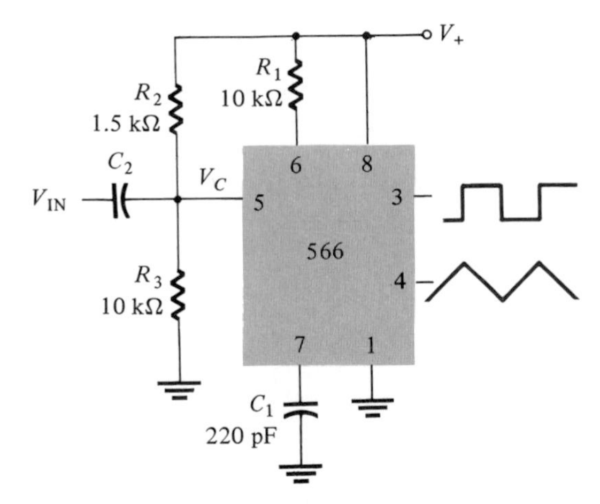

Figure 17.24 Operation of VCO with frequency modulating input.

17.6 PHASE-LOCKED LOOP

A phase-locked loop (PLL) is an electronic circuit that consists of a phase detector, a low-pass filter, and a voltage-controlled oscillator connected as shown in Fig. 17.25. Common applications of a PLL include: (1) frequency synthesizers that provide multiples of a reference signal frequency (for example, the carrier frequency for the multiple channels of a citizens band (CB) unit or marine-radio-band unit can be generated using a single-crystal-controlled frequency and its multiples generated using

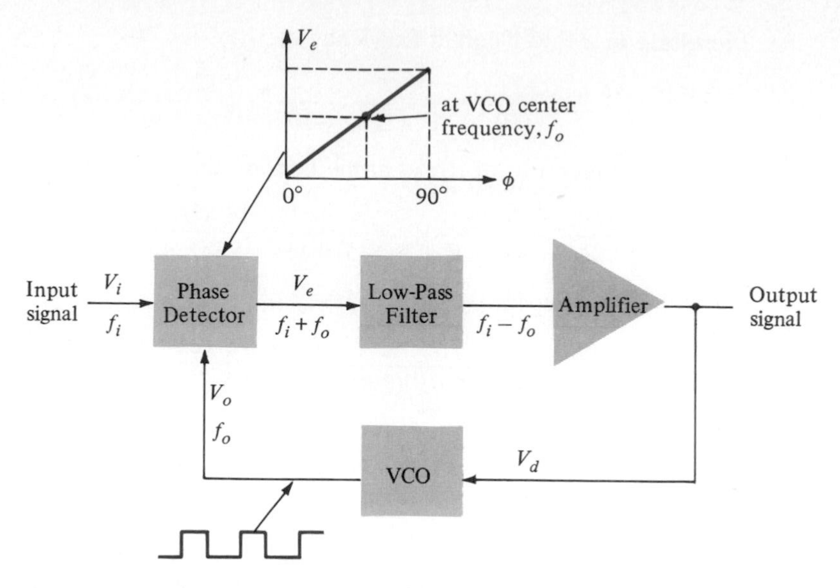

Figure 17.25 Block diagram of basic phase-locked loop (PLL).

a PLL); (2) FM demodulation networks for fm operation with excellent linearity between the input signal frequency and the PLL output voltage; (3) demodulation of the two data transmission or carrier frequencies in digital-data transmission used in frequency-shift keying (FSK) operation; and (4) a wide variety of areas including modems, telemetry receivers and transmitters, tone decoders, AM detectors, and tracking filters.

An input signal, V_i, and that from a VCO, V_o, are compared by a phase comparator (refer to Fig. 17.25) providing an output voltage, V_e, that represents the phase difference between the two signals. This voltage is then fed to a low-pass filter that provides an output voltage (amplified if necessary) that can be taken as the output voltage from the PLL and is used internally as the voltage to modulate the VCO's frequency. The closed-loop operation of the circuit is to maintain the VCO frequency locked to that of the input signal frequency.

Basic PLL Operation

The basic operation of a PLL circuit can be explained using the circuit of Fig. 17.25 as reference. We will first consider the operation of the various circuits in the phase-locked loop when the loop is operating in lock (input signal frequency and VCO frequency are the same). When the input signal frequency is the same as that from the VCO to the comparator, the voltage, V_d, taken as output is the value needed to hold the VCO in lock with the input signal. The VCO then provides output of a fixed-amplitude square-wave signal at the frequency of the input. Best operation is obtained if the VCO center frequency, f_o, is set with the dc bias voltage midway in its linear operating range. The amplifier allows this adjustment in dc voltage from that obtained as output of the filter circuit. When the loop is in lock, the two signals to the comparator are of the same frequency although not necessarily in phase. A fixed phase difference between the two signals to the comparator results in a fixed dc voltage to the VCO. Changes in the input signal frequency then result in change in the dc voltage to the VCO. Within a capture-and-lock frequency range, the dc voltage will drive the VCO frequency to match that of the input.

While the loop is trying to achieve lock, the output of the phase comparator contains frequency components at the sum and difference of the signals compared. A low-pass filter passes only the lower-frequency component of the signal so that the loop can obtain lock between input and VCO signals.

Owing to the limited operating range of the VCO and the feedback connection of the PLL circuit, there are two important frequency bands specified for a PLL. The *capture range* of a PLL is the frequency range centered about the VCO free-running frequency, f_o, over which the loop can *acquire* lock with the input signal. Once the PLL has achieved capture, it can maintain lock with the input signal over a somewhat wider frequency range called the *lock range*.

Applications

The PLL can be used in a wide variety of applications, including (1) frequency demodulation, (2) frequency synthesis, and (3) FSK decoders. Examples of each of these follow.

FREQUENCY DEMODULATION

FM demodulation or detection can be directly achieved using the PLL circuit. If the PLL center frequency is selected or designed at the FM carrier frequency, the filtered or output voltage in the circuit of Fig. 17.25 is the desired demodulated voltage, varying in value proportional to the variation of the signal frequency. The PLL circuit thus operates as a complete intermediate-frequency (IF) strip, limiter, and demodulator as used in FM receivers.

One popular PLL unit is the 565, shown in Fig. 17.26a. The 565 contains a phase detector, amplifier, and voltage-controlled oscillator, which are only partially connected internally. An external resistor and capacitor, R_1 and C_1, are used to set the free-running or center frequency of the VCO. Another external capacitor, C_2, is used to set the low-pass filter passband, and the VCO output must be connected back as input to the phase detector to close the PLL loop. The 565 typically uses two power supplies, V^+ and V^-.

Figure 17.26b shows the 565 PLL connected to work as an FM demodulator. Resistor R_1 and capacitor C_1 set the free-running frequency, f_o,

$$\boxed{f_o = \frac{0.3}{R_1 C_1}} \tag{17.9}$$

$$= \frac{0.3}{(10 \times 10^3)(220 \times 10^{-12})} = 136.36 \text{ kHz}$$

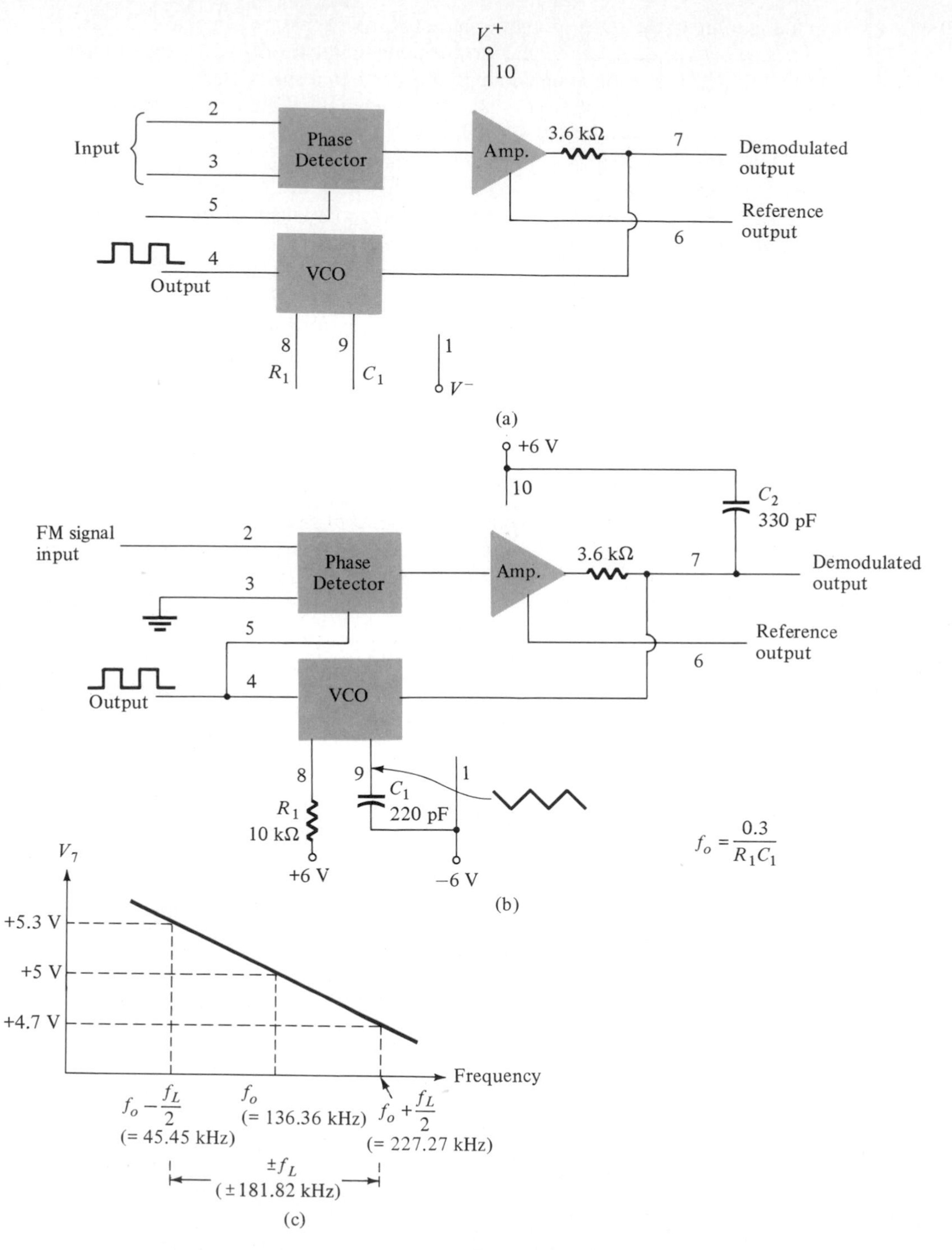

Figure 17.26 (a) Block diagram of 565 PLL unit; (b) connection as FM demodulator; (c) output voltage-frequency relation.

with limitation $2\ k\Omega \leq R_1 \leq 20\ k\Omega$. The lock range is

$$\boxed{f_L = \pm\frac{8f_o}{V}}$$

$$= \pm\frac{8(136.36 \times 10^3)}{6} = \pm 181.8\ \text{kHz}$$

for supply voltages V = $\pm$ 6 V. The capture range is

$$\boxed{f_C = \pm\frac{1}{2\pi}\sqrt{\frac{2\pi f_L}{R_2 C_2}}}$$

$$= \pm\frac{1}{2\pi}\sqrt{\frac{2\pi(181.8 \times 10^3)}{(3.6 \times 10^3)(330 \times 10^{-12})}} = 156.1\ \text{kHz}$$

The signal at pin 4 is a 136.36 kHz square wave. An input within the lock range of 181.8 kHz will result in the output voltage at pin 7 varying around its dc voltage level set with input signal at f_o. Figure 17.26c shows the output at pin 7 as a function of the input signal frequency. The dc voltage at pin 7 is linearly related to the input signal frequency within the frequency range f_L = 181.8 kHz around the center frequency 136.36 kHz. The output voltage is the demodulated signal that varies with frequency within the operating range specified.

FREQUENCY SYNTHESIS

A frequency synthesizer can be built around a PLL as shown in Fig. 17.27. A frequency divider is inserted between the VCO output and the phase comparator so that the loop signal to the comparator is at frequency f_o while the VCO output is Nf_o. This output is a multiple of the input frequency as long as the loop is in lock. The input signal can be crystal stabilized at f_1 with the resulting VCO output at Nf_1 if the loop is set up to lock at the fundamental frequency (when $f_o = f_1$). Figure 17.27b shows an example using a 565 PLL as frequency multiplier and 7490 as divider. The input V_i at frequency f_1 is compared to the input (frequency f_o) at pin 5. An output at Nf_o ($4f_o$ in the present example) is connected through an inverter circuit to provide an input at pin 14 of the 7490, which varies between 0 V and +5 V. Using the output at pin 9, which is divided by four from that at the input to the 7490, the signal at pin 4 of the PLL is four times the input frequency as long as the loop remains in lock. Since the VCO can only vary over a limited range from its center frequency, it may be necessary to change the VCO frequency whenever the divider value is changed. As long as the PLL circuit is in lock the VCO output frequency will be exactly N times the input frequency. It is only necessary to readjust f_o to be within the capture-and-lock range, the closed loop then resulting in the VCO output becomes exactly Nf_1 at lock.

FSK DECODERS

An FSK (frequency-shift keyed) signal decoder can be built as shown in Fig. 17.28. The decoder receives a signal at one of two distinct carrier frequencies,

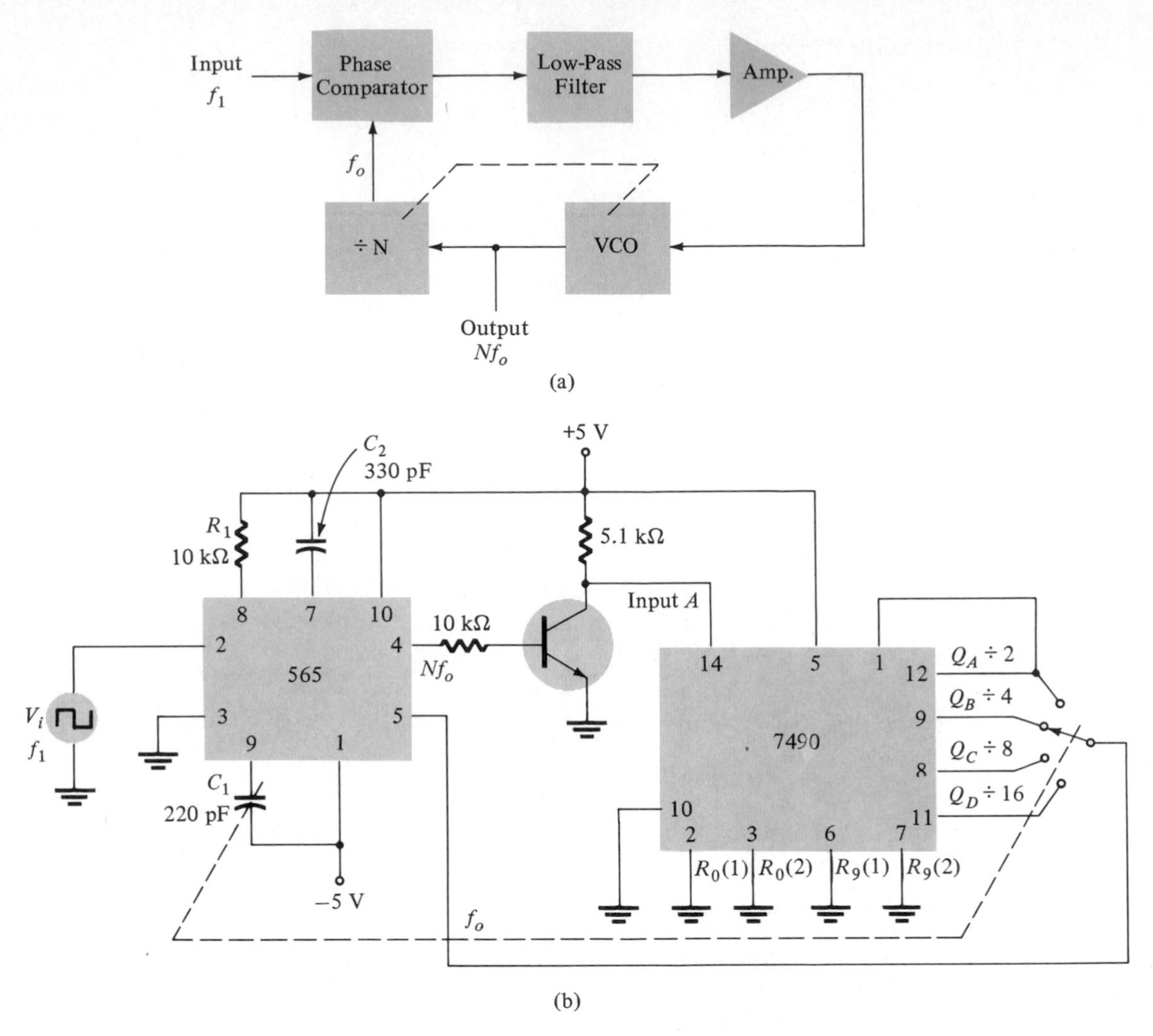

Figure 17.27 Frequency synthesizer: (a) block diagram; (b) implementation using 565 PLL unit.

1270 Hz or 1070 Hz, representing the RS-232C logic levels of mark (−5 V) or space (+14 V), respectively. As the signal appears at the input, the loop locks to the input frequency and tracks it between the two possible frequencies with a corresponding dc shift at the output.

The *RC* ladder filter (three sections of $C = 0.02\ \mu\text{F}$ and $R = 10\ \text{k}\Omega$) is used to remove the sum frequency component. The free-running frequency is adjusted with R_1 so that the dc voltage level at the output (pin 7) is the same as that at pin 6. Then, an input at frequency 1070 Hz will drive the decoder output voltage to a more positive voltage level, driving the digital output to the high level (space or +14 V). An input at 1270 Hz will correspondingly drive the 565 dc output less positive with the digital output, which then drops to the low level (mark or −5 V).

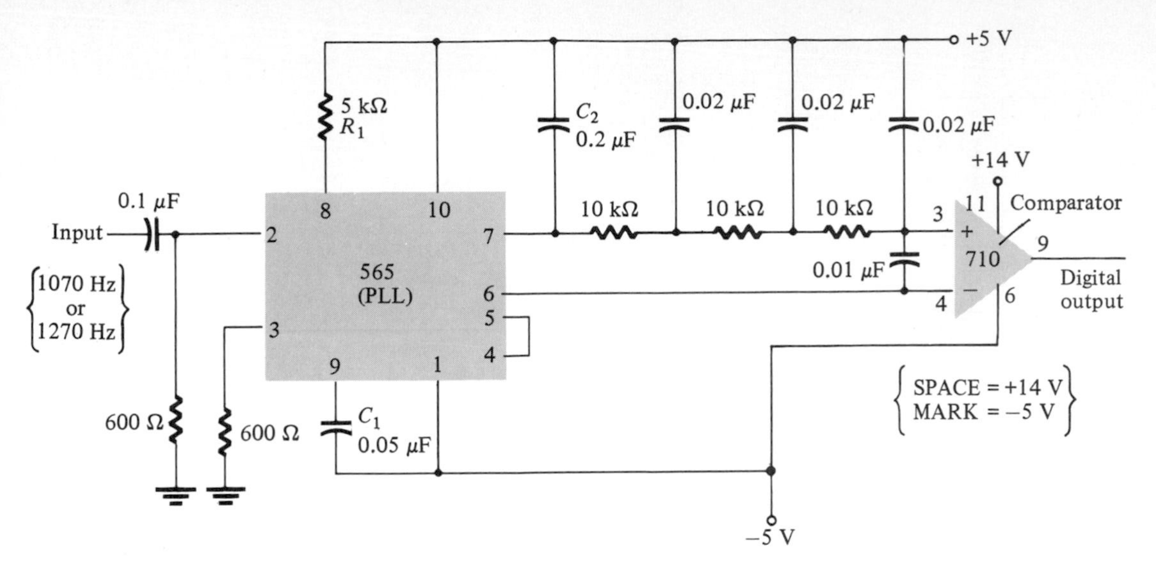

Figure 17.28 Connection of 565 as FSK decoder.

17.7 INTERFACING

Connecting different types of circuits, different analog or digital units, and inputs or loads to other electronics all require some sort of interfacing. Interface circuits may be categorized as either driver or receiver units. A receiver essentially accepts inputs, providing high input impedance to minimize loading of the input signal. A driver circuit provides the output signal at voltage or current levels suitable to operate a number of loads, or to operate such devices as relays, displays, and power units. Furthermore, these inputs or outputs may require strobing, which provides the interface signal connection during specific time intervals as established by the strobe.

Figure 17.29a shows a dual line driver, each driver having input operation capable of accepting TTL signals and output which can operate TTL, DTL, or MOS devices. Often, the interface circuit is needed to receive signals from one type of circuit (TTL, DTL, ECL, MOS) and transmit the signal to another circuit type. The relation between the input and output signal may be such that the interface unit can be used as either noninverting unit or inverting unit. Interface circuits or all these various configurations are necessary and exist as IC units.

The circuit of Fig. 17.29b shows a dual line receiver having both inverting and noninverting inputs so that either operating condition can be selected. As example, connection of an input signal to the inverting input would result in an inverted output from the receiver unit. Connecting the input to the noninverting input would provide the same interfacing except that the output obtained would have the same polarity as the received signal. In the case of both driver and receiver circuits shown in Fig. 17.29, the outputs will be present only when the strobe signal is present—high level in the present circuits.

Another type of interfacing that is quite important occurs when connecting signals between various terminals of a digital system. Signals from such devices as a teletype,

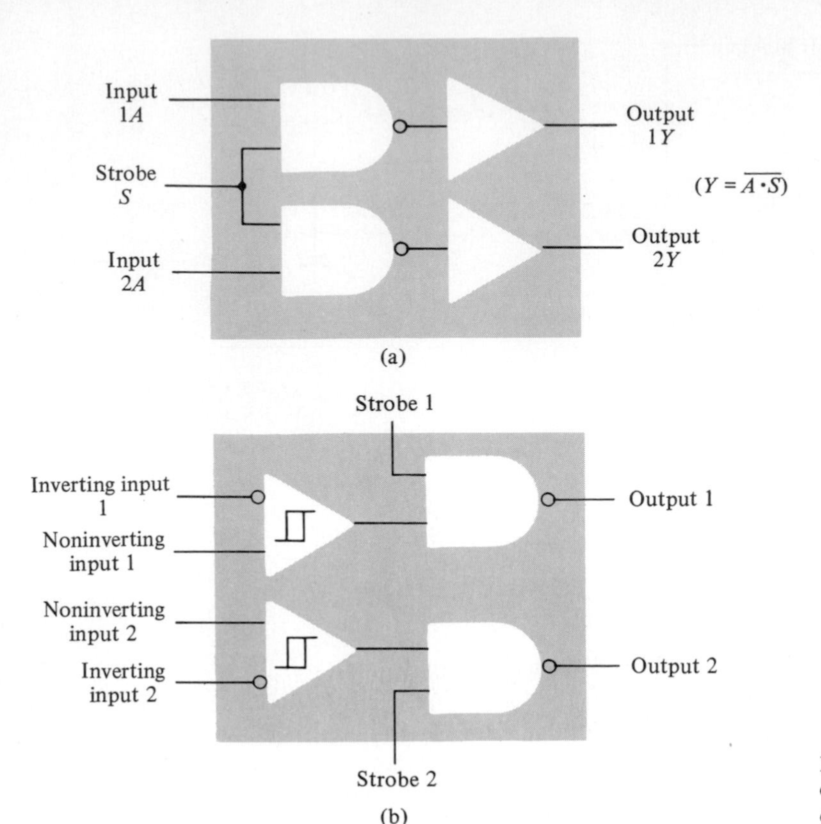

Figure 17.29 Interface units: (a) dual-line drivers (SN75150); (b) dual-line receivers (SN75152).

video terminal, card reader, or line printer are usually one of a number of signal forms. The EIA electronics industry's most popular standard is referred to as RS-232C. Complete details of the expected signal conditions for this standard can be stated simply here as binary signals representing mark (logic-1) and space (logic-0) corresponding to the voltage levels of −12 V and +12 V, respectively. TTL circuits operate with signals defined as +5 V as mark and 0 V as space. Teletype units are sometimes wired to operate with current-loop signals, for which 20 mA represents a mark with no current representing a space. These different types of signals may occur as either input or output of a particular terminal, so that a variety of interface circuitry is necessary to convert from one signal type to one of the other. Some popular examples of interfaces will be described next.

Figure 17.30a shows the defined mark and space conditions for current loop, RS-232C, and TTL signals.

RS-232C-to-TTL Converter

If a unit having output defined by RS-232C is to operate into another unit which operates with TTL-signal levels, the interface circuit of Fig. 17.30b could be used. A mark output from the driver (−12 V) would get clipped by the diode so that the input

	Current Loop	RS-232-C	TTL
MARK	20 mA	−12 V	+5 V
SPACE	0 mA	+12 V	0 V

(a)

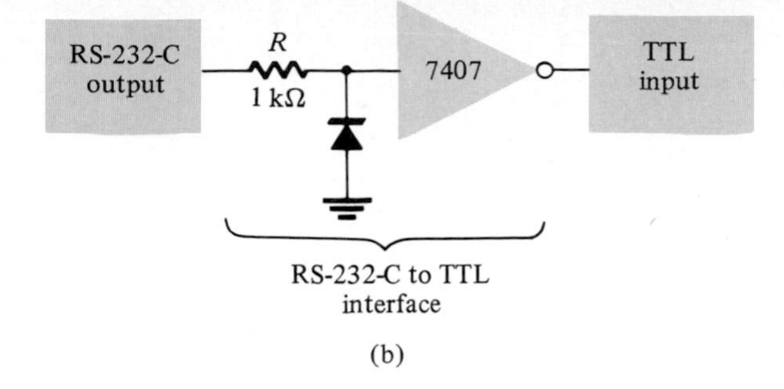

(b)

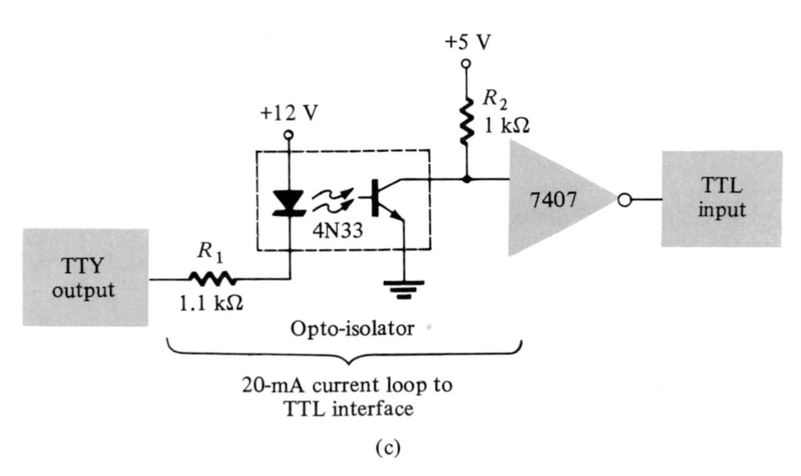

(c)

Figure 17.30 Interfacing signal standards and converter circuits.

to the inverter circuit is near 0 V, resulting in an output of +5 V or a TTL-level mark. A space output at +12 V would drive the inverter output low for a 0-V space (TTL).

Another example of interface is that between a current-loop input and TTL, as shown in Fig. 17.30c. An input mark results when 20 mA current is drawn through the output line of the Teletype (TTY). This current then goes through the diode element of an opto-isolator driving the output transistor on. The input to the inverter going low results in a +15-V signal to the TTL input, so that a mark from the Teletype results in a mark into the TTL input. A space from the Teletype current loop provides no current with opto-isolator transistor remaining off and inverter output then 0 V, which is a TTL space signal.

Other types of interface circuits can be considered, those of Fig. 17.31 being representative. Another means of interfacing digital signals is made using open-collector output and using tri-state buffer outputs. When a signal is output from a transistor collector (see Fig. 17.31) which is not connected to any other electronic components, the output is open-collector. This allows connecting a number of signals to the same signal wire or signal bus. Then any transistor going on provides a low output condition, while all transistors off provide a high output.

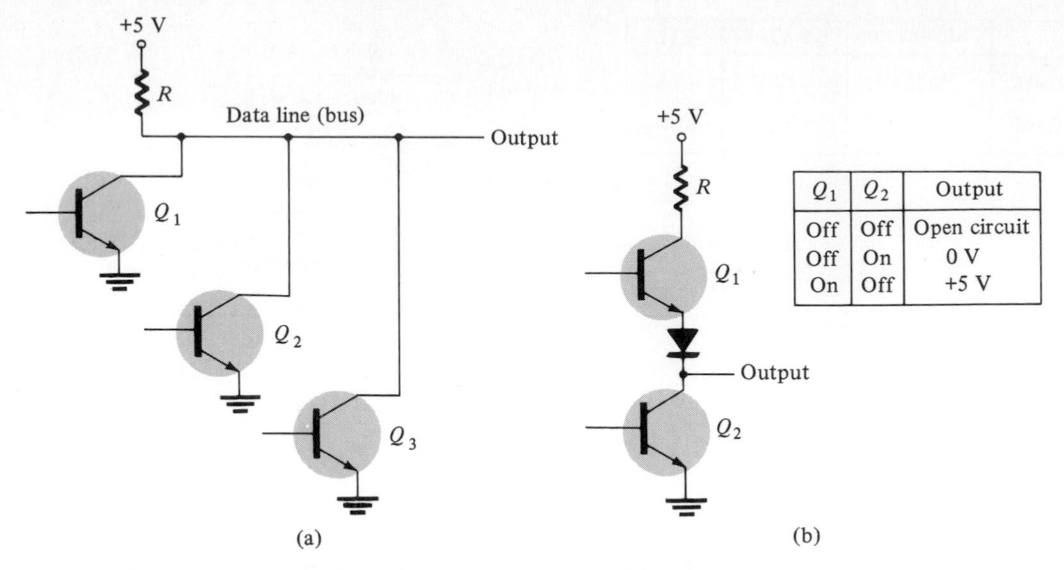

Q_1	Q_2	Output
Off	Off	Open circuit
Off	On	0 V
On	Off	+5 V

Figure 17.31 Connections to data lines: (a) open-collector output; (b) tri-state output.

PROBLEMS

§ 17.2

1. Draw the diagram of a 741 op-amp operated from ±15-V supplies with input to minus input terminal and plus input terminal connected to a +5-V reference voltage. Include terminal pin connections.

2. Sketch the voltage waveforms for the circuit of Problem 1, with input of 10 V rms.

3. Draw the connection diagram of a 311 op-amp showing 10 V rms input applied to pin 3 and ground to pin 2.

4. Using the ±12-V supply and output connected to positive supply through a 10-kΩ resistor. Sketch the input and resulting output waveforms.

5. Describe the operation of the circuit of Fig. 17.7 with input of 10 V rms applied.

6. Draw the circuit diagram of a zero-crossing detector using a 339 comparator stage with ±12-V supplies.

7. For the circuit of Problem 6, sketch the voltage waveforms with 10-V rms input applied to the minus input and plus input grounded.

8. Describe the operation of a window detector circuit as in Fig. 17.11 for resistor values of 7.5 kΩ and 8.2 kΩ changed to 6.2 kΩ.

§ 17.3

9. Sketch a three-input ladder network using 15-kΩ and 30-kΩ resistor values.

10. For a reference voltage of 16 V, calculate the output of the network in Problem 9 with an input of 110.

11. What voltage resolution is possible using a 12-stage ladder network with a 10-V reference voltage?

12. Describe what occurs during the fixed time interval and during the count interval of dual-slope conversion.

13. How many count steps occur using a 12-stage digital counter as the output of an A/D converter?

14. What is the maximum count interval resulting using a 12-stage counter operated at a clock rate of 2 MHz?

§ 17.4

15. Sketch a 555 timer connected as an astable multivibrator for operation at 100 kHz. Determine the value of capacitor C needed if $R_A = R_B = 7.5\ \text{k}\Omega$.

16. Draw a 555 timer used as a one-shot using $R_A = 7.5\ \text{k}\Omega$ for a time period of 25 μs. Determine the value of capacitor C needed.

17. Sketch the input and output waveforms for a one-shot as in Problem 16 for input triggered by a 10-kHz clock.

§ 17.5

18. Calculate the center frequency of a VCO using a 566 as in Fig. 17.22 for $R_1 = 4.7\ \text{k}\Omega$, $R_2 = 1.8\ \text{k}\Omega$, $R_3 = 11\ \text{k}\Omega$, and $C_1 = 0.001\ \mu\text{F}$.

19. What frequency range results in the circuit of Fig. 17.23 for $C_1 = 0.001\ \mu\text{F}$?

20. Determine the capacitor needed in the circuit of Fig. 17.22 to obtain a 100-kHz output.

§ 17.6

21. Calculate the VCO free-running frequency for the circuit of Fig. 17.26b with $R_1 = 4.7\ \text{k}\Omega$ and $C_1 = 0.001\ \mu\text{F}$.

22. What value capacitor, C_1, is required in the circuit of Fig. 17.26b to obtain a center frequency of 100 kHz?

23. What is the lock range of the PLL circuit in Fig. 17.26b for $R_1 = 4.7\ \text{k}\Omega$ and $C_1 = 0.001$ F?

§ 17.7

24. Describe the signal conditions for current-loop and RS-232C interfaces.

25. What is a data bus?

26. What is the difference between open-collector and tri-state outputs?

18

Feedback Amplifiers and Oscillator Circuits

18.1 FEEDBACK CONCEPTS

Feedback was mentioned when considering dc bias stabilization in Chapters 5 and 7. Amplifier gain was sacrificed in the circuit design for improvement in dc bias stability. We might say that a trade-off of gain for stability was made in the circuit design. Such trade-off is typical of engineering design compromises. If *negative* voltage feedback is used, for example, a circuit can be designed to couple some of the output voltage back to the input, reducing the overall voltage gain of the circuit. For this loss of gain, however, it is possible to obtain higher input impedance, lower output impedance, more stable amplifier gain, or higher cutoff frequency operation.

If the feedback signal is connected in order to aid or add to the input signal applied, however, *positive* feedback occurs, which could drive the circuit into operation as an oscillator.

A typical feedback connection is shown in Fig. 18.1. The input signal, V_s, is applied to a mixer network, where it is combined with a feedback signal, V_f. The difference of these signals, V_i, is then the input voltage to the amplifier. A portion of the amplifier output, V_o, is connected to the feedback network (β), which provides a reduced portion of the output as feedback signal to the input mixer network.

The overall gain of the amplifier is reduced by this *negative* feedback (feedback signal opposite in polarity to the input signal). In return for this gain reduction a number of improvements can be obtained, such as the following:

1. Higher input impedance
2. Better stabilized voltage gain

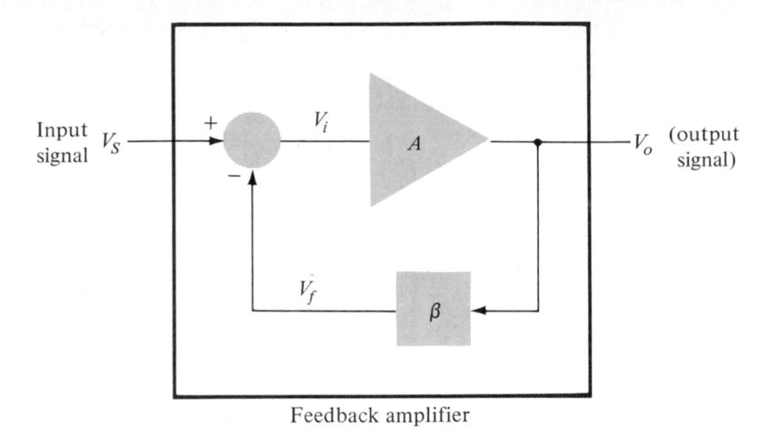

Figure 18.1 Simpler block diagram of feedback amplifier.

3. Improved frequency response
4. More linear operation
5. Lower output impedance
6. Reduced noise

18.2 FEEDBACK CONNECTION TYPES

There are four basic ways of connecting the feedback signal. Both *voltage* and *current* can be fed back to the input either in *series* or *parallel*. Specifically, there can be

1. Voltage-series feedback (Fig. 18.2a)
2. Voltage-shunt feedback (Fig. 18.2b)
3. Current-series feedback (Fig. 18.2c)
4. Current-shunt feedback (Fig. 18.2d)

In the list above, *voltage* refers to connecting the output voltage as input to the feedback network; *current* refers to tapping off some output current through the feedback network. *Series* refers to connecting the feedback signal is series with the input signal voltage; *shunt* refers to connecting the feedback signal in shunt (parallel) with an input current source.

Series feedback connections tend to *increase* the input resistance while shunt feedback connections tend to *decrease* the input resistance. Voltage feedback tends to *decrease* the output impedance while current feedback tends to *increase* the output impedance. Typically, higher input and lower output impedances are desired for most cascade amplifiers. Both of these are provided using the voltage-series feedback connection. We shall therefore concentrate first on this amplifier connection.

Gain with Feedback

In this section we examine the gain of each of the feedback circuit connections of Fig. 18.2. The gain without feedback, A, is that of the amplifier stage. With feedback,

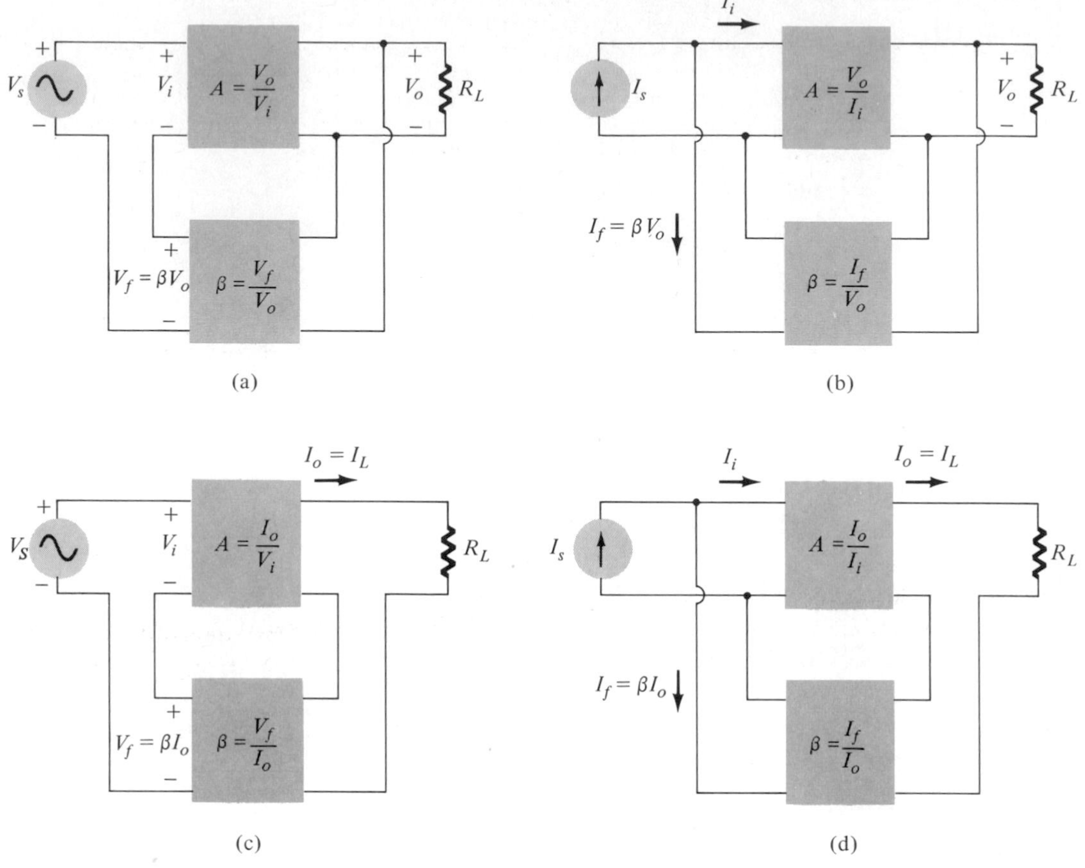

Figure 18.2 Feedback amplifier connection types: (a) voltage-series feedback, $A_f = V_o/V_s$; (b) voltage-shunt feedback, $A_f = V_o/I_s$; (c) current-series feedback, $A_f = I_o/V_s$; (d) current-shunt feedback, $A_f = I_o/I_s$.

β, the overall gain of the circuit is reduced by a factor $(1 + \beta A)$, as detailed below. A summary of the gain, feedback factor, and gain with feedback of Fig. 18.2 is provided for reference in Table 18.1.

VOLTAGE SERIES

Figure 18.2a shows the voltage-series feedback connection with a part of the output voltage fed back in series with the input signal, resulting in an overall gain reduction. If there is no feedback ($V_f = 0$), the voltage gain of the amplifier stage is

$$A = \frac{V_o}{V_s} = \frac{V_o}{V_i} \tag{18.1}$$

If a feedback signal, V_f, is connected in series with the input, then

$$V_i = V_s - V_f \tag{18.2}$$

TABLE 18.1 Summary of Gain, Feedback, and Gain with Feedback from Fig. 18.2

		Voltage-Series	*Voltage-Shunt*	*Current-Series*	*Current-Shunt*
Gain without feedback	A	$\frac{V_o}{V_i}$	$\frac{V_o}{I_i}$	$\frac{I_o}{V_i}$	$\frac{I_o}{I_i}$
Feedback	β	$\frac{V_f}{V_o}$	$\frac{I_f}{V_o}$	$\frac{V_f}{I_o}$	$\frac{I_f}{I_o}$
Gain with feedback	A_f	$\frac{V_o}{V_s}$	$\frac{V_o}{I_s}$	$\frac{I_o}{V_s}$	$\frac{I_o}{I_s}$

Since $$V_o = AV_i = A(V_s - V_f) = AV_s - AV_f = AV_s - A(\beta V_o)$$

then $$(1 + \beta A)V_o = AV_s$$

so that the overall voltage gain *with* feedback is

$$A_f = \frac{V_o}{V_s} = \frac{A}{1 + \beta A} \tag{18.3}$$

Equation (18.3) shows that the gain *with* feedback is the amplifier gain reduced by the factor $(1 + \beta A)$. This factor will be seen also to affect input and output impedance among other circuit features.

VOLTAGE SHUNT

The gain with feedback for the network of Fig. 18.2b is

$$A_f = \frac{V_o}{I_s} = \frac{AI_i}{I_i + I_f} = \frac{AI_i}{I_i + \beta V_o} = \frac{AI_i}{I_i + \beta AI_i}$$

$$= \frac{A}{1 + \beta A}$$

Input Impedance with Feedback

SERIES FEEDBACK

A more detailed series feedback connection is shown in Fig. 18.3. The input impedance can be determined as follows:

$$I_i = \frac{V_i}{Z_i} = \frac{V_s - V_f}{Z_i} = \frac{V_s - \beta V_o}{Z_i} = \frac{V_s - \beta AV_i}{Z_i}$$

$$I_i Z_i = V_s - \beta AV_i$$

$$V_s = I_i Z_i + \beta AV_i = I_i Z_i + \beta A I_i Z_i$$

$$Z_{if} = \frac{V_s}{I_i} = Z_i + (\beta A)Z_i = Z_i(1 + \beta A)$$

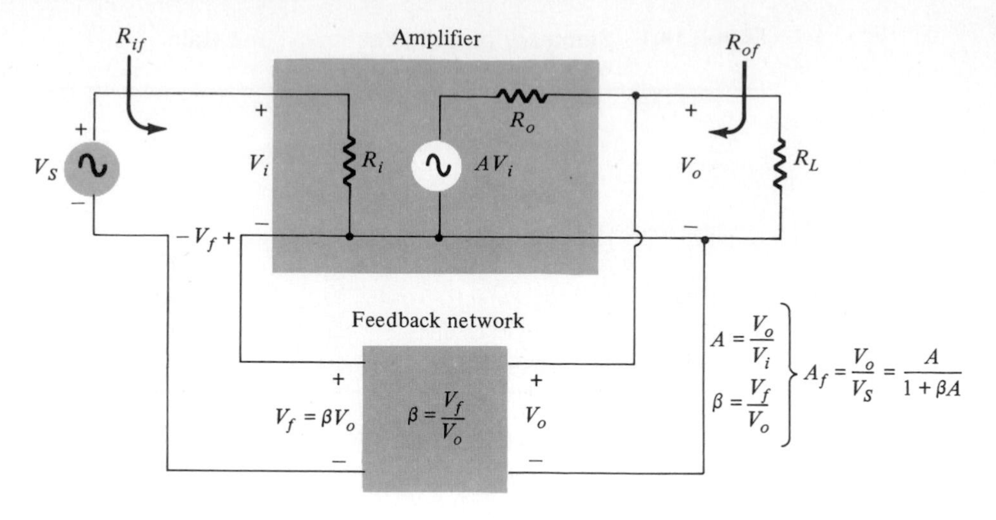

Figure 18.3 Voltage-series feedback connection.

The input impedance with series feedback is seen to be the value of the input impedance without feedback multiplied by the factor $(1 + \beta A)$ and applies to both voltage-series (Fig. 18.2a) and current-series (Fig. 18.2c) configurations.

SHUNT FEEDBACK

A more detailed shunt feedback connection is shown in Fig. 18.4. The input impedance can be determined to be

$$Z_{if} = \frac{V_i}{I_s} = \frac{V_i}{I_i + I_f} = \frac{V_i}{I_i + \beta V_o}$$

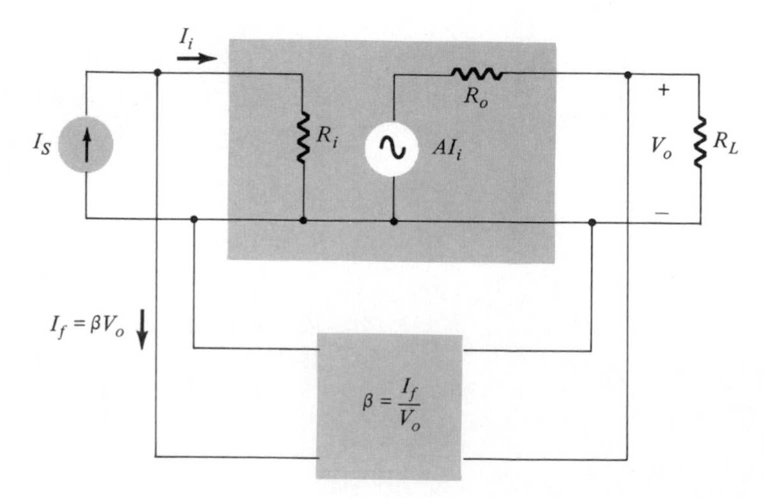

Figure 18.4 Voltage-shunt feedback connection.

$$= \frac{V_i/I_i}{I_i/I_i + \beta V_o/I_i}$$

$$= \frac{Z_i}{1 + \beta A} \qquad (18.5)$$

This reduced input impedance applies to both the voltage-shunt connection of Fig. 18.2b and the current-shunt connection of Fig. 18.2d.

Output Impedance with Feedback

The output impedance for the connections of Fig. 18.2 are dependent on whether voltage or current feedback is used. For voltage feedback the output impedance is decreased, while current feedback increases the output impedance.

VOLTAGE FEEDBACK

The voltage-series feedback circuit of Fig. 18.3 provides sufficient circuit detail to determine the output impedance with feedback. The output impedance is determined by applying a voltage, V, resulting in a current, I, with V_s shorted out ($V_s = 0$). The voltage V is then

$$V = IZ_o + AV_i$$

For $V_s = 0$,

$$V_i = -V_f$$

so that

$$V = IZ_o - AV_f = IZ_o - A(\beta V)$$

Rewriting the equation as

$$V + \beta AV = IZ_o$$

allows solving for the output resistance with feedback:

$$\boxed{Z_{of} = \frac{V}{I} = \frac{Z_o}{1 + \beta A}} \qquad (18.6)$$

Equation (18.6) shows that with voltage feedback the output impedance is reduced from that without feedback by the factor $(1 + \beta A)$.

CURRENT FEEDBACK

The output impedance with current feedback can be determined by applying a signal V to the output with V_s shorted out, resulting in a current I, the ratio of V to I being the output impedance. Figure 18.5 shows a more detailed connection with current-series feedback. For the output part of a current-series feedback connection shown in Fig. 18.5, the resulting output impedance is determined as follows. With $V_s = 0$,

$$V_i = V_f$$

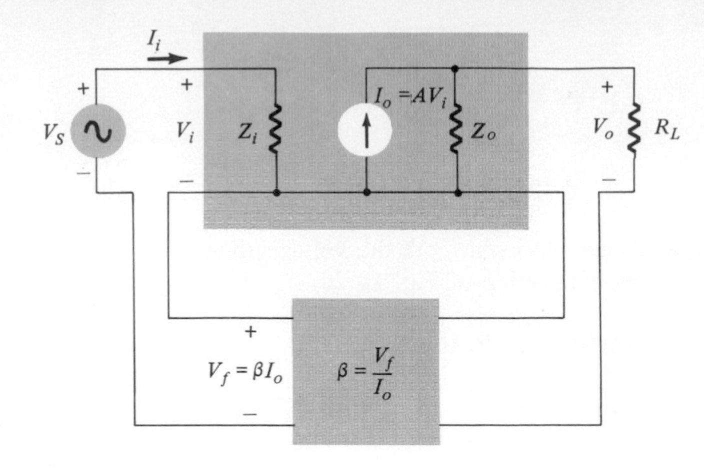

Figure 18.5 Current-series feedback connection.

$$I = \frac{V}{Z_o} - AV_i$$

$$= \frac{V}{Z_o} - AV_f = \frac{V}{Z_o} - A\beta I$$

$$Z_o(1 + \beta A)I = V$$

$$Z_{of} = \frac{V}{I} = Z_o(1 + \beta A) \tag{18.7}$$

A summary of the effect of feedback on input and output impedance is provided in Table 18.2.

TABLE 18.2 Effect of Feedback Connection on Input and Output Impedance

	Voltage-Series	*Current-Series*	*Voltage-Shunt*	*Current-Shunt*
Z_{if}	$Z_i(1 + \beta A)$ (increased)	$Z_i(1 + \beta A)$ (increased)	$\frac{Z_i}{1 + \beta A}$ (decreased)	$\frac{Z_i}{1 + \beta A}$ (decreased)
Z_{of}	$\frac{Z_o}{1 + \beta A}$ (decreased)	$Z_o(1 + \beta A)$ (increased)	$\frac{Z_o}{1 + \beta A}$ (decreased)	$Z_o(1 + \beta A)$ (increased)

most used

EXAMPLE 18.1

Determine the voltage gain, input, and output impedance with feedback for voltage-series feedback having $A = -100$, $R_i = 10\ \text{k}\Omega$, $R_o = 20\ \text{k}\Omega$ for feedback of (a) $\beta = -0.1$ and (b) $\beta = -0.5$.

Solution:

Using Eqs. (18.3), (18.5), and (18.7)

(a) $$A_f = \frac{A}{1 + \beta A} = \frac{-100}{1 + (-0.1)(-100)} = \frac{-100}{11} = \mathbf{-9.09}$$

$$Z_{if} = Z_i(1 + \beta A) = 10\ \text{k}\Omega\,(11) = \mathbf{110\ k\Omega}$$

$$Z_{of} = \frac{Z_o}{1 + \beta A} = \frac{20 \times 10^3}{11} = \mathbf{1.82\ k\Omega}$$

(b) $$A_f = \frac{A}{1 + \beta A} = \frac{-100}{1 + (0.5)(100)} = \frac{-100}{51} = \mathbf{-1.96}$$

$$Z_{if} = Z_i(1 + \beta A) = 10\ \text{k}\Omega(51) = \mathbf{510\ k\Omega}$$

$$Z_{of} = \frac{Z_o}{1 + \beta A} = \frac{20 \times 10^3}{51} = \mathbf{392.16\ \Omega}$$

Example 18.1 demonstrates the trade-off of gain for improved input and output resistance. Reducing the gain by a factor of 11 (from 100 to 9.09) is complimented by a reduced output resistance and increased input resistance by the same factor of 11. Reducing the gain by a factor of 51 provides a gain of only 2 but with input resistance increased by the factor of 51 (to over 500 kΩ) and output resistance reduced from 20 kΩ to under 400 Ω. Feedback offers the designer the choice of trading away some of the available amplifier gain for other improved circuit features.

Reduction in Frequency Distortion

For a negative-feedback amplifier having $\beta A \gg 1$ the gain with feedback is $A_f \cong 1/\beta$. It follows from this that if the feedback network is purely resistive, the gain with feedback is not dependent on frequency even though the basic amplifier gain is frequency dependent. Practically, the frequency distortion arising because of varying amplifier gain with frequency is considerably reduced in a negative-voltage feedback amplifier circuit.

Reduction in Noise and Nonlinear Distortion

lumpy out put of Am

Signal feedback tends to hold down the amount of noise signal (such as power-supply hum) and nonlinear distortion. The factor $(1 + \beta A)$ reduces both input noise and resulting nonlinear distortion for considerable improvement. However, it should be noted that there is a reduction in overall gain (the price required for the improvement in circuit performance). If additional stages are used to bring the overall gain up to the level without feedback, it should be noted that the extra stage(s) might introduce as much noise back into the system as that reduced by the feedback amplifier. This problem can be somewhat alleviated by readjusting the gain of the feedback-amplifier circuit to obtain higher gain while also providing reduced noise signal. Can reduce noise introduced by the amplifier

Effect of Negative Feedback on Gain and Bandwidth

In Eq. (18.3) the overall gain with negative feedback is shown to be

$$A_f = \frac{A}{1 + \beta A} \cong \frac{A}{\beta A} = \frac{1}{\beta} \qquad \text{for } \beta A \gg 1$$

As long as $\beta A \gg 1$ the overall gain is approximately $1/\beta$. We should realize that for a practical amplifier (for single low- and high-frequency breakpoints) the open-loop gain drops off at high frequencies due to the active device and circuit capacitances. Gain may also drop off at low frequencies for capacitively coupled amplifier stages. Once the open-loop gain A drops low enough and the factor βA is no longer much larger than 1, the conclusion of Eq. (18.3) that $A_f \cong 1/\beta$ no longer holds true.

Figure 18.6 shows that the amplifier with negative feedback has more bandwidth (B_f) than the amplifier without feedback (B). The feedback amplifier has a higher upper 3-dB frequency and smaller lower 3-dB frequency.

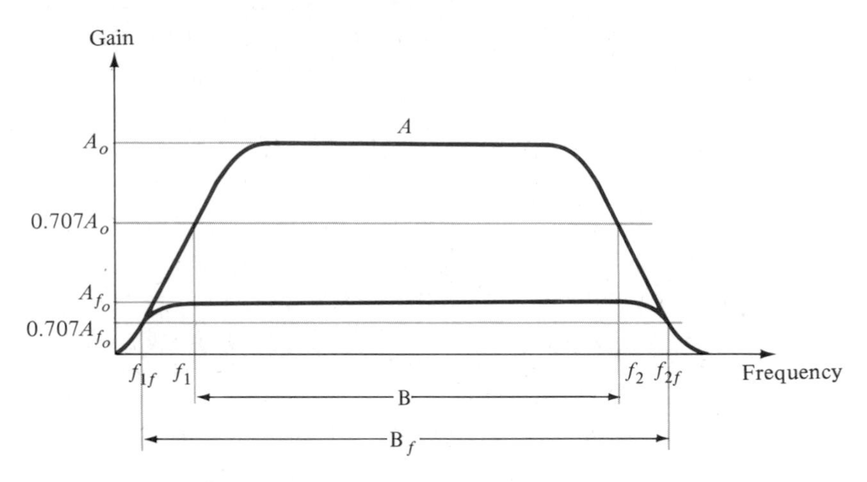

Figure 18.6 Effect of negative feedback on gain and bandwidth.

It is interesting to note that the use of feedback, while resulting in a lowering of voltage gain, has provided an increase in B and in the upper 3-dB frequency, particularly. In fact, the product of gain and frequency remains the same so that the gain-bandwidth product of the basic amplifier is the same value for the feedback amplifier. However, since the feedback amplifier has lower gain, the net operation was to *trade* gain for bandwidth (we use bandwidth for the upper 3-dB frequency since typically $f_2 \gg f_1$).

Gain Stability with Feedback

In addition to the β factor setting a precise gain value, we are also interested in how stable the feedback amplifier is compared to an amplifier without feedback. Differentiating Eq. (18.3) leads to

$$\frac{dA_f}{A_f} = \frac{1}{|1 + \beta A|}\frac{dA}{A} \tag{18.8}$$

$$\frac{dA_f}{A_f} \cong \frac{1}{\beta A}\frac{dA}{A} \quad \text{for } \beta A \gg 1 \tag{18.9}$$

This shows that the change in gain (dA) is reduced by the factor βA when feedback is employed.

EXAMPLE 8.2

If an amplifier with gain of -1000 and feedback of $\beta = -0.1$ has a gain change of 20% due to temperature, calculate the change in gain of the feedback amplifier.

Solution:

Using Eq. (18.9), we get

$$\frac{dA_f}{A_f} \cong \frac{1}{\beta A}\frac{dA}{A} = \frac{1}{-0.1(-1000)}(20\%) = \mathbf{0.2\%}$$

The improvement is 100 times. Thus, while the amplifier gain changes from $A = -1000$ by 20%, the gain with feedback changes from $A_f = -100$ by only 0.2%.

18.3 PRACTICAL FEEDBACK CIRCUITS

Examples of practical feedback circuits will provide a means of demonstrating the effect feedback has on the various connection types.

Voltage Series Feedback

Figure 18.7 shows a FET amplifier stage with voltage-series feedback. A part of the output signal (V_o) is obtained using a feedback network of resistors R_1 and R_2. The feedback voltage V_f is connected in series with the source signal V_s, their difference being the input signal V_i.

Without feedback the amplifier gain is

$$A = \frac{V_o}{V_i} = -g_m R_L \tag{18.10}$$

where R_L is the parallel combination of resistors

$$R_L = R_D \| R_o \| (R_1 + R_2) \tag{18.11}$$

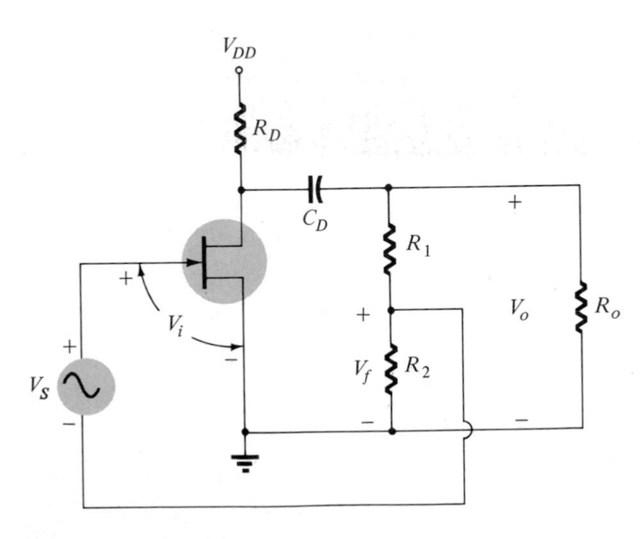

Figure 18.7 FET amplifier stage with voltage-series feedback.

The feedback network provides a feedback factor of

$$\beta = \frac{V_f}{V_o} = \frac{-R_2}{R_1 + R_2} \tag{18.12}$$

Using the values of A and β above in Eq. (18.3), we find the gain with negative feedback to be

$$A_f = \frac{A}{1 + \beta A} = \frac{-g_m R_L}{1 + [R_2 R_L/(R_1 + R_2)]g_m} \tag{18.13}$$

If $\beta A \gg 1$, we have

$$A_f \cong \frac{1}{\beta} = -\frac{R_1 + R_2}{R_2} \tag{18.14}$$

EXAMPLE 18.3

Calculate the gain without and with feedback for the FET amplifier circuit of Fig. 18.7 and the following circuit values: $R_1 = 80\ \text{k}\Omega$, $R_2 = 20\ \text{k}\Omega$, $R_o = 10\ \text{k}\Omega$, $R_D = 10\ \text{k}\Omega$, and $g_m = 4000\ \mu\text{S}$.

Solution:

$$R_L \cong \frac{R_o R_D}{R_o + R_D} = \frac{10\ \text{k}\Omega\ (10\ \text{k}\Omega)}{10\ \text{k}\Omega + 10\ \text{k}\Omega} = 5\ \text{k}\Omega$$

(neglecting 100 kΩ resistance of R_1 and R_2 in series)

$$A = -g_m R_L = -(4000 \times 10^{-6})(5\ \text{k}\Omega) = \mathbf{-20}$$

The feedback factor is

$$\beta = \frac{-R_2}{R_1 + R_2} = \frac{-20}{80 + 20} = -0.2$$

The gain with feedback is

$$A_f = \frac{A}{1 + \beta A} = \frac{-20}{1 + (-0.2)(-20)} = \frac{-20}{5} = \mathbf{-4}$$

Figure 18.8 shows a voltage-series feedback connection using an op-amp. The

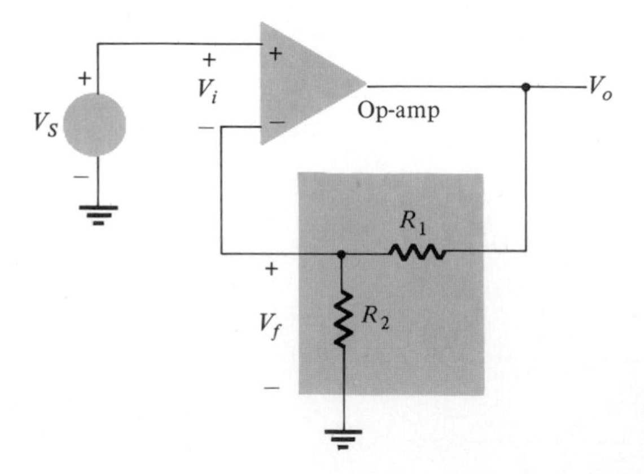

Figure 18.8 Voltage-series feedback in an op-amp connection.

gain of the op-amp, A, without feedback, is reduced by the feedback factor

$$\beta = \frac{-R_2}{R_1 + R_2} \tag{18.15}$$

EXAMPLE 18.4

Calculate the amplifier gain of the circuit of Fig. 18.8 for op-amp gain $A = -100{,}000$ and resistances $R_1 = 1.8\ \text{k}\Omega$ and $R_2 = 200\ \Omega$.

Solution:

$$\beta = \frac{-R_2}{R_1 + R_2} = \frac{-200}{200\ \Omega + 1.8\ k\Omega} = -0.1$$

$$A_f = \frac{A}{1 + \beta A} = \frac{-\ 100{,}000}{1 + (-0.1)(-100{,}000)}$$

$$= \frac{-100{,}000}{10{,}001} = -9.999$$

Note that since $\beta A \gg 1$,

$$A_f \cong \frac{1}{\beta} = \frac{1}{-0.1} = -10$$

The emitter-follower circuit of Fig. 18.9 provides voltage-series feedback. The signal voltage, V_s, is the input voltage, V_i. The output voltage, V_o, is also the voltage fed back in series with the input voltage. The amplifier, as shown in Fig. 18.9, provides the operation *with* feedback. The operation of the circuit without feedback provides $V_f = 0$, so that

$$A = \frac{V_o}{V_s} = \frac{h_{fe} I_b R_E}{V_s} = \frac{h_{fe} R_E \left(\dfrac{V_s}{h_{ie}}\right)}{V_s} = \frac{h_{fe} R_E}{h_{ie}}$$

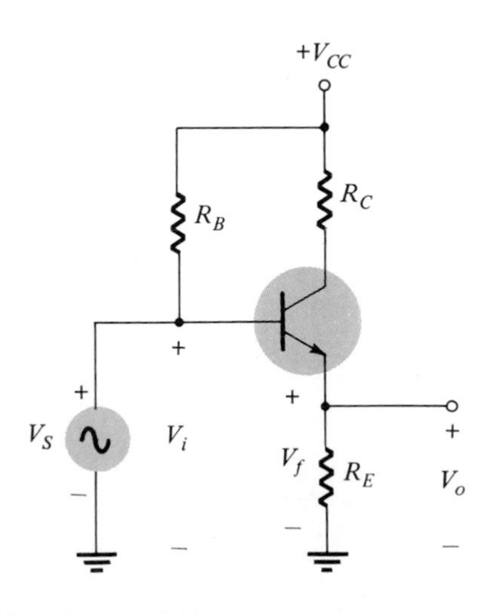

Figure 18.9 Voltage-series feedback circuit (emitter-follower).

and $$\beta = \frac{V_f}{V_o} = 1$$

The operation with feedback then provides that

$$A_f = \frac{V_o}{V_s} = \frac{A}{1 + \beta A} = \frac{\dfrac{h_{fe} R_E}{h_{ie}}}{1 + (1)\left(\dfrac{h_{fe} R_E}{h_{ie}}\right)}$$

$$= \frac{h_{fe} R_E}{h_{ie} + h_{fe} R_E}$$

For $h_{fe} R_E \gg h_{ie}$,

$$A_f \cong 1$$

Current-Series Feedback

Another feedback technique is to sample the output current (I_o) and return a proportional voltage in series with the input. While stabilizing the amplifier gain, the current-series feedback connection increases input resistance.

Figure 18.10 shows a single transistor amplifier stage. Since the emitter of this stage has an unbypassed emitter, it effectively has current-series feedback. The current through resistor R_E results in a feedback voltage that opposes the source signal applied so that the output voltage V_o is reduced. To remove the current-series feedback the emitter resistor must be either removed or bypassed by a capacitor (as is usually done).

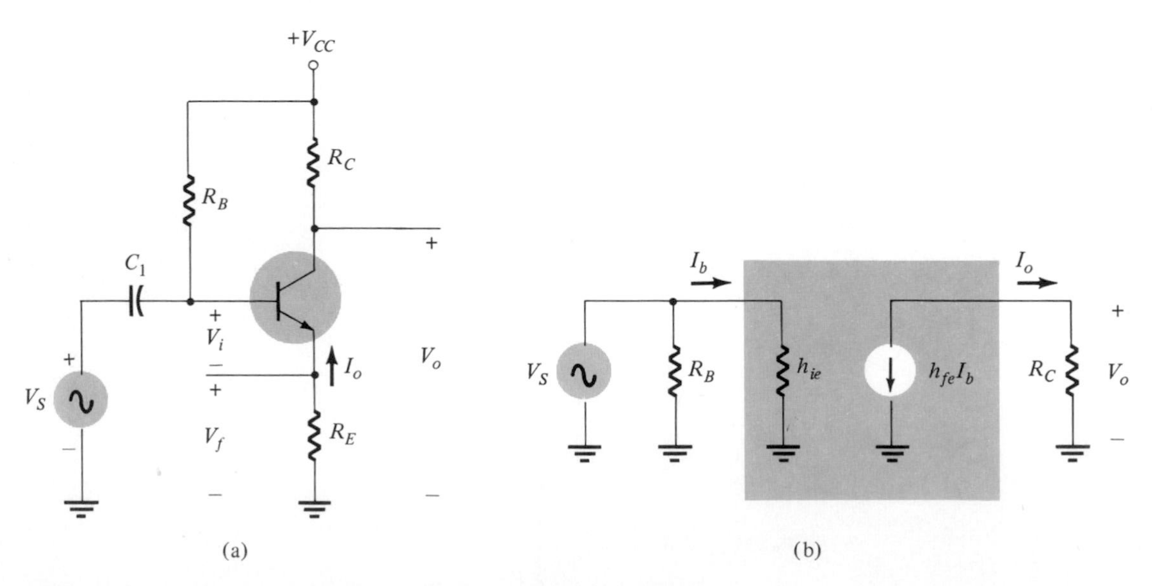

Figure 18.10 Transistor amplifier with unbypassed emitter resistor (R_E) for current-series feedback: (a) amplifier circuit; (b) ac equivalent circuit without feedback.

WITHOUT FEEDBACK

Referring to the basic format of Fig. 18.2a and summarized in Table 18.1, we have

$$A = \frac{I_o}{V_i} = \frac{-I_b h_{fe}}{I_b h_{ie}} = \frac{-h_{fe}}{h_{ie}} \tag{18.16}$$

$$\beta = \frac{V_f}{I_o} = \frac{-I_o R_E}{I_o} = -R_E \tag{18.17}$$

The input and output impedances are

$$Z_i = R_B \parallel h_{ie} \cong h_{ie} \tag{18.18}$$

$$Z_o = R_C \tag{18.19}$$

WITH FEEDBACK

$$A_f = \frac{I_o}{V_s} = \frac{A}{1 + \beta A} = \frac{\dfrac{-h_{fe}}{h_{ie}}}{1 + (-R_E)\left(\dfrac{-h_{fe}}{h_{ie}}\right)}$$

$$= \frac{-h_{fe}}{h_{ie} + h_{fe} R_E} \tag{18.20}$$

The input and output impedance is calculated as specified in Table 18.2.

$$Z_{if} = Z_i(1 + \beta A) = h_{ie}\left(1 + \frac{h_{fe} R_E}{h_{ie}}\right) = h_{ie} + h_{fe} R_E \tag{18.21}$$

$$Z_{of} = Z_o(1 + \beta A) = R_C\left(1 + \frac{h_{fe} R_E}{h_{ie}}\right) \tag{18.22}$$

The voltage gain with feedback is

$$A_f = \frac{V_o}{V_s} = \frac{I_o R_C}{V_s} = \left(\frac{I_o}{V_s}\right) R_C = A_f R_C = \frac{-h_{fe} R_C}{h_{ie} + h_{fe} R_E} \tag{18.23}$$

EXAMPLE 18.5

Calculate the voltage gain of the circuit of Fig. 18.11.

Solution:

Without feedback,

$$A = \frac{I_o}{V_i} = \frac{-h_{fe}}{h_{ie}} = \frac{-120}{900} = -0.1333$$

$$\beta = \frac{V_f}{I_o} = -R_E = -510$$

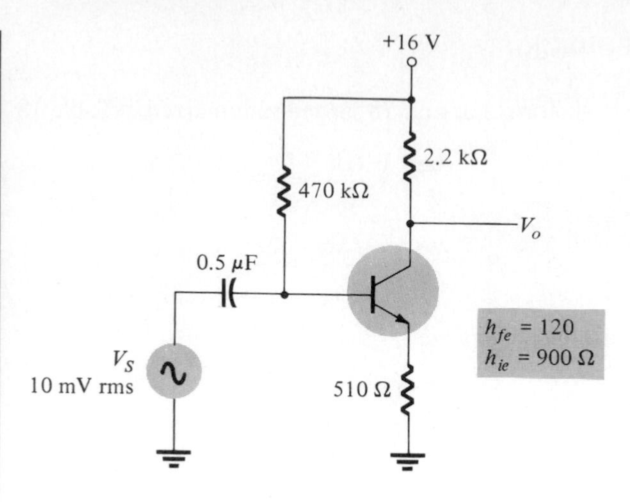

Figure 18.11 BJT amplifier with current-series feedback for Example 18.5.

The factor $(1 + \beta A)$ is then

$$1 + \beta A = 1 + (-0.1333)(-510) = 69$$

The gain with feedback is then

$$A_f = \frac{I_o}{V_s} = \frac{A}{1 + \beta A} = \frac{(-0.1333)}{69} = -1.93 \times 10^{-3}$$

and the voltage gain with feedback is

$$A_{vf} = \frac{V_o}{V_s} = A_f R_C = (-1.93 \times 10^{-3})(2.2 \times 10^3) = \mathbf{-4.25}$$

(Without feedback the magnitude of the voltage gain is

$$|A_v| = \frac{-R_C}{r_e} = \frac{-2.2 \times 10^3}{7.5} = -293.3$$

Voltage-Shunt Feedback

The constant-gain op-amp circuit of Fig. 18.12a provides voltage-shunt feedback. Referring to Fig. 18.2b and Table 18.1 and the op-amp ideal characteristics $I_i = 0$, $V_i = 0$, and voltage gain of infinity; we have

$$A = \frac{V_o}{I_i} = \text{infinity} \tag{18.24}$$

$$\beta = \frac{I_f}{V_o} = \frac{-1}{R_o} \tag{18.25}$$

The gain with feedback is then

$$A_f = \frac{V_o}{I_s} = \frac{V_o}{I_i} = \frac{A}{1 + \beta A} = \frac{1}{\beta} = -R_o \tag{18.26}$$

This is a transfer resistance gain. The more usual gain is the voltage gain with

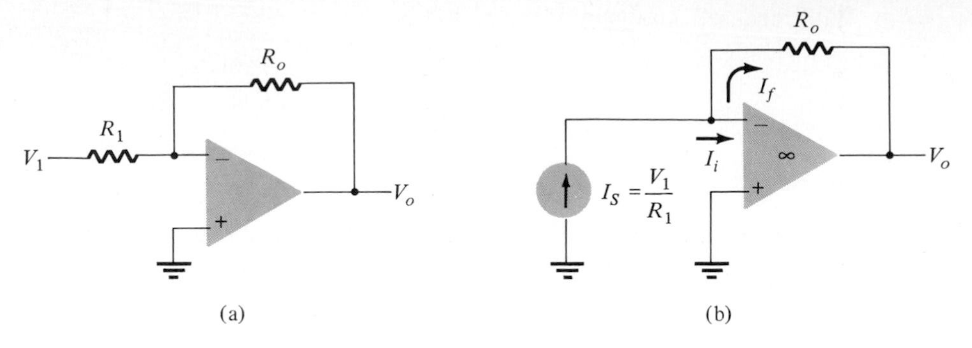

Figure 18.12 Voltage-shunt negative feedback amplifier: (a) constant-gain circuit; (b) equivalent circuit.

feedback,

$$A_{vf} = \left(\frac{V_o}{I_s}\right)\left(\frac{I_s}{V_1}\right) = (-R_o)\left(\frac{1}{R_1}\right) = \frac{-R_o}{R_1} \tag{18.27}$$

The circuit of Fig. 18.13 is a voltage-shunt feedback amplifier using an FET. With the feedback resistor not connected, the gain without feedback ($I_s = I_i$) is that of a common-source circuit times the source resistance.

$$A = \frac{V_o}{I_i} = \left(\frac{V_o}{V_i}\right)\left(\frac{V_i}{I_i}\right) = \left(\frac{V_o}{V_s}\right)\left(\frac{V_s}{I_s}\right) = (-g_m R_D)R_S \tag{18.28}$$

The feedback is

$$\beta = \frac{I_f}{V_o} = \frac{-1}{R_F} \tag{18.29}$$

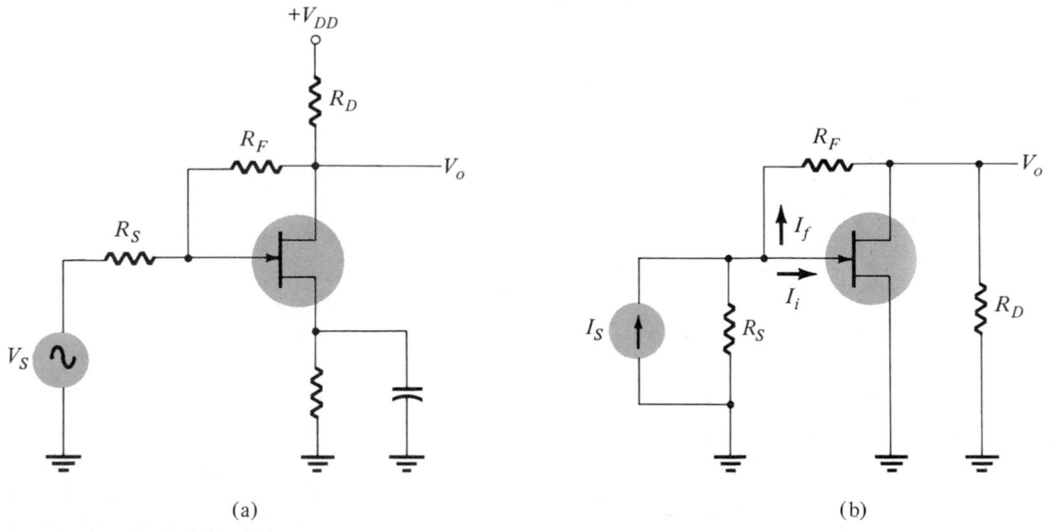

Figure 18.13 Voltage-shunt feedback amplifier using an FET: (a) circuit; (b) equivalent circuit.

With feedback, the gain of the circuit is

$$A_f = \frac{V_o}{I_s} = \frac{A}{1 + \beta A} = \frac{-g_m R_D R_S}{1 + (-1/R_f)(-g_m R_D R_S)}$$

$$= \frac{-g_m R_D R_S R_F}{R_F + g_m R_D R_S} \tag{18.30}$$

The voltage gain of the circuit with feedback is then

$$A_{vf} = \left(\frac{V_o}{I_s}\right)\left(\frac{I_s}{V_s}\right) = \frac{-g_m R_D R_S R_F}{R_F + g_m R_D R_S} \cdot \frac{1}{R_S}$$

$$= \frac{-g_m R_D R_F}{R_F + g_m R_D R_S} = (-g_m R_D)\frac{R_F}{R_F + g_m R_D R_S} \tag{18.31}$$

EXAMPLE 18.6

Calculate the voltage gain with and without feedback for the circuit of Fig. 18.13a with values of $g_m = 5$ mS, $R_D = 5.1$ kΩ, $R_S = 1$ kΩ, and $R_F = 20$ kΩ.

Solution:

Without feedback the voltage gain is

$$A_v = -g_m R_D = -(5 \times 10^{-3})(5.1 \times 10^3) = \mathbf{-25.5}$$

With feedback the gain is reduced to

$$A_{vf} = (-g_m R_D)\frac{R_F}{R_F + g_m R_D R_S}$$

$$= (-25.5)\frac{20 \times 10^3}{(20 \times 10^3) + (5 \times 10^{-3})(5.1 \times 10^3)(1 \times 10^3)}$$

$$= -25.5(0.44) = \mathbf{-11.2}$$

18.4 FEEDBACK AMPLIFIER—PHASE AND FREQUENCY CONSIDERATIONS

So far we have considered the operation of a feedback amplifier in which the feedback signal was *opposite* to the input signal—negative feedback. In any practical circuit this condition occurs only for some midfrequency range of operation. We know that an amplifier gain will change with frequency, dropping off at high frequencies from the midfrequency value. In addition, the phase shift of an amplifier will also change with frequency.

If, as the frequency increases, the phase shift changes then some of the feedback signal *adds* to the input signal. It is then possible for the amplifier to break into oscillations due to positive feedback. If the amplifier oscillates at some low or high frequency, it is no longer useful as an amplifier. Proper feedback-amplifier design requires that the circuit be stable at *all* frequencies, not merely those in the range of interest. Otherwise, a transient disturbance could cause a seemingly stable amplifier to suddenly start oscillating.

Nyquist Criterion

In judging the stability of a feedback amplifier, as a function of frequency, the βA product and the phase shift between input and output are the determining factors. One of the most popular techniques used to investigate stability is the Nyquist method. A Nyquist diagram is used to plot gain and phase shift as a function of frequency on a complex plane. The Nyquist plot, in effect, combines the two Bode plots of gain versus frequency and phase-shift versus frequency on a single plot. A Nyquist plot is used to quickly show whether an amplifier is stable for all frequencies and how stable the amplifier is relative to some gain or phase-shift criteria.

As a start, consider the *complex plane* shown in Fig. 18.14. A few points of various gain (βA) values are shown at a few different phase-shift angles. By using the positive real axis as reference (0°) a magnitude of $\beta A = 2$ is shown at a phase shift of 0° at point 1. Additionally, a magnitude of $\beta A = 3$ at a phase shift of $-135°$ is shown at point 2 and a magnitude/phase of $\beta A = 1$ at 180° is shown at point 3. Thus points on this plot can represent *both* gain magnitude of βA and phase shift. If the points representing gain and phase shift for an amplifier circuit are plotted at increasing frequency, then a Nyquist plot is obtained as shown by the plot in Fig. 18.15. At the origin the gain is 0 at a frequency of 0 (for *RC*-type coupling). At increasing frequency

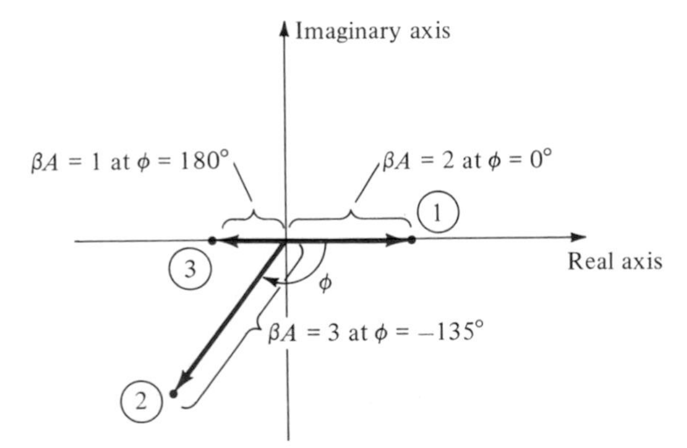

Figure 18.14 Complex plane showing typical gain-phase points.

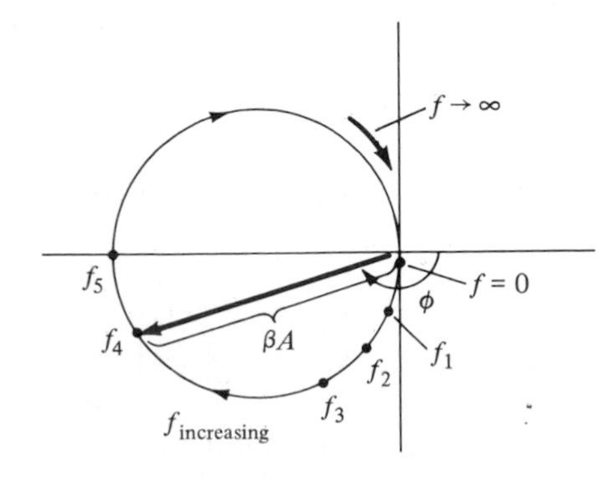

Figure 18.15 Nyquist plot.

points f_1, f_2, and f_3 and the phase shift increased as did the magnitude of βA. At a representative frequency f_4 the value of A is the vector length from the origin to point f_4 and the phase shift is the angle ϕ. At a frequency f_5 the phase shift is 180°. At higher frequencies the gain is shown to decrease back to 0.

The Nyquist criterion for stability can be stated as follows: *The amplifier is unstable if the Nyquist curve plotted encloses (encircles) the* -1 *point, and it is stable otherwise.*

An example of the Nyquist criterion is demonstrated by the curves in Fig. 18.16. The Nyquist plot in Fig. 18.16a is stable since it does not encircle the -1 point, whereas that shown in Fig. 18.16b is unstable since the curve does encircle the -1 point. Keep in mind that encircling the -1 point means that at a phase shift of 180° the loop gain (βA) is greater than 1; therefore, the feedback signal is in phase with the input and large enough to result in a larger input signal than that applied, with the result that oscillation occurs.

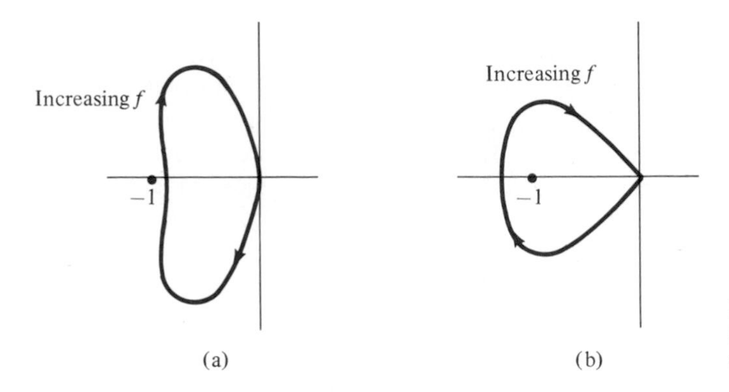

Figure 18.16 Nyquist plots showing stability conditions: (a) stable; (b) unstable.

Gain and Phase Margins

From the Nyquist criterion we know that a feedback amplifier is stable if the loop gain (βA) is less than unity (0 dB) when its phase angle is 180°. We can additionally determine some margins of stability to indicate how close to instability the unit is. That is, if the gain (βA) is less than unity but, say 0.95 in value, this would not be as relatively stable as another amplifier having, say, $(\beta A) = 0.7$ (both measured at 180°). Of course, amplifiers with loop gains 0.95 and 0.7 are both stable, but one is closer to instability, if the loop gain increases, than the other. We can define the following terms:

Gain margin (GM) is defined as the value of βA in decibels at the frequency at which the phase angle is 180°. Thus, 0 dB, equal to a value of $\beta A = 1$, is on the border of stability and any negative decibel value is stable. The more negative the decibed gain value the more stable the feedback circuit. The GM may be evaluated in decibels from the curve of Fig. 18.17.

Phase margin (PM) is defined as the angle of 180° minus the magnitude of the angle at which the value βA is unity, (0 dB). The PM may also be evaluated directly from the curve of Fig. 18.17.

An example of these two amplifier factors is shown on the Bode plots of Fig. 18.17. Instability occurs, therefore, with a positive GM and PM greater than 180°.

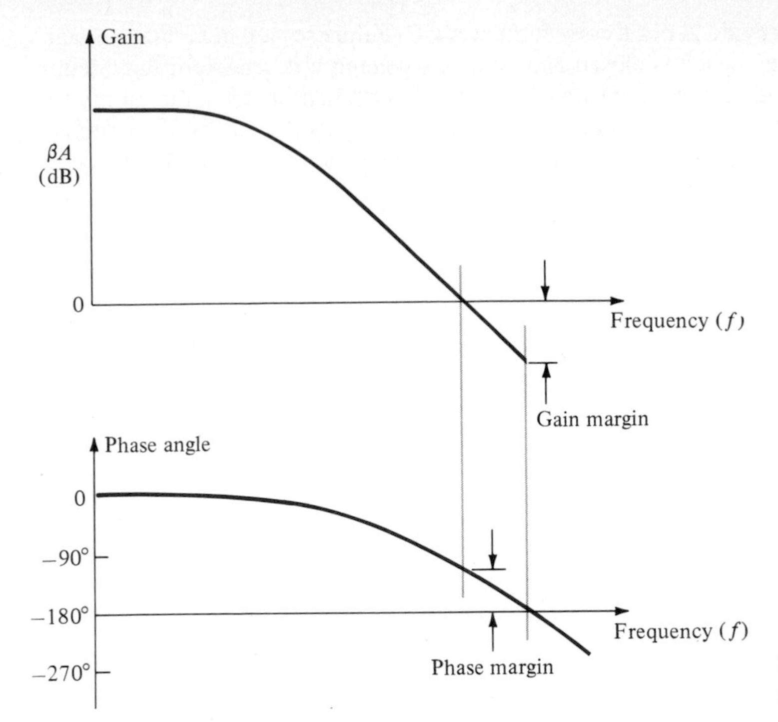

Figure 18.17 Bode plots showing gain and phase margins.

18.5 OSCILLATOR OPERATION

The use of positive feedback which results in a feedback amplifier having closed-loop gain A_f greater than 1 and satisfies the phase conditions will result in operation as an oscillator circuit. An oscillator circuit then provides a constantly varying output signal. If the output signal varies sinusoidally, the circuit is referred to as a *sinusoidal oscillator*. If the output voltage rises quickly to one voltage level and later drops quickly to another voltage level, the circuit is generally referred to as a *pulse* or *square-wave oscillator*.

To understand how a feedback circuit performs as an oscillator consider the feedback circuit of Fig. 18.18. When the switch at the amplifier input is open, no oscillation occurs. Consider that we have a *fictitious* voltage at the amplifier input (V_i). This results in an output voltage $V_o = AV_i$ after the base amplifier stage and in a voltage $V_f = \beta(AV_i)$ after the feedback stage. Thus, we have a feedback voltage $V_f = \beta AV_i$, where βA is referred to as the *loop gain*. If the circuits of the base amplifier and

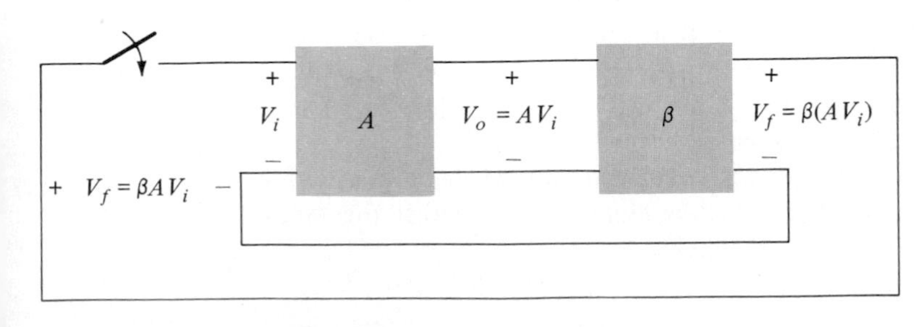

Figure 18.18 Feedback circuit used as an oscillator.

feedback network provide βA of a correct magnitude and phase, V_f can be made equal to V_i. Then, when the switch is closed and fictitious voltage V_i is removed, the circuit will continue operating since the feedback voltage is sufficient to drive the amplifier and feedback circuits resulting in a proper input voltage to sustain the loop operation. The output waveform will still exist after the switch is closed if the condition

$$\beta A = 1 \tag{18.32}$$

is met. This is known as the *Barkhausen criterion* for oscillation.

In reality, no input signal is needed to start the oscillator going. Only the condition $\beta A = 1$ must be satisfied for self-sustained oscillations to result. In practice βA is made greater than 1, and the system is started oscillating by amplifying noise voltage which is always present. Saturation factors in the practical circuit provide an "average" value of βA of 1. The resulting waveforms are never exactly sinusoidal. However, the closer the value βA is to exactly 1 the more nearly sinusoidal is the waveform. Figure 18.19 shows how the noise signal results in a buildup of a steady-state oscillation condition.

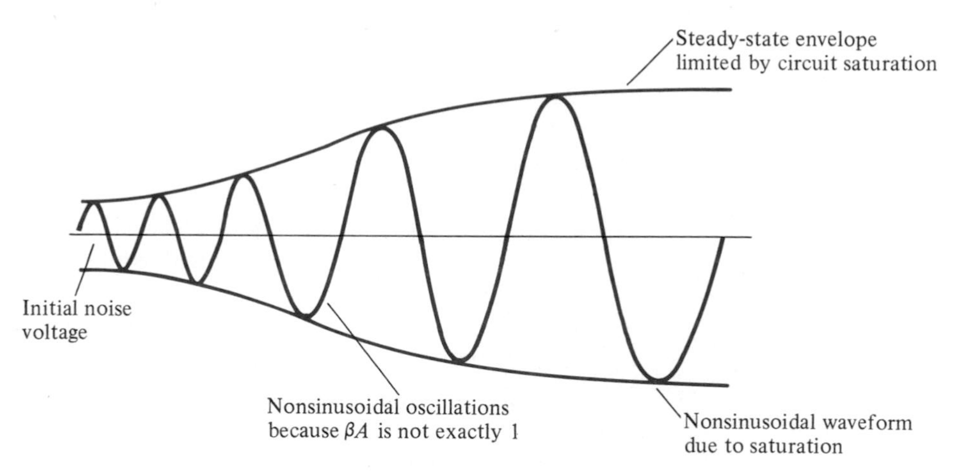

Figure 18.19 Buildup of steady-state oscillations.

Another way of seeing how the feedback circuit provides operation as an oscillator is obtained by noting the denominator in the basic feedback equation, (Eq. 18.3), $A_f = A/(1 + \beta A)$. When $\beta A = -1$ or magnitude 1 at a phase angle of 180°, the denominator becomes 0 and the gain with feedback, A_f, becomes infinite. Thus, an infinitesimal signal (noise voltage) can provide a measurable output voltage, and the circuit acts as an oscillator even without an input signal.

The remainder of this chapter is devoted to various oscillator circuits that use a variety of components. Practical considerations are included so that workable circuits in each of the various cases are discussed.

18.6 PHASE-SHIFT OSCILLATOR

An example of an oscillator circuit that follows the basic development of a feedback circuit is the *phase-shift oscillator*. An idealized version of this circuit is shown in Fig. 18.20. Recall that the requirements for oscillation are that the loop gain, βA, is greater than unity *and* that the phase shift around the feedback network is 180° (providing positive feedback). In the present idealization we are considering the feedback network to be driven by a perfect source (zero source impedance) and the output of the feedback network is connected into a perfect load (infinite load impedance). The idealized case will allow development of the theory behind the operation of the phase-shift oscillator. Practical circuit versions will then be considered.

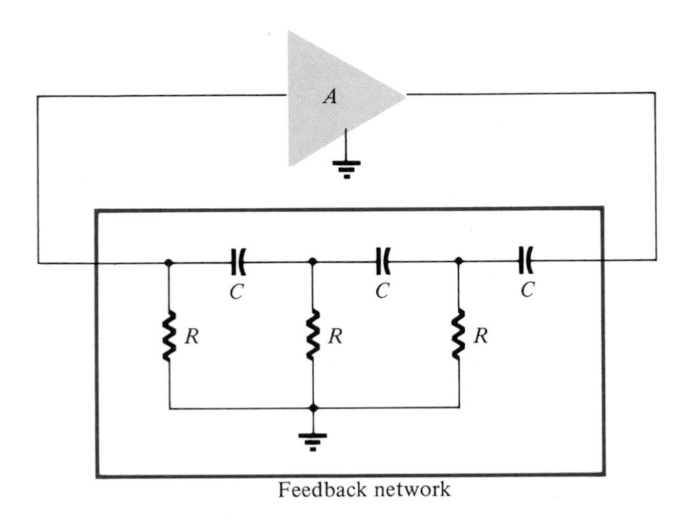

Figure 18.20 Idealized phase-shift oscillator.

Concentrating our attention on the phase-shift network we are interested in the attenuation of the network at the frequency at which the phase shift is exactly 180°. Using classical network analysis, we find that

$$\boxed{f = \frac{1}{2\pi RC\sqrt{6}}} \tag{18.33}$$

$$\beta = \frac{1}{29} \tag{18.34}$$

and the phase shift is 180°.

For the loop gain βA to be greater than unity the gain of the amplifier stage must be greater than $1/\beta$ or 29

$$A > 29 \tag{18.35}$$

When considering the operation of the feedback network one might naively select the values of R and C to provide (at a specific frequency) 60°-phase shift per section for three sections, resulting in 180°-phase shift as desired. This, however, is not the case, since each section of the RC in the feedback network loads down the previous one. The net result that the *total* phase shift be 180° is all that is important. The frequency given by Eq. (18.33) is that at which the *total* phase shift is 180°. If one measured the phase shift per RC section, each section would not provide the same phase shift (although the overall phase shift is 180°). If it were desired to obtain exactly 60°-phase shift for each of three stages, then emitter-follower stages would be needed for each RC section to prevent each from being loaded from the following circuit.

FET Phase-Shift Oscillator

A practical version of a phase-shift oscillator circuit is shown in Fig. 18.21a. The circuit is drawn to show clearly the amplifier and feedback network. The amplifier stage is self-biased with a capacitor bypassed source resistor R_S and a drain bias resistor R_D. The FET device parameters of interest are g_m and r_d. From FET amplifier theory the amplifier gain magnitude is calculated from

$$|A| = g_m R_L \tag{18.36}$$

where R_L in this case is the parallel resistance of R_D and r_d

$$R_L = \frac{R_D r_d}{R_D + r_d} \tag{18.37}$$

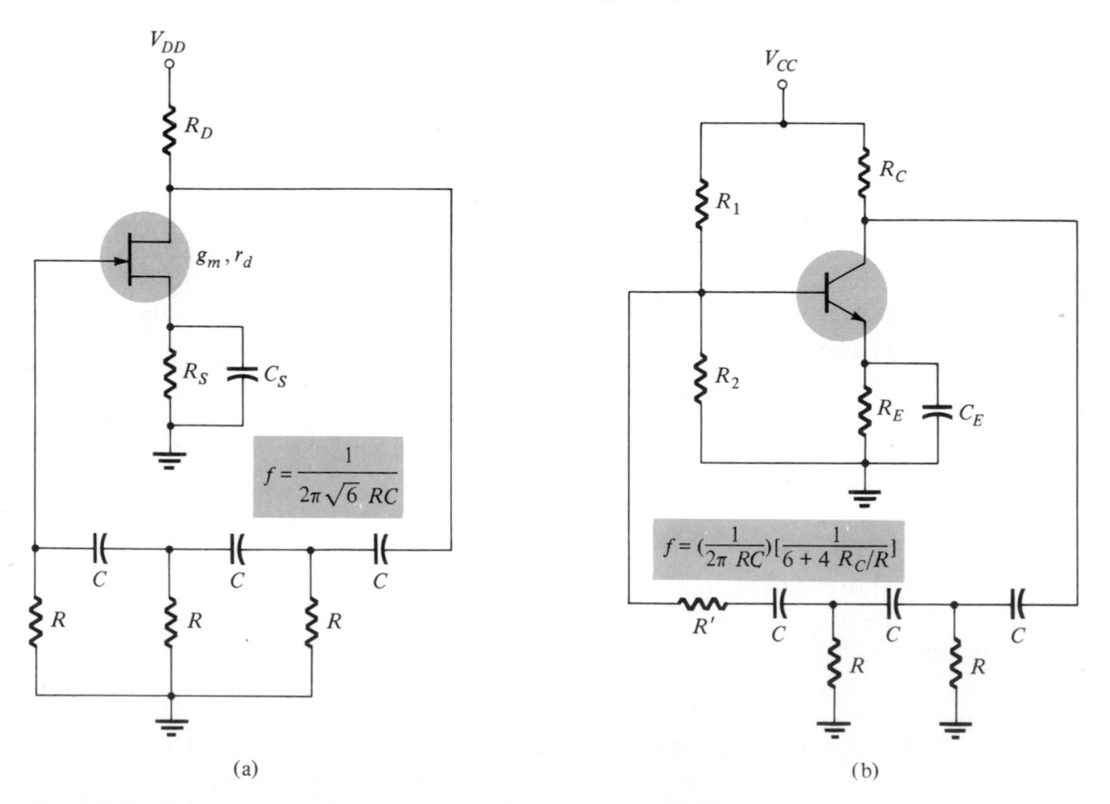

Figure 18.21 Practical phase-shift oscillator circuits: (a) FET version; (b) BJT version.

We shall assume as a very good approximation that the input impedance of the FET amplifier stage is infinite. This assumption is valid as long as the oscillator operating frequency is low enough so that FET capacitive impedances can be neglected. The output impedance of the amplifier stage given by R_L should also be small compared to the impedance seen looking into the feedback network so that no attenuation due to loading occurs. In practice, these considerations are not always negligible, and the amplifier stage gain is then selected somewhat larger than the needed factor of 29 to assure oscillator action.

EXAMPLE 18.7

It is desired to design a phase-shift oscillator (as in Fig. 18.21a) using an FET having $g_m = 5000\ \mu S$, $r_d = 40\ k\Omega$, and feedback circuit value of $R = 10\ k\Omega$. Select the value of C for oscillator operation at 1 kHz and R_D for $A > 29$ to ensure oscillator action.

Solution:

Equation (18.33) is used to solve for the capacitor value. Since $f = 1/2\pi RC\sqrt{6}$, we can solve for C:

$$C = \frac{1}{2\pi Rf\sqrt{6}} = \frac{1}{(6.28)(10 \times 10^3)(1 \times 10^3)(2.45)} = \mathbf{6.5\ nF}$$

Using Eq. (18.36), we solve for R_L to provide a gain of, say, $A = 40$ (this allows for some loading between R_L and the feedback network input impedance):

$$|A| = g_m R_L$$

$$R_L = \frac{|A|}{g_m} = \frac{40}{5000 \times 10^{-6}} = 8\ k\Omega$$

Using Eq. (18.37), we solve for $R_D = \mathbf{10\ k\Omega}$.

Transistor Phase-Shift Oscillator

If a transistor is used as the active element of the amplifier stage, the output of the feedback network is loaded appreciably by the relatively low input resistance (h_{ie}) of the transistor. Of course, an emitter follower input stage followed by a common-emitter amplifier stage could be used. If a single transistor stage is desired, however, the use of voltage-shunt feedback (as shown in Fig. 18.21b) is more suitable. In this connection, the feedback signal is coupled through the feedback resistor R_S in *series* with the amplifier stage input resistance (R_i).

Analysis of the ac circuit provides the following equation for the resulting oscillator frequency:

$$\boxed{f = \frac{1}{2\pi RC}\frac{1}{\sqrt{6 + 4(R_C/R)}}} \qquad (18.38)$$

For the loop gain to be greater than unity, the requirement on the current gain of the transistor is found to be

$$h_{fe} > 23 + 29\frac{R_C}{R} + 4\frac{R}{R_C} \qquad (18.39)$$

IC Phase-Shift Oscillator

As IC circuits have become more popular they have been adapted to operate in oscillator circuits. One need buy only an op-amp to obtain an amplifier circuit of stabilized gain setting and incorporate some means of signal feedback to produce an oscillator circuit. For example, a phase-shift oscillator is shown in Fig. 18.22. The output of the op-amp is fed to a three-stage RC network which provides the needed 180° of phase shift (at an attenuation factor of 1/29). If the op-amp provides gain (set by resistors R_i and R_f) of greater than 29, a loop gain greater than unity results and the circuit acts as an oscillator [oscillator frequency is given by Eq. (18.33)].

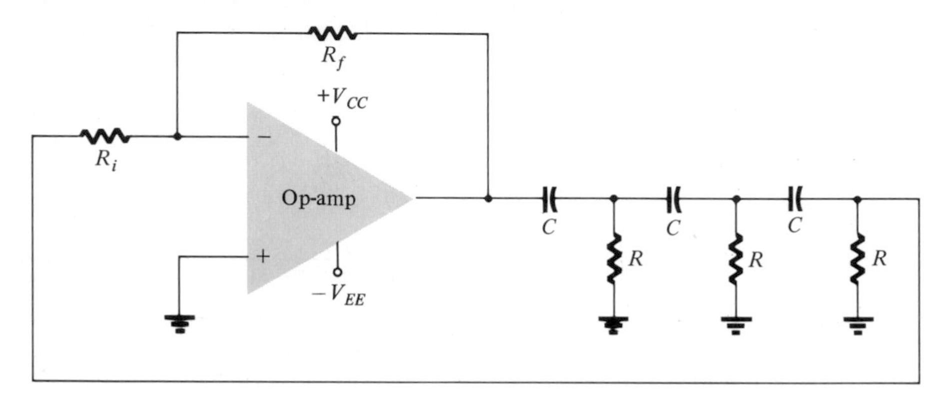

Figure 18.22 Phase-shift oscillator using op-amp.

18.7 WIEN BRIDGE OSCILLATOR

A practical oscillator circuit uses an op-amp and RC bridge circuit, with the oscillator frequency set by the R and C components. Figure 18.23 shows a basic version of a Wien bridge oscillator circuit. Note the basic bridge connection. Resistors R_1, R_2 and capacitors C_1, C_2 form the frequency-adjustment elements, while resistors R_3 and R_4 form part of the feedback path. The op-amp output is connected as the bridge input at points a and c. The bridge circuit output at points b and d is the input to the op-amp.

Neglecting loading effects of the op-amp input and output impedances, the analysis of the bridge circuit results in

$$\frac{R_3}{R_4} = \frac{R_1}{R_2} + \frac{C_2}{C_1} \tag{18.40}$$

and

$$\boxed{f_o = \frac{1}{2\pi\sqrt{R_1C_1R_2C_2}}} \tag{18.41}$$

If, in particular, the values are $R_1 = R_2 = R$ and $C_1 = C_2 = C$, the resulting oscillator frequency is

$$f_o = \frac{1}{2\pi RC} \tag{18.42}$$

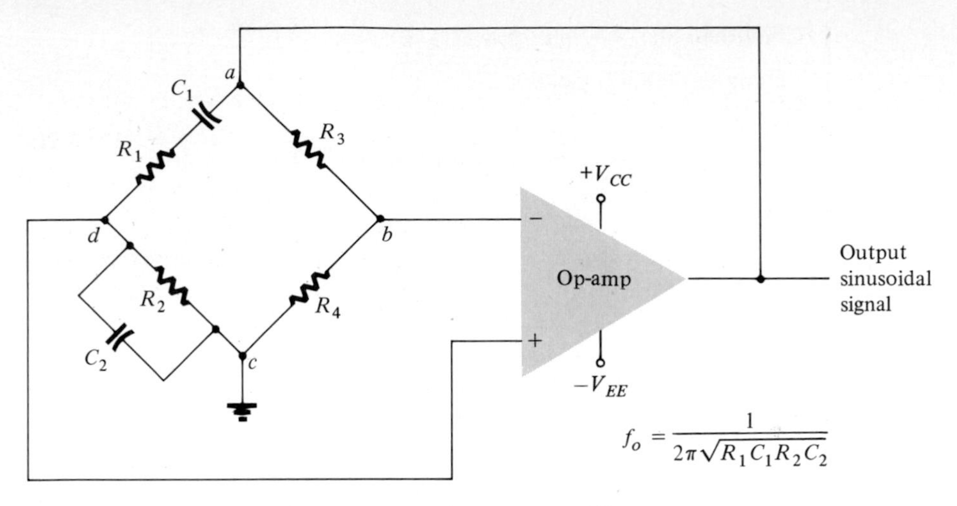

Figure 18.23 Wien bridge oscillator circuit using op-amp amplifier.

and

$$\frac{R_3}{R_4} = 2 \tag{18.43}$$

Thus a ratio of R_3 to R_4 greater than 2 will provide sufficient loop gain for the circuit to oscillate at the frequency calculated using Eq. (18.42).

EXAMPLE 18.8

Calculate the resonant frequency of the Wien bridge oscillator of Fig. 18.24.

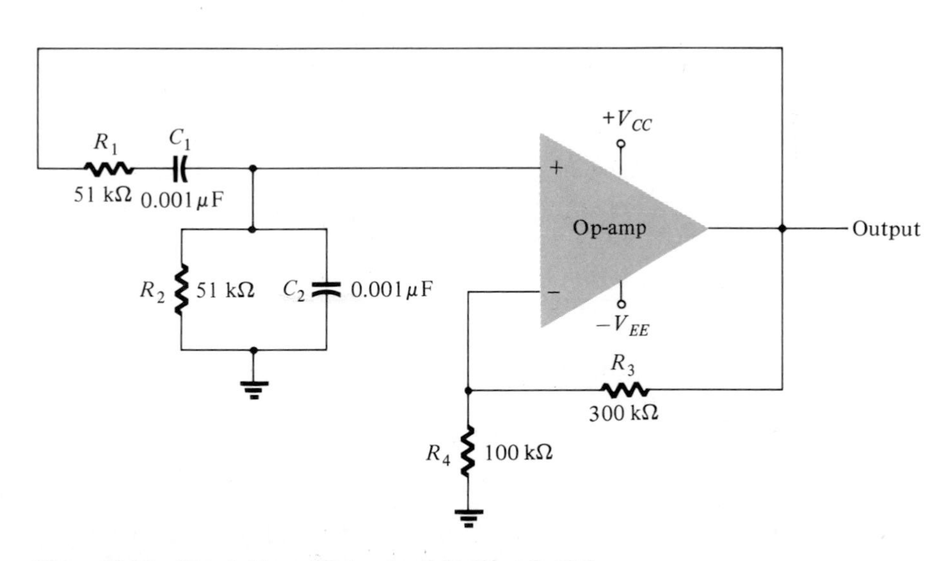

Figure 18.24 Wien bridge oscillator circuit for Example 18.8.

Solution:

Using Eq. (18.42) yields

$$f_o = \frac{1}{2\pi RC} = \frac{1}{2\pi(51 \times 10^3)(0.001 \times 10^{-6})} = \mathbf{3120.7\ Hz}$$

EXAMPLE 18.9

Design the *RC* elements of a Wien bridge oscillator as in Fig. 18.24 for operation at $f_o = 10$ kHz.

Solution:

Using equal values of *R* and *C* we can select $R = 100\ k\Omega$ and calculate the required value of *C* using Eq. (18.42):

$$C = \frac{1}{2\pi f_o R} = \frac{1}{6.28(10 \times 10^3)(100 \times 10^3)} = \frac{10^{-9}}{6.28} = \mathbf{159\ pF}$$

We can use $R_3 = 300\ k\Omega$ and $R_4 = 100\ k\Omega$ to provide a ratio R_3/R_4 greater than 2 for oscillation to take place.

18.8 TUNED OSCILLATOR CIRCUIT

Tuned-Input, Tuned-Output Oscillator Circuits

A variety of circuits can be built using that shown in Fig. 18.25 by providing tuning in both the input and output sections of the circuit. Analysis of the circuit of Fig. 18.25 reveals that the following types of oscillators are obtained when the reactance elements are as designated:

	Reactance Element		
Oscillator Type	X_1	X_2	X_3
Colpitts oscillator	*C*	*C*	*L*
Hartley oscillator	*L*	*L*	*C*
Tuned input, tuned output	*LC*	*LC*	—

Colpitts Oscillators

FET COLPITTS OSCILLATOR

A practical version of an FET Colpitts oscillator is shown in Fig. 18.26. The circuit is basically the same form as shown in Fig. 18.25 with the addition of the components needed for dc bias of the FET amplifier. The oscillator frequency can be

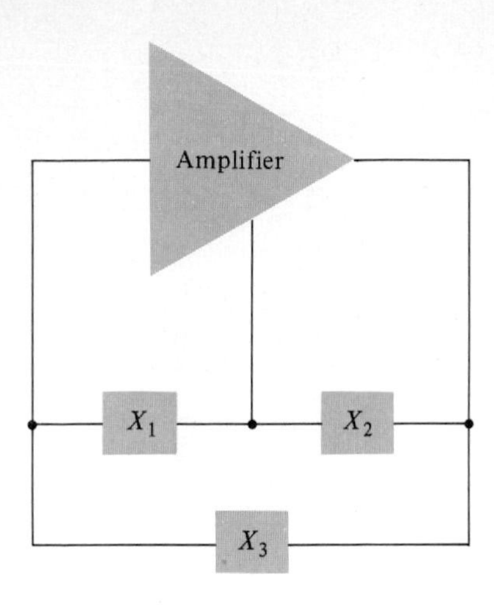

Figure 18.25 Basic configuration of resonant circuit oscillator.

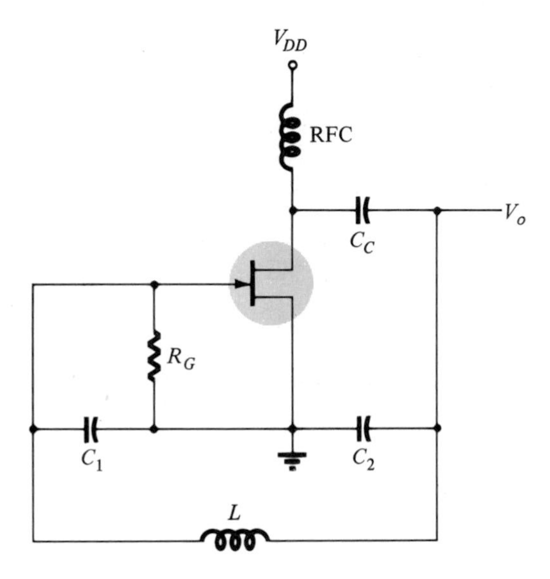

Figure 18.26 FET Colpitts oscillator.

found to be

$$f_o = \frac{1}{2\pi\sqrt{LC_{eq}}} \tag{18.44}$$

where

$$C_{eq} = \frac{C_1 C_2}{C_1 + C_2} \tag{18.45}$$

TRANSISTOR COLPITTS OSCILLATOR

A transistor Colpitts oscillator circuit can be made as shown in Fig. 18.27. The circuit frequency of oscillation is given by Eq. (18.44).

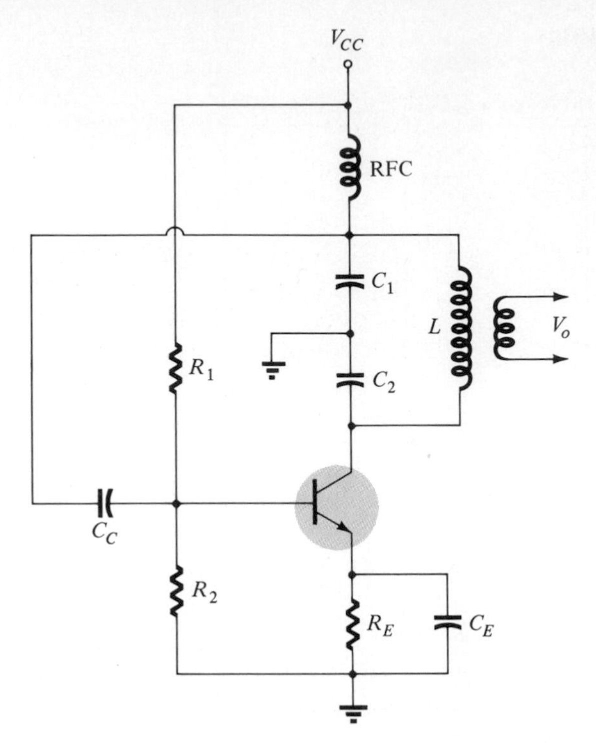

Figure 18.27 Transistor Colpitts oscillator.

IC COLPITTS OSCILLATOR

An op-amp Colpitts oscillator circuit is shown in Fig. 18.28. Again, the op-amp provides the basic amplification needed while the oscillator frequency is set by an *LC* feedback network of a Colpitt configuration. The oscillator frequency is given by Eq. (18.44).

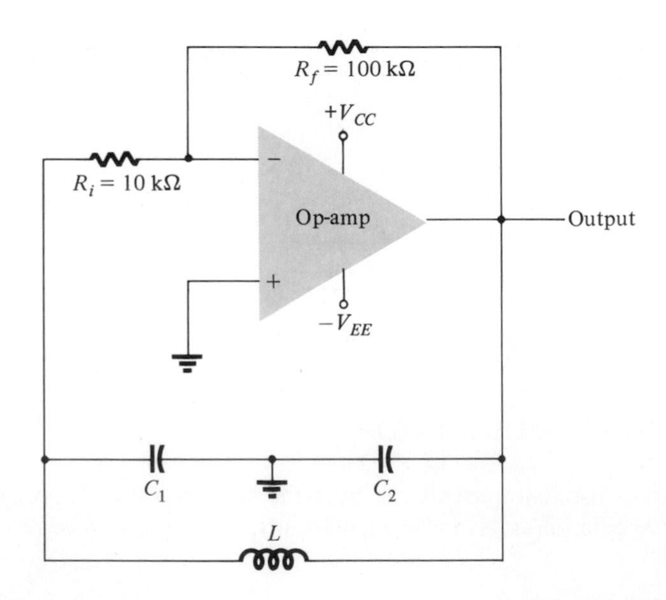

Figure 18.28 Op-amp Colpitts oscillator.

Hartley Oscillator

If the elements in the basic resonant circuit of Fig. 18.25 are X_1 and X_2 (inductors), and X_3 (capacitor), the circuit is a Hartley oscillator.

FET HARTLEY OSCILLATOR

An FET Hartley oscillator circuit is shown in Fig. 18.29. The circuit is drawn so that the feedback network conforms to the form shown in the basic resonant circuit (Fig. 18.25). Note, however, that inductors L_1 and L_2 have a mutual coupling, M, which must be taken into account in determining the equivalent inductance for the resonant tank circuit.

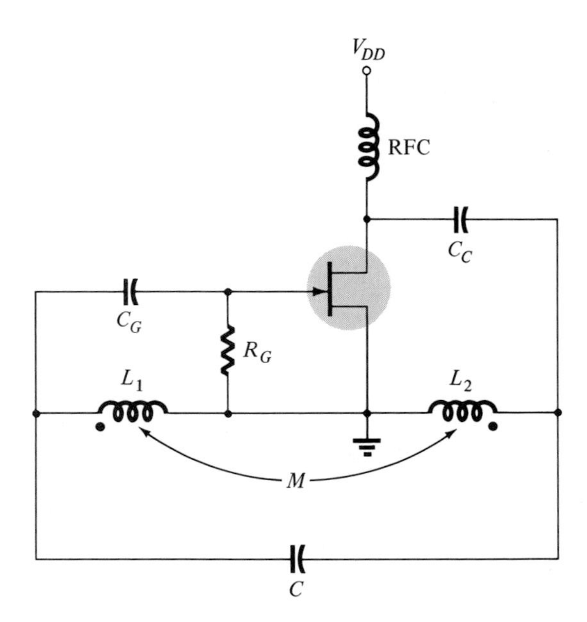

Figure 18.29 FET Hartley oscillator.

The circuit frequency of oscillation is then given approximately by

$$f_o = \frac{1}{2\pi\sqrt{L_{eq}C}} \tag{18.46}$$

with

$$L_{eq} = L_1 + L_2 + 2\,\text{M} \tag{18.47}$$

TRANSISTOR HARTLEY OSCILLATOR

Figure 18.30 shows a transistor Hartley oscillator circuit. The circuit operates at a frequency given by Eq. (18.46).

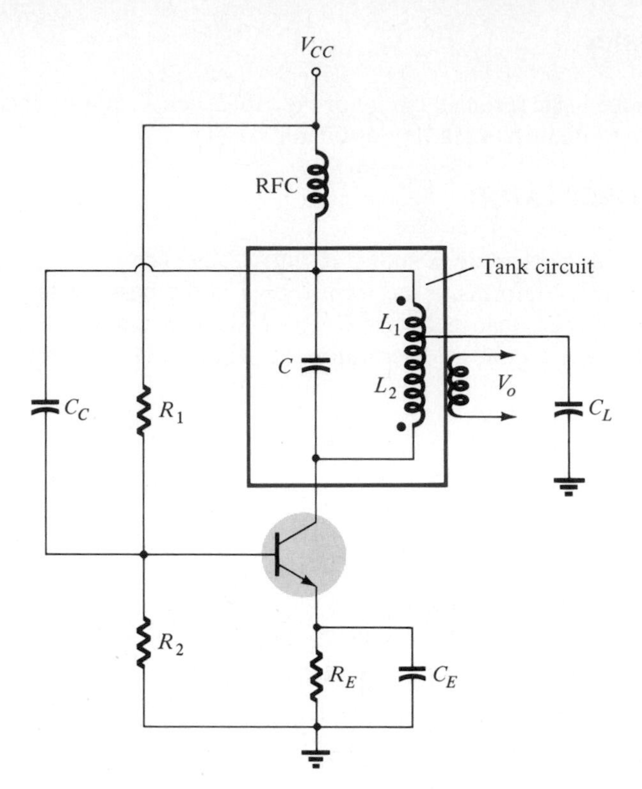

Figure 18.30 Transistor Hartley oscillator circuit.

18.9 CRYSTAL OSCILLATOR

A crystal oscillator is basically a tuned-circuit oscillator using a piezo-electric crystal as a resonant tank circuit. The crystal (usually quartz) has a greater stability in holding constant at whatever frequency the crystal is originally cut to operate. Crystal oscillators are used whenever great stability is required, for example, in communication transmitters and receivers.

Characteristics of a Quartz Crystal

A quartz crystal (one of a number of crystal types) exhibits the property that when mechanical stress is applied across the faces of the crystal, a difference of potential develops across opposite faces of the crystal. This property of a crystal is called the *piezoelectric effect*. Similarly, a voltage applied across one set of faces of the crystal causes mechanical distortion in the crystal shape.

When alternating voltage is applied to a crystal, mechanical vibrations are set up—these vibrations having a natural resonant frequency dependent on the crystal. Although the crystal has electromechanical resonance, we can represent the crystal action by an equivalent electrical resonant circuit as shown in Fig. 18.31. The inductor L and capacitor C represent electrical equivalents of crystal mass and compliance while resistance R is an electrical equivalent of the crystal structure's internal friction. The shunt capacitance C_M represents the capacitance due to mechanical mounting of

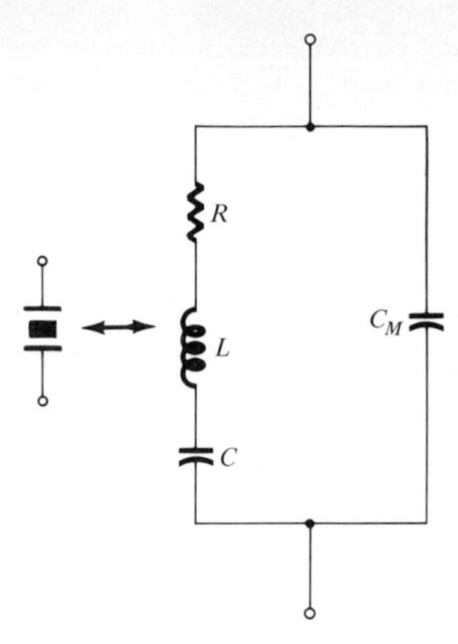

Figure 18.31 Electrical equivalent circuit of a crystal.

the crystal. Because the crystal losses, represented by R, are small, the equivalent crystal Q (quality factor) is high—typically 20,000. Values of Q up to almost 10^6 can be achieved by using crystals.

The crystal as represented by the equivalent electrical circuit of Fig. 18.31 can have two resonant frequencies. One resonant condition occurs when the reactances of the series RLC leg are equal (and opposite). For this condiiton the *series-resonant* impedance is very low (equal to R). The other resonant condition occurs at a higher frequency when the reactance of the series-resonant leg equals the reactance of capacitor C_M. This is a parallel resonance or antiresonance condition of the crystal. At this frequency the crystal offers a very high impedance to the external circuit. The impedance versus frequency of the crystal is shown in Fig. 18.32. In order to use the

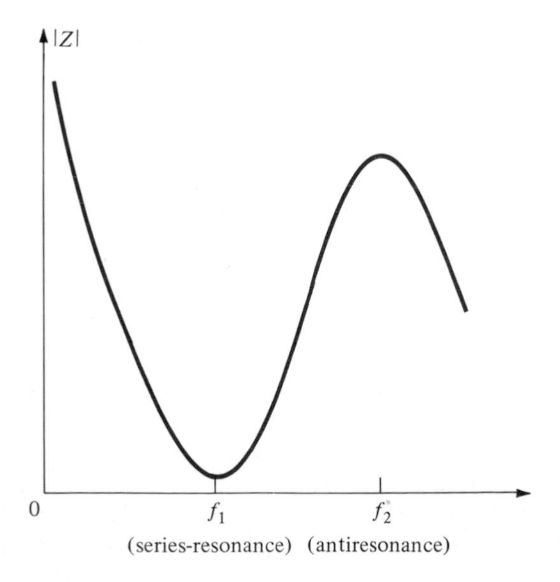

Figure 18.32 Crystal impedance versus frequency.

crystal properly it must be connected in a circuit so that its low impedance in the series-resonant operating mode or high impedance in the antiresonant operating mode is selected.

Series-Resonant Circuits

To excite a crystal for operation in the series-resonant mode it may be connected as a series element in a feedback path. At the series-resonant frequency of the crystal its impedance is smallest and the amount of (positive) feedback is largest. A typical transistor circuit is shown in Fig. 18.33. Resistors R_1, R_2, and R_E provide a voltage-divider stabilized dc bias circuit. Capacitor C_E provides ac bypass of the emitter resistor and the RFC coil provides for dc bias while decoupling any ac signal on the power lines from affecting the output signal. The voltage feedback from collector to base is a maximum when the crystal impedance is minimum (in series-resonant mode). The coupling capacitor C_C has negligible impedance at the circuit operating frequency but blocks any dc between collector and base.

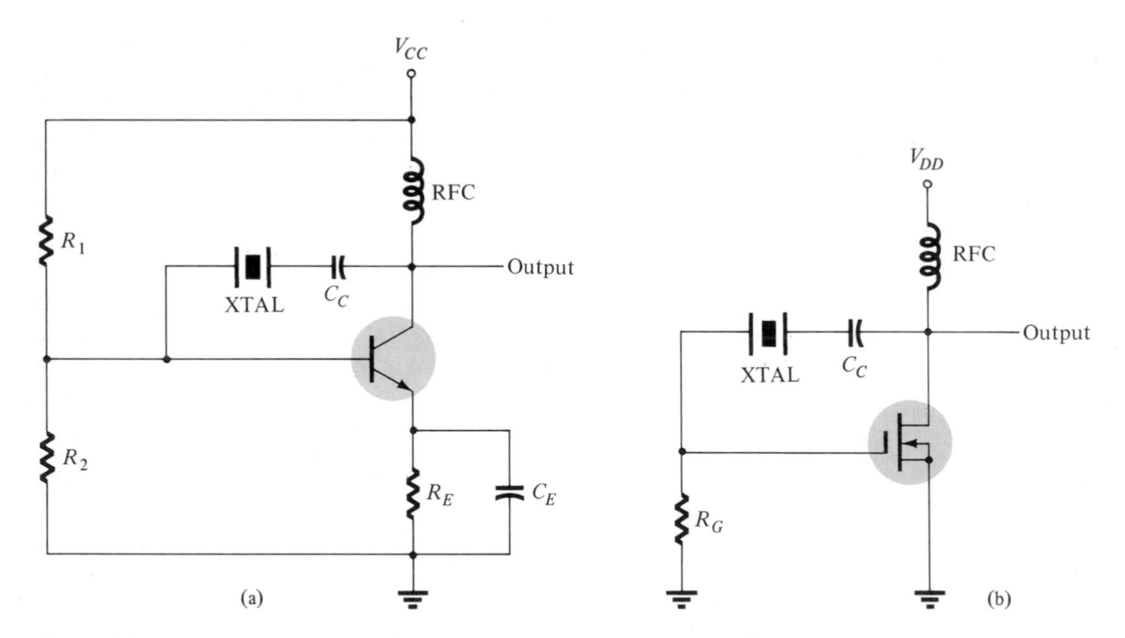

Figure 18.33 Crystal-controlled oscillator using crystal in series-feedback path: (a) BJT circuit; (b) FET circuit.

The resulting circuit frequency of oscillation is set, then, by the series-resonant frequency of the crystal. Changes in supply voltage, transistor device parameters, and so on, have no effect on the circuit operating frequency which is held stabilized by the crystal. The circuit frequency stability is set by the crystal frequency stability, which is good.

The circuits shown in Fig. 18.33a and 18.33b are generally called Pierce crystal-controlled oscillators.

Parallel-Resonant Circuits

Since the parallel-resonant impedance of a crystal is a maximum value, it is connected in shunt. At the parallel-resonant operating frequency a crystal appears as an inductive reactance of largest value. Figure 18.34 shows a crystal connected as the inductor element in a modified Colpitts circuit. The basic dc bias circuit should be evident. Maximum voltage is developed across the crystal at its parallel-resonant frequency. The voltage is coupled to the emitter by a capacitor voltage divider—capacitors C_1 and C_2.

A *Miller* crystal-controlled oscillator circuit is shown in Fig. 18.35. A tuned *LC* circuit in the drain section is adjusted near the crystal parallel-resonant frequency. The

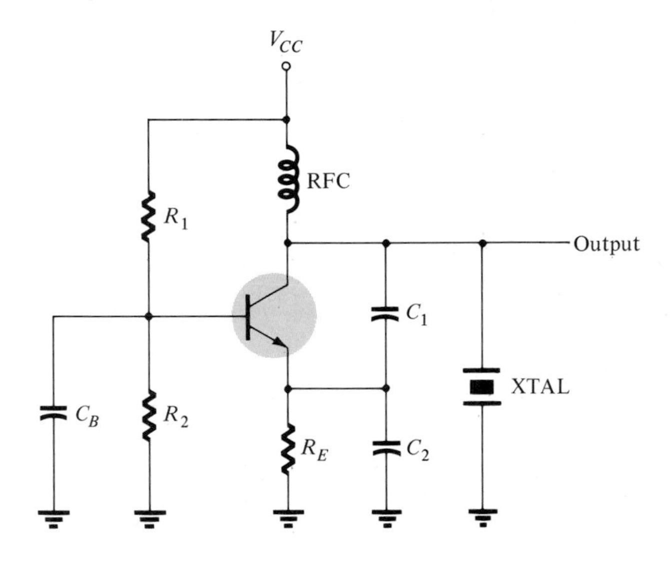

Figure 18.34 Crystal controlled oscillator operating in parallel-resonant triode.

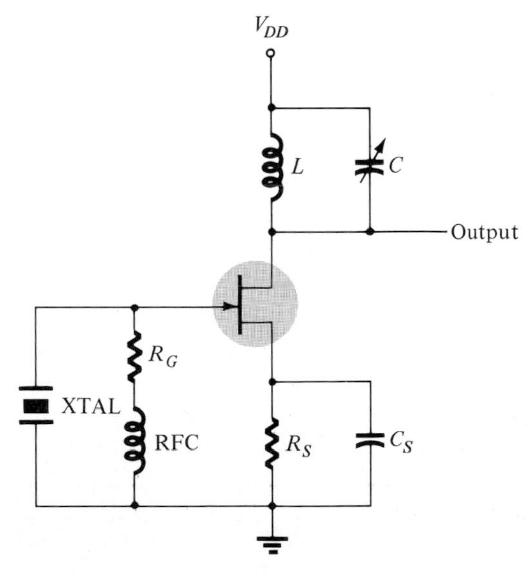

Figure 18.35 Miller crystal-controlled oscillator.

maximum gate-source signal occurs at the crystal antiresonant frequency controlling the circuit operating frequency.

Crystal Oscillator

An op-amp can be used in a crystal oscillator as shown in Fig. 18.36. The crystal is connected in the series-resonant path and operates at the crystal series-resonant frequency. The present circuit has a high gain so that an output square-wave signal results as shown in the figure. A pair of Zener diodes is shown at the output to provide output amplitude at exactly the Zener voltage (V_Z).

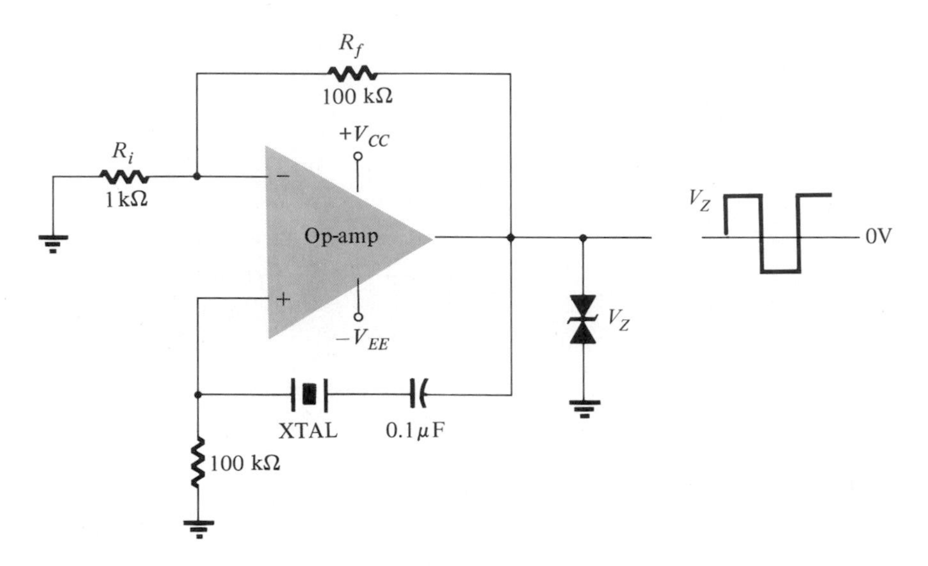

Figure 18.36 Crystal oscillator using op-amp.

18.10 UNIJUNCTION OSCILLATOR

A particular device, the unijunction transistor can be used in a single-stage oscillator circuit to provide a pulse signal suitable for digital-circuit applications. The unijunction transistor can be used in what is called a *relaxation oscillator* as shown by the basic circuit of Fig. 18.37. Resistor R_T and capacitor C_T are the timing components that set the circuit oscillating rate. The oscillating frequency may be calculated using Eq. (18.48) which includes the unijunction transistor *intrinsic stand-off ratio* η, as a factor (in addition to R_T and C_T) in the oscillator operating frequency.

$$f_o \cong \frac{1}{R_T C_T \ln [1/(1 - \eta)]} \tag{18.48}$$

Typically, a unijunction transistor has a stand-off ratio from 0.4 to 0.6. Using a value

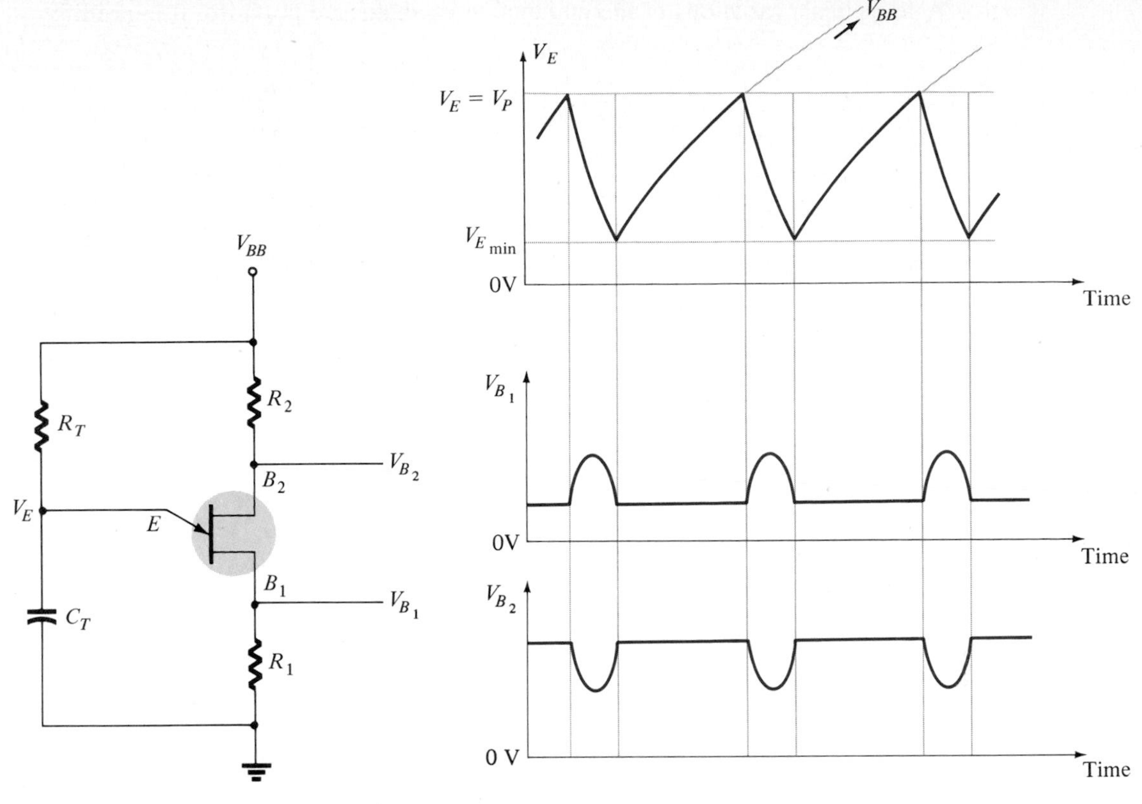

Figure 18.37 Basic unijunction oscillator circuit.

Figure 18.38 Unijunction oscillator waveforms.

of $\eta = 0.5$, we get

$$f_o \cong \frac{1}{R_T C_T \ln [1/(1 - 0.5)]} = \frac{1}{R_T C_T \ln 2} = \frac{1.44}{R_T C_T}$$

$$\cong \frac{1.5}{R_T C_T} \tag{18.49}$$

Capacitor C_T is charged through resistor R_T toward supply voltage V_{BB}. As long as the capacitor voltage V_E is below a stand-off voltage (V_P) set by the voltage across $B_1 - B_2$, and the transistor stand-off ratio η

$$V_P = \eta V_{B_1} V_{B_2} - V_D \tag{18.50}$$

the unijunction emitter lead appears as an open circuit. When the emitter voltage across capacitor C_T exceeds this value (V_P), the unijunction circuit fires, discharging the capacitor, after which a new charge cycle begins. When the unijunction fires, a voltage rise is developed across R_1 and a voltage drop is developed across R_2 as shown in Fig. 18.38. The signal at the emitter is a sawtooth voltage waveform, that at base 1 is a positive-going pulse, and that at base 2 is a negative-going pulse.

A few circuit variations of the unijunction oscillator are provided in Fig. 18.39.

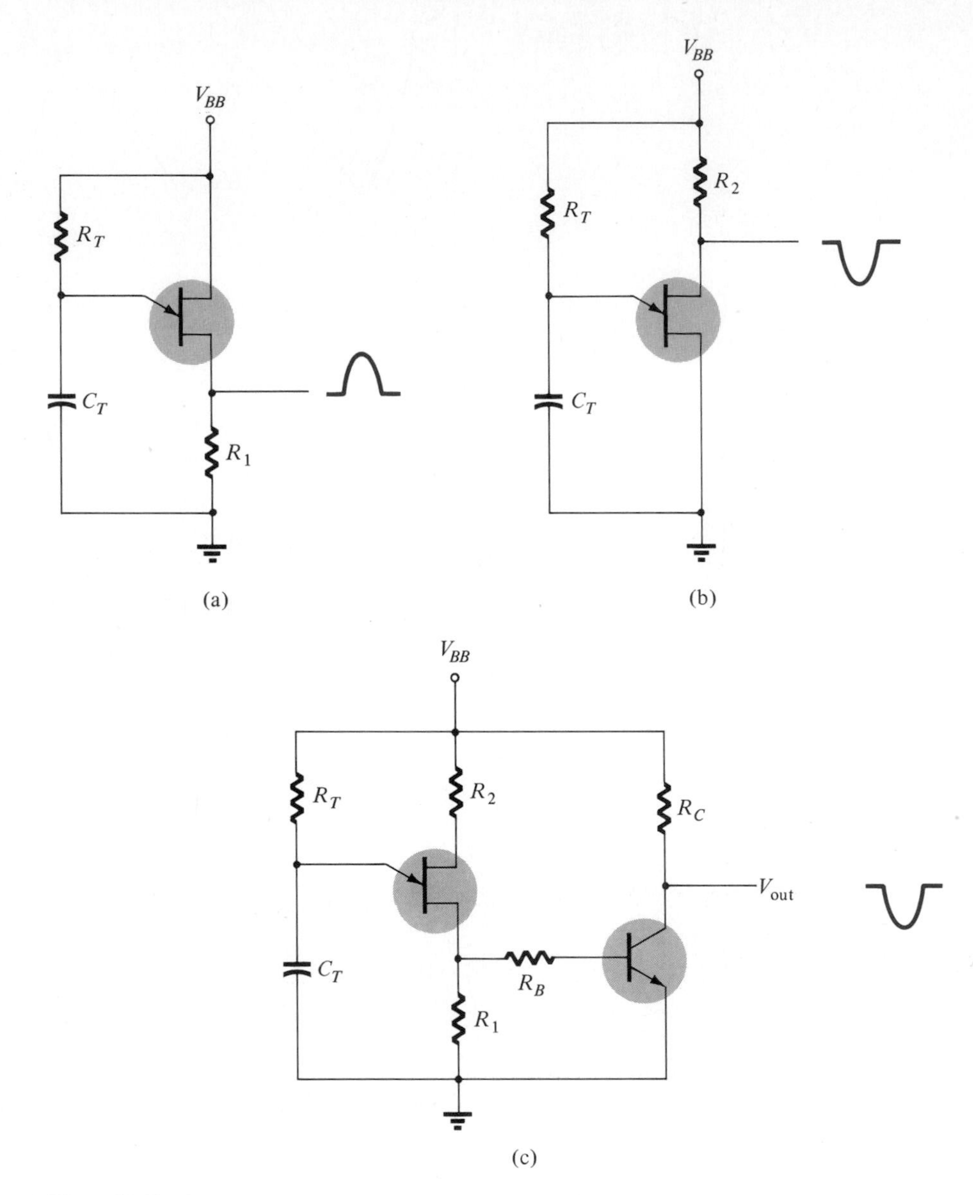

Figure 18.39 Some unijunction oscillator circuit configurations.

PROBLEMS

§ 18.2

1. Calculate the gain of a negative-feedback amplifier having $A = -2000$, $\beta = -1/10$.
2. If the gain of an amplifier changes from a value of 1000 by 10%, calculate the gain change if the amplifier is used in a feedback circuit having $\beta = -1/20$.

§ 18.3

3. Calculate the gain, input, and output impedances of a voltage-series feedback amplifier having $A = -300$, $R_i = 1.5$ kΩ, $R_o = 50$ kΩ, and $\beta = -1/15$.

4. Calculate the gain with and without feedback for an FET amplifier as in Fig. 18.5 for circuit values $R_1 = 200$ kΩ, $R_2 = 800$ Ω, $R_o = 40$ kΩ, $R_D = 8$ kΩ, and $g_m = 5000$ μS.

5. For a circuit as in Fig. 18.11 and the following circuit values, calculate the circuit gain and the input and output impedance with and without feedback: $R_B = 600$ kΩ, $R_E = 1.2$ kΩ, $R_c = 12$ kΩ, $h_{ie} = 2$ kΩ, and $h_{fe} = 75$. Use $V_{CC} = 16$ V.

§ 18.6

6. An FET phase-shift oscillator having $g_m = 6000$ μS, $r_d = 36$ kΩ, and feedback resistor $R = 12$ kΩ is to operate at 2.5 kHz. Select R_D and C for specified oscillator operation.

7. Select values of capacitor C and transistor gain h_{fe} to provide operation of a transistor phase-shift oscillator at 5 kHz for circuit values $R_1 = 24$ kΩ, $R_2 = 75$ kΩ, $R_C = 18$ kΩ, $R = 6$ kΩ, and $h_{ie} = 2$ kΩ.

§ 18.7

8. Design the RC elements of a Wien bridge oscillator circuit (as in Fig. 18.23) for operation at $f_o = 2$ kHz.

§ 18.8

9. For an FET colpitts oscillator as in Fig. 18.26 and the following circuit values determine the circuit oscillation frequency: $C_1 = 750$ pF, $C_2 = 2500$ pF, $L = 40$ μH, $R_G = 750$ kΩ, $L_{RFC} = 0.2$ mH, $C_C = 2000$ pF.

10. For the transistor Colpitts oscillator of Fig. 18.27 and the following circuit values calculate the oscillation frequency: $L = 100$ μH, $L_{RFC} = 0.5$ mH, $C_1 = 0.005$ μF, $C_2 = 0.01$ μF, $C_C = 10$ μF.

11. Calculate the oscillator frequency for an FET Hartley oscillator as in Fig. 18.29 for the following circuit values: $C = 250$ pF, $L_1 = 1.5$ mH, $L_2 = 1.5$ mH, $M = 0.5$ mH.

12. Calculate the oscillation frequency for the transistor Hartley circuit of Fig. 18.30 and the following circuit values: $L_{RFC} = 0.5$ mH, $L_1 = 750$ μH, $L_2 = 750$ μH, $M = 150$ μH, $C = 150$ pF.

§ 18.9

13. Draw circuit diagrams of (a) a series-operated crystal oscillator and (b) a shunt-excited crystal oscillator.

§ 18.10

14. Design a unijunction oscillator circuit for operation at (a) 1 kHz and (b) 150 kHz.

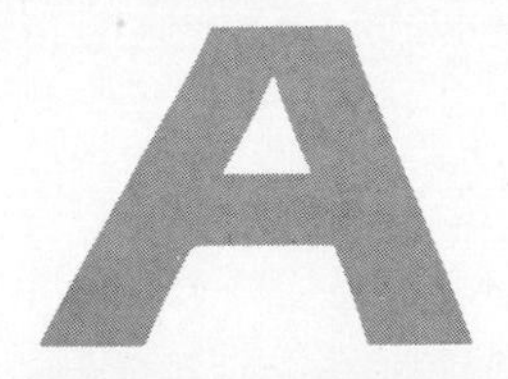

Hybrid Parameters—Conversion Equations (Exact and Approximate)

A.1 EXACT

Common-Emitter Configuration

$$h_{ie} = \frac{h_{ib}}{(1 + h_{fb})(1 - h_{rb}) + h_{ob}h_{ib}} = h_{ic}$$

$$h_{re} = \frac{h_{ib}h_{ob} - h_{rb}(1 + h_{fb})}{(1 + h_{fb})(1 - h_{rb}) + h_{ob}h_{ib}} = 1 - h_{rc}$$

$$h_{fe} = \frac{-h_{fb}(1 - h_{rb}) - h_{ob}h_{ib}}{(1 + h_{fb})(1 - h_{rb}) + h_{ob}h_{ib}} = -(1 + h_{fc})$$

$$h_{oe} = \frac{h_{ob}}{(1 + h_{fb})(1 - h_{rb}) + h_{ob}h_{ib}} = h_{oc}$$

Common-Base Configuration

$$h_{ib} = \frac{h_{ie}}{(1 + h_{fe})(1 - h_{re}) + h_{ie}h_{oe}} = \frac{h_{ic}}{h_{ic}h_{oc} - h_{fc}h_{rc}}$$

$$h_{rb} = \frac{h_{ie}h_{oe} - h_{re}(1 + h_{fe})}{(1 + h_{fe})(1 - h_{re}) + h_{ie}h_{oe}} = \frac{h_{fc}(1 - h_{rc}) + h_{ic}h_{oc}}{h_{ic}h_{oc} - h_{fc}h_{rc}}$$

$$h_{fb} = \frac{-h_{fe}(1 - h_{re}) - h_{ie}h_{oe}}{(1 + h_{fe})(1 - h_{re}) + h_{ie}h_{oe}} = \frac{h_{rc}(1 + h_{fc}) - h_{ic}h_{oc}}{h_{ic}h_{oc} - h_{fc}h_{rc}}$$

$$h_{ob} = \frac{h_{oe}}{(1 + h_{fe})(1 - h_{re}) + h_{ie}h_{oe}} = \frac{h_{oc}}{h_{ic}h_{oc} - h_{fc}h_{rc}}$$

Common-Collector Configuration

$$h_{ic} = \frac{h_{ib}}{(1 + h_{fb})(1 - h_{rb}) + h_{ob}h_{ib}} = h_{ie}$$

$$h_{rc} = \frac{1 + h_{fb}}{(1 + h_{fb})(1 - h_{rb}) + h_{ob}h_{ib}} = 1 - h_{re}$$

$$h_{fc} = \frac{h_{rb} - 1}{(1 + h_{fb})(1 - h_{rb}) + h_{ob}h_{ib}} = -(1 + h_{fe})$$

$$h_{oc} = \frac{h_{ob}}{(1 + h_{fb})(1 - h_{rb}) + h_{ob}h_{ib}} = h_{oe}$$

A.2 APPROXIMATE

Common-Emitter Configuration

$$h_{ie} \cong \frac{h_{ib}}{1 + h_{fb}} \cong \beta r_e$$

$$h_{re} \cong \frac{h_{ib}h_{ob}}{1 + h_{fb}} - h_{rb}$$

$$h_{fe} \cong \frac{-h_{fb}}{1 + h_{fb}} \cong \beta$$

$$h_{oe} \cong \frac{h_{ob}}{1 + h_{fb}}$$

Common-Base Configuration

$$h_{ib} \cong \frac{h_{ie}}{1 + h_{fe}} \cong \frac{-h_{ic}}{h_{fc}} \cong r_e$$

$$h_{rb} \cong \frac{h_{ie}h_{oe}}{1 + h_{fe}} - h_{re} \cong h_{rc} - 1 - \frac{h_{ic}h_{oc}}{h_{fc}}$$

$$h_{fc} \cong \frac{-h_{fe}}{1 + h_{fe}} \cong \frac{-(1 + h_{fc})}{h_{fc}} \cong -\alpha$$

$$h_{ob} \cong \frac{h_{oe}}{1 + h_{fe}} \cong \frac{-h_{oc}}{h_{fc}}$$

Common-Collector Configuration

$$h_{ic} \cong \frac{h_{ib}}{1 + h_{fb}} \cong \beta r_e$$

$$h_{rc} \cong 1$$

$$h_{fc} \cong \frac{-1}{1 + h_{fb}} \cong -\beta$$

$$h_{oc} \cong \frac{h_{ob}}{1 + h_{fb}}$$

Ripple Factor and Voltage Calculations

B.1 RIPPLE FACTOR OF RECTIFIER

The ripple factor of a voltage is defined by

$$r \equiv \frac{\text{rms value of ac component of signal}}{\text{average value of signal}}$$

which can be expressed as

$$r = \frac{V_r(\text{rms})}{V_{\text{dc}}}$$

Since the ac voltage component of a signal containing a dc level is

$$v_{\text{ac}} = v - V_{\text{dc}}$$

the rms value of the ac component is

$$\begin{aligned} V_r(\text{rms}) &= \left[\frac{1}{2\pi}\int_0^{2\pi} v_{\text{ac}}^2\, d\theta\right]^{1/2} = \left[\frac{1}{2\pi}\int_0^{2\pi} (v - V_{\text{dc}})^2\, d\theta\right]^{1/2} \\ &= \left[\frac{1}{2\pi}\int_0^{2\pi} (v^2 - 2vV_{\text{dc}} + V_{\text{dc}}^2)\, d\theta\right]^{1/2} \\ &= [V^2(\text{rms}) - 2V_{\text{dc}}^2 + V_{\text{dc}}^2]^{1/2} = [V^2(\text{rms}) - V_{\text{dc}}^2]^{1/2} \end{aligned}$$

where $V(\text{rms})$ is the rms value of the total voltage. For the half-wave-rectified signal,

$$V_r(\text{rms}) = [V^2(\text{rms}) - V_{dc}^2]^{1/2}$$
$$= \left[\left(\frac{V_m}{2}\right)^2 - \left(\frac{V_m}{\pi}\right)^2\right]^{1/2}$$
$$= V_m\left[\left(\frac{1}{2}\right)^2 - \left(\frac{1}{\pi}\right)^2\right]^{1/2}$$

$$\boxed{V_r(\text{rms}) = 0.385V_m \qquad \text{(half-wave)}} \tag{B.1}$$

For the full-wave-rectified signal,

$$V_r(\text{rms}) = [V^2(\text{rms}) - V_{dc}^2]^{1/2}$$
$$= \left[\left(\frac{V_m}{\sqrt{2}}\right)^2 - \left(\frac{2V_m}{\pi}\right)^2\right]^{1/2}$$
$$= V_m\left(\frac{1}{2} - \frac{4}{\pi^2}\right)^{1/2}$$

$$\boxed{V_r(\text{rms}) = 0.308V_m \qquad \text{(full-wave)}} \tag{B.2}$$

B.2 RIPPLE VOLTAGE OF CAPACITOR FILTER

Assuming a triangular ripple waveform approximation as shown in Fig. B.1, we can write (see Fig. B.2)

$$V_{dc} = V_m - \frac{V_r(p\text{-}p)}{2} \tag{B.3}$$

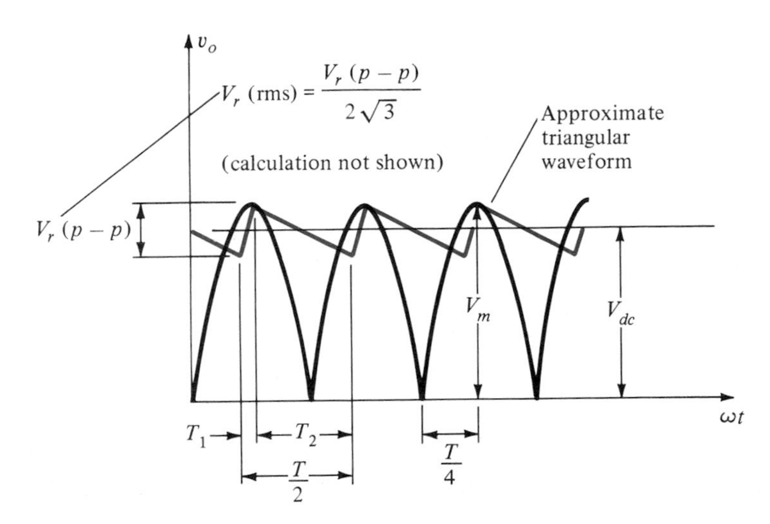

Figure B.1 Approximate triangular ripple voltage for capacitor filter.

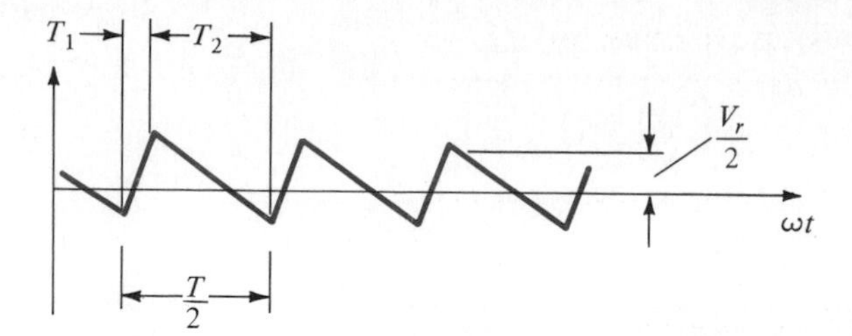

Figure B.2 Ripple voltage.

During capacitor-discharge the voltage change across C is

$$V_r(p\text{-}p) = \frac{I_{dc} T_2}{C} \tag{B.4}$$

From the triangular waveform in Fig. B.1

$$V_r(\text{rms}) = \frac{V_r(p\text{-}p)}{2\sqrt{3}} \tag{B.5}$$

(obtained by calculations, not shown).

Using the waveform details of Fig. B.1 results in

$$\frac{V_r(p\text{-}p)}{T_1} = \frac{V_m}{T/4}$$

$$T_1 = \frac{V_r(p\text{-}p)(T/4)}{V_m}$$

Also,

$$T_2 = \frac{T}{2} - T_1 = \frac{T}{2} - \frac{V_r(p\text{-}p)(T/4)}{V_m} = \frac{2TV_m - V_r(p\text{-}p)T}{4V_m}$$

$$T_2 = \frac{2V_m - V_r(p\text{-}p)}{V_m}\frac{T}{4} \tag{B.6}$$

Since Eq. (B.3) can be written as

$$V_{dc} = \frac{2V_m - V_r(p\text{-}p)}{2}$$

we can combine the last equation with Eq. (B.6):

$$T_2 = \frac{V_{dc}}{V_m}\frac{T}{2}$$

which, inserted into Eq. (B.4), gives

$$V_r(p\text{-}p) = \frac{I_{dc}}{C}\left(\frac{V_{dc}}{V_m}\frac{T}{2}\right)$$

$$T = \frac{1}{f}$$

$$V_r(p\text{-}p) = \frac{I_{dc}}{2fC}\frac{V_{dc}}{V_m} \tag{B.7}$$

Combining Eqs. (B.5) and (B.7), we solve for V_r(rms):

$$V_r(\text{rms}) = \frac{V_r(p\text{-}p)}{2\sqrt{3}} = \frac{I_{dc}}{4\sqrt{3}\,fC}\frac{V_{dc}}{V_m} \tag{B.8}$$

B.3 RELATION OF V_{dc} AND V_m TO RIPPLE, r

The dc voltage developed across a filter capacitor from a tranformer providing a peak voltage, V_m, can be related to the ripple as follows:

$$r = \frac{V_r(\text{rms})}{V_{dc}} = \frac{V_r(p\text{-}p)}{2\sqrt{3}\,V_{dc}}$$

$$V_{dc} = \frac{V_r(p\text{-}p)}{2\sqrt{3}r} = \frac{V_r(p\text{-}p)/2}{\sqrt{3}r} = \frac{V_r(\text{p})}{\sqrt{3}r} = \frac{V_m - V_{dc}}{\sqrt{3}r}$$

$$V_m - V_{dc} = \sqrt{3}rV_{dc}$$

$$V_m = (1 + \sqrt{3}r)V_{dc}$$

$$\frac{V_m}{V_{dc}} = 1 + \sqrt{3}r \tag{B.9}$$

The relation of Eq. (B.9) applies to both half- and full-wave rectifier-capacitor filter circuits and is plotted in Fig. B.3. As example, at a ripple of 5% the dc voltage is $V_{dc} = 0.92V_m$, or within 10% of the peak voltage, where as at 20% ripple the dc voltage drops to only $0.74V_m$ which is more than 25% less than the peak value. Note that V_{dc} is within 10% of V_m for ripple less than 6.5%. This amount of ripple represents the borderline of the light-load condition.

B.4 RELATION OF V_r (RMS) AND V_m TO RIPPLE, r

We can also obtain a relation between V_r(rms), V_m, and the amount of ripple for both half-wave and full-wave rectifier-capacitor filter circuits as follows:

$$\frac{V_r\,(\text{p-}p)}{2} = V_m - V_{dc}$$

$$\frac{V_r\,(\text{p-}p)/2}{V_m} = \frac{V_m - V_{dc}}{V_m} = 1 - \frac{V_{dc}}{V_m}$$

$$\frac{\sqrt{3}V_r\,(\text{rms})}{V_m} = 1 - \frac{V_{dc}}{V_m}$$

Using Eq. (B.9), we get

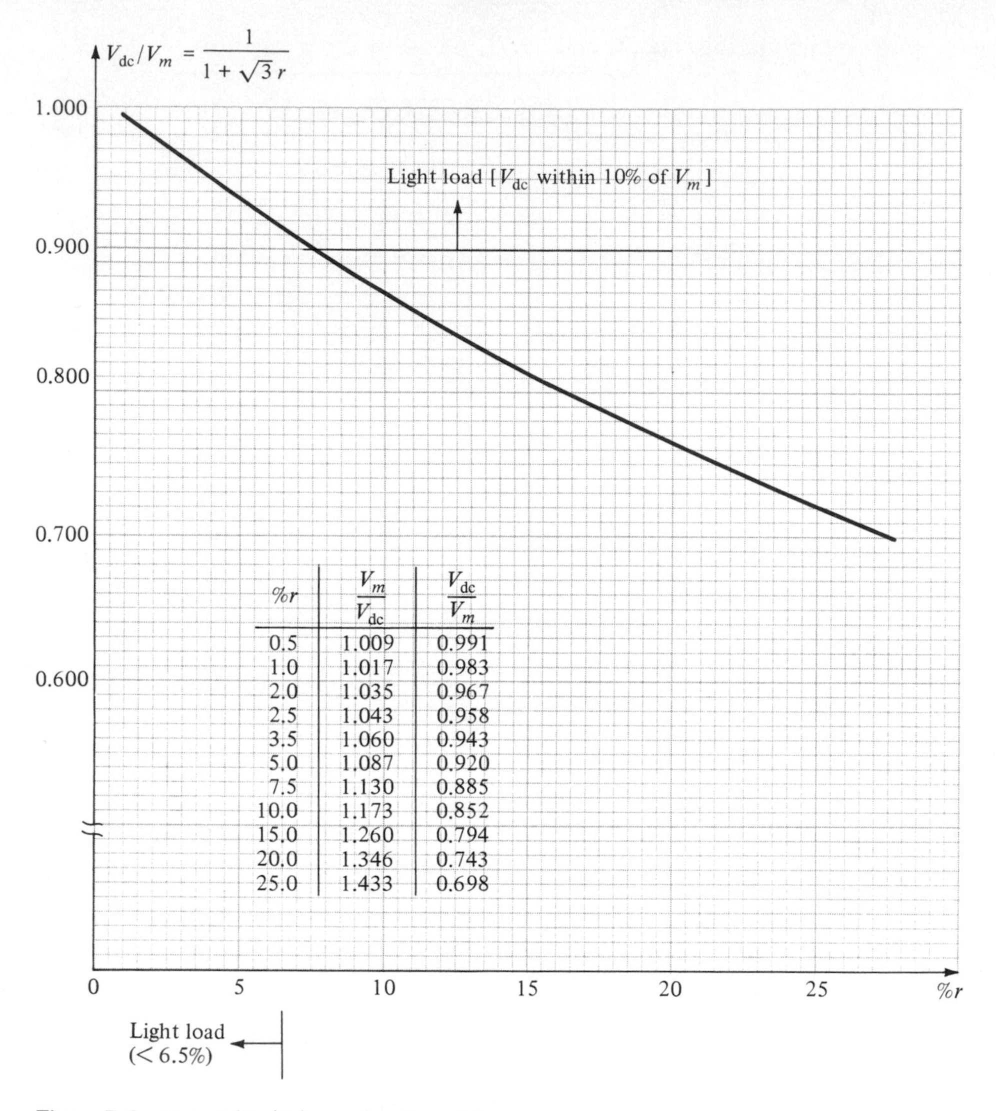

%r	$\dfrac{V_m}{V_{dc}}$	$\dfrac{V_{dc}}{V_m}$
0.5	1.009	0.991
1.0	1.017	0.983
2.0	1.035	0.967
2.5	1.043	0.958
3.5	1.060	0.943
5.0	1.087	0.920
7.5	1.130	0.885
10.0	1.173	0.852
15.0	1.260	0.794
20.0	1.346	0.743
25.0	1.433	0.698

Figure B.3 Plot of (V_{dc}/V_m) as a function of % r.

$$\frac{\sqrt{3}V_r\,(\text{rms})}{V_m} = 1 - \frac{1}{1+\sqrt{3}\,r}$$

$$\frac{V_r\,(\text{rms})}{V_m} = \frac{1}{\sqrt{3}}\left(1 - \frac{1}{1+\sqrt{3}r}\right) = \frac{1}{\sqrt{3}}\left(\frac{1+\sqrt{3}r-1}{1+\sqrt{3}r}\right)$$

$$\boxed{\frac{V_r\,(\text{rms})}{V_m} = \frac{r}{1+\sqrt{3}r}} \tag{B.10}$$

Equation (B.10) is plotted in Fig. B.4.

Since V_{dc} is within 10% of V_m for ripple $\leq 6.5\%$,

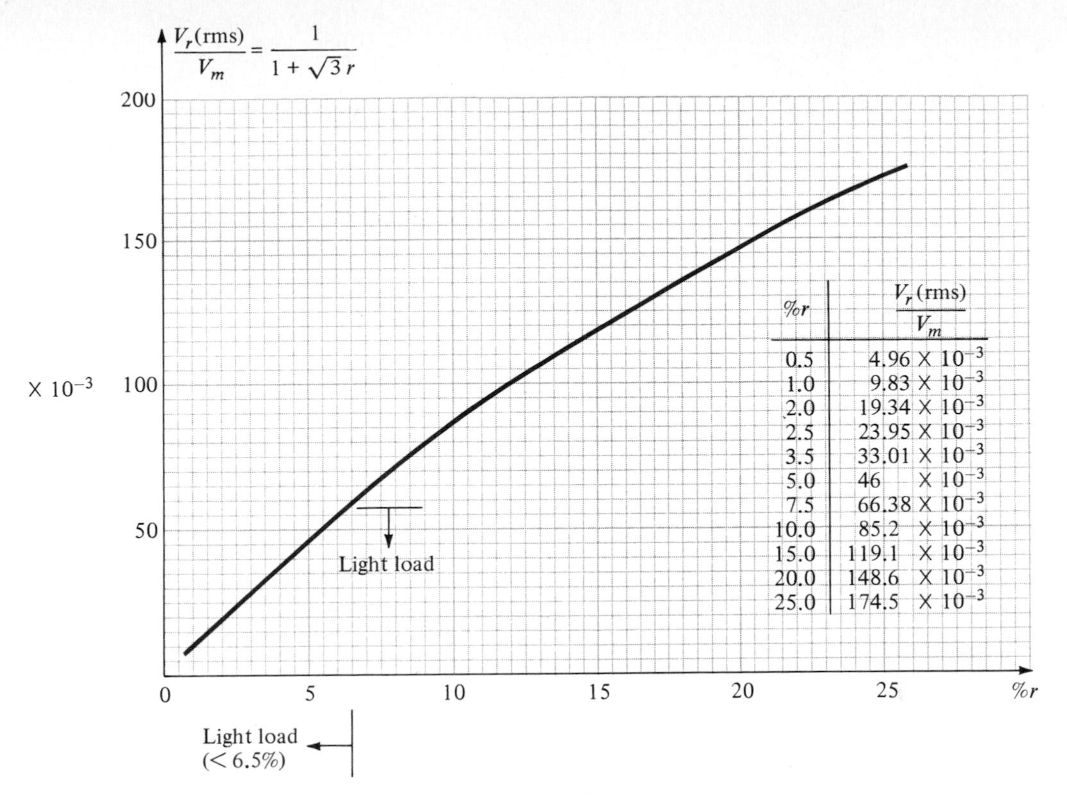

%r	$\frac{V_r(\text{rms})}{V_m}$
0.5	4.96 X 10^{-3}
1.0	9.83 X 10^{-3}
2.0	19.34 X 10^{-3}
2.5	23.95 X 10^{-3}
3.5	33.01 X 10^{-3}
5.0	46 X 10^{-3}
7.5	66.38 X 10^{-3}
10.0	85.2 X 10^{-3}
15.0	119.1 X 10^{-3}
20.0	148.6 X 10^{-3}
25.0	174.5 X 10^{-3}

Figure B.4 Plot of V_r (rms)/V_m as a function of % r.

$$\frac{V_r\text{ (rms)}}{V_m} \cong \frac{V_r\text{ (rms)}}{V_{\text{dc}}} = r \qquad \text{(light load)}$$

and we can use V_r (rms)/V_m = r for ripple $\leq$ 6.5%.

B.5 RELATION BETWEEN CONDUCTION ANGLE, % RIPPLE, AND $I_{\text{peak}}/I_{\text{dc}}$ FOR RECTIFIER-CAPACITOR FILTER CIRCUITS

In Fig. B.1 we can determine the angle at which the diode starts to conduct, θ, as follows: Since

$$v = V_m \sin\theta = V_m - V_r\,(p\text{-}p) \qquad \text{at} \qquad \theta = \theta_1$$

$$\theta_1 = \sin^{-1}\left[1 - \frac{V_r\,(p\text{-}p)}{V_m}\right]$$

Using Eq. (B.10) and V_r (rms) = $V_r\,(p\text{-}p)/2\sqrt{3}$ gives

$$\frac{V_r\ (p\text{-}p)}{V_m} = \frac{2\sqrt{3}V_r\ (\text{rms})}{V_m}$$

so that
$$1 - \frac{V_r\ (p\text{-}p)}{V_m} = 1 - \frac{2\sqrt{3}V_r\ (\text{rms})}{V_m} = 1 - 2\sqrt{3}\left(\frac{r}{1+\sqrt{3}r}\right)$$

$$= \frac{1-\sqrt{3}r}{1+\sqrt{3}r}$$

and
$$\boxed{\theta_1 = \sin^{-1}\frac{1-\sqrt{3}r}{1+\sqrt{3}r}} \tag{B.11}$$

where θ_1 is the angle at which conduction starts.

When the current becomes zero after charging the parallel impedances R_L and C, we can determine that

$$\theta_2 = \pi - \tan^{-1}\omega R_L C$$

An expression for $\omega R_L C$ can be obtained as follows:

$$r = \frac{V_r\ (\text{rms})}{V_{\text{dc}}} = \frac{(I_{\text{dc}}/4\sqrt{3}\,fC)(V_{\text{dc}}/V_m)}{V_{\text{dc}}} = \frac{V_{\text{dc}}/R_L}{4\sqrt{3}\,fC}\frac{1}{V_m}$$

$$= \frac{V_{\text{dc}}/V_m}{4\sqrt{3}\,fCR_L} = \frac{2\pi\left(\dfrac{1}{1+\sqrt{3}r}\right)}{4\sqrt{3}\ \omega CR_L}$$

so that
$$\omega R_L C = \frac{2\pi}{4\sqrt{3}\ (1+\sqrt{3}r)r} = \frac{0.907}{r(1+\sqrt{3}r)}$$

Thus conduction stops at an angle

$$\boxed{\theta_2 = \pi - \tan^{-1}\frac{0.907}{(1+\sqrt{3}r)r}} \tag{B.12}$$

From Eq. (16.10b) we can write

$$\frac{I_{\text{peak}}}{I_{\text{dc}}} = \frac{I_p}{I_{\text{dc}}} = \frac{T}{T_1} = \frac{180^\circ}{\theta} \quad \text{(full-wave)}$$

$$= \frac{360^\circ}{\theta} \quad \text{(half-wave)} \tag{B.13a}$$

A plot of I_p/I_{dc} as a function of ripple is provided in Fig. B.5 for both half- and full-wave operation.

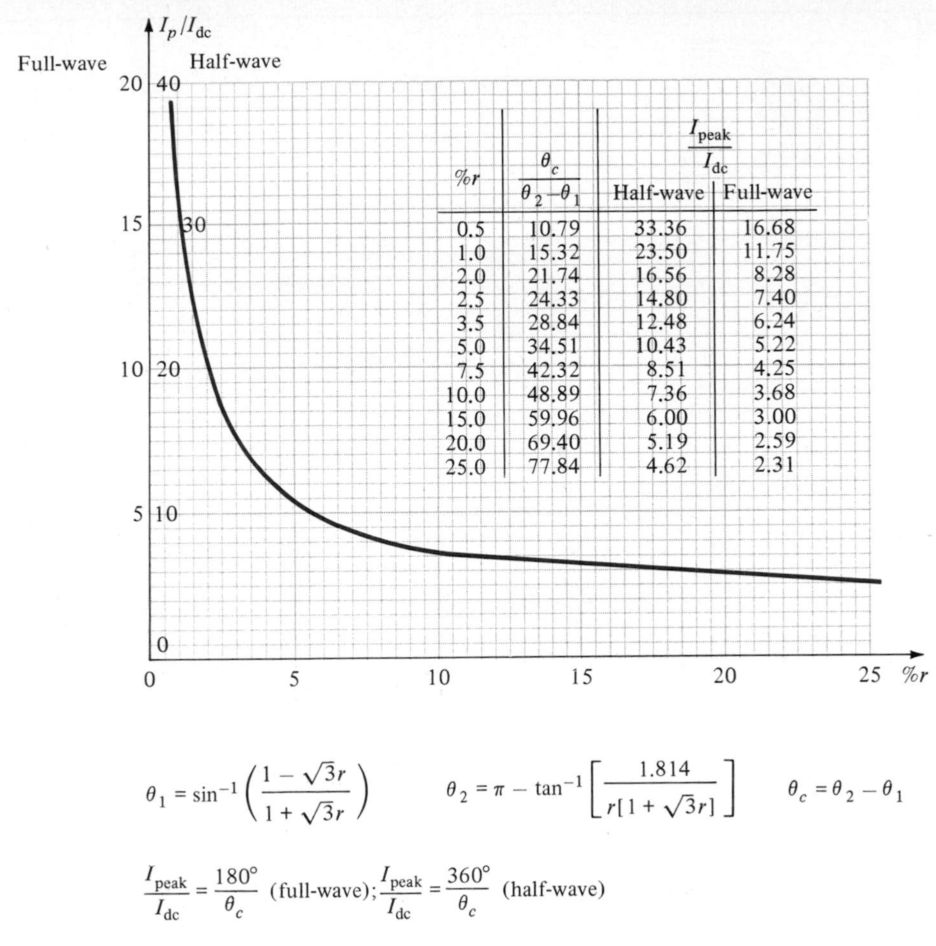

%r	$\frac{\theta_c}{\theta_2 - \theta_1}$	$\frac{I_{peak}}{I_{dc}}$ Half-wave	Full-wave
0.5	10.79	33.36	16.68
1.0	15.32	23.50	11.75
2.0	21.74	16.56	8.28
2.5	24.33	14.80	7.40
3.5	28.84	12.48	6.24
5.0	34.51	10.43	5.22
7.5	42.32	8.51	4.25
10.0	48.89	7.36	3.68
15.0	59.96	6.00	3.00
20.0	69.40	5.19	2.59
25.0	77.84	4.62	2.31

$$\theta_1 = \sin^{-1}\left(\frac{1 - \sqrt{3}r}{1 + \sqrt{3}r}\right) \qquad \theta_2 = \pi - \tan^{-1}\left[\frac{1.814}{r[1 + \sqrt{3}r]}\right] \qquad \theta_c = \theta_2 - \theta_1$$

$$\frac{I_{peak}}{I_{dc}} = \frac{180^\circ}{\theta_c} \text{ (full-wave)}; \frac{I_{peak}}{I_{dc}} = \frac{360^\circ}{\theta_c} \text{ (half-wave)}$$

Figure B.5 Plot of I_p/I_{dc} versus % r, half- and full-wave operation.

TABLE C.1 Greek Alphabet and Common Designations

Name	Capital	Lowercase	Used to Designate:
alpha	A	α	Angles, area, coefficients
beta	B	β	Angles, flux density, coefficients
gamma	Γ	γ	Conductivity, specific gravity
delta	Δ	δ	Variation, density
epsilon	E	ϵ	Base of natural logarithms
zeta	Z	ζ	Impedance, coefficients, coordinates
eta	H	η	Hysteresis coefficient, efficiency
theta	Θ	θ	Temperature, phase angle
iota	I	ι	
kappa	K	κ	Dielectric constant, susceptibility
lambda	Λ	λ	Wave length
mu	M	μ	Micro, amplification factor, permeability
nu	N	ν	Reluctivity
xi	Ξ	ξ	
omicron	O	o	
pi	Π	π	Ratio of circumference to diameter = 3.1416
rho	P	ρ	Resistivity
sigma	Σ	σ	Sign of summation
tau	T	τ	Time constant, time phase displacement
upsilon	Y	υ	
phi	Φ	ϕ	Magnetic flux, angles
chi	X	χ	
psi	Ψ	ψ	Dielectric flux, phase difference
omega	Ω	ω	Capital: ohms; lower case: angular velocity

Answers to Selected Odd-Numbered Exercises

CHAPTER 1

7. 6.4×10^{-19} C **19.** $\cong 0.3$ V **21.** 56.4 mA **25.** 30 Ω **27.** 10 Ω **29.** $R_{dc} = 40\ \Omega$, $r_{ac} = 2\ \Omega$ **31.** $R_{dc} = 800\ \Omega$, $r_{ac} = 333\ \Omega$ **33.** (a) V_T, r_{av}, ideal diode (b) $I_R = 17.1$ mA, $V_R = 3.762$ V **37.** (a) 0 V: 3.2 pF 0.25 V: 9 pF **39.** 0.2 V: 3.54 kΩ; −20 V: 44.25 kΩ **41.** −75°C: $V_D \cong 1.7$ V, $I_S \cong 0.1\ \mu$A, +25°C: $V_D \cong 1.3$ V, $I_S \cong 0.4\ \mu$A, +100°C: $V_D \cong 0.98$ V, $I_S \cong 1.0\ \mu$A, +200°C: $V_D \cong 0.64$ V, $I_S \cong 2.2\ \mu$A **43.** $P_{D_{max}} = 200$ mW, $P_{D_{max}}$ (reverse bias) = 10 μW **47.** 26.54 kΩ **49.** 628.93 mA **51.** (a) 2.3 mV/°C (b) 46.0 mV **53.** 493 Ω **57.** 240 mW **59.** 320 mA, 1 V

CHAPTER 2

1. (a) −4.3 V, 0.915 mA (b) 4.75 V, 1.039 mA **3.** (a) $V_{o_1} = 11.3$ V, $V_{o_2} = 0.3$ V (b) $V_{o_1} = -9$ V, $V_{o_2} = -5.4$ V **5.** (a) 9.7 V, 9.7 mA (b) 14.6 V, 1.893 mA **7.** 4.65 V, 2.325 mA **9.** 9.3 V **11.** 10 V **13.** −0.7 V **17.** $V_m = 155.54$ V, $V_{dc} = 49.462$ V **19.** (a) 20 mA (b) 36.74 mA (c) 18.37 mA (d) Less (e) 36.74 mA > 20 mA **23.** $V_m = 4.3$ V, $V_{dc} = 1.367$ V **25.** (a) 0 V, −3 V (b) 20 V, 5 V **27.** I_R (positive peak) = 0.4 mA I_R (negative peak) = 0.2 mA **29.** (a) 28 ms (b) $5\tau \gg 0.5$ ms (c) −1.3 V, −21.3 V

CHAPTER 3

3. 0.053%/°C **5.** ≅13 Ω **7.** (a) $V_L = 9$ V, $I_L = 50$ mA, $I_Z = 0$ mA, $I_R = 50$ mA (b) $V_L = 10$ V, $I_L = 21$ mA, $I_Z = 24$ mA, $I_R = 45$ mA

(c) 2 kΩ (d) 220 Ω **9.** $V_{i_{min}} = 11.309$ V, $V_{i_{max}} = 15.826$ V **15.** 47.5°C, lower current levels **17.** 0 → 2 V: 33%; 8 → 10 V: 5.4% **19.** (a) 27 pF (b) −8 V: 2 pF/V; −2 V: 9.25 pF/V **21.** 6.67 **23.** Lower levels **29.** 3.97×10^{-19} J = 2.48 eV **31.** ≅350 μA **33.** 42.5 V **39.** Yellow **41.** (a) 0.77 **45.** 2.3 V **47.** (a) 36 mA (b) 54 mA **59.** 20 kΩ **61.** 90 Ω

CHAPTER 4

9. 7.92 mA **11.** 25 **13.** (a) 4.95 mA (b) 3 mA (c) 800 mV **17.** (a) 114.3 (b) 0.991 (c) 300 μA (d) 2.62 μA **23.** 0.972 **27.** (a) 3.3 mA (b) 28 V (c) 25 μA **29.** (a) 4.57 mW/°C (b) 1.714 W (c) Good match **31.** (a) 7.5 nA (b) 1.5 μA (c) 0.267 nA/°C

CHAPTER 5

1. $V_C = 7$ V **3.** $I_C = 3.3$ mA, $V_{CE} = 6.06$ V **5.** $R_C = 3.19$ kΩ (use 3.3 kΩ) **7.** $V_{CE} = 4.77$ V **9.** $\beta = 19.49$ **11.** $V_{CE} = 13.88$ V **13.** $R_B = 445.16$ kΩ (use 430 kΩ) **15.** $V_E = 0.637$ V, $V_C = 6.47$ V, $V_B = 2.07$ V **17.** $I_B = 13.05$ μA, $I_C = 0.914$ mA **19.** $I_C = 0.383$ mA, $V_{CE} = 5$V **21.** $V_{CB} = 7.14$ V **23.** % change in $V_{CE} = 10.1\%$ **25.** $R_C = 5.9$ kΩ (use 6.2 kΩ) **27.** $V_C = 8.66$ V **29.** $V_C = 18.03$ V **31.** $R_C = 9.4$ kΩ (use 9.1 kΩ) **33.** % change in $V_C = 9.75\%$ **35.** $V_C = 7.42$ V **37.** % change in $I_C = 38.35\%$ **39.** $V_C = -11.44$ V **41.** $V_{CE} = 20.18$ V **43.** $V_C = -6.375$ V **47.** $R_C = 1.7$ kΩ (use 1.6 kΩ) **49.** $R_E = 360$ Ω, $R_C = 1.64$ kΩ (use 1.6 kΩ), $R_B = 310$ kΩ (use 300 kΩ) **51.** $R_E = 1.069$ kΩ (use 1.1 kΩ), $R_B = 1.66 \times 10^6$ (use 1.6 MΩ) **53.** $R_C = 4.8$ kΩ (use 4.7 kΩ), $R_{B_2} = 9.6$ kΩ (use 10 kΩ), $R_{B1} = 51$ kΩ

CHAPTER 6

1. (a) $I_D = 3.556$ mA, (b) $I_D = 8$ mA, (c) $I_D = 0.889$ mA **3.** $V_{GS} = 0.628$ V **5.** $V_P = -5$ V **7.** $V_{GS} = -1.5$ V **11.** $g_{mo} = 3.556$ mS **13.** $V_P = 3.692$ V **15.** $g_m = 2.5$ mS **17.** $V_P = -3.966$ V, $g_m = 2352$ μS **19.** $g_m = 4.53$ mS **21.** $g_m = 3150$ μS **23.** $g_{mo} = 6000$ μS **25.** $g_m = 8$ mS **27.** $V_T = 3$ V **29.** $V_T = 3.33$ V

CHAPTER 7

1. $V_D = 13.56$ V **3.** $R_D = 7.2$ kΩ (use 7.5 kΩ) **5.** $V_{GS} = V_{GG} = 1.32$ V **7.** $V_S = 2.8$ V **9.** $R_S = 562$ Ω (use 560 Ω) **11.** $V_S = 2.8$ V **13.** $I_D = 1.7$ mA **15.** $V_{DS} = 3.77$ V **17.** $R_S = 140.6$ Ω (use 150 Ω) **19.** $V_{DS} = 0.6$ V **21.** $V_{DS} = 2.2$ V **23.** $V_{DS} = 9.11$ V **25.** $R_D = 2.42$ kΩ (use 2.4 kΩ) **27.** Increase **29.** $V_{DS} = 3.36$ V **31.** $V_{DS} = 1.8$ V **33.** Decreases (to

6mA

I_D = 4.86 mA) **35.** V_{DS_Q} = 9.6 V **37.** K = 0.25 mA/V **39.** V_{DS_Q} = 8.1 V **41.** V_{DS} = 2.48 V **43.** I_{DSS} = 9.37 mA, V_P = −3 V **45.** R_S = 390 Ω, R_D = 3.61 kΩ (use 3.6 kΩ) **47.** R_D = 2.39 kΩ (use 2.4 kΩ) **49.** V_{GS_Q} = −2.61 V **51.** I_D = 1.32 mA **53.** V_{GS} = −2.43 V

CHAPTER 8

9. 4.227%, Yes **11.** 14.529%, No **13.** (a) 100 μS (b) 20 μS **15.** (a) 10×10^{-4} (b) 15×10^{-4} **17.** β = 100, r_e = 7.88 Ω **19.** (a) 66.7% (b) 50% (c) 150% **21.** (a) 2 μS (b) Yes **23.** (a) 5.4×10^{-4} (b) No

CHAPTER 9

1. (a) $Z_i \cong$ 0.5 kΩ, $Z_o \cong$ 5.1 kΩ, A_v = −612, A_i = 60 (b) Z_i = 0.5 kΩ, $Z_o \cong$ 4.628 kΩ, A_v = −555.36, A_i = 54.45 **3.** (a) I_B = 23.85 μA, I_C = 2.385 mA, r_e = 10.9 Ω (b) h_{fe} = 100, h_{ie} = 1.09 kΩ (c) $Z_i \cong$ 1.09 kΩ, $Z_o \cong$ 4.3 kΩ, A_v = −394.5, $A_i \cong$ 100 (d) A_v = −355.96, A_i = 90.29 **5.** V_{CC} = 30.66 V **7.** (a) $Z_i \cong$ 118.37 kΩ, $Z_o \cong$ 4.7 kΩ, A_v = −3.87, A_i = 97.51 (b) $Z_i \cong$ 118.37 kΩ, $Z_o \cong$ 4.296 kΩ, A_v = −3.54, A_i = 89.07 **9.** (a) r_e = 7.075 Ω (b) $Z_i \cong$ 74.71 kΩ, A_v = −371.17 (c) A_i = 61.89 **11.** (a) $Z_i \cong$ 231.4 kΩ, $Z_o \cong$ 33.5 Ω, A_v = 0.994, A_i = 41.074 (b) V_o = 0.994 mV **13.** (a) r_e = 33.376 Ω (b) $Z_i \cong$ 33.4 Ω, $Z_o \cong$ 4.7 kΩ, A_v = 140.72, A_i = 1 **15.** $Z_i \cong$ 0.632 kΩ, $Z_o \cong$ 1.664 kΩ, A_v = −207.35, A_i = 73.95 **17.** (a) r_e = 13.2 Ω (b) $Z_i \cong$ 0.511 kΩ, $Z_o \cong$ 3.83 kΩ, A_v = −29.25, A_i = 38.37 **19.** (d) A_v = −1.833, A_i = 25.354, $Z_i \cong$ 40.76 kΩ, $Z_o \cong$ 2.16 kΩ **21.** (a) A_{v_s} = −30.1, A_i = 25.67 (b) A_v = −121.88, A_i = 56.6 **23.** (a) A_{v_s} = 0.95, A_i = 35.243 (b) A_v = 0.997, A_i = 70.518 **25.** (a) $Z_i \cong$ 0.459 kΩ, $Z_o \cong$ 4.924 kΩ, A_v = −590, A_i = 53.22, A_p = 31.4×10^3 (b) $Z_i \cong$ 0.5 kΩ, $Z_o \cong$ 5.1 kΩ, A_v = −612, A_i = 60, A_p = 36.72×10^3 (c) A_{v_s} = −231.66, A_v = −342.69, $Z_i \cong$ 0.459 kΩ, $Z_o \cong$ 4.924 kΩ, A_i = 23.989, A_p = 8.22×10^3 (d) A_{v_s} = −242.68, A_v = −349.68, $Z_i \cong$ 0.5 kΩ, $Z_o \cong$ 5.1 kΩ, A_i = 25.74, A_p = 6.25×10^3 **27.** (a) A_v = 138.16, A_i = −1, $Z_i \cong$ 34.24 Ω, $Z_o \cong$ 4.56 kΩ, A_p = 138.16 (b) A_v = 140.72, A_i = −1, $Z_i \cong$ 33.4 Ω, $Z_o \cong$ 4.7 kΩ, A_p = 140.72 (c) A_{v_s} = 79.365, A_v = 81.735, A_i = −0.409, $Z_i \cong$ 33.4 Ω, $Z_o \cong$ 4.7 kΩ, A_p = 33.43 (d) A_{v_s} = 11.91, A_v = 83.204, A_i = −0.409, $Z_i \cong$ 33.4 Ω, $Z_o \cong$ 4.7 kΩ **29.** (a) R_i = 1.815 kΩ, A_v = −66.59, R_o = 2.131 kΩ (b) A_{v_s} = −30.1, same solution

CHAPTER 10

1. A_v = −4.392 **3.** A_v = −3.301 **5.** A_v = −6.761 **7.** V_o = −0.33 V rms **9.** V_o = 224 mV peak **11.** V_o = −353.25 mV peak **13.** V_o = −737.786 mV peak **15.** R_i = R_G = 750 kΩ, R_o = R_D = 3.3 kΩ, V_o = 344.5 mV peak

17. $A_v = -7.24$ **19.** $A_v \cong -8.2$ **21.** $V_o = 333.3$ mV **23.** $V_o = 311.6$ mV **25.** $R_{os} = 76.9\ \Omega$ **27.** $V_o = 72.4$ mV rms **29.** $A_v = 9.18$ **31.** $A_v = 6.89$ **33.** $V_o = 1.026$ V rms **35.** $R_S = 150\ \Omega$ **37.** $R_D = 6.8\ \text{k}\Omega$, $R_S = 1.5\ \text{k}\Omega$ **39.** $C_i = 40.185$ pF

CHAPTER 11

1. (a) $A_{i_T} = 80$, $A_{v_T} = 160$ (b) $A_v = 12.65$, $A_i = 8.94$ **3.** $Z_i \cong 1.31\ \text{k}\Omega$, $Z_o \cong 3.3\ \text{k}\Omega$, $A_{v_T} = 5968.4$, $A_{i_T} = 3553.91$, $A_{P_T} = 21.2 \times 10^6$ **5.** $Z_i \cong 13.85\ \text{k}\Omega$, $Z_o \cong 24.51$, $A_{v_1} = -2$, $A_{v_2} \cong 1$, $A_{v_T} = -2$, $A_{i_T} = 27.7$ **7.** $Z_i \cong 1.19\ \text{k}\Omega$, $Z_o \cong 10\ \text{k}\Omega$, $A_{v_1} = -0.779$, $A_{v_2} = -21.61$, $A_{v_T} = 16.83$, $A_{i_T} = 35.76$ $A_{P_T} = 601.84$ **9.** $A_{v_1} = -10.486$, $A_{v_2} = -2.318$, $A_{v_T} = 24.307$ **11.** $a = 20$ **13.** $V_{E_2} = 11$ V, $V_{B_2} = 11.7$ V, $V_{E_1} = 1.716$ V, $V_{B_1} = 2.416$ V **15.** (a) $r_{e_1} = 4.828\ \Omega$, $r_{e_2} = 6.976\ \Omega$ (b) $A_{v_T} = 455.71$, $V_o = 4.56$ V (c) $Z_i \cong 222\ \Omega$, $Z_o \cong 2.2\ \text{k}\Omega$ **17.** $A_i = 843.75$, $Z_i \cong 1.012\ \text{M}\Omega$ $Z_o \cong 2.2\ \text{k}\Omega$, $A_v = -1.834$ **19.** $A_i = 2006.98$, $Z_i \cong 301.1\ \text{k}\Omega$, $Z_o \cong 2.2\ \text{k}\Omega$, $A_v = -14.664$ **21.** 12 dB **23.** (a) $\text{dB}_1 = 21.05$, $\text{dB}_2 = 42.1$, $\text{dB}_3 = 56.84$ (b) $A_{v_1} = 11.29$, $A_{v_2} = 127.35$, $A_{v_3} = 691.83$ **25.** (a) $f_{L_S} = 85.61$ Hz, $f_{L_C} = 35.78$ Hz (b) $A_{v_{\text{mid}}} = -2.08$ (c) $f_{H_i} = 9.75$ MHz, $f_{H_o} = 8.52$ MHz **27.** $A_{v_t} = 160 \times 10^3$ **29.** $f_1 = 92.17$ Hz **31.** GBW $= -50$ **33.** $A_v \cong 4.445 \times 10^6$

CHAPTER 12

1. $P_i = 15.1875$ W, $P_o = 312.5$ mW **3.** max $P_o(\text{ac}) = 0.582$ W **5.** $R'_L = 2.5\ \text{k}\Omega$ **7.** $n = 44.72/1$ **9.** $n = 37.04\%$ **17.** (a) $P_D = 28.125$ W (b) $P_i = 50.643$ W (c) $n = 55.536$ % (d) $P_{ZQ} = 22.518$ W **19.** (a) $n = 35.8\%$ (b) $n = 64.2\%$ **21.** $P_D = 25$ W **23.** $P_{D,\max} = 3$ W

CHAPTER 13

5. (a) Yes, for $T = 25°\text{C}$, $I_G > \cong 40$ mA (b) No (c) No, minimum of 3 V (d) 6 V, 800 mA: excellent and 4 V, 1.6 A: no **11.** $\cong 0.7\ \text{mW/cm}^2$ (b) 81.25% reduction **17.** (a) $R_{B_2} = 1.08\ \text{k}\Omega$ (b) $R_{BB} = 3.08\ \text{k}\Omega$ (c) $V_{R_{B1}} = 13$ V (d) $V_P = 13.7$ V **19.** $I_B = 27\ \mu\text{A}$, $I_C = 1.08$ mA **21.** (a) 0.67%/°C (b) Yes **23.** Ratio = 0.471, yes **25.** (a) $I_C = 2$ mA (b) R_{L_1}: R_{L_2} = 10:1, t_{s1}: t_{s2} = 4.4:1 **27.** $Z_P \cong 87\ \text{k}\Omega$ (open-circuit), $Z_v \cong 181.8\ \Omega$ (short-circuit) **29.** (a) Yes, V_{BB} (min) = 8.18 V (b) $R < 2\ \text{k}\Omega$ (c) $R = 1.82\ \text{k}\Omega$

CHAPTER 15

5. $V_C = 7.84$ V **7.** $R_i = 7.8\ \text{k}\Omega$, $R_o = 15\ \text{k}\Omega$ **9.** $V_C = 11.438$ V **11.** $R_o = 3.496\ \text{M}\Omega$ **13.** CMRR = 51.126 dB **15.** $A_d = 172$,

$A_c = 0.355 \times 10^{-3}$, CMRR = 113.7 dB **17.** $R_i = 10.714$ kΩ **21.** $V_o = 0$ V **23.** $V_o = -12$ V **27.** $I_o = 1.667$ mA

CHAPTER 16

1. $r = 2.8\%$ **3.** $V_r = 22.4$ V **5.** $V_r = 1.233$ V **7.** $V_{de} = 16.958$ V **9.** $r = 0.96\%$ **11.** $V_r = 1.663$ V, $V_{dc} = 12$ V **13.** $V_{dc} = 22.22$ V, $r = 7.201\%$ **15.** $V_r = 1.6$ V **17.** $I_L = 73$ mA **19.** V.R. = 32.01% **23.** V.R. = 5.263% **25.** Current regulation = 1% **27.** $R_{s_{min}} = 307.692\ \Omega$ **29.** (a) $V_L = 10.7$ V (b) $T_L = 2.675$ mA (c) $I_{p_s} = 4.65$ mA (d) $I_z = 44.61\ \mu$A, $I_c = 2.23$ mA **33.** $V_{out} = 10.805$ V **35.** $v = 5.485\%$ **37.** (a) $V_{de} = 13.749$ V (b) V_r(peak) = 1.247 V

CHAPTER 17

11. Resolution = 2.44 mV **13.** 4096 Count steps **15.** $C = 640$ pF **19.** Frequency range = 42.47 kHz **21.** $f_o = 63.83$ kHz **23.** $f_L = \pm 85.11$ kHz

CHAPTER 18

1. $A_f = 9.95$ **3.** $A_f = 14.286$, $R_{if} = 31.5$ kΩ, $R_{of} = 2.381$ kΩ **5.** With R_e removed (bypassed): $A = 450$, $R_i = 2$ kΩ, $R_o = 12$ kΩ with R_e in circuit: $A_f = 0.998$, $R_{if} = 90.2$ kΩ, $R_{of} = 12$ kΩ **7.** $hfe \geq 44.67$ **9.** $f_o = 1.048$ MHz **11.** $f_o = 159.155$ kHz

Index

A

B

C

D

E

F

G

H

I

J

L

M

N

O

P

Q

R

S

T

U

V

W

Y

Z

11 Multistage Systems and Frequency Considerations - cont.
$Z_{o_1} \cong (R_{S_1} + h_{ie_1})/h_{fe_1}$, $Z_{o2} = [(Z_{o_1} \parallel 1/h_{oe_1}) + h_{ie_2}]/h_{fe_2}$, $A_v \cong 1/(1 + h_{ie_2}/h_{fe_2}R_E)$;
Decibels: $G_{dB} = 10 \log_{10} P_2/P_1 = 20 \log_{10} V_2/V_1$, $G_v = G_{v_1} + G_{v_2} + G_{v_3} + \cdots + G_{v_n}$,
$\log_b x = y$ is equivalent to $b^y = x$;
Frequency response: $P_{o_{HPF}} = 0.5\, P_{o_{mid}}$, $BW = f_2 - f_1$;
BJT low-frequency: $f_1 = 1/2\pi RC$, $A_v = 1/[1 - j(f_1/f)]$,
$A_v = 1/\sqrt{1 + (f_1/f)^2}\, \underline{/\tan^{-1}(f_1/f)}$, $A_v|_{dB} = -20 \log_{10} f_1/f$ for $f \ll f_1$,
$f_{L_S} = 1/2\pi(R_S + R_i)C_S$, $f_{L_C} = 1/2\pi(R_C + R_L)C_C$, $f_{L_E} = 1/2\pi R_e C_E$;
High frequency: $A_v = 1/[1 + j(f/f_2)]$, $f_{H_i} = 1/2\pi R_{Th_1} C_i$, $R_{Th_1} = R_S \parallel R_{B_1} \parallel R_{B_2} \parallel R_i$,
$C_i = C_{W_1} + C_{be} + C_M = C_{W_1} + C_{be} + (1 + |A_v|)C_{bc}$, $f_{H_o} = 1/2\pi R_{Th_2} C_o$, $R_{Th_2} = R_C \parallel R_L$,
$C_o = C_{W2} + C_{ce}$, $h_{fe} = h_{fe_{mid}}/(1 + jf/f_\beta)$, $f_\beta = g_{b'e}/2\pi(C_{b'e} + C_{b'c}) =$
$1/h_{fe_{mid}}[g_m/2\pi(C_{b'e} + C_{b'c})]$, $f_\beta = f_\alpha(1 - \alpha)$, $f_T \cong h_{fe_{mid}} f_\beta$, $f_\beta = f_T/\beta$,
$f_T \cong g_m/2\pi(C_{be} + C_{bc})$, $BW \cong f_{H_i} - f_{L_E} \cong f_{H_i}$;
Multistage effects: $f_1' = f_1/\sqrt{2^{1/n} - 1}$, $f_2' = \sqrt{2^{1/n} - 1}\, f_2$;
FET multistage: $f_S = [1 + R_S(1 + g_m r_d)/(rd + R_D \parallel R_L)]/2\pi C_S R_S$,
$f_1 = 1/2\pi C_C(R' + R_{G_2})$, $R' = r_{d_1} \parallel R_{D_1}$, $f_2 = 1/2\pi C_{high} R_{high}$, $C_{high} = C_W + C_{i_2}$,
$R_{high} = R = r_{d_1} \parallel R_{D_1} \parallel R_{G_2}$, $C_i = C_{gs} + C_{gd}(1 + |A_v|)$,
$GBW = g_m/2\pi(C_W + C_i) = f_T = A_{v_{mid}} f_2$.

12 Large-Signal Amplifiers Series-fed class-A amplifier: $I_B = (V_{CC} - V_{BE})/R_B$,
$I_C = \beta I_B = h_{fe} I_B$, $V_{CE} = V_{CC} - I_C R_C$, $P_{i_{dc}} = V_{CC} I_C$, $P_{o_{ac}} = I_C^2(p\text{-}p)R_C/2 =$
$V_{CC}^2(p\text{-}p)/2R_C = I_C\,(p\text{-}p)V_{CE}(p\text{-}p)/8$, $\%\eta = P_o\,(ac)/P_i\,(dc) \times 100\%$, $P_{transistor} = P_i\,(dc) -$
$P_o\,(ac)$, max $P_o\,(ac) = I_C\,(p\text{-}p)V_{CE}\,(p\text{-}p)/8 = V_{CC}^2/8R_C$, max$\%\eta = 25\%$;
Transformer coupled: $R_L' = (N_1/N_2)^2 R_L$, $\Delta I_C = \Delta V_{CE}/R_L'$,
$P_o\,(ac) = \dfrac{(V_{CE_{max}} - V_{CE_{min}})(I_{C_{max}} - I_{C_{min}})}{8}$, $V_2 = (N_2/N_1)V_1$, $P_L = V_L^2\,(rms)/R_L$, $I_L\,(rms) =$
(N_1/N_2), I_C (rms), $P_L = I_L^2\,(rms)\,R_L$,
Series-fed: $\eta = 25[(V_{CE_{max}} - V_{CE_{min}})/V_{CC}]\%$,
Transformer coupled: $\eta = 50[(V_{CE_{max}} - V_{CE_{min}})/(V_{CE_{max}} + V_{CE_{min}})]\%$;
Class B: $P_i\,(dc) = V_{CC} I_{dc}$, $I_{dc} = (2/\pi)I_{peak}$, $P_o\,(ac) = V_L^2\,(rms)/R_L$, $P_{2Q} = P_i - P_o$,
$P_Q = P_{2Q}/2$, max $P_o\,(ac) = V_{CC}^2/2R_L$, max $P_i\,(dc) = \dfrac{2V_{CC}^2}{\pi R_L}$, max $\eta = 78.54\%$,
(max) $P_{2Q} = \dfrac{2V_{CC}^2}{\pi^2 R_L}$;
Distortion $D_2 = |I_2/I_1| = |[\frac{1}{2}(I_{C_{max}} + I_{C_{min}}) - I_{CQ}]/[I_{C_{max}} - I_{C_{min}}]| \times 100\% =$
$|[\frac{1}{2} V_{CE_{max}} + V_{CE_{min}}) - V_{CE_Q}]/[V_{CE_{min}} - V_{CE_{min}}]| \times 100\%$, $D_n = |I_n/I_1|$,
$D = \sqrt{D_2^2 + D_3^2 + D_4^2 + \cdots}$, $P = (1 + D^2)P_1$;
Heat sinks: $P_D = V_{CE} I_C$, $P_{D_{t1}} = P_{D_{t0}} - (t_1 - t_0)$(Derating factor), $\theta_{JA} = \theta_{JC} + \theta_{CS} + \theta_{SA}$,
$T_J = P_D \theta_{JA} + T_A$.

13 *pnpn* and Other Devices SCR: $V_{(BR)F^*}$ forward breakover voltage-transition state voltage, I_H holding current-switching current; DIAC: $V_{BR_1} = V_{BR_2} \pm 10\%\, V_{BR_2}$;
Unijunction transistor: $R_{BB} = (R_{B_1} + R_{B_2})|I_E = 0$, $V_{RB_1} = R_{B1} V_{BB}/[R_{B_1} + R_{B_2}] = \eta V_{BB}$ with
$I_E = 0$, $\eta = R_{B_1}/(R_{B_1} + R_{B_2})$ with $I_E = 0$, $V_P = \eta V_{BB} + V_D$,
To insure firing $R_1 < (V - V_P)/I_P$,
To insure turning off $R_1 > (V - V_v)/I_V$, $V_{R_2} \cong R_2 V/(R_2 + R_{BB})$ with $I_E = 0$, charging
$v_C = V_v + (V - V_v)(1 - e^{-t/R_1C})$,
Discharging $v_C \cong V_P e^{-t/(R_{B1} + R_2)C}$, V_{R_2} (peak) $\cong R_2(V_p - 0.7)/(R_2 + R_{B_1})$,
Period $T = t_1 + t_2 = R_1 C \log_e (V - V_v)/V - V_P) + (R_{B_1} + R_2)C \log_e V_P/V_v$,
$f_{osc} = 1/T \cong 1/R_1 C \log_e (1/1 - \eta)$;